Y. Touitou    E. Haus    (Eds.)

# Biologic Rhythms in Clinical and Laboratory Medicine

With 347 Figures and 54 Tables

Springer-Verlag
Berlin  Heidelberg  New York
London  Paris  Tokyo
Hong Kong  Barcelona
Budapest

YVAN TOUITOU
Professor of Biochemistry
Faculté de Médecine Pitié-Salpétrière
Laboratoire de Biochimie Médicale
91, Boulevard de l'Hôpital
75634 Paris Cedex 13, France

ERHARD HAUS
Professor, Department of Laboratory Medicine and Pathology
University of Minnesota
Chairman, Department of Anatomic and Clinical Pathology
St. Paul-Ramsey Medical Center
640 Jackson Street
St. Paul, MN 55101-259, USA

*Cover picture:* Sculpture by Arman, gare Saint-Lazare, Paris

Second Printing 1994

ISBN 3-540-57592-8  Springer-Verlag Berlin Heidelberg New York
ISBN 0-387-57592-8  Springer-Verlag New York Berlin Heidelberg

CIP data applied for

© Springer-Verlag Berlin Heidelberg 1992
Printed in Germany

Production Editor: Isolde Scherich
Reproduction of the figures: Gustav Dreher GmbH, Stuttgart
Typesetting, printing, and binding: Appl, Wemding
27/3130 – 5 4 3 2 1 – Printed on acid-free paper

# Foreword

With the help of our European and American colleagues and friends, we have been able to prepare this book on *Biologic Rhythms in Clinical and Laboratory Medicine,* which includes 20 parts consisting of 47 chapters on different aspects of chronobiology and medicine.

Parts I and II present six chapters on the basic concepts and the mechanisms of biologic rhythms, followed in Part III by a review of the appropriate methodology for their study. Part IV summarizes the large amount of data accumulated on the time-dependent changes in drug effects and side effects and in toxicology. Part V presents in four chapters observations on biologic rhythms beginning in pregnancy and extending through childhood and development to old age. Part VI is dedicated to the rhythms in physical and mental performance and is followed in Part VII by the study of the adaptation of these rhythms during shiftwork and after transmeridian flights. In Part VIII is presented the physiology and pathophysiology of sleep. Parts IX–XIX discuss biologic rhythms in different organ systems in health and disease, including aspects of physiology, pharmacology, and toxicology. The clinically important rhythm alterations found in certain psychiatric conditions, i.e., in affective disorders are presented in Part IX. The eight chapters in Part X are dedicated to the rhythms of multiple frequencies found in endocrine and metabolic functions. In Part XI, the biologic rhythms in hepatic drug metabolism are discussed which determine, in part, circadian and other variations in drug effects. Part XII describes the rhythmicity of the gastrointestinal system with its implications for ulcerogenesis and timed treatment. In Part XIII are discussed rhythms of the cardiovascular system, including hypertension, and in Part XIV rhythms in the respiratory system with the clinically important application to the treatment of bronchial asthma. Part XV presents rhythms in renal function and pathology. The chronobiology of the inflammatory reaction is presented in Part XVI. Biologic rhythms in hematology and immunology are presented in Part XVII and include, in addition to the large-amplitude rhythms of some of the formed elements in the peripheral blood, new data on the rhythmicity of the human bone marrow, which will be of considerable importance for the effects and/or side effects of the numerous drugs and other agents acting upon this complex and vital tissue. This part also includes observations on biologic rhythms in cell migration, in the immune response, and of the blood coagulation system with its implications for transient risk states for thrombosis or hemorrhage. Part XVIII presents first basic data on rhythms in normal and abnormal cell proliferation, then the endocrine regulations related to the cancerogenesis in endocrine responsive tissues, i.e., the breast, the rhythms in the appearance of tumor markers in the circulation, possibly representing periodic variations in tumor metabolism and, finally, the opportunities provided by the rhythms of host and tumor to optimize chemotherapy by treatment timed according to the biologic rhythms of the host and/or the tumor. Epidemiologic aspects of chronobiology are presented in

Part XIX in regard to physiologic events like birth and to chronopathology and death from numerous causes. Finally, the implication of biologic rhythms for laboratory medicine are discussed in Part XX including time-qualified reference values for high-amplitude rhythms and the tools to improve laboratory diagnosis by taking into account the principles of chronobiology.

In all these chapters, our purpose is to present to the reader, whether experienced in chronobiology or not, a general review of the role of biologic rhythms in the normal and abnormal function of different organ systems.

Each of the contributors is a known specialist in his or her field, and we would like to express our great appreciation for their participation in this didactic teamwork. We hope that this book will interest the readers who wish to extend their familiarity with chronobiology and will provide the investigators in the study of biologic rhythms a general overview of the field.

March 1992                                          Yvan Touitou, Paris
                                                    Erhard Haus, St. Paul

# Contents

Part XVIII    **Cell Proliferation and Cancer**

Part XIX    **Epidemiology**

Part XX    **Laboratory Medicine**

# Contributors

ANGELI, A.
Cattedra di Medicina Interna, Università degli Studi di Torino,
Divisione di Clinica Medica Generale, Ospedale San Luigi Gonzaga,
10043 Orbassano, Italy

ARENDT, J.
Department of Biochemistry, University of Surrey, Guilford Surrey GU2 5XH,
Great Britain

BÉLANGER, P. M.
Université Laval, Ecole de Pharmacie, Cité Universitaire, Québec G1K 7P4,
Canada

BLOMQUIST, C. H.
Department of Obstetrics and Gynecology, Ramsey Clinic, St. Paul-Ramsey
Medical Center, 640 Jackson Street, St. Paul, MN 55101, USA

BOGDAN, A.
Faculté de Médecine Pitié-Salpétrière, Laboratoire de Biochimie Médicale,
91 Boulevard de l'Hôpital, 75634 Paris Cedex 13, France

BRUGUEROLLE, B.
Laboratoire de Pharmacologie Médicale et Clinique, Faculté de Médecine,
CHU de la Timone, 27 Boulevard Jean Moulin, 13385 Marseille, France

BUREAU, J. P.
CHR de Nîmes, Laboratoire de Cytologie Clinique et Cytogénétique,
Faculté de Médecine, Avenue Kennedy, 30000 Nîmes, France

CAL, J. C.
Groupe d'Etude de Physiologie et Physiopathologie Rénales,
Faculté de Médecine, 91 rue Leyteire, 33000 Bordeaux, France

CAMBAR, J.
Groupe d'Etude de Physiologie et Physiopathologie Rénales,
Faculté de Médecine, 91 rue Leyteire, 33000 Bordeaux, France

CANON, C.
I. C. I. G., Service des Maladies Sanguine et Tumorales, Hôpital Paul Brousse,
14–16 Avenue Paul Vaillant Couturier, 94800 Villejuif, France

CARIGNOLA, R.
Cattedra di Medicina Interna, Università degli Studi di Torino,
Divisione di Clinica Medica Generale, Ospedale San Luigi Gonzaga,
10043 Orbassano, Italy

CORNÉLISSEN, G.
Chronobiology Laboratories, University of Minnesota, Minneapolis, MN, USA

CZEISLER, C. A.
Harvard Medical School, Neuroendocrinology Laboratory, Brigham and
Women's Hospital, 221 Longwood Avenue, Boston, MA 02115, USA

D'ALONZO, G. E.
School of Public Health, University of Texas, P.O. Box 20186,
Astrodome Station East 442, Houston, TX 77025, USA

DECOUSUS, H.
Service de Médecine Interne et Thérapeutique, Hôpital de Bellevue,
42023 Saint-Etienne, France

DE PRINS, J.
Stanley 7, 3080 Tervuren, Belgium

DJAKOVIC, M.
Unité 70 INSERM, 388 rue du Mas de Prunet, 34070 Montpellier, France

DRENNAN, M. D.
Department of Psychiatry, Veterans Administration Hospital,
3350 La Jolla Village Drive, San Diego, CA 92161, USA

DROUIN, P.
Université de Nancy, Département de Nutrition et des Maladies Métaboliques,
INSERM U 59, 40 rue Lionnois, 54000 Nancy, France

EDMUNDS, L. N., Jr.
Division of Biological Sciences, Life Sci Bldg 370, State University of New York,
Stony Brook, New York 11794, USA

ELLIOTT, J. A.
Department of Psychiatry, Veterans Administration Hospital,
3350 La Jolla Village Drive, San Diego, CA 92161, USA

FERNANDES, G.
University of Texas, Health Science Center at San Antonio,
7703 Floyd Curl Drive, San Antonio, TX 78284-7874, USA

FEUERS, R. J.
Division of Genetic Toxicology, National Center for Toxicological Research,
Jefferson, Arkansas 72079, USA

FOCAN, C.
Service de Médecine interne, Clinique St. Joseph – Ste. Elisabeth, Liège, Belgique

GARRELLY, L.
CHR de Nîmes, Laboratoire de Cytologie Clinique et Cytogénétique,
Faculté de Médecine, Avenue Kennedy, 30000 Nîmes, France

GATTI, G.
Cattedra di Medicina Interna, Università degli Studi di Torino,
Divisione di Clinica Medica Generale, Ospedale San Luigi Gonzaga,
10043 Orbassano, Italy

HALBERG, F.
5-197 Lyons Lab, University of Minnesota, 420 Washington Avenue S.E.,
Minneapolis, MN 55455, USA

HAUS, E.
St. Paul-Ramsey Medical Center, Department of Anatomic and
Clinical Pathology, 640 Jackson Street, St. Paul, MN 55101, USA

HAYES, B.
Harvard Medical School, Neuroendocrinology Laboratory, Brigham and
Women's Hospital, 221 Longwood Avenue, Boston, MA 02115, USA

HECQUET, B.
Centre Oscar Lambret, B.P. 307, 59020 Lille, France

HOLLEY, D.C.
Biomedical Research Division, 240 A-3, NASA/Ames Research Center,
Moffett Field, CA 94035, USA

HOLT, J.P., Jr.
Reproductive Endocrinology, St. Paul-Ramsey Medical Center, Ramsey Clinic,
640 Jackson Street, St. Paul, MN 55101, USA

HRUSHESKY, W.J.M.
Division of Oncology, 47 New Scotland Avenue, Albany Medical College,
State University of New York, MSX-30 Albany, NY 12208, USA

IRANMANESH, A.
Division of Endocrinology and Metabolism, Department of Internal Medicine,
Salem Veterans Administration Hospital, Salem, Virgina 24153, USA

JOHNSON, M.L.
Division of Endocrinology and Metabolism, Department of Internal Medicine,
Box 202, Health Sciences Center, University of Virginia, Charlottesville, VI,
22908, USA

KANABROCKI, E.L.
Nuclear Medicine Service, Veterans Administration Hospital, Hines,
Illinois 60141, USA

KOLOPP, M.
Université de Nancy, Département de Nutrition et des Maladies Métaboliques,
INSERM U 59, 40 rue Lionnois, 54000 Nancy, France

KRIPKE, D. F.
Department of Psychiatry, Veterans Administration Hospital,
3350 La Jolla Village Drive, San Diego, CA 92161, USA

L'AZOU, B.
Groupe d'Etude de Physiologie et Physiopathologie Rénales,
Faculté de Médecine, 91 rue Leyteire, 33000 Bordeaux, France

LABRECQUE, G.
Ecole de Pharmacie, Université Laval, Pavillon Vachon, Québec G1K 7P4,
Canada

LAERUM, O. D.
Department of Oncology, Haukeland Hospital, University of Bergen,
5021 Bergen, Norway

LAKATUA, D. J.
Department of Laboratory Medicine and Clinical Pathology, St. Paul-Ramsey
Medical Center, 640 Jackson Street, St. Paul, MN 55101, USA

LEMMER, B.
Klinikum der Johann-Wolfgang-Goethe-Universität, Theodor-Stern-Kai 7,
6000 Frankfurt 70, Federal Republic of Germany

LÉVI, F.
I. C. I. G., Service des Maladies Sanguines et Tumorales, Hôpital Paul Brousse,
14–16 Avenue Paul Vaillant Couturier, 94800 Villejuif, France

LIZARRALDE, G.
Division of Endocrinology and Metabolism, Department of Internal Medicine,
Salem Veterans Administration Hospital, Salem, Virgina 24153, USA

MARKOWITZ, M. E.
Department of Pediatrics
Montefiore Medical Center, 111 E. 210th Street, Bronx, NY 10467, USA

MÄRZ, W. J.
Division of Oncology, 47 New Scotland Avenue, Albany Medical College,
State University of New York, MSX-30 Albany, NY 12208, USA

MASERA, R.
Cattedra di Medicina Interna, Università degli Studi di Torino,
Divisione di Clinica Medica Generale, Ospedale San Luigi Gonzaga,
10043 Orbassano, Italy

MEIS, P. J.
Department of Obstetrics, Bowman Gray School of Medicine,
300 South Hawthorne Road, Winston Salem, NC 27103, USA

MÉJEAN, L.
Université de Nancy, Département de Nutrition et des Maladies Métaboliques,
INSERM U 59, 40 rue Lionnois, 54000 Nancy, France

MEYLOR, J. S.
Biomedical Research Division, 240 A-3, NASA/Ames Research Center,
Moffett Field, CA 94035, USA

MONK, T. H.
Western Psychiatric Institute and Clinic, University of Pittsburg, Pittsburg, PA,
USA

MONTAGNER, H.
Unité 70 INSERM, 388 rue du Mas de Prunet, 34070 Montpellier, France

MOORE, J. G.
Utah School of Medicine, GI Section, Department of Veterans Affairs Medical
Center, 500 Foothill Blvd., Salt Lake City, Utah 84148, USA

NICOLAU, G. Y.
Endocrine Rhythms Laboratory, The "C. I. Parhon" Institute of Endocrinology,
Bd. Aviatorilor 34–36, 79600 Bucharest, Romania

QUEIROZ, O.
5 avenue des Bouvreuils, 78720 Cernay La Ville, France

QUEIROZ-CLARET, C.
Institut National Agronomique Paris-Grignon, Chaire de Biochimie,
78850 Thiverval Grignon, France

REINBERG, A. E.
URA 58 1 CNRS, Chronobiologie et Chronopharmacologie,
Fondation A. de Rothschild, 29 rue Manin, 75940 Paris, France

RIETVELD, W. J.
Department of Chronobiology, University of Leiden, Wassenaarseweg 62,
2333 AL Leiden, The Netherlands

VON ROEMELING, R.
Division of Medical Oncology, 47 New Scotland Avenue, Albany Medical
College, State University of New York, MSX-30 Albany, NY 12208, USA

ROGOL, A. D.
Department of Pediatrics, Box 386, University of Virginia,
Health Sciences Center, Charlottesville, VI 22908, USA

DE ROQUEFEUIL, G.
Unité 70 INSERM, 388 rue du Mas de Prunet, 34070 Montpellier, France

SAUERBIER, I.
Institut für Anatomie, Medizinische Hochschule Hannover,
Konstanty-Gutschow-Straße 8, 3000 Hannover 61, Federal Republic of Germany

SCHEVING, L. A.
Department of Medicine, Stanford University School of Medicine, Stanford,
CA 94305, USA

SCHEVING, L.E.
University of Arkansas for Medical Sciences, Department of Anatomy, Slot 510,
4301 W. Markham, Little Rock, AR 72201, USA

SMAALAND, R.
The Gade Institute, Department of Pathology, Haukeland Hospital,
University of Bergen, 5021 Bergen, Norway

SMOLENSKY, M. H.
School of Public Health, University of Texas, P. O. Box 20186,
Astrodome Station East 442, Houston, TX 77025, USA

SOLIMAN, M. R. I.
Biomedical Research Division, 240 A-3, NASA/Ames Research Center,
Moffett Field, CA 94035, USA

TOUITOU, Y.
Faculté de Médecine Pitié-Salpétrière, Laboratoire de Biochimie Médicale,
91 Boulevard de l'Hôpital, 75634 Paris Cedex 13, France

TRANCHOT, J.
Groupe d'Etude de Physiologie et Physiopathologie Rénales,
Faculté de Médecine, 91 rue Leyteire, 33000 Bordeaux, France

TSAI, T.-H.
University of Arkansas for Medical Sciences, Department of Anatomy, Slot 510,
4301 W. Markham, Little Rock, AR 72201, USA

VELDHUIS, J. D.
Division of Endocrinology and Metabolism, Department of Internal Medicine,
Box 202, Health Sciences Center, University of Virginia, Charlottesville, VI,
22908, USA

WINGET, C. M.
Biomedical Research Division, 240 A-3, NASA/Ames Research Center,
Moffett Field, CA 94035, USA

# Biologic Rhythms from Biblical to Modern Times. A Preface

Y. Touitou and E. Haus

## Introduction

The origins of chronobiology go back to the very beginning of life on this planet. Living matter and the evolving organisms were exposed to the earth's revolution around the sun with its periodicity of day and night, of light and darkness, with the periodic changes in the length of the daily light and dark span and with the climatic changes of the seasons. In addition some aquatic forms of life especially were exposed to the periodic input provided by the cycles of the moon with its influence on ocean tides. Adaptation to the periodically changing environment on our planet was a necessity for the earliest and all later forms of life. The related periodic functions – originally in response to the environmental stimuli – seem to have impressed themselves on the genetic makeup of living matter. Periodic variations, many but not all of which follow the frequencies of the periodic environmental input, are found in the most primitive and ancient forms of life presently available for study. Many other periodic functions, however, ranging in the length of their cycle from milliseconds (as in the activity of single neurons) to seconds (such as the heart and respiration rate) and to months (such as the menstrual cycle in sexually mature women) have no known environmental counterpart. Some biochemical and biophysical mechanisms creating or maintaining periodic functions at the cellular level are related to the genetic material in nuclear DNA, while others are apparently functioning apart from nuclear material in relation to membranes or to metabolic processes in the cytoplasm (Edmunds, this volume; Lakatua, this volume).

## Early Observation of Biologic Periodicity

The concept of time and the environmental periodicities related to its passage were familiar to prehistoric man as hunter and gatherer of food. Attempts to measure time most likely date back over more than 30 000 years. Some of these early attempts are expressed in archeological sites showing apparent relations in their construction to the cycles of the sun, the moon and the stars.

In biblical times the importance of temporal factors was recognized as stated in Ecclesiastes: "To everything there is a season and a time to every purpose under the Heaven: a time to be born and a time to die; a time to plant and a time to harvest." In Genesis, the first task of God was to create light and then the alternation of light and darkness. Also the 7-day cycle of activity and rest is part of the story of creation as presented in the Old Testament; the 6 days of work were followed by a day of rest.

The earliest recorded recognition of the importance of biologic rhythms in plants and animals dates back to at least 5000 BC. The Egyptian calender was invented around 4200 BC. The time and the periodic variations of biologic events in health and disease played a large role in the minds of the ancient physicians. The Egyptians had developed a concept of periodicity shrouded in a number magic and a doctrine of "critical days" on which certain symptoms were thought to appear or to become exacerbated. The number 7 played an important role in this concept, although we have no records of actual physiologic or clinical observations from that time.

The Greek naturalists and physicians Aristotle, Hippocrates, Diocles, and others adopted these ideas and added to them some of their own observations on the periodic course of disorders like intermittent fevers, which followed different periods, some of which nowadays could most likely be explained by the life cycle of the malarial parasite. For Diocles, the number 7 also seemed to play an important role in his thoughts on the periodicity of health and disease. The cause of the sometimes surprisingly regular periodic events observed was sought in astronomy. Aristotle believed that the moon influenced reproductive functions and various derangements peculiar to women. Hippocrates described the tendency of crises in disease to occur at regular intervals and felt that the pathologic phenomena seemed to obey the same mathematical regularity that was attributed to celestial bodies. Hippocrates recommended that no physician be permitted to treat patients unless he had learned the astronomical signs.

The capability for exact measurements of time was limited in the ancient days and early middle ages to water clocks, sand clocks and sun dials. Only in the 14th and 15th century AD did more accurate time keeping become possible with the invention of mechanical clocks. The pendulum clock was invented in 1656 by Christian Huygens and a chronometer, the action of which was based on the coiling and uncoiling of a spring, by John Harrison in 1761. A "physicians pulse watch" was introduced by John Florer in 1707.

The periodicity of sleep and wakefulness is discussed in the writings, among others, of Aristotle and later of Galen. While Aristotle still regarded the heart as the predominant organ related to sleep-wakefulness, Galen placed this function in the brain. Galen's concept that sleep is caused by increased pressure on the brain, which interferes with its activity (secondary to fumes ascending from the stomach), was challenged only 1300 years later by Joannes Argentinus in his book *On Sleep* (1555). Argentinus regarded sleep as a "faculty" of the spirits and an "instinct" and "desire." Apparently similar ideas were presented by Conrad Victor Schneider (1614–1680; cited after Lavie 1989). The passive explanations of sleep were conclusively disproved only during the 1950s by Aserinsky and Kleitman's discovery of REM sleep (Kleitman 1953).

The continuous recording of electric potentials in the EEG and eye movements during sleep led Kleitman in 1953 to the description of the electrical stages of sleep associated with the ultradian periodicity in the approximate 90-min range of episodes of rapid eye movement (REM), and led to the change of the concept of sleep from that of a condition of passive inactivity to our understanding as an active process forming an essential part of the circadian system.

Similarly, in exploring neuromuscular activity, both Fessart in 1936 and Cardot in 1933 found that in the ultradian frequency range "rhythmic activity is a basic property of excitable systems."

Although the appearance of elevated body temperature in diverse disease states in the afternoon was known in ancient times, objective measurements had to wait for the invention of a thermometer suitable for repeated and rapid in vivo measurements. In 1843, Chossat described a circadian rhythm in body temperature in pigeons which were maintained under conditions of complete inanition. The body temperature of the birds deprived of food and water not only did not lose its circadian periodicity, but showed a marked increase in amplitude. This observation, which later was extended to other animal species, was designated the "Chossat phenomenon."

Davy in 1845 reported both a circadian and a circannual rhythm in his own body core temperature which were closely related neither to physical activity nor to the environmental temperature. Morel in 1866 found the circadian body temperature rhythm to be independent of exercise, food, and ambient temperature and described a decrease of the circadian amplitude but not an abolishment of the rhythm as a result of reduced muscular activity.

The idea of the constancy of the internal milieu *(milieu interieur)* was originally proposed by Claude Bernard (1865, 1885, 1926) and was then expanded by Cannon (1929) to the theory of homeostasis. These investigators assumed that within a "biological equilibrium" or a "steady state" changes in the system would lead by the way of feedback regulations to counterreactions which attempt to reestablish the "constant level" of a function.

However, a study of the papers by Bernard indicates that, in spite of his assumption of a "constant internal environment" in the face of a changing external milieu, he was well aware of the great variability encountered in the physiologic condition of the objects of his studies. One finds the statement that "the physiologic conditions of the internal milieu manifest an extreme variability" and that "one must keep in mind not only the variations of the cosmic external milieu, but also the variations of the organic milieu, i.e., that of the actual state of the organism." He also lists the observation of so-called "oscillatory" life, although it appears unlikely that rhythms as such were clearly recognizable and even less measurable in his studies.

## Endogenous Nature of Biologic Rhythms

While it had been generally believed in the ancient and not too ancient days that cyclic changes in the organism represented exclusively the effects of cyclic changes in environmental factors, the first observations which led to the recognition of the endogenous nature of certain periodicities were made in 1729 when the French astronomer J.J.Ortons de Mairan reported that the daily changes in the position of the appendages of the heliotrope persisted in continuous darkness. This observation was followed in 1745 by the designation of a "flower clock" by K.Linné, who showed that at certain sun-related clock hours the flowers of some plant species are open while those of others are closed (Fig.1).

The observation that circadian rhythms persisting under as far as feasible constant conditions may show a period differing from 24 h, and thus be free running from the environmental 24-h cycle, was first reported by A. de Candolle (1832), who found that the leaf movement in *Mimosa pudica* persists in complete darkness with a period of 22–23 h. Similar findings under more controlled experimental conditions were presented for the endogenous origin of certain circadian rhythms in plants by Pfeffer (1875, 1915).

In human subjects, Thomas Lacock expressed in 1842 the concept that periodicity originated from within the system ("isoteric origin") or was caused by periodic factors acting from without ("exoteric origin") or by a combination of both ("endoexoteric").

This observation has been extended and confirmed under more precise experimental conditions for many species from eukaryote to man. It became obvious that organisms are not passive responders to environmental changes but have internal accurate time-measuring systems or "clocks." This concept of biologic clocks was developed during the 1950s and 1960s by Bünning, Hastings, Schweiger, Aschoff, Richter, Pittendrigh, Menaker and others.

The investigation of the endogenous nature of this "clock" mechanism led to diverse hypotheses and theories postulating a population of interacting oscillators synchronized by structures acting as pacemakers, which are subject to exogenous phase information (Fig.2). Alternatively, theories were developed assuming one or two central oscillators. The environment then acts on an organism to keep the clock set to a "correct" time. The information now available on rhythmic functions clearly does not allow the simple hypothesis of a single central "mas-

**Fig.1.** Linné's flower clock. Opening and closing of different species of flowers occurs at different times allowing the botanist to estimate the clock hour on the basis of the observation of his plants

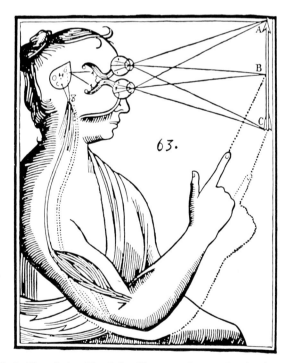

**Fig.2.** The relationship of visual factors to the neural and endocrine system was suggested by Rene Descartes in his *Traité de l'Homme* in 1664

ter clock." Even in unicellular organisms like acetabularia, both the nucleus and anucleated fragments of cytoplasm maintain the circadian periodicity even in the presence of inhibitors of the synthesis of chloroplastic or mitochondrial RNA. The studies in acetabularia showed that in a single cell more than one oscillator may produce independent periodic activities. The now evolving concept is the existence in the cell as well as in the multicellular organism of interconnected oscillators, some of which are sensitive to information provided by environmental factors. It furthermore appears that more than one factor controls each of these biochemical rhythm components. Some of the structures acting as pacemakers for numerous endogenous rhythms show their own endogenous periodicity as, e.g., the suprachiasmatic nuclei of the hypothalamus.

In mammals, the suprachiasmatic nucleus (SCN) is likely to be the circadian oscillator system, which as pacemaker drives certain neuroendocrine circadian rhythms, such as those of ACTH, TSH and prolactin. Physical destruction or neurotoxic inhibition of the SCN results in the cessation of these neurohormonal circadian rhythms.

However, also after destruction of the SCN, certain circadian rhythms of the organism persist, including rest and activity, eating and drinking, body temperature and corticosterone secretion from the adrenal cortex as shown by Moore in the United States, Assenmacher in France and others.

Most recently, Ralph et al. (1988) showed that the transplantation of suprachiasmatic nuclei between animals showing genetically determined differences in their endogenous circadian cycle length and different circadian timing of their synchronization will impress the period as well as the phase of the donor to the recipient.

## Genetics

The genetic origin of circadian rhythm characteristics was shown first by Bünning in 1935 in the bean plant *Phaseolus*. The circadian rhythm of stem and leaf movements in this plant was found to be different between two genetically distinct stalks, exhibiting periodicities of about 23 and 27 h respectively. Hybridization experiments produced plants with hybrid circadian rhythm characteristics and an intermediate period of about 25 h. These studies showed that biologic rhythm characteristics are transmitted from

generation to generation according to genetic rules. In drosophila, several genes have been identified responsible for certain biologic rhythm characteristics.

Konopka and Benzer in 1971 identified on the X chromosome of drosophila a region controlling the period of a circadian rhythm with three mutations, one of which led to a longer period than 24 h, the other to a shorter period, and the third to an aperiodic behavior of the animal. In 1984, Bargiello and colleagues showed that a fragment of 7100 base pairs corresponding to the periodicity gene *(per)* injected in embryos of arrhythmic drosophilas was capable of restoring the circadian rhythm of activity and of eclosion. Of special importance for the expression of the genetic information transmitted seems to be the synchronization between cells. The synchronization of the apparently individually cycling cells leads to the grossly manifest rhythms, the better the synchronization observed the higher the amplitude. The *per* mutations were shown to alter this intercellular communication. For a more detailed discussion of these mechanisms, see the chapters by Edmunds and Lakatua in this book.

## Biologic Rhythms and Environmental Factors

The relations between endogenous apparently genetically controlled biologic rhythms and environmental factors were explored in the early 1950s, and it became obvious that environmental factors, and especially the lighting regimen, were capable of determining the timing of circadian rhythms and could act as synchronizer (Halberg), entraining agent (Pittendrigh), or zeitgeber (Aschoff). These terms, which were originally coined at about the same time, are now used synonymously. While well documented for circadian rhythms, synchronizers for rhythms of other frequencies are less well understood. It seems likely that weekend and work week may synchronize circaseptan periodicities in man, and that the length of the daily light span as well as environmental temperature are likely synchronizers for circannual periods. It has to be understood that synchronizers do not create rhythms, but do determine their placement in time.

A susceptibility cycle to the influence of light upon circadian rhythms was shown by Bünning in 1963, who found that a time-restricted short exposure to light induced flowering when given at a certain critical clock hour but that the same quantity and quality of light when given at another clock hour was ineffective.

In the late 1950s, it was first demonstrated by Pittendrigh and Bruce that single short light exposures could modify the phase of circadian rhythms in experimental animals and it was found that the magnitude and direction of these phase shifts were dependent, not only on the intensity and the duration of the light exposure, but also on the circadian phase at the time of light exposure (Hastings and Sweeney 1958). In mammals, the phase information from the environment is relayed from specialized retinal photoreceptors to a central circadian pacemaker in the hypothalamus (the suprachiasmatic nuclei; SCN) via a monosynoptic retinohypothalamic tract (RHT) (Fig. 2). This function can still be observed in the animals after induction of behavioral blindness by lateral transection of the primary optic tracts.

In humans, cyclic changes of socioecologic factors are a powerful synchronizer of circadian rhythms. Recently, however, it has been shown that also in the human the exposure to bright light can play a significant role as synchronizer and that by appropriately placed bright light the circadian system can be phase shifted more effectively than by social factors (see chapter by Czeisler, this volume). Bright light (exceeding a minimum threshold of 2500 lux), appropriately placed, appears to overide the effects of the social environment as entraining agent.

## Biologic Rhythms in Disease

Virey (1814) explored human diurnal rhythms in relation to health and disease. In the first thesis in chronobiology for the doctorate of medicine presented at the University of Paris in 1814, he emphasized the importance of timing in therapeutic interventions.

The first mention of treatment timed prospectively, according to the stages of a rhythm, was presented by Balfour (1815), who tried to predict fevers according to the lunar cycle and according to the timing of previous attacks. Balfour suggested that treatment should be timed according to the expected occurrence of the next attack of the disorder.

Edward Smith, in 1861, reported a higher toxicity of ethanol in the morning and stressed that drug effects are cycle stage dependent. Smith also reported a circadian periodicity of time of death: "death occurs much more frequently from 1 to 5 a.m. than at other hours of the day or night", which has since been confirmed numerous times by many investigators. He drew the conclusion from this observation that "patients should be carefully watched at these vulnerable times." As a forerunner of contemporary chronopharmacology, Smith stated that "the efficiency of the remedy will as much depend upon the right period being chosen for its administration as on its own properties."

These early reports went largely unnoticed until almost the middle of the 20th century when the importance of timing of treatment according to the rhythmic course of disease was realized, e. g., the time insulin has to be given to diabetic patients (von Möllersdorf). In the 1950s to 1970s, extensive animal experimentation in rodent models showed that sensitivity and resistance of the animals to a wide variety of noxious agents, including many drugs used in clinical medicine, was critically dependent on the circadian system phase at the time of administration.

The concept of chronopharmacology has been extended to human subjects in studies by Reinberg, Halberg and many others since the early 1960s and has developed rapidly during the last 2 decades, and a large amount of pertinent literature has accumulated, which is reviewed in the chapter by Bruguerolle in this book.

During the last three decades the development of more advanced measurement and monitoring techniques of physiologic and laboratory variables has greatly facilitated the introduction of chronobiology into clinical medicine. The advent of electronic computers during this same time has allowed the use of inferential statistical methods for the quantitative measurement of time-related changes in body functions and their evaluation with statistically meaningful endpoints. These developments have opened new avenues in the recognition and understanding of abnormality and in the treatment of disease, and allow a meaningful and cost-effective application of chronobiologic concepts and methods to clinical medicine. Together with the recent advances in chronopharmacology, such an application appears now timely and in some areas urgent.

It is the purpose of this book to provide the background for the application of chronobiology to clinical medicine and to outline some present and some potential future applications.

*Acknowledgements.* We wish to express our gratitude to Professor P. Lavie (Israel) for providing us with a draft of a historic review entitled "From intermittent fever to chronobiology."

# Principles of Clinical Chronobiology

E. Haus and Y. Touitou

## Definitions and Basic Principles

If one examines living matter as a function of time under appropriate experimental conditions, on the cellular level, in tissue culture or in multicellular organisms, including man, at different levels of physiologic organization one invariably finds nonrandom variations of the variables examined (Aschoff and Wever 1976; Bünning 1973; Conroy and Mills 1970; Halberg 1959; Haus et al. 1980, 1988; Haus and Halberg 1980; Reinberg and Ghata 1964). Many of these time-dependent changes recur in regular intervals and thus represent rhythms, which are to a certain degree predictable in time. With the use of statistical procedures of rhythmometry, a large proportion of

**MESOR**

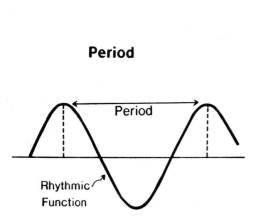

**Amplitude**

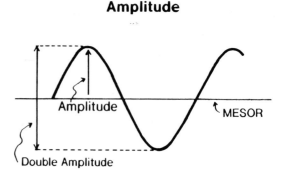

**Period**

**Acrophase**

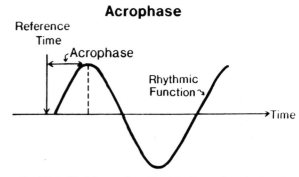

**Fig. 1.** Definition of parameters of a rhythmic function (e. g., by a cosine curve) fitted to the data. MESOR *(M):* rhythm-adjusted mean. The average value of the fitted rhythmic function. Expressed in the same units as original data. The MESOR will be equal to the arithmetic mean only if the data are obtained at equidistant intervals over one or more full cycles. Period (τ): duration of one complete cycle. Expressed in time units. Amplitude *(A):* half of the total predictable change in a rhythm described by a mathematical model (half the peak-trough distance of the fitted function). Expressed in original or "relative" units, e. g., as percentage of MESOR. Acrophase (Φ): crest time of the fitted rhythmic function expressed as lag from reference time as (negative) degrees with 360° = period, 0° = reference time or in customary time units (e. g., clock-hours, days, months, etc.)

the variability encountered in most series of measurements of biologic variables can be shown to be due to a multitude of rhythms in different frequency ranges (Halberg et al. 1965 a, b; Halberg and Panofsky 1961; Halberg and Reinberg 1967; Haus et al. 1980, 1981; Panofsky and Halberg 1961), which may be superimposed upon each other and upon trends, e. g., as a function of aging. Chronobiology is the science investigating and objectively quantifying the mechanisms of this biologic time structure including the rhythmic manifestations of life (Halberg et al. 1977).

## Rhythm Parameters

A rhythm represents a regularly recurring oscillation, the repeating unit of which is referred to as a cycle. The time required to complete one cycle (Fig. 1) (i. e., the time interval after which a distinct phase of the oscillation recurs) is referred to as the period (which is the reciprocal of the rhythm's frequency).

### Periods of Biologic Rhythms

The periods of the rhythms encountered in biology and medicine may range from a fraction of a second as in single neurons to seconds as in the cardiac and respiratory cycles, a few hours as they occur in certain endocrine parameters, to about 24 h as seen in the prominent circadian frequency domain which is found almost ubiquitously in all metabolizing structures, and physiologic parameters. The periodic variations with shorter periods (higher frequencies) than circadian, the so-called ultradian rhythms are superimposed upon the circadian rhythms of the same parameter. The circadian rhythms in turn are superimposed upon rhythms with longer periods (or lower frequencies), the so-called infradian rhythms, which

include, among others, rhythms with a period of about 1 week (circaseptan rhythms), rhythms with a period of about 30 days (circatrigintan rhythms) and rhythms with a period of about 1 year (circannual rhythms and/or seasonal variations; see Table 1) (Haus and Halberg 1970; Nicolau et al. 1983; Reinberg 1974).

### Rhythm-Adjusted Mean: MESOR

The mean value of a rhythm would ideally be represented by the mean of all instantaneous values of the oscillating variable within one period. However, in biologic time series, i.e., in clinical medicine, quasi-continuous measurements will seldom be feasible. The data available will more often be spot checks of the rhythmic variable obtained in larger discrete and often irregular intervals. In the latter case, the arithmetic mean will not correctly represent the mean of the rhythm, but will be biased in the direction of the higher sampling density (Fig. 2). If the rhythm under study can be approximated and defined by a mathematical model, e. g., a cosine curve, a rhythm-adjusted mean or Midline Estimating Statistic Of Rhythm, a so-called MESOR (Halberg et al. 1977), may be preferable. The MESOR then represents the value midway between the highest and the lowest values of the (sinusoidal or other) function used to approximate the rhythm. In fitting of a sinusoidal model the MESOR will be equal to the arithmetic mean of the data only if the data were obtained at equal intervals over an entire cycle of the rhythm (or multiples thereof).

### Extent of a Rhythm: Amplitude

The extent of an oscillation may be expressed by its range (i. e., the difference between the maximum and the minimal value within one period) or if a rhythm can be defined by a sinusoidal mathematical model by the amplitude. The amplitude is defined as one-half the difference between the highest and the lowest point of the mathematical model (Fig. 1) and can be considerably different from the overall range of the data (Fig. 2). This difference may be due to nonsinusoidality of the variable measured (in which case the fitting of a sinusoidal model may be inappropriate) or due to outliers in the measurements in which case the amplitude of the model may be more representative of the rhythm than the apparent peak-trough difference in the data set, i. e., if the measurements are limited to a single cycle.

**Table 1.** Frequency ranges frequently encountered in biologic rhythms

| Domain; region | Range |
|----------------|-------|
| Ultradian | $t < 20$ h |
| Circadian | $20$ h $\leq t \leq 28$ h |
| Infradian | $t > 28$ h |
|   Circaseptan | $t = 7 \pm 3$ days |
|   Circadiseptan | $t = 14 \pm 3$ days |
|   Circavigintan | $t = 21 \pm 3$ days |
|   Circatrigintan | $t = 30 \pm 5$ days |
|   Circannual | $t = 1$ year $\pm 3$ months |

$t$, period

**MESOR vs Arithmetic Mean**

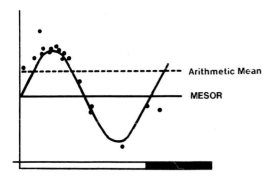

**Amplitude vs Range**

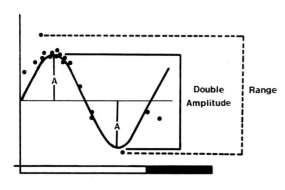

**Acrophase (∅) vs Actual Peak Time**

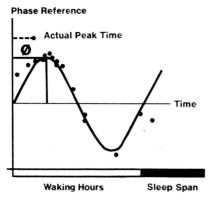

**Fig. 2.** Comparison of some rhythmometric end points. MESOR versus arithmetic mean: unequal sampling density and/or incomplete cycle(s) sampled lead to differences. Amplitude versus range: difference between double amplitude and range, may be due to shape of rhythm not fitting to the model (e.g., nonsinusoidal rhythm) or due to outliers. Acrophase versus actual peak time: differences due to same factors as for amplitude

Timing of a Rhythm: Phase

The adjustment of a rhythm in time can be expressed by the time location of its measured peak and trough. However, this approach is sensitive to outliers and the location of a peak or trough has to be confirmed by the study of multiple cycles, which then may also provide a variance estimate for these parameters. Alternatively, the mathematical-statistical determination of the phase of the rhythm by fitting an appropriate model to the data may often be preferable. The location in time of the rhythm is then defined by the highest point (acrophase) or the lowest point (bathyphase) of the fitted model (e.g., of a cosine curve) in relation to a phase reference chosen by the investigator (e.g., local midnight for circadian or January 1 for circannual rhythms) (Figs. 1, 2). The timing of the phase (e.g., the acrophase) of the rhythm in relation to the phase reference is called the phase angle and is expressed in units of time or in angular degrees (one period = 360°) in a clockwise direction as lag from zero time (0° = the phase reference). In the latter case, the phase angle is customarily expressed in negative degrees, e.g., − X°. The phase relation between two rhythms of the same period is described by the phase angle difference between corresponding phases; e.g., the acrophase of plasma cortisol at 0700 hours (− 105°) shows a phase angle difference of 4 h (60°) to the acrophase of urinary cortisol at 1100 hours. The phase angle difference has a positive sign if one rhythm leads the other in phase (its corresponding phase occurs earlier) or has a negative sign if a rhythm's phase lags behind that of the rhythm it is to be compared with; e.g., the urinary cortisol rhythm lags behind that of plasma cortisol by 4 h (− 60°). A shift in phase toward an earlier time in the cycle measured is designated as phase advance and a shift to a later time as phase delay; e.g., several days after earlier rising, the acrophase of plasma cortisol may be found at 0500 hours, a 2-h (+ 30°) phase advance or after several days of longer morning sleep may be found at 0900 hours, a 2-h (− 30°) phase delay.

Actual time series of measurements, however, are often rather complex due to the superimposition of rhythms of different frequencies and the response to environmental stimuli ("noise") (Fig. 3). The characterization of the rhythm parameters often requires mathematical statistical procedures, which may be quite different for different frequencies and different questions to be answered. Also each procedure of analysis has to be critically evaluated for its applicability to a given series of measurements and the results of single or multiple measurements have to be

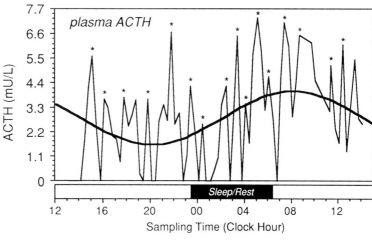

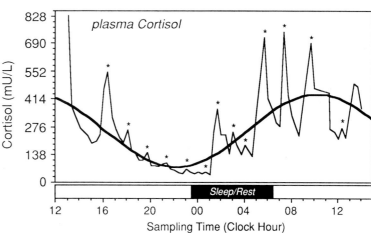

**Fig. 3.** Measurements of ACTH and cortisol in clinically healthy adult women sampled at 20-min intervals over a 24-h span. Episodic (pulsatile) secretion or ultradian rhythm superimposed upon and modulated by circadian rhythm. Increased peak height during late night and early morning hours leads to circadian variations in hormone concentration. Peaks identified by cluster program of Veldhuis and Johnson (1986). Circadian rhythm described by least squares fit of cosine function (Nelson et al. 1979). Not all peaks in ACTH secretion are followed by a comparable rise in plasma cortisol suggesting circadian and probably ultradian susceptibility cycles of the adrenal to ACTH leading to complex rhythmic interaction. The pitfalls of infrequent sampling are obvious

seen within the context of the often complex multifrequency time structure of a given variable. Curve-fitting procedures like the cosinor (Nelson et al. 1979) provide numerical end points for rhythm-adjusted mean (MESOR), amplitude and acrophase with their variance estimates, but have to be qualified because of the often less than ideal fit of the model (i.e., the "best-fitting" cosine curve in the cosinor analysis). Fig. 4 shows the chronogram of plasma cortisol, the superimposed "best-fitting" cosine curve and the polar cosinor plot often used for the graphic presentation of phase, amplitude and their confidence region.

## Endogenous or Exogenous Origin of Rhythms

Life on earth has developed in a rhythmic surrounding. The earth's rotation around its axis and around the sun, with its changes of light and darkness and warm and cold, has from earliest days impressed its timing upon living matter.

Thus, rhythmic changes observed in biology and medicine may in some instances simply be a response to the rhythmic changes in our environment. However, many of the rhythms in the circadian, and in some infradian, frequency ranges usually reflect only in part the organism's reaction to a periodic input from the environment; they often are related to genetically fixed periodic processes. These genetically determined ("endogenous") rhythms continue after removal of all periodic environmental input as self-sustaining oscillations. The recognition of an en-

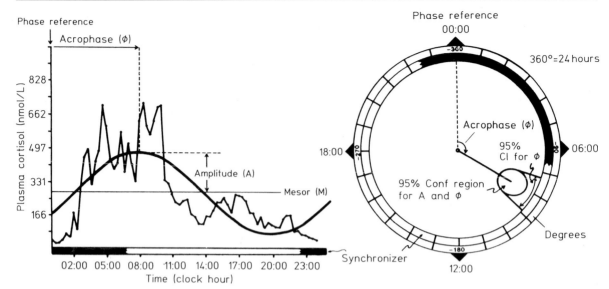

**Fig. 4.** Description of rhythm parameters by the cosinor procedure (Nelson et al. 1979). *Left,* a cosine curve is fitted to the raw data by the least squares technique and its rhythm parameters are determined, i.e., the "MESOR" (rhythm-adjusted mean), "amplitude," the distance from the MESOR to the peak (or trough) of the cosine curve best fitting to the data and "acrophase," the timing of the peak of the cosine curve best fitting to the data in relation to a phase reference chosen by the investigator, e.g., local midnight. The less than ideal sinusoidality of the data measured and the potential pitfalls of infrequent sampling are obvious. *Right,* polar cosinor plot. The period length is shown as a *full circle,* local midnight as a phase reference *on top of the circle.* The rest span is indicated in the *inner circle* of the polar display. The direction of the vector expressed in negative degrees from the phase reference indicates the acrophase and, in its length, the amplitude. The error ellipse indicates the 95 % confidence region for amplitude and acrophase

dogenous component of a rhythm requires the persistence of a rhythm under – as far as feasible – constant conditions removed from any known environmental time cues as has been shown in human subjects for circadian rhythms during isolation in natural caves (Siffre et al. 1966; Halberg et al. 1970; Reinberg et al. 1966) or other isolation facilities (Haus et al. 1968; Weitzman et al. 1979; Wever 1979). Under those conditions, the period of the rhythm usually deviates slightly but consistently from the environmental cycle to which it is normally entrained (synchronized); it thus "free runs" from the synchronizer cycle. In contrast, rhythmic variations directly imposed upon the organism by exogenous factors (e. g., environmental temperature) will disappear when the driving force has been removed.

The term "circa" (about) is used in connection with the designation of a certain frequency to convey the likelihood of such genetic programming with the potential to develop periods different from any environmental counterpart. Since rhythmic variations in biologic systems are not as precise in their period as their counterparts in physics, the circa also serves to indicate a statistical scatter, and a limitation to resolution due to the finiteness of the observation span and the variability encountered. The term is used,

therefore, in this presentation and by most (but not all) investigators in a broader sense and not limited to those rhythms in which a self-sustaining endogenous oscillation could be conclusively proven. The use of the term "circatrigintan" rather than "menstrual" for the approximate 30-day frequency range (which in sexually mature women includes the menstrual cycle) is prompted by the fact that circatrigintan changes can be found also in the male and in premenarchal and postmenopausal women.

## Synchronization (Entrainment) of Rhythms

Although many of the rhythms observed in different frequency ranges seem to be genetically determined, they are continuously modulated, modified and adjusted in time (entrained or synchronized) by periodic events in the environment.

Environmental geophysical cycles like the astronomic day/night (light/dark) cycle or our social surrounding with its activity-rest or sleep-wakefulness pattern, the exposure to bright light stimuli or the time of food uptake, or seasonally changing factors like the length of the daily light span or the environmental temperature, may induce periodic responses but in many instances rather serve as entraining agents (synchronizers) for endogenous rhythms and determine their timing (Aschoff 1978a; Halberg 1959, 1960; Pittendrigh and Daan 1976; Czeisler et al. 1985, 1986).

Entrainment of a self-sustained (endogenous) oscillation implies that an entraining oscillation (e.g., the light-dark cycle) exerts phase control over the entrained oscillation (the biologic rhythm) with a tendency of the organism to maintain a certain phase angle difference with this environmental oscillation. The rhythmic function exerting phase control over an endogenous rhythm is designated almost synonymously as "synchronizer" (Halberg et al. 1977), "entraining agent" (Pittendrigh and Daan 1976) or "zeitgeber" (Aschoff 1978a). An endogenous rhythm (representing a self-sustained oscillation) can be entrained only to frequencies that do not deviate too much from its own natural frequency. Outside this "range of entrainment" during exposure to environmental periods unacceptable to the endogenous oscillator (e.g., 8-h or 30-h light-dark cycles acting upon a circadian rhythm) the rhythm will no longer be synchronized but will be free running with its own frequency, which, however, may still be modulated by the environmental stimulus.

## Synchronizers

The effect of an entraining agent (synchronizer) upon an endogenous rhythm depends upon the stage of the rhythm when the stimulus is applied. Endogenous rhythms exhibit rhythms of sensitivity and responsiveness to entraining agents and their phase-shifting effects, e.g., pulses of bright light have different effects upon human circadian rhythms when applied at different circadian rhythm stages (Wever 1985b; Czeisler et al. 1986). A graph of those effects (phase advances or phase delays) versus the time of application of the entraining signal or other stimuli is called a "phase response curve" (Pittendrigh 1981). The phase response to different entraining agents, the length of the free running periods, and the range of periods over which entrainment can be obtained vary among individuals of the same species. Thus, the time structure of different individuals in a free living population may vary sometimes considerably from one subject to the other (Halberg et al. 1981; Lakatua et al. 1984; Nicolau et al. 1984; Touitou et al. 1981, 1982, 1983a,b, 1989; Haus et al. 1984a, 1988, 1989, 1990b).

A certain clock hour, a certain day of the week or time of the year may in a given individual not always be representative of a certain stage of its biologic time structure as compared to time-qualified reference values.

The influence and importance of different environmental synchronizers like light, social environment or time of food uptake, acting upon endogenous rhythms, will vary from parameter to parameter. Several environmental synchronizers may in their interaction determine the temporal placement of a rhythmic function in relation to its environment and to other rhythms of the body (Goetz et al. 1976; Halberg et al. 1959; Lakatua et al. 1983a,b; Nelson et al. 1975; Scheving et al. 1976), e.g., the relative length, or the timing of the daily light and dark spans, temperature variations, the main time of food uptake, the activity-rest schedule and social routine may all act upon the same circadian rhythmic parameter and in their ensemble determine its adjustment in time. Some synchronizers may be dominant in their effects upon certain rhythms, e.g., the time of food uptake determines the timing of the circadian rhythms in intestinal cell proliferation and of certain metabolic parameters but not or much less the circadian rhythms in the number of circulating lymphocytes (Lakatua et al. 1983a,b; Haus et al. 1984a,b, 1988). Bright-light stimuli, appropriately timed, may override the synchronizer effect of human social routine (Czeisler et al. 1986). Changes in one or the other of these different synchronizers may lead to changes in the temporal relation of rhythms to each other. Also, several synchronizers acting upon the same rhythmic parameter ("competing synchronizers") may lead to other than usual time relations between this and other rhythmic body functions, and their environmental synchronizers (Goetz et al. 1976; Haus et al. 1984a, 1988; Lakatua et al. 1983a,b). Under the influence of conflicting synchronizers one component of the circadian system may remain entrained, while another may become free running ("partial entrainment") (Aschoff 1978a,b). The human time structure, therefore, is not necessarily the same in subjects living under different environmental conditions and following different habits and/or work schedules. Environmental stimuli may also lead to short-term alterations of the parameters measured, lasting not much longer than the stimulus persists ("masking" of the rhythm; see below). Any measurement of a physiologic variable is characterized by the ensemble of the rhythms of many frequencies, which this function exhibits. These rhythms are at any one time subject to numerous, and sometimes competing, environmental synchronizers and masking stimuli. Thus a multitude of factors determine at any given moment the organism's stage of cycle in a parameter measured, and with it its functional state and susceptibility to environmental agents (chronesthesy).

## Free Running of Rhythms: Phase Drift

If under as far as feasible constant environmental conditions various endogenous circadian rhythms are measured simultaneously in the same clinically healthy subject, they usually remain synchronized to each other with a common period of about 24 h. However, during prolonged exposure to such conditions, the sleep-wake cycle may suddenly become substantially longer (e. g., 28–30 h or more) or shorter (e. g., less than 22 h) and desynchronize from other

rhythms which continue to free run with a period close to 24 h (Aschoff and Wever 1976; Haus 1970). Such a "spontaneous internal desynchronization" was observed by Wever (1979) in 24 % of the subjects examined and the proportion was greater for subjects older than 40 years (70 %) than for younger individuals (22 %), suggesting a role for the maintenance of internal synchronization as part of the adaptive processes which may become deficient during aging. There was also a significantly higher incidence of desynchronization in subjects with relatively high scores for neuroticism (Lund 1974), suggesting that the capability to maintain internal synchronization may also be of importance in psychology and psychiatry.

Free-running rhythms, however, are not only found under artificial experimental conditions and in the circadian frequency range, but may occur under diverse clinical conditions and in different usually en-

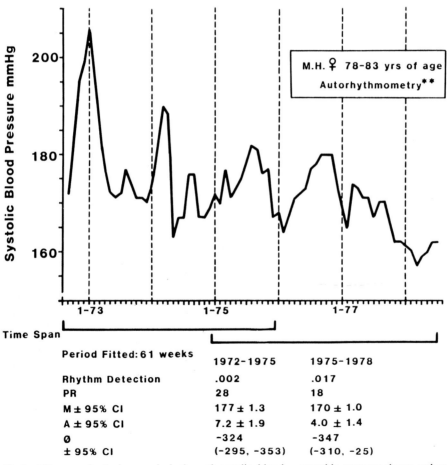

| Period Fitted: 61 weeks | 1972–1975 | 1975–1978 |
|---|---|---|
| Rhythm Detection | .002 | .017 |
| PR | 28 | 18 |
| M ± 95% CI | 177 ± 1.3 | 170 ± 1.0 |
| A ± 95% CI | 7.2 ± 1.9 | 4.0 ± 1.4 |
| Ø | -324 | -347 |
| ± 95% CI | (-295, -353) | (-310, -25) |

**Fig. 5.** "Free-running" circannual rhythm of systolic blood pressure in an elderly women measuring her blood pressure 5–7 times/day over a 7-year span. Two 3$^1/_2$-year spans show a "circannual" period of 61 weeks (analysis by least squares fits to monthly averages shown *at bottom of figure*). The circannual peak and trough in systolic blood pressure vary from one year to the other in their relation to the calendar year

vironmentally synchronized frequency ranges (Charyulu et al. 1974; Halberg et al. 1965a,b). Figure 5 shows the example of a free-running circannual rhythm in systolic blood pressure in a moderately hypertensive but otherwise healthy elderly women.

## Phase Shift

Endogenously determined but environmentally synchronized rhythms, especially in the circadian frequency domain, follow a shift of their dominant synchronizer not abruptly but over several and sometimes numerous transient cycles. If several synchronizers act upon a circadian periodic function, its rhythm will tend to follow its dominant synchronizer (e.g., the lighting regimen), although it may be modified in its temporal adjustment by secondary synchronizers (e.g., the time of food uptake). The rate of phase adaptation (re-entrainment) of a rhythm after an acute shift in the phase of the synchronizing oscillation varies greatly with (1) the nature of the rhythm

concerned, e.g., the circadian rhythm in body temperature or the rhythm in plasma corticosteroids will adapt to a sudden shift in synchronizer phase of several hours faster and with fewer transient cycles than the circadian rhythms in some excretory parameters (Haus 1970); (2) with the extent of the phase shift; and (3) with its direction. Phase adaptation after phase advance, e.g., after an eastward flight over several time zones usually (but not invariably) takes longer than after phase delay, e.g., after a westward flight. Actually slow- and fast-adapting rhythm components appear to be influenced differently by the direction of the shift as described for the rhythm in plasma cortisol when measured by frequent sampling (Van Cauter et al. 1984; Van Cauter and Honinckx (1985). (4) The speed of phase adaptation is also at least partially determined by the strength of the synchronizers (e.g., appropriately timed bright light stimuli of 3000–10000 lux or more will phase shift circadian rhythms substantially faster than social routine and/or a lighting regimen of usual indoor room light with an intensity of less than 1000 lux). For de-

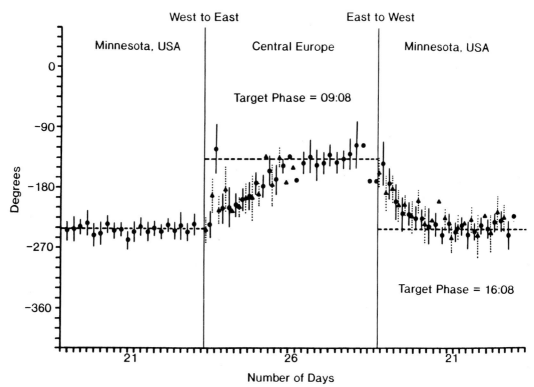

**Fig. 6.** Acrophase diagram of oral temperature summarizing shift in acrophase in pooled data of 12 transmeridian flights west-east and east-west over 7 time zones in the same subject (age 38–53 years). Acrophases determined by cosinor for non-overlapping 3-day time spans. Note time required for phase adaptation is shorter after east-west than after west-east flights (asymmetry in phase adaptation). (Haus et al. 1981). ●, acrophase ±95% CI for stay in Europe of over 11 days. ▲, acrophase ±95% CI for stay in Europe of 11 days or less

tails on observations on transmeridian flights and shift work see the chapters by Reinberg and Smolensky, and by Winget et al. During phase adaptation the usual phase relation between fast- and slow-adapting rhythms will be altered leading to an "internal desynchronization" of rhythms. The usual phase relation between rhythms is thought to be of importance for certain body functions. If a disruption of this relation occurs during a rapid phase shift forced upon the organism by a sudden change in the phase of the environmental synchronizer, a transient functional impairment may result which may become clinically manifest as "jet lag" with performance decrements, sleep disorders, gastrointestinal disorders, etc. A similar functional disorder may occur in shift workers. The phase adaptation found in the pooled data of self-measured body temperature during 12 transmeridian flights of a clinically healthy man is shown as an acrophase diagram in Fig. 6. Attempts to facilitate or accelerate phase adaptation have been made with some success in animal experiments with drugs such as benzodiazepines (Turek and Losee-Olson 1986) and in human subjects with melatonin (Arendt et al. 1986, 1987) and with bright light stimuli applied at appropriate circadian phases (Daan and Lewy 1984; Czeisler et al. 1986; Wever 1985b).

Shift workers are frequently exposed to a situation in which some synchronizer cycles are shifted (e. g., the activity pattern during working days, and the artificial light-dark cycle) while others remain unaltered (e. g., the rest-activity pattern on weekends, the time of the main meals, the natural light-dark cycle and the routines of family life). Under the effect of such "conflicting synchronizers" internal desynchronization with only partial and/or selective entrainment of some functions may occur. This may be compounded by alterations of some rhythms by environmental factors ("masking"). The findings in shift workers thus may be complex and variable and their interpretation may be difficult (see the chapter by Reinberg and Smolensky).

## "Masking" of Rhythms

Environmental stimuli may act upon a rhythmic variable and may lead to short-term alterations of its rhythm parameters, lasting not much longer than the stimulus persists ("masking" of the rhythm), e. g., a hot bath can substantially alter the body core temperature without changing the endogenous circadian rhythm of this variable. The effect of external stimuli upon the parameters of an endogenous rhythm depends not only upon their nature and intensity, but also upon the time when a stimulus is applied. Masking may affect ultradian as well as circadian, circaseptan and other infradian rhythms and can severely modify or even suppress the expression of the endogenous rhythm. In the study of human subjects living freely in their usual environment, the endogenous and exogenous rhythm components cannot easily be separated and the rhythm actually observed usually consists of both components. Attempts can be made to minimize the masking effects by the study design, as for circadian rhythms complete bedrest, distribution of caloric intake to numerous equally spaced small meals or intravenous feeding, isolation from outside time cues, exposure to continuous dim light and various combinations of these methods. It is obvious that experimental designs which attempt an approximation of the endogenous rhythm by removal of environmental stimuli are quite artificial and not applicable to everyday clinical practice. The values measured in the clinical examination of a patient will, usually include both the endogenous rhythm component and the masking of this rhythm by environmental factors. It is essential, therefore, to define in every instance when rhythms are measured the circumstances under which the measurements were obtained. Different study designs may lead to differences in rhythm parameters, e. g., the study of the circadian rhythm in body temperature, under sleep deprivation, which is felt to "unmask" this rhythm will lead to a more sinusoidal appearance of the temperature curve but with significantly decreased amplitude (Folkard 1989). However, at the same time sleep deprivation leads to phase alterations by masking of numerous other circadian periodic parameters (Haus et al. 1988).

A circadian rhythm can be masked by any environmental signal to which the organism is sensitive, independent of whether or not the signal entrains the rhythm when given periodically. However, many environmental stimuli may not only exert a direct effect upon the rhythmic variable under study (a masking effect), but may also act as synchronizer and adjust the timing of the endogenous rhythm. Vice versa a stimulus, which serves as synchronizer, may also exert a "masking effect" upon the same rhythm and in addition to phase adjustments may lead, e. g., to an amplification of the circadian amplitude in the exposed individual due to masking. Accordingly, the amplitude of the circadian rhythms in body temperature, blood pressure, catecholamines, corticosteroids, etc., is larger in a subject following its usual daily activity than in the same subject during bedrest. Whenever a synchronizer also exerts masking effects, it is difficult to decide where "masking" ends and where synchronization begins.

The human environment in our so-called civilized society is an artificial one – with respect to lighting, food availability, heating and/or air conditioning, activity and rest spans, and social activity and noise. Since human subjects can modify their environment, masking effects can be self selected by individuals rather than imposed upon them from outside, and thus may vary in the same subject during different times and/or may vary from one subject to the other.

The degree of masking of circadian rhythms by external loads varies for different circadian periodic functions. For example, catecholamines respond rapidly and strongly to arousal, psychologic and physical loads (Akerstedt et al. 1983), while the response of, e.g., plasma cortisol is considerably slower in appearance and less in extent. Certain circadian rhythms are affected directly and strongly by sleep, e.g., sleep promotes the secretion of growth hormone while some (e.g., cortisol) are affected very little and others (e.g., prolactin) to an intermediate degree (Krieger 1979; Parker et al. 1987).

Pineal melatonin may play a role in the translation of environmental changes in lighting to circadian rhythmicity. Although an endogenous circadian pineal rhythm can be demonstrated, melatonin production is strongly and predominantly masked negatively by natural or by bright artificial light (Lewy et al. 1980). By measuring the onset of melatonin production during bedrest under dim light conditions, Lewy and Sack (1989) reported being able to assess the true circadian phase position of this endogenous rhythm and to obtain phase information on related rhythms of an individual's circadian system.

It has been tried to estimate the exogenous or "masking" component of a circadian rhythm and then subtract it from the observed rhythm "to reveal" the endogenous component (Wever 1979, 1985 a; Folkard 1989). Regression models for the estimation of circadian rhythms in the presence of masking have been developed (Spencer 1989). The interpretation of the results, however, is often difficult and the separation of evoked effects from the endogenous rhythm is often uncertain.

# Mechanisms of Rhythmicity: "Biologic Clocks"

Circadian rhythms represent an ubiquitous regulating mechanism found in all eukaryotic cells and in multicellular organisms including man. In the search of mechanisms for the origin and regulation of these rhythms the idea of a "biologic clock" has been developed (Hastings 1959, 1986; Bünning 1964; Aschoff 1965 a; Richter 1965; Menaker 1969, 1982; Edmunds 1987). The investigation of the endogenous nature of this "clock" mechanism led to diverse hypotheses and theories postulating a population of interacting oscillators synchronized by structures acting as pacemakers which are subject to exogenous phase information. Alternatively, theories were developed assuming one or two central oscillators.

## Genetic Origin of Biologic Rhythms

The genetic origin of circadian rhythm characteristics was shown for the first time in plants. In the bean plant *Phaseolus*, Bünning (1935) found the circadian rhythm of stem and leaf movements to be different between two genetically distinct stalks exhibiting periodicities of about 23 and 27 h. Hybridization experiments produced plants with hybrid circadian rhythm characteristics and an intermediate period of about 25 h. These studies showed that biologic rhythm characteristics are transmitted from generation to generation according to genetic rules. The hereditary and genetically transmitted nature of circadian rhythms was also obvious in different strains of *Drosophila* (Konopka and Benzer 1971; Rensing 1973), in mice (Bean 1988; Possidente and Stephan 1988) and in hamsters (Ralph and Menaker 1988) and was supported by twin studies in human subjects (Reinberg et al. 1985; Hanson et al. 1984).

Studies in the unicellular alga *Acetabularia* indicated that the nucleus was essential for the maintenance of the circadian rhythm and for the position of its phase in the 24-h cycle. Cyclic protein synthesis, however, continues to oscillate also in anucleate cells (Hartwig et al. 1986). Schweiger and Schweiger (1977) proposed on the basis of these observations a "coupled translation membrane model" which envisions rhythmicity as the result of periodic synthesis of a polypeptide in the cytosol which inhibits its own synthesis by a feedback regulation.

Studies in *Drosophila* provide some insight into the genetic mechanisms involved in circadian rhythmicity. In *D. melanogaster* a chromosome segment, the *per* region has been identified and mapped on the X chromosome which is involved in the control of the expression of the circadian rhythm (Konopka and Benzer 1971). Three mutants have been isolated, one of which leads to a longer period than 24 h, the other to a shorter period, and the third to an aperiodic behavior of the animals. From the *per* region Bargiello and Young (1984) isolated a 90-kb pair DNA from which a 7.1-kb fragment, affecting the circadian rhythms, was obtained. When injected into embryos of arrhythmic *Drosophila*, this fragment was capable of restoring both the circadian rhythm in activity (an individual rhythm) and in eclosion (a population rhythm) (Bargiello et al. 1984). The gene product has been shown to be a proteoglycan (Jackson et al. 1986; Reddy et al. 1986). It appears that the *per* mutations by way of their gene product modulate the intercellular junctional communication (Bargiello et al. 1987; Young et al. 1988) and direct the synchronization between cells. The synchronization of the apparently individually cycling cells leads to the grossly manifest circadian rhythms with a higher amplitude the better the synchronization.

The mutations of *per* appear to have a fundamental effect on rhythmicity in *Drosophila* beyond the circadian frequency range, since they also affect ultradian behavioral rhythms (Kyriacou and Hall 1980).

In the study of the basic mechanisms of circadian rhythmicity it became obvious that a single protein cannot alone be responsible for circadian periodicity (Vanden Driessche 1989 a, b). Even in a single cell more than one oscillator may produce independent periodic activities and dissociation of some of these oscillations can be obtained experimentally. Dissociation between circadian rhythms of different parameters was found in the unicellulars *Acetabularia* (Schweiger et al. 1986; Vanden Driessche et al. 1988 a, b) and *Euglena* (Lonergan 1986) and some periodic behavior remains in *Drosophila* in the *per°* mutant (Weitzel and Rensing 1981). These observations as well as the internal desynchronization of circadian rhythms found in multicellulars, including man, do not allow the hypothesis of a single "master clock." The concept, which on the basis of these observations seems to evolve, is the existence in the cell as well as in the multicellular organism of interconnected oscillators, some of which are sensitive to information provided by environmental factors and may act as pacemakers for numerous endogenous rhythms.

## Pacemakers

Pacemakers are primary oscillators, which exhibit a genetically determined endogenous self-sustained oscillation in the absence of external time cues and which provide timing signals to the organism that synchronize a multitude of rhythms in the same frequency range.

In most mammalian species, including man, the dominant pacemaker for many of the circadian rhythms is the paired suprachiasmatic nuclei (SCN) located in the anterior hypothalamus immediately dorsal to the optic chiasm. Circadian oscillation intrinsic to the SCN has been demonstrated (Inouye and Kawamura 1982), even in individual neurons (Green and Gillette 1982), and the period is apparently genetically determined (Ralph et al. 1990). Total bilateral destruction of the SCN does disrupt several but not all circadian rhythms.

Ralph et al. (1990) studied the pacemaker role of the SCN by neural transplantation in a mutant strain of hamster that shows a short (20-h) circadian period. Small neural grafts from the suprachiasmatic region restored circadian activity rhythms to arrhythmic animals whose own nucleus had been ablated. The restored rhythms always exhibited the period of the donor genotype regardless of the direction of the transplant or genotype of the host. These studies indicate that the basic period of the overt circadian rhythm in the animal's activity is determined by cells of the suprachiasmatic region.

Pacemakers are amenable to receive environmental timing information and may adjust the timing of their oscillation accordingly.

Environmental phase information is provided to the SCN from specialized retinal photoreceptors via a normosynaptic retinohypothalamic tract (Moore 1973; Johnson et al. 1988) and a tract originating in the retinorecipient area of the lateral geniculate nuclei and in parts of the ventral-lateral geniculate nuclei (Card and Moore 1982; Harrington et al. 1987; Pickard 1987). Neural transmitters appear to be involved in conveying photic information to the SCN (Card and Moore 1982; Harrington et al. 1987; Rusak and Bina 1990; Shibata et al. 1986) and alterations in c-*fos* proto-oncogene expression in SCN neurons was reported in response to retinal illumination, but only at times in the circadian cycle when light is capable of influencing entrainment. The phase-setting capability of light stimulation is independent from vision since phase resetting can be accomplished even after induction of behavioral blindness by bilateral transection of the primary optic tracts (Rusak 1979).

Beyond the retina, the part of the optic system regulating the circadian pacemaker appears to be distinct from the visual system and it appears that the nonvisual photoreceptive system may serve the entrainment process for circadian rhythms in mammals (Takahashi et al. 1984). In mammals, the SCN is at this time the only structure that has been clearly identified as a central circadian pacemaker (Turek 1985). Extensive experimentation with SCN-lesioned animals, however, and the observation of internal desynchronization in animals and in man strongly suggest the existence of other circadian pacemakers than the SCN.

By way of neural or neurohormonal mediation, the time information is transmitted from the pacemaker to secondary oscillators (e.g., from the SCN to other hypothalamic centers and/or the pituitary gland). The secondary oscillators are capable of self-sustained oscillations but are adjusted in their timing (synchronized) by the pacemaker. The secondary oscillators and their mediators (hormonal or others) may then either synchronize tertiary oscillators, e.g., in peripheral tissues or impress rhythmic variations in systems that are not capable of self-sustained oscillations. Under some circumstances, the secondary oscillator, if isolated from the superimposed control, may show a different free-running period than the pacemaker.

A model of a circadian oscillating and time-keeping system is shown in Fig. 7.

In the mammalian organism, phase information can be provided in different ways (e.g., light, social routine, time of food uptake) and not all circadian rhythms follow the same pacemaker. Some investigators have postulated two interacting pacemakers as governing the human circadian system. Studies in isolation of human subjects from outside time cues have shown a desynchronization among circadian rhythmic functions, one group of which follows the sleep-wake cycle (including among others plasma growth hormone, subjective and objective alertness and cognitive performance) and another group coupled with the circadian rhythm in body core temperature (which includes, e.g., plasma cortisol, manual dexterity) (Aschoff 1965a,b; Wever 1979; Czeisler et al. 1980; Kronauer et al. 1982; Borbely 1982).

Pacemakers for other than circadian endogenous rhythms, i.e., in the circaseptan, circatrigintan and circannual frequency range also have to be postulated, but have thus far not been identified.

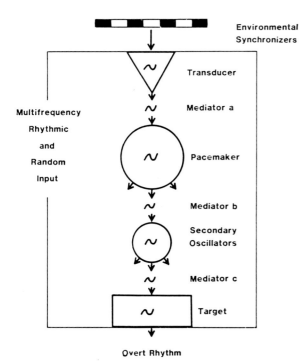

**Fig. 7.** Simplified diagram of circadian timing mechanism: self-sustained genetically determined circadian oscillations characterize the circadian pacemaker (e.g., the suprachiasmatic nucleus) but may also be found in transducers (e.g., the retina), secondary oscillators (e.g., pituitary and/or adrenal) and in target tissues (e.g., mesenchymal tissues, lymphocytes). The environmental circadian synchronizer acting upon the transducer provides phase information to the pacemaker. This time information may be modified by multifrequency rhythmic interactions and by random or nonrandom environmental input, potentially leading to alterations and/or masking of the rhythm measured (e.g., plasma cortisol, rhythm in cell division)

## Circadian Rhythms

Rhythms of the length of about 1 day, the so-called circadian rhythms are ubiquitous phenomena in living matter and are found at all levels of organization from subcellular particles to the mammalian including human organism as a whole.

The name circadian (*circa dian* = about 1 day) is used rather than 24-h to indicate the statistical nature of the biologic rhythm, which is continuously adjusted to its surroundings by external synchronizers, and to indicate its capability to free run with a frequency slightly but consistently different from 24 h. The name "diurnal" is not recommended for the designation of rhythms of this frequency since this term is commonly used in contrast to "nocturnal", which may lead to confusion.

A large number of these rhythms are genetically

fixed and continue free running with a period slightly different from 24 h under as far as feasible constant environmental conditions. Circadian rhythms are usually synchronized with our 24-h periodic surrounding by environmental synchronizers and follow a sudden synchronizer shift (e.g., through a transmeridian flight) slowly over several transient (slightly different from 24-h) cycles. The circadian rhythms are most extensively studied and for many physiologic variables show high amplitudes.

Although a large portion of the material presented in this book will be centered around circadian rhythmicity, it has to be kept in mind that circadian rhythms are only one frequency in the complex multifrequency human time structure.

## Ultradian Rhythms

Ultradian rhythms are found in biologic systems as ubiquitous phenomena over a wide range of frequencies ranging from milliseconds to a few hours. Ultradian rhythms include rhythms in the EEG, the heart and respiration rate, rhythms in blood pressure, the alteration between rapid eye movement (REM) and non-REM sleep and a multitude of others, a discussion of which is far beyond the scope of this review. Many of these are clearly unrelated in frequency, causal mechanisms and functional significance. Others may be related to the ultradian rhythms of other variables in the same frequency range suggesting some common oscillators (Pavlou et al. 1990). Some ultradian rhythms, e.g., in the secretion and the plasma concentration of certain polypeptide and steroid hormones, are less regular and less reproducible than the circadian rhythms of the same variables, and may be regarded as pseudoperiodic, pulsatile or episodic. Although random distribution of the secretory peaks of these hormones is frequently assumed, closer observation and appropriate means of statistical analysis (Veldhuis and Johnson 1986) often reveal patterns with repetitive (i.e., rhythmic) changes in frequency and amplitude over the 24-h scale. Interactions between the circadian system and certain ultradian rhythms are observed in the endocrine system. In some hormonal parameters, the circadian oscillator seems to modulate both the magnitude and the frequency of the hormonal pulses (see Fig. 3) (Van Cauter and Honinckx 1985).

The repetition of secretory episodes in the endocrine system varies from less than 1 to over 4 h. The frequency, amplitude and distribution over the 24-h span are characteristic for each hormone. In some variables, the secretory pulses vary in amplitude and frequency as a function of the circadian phase. In human subjects, luteinizing hormone (LH) shows the fastest oscillations with secretory episodes at about hourly intervals (Veldhuis et al. 1984; Reame et al. 1984; see the chapter by Blomquist and Holt) with little circadian modulation in the adult (Touitou et al. 1981, 1983 a,b) but apparently some circadian modulation in children, which allows the recognition of a circadian rhythm of this parameter in the younger age group (Haus et al. 1988). The average number of pulses reported over a 24-h span was around 15 for ACTH, 12 for prolactin, 9 for TSH, and 4 for growth hormone (Van Cauter and Honinckx 1985; see the chapter by Veldhuis et al.). The temporal placement of those pulses over the 24-h span varies with the variable with, e.g., the growth hormone pulses occurring shortly after sleep onset, the highest TSH pulses during the evening and night hours, but apparently unrelated to sleep and the largest and most frequent ACTH and cortisol pulses during the late night and early morning hours (Desir et al. 1986; Wehrenberg et al. 1982; Pohl and Knobil 1982; Sassin et al. 1969). The physiologic effect of some hormones upon their target organs was shown to depend critically upon their pulsatile rather than continuous secretion or therapeutic dosing, e.g., in LH. But also in several others such as TSH and insulin, their action appears to be augmented by this form of secretion, which seems to be optimal in eliciting the appropriate physiologic response (Matthews et al. 1983; Bratusch-Marrain et al. 1986; Pohl and Knobil 1982). There is substantial evidence for a central nervous system origin of hormonal pulsatility, although relatively little is known at this time concerning the single or multiple nature of the pacemakers involved, their interaction, feedback regulation and modulation by other rhythms of the same or of related parameters. The topic of endocrine pulsatility is discussed in more detail in the chapters by Veldhuis et al., Angeli et al. (on adrenal periodicity), Nicolau and Haus (on thyroid periodicity), and Blomquist and Holt (on gonadal axis periodicity).

## Infradian Rhythms

### Circaseptan Rhythms

Rhythms of about 7-day duration (circaseptan) are widespread in nature and are found in unicellulars (Schweiger et al. 1986; Cornelissen et al. 1986), in insects (Hayes et al. 1985; Marques et al. 1987), in rodents (Uezono et al. 1987; Sanchez de la Pena et al. 1984) as well as in human subjects (Halberg et al. 1965 a, b; Haus et al. 1984 a, 1988; Uezono et al. 1984). Circaseptan variations are regular features of human body functions. They can be found in single subjects studied over prolonged timespans (Halberg et al. 1965 a, b) and are reproducible in groups of subjects studied years apart (Haus et al. 1981, 1984 a) (Fig. 8). If multiple parameters are studied in the same subject, each parameter seems to have its own characteristic acrophase and phase relation to other circaseptan rhythms. It is likely that some circaseptan variations are induced by our societal habits. However, circaseptan rhythms have been found in forms of life other than human and under schedules designed to eliminate any circaseptan environmental input (Uezono et al. 1987). Free-running circaseptan rhythms have been described for urinary volume and 17-ketosteroid excretion in a man while living on his usual weekly work-rest schedule (Halberg et al. 1965 a, b) and have been found for blood pressure and heart rate in human subjects kept for several months, under as far as feasible constant environ-

mental conditions, in temporal isolation, on a self-selected schedule (Halberg 1989). Circaseptan phase shifts have been described after transmeridian flights with a new circaseptan phase location, e.g., of diastolic blood pressure found for spans as long as 1 year (Halberg et al. 1988). These observations suggest that some circaseptan rhythms are the expression of innate rhythmic body functions similar to many circadian rhythms.

The mammalian organism shows overt circaseptan periodicity in a wide variety of functions ranging from biochemical variables (Haus et al. 1988) to the occurrence of sudden death (Halberg et al. 1984; Nicolau et al. 1991) and seems to be endowed with a circaseptan response pattern to environmental stimuli (Gehlken et al. 1961; Hildebrandt et al. 1980; Hübner 1967; Levi and Halberg 1982; Uezono et al. 1987; Weigle 1975). In the organism "at rest" circaseptan bioperiodicity usually shows a small amplitude. After stimulation which may be physical, chemical or antigenic, the organism often reacts with a marked infradian, i.e., circaseptan, periodicity resembling single stimulus induction. The single stimulus does, of course, not contain any circaseptan information but either elicits or synchronizes a circaseptan response of the organism or amplifies a preexisting circaseptan variation so that it becomes large enough to be detected in a chronogram or by appropriate rhythmometric analyses. The responses to such stimulation are not always in exactly the 6- to 7-day range, but may vary depending upon the oscillating physiologic system, the type of stimulation, its severity and pos-

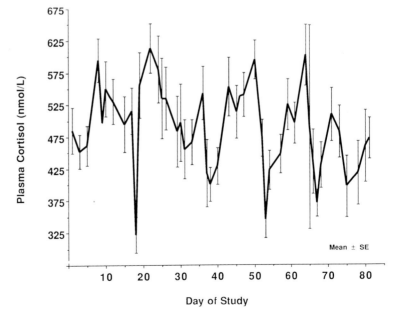

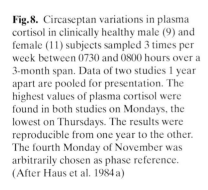

**Fig. 8.** Circaseptan variations in plasma cortisol in clinically healthy male (9) and female (11) subjects sampled 3 times per week between 0730 and 0800 hours over a 3-month span. Data of two studies 1 year apart are pooled for presentation. The highest values of plasma cortisol were found in both studies on Mondays, the lowest on Thursdays. The results were reproducible from one year to the other. The fourth Monday of November was arbitrarily chosen as phase reference. (After Haus et al. 1984 a)

sibly its timing (Romball and Weigle 1973; Levi and Halberg 1982).

Circaseptan rhythms in sensitivity and resistance have been observed in experimental animals and in human subjects (Derer 1956, 1960) and may play a role in cancer chemotherapy (Bixby et al. 1979; Levi et al. 1981). A circaseptan rhythm in an experimental tumor was shown by Moore et al. (1979).

Circaseptan and infradian rhythms of similar or related frequencies have been described in immunology for the development of humoral (Romball and Weigle 1973; Weigle 1975) and for cellular immunity (Barrett and Hansen 1957; Denham et al. 1970). The periods of these immunologic rhythms vary in the infradian range between 5 and 14 days. Most frequently, however, approximately 7- to 8-day periods are found. Also the cyclic variations in immune response do not depend upon the calendar week but take their apparent "origin" from the time of the introduction of the antigen which is also manifest in experimental (Levi et al. 1981; Ratte et al. 1977) and human transplantation biology. The incidence of rejection of renal, cardiac and pancreatic allografts in the untreated rat model follows a circaseptan bioperiodicity which appears to be triggered by the time of introduction of the experimental allograft, and, if delayed by partial immunosuppression, tends to occur at multiples of circaseptan periods. Similar observations have been made in human subjects (DeVecchi et al. 1978, 1979; Levi et al. 1981). The possibility of inducing circaseptan periodicity by environmental factors, like exposure to, e. g., high-salt diet (Uezono et al. 1987) or other forms of stimulation (Hildebrandt and Nunhöfer 1977) are of interest and may have practical application in chronopharmacology.

A circaseptan response to physical therapy was reported by Hildebrandt (1972) and Hildebrandt et al. (1980) with apparent individual and group synchronization due to the physical therapeutic exposure. Also the appearance of odontogenic suppurative processes in the course of physical therapy was found to follow a circaseptan periodicity with peaks around the 7th, 14th, 21st and 28th day after the arrival of the patients at the treatment centers (Tollman and Heller 1984; Hildebrandt et al. 1980).

Circaseptan chronopathology has been described for sudden cardiac death in adults with peak mortality on Mondays (Rabkin et al. 1980; Nicolau et al. 1991) and in sudden infant death syndrome in older infants (3–12 months of age) with peak mortality on weekends (Murphy et al. 1986).

Circaseptan rhythms are prominent in the blood pressure of newborn babies at term (Halberg et al. 1989, 1990). Circadian and circaseptan changes in vital signs of premature babies assessed during the first 3 weeks postpartum showed a circaseptan over circadian prominence (Halberg et al. 1989, 1990). Circaseptan rhythmicity may be reset and synchronized by rhythms with lower frequency, e. g., by circatrigintan (notably menstrual) cycles (Murphy 1987; Simpson et al. 1989). The accumulating evidence on circaseptan periodicity suggests that human custom and culture "recognized" nature in adopting 7-day social periods and periods of activity and rest rather than circaseptan rhythms appearing as response to environmental stimulation. It appears likely that our societal circaseptan periodic habits, which are found in many ancient and modern civilizations, may be the adaptation to an innate biologic periodicity, which in some rhythms is adjusted in its timing and synchronized within groups of subjects by our weekly societal activity pattern.

## Circatrigintan and Circavigintan Rhythms

In the analysis of time series of biologic functions obtained over several months duration, rhythms are found with a period of about 20 days (circavigintan) and about 30 days (circatrigintan) (Halberg et al. 1965b; Halberg and Reinberg 1967; Reinberg and Smolensky 1974; Haus et al. 1981, 1984a, 1988). The latter include in sexually mature women during the reproductive age most prominently the menstrual cycle and related secondary rhythms some of which may show high amplitudes. However, the term "circatrigintan" is preferred for this frequency range to "menstrual" since similar but apparently low amplitude rhythms of some functions were found in postmenopausal women, premenopausal girls and men (Halberg et al. 1965b, 1980; Hansen et al. 1975; Haus et al. 1981). Circavigintan and circatrigintan rhythms were found, among others, in plasma cortisol concentration, in both men and women, and were reproducible as a group phenomenon from year to year (Haus et al. 1981, 1984a) and in urinary ketosteroid excretion in an adult man studied over a 15-year span (Halberg et al. 1965a,b). However, the amplitudes of the circaseptan, circavigintan and circatrigintan rhythms in plasma cortisol were considerably smaller than the circadian amplitudes measured in the same subjects.

Circatrigintan secondary rhythms related to the hormonal changes in the menstrual cycle have been observed and rhythmometrically analyzed for many variables (Reinberg and Smolensky 1974). The changes observed include the circadian means as well as other circadian rhythm parameters. There is a need for adequate sampling in the evaluation of such frequencies. With daily sampling limited to a single fixed clock hour changes in the circadian acrophase may lead to apparent changes in the concentration of an analyte which could be misinterpreted as indicating an infradian change in MESOR when actually only a change in timing has occurred (Reinberg and Smolensky 1974; Simpson and Halberg 1974). Circavigintan changes in hematopoietic elements have been postulated in clinically healthy subjects (Morley 1966; Morley et al. 1970) and are found in the clinical syndrome of cyclic neutropenia (see the chapters on hematologic periodicities). Also of interest are low-frequency infradian variations in the number of circulating leukemic cells which have been found in some forms of chronic granulocytic leukemia and which in some cases were rather dramatic in extent and persisted irrespective of therapy (Chikkappa et al. 1976; Gatti et al. 1973; Kennedy 1970).

Discussion of the physiology and pathophysiology of the reproductive cycle is beyond the scope of this review (see the chapter by Blomquist and Holt).

## Circannual Rhythms and Seasonal Variations

Seasonal variations of numerous body functions occur in many if not in most species of free-living animals in the northern as well as the southern hemisphere and include among others migration patterns, reproduction and metabolism, and are also found in human subjects. Seasonal variations may be induced environmentally and may represent the organism's response to changes in temperature and to the relative length of the daily light and dark span and to the seasonal availability of certain food materials. For example, the seasonal changes in thyroid function show a close inverse relation with the environmental temperature (see the chapter by Nicolau and Haus). However, such changes persist in many parameters also in the controlled environment of the laboratory under as far as feasible constant conditions of temperature, lighting regimen and food uptake (Haus and Halberg 1970; Wallace 1979). Also exposure of animals for 3 weeks prior to study to prolonged (16 h) or shortened (8 h) daily light spans does not abolish or substantially change the seasonal variations in cell proliferation in numerous organs (Haus et al. 1984b). A circannual rhythm free running from the calendar year has been shown for 17-ketosteroids (Halberg et al. 1965b) and for human blood pressure (see Fig. 5) (Engel et al. 1985; Haus et al. 1981) and a circannual

rhythm established in one hemisphere was found to persist with the same timing several years after the subject changed from the southern to the northern hemisphere (Manton 1977). Therefore, it appears likely that many of the "seasonal variations" observed are based upon endogenous circannual rhythms which may be determined in their timing (synchronized) and possibly in their amplitude by environmental factors but are not directly induced by them.

For the purpose of this discussion, the term "seasonal variation" is used whenever environmental factors like light and/or temperature seem to determine the changes observed, and the term "circannual rhythm" is reserved for variations with a period of about 1 year for which an endogenous rhythm component has been shown, irrespective of whether it may be modualted and/or synchronized by the seasonal changes in the environment.

Circannual rhythms can be shown in longitudinal studies of the same subjects by repeatedly sampling over a timespan of one or several years or by the study of different subjects only once or a few times each during different seasons. To effectively characterize a seasonal variation or a circannual rhythm the interaction of rhythms of different frequencies shown by the same parameter have to be taken into account. Single sampling at a fixed clock hour throughout the year may be misleading since circannual changes in circadian amplitude or acrophase would not be recognized which could lead to misinterpretation of the results.

Seasonal variations may also be the result of a circannual rhythm in sensitivity of an organ or of an organ system to environmental stimuli. Rhythms of the activity of hormone receptors and of the responsiveness of metabolizing structures of target tissues have been described for several, including circannual, frequency ranges (Halberg and Howard 1958; Halberg et al. 1983; Hrushesky et al. 1979; Hughes et al. 1976; Lakatua et al. 1979, 1986; Spelsberg et al. 1979; Haus et al. 1988; Sensi et al. 1984; Touitou et al. 1983a,b, 1984, 1986; Holdaway et al. 1990).

## Time-Dependent Changes in Sensitivity, Resistance and Responsiveness

The periodically changing functional state of the organism and of its subsystems leads to time-dependent differences in the handling of metabolic loads

## Lethality

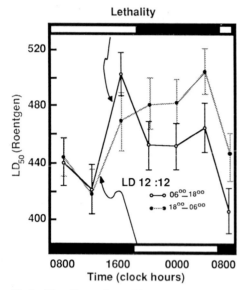

## Bone Marrow Depression

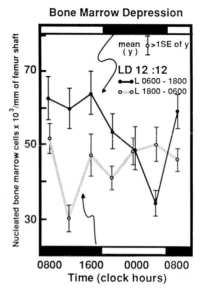

**Fig. 9.** Circadian radiosensitivity rhythm of adult male Balb/C mice after standardization for 14 days on two lighting regimens differing 180° (12 h) in phase. Radiation effect of single dose of whole-body X-irradiation in relation to circadian system stage at time of exposure. *Left,* Circadian change in $LD_{50}$; *right,* bone marrow depression (number of surviving nucleated cells in bone marrow per millimeter of femur shaft, 4 days after exposure to single dose of 350 r at different circadian system stages. (Haus et al. 1974 a, b)

and to differences in response to endogenous and exogenous stimuli of all kinds. This includes rhythmic changes in the response of secondary or tertiary oscillators to their superimposed controls (e.g., the difference in responsiveness of the adrenal to endogenous and exogenous ACTH) and applies to the metabolic handling of chemical agents introduced into the body from the outside, including drugs used in clinical medicine for the treatment of a wide variety of conditions. Depending upon the functional state of the multifrequency rhythmic systems at the time of exposure, there are temporal differences in absorption, distribution, metabolism and elimination rate of food materials, chemical agents, and drugs. Administration of drugs at different circadian times will lead to differences in the pharmacokinetic parameters (chronopharmacokinetics) and the constant infusion of many pharmacologic agents over a 24-h span does not lead to equally constant plasma concentrations.

The response and sensitivity of the organism to pharmacologic agents varies rhythmically and thus predictably in time (chronesthesy). The mechanisms involved include rhythmic differences at the receptor level, in enzyme activities or in the proliferative state of a tissue, e.g., the bone marrow. The time-dependent differences in drug handling (chronopharmacokinetics) and in drug susceptibility (chronesthesy)

lead to time-dependent differences in drug effects upon the organism (chronergy). The latter includes the rhythmic variation of desired effects (chronoefficiency) as well as the equally periodic variations of undesired side effects (chronotoxicity) and the tolerance of the organism toward potential side effects of the drug (chronotolerance). The recognition of biologic rhythms with high amplitude in drug handling and drug response have led to the development of the rapidly expanding fields of chronopharmacology and chronotoxicology.

Although susceptibility-resistance cycles can be observed in several frequency ranges, the circadian rhythms are most widely explored. By animal experimentation, statistically highly significant reproducible rhythms in responsiveness and in susceptibility or resistance were demonstrated in separate groups of experimental animals, comparable as for genetic background, history, age and sex, and kept under conditions suitable for light-synchronized periodicity analysis (Halberg 1959) when these animals were exposed to identical stimuli, e. g., at 4-h intervals along the 24-h scale. Life or death from a given toxic agent may then experimentally be made a function of the circadian system stage at the time of exposure. By the use of such procedures of indirect periodicity analysis, reproducible susceptibility-resistance cycles were ascertained in rodents to numerous potentially

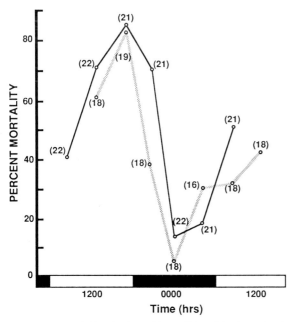

**Fig. 10.** Susceptibility rhythm of Bagg albino *(c)* mice to i. p. injection of *E. coli* endotoxin. A dose of endotoxin, which is compatible with survival in most animals if given at one time, is highly lethal when it is injected into comparable animals at a different circadian phase. (After Halberg et al. 1960; Haus et al. 1974a)

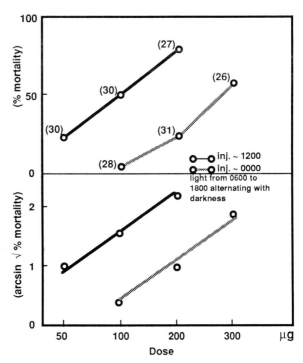

**Fig. 11.** Bioassay of the same preparation of *E. coli* endotoxin at the times of high and of low susceptibility of the animals. The striking difference in response is not due to a change in potency of the drug, but only to the circadian periodic change in sensitivity of the animals, which under standardized conditions is predictable in time and extent. (After Halberg et al. 1960; Haus et al. 1974a)

harmful agents. Some of these stimuli did exert a noxious or even lethal effect within widely different timespans. Circadian cycles of responsiveness characterize agents such as noise which acts within seconds to induce audiogenic convulsions in susceptible strains of inbred mice (Halberg et al. 1958) or agents which act within minutes, such as acetylcholine (Jones et al. 1963), pentobarbital (Davis 1962) or ouabain (Halberg and Stephens 1959); or within hours such as ethanol (Haus 1964; Haus et al. 1974a). But also the effect of ionizing radiation (Fig. 9) (Haus et al. 1974a, b, 1983) or of bacterial endotoxins (Halberg et al. 1960) which act within days (Fig. 10) and of certain chemical carcinogens, which manifest their effects only after several months, depends predictably upon the circadian system stage of the animals at the moment when a given agent is administered. Applied to drug testing in animal models, these often dramatic changes in susceptibility may lead to marked (and misleading) differences in the apparent potency of the same agent when given at different circadian phases (Fig. 11) and/or may miss the toxicity of an agent at one time and the organisms' tolerance at another. It appears imperative that chronobiologic concepts and methods be applied to drug testing in the future, which can lead to a marked improvement in the sensitivity and validity of the results. It also appears somewhat disconcerting that much of the drug-testing done without such considerations in the past may have to be qualified and for some agents probably should be redone. Since these early studies, a large amount of information has accumulated on various aspects of experimental and clinical chronopharmacology (for review see Reinberg and Smolensky 1983; Reinberg et al. 1984, 1986a, b, 1988a, b, 1990a, b; Lemmer 1989) and the chapters by Bruguerolle and by Cambar et al. The addition of the time factor adds a new dimension to pharmacology and in the application of these findings to human subjects can in many instances lead to an improvement of the therapeutic effect and/or decrease of undesired side effects.

## Chronobiologic Sampling Schedules

In chronobiologic studies, it is essential to meet the sampling requirements both in sampling density and in the length of the sampling span necessary to obtain a statistically and biologically (and, i.e., clinically) meaningful result. This aspect will be discussed in more detail in the chapter "Chronobiology in Laboratory Medicine." In the collection of data for rhythmometric evaluation, one can either follow an individual subject by frequent sampling over several cycles (longitudinal study) or a group of subjects over a single period (transverse approach). Hybrid solutions of the study of few individuals over a limited number of cycles are also appropriate for certain questions.

In the application of chronobiology to individual subjects a longitudinal approach is desirable. In such an instance the sampling density has to be adequate to provide statistically and methodologically valid information. Thus in the study of ultradian rhythms, the sampling density will have to allow the measurement and mathematical-statistical definition of an oscillation, and will have to take into account factors like the recognition of a secretory episode of a hormone, the duration of the secretory episode, and the half-life of the compound to be measured. Appropriate statistical methods of rhythm detection and quantitation should be used wherever feasible to obtain objective quantitative end points and rhythm parameters. The sampling requirements of some of these methods will vary and are discussed in the chapter by DePrins and Hecquet. High-amplitude rhythms of any variable of clinical importance require sampling at defined stages of the rhythm, or preferably measurement of its rhythm parameters. The latter may be possible with automatic monitoring devices; for example, in the assessment of blood pressure and in the treatment of hypertension, this approach is widely used and recommended (see the chapter by Cornelissen et al.).

In sampling at a "defined stage" of a high-amplitude rhythm, it has to be emphasized that a certain clock hour, a certain day of the week, or time of the year do not necessarily represent a defined stage of a rhythm, e.g., circadian periodic changes must not be misinterpreted as time of day effects. The same time of day may have an entirely different meaning for two individuals on two different living schedules. Phase shifts, phase drift and within group desynchronization have to be kept in mind in the interpretation of experimental or laboratory results. Such rhythm disturbances may occur during an ongoing study or during the course of treatment of a patient (Charyulu et al. 1974). In the study of human subjects or populations, chronobiologic sampling should be accomplished against the background of the subject's rest-activity and sleep-wakefulness patterns, working schedules and dietary habits, including the times of the main meals and possible interference by other environmental, including iatrogenic stimuli. Inquiring into some or all of these simple rhythmic biologic functions provides information on the subject's synchronization and may alert about possible rhythm alterations which may be critical if sampling has to be limited to a single or to a few time points (see the chapter "Chronobiology in Laboratory Medicine").

## Marker Rhythms

A "marker rhythm" or "reference function" is a rhythmic variable which is relatively easily measurable and characterizes the timing of the endogenous rhythmic time structure or more frequently shows a fixed phase relation to a certain rhythm to be studied, which in itself would be difficult to measure by direct means. Some rather elaborate schemes have been derived for specialized laboratories to detect the timing of the endogenous circadian time structure (Monk et al. 1985; Lewy and Sack 1989; Folkard et al. 1985). For example, the circadian rhythm in melatonin under constant dim light has been used as an indicator of an organism's biologic time (Lewy and Sack 1989). Such settings, however, have to remain limited to specialized laboratories and are not amenable to everyday clinical medicine, and may even lead to a phase estimation which may be substantially different from that observed during a subject's exposure to its everyday environment. A reference function (e.g., for the susceptibility cycle to a chemotherapeutic agent) has to be not only periodic in the same frequency and easily measurable, but has to maintain in the disease state to be treated, and during the treatment a fixed phase relation to the susceptibility cycle which needs to be monitored. Whether a potential marker rhythm meets these conditions will have to be explored and documented in each instance. After a potential reference rhythm has been found to be pertinent one will have to establish the sampling requirements to obtain statistically valid results on the parameters of this rhythm in a single subject in an ongoing and timely manner. In the circadian domain, body temperature, urinary volume, some urinary and salivary functions, notably electrolytes and corticosteroids, melatonin, sweat electrolytes, etc., have been used. Preferably a reference function should be longitudinally measurable at frequent intervals by noninvasive methods. Physical or chemical end points that do not involve expensive and/or lengthy chemical analysis and which can be recorded automatically in a computer-compatible form are particularly attractive.

The detection and characterization of pertinent reference functions may be essential for the choice of the right time for treatment to obtain optimal results.

## Role of Chronobiology in Clinical Medicine

A critical amount of information on human chronobiology has accumulated over the last 3 decades to lead to practical applications in human physiology, pathology and in the treatment of disease. The time domain has become an essential part in the analysis of any physiologic process. The recognition of the periodic nature of biologic events requires a reassessment of many earlier results obtained without regard to the human time structure. In its application to clinical medicine, chronobiology faces a number of major tasks. Among these are:

### Establishment of Chronobiologic Reference Standards

A considerable effort will still be required in the further exploration and quantitative measurement of the human time structure in subjects of different ethnic-geographic background, age and sex, and in the establishment of appropriate reference standards as new additional end points to measure the so-called "normalcy." Chronobiology adds to the observation of where a physiologic process takes place (the anatomy) and to the question of what happens in a given process (the biochemistry), the question of when this process occurs.

The application of chronobiology to clinical medicine has led to a breaking up of the "normal range" into cyclic patterns with the timing of the biologic rhythms, their amplitude and their internal and external phase relation as new (additional) end points. The usual values expected in clinically healthy subjects have to be defined now in terms of the human time

structure, taking into account the timing, amplitude and phase relations of rhythmic variables wherever feasible in several frequencies.

Extensive mapping to establish the usual ranges for rhythm parameters in different frequency ranges and their time relations is presently underway and preliminary information providing some chronobiologic reference values has been published (Haus et al. 1984a, 1988, 1990b; Haus and Halberg 1980; Pocock et al. 1989; Kanabrocki et al. 1987, 1990; Scheving et al. 1977; Touitou et al. 1979, 1986, 1989). Much additional work, however, still has to be done, which requires worldwide cooperation between numerous laboratories with standardization of techniques of sampling and measurement, and establishment of a generally accessible computerized data bank. This task is compounded by the difficulty of the determination of a "health" status of the reference population, especially, in view of our present knowledge on the genetic determination of "risk states" and their early manifestation in the form of subtle changes in the human time structure.

### Detection of Earliest Deviations in the Organism's Time Structure: Detection of Risk States

A promising task for the application of chronobiology to clinical medicine is the detection of earliest deviations from conventional and chronobiologic reference standards, early changes in time structure and recognition of risk states for common diseases. Recognition of earliest changes in biologic functions by chronobiologic investigation may occur in instances where non-time-qualified measurements do not allow the recognition of abnormality. In its earliest stages this approach aims at the risk assessment for the development of disease. Assuming that early changes in the human time structure may precede open disease and may be related to genetically determined risk states to develop certain common diseases, it appears of interest to compare and correlate ultradian, circadian and circannual rhythm parameters with epidemiologically assessed risk states.

Such correlations were reported between the risk of developing breast cancer and the circannual amplitudes of prolactin and TSH (Halberg et al. 1981; Hermida et al. 1984). In these studies, the circannual amplitude of plasma prolactin was found to be lower, the higher the familiar risk of developing breast cancer while the circannual amplitude of plasma TSH did correlate positively with breast cancer risk. A rhythm alteration was found in breast-specific skin temperature in relation to breast cancer risk (Simpson et al. 1989). The menstrual cycle related temperature rhythm, present in the low-risk controls, was not detected in the high-risk cases; instead a circaseptan rhythm was found in many high-risk patients and appears to characterize the high-risk population as a whole.

Alteration in circadian rhythm parameters of some endocrine functions were reported for the risk to develop elevated blood pressure (Hermida and Halberg 1986; Hermida et al. 1987) and characterized the Minnesota Multiphasic Personality Inventory (MMPI) scale indicating subjects prone to drug addiction and alcoholism (Hermida et al. 1982). The problem with these studies especially those using infradian rhythms are the large sampling requirements to obtain statistically meaningful results and often the absence of adequate reference ranges. These findings were frequently obtained in groups of human subjects and the sampling requirements to obtain an estimate of, e.g., the circannual amplitude of TSH or prolactin can seldom be met in an individual patient.

Prominent among risk state assessment is the recognition of transient risk states for cardiovascular and cerebrovascular disease (Master and Jaffe 1952; Muller et al. 1985; Nicolau et al. 1991; Johansson et al. 1990; Marshall 1977; Ramirez-Lassepas et al. 1991). These transient risk states appear to be a function of cardiovascular factors like the circadian rhythmicity of blood pressure variations, sympathetic activation in the early morning hours and changes in coagulation parameters favoring hypercoagulability at certain circadian phases but not at others (Tofler et al. 1987; Conchonnet et al. 1990; Haus et al. 1990a; see the chapter by Decousus).

An increased incidence of premature ventricular beats combined with a phase shift in their circadian rhythm in a 24-h electrocardiographic record obtained following a myocardial infarction characterizes patients who are likely to suffer sudden cardiac death within 5 years as compared with those who survive. The groups can be separated by considering the features of the circadian rhythm but not when only the average 24-h incidence of premature ventricular beats is used (Orth-Gomer et al. 1986).

The circadian amplitude of blood pressure during pregnancy correlates positively with high statistical significance with the cardiovascular risks which are determined on the basis of a questionnaire inquiring about the personal and familial history of high blood pressure and cardiovascular disease (Cornelissen et al. 1989).

In clinically healthy full-term newborns studied during the 1st week of life, ultradian and circadian amplitudes separate on a group basis newborns at low or at high risk of developing high blood pressure later in life (Halberg et al. 1986, 1988). The same applies to 9-year-old children (Scarpelli et al. 1986), and in 14-year-old children the circadian amplitude of the diastolic blood pressure correlates positively with the interventricular septal thickness determined by M-mode echocardiography (Halberg et al. 1988; Anderson et al. 1989). Risk state assessment for high blood pressure is discussed in detail in the chapter by Cornelissen et al. Recognition of changes in the human time structure related to permanent or temporary risk states carries with it the possibility of preventive action by timed intervention and/or by attempts of adjusting the risk-related temporal abnormalities.

## Detection of Temporal Abnormalities as Cause or Consequence of Disease (Chronopathology)

The recognition of temporal abnormalities as cause or consequences of disease represents a challenge and an opportunity for clinical chronobiology. The importance of the human time structure for normal function has been shown in several frequency ranges, e.g., the dependence of normal reproductive function on the pulsatile nature of the secretion of LH has been widely shown (Knobil 1980) and has become common knowledge for every reproductive physiologist and physician concerned with (male and female) fertility problems and will be discussed in detail in the chapter by Blomquist and Holt.

Changes of the temporal parameters of biologic functions related to disease can either represent an abnormality occurring as a result of clinical disease or alternatively the alteration of an organism's time structure can be involved in the etiology and/or pathogenesis of the pathologic state. A very substantial amount of literature on rhythm alterations in disease has appeared over the past few decades (Haus et al. 1984a). The recent developments in chronopathology of heart disease, blood pressure, gastrointestinal and renal disease, pulmonary disease, i.e., bronchial asthma, endocrine disorders, immunology, malignant growth, affective disorders, etc., are presented in the appropriate chapters of this book and go beyond the scope of this introductory review.

## Work Hygiene (Work Schedules, Shift Work, Transmeridian and Space Flight)

Chronobiology is essential in the design of work schedules for the ever increasing numbers of night and shift workers, and subjects exposed to transmeridian flights and during spaceflight and prolonged stay in extraterrestrial space. Maintenance of optimal performance and with it health and work safety require familiarity with chronobiologic principles and the application of chronobiologic schedules (see the chapter by Reinberg and Smolensky).

## Time-Dependent Therapeutic Intervention: Chronopharmacology

Chronobiology provides the capability of therapeutic intervention at a time when this intervention is useful and best tolerated and avoidance when it is not. The chronobiologic approach to treatment is especially critical when potentially damaging or toxic agents have to be used as is the case, e.g., in cancer chemotherapy (see the chapter by Hrushesky). But far beyond this application, the time factor has to be introduced in just about all aspects of clinical pharmacology and many "time-honored" customs like "three times a day" and "four times a day" medications will be replaced by more meaningful, and often more effective and less toxic, chronobiologic treatment schedules. The choice of the "right time," however, will require chronobiologic knowledge and experience since treatment at the "wrong time" can be potentially harmful (Haus and Halberg 1978; see the chapter by Haus).

## Chronotherapy

Some pharmacologic agents including hormones like corticosteroids or melatonin, drugs like benzodiazepines or stimuli like bright light, appear to influence the phase setting of some, i.e., circadian rhythms (chronopharmacologic agents). The use of such agents may be of interest if temporal changes are documented as causative or participating factors in human organic disease or functional disturbances (including shift work intolerance, jet lag, etc.) or if a certain synchronization is desired for maximal performance at a certain time (e.g., during space-flight) or for maximal resistance (e.g., to chemotherapeutic agents).

Developments in diagnoses and treatment may evolve from the combination of several emerging technologies like the availability of portable long-term ambulatory monitors for physical and biochemical monitoring of biologic functions, together with the availability of procedures to rapidly analyze biologic rhythms arrived at by such a series of measurements with rapid turnaround of the analysis in time to devise optimal dosage time patterns for specific individuals.

Automatic measurement and recording and possibly even telemetering of some biochemical or physical reference functions will be of crucial importance, and will in the long run greatly reduce the cost of chronobiologic studies. The practical application of human chronobiology to environmental and work physiology and to clinical medicine will depend critically on the availability of automated instrumentation for the study of low-amplitude rhythms and of marker rhythms. Chronopharmacology will be greatly aided by the development of portable programmable devices or other means of timed drug delivery to take advantage of these rhythmic variations.

## Conclusion

The human time structure is a basic fact of our existence, no matter if one wants to study it or not. The time-dependent, mostly rhythmic, and thus to a certain degree predictable, variations of physiologic functions and of sensitivity and resistance to many environmental agents are often quite large and offer not only new insight into human physiology and pathology but also diagnostic possibilities and therapeutic advantages. Chronobiology and its subspecialties, like chronopharmacology, will certainly play an important role in the clinical medicine of the future. Successful application of chronobiology to clinical medicine, however, depends critically on a thorough knowledge of its basic principles.

## References

Akerstedt TB, Gillberg M, Hjemdahl P. Sigurdson K, Gustavsson I, Daleskog M, Pollare T (1983) Comparison of urinary and plasma catecholamine responses to mental stress. Acta Physiol Scand 117 (1): 19–26

Anderson S, Cornelissen G, Halberg F, Scarpelli PT, Cagnoni S, Germano G, Livi R, Scarpelli L, Cagnoni M, Holte JE (1989) Age effects upon the harmonic structure of human blood pressure in clinical health. In: IEEE (ed) Proceedings of the 2nd annual IEEE symposium on computer-based medical systems, June 26–27, Minneapolis. Computer Society, Washington, pp 238–243

Arendt J, Aldhous M, Marks V (1986) Alleviation of jet-lag by melatonin: preliminary results of controlled double-blind trial. Br Med J 292: 1170

Arendt J, Aldhous M, Marks M, Folkard S, English J, Marks V, Arendt J (1987) Some effects of jet-lag and its treatment by melatonin. Ergonomics 30: 1379–1393

Aschoff J (1965a) Cell division rhythms and the circadian clock. In: Aschoff J (ed) Circadian clocks. North-Holland, Amsterdam, pp 125–138

Aschoff J (1965b) Circadian rhythms in man. Science 148: 1427–1432

Aschoff J (1978a) Circadian rhythms within and outside their ranges of entrainment. In: Assenmacher I, Farner DS (eds) Environmental endocrinology. Springer, Berlin Heidelberg New York, pp 172–181

Aschoff J (1978b) Features of circadian rhythms relevant for the design of shift schedules. Ergonomics 21: 739–754

Aschoff J, Wever R (1976) Human circadian rhythms: a multioscillator system. Fed Proc 35: 2326–2332

Bargiello TA, Young MW (1984) Molecular genetics of a biological clock in *Drosophila*. Proc Natl Acad Sci USA 81: 2142–2146

Bargiello TA, Jackson FR, Young MW (1984) Restoration of circadian behavioural rhythms by gene transfer in *Drosophila*. Nature 312: 752–754

Bargiello TA, Saez L, Baylies MK, Gasic G, Young MW, Spray DC (1987) The *Drosophila* clock gene *per* affects intracellular junctional communication. Nature 328: 686–691

Barrett MK, Hansen WH (1957) Undulations in the time-response curve for tumor immunity after primary immunization with wash erythrocytes. JNCI 18: 57–63

Bean J (1988) Polygenic correlates of activity rhythm in mammals: analysis of two inbred mouse strains C57BL/6By and BALB/cBy. C R Acad Sci [III] 307: 37–40

Bixby EK, Levi F, Haus R, Sackett LL, Haus E, Halberg F, Hrushesky W (1979) Circadian aspects of cisplatin (CP) in urine nephrotoxicity (Abstr). Chronobiologia 6: 80

Borbely AA (1982) A two process model of sleep regulation. Hum Neurobiol 1: 195–204

Bratusch-Marrain PR, Komjati M, Waldhause WK (1986) Efficacy of pulsatile versus continuous insulin administration on hepatic glucose production and glucose utilization in type I diabetic humans. Diabetes 35: 922–926

Bünning E (1935) Zur Kenntnis der erblichen Tagesperiodizität bei den Primärblättern von *Phaseolus multiflorus*. Jahrb Wiss Bot 81: 411–418

Bünning E (1964) The physiological clock: endogenesis, diurnal rhythms, and biological chronometry. Academic, New York

Bünning E (1973) The physiological clock: circadian rhythms and biological chronometry, 3rd edn. Springer, Berlin Heidelberg New York

Card JP, Moore RY (1982) Ventral lateral geniculate nucleus efferents to the rat suprachiasmatic nucleus exhibit avian pancreatic polypeptide-like immunoreactivity. J Comp Neurol 206: 390–396

Charyulu K, Halberg F, Reeker E, Haus E, Buchwald H (1974) Autorhythmometry in relation to chemotherapy: case report

as tentative feasibility check. In: Scheving LE, Halberg F, Pauly JE (eds) Chronobiology. Igaku-Shoin, Tokyo, pp 265–272

Chikkappa G, Borner G, Burlington H, Chanana AD, Cronkite EP, Ohl S, Pavelec M, Robertson JS (1976) Periodic oscillation of blood leukocytes, platelets or reticulocytes in a patient with chronic myelocytic leukemia. Blood 47: 1023–1030

Conchonnet P, Decousus H, Boissier C, Perpoint B, Raynaud J, Mismetti P, Tardy B, Queneau P (1990) Morning hypercoagulability in man. Annu Rev Chronopharmacol 7: 165–168

Conroy RTWL, Mills JN (1970) Human circadian rhythms. Churchill, London

Cornelissen G, Broda H, Halberg F (1986) Does *Gonyaulax polyedra* measure a week? Cell Biophys 8: 69–85

Cornelissen G, Kopher R, Brat P, Rigatuso J, Work B, Eggen D, Einzig S, Vernier R, Halberg F (1989) Chronobiologic ambulatory cardiovascular monitoring during pregnancy in Group Health of Minnesota. In: IEEE (ed) Proceedings of the 2nd annual IEEE symposium on computer-based medical systems, June 26–27, Minneapolis. Computer Society, Washington, pp 226–237

Czeisler CA, Weitzman ED, Moore-Ede MC, Zimmerman JC, Knauer RS (1980) Human sleep: its duration and organization depend on its circadian phase. Science 210: 1264–1267

Czeisler CA, Brown EN, Ronda JM, Kronauer RE, Richardson GS, Freitag WO (1985) A clinical method to assess the endogenous circadian phase (ECP) of the deep circadian oscillator in man. Sleep Res 14: 295

Czeisler CA, Allan JS, Strogatz SH, Ronda JM, Sanchez R, Rios CD, Freitag WO, Richardson GS, Kronauer RE (1986) Bright light resets the human circadian pacemaker independent of the timing of the sleep-wake cycle. Science 233: 667–671

Daan S, Lewy AJ (1984) Scheduled exposure to daylight. A potential strategy to reduce "jet lag" following transmeridian flight. Psychopharmacol Bull 20: 566–568

Davis WM (1962) Day-night periodicity in pentobarbital response of mice and the influence of socio-psychological conditions. Experientia 18: 235–237

Denham S, Grant CK, Hall JG, Alexander P (1970) The occurrence of two types of cytotoxic lymphoid cells in mice immunized with allogeneic tumor cells. Transplantation 9: 366–382

Derer L (1956) Concealed macroperiodicity in the reactions of the human organism. Rev Czech Med 2: 277–287

Derer L (1960) Rhythm and proliferation with special reference to the 6-day rhythms of blood leukocyte count. Neoplasma 7: 117–133

Desir D, van Cauter E, Beyloos M, Bosson D, Golstein J, Copinschi G (1986) Prolonged pulsatile administration of ovine corticotropin-releasing hormone in normal man. J Clin Endocrinol Metab 63: 1292–1299

DeVecchi A, Halberg F, Sothern RB, Cantaluppi A, Ponticelli C (1978) Circaseptan rhythmic aspects of rejection in treated patients with kidney transplants (Abstr). Proc int symp clinic chronopharmacol, chronotherapeutics and chronopharmacy, Tallahassee, pp 23–25

DeVecchi A, Carandente F, Fryd DS, Halberg F, Sutherland DE, Howard RJ, Simmons RL, Najarian JS (1979) Circaseptan (about 7-day) rhythms in human kidney allograft rejection in different geographic locations. Chronopharmacol Adv Biosci 19: 193–202

Edmunds LN Jr (1987) Cellular and molecular basis of biological clocks. Springer, Berlin Heidelberg New York

Engel R, Sothern RB, Halberg F (1985) Circadian and infradian aspects of blood pressure in a treated elderly mesorhypertensive physician (Abstr). Chronobiologia 12: 243

Folkard S (1989) The pragmatic approach to masking. Chronobiol Int 6: 55–64

Folkard S, Hume KI, Minors DS, Waterhouse JM, Watson FL (1985) Independence of the circadian rhythm in alertness from the sleep/wake cycle. Nature 313: 678–679

Gatti RA, Robinson WA, Deinard AS, Nesbit M, McCullough JJ, Ballow M, Good RA (1973) Cyclic leukocytosis in chronic myelogenous leukemia: new perspectives on pathogenesis and therapy. Blood 41: 771–782

Gehlken K, Hildebrandt G, Franke M (1961) Psychophysical correlations in the course of treatment. Arch Phys Ther 13: 171–175

⤬ Goetz F, Bishop J, Halberg F, Sothern R, Brunning R, Senske B, Greenberg B, Minors D, Stoney P, Smith ID, Rosen GD, Cressey D, Haus E, Apfelbaum M (1976) Timing of single daily meal influences relations among human circadian rhythms in urinary cyclic AMP and hemic glucagon, insulin and iron. Experientia 32: 1081–1084

Green DJ, Gillette R (1982) Circadian rhythm of firing rate recorded for single cells in the rat suprachiasmatic brain slice. Brain Res 245: 198–200

Halberg E, Halberg J, Halberg F, Halberg F (1980) Infradian rhythms in oral temperature before human menarche. In: Dan A, Graham E, Beecher C (eds) Menstrual cycle: a synthesis of interdisciplinary research. Springer, Berlin Heidelberg New York, pp 99–106

Halberg F (1959) Physiologic 24-hour periodicity; general and procedural considerations with reference to the adrenal cycle. Z Vitam Horm Fermentforsch 10: 225–296

Halberg F (1960) Temporal coordination of physiologic function. Cold Spring Harbor Symp Quant Biol 25: 289–310

Halberg F (1989) Chronobiologic engineering. In: Ensminger WD, Selam JL (eds) Infusion systems in medicine. Futura, Mount Kisco, pp 263–297

Halberg F, Howard RB (1958) 24-hour periodicity and experimental medicine. Example and interpretations. Postgrad Med 24: 349–358

Halberg F, Panofsky HJ (1961) I. Thermo-variance spectra; method and clinical illustrations. Exp Med Surg 19: 284–309

Halberg F, Reinberg A (1967) Rythmes circadiens et rythmes de basses fréquences en physiologie humaine. J Physiol (Paris) [Suppl] 59: 117–200

Halberg F, Stephens AN (1959) Susceptibility to ouabain and physiologic circadian periodicity. Proc Minn Acad Sci 27: 139–143

Halberg F, Jacobson E, Wadsworth G, Bittner JJ (1958) Audiogenic abnormality spectra, 24-hour periodicity and lighting. Science 128: 657–658

Halberg F, Halberg E, Barnum CP, Bittner JJ (1959) Physiologic 24-hour periodicity in human beings and mice, the lighting regimen and daily routine. In: Withrow RB (ed) Photoperiodism and related phenomena in plants and animals. American Association for the Advances in Science, Washington, pp 803–878 (Publication no 55)

Halberg F, Johnson EA, Brown BW (1960) Susceptibility rhythm to E. coli endotoxin and bioassay. Proc Soc Exp Biol Med 103: 142–144

Halberg F, Engeli M, Hamburger C (1965 a) The 17-ketosteroid excretion of a healthy man on weekdays and weekends. Exp Med Surg 23: 61–69

Halberg F, Engeli M, Hamburger C, Hillman D (1965 b) Spectral resolution of low-frequency, small amplitude rhythms in excreted 17-ketosteroids; probable androgen-induced circaseptan desynchronization. Acta Endocrinol [Suppl] (Copenh) 103: 5–54

Halberg F, Reinberg A, Haus E, Ghata J, Siffre M (1970) Human biological rhythms during and after several months of isolation in underground natural caves. Natl Speleol Soc Bull 32: 89–115

Halberg F, Carandente F, Cornelissen G, Katinas GS (1977) Glossary of chronobiology. Chronobiologia [Suppl] 4: 1–189

Halberg F, Cornelissen G, Sothern RB, Wallach LA, Halberg E, Ahlgren A, Kuzel M, Radke A, Barbosa J, Goetz F, Buckley J, Mandel J, Schuman L, Haus E, Lakatua D, Sackett L, Berg H, Wendt HW, Kawasaki T, Ueno M, Uezono K, Matsuoka M, Omae T, Tarquini B, Cagnoni M, Garcia Sainz M, Perez Vega E, Wilson D, Griffiths K, Donati L, Tatti P, Vasta M, Locatelli J, Camagna A, Lauro R, Tritsch G, Wetterberg L (1981) International geographic studies of oncological interest on chronobiological variables. In: Kaiser H (ed) Neoplasms – comparative pathology of growth in animals, plants and man. Williams and Wilkins, Baltimore, pp 553–596

Halberg F, Lagoguey M, Reinberg A (1983) Human circannual rhythms over a broad spectrum of physiological processes. Int J Chronobiol 8: 225–268

Halberg F, Drayer JIM, Cornelissen G, Weber MA (1984) Cardiovascular reference data base for recognizing circadian mesor – and amplitude – hypertension in apparently healthy men. Chronobiologia 11: 275–298

Halberg F, Cornelissen G, Bingham C, Tarquini B, Mainardi G, Cagnoni M, Panero C, Scarpelli P, Romano S, Marz W, Hellbrugge T, Shinoda M, Kawabata Y (1986) Neonatal monitoring to assess risk for hypertension. Postgrad Med 79: 44–46

Halberg F, Cornelissen G, Halberg E, Halberg J, Delmore P, Bakken E, Shinoda M (1988) Chronobiology of human blood pressure, 4th edn. The sphygmochron. A Medtronic Seminar, Medtronic Corp, MN, pp 242

Halberg F, Cornelissen G, Kopher R, Choromanski L, Eggen D, Otsuka K, Bakken E, Tarquini B, Hillman DC, Demore P, Kawabata Y, Shinoda M, Vernier R, Work B, Cagnoni M, Cugini P, Ferrazzani S, Sitka U, Weinert D, Schuh J, Kato J, Kato K, Tamura K (1989) Chronobiologic blood pressure and ECG assessment by computer in obstetrics, neonatology, cardiology and family practice. In: Maeda K, Hogaki M, Nakano H (eds) Proceedings of the Proc 2nd world symposium computers in the care of the mother, fetus and newborn, Oct 23–26, Kyoto, Elsevier, Amsterdam

Halberg F, Cornelissen G, Halberg J, Bakken E, Delmore P, Wu J, Sanchez de la Pena S, Halberg E (1990) The sphygmochron for blood pressure and heart rate assessment: a chronobiologic approach. In: Miles LE, Broughton RJ (eds) Medical monitoring in the home and work environment. Raven, New York, pp 85–98

Hansen JW, Hoffman HJ, Ross GT (1975) Monthly gonadotropin cycles in premenstrual girls (Abstr). Chronobiologia [Suppl 1] 2: 27

Hanson BR, Halberg F, Tuna N, Bouchard TJ, Lykken DT, Cornelissen G, Heston LL (1984) Rhythmometry reveals heritability of circadian characteristics of heart rate of human twins reared apart. Cardiologia 29: 267–282

Harrington ME, Nance DM, Rusak B (1987) Double labeling of neuropeptide Y-immunoreactive neurons which project from the geniculate to the suprachiasmatic nuclei. Brain Res 410: 275–282

Hartwig R, Schweiger R, Schweiger HG (1986) Circadian

rhythm of the synthesis of a high molecular weight protein in anucleate cells of the green alga *Acetabularia.* Eur J Cell Biol 41: 139–141

Hastings JW (1959) Unicellular clocks. Annu Rev Microbiol 13: 297–312

Hastings JW (1986) The elusive mechanism of the circadian clock. Am Sci 74: 29–36

Haus E (1964) Periodicity in response and susceptibility to environmental stimuli. Ann NY Acad Sci 117: 281–291

Haus E (1970) Biologic aspects of a chronopathology. PhD Thesis, University of Minnesota

Haus E, Halberg F (1970) Circannual rhythm in level and timing of serum corticosterone in standardized inbred mature C-mice. Environ Res 3: 81–106

Haus E, Halberg F (1978) Cronofarmacologia della neoplasia con speciale riferimento alla leucemia. In: Bertelli A (ed) Farmacologia clinica e terapia. CG Edizioni Medico-Scientifiche, Turin, pp 29–85

Haus E, Halberg F (1980) The circadian time structure. NATO Adv Study Inst [D] 47–94

Haus E, Halberg F, Nelson W, Hillman D (1968) Shifts and drifts of phase of human circadian system following intercontinental flights and in isolation (Abstr). Fed Proc 27: 224

Haus E, Halberg F, Kuhl JFW, Lakatua DJ (1974a) Chronopharmacology in animals. Chronobiologia [Suppl 1] 1: 122–156

Haus E, Halberg F, Loken MK (1974b) Circadian susceptibility-resistance cycle of bone marrow cells to whole body x-irradiation on Balb/C mice. In: Scheving LE, Halberg F, Pauly JE (eds) Chronobiology. Igaku-Shoin, Tokyo, pp 115–122

Haus E, Cornelissen G, Halberg F (1980) Introduction to chronobiology. NATO Adv Study Inst [D] 1–32

Haus E, Sackett LL, Haus M Sr, Babb WK, Bixby EK (1981) Cardiovascular and temperature adaptation to phase shift by intercontinental flights – longitudinal observations. Adv Biosci 30: 375–390

Haus E, Lakatua D, Swoyer J, Sackett-Lundeen L (1983) Chronobiology in hematology and immunology. Am J Anat 168: 467–517

Haus E, Lakatua DJ, Sackett-Lundeen L, Swoyer J (1984a) Chronobiology in laboratory medicine. In: Reitveld WT (ed) Clinical aspects of chronobiology. Bakker, Baarn, pp 13–82

Haus E, Lakatua DJ, Sackett-Lundeen L, White M (1984b) Circannual variation of intestinal cell proliferation in $BDF_1$ male mice on three lighting regimens. Chronobiol Int 1: 185–194

Haus E, Nicolau GY, Lakatua DJ, Sackett-Lundeen L (1988) Reference values for chronopharmacology. Annu Rev Chronopharmacol 4: 333–424

Haus E, Nicolau G, Lakatua DJ, Sackett-Lundeen L, Petrescu E (1989) Circadian rhythm parameters of endocrine functions in elderly subjects during the seventh decade to the ninth decade of life. Chronobiologia 16: 331–352

Haus E, Cusulos M, Sackett-Lundeen L, Swoyer J (1990a) Circadian variations in blood coagulation parameters, alpha-antitrypsin antigen and platelet aggregation and retention in clinically healthy subjects. Chronobiol Int 7: 203–216

Haus E, Nicolau GY, Lakatua DJ, Sackett-Lundeen L, Swoyer J (1990b) Circadian rhythms in laboratory medicine. In: Fanfani M, Tarquini B (eds) Reference values and chronobiology. Arand and Brent, Florence, pp 21–33

Hayes DK, Shade L, Cornelissen G, Halberg E, Miller RW, Halberg F (1985) Chronomodulatory infradian synchroniza-

tion by placebo or ACTH 1–17 of *Musca autumnalis* mortality on shifted lighting regimens. Chronobiologia 12: 361–365

Hermida RC, Halberg F (1986) Bootstrapping and added data discriminate, at low blood pressures, neuroendocrine risk of developing mesor hypertension. Chronobiologia 13: 29–36

Hermida RC, Halberg F, del Pozo F, Haus E (1982) Toward a chronobiologic pattern of the risk of breast cancer and other diseases. Rev Esp Oncol 29: 199–267

Hermida RC, Halberg F, del Pozo F, Chavarria F (1984) Pattern discrimination and the risk to develop breast cancer. In: Haus E, Kabat H (eds) Chronobiology 1982–1983. Karger, New York, pp 399–412

Hermida RC, Bingham C, Halberg F, del Pozo F (1987) Bootstrapped potential circadian harbingers if not determinants of cardiovascular risk. Prog Clin Biol Res 227B: 571–583

Hildebrandt G (1972) Therapeutische Zeitordnung und Kurerfolg. Z Angew Bäder Klimaheilkd 19: 219

Hildebrandt G, Nunhöfer C (1977) Zur Frage der reaktiven Periodik des erythropoetischen Systems. Untersuchungen in 1200 m Höhe und nach Aderlaß. Z Phys Med [Suppl] 186

Hildebrandt G, Emde L, Geyer F, Weimann H (1980) Zur Frage der periodischen Gliederung adaptiver Prozesse. Z Phys Med 9: 90–92

Holdaway IM, Mason BH, Marshall RJ, Neave LM, Kay RG (1990) Seasonal change in the concentration of progesterone receptors in breast cancer. Cancer Res 50: 5883–5886

Hrushesky W, Teslow T, Halberg F, Kiang D, Kennedy BJ (1979) Temporal components of predictable variability along the 1-year scale in estrogen receptor concentration of primary human breast cancer (Abstr). Proc Am Soc Clin Oncol 331: 165

Hübner K (1967) Kompensatorische Hypertrophie, Wachstum, und Regeneration der Rattenniere. Ergeb Allg Pathol Anat 109: 1–80

Hughes A, Jacobson HI, Wanger RK, Jungblut PW (1976) Ovarian independent fluctuations of estradiol receptor levels in mammalian tissues. Mol Cell Endocrinol 5: 379–388

Inouye ST, Kawamura H (1982) Characteristics of a circadian pacemaker in the suprachiasmatic nucleus. J Comp Physiol 146: 153–160

Jackson FR, Bargiello TA, Yun SH, Young MW (1986) Product of the *per* locus of *Drosophila* shares homology with proteoglycans. Nature 320: 185–188

Johansson BB, Norrving B, Widner H, Wu J, Halberg F (1990) Stroke incidence: Circadian and circaseptan (about-weekly) variations in onset. In: Hayes DR, Pauly JE, Reiter RJ (eds) Chronobiology: its role in clinical medicine, general biology, and agriculture, part A. Wiley, New York, pp 427–436

Johnson RF, Morin LP, Moore RY (1988) Retinohypothalamic projections in the hamster and rat, demonstrated using cholera toxin. Brain Res 462: 301–312

Jones F, Haus E, Halberg F (1963) Murine circadian susceptibility-resistance cycle to acetylcholine. Proc Minn Acad Sci 31: 61

Kanabrocki EL, Graham R, Veatch R, Greco J, Kaplan E, Nemchausky BM, Halberg F, Sothern RB, Scheving LE, Pauly JE, Wetterberg L, Olwin J, Marks GE (1987) Circadian variations in eleven radioimmunoassay variables in the serum of clinically healthy men. Prog Clin Biol Res 227A: 317–327

Kanabrocki EL, Sothern RB, Scheving LE, Vesely DL, Tsai TH, Shelstad J, Cournoyer C, Greco J, Mermall H, Ferlin H, Nemchausky BM, Bushnell DL, Kaplan E, Kahn S, Augustine G, Holmes E, Rumbyrt J, Sturtevant RP, Sturtevant F,

Bremner F, Third JLHC, McCormick JB, Dawson S, Sackett-Lundeen L, Haus E, Halberg F, Pauly JE, Olwin JH (1990) Reference values for circadian rhythms of 98 variables in clinically healthy men in the fifth decade of life. Chronobiol Int 7: 445–461

Kennedy BJ (1970) Cyclic leukocyte oscillations in chronic myelogenous leukemia during hydroxy-urea therapy. Blood 35: 751–760

Knobil E (1980) The neuroendocrine control of the menstrual cycle. Recent Prog Horm Res 36: 53–88

Konopka RJ, Benzer S (1971) Clock mutants of *Drosophila melanogaster*. Proc Natl Acad Sci USA 68: 2112–2116

Krieger DT (1979) Rhythms in CRF, ACTH and corticosteroids. In: Krieger DT (ed) Endocrine rhythms. Raven, New York, pp 123–142

Kronauer RE, Czeisler CA, Pilato SF, Moore-Ede MC, Weitzman ED (1982) Mathematical model of the human circadian system with two interacting oscillators. Am J Physiol 242: R3–R17

Kyriacou CP, Hall JC (1980) Circadian rhythm mutations in *Drosophila melanogaster* affect short-term fluctuations in the male's courtship song. Proc Natl Acad Sci USA 77: 6729–6733

Lakatua DJ, Labrosse K, Haus E, Sackett L, Halberg F (1979) Circadian periodicity in sex hormone receptor activity in Balb/c mice. Chronobiologia 6: 267

Lakatua DJ, Haus E, Sackett-Lundeen L (1983a) Thyroid and adrenal synchronization in Swiss-Webster female mice under the effect of "competing" synchronizers. Chronobiologia 10: 137

Lakatua DJ, White M, Sackett-Lundeen LL, Haus E (1983b) Change in phase relations of circadian rhythms in cell proliferation induced by time-limited feeding in Balb/c x D8A/2F$_1$ mice bearing a transplantable Harding-Passey tumor. Cancer Res 43: 4068–4072

Lakatua DJ, Nicolau GY, Bogdan C, Petrescu E, Sackett-Lundeen L, Irvine P, Haus E (1984) Circadian endocrine time structure in clinically healthy human subjects above 80 years of age. J Gerontol 6: 648–654

Lakatua DJ, Haus E, Labrosse K, Veit C, Sackett-Lundeen L (1986) Circadian rhythm in mammary cytoplasmic estrogen receptor content of Balb/c female mice with and without pituitary isografts. Chronobiol Int 3: 213–219

Lemmer B (ed) (1989) Chronopharmacology. Dekker, New York

Levi F, Halberg F (1982) Circaseptan (about 7-day) bioperiodicity – spontaneous and reactive – and the search for pacemakers. Ric Clin Lab 12: 323–370

Levi F, Halberg F, Nesbit M, Haus E, Levine H (1981) Chrono-oncology. In: Kaiser HE (ed) Neoplasms – comparative pathology of growth in animals, plants and man. Williams and Wilkins, Baltimore, pp 267–316

Lewy AJ, Sack RL (1989) The dim light melatonin onset as a marker for circadian phase position. Chronobiol Int 6: 93–102

Lewy AJ, Wehr TA, Goodwin FK, Newsome DA, Markey SP (1980) Light suppresses melatonin secretion in humans. Science 210: 1267–1269

Lonergan TA (1986) A possible second role for calmodulin in biological clock-controlled processes of *Euglena*. Plant Physiol 82: 226–229

Lund R (1974) Personality factors and desynchronization of circadian rhythms. Psychosom Med 36: 224–228

Manton WI (1977) Sources of lead in blood. Arch Environ Health 32: 149–159

Marques N, Edwards SW, Fry JC, Halberg F, Lloyd D (1987) Temperature-compensated ultradian variation in cellular protein content of Acanthamseba castellani revisited. Prog Clin Biol Res 22717: 257–264

Marshall J (1977) Diurnal variation in the occurrence of strokes. Stroke 8: 230–231

Master AM, Jaffe HL (1952) Factors in the onset of coronary occlusion and coronary insufficiency: effort, occupation, trauma, and emotion. JAMA 148: 794–798

Matthews DR, Naylor BA, Jones RG, Ward GM, Turner RC (1983) Pulsatile insulin has greater hypoglycemic effect than continuous delivery. Diabetes 32: 617–621

Menaker M (1969) Biological clocks. Bioscience 19: 681–689

Menaker M (1982) The search for principles of physiological organization in vertebrate circadian systems. In: Aschoff J, Daan S, Gross G (eds) Vertebrate circadian systems. Springer, Berlin Heidelberg New York, pp 1–12

Monk TH, Fookson JE, Moline ML, Pollak CP (1985) Diurnal variation in mood and performance in a time-isolated environment. Chronobiol 2: 185–193

Moore J, Rawley R, Hopkins HA, Ritenour RE, Looney WB (1979) Cyclophosphamide as an adjuvant to x-rays in treatment of a radio-resistant solid tumor of the rat, hepatoma H-4-11-E. Int J Radiat Oncol Biol Phys 5: 1471–1474

Moore RY (1973) Retinohypothalamic projections in mammals: a comparative study. Brain Res 49: 403–409

Morley AA (1966) A neutrophil cycle in healthy individuals. Lancet 2: 1220–1222

Morley AA, King-Smith EA, Stohlman F Jr (1970) The oscillatory nature of hemopoiesis. In: Stohlman F Jr (ed) Symposium on hemopoietic cellular proliferation. Grune and Stratton, New York, pp 3–14

Muller JE, Stone PH, Turi ZG, Rutherford JD, Czeisler CA, Parker C, Poole WK, Passamani E, Roberts R, Robertson T, Sobel BE, Willerson JT, Braunwald E, MILIS Study Group (1985) Circadian variation in the frequency of onset of acute myocardial infarction. N Engl J Med 313: 1315–1322

Murphy MF, Campbell MJ, Jones DR (1986) Increased risk of sudden infant death syndrome in older infants at weekends. Br Med J [Clin Res] 293: 364–365

Murphy N (1987) Phase relationships between female menstrual and cognitive cycles. PhD Thesis, Boston University

Nelson W, Scheving L, Halberg F (1975) Circadian rhythms in mice fed a single daily "meal" at different stages of LD (12:12) lighting regimen. J Nutr 105: 171–184

Nelson W, Tong YL, Lee JK, Halberg F (1979) Methods for cosinor rhythmometry. Chronobiologia 6: 305–323

Nicolau GY, Lakatua D, Sackett-Lundeen L, Haus E (1983) Circadian and circannual rhythms of hormonal variables in clinically healthy elderly men and women (Abstr). Chronobiologia 10: 144

Nicolau GY, Lakatua DJ, Sackett-Lundeen L, Haus E (1984) Circadian and circannual rhythms of hormonal variables in clinically healthy elderly men and women. Chronobiology Int 1: 301–319

Nicolau GY, Haus E, Popescu M, Sackett-Lundeen L, Petrescu E (1991) Circadian, weekly and seasonal variations in cardiac mortality, blood pressure and catecholamine excretion. Chronobiol Int 8: 149–159

Orth-Gomer K, Cornelissen G, Halberg F, Sothern R, Akerstedt T (1986) Relative merits of chronobiologic vs conventional monitoring of ventricular ectopic beats (VEB) In: Halberg F, Reale L, Tarquini B (eds) 2nd International conference on medico-social aspects of chronobiology, Oct 2,

1984, Florence. Istituo Italiano di Medicina Sociale, Rome, pp 767–770

Panofsky H, Halberg F (1961) II. Thermo-variance spectra; simplified computational example and other methodology. Exp Med Surg 19: 323–338

Parker DC, Rossman LG, Pekary AE, Hershman JM (1987) Effect of 64-hour sleep deprivation on the circadian waveform of thyrotropin (TSH): further evidence of sleep-related inhibition of TSH release. J Clin Endocrinol Metab 64: 157–161

Pavlou SN, Veldhuis JD, Lindner J, Souza KH, Urban RJ, Rivier JE, Vale WW, Stallard DJ (1990) Persistence of concordant luteinising hormone (LH), testosterone, and alpha-subunit pulses after LH releasing hormone antagonist administration in normal men. J Clin Endocrinol Metab 70: 1472–1478

Pickard GE (1987) Circadian rhythm of nociception in the golden hamster. Brain Res 425: 395–400

Pittendrigh CS (1981) Circadian systems. Entrainment. In: Aschoff J (ed) Biological rhythms. Plenum, New York, pp 95–124 (Handbook of behavioral neurobiology, vol 4)

Pittendrigh CS, Daan S (1976) A functional analysis of circadian pacemakers in nocturnal rodents. V. Pacemaker structure: a clock for all seasons. J Comp Physiol 106: 333

Pocock SJ, Ashby D, Shaper AG, Walker M, Broughton PMG (1989) Diurnal variations in serum biochemical and haematological measurements. J Clin Pathol 42: 172–179

Pohl CR, Knobil E (1982) The role of the central nervous system in the control of ovarian function in higher primates. Annu Rev Physiol 44: 583–593

Possidente B, Stephan FK (1988) Circadian period in mice: analysis of genetic and material contributions to inbred strain differences. Behav Genet 18: 109–117

Rabkin SW, Mathewson FAL, Tate RB (1980) Chronobiology of cardiac sudden death in men. JAMA 244: 1357–1358

Ralph MR, Menaker M (1988) A mutation of the circadian system in golden hamsters. Science 241: 1225–1227

Ralph MR, Foster RG, Davis FC, Menaker M (1990) Transplanted suprachiasmatic nucleus determines circadian period. Science 247: 975–978

Ramirez-Lassepas M, Haus E, Lakatua DJ, Sackett-Lundeen L, Swoyer J (1992) Circadian variation in time of onset of acute intracerebral hemorrhage. Stroke (in press)

Ratte J, Halberg F, Kuhl JFW, Najarian JS (1977) Circadian and circaseptan variations in rat kidney allograft rejection. In: McGovern J, Reinberg A, Smolensky MH (eds) Chronobiology in allergy and immunology. Thomas, Springfield, pp 250–257

Reame N, Sauder SE, Kelch RP, Marshall JC (1984) Pulsatile gonadotropin secretion during the human menstrual cycle. Evidence for altered frequency of gonadotropin releasing hormone secretion. J Clin Endocrinol Metab 59: 328–337

Reddy P, Jacquier AC, Abovich N, Petersen G, Rosbach M (1986) The period clock locus of Drosophila melanogaster codes for a proteoglycan. Cell 46: 53–61

Reinberg A (1974) Aspects of circannual rhythms in man. In: Pengelley ET (ed) Circannual clocks annual biologic rhythms. Academic, New York, p 423

Reinberg A, Ghata J (1964) Biological rhythms. Walker, New York

Reinberg A, Smolensky M (1974) Circatrigintan secondary rhythms related to hormonal changes in the menstrual cycle: general considerations. In: Anderson JA (ed) Biorhythms and human reproduction – seminars in human reproduction. Wiley, New York, pp 241–258

Reinberg A, Smolensky M (1983) Biological rhythms and medicine. Cellular, metabolic, physiopathologic and pharmacologic aspects. Springer, Berlin Heidelberg New York

Reinberg A, Halberg F, Ghata J, Siffre M (1966) Spectre thermique (rhythmes de la température rectale) d'une femme adulte avant, pendant et après son isolement souterrain de trois mois. C R Acad Sci [D] (Paris) 262: 782–785

Reinberg A, Smolensky M, Labrecque G (eds) (1984) Biological rhythms and medications. Pergamon, New York (Annual review of chronopharmacology, vol 1)

Reinberg A, Touitou Y, Restoin A, Migraine C, Levi F, Montagner H (1985) The genetic background of circadian and ultradian rhythm patterns of 17-hydroxycorticosteroids: a cross-twin study. J Endocrinol 105: 247–253

Reinberg A, Smolensky M, Labrecque G (eds) (1986a) Annual review of chronopharmacology, vol 2. Pergamon, New York

Reinberg A, Smolensky M, Labrecque G (eds) (1986b) Biological rhythms and medications. Pergamon, New York (Annual review of chronopharmacology, vol 3)

Reinberg A, Smolensky M, Labrecque G (eds) (1988a) Annual review of chronopharmacology, vol 4. Pergamon, New York

Reinberg A, Smolensky M, Labrecque G (eds) (1988b) Biological rhythms and medications. Pergamon, New York (Annual review of chronopharmacology, vol 5)

Reinberg A, Smolensky M, Labrecque G (eds) (1990a) Annual review of chronopharmacology, vol 6. Pergamon, New York

Reinberg A, Smolensky M, Labrecque G (eds) (1990b) Biological rhythms and medications. Pergamon, New York (Annual review of chronopharmacology, vol 7)

Rensing L (1973) Biologische Rhythmen und Regulation. Fischer, Stuttgart

Richter CP (1965) Biological clocks in medicine and psychiatry. Thomas, Springfield

Romball CG, Weigle WO (1973) A cyclical appearance of antibody-producing cells after a single injection of serum protein antigen. J Exp Med 138: 1426–1442

Rusak B (1979) Neural mechanisms for entrainment and generation of mammalian circadian rhythms. Fed Proc 38: 2589–2595

Rusak B, Bina KH (1990) Neurotransmitters in the mammalian circadian system. Annu Rev Neurosci 13: 387–401

Sanchez de la Pena S, Halberg F, Schweiger HG, Eaton J, Sheppard J (1984) Circadian temperature rhythm and circadian-circaseptan (about 7-day) aspects of murine death from malaria. Proc Soc Exp Biol Med 175: 196–204

Sassin JF, Parker DS, Mace JW (1969) Human growth hormone release: relation to slow-wave sleep and sleep-waking cycles. Science 165: 513–515

Scarpelli PT, Romano S, Cagnoni M, Livi R, Scarpelli L, Croppi E, Bigiolo F, Marz W, Halberg F (1986) Blood pressure self-measurement as part of instruction in the Regione Toscana In: Halberg F, Reale L, Tarquini B (eds) 2nd International conference on medico-social aspects of chronobiology, Oct 2, 1984, Florence. Istituto Italiano di Medicine Sociale, Rome, pp 345–366

Scheving LE, Burns E, Pauly JE, Tsai S, Halberg F (1976) Meal scheduling, cellular rhythms and the chronotherapy of cancer. Chronobiologia 3: 80

Scheving LE, Halberg F, Kanabrocki EL (1977) Circadian rhythmometry on forty two variables of thirteen presumably healthy young men. In: 12th international conference of the International Society for Chronobiology, Washington D.C., Il Ponte, Italy, pp 47–71

Schweiger HG, Schweiger M (1977) Circadian rhythms in unicellular organisms: an endeavor to explain the molecular mechanism. Int Rev Cytol 51: 315–342

Schweiger HG, Hartwig R, Schweiger M (1986) Cellular aspects of circadian rhythms. J Cell Sci [Suppl] 4: 181–200

Sensi S, Haus E, Nicolau GY, Halberg F, Lakatua DJ, del Ponte A, Guagnano MT (1984) Circannual variations of insulin secretions in clinically healthy subjects in Italy, Romania and the USA. Riv Ital Med 4: 1–8

Shibata S, Liou SY, Ueki S (1986) Influence of excitatory amino acid receptor antagonists and of baclofen on synaptic transmission in the optic nerve to the suprachiasmatic nucleus in slices of rat hypothalamus. Neuropharmacology 25: 403–409

Siffre M, Reinberg A, Halberg F, Ghata J, Perdriel G, Slind R (1966) L'isolement souterrain prolongé. Etude de deux sujets adultes sains avant, pendant et après cet isolement. Presse Med 74: 915–919

Simpson HW, Halberg EA (1974) Menstrual changes of the circadian temperature rhythm in women. In: Anderson JA (ed) Biorhythms and human reproduction. Wiley, New York, pp 549–556

Simpson HW, Paulson A, Cornelissen G (1989) The chronopathology of breast pre-cancer. Chronobiologia 16: 365–372

Spelsberg TC, Boyd PO, Halberg FC (1979) Circannual rhythms in chick oviduct progesterone receptor and nuclear acceptor. In: Reinberg A, Halberg F (eds) Chronopharmacology. Advances in the biosciences. Pergamon, New York, pp 85–88

Spencer MB (1989) Regression models for the estimation of circadian rhythms in the presence of effects due to masking. Chronobiol Int 6: 77–91

Takahashi JS, DeCoursey PJ, Bauman L, Menaker M (1984) Spectral sensitivity of a novel photoreceptive system mediating entrainment of mammalian circadian rhythms. Nature 308: 186–188

Tofler GH, Brezinski D, Schafter AI, Czeisler CA, Rutherford JD, Willich SN, Gleason RE, Williams GH, Muller JE (1987) Concurrent morning increase in platelet aggregability and the risk of myocardial infarction and sudden cardiac death. N Engl J Med 316: 1514–1518

Tollman L, Heller M (1984) In: Rettig HM (ed) Biomaterialien und Nahtmaterial. Springer, Berlin Heidelberg New York, pp 328–330

Touitou Y, Touitou C, Bogdan A, Chasselut J, Beck H, Reinberg A (1979) Circadian rhythm in blood variables of elderly subjects. In: Reinberg A, Halberg F (eds) Chronopharmacology. Pergamon, Oxford, pp 283–290

Touitou Y, Fevre M, Lagoguey M, Carayon A, Bogdan A, Reinberg A, Beck H, Cesselin F, Touitou C (1981) Age and mental health-related circadian rhythm of plasma levels of melatonin, prolactin, luteinizing hormone and follicle-stimulating hormone in man. J Endocrinol 91: 467–475

Touitou Y, Sulon J, Bogdan A, Touitou C, Reinberg A, Beck H, Sodoyez JC, van Cauwenberge H (1982) Adrenal circadian system in young and elderly human subjects: a comparative study. J Endocrinol 93: 201–210

Touitou Y, Lagoguey M, Bogdan A (1983a) Seasonal rhythms of plasma gonadotropins: their persistence in elderly men and women. J Endocrinol 96: 15–21

Touitou Y, Sulon J, Bogdan A, Reinberg A, Sodoyez JC, Demey-Ponsart E (1983b) Adrenocortical hormones, ageing and mental condition: seasonal and circadian rhythms of plasma 18-hydroxy-11-deoxycorticosterone, total and free cortisol and urinary corticosteroids. J Endocrinol 96: 53–64

Touitou Y, Fevre M, Bogdan A, Reinberg A, de Prins J, Beck H, Touitou C (1984) Patterns of plasma melatonin with ageing and mental condition: stability of nyctohemeral rhythms and differences in seasonal variations. Acta Endocrinol (Copenh) 106: 145–151

Touitou Y, Touitou C, Bogdan A, Reinberg A, Auzeby A, Beck H, Guillet P (1986) Differences in seasonal and circadian variations of total plasma proteins and blood volume as reflected by hemoglobin, hematocrit and erythrocyte counts. A transverse study in young and elderly subjects. Clin Chem 32: 801–804

Touitou Y, Touitou C, Bogdan A, Reinberg A, Motohashi Y, Auzeby A, Beck H (1989) Circadian and seasonal variations of electrolytes in aging humans. Clin Chim Acta 180: 245–254

Turek FW (1985) Circadian neural rhythms in mammals. Annu Rev Physiol 47: 49–64

Turek FW, Losee-Olson S (1986) A benzodiazepine used in the treatment of insomnia phase-shifts in the mammalian circadian clock. Nature 321: 167–168

Uezono K, Haus E, Swoyer J, Kawasaki T (1984) Circaseptan rhythms in clinically healthy subjects. In: Haus E, Kabat H (eds) Chronobiology 1981–1983. Karger, New York, pp 257–262

Uezono K, Sackett-Lundeen L, Kawasaki T, Omae T, Haus E (1987) Circaseptan rhythm in sodium and potassium excretion in salt sensitive and salt resistant Dahl rats. Prog Clin Biol Res 227A: 297–307

Van Cauter E, Honinckx E (1985) The pulsatility of pituitary hormones. In: Schulz H, Lavie P (eds) Ultradian rhythms in physiology and behavior. Springer, Berlin Heidelberg New York, pp 41–60

Van Cauter E, Desir D, Copinschi G, Refetoff S (1984) Sleep and hormone secretion: lessons from the "jet lag" study. In: Koella WP, Schulz H, Ruther E (eds) Sleep. Fischer, Stuttgart

Vanden Driessche T (1989a) Evolution des concepts généraux qui ont soustendu les recherches sur les mécanismes de la rythmicité circadienne ces vingt dernières années. Bull Group Etude Rythmes Biol 21: 59–62

Vanden Driessche T (1989b) The molecular mechanism of circadian rhythms. Arch Int Physiol Biochim 97: 1–11

Vanden Driessche T, Jerebzoff S, Jerebzoff-Quintin S (1988a) Phase-shifting effects of indole-3-acetic acid and the synchronization of three circadian metabolic rhythms in Acetabularia. J Interdiscip Cycle Res 19: 81–87

Vanden Driessche T, Lateur L, Rzemieniewski P, Guisset JL (1988b) Natural seawater may be unreliable for the culture of Acetabularia. Chronobiol Int 5: 1–3

Veldhuis JD, Johnson ML (1986) Cluster analysis: a simple, versatile and robust algorithm for endocrine pulse detection. Am J Physiol 250: E486–E496

Veldhuis JD, Evans WS, Rogel AD, Drake CR, Thorner MO, Merriam GR, Johnson ML (1984) Intensified rates of venous sampling unmask the presence of spontaneous high-frequency pulsations of luteinizing hormone in man. J Clin Endocrinol Metab 59: 96–102

Wallace ALC (1979) Variations in plasma thyroxin concentrations throughout one year in penned sheep on a uniform food intake. Aust J Biol Sci 32: 371–374

Wehrenberg WB, Brazeau P, Luben R (1982) Inhibition of the pulsatile secretion of growth hormone by monoclonal antibodies to the hypothalamic growth hormone-releasing factor. Endocrinology 111: 2147–2148

Weigle WD (1975) Cyclical production of antibody as a regulatory mechanism in the immune response. Adv Immunol 21: 87–111

Weitzel G, Rensing L (1981) Evidence for cellular circadian rhythms in isolated fluorescent dye labelled salivary glands of wild type and an arrhythmic mutant of *Drosophila melanogaster.* J Comp Physiol 143: 229–235

Weitzman ED, Czeisler CA, Moore-Ede MC (1979) Sleepwake, neuroendocrine and body temperature circadian rhythms under entrained and non-entrained (free-running) conditions in man. In: Suda M, Hayaishi O, Nakagawa H (eds) Biological rhythms and their central mechanism. Elsevier/North-Holland, Amsterdam, pp 199–227

Wever RA (1979) The circadian system of man. Results of experiments under temporal isolation. Springer, Berlin Heidelberg New York

Wever RA (1985a) Internal interactions within the circadian system: the masking effect. Experientia 41: 332–342

Wever RA (1985b) Use of bright light to treat jet lag: differential effects of normal and bright artificial light on human circadian rhythms. Ann NY Acad Sci 453: 282–304

Young MW, Bargiello TA, Baylies MK, Saez L, Spray DC, Jackson FR (1988) The molecular genetic approach to a study of biologic rhythms in *Drosophila.* Adv Biosci 73: 43–53

# Cellular and Molecular Aspects of Circadian Oscillators: Models and Mechanisms for Biological Timekeeping

L. N. Edmunds, Jr.

## Introduction

Several approaches to elucidate the nature of biological clocks, particularly circadian oscillators, have emerged over the years (Hastings and Schweiger 1976; Edmunds 1988). These include the attempt to locate the anatomical loci responsible for generating these periodicities, efforts to trace the entrainment pathway for light signals (and other zeitgebers) from the photoreceptor(s) to the clock itself, the experimental dissection of the clock using chemicals and metabolic inhibitors and employing the exciting new techniques of molecular genetics, and the characterization of the coupling pathways and the transducing mechanisms between the clock(s) and the overt rhythmicities (hands) it drives. The results obtained by these experimental lines of attack, in turn, have provided the grist for several classes of biochemical and molecular model for autonomous circadian oscillators (COs).

In the following sections, some recent experimental results obtained by each of these strategies are summarized. It should be readily apparent that what constitutes "mechanism" for one investigator is merely "descriptive phenomenology" for another working at a lower level of organization! Finally, given the didactic nature of this brief overview, no attempt has been made to furnish all of the relevant primary references; these are available in other extended monographs (see Edmunds 1988) and in the cited review articles.

This study was supported in part by grants to L. N. E. from the National Science Foundation (DCB-9105752 and DCB-8901944). It was adapted and updated, by permission, from an overview in *Trends in Chronobiology* (Hekkens WThJM, Kerkhof GA, Rietveld WJ, eds), Pergamon Press, Oxford, pp 1–17, 1988. The study is dedicated to the memory of Dr. Beatrice M. Sweeney (1914–1989) and Dr. Ruth L. Satter (1923–1989), whose recent deaths deprive the fields of biological rhythms and plant physiology of two of their most remarkable and valuable workers.

## Quest for an Anatomical Locus

One obvious approach to discovering the clock is to attempt to localize it within the organism, assuming, of course, that it has an anatomically defined locus.

## Circadian Pacemakers at the Organ and Tissue Levels

At higher levels of biological organization, circadian pacemakers have been discovered in the suprachiasmatic nuclei (SCN) of the vertebrate hypothalamus, the optic lobes or median neurosecretory cells of the brains of certain insects, the eyes of several marine gastropods (*Aplysia* spp., *Bulla* spp.), the avian pineal gland, and the pulvini of leaves (Jacklet 1984, 1989a, b; Moore and Card 1986; Satter et al. 1988; Takahashi et al. 1989). In most cases, the entire organ or tissue is not necessary for the maintenance of circadian rhythmicity, as has been revealed in tissue reduction experiments in vitro, and it may be that each cell of a pacemaker is capable of circadian outputs under appropriate conditions. Indeed, Block and McMahon (1984) have shown that the rhythm in compound action potential of the isolated *Bulla* eye persists in constant darkness (DD) after all photoreceptors have been removed and as few as six pacemaker neuron cell bodies remain. Thus, it is likely that the entire circadian pacemaker, including the photoreceptor required for light-mediated phase shifts ($\Delta\Phi$s), is contained within the circuitry of the basal retinal pacemaker neurons. Similarly, the circadian rhythm (CR) of electrical activity of SCN neurons in cultured rat brain slices continues for up to 60 h *in vitro* and can be reset by 1-h pulses of drugs that alter endogenous levels of cAMP (Prosser and Gillette 1989) or cGMP (Prosser et al. 1989). Finally, dispersed cell cultures of the chick pineal gland continue to release melatonin rhythmically for several cycles in DD (Robertson and

Takahashi 1988a) and to retain their in vivo photosensitivity (Robertson and Takahashi 1988b; Zatz et al. 1988) – unlike the mammalian pineal, in which the components of the circadian system (photoreceptor, pacemaker, and melatonin production) are anatomically distinct (located, respectively, in the retina, SCN, and pineal) but interconnected by multisynaptic neuronal pathways (Moore 1983).

Transplantation experiments underscore the autonomy of these circadian pacemakers. For example, if hypothalamic tissue containing the SCN is taken from neonatal rats and placed into the third ventricle of young adult rats that previously had received bilateral electrolytic lesions of their SCN (and whose rhythmicity of wheel-running activity thus had been disrupted), circadian rhythmicity reappeared from 2 to 8 weeks after transplantation (Sawaki et al. 1984). DeCoursey and Buggy (1988) reported similar findings in SCN-lesioned golden hamsters into the ventricles of which fetal hypothalamic tissue containing SCN anlagen had been implanted. Such grafts contained an organization of peptidergic cells similar to that of the intact SCN and appeared to establish both afferent and efferent connections with the host brain (Lehman et al. 1987). Relatively few peptidergic cells and fibers needed to be present, however, for the restoration of circadian rhythmicity although light entrainability and gonadal photoperiodic responses were lacking. Reappearance of rhythmicity in DD was correlated with the presence in the graft of neuropeptides normally present in the SCN of unlesioned hamsters (Lehman et al. 1987). Further, there is a specificity of circadian function in these transplants of the fetal SCN: neurons transplanted from the presumptive paraventricular nucleus (PVN) or other non-SCN neural tissue of rats or hamsters to SCN-lesioned hosts did not restore the CR of locomotor activity or of VP level in the cerebrospinal fluid, although transplanted precursor neurons of the SCN did (DeCoursey and Buggy 1989; Earnest et al. 1989). Finally, Ralph et al. (1990) have reported that small neural grafts from the SCN of a mutant strain of hamster that shows a short (20–22 h) free-running period under constant conditions ($\tau$) restored rhythmicity to arrhythmic animals whose own SCN had been ablated. The period of the restored rhythm always reflected that of the donor and not that of the lesioned host, indicating that the fundamental $\tau$ of the overt CR was determined by cells of the suprachiasmatic region.

## Circadian Rhythms in Isolated Cells

One can conclude from the preceding section that isolated organs and tissues can oscillate autonomously, but what of lower levels of anatomical complexity? That populations of eukaryotic microorganisms exhibit circadian organization is well known (Edmunds 1984) although there is always the formal possibility that rhythmicity is generated by a network of intercommunicating cells. Mammalian cell cultures also display CRs.

The situation for isolated microorganisms is less ambiguous: CRs have been recorded in single cells of *Acetabularia, Gonyaulax,* and *Pyrocystis,* to name a few. Recently, a rhythm of mating reactivity has been demonstrated within single cells of *Paramecium,* even in individuals taken from arrhythmic populations in which CRs had damped out after 2 weeks in dim constant light of 1000 lux (Miwa et al. 1987). Population arrhythmicity in this case appears to result from desynchronization among the constituent oscillators. A single dark pulse restored rhythmicity in a popultion, probably by its differential phase-shifting action on individual cells, whose phases were randomly distributed at the time of the signal. A dark-pulse phase-response curve (PRC) was derived, in which the $\Delta\Phi$ of the free-running rhythm engendered by the perturbation was plotted as a function of the circadian time (CT) at which it was applied (CT 0 indicates the phase point of a rhythm, which has been normalized to 24 h, that corresponds to the phase points occurring at the onset of light in an LD:12,12 reference cycle).

Although it has been an article of faith that CRs are manifested only in eukaryotic organisms [a tenet incorporated into several models for circadian clocks, such as the chronon of Ehret and Trucco (see "Biochemical and Molecular Models for Circadian Clocks" below)], this central dogma may have to yield to experimental advances. There have been recent reports of CRs in nitrogen-fixing cyanobacteria, which are unique in their ability to carry out photosynthetic $O_2$ evolution and $O_2$-labile nitrogen fixation within the same organism (Grobbelaar et al. 1986; Mitsui et al. 1986). These seemingly incompatible reactions take place in heterocystous cyanobacteria by spatial separation of the sites for the two processes. In nonheterocystous species, however, other mechanisms must be invoked. For example, in diurnal light-dark (LD) cycles, temporal separation of photosynthesis and nitrogen fixation into the light and dark intervals of growth, respectively, may occur. Mitsui et al. (1986) have found, however, that the two processes can continue in constant illumination (LL) for at least 3 days in certain species of *Synechococcus*, the growth of which had been synchronized by LD:12,12. The CR in $O_2$ evolution and nitrogenase activity were approximately 180° out of phase. Maximum activity in net $O_2$ evolution occurred during the subjective day, just before cell division, whereas peak activity of nitrogenase was found at night. If $O_2$ evolution was inhibited with diuron (DCMU), acetylene reduction was not affected, although in nonheterocystous species blockage enhances reduction (indicating that photosynthetic $O_2$ evolution has adverse effects on nitrogen fixation). Thus, a circadian oscillator has enabled *Synechococcus* to achieve a functional separation of two otherwise incompatible reactions during the cell division cycle, with the attendant implication that CRs (in this case of all three processes) can be expressed in a prokaryote. Similar findings have been reported for the filamentous, nonheterocystous cyanobacterium *Oscillatoria* (Stal and Krumbein 1987). That these prokaryotic rhythmicities are truly circadian in nature has been further supported by the recent report (Sweeney and Borgese 1989) of a clear-cut CR of cell division in *Synechococcus*, under conditions in which the generation (doubling) time *(g)* was longer than 1 day, that was LD-entrainable, that persisted for at least four cycles in LL, and that displayed a temperature-constant $\tau$.

Finally, a basic but unanswered question in the field of CRs concerns the number of clocks that might exist in a single cell or microorganism. Although the prevailing opinion for many years probably has been that there is but one central oscillator within the cell – or that it is the cell itself or most of it – and that the numerous overt rhythms observed as circadian outputs (in *Acetabularia*, *Euglena*, or *Gonyaulax*, for example) are merely different hands of this single driving entity, this dogma too is challenged by recent observations. Although simultaneous recording of several CRs in a single unicell usually reveals identical, stable $\tau$s, the phase relation among which is retained after a light or dark pulse, Schweiger et al. (1986) have noted that occasionally a dark pulse shifted only one of two CRs in an individual *Acetabularia* cell, a finding that suggests that one is dealing with at least two different clocks the phases of which are closely coupled. That an ultradian ($\tau$ ~55 m) rhythm and a circadian rhythm of the same variable can occur simultaneously in the same unicell has been demonstrated for chlorophyll *a* levels in *Chlamydomonas* during LD-synchronized growth, with the attendant hypothesis that the underlying ultradian os-

cillator may be a common element in both circadian and cell cycle timing (Jenkins et al. 1989).

## Subcellular Circadian Rhythmicity

We now turn to the possibility that circadian organization may reside at the subcellular level. Indeed, it is well known that CRs are observed routinely in isolated, enucleated *Acetabularia* cells and cell fragments. Human erythrocytes (RBCs) cultured in vitro, which constitute minimal systems in that they lack nuclei, mitochondria, ribosomes, endoplasmic reticula, and vacuoles, appear to exhibit ultradian (but not circadian) fluctuations (oscillations?) in enzyme activity (GAPD, G6PD) and $^{45}Ca^{2+}$-binding capacity (Edmunds 1988). Perhaps this is not surprising given the fact that during ontogeny RBCs have lost their capacity for protein synthesis. RBC hemolysates prepared by sonication (as opposed to hypotonically produced hemolysates) do not exhibit significant oscillations, a finding that suggests the observed rhythms may reflect a time-dependent attachment and detachment of enzyme molecules from the cell membrane and that in the bound state the enzyme may be inactive (Peleg et al. 1990).

In this regard, Radha et al. (1985) have reported that glutathione (GSH) levels in human platelets (but not in RBCs) display a CR *in vitro* in concentrates stored in LD, LL, or DD. Platelets are not entire cells, but represent small cell fragments that originate as vesicles that detach in great numbers from the outer regions of large megakaryocytes. The phasing of the CR appeared to be independent of the LD cycle and of the sleep-wake cycle of the donor. Stored platelets actively incorporated [$^{14}$C]glutamine into a small peptide. The rhythm in GSH levels was abolished by incubation with buthionine sulfoximine (an inhibitor of glutamyl cysteine synthetase), a finding that suggested that *de novo* synthesis was responsible for the periodicity. Lastly, the presence of a CR in the proportion of E-rosette-forming lymphocytes has been observed in cultured lymphocytes (Gamalega et al. 1988).

It would be most provocative if circadian periodicity could be demonstrated in an isolated subcellular component, such as an *in vitro* chloroplastic or mitochondrial preparation. In such a case, an anatomical locus for a CO would have been shown to exist at the subcellular level – perhaps not impossible if one considers these structures to have had an ancestry as an invading endosymbiont, replete with genome and membranes.

## Intercellular Communication and Coupled Cellular Oscillators

The notion of intercellular cross-talk within a population of interacting oscillators is important for circadian clock models in which the long $\tau$ derives from the coupling of ultradian oscillators (see Edmunds 1988, p.372). It is noteworthy that the product of the *per* clock gene in *Drosophila* (see "Recombinant DNA Studies: Cloning Clock Genes" below) appears to regulate the degree of intercellular communication by altering the number and organization of functional gap junctions (Bargiello et al. 1987). Direct measurement of the coupling coefficient, apparent space constant, and junctional conductance in the larval salivary glands of $per^o$, $per^+$ and $per^s$ demonstrated that the extent of coupling was correlated inversely with $\tau$, being weakest in $per^o$. These provocative results are consistent with the hypothesis (Dowse and Ringo 1987; Dowse et al. 1987) that the $per^+$ gene product mediates the coupling of multiple ultradian oscillators (which predominate in the "arrhythmic" mutant $per^o$) to produce wild-type CRs. In this scheme, $\tau$ would be a function of the coupling tightness among ultradian oscillators, increasing as coupling loosens. Ultradian rhythms would become apparent under weak coupling or in its absence. This suggestion also might explain weak or erratic rhythms of membrane potential in isolated salivary glands and larval heartbeat in the $per^o$ mutant (see Bargiello et al. 1987).

Further support for this hypothesis has been obtained by Dowse et al. (1987). These workers, using digital techniques for signal analysis, have detected weak, shorter CRs and ultradian periodicities in the locomotor activity of supposedly arrhythmic $per^o$ males and of females lacking the *per* locus ($per^-$; heterozygous for two deficiencies, each of which deletes the gene). The dominant periods ranged from 4 to 22 h, and most of the significantly rhythmic flies exhibited multiple periodicities, unlike the wild type. Thus, the $per^+$ gene product might well serve to mediate the coupling of multiple ultradian oscillators to produce wild-type CRs. Dowse and Ringo (1987) have obtained additional data that are consistent with this notion. Signal-to-noise ratios (SNRs) were computed for the *per* mutants (in order to characterize the precision of their rhythms of locomotor activity), and the following progression was obtained: *pero* (noisiest) $> per^L > per^+ > per^s$. The SNR was found to decrease as $\tau$ increased in *pers*, $per^+$ and $per^L$; $per^o$ typically had multiple ultradian periodicities and the lowest SNR. At least 70% of $per^L$ individuals also

had ultradian rhythms. Thus, ultradian rhythms, the periods of which would be inversely correlated with coupling strength, would become apparent under weak coupling, or in the absence of coupling, as Bargiello et al. (1987) indeed found to be the case by direct measurement. Finally, Dowse and Ringo (1989) have demonstrated that rearing *Drosophila* in DD produced predominantly phenocopies of the $per^o$ and $per^L$ mutants (the latter having relatively weak long-period rhythms and multiple ultradian rhythms). These authors hypothesize that exposure to light at some point in development acts to couple a population of ultradian oscillators into a composite clock, and that the disruption of the rhythms seen in DD-reared flies is functionally identical to that found in the *per* mutants.

The very slow decay of synchronization of CRs in populations of microorganisms might also be attributed to cell-cell communication, although earlier tests of this hypothesis in *Acetabularia, Euglena, Gonyaulax,* and *Neurospora* were inconclusive or negative. Nevertheless, a reexamination of the question in *Gonyaulax* (Broda et al. 1986), utilizing mixing experiments of out-of-phase cultures, has indicated an affect on $\tau$ by cell "conditioning" of the medium. It is provocative that a substance (gonyauline) that shortens the $\tau$ of the glow rhythm by as much as 4 h in a dose-dependent manner is present in extracts of *Gonyaulax* and has been isolated and characterized as a novel, low molecular weight cyclopropanecarboxylic acid (Roenneberg et al. 1991).

## Tracing the Entrainment Pathway for Light Signals

Another strategy for discovering the mechanism of a circadian clock is to delineate the pathway by which information regarding the LD cycle (and other environmental cues) is received, transduced, coded and propagated to the pacemaker.

### Nature and Localization of the Photoreceptor

As one of the first steps in the entrainment pathway, a photopigment in an organized receptor or receptor cell must absorb the photons of the light signal. A classic approach to the identification of a photoreceptor pigment is the determination of an action spectrum for a given process in the hope that it will match the absorption spectrum for some putative substance present in the cell or tissue. In her comprehensive review, Ninnemann (1979) concluded that there is no single, specific light-absorbing pigment in plants and animals that is effective for initiating, phase-shifting, or inhibiting CRs. Thus, action spectra for resetting rhyhtms in *Neurospora, Gonyaulax, Drosophila,* the moth *Pectinophora,* and the hamster *Mesocricetus* are all different. Just as no universal structure has been developed as the anatomical locus for COs (see "Quest for an Anatomical Locus" above), no universal photoreceptor or pigment has evolved. Nevertheless, some generalizations can be made (Edmunds 1988).

Lower plants appear to use blue light-absorbing photoreceptors for CRs, such as flavoproteins and carotenoids. Recently, for example, Fritz et al. (1989) have demonstrated by using riboflavin-deficient mutants that free cellular riboflavin mediates phase-shifting by light signals in *Neurospora.* Higher plants use phytochrome and, perhaps, chlorophyll as well. Less is known about the identity of the absorbing pigments for circadian systems in animals. Insects appear to use flavoproteins (or carotenoids), whereas retinal rhodopsin appears to be the clock photopigment in mammals. There is a large body of evidence also for light reception by extraretinal photoreceptors in birds and mammals, with the possibility that other pigments, such as the red protoporphyrins found in the harderian glands, play a role.

It may be that more than one photoreceptor participates in light mediation by circadian systems. Thus, blue (470 nm) and red (660 nm) light pulses had qualitatively different phase-shifting effects on the free-running phototaxis rhythm in *Chlamydomonas:* blue generated advances ($+ \Delta \Phi s$), red produced delays ($- \Delta \Phi s$). Similar differences between the effects of the two wavelengths were recorded as a function of fluence and in the measurement of PRCs (they were displaced by 24 h). Roenneberg and Hastings (1988) have made similar findings for the effects of constant blue or red light of different intensities on the rhythm of glow bioluminescence in *Gonyaulax.* For blue light, $\tau$ became shorter the higher the intensity, whereas with red light $\tau$ increased with intensity.

## Coupling Links Between Photoreceptor and Clock

The next steps in the entrainment pathway(s) that follow the primary photochemical reactions that must occur have been dissected pharmacologically in the isolated eye of the sea hare, *Aplysia californica,* and the cloudy bubble snail, *Bulla gouldiana,* in which a circadian pacemaker that is responsible for the rhythm of compound action potential (CAP) resides in one or a few of the basal retinal neurons (BRNs) of the photoreceptor layer (see Jacklet 1984, 1989 b). In the *Bulla* retina, light appears to directly depolarize the BRNs. Depolarization, whether achieved by light, direct intracellular current injection into BRNs, or elevated potassium concentration, can phase-shift the CAP rhythm, chronic changes in membrane potential lengthen $\tau$, and light-induced $\Delta \Phi s$ are blocked if the membrane potential is returned to rest by injecting hyperpolarizing current (McMahon and Block 1987 a, b). These findings have led to the conclusion that membrane depolarization is a critical step in the light-entrainment pathway.

A transmembrane calcium flux also appears to be involved in phase-shifting by depolarizing treatments and may represent the next critical step in the light-entrainment pathway (see "Role of Ions and Second Messengers: Calcium" below). Thus, reducing extracellular free $Ca^{2+}$ inhibited light-induced $\Delta \Phi s$ without blocking the photic response of *Bulla* BRNs (Khalsa and Block 1988 a; McMahon and Block 1987 a), and pulses of low-calcium EGTA solution yielded a PRC similar to that for pulses of a hyperpolarizing low potassium-low sodium solution (Khalsa and Block 1990). Calcium, therefore, may act as a second messenger transducing the effects of light on the circadian pacemaker.

How, then, does calcium act to modify the elements that generate the CR in the BRNs? Although calcium can directly affect neuronal excitability via calcium-dependent potassium channels, its long-term effects are often mediated by protein kinases, which serve to phosphorylate membrane channels and change neuronal excitability. That this may be the case for the *Bulla* system is suggested by the recent findings of Roberts et al. (1989) that the highly specific inhibition of kinase activity by isoquinoline sulfonamide (H-8) lengthens the $\tau$ of the CR in a dose-dependent manner. In addition, the phosphorylation of five proteins was markedly affected by H-8, an observation which suggests that these proteins are putative elements for regulation of the period of the pacemaker.

Recent results have implicated cyclic AMP (cAMP) in the mediation of the phase-shifting effects of serotonin in the serotonergic entrainment pathway of *Aplysia* (see Edmunds 1988; Jacklet 1989 b). Thus, cAMP either is part of the input pathway by which serotonin entrains the CO, or it (along with the cAMP generating and degrading system) is part of the clock mechanism itself (see "Role of Ions and Second Messengers: Cyclic AMP" below).

The next step of the pathway may entail protein phosphorylation through activation of a kinase by cAMP (Lotshaw and Jacklet 1987), just as was hypothesized for the *Bulla* circadian system. Indeed, there is a requirement of protein synthesis (see "Clock-Controlled Genes: Transcriptional and Translational Control" below) in the regulation of the CAP rhythm by the serotonergic pathway: serotonin pulses given at those CTs known to produce $\Delta\Phi$s significantly increased the synthesis of a specific 34-kDa protein but did not do so at those times when serotonin itself was ineffective in phase resetting (Yeung and Eskin 1987). More recently, Raju et al. (1990) have examined other proteins separated by two-dimensional polyacrylamide gel electrophoresis that may be involved in light-resetting of the *Aplysia* ocular pacemaker and have discovered three (30, 31, and 42 kDa) that were all affected similarly by light, elevated potassium, 8-bromo-cGMP, and serotonin (all of which generate comparable $\Delta\Phi$s of the CAP rhythm). These proteins, therefore, are likely candidates for elements of either the entrainment pathway or the oscillator mechanism. At least some of the newly appearing protein species are phosphorylated, as indicated by the fact that some 14 proteins showed increased incorporation of $^{32}$P when *Aplysia* eyes were exposed to serotonin pulses between CT 06 and CT 12, although light pulses (which have little or no effect on the CAP rhythm at this phase) affected only one protein (Zwartjes and Eskin 1990).

## Dissection of the Clock: Biochemical and Molecular Analysis

We now have arrived at the oscillator itself. Recent areas of emphasis center on the role of protein synthesis and "clock proteins" in the timekeeping mechanism(s), transcriptional and posttranscriptional control of clock genes, and the involvement of ions and second messengers – particularly calcium and the calcium-calmodulin complex, and cAMP – in circadian rhythmicity. (Comprehensive treatments and recent reviews may be found in: Edmunds 1988; Hall and Rosbash 1987, 1988; Hastings and Schweiger 1976; Johnson and Hastings 1986; Rosbash and Hall 1989; Vanden Driessche 1989.)

## Protein Synthesis and Clock Proteins

Results obtained from experimental perturbation of the clock with inhibitors of protein synthesis (such as cycloheximide, puromycin and anisomycin) in *Acetabularia*, *Aplysia*, *Euglena*, *Gonyaulax*, *Neurospora* and a number of other organisms have implicated the synthesis of proteins on 80S (but not 70S) ribosomes in the operation of the CO. In all systems studied, short (even as little as 1 min) inhibitor pulses induce phase-dependent shifts (both $+\Delta\Phi$s and $-\Delta\Phi$s) of CRs. Extension of these studies to vertebrates has been undertaken recently. Thus, microinjections of anisomycin directly into the SCN of free-running golden hamsters induce $\Delta\Phi$s of the activity rhythm (Inouye et al. 1988) similar to those obtained after subcutaneous injections (Takahashi and Turek 1987), and systemic administration of cycloheximide (CHX) yielded comparable PRCs (Wollnik et al. 1989).

A degree of caution must be exercised, however, in inferring from the PRC for a given drug when protein synthesis is important, due to the slow recovery times of the system from the inhibition of protein synthesis (Olesiak et al. 1987; Yeung and Eskin 1988). Further, protein synthesis per se may not be part of the oscillatory feedback loop ( = oscillator) since the clock runs relatively unimpeded in the *frq-7* mutant of *Drosophila* (see "Isolation of Clock Mutants" below) even if protein synthesis is inhibited by CHX. Rather, daily resynthesis during specific CTs of certain protein(s) with a high turnover rate probably is required for normal operation of the *Neurospora* clock (Dunlap and Feldman 1988).

But which protein species are important? [Measuring the *overall* rate of protein synthesis in the SCN, for example, may provide no evidence of a CR, suggesting that the bulk of synthesized proteins are involved in non-circadian "housekeeping" functions (Scammell et al. 1989).] Hartwig et al. (1985, 1986) have detected a single, high-molecular-weight (230 kDa) protein in the chloroplast fraction of both nucleate and anucleate *Acetabularia*, p230, that fulfills three requirements of an essential clock protein: its rate of synthesis in LL exhibited circadian rhythmicity, its synthesis was inhibited by CHX, and pulses of CHX phase-shifted the rhythm of synthesis. *In vivo* labeling of the unicellular green alga *Chlorella fusca* with [$^{35}$S]methionine also revealed the circadian ($\tau$ = 21 h) synthesis of a 41-kDa polypeptide that met these criteria (Walla et al. 1989). Similarly, the 30-, 31-, 34-, and 42-kDa proteins discussed previously ("Coupling Links Between Photoreceptor and Clock" above) may play an analogous role in the mediation of light-induced $\Delta\Phi$s of the CAP rhythm in *Aplysia* (Raju et al. 1990; Yeung and Eskin 1987), as might the phosphoproteins implicated in the light or serotonin entrainment pathway of *Bulla* (Roberts et al. 1989) or *Aplysia* (Lotshaw and Jacklet 1987). Finally, the products of the *per* gene in *Drosophila* and the *frq* locus in *Neurospora* clearly play a key role in circadian timekeeping (see "Recombinant DNA Studies: Cloning Clock Genes" below).

## Clock-Controlled Genes:
## Transcriptional and Translational Control

If circadian fluctuations in specific, essential "clock" proteins are important for circadian timekeeping, the question now arises as to how they are controlled. Gene expression may be regulated at different levels, including the abundance of the gene transcript, the amount of the translation product, and the biological activity of the protein itself. Thus, if the control is transcriptional, then the level of the relevant mRNA(s) should cycle, but if it is posttranscriptional, then the level of the transcript should remain constant but the transcript should be translated rhythmically. Finally, if oscillations in the protein or transcript, or both, are part of the clock itself, then their elimination should alter or suppress output circadian rhythmicity, but if they are not then there should be no effect on the clock in the absence of the cycling transcript or protein (assuming that the course of the oscillator could be measured independently). Recent progress has been made on this front (Edmunds 1988; Rosbash and Hall 1989).

Circadian oscillations in mRNA levels have been reported in several animal and plant systems. For example, the level of the *per* gene product of *Drosophila* (see "*Drosophila:* the *per* locus" below) undergoes a circadian oscillation, which is accompanied by an underlying cycling of *per* RNA in both LD and DD (Hardin et al. 1990). This cycling was not noted in earlier studies in DD, in part because the RNA oscillation is less dramatic than the 10-fold fluctuation in abundance observed in LD, and in part because it is obscured somewhat because the amount of another, 0.9-kb RNA transcript fluctuates more than 20-fold over the course of a day, although this latter oscillation has now been attributed to developmental gating during eclosion and not to a true cycling of transcript as a function of circadian time (Lorenz et al. 1989). Hardin et al. (1990) interpret these results to indicate that the cycling of *per*-encoded protein results from *per* RNA cycling, and that there is a

feedback loop through which the activity of the *per* product causes cycling of its own mRNA levels. This regulation could be either transcriptional or posttranscriptional. Similarly, Loros et al. (1989) have identified at least two RNA transcripts that cycle with circadian periods of 21.5 h or 29 h, respectively, in the wild-type *frq*$^+$ strain or the long-$\tau$ *frq-7* clock mutant of *Neurospora* (see "*Neurospora:* the *frq* Locus" below), and in rats, vasopressin mRNA levels show circadian rhythmicity in the SCN, but not in other hypothalamic nuclei (PVN, SON) where this transcript is also found (Reppert and Uhl 1987).

Plant systems, too, exhibit circadian cycling of RNA transcripts (see "Other Systems" below). For example, the mRNAs coding for some of the major nuclear-encoded, chloroplast gene products, such as the light-harvesting, chlorophyll *a/b* binding *(cab)* protein of tomato (Giuliano et al. 1988; Piechulla 1988, 1989), tobacco (Paulsen and Bogorad 1988), pea (Otto et al. 1988), wheat (Nagy et al. 1988a), and maize (Taylor 1989), the precursor apoprotein of the LHC-I and LHC-II light-harvesting complexes of photosystems I (Tavladoraki and Argyroudi-Akoyunoglou, 1989) and II (Tavladoraki et al. 1989) of red kidney bean leaves, and the small subunit of ribulose-1,5-biphosphate carboxylase/oxygenase of the pea (Otto et al. 1988) all have been shown to display circadian rhythmicity in DD. Likewise, the level of mRNA coding for nitrate reductase (NR) in the leaves of tobacco is light-inducible and cycles with a circadian $\tau$ in DD (Deng et al. 1990). In fact, exposure of corn seedlings to light results in a rapid ( < 3 h) pulse of NR mRNA, which then leads to an increase in NR activity, an observation that suggests that the latter is driven due to positive feedback kinetics at the nucleic acid level (Lillo 1989). Thus, one can anticipate that light regulation of gene expression in plants may act through the CO, with the mediation of phytochrome in the entrainment pathway (Nagy et al. 1988b). Nevertheless, it is important to note that in some systems (such as *Gonyaulax*) circadian control over synthesis of at least some proteins is at a translational level (see "Characterizing the Coupling Pathway" below).

## Role of Ions and Second Messengers

The fact that single cells can exhibit circadian rhythmicity simultaneously in quite different processes, such as those of photosynthesis and bioluminescence in *Gonyaulax,* suggests that membrane-bound compartmentalization is important for temporal organization. Since both of these rhythms, as well as others, are known to be affected by changes in the ionic environment and are probably membrane-bound systems, it is not surprising that transmembrane ion transport or flux has been proposed to be a key feature of the CO (see "Classes of Model for Autonomous Oscillators" below). Let us examine some current research trends along these lines of attack.

Calcium

Since intracellular free calcium acts as a "second messenger" and is well known as a cell regulator, coordinating many kinds of intracellular reactions and even its own concentration, it is quite possible that it may also play a significant role in the functioning of circadian clocks (Edmunds 1988; Edmunds and Tamponnet 1990; Techel et al. 1990). For example, the calcium ionophore A23187 is quite effective in phase-shifting the CR in compound action potential of the isolated eye of *Aplysia* (see "Coupling Links Between Photoreceptor and Clock" above). Similarly, light signals impinging on the *Bulla* eye preparation cause membrane depolarization, which can be mimicked by direct electrical stimulation of a single basal retinal neuron. Depolarization may be causing changes in the ionic flux of $Ca^{2+}$, allowing for its entry, perhaps by opening and closing $Ca^{2+}$ channels (Khalsa and Block 1988a, 1990). If the extracellular $Ca^{2+}$ concentration was reduced with the chelator EGTA, light-induced $\Delta\Phi$s of the *Bulla* ocular rhythm were blocked (McMahon and Block 1987a). In contrast, a 2-h pulse (CT 14–16) of the convulsant agent pentylenetetrazole, which is known to directly modulate $Ca^{2+}$ levels by releasing $Ca^{2+}$ from intracellular stores, generated phase delays in the *Bulla* eye comparable to those produced by light pulses (Khalsa and Block 1986). These results, therefore, indicate that $\Delta\Phi$s can be generated despite reduced extracellular $Ca^{2+}$. Mediation by a $Ca^{2+}$ mechanism has been implicated (Earnest and Sladek 1987) in the CR of vasopressin release in response to membrane polarization that has been reported for perfused rat SCN explants *in vitro*, and the concentration of $Ca^{2+}$ ion may be important (Shibata et al. 1987) for maintaining and regulating the rhythm of metabolic activity in rat hypothalamic-slice preparations. Phase-shifting by agents that cause transitory perturbations (increases) of intracellular $Ca^{2+}$ has been reported also for the CRs of cell shape in *Euglena,* leaf movement in *Trifolium* and *Cassia,* phototaxis in *Chlamydomonas,* and conidiation in *Neurospora* (see Edmunds and Tamponnet 1990). Recently, Tamponnet and Edmunds (1990) have found for the cell division rhythm in *Euglena* that the PRC for pulses of LoCa is a mirror image of that for HiCa but is virtually identical to that for light signals. It is interesting to note also that the phosphatidylinositol cycle has been hypothesized to mediate the effects of light on leaflet movement in *Samanea saman* pulvini. Hydrolysis of membrane-localized phosphoinositides, accompanied by an increase in cytosolic free $Ca^{2+}$, would provide a mechanism for phototransduction in the motor cells (Morse et al. 1987). Finally, the product of the *frq* gene may be related to calcium metabolism: in mutant strains of *Neurospora* that have a $Ca^{2+}$ dependency for growth, $\tau$ is also affected by the concentration of external $Ca^{2+}$ and even entrainability and rhythmicity itself (Nakashima 1985, personal communication).

There also is considerable evidence that oscillations in $Ca^{2+}$ levels may play an important role in the generation of ultradian rhythms. For example, Yada et al. (1986) have investigated the higher-frequency oscillations in membrane potential with repeated hyperpolarizations that occur in cultured, secretory epithelial intestine 407 (I-407) cells (as well as in fibroblastic L cells, macrophages, sympathetic neurons and hamster eggs). Periodic activation of $Ca^{2+}$-dependent $K^+$ conductance has been shown to be a common mechanism. The cytosolic $Ca^{2+}$ oscillation results from cyclic release of $Ca^{2+}$ from an intracellular storage site, which depends in turn on mitochondrial activity.

Calcium can bind with calcium-binding proteins such as calmodulin, which, in conjunction with mitochondrial $Ca^{2+}$ transport, plays a pivotal role in cellular regulation. Nakashima (1986) has examined the effects of pulses of several calmodulin antagonists on the CR of conidiation in *Neurospora*. The order of effectiveness in their phase-shifting activity was qualitatively paralleled by the efficacy of the drugs in inhibiting calmodulin-induced activation of phosphodiesterase, a finding that suggests that calmodulin antagonists cause $\Delta\Phi$s by affecting calmodulin-dependent reactions. This conclusion was buttressed by the observation that pulses of W7 or W13, both specific calmodulin antagonists, resulted in large $\Delta\Phi$s, while their ineffective, respective chlorinated analogues, W5 and W12, did not. The amount of calmodulin itself, however, extracted at different phases in DD and assayed by its stimulation of phosphodiesterase activity, showed no circadian variation, a fact which suggested that the cyclic changes in sensitivity to calmodulin antagonists might be due to variation in the synthesis and levels of calmodulin-binding proteins. Similarly, pulses of W7, TFP, and chlorpromazine generated CT-specific $+\Delta\Phi$s and $-\Delta\Phi$s in rat hypothalamic brain slices (Moore 1987, personal communication). Khalsa and Block (1988b), however, found that pulses of W7, TFP, and a third calmodulin antagonist, calmidazolium, did not block light-induced $\Delta\Phi$s of the *Bulla* ocular rhythm – results that indicate it is unlikely that calmodulin itself mediates these $\Delta\Phi$s in this system.

The foregoing experimental findings suggesting that $Ca^{2+}$, $Ca^{2+}$ transport, and calmodulin may play a key role in the clock mechanism have been incorporated into a model for a CO in what is most probably a cellular clock shop, or network (Goto et al. 1985). This regulatory scheme includes the following three steps: (1) $NAD^+$ (or a stimulated photoreceptor, such as phytochrome or a blue-light photopigment) would enhance the rate of net $Ca^{2+}$ efflux from the mitochondria (or other compartment) or net $Ca^{2+}$ influx across the plasmalemma into the cytoplasm, resulting 6 h ($90°$) later in a maximal concentration of cytosolic $Ca^{2+}$; (2) $Ca^{2+}$ would immediately form an activated $Ca^{2+}$-calmodulin complex in the cytoplasm (maximal level at $90°$); (3) this active from of $Ca^{2+}$-calmodulin would decrease the rate of net production of $NAD^+$ by both activating NAD kinase and inhibiting NADP phosphatase in the cytoplasm so that the rate would become maximal 12 h later (at $270°$) when $Ca^{2+}$-calmodulin reaches its minimum level. After 6 h more when the *in vivo* level of $NAD^+$ becomes lowest (at $0°$), the regulatory sequence would be closed, and the cycle would repeat. For the moment, the site of the sequestered calcium pool remains uncertain. Should additional experimental data demand it, other elements (for example, cAMP) could be inserted into the proposed loop and the oscillator expanded.

The model shares several aspects in common with the calcium cycle model of Kippert (1987), in which the $Ca^{2+}$ concentration gradient between cytoplasm and mitochondrial matrix alters cyclically (attributed simply to their counteracting $Ca^{2+}$ transport pathways) and to that of Lakin-Thomas (1985), based on oscillations in intracellular $Ca^{2+}$ compartmentation (see "Classes of Model for Autonomous Oscillators" below). Kippert (1987) provocatively suggests that endogenous rhythms and the calcium system of intracellular signaling coevolved as a consequence of mutual interaction among partners in a cellular symbiotic consortium developing toward the eukaryotic cell, with its defined nucleoplasm and mitochondria. Thus, the early, endogenous, self-sustaining clock would have provided a device for internal timekeeping and temporal coordination between still largely autonomous compartments (host and endosymbionts). A homeostatic, self-regulated, oscillatory regimen would have helped preserve this developing association from destabilizing, chaotic fluctuations that otherwise might have developed in the complex network of feedback mechanisms.

## Cyclic AMP

Adenosine $3',5'$-cyclic monophosphate (cAMP), which plays a pivotal role in the regulation of a number of cellular functions, could be a possible second messenger facilitating the regulation of rhythmic processes by a circadian clock(s). Oscillations in cAMP level have also been proposed to be part of the biochemical feedback loop itself that is believed to underlie circadian rhythmicity (see "Classes of Model for Autonomous Oscillators" below). Circadian variations in cellular cAMP level indeed have been observed in *Neurospora*, *Tetrahymena pyriformis*, *Acetabularia*, *Euglena*, and rat (Edmunds 1988; Techel et al. 1990).

Genetic experiments on the budding yeast *Saccharomyces*, as well as physiological studies in mammalian cells, have shown that transient changes in cAMP level are necessary for the transit of these cell types through the different phases of the cell division cycle (CDC) (Whitfield et al. 1987). Indeed, there is increasing evidence that both a transient rise and the ensuing fall in the level of cAMP are required for the initiation of DNA synthesis. A second cAMP surge is observed during $G_2$, which may be correlated with the onset of mitosis. A clock-controlled variation of cAMP level – that is, the periodic repetition of a cAMP signal – thus may participate in the "gating" of DNA synthesis and cell division to a certain phase of the circadian cycle.

This possibility has been examined for the cellular CO of *Euglena*, which has been shown to modulate the progression of cells through the different phases of their CDC (Carré et al. 1989). We have demonstrated a bimodal, circadian variation of cAMP content in the photosynthesis-deficient ZC mutant of this unicell in both LD and DD. Rhythmic changes of cAMP level, which may reflect the transition of cell cycle transit in division-phased cultures, also persisted after the culture medium had become limiting and the cells had stopped dividing. The cAMP rhythm, free-running in DD, could be phase-shifted by a light signal in a manner that could be predicted from the

PRC previously obtained for the cell division rhythm in the ZC mutant. These results suggest a possible role for cAMP, either as an element of the coupling pathway for the control of the CDC by the CO, or as a "gear" of the block itself. The latter possibility seems to be excluded by the fact that cAMP pulses, though transiently resetting the division cycle immediately following the pulse, yielded no final steady-state $\Delta\Phi$ of the rhythm (Carré and Edmunds, unpublished results). The inability of cAMP to phase-shift the rhythm of melatonin prodution in chick pineal cells likewise strongly suggests that cAMP acts as an output signal of the CO (Nikaido and Takahashi 1989).

## Molecular Genetics of the Clock

Just as dissection of the clock with inhibitors of macromolecular synthesis has implicated one or more key proteins in circadian timekeeping, so the techniques of modern molecular genetics have provided a powerful approach toward the elucidation of the protein products of clock genes and their regulation by the oscillator. Genetic approaches have been along two major lines: (1) the isolation of mutants that have alterations in $\tau$ or other clock properties, such as phase, sensitivity to light, or temperature compensation; and (2) the isolation of biochemical mutants with known metabolic lesions and their subsequent assay to determine the effect of such lesions on the functioning of their clocks (Edmunds 1988; Feldman 1982, 1988; Feldman and Dunlap 1983; Hall and Rosbash 1987, 1988; Rosbash and Hall 1989).

### Isolation of Clock Mutants

In both *Neurospora* and *Drosophila*, at least 15 clock mutants have been isolated that display altered $\tau$s (Feldman 1982, 1988). For example, approximately half of the *Neurospora* mutants map to a single locus (frequency, *frq* ) and have $\tau$s of the conidiation (and other) rhythms ranging from 16.5 h *(frq-1)* to about 34 h *(frq-9)* (the *frq*$^+$ wild type has a $\tau$ of 21.5 h). The mutant *frq-9* shows a complete lack of temperature compensation. Analysis of heterokaryons containing different ratios of the mutant *frq* and wild-type *frq*$^+$ alleles revealed a gene dosage effect, in which the effect on $\tau$ was proportional to the fraction of *frq* nuclei.

In *Drosophila*, perhaps the most interesting set of mutants maps to a single locus (period, *per* ) on the X chromosome. These *per* mutants exhibit one of three main clock phenotypes, in which both the CRs of eclosion and locomotor activity are similarly altered: *per*$^s$ (short-period of 19 h), *per*$^L$ (long period, 29 h), and *per*$^o$ (arrhythmic). The tissues that determine the *per* gene-controlled behavior have been localized by fate-mapping in genetic mosaics to the brain area of the fly. Indeed, if the brain of a *per*$^s$ mutant is transplanted into the abdomen of an arrhythmic *per*$^o$ fly, the host recipient expressed a short-period rhythm, an observation suggesting that clock information is mediated via a diffusible substance. The *per* locus does not seem to be causally involved in photoperiodic time measurement (contrary to what might have been anticipated by the Bünning hypothesis): induction of ovarian diapause by short days still occurs in the *per*$^o$ mutant (Saunders et al. 1989).

Provocatively, the *per* mutations also altered the $\tau$ of the short-term, ultradian fluctuations in the interpulse interval of the male fly's courtship "love song" in exactly the same fashion as they did the circadian behaviorial rhythms (the mutational site of action seems to be the thoracic ganglion cells instead of the brain). Thus, the $\tau$s for *per*$^+$, *per*$^s$, and *per*$^L$ were approximately 56, 40, and 76 s, while *per*$^o$ was arrhythmic. The obvious inference is that the *per* gene product modulates the $\tau$s of both the circadian and ultradian oscillators.

### Recombinant DNA Studies: Cloning Clock Genes

In the preceding section we have seen that the *per* and *frq* loci play a fundamental role in the construction or maintenance of both a circadian and an ultradian oscillator. Now let us turn to the isolation and molecular analysis of these loci themselves (see reviews by Edmunds 1988; Feldman 1988; Hall and Rosbash 1987, 1988; Rosbash and Hall 1989).

#### *Drosophila: the per Locus*

*Restoration of Biological Rhythms in Transgenic Drosophila.* Germ-line transformation (mutant-rescue) experiments of arrhythmic *per*$^o$ fruit flies have indicated that the genomic region encoding the 4.5-kb transcript, which has now been shown to undergo circadian cycling (Hardin et al. 1990), is the core of the *per* locus, even though (1) the entirety of the transcript is not necessary for rather strong rhythmicity and (2) it does not seem to be completely sufficient, in transformants, to produce wild-type behavioral phenotypes. In contrast, the 0.9-kb RNA species, which oscillates as a result of developmental gating during eclosion and not, as first thought, as a function of CT (Lorenz et al. 1989), is neither necessary nor sufficient for rhythmicity. Therefore, the effects of the $per^o$ mutation on the molecular oscillation must be indirect (see Hamblen et al. 1986).

Recently, Yu et al. (1987a) mapped point mutations in the $per^{ol}$ and $per^s$ loci to single nucleotides. Chimeric DNA fragments consisting of well-defined wild-type and mutant DNA subsegments were constructed, introduced into flies by germ-line transformation, and assayed for their biological activity (circadian eclosion rhythm). These experiments further localized both $per^{ol}$ and $per^s$ to a 1.7-kb DNA fragment that was mostly coding DNA. Sequencing of this subsegment from each mutant showed that each reflected a single base-pair substitution: $per^{ol}$ was completely accounted for by a nonsense mutation (a stop codon) in the third coding exon of the 4.5-kb RNA transcript, whereas $per^s$ was a missense mutation in the fourth coding exon. Similar results have been obtained by Baylies et al. (1987).

Other interesting discoveries have been made during the course of these transformation-rescue experiments. Baylies et al. (1987) observed that $\tau$ ranged from about 25 to 40 h in various transformed lines, even though the arrhythmic ($per^-$) Drosophila had received identical transforming $per$-locus DNA fragments. Transcription studies revealed a tenfold variation in the level of $per$ RNA among transformed flies. These levels were inversely correlated with $\tau$, so that flies with lowest levels of $per$ product had the most slowly running clocks. Thus, $\tau$ would appear to be set by the level of gene product. But, because $per^s$ and $per^L$ both produce wild-type levels of $per$ mRNA, Baylies et al. (1987) suggest that the amino acid substitutions in these two mutants respectively increase or decrease the stability of the $per$ product; or, alternatively, $per^s$ protein would be hyperactive, and $per^L$ protein hypoactive.

In the same vein, Yu et al. (1987b) have found that the length of the Gly-Thr run of the $per$ gene (see next section) varies between 17 and 23 repeats in $per^+$ alleles. In order to ascertain if this variable region of the $per$ locus is functionally significant, these workers constructed an in-frame deletion of this region that removed the entire Gly-Thr repeat and used it to transform $per^{ol}$ flies. Surprisingly, this mutant construct rescued the CR phenotype but yielded a short love-song rhythm of only 40 s; this restored ultradian $\tau$ is more characteristic of $per^s$, contrasting with that (60 s) of both $per^+$ and of $per^{ol}$ flies transformed with a normal $per$ gene having an intact Gly-Thr repeat region. Thus, the effects of $per$ on the circadian and the male courtship-song rhythms could be dissociated. Perhaps this polymorphism of the $per$ gene within the Gly-Thr tract could explain how one gene product can differentially affect two different classes of rhythms. The species-constant circadian period could be conserved if the oscillation depended on the absolute amount of $per$ product (which might oscillate if the protein were unstable and blocked translation of its own mRNA), whereas the highly species-specific, ultradian song period might derive from variations in the concentration of $per$ product. If this reasoning is valid, a molecular basis for a temporal sexual isolating mechanism in courtship behavior would have been provided within the fine structure of the $per$ gene.

*The Product of the per Locus of Drosophila.* Recent work based on transcript mapping and DNA sequencing of the 7217 bases in the biologically active segment of $per$ locus DNA that encodes for the 4.5-kb RNA species supports the notion of a long transcript required for clock function. Using a complementary DNA clone that hybridized to a specific interval of the $per$ gene, Jackson et al. (1986) were able to place the transcript map onto the DNA restriction map. None of the fragments used for transformation could have been expected to encode this full-length transcript, all (or a large part) of which probably alone constitutes the $per$ gene.

The conceptual translation of the open reading frame determined from the $per$ DNA sequence yielded a polypeptide of 1127 amino acids (Jackson et al. 1986; Yu et al. 1987a; Citri et al. 1987). Several abnormal phenotypes displayed by some transformed flies, characterized by long-period rhythms, could be associated with changes in the sequence of untranslated portions of the transcription unit. In each case, the change in DNA sequence failed to affect the structure of the $per$ protein; therefore, the long $\tau$s were probably generated by altered regulation of $per$ protein synthesis. Surprisingly, nearly half of the predicted gene product comprised only five amino acids (serine, threonine, glycine, alanine, proline), and these often formed simple repeats, such as polyalanine and polyglycine tracts up to 17 residues in length. The hypothesized $per$ protein also included a potential site for phosphorylation by cAMP-dependent protein kinase. The putative $per$ protein sequence was compared to all available sequences in the Dayhoff and Doolittle database library (Jackson et al. 1986). Extensive regions of homology to the core protein of a rat chondroitin sulfate proteoglycan were detected, which is known to play a role in the glycosylation of the core protein. Thus, this unusual repeat sequence, together with the Thr-Gly repeat tracts that were also found in the *per* protein, might form a site serving for glycosylation, a process with an unknown role in clock function.

A similar DNA sequencing of a portion of the 4.5-kb transcript's source (Reddy et al. 1986) also implicated an unusual, multiresidue, Gly-Thr repeat in the protein that was inferred to be encoded within the *per* locus. By recombinant DNA techniques, a small subregion of the coding sequence was cloned and expressed in bacteria as part of a fusion protein, which was then used to immunize rabbits. When the resultant antisera were characterized and used to probe protein preparations from *Drosophila*, an antigen was detected in wild-type flies but not in a *per*$^-$ mutant. Biochemical characterization of this antigen indicated that it indeed was a proteoglycan as Shin et al. (1985) and Jackson et al. (1986) also independently had concluded.

Although not much is known about the function of proteoglycans, their distribution has been studied intensively. They typically are found in extracellular locations or in association with cell surfaces. James et al. (1986) have examined the temporal and spatial expression of the 4.5-kb mRNA that is transcribed from the *per* locus and that presumably codes for a proteoglycan-like clock protein(s). Both Northern blot analyses and *in situ* hybridizations to tissue sections revealed significant expression of this transcript in *Drosophila* embryos of different developmental stages. Although none was detectable in embryos from 0 to 6 h after egg laying, significant amounts were present thereafter, from gastrulation until the embryos hatched some 22–24 h later. Expression of the 4.5-kb mRNA then was undetectable in all larval instars and through the first half of the pupal stage, whereupon *per* locus activity recommenced. This expression of the *per* clock gene was limited to the central nervous system of the developing embryo and was localized within the brain and ventral ganglia. It is significant that these sites correspond, respectively, to the foci for the eclosion and courtship song oscillators localized by fate-map, genetic mosaic analysis. In addition to the physiological role that the 4.5-kb mRNA species and the nervous system-specific proteoglycan for which it codes might have in maintaining biological rhythms, James et al. (1986) suggest that they participate in development by establishing mechanisms necessary for eventual expression of

clock functions. The *per* gene product also appears to regulate the degree of intercellular communication by altering the number and organization of functional gap junctions in *Drosophila* (Bargiello et al. 1987). More recent studies have shown that the *per* product is expressed in an even greater variety of tissues of *Drosophila*, which perhaps contain their own intrinsic oscillator activity (Liu et al. 1988; Saez and Young 1988).

DNA homologous to the *per* locus has been found in a number of vertebrates, and the unusual coding sequence from this *Drosophila* clock gene appears to have been conserved. The peculiar, tandemly repeated sequence forming a portion of the 4.5-kb *per* transcript was found to be strongly homologous to mouse DNA and, to a lesser extent, to DNA in chicken and man (Shin et al. 1985). Sequences homologous to those in *per* probes from *Drosophila* recently have been reported in spinach, rape, and *Acetabularia*, although in *Acetabularia* they were found not in nuclear DNA but rather in the chloroplast genome (Li-Weber et al. 1987). More recently, an antibody to the *per* protein has been shown to label putative circadian pacemaker neurons and fibers in the eyes of *Aplysia* and *Bulla* (Siwicki et al. 1989). Finally, the *Neurospora frq* gene (see next section) has been found to share a sequence element with the *per* clock gene, although these homologies do not correspond to the region of the Gly-Thr repeat (McClung et al. 1989). These findings, then, raise the attractive possibility that homologous DNA sequences play a role in the generation of both ultradian and circadian biological rhythms in many species.

## *Neurospora: the frq Locus*

The techniques of molecular genetics also have been used to determine how the circadian clock controls the expression of the *frq* gene in *Neurospora*. The first step was the characterization of the *frq* locus itself. A set of contiguous, overlapping cosmid and phage clones that generated a physical map extending 130 kb in either direction from the *oli* marker (a total of more than 8 map units) on linkage group VIIR was identified by using chromosome-walk methodology (McClung et al. 1989). Transformation and phenotypic rescue of recessive mutants further identified the *frq* and *for* (formate) genes; the physical map for 200 kbp in the *oli* region that included these loci agreed with the previously determined genetic map. Transformation experiments using successively smaller subclones allowed localization of *frq* to an 8-kbp region of DNA. Phenotypic rescue of *frq-9* restored not only rhythmic conidial banding with a wild-type $\tau$, but also temperature compensation of $\tau$ over a 20° to 30°C temperature range and wild-type carotenogenesis.

The next step was to isolate "timed target" (clock-controlled) genes whose level of expression is regulated by the CO. The *frq* gene was cloned (Loros et al. 1989) by using complementary DNAs corresponding to mRNAs present at different times (CT 1, CT 13) to isolate cDNAs by subtractive and differential hybridization. These time-specific cDNA populations then were used to probe cDNA and genomic libraries. A putative clock-controlled gene identified by this protocol was verified by using the cloned DNA to probe Northern blots of RNA that had been isolated every 4 h from *frq*$^+$ ($\tau$ = 21.5 h) and from *frq-7* ($\tau$ = 29 h) strains maintained for 12–56 h in DD. In these tests, the mRNAs derived from wild-type clock-controlled genes was found to cycle 2.5 times from maximum to minimum concentration, whereas mRNAs from the same genes in *frq-7* cycled only twice; consequently, the two mRNA species were out of phase over a 40-h time span. Thus, these cDNA clones clearly represented clock-controlled genes rather than merely developmentally controlled genes that were responding to culture conditions.

Are *Neurospora* clock-gene components similar to those of *Drosophila*? McClung et al. (1989) have found that DNA sequences of *Neurospora* hybridize with the 8.0-kb rescuing fragment from the *per* locus of *Drosophila* (see preceding section) and that these sequences are homologous. These homologies, however, do not correspond to the region of the Gly-Thr repeat (although they do correspond to a region where at least two of the known *per* mutations are localized. Similarities between the *frq* and *per* mutants suggest that the two genes might code for similar proteins, but none of the *frq* sequences that have been isolated as homologous to the *per* gene map to the *frq* locus. One must now identify the *frq* gene product and determine if the biochemical functions encoded in these *Neurospora* genomic fragments are similar to those encoded in those of *Drosophila*.

*Other Systems*

One of the most sophisticated recent studies of the molecular genetics of circadian systems has concerned the clock regulation of the transcription of the wheat *Cab-1* gene (Nagy et al. 1988a,b), which encodes the major light-harvesting chlorophyll-binding protein protein of the chloroplast (see "Clock-Controlled Genes: Transcriptional and Translational Control" above). In plants grown in LD:12,12, the expression of this gene continued to cycle upon transfer of the plant of LL or DD. The plant photoreceptor, phytochrome, interacts with the clock to govern the level of *Cab-1* RNA. Furthermore, when the wheat gene was transferred to tobacco, the circadian regulation was maintained. Fusion of the upstream region of the *Cab-1* gene, containing the *cis*-acting element responsible for its regulation by light, to be bacterial chloramphenicol acetyltransferase (CAT) gene resulted in the circadian fluctuation of CAT mRNA in transgenic plants. These results indicate, therefore, that the clock is acting at the transcriptional level and identifies the *cis*-acting element that mediates this response.

Another organism that seems promising for the molecular genetic analysis of clocks is *Arabidopsis thaliana* (L.), which is already serving as a model system for the study of plant genetics. This plant is small, has a rapid (5- to 6-week) generation time, and possesses a small and simple genome. The genetic map comprises more than 80 loci on 5 chromosomes, and a complete physical map is being assembled. These maps will facilitate gene cloning by chromosome walking from linked genes that have been cloned. The circadian system is only now being explored, but rhythms in the abundance of the mRNAs for several genes have been observed (McClung, personal communication).

Alteration of Clock Properties
in Biochemical Mutants

In addition to the isolation and characterization of clock mutants discussed earlier, another related approach in the genetic dissection of clock mechanisms is the isolation of biochemical mutants with known metabolic lesions and the determination of the effects of such mutants on the functioning of the clock (reviewed by Feldman 1982). One of the best systems for illustrating this line of attack is afforded by *Neurospora* (Feldman and Dunlap 1983).

Although the great majority of auxotrophic and morphological mutations that have been characterized in *Neurospora* do not affect the circadian clock, several mutants with various biochemical lesions have been discovered that exhibit significantly altered clock properties. Conversely, certain biochemical mutants have been found in which clock characteristics have not been changes, thereby ex-

**Table 1.** Some biochemical mutants in *Neurospora crassa* in which clock properties (rhythm of conidiation) have been examined

| Strain | Mutation |
| --- | --- |
| **Mutations affecting respiratory and photosensitive pigments** | |
| *al-1; al-2* (albino) | No detectable carotenoids; damping of rhythm in LL unaltered |
| *poky* | Respiratory mutant; reduction in nonmitochondrial cytochrome; higher threshold intensity for inhibition of banding in LL |
| *rib-1; rib-2* | Riboflavin auxotrophs; reduction in levels of FAD and FMN in mycelia; reduction in light-sensitivity of clock for phase-shifting and damping |
| *nit-1; nit-2; nit-3* | Reduction in activity of nitrate reductase; no effect on photosuppression or phase-shifting of conidiation rhythm |
| **Mutations affecting cyclic 3′,5′-AMP** | |
| NG 6–3; NG 6–11 (revertants of *crisp-1*) | No colonial morphology; reduced levels of adenylate cyclase and cAMP; entrainment and free-running period unaffected |
| *cpd-1; cpd-2* | Reduced levels of cAMP caused by reduction in activity of $Mg^{2+}$-stimulated cyclic PDE or of AC; $\tau$ unaffected; rhythm in cAMP |
| **Mutations affecting $Ca^{2+}$ metabolism** | |
| *Ca4; Ca23* | $Ca^{2+}$ dependency for growth; growth rate slower than wild type; $\tau$ in *Ca4* affected by changes in $Ca^{2+}$ concentration; *Ca23* arrhythmic, does not entrain to LD or temperature cycles |
| **Mutations affecting cysteine biosynthesis** | |
| *cys-X; cys-4; cys-12* | Cysteine auxotrophs; shortened period |
| **Mutations affecting fatty acid synthesis** | |
| *cel* | Defective fatty acid synthetase complex; deficient in synthesis of palmitic acid (16:0); addition of unsaturated or short-chain saturated fatty acids lengthened period; loss of temperature compensation below 22°C, exacerbated by supplementation with linoleic acid (18:2), but restored in part by addition of 16:0 |
| **Ergosterol-deficient mutants** | |
| *erg-1; erg-3* | Steroid deficiency in plasma membrane; normal clock; growth rate resistant to nystatin, but phase shifting by nystatin pulses attenuated |
| **Oligomycin-resistant mutants** | |
| *oli$^r$* | Resistance to drug oligomycin due to mutations in the DCCD binding protein of mitochondrial ATPase; $\tau$ shortened in proportion to degree of resistance; introduction into *cel* restores temperature compensation below 22°C and negates effect of linoleic acid supplementation |
| **Cycloheximide-resistant mutants** | |
| *cyh-1; cyh-2* | Cycloheximide-resistant 80S ribosomes; clock unaltered, but pulses of CEX could no longer phaseshift rhythm |

For further details and references, see Feldman (1982), Feldman and Dunlap (1983), and Edmunds (1988)

cluding a particular pathway of sequence as a key part of the oscillator itself. Particularly interesting are those mutations (Table 1) affecting respiratory and photosensitive pigments, cAMP levels, fatty acid metabolism, and drug resistance (such as to oligomycin and cycloheximide).

## Characterizing the Coupling Pathway: Transducing Mechanisms Between Clocks and Their Hands

Another time-honored approach toward elucidating the mechanism underlying an overt, physiological rhythm, or expression ("hand") of the clock, is to attempt to thread one's way back through the biochemical pathways mediating the rhythm – the so-called transducing mechanisms – until one arrives at their point of coupling to the oscillator. The CR of glow bioluminescence of *Gonyaulax* will serve here as an excellent minicase history. (Another illustrative system is that of the compound action potentials produced by the eye in *Aplysia* and *Bulla*, whose entrainment pathways leading to the input side of the clock were discussed above.)

The reaction responsible for light production involves the oxidation of dinoflagellate luciferin molecular oxygen, catalyzed by a specific luciferase enzyme:

$$\text{Luciferin} + O_2 \xrightarrow{\text{luciferase}} \text{light } (\lambda = 475 \text{ nm}) + \text{products}$$

The absolute level of luciferin, the binding capacity of its specific binding protein, and the activity of luciferase have all been shown to be under circadian clock control (by Johnson and Hastings 1986).

Earlier work on the cause of rhythmic luciferase activity ruled out simple explanations involving differences in enzyme extractability or extractable inhibitors or activators, leaving open the alternatives of cyclic synthesis and degradation of enzyme (constant specific activity) or cyclic covalent modification of the polypeptide, thus altering its activity (cyclic specific activity). Unproteolyzed, higher-molecular-weight luciferase from both day- and night-phase cells has been purified and the two preparations compared with respect to several physicochemical, enzymatic, and immunological criteria. A given amount of antiluciferase inactivated the same amount of luciferase activity in both extracts, indicating that their specific activities were the same, and suggesting that the luciferase was the same polypeptide in day and night preparations but that there were different amounts of the enzyme in each. These findings were confirmed by direct measurement of luciferase protein with antibody to luciferase. The cyclic activity of the enzyme corresponded to a rhythm in the concentration of immunologically reactive luciferase protein. Thus, the CR of luciferase activity could be attributed to circadian clock-modulated synthesis or degradation, or both, of the luciferase polypeptides.

Of course, the question now arises (as it always does) as to the next step in the quest for the elusive clock. In particular, is the control of the synthesis of luciferase and luciferin-binding protein (LBP) transcriptional (in which case their respective concentrations should be rhythmic), or translational (where constant mRNA concentrations would be translated rhythmically? A molecular genetic attack on this problem comprises the cloning of luciferase and LBP cDNAs and using them as probes to measure their luciferase mRNAs as a function of circadian time. Just such an approach has been taken in the *Gonyaulax* system (Morse et al. 1989). The LBP cDNA was isolated by immunological screening of a cDNA library that had been subcloned into an expression vector. The identity of the cDNA was confirmed by *in vitro* translation of a mRNA hybrid selected from total RNA by the LBP alone. Northern hybridization of the cDNA to mRNA, isolated at different CTs, showed that the amounts of the LBP mRNA were invariant. A putative luciferase cDNA also was isolated, and similar experiments demonstrated constant levels of the corresponding luciferase mRNA over a 24-h time span. In a complementary approach, *in vitro* translation of the mRNA that was used in the Northern blots showed that the synthesis of LBP at all CTs was identical. Thus, the regulation of the CRs in the amounts of LBP and luciferase is exerted at the translational level. Indeed, Milos et al. (1990) recently have demonstrated circadian translational control over the synthesis of many other *Gonyaulax* proteins as well.

## Biochemical and Molecular Models for Circadian Clocks

Given the experimental results obtained from the various lines of experimental attack on circadian clock mechanisms, it is not surprising that a number of models for the biochemical and molecular bases of COs have been proposed over the years. Despite the fact that models in this field are plentiful, they do serve as useful foci, often leading to important experimental advances. If these results than falsify the construct, so much the better: a good model sows the seeds of its own destruction! Furthermore, it is perhaps useful to introduce the potential initiate to this field to some of the extant notions, if for no other reason than to provide reassurance that it is still virgin territory. We have a long way to go to placate the editorial writer who plaintively queried, "Why is so little known about the biological clock?" and expressed hope that it soon would be "wound up" (Editorial 1971). Indeed, the very fact that so many models have been put foward should serve as a warning that the biological clock possibly is much more complex than we earlier had envisioned.

## Classes of Model of Autonomous Oscillators

There are several different classes of model for endogenous, self-sustaining circadian clocks (Table 2), which are neither mutually exclusive nor jointly exhaustive. They can be grouped into several main categories (Edmunds 1988): (1) strictly molecular models, which rely on the properties of molecules themselves for generating persisting 24-h rhythms; (2) feedback-loop, network models for oscillations in energy metabolism and in other biosynthetic pathways, in which longer periods would be generated (frequency demultiplication) by energy reservoirs or depots, appropriate allosteric constants and turnover numbers of key enzymes, or by negative cross-coupling among individual oscillators (or even among prokaryotic endosymbionts) within a cell; (3) transcriptional (tape-reading) models, wherein the tran-

**Table 2.** Some molecular models for circadian clocks

| Model | Key elements |
| --- | --- |
| Molecular *(in vitro)* | Periodicity in the structure and the properties of molecules themselves (e. g., alternation between two conformational subunit states resulting from posttranscriptional phosphorylation and dephosphorylation) |
| Network (biochemical feedback loops) | Glycolytic oscillator: by suitable selection of allosteric constants and turnover numbers of key enzymes, frequency can be controlled over a large range (in principle, even 24 h) |
| | Cell energy metabolism = the clock: appropriate choice of depot (deposition effect) would allow self-oscillatory reactions on a circadian time scale |
| | Coupled oscillators: cross-coupling among high-frequency oscillations in energy metabolism could generate circadian rhythmicity in energy transduction |
| | Cyclic AMP model: cAMP, ATP, AC and PDE are oscillating variables which would exhibit limit-cycle behavior by allosteric feedback of AMP on AC and PDE |
| | Mitochondrial $Ca^{2+}$ Cycle: $NAD^+$, NAD kinase and NADP phosphatase, $Ca^{2+}$, the mitochondrial $Ca^{2+}$ transport system, and calmodulin represent clock gears, which, in ensemble, would constitute a self-sustained, negatively cross-coupled, circadian oscillator |
| | Heterodyne endosymbiont hypothesis: two prokaryotic colonists of a putative ancestral eukaryotic cell emitted chemical pulses with slightly different short periods; their coincidence would have yielded a longer, circadian period |
| Transcriptional (tape-reading) | Chronon model: sequential transcription of long, polycistronic, DNA complexes in eukaryotic chromosomes, coupled to rate-limiting, time-consuming, temperature-independent diffusion steps by mRNA to the ribosomes for translation would yield circadian periods |
| | Chronogene-cytochron Model for cell cycle clocks: programmable, sequential transcription of segment of chromosomal DNA without requirement for translation |
| Membrane | Molecule X actively transported into organelles, changing configuration and transport capacity of membranes; passive diffusion then occurs until X is evenly redistributed |
| | Limit cycle behavior in which an ion concentration gradient and membrane transport activity are the oscillating variables; slow translational diffusion of membrane proteins; cross-coupling; temperature compensation by changes in membrane lipid saturation |
| | Membrane oscillator hypothesis: membrane-bound photoreceptors modulate membrane-bound energy transduction; energetic state of membranes, in turn, determines photoreceptor sensitivity |
| | Membrane transport: rhythmic interplay of membrane pumps, leaks and porters; role of cell division cycle |
| | Proton gradient model: membrane-bound carrier actively transports $H^+$ to the outside; key enzyme having pH-dependent activity profile translocates $H^+$ from exterior to interior (or produces $H^+$ as a product); sustained oscillations occur when the two competing processes balance |
| | Coupled translation-membrane model: assembly, transport and insertion (loading) of essential proteins into membranes |
| | Monovalent ion-mediated translational control model: intracellular monovalent ion concentration feedback-regulates the synthesis and insertion of membrane proteins, resulting in changes in ion concentration |

See Edmunds (1988) for further details and references

scription of a long, polycistronic piece(s) of DNA with associated rate-limiting diffusion steps leading to translation would meter circadian time; and, (4) membrane models, in which the transport activity and other properties of various membranes in the cell (intimately related to state transitions in the fluid mosaic membrane, which in turn affect membrane structure) would constitute, in ensemble, a stable limit cycle oscillator that ultimately would account for the properties of CRs.

## Problems and Prospects

The various models proposed for circadian clocks (Table 2) sometimes overlap, for they often incorporate several different notions from each other. This is not at all surprising, since each has strengths and shortcomings for which it attempts to compensate by hybridization (Edmunds 1988). For example, although the network models deal with known biochemical oscillations, they have difficulty in accounting for the long $\tau$ of CRs and their temperature compensation (notwithstanding appeals to deposition effects and the fiddling with parameter values). Similarly, the chronon hypothesis and the chronogene model combine gene action and cell cycle controls but are weakened by the demonstration that DNA-dependent RNA synthesis does not seem to be important in the clock mechanism of *Acetabularia*. Enucleated cells continue to exhibit a rhythm of photosynthesis, and inhibiting extranuclear RNA synthesis with rifampicin does not stop the clock. Finally, the earlier membrane models, although neatly accounting for temperature compensation, had difficulty in accounting for the circadian period. This deficiency was addressed by the coupled translation-membrane model, which invoked the time-consuming processes of assembly, transport, and loading of essential proteins into membranes. Further, it incorporated the recent reemphasis on the necessary role of protein synthesis in the maintenance (if not the generation) of persisting CRs in *Acetabularia*, *Gonyaulax* and *Neurospora* (see "Protein Synthesis and Clock Proteins" above).

Although the search for essential proteins may be very time-consuming with no guarantee of immediate success, it seems to have been rewarded by the identification of a cyclically appearing clock species (p 230) in *Acetabularia* (Hartwig et al. 1985) and of a 34-kDa protein in *Aplysia* that Yeung and Eskin (1987) have nominated as a worthy candidate for a component of the CO (see "Protein Synthesis and

Clock Proteins" above). The advent of the powerful techniques of recombinant DNA research and gene cloning may offer some solace, as indeed they have in the case of the identification of the gene product (a proteoglycan) of the *per* locus in *Drosophila* (Jackson et al. 1986; Reddy et al. 1986; see "*Drosophila*: The *per* Locus" above). Li-Weber et al. (1987) have reported provocatively that the chloroplast genome of *Acetabularia* (which is assumed to encode the p230 clock protein) shares a homology with the *per* locus of *Drosophila*. But what does the *per* protein do? Does it only secondarily affect the rhythmicity generated by some other oscillator, or is it a component of an intercellular clock, forming or maintaining communication (see "Intercellular Communication: Coupled Cellular Oscillators" above) among rhythmic cells that would lead to mutual synchronization (Bargiello et al. 1987; Dowse and Ringo 1987). Perhaps the *per* gene product is essential for the modulation (frequency demultiplication) of another ultradian oscillator located in a membrane or neural net.

One must note that the search for the elusive clock per se could be doomed to failure if circadian timekeeping is not attributable to any one entity or subset of reactions in a cell; or, exceedingly difficult, at least, if "all these aspects of cell chemistry [soluble enzyme kinetics, nuclear message transcription, membranes, and second messengers] are susceptible to circadian modification of their regulatory dynamics, and that the clock, volatile as a ghost, lurks now in one room, now in another, in different cell types" (Hastings and Schweiger 1976, pp. 54–55).

No matter whether our taste runs to ghosts or not, let us end on a note of cautious optimism: "... We will all have some fun in any case" (Pittendrigh 1960).

## References

Bargiello TA, Saez L, Baylies MK, Gasic G, Young MW, Spray DC (1987) The *Drosophila* clock gene *per* affects intercellular junctional communication. Nature 328: 686–691

Baylies MK, Bargiello TA, Jackson FR, Young MW (1987) Changes in abundance or structure of the *per* gene product can alter periodicity of the *Drosophila* clock. Nature 326: 390–392

Block GD, McMahon DG (1984) Cellular analysis of the *Bulla* ocular circadian pacemaker system. III. Localization of the circadian pacemaker. J Comp Physiol [A] 155: 387–395

Broda H, Brugge D, Homma K, Hastings JW (1986) Circadian communication between unicells? Effects on period by cell-conditioning of medium. Cell Biophys 8: 47–67

Carré IA, Laval-Martin DL, Edmunds LN Jr (1989) Circadian changes in cyclic AMP levels in synchronously dividing and stationary-phase cultures of the achlorophyllous ZC mutant fo *Euglena gracilis*. J Cell Sci 94: 267–272

Citri Y, Colot HV, Jacquier AC, Yu Q, Hall JC, Baltimore D, Rosbash M (1987) A family of unusually spliced biologically active transcripts encoded by a *Drosophila* clock gene. Nature 326: 42–47

DeCoursey PJ, Buggy J (1988) Restoration of circadian locomotor activity in arrhythmic hamsters by fetal SCN transplants. Comp Endocrinol 7: 49–64

DeCoursey PJ, Buggy J (1989) Circadian rhythmicity after neural transplant to hamster third ventricle: specificity of suprachiasmatic nuclei. Brain Res 500: 263–275

Deng MD, Moureaux T, Leydecker MT, Caboche M (1990) Nitrate-reductase expression is under the control of a circadian rhythm and is light inducible in *Nicotiana tabacum* leaves. Planta 180: 257–261

Dowse HB, Ringo JM (1987) Further evidence that the circadian clock in *Drosophila* is a population of coupled ultradian oscillators. J Biol Rhythms 2: 65–76

Dowse HB, Ringo JM (1989) Rearing *Drosophila* in constant darkness produces phenocopies of *period* circadian clock mutants. Physiol Zool 62: 785–803

Dowse HB, Hall JC, Ringo JM (1987) Circadian and ultradian rhythms in *period* mutants of *Drosophila melanogaster*. Behav Genet 17: 19–35

Dunlap JC, Feldman JF (1988) On the role of protein synthesis in the circadian clock of *Neurospora crassa*. Proc Natl Acad Sci USA 85: 1096–1100

Earnest DJ, Sladek CD (1987) Circadian vasopressin release from perifused rat suprachiasmatic explants *in vitro*: effects of acute stimulation. Brain Res 422: 398–402

Earnest DJ, Sladek CD, Gash DM, Wiegand SJ (1989) Specificity of circadian function in transplants of the fetal suprachiasmatic nucleus. J Neurosci 9: 2671–2677

Editorial (1971) Is it time to wind up the biological clock? Nature [New Biol] 231: 97–98

Edmunds LN Jr (1984) Physiology of circadian rhythms in microorganisms. Adv Microb Physiol 25: 61–148

Edmunds LN Jr (1988) Cellular and molecular bases of biological clocks. Springer, Berlin Heidelberg New York

Edmunds LN Jr, Tamponnet C (1990) Oscillator control of cell division cycles in *Euglena*: role of calcium in circadian timekeeping. In: O'Day DH (ed) Calcium as an intracellular messenger in eucaryotic microbes. American Society for Microbiology, Washington, pp 97–123

Feldman J (1982) Genetic approaches to circadian clocks. Annu Rev Plant Physiol 33: 583–608

Feldman JF (1988) Genetics of circadian clocks. Bot Acta 101: 128–132

Feldman J, Dunlap JC (1983) *Neurospora crassa*: a unique system for studying circadian rhythms. Photochem Photobiol Rev 7: 319–368

Fritz BJ, Kasai S, Matsui K (1989) Free cellular riboflavin is involved in phase shifting by light of the circadian clock in *Neurospora crassa*. Plant Cell Physiol 30: 557–564

Gamalega NF, Shishko ED, Chyorny AP (1988) Preservation of circadian rhythms by human lymphocytes *in vitro* (in Russian). Bull Eksp Biol Med 106: 598–600

Giuliano G, Hoffman NE, Ko K, Scolnik PA, Cashmore AR (1988) A light-entrained circadian clock controls transcription of several plant genes. EMBO J 4: 3635–3642

Goto K, Laval-Martin D, Edmunds LN Jr (1985) Biochemical modeling of an autonomously oscillatory circadian clock in *Euglena*. Science 228: 1284–1288

Grobbelaar N, Huang TC, Lin HY, Chow TJ (1986) Dinitrogen-fixing endogenous rhythm in *Synechococcus* RF-1. FEMS Microbiol Lett 37: 173–178

Hall JC, Rosbash M (1987) Genetic and molecular analysis of biological rhythms. J Biol Rhythms 2: 152–178

Hall JC, Rosbash M (1988) Mutations and molecules influencing biological rhythms. Annu Rev Neurosci 11: 373–393

Hamblen M, Zehring WA, Kyriacou CP, Reddy P, Yu Q, Wheeler DA, Zwiebel LJ, Konopka RJ, Robash M, Hall JC (1986) Germ-line transformation involving DNA from the *period* locus in *Drosophila melanogaster*: overlapping genomic fragments that restore circadian and ultradian rhythmicity to $per^o$ and $per^-$ mutants. J Neurogenet 3: 249–291

Hardin PE, Hall JC, Rosbash M (1990) Feedback of the *Drosophila period* gene product on circadian cycling of its messenger RNA levels. Nature 343: 536–540

Hartwig R, Schweiger M, Schweiger R, Schweiger HG (1985) Identification of a high molecular weight polypeptide that may be part of the circadian clockwork in *Acetabularia*. Proc Natl Acad Sci 82: 6899–6902

Hartwig R, Schweiger R, Schweiger HG (1986) Circadian rhythm of the synthesis of a high molecular weight protein in anucleate cells of the green alga *Acetabularia*. Eur J Cell Biol 41: 139–141

Hastings JW, Schweiger HG (eds) (1976) The molecular basis of circadian rhythms. Abakon, Berlin (Dahlem Konferenzen)

Hastings JW, Johnson C, Kondo T (1987) Action spectrum for phase shifting of the circadian rhythm of phototaxis in *Chlamydomonas*. Photochem Photobiol [Suppl] 45: 86S

Inouye ST, Takahashi JS, Wollnik F, Turek FW (1988) Inhibitor of protein synthesis phase shifts a circadian pacemaker in mammalian SCN. Am J Physiol 255: R1055–R1058

Jacklet JW (1984) Neural organization and cellular mechanisms of circadian pacemakers. Int Rev Cytol 89: 252–294

Jacklet JW (ed) (1989a) Neuronal and cellular oscillators. Dekker, New York

Jacklet JW (1989b) Circadian neuronal oscillators. In: Jacklet JW (ed) Neuronal and cellular oscillators. Dekker, New York, pp 483–527

Jackson FR, Bargiello TA, Yun SH, Young MW (1986) Product of *per* locus of *Drosophila* shares homology with proteoglycans. Nature 320: 185–187

James AA, Ewer J, Reddy P, Hakk JC, Rosbash M (1986) Embryonic expression of the *period* clock gene in the central nervous system of *Drosophila melanogaster*. EMBO J 5: 2313–2320

Jenkins HA, Griffiths AJ, Lloyd D (1989) Simultaneous operation of ultradian and circadian rhythms in *Chlamydomonas reinhardii*. J Interdiscip Cycle Res 20: 257–264

Johnson CH, Hastings JW (1986) The elusive mechanism of the circadian clock. Am Sci 74: 29–36

Khalsà SBS, Block GD (1986) The *Bulla* ocular circadian pacemaker is phase shifted by pentylenetetrazole independently of extracellular calcium concentration. Soc Neurosci Abstr 12: 596

Khalsa SBS, Block GD (1988a) Calcium channels mediate phase shifts of the *Bulla* circadian pacemaker. J Comp Physiol [A] 164: 195–206

Khalsa SBS, Block GD (1988b) Phase-shifts of the *Bulla* ocular circadian pacemaker in the presence of calmodulin antagonists. Life Sci 43: 1551–1556

Khalsa SBS, Block GD (1990) Calcium in phase control of the *Bulla* circadian pacemaker. Brain Res 506: 40–45

Kippert F (1987) Endocytobiotic coordination: intracellular calcium signalling and the origin of endogenous rhythms. Ann NY Acad Sci 503: 476–495

Lakin-Thomas PL (1985) Biochemical genetics of the circadian rhythm in *Neurospora crassa:* studies on the *cel* strain. Thesis, University of California, San Diego

Lehman MN, Silver R, Gladstone WR, Kahn RM, Gibson M, Bittman EL (1987) Circadian rhythmicity restored by neural transplant. Immunocytochemical characterization of the graft and its integration with the host brain. J Neurosci 7: 1626–1638

Lillo C (1989) An unusually rapid light-induced nitrate reductase mRNA pulse and circadian oscillations. Naturwissenschaften 76: 526–528

Liu X, Lorenz L, Yu Q, Hall JC, Rosbash M (1988) Spatial and temporal expression of the *period* gene in *Drosophila melanogaster.* Genes Dev 2: 228–238

Li-Weber M, de Groot EJ, Schweiger HG (1987) Sequence homology to the *Drosophila per* locus in higher plant nuclear DNA and in *Acetabularia* chloroplast DNA. Mol Gen Genet 209: 1–7

Lorenz LJ, Hall JC, Rosbash M (1989) Expression of a *Drosophila* mRNA is under circadian clock control during pupation. Development 107: 869–880

Loros JJ, Denome SA, Dunlap JC (1989) Molecular cloning of genes under control of the circadian clock in *Neurospora.* Science 243: 385–388

Lotshaw DP, Jacklet JW (1987) Serotonin induced protein phosphorylation in the *Aplysia* eye. Comp Biochem Physiol 86C: 27–32

McClung CR, Fox BA, Dunlap JC (1989) The *Neurospora* clock gene *frequency* shares a sequence element with the *Drosophila* clock gene *period.* Nature 339: 558–562

McMahon DG, Block GD (1987a) The *Bulla* ocular circadian pacemaker. I. Pacemaker neuron membrane potential controls phase through a calcium-dependent mechanism. J Comp Physiol [A] 161: 335–346

McMahon DG, Block GD (1987b) The *Bulla* ocular circadian pacemaker. II. Chronic changes in membrane potential lengthen free running period. J Comp Physiol [A] 161: 347–354

Milos P, Morse D, Hastings JW (1990) Circadian control over synthesis of many *Gonyaulax* proteins is at a translational level. Naturwissenschaften 77: 87–89

Mitsui A, Kumazawa S, Takahashi A, Ikemoto H, Cao S, Arai T (1986) Strategy by which nitrogen-fixing unicellular cyanobacteria grow photoautotrophically. Nature 323: 720–722

Miwa I, Nagatoshi H, Horie T (1987) Circadian rhythmicity within single cells of *Paramecium bursaria.* J Biol Rhythms 2: 57–64

Moore RY (1983) Organization and function of a central nervous system circadian oscillator: the suprachiasmatic hypothalamic nucleus. Fed Proc 42: 2783–2789

Moore RY, Card JP (1986) Visual pathways and the entrainment of circadian rhythms. Ann NY Acad Sci 453: 123–133

Morse D, Milos PM, Roux E, Hastings JW (1989) Circadian regulation of bioluminescence in *Gonyaulax* involves translational control. Proc Natl Acad Sci USA 86: 172–176

Morse MJ, Crain RC, Satter RL (1987) Phosphatidylinositol cycle metabolites in *Samanea saman* pulvini. Plant Physiol 83: 640–644

Nagy F, Kay SA, Chua NH (1988a) A circadian clock regulates transcription of the wheat *Cab-1* gene. Genes Dev 2: 376–382

Nagy F, Kay SA, Chua NH (1988b) Gene regulation by phytochrome. Trends Genet 4: 37–42

Nakashima H (1986) Phase shifting of the circadian conidiation rhythm in *Neurospora crassa* by calmodulin antagonists. J Biol Rhythms 1: 163–169

Nikaido SS, Takahashi JS (1989) Twenty-four hour oscillation of cAMP in chick pineal cells: role of cAMP in the acute and circadian regulation of melatonin production. Neuron 3: 609–619

Ninnemann H (1979) Photoreceptors for circadian rhythms. Photochem Photobiol Rev 4: 207–265

Olesiak W, Ungar A, Johnson CH, Hastings JW (1987) Are protein synthesis inhibition and phase shifting of the circadian clock in *Gonyaulax* correlated? J Biol Rhythms 2: 121–138

Otto B, Grimm B, Ottersbach P, Kloppstech K (1988) Circadian control of the accumulation of mRNAs for light- and heat-inducible chloroplast proteins in pea (*Pisum sativum* L.). Plant Physiol 88: 21–25

Paulsen H, Bogorad L (1988) Diurnal and circadian rhythms in the accumulation and synthesis of mRNA for the light-harvesting chlorophyll *a/b*-binding protein in tobacco. Plant Physiol 88: 1104–1109

Peleg L, Dotan A, Luzato P, Ashkenazi IE (1990) "Long ultradian" rhythms in red blood cells and ghost suspensions: possible involvement of cell membrane. In Vitro Cell Dev Biol 26: 978–982

Piechulla B (1988) Plastid and nuclear mRNA fluctuations in tomato leaves – diurnal and circadian rhythms during extended dark and light periods. Plant Mol Biol 11: 345–353

Piechulla B (1989) Changes of the diurnal and circadian (endogenous) mRNA oscillations of the chlorophyll *a/b* binding protein in tomato leaves during altered day/night (light/dark) regimes. Plant Mol Biol 12: 317–327

Pittendrigh CS (1960) In "Discussion" following his article "Circadian rhythms and the Circadian Organization of Living Systems." Cold Spring Harbor Symp Quant Biol 25: 183

Prosser RA, Gillette MU (1989) The mammalian circadian clock in the suprachiasmatic nuclei is reset *in vitro* by cAMP. J Neurosci 9: 1073–1081

Prosser RA, McArthur AJ, Gillette MU (1989) cGMP induces phase shifts of a mammalian circadian pacemaker at night, in antiphase to cAMP effects. Proc Natl Acad Sci USA 86: 6812–6815

Radha E, Hill TD, Rao GHR, White JG (1985) Glutathione levels in human platelets display a circadian rhythm *in vitro.* Thromb Res 40: 823–831

Raju U, Yeung SJ, Eskin A (1990) Involvement of proteins in light resetting ocular circadian oscillators in *Aplysia.* Am J Physiol 258: R256–R262

Ralph MR, Foster RG, Davis FC, Menaker M (1990) Transplanted suprachiasmatic nucleus determines circadian period. Science 247: 975–978

Reddy P, Jacquier AC, Abovich N, Petersen G, Rosbash G (1986) The *period* clock locus of *D. melanogaster* codes for a proteoglycan. Cell 46: 53–61

Reppert SM, Uhl GR (1987) Vasopressin messenger ribonucleic acid in supraoptic and suprachiasmatic nuclei: appearance and circadian regulation during development. J Endocrinol 120: 2483–2487

Roberts MH, Bedian V, Chen Y (1989) Kinase inhibition

lengthens the period of the circadian pacemaker in the eye of *Bulla gouldiana.* Brain Res 504: 211–215

Robertson LM, Takahashi JS (1988a) Circadian clock in cell culture. I. Oscillation of melatonin release from dissociated chick pineal cells in flow-through microcarrier culture. J Neurosci 8: 12–21

Robertson LM, Takahashi JS (1988b) Circadian clock in cell culture. II. *In vitro* photic entrainment of melatonin oscillation from dissociated chick pineal cells. J Neurosci 8: 22–30

Roenneberg T, Hastings JW (1988) Two photoreceptors control the circadian clock of a unicellular alga. Naturwissenschaften 75: 206–207

Roenneberg T, Nakamura H, Cranmer LD III, Ryan K, Kishi Y, Hastings JW (1991) Gonyauline: a novel endogenous substance shortening the period of the circadian clock of a unicellular alga. Experientia 47: 103–106

Rosbash M, Hall JC (1989) The molecular biology of circadian rhythms. Neuron 3: 387–398

Saez L, Young M (1988) *In situ* localization of the per clock protein during development of *Drosophila melanogaster.* Mol Cell Biol 8: 5378–5385

Satter RL, Morse MJ, Lee Y, Crain RC, Coté GG, Moran N (1988) Light- and clock-controlled leaflet movements in *Samanea saman:* a physiological, biophysical and biochemical analysis. Bot Acta 101: 205–213

Saunders DS, Henrich VC, Gilbert LI (1989) Induction of diapause in *Drosophila melanogaster:* photoperiodic regulation and the impact of arrhythmic clock mutations on time measurement. Proc Natl Acad Sci USA 86: 3748–3752

Sawaki Y, Nihonmatsu I, Kawamura H (1984) Transplantation of the neonatal suprachiasmatic nuclei into rats with complete bilateral suprachiasmatic lesions. Neurosci Res 1: 67–72

Scammell TE, Schwartz WJ, Smith CB (1989) No evidence for a circadian rhythm of protein synthesis in the rat suprachiasmatic nuclei. Brain Res 494: 155–158

Schweiger HG, Hartwig R, Schweiger M (1986) Cellular aspects of circadian rhythms. J Cell Sci [Suppl] 4: 181–200

Shibata S, Newman GC, Moore RY (1987) Effects of calcium ions on 2-deoxyglucose uptake in the suprachiasmatic nucleus *in vitro.* Brain Res 426: 332–338

Shin HS, Bargiello TA, Clark BT, Jackson RJ, Young MW (1985) An unusual coding sequence from *Drosophila* clock gene is conserved in vertebrates. Nature 317: 445–448

Siwicki KK, Strack S, Rosbash M, Hall JC, Jacklet JW (1989) An antibody to the *Drosophila period* protein recognizes circadian pacemaker neurons in *Aplysia* and *Bulla.* Neuron 3: 51–58

Stal LJ, Krumbein WE (1987) Temporal separation of nitrogen fixation and photosynthesis in the filamentous, non-heterocystous cyanobacterium *Oscillatoria* sp. Arch Microbiol 149: 76–80

Sweeney BM, Borgese MB (1989) A circadian rhythm in cell division in a prokaryote, the cyanobacterium *Synechococcus* WH7803. J Phycol 25: 183–186

Takahashi JS, Turek FW (1987) Anisomycin, an inhibitor of protein synthesis, perturbs the phase of a mammalian circadian pacemaker. Brain Res 405: 199–203

Takahashi JS, Murakami N, Nikaido SS, Pratt BL, Robertson LM (1989) The avian pineal, a vertebrate model system of the circadian oscillator: cellular regulation of circadian rhythms by light, second messengers, and macromolecular synthesis. Recent Prog Horm Res 45: 279–352

Tamponnet C, Edmunds LN Jr (1990) Entrainment and phase-shifting of the circadian rhythm of cell division by calcium in synchronous cultures of the wild-type Z strain and of the ZC achlorophyllous mutant of *Euglena gracilis.* Plant Physiol 93: 425–431

Tavladoraki P, Argyroudi-Akoyunoglou J (1989) Circadian rhythm and phytochrome control of LHC-I gene transcription. FEBS Lett 255: 305–308

Tavladoraki P, Kloppstech K, Argyroudi-Akoyunoglou J (1989) Circadian rhythm in the expression of the mRNA coding for the apoprotein of the light-harvesting complex of photosystem II. Plant Physiol 90: 665–672

Taylor WC (1989) Transcriptional regulation by a circadian rhythm. Plant Cell 1: 259–264

Techel D, Gebauer G, Kohler W, Braumann T, Jastorff B, Rensing L (1990) On the role of $Ca^{2+}$-calmodulin-dependent and cAMP-dependent protein phosphorylation in the circadian rhythm of *Neurospora crassa.* J Comp Physiol [B] 159: 695–706

Vanden Driessche T (1989) The molecular mechanism of circadian rhythms. Arch Int Physiol Biochim 97: 1–11

Walla OJ, de Groot EJ, Schweiger M (1989) Identification of a polypeptide in *Chlorella* that apparently is involved in circadian rhythm. Eur J Cell Biol 50: 181–186

Whitfield JF, Durkin JP, Kleine LP, Raptis L, Rixon RH, Sikorska M, Roy Walher P (1987) Calcium, cyclic AMP and protein kinase C – partners in mitogenesis. Cancer Metastasis Rev 5: 205–250

Wollnik F, Turek FW, Majewski P, Takahashi JS (1989) Phase shifting the circadian clock with cycloheximide: response of hamsters with an intact or a split rhythm of locomotor activity. Brain Res 496: 82–88

Yada T, Oiki S, Ueda S, Okada Y (1986) Synchronous oscillation of the cytoplasmic $Ca^{2+}$ concentration and membrane potential in cultured epithelial cells (intestine 407). Biochim Biophys Acta 887: 105–112

Yeung SJ, Eskin A (1987) Involvement of a specific protein in the regulation of a circadian rhythm in *Aplysia* eye. Proc Natl Acad Sci USA 84: 279–283

Yeung SJ, Eskin A (1988) Responses of the circadian system in the *Aplysia* eye to inhibitors of protein synthesis. J Biol Rhythms 3: 225–236

Yu Q, Jacquier AC, Citri Y, Hamblen M, Hall JC, Rosbash M (1987a) Molecular mapping of point mutations in the period gene that stop or speed up biological clocks in *Drosophila melanogaster.* Proc Natl Acad Sci USA 84: 784–788

Yu Q, Colot HV, Kyriacou P, Hall JC, Rosbash M (1987b) Behaviour modification by *in vitro* mutagenesis of a variable region within the *period* gene of *Drosophila.* Nature 326: 765–769

Zatz M, Mullen DA, Moskal JR (1988) Photoendocrine transduction in cultured chick pineal cells: effects of light, dark, and potassium on the melatonin rhythm. Brain Res 450: 199–215

Zwartjes RE, Eskin A (1990) Changes in protein phosphorylation in the eye of *Aplysia* associated with circadian rhythm regulation by serotonin. J Neurobiol 21: 376–383

# The Suprachiasmatic Nucleus and Other Pacemakers

W. J. Rietveld

## Introduction

A host of experiments involving mainly rats and hamsters have led to the recognition of the suprachiasmatic nuclei (SCN) of the hypothalamus as the site of an endogenous circadian oscillator in mammals (Rietveld and Groos 1980; Meijer and Rietveld 1989). Using an autoradiographic tracing method Moore (1973) demonstrated a direct neuronal connection between the retina and the SCN in the rat, the retinohypothalamic projection (RHP). In addition to this anatomical finding there is other empirical support for the assumption that the SCN are a major pacemaker. Many behavioural circadian rhythms are abolished by complete bilateral SCN lesions or surgical islation (Rusak and Zucker 1979). Electrical stimulation of the SCN alters the phase of circadian rhythms in locomotor activity in rodents (Rusak and Groos 1982). With the aid of the 2-DG method, Schwartz et al. (1980) demonstrated a circadian rhythm in metabolic activity in the SCN, glucose utilization being high during the light period. No other brain area exhibits a similar rhythm. In accordance with this are electrophysiological studies (Inouye and Kawamura 1979) showing that in vivo and in vitro the multiunit activity within the SCN is high during the light period and low during darkness.

From the anatomical point of view the SCN are two small nuclei lying immediately above the optic chiasm. Each nucleus contains about 10 000 neurones. There is also anatomical evidence that the SCN in the rat has at least three subdivisions. A rostral part, about one-fourth of the total nucleus, contains small neurones with a scant cytoplasm and relatively few organelles. The few dendrites have a limited arborization (Moore et al. 1980). The caudal part consists of a dorsomedial part, quite similar to the rostral part. The neurones of the ventrolateral part are larger with more cytoplasm and a more extensive dendritic arborization. This latter region is

characterized by the presence of the terminals of the RHP. Using classical neuroanatomical techniques, van den Poll (1980) described that SCN neurones have relatively simple dendritic arbors. He identified simple bipolar, curly bipolar, radial monopolar and spinous multipolar cells. At the ultrastructural level Gueldner (1976) described two Gray I type synapses and three Gray II types. The compartmentalization of the SCN is further supported by histochemical studies. Three groups of neuropeptide-containing cells are found whithin the nucleus. Vasopressin-containing neurones are located exclusively in the rostral and dorsomedial part. Vasoactive intestinal peptide is present in neurones of the ventrolateral part, whereas neurones containing somatostatin, substance P or avian pancreatic polypeptide are present throughout the whole SCN. In addition a number of other peptides are found in terminals. Serotonin and molluscan cardioexcitatory peptide containing fibres are found in the ventral portion of the SCN, whereas fibres containing leu-enkephalin and cholecystokinin are found in the area immediately surrounding the border of the SCN. Terminals in this region probably innervate dendrites extending outside the cellular area of the SCN. It is an intriguing but as yet unsolved problem why so many neuropeptides are localized within the SCN.

## The SCN As an Endogenous Oscillator

A pacemaker function of the SCN has been suggested in many studies in which complete bilateral lesions or surgical isolation of the SCN abolished various circadian rhythms in rodents (Rusak and Zucker 1979). These experiments, however, are not conclusive in addressing the question whether the SCN is a pacemaker of circadian rhythmicity. First, the SCN may function as a relay station for those particular rhythms that are abolished after SCN lesions. The demonstration that disruption of several afferent connections (e.g. from the eyes, the ventral lateral geniculate nucleus and the raphe nuclei) fails to abolish free-running activity patterns does not solve this problem since other, undisrupted, inputs may be responsible for maintaining the activity rhythm. Second, physical injury that causes neurones to die will bring about permanent changes in the structure of the nervous system and this structural change is usually accompanied by long-lasting alterations in the functions of the affected areas (Kelly 1981). Such secondary effects of SCN lesions may not be restricted to the SCN.

In several studies, a non-destructive technique was used to investigate the pacemaker function of the SCN (Albers et al. 1984a; Albers and Ferris 1984; Meijer et al. 1984; Meijer and Groos 1988; Rusak and Groos 1982; Zatz and Brownstein 1981; Takahashi and Turek 1987). Electrical stimulation of the SCN in hamsters and rats resulted in phase-dependent phase shifts of the free-running activity cycle (Rusak and Groos 1982). Phase-shifting effects are also observed after local stimulation of the SCN with neuropeptide Y (Albers and Ferris 1984), with the cholinergic agonist carbachol (Zatz and Brownstein 1979) and with glutamate (Meijer et al. 1988). Local application of the protein synthesis inhibitor anisomysin also induces phase-dependent phase shifts (Takahashi and Turek 1987). In all these cases the demonstration that stimulation of the SCN phase shifts circadian rhythms implies that the SCN drives rhythmicity by imposing its oscillation on structures located elsewhere in the animal.

When suprachiasmatic tissue of young embryos is transplanted into the anterior chamber of the eye or in the lateral, third or fourth ventricle, differentiation and growth occurs (Boer et al. 1985; Roberts et al. 1987; Wiegand and Gash 1988). Implantation of suprachiasmatic tissue of rat fetuses in the third ventricle results in a number of efferent connections. Vasopressinergic connections have been observed to the medial preoptic area, the periventricular and dorsomedial hypothalamic nuclei, the paraventricular nucleus of the thalamus and hypothalamus, the retrochiasmatic area, the arcuate nucleus and to the SCN of the horst brain, 4–6 weeks after the implantation (Wiegand and Gash 1988). Iontophoretic application of the orthograde tracer *Phaseolus vulgaris* leucoagglutinin in the graft (after at least 14 weeks following the transplantation) resulted only in a few labelled fibres in the adjacent host hypothalamus (Lehman et al. 1987). These results show that transplanted SCN tissue grows with (few) efferents into the surrounding brain areas.

Fetal or neonatal suprachiasmatic tissue has also been transplanted in host animals that had previously received a SCN lesion. Such transplantations could restore rhythmicity of the behavioural activity and drinking rhythm (Aguilar-Roblero et al. 1986; Drucker-Colin et al. 1984; DeCoursey and Buggy 1986; Lehman et al. 1987; Sawaki et al. 1984). Rhythmicity was restored in blinded host animals as well as in rats that were exposed to a light-dark cycle or constant darkness. In one of these studies (Lehman et al. 1987) extensive histological inspection afterwards revealed that unsuccessful grafts were characterized by an incomplete immunostaining for several neuropeptides (vasoactive intestinal polypeptide, neuropeptide Y, somatostatin and neurophysin or vasopressin). This may suggest that the organization of peptides within the SCN is critical for locomotor rhythmicity (Lehman et al. 1987). On the other hand, a well-organized peptidergic structure of the transplanted SCN may reflect an overall well-developed structure of this tissue.

In summary, it can tentatively be concluded that the SCN is not merely a circadian oscillator but also functions as a circadian pacemaker. As yet, we lack criteria by which SCN neurones that are part of the rhythm-generating mechanism can be recognized. As a result, the distinction between the afferents of the circadian pacemaker and the pacemaker of the SCN itself is arbitrary. For instance, visual responsive cells of the mammalian SCN can be considered as input to the pacemaker. However, they could just as well be part of the rhythm-generating mechanism. The same is true for the neurotransmitters that are present within the SCN. The phase-shifting effects of these transmitters may indicate that they are of importance for photic entrainment (or entrainment to other physiological processes). Alternatively the induced phase shifts could also reflect that these substances are involved in mutual entrainment of groups of pacemakers (or pacemaker cells) inside the SCN. This latter possibility holds especially for transmitters that are implicated in suprachiasmatic interneurones (such as GABA).

On the other hand, the properties of suprachiasmatic cells, as they are determined by electrophysiological experiments, may be primarily of importance for entrainment of the pacemaker and may not be relevant for the generation of rhythms.

According to Moore (1982), the SCN neurones are initially produced as a set of genetically determined, independent oscillators that become interconnected during development so that individual neuronal function now becomes a network function. Firstly, each SCN is interconnected with the contralateral SCN by a highly topographically ordered fibre system. Neurones of an individual SCN subsequently differentiate probably into at least two neuronal networks within the SCN, a ventrolateral and a dorsomedial group including a rostral component. This fits with the results of experiments described by Rietveld (1984). Electrolytic lesions of the rostral part of both SCN in blinded rats alter the period of their free-running behaviour. After such partial lesions the rhythms in locomotor activity, food and water intake, in body temperature as well as in urine corticosterone return within a period of 30–60 days but now with a shorter free-running period than before the lesion. Complete lesions of the whole SCN as well as lesions of the caudal part completely disrupt all circadian rhythmicity (suggesting a dual oscillator system). As for the generation of circadian rhythms synaptic interactions between SCN cells are important. Infusion of tetrodotoxin (TTX) into the SCN of unanaesthetized and unrestrained rats blocks the function of input and output pathways without affecting the actual oscillatory mechanism of the nuclei themselves (Schwartz et al. 1987). Drinking activity disappeared during infusion of TTX during 14 days, but reappeared with a phase that could be predicted by extrapolation of the period length before infusion. This suggests that intercellular communication plays a role in the synthesis of the oscillation. Ca-dependent spike activity or graded Ca-dependent release of neurotransmitters play a greater role than firing all-or-none spikes. Also glia-neuronal interactions may play a role in circadian rhythm generation. Morin et al. (1989) describe a dense GFAP-like immunoreactivity in the SCN suggesting that astrocytes are involved, either by producing a specific trophic substance required for maintenance of the clock function or by enhancing the neuronal communication. SCN astrocytes have extensive gap junctions which could facilitate intranuclear communication. In addition it is noteworthy that astrocytes contain numerous specific receptors (Murphy and Pearce 1987). So neuropharmacological effects might be due to an effect on the glial-neurone interaction.

For most of the neuropeptides present in the SCN cells it is still unknown how far they are involved in circadian time-keeping. Local injection of $\alpha$-bungarotoxin, an irreversible cholinergic antagonist, into the SCN does not affect circadian rhythmicity in pineal activity (Zatz and Brownstein 1981). Brattleboro rats that lack vasopressin in the SCN show undisturbed circadian rhythms. Injection of vasopressin in the rat SCN does not change the period of free-running activity rhythms. The serotonin (5-HT) in the SCN is located in terminals, the perikarya of which are located in the midbrain raphe nuclei. 5-HT may act as a transmitter between these nuclei and the SCN. Electrophysiological studies reveal a response of SCN neurones to iontophoretic application of serotonin. In spite of this there is no effect on free-running period after local application of 5-HT into the SCN.

One of the first agents known to affect circadian rhythms is lithium. In most of the earlier experiments on plants, on insects, on mammals including man, it has been described that it lengthens the period of free-running rhythms (Wirz-Justice et al. 1982; Engelmann 1973; Hofmann et al. 1978; Kavaliers 1981; Kripke et al. 1978; Kripke and Wyborney 1980; Johnsson et al. 1980). Others like Delius et al. (1984) studying locomotor activity of hamsters could not find any consistent effect of lithium added to the drinking water. Infusion of lithium into rats using Alzet miniosmic pumps did not show a difference in period length between the lithium group and control rats. So any consistent effect on the pacemaker is not easy to understand. Delius et al. (1984) therefore propose a model in which lithium alters the coupling between circadian oscillators equivalent to the model developed by Kronauer et al. (1982). The fact that gene dependency has been shown in mice by a differential lengthening in different strains (Possidente and Hegman 1982; Possidente and Exner 1986) supports the view of a more complicated action of lithium on the central control of behaviour.

Similar rather inconsistent data have been obtained from experiments with chronic application of the MAO-A inhibitor clorgyline. Administration of the drug by means of miniosmic pumps increases the circadian period of locomotor activity of female hamsters. Similar experiments done in rats do not show any effect (Rietveld et al. 1986). Clorgyline applied for 2 weeks by means of osmotic minipumps reduces significantly 5-HT and 5-HIAA levels of brain tissue as compared to control rats. The amplitude of free-running food intake is reduced for about 3 weeks after implantation. In none of the animals was there any change in period value. However, in the case of local implantation of clorgyline near the SCN there is some evidence of an increase in period length (Wirz-Justice et al. 1982).

Another antidepressant drug, imipramine, seems to have an effect on the period length (Wirz-Justice and Campbell 1982). They claim that in rats as well as in hamsters free-running rhythms in behaviour are slowed down after continuous infusion of the drug. A similar slowing of the pacemaker has been described after application of deuterium oxide and lithium (Daan and Pittendrigh 1976).


## Entrainment of the Oscillator to the Environment


In addition to the free-running period, the phase-dependent phase shifts of a rhythm exposed to stimuli at various time points is another parameter of an oscillator that can be measured. However, little is known about the physiological mechanisms that mediate the entrainment of SCN neurones to the L/D cycle or communicate circadian information to other physiological systems. Recent studies have provided information on the morphology and neurochemistry of afferent projections to the SCN. Although the RHP is the main pathway mediating the effects of light on circadian rhythms, it does not imply that it is the only one. Later anatomical data proved the existence of a secondary projection from the ventral geniculate nucleus to the SCN. The terminal distribution of both tracts overlap. Cells of the SCN that are light-responsive change their firing rate tonically either by an increase or by a decrease when the ambient luminance level is changed (Groos and Hendriks 1979). Similar cell types are found in the VLGN. Photic entrainment may involve acetylcholine and its nicotinic cholinergic receptor in the SCN. Injections of the cholinergic agonist carbachol into the SCN phase-shifts locomotor activity in rodents (Zatz and Herkenham 1981). It is, however, unknown whether acetylcholine is a transmitter between RHP terminals and SCN cells or between intrinsic cells of the SCN itself. Recent experiments described by Liou et al. (1986) suggest a role of glutamate and of aspartate in the transmission of signals from the RHP into the SCN. There is evidence that the ventral part of the lateral geniculate nucleus (VLGN) projects to the SCN with avian pancreatic polypeptide (APP)-containing fibres (Card and Moore 1982). Recent experiments (Moore et al. 1984) have demonstrated that this is probably neuropeptide Y (NPY), another 36 amino acid peptide which has 20 amino acids homologous with APP. Microinjections of APP into the SCN shift hamster circadian activity rhythms in a similar way to dark pulses during constant light (Albers et al. 1984a). However, lesions of the VLGN have no effect on the

rate of reentrainment after a phase shift. So the function of this afferent connexion remains unclear for the rat. Unilateral lesions of the SCN shorten the period of locomotor activity in the hamster in constant light. In the rat no change in period can be observed; there is a slight increase in reentrainment after a phase shift in this animal. Unilateral blinding, however, retards the rate of phase shifting by 2 days and decreases the period of free-running in constant light, suggesting an effect of asymmetrical innervation of the two SCN by the RHP. This view is supported by the fact that this phenomenon is not present in the hamster, in which the RHP innervates the SCN symmetrically (Donaldson and Stephan 1981).

Slowing down a circadian pacemaker will delay the phase position of entrained rhythms (Aschoff 1965). This is the case for lithium, clorgyline and pargyline in rats and in hamsters (von Rommelspacher et al. 1976; Craig et al. 1981). This does not hold for all rhythms. McEachron et al. (1982; McEachron 1984) showed that in lithium-fed rats that were synchronized to a light-dark cycle the plasma prolactin rhythm was phase delayed, whereas the pineal serotonin rhythm was not. A similar delay was described for locomotor activity. It is noteworthy that the C14-2-deoxyglucose rhythm of the SCN was undisturbed (McEachron 1984). Aschoff (1986) described that adding imipramine to the drinking water over 1 month did not affect the circadian activity rhythm of hamsters. There were no changes in phase position nor differences in reentrainment after a 6-h phase shift. Similar negative results of effect of imipramine on entrained rhythms are described by Fowler et al. (1985). After oral administration for 21 days there was no effect on the pineal melatonin rhythm in the rat. Haeusler et al. (1985) describe a lack of effect on rat urinary corticosterone rhythm after repeated intraperitoneal injections of imipramine.

In addition to an effect of species differences, these results suggest a selective effect on different pacemakers of the circadian oscillator networks. In man Kripke et al. (1979) described a slight delay of the phase of midpoint sleep.

Not only the phase position of a rhythm in respect to the entrainment signal, but also the shape of the phase response curve itself seems to be affected. Han (1984) demonstrated an effect of lithium on the phase response curve of hamster locomotor activity, reducing the phase advance area, whereas the phase delay area increased. This might explain the finding of Reinhard (1983, 1985) that during application of lithium hamsters show an increase of the upper range of entrainment.

## Control of Subordinate Structures

The SCN drive a great number of behavioural and physiological rhythms. SCN lesions affect the rhythmicity in locomotor activity (Rusak 1977), food intake (Boulos et al. 1980; Stoynev et al. 1982), water intake (Boulos et al. 1980), sexual behaviour (Eskes 1984; Sodersten et al. 1981) as well as deep body temperature (Eastman et al. 1983; Ruis et al. 1987; Saleh et al. 1977) and sleep wakefulness cycle (Eastman et al. 1983; Hanada and Kawamura 1981; Ibuka et al. 1977). In addition, several hormone levels are under the influence of the SCN. This holds in particular for the synthesis and/or secretion of adrenocorticotropic hormone (Assenmacher 1982), adrenal corticosterone (Moore and Eichler 1972), pituitary prolactin (Bethea and Neill 1980; Kawakami et al. 1980), pineal melatonin (Klein and Moore 1979) and gonadotropin (Wiegand and Terasawa 1982).

The involvement of neuronal pathways in the control of these functions becomes evident as effects of SCN lesions can be mimicked by surgical isolation of the SCN (Eskes and Rusak 1985). Moreover, TTX infusion in the SCN, which blocks sodium-dependent spikes, produces arrhythmicity in activity and food intake during the infusion (Schwartz et al. 1987). This result also indicates neuronal control of these rhythms. Transplantation of the SCN into an arrhythmic animal can restore circadian rhythmicity in activity and drinking. In these experiments only a few efferents of the graft were detected (Lehman et al. 1987). The transplantation experiments therefore contradict the importance of neuronal efferents for rhythmicity outside the SCN. Instead these results suggest humoral control of circadian rhythms (Lehman et al. 1987). Although this matter remains as yet unresolved, it is conceivable that a small number of efferent projections is sufficient to impose rhythmicity on other functions.

Evidence on the distribution of projections from the SCN is based on numerous studies, using anterograde as well as retrograde labelling techniques. Autoradiographic experiments using [3H]-amino acids (Swanson and Cowan 1975; Stephan et al. 1981; Berk and Finkelstein 1981) as well as experiments using *Phaseolus vulgaris* leucoagglutin (PHA-L) (Watts et al. 1987) reveal six main efferent pathways to areas in which fibre terminals can be identified (see Watts et al. 1987).

A dense plexus of fibres originate in the SCN just dorsal and caudal to the nucleus between the periventricular nucleus and the anterior hypothalamic area. Fibre endings are mostly ipsilateral in a region ventral to the posterior part of the paraventricular nucleus, the subparaventricular zone. A few fibres continue dorsally from this zone, pass through the parvocellular parts of the paraventricular nucleus and midline thalamic nuclei, to end in the midrostal parts of the paraventricular nucleus of the thalamus, the dorsomedial nucleus and the area around the ventromedial nucleus as well as the posterior hypothalamic area.

The other pathways consist of relatively smaller amounts of fibres. A second bundle runs rostrally and ends in the ventral parts of the medial preoptic area and anteroventral periventricular nucleus. Anterodorsally fibres pass through the medial preoptic nucleus to end in the intermediate lateral septal nucleus. Caudally to this group, fibres have been traced that end in the preoptic continuation of the bed nucleus of the stria terminalis, in the parataenial nucleus, in the rostral part of the paraventricular nucleus of the thalamus as well as in the midbrain central grey and the raphe nuclei (Bons et al. 1983; Kucera and Favrod 1979; Stephan et al. 1981; Berk and Finkelstein 1981; Watts et al. 1987).

Laterally directed fibres end in the ventral lateral geniculate nucleus. Finally, descending fibres connect the SCN with the zone between the arcuate nucleus, the ventral part of the ventromedial nucleus as well as parts of the lateral hypothalamic area. All these pathways have been confirmed by retrograde labelling with fluorescent dyes (Watts and Swanson 1987).

A striking observation from these studies is the finding that retrograde labelling was never restricted to the SCN, but was always accompanied by a marked labelling of the area around the SCN as well as of the subparaventricular zone (Watts et al. 1987). Implants of the dye true blue in the zona incerta, the dorsomedial nucleus and the ventromedial nucleus labelled even more neurones in the area immediately surrounding the SCN than within the nucleus (Watts et al. 1987). Two efferent projections of the SCN terminate in areas from which in turn afferents to the SCN arise. One is the IGL (Watts et al. 1987) which projects with neuropeptide Y containing fibres to the SCN. The second is the raphe (Bons et al. 1983) from which serotonergic fibres arise. Both feedback loops could allow the SCN to modulate its own input.

Several attempts have been made to identify transmitter substances of these efferent systems (Card et al. 1981; Moore 1983; Sofroniew and Weindl 1982; Ueda et al. 1983; Watts and Swanson 1987). Neurochemically, cells are organized in several subfields within the SCN in a heterogeneous manner (Card and Moore 1984; van den Pol and Tsujimoto 1985; Watts et al. 1987). Immunocytochemistry has been used to trace the efferent fibres from the SCN. However, as the origin of immunoreactive fibres is difficult to ascertain, other tracing techniques are required.

Fibres from the SCN to the paraventricular nucleus have been described containing vasopressin, neurophysin and VIP (Sofroniew and Weindl 1978; Sims et al. 1980). Combining immunohistochemistry and retrograde transport of fluorescent dyes (Watts et al. 1987) shows that projections from the SCN to the midline thalamus, the subparaventricular zone and the dorsomedial nucleus contain vasopressin, VIP and neurotensin-stained fibres. These fibres follow separate pathways. Vasopressin axons run laterally around the lateral edge of the paraventricular nucleus into the dorsomedial nucleus, whereas VIP neurones do not project as far laterally but terminate caudally to the paraventricular zone (Watts et al. 1987). For a detailed study and review on the neuropharmacology of SCN efferents we refer to Watts and Swanson (1987).

An important finding is that the efferent fibre system of the SCN projects to the same area that receives an input from the peri-SCN as well as from the supraventricular area. This suggests some integrative or amplifying function of the latter (Watts and Swanson 1987). However, the question still remains to be solved whether this only involves the transmission of the timing signal from the SCN pacemaker already controlled by the photic signal that is presented via the RHT (Rusak 1982).

The question arises to what degree different homeostatic control systems are affected by the efferents of the SCN. The rostral fibres end diffusely in the medial preoptic area. This connection allows the SCN to control the intake of water, the deep body temperature as well as the reproductive behaviour. In the female rat the cyclic control of gonadotropins is exerted by the anteroventral periventricular nucleus, which also contains endings of the SCN (Simerly and Swanson 1986; Wiegand and Terasawa 1982). However, the control of neuroendocrine rhythms probably takes place by way of multisynaptic pathways as there are no endings at the level of the median eminence (Watt et al. 1987). This is in contrast with earlier reports (Swanson and Cowan 1975; Stephan et al. 1981). In these studies probably neurons in the neighbourhood of the SCN have been labelled (Berk and Finkelstein 1981; Watts et al. 1987). This widespread distribution might explain why knife cuts posterior to the SCN abolish adrenal corticosterone and

pineal N-transferase rhythms as well as the oestrous cycle, whereas food intake and drinking are abolished only by larger semicircular knife cuts around the SCN (Dark 1980; Honma et al. 1984; Nishio et al. 1979; Nunez and Stephan 1977; Nunez and Casati 1979; Stephan and Nunez 1977; Wiegand and Terasawa 1982).

The caudal fibres reach the anterior hypothalamus including the retrochiasmatic zone, diverge and end in zones immediately surrounding the arcuate nucleus, the ventromedial nucleus as well as the dorsomedial nucleus. Luiten and Room (1980) have shown that neuronal connections between the ventromedial and the lateral hypothalamic area all pass through the dorsomedial nucleus. This explains the finding that selective lesioning of the dorsomedial nucleus interrupts the circadian control of food intake without affecting the rhythm in body temperature, drinking and locomotor activity (Rietveld et al. 1983).

An important pathway is the one involved in the generation by the SCN of the circadian rhythm in pineal N-acetyltransferase. This rhythm is controlled by a release of noradrenaline from postganglionic fibres of the superior cervical ganglion. Preganglionic fibres originate in the intermediolateral cell column of the spinal cord. Monosynaptic projections to the spinal cord have been described from the dorsal and medial parvocellular area of the paraventricular hypothalamic nucleus as well as from the retrochiasmatic area (Swanson and Kuypers 1980a,b). Both areas receive efferents from the SCN.

Although the present studies provide some evidence that different circadian rhythms are controlled by separate neuronal pathways, it remains to be solved whether or not there is any interaction between these pathways. Such interactions could explain complex behaviour in which several functions are mutually integrated.

Dissociation of a rhythm can be considered as a modification of the coupling between the driving oscillator and subordinate control centres. Sometimes this can be induced by constant light as well as by administration of drugs. A similar effect has been described after continuous infusion of clorgyline and imipramine by means of osmic mini-pumps during 4–28 days (Wirz-Justice and Campbell 1982). This is specific drug effect, as in the case of application of Deuterium oxide or lithium, also drugs that lengthen the free-running period too, such an effect has never been seen.

The most dramatic internal desynchronization has been observed after application of methamphetamine, a stimulant with a mild antidepressive action (Rietveld et al. 1985; Honma et al. 1986; Honma and Honma 1986). Oral application of the drug in blinded rats while recording free-running food intake, water intake as well as wheel-running activity results in a delay of part of the activity sometimes in the range of circabidian days. Other components free run with the same period length as before the drug, suggesting an uncoupling of the SCN from other lower order control centres. This is supported by the fact that similar circabidian activity can be induced by methamphetamine in rats that have received a bilateral lesion of the SCN (Rietveld et al. 1986).

Administration of methamphetamine by means of osmotic minipumps seems to be ineffective as was described by Kraeuchi et al. (1986).

A final question to ask is whether there are other circadian oscillators outside the SCN. There are conflicting data about the presence of body temperature rhythms in SCN-lesioned animals. Stephan et al. (1981) describe an anticipatory response in animals that have received complete lesioning of both SCN; however, both location and nature of such an oscillator remain to be elucidated.

# References

Aguilar-Roblero R, Garcia-Hernandez F, Aguilar R, Arankowsky-Sandoval G, Drucker-Colin R (1986) Suprachiasmatic nucleus transplants function as an endogenous oscillator only in constant darkness. Neurosci Lett 69: 47–52

Albers HE, Ferris CF (1984) Neuropeptide Y: role in light-dark cycle entrainment of hamster circadian rhythms. Neurosci Lett 50: 163–168

Albers HE, Ferris CF, Leeman SE, Goldman BD (1984a) Avian pancreatic polypeptide phase shifts hamster circadian rhythms when injected into the suprachiasmatic region. Science 223: 833–835

Albers HE, Lydic R, Gander PH, Moore-Ede MC (1984b) Role of the suprachiasmatic nuclei in the circadian timing system of the squirrel monkey. I. The generation of rhythmicity. Brain Res 300: 275–284

Aschoff J (1965) The phase-angle difference in circadian periodicity. In: Aschoff J (ed) Circadian clocks. North-Holland, Amsterdam, pp 263–276

Aschoff J (1986) Circadian activity rhythm in hamsters unaffected by imipramine. In: Hildebrandt G, Moog An R, Raschke F (eds) Chronobiology and chronomedicine. Lang, Frankfurt, pp 243–247

Assenmacher I (1982) CNS structures controlling circadian neuroendocrine and activity rhythms in rats. In: Aschoff J, Daan S, Groos GA (eds) Vertebrate circadian system. Springer, Berlin Heidelberg New York, pp 87–96

Berk ML, Finkelstein JA (1981) An autoradiographic determination of the efferent projections of the suprachiasmatic nucleus of the hypothalamus. Brain Res 226: 1–13

Bethea CL, Neill J (1980) Lesions of the suprachiasmatic nuclei abolish the cervically stimulated prolactin surges in the rat. Endocrinology 107: 1–5

Boer GJ, Gash DM, Dick L, Schluter N (1985) Vasopressin neuron survival in neonatal Brattleboro rats; critical factors in graft development and innervation of the host brain. Neuroscience 15: 1087–1109

Bons N, Combes A, Szafarczyk A, Assenmacher I (1983) Efférences extrahypothalamiques du noyau suprachiasmatique chez le rat. CR Acad Sci [III] 297 (111): 347–350

Boulos Z, Rosenwasser AM, Terman M (1980) Feeding schedules and the circadian organization of behavior in the rat. Behav Brain Res 1: 39–65

Branchey L, Weinberg U, Branchey M, Linkowsky P, Mendelwicz J (1982) Simultaneous study of 24-hour patterns of melatonin and cortisol secretion in depressed patients. Neuropsychobiology 8: 225–232

Brown R, Kocsis JH, Caroff S, Amsterdam J, Winokur A, Stokes PE, Frazer A (1985) Differences in nocturnal melatonin secretion between melancholic depressed patients and controls. Am J Psychiatry 142: 811–816

Card JP, Moore RY (1982) Ventral lateral geniculate nucleus efferents to the rat suprachiasmatic nucleus exhibit avian pancreatic polypeptide-like immunoreactivity. J Comp Neurol 206: 390–396

Card JP, Moore RY (1984) The suprachiasmatic nucleus of the golden hamster: immunohistochemical analysis of cell and fiber distribution. Neuroscience 13: 415–431

Card JP, Brecha N, Karten HJ, Moore RY (1981) Immonocytochemical localization of vasoactive intestinal polypeptide containing cells and processes in the suprachiasmatic nucleus of the rat: light and electron microscopic analysis. J Neurosci 1: 1289–1303

Craig C, Tamarkin L, Garrick N, Wehr TA (1981) Long-term and short-term effects of clorgyline (a monoamineoxidase type A inhibitor) on locomotor activity and on pineal melatonin in the hamster. Abstract no. 2294.14, Society for Neuroscience 11th Annual Meeting

Daan S, Pittendrigh CS (1976) A functional analysis of circadian pacemakers in nocturnal rodents. J Comp Physiol 106: 267–290

Dark J (1980) Partial isolation of the suprachiasmatic nuclei: effects in the circadian rhythms of rat drinking behavior. Physiol Behav 25: 863–873

Decoursey PJ, Buggy J (1986) Restoration of locomotor rhythmicity in SCN-lesioned golden hamsters by transplantation of fetal SCN. Neurosci Abstr 12: 210

Delius K, Gunderoth-Palmowski M, Krause I, Engelmann W (1984) Effects of lithium salts on the behaviour and the circadian system of Mesocricetus auratus W. J Interdisc Cycle Res 15: 289–299

Donaldson JA, Stephan FK (1981) Entrainment of circadian rhythms: retinofugal pathways and unilateral suprachiasmatic nucleus lesions. Physiol Behav 29: 1161–1169

Drucker-Colin R, Aguilar-Roblero R, Garcia-Hernandez F, Fernandez-Cancinoa F, Rattoni FB (1984) Fetal suprachiasmatic nucleus transplants: diurnal rhythm recovery of lesioned rats. Brain Res 311: 353–357

Eastman CI, Mistlberger RE, Rechtschaffen A (1983) Suprachiasmatic nuclei lesions eliminate circadian temperature and sleep rhythms in the rat. Physiol Behav 32: 357–368

Engelmann W (1973) A slowing down of circadian rhythms by lithium ions. Z Naturforsch [C] 28: 733–736

Eskes GA (1984) Neural control of the daily rhythm of sexual behavior in the male golden hamster. Brain Res 293: 127–141

Eskes GA, Rusak B (1985) Horizontal knife cuts in the suprachiasmatic area prevent hamster gonadal responses to photoperiod. Neurosci Lett 61: 261–266

Fowler CJ, Hall H, Saaf J, Ask A-L, Ross SB (1985) Pharmacological manipulation of biochemical measured rhythms in the mammalian central nervous system. In: Redfern PH, Campbell IC, Davies TA (eds) Circadian rhythms in the central nervous system. MacMillan, London, pp 111–121

Groos G, Hendriks J (1979) Regularly firing neurones in the rat suprachiasmatic nucleus. Experientia 35: 1597–1598

Gueldner FH (1976) Synaptology of the rat suprachiasmatic nucleus. Cell Tissue Res 165: 509–544

Haeusler A, Hauser K, Meeker JB (1985) Effects of subchronic administration of psychoactive substances on the circadian rhythm of urinary corticosterone excretion in rats. Psychoneuroendocrinology 10: 421–429

Han SZ (1984) Lithium chloride changes sensitivity of hamster rhythm to light pulses. J Interdiscip Cycle Res 15: 139–146

Hanada Y, Kawamura H (1981) Sleep-waking electrocorticographic rhythms in chronic cerveau isolé rats. Physiol Behav 26: 725–728

Hofmann K, Gunderoth-Palmowski M, Wiedenmann G, Engelmann W (1978) Further evidence for period lengthening effect of Li + on circadian rhythms. Z Naturforsch [C] 33: 231–234

Honma K, Honma S (1986) Effects of methamphetamine on development of circadian rhythms in rats. Brain Dev 8: 397–401

Honma K, Honma S, Hiroshige T (1986) Disorganization of the rat activity rhythm by chronic treatment with methamphetamine. Physiol Behav 38: 687–695

Honma S, Honma K, Hiroshige T (1984) Dissociation of circadian rhythms in rats with a hypothalamic island. Am J Physiol 246: R949–R954

Ibuka M, Inouye S, Kawamura H (1977) Analysis of sleep-wakefulness rhythms in male rats after suprachiasmatic nucleus lesions and ocular enucleation. Brain Res 122: 33–47

Inouye SI, Kawamura H (1979) Persistence of circadian rhythmicity in a mammalian hypothalamic "island" containing the suprachiasmatic nucleus. Proc Natl Acad Sci USA 76: 5962–5966

Johnsson A, Engelmann W, Pflug B, Klempke W (1980) Influence of lithium ions on human circadian rhythms. Z Naturforsch [C] 35: 503–507

Kavaliers M (1981) Period lengthening and disruption of socially facilitated circadian activity rhythms of goldfish by lithium. Physiol Behav 27: 625–628

Kawakami M, Arita J, Yoshioka E (1980) Loss of estrogen induced daily surges of prolactin and gonadotropin by suprachiasmatic nucleus lesions in ovariectomized rats. Endocrinology 106: 1087–1092

Kelly JP (1981) Reactions of neurons to injury. In: Kandell ER, Schwartz JH (eds) Principles of neural science. Elsevier, New York, pp 138–146

Klein DC, Moore RY (1979) Pineal-N-acetyltransferase and hydroxyindole-o-methyltransferase. Control by the retino hypothalamic tract and the suprachiasmatic nucleus. Brain Res 174: 245–262

Kraeuch F, Wirz-Justice A, Feer H (1986) Food selection in methamphetamine treated rats: dependence on circadian period. In: Hildebrandt G, Moog An R, Raschke F (eds) Chronobiology and chronomedicine. Lang, Frankfort, pp 247–252

Krieger DT (1980) Ventromedial hypothalamic nucleus lesions abolish food-shifted circadian adrenal and temperature rhythmicity. Endocrinology 106: 649–654

Kripke DF (1983) Phase advance theories for affective illnesses. In: Wehr TA, Goodwin FK (eds) Circadian rhythms in psychiatry. Boxwood, Pacific Grove, pp 41–69

Kripke DF, Wyborney VW (1980) Lithium slows at circadian rhythms. Life Sci 26: 1319–1321

Kripke DF, Mullaney DJ, Atkinson M, Wolfe S (1978) Circadian rhythm disorders in manic-depressives. Biol Psychiatry 13: 335–348

Kripke DF, Judd LL, Hubbard B, Janowsky DS, Huey LY (1979) The effect of lithium carbonate on the circadian rhythm of sleep in normal human subjects. Biol Psychiatry 14: 545–548

Kronauer RE, Czeisler CA, Pilato SF, Moore-Ede C, Weitzman ED (1982) Mathematical model of the human circadian system with two interacting oscillators. Am J Physiol 242: R3–R17

Kucera P, Favrod P (1979) Suprachiasmatic nucleus projection to the mesencephalic grey in the woodmouse (Apodemus sylvaticus L.). Neuroscience 4: 1705–1715

Lehman MN, Silver R, Gladstone WR, Kahn RM, Gibson M, Bittman EL (1987) Circadian rhythmicity restored by neural transplant. Immunocytochemical characterization with the host brain. J. Neurosci 7: 1626–1638

Lewy AJ, Sack RA, Singer CL (1984) Assessment and treatment of chronobiologic disorders using plasma melatonin levels and bright light exposure: the clock-gate model and the phase response curve. Psychopharmacol Bull 20: 561–565

Lewy AJ, Sack RL, Singer CL (1985) Treating phase typed chronobiological sleep and mood disorders using appropriately time bright artificial light. Psychopharmacol Bull 21: 368–372

Liou SY, Shibata S, Iwasaki K, Ueki S (1986) Optic nerve stimulation-induced increase of release of 3H-glutamate and 3H aspartate but not of 3H-GABA from the suprachiasmatic nucleus in slices or rat hypothalamus. Brain Res Bull 16: 527–531

Luiten PG, Room P (1980) Interrelations between lateral, dorsomedial and ventromedial hypothalamic nuclei. Brain Res 190: 321–332

McEachron DK (1984) Testing the circadian hypothesis of affective disorders: lithium's effect on selected circadian rhythms in rats. Thesis, University of California, San Diego

McEachron DL, Kripke DF, Hawkins R, Haus E, Pavlinac D, Deftos L (1982) Lithium delays biochemical circadian rhythms in rats. Neuropsychobiology 8: 12–29

Meijer JH, Groos GA (1988) Responsiveness of suprachiasmatic and ventral lateral geniculate neurons to serotonin and imipramine: a micro-iontophoretic study in normal and imipramine-treated rats. Brain Res Bull 20: 89–96

Meijer JH, Rietveld WJ (1989) Neurophysiology of the Suprachiasmatic Circadian Pacemaker in rodents. Physiological Reviews 69: 671–707

Meijer JH, Rusak B, Harrington ME (1984) Geniculate stimulation phase shifts hamster circadian rhythms. Soc Neurosci Abstr 10: 502

Moore RY (1973) Retino-hypothalamic projection in mammals. A comparative study. Brain Res 49: 403–409

Moore RY (1982) The suprachiasmatic nucleus and the organization of a circadian system. Trends Neurosci 5: 404–407

Moore RY (1983) Organization of function of a central nervous system circadian oscillator: the suprachiasmiatic hypothalamic nucleus. Fed Proc 42: 2783–2789

Moore RY, Eichler VB (1972) Loss of circadian adrenal corticosterone rhythm following suprachiasmatic nucleus lesions in the rat. Brain Res 42: 201–206

Moore RY, Card JP, Riley JN (1980) The suprachiasmatic nucleus: neuronal ultrastructure. Neurosci Abstr 6: 758

Moore RY, Gustafson EL, Card JP (1984) Identical immunoreactivity of afferents to the rat suprachiasmatic nucleus with antisera against avian polypeptide, molluscan cardioexcitatory peptide and neuropeptide Y. Cell Tissue Res 236: 41–46

Morin LP, Johnson RF, Moore RY (1989) Two brain nuclei controlling circadian rhythms are identified by GFAP immunoreactivity in hamsters and rats. Neurosci Lett 99: 55–60

Murphy S, Pearce B (1987) Functional receptors for neurotransmitters on astroglial cells. Neuroscience 22: 381–394

Nishio T, Shiosaka S, Nakaga H, Sakumoto T, Satoh K (1979) Circadian feeding rhythm after hypothalamic knife cut isolating suprachiasmatic nucleus. Physiol Behav 23: 763–769

Nunez AA, Casati MJ (1979) The role of the efferent connections of the suprachiasmatic nucleus in the control of circadian rhythms. Behav Neurol Biol 25: 263–267

Nunez AA, Stephan FK (1977) The effects of hypothalamic knife cuts on drinking rhythms and on the estrous cycle of the rat. Behav Biol 20: 224–234

Possidente B, Exner RH (1986) Gene-dependent effect of lithium on circadian rhythms in mice (Mus musculus). Chronobiol Int 3: 17–21

Possidente B, Hegmann JP (1982) Gene differences modify Aschoff's rule in mice. Physiol Behav 28: 199–200

Reinhard P (1983) Die Wirkung von Lithium auf das circadiane Verhalten von Schaben und Hamstern in Lichtprogrammen verschiedener Periodenlänge. Thesis, University of Tübingen

Reinhard P (1985) Effect of lithium chloride on the lower range of entrainment in Syrian hamsters. J Interdiscip Cycle Res 16: 227–237

Rietveld WJ (1981) The functional significance of a retino-hypothalamo-retinal loop for the circadian eating activity of rabbits. Behav Brain Res 2: 274–275

Rietveld WJ (1984) The effect of partial lesions of the hypothalamic suprachiasmatic nucleus on the circadian control of behaviour. In: Reinberg A, Smolensky M, Labreque G (eds) Annual review of chronopharmacology. Pergamon, Oxford

Rietveld WJ, Groos GA (1980) The central neural regulation of circadian rhythms. NATO Adv Study Inst [D] 3: 189–204

Rietveld WJ, Groos GA (1981) The role of the suprachiasmatic nucleus afferents in the central regulation of circadian rhythms. In: von Mayersbach H, Scheving LE, Pauly JE (eds) Biological rhythms in structure and function. Liss, New York

Rietveld WJ, Wirz-Justice A (1986) The effect of chronic application of methamphetamine in suprachiasmatic lesioned rats. In: Hildebrandt G, Moog An R, Raschke F (eds) Chronobiology and chronomedicine. Lang, Frankfort, pp 252–257

Rietveld WJ, ten Hoor F, Kooij M, Flory W (1978) Changes in 24-hours fluctuations of feeding behaviour during hypothalamic hyperphagia in rats. Physiol Behav 21: 15–622

Rietveld WJ, Kooij M, Aardoom OR, Boon ME (1983) The role of the dorsomedial hypothalamic nucleus in circadian control of food intake in rats. Neurosci Lett [Suppl] 14: 310–311

Rietveld WJ, Korving J, Wirz-Justice A (1985) The effect of chronic methamphetamine application on the circadian con-

trol of behaviour in the rat. J Interdiscip Cycle Res 16: 155–156

Rietveld WJ, Hekkens WTJM, Groos GA (1986) Changes in the rat circadian rhythm of food intake after long term application of clorgyline. In: Hildebrandt G, Moog An R, Raschke F (eds) Chronobiology and chronomedicine. Lang, Frankfort, pp 238–243

Roberts MH, Bernstein MF, Moore RY (1987) Differentiation of the suprachiasmatic nucleus in fetal rat anterior hypothalamic transplants in oculo. Dev Brain Res 32: 59–66

Ruis JF, Rietveld WJ, Buys PJ (1987) Effects of suprachiasmatic nuclei lesions on circadian and ultradian rhythms in body temperature in ocular enucleated rats. J Interdiscip Cycle Res 18: 259–273

Rusak B (1977) The role of the suprachiasmatic nuclei in the generation of the circadian rhythms in the golden hamster Mesocricetus auratus. J Comp Physiol 118: 145–164

Rusak B (1982) Physiological models of the rodents circadian system. In: Aschoff J, Daan S, Groos GA (eds) Vertebrate circadian systems. Springer, Berlin Heidelberg New York, pp 62–74

Rusak B, Groos GA (1982) Suprachiasmatic stimulation phase shifts rodent circadian rhythms. Science 215: 1407–1409

Rusak B, Zucker I (1979) Neural regulation of circadian rhythms. Physiol Rev 59: 449–526

Saleh MA, Haro PJ, Winget CM (1977) Loss of circadian rhythmicity in body temperature and locomotor activity following suprachiasmatic lesions in the rat. J Interdiscip Cycle Res 8: 341–346

Sawaki Y, Nihonmatsu I, Kawamura H (1984) Transplantation of the neonatal suprachiasmatic nuclei into rat: with complete bilateral suprachiasmatic lesions. Neurosci Res 1: 67–72

Schwartz WJ, Davidsen LC, Smith CB (1980) In vivo metabolic activity of a putative circadian oscillator, the rat suprachiasmatic nucleus. J Comp Neurol 189: 157–167

Schwartz WJ, Gross RA, Morton MT (1987) The suprachiasmatic nuclei contain a tetrodotoxin-resistant circadian pacemaker. Proc Natl Acad Sci USA 84: 1694–1698

Simerly RB, Swanson LW (1986) The organization of neural inputs to the medial preoptiv nucleus of the rat. J Comp Neurol 246: 312–342

Sims KB, Hoffman DL, Said SI, Zimmerman EA (1980) Vasoactive intestinal polypeptide (VIP) in mouse and rat brain: an immunocytochemical study. Brain Res 186: 165–183

Sodersten P, Hansen S, Srebro B (1981) Suprachiasmatic lesions disrupt the daily rhythmicity in the sexual behavior of normal male rats and of male rats treated neonatally with antioestrogen. J Endocrinol 88: 125–130

Sofroniew MV, Weindl A (1978) Projections from the parvocellular vasopressin- and neurophysin-containing neurons of the suprachiasmatic nucleus. Am J Anat 153: 391–430

Sofroniew MV, Weindl A (1982) Neuroanatomical organization and connections of the suprachiasmatic nucleus. In: Aschoff J, Daan S, Groos GA (eds) Vertebrate circadian systems. Springer, Berlin Heidelberg New York, pp 75–87

Stephan FK, Nunez AA (1977) Elimination of circadian rhythms in drinking, activity, sleep and temperature by isolation of the suprachiasmatic nuclei. Behav Biol 20: 1–16

Stephan FK, Berkley M, Moss R (1981) Efferent connections of the rat suprachiasmatic nucleus. Neuroscience 6: 2625–2641

Stoynev AG, Ikonomov OC, Usonoff KG (1982) Feeding pattern and light-dark variations in water intake and renal excretion after suprachiasmatic nuclei lesions in rats. Physiol Behav 29: 35–40

Swanson LW, Cowan WM (1975) The efferent connections of the suprachiasmatic nucleus of the hypothalamus. J Comp Neurol 160: 1–12

Swanson LW, Kuypers HGJM (1980a) A direct projection from the ventromedial nucleus and the retrochiasmatic area of the hypothalamus to the medulla and spinal cord of the rat. Neurosci Lett 17: 307–312

Swanson LW, Kuypers HGJM (1980b) The paraventricular nucleus of the hypothalamus: cytoarchitectonic subdivisions and organization of projections to the pituitary, dorsal vagal complex and spinal cord as demonstrated by retrograde double-labelling methods. J Comp Neurol 194: 555–570

Takahashi JS, Turek FW (1987) Anisomysin, an inhibitor of protein synthesis, perturbs the phase of a mammalian circadian pacemaker. Brain Res 405–199–203

Ueda S, Kawata M, Sano Y (1983) Identification of serotonin- and vasopressin immunoreactivities in the suprachiasmatic nucleus of four mammalian species. Cell Tissue Res 234: 237–248

Van den Poll AN (1980) The hypothalamic suprachiasmatic nucleus of rat: intrinsic anatomy. J Comp Neurol 191: 661–702

Van den Poll AN, Tsujimoto KL (1985) Neurotransmitters of the hypothalamic suprachiasmatic nucleus: immunocytochemical analysis of 25 neuronal antigens. Neuroscience 15: 1049–1086

Von Rommelspacher H, Bade P, Bludau J, Strauss S (1976) Hemmung der Monoaineoxidase und Tag-Nacht-Rhythus: Korrelation zwischen physiologischen und biochemischen Parametern. Arzneimittelforschung 26: 1078–1080

Watts AG, Swanson LW (1987) Efferent projection of the suprachiasmatic nucleus. II. Studies using retrograde transport of fluorescent dyes and simultaneous peptide immunohistochemistry in the rat. J Comp Neurol 258: 230–252

Watts AG, Swanson LW, Sanchez-Watts G (1987) Efferent projections of the suprachiasmatic nucleus. I. Studies using anterograde transport of phaseolus vulgaris leucoagglutin in the rat. J Comp Neurol 258: 204–229

Wiegand SJ, Gash DM (1988) Organization and efferent connections of transplanted suprachiasmatic nuclei. J Comp Neurol 267: 562–579

Wiegand SJ, Terasawa E (1982) Discrete lesions reveal functional heterogeneity of suprachiasmatic structures in regulation of gonadotropin secretion in the female rat. Neuroendocrinology 34: 395–404

Wirz-Justice A, Campbell IC (1982) Antidepressant drugs can slow or dissociate circadian rhythms. Experientia 38: 1301–1309

Wirz-Justice A, Groos GA, Wehr TA (1982) The neuropharmacology of circadian timekeeping in mammals. In: Aschoff J, Daan S, Groos G (eds) Vertebrate circadian systems. Springer, Berlin Heidelberg New York, pp 183–193

Zatz M, Brownstein MJ (1981) Injection of alpha-bungarotoxine near the suprachiasmatic nucleus blocks the effects of light on nocturnal pineal enzyme activity. Brain Res 213: 438–442

Zatz M, Herkenham MA (1981) Intraventricular carbachol mimics the phase shifting effects of light on the circadian rhythm of wheel-running activity. Brain Res 212: 234–238

Zimmerman NH, Menaker M (1979) Neural connections of the sparrow pineal: role in circadian control of activity. Science 190: 477–479

# Molecular and Genetic Aspects of Chronobiology

D. J. Lakatua

The endogenous nature of a biological rhythm in the circadian, ultradian as well as infradian frequency range is strongly suggested by their documented persistence under constant environmental conditions [1-5]. These findings implicate molecular events at the gene level that seem to be required for the control of the underlying maintenance and continued function of behavioral, physiological and biochemical rhythms [6]. In mammalian systems, the control and timing of developmental and behavioral cycles, such as the hormonal and sleep-wake cycles, appear to be localized in discrete regions of the brain [7]. On the other hand, single chick pinealocytes, when cultured in vitro, maintain circadian rhythmicity of melatonin synthesis and release and remain amenable to phase shifts by manipulation of the light-dark cycle [8 a, b, 9] suggesting that circadian periodicity at least is a genetically inherent property of the individual cell. Although significant advances have been made in the isolation of gene sequences controlling bioperiodicity in *Drosophila*, little is known about how transcriptional or translational events at the genomic DNA level lead to and maintain biorhythmicity [10, 11].

That the nucleus is essential for the maintenance of circadian periodicity was shown in *Acetabularia* [10, 12]. The circadian rhythm of its physiological functions as well as morphogenesis, once transcription has occurred, persists following enucleation [13] and can be phase-shifted by cycloheximide, an inhibitor of protein synthesis on the 80s ribosomes. The cycloheximide effect has also been shown to phase shift the mammalian SCN [14] and is highly circadian stage dependent. Synthesis of a high molecular weight protein product of about 230000 [15] remains rhythmic in acetabularia, in whole as well as anucleate cells [16].

Genetic mutational variants affecting biorhythmicity have been well studied in *Neurospora* [17, 18], *Chlamydomonas* [19, 20], *Drosophila melanogaster* [21–23] and *Pseudoobscura* [22, 24]. In *Drosophila melanogaster*, five genes have been defined to affect circadian locomotor and ultradian eclosion rhythmicity [25]. The first *per* or period gene to be recognized in *Drosophila melanogaster* controls circadian or 24-h periodicity and is X chromosome linked [21]. Several mutant alleles have been recovered; the *per*$^o$ allele abolishes rhythmicity of locomotor activity and eclosions, while *per*$^1$ mutants maintain rhythmicity, but with average period lengths of 29 h. The *per*$^s$ mutants on the other hand have a shorter period rhythmicity of 19 h. In *Drosophila* the *per*$^s$ control has been localized in the "brain," with restoration of the short period locomotor activity rhythm following abdominal transfer of *per*$^s$ brain into *per*$^o$ recipients [22, 26]. These findings have been confirmed in genetic mosaics for the *per* mutations at the *per* locus and also found to affect ultradian rhythmicity, i.e., the courtship song in *Drosophila melanogaster*, which in the wild type has a periodicity of 55 s, but in those that carry the *per*$^1$ mutation is lengthened to 80 s and in the *per*$^s$ mutant shortened to 40 s [27]. The *per*$^o$ flies sing an arrhythmic courtship song [28]. The tissue controlling these rhythms appears to be localized to neural tissue in the thorax [29].

The gene of the *per* locus on the X chromosome has been mapped to a 7.1-kb DNA segment. This DNA segment contains a single transcription unit that transcribes a 4.5-kb poly (A)$^+$ RNA [26]. There is no rhythm detectable in whole body RNA transcription when *per* expression is studied over a 24-h cycle on an LD 12:12 lighting schedule [25]. *Per* gene mRNA transcription and its protein product isolated from the adult heads of *Drosophila melanogaster*, on the other hand, show circadian rhythmicity and peak at the onset of darkness. Under DD there is a dampening of the amplitude of the mRNA cycling [30]. These authors also propose that the *per* protein product influences mRNA circadian rhythmicity through a feedback loop. When wild-type DNA, containing this transcription unit is transferred to the genome of the *per*$^o$ fly, the 4.5-kb RNA is transcribed and rhythmic behavior restored. This transforming DNA also complements chromosomal deletions that include the *per* locus. The *per* DNA was isolated by rescue studies of *per* mutants and is localized within band 3B 1–2 of the zesty white region on the X chromosome of *Drosophila melanogaster* [26]. Through well-characterized deletion (Df) and rearrangment breakpoint studies in this genomic interval the *per* locus has been precisely localized and using cloned restriction fragments cytogenetically mapped [31]. The *per* region was also isolated using chromosome walking/jumping and microexcision techniques. From these studies, it became clear that deletion of up to 15 kb of this DNA region results in apparent absence of circadian rhythmicity. At least three transcripts complementary to this 15-kb DNA region are absent in homozygous Df adults [32, 33].

A translocation breakpoint, JC43, is in the center of the *per* DNA region. Flies, hemizygous for JC43/Df, have weak, long-period rhythms. *Per* activity by conventional designation is restricted to the left of the JC43 breakpoint. The RNA complementary to this region, left of the JC43 breakpoint, is the 4.5-kb RNA and is presumed to code for a product with *per* activity. The periodicity of circadian rhythms, at least in *Drosophila melanogaster,* is determined by the level of expression of *per* locus proteins encoded by the 4.5-kb transcript [26].

In several vertebrate species DNA homologous to the *per* locus has been found and mouse DNA of high homology to portions of the *per* locus has been cloned and the DNA sequence of the conserved segment determined. Both mouse and fly DNAs appear to code for long-protein segments composed of alternating threonine and glycine or serine and glycine residues [34]. DNA to the right of the JC43 breakpoint transcribes an about 1-kb RNA transcript which fluctuates during an LD 12:12 light-dark cycle. This periodicity persists in constant darkness, but in *per*° mutants cycling is abolished and the RNA maintained at the dark (night time) level [25]. The *per* gene has been cloned and identified by transformation and functional in vitro assays and sequenced [26]. Analysis of the transcript and cDNA sequencing incidates that the *per* gene codes for a primary translation product of 1218 amino acids as well as two minor variants [35, 36]. Biochemical analysis have suggested that the *per* gene codes for a proteoglycan. The *per* mutations have been found to be single nucleotide substitutions [37]. Northern blot and in situ hybridization experiments to embryo sections and RNA blot analyses show that expression of the *per* gene begins during mid-late embryogenesis in the nervous system. Expression then diminishes until the mid-pupal [38] stage when the 4.5-kb transcript increases substantially and remains at a high level more in the heads than in the body. Experiments with the *per* gene fused to $\beta$-galactosidase [39] demonstrated by in situ hybridization positively stained tissues from the antennae, proboscis, eye, optic lobes, central brain, thoracic ganglia, gut, malpighian tubules and ovarian follicles [40].

Mutations have also been isolated from circadian rhythm variants on autosomes in *Drosophila melanogaster* [21, 41]. Three mutant strains established by mutagen-treated autosomes have been shown to exhibit altered synchronization to entraining environmental cycles. The circadian peaks of adult emergence rhythms occur earlier than in the wild type, when the strains are entrained to either light-dark or high- and low-temperature cycles. These psi strains show lengthened adult emergence circadian rhythm periods. Abnormal synchronization of locomotor activity to light-dark cycles resulting in lengthened periods of the activity rhythm has been found in *Drosophila melanogaster.* Emergence rhythms in a mutant strain, *gat,* becomes nearly aperiodic under constant light [42].

In *Neurospora crassa* the circadian conidiation rhythms are controlled by the *frq* locus [17, 43]. Mutants of the *frq* gene have been studied to isolate linkage groups to identify the *frq* gene and *for* gene by using chromosome walking methodology and transformation assays of recessive mutants showing segregation of period length and growth rate. They have been associated with single nuclear genes, but are not necessarily linked mutations [44]. Altered period lengths of the *Neurospora* circadian conidiation rhythm appears to be additive in double mutant experiments, which exhibit short and long periodicities with the same slow growth rate characteristic of their mutant *frq-5* [45]. Segregations between period length and growth rate were also observed under widely varying conditions of temperature and nutrition. Other alterations observed in the *frq-5* mutant, but not well studied, show differences in hyphal branching patterns, reduced formation of protoperithecia and germination of ascospores. Gene walking experiments also reveal similarities in DNA sequences of the *frq* gene in *Neurospora* and *per* gene in *Drosophila* [46]. This sequence similarity is not extensive and the significance of evolutionary conservation of such small regions of the gene remains speculative. Regulation of mRNA transcripts and their circadian control have been studied by screening genomic DNA with cDNA probes by subtractive hybridization [47].

In *Gonyaulax polyedra* bioluminescence of this marine dinoflagellate shows a circadian rhythm of luciferin-binding protein – (LBP), a dimer of two 72-kDa subunits [48]. This protein stabilizes the bioluminescence substrate and shows a tenfold circadian variation with a peak during the dark and trough during the light period [49, 50]. Pulse-labeling experiments have demonstrated that LBP is rapidly synthesized in vivo during the early dark phase while at other circadian system phases LBP synthesis is at least 50 times lower. LBP mRNA levels by in vitro translation and northern hybridization studies do not vary over the 24-h period, suggesting that the translational but not transcriptional process of bioluminescence is under cirdadian control [48, 51]. Bovine creatine shortens the period of free-running circadian rhythms in this marine alga *Gonyaulax polyedra*. This effect is dose related and shortens the period by as much as 4 h. Extracts of *Gonyaulax* have been shown to have a similar effect on the circadian period [52].

A circadian rhythm has also been shown in rod photoreceptor renewal of the outer disk membrane in lower vertebrates with 2%–4% of the outer segment produced daily. In toad and fish retina mRNA for opsin, the rod disks protein fluctuates with a circadian frequency, synchronized by light [53]. The rod opsin mRNA level rises before the onset of light, remains elevated during the light period and decreases during the dark period four to ten times. In constant darkness, however, the elevation in mRNA is maintained and occurs during the subjective daytime. Rod-opsin mRNA can be stimulated by light exposure during the dark span. A 4- to 5-day rhythmicity has also been described for disk shedding in *Xenopus* and *Rana* [54, 55].

## Molecular Chronobiology of Mammalian Systems

The concept of a genetic control of circadian rhythms in mammalian organisms is more complex at the molecular level. A dominant "pacemaker" controlling many circadian rhythms in certain species has been localized in the suprachiasmatic nucleus (SCN), located in the anterior hypothalamus, immediately dorsal to the optic chiasm [56]. At the molecular level, in hamsters and rats a protooncogene c-*fos* is expressed following photic stimulation or activation of the SCN cells. In both species, using c-*fos-lir* (c-*fos*-like immunoreactive antibodies) 30- to 60-min light

pulses (~30 lux) caused increased c-*fos-lir*-labeling in the region of the SCN that receives retinal fibers (RHT). This effect is highly circadian stage dependent to those phases of the circadian rhythm amenable to phase-shifting by light [57]. These light pulses not only increased c-*fos-lir* but also mRNA for the *fos*-protein and the immediate-early protein NGFI-A in the rat and hamster SCN. These results are consistent with changes in gene expression in response to retinal light exposure. A circadian variation in the expression of the c-*fos* protooncogene, as measured by c-*fos* protein immunoreactivity of adrenocortical cells, has been reported in adult male rats [58]. On an LD 12:12 with lights on from 07.00 to 19.00 hours, the number of c-*fos* IR cells in the zona fasciculata and reticularis was reduced to 53% at 08.00, to 18% at 12.00, and to 63% at 16.00 hours compared to the number at 20.00, 24.00 and 04.00 hours, which remained relatively constant. There is a rapid decrease at the end of the dark span at 08.00 hours to 52%. C-*fos* IR was detectable only at 20.00 and 24.00 hours in the zona glomerulosa. Dexamethasone significantly reduced the number of c-*fos* IR cells, suggesting that c-*fos* gene activation is under control of pituitary ACTH, and may be a physiological mediator of ACTH action.

Expression of mRNA in rat brain and hippocampus has also been studied with molecular probes for glucocorticoid (GR) and mineralocorticoid (MR) receptor mRNA [59]. Glucocorticoid receptors (type II) are uniformly distributed in several brain regions as well as the GR mRNA.

Mineralocorticoid receptors (type I) are predominantly concentrated in the hippocampus and MR mRNA shows a higher level of expression in the hippocampus. There is a corresponding inverse relation of type II receptor binding and peripheral steroid hormone levels, but not in the mRNA levels in the hippocampus. The circadian rhythm of serum corticosterone in intact rats also regulates type I receptor binding in the hippocampus but does not affect mRNA expression [60]. Expression of mRNA for both receptors does not appear to be modulated by the peripheral level of corticosterone.

Prolactin (PRL) mRNA levels in the rat pituitary maintain circadian rhythmicity, but during proestrus and less pronounced during the estrus phase the mesor and amplitude show increases during the major prolactin surge, i.e., toward the late light span of the lighting regimen (17.00 hours), when the animals are kept on a lighting regimen of LD 14:10, with lights on at 06.00 hours [61]. The circadian PRL mRNA levels show an abrupt decrease at

11.00 hours, and a sevenfold rise at the end of the light and early dark span (between 17.00 and 20.00 hours). There is a phase advance of the circadian rhythm of PRL mRNA of about 12 hours during metestrous and diestrous which is not associated with increased serum prolactin levels. Bromocriptine abolishes these changes in serum and pituitary prolactin and PRL mRNA [62, 63].

In sheep, the pituitary content of the beta-subunit mRNA of FSH declines during the preovulatory stage of the estrus cycle and reaches a trough when the alpha subunit and the pituitary content of LH mRNA beta subunit are at their maximum. This decrease in beta-subunit of FSH mRNA is accompanined by a rise in serum FSH, which reaches a maximum when the FSH mRNA for the beta-subunit reaches its minimum. In contrast serum LH and the beta-subunit of LH mRNA concentrations rise and fall concomitantly. These findings suggest that the regulatory control of pituitary FSH and LH transcription and translation is in opposite direction and estrus cycle dependent [64].

Hypothalamic mRNA, coding for preprocorticotrophin-releasing (ppCRH) hormone, a 41-amino acid peptide, was studied in Sprague-Dawley rats by in situ hybridization of the paraventricular nucleus. A circadian variation in ppCRH-mRNA content in male and ovariectomized estrogen-treated female rats was found to show a significant decrease during the beginning of the dark span on an LD 12:12 light-dark regimen with lights on at 06.00 hours [65]. The amplitude of the mRNA content was greater in female than in male rats, and the circadian variation in plasma corticosterone, sampled concomitantly, appears about 6 hours out of phase with the ppCRH-mRNA content of the hypothalamus [66].

Thyroid hormone action at the hepatocellular level stimulates S14 gene transcription of the mRNA-S14, which encodes for a protein of 17000 mol weight and p$I$ of 4.9. This increased expression of the S14 gene has also been observed following fasting and experimental diabetes mellitus. Glucagon administration leads to decreased levels of mRNA-S14, high-carbohydrate and low-fat diets augment expression. The S14 gene is associated with fatty acid synthesis, and most abundantly expressed in fat tissue, lactating mammary gland and liver. The mRNA-S14 in intact rats shows a circadian variation with an amplitude about threefold from the base value at 08.00 hours and observed peak levels at 20.00 hours on a lighting regimen of LD 12:12 with lights on at 07.00 hours and ad libitum feeding. This rhythm in S14 gene expression can be phase shifted by inverting the lighting regimen for 15 days, but not by restricting the feeding regimen to 12 h given, either during the light or dark span. Administration of a single receptor-saturating dose of triiodothyronine ($T_3$) increased the mesor of the rhythm of mRNA-S14 gene expression, and shifted the timing of the rhythm into the late dark span. Hypophysectomy with $T_3$ substitution and exposure to constant light for 15 days had no effect on the timing of the rhythm but the mesor was increased. In brown adipose tissue, mRNA-S14 concentrations show no circadian variations [67]. Of the four DNase I hypersensitive sites (HS-1 to 4) in the 5′-flanking regions of the S14 gene, only one, the HS-1, which is located immediately adjacent to the transcription initiation site, is consistently present in each tissue, in which the S14 gene is actively expressed, and hormonally regulated. Gel shift analysis with a 111-bp fragment within the HS-1 region revealed the presence of an exclusive hepatic nuclear protein PI which binds to this specific DNA segment. The level of the PI nuclear protein is responsive to hormonal regulation. The PI protein is not present in other tissues other than the liver. The binding activity of the PI protein correlates with the circadian rhythm of the hepatic mRNA-S14. These findings corroborate the apparent absence of a circadian rhythm in mRNA-S14 in tissues other than the liver. The PI protein may therefore be considered a circadian modulator of hepatic S14 gene expression and a defined nucleotide sequence that binds specifically the PI protein in the S14 gene, the apparent gene region regulating circadian rhythmicity [68].

The thyroid content of immunoassayable calcitonin and calcitonin-specific RNA, determined by dot blot hybridization of mRNA with [32]P-labeled human calcitonin-specific cDNA, was studied at seven different time points in male CF Wistar rats kept on an LD 12:12 lighting regimen [69]. There were significant circadian variations of high amplitude(A) detected in immunoassayable thyroid calcitonin ($A = 67\%$) and calcitonin hybridizable RNA ($A = 113\%$) which appear in phase also with plasma calcitonin levels, but apparently not with plasma calcium and phosphate. The apparent peak concentrations of thyroid calcitonin and hybridizable calcitonin RNA or calcitonin gene expression occurs 4 h before the onset of the dark span and reaches the lowest levels 4 h thereafter. By third-order polynomial regression curve fitting procedure a correct fit was detected for all circadian variations, except for the plasma calcitonin data. By increasing the sample size of plasma calcitonin to 15 per time point, a significant circadian rhythm was detected with a phase advance of about 1 h compared to all other circadian rhythms. Positive correlations were found between plasma calcium and thyroid calcitonin [70] and between plasma phosphate and hybridizable calcitonin RNA [71].

A pretranslational, i.e., mRNA circadian regulation of cholesterol $7\alpha$-hydroxylase activity has also been described in 2-months-old Wistar rats [72]. This circadian rhythm has been found for the enzyme protein as well as mRNA. In animals fed ad lib and on a natural lighting regimen the maximum level of enzyme activity, enzyme protein and mRNA occurs during the mid-dark span (22.00 hours) and the minimum at 10.00 hours. There were no significant sex differences. Cholestyramine feeding showed a significant increase in all three parameters only at 10.00 hours but not at 22.00 hours, while starvation decreased enzyme activity, enzyme protein and mRNA at both time points, while maintaining circadian rhythmicity. The timing of these rhythms may be synchronized by the feeding schedule since enzyme synthesis is protein and mRNA dependent [73].

## Mammalian Genetic Studies

Inbred strains of hamsters and other rodents are uniquely amenable to genetic studies, since each member of an inbred strain carries the same unique genetic fingerprint in behavioral and biochemical functions. Genetic mutations by hybrid breeding can be studied by the pattern of inheritance of the hybrid mutation and a locus assigned tentatively to an autosomal gene through linkage studies. A single gene mutation that affects the period of a circadian rhythm has been found in a strain of male golden hamsters from the Charles River Breeding laboratories [74, 75]. This gene mutation affected the circadian locomotor activity rhythm by shortening the period length in constant darkness (DD). On a lighting regimen of LD 14:10, the onset of activity occurred about 4 h earlier than in nonmutant male golden hamsters. When a male golden hamster of this strain was bred with 3 female hamsters with normal free-running circadian periodicities, 20 of 21 animals in the $F_1$ hybrid generation could be divided into two groups with different circadian periods: a Ts group, with a mean $\tau$ of $22.31 \pm 0.15$ hours and a Tn group with a mean $\tau$ of $24.03 \pm 0.07$ hours. The $F_1$ ratio of 50% is consistent with a mutation at a single autosomal locus. The Ts animals in the $F_1$ generation are heterozygous for this trait and the Tn animals can be considered the wild type. When Ts animals were crossbred with Ts animals, the offspring showed three distinct phenotypic groups with period lengths of their circadian locomotor activity rhythms of about 24, 22 and 20 h [74]. These studies suggest semidominance of a single autosomal mutation at a single locus with the homozygous having a period length of 20 h. Of interest is the increase in amplitude following phase-shifting by lighting scheduling experiments in the heterozygous animals, which entrain abnormally or are unable to entrain to 24-h light-dark schedules. Transplantation and ablation studies in golden hamsters carrying the mutation controlling the altered circadian period, Ts, have been further localized to the cells in the suprachiasmatic nucleus.

Differences among inbred strains of mice [76] and rats and crossbreeding experiments have identified the genetic contribution to the circadian rhythmicities in locomotor activity [77–79], food and water consumption [80] and body temperature in mice [81], as well as oxygen consumption in rats [82]. Strain-dependent differences have also been reported in response to phase shifts in lighting regimen, or following lithium administration [83]. In C57BL/6J and A/J male mice exposure to continuous light for different lengths of time results in free-running circadian rhythmicities of certain enzyme activities in the thymus, kidney, gonads, spleen and brain and show genetic differences in rhythm characteristics, including acrophases [84]. Lengthening of the free-running period of locomotor activity by lithium [85] and differences in sensitivity to maternal crossfostering effects on the circadian locomotor activity rhythm have been reported in Balb and C57 mice [86]. These studies also showed differences in the inheritance pattern in $F_1$ hybrids and dependence on age and maternal fostering, consistent with epigenetic modulation of the circadian locomotor activity. Similar genetic differences in the threshold of pain and morphine analgesia have been described to be circadian periodic, with decreased pain sensitivity and increased latency to morphine analgesia in nocturnal rodents occurring during the dark span [87]. In C57BL/6 and Swiss mice, kept on an LD 12:12 regimen, with lights on at 08.00 hours, tailflick pain shows a circadian periodic response. The highest latency values at 02.00 hours and the lowest at 14.00 hours were observed in Swiss mice, while in C57BL/6 mice the response is the opposite with the highest latency values at 14.00 hours and the lowest at 02.00 hours. A circadian rhythm in the number of brain-opiate binding sites have also been described and may underly these differences in nociception.

Light exposure during the dark span apparently through inhibition of pineal melatonin secretion appears to attenuate the increase in latency response [88]. Those findings suggest a difference in pineal response to light between the dark-coated, dark-eyed C57 mouse and Swiss albino [89]. Although these explanations remain speculative, the C57 mouse is characterized by a higher serotonin turnover time than the Swiss albino and Balb/c mouse [90]. The circadian rhythm in locomotor activity has also been studied in inbred strains of rats, which by cosinor and variance spectrum analysis show differences in amplitude, acrophase and period. The male LEW/Ztm rat shows by variance spectrum analysis a prominent circadian or approximate 24-h component of its locomotor activity rhythm and two minor periods of 4 and 4.8 h [91]. The female LEW/Ztm rat did not display such ultradian components, which in the male rat appears to be counteracted by estradiol administration. Under continuous darkness, these strain characteristic rhythms persist. In the genetic analysis of these circadian and ultradian rhythms, inbred strains of rats were studied by crossbreeding between ACI/Ztm female rats, which showed only a 24-h locomotor activity rhythm, and LEW/Ztm male rats. The acrophase by cosinor analysis of the locomotor activity rhythm of the ACI/Ztm strain is at $-84.7$ (05:39 hours), which is about 1 h later than in the LEW/Ztm strain, where it occurs at $-66.10$ (04:24 hours). The $F_1$ hybrid between these two strains showed an acrophase at $-79.99$ (05:19 hours), which is in between the two parental strains. The amplitude of the 4-h and 4.8-h spectral components was significant only in the LEW/Ztm strain. The $F_1$ generation showed dominance toward the circadian frequency component and was not influenced by sex or maternal effects. Backcross generations showed dominance toward low-amplitude, i.e., LEW/Ztm strain and nonultradian frequencies, i.e., ACI/Ztm strain. These studies, conducted under an LD 12:12 lighting regimen, did not allow distinction of genetic differences in the observed rhythms from the genetic differences in the "pacemaker" driving these rhythms. Under constant darkness, genetic studies of free-running periodicities and phase response curves of locomotor activity might be helpful to determine whether the genetic differences are in the "circadian pacemaker" [92]. Genetic differences in locomotor activity have also been studied in crossbreeding experiments of C57BL/6J with DBA mouse strains and in BALB/CByJ with C57BL/10Sn strains with similar results. The results indicate that dominance of short period rhythms is variable with resulting period lengths, which may be intermediate. Under conditions of constant darkness a differential aftereffect in response to a change in lighting regimen suggests genetic differences in free-running periodicity and interactions with the endogenous circadian "pacemaker." Heritable genetic differences in this mouse model of the gene effects on the circadian period are to a certain extent expressed independently of maternal environmental influences with the genetic allelic autosomal effects being additive and predominate but with the dominance interactions being dependent on each individual strain [93].

In comparing rectal body temperature rhythms between C57BL/6j and C3H/2Ibg mice, differences in the circadian rhythm parameters have been studied and genetically analyzed by cross-breeding experiments. In these studies rectal body temperature rhythms in the two strains of mice and their $F_1$ and $F_2$ hybrids were analyzed with the cosinor method by least squares analysis and the timing of the rhythm expressed in relation to the time of lights off or by their phase-angle differences with the animals standardized on an LD 12:12 regimen, with lights on from 07.00 to 23.00 hours. In order to obtain an estimate of the light-off anticipatory time, the rectal temperature, measured at 22.00 hours, was expressed as a percentage of the rhythm-adjusted mean or mesor. Genetic analysis was carried out using the method of Mather and Jinks. The amplitude $A$ of the C3H mice was greater than that of the C57BL mice. The $F_1$ hybrids showed a strong dominance toward the C3H strain. The acrophase of the C57BL strain showed a phase delay compared to the C3H, while the $F_1$ hybrid generation was intermediate. The same relation was seen in anticipation of the lights-off parameter. There were no significant differences in mesor of the parental strains. In the $F_1$ hybrids the mesor was higher than in either one of the parents and presumably dominantly inherited. The genetically influenced differences in the mesor of body temperature are specifically noted during the dark span and also in wheel-running activity [81].

EEG-recorded and -defined sleep was also studied in C57BL/6 and SEC/1RE mice under conditions of LD 12:12, 6:6, 3:3, and 1:1 and under constant light [77]. In the C57BL strain the LD-synchronized differences in locomotor activity rhythm and sleep were less pronounced and superimposed on a circadian rhythm, with wheel-running activity occurring during the dark period. Under constant light circadian rhythmicity was maintained in C57BL, but not in the SEC strain. Sleep was enhanced by light and superimposed on the circadian rhythm in both strains. The circadian amplitude of the activity rhythm in the C57BL was greater than in the SEC strain and the free-running circadian rhythmicity was more pronounced in the C57BL/6 than in the SEC/1RE mice under conditions of continuous light. These findings provide evidence for a genetic difference in response and expression of circadian rhythmicity under conditions of free-running as well as light-dark synchronized schedules shorter than 24 h.

## Plant Rhythms

The effect of light on the expression of plastid and nuclear-encoded genes specific for photosynthesis have been studied in tomato plants. Under natural light-dark cycles and greenhouse conditions RNA was measured in 14- to 15-day-old tomato fruits harvested at ten time points over a 38-h period. Relative transcript levels as a percentage of maximum level for mitochondrial ATPase, the $\beta$-subunit of tubulin, the small and large subunits of RuBPC/Oase, LHCPII, the $Q_B$-binding protein *(psbA)*, the P680 reaction of PSII *(psbB)* and the p700 reaction center of PSI *(psaA)* were determined. Only LHCP mRNA levels showed significant circadian alterations with high levels during the light period, while transcription levels for the other genes encoding for photosynthesis specific, as well as nonphotosynthesis specific genes ($\beta$-subunit of mitochondrial ATPase and tubulin), showed circadian fluctuations of much lower amplitude [94]. Differences in timing including the phase of the mRNA rhythm are circadian stage dependent and are related to the time of sunrise. Sunlight has been demonstrated to modulate *rbcS* gene expression in tomato leaves and other plants including transgenic plants. The mechanisms involved in the control of circadian rhythms at the molecular level remain at present unknown. Transcription rates, measured by nuclear run on experiments, however, did show circadian variations with a peak at 08.00 hours for CAB/II and *rbcS* while that for a biotin clone showed a peak around 12.00 hours. Constant light or darkness alters the amplitude of transcriptions, but does not abolish it, while maintaining 24-h periodicity [95]. In wheat, genes encoding for the light-harvesting chlorophyll *a/b*-binding protein of the photosystem II *(Cab)* appear to be circadian periodic and regulated by phytochrome. This regulation occurs at the transcriptional level. The timing of maximal *cab* gene expression is synchronized by the time of light to dark transition even under free-running conditions of constant darkness. Those findings suggest that the onset of darkness sychronizes the phase of the circadian timing of *Cab* gene expression as if it were to anticipate the time of sunrise the following day [96].

## Human Studies

The genetics of human circadian rhythms have been studied in twins. In some studies rhythm parameters have been quantitatively analyzed by statistical means, and their error estimates quantified at the 95% confidence level, in order to compare differences and similarities in rhythm parameters between individual subjects of a twin pair. In other studies, such parameters as phase, amplitude and rhythm-adjusted mean or mesor are not evaluated and quantified with the same statistical methods of analyses to allow comparison. In one of the earliest investigations [97], eight pairs of human male twins between the age of 20 and 21 years, of which 4 were monozygotic and 4 dizygotic, were studied. Zygosity was determined by blood group antigens (ABO, Rh/Hr, MN, P and Lewis) and serum proteins (haptoglobin, and group-specific gamma globulins). This study was conducted for 4–5 weeks under standardized conditions in a hospital setting following a 10-day period of adaptation. At 4-hourly intervals from 04.00 to 12.00 hours, pulse and respiration rate, axillary body temperature, first, fourth and fifth phases of arterial blood pressure, P and T waves in the second bipolar extremity lead of the ECG were measured and blood samples for histamine, serotonin, uric acid, lipids, plasma euglobin fibrinolysis time and in the urine 17-hydroxycorticosteroid and 3-methoxy-4 hydroxy mandelic acid were collected every 4 h. Sampling times were extended every 4 h the following day from 12.00 hours and 16.00 hours until 08.00 hours. There were a total of seven time points of sampling at equal intervals. Curve-fitting procedures were done mathematically by least squares of polynomials and derivatives of all degrees to obtain smoothing of the curves with derivatives of the 5th degree being the most optimal for the best-fitting curve. No correlation coefficients were calculated, but instead, by calculation and comparison of the means of the seven time points, a quotient and difference between nominator and denominator were used to draw conclusions. This study finds only genetic effects on the circadian variation of pulse rate and systolic blood pressure, but not in the urinary excretion of 3-methoxy-4-hydroxymandelic acid. The mathematical analyses in this study did not allow comparisons of phase or amplitude, but were based strictly on the comparison of curve fit and circadian means. A more recent study [98] was conducted in 16 monozygotic twins and 1 triplet ranging in age from 23–52 years including a pair of boys age 11. They were continuously monitored covering at least 1 week carrying a Holter monitor, i.e., an ambulatory electrocardiocorder model 445B for spans of about 24 h. Continuous data were analyzed by 24-h cosinor with the method of least squares. Intraclass correlations were estimated for the mesors, amplitudes and acrophases of the 16 monozygotic twin pairs and 1 triplet and also for 4 dizygotic twin pairs (2 males and 2 females); computations were also carried out for monozygotic pairs of men and women separately studied either on the same day or on different days and for the dizygotic twins of the same sex. In this study the twins were reared apart from early life, average age 2.9 months and subsequently reunited after an average 16.1 years and thus allowing estimation of the variability due to genetic relative to environmental factors. Under such circumstances the quantification of intraclass correlations for the mesor, amplitude and acrophase of monozygotic twins showed heritability of the circadian rhythm of heart rate.

An endogenous or genetic component of the circadian and ultradian rhythm in the urinary excretion of 17-hydroxycorticosteroids has been studied in one monozygotic and one dizygotic twin pair [99]. In this study urinary 17-hydroxycorticosteroids (17-OH-CS) were determined in urine samples voluntarily voided during the day and night at random, i.e., at unequal time intervals over an 8-day time span. The data included urine volume and 17-OH-CS and creatinine excretion expressed in their respective units per hour calculated at the time of the mid-collection period. In order to maintain continuity the studies were scheduled during school vacations at the same time of the year (summer) and lights-on (08.30 hours) and lights-off (20.00 hours) were standardized in each of the studies. The monozygotic twins were studied in 1977 at age 4.3 years and in 1983 at age 10.3 years June and July, respectively; the dizygotic twins were studied at age 6 years in September 1983. The time series were analyzed by four different methods, initially by inspection of chronograms then by cosinor analyses using the best-fitting cosine functions by the method of least squares, yielding estimates of acrophase, amplitude and mesor with their respective confidence intervals. Power spectrum analysis to detect prominent period(s) was also used, adapted to nonequidistant intervals. The rhythmometric end points were used to estimate intraclass correlation coefficient($r$) for the monozygotic versus dizygotic twin pairs, by linear regression analysis. The monozygotic twins at 4.3 and 10.3 years showed prominent periods at 24 and 8 h, with the acrophase at age 4.3 years at 12.04 and 12.94 hours for the 24-h and at 01.46 and 01.79 hours for the 8-h periodicity. At age 10.6 years there was a twofold increase in mesor of 17-OH-CS and a phase advance in acrophases to 10.73 and 11.52 hours for the 24-h versus 00.72 and 00.73 hours for the 8-h periodicity. There were no statistically significant differences between this monozygotic twin pair in rhythm parameters. The dizygotic twin pair showed differences in mesor and acrophases (11.00 vs. 14.00 hours) and by harmonic analysis large-amplitude periods of 24 and 12 h were detected in one and 24 h and 8 h in the other member of the dizygotic twin pair. Intraclass correlation for the timing of the rhythm or acrophases in the monozygotic twin pair was $r = 0.80$ ($P < 0.001$), while the correlation of the dizygotic twins was similar to that of nontwin correlations. The circadian rhythms for water excretion and urinary creatinine were strikingly similar in monozygotic twins for acrophase, amplitude and mesor as well as other harmonic components, i.e., period of 8 h at age 4.3 years of age but not 10 years of age for water excretion, but a 12-h component in urinary creatinine at 10 years but not at 4.3 years of age. The circadian rhythms in urinary excretion of water in the dizygotic twins had a prominent 8-h period with a larger amplitude than the circadian component. The circadian acrophases of this variable differed by about 3 h. No circadian rhythmicity was detectable in urinary creatinine of the dizygotic twins.

In single circadian stage (three samples at 20-min intervals from 08.00 to 09.30 hours) collection studies on clinically healthy monozygotic ($N = 15$) and dizygotic ($N = 15$) male twins a genetic and environmental component underlying variation in thyroid hormone plasma concentrations has been reported [100]. Intraclass correlations were significantly greater in monozygotic twins for total and free $T_4$ with a heritability index [$H = 2 ({}^tMZ - {}^tDZ)$] of greater than 28% ($P < 0.05$). The variability in TBG concentrations accounted for less than 20%, while those in TSH and $T_3$ concentrations were influenced by both genetic and environmental factors.

The heritability of adrenocortical functions in 20 monozygotic and 20 dizygotic male twin pairs has also been studied at a single circadian stage on 3 equal blood volume collections from 08.00 to 09.30 hours [101]. They ranged in age from 20 to 60 years. By intraclass correlation a heritability index was calculated, which was 45.4% ($P < 0.05$) for total cortisol, 50.6% ($P < 0.05$) for plasma-free cortisol, and 57.8% ($P < 0.05$) for DHEA-S, but not for CBG ($P \geqslant 0.05$). Environmental influences such as age, smoking, drinking, exercise, and degree of obesity were corrected for by factor analysis.

The study of monozygotic twins lends itself also to identification of genetic components in the control of serum high-density lipoprotein levels [102] with heritability indices for apo A-II of 0.35 and 0.30 for males and females respectively, although this study was not circadian stage controlled.

Genetic and environmental factors regulating cytosolic epoxide hydrolase activity (cEH) in human lymphocytes of clinically healthy subjects have been studied in 6 pairs of monozygotic and 6 pairs of dizygotic twins, revealing significantly less intrapair variance for monozygotic than dizygotic twins, while in 100 unrelated male subjects the extent of interindividual variation was 11-fold [103]. By pedigree analysis an autosomal dominant either monogenic or polygenic inheritance pattern controls the variations in lymphocyte cEH activity. There was no significant circadian variation by A.M. and P.M. sampling only and menstrual cycle variations were of high amplitude, but irregular and not further analyzed by time-series methods.

Cyclic neutropenia has also been described in monozygotic twin girls, but the onset appeared 3 years apart and the symptoms decreased initially during a 5-year follow-up [104]. No rhythmometric or time-series analysis were applied to the data to allow intraclass correlations of stage-dependent rhythm parameters. The chronograms appear on inspection entirely different for either member of the twin pair.

A genetic component in sleep disorders in humans, such as narcolepsy, sleep walking, periodic movements in sleep, circadian delay syndrome and familial insomnia has long been recognized but shows differences in control mechanisms of the circadian sleep and wakefulness pattern, and the prevalence and onset of non-REM versus REM activity [105, 106]. A genetic defect in narcolepsy has been tentatively localized on chromosome 6, but a specific defect in neutrotransmission, underlying these disorders, such as somatomotor inhibition and excitation, autonomic discharge as well as circadian regulation remains unknown. In infants and children with corticoreticular seizures including febrile convulsions and

ultradian rhythmic theta activity, a characteristic interictal EEG pattern appears as an age-dependent genetic predisposition to convulsions.

At the chromosomal level, genetic damage, as measured by an increased rate of sister chromatid exchanges (SCES) and other chromosome aberrations, appears to be menstrual and circadian stage dependent with higher rates of damage occurring during estrogenic and ovulatory stages, compared to progestrogenic stages [107] and higher rates at 21.00 compared to 09.00 hours. Variations in frequency have also been observed in men, and may be due to other factors so far not yet elucidated [108]. The number and subpopulations of lymphocytes vary as a function of circadian stage as well as their responsiveness to in vitro mitogen stimulation. These early findings may well explain the differential sensitivity as a function of subpopulation of circulating lymphocytes, which may be more or less sensitive to chromosomal or genetic damage [107].

Sex chromosomal abnormalities have been reported to affect the circadian rhythm of sleep in one XXY and one XYY infant, from birth to 8 months and 3 years of age respectively. There is no delay in CNS establishment of the circadian sleep-wakefulness pattern. The XXY infant slept more in short naps throughout both day and night [109].

## Conclusions

A gene sequence controlling circadian rhythmicity has been isolated in *Drosophilia*. This 7.1-kb DNA segment has been sequenced and mapped to the X-chromosome and a transcriptional product of 4.5 kb (poly(A)$^+$ RNA identified. This transcriptional product translates to a 1218 amino acid glycosylated proteoglycan, which is apparently involved in a feedback control of mRNA circadian rhythmicity in *Drosophila* brain. Despite the elucidation of a number of molecular events controlling this circadian rhythmicity, the actual mechanism of action involved in these periodic phenomena remains unknown and is proposed to result from the interaction of the *per* protein with the *per* gene. A DNA region in the *per* gene of *Drosophila* shows sequence similarity with a region within the *frq* gene of *neurospora,* which may be statistically significant [46]. This similarity extends to the protein sequence product, since this similarity lies within the open reading frame (ORF) of the *frq* gene and this homology to the *per* gene includes several proteoglycans, i. e., threonine-glycine repeats (TG repeats). This similarity between the *per* and *frq* gene extends both 3′ and 5′ beyond the direct TG repeat regions. The TG repeat regions may be the sites for some of the extensive glycosylations, thought to be the protein products of the *per* gene. This sequence similarity is not extensive and may be incidental, although the similarity, including glycosylation sites suggests evolutionary conservation. Hybridization at moderate stringency of DNA from the *frq* and *per* loci has been used to screen genomic DNA from a number of vertebrate, invertebrate, fungal and plant systems. Potential homologous sequences have been identified in several mammalian species [34], yeast and soybean, and promise further exploration in higher mammalian systems including humans.

Mammalian molecular studies on the control of rhythmicity have been limited to the transcriptional and translational events at the gene level. A better worked out molecular mechanism is the circadian pattern of expression of the S14 gene in rat hepatic nuclei, an example of a gene, which is responsive to and regulated by triiodothyronine ($T_3$) and dietary factors. The rhythm persists after hypophysectomy and a saturating dose of $T_3$ and remains synchronized by the photoperiod. Activation of this gene and circadian rhythmicity of gene expression has been shown to be modulated by a PI protein, which specifically binds to a hypersensitive region 5′ to the transcription initiation site of the S14 gene. This protein is not present in other lipogenic tissues other than the liver and is a unique example of a circadian periodic *trans*-acting binding factor, which may be responsible for the periodicity in mRNA-S14.

The human studies are limited to studies on hormonal and cardiovascular rhythms in identical or monozygotic twins, but differences in rhythmometric analyses among the studies do not allow comparisons of quantitative rhythm parameters, such as phase or timing, amplitude or extent of change and mesor or rhythm-adjusted level. By comparing heritability indices, monozygotic twins show greater intraclass correlations for periods and mesors of several functions studied, and also for the timing of their rhythms. These findings strongly suggest that in humans bioperiodicity and its parameters have strong genetic determinants and may be regulated by or share genomic sequences similar to the *per* gene in *Drosophila*.

*Acknowledgement.* I thank Ms. Karen Cloutier for her excellent and dedicated assistance in the preparation of the manuscript.

# References

1. Halberg F, Siffre M, Engeli M, Hillman D, Reinberg A (1965) Etude en libre cours des rythmes circadiens du pouls, de l'alternance veille-sommeil et de l'estimation du temps pendant les deux mois de séjour souterrain d'un homme adulte jeune. C R Acad Sci (Paris) 260: 1259–1262
2. Ghata J, Halberg F, Reinberg A, Siffre M (1968) Rythmes circadiens désynchronisés (17-hydroxycorticosteroides, température rectale, veille-sommeil) chez deux sujets adultes sains. Ann Endocrinol (Paris) 29: 269–270
3. Halberg F, Reinberg A, Haus E, Ghata J, Siffre M (1970) Human biological rhythms during and after several months of isolation underground in natural caves. Natl Speleol Soc Bull 32: 89–115
4. Reinberg A, Halberg F, Ghata J, Siffre M (1966) Spectre thermique (rythmes de la température rectale) d'une femme adulte avant, pendant et après son isolement souterrain de trois mois. C R Acad Sci (Paris) 262: 782–785
5. Siffre M, Reinberg A, Halberg F, Ghata J, Perdriel G, Slind R (1966) L'isolement souterrain prolongé. Etude de deux sujets adultes sains avant, pendant et après cet isolement. Presse Med 74: 915–919
6. Mergenhagen D (1986) The circadian rhythm in *Chlamydomonas renhardi* in a Zeitgeber free environment. Naturwissenschaften 73: 410–412
7. Ralph MR, Foster RG, Davis FC, Menaker M (1990) Transplanted suprachiasmatic nucleus determines circadian period. Science 247: 975–978
8a. Robertson LM, Takahashi JS (1988) Circadian clock in cell culture. I. Oscillation of melatonin release from dissociated chick pineal cells in flow-through microcarrier culture. J Neurosci 8: 12–21
8b. Robertson LM, Takahashi JS (1988) Circadian clock in cell culture. II. In vitro photic entrainment of melatonin oscillation from dissociated chick pineal cells. J Neurosci 8: 22–30
9. Takahashi JS, Murakami N, Nikaido SS, Pratt BL, Robertson LM (1989) The avian pineal, a vertebrate model system of the circadian oscillator: cellular regulation of circadian rhythms by light, second messengers, and macromolecular synthesis. Recent Prog Horm Res 45: 279–348; discussion 348–352
10. Vanden Driessche T (1989) The molecular mechanism of circadian rhythms. Arch Int Physiol Biochim 97: 1–11
11. Schweiger HG, Schweiger M (1977) Circadian rhythms in unicellular organisms: an endeavour to explain the molecular mechanism. Int Rev Cytol 51: 315–342
12. Schweiger HG, Hartwig R, Schweiger M (1986) Cellular aspects of circadian rhythms. J Cell Sci [Suppl] 4: 181–200
13. Hartwig R, Schweiger R, Schweiger HG (1986) Circadian rhythms of the synthesis of a high molecular weight protein in anucleate cells of the green alga *acetabularia*. Eur J Cell Biol 41: 139–141
14. Wollnik F, Turek FW, Majewski P, Takahashi JS (1989) Phase shifting the circadian clock with cycloheximide: response of hamsters with an intact or a split rhythm of locomotor activity. Brain Res 469: 82–88
15. Hartwig R, Schweiger M, Schweiger R, Schweiger HG (1985) Identification of a high molecular weight polypeptide that may be part of the circadian clockwork in *Acetabularia*. Proc Natl Acad Sci USA 82: 6899–6902
16. Karakashian MW, Schweiger HG (1976) Evidence for a cy-

cloheximide-sensitive component in the biological clock of *Acetabularia.* Exp Cell Res 98: 303–312

17. Feldman JF (1985) Genetic and physiological analysis of a clock gene in *Neurospora crassa.* In: Rensing L, Jaeger NI (eds) Temporal order. Springer, Berlin Heidelberg New York, pp 238–245

18. Gardner GF, Feldman JF (1980) The *frq* locus in *Neurospora crassa:* a key element in circadian clock organization. Genetics 96: 877–886

19. Bruce VG (1972) Mutants of the biological clock in *Chlamydomonas reinhardi.* Genetics 70: 537–548

20. Bruce VG (1974) Recombinants between clock mutants of *Chlamydomonas reinhardi.* Genetics 77: 221–230

21. Konopka R, Benzer S (1971) Clock mutants of *Drosophila melanogaster.* Proc Natl Acad Sci USA 69: 2112–2116

22. Konopka RJ (1979) Genetic dissection of the *Drosophila* circadian system. Fed Proc 38: 2602–2605

23. Rosbash M, Hall JC (1985) Biological clocks in *Drosophila:* finding the molecules that make them tick. Cell 43: 3–4

24. Pittendrigh CS (1974) Circadian oscillations in cells and the circadian organization of multicellular systems. In: Schmitt FO, Worden FG (eds) The Neurosciences, Third Study Program, MIT Press 437–458

25. Young MW, Jackson FR, Shin HS, Bargiello TA (1985) A biological clock in *Drosophila.* Cold Spring Harbor Symp Quant Biol 50: 865–875

26. Bargiello TA, Young MW (1984) Molecular genetics of a biological clock in *Drosophila.* Proc Natl Acad Sci USA 81: 2142–2146

27. Hamblen-Coyle M, Konopka RJ, Zwiebel LJ, Colot HV (1989) A new mutation at the period locus of *Drosophila melanogaster* with some level effects on circadian rhythms. J Neurogenet 5: 229–256

28. Kyriacou CP, Hall JC (1980) Circadian rhythm mutations in *Drosophila melanogaster* affect short-term fluctuations in the male's courtship song. Proc Natl Acad Sci USA 77: 6929

29. Hall JC (1984) Complex brain and behavioral functions disrupted by mutations in *Drosophila.* Dev Genet 4: 355

30. Hardin PE, Hall JC, Rosbash M (1990) Feedback of the *Drosophila* period gene product on circadian cycling of its messenger RNA levels. Nature 343: 536–554

31. Rosbash M, Hall JC (1989) The molecular biology of circadian rhythms. Neuron 3: 387–398

32. Hall JC, Roshbash M (1987) Genetics and molecular biology of rhythms. Bioassays 7: 108–112

33. Ewer J, Rosbash M, Hall JC (1988) An inducible promotor fused to the period gene in *Drosophila* conditionally rescues adult *per*-mutant arrhythmicity. Nature 333: 82–84

34. Shin HS, Bargiello TA, Clark BT, Jackson FR, Young MW (1985) An unusual coding sequence from a *Drosophila* clock gene is conserved in vertebrates. Nature 317: 445–448

35. Jackson FR, Bargielle TA, Yun SH, Young MW (1986) Product of *per* locus of *Drosophila* shares homology with proteoglycans. Nature 320: 185–188

36. Citri Y, Colot HV, Jacquier AC, Yu Q, Hall JC, Baltimore D, Rosbash M (1987) A family of unusually spliced biologically active transcripts encoded by a *Drosophila* clock gene. Nature 326: 42–47

37. Yu Q, Jacquier AC, Citri Y, Hamblen M, Hall JC, Rosbash M (1987) Molecular mapping of point mutations in the period gene that stop or speed up biological clocks in *Drosophila melanogaster.* Proc Natl Acad Sci USA 84: 784–788

38. James AA, Ewer J, Reddy P, Hall JC, Rosbash M (1986) Embryonic expression of the period clock gene in the central nervous system of *Drosophila melanogaster.* EMBO J 5: 2313–2320

39. Mismer D, Rubin GM (1987) Analysis of the promoter of the nina opsin gene in *Drosophila melanogaster.* Genetics 116: 565–578

40. Liu X, Lorenz L, Yu Q, Hall JC, Rosbash M (1988) Spatial and temporal expression of the period gene in *Drosophila melanogaster.* Genes Dev 2: 228–238

41. Lindsley D, Grell E (1986) Genetic variations of *Drosophila melanogaster.* Carnegie Inst Wash Publ 627

42. Jackson FR (1983) The isolation of biological rhythm mutations on the autosomes of *Drosophila melanogaster.* J Neurogenet 1: 3–15

43. Feldman JF, Dunlap JC (1983) *Neurospora crassa*: a unique system for studying circadian rhythms. Photochem Photobiol Rev 7: 319–368

44. Dunlap JC (1990) Closely watched clocks: molecular analysis of circadian rhythms in *Neurospora* and *Drosophila.* Trends Genet 6: 159–165

45. Feldman JF, Atkinson CA (1978) Genetic and physiological characteristics of a slow-growing circadian clock mutant of *Neurospora crassa.* Genetics 88: 255–265

46. McClung DR, Fox BA, Dunlap JC (1989) The *Neurospora* clock gene frequency shares a sequence element with the *Drosophila* clock gene period. Nature 339: 558–562

47. Loros JJ, Denome SA, Dunlap JC (1989) Molecular cloning of genes under control of the circadian clock in *Neurospora.* Science 243: 385–388

48. Hastings JW (1989) Chemistry, clones, and circadian control of the dinoflagellate bioluminescent system. J Biolumin Chemilumin 4: 12–19

49. Morse D, Milos PM, Roux E, Hastings JW (1989) Circadian regulation of bioluminescence in *Gonyaulax* involves translational control. Proc Natl Acad Sci USA 86: 172–176

50. Morse D, Pappenheimer AM Jr, Hastings JW (1989) Role of a luciferin-binding protein in the circadian bioluminescent reaction of *Gonyaulax polyedra.* J Biol Chem 264: 11822–11826

51. Milos P, Morse D, Hastings JW (1990) Circadian control over synthesis of many *Gonyaulax* proteins is at a translational level. Naturwissenschaften 77: 87–89

52. Roenneberg T, Nakamura H, Hastings JW (1988) Creatine accelerates the circadian clock in a unicellular alga. Nature 334: 432–434

53. Korenbrot JI, Fernald RD (1989) Circadian rhythm and light regulate opsin mRNA in rod photoreceptors. Nature 337: 454–457

54. Besharse JC, Iuvone PM (1983) Circadian clock in *Xenopus* eye controlling retinal serotonin *N*-acetyltransferase. Nature 305: 133–135

55. Basinger SF, Hoffman R, Matthes M (1976) Photoreceptor shedding is initiated by light in the frog retina. Science 194: 1074–1076

56. Kornhauser JM, Nelson DE, Mayo KE, Takahashi JS (1990) Photic and circadian regulation of c-*fos* gene expression in the hamster suprachiasmatic nucleus. Neuron 5: 127–134

57. Rusak B, Robertson HA, Wisden W, Hunt SP (1990) Light pulses that shift rhythms induce gene expression in the suprachiasmatic nucleus. Science 248: 1237–1240

58. Koistinaho J, Roivainen R, Yang G (1990) Circadian

rhythm in c-*fos* protein expression in the rat adrenal cortex. Mol Cell Endocrinol 71: R1–R6

59. Chao HM, Choo PH, McEwen BS (1989) Glucocorticoid and mineralocorticoid receptor mRNA expression in rat brain. Neurendocrinology 50: 365–371

60. Van Eekelen JA, Jiang W, de Kloet ER, Bohn MC (1988) Distribution of the mineralocorticoid and the glucocorticoid receptor mRNAs in the rat hippocampus. J Neurosci Res 21: 88–94

61. Haisenleder DJ, Ortolano GA, Landefeld TD, Zmeili SM, Marshall JC (1989) Prolactin messenger ribonucleic acid concentrations in 4-day cycling rats and during the prolactin surge. Endocrinology 124: 2023–2028

62. Maurer RA (1981) Transcriptional regulation of the prolactin gene by ergocryptine and cyclic AMP. Nature 294: 94

63. Carrillo AJ, Sharp ZD, DePaolo LV (1987) Correlation of rat pituitary prolactin mRNA and hormone content with serum levels during the estrogen-induced surge. Endocrinology 121: 1993

64. Leung K, Kim KE, Maurer RA, Landefeld TD (1988) Divergent changes in the concentrations of gonadotropin beta-subunit messenger ribonucleic acid during the estrous cycle of sheep. Mol Endocrinol 2: 272–276

65. Swanson LW, Simmons DM (1989) Differential steroid hormone and neural influences on peptide mRNA levels in CRH cells of the paraventricular nucleus: a hybridization histochemical study in the rat. J Comp Neurol 285: 413–435

66. Watts AG, Swanson LW (1989) Diurnal variations in the content of preprocorticotropin-releasing hormone messenger ribonucleic acids in the hypothalamic paraventricular nucleus of rats of both sexes as measured by in situ hybridization. Endocrinology 125: 1734–1738

67. Kinlaw WB, Fish LH, Schwartz HL, Oppenheimer JH (1987) Diurnal variation in hepatic expression of the rat S14 gene is synchronized by the photoperiod. Endocrinology 120: 1563–1567

68. Wong NC, Perez-Castillo AM, Sanders MM, Schwartz HL, Oppenheimer JH (1989) Thyroid hormone and circadian regulation of the binding activity of a liver-specific protein associated with the 5'-flanking region of the S14 gene. J Biol Chem 264: 4466–4470

69. Lausson S, Segond N, Milhaud G, Staub JF (1989) Circadian rhythms of calcitonin gene expression in the rat. J Endocrinol 122: 527–534

70. Segond N, Jullienne A, Lasmoles F, Desplan C, Milhaud G, Moukhtar MS (1984) Rapid increase of calcitonin-specific mRNA after acute hypercalcemia. Eur J Biochem 139: 209–215

71. Segond N, Legendre B, Tahri EH, Besnard P, Jullienne A, Moukhtar MS, Garel JM (1985) Increased level of melatonin mRNA after 1,25-dihydroxyvitamin D3 injection in the rat. FEBS 184: 268–272

72. Noshiro M, Nishimoto M, Okuda K (1990) Rat liver cholesterol 7 alpha-hydroxylase. Pretranslational regulation for circadian rhythm. J Biol Chem 265: 10036–10041

73. Gielen J, van Cantfort J, Robaye B, Renson J (1975) Rat liver cholesterol 7 alpha-hydroxylase-3. New results about its circadian rhythm. Eur J Biochem 55: 41–48

74. Ralph MR, Menaker M (1988) A mutation of the circadian system in golden hamsters. Science 241: 1225–1227

75. Mrosovsky N (1989) Circadian rhythms: mutant hamster in a hurry [news]. Nature 337: 213–214

76. Beau J (1988) Polygenic correlates of activity rhythm in mammals: analysis of two inbred mouse strains C57BL/6By and BALB/cBy. C R Acad Sci [III] 307: 37–40

77. Oliverio A, Malorni W (1979) Wheel running and sleep in two strains of mice. Plasticity and rigidity in the expression of circadian rhythmicity. Brain Res 163: 121–133

78. Diez-Noguera A, Cambras T, Ribot M, Torralba A (1988) Detection of changes in the mouse circadian rhythm induced by stress. Rev Esp Fisiol 44: 247–255

79. Cambras T, Diez-Noguera A (1988) Generational variability in the patterns of motor activity circadian rhythm in the rats. Rev Esp Fisiol 44: 243–246

80. Possidente B, Hegmann JP (1980) Circadian complexes: circadian rhythms under common gene control. J Comp Physiol 139: 121–125

81. Connolly MS, Lynch CB (1983) Classical genetic analysis of circadian body temperature rhythms in mice. Behav Genet 13: 491–500

82. Buttner D (1987) Influence of strain specific patterns of locomotor activity on the daily pattern, minimal, means and maximal oxygen consumption in the rat. Z. Vertuchstierkd 29: 121–128

83. Kripke DF, Wyborney VW (1980) Lithium slows rat circadian rhythms. Life Sci 26: 1319–1321

84. Peleg L, Nesbitt MN, Ashkenazi IE (1989) Strain dependent response of circadian rhythms during exposure to continuous illumination. Life Sci 44: 893–900

85. Possidente B, Exner RH (1986) Gene-dependent effect of lithium on circadian rhythms in mice (*Mus musculus*). Chronobiol Int 3: 17–21

86. LePape G, Lassalle JM (1986) Behavioral development in mice: effects of maternal environment and the albino locus. Behav Genet 16: 531–541

87. Castellano C, Puglisi-Allegra S, Renzi P, Oliverio A (1985) Genetic differences in daily rhythms of pain sensitivity in mice. Pharmacol Biochem Behav 23: 91–92

88. Kavaliero M, Hirst M, Reskey GC (1983) Aging, opioid analgesia and the pineal gland. Life Sci 32: 2279–2287

89. Oliverio A, Castellano C, Puglisi-Allegra (1982) Opiate analgesia: evidence for circadian rhythms in mice. Brain Res 249: 265–270

90. Kempf E, Mandel P, Oliverio A, Puglisi-Allegra S (1982) Circadian variations of noradrenaline, 5-hydroxytryptamine and dopamine in specific brain areas of C57BL/6 and BALB/c mice. Brain Res 232: 472–478

91. Buttner D, Wollnik F (1984) Strain-differentiated circadian and ultradian rhythms in locomotor activity of the laboratory rat. Behav Genet 14: 137–152

92. Wollnik F, Gartner K, Buttner D (1987) Genetic analysis of circadian and ultradian locomotor activity rhythms in laboratory rats. Behav Genet 17: 167–178

93. Possidente B, Stephan FK (1988) Circadian period in mice: analysis of genetic and maternal contributions to inbred strain differences. Behav Genet 18: 109–117

94. Piechulla B, Gruissen W (1987) Diurnal mRNA fluctuations of nuclear and plastid genes in developing tomato fruits. EMBO J 6: 3593–3599

95. Giuliano G, Hoffman NE, Ko K, Scolnik RA, Cashmore AR (1988) A light-entrained circadian clock controls transcription of several plant genes. EMBO J 7: 3635–3642

96. Lam E, Chua NH (1989) Light to dark transition modulates the phase of antenna chlorophyll protein gene expression. J Biol Chem 264: 20175–20176

97. Barcal R, Sova J, Krizanovska M, Levy J, Matousek J

(1968) Genetic background of circadian rhythms. Nature 220: 1128–1131

98. Hanson BR, Halberg F, Tuna N, Bouchard TJ, Lykken DT, Cornelissen G, Heston LL (1984) Rhythmometry reveals heritability of circadian characteristics of heart rate of human twins reared apart. Cardiologia 29: 267–282

99. Reinberg A, Touitou Y, Restoin A, Migraine C, Levi F, Montagner H (1985) The genetic background of circadian and ultradian rhythm patterns of 17-hydroxycorticosteroids: a cross-twin study. J Endocrinol 105: 247–253

100. Meikle AW, Stringham JD, Woodward MG, Nelson JC (1988) Hereditary and environmental influences on the variation of thyroid hormones in normal male twins. J Clin Endocrinol Metab 66: 588

101. Meikle AW, Stringham JD, Woodward MG, Bishop DT (1988) Heritability of variation of plasma cortisol levels. Metabolism 37: 514–517

102. Sistonen P, Ehnholm C (1980) On the heritability of serum high density lipoprotein in twins. Am J Hum Genet 32: 1–7

103. Norris KK, DeAngelo TM, Vesell ES (1989) Genetic and environmental factors that regulate cytosolic epoxide hydrolase activity in normal human lymphocytes. J Clin Ivest 84: 1749–1756

104. Chusid MJ, Casper JT, Camitta BM, McCreadie SR (1986) Cyclic neutropenia in identical twins. Am J Med 80: 994–996

105. Doose H, Baier W (1988) Theta rhythms in the EEG: a genetic trait in childhood epilepsy. Brain Dev 10: 347–354

106. Parkes JD, Lock CB (1989) Genetic factors in sleep disorders. J Neurol Neurosurg Psychiatry 52 [Suppl]: 101–108

107. D'Souza D, Thomas IM, Das BC (1988) Variation in spontaneous chromosomal damage as a function of biologic rhythms in women. Hum Genet 79: 83–85

108. Slozina NM, Golovachev GD (1986) The frequency of sister chromatid exchanges in human lymphocytes determined at different time within 24 hours. Tsitologiia 28: 127–129

109. Higurashi M, Kawai H, Segawa M, Iijima K, Ikeda Y, Tanaka F, Egi S, Kamashita S (1986) Growth, psychologic characteristics, and sleep-wakefulness cycle of children with sex chromosomal abnormalities. Birth Defects 22: 251–275

# Seasonal and Daily Control of Enzyme Synthesis and Activity by Circadian Clocks

O. Queiroz and C. Queiroz-Claret

## Introduction

The ability *in due time* to turn on and off gene activity, metabolic pathways and regulation mechanisms is essential for the efficient adaptability of living systems. Free-living organisms in natural conditions are subjected to the action of periodic changes in environmental constraints, in particular those resulting from the earth's 24-h rotation and its orbit around the sun, and tend to adapt to these variations by readjusting periodically their metabolic functions and behaviour in order to match the course of the seasons and, for each seasonal pattern, the day/night alternation. More precisely, the term "seasonal cycles" includes (1) the variation in the relative lengths of day and night in the 24-h cycle, a variation which is strictly stable from year to year and thus predictable, and (2) a set of other factors, in particular temperature and water availability, showing only probable seasonal variations and irregular fluctuations. It seems therefore reasonable to assume that the necessity of shifting from one metabolic strategy to another, in order to adapt to environmental changes and to trigger growth and reproduction only when the favourable season arrives, is led by selective advantage to utilize some mode of measuring the daily duration of darkness or light as the only reliable indicator of the course of the seasons.

The first important discovery in this area was by Garner and Allard (1920): they established that in plants the shift from vegetative growth to flowering occurs when a critical duration of day is attained in the course of the year, this *critical photoperiod* depending on the species. Bünning (1936) proposed that circadian rhythmicity was involved in *photoperiodism*, the mechanisms detecting changes in photoperiod. Since then, results have accumulated continuously on the relationships between circadian rhythms, photoperiodism and physiological functions in plants and animals (see, e.g. Aschoff 1965; Bün-ning 1973; Queiroz 1974; Hillman 1976; Brady 1982). In contrast less attention has been devoted to the role of these interactions in the resistance to stress caused by extreme climatic factors.

The present chapter discusses conceptual and experimental bases for the study of the mechanisms through which enzyme synthesis and rhythmicity are controlled by the coupling between circadian clocks and photoperiodism to achieve resistance to stressing factors. Rather than listing a multiplicity of examples, the discussion is illustrated more particularly by one system (the CAM-mode of photosynthesis) because it has been the object of extensive and systematic studies, ranging from clock control of gene expression to posttranslational enzyme regulation. It thus provides elements for a coherent model attempting to relate different modes by which seasonal and daily changes in metabolism can be driven by the circadian clock and combined into an integrated adaptive process.

## Circadian Organization and Adaptive Mechanisms

### Rhythms for Time Measurement and Rhythms for Operational Efficiency

Two main categories of endogenous rhythms occur in plant and animal organisms: (1) rhythms which underlie time measurement and have adaptive potentialities. We will refer more particularly to *circadian* rhythms which can be said (utilizing a modern and somewhat teleological terminology) to be a basis for adaptive "*strategy*" by coordinating seasonal and diurnal metabolic organization. (2) Biochemical *short-period oscillations* such as those resulting from alternated action and release from feedback or feedforward mechanisms. The role of these oscillations is not to measure time but obviously to *optimize* metabolic operation by enabling the regulated enzyme to receive information on the metabolic flux through its own pathway or connected sequences (Hess and Boiteux 1971; Queiroz 1979, 1983b; Crabtree and Newsholme 1987). It should be kept in mind that type (1) and (2) oscillations coexist and may exert their specific functional roles in the same metabolic sequence (e.g., glycolysis).

## Circadian Rhythmicity Versus Stimulus-Response Systems: Adaptive Value for Metabolic Anticipation

Probably the more fundamental feature underlying seasonal and daily adaptation is that in most cases the basic mechanism is endogenous circadian rhythmicity and not direct stimulus-response effects. Starting the day's or season's adaptive sequences of metabolic events by a specific stimulus from the environment (e.g., dawn, dusk, jumps in the temperature or humidity regimes) would risk adaptive inefficiency because a number of relevant seasonal responses may require days or weeks to be effective (e.g., endocrine changes or synthesis of a sufficient amount of enzyme to attain the necessary level of activity). Moreover, out-of-season momentary fluctuations in the external factors would needlessly trigger seasonal adaptive mechanisms. Also in night/day changes, lags in metabolic activity are not compatible with efficient performance.

In contrast, the properties of circadian rhythms (Table 1) involve clear adaptive advantages by providing a temporal metabolic structure which can easily couple to 24-h periodicities (entrainment) and secure the timely reinitiation of metabolic sequences.

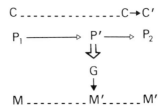

**Fig. 1.** A model for seasonal anticipation. Critical photoperiod *P'* would induce in advance the expression of genetic programme *G* required for the synthesis of the metabolic machinery *M'* which enables the organism to respond to seasonally probable changes in the environmental factor *C*

In fact, the main adaptive value of endogenous circadian rhythmicity is that, by proper phase adjustment under entrainment, metabolism anticipates the external periodic changes, i.e. *prepares in advance* (Hoffmann 1976; Queiroz 1983a). Thus the biochemical machinery is ready for optimal adaptive performance when the expected (cyclic) changes in environmental conditions occur. Anticipation is a well-established feature in night/day physiological changes (e.g. in body temperature) but also at the metabolic and enzymatic levels. At the seasonal scale, combination of circadian rhythmicity with photoperiodism confers on the organism the possibility of timely preparation of seasonal adaptive response by restricting a particular metabolic activity to a certain time of the year.

The basis for *seasonal anticipation* can be schematized as follows (Fig. 1): let *M'* be a metabolic programme which permits resistance to seasonal values *C'* of a factor *C* (e.g. drought); selective advantage is obvious if the enzymatic machinery required to perform *M'* can be synthesized in advance so that maximum efficiency is immediately attained when *C'* occurs. This "anticipation" can be achieved through timely induction of *M'* by a critical photoperiod "announcing" the coming on of the season during which occurrence of *C'* is probable.

## Clocks for Daily and Seasonal Organization

### Features in Circadian Time Measurement

The ability to measure time by photoperiodism implies the existence of systems capable of *(a)* transducing information on periodic light signals obtained via flavoproteins or phytochrome in plants, retinal or non-retinal systems in animals; *(b)* detecting changes in the day length and/or night length; and *(c)* distinguishing between "short" and "long"; for instance, reproduction programs in plants and animals are trig-

**Table 1.** Characteristics of circadian rhythms and functional significance for the circadian organization in response to external perturbations. Damping may occur after a few cycles owing to poisoning or starvation effects in the absence of the necessary complementarity between day and night metabolic activity

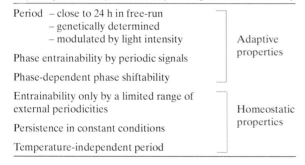

| | |
|---|---|
| Period – close to 24 h in free-run<br> – genetically determined<br> – modulated by light intensity<br>Phase entrainability by periodic signals<br>Phase-dependent phase shiftability | Adaptive properties |
| Entrainability only by a limited range of external periodicities<br>Persistence in constant conditions<br>Temperature-independent period | Homeostatic properties |

gered for some species when night becomes short, for others when it becomes long, the absolute values for "short" or "long" being specific to a species. Properties *b* and *c* are the basis of photoperiodism. The modified metabolic programe is then driven at the day/night time scale.

Phase response curves showed that photoperiodic entrainment is achieved by the interaction of a phase advance at dawn and a phase delay at dusk (Pittendrigh 1976). The resulting abrupt phase shifts could be observed for instance in the levels of glycolytic enzymes (Pierre et al. 1984). Also 24-h period temperature signals or appropriate chemicals can entrain the circadian oscillator (see Edmunds 1988 for discussion and references) but under natural conditions the photoperiod is the predominant entraining agent (Zeitgeber), probably because of its reliability, except in the case of strong social cues.

## Basic Models: Hourglass or Circadian Oscillators

Locking the day/night organization to photoperiod as a means of measuring time can be achieved through various mechanisms which are basically similar across the plant and animal kingdoms. At the present state of knowledge three main models account for most cases:

1. The *hourglass model* assumes that a precursor molecule accumulates steadily (or an inhibitor is steadily depleted) by a process set in motion at dusk. If night duration is long enough ("critical night-length") to allow the concentration of that substance to attain a critical threshold, then a metabolic sequence can be started and long-night (short-day) effects are observed. In this model extending darkness results in increasing response: the hourglass does not reset itself in the absence of light cycles, which means that it does not contain circadian rhythmicity by itself. First observed in plants, the hourglass process has also been shown to account for seasonal time measurement in number of animal organisms.

2. The *"Bünning hypothesis" ("external coincidence" model)*, proposed by Bünning (1936) and reformulated by Pittendrigh (1972), involves an oscillator, assumed to be composed of two half-cycles of about 12 h each differing in sensitivity to light, and to restart its motion at dusk. Photoperiodic induction only occurs when light coincides with a particular phase of the rhythm of sensitivity. For instance, short-day responses are triggered only under photoperiods for which dark extends beyond the light-sensitive phase of the rhythm (located at the end of the critical night-length); when in the course of seasons night becomes shorter so that the light-sensitive phase receives light, long-day effects are produced. In contrast with an hourglass model the circadian sensitivity rhythm continues in extended darkness alternating phases (corresponding to "subjective night") during which night length is effectively measured, and phases (corresponding to "subjective day") during which darkness is not measured. To sum up, light has two

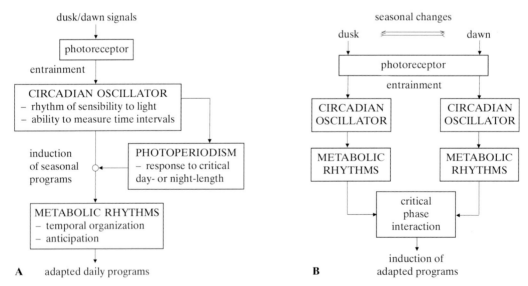

**Fig. 2 A, B.** Conceptual outline of seasonal control of metabolic programs by photoperiodism via the "external coincidence" (**A**) or "internal coincidence" (**B**) mechanisms

roles: entrainment and induction (Fig. 2 A). Bünning's hypothesis accounts for a very large number of circadian rhythms in plants, insects, birds and mammals (see, e.g. Aschoff 1965; Bünning 1973; Queiroz 1974; Brady 1982).

3. The *"internal coincidence"* model was proposed by Pittendrigh (1972) to account for a number of results obtained with plants and animals suggesting that an organism or a cell could have not one but several circadian oscillators. In particular, if one oscillator is locked to dusk and another to dawn, changes in photoperiod would modify the phase relation between the two oscillators (and consequently between the metabolic rhythms they drive) thus inducing changes in metabolic programs (Fig. 2 B). In the internal coincidence model light has only a single role, that of entrainment; induction would result from a particular entrained steady state of the circadian organization in the multioscillator system.

An important consequence of the model is that "photoperiodic" induction should also be obtained in the absence of light/dark signals if the oscillators can be conveniently phased by temperature or chemical pulses. This conclusion is supported for instance by results showing that injections of corticosterone and prolactin differently timed in order to change the phase of the rhythms of these hormones in birds mimicked the effects of long days or short days (Meier 1973); similar effects were obtained with hamsters by daily timed injections of 5-HTP and DOPA (Wilson and Meier 1989). The internal coincidence model was also supported by extensive research on plants (e.g. Papenfuss and Salisbury 1967), insects (Saunders 1974; Pittendrigh and Takamura 1987), rodents (Pittendrigh and Daan 1976) and tadpole epidermal cell proliferation (Wright et al. 1988), to quote only a few references in a rich literature.

From the point of view developed in the present article an important feature in the internal coincidence model is that it goes deeply into the notion that circadian organization arises from the interactive character of the metabolic network (see the next section).

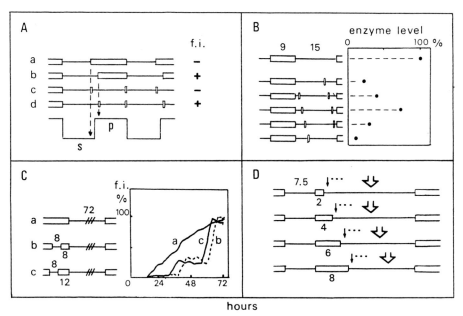

**Fig. 3 A–D.** Studies of clock/photoperiod coupling: illustrative early examples of still usual methodology (light: *white bars*). **A** Skeleton photoperiods *(c, d)* mimicked respectively long days *(a)* or short days *(b)* for flower induction *(f. i.)* in *Lemna*. The postulated skotophil *(s)* and photophil *(p)* phases of the "Bünning oscillator" are shown (modified from Hillman 1964). **B** Effect on the level of the enzyme PEP carboxylase of 10 h darkness (critical night length) timed by two light flashes at different moments of 15-h nights (modified from Queiroz 1979). **C** *a*, Different durations of extended darkness (up to 72 h) showed the existence in *Pharbitis* of an hourglass process to control flower induction. Interpolation of a single 8-h or 12-h light span *(b, c)* triggered an oscillator superimposed on the hourglass and measuring time from lights-on (modified from Takimoto and Hamner 1964). **D** In *Xanthium* subjected to LL then to phasing 7.5 h darkness, light spans up to 4 h triggered time measurement from lights-on as shown by time of optimum effect *(white arrows)* of one light flash *(black arrows)* given at different times of the following night; for light spans longer than 6 h time measurement started from lights off. (Drawn from results by Papenfuss and Salisbury 1967)

## Experimental Methods

The existence of the three main types of models delineated above and their variants reflects the fact that in the course of evolution photoperiodism appeared in different forms, involving diverse mechanisms as a means to couple internal coordination to the daily cycles. Moreover it has been found that the same species or even the same organism can switch from one to another of these mechanisms or combine them, depending on the prevailing environmental conditions (Claret 1985; Takeda and Skopik 1985; Dumortier and Brunnarius 1989). Experimental protocols to distinguish hourglass from circadian sensitivity can utilize more particularly three techniques (illustrated in Fig. 3 by classic examples): "skeleton photoperiods", i.e. appropriately timed light flashes mimicking photoperiods in constant darkness; night interruption by short (minutes or even less) light flashes applied at different times of the night in a regime of 24-h day/night cycles; and "resonance" experiments, in which short days (less than 12 h) are combined with different lengths of prolonged darkness, or darkness interrupted at different times by a light flash. Distinction between external or internal coincidence may rely on detecting whether metabolic pulses can replace light signals for phase entrainment of the rhythms assumed to be involved in the photoperiodic response.

It should also be realized that photoperiodic induction has a cumulative component: in most cases a minimum number of photoperiodic cycles (lag time) is necessary to induce a detectable response, and maximum response levels are attained only by cumulating the effect of several cycles.

# Biochemical Bases of Adaptation to Environmental Periodic Constraints

## Readjusting the Metabolic Machinery

### Functional Coordination

Metabolism in any tissue or cell should be viewed as an interactive, highly coordinated network of biochemical reactions, continuously readjusting its functional patterns in order to counteract the influences from external inputs and their variations. The coordinated character of metabolism implies that several metabolic functions (which means several enzymes

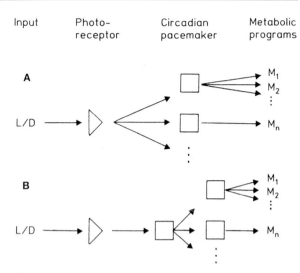

**Fig. 4 A, B.** Block models of circadian organization (see text)

or groups of enzymes, hence several genes) should be expected to intervene in a particular response to external modifications. Therefore identification of the primary target of the external signal (necessary for the interpretation and application of the results) is often difficult.

In a more general form, the question is to distinguish between two modes of response: *(a)* The external signal could trigger simultaneously different components of a function, or different functions. For example, it was shown with some plant species that photoperiodism controls simultaneously the change from the vegetative to reproductive programs and the induction of an enzymatic pathway required for resistance to drought (next section). In terms of chronobiology, one would suppose that the external signal either entrains one clock governing several functions or several clocks (Fig. 4 A). *(b)* The external signal could trigger a main rhythmic pathway which in turn would act as internal pacemaker to other pathways or functions. In the above example this would imply an alternative hypothesis, viz. as flowering in those species occurs during the dry season photoperiodism would trigger the onset of a drought-resistance mechanism which in turn would act as an internal signal unlocking the genetic program for reproduction; this latter would thus be carried in an organism previously capable of resisting seasonal stress. In terms of chronobiology hierarchic clock organization would fulfil this type of control (Fig. 4 B).

At the present time the major hindrances to answering the question whether multiple switches or cascade switches are involved in clock control are ignorance of the secondary messenger molecules carrying the intermediary information and in most cases insufficient knowledge of the molecular rhythmic mechanisms modulating the different steps of gene expression.

## Synergistic Oscillatory Mechanisms

Let us suppose that the concentration of an inhibitor increases so that its regulatory effect becomes operative at time $t$ of the diurnal cycle: this effect would be more efficient and clearcut if by that time the amount of active enzyme is decreased, or its affinity to the substrate decreases and/or its sensitivity to the inhibitor increases. This example illustrates the general assumption that more efficient homeostasis and selective advantage should be achieved if several mechanisms rather than only one are involved in the adaptive metabolic readjustment, and if these concurrent effects arise from independent processes rather than from a sequential process (in which for example the inhibitor would itself bring up the sensitivity of the enzyme to its own effect). This assumption, more generally discussed by Queiroz-Claret and Queiroz (1990), will be illustrated in the next section.

## Rhythms of Enzyme Capacity and Enzyme Activity

In vivo oscillations in enzyme activity can proceed from (a) feedback or feedforward regulation, or (b) oscillations in the concentration and/or the catalytic properties of the enzyme, i.e. the actual quantitative or qualitative status of the molecular population of the enzyme. The term *"enzyme capacity"* will refer to the catalytic potentiality of the cells to perform a given enzyme reaction (Queiroz 1983b). Enzyme capacity is evaluated experimentally by the maximum activity of the total extractable enzyme from a tissue, or if possible from the corresponding cell compartment, measured at substrate saturation in non-limiting, well-defined reaction medium. It should be noted that the maximum activity thus obtained approaches asymptotically the theoretical $V_{max}$, which can be computed from the kinetic data. Rhythmic variations in enzyme capacity indicate that quantitative and/or qualitative rhythmic modifications of the enzymatic machinery should be sought (Table 2).

A basic point in metabolic organization is that under the usual physiological conditions in vivo enzyme capacity exceeds largely the actual enzyme activity, i.e. the rate of substrate transformation (measured, e.g. by incorporating radioactive sub-

**Table 2.** Analytical indications for the mechanisms assumed to underlie rhythmic changes in enzymatic capacity

| Enzyme concentration | Catalytic properties[a] | Postulated mechanism |
|---|---|---|
| Rhythmic | Non-rhythmic | Rhythmic synthesis[b] |
| Rhythmic | Rhythmic | Rhythmic synthesis of an isoenzyme |
| Non-rhythmic | Rhythmic | Posttranslational rhythmic molecular modifications |

[a] E.g. $V_{max}$, $K_m$, affinity for inhibitors or activators
[b] Rhythms of degradation have also been reported

strate). In other words, enzymes in vivo utilize only a fraction of their catalytic potentiality. In fact they work at about $v = 0.5\ V_{max}$, which corresponds to an intracellular level of substrate concentration $(S)$ close to $K_m$. If in the course of metabolic rhythmicity a large amplitude in the production of $S$ leads to $v$ tending to $V_{max}$, maintenance of $(S)$ close to $K_m$ together with high activity rates can be obtained by a rhythm in enzyme capacity, for instance a rhythm in enzyme synthesis or a rhythm in affinity for the substrate (Queiroz 1983b; Queiroz-Claret and Queiroz 1990).

## General Categories of Response to Stress

When external variations are too steep or have too large an amplitude to be counteracted by the readjusting capacity of the current metabolic program they act as a stress. This term often refers to the effect of sudden, unexpected aggressions. However, stress effects can occur as part of normal environmental periodicities. For instance, in hot habitats the rapid increase in temperature following sunrise is responded to by the synthesis of "heat-shock" proteins, with involvement of clock-controlled translation (Rensing et al. 1987) or transcription (Otto et al. 1988).

Two main strategies can be developed by the organisms as a response to stress situations (Levitt 1972): *stress avoidance* through essentially conservative mechanisms tending to escape the effects of the stressing factor, for example by migration or by adequately timing parts of the life cycle (insect diapause or plant bud dormancy to survive winter, seed dormancy during the cold or the dry season); or *stress tolerance* by changing metabolic functioning into a different metabolic program capable of integrating or counterbalancing the effects of stress, e.g. poikilothermic responses, synthesis of hormones, "stress proteins" or specific enzymes. Conversely growth

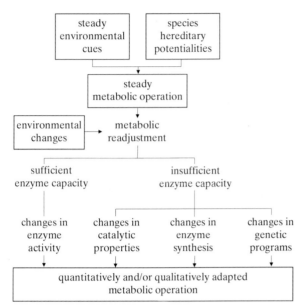

**Fig. 5.** Various ways to achieve stress tolerance by modifying the enzymatic machinery. The term "changes in genetic program" refers to production of isoenzymes, enzymes catalysing previously absent reactions (e.g. anaplerotic enzymes) or stress proteins. See also Fig. 6

and reproduction processes in animals and plants occur in seasons offering favourable conditions or are concomitant with the development of stress tolerance metabolisms.

Figure 5 outlines how the diverse modes of metabolic flexibility discussed in the preceding sections might contribute to stress tolerance. Summing up, since only a fraction of the existing enzyme capacity is utilized in a prevailing environmental situation, small-amplitude changes either periodic or accidental can be responded to by utilizing the safety margin in enzyme capacities; if the external changes have amplitudes too large for the existing enzyme capacities, or require bringing into action other metabolic pathways (this is for instance the case at the onset of the day in plants and day-active animals) then one should expect to observe de novo enzyme synthesis and/or posttranslational modifications of enzyme properties.

## Multilevel Study on the Control by Circadian Clocks of Enzyme Rhythms

Figure 6 summarizes the process leading from gene activity to operational enzyme forms: it is known from work with a number of organisms that each of the different steps may be sensitive to control by cir-

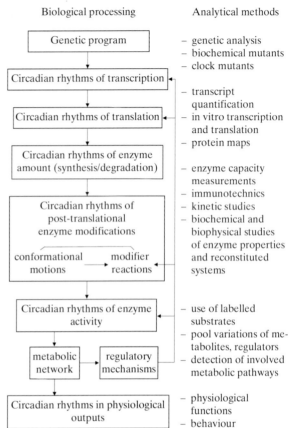

**Fig. 6.** Multilevel study of enzyme rhythms. Note that (as shown by a number of examples) clock control at one level does not necessarily imply rhythmicity in the preceding or subsequent steps. Research in most cases starts from rhythms observed at physiological levels then proceeds by dissecting the underlying biochemical mechanisms at progressively lower levels of the biological organization. As documented by recent literature, an increasingly wider range of analytical approaches and techniques, including protein dynamics and molecular biology tools, is now applied to the study of the control of enzyme rhythmicity by circadian clocks

cadian clocks, particularly via photoperiodism. Research usually starts by characterizing a rhythmic physiological output, then attempts to trace its causality upward to the involved metabolic pathways and enzymes, more precisely the key-enzyme (i.e. the enzyme with the lowest catalytic capacity in the pathway), in order to detect whether its rhythm accounts for the pathway's rhythm. Early research on enzyme rhythmicity centered mostly on regulation mechanisms. More recently attention was drawn to circadian rhythms in posttranslation enzyme modifications, translation and transcription (for reference see Rensing et al. 1987; Edmunds 1988; Queiroz-Claret and Queiroz 1990).

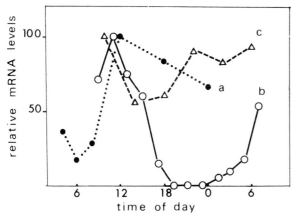

**Fig. 7.** Examples of recent results obtained by mRNA blot analysis indicating control by circadian clocks of the transcription ability for various plant genes. Data were confirmed by in vitro transcription experiments with isolated nuclei. Results expressed as relative mRNA abundance (percentage of maximum) for: *(a)* a biotin-binding protein in tomato seedlings in continuous darkness (modified from Giuliano et al. 1988); *(b)* the light-harvesting chlorophyll-binding protein from tobacco leaves in continuous light (modified from Paulsen and Bogorad 1988); *(c)* a 24-kD heat-shock protein in pea chloroplasts induced by 2 h at 42°C immediately before sampling in continuous light (modified from Otto et al. 1988). All plants were previously under photoperiodic conditions

A promising field concerning the study of clock-controlled circadian transcription rhythms develops rapidly through the utilization of molecular biology tools. This is at present better illustrated by the very recent studies progressing in several laboratories on chloroplast enzymes and proteins (Fig. 7). In the case of the light-harvesting chlorophyll-binding protein the established circadian rhythm of transcription was shown to be entrained by nightfall so that the working capacity of the light energy-harvesting machinery is primed before the onset of solar radiation; a *cis-acting* element mediating the clock-gene interaction has been identified. Entrainment by dusk would achieve connection with the photoperiod throughout the year (Lam and Chua 1989 and references therein).

Thus it can be said that, at present, circadian rhythms at one or the other of these different steps are separately well documented in the literature. In contrast, comprehensive studies covering the complete sequence from gene control to the physiological rhythmic output are lacking, except in very few cases as examplified in the following section.

# Seasonal and Daily Control of Enzyme Rhythms by Photoperiodism: An Integrated Adaptive Process

The purpose of this section is to emphasize through a particularly comprehensive example that achievement and coordination of the two major roles of circadian time measurement, viz. seasonal induction of gene expression and daily entrainment of enzyme rhythms, may require the conjunction of diversified rhythmic mechanisms intervening at the different steps of the gene → active enzyme process.

## Seasonal Modulation of Gene Activity by Photoperiodism

Season-dependent patterns of physiological functions and behaviour have been widely reported in the literature for higher plants and animal organisms. In contrast, data on the corresponding changes in gene expression are generally lacking. To our knowledge, the few data available on the control by photoperiodism of gene activity on a seasonal time scale come mostly from studies on plant enzymes.

The CAM mode of $CO_2$ incorporation is present in plant families adapted to arid and hot habitats. Its adaptive value derives from the ability to fix external $CO_2$ during the night (by the enzyme PEP carboxylase) and to store it as malate within the cell vacuole until the following morning (Fig. 8 A); after dawn the malate becomes accessible to decarboxylation providing an internal source of $CO_2$ for carbohydrate photosynthesis, thus enabling it to proceed even if external $CO_2$ is no longer available because of stomata closure in order to prevent tissue dehydration during the hot and dry daytime. Short-day photoperiods induce (by "external coincidence", Fig. 3 B) the activity of a gene coding for a specific isoform of PEP carboxylase (Fig. 8 B). The induction of the CAM pathway obviously implies readjusting the metabolic flows and the enzymatic capacities of the connected pathways (Fig. 8 A; see Queiroz 1983 a, b; Queiroz-Claret and Queiroz 1990 and references therein). Kinetic similarities and coherence in phase shifts suggested that the same timing mechanism would coordinate the instauration of the "CAM program" over this entire network.

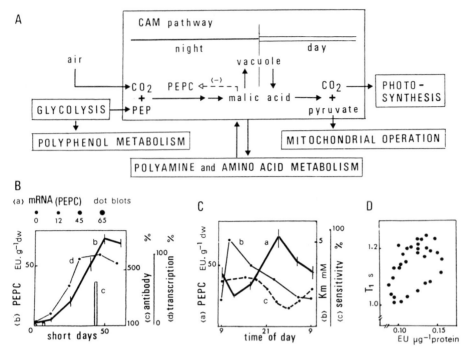

**Fig. 8. A** Plant CAM pathway for primary $CO_2$ fixation and connected pathways. Seasonal and diurnal variations of the entire network are under control by photoperiodism. **B** Seasonal control: short days induce the production of increasing levels of PEPC-specific mRNA (*a* schematic representation of dot blots) and increasing enzyme capacity (*b*) due to increased enzyme synthesis as established by immunotitration (*c*) relative to control in long days. Isolated nuclei showed increased in vitro transcription capacity (*d*). **C** Circadian rhythms in PEP carboxylase capacity (*a*) are related to rhythms in enzyme properties, e.g. rhythms in $Km_{PEP}$ (*b*) and in sensitivity to inhibition by malate (*c*). **D** Example of circadian oscillations of enzyme activity arising spontaneously in a solution (purified malate dehydrogenase) and concomitant oscillations in spin-lattice relaxation time $T_1$ of enzyme-water interactions (measured by NMR). See text for references

## Concurrent Rhythmic Mechanisms for Circadian Cycling

Circadian rhythmicity of the CAM pathway was first observed at the level of its physiological function, namely the typical $CO_2$ exchange pattern, which can proceed in either prolonged darkness or dim light, and be phase-shifted by light or temperature pulses. The pathway alternates (a) a night phase of $CO_2$ incorporation into malate; when the storage pool is filled up feedback inhibition of PEP carboxylase by excess malate stops the process acting as a "phase-delay" step until (b) the dawn signal entrains a circadian rhythm (Morel and Queiroz 1974) governing the vacuole's permeability to malate, which then leaks into the cytosol and is decarboxylated. It is obvious that for efficient CAM operation PEP carboxylase activity must be switched off during the day so that the internally produced $CO_2$ can be utilized for carbohydrate synthesis. Since earlier work (Queiroz and Morel 1974) indicating that PEP carboxylase activity rhythm was governed by a capacity rhythm (Fig. 8C),

the causality of the latter has been sought in several laboratories. Results established that the capacity rhythm of PEP carboxylase was not due to a rhythm of enzyme synthesis (Brulfert et al. 1982) but rahter to rhythmic posttranslational modifications of the enzyme molecule (Kluge et al. 1981; Fig. 8C). More recently work by Brulfert et al. (1986) and Nimmo et al. (1987) established that the enzyme undergoes a circadian rhythm alternating phosphorylated (active) and dephosphorylated states of the molecule.

## Spontaneous Conformational Oscillators

Research on the posttranslational bases for enzyme rhythms, summarized above, has looked so far for mechanisms in which the enzyme is a target molecule for the periodic action of effectors or modifying reactions. A less conventional approach is to consider that an enzyme molecule could behave by itself as an oscillator shifting between different conformations. This assumption was grounded on the new well-established evidence that proteins are dynamic systems undergoing continuous reversible shifting from one configuration to another (Karplus and McCammon 1983; Wagner and Wüthrich 1986), e. g. cyclic subunit assembly in porcine lactate dehydrogenase (King and Weber 1986), conformational interconversions shown by Domenech et al. (1987) for malate dehydrogenase from various animal sources or by Mandell and Russo (1981) with rat striated tyrosine hydroxylase. Systematic research was performed with crude or purified preparations of CAM PEP carboxylase and malate dehydrogenase, combining different techniques, e. g. kinetic studies, active site accessibility, proton exchangeability, chromatofocusing and proton NMR (Queiroz-Claret et al. 1985, 1988a,b, Queiroz-Claret and Queiroz 1991). Results (Fig.8D) yielded evidence that spontaneous reversible transformation between conformational substates would generate circadian oscillations in the catalytic capacity of the enzyme and its sensitivity to ligands.

## Conclusions

As schematized in Fig.9, the study of the control of the CAM system by photoperiodism is illustrative of (1) the complementarity of the seasonal and daily roles of circadian clocks in the timely organization of adaptive responses, through which the organism can *prepare* (on the seasonal time scale) its metabolic ma-

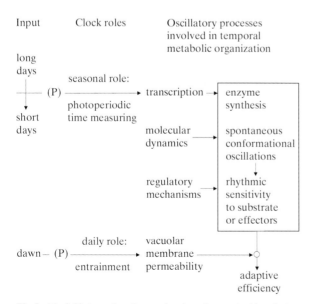

**Fig.9.** Model integrating the mechanisms for control by photoperiodism *(P)* of seasonal induction of the CAM mode of photosynthesis and its circadian rhythmicity (see text)

chinery in order to *tolerate* (through adequate daily rhythmic metabolic tools) the effects of seasonal stressing factors; and (2) the achievement of this adaptive efficiency through the combination of circadian mechanisms intervening at different levels of the epigenetic and metabolic organization, viz. clock-controlled gene expression, feedback regulation as a delaying step, daily entrainment of the vacuolar membrane permeability, circadian conformational oscillators originating or contributing to circadian rhythms in enzymatic properties, phosphorylated state and sensitivity to regulators.

## Prospective

Experimental reasearch on the modalities of the clock photoperiodism interaction was first conducted on development and physiological rhythms in plants (e. g., Fig.3) and later in animal and human organisms. Data describing metabolic and enzymatic circadian rhythms have accumulated progressively in the literature during the last 20 years. More recently, several laboratories investigated the relationships between clocks and protein synthesis, and results were published on circadian rhythms of total or particular RNA and protein species (for references see Edmunds 1988). Progress in molecular biology techniques led particularly in higher plants to a very recent burst of results dealing more precisely with clock control at the transcriptional level for genes encoding identified enzyme species (see references in Queiroz-Claret and Queiroz 1990). The adaptive value of the clock-photoperiodism relationship at this level was emphasized by the study of enzymes of photosynthesis regarding either seasonal induction of gene expression or circadian rhythms of transcription. These results show that technology is now available for developing the study of clock-gene interactions and upward for the identification of the *cis*- and *trans*-acting elements that mediate the rhythmic process.

Adaptive programs imply selective recruitment of sets of genes or elements of gene families brought to activity or arrest as a response to environmental changes. Developing work to determine whether the circadian clock coordinates gene expression via multiple switches or a single main switch is a key to an understanding of its role in physiological adaptation.

At the terminal step of the gene → protein process, making the enzyme molecule operative may involve various posttranslational mechanisms some of which, e.g. periodic phosphorylation, received particular attention in the last few years. But further progress implies taking into account that the protein molecule contains in its structure all the information required for shifting reversibly between conformational substates (macromolecule/solvent oscillators). This feature opens novel avenues to understanding the molecular mechanisms that concur to achieve rhythmicity in enzyme activity and, further, rhythmic enzyme-enzyme interactions and metabolic channeling (Queiroz-Claret and Queiroz 1991).

It is obvious that such a wide range of complementary approaches requires research policies favouring interdisciplinary cooperation on selected crucial problems.

# References

Aschoff J (ed) (1965) Circadian clocks. North-Holland, Amsterdam

Brady J (ed) (1982) Biological timekeeping. Cambridge University Press, Cambridge

Brulfert J, Vidal J, Gadal P, Queiroz O (1982) Daily rhythm of phosphoenol-pyruvate carboxylase in crassulacean acid metabolism plants. Immunological evidence for the absence of a rhythm in protein synthesis. Planta 156: 92–94

Brulfert J, Vidal J, Le Maréchal P, Gadal P, Queiroz O, Kluge M, Kruger I (1986) Phosphorylation-dephosphorylation process as a probable mechanism for the diurnal regulatory changes of PEP carboxylase in CAM plants. Biochem Biophys Res Comm 136: 151–159

Bünning E (1936) Die endonome Tagesrhythmik als Grundlage der photoperiodischen Reaktion. Ber Dtsch Bot Ges 54: 590–607

Bünning E (1973) The physiological clock. Springer, Berlin Heidelberg New York

Claret J (1985) Two mechanisms in the biological clock of Pieris brassicae: an oscillator for diapause induction, an hourglass for diapause termination. Experientia 41: 1613–1615

Crabtree B, Newsholme EA (1987) A systematic approach to describing and analysing metabolic control systems. Trends Biochem Sci 12: 4–12

Domenech C, Abante J, Bozal FX, Mazo A, Cortes A, Bozal J (1987) Micro-heterogeneity of malate dehydrogenase from several sources. Biochem Biophys Res Commun 147: 753–757

Dumortier B, Brunnarius J (1989) Diet-dependent switch from circadian to hourglass-like operation of an insect photoperiodic clock. J Biol Rhythms 4: 481–490

Edmunds LN Jr (1988) Cellular and molecular bases of biological clocks. Springer, Berlin Heidelberg New York

Garner WW, Allard HA (1920) Effect of the relative length of the day and night and other factors of the environment on growth and reproduction in plants. J Agric Res 18: 553–606

Giuliano G, Hoffman NE, Ko K, Scolnik PA, Cashmore AR (1988) A light-entrained circadian clock controls transcription of several plant genes. EMBO J 7: 3635–3642

Hess B, Boiteux A (1971) Oscillatory phenomena in biochemistry. Annu Rev Biochem 40: 237–258

Hillman WS (1964) Endogenous circadian rhythms and the response of Lemna perpusilla to skeleton photoperiods. Am Nat 98: 323–328

Hillman WS (1976) Biological rhythms and physiological timing. Annu Rev Plant Physiol 26: 159–179

Hoffman K (1976) The adaptive significance of biological rhythms corresponding to geophysical cycles. In: Hastings JW, Schweiger HG (eds) The molecular basis of circadian clocks. Abakon, Berlin, pp 63–75

Karplus M, McCammon JA (1983) Dynamics of proteins: elements and function. Annu Rev Biochem 53: 263–300

King L, Weber G (1986) Conformational drift and cryoinactivation of lactate dehydrogenase. Biochemistry 25: 3637–3640

Kluge M, Brulfert J, Queiroz O (1981) Diurnal changes in the regulatory properties of PEP carboxylase in CAM. Plant Cell Environ 4: 251–256

Lam E, Chua NH (1989) Light to dark transition modulates the phase of antenna chlorophyll protein gene expression. J Biol Chem 264: 20175–20176

Levitt J (1972) Response of plants to environmental stresses. Academic, New York

Mandell AJ, Russo PV (1981) Striatal tyrosine hydroxylase activity: multiple conformational kinetic oscillators and product concentration frequencies. J Neurosci 1: 380–389

Meier AH (1973) Daily hormone rhythms in the white-throated sparrow. Am Sci 61: 184–187

Morel C, Queiroz O (1974) Dawn signal as a rhythmical timer for the seasonal adaptive variation of CAM: a model. Plant Cell Environ 1: 141–149

Nimmo GA, Wilkins MB, Fewson CA, Nimmo HG (1987) Persistent circadian rhythms in the phosphorylation state of PEPC from Bryophyllum fedtschenkoi leaves and its sensitivity to inhibition by malate. Planta 170: 408–415

Otto B, Grimm B, Ottersbach P, Kloppstech K (1988) Circadian control of the accumulation of mRNAs for light- and heat-inducible chloroplast proteins in pea. Plant Physiol 88: 21–25

Papenfuss HD, Salisbury FB (1967) Properties of clock resetting in flowering of Xanthium. Plant Physiol 42: 1562–1568

Paulsen H, Bogorad L (1988) Diurnal and circadian rhythms in the accumulation and synthesis of mRNA for the light-harvesting chlorophyll a/b binding protein in tobacco. Plant Physiol 88: 1104–1109

Pierre JN, Celati C, Queiroz O (1984) Control, entrainment and co-ordination by photoperiodism of 24h-period enzyme rhythms (phosphofructokinase and PEP carboxylase) in Kalanchoe. J Interdisc Cycle Res 15: 267–279

Pittendrigh CS (1972) Circadian surfaces and the diversity of possible roles of circadian organization in photoperiodic induction. Proc Natl Acad Sci USA 69: 2734–2737

Pittendrigh CS (1976) Circadian clocks: what are they? In: Hastings JW, Schweiger HG (eds) Molecular basis of circadian rhythms. Abakon, Berlin, pp 11–48

Pittendrigh CS, Daan S (1976) A functional analysis of circadian pacemakers in nocturnal rodents. I–V. J Comp Physiol 106: 223–355

Pittendrigh CS, Takamura T (1987) Temperature dependence and evolutionary adjustment of critical night-length in insect photoperiodism. Proc Natl Acad Sci USA 81: 7169–7173

Queiroz O (1974) Circadian rhythms and metabolic patterns. Annu Rev Plant Physiol 25: 115–134

Queiroz O (1979) CAM: rhythms of enzyme capacity and activity as adaptive mechanisms. In: Gibbs M, Latzko E (eds)

Photosynthesis II. Springer, Berlin Heidelberg New York, pp 126–139 (Encyclopedia of plant physiology, new series, vol 6)

Queiroz O (1983a) An hypothesis on the role of photoperiodism in the metabolic adaptation to drought. Physiol Veg 23: 577–588

Queiroz O (1983b) Interaction between external and internal factors affecting the operation of phosphoenolpyruvate carboxylase. Physiol Veg 23: 963–975

Queiroz O, Morel C (1974) Metabolic feedback and endogenous circadian rhythmicity: the case of the enzyme PEP carboxylase: J Interdisc Cycle Res 5: 217–222

Queiroz-Claret C, Queiroz O (1990) Multiple levels in the control of rhythms of enzyme synthesis and activity by circadian clocks: recent trends. Chronobiol Int 5: 25–33

Queiroz-Claret C, Queiroz O (1991) Enzyme circadian rhythms and conformational oscillators. Survey and prospects. J Interdiscip Cycle Res 22: 41–45

Queiroz-Claret C, Girard Y, Girard B, Queiroz O (1985) Spontaneous long-period oscillations in the catalytic capacity of enzymes in solution. J Interdiscip Cycle Res 16: 1–9

Queiroz-Claret C, Lenk R, Queiroz O, Greppin H (1988a) NMR studies of in vitro oscillations in enzyme properties and dissipative structures. Plant Physiol Biochem 26: 333–338

Queiroz-Claret C, Valon C, Queiroz O (1988b) Are spontaneous conformational interconversions a molecular basis for long-period oscillations in enzyme activity? Chronobiol Int 5: 301–309

Rensing L, Techel D, Schroeder-Lorenz A (1987) Protein phosphorylation and circadian clock mechanisms. In: Hildebrand G, Moog R, Raschke K (eds) Chronobiology and chronomedicine. Lang, Frankfurt, pp 39–48

Saunders DS (1974) Evidence for "dawn" and "dusk" oscillators in the Nasonia photoperiodic clock. J Insect Physiol 20: 77–88

Takeda M, Skopik SD (1985) Geographic variation in the circadian system controlling photoperiodism in Ostrinia nubilalis. J Comp Physiol 156: 653–658

Takimoto A, Hamner KC (1964) Effect of temperature and pre-conditioning on photoperiodic response in Pharbitis nil. Plant Physiol 39: 1024–1030

Wagner G, Wüthrich K (1986) Observation of internal motility of protein by NMR in solution. Methods Enzymol 131: 307–326

Wilson JM, Meier AH (1989) Resetting the annual cycle with timed daily injections of 5-hydroxytryptophan and L-dihydroxyphenylalanine in Syrian hamster. Chronobiol Int 6: 113–121

Wright ML, Myers YM, Karpells ST, Skibel CA, Clark MB, Fieldstad ML, Driscoll IJ (1988) Effect of changing the light/dark schedule, the time of onset of the light or dark period, or the daylength, on rhythms of epidermal cell proliferation. Chronobiol Int 5: 317–330

# Data Processing in Chronobiological Studies

J. De Prins and B. Hecquet

## Introduction

Observations of plants and animals show that their activities are time dependent. Measurements of physiological quantities as a function of time often display the presence of more or less repetitive patterns in living organisms, including man. Such observations raise many questions. Is a pattern, i.e. a rhythm, present or do we only observe random fluctuations? How are we able to characterize what we see? Precise answers to these questions require the choice of procedures, giving us pertinent parameters, to qualify and quantify our observations.

This is done by data processing, which is certainly a very important step in experimental study. The aim of processing is to extract as much information as the data potentially contain. Ill-performed data analysis will thus result either in a decrease of the available information or in an excess of information incorrectly deduced from experimental data. The second risk is probably the more pernicious, especially when the conclusions obtained are in accordance with the expected ones.

Data processing is limited by the experiment studies. It cannot compensate for lack of design accuracy, or imprecision in measurements. In other words, only a better design or more careful measurement could have produced a better experiment. Concerning the design, this also means that the experiment should deal as directly as possible with the ideas one is investigating, whereas all other influences are minimized.

Chronobiological experiments generally provide one or several sequences of numbers representing a variable (temperature, blood pressure, activity, etc.) with respect to time. In this case, data processing is mainly concerned with the extraction of some characteristic parameters, and possibly their confidence limits. Data processing implies the use of several techniques, including data acquisition, computation, modelization and statistics. The availability of computers has fostered the development of increasingly complex and sophisticated processing. Furthermore these procedures are available on every personal computer. The simplicity of use is such that anyone is able to apply them. Nevertheless, to give good conclusions, the user should have a full and deep understanding of the processing principles.

In this paper, we wish to clarify the underlying principles of chronobiological data processing. This implies the study of the relations between mathematical analysis and experimental measurements. In fact, we will never be able to *demonstrate* some experimental properties or facts. At best, we may give some *plausible* arguments based on theory *and* experiment. Due to this fuzziness, we need more than ever to keep a rational behaviour, and to clearly understand the limits of our arguments. We will try to explain those ideas with the help of several graphic examples, avoiding mathematical expressions as much as possible. Readers wanting more practical or precise considerations are referred to previous publications [1, 2].

## Data Acquisition

### Quantification and Sampling

Measurements imply quantification and sampling. Quantification is an operation assigning a numerical value to a "state" of a phenomenon. It implies adherence to a theoretical model. For example the measurement of a temperature relies at least on thermodynamics. When semiconductors are used for this measurement, electromagnetism also has to be considered. In practice, quantification implies the use of values characterized by a finite number of digits, resulting in rounding errors.

Moreover, during an experiment, we are able to only make a finite number of measurements. In other words, we have to make quantification at some selected or defined moments. This is sampling. When measurements are taken at (reasonably) equal time intervals, we speak of "equidistant data". In this case and *if* the time interval (or its inverse: the sampling rate) is well chosen, the sampling theory shows that discrete data may give a full knowledge of the studied phenomenon [3, 4].

We will consider, in this paper, the easiest case, i. e. an experiment giving $N$ couples of values $y_i$, $t_i$. Conventionally, $y_i$ values are associated with the measurements of a physiological quantity (for example, temperature, blood pressure, activity) and $t_i$ values are the times of the $y_i$ measurements.

## Data Characteristics

Firstly, it is useful to clearly evaluate the characteristics of the data. Characteristics are related to both criteria of data acquisition: quantification and sampling.

### Quantification, Errors and Variability

Usually, in biology, physiological measurements are not very precise. In other words, these quantities are subject to considerable experimental errors and are subject to variations due to the inherent biological variability. It is convenient to write that:

$$y_i = Y_i + e_i \tag{1}$$

where $Y_i$ is the "true value". This value is given by an ideal instrument applied to an ideal subject. "Error" $e_i$ is by definition equal to the difference between measured value $y_i$ and true value $Y_i$.

Errors are due to several causes. The type of error can be illustrated taking as an example the measurement of the core temperature of a healthy young man:

### Instrumental Errors

Let us consider a temperature measurement performed with the aid of a semiconductor device. Its resistance is a function of its temperature. The final value is influenced by instrumental imperfections: fluctuations of the applied voltages or currents, digital conversion rounding-off, etc. For a well-conceived apparatus these errors should be negligible and independent for each measurement. Nevertheless, even for a well-studied instrument, there will be drift of the temperature scale due to the aging of the components. Worse, there may be some *transcription errors*. For example, a measurement of 37.2°C is written as 27.2°C, indubitably giving a very false value (outlier).

### Transposition Errors

We wish to measure the "core temperature". This is a conceptual temperature, and we have to carefully study the way to measure it. We have to make the measurements at a fixed spot, and associate them with our concept of "core temperature". This transposition is not necessarily adequate and may give considerable random or systematic errors.

### Biological Variability

It can be divided into two parts:

*"Intraindividual Variability"*. It is quite impossible, in biology, to maintain the experimental conditions constant in order to study the variation of a single one (time in our case). In the previous example, during the measurement, stress in our subject can temporarily raise the core temperature. If the aim of our experiment is the study of a man in normal conditions, we may consider that this variation is abnormal and is thus an experimental error. This type of decision is always arduous to make, due to its subjectivity. However, we have to come to a decision! Let us also note that it is always our decision to consider some unwanted variations and facts as errors.

*"Interindividual Variability"*. The same experimental conditions in two subjects considered as identical (normal, for example) will result in different values of the parameters studied.

Those reflections show that neither the data nor their errors are evidence. Therefore, the first step in data processing is to make a critical study of the crude values and to reach a good understanding of their biological meaning.

### Sampling Characteristics (Number, Interval)

The number of measurements is a first important characteristic of the data. In chronobiology the range of number measurements is quite extensive. It depends on the type of experiment (in vitro, in vivo), on the species studied (animal, man) and also on the parameters measured (chemical measurement, physical measurement, etc.). The data size will greatly influence the result analysis. Small data sets only allow the use of specific mathematical and statistical methods. On the contrary large data sets could be analysed by powerful methods usually devoted to physical sciences.

The type of interval sampling also determines the analysis procedure. Without developing this point it can be considered that, in chronobiology, only time equidistant measurements can be correctly analysed by most of the mathematical methods. Thus the design of experiments has to comply with equidistant sampling. As previously noted, in this case and *if* the time interval is well chosen, the sampling theory shows that discrete data may give a full knowledge of the phenomenon studied [3, 4].

## Display of Experimental Data and Preliminary Estimations

Plotting $y_i$ versus $t_i$ immediately shows us the time function. It is often convenient to connect adjacent data with lines. Figure 1 gives us an example of such a representation for a "real alike" sequence. It seems obvious that a rhythm is present. Do not be afraid to make several representations with some scale changes to give another picture. It is surprising to "see" these difference. Fortunately these effects are well studied [5, 6], and graphical tools on personal computers enable such representations to be performed rapidly.

In this example, for preliminary estimations, it is useful to produce several graphics representing two cycles with a larger time scale. Figure 2 gives such a representation. Taking a black pencil and an eraser, we are able with a lot of common sense to pin point

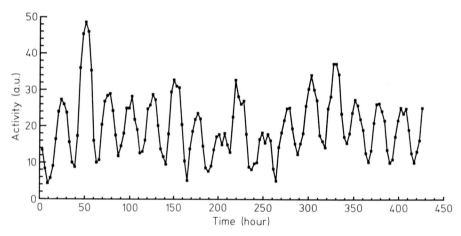

**Fig. 1.** Activity of mice as a function of time. Simulated data from graphical display of ASCHOFF experiments. [35]

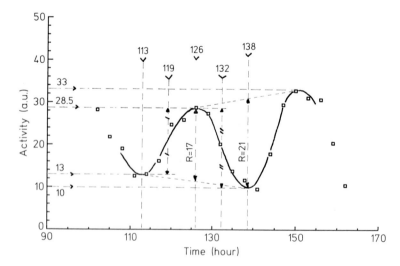

**Fig. 2.** Larger time scale representation of Fig. 1 and "manual evaluation" of maximum, minimum and half-range crossing points. All these points are indicated by their values, moments and a rough estimation of their uncertainties. It is to be noted that half-range moments are much more precise than maximum and minimum moments

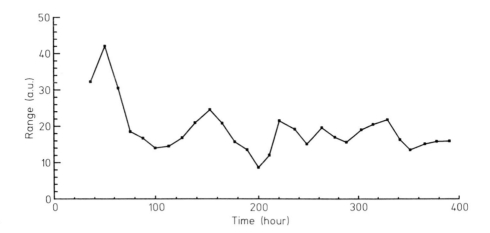

**Fig. 3.** Ranges of successive cycles in Fig. 1

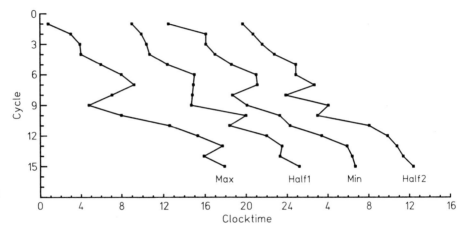

**Fig. 4.** Characteristics of successive cycles in Fig. 1. *Max*, maximum; *Half1*, first half-range; *Min*, minimum; *Half2*, second half-range

maximum and minimum values, to evaluate the corresponding times and possibly we are able to roughly estimate the uncertainty of these values (Fig. 2). We are then able to evaluate the "range" of each cycle. The range of a cycle is defined as the difference between its maximum and minimum values. This quantity may vary from cycle to cycle, so that we may proceed to more than one estimation during a cycle duration (Fig. 2). Now we can measure the time of the "half-range crossing point", i.e. the moment when the value of our variable is just midway between the adjacent maximum and minimum (Fig. 2).

Having done this work, it is time to make a series of graphics representing those values and to think intensively, in order to extract the suspected (and unsuspected) facts which may arise. Each experiment being a particular one, it is impossible to give a general procedure or even good general advice. Nevertheless the following example will illustrate such procedures and help understand the next sections of this paper.

Figure 3 gives the ranges of the successives cycles obtained for the data represented in Fig. 1. The higher range (42 a. u.) is nearly five times bigger than the lower range (8.6 a. u.). Was that your first impression looking at Fig. 1? Though evaluations are *very* criticizable, we have to admit that the range has no constant value. Nevertheless, we are unable to prevent somebody from calculating the "mean range" and its standard deviation (in this case $18.9 \pm 6.7$ a. u.), or worse the mean standard error ($18.9 \pm 1.3$ a. u.) and giving those figures as the final result! The latter gives great respectability to the stability of this rhythm, but what is the meaning of such results?

Figure 4 shows the times of successive maxima and minima on a 24-h time scale for the previous example. Visibly, there is a trend indicating that globally the duration of a cycle is longer than 24 h. It can also be deduced from Fig. 4 that the *estimated* duration of *a* cycle varies from $\simeq 22$ h to $\simeq 29$ h. We may also measure that, from the 1st to the 15th minimum,

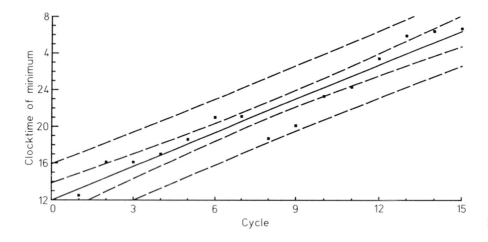

Fig. 5.

the shift is $\simeq 18$ h. The mean duration of a cycle is thus $\simeq 25.3$ h, as given by $24 + 18/(15-1)$.

We can now look for a deeper mathematical method to analyse these results. Linear regression being implemented on every calculator, it seems sound to use this method. Figure 5 gives the results as obtained using a conventional statistical package. The results concerning the slope imply a mean adjusted cycle duration of $25.23 \pm 0.1$ h. Due to the use of a sophisticated tool some people will find this approach much more "serious". It is important to note that it is not! Linear regressions are based on the hypothesis that errors are independent. This may be the case for the (unknown) errors of our estimations, but certainly not for the "biological errors". As a matter of fact, if one cycle was really shorter, there is no reason to believe that, in the absence of synchronization, the next cycle will be longer, thus allowing the next value to be centered in our adjustment line. In other words, in free run, the biological errors are given by the summation of the preceding ones. Even if the line seems to be well adjusted, the standard error on the slope is certainly false!

In conclusion we strongly believe from such examples that the advised heuristic approach, privileging graphical representations and common sense, is essential for a first understanding of the phenomena studied. This method also leads to some quantitative values.

If some regular patterns appear, we may believe we have detected a rhythm. We have to confirm it using adequate mathematical methods.

On the other hand, if we detect only irregularities, it is not exceptional for powerful mathematical methods to apparently detect regular patterns. We have seen such an example with linear regression. This contradiction has to be eliminated. Usually, a deep understanding of the mathematical method will show its inadequacy. Nevertheless, in a few situations, we may conclude a hidden cycle is present, "masked" by several unwanted effects. We have to be very cautious before taking a decision. It is imperative to reach the final decision on the basis of several acceptable methods.

Finally let us note that it is possible to make this heuristic approach more "impersonal" by using a specially designed computer program. This is the way we obtained the figures in this section. It is certainly more convenient for the user, but it is certainly not safer. The computer will do exactly what is written in its program, remaining unaffected by its more stupid unforeseen consequences. It is imperative, to avoid major errors, to display several graphical representations allowing rapid and adequate control.

## Mathematical Tools and Models

Further investigations require modelization. We have to consider three steps:
1. Description of mathematical tools and models
2. Estimations based on those tools
3. Fitting models to the data

### Mathematical Tools

Formulation

Taking only the case of equidistant data, the successive $Y_i$ measurements constitute a sequence of real numbers.

*Definition.* A sequence of reals $Y_1$, $Y_2$, $Y_3$, ... represents a function $Y(n)$ defined on the set of non-negative integers $n$ [7].

The notation $f[n]$ will mean a sequence of numbers. Our data may be considered as the equidistant sampling of a function $Y(t)$ so that:

$$Y[n] = Y(t_0 + n\theta) \tag{2}$$

where $\theta$ ist the value of the sampling interval, and $t_0$ is the initial time. Let us note that a time scale is very specific. Here $\theta$ is a time interval or duration. On the other hand, $t_0$ is a reference, i.e. a specific data: for example 31 December 1988, 1100 a.m. This data or reference is often very important from a biological point of view. Nevertheless, in mathematical expressions, for our convenience we will take $t_0$ as equal to zero so that we may write:

$$Y[n] = Y(n\theta) \tag{3}$$

In applications, we indeed have to remember the date corresponding to $t_0 = 0$.

For example, the sampling of a sinusoidal function will lead to:

$$f[n] = A \sin(n\theta/\tau + \varphi) \tag{4}$$

where $\tau$ is the period and $\varphi$ is the phase at $t = t_0$ (here $t_0 = 0$).

Let us insist on the fact that sampling conditions have to be fulfilled so that $Y[n]$ constitutes a good representation of $Y(t)$. This will not be the case if the sampling interval is too big. As a rule of thumb, in chronobiology, the difference between two successive values of the sequence should never exeed half the range.

Having discrete signals, we will use discrete mathematics or systems.

*Definition.* A discrete system is a rule for assigning to the sequence $f[n]$ another sequence $g[n]$ [3].

For example:

$$g[n] = 1/3 \{f[n-1] + f[n] + f[n+1]\} \tag{5}$$

where the sequence $g[n]$ is a smoothing of $f[n]$. Another example is:

$$g[n] - 1.648 \, g[n-1] + 0.848 \, g[n-2] = f[n] \tag{6}$$

In this case, to find $g[n]$ we need to know not only $f[n]$ but also $g[n-1]$ and $g[n-2]$. Thus $g[n]$ is obtained by solving a recursion equation. In this simple case, we obtain:

$$g[n] = f[n] + 1.648 \, g[n-1] - 0.848 \, g[n-2] \tag{7}$$

## Computation

Usually, calculations are done mainly on computers with the help of algorithms and it seems important to clarify the definition and properties of algorithms.

*Definition.* An algorithm is any method of computing consisting of a finite set of unambiguous rules which specify a finite sequence that provides a solution to a problem [7].

Every meaningful algorithm is defined by:
1. A set of *rules*, expressed by mathematical and logical expressions
2. One or several *outputs* (the solution)
3. Zero or several *inputs* (data)

Let us consider what seems to be a very simple algorithm. The value $C$ (solution) has to be given by the addition (rule) of values $A$ and $B$ (inputs). Addition is defined by the well-known mathematical properties. In fact, the computer processor executes several tens of elementary operations to perform the addition. Even in this simple algorithm, the rules implemented by the computer are manifold.

It is imperative to realize that the algorithm results, as given by a computer, are affected by several errors and are thus, strictly speaking, inexact. This is due to number representation and round-off errors in the basic operations [8]. The severity of those errors depends on the computer hardware, the compiler and the user. In well-studied situations, those errors may be negligible.

Moreover attractive algorithms can be unstable. This means that very small errors in the initial data are magnified by the algorithm until they finally swamp the true theoretical value.

Usually, it is imperative that we avoid unstable algorithms. Indeed, our calculations are necessarily justified by a theory but this justification is devoid of reason if our results are false! Hopefully, in current stituations some small modifications change an unstable algorithm into a stable one.

The choice of hardware and software is thus crucial for the quality of the computed results. The chronobiologist generally has to be helped by an expert in computation to perform advisable choices.

## Mathematical Models

The term "mathematical model" may be used for any complete and consistent set of mathematical equations which is thought to correspond to some biological or physical entities [9].

Models may be based on the deep understanding of the properties of the biological phenomena studied. Without doubt, modelization (in its general meaning) is the best scientific approach and the most promising from an epistemological point of view.

Unfortunately, this is rarely the case in chronobiology, where we face a much more empirical stituation. In a first step, we will try to imagine some "general algorithms" able to more or less simulate the real data. Taken as models, they generally allow the estimation of several characteristics of the phenomena studied. A further step is to adjust (or fit) those simple models to the data. This procedure allows possible rejection of the model.

By "general algorithm" we mean a procedure justified by theoretical arguments and by plausible practical considerations, and certainly not an ad hoc adjustment especially conceived for a specific experiment.

General algorithms usually have adjustable parameters, able to allow a reasonable fitting of the results with the observed data. It is always amusing to see the apparent diversity of patterns obtainable from the adjustment of very few parameters. It is thus not surprising to find very good fitting when an algorithm possesses a great number of adjustable parameters. This fitting does not necessarily imply the adequacy of the model!

Let us have a pragmatic look at the favourite models in chronobiology.

## Deterministic Models

### Periodic Functions, Fourier Series

These models rely on the concept of periodic function. The function $f(t)$ is said to be periodic with period $\tau$, if it assumes the same value for times $t$ and $t + \tau$, i.e. if:

$$f(t) = f(t + \tau) \quad \text{for all values of } t \tag{8}$$

If $\tau$ is a period of the function, any value equal to the product of $\tau$ and an integer is also a period. To avoid any ambiguity, the "period of the function" is defined as the smallest value of $\tau$ satisfying Eq. 8. It is agreed to call "frequency" of the same function the inverse of $\tau$. It will be denoted $f$, with $f = 1/\tau$ [2].

Usually, a periodic function $f(t)$, with period $\tau$, can be considered as the sum of a constant and sinusoidal functions with periods $\tau$, $\tau/2$, $\tau/3$, ..., $\tau/n$, ..., with given amplitudes $(A_i)$ and phases $(\varphi_i)$:

$$f(t) = A_0 + A_1 \sin(2\pi t/\tau + \varphi_1) + A_2 \sin(4\pi t/\tau + \varphi_2) + \dots + A_n \sin(2\pi n t/\tau + \varphi_n) + \dots \tag{9}$$

or:

$$f(t) = A_0 + A_1 \sin(2\pi f t + \varphi_1) + A_2 \sin(4\pi f t + \varphi_2) + \dots + A_n \sin(2\pi n f t + \varphi_n) + \dots \tag{10}$$

The term in $2\pi f$ corresponds to the fundamental (or first harmonic), and the $n$th term in $2\pi nf$ to the $n$th harmonic [10].

Similarly, in discrete mathematics, it is possible to obtain a relatively simple relation between the $N$ values of a sequence $f[n]$, and $N/2$ couples of values $A_i$ and $\varphi_i$. Furthermore, if the sequence $f[n]$ constitutes a "good sampling" [2] of the function $f(t)$, then the values obtained are the same as the ones given by Eq. 10! This relation is expressed with the help of complex numbers. As we have aimed in this paper to keep things simple, this equation should not be shown. Nevertheless, we have included it here, without explanation, for its esthetic simplicity [3]:

$$c_l = A_l \cdot e^{j\varphi_l} = \sum_{m=0}^{N-1} f_m \cdot W_N^{lm} \text{ for } l = 0, \dots, N-1 \tag{11}$$

$$\text{where: } j = \sqrt{-1}$$
$$W_N = e^{j2\pi/N}$$

The $c_l$ coefficients represent a new sequence of complex numbers $c[n]$, constituting the "spectrum". Despite the importance of the phase $(\varphi_l)$ it is usual, in a graphic representation of the spectrum, to display the amplitude $(A_l)$ values only.

Equation 11 defines the "Fourier transform" (FT). A symmetrical equation allows calculation of the $f[n]$ sequence from the $c[n]$, corresponding to an inverse Fourier transform (IFT).

Despite its formal simplicity, such calculations become laborious for increasing $N$. However, some clever rearrangement of those equations allows faster calculations. The best known is the so-called "Fast Fourier Transform" (FFT). In its implest form, it allows only calculations with $N = 2^p$, where $p$ is an integer (for example $N = 2, 4, 8, 16, 32$, etc.). For this reason, despite some disadvantages, we prefer and use the "Goertzel algorithm", described elsewhere [2, 8].

This mathematical method, despite its cleverness, leads to artificial and implausible explanations. Consider Fig. 6, which shows a time function and its spectrum. The time function is an interrupted sinusoidal function. It is clear that during the interruption (or silence) no signal is present (or emitted). Interpretation of the spectrum implies that the addition of the

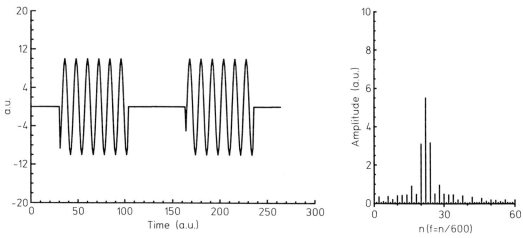

**Fig. 6.** Interrupted sinus *(left)* and display of the $A_1$ values, i.e. discrete Fourier spectrum (right)

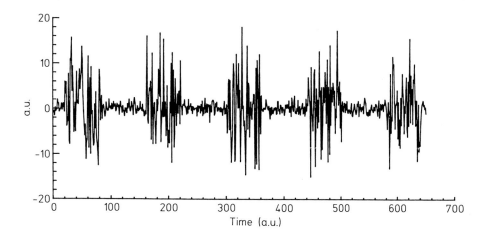

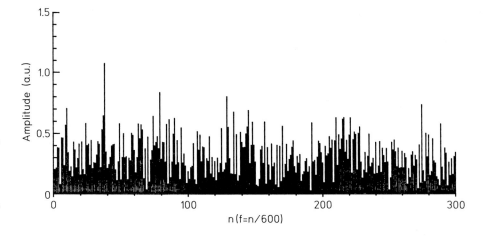

**Fig. 7.** Pseudo-random sequence multiplied periodically by a constant factor, here 6 *(up)*. The periodicity of multiplication should correspond to harmonic number 4 in the spectrum *(below)*

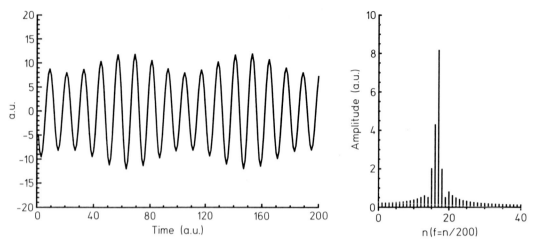

**Fig. 8.** Amplitude modulation of a sinusoidal signal *(left)* and its spectrum *(right)*

different terms of the Fourier series gives a nil total result during the "silence". In other words, the absence of signal is explained by a destructive interference between several signals!

On the other hand, Fig. 7 shows a time function, where you would believe that a periodicity is clearly present. Unfortunately there is no periodicity at all, as shown by the spectrum. Considering equation 8, it is comprehensible. As a matter of fact, every value of the sequence being random, this equation is certainly *not* fulfilled.

Another related approach, giving some information, is the estimation of the autocovariance function $\gamma(k)$ [11], which may be taken as:

$$\gamma(k) = 1/N \sum_{i=1}^{N-k} (f_i - \mu) \cdot (f_{i+k} - \mu) \qquad (12)$$

$$\text{where} \quad \mu = 1/N \sum_{i=1}^{N} f_i$$

This method is usually efficient for numerous data distributed over a great number of periods, which is seldom the case in chronobiology. The Fourier transform of the autocovariance function leads to knowledge of the $A_i^2$ (power spectrum). Unfortunately, phases $\varphi_i$ remain unknown. Nevertheless, this approach was adapted for non-equidistant data [2]. It is usual to speak of the autocorrelation function $\rho(k)$, which is given by:

$$\rho(k) = \frac{\gamma(k)}{\gamma(0)} \qquad (13)$$

*Quasi-Periodic Functions, Modulation*

When a function is obtained by the sum of several sinusoidal functions, i.e. when $f(t)$ can be expressed as:

$$f(t) = \Sigma_i A_i \sin (2\pi f_i t + \varphi_i) \qquad (14)$$

and at least one $f_k/f_1$ ratio is irrational, it can easily be understood that $f(t)$ is no longer periodic as defined above (8). This situation is conceptually disturbing, since a display of such a function over a finite time interval often presents as being rhythmic. Moreover, it is experimentally impossible to decide whether such a ratio is irrational or not. For those reasons, experimenters define such functions as quasi-periodic.

Special cases are given by the modulation of a sinusoidal signal. They are, in general, described by:

$$f(t) = A(t) \cos\{2\pi ft + \varphi(t)\} \qquad (15)$$

where $A(t)$ and $\varphi(t)$ vary slowly with respect to the cosine function.

When only $A(t)$ varies, we have an "amplitude modulation" (Fig. 8). We speak of "phase modulation" when only $\varphi(t)$ varies (Fig. 9). Those modulations will allow us to simulate varying signals, as shown by the two preceding figures. Those figures also show the $A_i$ of the related spectra. The amplitude modulation gives sidelines related to the spectrum of $A(t)$. They are important for the spectral interpretation. On the other hand, phase modulation gives more complicated spectra, much more difficult to interpret.

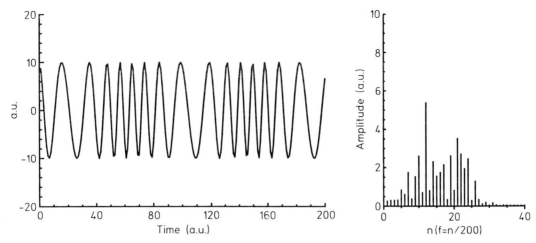

**Fig. 9.** Phase modulation of a sinusoidal signal *(left)* and its spectrum *(right)*

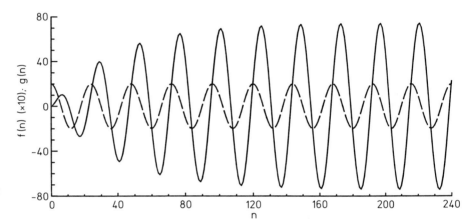

**Fig. 10.** ARMA filtering *(unbroken line)* of a sinusoidal function *(dotted line)*

*ARMA Filters*

In modelization we often want *linear* discrete systems for the sake of simplicity. The general linear relation, defining a sequence $g[n]$ considered as output, from another sequence $f[n]$ taken as input, is given by the equation:

$$g[n] = b_0 f[n] + b_1 f[n-1] + \ldots + b_h f[n-h] \\ - a_1 g[n-1] - \ldots - a_k g[n-k] \quad (16)$$

where $b_0, b_1, \ldots, b_h, a_1, \ldots, a_k$ are constants. It is usual to write that this equation defines a "linear filter", $f[n]$ being the input signal and $g[n]$ the "filtered" output.

If at least one $a_i$ is different from zero, we have a recursion equation, as defined previously. We may define two particular cases:

− AR filters of order $k$:

$$g[n] = f[n] - a_1 g[n-1] - \ldots - a_k g[n-k] \quad (17)$$

− MA filters of order $h$

$$g[n] = b_0 f[n] + b_1 f[n-1] + \ldots + b_h f[n-h] \quad (18)$$

For these three equations, with well-chosen coefficients, the $g[n]$ sequence will constitute a linear filtering of sequence $f[n]$.

As an example let us consider the sequence:

$$f[n] = \sin(2\pi n/24) \quad (19)$$

which corresponds to a sinusoidal function of period equal to 24.

Let us use the AR filter of order 2 defined by:

$$g[n] = f[n] + 1.9217 \, g[n-1] - 0.98956 \, g[n-2] (20)$$

The sequence illustrated in Fig. 10 is obtained. We clearly observe a transient state, corresponding to an increase in the output signal amplitude. After about eight cycles we note a stable periodic sequence (permanent state). Furthermore, the analysis shows that we have a new sinusoidal sequence, with well-defined amplitude and phase. This is a general property of

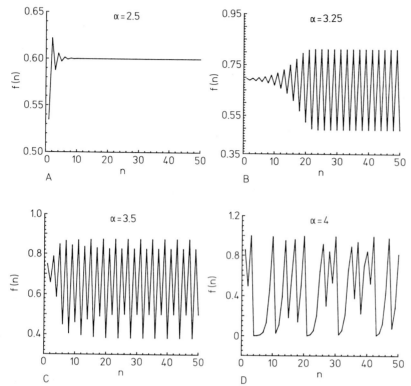

**Fig. 11 A-D.** Plotting of a chaotic function (Eq. 21) for different values of $\alpha$

linear filters. Applied to a sinusoidal sequence, after a transient state, we obtain a new sinusoidal sequence (permanent state). This is the fact that justifies, *in mathematics,* the intense use of sinusoidal functions. This property, used jointly with the Fourier series, allows easy calculations. However, in the "real world", it is not easy to generate a sinusoidal function! In fact, it is seldom, or even never, found in "biological emitters".

The theory gives, without difficulty, the ratio of the amplitude of both the sinusoids $(G(f))$, as well as the expected phase shift $(\Phi(f))$ between the two sequences with respect to the frequency $f$ [2, 3]. Those functions $(G(f)$ and $\Phi(f))$ give a full knowledge of the linear filters. They correspond to the two components of the complex function $S(t)$, called "transfer function of the filter".

## Pseudo-Random Models

### Chaos

Definitions of chaos require elaborate mathematical notions. Furthermore, there is no complete agreement on this concept. We will try to give some (imprecise) comprehension of this matter with the help of simple examples. Let us take the well-studied example of the sequence defined by:

$$f[n] = \alpha f[n-1] (1 - f[n-1])  \qquad (21)$$

where $\alpha$ is a constant such that $0 < \alpha \le 4$.

Once $f[1]$ known, the sequence is defined. We are able to introduce a seeding value $f[1]$ such as $0 \le f[1] \le 1$. When $\alpha < 3$ we observe that, whatever the value of $f[1]$, quite rapidly the $f[n]$ values converge to a value $\beta$ given by $\beta = 1 - 1/\alpha$ (Fig. 11 a). When $3 \le \alpha \le 3.44$, and if we take a seeding value different from $\beta$, we observe a periodic function with $\tau = 2$ (Fig. 11 B). For a slight increase in $\alpha$ we observe the appearance of longer periodicities (Fig. 11 C). In each case, for a given value of $\alpha$, the final pattern will be the same whatever the value of the seeding $f[1]$. On the contrary, when $\alpha$ becomes greater than 3.57, we *may* observe the disappearance of "simple" periodicities, and obtain apparently random patterns (Fig. 11 D). Furthermore, changing the $f[1]$ value will lead to quite different sequences. We have obtained chaotic sequences [12].

We may argue that a necessary condition for having a chaotic sequence is that a small change in the seeding number yields a new sequence which differs

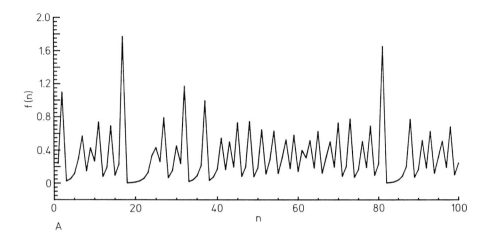

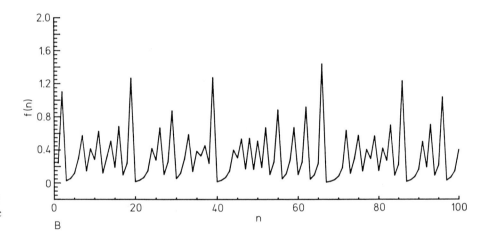

**Fig. 12 A, B.** Influence of small errors on a "chaotic function" (Eq. 22)

notably from the original one. This property renders numerical studies necessarily inaccurate, due to inevitable rounding-off errors.

Consider the sequence defined by:

$$f[n] = -1/3 \ln\{\,|\,1 - 2 \exp(-3\,f[n-1])\,|\,\} \qquad (22)$$

where the notation $|A|$ means the absolute value of $A$ [13]. The seeding value is such that $f[1] > 0$. Figure 12 A gives the result obtained with our computer installation (hard- and software) with $f[1] = 0.1$. However, using other hard- and software will lead to another sequence. To show the effects of small errors, let us take:

$$f[n] = -1/3 \ln\{\,|\,1 - 2 \exp(-3\,f[n-1])\,|\,\} + \varepsilon[n] \qquad (23)$$

where $\varepsilon[n]$ is a sequence of small errors, defined by an uniform distribution between 0 and $10^{-5}$. Figure 12 B gives the results, and compared to Fig. 12 A clearly shows the difference. In other words, any cal-

culation of the sequence defined by Eq. 22 is necessarily, sooner or later, inaccurate. Moreover, the calculations may always exhibit recurrence. That means that $f[n]$ is equal to a previously obtained value $f[n-j]$, necessarily involving a periodicity of $j$.

Those remarks lead to the popular property of chaos. Chaotic sequences are so sensitive to initial conditions that the combination of measurement errors and round-off errors make the long-term prediction of the sequence behaviour impossible. Nevertheless, some global properties may be found from a chaotic sequence. As an example, with the very simple sequence defined by Eq. 23, Fig. 13, displaying the $f[n+1]$ with respect to the $f[n]$, immediately shows the dependence underlying this sequence.

Some procedures related to those chaotic sequences are welcomed in simulations where they generate what are called "pseudo-random sequences". Those sequences, if well studied, have some important properties of random sequences.

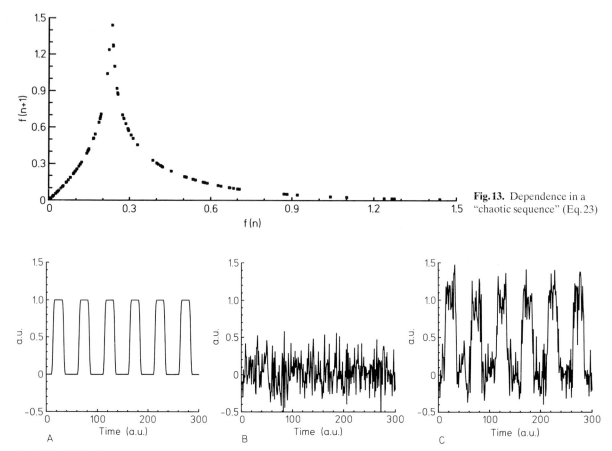

**Fig. 13.** Dependence in a "chaotic sequence" (Eq. 23)

**Fig. 14 A–C.** Simulation of experimental errors: **A** true periodic function; **B** pseudo-random sequence; **C** Sequence obtained by addition of **A** and **B**

Ideally, they consist in mutually independent, identically distributed variables [14]. Nevertheless, users always have to be very cautious [15]. Those sequences are also called "white noise" sequences. This refers to the fact that, theoretically, the power density is constant with respect to the frequency. This means that, taking several sequences, the mean value of the $A_i$ coefficients converges to the same constant value. This implies, for example, that the mean values of the $A_i$ at low and at high frequencies are equivalent.

Those white noise sequences may be useful to simulate experimental errors. As an example, Fig. 14 gives the addition of a periodical signal and white noise.

### ARMA Process

If we take a pseudo-random sequence $f[n]$ (Fig. 15 A), and filter it with the help of an ARMA filter, we obtain quite interesting $g[n]$ sequences. This

will be named the ARMA process. Figure 15 B–D shows several examples. Figure 15 B resembles often-encountered noises following a normal distribution; some observers will even be convinced of seeing irregular cycles. However, successive values are not independent. On the other hand, Fig. 15 C may simulate an instrument drift.

Much more puzzling to the chronobiologist is Fig. 15 D, representing the results of white noise filtered by very simple, i.e. low-order, AR filters. It presents some features of the free-run cycles. Even if this sequence appears periodic or at least rhythmic, from a mathematical viewpoint it is purely random. Indeed, knowing only the beginning of the sequence $g[n]$, we are unable to foresee over long intervals both the range and the maximum or minimum positions of the following cycles. From a mathematical viewpoint it is a random process. Despite this fact, and due to the recursion equation which gives a dependence between successive values, we are often able to give some efficient prediction for a relatively short future.

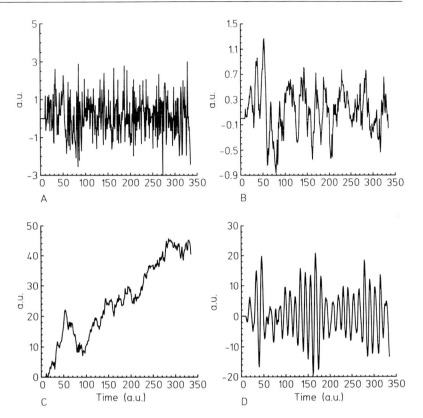

**Fig. 15 A–D.** ARMA processing of a pseudo-random sequence: **A** pseudo-random sequence $f[n]$ as input; **B** filtered sequence $g[n]$ with MA 8 ($b_0 = \ldots = b_7 = 0.125$); **C** filtered sequence $g[n]$ with AR 1 ($a_1 = 1$, equivalent integrator); **D** filtered sequence $g[n]$ with AR 2 ($a_1 = 1.7139$, $a_2 = -0.97917$)

## Statistical Tools

The mathematical model and related algorithms allow us to define and calculate one or several parameters. For each, we want to determine at least two values:
1. The best estimation of the parameter
2. A measurement of the uncertainty for the estimated value

The determination of both the values is influenced, not only by the assumptions of the mathematical model, but also by the choice of the statistical theory. For the latter, we will distinguish four possibilities: parametric methods, nonparametric tests, robust and resampling methods.

The theory of parametric estimations relies on the (a priori) knowledge of the error distribution around the "true value". For each "reasonable" distribution there exists an asymptotically efficient method. The methods of estimation and testing with a known distribution are the most powerful. For the sake of ease, it is often postulated that the error distribution is gaussian (or "normal distribution").

There is no agreement among statisticians concerning the definition of nonparametric tests. Let us simply state that a test is nonparametric if we do not have to specify the error distributions. This is, for example, the case for the Mann-Whitney test [16].

Robust methods are relatively insensitive to deviations from the theoretical assumptions on error distribution. They allow estimations of parameters and testing. By compromising between efficiency and risk of error, robust estimation can be optimized [17]. Most often, robust methods consist of assigning different weights to extreme values, although this is not their only aspect. As an example, the median of several values is more robust than the mean. The latter is much more sensitive to outliers than the former.

Based on a *very* large number of calculations, the resampling methods provide unbiased estimations and uncertainties. About 10 years ago, Efron introduced this new approach [18, 19]. He considers that the sample of measurements is representative of its theoretical distribution. From the initial set of data, several hundreds to several tens of thousands of samples are prepared by random sampling with replacement among the original data. For each such sample the value of the estimator is calculated. The large number of estimations allows the use of conventional methods to derive an unbiased value for the estimator and its uncertainty.

Nevertheless, all the above methods usually postulate independence of the errors. In most lecture and books, this seems the most obvious assumption. However, an adequate sampling rate of observations necessarily implies error *dependence* [1]. Any estimation of uncertainties or confidence limits, not accounting for this dependence, will be grossly erroneuous. Corrections must be applied by specific methods: estimating correlations [20] or decimation.

We would like to present another comment. The aims of mathematicians and experimenters are basically not the same. The mathematician wants to create the most powerful methods based on a well-defined axiomatic. This means in practice that he assumes the truth of several conditions. He is then in a position to *demonstrate* his conclusions. Furthermore, to be able to proceed to complete calculations, his interest is to take the easier hypotheses.

On the other hand, the experimenter is confronted with very complex situations. Some mathematical hypotheses are very difficult to assess. Let us think of hypotheses on error distribution. Others, like stationarity or randomness, are simply not operational, i.e. not verifiable for a finite data series. Taking into account that the failure of postulated hypotheses leads to fallacious results, the experimenter should thus prefer to use methods less dependent on the hypotheses, even if they are less powerful or more time consuming. It is better to admit one's ignorance than to derive erroneous conclusions, based on assumptions known to be violated.

Everyone is confused when confronted with the large choice of methods. Each of them has a set of advantages and limitations, from both a mathematical and a biological point of view. Often, there are contradictions between those standpoints. It is thus always advisable to select several acceptable procedures. Then, one has to compare with common sense the results obtained by each of them. Once the agreements and divergences are understood, reliable results may be given.

## Data Processing

### Pertinent Estimator Choice

Discrete Fourier Spectral Analysis

As seen before, direct calculations lead us to the $A_i$ and $\varphi_i$ coefficients characterizing a spectrum (Eq. 11). Unfortunately, in practice, many difficulties arise. As a matter of fact, spectral theory is based on strong assumptions. Calculations lead to correct interpretations only if the analysis is performed on an integer number of periods.

As an example, consider the sampling of a sinusoidal function (amplitude = 10), with a sampling rate of 12 measurements per cycle. Let us suppose we have 99 data, corresponding to the observation of 8.25 periods. If we take only 96 data, the spectral analysis will be done on exactly 8 periods. Figure 16 A shows the spectrum obtained. Figure 16 B shows the results when taking 8.25 periods (99 data). Not so nice! Figure 16 C will show another situation. Several algorithms, like the well-known FFT, require a number of data equal to $2^n$, where $n$ is an integer. Several authors advice one

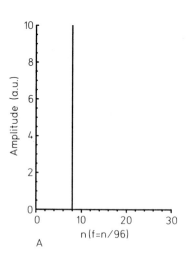

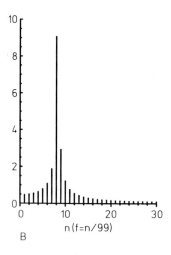

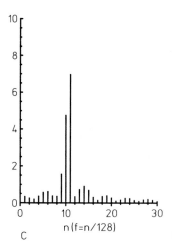

A    n (f=n/96)

B    n (f=n/99)

C    n (f=n/128)

**Fig. 16 A–C.**

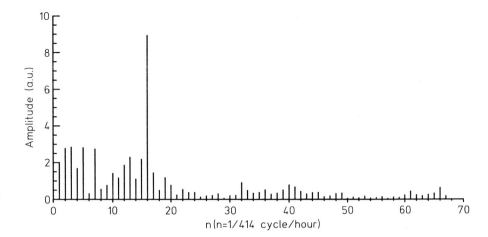

**Fig. 17.** Discrete Fourier spectrum for the series respresented in Fig. 1

to reach this number, by adding zeros. In our case we have to add 29 zeros. Looking at Fig. 16 C, one immediately realizes that his is clearly not a good advice!

Some theoretician readers may be scandalized by our approach. Often, spectral analysis theory is used for a stationary random process. In this context, the use of "windows" is often advised [21]. Well-studied functions are applied in the time or frequency domains, improving the smoothness of the spectrum and making boundary problems easier. Unfortunately, interesting lines are also smoothed, i.e. widened or shortened, increasing difficulties of interpretation. Knowing that in chronobiology we generally have short sequences of non-random rhythms, we avoid this approach. On the contrary, we deliberately take an "open window" (or no window) approach, as initially described by Burgess [22].

Mathematically, the Fourier Transform gives "exact" results. Practical calculations give negligible round-off errors. Nevertheless, as pointed out above, the amplitude related to a given frequency will strongly depend on the number of data *(N)* and on the global characteristics of the noise in its frequency neighbourhood. Evaluating valuable uncertainty limits is a very difficult problem, requiring deep theoretical knowledge. Some acceptable approximate methods are given in the literature [2, 23]. However, once again, one has to be very cautious. As an example, in Fig. 7 (below), it may be shown that the highest amplitude is still within the 95% confidence limit for the *whole* spectrum (random sequence). This is a consequence of the Rayleigh distribution of the spectral amplitudes, propitious to some "higher" values. Let this be a warning to the "cycle hunters"!

We must also make some comments on the vicious circle encountered. On the one hand, we want to use spectral analysis to evaluate the period of a cycle, but

on the other hand we need to make the analysis on an integer number of this unknown period! Nevertheless, this can be done thanks to our preliminary estimations, giving us a very approximate period. This knowledge may be clarified with the help of the Goertzel algorithm, which allows the quick evaluation of some lines. By changing the limits of the time interval considered for the analysis, one can optimize the dominant spectral amplitude in the suspected frequency domain. The condition of an integer number of periods is then fulfilled with an excellent approximation. This method was applied to the sequence of Fig. 1, giving the spectrum of Fig. 17.

### Filtering

Filtering consists in considering the data values as the sum of a "signal" (the biological rhythm) and a "noise" (experimental errors and biological variability). According to these assumptions biological data can be treated like any physical wave and, especially, can be filtered through digital filters.

Many methods are available. They consider that signal frequency is not in the same range as noise frequency. By amplifying preferentially the signal part, one obtains a strong attenuation of the noise.

As an example, if the noise consists of rapid fluctuations with respect to the signal variations, the simplest smoothing technique is the moving average. This method takes the mean of the points surrounding selected target values. This leads to "smoothed" values of the original data.

As another example, consider that an important trend is present in the data, and that its variations are slow with respect to the rhythm (signal). This trend may be removed by fitting the experimental points to

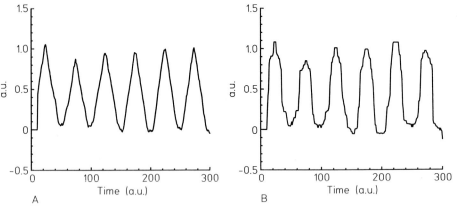

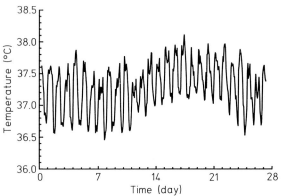

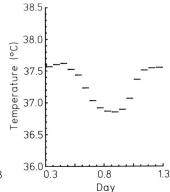

**Fig. 18 A, B.** Digital filtering of signal represented in Fig. 14 C: **A** moving average; **B** moving median

**Fig. 19.** Signal averaging *(right)* a simulated data *(left)* from a graphical display of body temperature in a woman. (Courtesy of Zimmermann, Montefiore hospital, New York)

an *n*-order polynomial. The polynomial used has no periodical meaning and is only a "passive" tool to represent a continuous curve. Fitting can be performed by a least-squares procedure.

Much more sophisticated linear filters are available. Essentially based on the ARMA filters theory, several powerful methods for the synthesis of digital filters are available [24]. Unfortunately, the frequency ranges of signal and noise usually overlap one another. Filtering will then necessarily alter the shape of the rhythm. A better solution may be the use of non-linear filters [25]. The easiest is the "moving median", which is the non-linear equivalent of the moving average. Here, the calculation of a weighted mean (Eq. 18) is replaced by the determination of the median. As suggested by Fig. 18 A, rapid transition may be respected by the moving median.

Let us also note the possibilities of optimal filters. They minimize a criterion on the basis of the characteristics to be estimated and of the available information regarding both the noise and the biological signal. The best-known applications are the Wiener and Kalman filters [26, 27]. For this matter, the help of an expert is essential.

Another method related to filtering is "signal averaging" used in instrumentation or in data analysis to recover repetitive signals. If we proceed to perform the same sampling from the same beginning of the signal, we are able to average all the measurements having the same relative sampling position. If the mean value of the noise is nil, its influence will decrease as the number of measurements increases [2, 28]. This method may be used for cycles synchronized for example by a D-L cycle (Fig. 19).

Demodulation

If we believe that our data correspond to the modulation of a sinusoidal signal of known frequency, we may extract interesting characteristics using the demodulation. To understand the procedure, we start from Eq. 15. We first calculate two new sequences corresponding to:

$$D_c(t) = f(t) \cdot \cos\{2\pi f t\}$$
$$D_s(t) = f(t) \cdot \sin\{2\pi f t\} \tag{24}$$

It is easy to show, with the help of trigonometry, that:

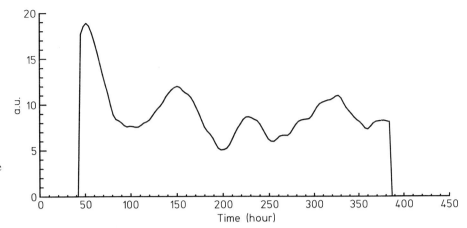

**Fig. 20.** Estimation of the amplitude as given by double demodulation. Due to the width of the filter the first and last values are unknown and are arbitrarily put as equal to zero

$D_c(t) = A(t) \cdot \cos\{2\pi ft + \varphi(t)\} \cdot \cos(2\pi ft)$
$D_c(t) = 1/2 \cdot A(t) \cdot \cos\{\varphi(t)\}$
$\qquad + 1/2 \cdot A(t) \cdot \cos\{\varphi(t)\} \cdot \underline{\cos\{2\pi(2f)t\}}$
$\qquad + 1/2 \cdot A(t) \cdot \sin\{\varphi(t)\} \cdot \underline{\sin\{2\pi(2f)t\}}$

The second and third terms are clearly signals centered on frequency 2*f*, whereas the first is a slow varying signal in the very low frequencies. If we now use digital filtering, rejecting the higher frequencies, we may obtain a new sequence corresponding to:

$$F_c(t) \simeq 1/2 \cdot A(t) \cdot \cos\{\varphi(t)\} \qquad (25)$$

Similarly, with the filtering of $D_s(t)$, we obtain

$$F_s(t) \simeq 1/2 \cdot (A(t) \cdot \sin\{\varphi(t)\} \qquad (26)$$

Those two functions give us a full knowledge of $A(t)$ and $\varphi(t)$. As a matter of fact, we obtain:

$$A(t) \simeq \sqrt{F_s^2(t) + F_c^2(t)}$$
$$\varphi(t) \simeq tg^{-1}\left\{\frac{-F_s(t)}{F_c(t)}\right\} \qquad (27)$$

Figure 9 has shown us that a phase modulation corresponds to an apparent modification of the period. The knowledge of $\varphi(t)$ allows us to calculate the apparent period and to refine the determination of $A(t)$. This is realize by the procedure of "double modulation" extensively described in another paper [29]. Figure 20 shows the result obtained with the data of Fig. 1.

This approach is specially recommended for the study of biological phenomena undergoing a phase shift. Nevertheless, it has to be noted that this method only considers the fundamental of the rhythm and that any variation in the rhythm's waveform may introduce phase fluctuations.

## Model Fitting

Partial Fourier series

Let us note that we have restricted ourselves in this apper to equidistant data. Nevertheless, most of the conclusions in this section may be extended to non-equidistant data. We will consider, from a very general point of view, the fitting of one or several sinusoidal functions of known period (or frequency). The well-known "cosinor" is an example of such a fitting based on the least squares method. It leads to the determination of three quantities: $A_0$ (called the "mesor"), the amplitude and the phase (acrophase) of the fitted cosinusoidal function, and their confidence limits [30].

In fact, it is theoretically shown that, for a well-sampled function, the Fourier coefficients correspond to a least squares fit. We may thus conclude, for pedagogical purposes, that selecting a few Fourier components of a spectrum will roughly correspond to a classical least squares fitting.

The main problem of interpretation is to decide which spectral components may belong to the cycles and which may be considered as noise. Once more we have to carefully study the characteristics of signal and of noise.

To understand the interpretation problems, let us take a very simple situation. Figure 14 A represents a sequence, corresponding to a true periodical function. The preliminary estimation is particularly easy. In the arbitrary units of this figure, the period is equal to 50, the range is equal to 1 and the signal is asymmetrical, i.e. the duration of the high level is shorter. The spectra are calculated for 250 data, i.e. exactly 5 periods. The $A_i$ values are shown in Fig. 21. Nine lines clearly emerge, with four others also being visible.

A cosinor fit imposing a period of 50 corresponds to taking $A_0$ and the fifth term of the Fourier series. Figure 22 shows the sinusoidal function obtained, with respect to the sequence studied.

Several remarks have to be made:

– The range of a sinus is equal to twice its amplitude. The latter being equal to 0.6, the sinus range is equal to 1.2. It is thus 20% higher than the real one.

– The mesor is given by $A_0$, here 0.4. This effect is related to the asymmetry of the pattern.

– Whereas the maximum values indicated by the sinusoidal function are nearly correct, the minimum ones are around $-0.2$, an error equal to 20% of the exact range. Those effects, anticipated by the theory, are related to the neglected harmonics of our periodic function. Unfortunately, they are *not*

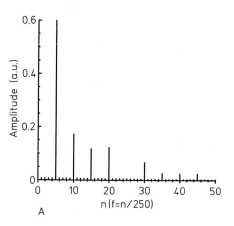

A

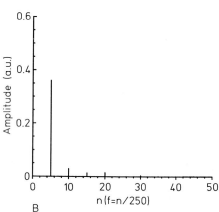

B

**Fig.21 A,B.** Spectral analysis of a signal represented in Fig. 14 A: **A** discrete Fourier spectrum; **B** power spectrum

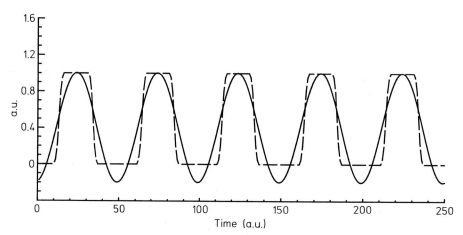

**Fig.22.**

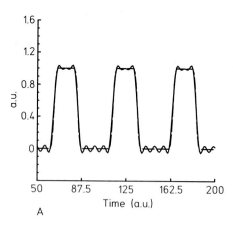

A

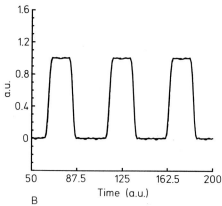

B

**Fig.23 A,B.** Reconstruction of signal represented in Fig. 14 A using spectrum display in Fig. 21: **A** with five harmonics; **B** with nine harmonics

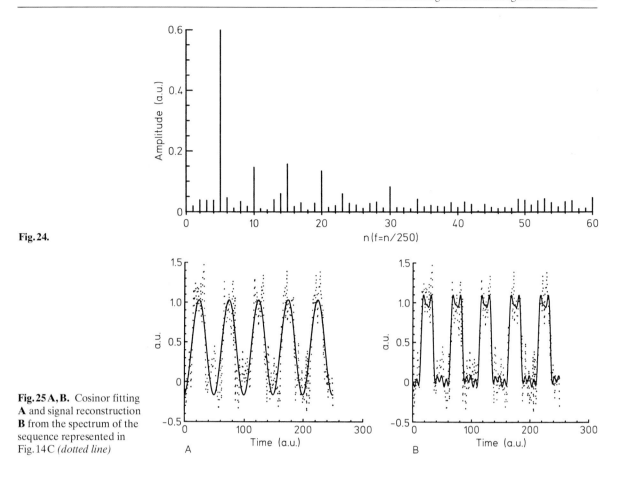

**Fig. 24.**

**Fig. 25 A, B.** Cosinor fitting **A** and signal reconstruction **B** from the spectrum of the sequence represented in Fig. 14 C *(dotted line)*

taken into account in the cosinor confidence limits, rendering them illusory. Figure 22 shows the poor approximation given by a simple sinusoidal function for a non-sinusoidal pattern, and its misleading appearance. For example, the adjusted sinus has sharp optima with respect to the broad ones of the sequence studied.

Figure 23 A shows the reconstruction obtained with five harmonics ($A_0$ + lines 5, 10, 15, 20 and 30), whereas Fig. 23 B gives the reconstruction with nine harmonics (same lines + lines 35, 45, 45, 55). Those two graphics clearly show the potential importance of the smallest harmonics in the reconstruction process. This importance is due to phase relations between harmonics. Unfortunately it is quite difficult to detect and to show those phase relations.

It is essential to remember that, as defined here, the spectrum is given by two sequences, related respectively to amplitudes and to phases. On the other hand, a power spectrum corresponds to one sequence related to the amplitude squared. In other words,

phases are unknown and no reconstructions are possible. Furthermore, due to the squaring, graphic representation may be very deceptive, as seen in Fig. 21 B. Here the harmonics appear negligible, despite their known importance.

If we add a pseudo-random white noise to our signal, we obtain the sequence previously presented in Fig. 14 C. This is quite a favourable case, because we have both an ideal signal and an ideal noise. Nevertheless, the interpretation becomes harder and the preliminary estimations less precise.

Let us first suppose that we know the period should be 50 arbitrary units, for example because they correspond to 24 h of a synchronized rhythm. This allows us to calculate the spectrum on 250 data (Fig. 24). This time only four lines clearly emerge (lines 5, 10, 15, 20). The cosinor fitting (Fig. 25 A) obviously has the same drawbacks as before, but, due to the noise, they are less evident. Figure 25 B shows the reconstruction obtained with the four emerging lines. It is more satisfactory, but the sensitivity of the spectral methods on noise with a small number of cycles is

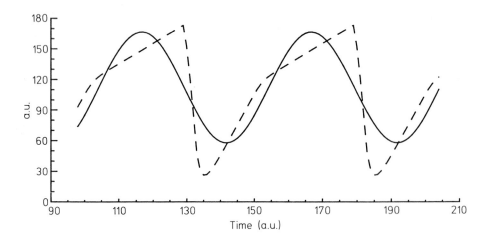

**Fig.26.** True periodic function *(dotted line)* and its adjusted cosinusoidal function *(unbroken line)*. The phase differences between maximums and minimums, expressed in degrees, are equal to 87 and 46, respectively

evident. The reader will certainly be interested in comparing those results with the ones obtained by non-linear filtering (Fig.18).

If we leave the simulations, going to real data, many more problems arise. Studying noise with the discrete Fourier approach is certainly not the easiest way. In theory, the phases of the noise spectral components should have a random distribution. However, this study is rather cumbersome. Signal reconstruction, using discrete Fourier transformations, may be helpful. As an example, in Fig.17 the low part of the spectrum (lines < 11) explains the variability of the mean value of successive cycles.

If the noise is considerable, the spectral analysis may give insufficient results. Quite rapidly, only the fundamental will emerge from the spectra, leaving us with a very poor knowledge of the pattern. We may insist on this point, looking at the simulation shown by Fig.26. The signal is quite asymmetrical, with a quick transition from higher to lower values. If we fit to this signal a sinusoidal function, we clearly see its inadequacy on the same figure.

Furthermore, concerning the cosinor, it is important to realize its confidence limits reflect the properties of the adjustment chosen, and not necessarily the properties of the studied function. Similarly, as seen in Fig.26, amplitude and phase are related to the fitted function and are *not* simply related to the studied phenomena.

ARMA Process

This method is related to the concept of "modern" spectral analysis, also characterized as "maximum entropy", "maximum likelihood" or "parametric" methods. The first papers on those methods appeared around 1967, and were justified by many philosophical considerations [2, 31]. One way to find the spectrum is to consider that the measured sequence results from an ARMA process. The transfer function $S(f)$ of the ARMA filter corresponds to the spectrum to be determined.

In order to clarify the method, let us consider a relatively simple approach, although it is not the most efficient in practice:

1. An AR model of order $k$ is considered.
2. This model is fitted by the least squares method to the data. This fit is performed with the help of Eq.17. For this analysis we have full knowledge of sequence $g[n]$, we ignore sequence $f[n]$, and we want to estimate the $a_i$ coefficients. Knowing that $f[n]$ is a white noise, with a mean equal to zero, from the knowledge of the $k$ preceding values of the $g$ sequence, the best guess $(\vec{g}[n])$ on $g[n]$ is given by:

$$\vec{g}[n] = -a_1 g[n-1] - \ldots - a_k g[n-k] \qquad (28)$$

With a reasonable number of values, we have a great many such equations. The coefficients $a_i$ are now estimated using the classical least squares method, by minimizing the sum of the differences $g[n] - \vec{g}[n]$.

3. By using this procedure, models of order $p = 1$ to $k$ can be fitted to the data, evidently giving different spectra. The important problem is to select a "good" value for $p!$ Several criteria are proposed [2, 32]. Based on different hypotheses, they also give quite different filters. To facilitate the choice, it is always advisable to represent the transfer function components graphically.

This approach, despite its advantages, also has its shortcomings. First, in an excellent article, entitled

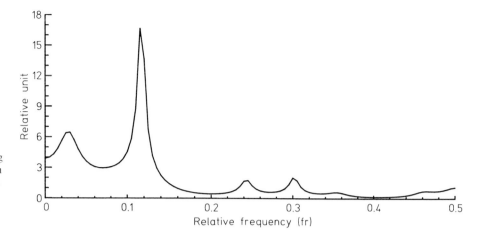

**Fig. 27.** ARMA processing of sequence represented in Fig. 1. The most important peak corresponds to a period equal to 25.6 h ($P = 0.9375$ cycles/day)

"Spectrum analysis: fact or fiction" [33], it is clearly shown that the estimated spectrum also greatly depends on the chosen filter (AR, MA or ARMA). As an example, for a sequence obtained by an MA process, the spectral analysis based on an AR filter is rather disappointing. As in reality the observed sequence has no reasons to be generated by such processes, one sees immediately the limits of those methods. Nevertheless, their use, even for synchronized cycles, may be helpful. It is always advisable to use them, when possible, jointly with the older spectral analysis methods. They are specially interesting when the number of data as well as the number of cycles are both small. Their computation time is comparable with that of more conventional methods.

As an example, Fig. 27 gives the result obtained with our favourite set of data.

## Conclusions

We have focused our paper chiefly on the understanding of rhythms and the estimation of some of their characteristics. Some important data processing steps, concerning measurement procedures and statistics, were not taken into account.

As an example, consider confidence limits and the "greatly favoured" $P$ values. The complexity of the problem is great. In several experiments one has some good information and estimations concerning the errors encountered. It is obvious that they have to be taken in account for the final conclusions. Unfortunately this knowledge is often neglected. Let us take as an example the often used "single cosinor". Some may argue that its 95% confidence limits con-

cern only the fitting. Nevertheless, this is often not the way a chronobiologist interprets the method. This is particularly harmful when the number of data is small. Concerning $P$ values, their use requires the greatest care. Rejecting an alternative hypothesis does not necessarily confirm another one.

As stated by many authors "scientists presuppose that nature is not capricious and that there exist in nature regularities of sufficiently low complexity to be accessible to the investigator" [34].

The research into those regularities is not an easy task. For instance go back to Fig. 12 A where the simple underlying law is evident from Fig. 13. Another example of such simplicity is given by Fig. 15 D. Those examples also show that knowledge of the law does not necessarily lead to long-range predictions.

Furthermore, those are "theoretical" examples. In real life things turn out to be less simple. We thus have to accept some approximations in the correspondence between our data and mathematical theories.

Tentative research into regular patterns thus requires the critical examination of several plausible theories. In the empirical approach, practically imposed in the majority of the chronobiological experimental studies, several mathematical approaches are possible. Due to their conceptual simplicity, generally no theory will perfectly fit the results. In fact, these hypotheses are much too restrictive: perfect regularity or, on the contrary, pure randomness. Nevertheless we have to live with those limits, trying to have some knowledge and understanding of our experimental data.

We advise the use of several methods and plotting of the results. As an example let us have a look at the data of Fig. 1. Different plottings are given in Fig. 3, 4, 17, 20 and 27. Some results may be contradictory, but most of them agree, leading to common properties regarding amplitudes, phases and periods. Each reader, on the basis of his or her knowledge and interest, is able to extract pertinent characteristics.

Considering the number of methods devoted to data processing, the chronobiologist is often hampered in choosing the optimal method of analysis. As in the conception and realization steps he has to learn data analysis with a more experienced expert. Although no systematic rules can be proposed, some general remarks can be suggested.

One must first take into account the data size.

Analysis of large samples allows the potential use of all the methods described. The choice will thus be governed by display and preliminary estimation. If we suspect the presence of noise in the represented data we can apply filtering. The "filtered data" will thus be treated as the crude one. We must then determine if a periodicity can be suspected. If not we will try to detect one, bearing in mind that the result obtained will have to be analysed cautiously (using different methods, for example). In the case of apparent periodicity we can look for its quantification by the methods involving a previous hypothesis of the expected value. Finally we will try to fit the data to a model. It is important to note that data fitting is the highest level of analysis. It allows some predictions on the rhythm. Thus, data fitting leads to hypotheses on further experiments. The results of these experiments can then make plausible the conclusion of our analysis.

Analysis of small samples is much more difficult. We can distinguish between two cases. In the first case data comprises few measurements for many subjects. The aim of data processing is thus to decrease interindividual biological variability to allow treatment by conventional methods. One of the powerful methods of doing so is to perform normalization of the values [1]. A more sophisticated way is to use non-parametric statistical analysis during data treatment. Non-parametric analysis substitutes the value magnitude by their ranks in the data so that the absolute values do not influence the results.

In the case of few measurements with few subjects one must be very cautious about every method used. Methods implying a normal distribution of errors should be avoided, this hypothesis seldom being verified in chronobiological practice. Non-parametric tests, robust methods and resampling should be preferred. For example it is possible, by non-parametric calculations to determine if, in a range of time, measurements gathered in time classes have different values according to these classes. If the result is positive we must then design experiments to confirm and clarify the phenomenon on a larger sample.

In conclusion, we hope to have demonstrated that data processing is never the simple application of formulas and algorithms, but refers to one of the deepest problems in science: interpretation of the "reality". The fact that "reality" appears in arithmetical form (numerical values) does not simplify the problem but only allows the use of specific methods, briefly described here.

# References

1. Hecquet B, De Prins J (1992) Statistical procedures in chronobiology. Scope and application. A critical survey. Chronobiol Int (in press)
2. De Prins J, Cornélissen G, Malbecq W (1986) Statistical procedures in chronobiology and chronopharmacology. Annu Rev Chronopharmacol 2: 27–141
3. Papoulis A (1977) Signal analysis. McGraw-Hill, New York
4. De Prins J, Lechien JP (1976) Remarques sur la pratique de l'echantillonnage et de l'analyse spectrale. Bull Cl Sci Acad Belg 62: 620–645
5. Cleveland J, McGill ME, McGill R (1988) The shape parameter of a two-variable graph. JASA 83: 289–300
6. Cleveland J, McGill R (1985) Graphical perception and graphical methods for analysing scientific data. Science 229: 828–833
7. Kronsjo LI (1979) Algorithms: their complexity and efficiency. Wiley, New York
8. Stoer J, Bulirsch R (1980) Introduction to numerical analysis. Springer, Berlin Heidelberg New York
9. Aris R (1978) Mathematical modelling techniques. Pitman, London
10. Bloomfield (1976) Fourier analysis of time series: an introduction. Wiley, New York
11. Schwartz M, Shaw L (1975) Signal processing: discrete spectral analysis, detection and estimation. McGraw-Hill, New York
12. Croquette V (1982) Déterminisme et chaos. Pour la Science 62: 62–77
13. Glass L, Mackey MC (1988) From clocks to chaos. The rhythms of life. Princeton University Press, Princeton
14. Chatfield C (1975) The analysis of time series. Theory and practice. Chapman and Hall, London
15. Press WH, Flannery BP, Teukolsky SA, Vetterling WV (1986) Numerical recipes. Cambridge University Press, Cambridge
16. Sokal RR, Rohlf RG (1981) Biometry, 2nd ed. Freeman, San Francisco
17. Rey W (1978) Robust statistical methods. Springer, Berlin Heidelberg New York
18. Diaconis P, Efron B (1983) Méthodes de calculs statistiques intensifs sur ordinateurs. Pour la Science 69: 46–58. English version in Scientific American. 96–108. May 1983
19. Efron B (1982) The jackknife, the bootstrap and other resampling plans. CBM-SNSF Regional Conference series in applied mathematics SIAM, Philadelphia

20. Malbecq W, De Prins J (1981) Applications of maximum entropy methods to cosinor analysis. J Interdiscip Cycle Res 12: 97–107
21. Blackman RB, Tukey JW (1958) The measurement of power spectra. Dover, New York
22. Burgess JC (1975) On digital spectrum analysis of periodic signals. Acoust Soc Am 58: 556–567
23. Granger CWJ (1964) Spectral analysis of economic time series. Princeton University Press, Princeton
24. Bogner RE, Constantinides AG (eds) (1975) Introduction to digital filtering. Wiley, London
25. Lee YH, Kassam SA (1985) Generalized median filtering and related nonlinear filtering techniques. IEEE Trans Acoustics Speech Signal Processing 33: 672–683
26. Bozic SM (1979) Digital and Kalman filtering. Arnold, London
27. Sorenson HW (1970) Least-squares estimation from Gauss to Kalman. IEEE Spectrum 7: 63–68
28. De Prins J, Malbecq W (1984) Signal averaging for the analysis of irregularly spaced data. J Interdiscip Cycle Res 15: 179–188
29. De Prins J, Cornélissen G (1979) Complex demodulation and double demodulation. Bull Cl Acad Belg 65: 445–455
30. Bingham C, Arbogast B, Cornélissen G, Lee JK, Halberg F (1982) Inferential statistical methods for estimating and comparing cosinor parameters. Chronobiologia 9: 397–439
31. Childer DG (ed) (1978) Modern spectrum analysis. IEEE, New York
32. Ulrych TJ, Bishop TN (1975) Maximum entropy spectral analysis and autoregressive decomposition. Rev Geophys Space Phys 13: 183–200
33. Gutowski PR, Robinson EA, Treitel S (1978) Spectral analysis estimation: fact or fiction. IEEE G.E. 16: 80–84
34. Losee J (1980) A historical introduction to the philosophy of science. Oxford University Press, Oxford
35. Bünning E (1973) The physiological clock. Springer, Berlin Heidelberg New York

# Chronopharmacology

B. Bruguerolle

## Introduction: Definitions and Aims

Chronobiological observations lead to the hypothesis that the response of an organism to a drug may depend on the hour of administration. Virey was one of the first who suggested in 1814 that desired and undesired effects of a drug may vary according to the time of its administration (Reinberg 1979). Later this hypothesis was validated and the toxicity and efficiency of many drugs have been shown to depend on their time of administration. Some of the mechanisms involved in the time-dependent (chronopharmacological) responses are documented at the present time.

The term "chronopharmacology" was introduced by Halberg in the 1960s and in 1971 Reinberg and Halberg reviewed the early studies on this subject.

Chronobiology makes it necessary to reevaluate numerous data according to the time they were obtained. Since an organism has a temporal structure which is composed of a multiplicity of biological rhythms, the effect and the fate of a drug may depend on its time of administration in relation to several frequences. Thus, chronopharmacology may involve qualitative and/or quantitative modifications of the efficiency of a drug according to the hour, the day or the month of administration. However, chronopharmacology not only consists of the study of temporal variations of the activity, toxicity and kinetics of drugs according to their time of administration but also concerns the study of possible alterations of the temporal structure of the organism receiving the drug (Reinberg 1974, 1976, 1978, 1982; Bruguerolle 1984a). Possible modifications of the biological rhythms of the organism by the drug constitute another aspect of chronopharmacology. According to the sum of the biological rhythms of the organism receiving the drug (temporal structure) the response of the organism will vary along the time scale.

Pharmacology is concerned with the effectiveness of drugs and thus can be divided into pharmacodynamics (study of the mode of action of a drug) and pharmacokinetics (study of the fate of drugs in the organism). These two aspects interfere since the response at the receptor site depends on the concentration of the drug at this site (involving pharmacokinetic processes) and the fate of the drug (i.e., absorption, distribution, metabolism and elimination). Temporal variations may act at all of these different stages and Reinberg (1974) has proposed three concepts and terms involving these interdependent aspects of chronopharmacology: chronesthesy, chronokinetics and chronergy.

*Chronesthesy* is defined as the temporal changes in biological susceptibility including temporal changes in receptors of target cells or organs, membrane permeability, etc., in other words chronesthesy concerns the temporal variations of the mode of action of the drug, i.e., its pharmacodynamics.

*Chronopharmacokinetics* involves the study of the temporal changes in absorption, distribution, metabolism and elimination of a drug and describes the influence of the time of administration on the mathematical parameters which describe these different stages.

Finally, *chronergy* represents the rhythmic change of the response of the organism to a drug (its total effect) according to its chronesthesy and its chronokinetics.

Thus the aim of chronopharmacology, which studies the time-dependent variations in pharmacology, is to document qualitative and/or quantitative temporal changes of the efficiency of drugs which lead to a better use of drugs in the treatment of diseases.

# Basis for Chronopharmacology: Chronotoxicology

Any pharmacological study begins with the study of toxicity. As first described by Halberg in the 1960s, the susceptibility of an organism to a physical, chemical or toxic agent varies according to the time of its administration (Halberg et al. 1955).

Chronotoxicology will be discussed elsewhere in this book but we only would like to underline some aspects of the influence of time of administration on the acute or chronic toxicity as well as on teratogenicity or tolerance to drugs.

## Temporal Variations of Acute Toxicity

The first step in the pharmacological development of a drug is the determination of its acute toxicity. The most used parameter to characterize this toxicity is the $LD_{50}$ (even if this concept tends to be more and more under discussion); increasing doses of the tested drug are administered to several species of animals by different routes, the end points studied being the death or the survival of the animals. Many physical, chemical or physiopathological factors are known to influence this acute toxicity and thus are often controlled during these studies. On the contrary the time of administration (hour of the day, day of the month, month of the year) is usually not systematically controlled. From studies such as those by Halberg et al. (1959) we know that the acute toxicity of drugs may critically depend on the time of administration, e.g., one of the first chronotoxicological studies concerned the temporal variation of the acute toxicity of ouabain in mice: the same dosage of this agent was administered every 4 h over 24 h to different groups of mice under controlled environmental conditions (i.e., temperature, humidity, lighting). The mortality was 80% when the drug was administered at 0900 hours and 20% at 2100 hours; thus a difference of 60% in ouabain-induced mortality was observed during a 12-h interval.

Many other drugs were shown to exhibit a time-dependent toxicity. A practical application of such a phenomenon is that in the determination of the acute toxicity of a drug. The investigator must now take into account the time of administration. Concerning chronotoxicity, several types of studies have been carried out:

– In some studies, the same dose is applied at different times.

– In others, the $LD_{50}$ (i.e., several doses given at the same time) is determined at different time points (stages) of an expected rhythmic variation.

Most of these studies reported the maximal toxicity occurring at different clock hours of the 24-h scale, even for drugs of the same pharmacological or chemical family. As an example, cholinergic substances such as the cholinomimetics acetylcholine, oxotremorine and physostigmine are in rodent more toxic at 2400 hours during the middle of their daily activity span, while cholinolytic agents such as atropine or scopolamine are more toxic at 1200 hours, i.e., in phase opposition (Friedman and Walker 1972).

Most of the drugs examined seem in nocturnal rodents to be more toxic during the dark phase (activity span).

The chronotoxicity of other periods than the circadian have also been studied for antibiotics by Pariat et al. (1984) or for phenobarbital by Bruguerolle et al. (1988c); e.g., we demonstrated that the acute toxicity of phenobarbital in mice varies both along the 24-h scale and during the year: phenobarbital is more toxic in January than in July (Bruguerolle et al. 1988).

## Influence of Time of Administration on Teratogenicity

The effect of a drug treatment during pregnancy is determined during its pharmacological development study by the determination of possible teratogenicity. Also in this field the hour of administration of the drug may play a role and the few investigations carried out in that field have shown that the type of drug-induced malformations may depend on the time of administration. Thus Isaacson (1959) has documented an increased frequency of the corticoid-induced deficit of the closure of the secondary palate when the drug is administered during the daily dark span. Sauerbier (1979, 1980) showed that cyclophosphamide is most toxic on the 12th day of gestation in mice when administered between 1300 hours and 1900 hours.

## Temporal Variations of Undesired Side effects of Drugs

The end points chosen in toxicology may also be a particular toxicity such as hepatic, digestive, and hematologic toxicity. This kind of toxicity may depend also on the time of administration, e.g., the hepatic toxicity of chloroform evaluated by SGPT, SGOT, and LDH is time dependent with a maximum when chloroform is administered to rats at 2100 hours (Lavigne et al. 1983). Levi et al. (1980a, 1981) have documented the maximal toxicity of *cis*-platinum in rats when this agent is administered during the light phase (resting period).

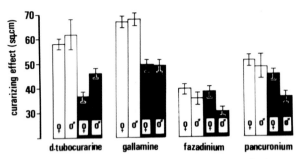

**Fig. 1.** Day-night differences in male and female rats in the curarizing effect of four curarizing agents; differences related to sex demonstrated during night time but not during day time (as usually studied). (From Bruguerolle et al. 1976)

## Experimental Methodology

Chronopharmacology which takes into account the time of administration of a drug requires particular and specific methodology concerning the experimental or clinical conditions and models, data collection or calculations.

The control of factors such as age, gender, species, feeding, lighting and manipulation becomes of particular importance in chronopharmacological studies.

### Age

The influence of age on biological rhythms has been studied in many biological phenomena. For example the development of some endocrinological phenomena have been studied in children and their evolution has been followed in the young adult and in the elderly. But in chronopharmacology little work has been done on the development (in children or young animals) or the evolution (in elderly subjects or in old animals) of chronopharmacological effects (Touitou 1982, 1987). From a methodological point of view, Jenni-Eiermann et al. (1985), in a study reporting on possible alterations of receptor rhythms during the ageing process, underline the necessity of pharmacological studies on drug effects in ageing (including different age groups). It appears obvious that the age must be taken into account in chronopharmacological studies just as much as in usual pharmacological studies.

### Gender

The study of time-dependent phenomena in pharmacology must take into account the influence of gender. Indeed the time of administration of a drug may reveal on some occasions a different pharmacological response related to gender, e.g., we have documented a daily rhythm of the curarizing effect of four curari-

mimetic agents (*d*-tubocurarine, gallamine, fazadinium and pancuronium) in rats with the maximal curarizing effect occurring during the day (Bruguerolle et al. 1976, 1977, 1978b). However, as shown in Fig. 1, a difference between male and female rats was not significant during the light span but was only detected during the dark span: this indicates that differences in pharmacological effects related to sex may depend on the time of administration. This observation was confirmed in man by Descorps-Declere et al. (1983), who have demonstrated a circannual difference between men and women in the curarizing effect of pancuronium.

### Species and Strains

We mentioned earlier the necessity in toxicology of using different species of animals. As far as chronopharmacology is concerned the choice of the species may be of particular importance; indeed it is well known that rodents have a circadian mode of synchronization opposite to man: rats or mice are nocturnal animals and their daily activity span (dark phase) corresponds to the rest span in man. Thus it appears to be feasible to extrapolate chronopharmacological results from rodents to men by inversing the phase. However, this is not always true and, in some cases, chronopharmacological effects are described at the same clock hour irrespective of the resting (or activity) period in man and in rodents (e.g., for the acrophase of melatonin and the maximal effect of oral anticoagulants). Numerous investigators describe the acrophase of a periodic phenomenon by referring to the "hour after the light on" (HALO).

Thus as in "classical" pharmacological studies, the extrapolation of the results from rodents to man is sometimes difficult; however, the pharmacological study of a drug in animals is a necessary and obligatory step in its development. The choice of appropriate hours of administration in clinical chronopharmacological studies requires previous chronopharmacological studies in animals.

As in "usual" pharmacological studies, more than one animal strain must be used for chronotoxicity studies prior to the extrapolation of the results to clinical trials. Carlebach et al. (1990) have recently reported strain differences in the chronotoxicity of three anticancerous agents. These results are of particular importance since many different strains of rodents are used in toxicological studies in cancerology.

## Feeding

Feeding conditions are of course of particular importance not only in nutritional studies; but also in pharmacology, particularly in the area of drug kinetics where food is a determinant for the resorption and influences the bioavailability of the drug and hence its efficiency. In experimental pharmacology the access to food is generally free. As mentioned in another chapter of this book, we have to emphasize that in rodents, contrary to man, feeding habits may be strong synchronizers while in nocturnal rodents the maximal food uptake normally occurs during the daily dark span. Time-limited access to food may be used to reinforce the mode of synchronization. Alternatively, in chronokinetic studies fasting is quite usual, in order to avoid interferences between food and drug.

## Lighting

The dark-light alternation is one of the strongest synchronizers in animals. Most animal laboratories are equipped with artificial light-dark schedules. All kinds of schedules may be chosen but a 12-h light and 12-h dark scale is widely used. As for biological rhythms, the chronopharmacological effects may depend on the lighting synchronization, and changing of the lighting schedule (suppression of light or dark periods, inversion of the light-dark cycle, modifications of their respective durations, etc.) may lead to a similar shift in the observed effects (Scheving et al. 1974, 1976).

The manipulation of animals during the dark period must be done under minimal light intensity (a red light is often chosen) since it has been shown that even a short period of light during the dark phase may induce alterations of some biological rhythms.

Thus synchronization by light is of great importance and the lighting regimen must be systematically described in the material and methods section of any publication; also, for the previously mentioned reasons, the adaptative period of animals in the laboratory rooms must be 2–3 weeks before starting an experiment.

Also the control of parameters such as temperature, humidity and lighting may be provided in special rooms.

## Stress and Effects of Manipulation

Halberg et al. (1955) demonstrated that the hour of application of white noise may influence its toxicity. It is well known that noise as well as other kinds of stress conditions may affect the animal's response to drugs and that their effect may be different at different stages of the cycle. Also animal handling or injections must be as gentle as possible.

## Data Collection

Since chronopharmacological studies document the influence of the time of administration of a drug on the organism's response, many measurements are carried out and the repetition of such measurements involves a higher variability than in conventional pharmacological studies. Thus special tools are needed to avoid disturbances leading to statistical variations: monitoring and data collection systems for unanesthetized and conscious freely moving laboratory animals are available. Continuous measurement of blood pressure, temperature, heart rate, ECG, EEG, EMG, activity, etc., may be performed by radiotelemetric procedures; also in humans continuous registration of ECG (Holter 1961), EEG, blood pressure, heart rate, etc., e. g. is possible and leads to a better knowledge of the efficiency of a drug along the 24-h scale. Gautherie and Gros (1977) have described procedures for registration of the skin temperature in breast cancer or in arthritis and have developed this methodology for chronopharmacological purposes in order to monitor the effects of drugs.

## Technical and Experimental Models

In order to demonstrate the efficiency of a drug the pharmacologist works on animal models which aim to mimic as closely as possible the conditions under which the drug is to be used in human pathology. In the first steps of a pharmacological study of drugs some of these animal models are used to "screen" the drugs ("screening") before a more extensive study. Those drugs which do not succeed in demonstrating their efficiency in these tests are not studied any further. The hour of administration of a drug is at the present time not generally taken into account during these studies. However, since chronopharmacological effects may be of great amplitude, it appears that in such studies ignoring chronopharmacological effects may lead to rejection of a drug which would have been active if given at the right time.

The time factor may be of importance in numerous technical or experimental models used in pharmacology, as indicated by the following examples.

Biological rhythms may play a role in the interpretation of data from cells in culture (Kadle and Folk 1983). Since the temporal variations observed on unicellular organisms cannot be extended easily to isolated cells derived from more complex multicellular animals, individual isolated cells may be considered in part as a basic unit for circadian clock studies (Kasal and Perez-Polo 1982). Circadian rhythms in tissue cultures have been documented by several authors and were reviewed by Kadle and Folk (1983).

Some experimental models study the efficiency of drugs and the biochemical mechanisms of their actions in vitro on isolated organs (Ungar and Reinberg 1962; Spoor and Jackson 1966); such models involve the killing of animals in order to sample the organ: thus the hour of sampling may have an influence on the pharmacological response. For instance, Spoor and Jackson (1966) reported on circadian variations of acetylcholine effects on isolated hearts of rats: the acetylcholine induced-heart rate decrease is maximal at 2300 hours.

Studies of "binding" are also often used in pharmacology in order to characterize the action of drugs (agonists and antagonists). As an example Lemmer and Lang (1984) reported differences in antagonist binding of [$^3$H]dihydropropranolol in heart ventricular membranes of rats when they were killed at 0800 hours or at 2000 hours. An approximately 40% significantly higher number of binding sites was observed at 2000 hours than at 0800 hours.

The study of antiinflammatory drugs involves several experimental models reproducing inflammatory processes; one of the more widely used tests in this area is the carrageenan-induced paw edema test: this test consists of the injection of carrageenan into the plantar tissue of the hind paw of rats. The volume of the paw edema is then measured before and after injection of the antiinflammatory drug tested. Labrecque et al. (1979, 1981, 1982) have demonstrated a circadian and a seasonal variation of the induced inflammation: the time required to obtain a maximal edema was much shorter after injection at 2000 hours and the maximal induced edema was observed in animals treated during the month of May.

Psychotropic drugs are "screened" with several pharmacological tests involving motor activity, apomorphine-induced stereotypies, etc., which all vary along the 24-h scale. Many other examples could be cited, indicating that in the study of a drug the time of the day in relation to the experimental conditions (i.e., the lighting regimen and the months in which the test is performed) must be taken into account to be able draw conclusions about the efficiency of the tested drug.

## Temporal Variations of the Dose-Response Effect

The pharmacodynamic study of a drug involves the evaluation of its dose-response effect. The quantitative aspects of drug potency and efficacy are generally studied on a log dose-response curve: the dose of a drug is plotted (on a log scale) against the corresponding measured effect and a sigmoid curve is usually obtained (log dose-response curve). For comparative purposes, the dose giving 50% of the maximal effect is often chosen: it is called the effective dose 50 ($ED_{50}$). Potency and efficacy of different drugs may be compared on such dose-response curves. To be accurate such studies should take into

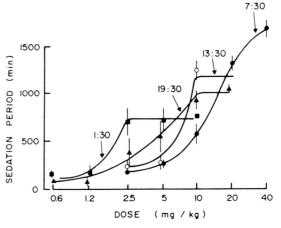

**Fig. 2.** Example of temporal variations in dose-response curves: different dose-response curves for the sedative effect of chlorpromazine in rats as function of time of administration. (From Nagayama et al. 1978)

account the time of administration of the drug. Unfortunately this factor is not usually considered. An example of the time-dependent differences encountered is given by data from Nagayama et al. (1978) as shown in Fig. 2: the dose-response curves for the sedative effect of chlorpromazine are significantly different according to the hour of administration of the drug.

Also Lemmer and Neumann (1987) reported on the circadian variation of seven $\beta$-receptor blocking drugs on the motor activity of rats; during the dark period, all $\beta$-receptor blocking drugs dose dependently decreased the spontaneous motor activity of rats while in contrast during the light span these drugs either did not affect motor activity compared with saline or even increased spontaneous motility. As these examples indicate, the dose-response curve studies of a drug do depend on the hour of administration of this drug.

## Data Calculation

Time series analysis in chronopharmacology as in chronobiological studies in general may be carried out by normal statistical methods (ANOVA for instance) or by specialized rhythmometric methods, e.g., cosinor (Nelson et al. 1979). These statistical methods are described elsewhere in this book (De Prins et al.).

# Methodology in Clinical Chronopharmacology

Some particular aspects of methodology must be emphasized in the study of drugs in man.

## Implications of Chronopharmacology in Clinical Trials

Growing interest in clinical pharmacology studies has led to the development of a particular methodology for clinical trials. Nevertheless only very few clinical trials have taken into account the time of administration.

A clinical chronopharmacological study requires:
- Chronopharmacological studies in animals as a background

- Chronopharmacokinetics and chronopharmacodynamics studies in healthy subjects
- Chronobiological studies of the pathology concerned
- Chronobiological markers of the pathology concerned
- Specific statistical methodology

At each stage of phase I, II or III trials chronopharmacology may be involved:
- Phase I trials must study the influence of the time of administration on drug tolerance in healthy subjects and have to provide chronokinetic information.
- Phase II trials study the implication of the hour of administration on the dose-response effects in patients.
- Finally phase III trials which represent the phase of comparative trials need to demonstrate the chronoefficiency of the drug studied.

We will develop these concepts further in the chapter on chronotherapeutics.

# A Particular Aspects of Clinical Chronopharmacology: The Placebo

Placebo is used in comparative clinical trials in order to demonstrate the efficiency of the tested drug. As we mentioned before a clinical trial in a given disease state must take its chronobiological aspects into account. Also if the tested drug is compared to a placebo, the possible chronopharmacological effect of this placebo must be considered. Pollmann (1981) for instance demonstrated that the analgesic effect of a placebo varies along the 24-h scale. Thus the analgesic effect of a drug tested and compared to a placebo must take this temporal "basal" variation into account.

# Examples of Chronopharmacological Effects

Even if most chronopharmacological studies have been carried out in rodents and humans, chronopharmacological effects may be observed in several other species for numerous periodicities (circadian and circannual periods are the most often studied) and are observed at the cellular level as well as at the organ level (isolated organs).

In terms of pharmacology the response of the organism to the drug may be qualitative (i.e., may concern the type of action of the drug: induction, inhibition, stimulation, etc.) or may be quantitative (i.e., may be expressed as an intensity, a duration of effect, etc.).

### Examples of Temporal Variations in Qualitative Effects

Beta-receptor blocking agents decrease dose dependently the motor activity of rats when administered during the dark phase; inversely during daytime they do not affect motor activity or may even increase spontaneous motility (Lemmer and Neumann 1987). The type of activity of $\beta$-receptor blocking drugs depends on the hour of their administration.

The same dosage of nalorphine has an opposite effect on the total brain levels of acetylcholine in the rat: at 0700 hours nalorphine induces a decrease of this neurotransmiter while at 1900 hours leads to an increase (Labrecque and Domino 1982).

Epinephrine injected into the lateral hypothalamus of rats may stimulate or inhibit eating and satiety centers depending on the time of injection (Margules et al. 1972).

### Examples of Temporal Variations in Quantitative Effects

We reported on the temporal variations of the curarizing effect of curarimimetic agents such as $d$-tubocurarine, fazadinium, pancuronium and gallamine in rats. As shown in Fig.3 the curarizing effect was demonstrated to vary along the 24-h scale with a

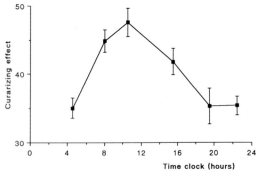

**Fig.3.** Circadian variation of the curarizing effect of pancuronium bromide, a curarizing agent, in rats. (From Bruguerolle et al. 1977)

minimal value during the dark phase; for pancuronium we have demonstrated a circadian rhythm in the curarizing effect with an acrophase located at 1030 hours (Bruguerolle et al. 1976, 1977, 1978b, 1979). Data from Scheving and Vedral (1966) have clearly demonstrated that the pentobarbital-induced duration of anesthesia varies in rats according to the time of injection; a 35-mg/kg dosage induces a 50-min duration of anesthesia at 0900 hours while this duration is about 90 min when the drug is injected at 2100 hours. We have also reported on time dependency of althesin in rats: for two different dosages of althesin we have demonstated a circadian rhythm of the duration of anesthesia with a peak at 1000 hours (Bruguerolle et al. 1977). Many other examples could be given to illustrate the importance of the hour of administration of a drug in the measurement of its pharmacological effects.

### Chronopharmacological Mechanisms

Many chronopharmacological studies have documented temporal variations of drug toxicity or efficiency but most were only descriptive. Underlying mechanisms of temporal variations in drug effects are less well described but there is at the present time a growing interest in such studies. The time-dependent efficiency of a drug may be explained either by temporal variations in the mode of action of this drug (i.e., chronopharmacodynamics or chronesthesy) or by variations in its pharmacokinetics (i.e., chronokinetics).

### Temporal Variations in the Mode of Action of Drugs: Chronopharmacodynamics

Drugs may be separated into different groups regarding their mechanism of action:
– Some drugs do not possess specific affinity for cellular or receptor sites: their action simply consists of a physicochemical reaction (local anesthetics, antacids, osmotic diuretics, etc.)
– Most other drugs have a specific action and have a particular affinity for molecular sites called receptors.

Both of these groups of drugs may be considered from a chronopharmacological point of view.

## Chronopharmacodynamics of Drugs Having a Nonspecific Mode of Action

Local anesthetics are known to modify cellular membrane permeability to ions such as Ca, Na and Cl. Reinberg and Reinberg (1977) reported on chronoefficiency of local anesthetics in dental practice: the lidocaine-induced duration of anesthesia lasts longer at 1500 hours than at 0900 hours. Among different hypotheses these authors have suggested the possibility of a temporal variation of membrane permeability. Such a variation had been proposed by Njus et al. (1974) and was demonstrated by Scott and Gulline (1975). We have found temporal changes in drug membrane permeability of several local anaesthetics (LAs) (Bruguerolle and Prat 1989). Since erythrocytes are usually chosen as a model for membrane permeability studies, the possible temporal changes of the penetration of LAs through the erythrocyte membrane were investigated by the determination of their respective erythrocyte concentrations, according to the hour of administration. Our data have revealed that the better penetration occurs at the end of the dark phase (bupivacaine and mepivacaine) or at the beginning of the light phase (etidocaine) (Fig. 4). A previously reported study had revealed that lidocaine penetration into erythrocytes was higher during the dark phase (2200 hours) (Bruguerolle et al. 1983).

The previously mentioned circadian variations of the curarizing effect of pancuronium in rats (Bruguerolle et al. 1976) may be explained by temporal changes in ionic permeability at the level of the neuromuscular junction.

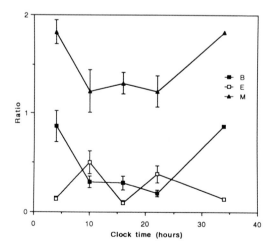

**Fig. 4.** Circadian variations in the erythrocytic permeability to three local anesthetics bupivacaine *(b)*, etidocaine *(E)*, and mepivacaine *(M)* in mice. (From Bruguerolle and Prat 1989)

## Chronopharmacodynamics of Drugs Having a Specific Mode of Action

The pharmacodynamic study of drugs involves the study of their mode of action at the molecular level, i.e., at the receptor site by determining the "binding studies" of these drugs or of their agonists. These binding studies are characterized by two parameters: the number of sites (expressed by the index of the maximal binding capacity, i.e., $B$max.) and the affinity of drugs to the binding site. Thus the chronopharmacological approach involves the study of temporal variations in the number and/or the affinity of binding sites. The characteristics of some studies carried out in animals or in human subjects for various drugs (opiates, benzodiazepines, insulin, imipramine, $\alpha$- or $\beta$-receptor blocking drugs, oestrogens, etc.) for several periodicities (circadian, mensual, circannual) and at a central or a peripheral level have been reviewed (Bruguerolle 1984a).

Wirz-Justice et al. (1980, 1981, 1983) and Naber et al. (1980, 1981 a, b) demonstrated circadian rhythms of rat brain neurotransmitter receptors, e.g., the number of binding sites for [$^3$H]dihydroalprenolol (a $\beta$-receptor blocking drug) in rat brain was maximal at 1000 hours during the month of February but was located at 0600 hours in April and June. Wirz-Justice (1987) reviewed the main studies concerning circadian rhythms in mamalian neurotransmitter receptors: $\alpha_1$, $\alpha_2$, $\beta$-adrenergic, muscarinic, cholinergic, dopaminergic, 5-HT1, 5-HT2, adenosine, opiates, benzodiazepines, GABA. Receptors isolated from rat forebrain homogenates varied over 24 h and during the season; these rhythms persist in the absence of time cues and may have different patterns according to the brain regions. As mentioned by Wirz-Justice (1987), mechanisms of membrane receptor regulation (extensively investigated) may provide some possible explanations for the mechanisms of receptor rhythms (internalization, desensitization, etc.).

Lemmer and Lang (1984) reported differences of antagonist binding of [$^3$H]dihydropropranolol on heart ventricular membranes of rats killed at 0800 hours or at 2000 hours. An approximately 40% statistically significantly higher number of binding sites was observed at 2000 hours than at 0800 hours. These findings may explain, in part at least, the higher efficiency of $\beta$-receptor blocking agent during the dark phase.

Schulz et al. (1983) have demonstrated a significant decrease of erythrocyte binding of insulin at 2400 hours compared to 0800 hours or 1600 hours. These findings concern the temporal variations of the affinity of the receptors but not their number and seem to be dependent on the circadian variations of insulin.

As for many chronopharmacological phenomena the manipulation of synchronizers may influence the circadian rhythms of receptors. The persistence of these rhythms under conditions of synchronizer manipulation indicates their endogenous nature. The development of such rhythms has been studied: Bruinink et al. (1983) documented the ontogeny of the circadian rhythms of central dopamine, serotonin, and spirodecanone in the rat; the rhythms of the dopaminergic D2 and serotoninergic S2 sites (binding of [³H]spiperone) differed between immature and adult animals. Possible alterations of receptor rhythms during the ageing process have been studied by Jenni-Eiermann et al. (1985). For example, circadian variations of neurotransmitter binding were evaluated in three age groups of rats: the phases were different between the young and adult group.

All these studies clearly indicate that the number of available sites on a receptor or its affinity sites may vary according to time and are not constant throughout the day, the month or the year. The changes at the receptor level may explain in part some of the chronopharmacological effects.

## Temporal Variations in Drug Pharmacokinetics: Chronokinetics

Some drugs have a narrow therapeutic range and it has been shown for these drugs that therapeutic monitoring is necessary since there is a good relationship between the pharmacological effect and the blood level. This leads to pharmacokinetic studies concerning the fate of drugs in the organism related to time after administration, i.e., the absorption, distribution, metabolism and elimination processes. Mathematical models are used to simplify and summarize all these processes and simulate the concentrations of the drug in all parts of the organism. However, these studies were based in the past upon the search of an as far as possible constant blood level of the drug which was thought to be necessary to obtain an effect as constant and reliable as possible. Obviously, all the chronopharmacological phenomena we have mentioned before deny this concept. We will develop the notion that the time of administration of a drug may be responsible for nonlinearity in pharmacokinetics.

Chronopharmacokinetics concerns the study of the temporal changes in absorption, distribution, me-

tabolism and elimination of a drug and thus the influence of time of administration on the mathematical parameters that describe these different stages. The chronokinetics of more than 100 drugs have been reported in animals or in man as reviewed by Reinberg and Smolensky (1982), Lemmer (1981b), Bruguerolle (1983a,b, 1987c) and Levi et al. (1989) and may explain in part at least the chronopharmacological data.

### Temporal Variations in Drug Absorption

Many locations are used for the administration of drugs: intravascular or extravascular sites including the oral, sublingual, buccal, intramuscular, subcutaneous, pulmonary and rectal routes. It is assumed that most of the drugs pass through membranes by passive diffusion. Many factors may change drug absorption processes, such as the structure of the membranes, the chemicophysical properties of the drugs, blood flow, pH and the degree of gastrointestinal tract emptying. Thus circadian variations in gastrointestinal pH or motility, digestive secretions, intestinal blood flow and membrane permeability may be involved in temporal variations of drug resorption.

The chronopharmacokinetics effects have often been considered to depend only on oral absorption and thus have been considered to be strictly related to gastric or intestinal emptying. Obviously, food influences the rate of absorption of a drug: in the presence of food, a drug passes less rapidly through intestinal membranes than under fasting conditions. Galenic considerations may also be involved in drug absorption. Smolensky (1989) recently reviewed the temporal variations in theophylline disposition and described differences in the temporal changes of the absorption of the slow-release forms according to the galenic formulation. Circadian-phase-dependent differences in the pharmacokinetics were also reported for an immediate release and a sustained release form of isosorbide-5-mononitrate (Lemmer et al. 1989).

Several clinical or experimental studies have reported temporal variations of drug absorption (Reinberg and Smolensky 1982). As an example we reported (Bruguerolle et al. 1984) on temporal variations in kinetic parameters characterizing lorazepam absorption in man under fasting and posture controlled conditions; absorption constant rate and absorption half-life were significantly modified by the hour of administration, the drug being more rapidly absorbed after the morning than after the evening administration.

Data from Smith et al. (1986) with another benzodiazepine, triazolam, agree with our results since they have demonstrated that the absorption half-life is twice as high in the evening than when the drug is taken in the morning. In contrast a more rapid absorption after morning drug administration than after evening intake was recently reported for propranolol (Langner and Lemmer 1988).

Most of the lipophilic drugs seem to be more rapidly absorbed in man when the drug is taken in the morning. Differences in the physiochemical properties of drugs have been proposed as an explanation of the differences in these circadian patterns. Thus, Belanger et al. (1981) studied the chronokinetics of the following drugs in the rat: acetaminophen, antipyrine, furosemide, hydrochlorothiazide, indomethacin, and phenylbutazone; they concluded that temporal variations in the rate of absorption of a drug seem to depend on its solubility: the absorption of highly water soluble drugs such as antipyrine does not seem to vary with time but the rate of absorption of lipid-soluble drugs is faster in the evening. These findings in night-active rats agree well with the more rapid absorption in day-active man at the onset of the activity period.

We also have demonstrated temporal variations of carbamazepine kinetics in rats showing a better absorption of the drug in the dark period (Bruguerolle et al. 1981); under comparable experimental conditions, the temporal variation of carbamazepine absorption persisted in fasting rats, although the rhythm was phase shifted.

The influence of posture is often evoked to explain temporal variations of the oral absorption of a drug. This factor must obviously be strictly controlled in chronokinetic studies, since posture is a factor which may modify the pharmacokinetics of a drug by modifying local blood flow.

Other than circadian rhythms periodicities such as circamensual rhythms (or the estrus cycle in rodents) have been reported to modify the absorption of drugs (Bruguerolle 1983b). For example sodium salicylate absorption varied with the menstrual cycle with a maximum in midcycle. We recently reported in the rat that theophylline absorption was five times higher during diestrous (final phase of the estrus cycle) than during proestrus (beginning of the cycle) (Bruguerolle 1987a). The same problem is under investigation in asthmatic women treated with theophylline.

In conclusion, it appears that temporal variations of drug absorption exist but may be influenced by many factors such as food, posture and galenic form of the drug (Bruguerolle 1989). Thus, future chronokinetic studies must consider and control all of these factors.

## Temporal Variations in Drug Distribution

The process of distribution of a drug involves its passage through biological membranes, its transport by plasma proteins or red blood cells and its tissue binding.

### Temporal Changes in Binding to Plasma Proteins

Drugs are transported from their site of administration to receptor sites by plasma proteins such as serum albumin, $\alpha_1$ acid glycoprotein, globulins and lipoproteins. Since only the unbound drug can diffuse through membranes and tissues (representing the active fraction of the drug), protein binding of drugs has important pharmacological implications.

Many factors may influence drug protein binding: temperature, pH, physicochemical properties of the drug, plasma concentration of the protein involved, etc. Each of these factors could theoretically be subject to temporal variations.

We reported in our laboratory circadian variations for the free plasma drug levels of acidic drugs (carbamazepine) or basic drugs (lidocaine, disopyramide). Thus, in rodents the lowest free drug levels (i.e., the highest protein binding) were located at 0400 hours for carbamazepine (Bruguerolle et al. 1981) and at 2200 hours for lidocaine (Bruguerolle et al. 1982) or

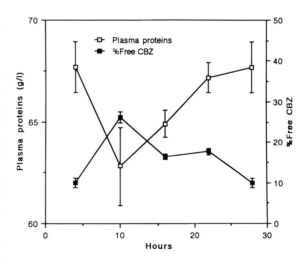

**Fig.5.** Circadian variation in the free carbamazepine (CBZ) percentage and total plasma proteins in the rat. (From Bruguerolle et al. 1981)

disopyramide (Bruguerolle 1984b). Such variations may in part at least be the mechanism explaining chronokinetics. It must be pointed out that total proteins and albumin (the protein specifically responsible for plasma binding of carbamazepine) also varied with a maximum during the dark phase of the light/dark cycle, the temporal variation being inversely correlated to those of the carbamazepine-free fraction (Fig. 5). Haen et al. (1985) reported circadian variations in propranolol plasma protein binding in rats with peaks at 1600 hours and 2400 hours.

In man until now few studies have been devoted to this phenomenon. Highest plasma levels of free phenytoin or valproic acid are observed in man between 0200 hours and 0600 hours (Patel et al. 1982, Lockard et al. 1985). Lowest levels of free diazepam (Naranjo et al. 1980) and carbamazepine (Riva et al. 1984) were reported to occur in the morning. Finally, a circadian rhythm in *cis*-platin binding on plasma proteins was described by Hecquet et al. (1984) (maximum in the afternoon and minimum in the morning).

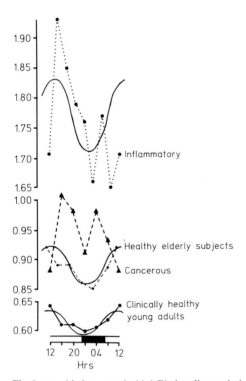

**Fig. 6.** $\alpha_1$ acid glycoprotein (AAG) circadian variations in clinically healthy young adults and elderly subjects and in patients with cancer or inflammatory diseases; significant differences in respective 24-h means were documented. In cancerous patients a circadian rhythm was not documented but an ultradian rhythm (12 h) was found. (From Bruguerolle et al. 1976; Focan et al. 1988)

Bauer et al. (1985) reported on circadian variations of valproic acid protein binding in young and elderly subjects: the clearance of the unbound drug was about 15% higher during the evening.

All these findings may depend on temporal variations of different plasma proteins as reported by Reinberg et al. (1977), Bruguerolle et al. (1986b) and Touitou et al. (1986). We studied in our group in collaboration with Yvan Touitou, Francis Levi and Christian Focan the physiological circadian time structure of some plasma proteins in man and its alterations by age or by different disease states such as cancer (Bruguerolle et al. 1986b) or inflammatory diseases (Focan et al. 1988) (Fig. 6).

From a methodological point of view, most of the previously described studies have reported temporal variations of plasma-free drug levels by direct measurement in the plasma and calculating plasma protein binding by the difference in the total plasma concentration of the drug. Temporal changes are usually reported to be dependent on the amount of plasma proteins but not on possible temporal changes in the affinity of the proteins concerned. Temporal changes of the protein-binding affinity characteristics of the drug should also be assessed in future studies.

Such changes are important concerning the possible explanations of chronopharmacological effects or of chronokinetic mechanisms. But the practical question raised by temporal variations of plasma protein drug binding are their clinical implications. It is generally admitted by pharmacologists that clinically significant consequences of changes in drug binding are observed only for drugs which are highly bound (more than 80%). Thus, temporal variations in plasma drug binding may have clinical implications only for drugs characterized by a high protein binding and a small apparent volume of distribution.

## Temporal Aspects in Drug Binding to Erythrocytes and Membrane Permeability

Drugs may also be transported by red blood cells. This binding depends on plasma protein binding of drugs and on their physicochemical properties. Time dependency of drug binding to erythrocytes has not been reported to our knowledge with the exception of that of local anesthetics (lidocaine, bupivacaine, etidocaine and mepivacaine) and of indomethacin.

We reported in rats the chronokinetics of lidocaine as well as the circadian variations of its passage into red blood cells (Bruguerolle et al. 1983). To assess the circadian time dependent passage into the erythrocytes, the ratio of drug concentration in erythrocytes to plasma concentrations was calculated; when the drug was given at 2200 hours the ratio was 0.74 versus 0.48 at 1000 hours (Bruguerolle et al. 1983).

We recently found that in female rats the kinetics of theophylline and also the passage of this drug into red blood cells were dependent on the phase of the estrous cycle (Bruguerolle 1987 b). The time dependency of the passage of drugs into red blood cells provides a strong argument for the existence of temporal variations in the passage of drugs through biological membranes for which red blood cells are often used as a model.

Finally, a particular membrane, namely the blood-brain barrier, must also be considered because centrally acting drugs must pass through this membrane.

To our knowledge, there are only two studies on temporal variations of blood-brain barrier permeability: an experimental study by Mato et al. (1981), who assessed temporal changes in the passage of horseradish peroxidase in rats, and a study by Lockard et al. (1985), who established circadian variations of valproic acid concentration in the CSF of monkeys correlated to the plasma-free level of the drug.

## Temporal Variations in Drug Metabolism

The main site of drug metabolism is the liver but many other tissue may also be implicated such as lungs, kidneys, blood and digestive tract. A general rule is that lipid-soluble drugs must be metabolized to more polar compounds before being excreted in the urine. Drugs undergo four main types of reaction in the body: oxidation (hydroxylation, N- or O-dealkylation, deamination) generally mediated by enzymes of the microsomal fraction of the liver (cytochrome P450), reduction, hydrolysis and conjugation such as glucuronoconjugation, sulfoconjugation, methylation, acetylation, etc.

Drug metabolism is generally assumed to depend on liver enzyme activity or hepatic blood flow: for some drugs such as lidocaine and propranolol (high hepatic extraction coefficient) the metabolism depends on the hepatic blood flow. For others the biotransformations essentially depend on the enzymatic activity of the hepatocytes.

Many factors may affect these biotransformations: temporal variations are of particular interest since they have been described for many years and used to explain chronopharmacokinetic changes. Several chronobiological studies were devoted to the tempo-

ral changes in liver enzyme activity by direct measurement or indirect evidence by measuring the chronokinetics of drugs and their metabolites.

### Temporal Variations in Enzyme Activity

Circadian or ultradian rhythms of liver drug metabolism activity were shown in vitro for the oxidative metabolism of drugs such as aminopyrine, $p$-nitro-anisol, hexobarbital (Radzialowsky and Bousquet 1968; Jori et al. 1971; Nair and Casper 1969) or 4-dimethylaminoazobenzene (Radzialowsky and Bousquet 1968) with a peak located during the dark (activity) period. The different enzymatic reactions such as $N$-demethylation (aminopyrine), hydroxylation (hexobarbital) or demethylation ($p$-nitroanisol) have been shown to be in phase.

The circadian rhythm of hexobarbital oxidase activity demonstrated by Nair (1974) was light/dark cycle dependent since abolition of the rhythmicity was observed in constant light or darkness. After adrenalectomy, Radzialowsky and Bousquet (1978) documented a disappearance of the circadian rhythm in drug metabolism of the previously cited drugs. Thus the daily rhythm of oxidative metabolism seems to be partially regulated by the adrenal gland. However, Jori et al. (1971) suggested the intervention of a central hormonal factor since Nair and Casper (1969) also found that hexobarbital oxidase activity is reduced by hypophysectomy or hypothalamus lesions in rats.

The oxidative reactions catalyzed by the cytochrome P450 system are the major pathway involved in drug metabolism: the temporal changes in the hepatic P450 system have been investigated by many authors and were reviewed by Belanger (1988).

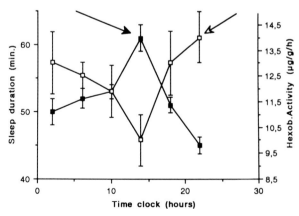

**Fig. 7.** Inverse relationship between temporal variations of the hexobarbital-induced sleep duration (*left*) and the hepatic hexobarbital oxidase activity (*right*) in rats. (From Nair 1974)

Many other rhythmic variations in enzymatic activity were documented and were reviewed by Belanger (1988) and by Feuers and Scheving (1988).

These circadian changes in metabolic pathways may be responsible for many instances of variation in drug response. Thus an interesting inverse relationship was found between the hepatic hexobarbital oxidase activity and the hexobarbital-induced sleep duration; as shown in Fig. 7 the maximal hepatic hexobarbital activity (2200 hours) corresponds to the minimal sleep duration in rats. This example is of particular interest since it underlines the temporal relationships between pharmacodynamic and pharmacokinetic parameters.

### Temporal Variations in Hepatic Blood Flow

For drugs with a high extraction ratio (lidocaine, propranolol, etc.) the hepatic metabolism depends on the hepatic blood flow. Circadian variations in hepatic blood flow induce changes in liver perfusion and thus temporal variations in the clearance of drugs. To our knowledge circadian variations of hepatic blood flow have not yet been documented in man; in rats, Dore et al. (1984) demonstrated circadian variations of hepatic blood flow, which is maximal during the dark phase. The temporal variations

of lidocaine kinetics (Bruguerolle et al. 1984) or bupivacaine kinetics (Fig. 8) were thought to be caused by circadian variations in hepatic blood flow.

Klotz and Ziegler (1982) reported on circadian variations in hepatic clearance of a benzodiazepine with a high extraction ratio, midazolam, in man: the plasma clearance of this drug was reported to be higher during the morning.

Lemmer (1981 a) demonstrated in the rat temporal variations in propranolol plasma clearance (higher values at the end of the activity span). The clearance of propranolol depends upon the liver blood flow which has been shown to be circadian periodic in these animals.

Finally, it appears from studies by Belanger et al. (1981) that temporal variations in hepatic clearance may be detected for drugs having a high hepatic extraction ratio (paracetamol, antipyrine, etc.) but not for those with a low ratio.

### Indirect Evidence of Temporal Variations in Drug Metabolism

Numerous experimental and clinical chronopharmacological studies have documented chronokinetics. Some of them have indirectly investigated temporal variations in hepatic drug metabolism capacity by

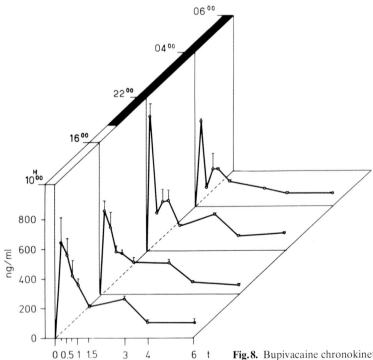

**Fig. 8.** Bupivacaine chronokinetics in mice. (From Bruguerolle and Prat 1987)

demonstrating chronokinetics of drugs and their metabolites; even if a metabolite is not pharmacologically active, its measurement indicates the rate of metabolizing capacity.

*Studies on Conjugations.* We reported chronokinetics of procainamide and its main metabolite, *N*-acetylprocainamide, in rats (Bruguerolle and Jadot 1985); the elimination of these two compounds is circadian time dependent with maximal elimination during the dark (activity) phase. These data have indicated a circadian rhythm of acetylation which is maximal in the rat at 0400 hours. Also glycuronoconjugation in man seems to be time dependent since ketoprofen and its glucuronoconjugate metabolite kinetics vary with time of day as demonstrated by Queneau et al. (1984).

*Hydrolysis.* Decarboxylation and hydrolysis of dipotassic chlorazepate leads to *N*-desmethyldiazepam, which is also the main metabolite of diazepam; several chronokinetics studies have been carried out with this drug which are in agreement. In man, Naranjo et al. (1980) demonstrated higher diazepam and *N*-desmethyldiazepam levels in the morning (0900 hours). Nakano et al. (1984) confirmed these data in man by demonstrating higher diazepam concentrations after morning administration.

*Oxidations.* Circadian variations of hydroxylation reactions in man were illustrated by Nakano and Hollister (1978), who measured nortriptiline (NT) and its major metabolite, 10-hydroxynortriptiline (10-NT) the plasma concentrations of which were found to be higher after morning intake.

*Demethylations.* Studies on chronokinetics of a sustained release form of indomethacin have led to similar results in man (Guissou et al. 1983; Bruguerolle et al. 1983): when given at 2000 hours indomethacin plasma levels remain much more stable than at 0800 hours or 1200 hours when plasma levels were found to be higher. Plasma *O*-desmethylindomethacin levels were found to be significantly higher after evening administration, indicating daily changes in the rat of hepatic demethylation of the parent drug.

Temporal Variations in Drug Excretion

Most drugs are eliminated by the renal route. Circadian rhythmicity of major renal functions, i.e., glomerular filtration, renal blood flow, urinary pH, tubular resorption, etc., has been documented by Cambar et al. (1979). Thus the urinary excretion of many drugs may depend on these rhythmic variations (Jones 1845).

One of the first studies in man concerning the renal excretion of drugs was reported by Beckett and Rowland (1964), who described the rhythmic renal excretion of amphetamine with a maximum at the beginning of the day dependent on pH variations (the acrophase of urinary pH in man occurs in the morning).

The physicochemical properties of drugs are of particular importance in this field since renal elimination depends partially on the ionization of drugs and thus may be modified by temporal changes in urinary pH. Related to these variations, acidic drugs such as sodium salicylate (Reinberg et al. 1975) and sulfasymazine (Dettli and Spring 1966) are excreted faster after an evening than a morning administration. In contrast, the renal excretion of sulfanilamide, another basic sulfamide ($pK_a = 10.5$), does not vary along the 24-h scale (Dettli and Spring 1966). The renal excretion of many other drugs has been shown to vary with time: as an example of the relationships between chronokinetics and chronotoxicity of drugs, studies by Levi et al. (1981) on an anticancer drug, *cis*-DDP, have shown an increased elimination of this drug when administered at 0600 hours in man corresponding to the time of higher nephrotoxicity of *cis*-DDP (Hrushesky et al. 1982).

Specific Methodological Aspects of Chronokinetics

All the previously mentioned studies clearly demonstrate that the time of administration of a drug is an important source of variation which must be taken into account in kinetic studies. The main findings of chronokinetic studies in man or animals for different groups of drugs, involving temporal variations in one or several stages of the fate of these drugs in the organism, were reviewed previously (Reinberg and Smolensky 1982; Lemmer 1981b; Bruguerolle 1983a, 1987c; Levi et al. 1989).

As an example Tables 1–3 illustrate the chronokinetics of anesthetic agents, drugs acting upon the cardiovascular system and antiinflammatory drugs.

*Predictability of a Temporal Kinetic Variation*

From the results of all these investigations it is not easy to define a specific chronokinetic pattern for drugs of the same chemical group. The predictability

**Table 1.** Chronopharmacokinetic changes of some general or local anesthetics in animals or in man

| Drugs | Species | Dosing/route | Major observation | Refs. |
|---|---|---|---|---|
| Bupivacaine | Mice | 20 mg/kg i.p. Sd | Highest $C_{max}$, $C_{max}/T_{max}$ ratio at 2200 hours, longest $T_{1/2}\beta$ at 2200 hours | Bruguerolle and Prat (1987) |
| Bupivacaine | Man | Peridural constant rate infusion during 36 h. 0.25 mg/kg/h | Circadian variations of bupivacaine plasma level in spite of a constant rate of infusion. Max. Cl at 0630 hours | Bruguerolle et al. (1988b) |
| Etidocaine | Mice | 40 mg/kg/i.p. Sd | Highest $C_{max}$, $C_{max}/T_{max}$ ratio at 0400 hours | Bruguerolle and Prat (1990) |
| Halothane | Rat | Inhalation of 28% cyclopropane | Minimum alveolar concentration highest during 10 min at 2000 hours | Munson et al. (1970) |
| Hexobarbital | Man | 500 mg p.o. | Highest $C_{max}$ and shortest $T_{max}$ at 0200 hours | Altmayer et al. (1979) |
| Lidocaine | Rat | 50 mg/kg i.p. Sd | Highest plasma levels at 1600 hours. Shortest $T_{1/2}\beta$ at 1600 hours | Bruguerolle et al. (1982) |
| Lidocaine | Man | 0.65 mg/kg Inj. Sd | Highest area under plasma concentration curve at 1530 hours | Bruguerolle and Isnardon (1985) |
| Mepivacaine | Mice | 60 mg/kg i.p. Sd | Highest $C_{max}/T_{max}$ ratio and $V_d$ at 2200 hours Longest $T_{1/2}\beta$ at 2200 hours | Bruguerolle and Prat (1988) |
| Pentobarbital | Mice | 86.5 / 78.7 / 71.5 / 65 mg/kg i.p. Sd | Higher brain pentobarbital concentration at 2000 hours | Nelson and Halberg (1973) |

Sd, single dose; i.p., intraperitoneal route; p.o., per os; $C_{max}$, maximal plasma concentration; $T_{max}$, maximal time to reach $C_{max}$; $T_{1/2}\beta$, elimination half-life; Cl, plasmatic clearance; $V_d$, volume of distribution

of the amplitude and the pattern of temporal variations in drug kinetics has been looked at by several authors.

Belanger et al. (1981) have demonstrated in rats that some physicochemical properties such as liposolubility or hydrosolubility may influence the chronokinetic pattern of a given drug. They have shown that drugs with low water solubility such as indomethacine, furosemide and phenylbutazone exhibit a circadian variation of absorption while this is not the case for water-soluble drugs such as antipyrine, hydrochlorothiazide or paracetamol. In contrast, these drugs have been shown to have a time-dependent clearance.

Lemmer and Neuman (1987) have pointed out the importance of lipophilicity for the circadian phase dependency of the kinetics of different $\beta$-receptor blocking drugs (e.g., there is a positive correlation between drug accumulation in rat brain and the lipophilicity of three different $\beta$-blocking agents).

Finally, we recently emphasized the role of lipophilicity, the degree of protein binding and the molecular weight of three amide-type local anesthetics in determining their respective plasma, heart and brain levels (Prat and Bruguerolle 1988, Bruguerolle and Prat 1990).

*Methodological Aspects and Chronokinetic Studies*

Certain specific methodological points must be emphasized before starting a chronokinetic study.

*Subjects.* Obviously, as in any other chronobiological work the synchronization of the subjects must be controlled.

The state of the subjects or patients (healthy or ill) participating in a chronokinetic study may influence its conclusions; temporal variations of biological rhythms in the organism are known to be modified in illness and thus may interfere with drug chronokinetics. Thus the kinetics of diphenylhydantoin are different in epileptic women and varies in catamenial epileptic women according to the stage of the menstrual cycle.

We reported recently (Bruguerolle et al. 1986b; Focan et al. 1988) alterations of the circadian time structure of plasma proteins in patients with cancer and with inflammatory disorders; such variations may modify drug protein binding and thus have kinetic implications.

*Necessity for Several Time Points.* To study the chronokinetics of a drug, several time points (times of administration) are needed. Limiting a study to only

**Table 2.** Chronopharmacokinetic changes of cardiovascular drugs in animals or in man

| Drugs | Species | Variable | Major observation | Refs. |
|---|---|---|---|---|
| Digoxin | Elderly men | $C_{max}$, $T_{max}$, AUC | $C_{max}$ higher at 0900 hours | Bruguerolle et al. (1988a) |
| Dipyridamole | Adult healthy men | Plasma levels Urinary excretion | Maxi bioavailability at 0600 hours Mini urinary excretion between 1800 and 0200 hours | Markiewicz and Semenowicz (1980) |
| Disopyramide | Mice | Cl, Vd, AUC $T_{1/2}\beta$ | Highest $V_d$, Cl at 2200 hours highest AUC at 0400 hours | Bruguerolle (1984b) |
| Isosorbide dinitrate | Adult healthy men | $C_{max}$, $T_{max}$, AUC, $C_{max}/T_{max}$ | Highest AUC at 0200 hours No diff. with Sr form | Lemmer et al. (1989) |
| Lidocaine | Rat | Plasma and RBC levels | Maxi $T_{1/2}\beta$:1000 hours Maxi AUC:1600 hours Maxi Vd:0400 hours | Bruguerolle et al. (1982) |
| Methyldigoxin | Adult patients | Plasma levels, AUC | Maxi AUC at 1600 hours second peak AM | Carosella et al. (1979) |
| Nitrindipine | Adult healthy subjects | $C_{max}$, $T_{max}$, $T_{1/2}\beta$, AUC Sd | $C_{max}$ higher at 0900 hours | Fujimura et al. (1989b) |
| Nitrindipine | Adult healthy subjects. 8 days treatment | | No significant differences | Lecocq et al. (1991) |
| Procainamide NAPA | Rat | Plasma levels | $T_{1/2}\beta$ shorter at 2200 hours N-acetylation maxi at 0400 hours | Bruguerolle (1984) |
| Procainamide | Adult men | Plasma levels | No significant differences in women | Fujimura et al. (1989a) |
| Propranolol | Rat | Plasma, heart, brain levels | $T_{1/2}\beta$ shortest at night | Lemmer et al. (1981a) |
| Metoprolol, Sotalol, Atenolol | Rat | Plasma, heart, brain levels Sd and repeated dosing | $T_{1/2}\beta$ shortest at night No significant differences after Rd | Lemmer et al. (1985) |
| Propranolol | Adult subjects | $C_{max}$, $T_{max}$, AUC $C_{max}/T_{max}$ $T_{1/2}$ | $C_{max}$, AUC and $C_{max}/T_{max}$ highest at 0800 hours $T_{max}$ and $T_{1/2}$ lowest at 0800 hours | Langner and Lemmer (1988) |

$C_{max}$, maximal plasma concentration; $T_{max}$, maximal time to reach $C_{max}$; $T_{1/2}\beta$, elimination half-life; Cl, plasmatic clearance; $V_d$, volume of distribution; AUC, area under plasma concentration curve; Sd, single dose; i.p., intraperitoneal route; p.o., per os; Sr, sustained release form

two time points may lead to missing a temporal variation which may be detected only by choosing more time points (Fig. 9).

If only two times of administration are possible, the choice of these time points must be determined according to a preliminary experiment or according to pertinent information such as a certain phase relation to the peak or trough time of a biological marker rhythm (Reinberg et al. 1990).

*Conditions of Food Intake and Posture.* The importance of food intake on temporal variations of drug absorption have been discussed before and were recently emphasized by Reinberg et al. (1990). We would also like to underline the necessity of controlling feeding habits in chronokinetics studies.

Posture has also been found to be of a great importance in chronokinetic studies the results of which are often very different between resting and activity

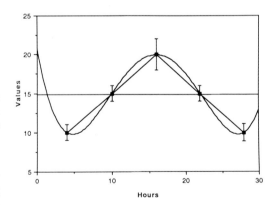

**Fig. 9.** Theoretical example of a circadian variation: choosing two time points (in the morning and in the evening, i.e., 0900 hours and 2100 hours) will show no significant difference; in contrast, choosing three or four time points (e.g., 0900, 1500 and 2100 or 0900, 1500, 2100 and 0300 hours) will show a significant difference

**Table 3.** Chronopharmacokinetic changes of antiinflammatory drugs in man

| Drug | Subjects | Hours of administration | Major findings | Refs. |
|------|----------|------------------------|----------------|-------|
| Indomethacin | $n = 9$ Healthy | 0700, 1100, 1500, 1900, 2300 | Highest $C_{max}$ at 0700–1100 hours longest $T_{1/2}\beta$ at 0900 hours | Clench et al. (1981) |
| Indomethacin SR | $n = 16$ Osteoarthritis | 0800, 1200, 2000 | Highest $C_{max}$ at 1200 hours, highest AUC and longest $T_{1/2}\beta$ at 2000 hours | Guissou et al. (1983) Bruguerolle et al. (1983) |
| Indomethacin SR | $n = 10$ Elderly | 0900, 1200, 2100 | $T_{max}$ higher at 2100 hours | Bruguerolle et al. (1986a) |
| Ketoprofen | $n = 10$ Healthy | 0700, 1300, 1900, 0100 | Highest $C_{max}$ and AUC at 0700 hours, longest $T_{1/2}\beta$ at 0100 hours | Queneau et al. (1984) |
| Ketoprofen SR | $n = 10$ Healthy | 0700, 1900 | Shortest $T_{max}$ at 0700 hours | Reinberg et al. (1986) |
| Ketoprofen constant infusion | $n = 8$ hospitalized for sciatica | Continuous i.v. during 24 h, sampling every 2 h | Variations of plasma levels with a peak at 2100 hours | Decousus et al. (1987b) |
| Pranoprofen | $n = 7$ Healthy | 1000, 2200 | $T_{max}$ shorter at 1000 hours | Fujimura et al. (1989a) |
| Salicylic acid | $n = 6$ Healthy | 0600, 1000, 1800, 2200 | Highest $C_{max}$ AUC and $T_{1/2}\beta$ at 0600 hours | Markiewicz and Semenowicz (1979) |
| Salicylic acid | $n = 6$ Healthy | 0700, 1100, 1500, 1900, 2300 | Urinary elimination longest at 0700 hours | Reinberg et al. (1975) |
| Sulindac | $n = 10$ Healthy | 0800, 2000 | No significant difference for the parent drug nor for its metabolites | Couet et al. (1986) |

$C_{max}$, maximal plasma concentration; $T_{max}$, maximal time to reach $C_{max}$; $T_{1/2}\beta$, elimination half-life; Cl, plasmatic clearance; Vd, volume of distribution; AUC, area under plasma concentrations curve; Sr, sustained release form

and which may influence some factors involved in kinetic processes. As an example, a 60% difference in hepatic blood flow has been found in man between standing and recumbent positions. Thus, posture must be also strictly controlled in kinetic studies.

*Single or Repeated Dosing and Constant Rate Delivery of Drugs.* Most chronokinetic studies have been carried out by comparing the kinetics of a drug taken at different time points and after a single dose intake. One can argue that such variations may disappear when repeated doses are applied; some chronokinetic investigations have been carried out with repeated doses (theophylline, diazepam, sodium valproate, etc.) and have nevertheless demonstrated significant temporal variations.

Drug delivery at a constant rate is supposed to produce constant blood levels. Recent chronokinetic studies on anticancer agents (5-fluorouracil, doxorubicin, vindesine) (Levi et al. 1986; Focan et al. 1989; Petit et al. 1988), antiinflammatory agents (ketoprofen, Decousus et al. 1987b) heparin (Decousus et al. 1985) and local anesthetics (bupivacaine, Bruguerolle et al. 1988b) demonstrated that continuous intravenous infusion does not lead to constant plasma levels but results in large-amplitude circadian changes. In man, in spite of a continuous (36 h) constant rate infusion by the peridural route, the bupivacaine plasma clearance varied with the 24-h scale with a maximum at 0600 hours (Bruguerolle et al. 1988).

Programmable implanted pumps in contrast allow the infusion rate to be varied in a sinusoidal fashion. Such findings have been applied to anticancer drugs. Levi et al. (1986) for instance demonstrated that a constant delivery rate of Adriamycin (doxorubicin hydrochloride) in patients suffering from advanced breast cancer resulted in temporal variations in Adriamycin plasma concentrations. In contrast when the same 24-h dosage was delivered by an infusion rate varying in a sinusoidal fashion, the temporal changes in plasma concentrations followed the pump delivery pattern. Thus it was possible to modulate the circadian infusion rate in order to give more drug when it was best tolerated.

*Sustained Release Forms.* Recent findings by Lemmer et al. (1989) on chronokinetics of nitrates and nifedipine (Lemmer et al. 1990) have demonstrated the influence of the galenic presentation on the detection and the amplitude of chronokinetic changes. Also, Smolensky (1989) reviewed recently the main studies concerning chronokinetics of theophylline and particularly the influence of the galenic form (sustained release forms).

*Kinetic Modelization.* As mentioned before the different steps in the fate of a drug in the organism are expressed as kinetic parameters by calculation of a theoretical mathematical model. The time of administration of a drug in such models is generally not taken into account. Hecquet and Sucche (1986) have proposed different models integrating the time of administration as a supplementary factor involved in kinetic processes.

Temporal variations are responsible for the nonlinearity of kinetics and must thus be taken into account (as many other factors contributing to nonlinearity).

The interesting point of chronokinetic studies is that they control a factor (the time of administration) which, among others, is responsible for variations in drug kinetics but they also try to explain some chronopharmacological effects. In most cases chronopharmacological effects are explained at least in part by chronokinetics but there are also some examples describing the opposition in phase of chronopharmacodynamic and chronokinetic findings.

To whatever extent correlations between effects and kinetics are based on a relationship between blood levels and pharmacological effect, this relationship depends on the time of administration of drugs; as an example theophylline plasma levels correlate well with the bronchodilatatory effects in asthmatic patients only when this drug is taken in the evening (Bruguerolle et al. 1987).

## Therapeutic Applications of Chronopharmacology: Chronotherapy

The application of chronobiological and chronopharmacological data to the treatment of illness leads to chronotherapy.

### Definition and Aims

If chronopharmacology studies the influence of the time of administration of a drug on its kinetics and its mode of action, chronotherapy is concerned with the influence of the timing of a treatment on its efficiency in patients. Obviously such an approach must integrate chronopharmacological and chronopathological information. Chronotherapy leads to a better use of drugs in the treatment of diseases and aims to improve the efficiency and/or the tolerance of drugs by the timing of their administration. This involves specific methodological points (clinical trials).

Chronotherapeutic studies are numerous at present and will be discussed in other chapters of this book together with the different areas of pathology such as corticotherapy, cancerology, rheumatology, hematology and pneumology. Some examples are presented here in order to illustrate the present state and the future of chronotherapeutics.

### Examples

#### Corticotherapy

The timing of corticosteroids therapy has been studied by Reinberg et al. (1990), Ceresa and Angeli (1977) and others. In adrenal insufficiency Reinberg et al. (1971) demonstrated 30 years ago that the best tolerance is obtained when two-thirds of the daily dose of steroids are taken in the morning in phase with the physiological secretion of the endogenous steroids; such a temporal distribution respects the temporal structure of the organism and thus avoids undesired side effects such as asthenia. In contrast when one wants to inhibit adrenal secretions for diagnostic purposes, steroids (i.e., dexamethasone) must be administered at 2300 hours. When corticosteroids are used as therapeutic agents in asthma for example, it has been shown that the best temporal distribution of steroids (as judged by the improvement of the peak expiratory flow) consists of two-thirds of the daily dosage in the morning and the remaining one-third in the afternoon (Reinberg et al. 1983).

#### Cancerology

One of the most impressive examples of chronotherapy concerns the chronotherapy of cancer. Anticancerous drugs are generally very potent and also very toxic, with a narrow therapeutic range. Thus, in cancerology, one of the first objectives of therapeutics is to ameliorate the tolerance of such toxic drugs. Obviously the choice of the time of administration of these drugs may modulate tolerance. Cancer chronotherapy well illustrates the importance and the perspectives of chronotherapeutics. The toxicity and efficiency of numerous anticancerous drugs have been shown to be time dependent (Levi et al. 1980 a, b; Hrusheski et al. 1980) both in animals and in man. For instance, Levi et al. (1981) have demonstrated that *cis*-platinum is best tolerated (gastrointestinal, renal and hematologic tolerance) in patients suffering from advanced cancers of either the bladder or ovary when administered in the evening compared to a morning administration.

A drug delivered at a constant rate over 24 h may cause wide variations in kinetics and effects; thus, drug delivery may be programmed in time in order to ameliorate tolerance and efficiency by enhancing the desired effects and reducing the undesired ones. Such a chronoptimization is now possible with chronobiological tools such as programmable infusion pump devices which allow the drug delivery to be programmed in time. Multicenter clinical trials with *cis*-platinum, Adriamycin, and other agents are presently under investigation.

Hrushesky discusses in a chapter of this book many chronopharmacological and chronotherapeutic examples in cancer chronotherapy.

### Rheumatology

Taking into account chronokinetic and chronopharmacological data on anti-inflammatory drugs, some clinical trials have been carried out on rheumatic patients.

Four multicenter clinical trials by Levi et al. (1985) were carried out in more than 500 osteoarthritis patients with a sustained release form of indomethacin given at 0800, 1200, or 2000 hours. Circadian variations were observed in the side effects of indomethacin with 32% of undesired side effects related to the morning ingestion compared to 7% when indomethacin was ingested in the evening. The evening administration was most effective in subjects with predominant nocturnal or morning pain and morning or noon ingestion was more effective in patients with pain predominantly during afternoon or evening.

More recently, Decousus et al. (1990) carried out a double-blind clinical trial in 114 outpatients with osteoarthritis of the hip or knee treated for 14 days with ketoprofen: patients stratified according to the circadian variation of self-rated pain intensity received ketoprofen in the morning or in the evening. No statistically significant differences between morning and evening intake were found concerning the efficiency

of this drug, but the tolerance was twice as high for the evening group as compared to the morning group. Thus evening administration of nonsteroidal antiinflammatory drugs in rheumatic patients seems to be better tolerated.

### Hematology

Platelet aggregation and the tendency for blood coagulation in man has been found to be increased in the morning when the fibrinolytic activity is lower. Decousus et al. (1985) demonstrated that the anticoagulant effect of heparin is at its minimum in the morning following the physiological temporal pattern of blood coagulation. These authors have suggested that the heparin doses should be modulated as a function of administration times in order to increase their efficiency and to minimize both bleeding risk and thrombosis (Decousus et al. 1987 a).

### Pneumology

For diseases whose symptoms vary over the 24-h span, the requirement for medications may be totally different during the daytime versus the nighttime.

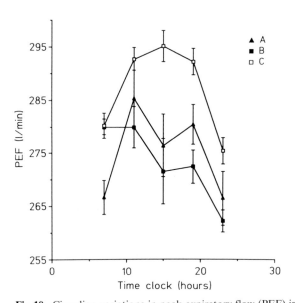

**Fig. 10.** Circadian variations in peak expiratory flow (PEF) in chronic obstructive pulmonary disease. The patients were treated by unequal twice daily sustained release theophylline administration; when two-thirds the total daily dose (C) was taken in the evening, PEF values were higher compared to one-third the total daily dose in the evening (A) or one-half the total daily dose in the evening (B). (From Bruguerolle et al. 1987)

Due to the nocturnal fall in bronchial patency, an unequal dosing schedule consisting of a greater intake of bronchopulmonary drugs and bronchodilators in the evening may be more beneficial than the conventional equal-interval, equal-dosing pattern.

Among other factors (a special issue of *Chronobiology International* was devoted to this topic) we have shown that theophylline given twice daily in unequal doses as sustained release preparation may be most beneficial in chronic obstructive pulmonary disease when two-thirds of the total daily dose is taken in the evening (Bruguerolle et al. 1987). With this dosage schedule peak expiratory flow values are significantly higher than in other regimens and the acrophase of the rhythms is maintained (Fig. 10).

## Conclusions

Future progress in chronopharmacology may be considered to depend on two different aspects:
- Advances in chronopharmacology and chronotherapeutics will obviously depend on the progress in medicine and biology (i. e., new drugs, new drug delivery devices, more specific and precise drug assay techniques, better understanding of drug kinetics and dynamics at the cellular and molecular levels, etc.).
- Future developments will also directly depend on specific progress in the field involving research of chronokinetic and chronodynamic mechanisms, development of multicenter clinical trials with specific and strict methodology, etc. Practical application of chronotherapeutic findings is at present possible with tools such as programmable infusion pump devices which allow drug delivery to be programmed in time: such applications are already used in the treatment of cancers, diabetes, sterility, pain, etc.

Chronopharmacology and chronotherapeutics should lead to a better use of drugs in the future through the choice of the best time of administration in order to optimize therapy, i.e., to enhance the desired effects and decrease the undesired side effects.

## References

Altmayer P, Groterath E, Lucker PW, Mayer D, von Mayersbach H (1979) Zirkadiane Schwankungen pharmakologischer Parameter nach oraler Gabe von Hexobarbital. Arzneimittelforschung 29: 1422–1428

Bauer LA, Davis R, Wilensky A, Raisys V, Levy RH (1985) Valproic acid clearance: unbound fraction and diurnal variation in young and elderly adults. Clin Pharmacol Ther 37: 697–700

Beckett AH, Rowland M (1964) Rhythmic urinary excretion of amphetamine in man. Nature 204: 1203–1204

Belanger PM (1988) Chronobiological variation in the hepatic elimination of drugs and toxic chemical agents. Annu Rev Chronopharmacol 4: 1–46

Belanger PM, Labrecque G, Dore F (1981) Rate limiting steps in the temporal variations in the metabolism of selected drugs. Int J Chronobiol 7: 208–215

Bruguerolle B (1976) Chronopharmacologie des substances curarisantes. Thesis, University of Marseille

Bruguerolle B (1983a) Influence de l'heure d'administration d'un médicament sur sa pharmacocinétique. Therapie 38: 223–235

Bruguerolle B (1983b) Cycle menstruel et pharmacocinétique des médicaments. J Gynecol Obstet Biol Reprod (Paris) 12: 825–827

Bruguerolle B (1984a) La chronopharmacologie. Ellipses, Paris

Bruguerolle B (1984b) Circadian phase dependent pharmacokinetics of disopyramide in mice. Chronobiol Int 1: 267–271

Bruguerolle B (1987a) Modifications de la pharmacocinétique de la theophylline au cours du cycle oestral chez la ratte. Pathol Biol 35: 181–183

Bruguerolle B (1987b) Erythrocyte concentrations of theophylline: influence of the oestrous cycle. Med Sci Res (Biochem) 15: 263–264

Bruguerolle B (1987c) Données récentes en chronopharmacocinétique. Pathol Biol 35: 925–934

Bruguerolle B (1989) Temporal aspects of drug absorption. In: Lemmer B (ed) Chronopharmacology cellular and biochemical interactions. Dekker, New York, pp 3–13

Bruguerolle B, Isnardon R (1985) Daily variations in plasma levels of lidocaine during local anaesthesia in dental practice. Ther Drug Monit 7: 369–370

Bruguerolle B, Jadot G (1983) Influence of the hour of administration of lidocaïne on its intraerythrocytic passage in the rat. Chronobiologia 10: 295–297

Bruguerolle B, Jadot G (1985) Circadian changes in procainamide and *N*-acetylprocainamide kinetics in the rat. J Pharm Pharmacol 37: 654–656

Bruguerolle B, Prat M (1987) Temporal changes in bupivacaine kinetics. J Pharm Pharmacol 39: 148–149

Bruguerolle B, Prat M (1988) Circadian phase dependent pharmacokinetics and acute toxicity of mepivacaine. J Pharm Pharmacol 40: 592–594

Bruguerolle B, Prat M (1989) Temporal variations in the erythrocyte permeability to bupivacaine, etidocaine and mepivacaine in mice. Life Sci 45: 2587–259

Bruguerolle B, Prat M (1990) Circadian phase dependent acute toxicity and pharmacokinetics of etidocaine in serum and brain of mice. J Pharm Pharmacol 42: 201–202

Bruguerolle B, Mesdjian E, Jadot G, Valli M, Agopian B, Bouyard P (1976) Variations de l'activité de diverses substances curarisantes en fonction de l'heure d'administration. Ann Anesthesiol Fr 16: 349–353

Bruguerolle B, Mesdjian E, Valli M, Jadot G, Agopian B, Bouyard P (1977) Etude chronopharmacologique du bromure de pancuronium chez le rat. J Pharmacol 8: 49–58

Bruguerolle B, Mesdjian E, Valli M, Blanc MC, Jadot G, Vignon E, Bouyard P (1978a) Etude chronopharmacologique de l'alfatésine chez le rat mâle. J Pharmacol 9: 53–64

Bruguerolle B, Valli M, Jadot G, Rakoto IC, Bouyard P (1978b) Chronopharmacologie du pancuronium chez le rat anesthésié à l'aide de CT1341(alfatésine). C R Soc Biol [D] (Paris) 172: 498–504

Bruguerolle B, Valli M, Rakoto JC, Jadot G, Bouyard P, Reinberg A (1979) Chronopharmacology of pancuronium in the rat: anaesthesia or seasonal influence? In: Reinberg A, Halberg F (eds) Chronopharmacology. Pergamon, Oxford, pp 117

Bruguerolle B, Valli M, Jadot G, Bouyard L, Bouyard P (1981) Circadian effect on carbamazepine kinetics in the rat. Eur J Drug Metab Pharmacokinet 6: 189–194

Bruguerolle B, Jadot G, Valli M, Bouyard L, Bouyard P (1982) Etude chronocinétique de la lidocaïne chez le rat. J Pharmacol 13: 65–76

Bruguerolle B, Desnuelle C, Jadot G, Valli M, Acquaviva PC (1983) Chronopharmacocinétique de l'indométhacine à effet prolongé en pathologie rhumatismale. Rev Int Rhumatisme 13: 263–267

Bruguerolle B, Bouvenot G, Bartolin R (1984) Temporal and sex-related variations in lorazepam kinetics. Annu Rev Chronopharmacol 1: 21–24

Bruguerolle B, Barbeau G, Belanger P, Labrecque G (1986a) Chronokinetics of indomethacin in elderly subjects. Annu Rev Chronopharmacol 3: 425–428

Bruguerolle B, Levi F, Arnaud C, Bouvenot G, Mechkouri M, Vannetzel J, Touitou Y (1986b) Alteration of physiologic circadian time structure of six plasma proteins in patients with advanced cancer. Annu Rev Chronopharmacol 3: 207–210

Bruguerolle B, Philip-Joet F, Parrel M, Arnaud A (1987) Unequal twice daily sustained release theophylline dosing in chronic obstructive pulmonary disease. Chronobiol Int 4: 381–385

Bruguerolle B, Bouvenot G, Bartolin R, Manolis J (1988a) Chronopharmacocinétique de la digoxine chez le sujet de plus de soixante dix ans. Therapie 43: 251–253

Bruguerolle B, Dupont M, Lebre P, Legre G (1988b) Bupivacaine chronokinetics in man after a peridural constant rate infusion. Annu Rev Chronopharmacol 5: 223–226

Bruguerolle B, Prat M, Douylliez C, Dorfman P (1988c) Are there circadian and circannual variations in acute toxicity of phenobarbital in mice? Fundam Clin Pharmacol 2: 301–304

Bruinink A, Lichtensteiger W, Schlumpf M (1983) Ontogeny of diurnal rhythms of central dopamine, serotonin and spirodecanone binding sites and of motor activity in the rat. Life Sci 33: 31–38

Cambar J, Lemoigne F, Toussaint C (1979) Diurnal variations: evidence of glomerular filtration in the rat. Experientia 35: 1607–1608

Carlebach R, Cohen-Kitay Y, Peleg L, Ashkenazi IE (1990) Chronotoxicity and genetic variability. Annu Rev Chronopharmacol 7: 241–244

Carosella L, Dinardo P, Bernabei R, Cocchi A, Carbonin P (1979) Chronopharmacokinetics of digitalis. Circadian variations of β-methyldigoxin serum levels after oral administration. In: Reinberg A, Halberg F (eds) Chronopharmacology. Pergamon, Oxford, p 125

Ceresa F, Angeli A (1977) Chronotherapie corticoide. In: Chronothérapeutique. Expansion Scientifique Française (ed.), Paris, pp 211–223

Clench J, Reinberg A, Dziewanowska Z, Ghata J, Smolensky M (1981) Circadian changes in the bioavailability and effects of indomethacin in healthy subjects. Eur J Clin Pharmacol 20: 359–369

Couet W, Ingrand I, Millerioux L, Lefebvre MA, Mainguy Y, Fourtillan JB (1986) Etude chronopharmacocinétique du sulindac et de ses métabolites. Sem Hop Paris 62: 2677–2682

Cuisinaud G, Guissou P, Sassard J (1984) Chronopharmacokinetic study of intravenous indomethacin in the rat. Annu Rev Chronopharmacol 1: 341–344

Decousus H, Croze M, Levi F, Perpoint B, Jaubert J, Bonnadona JF, Reinberg A, Queneau P (1985) Circadian changes in anticoagulant effect of heparin infused at a constant rate. Br Med J 290: 341–344

Decousus H, Ollagnier M, Jaubert J, Perpoint B, Hocquart J, Queneau P (1987a) Biological rhythms and thromboembolic disease. Physiologic, epidemiologic and pharmacologic aspects. Pathol Biol 35: 985–990

Decousus H, Ollagnier M, Cherrah Y, Perpoint B, Hocquart J, Queneau P (1987b) Chronokinetics of ketoprofen infused intravenously at a constant rate. Annu Rev Chronopharmacol 3: 321–324

Decousus H, Perpoint B, Boissier C, Ollagnier M, Mismetti P, Hocquart J, Queneau P (1990) Timing optimizes sustained-release ketoprofen treatment of osteoarthritis. Annu Rev Chronopharmacol 7: 289–292

Descorps-Declere A, Bonnafous M, Reinberg A, Begon C (1983) Sex-related differences in diurnal and seasonal changes in effectiveness of pancuronium bromide in man. Chronobiologia 10: 121–122

Dettli L, Spring P (1966) Diurnal variations in the elimination rate of a sulfonamide in man. Helv Med Acta 4: 921–926

Dore F, Belanger P, Labrecque G (1984) Distribution tissulaire de microsphères radioactives en fonction de l'heure du jour et de l'état nutritionel chez le rat. Union Med Can 38: 964–966

English J, Biol LI, Dunne M, Marks V (1983) Diurnal variation in prednisolone kinetics. Clin Pharmacol Ther 33: 381–385

Feuers RJ, Scheving LE (1988) Chronobiology of hepatic enzymes. Annu Rev Chronopharmacol 4: 209–254

Focan C, Bruguerolle B, Arnaud C, Levi F, Mazy V, Focan-Henrard D, Bouvenot G (1988) Alteration of circadian time structure of plasma proteins in patients with inflammation. Annu Rev Chronopharmacol 5: 21–24

Focan C, Doalto L, Mazy V, Levi F, Bruguerolle B, Cano JP, Rahmani R, Hecquet B (1989) Vindesine en perfusion continue de 48 heures (suivie de cisplatine) dans le cancer pulmonaire avancé. Données chronopharmacocinétiques et efficacité clinique. Bull Cancer (Paris) 76: 909–912

Friedman AH, Walker CA (1972) The acute toxicity of drugs acting at cholinoceptive sites and twenty four hour rhythms in brain acetylcholine. Arch Toxicol 39: 39–50

Fujimura A, Kajiyama H, Kumagai Y, Nakashima H, Sugimoto K, Ebihara A (1989a) Chronopharmacokinetic studies of propanofen and procainamide. J Clin Pharmacol 29: 786–790

Fujimura A, Ohashi K, Sugimoto K, Kumagai Y, Ebihara A (1989b) Chronopharmacokinetic study of nitrendipine in healthy subjects. J Clin Pharmacol 29: 909–915

Gautherie M, Gros CH (1977) Circadian rhythm alteration of skin temperature in breast cancer. Chronobiologia 4: 1–17

Guissou P, Cuisinaud G, Llorca G, Lejeune E, Sassard J (1983)

Chronopharmacokinetic study of a prolonged release form of indomethacin. Eur J Clin Pharmacol 24: 667–670

Haen E, Gerdsmeier W, Arbogast B (1985) Circadian variation in propranolol protein binding. Naunyn Schmiedebergs Arch Pharmacol 329: 393

Halberg F (1969) Chronobiology. Annu Rev Physiol 3: 20–25

Halberg F, Stephens A (1959) Susceptibility to ouabain and physiologic circadian periodicity. Proc Minn Acad Sci 27: 139–143

Halberg F, Bettner J, Gully R (1955) Twenty four hour periodic susceptibility of audiogenic convulsions in mice. Fed Proc 14: 67–71

Haus E, Halberg F (1959) 24 hours rhythm in susceptibility of C mice to toxic dose of ethanol. J Appl Physiol 14: 878–882

Hecquet B (1986) Constant pharmacologic effect and chronopharmacology: theoretical aspects. Chronobiol Int 3: 149–154

Hecquet B, Sucche M (1986) Theoretical study of the influence of the circadian rhythm of plasma protein binding on cisplatin area under the curve. J Pharmacokinet Biopharm 14: 79–93

Hecquet B, Meynadier J, Bonneterre J, Adenis L (1984) Circadian rhythm in cisplatin binding on plasma proteins. Annu Rev Chronopharmacol 3: 115–118

Holter NJ (1961) New method for heart studies: continuous electrocardiography of active subjects over long periods is now practical. Sciences 134: 1214

Hrushesky W, Levi F, Kennedy BJ (1980) Cis-diaminedichloroplatinum toxicity to the human kidney reduced by circadian timing. Proc Am Soc Clin Oncol 21: 45

Hrushesky W, Borch R, Levi F (1982) Circadian time dependence of cisplatin urinary kinetics. Clin Pharmacol Ther 32: 330–339

Isaacson RJ (1959) An investigation of some of the factors involved in the closure of the secondary palate. PhD Thesis, University of Minnesota, Minneapolis

Jenni-Eiermann S, von Han HP, Honegger CG (1985) Circadian variations of neurotransmitter binding in three age groups of rats. Gerontology 31: 138–149

Jones F, Haus E, Halberg F (1963) Murine circadian susceptibility resistance cycle to acetylcholine. Proc Minn Acad Sci 31: 61–62

Jones HP (1845) On the variations of the acidity of the urine in the state of health. Philos Trans R Soc Lond 135: 335–349

Jori A, di Salle E, Santini V (1971) Daily rythmic variation and liver drug metabolism in rats. Biochem Pharmacol 20: 2965–2969

Kadle R, Folk GE (1983) Importance of circadian rhythms in animal cell cultures. Comp Biochem Physiol 76: 773–776

Kasal C, Perez-Polo JR (1982) Circadian rhythms in vitro. Trends Biochem Sci 7: 59–61

Klotz U, Ziegler G (1982) Physiologic and temporal variation in hepatic elimination of midazolam. Clin Pharmacol Ther 32: 107–112

Labrecque G, Domino EF (1982) Temporal variations in the effect of morphine and nalorphine on total brain acetylcholine content in the rat. Arch Int Pharmacodyn Ther 259: 244–249

Labrecque G, Dore F, Belanger PM (1981) Circadian variation of carrageenan-paw edema in the rat. Life Sci 28: 1337–1343

Labrecque G, Dore F, Laperriere A, Perusse F, Belanger PM (1979) Chronopharmacology. II. Variations in the carrageenan induced edema, in the action and the plasma levels of indomethacin. In: Reinberg A, Halberg F (eds) Chronopharmacology. Pergamon, Oxford, pp 231–238

Labrecque G, Belanger PM, Dore F (1982) Circannual variations in carrageenan-induced paw edema and the anti-inflammatory effect of phenylbutazone in the rat. Pharmacology 24: 169–174

Langner B, Lemmer B (1988) Circadian phase dependency in pharmacokinetics and cardiovascular effects of oral propranolol in man. Annu Rev Chronopharmacol 5: 335–338

Lavigne JG, Belanger PM, Dore F, Labrecque G (1983) Temporal variations in chloroform-induced hepato-toxicity in rats. Toxicology 26: 267–273

Lecocq B, Midavaine M, Lecocq V, Jaillon P (1992) Chronopharmacocinetique de la nitrindipine chez le sujet sain. Therapie (in press)

Lemmer B (1981a) Pharmacokinetics of beta-adrenoceptors blocking drugs of different polarity (propranolol, metoprolol, atenolol) in plasma and various organs of the light-dark synchronised rat. Naunyn Schmiedebergs Arch Pharmacol 316: R 60

Lemmer B (1981b) Chronopharmacokinetics. In: Breimer D, Speiser P (ed) Topics in pharmaceutical sciences. Elsevier/North-Holland, Amsterdam, p 49–68

Lemmer B, Lang PH (1984) Circadian stage dependency in antagonist binding of 3H-dihydroalprenolol to rat heart ventricular membranes. Annu Rev Chronopharmacol 1: 335–338

Lemmer B, Neumann G (1987) Circadian phase dependency of the effects of different β-receptor blocking drugs on motor activity of rats. Arzneimittelforschung 37: 321–325

Lemmer B, Winkler H, Ohm T, Fink M (1985) Chronopharmacokinetics of β-receptor blocking drugs of different lipophilicity (propranolol, metoprolol, sotalol, atenolol) in plasma and tissues after single and multiple dosing in rats. Naunyn Schmiedebergs Arch Pharmacol 330: 42–49

Lemmer B, Scheidel B, Stenzhorn G, Blume H, Lenhard G, Grether D, Renczes J, Becker HJ (1989) Clinical chronopharmacology of oral nitrates. Z Kardiol 78: 61–63

Lemmer B, Nold G, Behne S, Becker HJ, Liefhold J, Kaiser R (1990) Chronopharmacokinetics and hemodynamic effects of oral nifedipine in healthy subjects and in hypertensive patients. Annu Rev Chronopharmacol 7: 121–124

Levi F (1982) Chronopharmacologie de trois agents doués d'activité anticancéreuse chez le rat et la souris. Chronoefficacité et chronotolérance. PhD Thesis, University of Paris

Levi F, Hrushesky YW, Haus E, Halberg F, Scheving LE, Kennedy BJ (1980a) Experimental chrono-oncology. NATO Adv Study Inst 3: 481–511

Levi F, Lakatua D, Haus E, Hrushesky W, Halberg F, Schwartz S, Kennedy BJ (1980b) Circadian urinary N-acetylglucoseaminidase (NAG) excretion gauges murine cisdiaminedichloroplatinum (DDP) nephrotoxicity. Am Assoc Cancer Res 21: 1232

Levi F, Hrushesky W, Kennedy BJ (1981) Chronotolerance for cytostatic drugs in cancer treatment. 15th ICAR International Congress of Rheumatology, Paris

Levi F, Lelouarn C, Reinberg A (1985) Timing optimizes sustained release indomethacin treatment of osteoarthritis. Clin Pharmacol Ther 37: 77–84

Levi F, Metzger G, Bailleul F, Reinberg A, Mathe G (1986) Circadian varying plasma pharmacokinetics of doxorubicin (DOX) despite continuous infusion at a constant rate. Proc Am Assoc Cancer Res 27: 175

Levi F, Bruguerolle B, Hecquet B (1989) Mecanismes et perspectives en chronopharmacocinétique clinique. Therapie 44: 313–321

Lockard JS, Levy RH, Ducharme LL, Congdom WC, Patel IH

(1979) Carbamazepine revisited in a monkey model. Epilepsia 20: 169–173

Lockard JS, Viswanathan CT, Levy RH (1985) Diurnal oscillations of CSF valproate in monkey. Life Sci 36: 1281–1285

Margules DL, Lewis MJ, Dragovitch JA, Margues A (1972) Hypothalamic norepinephrine: circadian rhythms and the control of feeding behavior. Science 178: 640–647

Markiewicz A, Semenowicz K (1979) Time dependent change in the pharmacokinetics of aspirin. Int J Clin Pharmacol Biopharm 17: 409–411

Markiewicz A, Semenowicz K (1980) Does a rhythmicity of serum concentrations and urinary excretion of dipyridamole exist during long term treatment? Pol J Pharmacol Pharm 32: 289–295

Marte E, Halberg F (1961) Circadian susceptibility rhythm of mice to librium. Fed Proc 20: 305

Marte E, Nelson DO, Matthews JH, Halberg F (1978) Circadian rhythm in murine susceptibility to anesthetics halothane and methohexital. Int J Chronobiol 5: 425–426

Mato M, Ookawara S, Tooyama K, Ishizaki T (1981) Chronobiological studies on the blood brain barrier. Experientia 37: 1013–1015

Maxeykyle G, Smolensky MH, McGovern JP (1979) Circadian variation in the susceptibility of rodents to the toxic effects of theophylline. In: Reinberg A, Halberg F (eds) Chronobiology. Pergamon, Oxford, p 239

Munson ES, Marucci RW, Smith RE (1970) Circadian variations in anaesthetic requirement and toxicity in rats. Anesthesiology 32: 507–514

Naber D, Wirz-Justice A, Kafka MS, Wehr TA (1980) Dopamine receptor binding in rat striatum: ultradian rhythm and its modification by chronic imipramine. Psychopharmacology 68: 1–5

Naber D, Wirz-Justice A, Kafka MS (1981a) Circadian rhythm in rat brain opiate receptor. Neurosci Lett 21: 45–50

Naber D, Wirz-Justice A, Kafka MS, Tobler I, Borbely A (1981b) Seasonal variation in the endogenous rhythm of dopamine receptor binding in rat striatum. Biol Psychiatry 16: 831–835

Nagayama H, Takagi A, Sakurai I, Nishiwaki K, Takahasi RT (1978) Chronopharmacological study of neuroleptics. Psychopharmacology 58: 49–53

Nair V (1974) Circadian rhythm in drug action: a pharmacologic, biochemical and electron-microscopic study: In: Scheving LE, et al. (eds) Chronobiology. Igaku-Shoin, Tokyo, p 182

Nair V, Casper R (1969) The influence of light on daily rhythm in hepatic drug metabolising enzymes in rat. Life Sci 8: 1291–1298

Nakano S, Hollister LE (1978) No circadian effect on nortriptyline kinetics in man. Clin Pharmacol Ther 23: 199–203

Nakano S, Watanabe H, Nagai K, Ogawa N (1984) Circadian stage dependent changes in diazepam kinetics. Clin Pharmacol Ther 36: 271–277

Naranjo CA, Sellers SEM, Giles HG, Abel JG (1980) Diurnal variations in plasma diazepam concentrations associated with reciprocal changes in free fraction. Br J Clin Pharmacol 9: 265–272

Nelson W, Halberg F (1973) An evaluation of time-dependent changes in susceptibility of mice to pentobarbital injection. Neuropharmacology 12: 509–524

Nelson W, Tong YL, Lee YK, Halberg F (1979) Methods for Cosinor-rhythmometry. Chronobiologia 6: 305–323

Njus D, Sulzman FM, Hastings TW (1974) Membrane model for the circadian clock. Nature 248: 116–120

Patel IH, Venkataramanan R, Levy RH, Viswanathan CT, Ojemann LM (1982) Diurnal oscillations in plasma protein binding of valproic acid. Epilepsia 23: 282–290

Pariat C, Cambar J, Courtois P (1984) Circadian variations in the acute toxicity of three aminoglycoside: gentamicin dibekacin and netilmicin in mice. Annu Rev Chronopharmacol 1: 381–384

Petit E, Milano G, Levi F, Thyss A, Bailleul F, Schneider M (1988) Circadian rhythm-varying plasma concentration of 5-fluorouracil during a five day continuous venous infusion at a constant rate in cancer patients. Cancer Res 48: 1676–1679

Pollmann L (1981) Circadian changes in the duration of local anesthesia. J Interdisc Cycle Res 12: 187–192

Prat M, Bruguerolle B (1988) Chronotoxicity and chronokinetics of two local anaesthetic agents, bupivacaine and mepivacaine, in mice. Annu Rev Chronopharmacol 5: 263–266

Queneau P, Ollagnier M, Decousus H, Cherrah Y (1984) Ketoprofen chronokinetics in human volunteers. Annu Rev Chronopharmacol 1: 353–356

Radzialowski FM, Bousquet WF (1968) Daily rhythmic variation in hepatic drug metabolism in the rat and mouse. J Pharmacol Exp Ther 163: 229–238

Reinberg A (1974) Chronopharmacology in man. Chronobiologia 1: 157–185

Reinberg A (1976) Advances in human chronopharmacology. Chronobiologia 3: 151–166

Reinberg A (1978) Clinical chronopharmacology, an experimental basis for chronotherapy. Arzneimittelforschung 28: 1861–1867

Reinberg A (1979) Des rythmes biologiques à la chronobiologie, 3rd edn. Gauthier-Villars, Paris

Reinberg A (1982) La chronopharmacologie. Recherche 13: 478–489

Reinberg A, Ghata J (1990) Les rythmes biologiques, 5th edn. Presses Universitaires de France, Paris

Reinberg A, Halberg F (1971) Circadian chronopharmacology. Annu Rev Pharmacol 11: 455–492

Reinberg A, Reinberg MA (1977) Circadian changes of duration of action of local anaesthetic agents. Naunyn Schmiedebergs Arch Pharmacol 297: 149–152

Reinberg A, Smolensky MH (1982) Circadian changes of drug disposition in man. Clin Pharmacokinet 7: 401–420

Reinberg A, Ghata J, Halberg F, Apfelbaum M, Gervais P, Bourdon P, Aboulker C, Dupont J (1971) Distribution temporelle du traitement de l'insuffisance corticosurrénalienne. Essai de chronothérapeutique. Ann Endocrinol (Paris) 32: 566–573

Reinberg A, Clench J, Ghata J, Halberg F, Abulker C, Dupont J, Zagula-Mally Z (1975) Rythmes circadiens des paramètres de l'excrétion urinaire du salicylate (chronopharmacocinétique) chez l'homme adulte sain. C R Acad Sci [D] (Paris) 280: 1697–1700

Reinberg A, Schuller E, Deslanerie N, Clench J, Helary M (1977) Rythmes circadiens et circannuels des leucocytes, proteines totales, immunoglobulines A, G et M. Etude chez neuf adultes jeunes et sains. Nouv Presse Med 6: 3819–3823

Reinberg A, Gervais P, Chaussade M, Fraboulet G, Duburque B (1983) Circadian changes in effectiveness of corticosteroids in eight patients with allergic asthma. Chronobiologia 1: 333–347

Reinberg A, Levi F, Touitou Y, Le Liboux A, Simon J, Frydman A, Bicakova-Rocher A, Bruguerolle B (1986) Clinical chronokinetic changes in a sustained release preparation of keto-profen. Annu Rev Chronopharmacol 3: 317–320

Reinberg A, Levi F, Smolensky M, Labrecque G, Ollagnier M, Decousus H, Bruguerolle B (1990) Chronokinetics. In: Hansch C (ed) Comprehensive medicinal chemistry, vol 5. Pergamon, Oxford, pp 279–296

Riva R, Albani F, Ambrosetto G, Contin M, Cortelli P, Perucca E, Baruzzi A (1984) Diurnal fluctuations in free and total steady state plasma levels of carbamazepine and correlation with intermittent side effects. Epilepsia 25: 476–481

Ross FH, Graig NL, Sermons Al, Walker CA (1981) Chronotoxicity of antidepressant and psychomotor stimulants in mice. In: Halberg F et al. (Eds.) 13th International Conference of the International Society of Chronobiology, p 125

Sauerbier R (1979) Circadian aspects of the teratogenicity of cytostatic drugs. Chronobiologia 6: 152

Sauerbier I (1980) Recent findings relative to teratology and chronobiology. NATO Adv Study Inst 3: 535

Scheving LE (1980) Chronotoxicology in general and experimental chronotherapeutics of cancer. NATO Adv Study Inst 3: 455

Scheving LE, Pauly JE (1976) Chronopharmacology; its implication for clinical medicine. Annu Rep Med Chem 11: 251–260

Scheving LE; Vedral D (1966) Circadian variation in susceptibility of the rat to several different pharmacological agents. Anat Rec 154: 417–422

Scheving LE, von Mayersbach H, Pauly JE (1974) An overiew of chronopharmacology. J Eur Toxicol 7: 203–227

Schulz B, Greenfield M, Reaven GM (1983) Diurnal variation in specific insulin binding to erythrocytes. Exp Clin Endocrinol 81: 273–279

Scott B, Gulline H (1975) Membrane changes in a circadian system. Nature 254: 69–70

Sermons AL, Ross FH, Walker CA (1980) 24 hours toxicity rhythms of sedative hypnotic drugs in mice. Arch Toxicol 45: 9–14

Shively CA, Vesell ES (1975) Temporal variations in acetaminophen and phenacetin half-life in man. Clin Pharmacol Ther 18: 413–424

Shively CA, Simons RJ, Passananti GT, Dvorchik BH, Vesell ES (1981) Dietary patterns and diurnal variations in aminopyrine disposition. Clin Pharmacol Ther 29: 65–73

Smith RB, Kroboth PD, Phillips JP (1986) Temporal variation in triazolam pharmacokinetics and pharmacodynamics after oral administration. J Clin Pharmacol 26: 120–124

Smolensky MH (1989) Chronopharmacology of theophylline and $\beta$-sympathomimetics. In: Lemmer B (ed) Chronopharmacology cellular and biochemical interactions. Dekker, New York, pp 65–114

Spoor RP, Jackson DB (1966) Circadian rhythms: variation in sensitivity of isolated rat atria to acetylcholine. Science 154: 782–789

Touitou Y (1982) Some aspects of the circadian time structure in the elderly. Gerontology 28: 53–67

Touitou Y (1987) Le vieillissement des rythmes biologiques chez l'homme. Pathol Biol 35: 1005–1012

Touitou Y, Touitou C, Bogdan A, Reinberg A, Auzeby A, Beck H, Guillet PH (1986) Differences between young and elderly subjects in seasonal and circadian variations of total plasma proteins and blood volume as reflected by hemoglobin, hematocrit and erythrocyte counts. Clin Chem 32: 801–804

Ungar F, Reinberg A (1962) Circadian rhythm in the in vitro response of mouse adrenal to adrenocorticotropic. Science 137: 1058–1060

Vesell ES, Shively CA, Passananti GT (1977) Temporal variations of antipyrine half-life in man. Clin Pharmacol Ther 22: 843–852

Wirz-Justice A (1987) Circadian rhythms in mammalian neurotransmitter receptors. Prog Neurobiol 29: 219–259

Wirz-Justice A, Kafka MS, Naber D, Wehr TA (1980) Circadian rhythms in rat brain $\alpha$ and $\beta$-adrenergic receptors are modified by chronic imipramine. Life Sci 27: 341–347

Wirz-Justice A, Tobler I, Kafka MS, Naber D, Marangos PJ, Borbely A, Wehr TA (1981) Sleep deprivation: effects on circadian rhythms of rat brain neurotransmitter receptors. Psychiat Res 5: 67–76

Wirz-Justice A, Krauchi K, Morima SAT, Willener R, Feer H (1983) Circadian rhythm of $^3$H-Imipramine binding in the rat suprachiasmatic nuclei. Eur J Pharmacol 87: 331–333

# Chronotoxicology

J. Cambar, B. L'Azou, and J. C. Cal

## Introduction

Biorhythmicity in living organisms has been observed empirically for many centuries, but has been measured precisely and objectively only during the last decades.

Structural and functional temporal variations can be shown in unicellular or multicellular organisms, in all physiological systems. In the mammalian organism these include the cardiovascular and nervous system and organs such as the liver, heart, brain, kidney, and intestine. These temporal changes in structure and function allow us to understand why the same exogenous agent, whatever its origin, will not cause the same effects in a living host if administered at different times of day or during different months in the year.

The nature of an exogenous agent acting upon a rhythmically changing system can cause changes of three major kinds: chronopathology (caused by pathological agents, e.g., microorganisms), chronopharmacology (in response to drugs or pharmacological agents), and chronotoxicology (elicited by toxic agents). Chronotherapeutics, based upon these periodic changes, contributes to an optimization of drug use by timed treatment in an attempt to increase favorable pharmacological effects and by decreasing undesired side effects.

Many reviews have reported circadian and circannual variations in toxic and pharmacological effects of physical and chemical agents, including many drugs used in clinical therapy (Reinberg and Halberg 1971; Moore-Ede 1973; Scheving et al. 1974; von Mayersbach 1976; Cambar and Cal 1985; Cal et al. 1989). A chronobiological approach to pharmacology and toxicology can add new and relevant information for a better knowledge of the precise effects of drugs in living organisms and can also open the way for a better clinical use of important drugs, such as anticancerous or antibiotic agents.

The present review about the temporal approach to toxicity cannot be exhaustive. It will cover acute and subacute circadian chronotoxicology in animals, particularly in rodents, but also human chronotolerance for important drugs; moreover, it will point out the importance of taking seasonal (circannual) changes in toxicity into account. We focus on the temporal changes in experimental and human nephrotoxicity, as discussed previously in more detail (Cal et al. 1985, 1989).

## Circadian Acute Chronotoxicology

### Historical Review

Most studies on the chronosusceptibility of living organisms have used acutely toxic doses.

The first experiment on circadian variations in murine tolerance to a drug was reported by Carlson and Serin (1950), who showed that nikethamide-induced mortality (assessed by the classical median lethal dose ($LD_{50}$) test) ranged from 33 % at 0200 hours to 67 % at 1400. These data, at that time very surprising and exciting, were obtained without any consideration of chronobiology (chronotoxicology as a branch of science did not yet exist!) but showed that temporal considerations in experimental medicine are as important as all the other numerous factors of variability.

As reported in an exhaustive review (Scheving et al. 1974), F. Halberg and E. Haus were the first to systematically and intensively demonstrate the importance of circadian changes in resistance and susceptibility of animals to a variety of different agents.

The general aim of all these early studies was to study in groups of comparable animals synchronized mainly by the lighting regimen the influence of the time of administration on the mortality rate induced by noxious agents. So, in experiments on groups of mice where a fixed dose of ouabain (a specific Na/K ATPase inhibitor) was given at 4-h intervals during a 24-h span, the number of deaths, recorded 10 min after injection, was circadian-dependent. A greater percentage of animals (75%) died at the beginning of the light phase (Halberg and Stephens 1959) and a smaller percentage at the beginning of the dark phase (15%). Numerous additional studies confirmed that the toxicity of the drug was circadian-phase dependent. Similar conclusions were reached with nicotine where the same dose lead to 80% dead animals at 1900 and only 10% at 1400 (Scheving and Vedral 1966). Recently, it was reported that 125 mg/kg propranolol induces death in 50% of the animals at 1100 hours and more than 80% at other times of the day (Lemmer et al. 1980). A similar study has been reported recently in spontaneously hypertensive rats (Fujimura and Ebihana 1987).

A large number of similar observations have been reported for many types of toxic chemicals such as ethanol (Haus et Halberg 1959), local anesthetics, such as lidocaine (Lutsch and Morris 1967), anti-cancer agents, such as cytosine arabinoside (Cardoso et al. 1970), or adriamycin (Kuhl et al. 1973); the latter induced no death at 0800 and 80% at other times of the day.

## Circadian Median Lethal Dose Changes

While circadian changes in mortality rate (expressed as a percentage of the animals dying of a given dose of a toxic agent) have often been reported, temporal changes in the toxicity of drugs assessed by $LD_{50}$ (which requires four or five different doses) are infrequent. Some recent papers have nevertheless shown in rodents (particularly in mice) $LD_{50}$ circadian

changes for diazepam (Ross et al. 1981), gentamicin (Nakano and Ogawa 1982), mercuric chloride (Cambar et Cal 1982), and procainamide (Bruguerolle 1984).

Very recently, $LD_{50}$ circadian changes have been compared for three different local anesthetic agents (bupivacaine, mepivacaine, and etidocaine) in animals synchronized for periodicity analysis (Prat and Bruguerolle 1988; Prat 1990). For bupivacaine, the $LD_{50}$ showed a maximum at 0100 ($62 \pm 2.9$ mg/kg) and a minimum at 2200 ($52.4 \pm 0.5$ mg/kg). For mepivacaine, the $LD_{50}$ showed a maximum at 1000 ($130 \pm 5$ mg/kg) and minimum (maximal toxicity) at

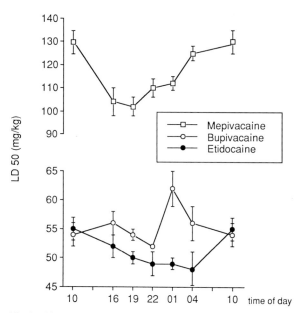

**Fig. 1.** Circadian changes in the median lethal dose (LD 50) of three anesthetic agents, mepivacaine, bupivacaine and etidocaine in mice. (Redrawn from Prat 1990)

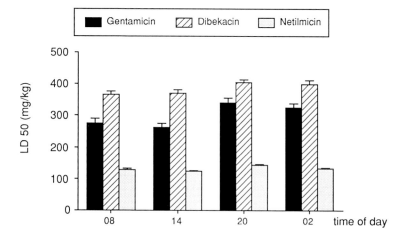

**Fig. 2.** Circadian changes in the median lethal dose (LD 50) of three antibiotics, gentamicin, dibekacin, and netilmicin in mice. (Redrawn from Pariat 1986)

1900 ($102 \pm 3$ mg/kg). For etidocaine, the minimum $LD_{50}$ occurred during the middle of the night ($47.5 \pm 2.8$ mg/kg) and the maximum at 1000 ($55 \pm 2.3$ mg/kg). It is noteworthy that, although the lowest $LD_{50}$ (the maximal toxicity) can be observed for all three anesthetic agents during the acitivity span of the mice, the maximum $LD_{50}$ occurs at 1000 for bupivacaine, but at 0100 for etidocaine (Fig. 1).

Similarly, circadian changes of the $LD_{50}$ of three aminoside antibiotics were reported in mice (Pariat 1986), showing a maximum in mortality in the middle of the rest span and a maximum in tolerance at 2000. At 1400, the $LD_{50}$ was 252 mg/kg for gentamicin, 370 for dibekacin, and 123 for netilmicin. At 2000, the $LD_{50}$ was increased to 340 mg/kg, 405, and 143 respectively (Fig. 2).

## Nonchemical Toxic Agents

Nonchemical toxic agents, such as bacterial or viral toxins or physical agents, can also induce death in experimental animals; the damage produced by these agents may vary greatly as a function of the time of administration. After Brucella endotoxin administration, mice injected at night lived for $75 \pm 15$ h while those injected during the daytime survived only $37 \pm 7$ h (Halberg et al. 1955). Similar studies with *Escherichia coli* endotoxin (Halberg et al. 1980) showed that mortality is 80% when mice are injected at 1600 and only 15% at 0000. Likewise, when the immune system of the mice was stimulated with Calmette-Guérin bacillus (BCG), it was found that 75% of the animals challenged during the middle of the light period survived, whereas only 50% of those challenged during the later part of the dark phase survived (Tsai et al. 1974).

Although only indirectly related to toxicology, the susceptibility to physical agents may also be circadian-phase dependent, e.g., the induction of seizures by white noise (about 100 dB during 60 s) in a susceptible strain of mice (Halberg et al. 1955b). X-irradiation showed circadian rhythms in mortality in *Drosophila* and in mice. In the latter, exposure to 555 R at night induced 100% death while during the day it was not lethal.

## Chronotoxicological Examples in Nonrodents

In most of the studies described above, rodents were used, for obvious practical reasons. The choice of such animals is justified by an easy synchronization under well-defined standardized conditions and also by a relatively moderate cost, allowing the use of a great number of animals for toxicological experimentation. Nevertheless, chronotoxicological studies have also been carried out in other species, including fish and insects. Insects have been recently proposed as simple and inexpensive animals models for chronopharmacological research (Hayes and Morgan 1988).

In fish, a paper reported circadian variations in zinc tolerance in rainbow trout (MacLeay and Munro 1979), and a study in the same species described the influence of the season on the lethal toxicity of cyanide (McGeachy and Leduc 1988). However, such observations seem too far from mammalian and human reality to be discussed here in detail.

Another interesting chronotoxicological approach may be to optimize in insects the use of pesticides in order to avoid unnecessarily large doses. A chronotoxicity (or "greatest kill rate") has been shown, e.g., for malathion in house flies (Frudden and Wellso 1968) and for Dursban in *Aedes aegypti* larvae (Roberts et al. 1974). All these investigations showed that the same concentration of a pesticide may have a different lethal potency on insects when the time of testing is considered. It has been reported that the same dose of pesticide can kill 90% of insect larvae if they are exposed during the second part of the night and only 20% at the end of the day. Such chronotoxicological considerations may reduce the amounts of pesticides required, having important ecological and economical consequences. Recent evidence of circadian changes in avian esterases can be used as indicators for exposure of birds to insecticides, which represent very dangerous pollutants for all nontarget organisms in the ecosystem (Thompson et al. 1988).

Such considerations make it necessary to review critically the significance of $LD_{50}$ values when the time of day of their determination is not known. The time has to be taken into account for mortality test evaluations which may have to be performed at two, three, or four different times if the toxin or the drug has a high toxicity. An improvement in the use of $LD_{50}$ testing may result from documenting the optimal (highest tolerance) and worst (highest toxicity) times of administration. Moreover, such observations point to the relative value of certain toxicological investigations in which differences in $LD_{50}$ values are often used to compare laboratory capacity or pertinence. It is fundamental in toxicology to take into account the exact hour of the day and even the month of the year, and the synchronizing schedule, e.g., the lighting regimen in rodents, to be able to compare possible conflicting results!

# Circadian Subacute Chronotoxicology

In spite of their great interest, acute toxicological studies have certain limits because the high lethal doses lead to a rapid and often atypical death (e.g., in 15 min for antibiotics) which does not allow any comparison with therapeutic conditions. Moreover, death is the result of damage to different physiological functions and numerous tissues. Therefore, a number of recent studies have used weaker nonlethal doses to document more accurately the subacute chronotoxicology of various noxious agents in different tissues. As we cannot be exhaustive, we will review only the reports dealing with the chronosusceptibility or chronotolerance of some organs frequently damaged by certain classes of therapeutic agents, i.e., the kidneys, liver, digestive tract, ear, and bone marrow.

The choice of the examples is in part suggested by another chapter of this book about the chronotoxicity and chronotolerance of anticancerous agents. As pointed out by a recent review, anticancerous drugs cause numerous side effects that can injure nearly all organs (Lévi et al. 1987). We will thus not approach the chronotoxicology of anticancerous agents except their nephrotoxicity, which our laboratory has studied for many years.

We will end this chapter by referring to some recent clinical studies about chronotolerance of several widely used drugs, to point out some practical clinical applications.

## Liver Chronotoxicity (Chronohepatotoxicity)

Some reports have been published on temporal variations in the effects of hepatotoxins such as 1,1-dichloroethylene (Jaeger et al. 1973), acetaminophen (Schnell et al. 1983), chloroform (Lavigne et al. 1983; Desgagne et al. 1988), or carbon tetrachloride (Harris and Anders 1980; Bruckner et al. 1984). The latter is a common contaminant of water supplies and a potent hepatotoxin.

Lavigne et al. (1983) reported that rats were most susceptible to chloroform (CHCl$_3$)-induced hepatotoxicity when treated 2 h after the beginning of their activity span (2100). The lowest susceptibility was noted at 0900. A more recent study approaches the mechanism of chloroform chronohepatotoxicity in rats (Desgagne et al. 1988). The administration of chloroform significantly depressed the hepatic concentration of glutathione both at 2100 and 0900, but more at 2100 than at 0900. Moreover, electron micro-

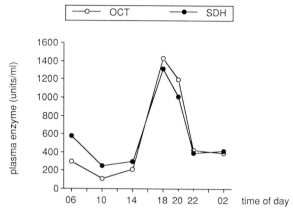

**Fig. 3.** Circadian changes in carbon tetrachloride-induced hepatotoxicity in rats, assessed by two plasma enzymes, ornithine carbamyl transferase (OCT) and sorbitol deshydrogenase (SDH). (Redrawn from Bruckner et al. 1984)

graphs of liver tissue showed that the necrotizing effect of chloroform on the mitochondria, nuclear membrane, and other components was much more pronounced at 2100 than at 0900. The authors concluded that a lower concentration of glutathione in rat liver at night would result in a greater susceptibility to cellular necrosis induced by metabolites generated from toxic chemicals.

Bruckner et al. (1984) reported similar chronohepatotoxicological investigations with carbon tetrachloride (CCl$_4$) in rats. The serum activity of four enzymes, glutamic-pyruvic-transaminase (GPT), sorbitol dehydrogenase (SDH), isocitrate dehydrogenase (ICDH), and ornithine-carbamyl-transferase (OCT), was dramatically increased when CCl$_4$ was given at 1800 or 2000 but remained unchanged at 1000. These results demonstrate that the rat is most sensitive to CCl$_4$ during the initial portion of its activity span (Fig. 3).

It is very interesting to note the similarity in liver chronosusceptibility for the hepatotoxins CCl$_4$ and CHCl$_3$.

## Ear Chronotoxicity (Chronoototoxicity)

A subacute chronotoxicological study of the auditory level has been conducted with kanamycin in rats (Fish et al. 1984). Female Sprague-Dawley rats under light-synchronized conditions, received (for 6 weeks) a daily subcutaneous injection of kanamycin sulfate (225 mg/kg) at four times (0800, 1400, 2000 and 1200). After 2 weeks, the 0800 group showed an average hearing loss of 11.5 dB at 32 kHz. The other groups had only minimal changes. After 6 weeks, the 0800 and 1400 groups (diurnal rest span groups) had a similar dramatic hearing loss (34 dB), while the two other groups had only slight losses. Such considerations can be of value to optimize the use of aminosides in clinical medicine by timing of their administration.

## Renal Chronotoxicity (Chrononephrotoxicity)

Renal damage induced by drugs or toxic agents is very frequent, since the kidney is the major route of elimination of xenobiotics and receives nearly 25 % of the cardiac output. Most of the drugs are filtered through the glomerular barrier and enter the tubular lumen in the tubular fluid. The fluid composition is continuously changed along the nephron by reabsorption-secretion processes. When this fluid contains a high concentration of a toxic substance, it can enter the tubular cells and induce severe intracellular damage. Thus, many water-soluble substances such as heavy metals, aminosides, or cyclosporin can dramatically injure the proximal part of the nephron and alter membrane exchange processes. In urine, a noninvasive liquid biopsy, the presence of some tubular components like proteins or enzymes makes it possible to quantify the renal damage. We use the change (increase in percentage) in urinary tubular enzymes released from the brush border membrane or from lysosomes to evaluate the renal damage caused by a drug administered at different times (Cal et al. 1985).

Heavy Metals Chrononephrotoxicity

We have previously mentioned the numerous papers describing circadian rhythms in lethal toxicity due to heavy metals. For a decade, sublethal chronotoxico-

logical investigations with heavy metals have been carried out for mercury (Cal 1983), cadmium (Cambar et al. 1983), and especially platinum derivatives, both in rodents (Lévi 1982; Boughattas 1989) and in human subjects (Caussanel 1989). These studies have been exhaustively reviewed (Lévi et al. 1987; Mormont et al. 1989).

Lévi et al. (1982 a, b) investigated for the first time heavy metal chrononephrotoxicity in cisplatin treated rats by monitoring serum urea concentration, urinary N-acetyl-glucosaminidase (NAG) activity, and renal histology. When cisplatin was given at the time of greater toxicity (as shown in studies of the lethal toxicity), a 3.8 times increase in the 24-h MESOR of urinary NAG was observed, in direct relation to the subsequent rise in serum urea. When cisplatin was given at a favorable circadian stage (in the middle of the night), however, the rats showed a small NAG rise (only 1.5-fold) and had little renal damage with a small rise in plasma urea.

We have performed similar studies with sublethal doses of mercuric chloride, the most nephrotoxic of the heavy metals (Cal et al. 1985, 1989 a). As explained above, circadian periodicity in heavy metal nephrotoxicity induced by a single sublethal injection at four different points in time was assessed in rats by the increase in the urinary activity of three tubular nephrotoxicity marker enzymes. Single cosinor analysis showed that the acrophase in toxicity-induced increase in enzyme excretion is located at 1200 with a MESOR of 1485 % (about 15-fold the control value). However, this increase reached about 1800 % when $HgCl_2$ was given in the middle of the light span and only 1000 % when given in the middle of the dark span, i.e., during the time of activity of rats (Fig. 4). The study of the mercury distribution revealed that when the rats were treated at the end of the light span, the Hg concentration was highest in the kidney

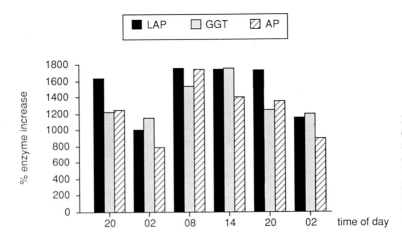

**Fig. 4.** Circadian changes in mercuric chloride induced nephrotoxicity in rats, assessed by increase in urinary enzyme excretion (in percentage of the control level) of three tubular enzymes, gamma glutamyl transferase (GGT), alkaline phosphatase (AP), and leucine aminopeptidase (LAP). (Redrawn from Cal et al. 1985)

and lowest in urine (Cal 1983; Cal et al. 1985). Similar results were obtained in cisplatin-treated rats by Lévi (1982a, b), who found that cisplatin was excreted in larger amonts in animals injected during the activity phase than in those injected during the rest phase.

Such interesting results in murine studies open the way to chronotherapy with anticancerous agents. A recent review reported investigations in patients with several widely used anticancerous agents, such as adriamycin, 5-fluorouracil, vindesine, or cisplatin (Lévi et al. 1987). These drugs exhibit circadian stage-dependent pharmacokinetics or pharmacodynamics in cancer patients. Moreover, it seems that these preclinical observations in rats or mice might be used (taking into account the phase difference in the activity/rest rhythm between rodents and humans) to design clinical chronobiological protocols. For example, murine experiments have shown that adriamycin was better tolerated in the second half of the host rest span and that cisplatin was least toxic when given in the second half of the host activity span. Patients were therefore randomized to receive either adriamycin in the morning and cisplatin in the evening (schedule A) or adriamycin in the evening and cisplatin in the morning (schedule B). In two independent studies, necessity of dose reduction, delay in treatment, bleeding, and infection were more often observed in patients on schedule B than in those on schedule A (Hrushesky 1985; Mormont et al. 1989).

It is interesting to point out the similarity in circadian timing of the lethal chronotoxicity of heavy metals with that of circadian sublethal chrononephrotoxicity. Moreover, mercury, cadmium, and cisplatinum exhibit very similar circadian bioperiodicity in heavy metal-induced toxicity.

Antibiotics Chrononephrotoxicity

Wachsmuth (1982) showed, for the first time, a circadian rhythm in the susceptibility of the rat kidney to an antibiotic (cephaloridine) as evaluated by histochemical staining of injured tubules and by the assay of enzymuria and proteinuria. The smallest response was observed when the animals were injected at the onset of the light span and the greatest after the injection at 1900.

We have also demonstrated the circadian biosusceptibility rhythms of four aminoglycoside antibiotics, which also are major nephrotoxic agents (Cal et al. 1985, 1989a; Dorian et al. 1988). For example, in November, the same amikacin dose (1.2 g/kg) in rat increased by nearly five the urinary tubular enzyme

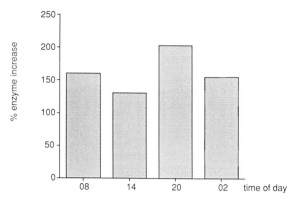

**Fig. 5.** Circadian changes in amikacin-induced nephrotoxicity in rats, assessed by increase in urinary gamma glutamyl transferase excretion (in percentage of the control level) 48 h after a single administration. (Redrawn from Dorian and Cambar 1986)

activity when injected at 2000 (Fig. 5), by at 0800, and only by 1.2 at 1400 (Dorian et al. 1986). Likewise, in June, the same 200 mg/kg gentamicin dose increased the urinary NAG activity in rats by +55% at 0200 and +180% at 1400, and the urinary AAP activity by +140% at 0200 and +22% at 1400. Moreover, at 0200, when the enzymuria increase was low, renal gentamicin accumulation was mild (345 mg/g kidney weight) and urinary gentamicin concentration (3900 µg/ml) and excretion (43300 µg/24 h) were very high. In contrast, at 1400 when increase in enzymuria was high, renal accumulation was high (442 µg/g kidney weight), and urinary concentration (2517 µg/ml) and excretion 30926 µg/24 h) were very low (Pariat 1986; Pariat et al. 1988).

**Cyclosporin Chronotoxicity**

Cyclosporin (CsA), a cyclic undecapeptide with unique marked immunosuppressive properties, has had a major impact in the field of organ transplantation. It has also been proposed as a chronic medication in the treatment of autoimmune diseases. Unfortunately, CsA has numerous side effects, the most serious of which is nephrotoxicity, which in transplant and nontransplant patients is correlated with the CsA blood level. We will review the main recent chronotoxicological studies on cyclosporin, including acute toxicology, sublethal toxicity, and pharmacokinetics in experimental animals. We will end this review with recent clinical chronotherapeutic considerations to optimize the use of this remarkable drug, which is so difficult to achieve in chronic treatment.

In Lewis rats treated daily IP with CsA

(20 mg/kg), body temperature decrease and weight loss were greater when the toxin was given during the dark span (Cavallini 1983). In 60 male inbred Lewis rats, a toxic dose of CsA (60 mg/kg) was given at one of six time points. Subsets of the animals were killed dat 2, 6, 12, and 24 h post-treatment. Whole blood samples were obtained at that time. The toxic effects of CsA in rats, as judged by weight loss and survival time, were significantly more severe when CsA was administered during the activity (dark) span than during the rest (light) span ($-20\%$ vs $-11\%$ for weight loss and 26 days vs 42 days for survival time). Moreover, the CsA plasma levels were highly correlated with the toxicological parameters. Thus, the plasma level was 2.3 times higher when CsA is given during the activity (dark) span (Bowers et al. 1986).

These observations were confirmed by a recent study, which also showed that the toxic effect and the blood levels were greater when the drug was administered during the active period of the rodent (Kabbaj 1989).

Finally, a recent chronopharmacological study in mice seems to indicate that the administration of oral CsA at the very end of the nocturnal activity span can be used to minimize renal toxicity (assessed by blood urea nitrogen) and achieve a satisfactory immunosuppressive effect (Pati et al. 1988).

All these observations indicate that nephrotoxicity, the major side effect of this immunosuppressive agent, is circadian stage dependent (Kabbaj 1989). Indeed, the serum creatinine increase (a classic test of renal function) was large on the third day of daily chronic injection, when the drug was administered at the onset of the light phase and at the end of the dark phase (0500 and 0900) with a plasma concentration of 146–148 $\mu M$ vs 92–95 in the controls. In contrast, this increase was small when the drug was administered during the light cycle (1300 and 1700) with 104–120 $\mu M$ vs 93–97 in the controls. After 16 and 21 days of treatment, serum creatinine was dramatically increased when cyclosporin was administered at the start of the dark cycle (2100) with 184–189$\mu M$ and mild when given at 0500 with only 108 $\mu M$. The glomerular filtration rate, assessed by the classic insulin clearance test, was also markedly reduced in lean Zucker rats when cyclosporin was administered at the beginning of the dark phase (Luke et al. 1988).

Glucose intolerance has been described among the adverse metabolic effects of chronic cyclosporin treatment. Malmary et al. (1988) showed the rapid damping, and even the disappearance, of glucose and insulin circadian rhythms during the first day of treatment.

Pharmacokinetic data make it possible to explain the circadian variation in cyclosporin tolerance in rodents. Kabbaj (1989) compared the main pharmacokinetic parameters when the same dose was given at different times and showed that when cyclosporin was given at 2100 many pharmacokinetic parameters, i.e., the plasma maximal concentration ($C_{max}$) and the area under the curve (AUC), were at their maximum. Plasma cyclosporin concentration varied dramatically according to injection time. The concentrations observed 1 h after injection at 0900, 1500, 2100, ad 0500 were 15700, 1000, 6380, and 3060 ng/ml respectively (Fig. 6; for legibility, only data for 1500 and 2100 are shown). The $C_{max}$ was maximal at 0900 (16240 ng/ml) and at 2100 (17450 ng/ml) and minimal at 1500 (5200 ng/ml). In the same way, the AUC was twofold greater at 2100 than at 0900.

Similar chronopharmacokinetic studies have been reported recently in human subjects. The concentration of CsA in the blood of transplant patients also varied with administration time (Bowers (1986; Alli-

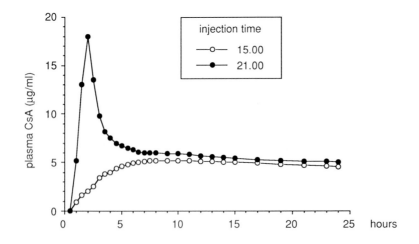

**Fig. 6.** Kinetics of plasma cyclosporin in rats during the first 24 h after a single injection at two different times. (Redrawn from Kabbaj 1989)

son and Trenton 1987). Evening administration induced higher plasma levels. Cipolle (1988) compared CsA pharmacokinetics in pancreatic transplant patients receiving 300 mg CsA at 0900 or at 2100. The pharmacokinetic parameters were higher with evening administration.

Seven adult pancreas transplant patients received CsA orally in two equally divided doses every 12 h at 0800 and 2000 (Canafax et al. 1988). The AUC during rest in evenig and during the night was greater than during the activity phase. The mean residence time, as an index of the average time the CsA molecule remained in the body, was greater during the resting phase (5.6 ± 0.4 h) than the activity phase (4.7 ± 0.6 h). These data indicate an increased exposure to CsA following the evening dose. These data were confirmed by Ramon et al. (1989), who showed that after administration at 0900 the CsA plasma concentration was two times higher than after administration at 2100.

Such data open the way for the optimization of CsA use in transplant patients by introducing chronotherapeutics to reduce the numerous side effects, such as nephrotoxicity. Similar approaches with anticancerous drugs are being initiated, with a chronobiological choice of a preferential administration time or by using peristaltic programmable computerized administration systems. Recently, programmable pump catheter systems were designed which infuse cyclosporin directly into the renal artery of the transplant (Gruber et al. 1988).

An experimental paper about segmental pancreatic transplant rejection in rodents points out that circadian rhythms, and also circaseptan and circannual rhythms, must be taken into consideration to optimize cyclosporin therapy (Liu et al. 1986).

## Chronesthesy as Mechanistic Approach to Chronotoxicology

The numerous examples of temporal changes in acute or subacute toxicity of drugs raise the question of the mechanism through which a drug can be less (or more) toxic at one time than at another. All living organisms present bioperiodicity, i.e., temporal changes in their structure and function. An organ, such as the liver or kidney, undergoes a circadian stage-dependent synthesis of its enzymes and even of its nuclear components and presents subsequent changes in membrane structure and function. Knowl-

edge of renal or hepatic chronophysiology makes it possible to explain the chronopharmacological variations in hepatic or renal elimination and metabolism of drugs (Belanger 1987; Feuers and Scheving 1987; Cal et al. 1985, 1989a).

It is very interesting to consider that chronograms of glycogen synthetase and glycogen phosphorylase are "in mirror image" with those of hexobarbital oxidase and barbiturate-induced sleep duration.

Acute and subacute chronotoxicity of acetaminophen have been compared in mice (Schnell et al. 1983). Acetaminophen (600 mg/kg) killed 70% of mice at 1800 and only 10% at 1000. Hepatic concentrations of nonprotein sulfhydril components exhibited circadian variations with a maximum at 1000 and a trough at 1800. Acetaminophen toxicity was minimal at 1000 when these components, which play a protective effect against the toxicity, were present in higher concentrations in the liver.

The general concept of chronesthesy was introduced by Reinberg (1976), to summarize all these considerations. Chronesthesy can be defined as rhythmic changes in susceptibility or tolerance of a biosystem. This biosystem can be a whole living organism, but also an organ, a tissue, or even a single cell. The chronesthesy of a biosystem to a drug or a toxin is all the more precise and well defined when its structure is well-known and space-limited.

We have introduced and discussed the chronesthesy of the renal proximal tubules (Cal et al. 1985, 1989a), which depends on glomerular filtration intensity, tubular membrane exchange, intracellular enzyme activity, and intraluminal toxic concentration. The tubular chronesthesy observed is dependent on the circadian state of all these factors.

## Circannual Chronotoxicology

Circadian changes in susceptibility to toxic agents must be examined during different months (or seasons) of the year (Cal et al. 1989b). Indeed, in many cases, it is very surprising to note that the circadian peaktime of tolerance found during one season may be totally different from that during another. We will give some examples to show the importance of studying the circannual changes of the circadian variations in drug or toxin susceptibility. The size and the high cost of such experiments can explain the scarcety of papers about circannual chronotoxicology.

Indeed, since the early reports of seasonal variations on the acute toxicity of X-rays both in the rat and in the mouse (Fochem et al. 1967), of arabinosylcytosine (Scheving et al. 1974), and of cyclophosphamide in the chinese hamster and the mouse (Pericin 1981), only a few recent papers have been concerned with this topic.

## Phenobarbital

A recent paper examined the circadian changes in acute toxicity of phenobarbital in mice during four different months in the year (Bruguerolle et al. 1988). It was found that the same toxic dose (190 mg/kg) given at 1600 killed no animal in July but killed 90 % of them in January. Similarly, 270 mg/kg of phenobarbital given at 0400 killed all the animals (100 %) in October and only 40 % in July. The $LD_{50}$ for drug administration at 1600 was $204 \pm 6.5$ mg/kg in January and $254 \pm 5.2$ in July whereas for a 0400 administration it was $185 \pm 5.5$ mg/kg in January and $270 \pm 7.7$ mg/kg in July. A circadian peak toxicity was found at 1000 in July and at 2000 in April or in October. Such observations clearly show that in toxicological experiments not only the time of day of drug administration, but also the month of the year have to be considered.

## Platinum Derivatives

A recent review of murine chronotoxicity of anticancer agents (Lévi et al. 1987) reported similar seasonal changes in the tolerance of mice for two anthracyclines, doxorubicin and donomycin. The tolerance was optimal in autumn and poorest in spring or in summer. Lévi (1982a) had already reported that CDDP tolerance in rats was twice higher in winter than in summer. Moreover, it was shown that the circadian tolerance peak occurred in winter at 0900 (40 % survival rate) and during the other seasons at 1700 (26 %). Documented at the same hour (0900), the tolerance was found to be maximal in winter (40 %) and low (11 %) during the other seasons.

A recent study in rodents compared the circannual chronotolerance of two platinum derivates, CDDP and I-OHP (Boughattas 1989). The circadian MESOR of total body weight in intoxicated animals presented a seasonal change with a circannual peak (minimal body weight loss) during the beginning of October for CDPP and during the beginning of February for I-OHP. The survival time for 14 mg/kg

CDDP was maximal in autumn and minimal in winter and for 17 mg/kg I-OHP was maximal in winter (35 days) and minimal during summer (20 days).

## Mercuric Chloride

For several years, we have performed investigations on the circadian chronotoxicology of mercuric chloride. To show circannual changes in circadian chronotoxicity, mice received a single IP injection of mercuric chloride at four different times during the day in 6 months of the year. It has been shown (Cal et al. 1984, 1986) that the same dose of this substance (5 mg/kg) induced a different mortality rate when given in September (53 %–86 %), in June (53 %–93 %), or in January (60 %–100 %). The mean mortality rate per month, with all times of day and doses pooled, was maximal in November (82 %) and minimal in February (69 %).

## Antibiotics

Another similar experiment, performed with amikacin, showed the nonreproducibility of the circadian chronotolerance (MESOR and acrophase) during two different seasons of the year (Dorian et al. 1986). The mortality rate in winter was 48 % at 1400 and 60 % at 0200 whereas in spring it was 37 % at 1400 and 23 % at 0200. Such observations show that the same toxic doses can reveal a circadian tolerance rhythm with apparently completely reversed acrophases at two different occasions. The circadian tolerance peak time at one occasion can become the trough at another one!

Similar marked phase differences were obtained in the study of subacute chronotoxicity of amikacin. In autumn urinary enzyme activity is increased 4.8 times when amikacin is given at 2000 and only 1.3 times at 1400; in contrast, during spring this increase is dramatic at 1400 (3.5 to 6.5 times) and weak at 2000. As for the mortality rate, during two different seasons the circadian acrophase of nephrotoxicity becomes the batyphase (Fig. 7)!

We have recently reported a comparable influence of the seasons on the severity of kidney damage due to gentamicin (Pariat 1986; Pariat et al. 1990). We have compared the MESOR and the acrophase of the gentamicin-induced circadian nephrotoxicity during the four seasons of the year and noted dramatic differences. In winter (January–February), urinary excretion of all enzymes studied increased with a maxi-

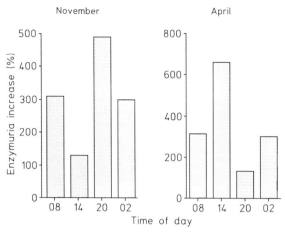

**Fig.7.** Seasonal differences of the circadian variations in gamma glutamyl transferase urinary excretion (expressed as a percentage of the control value) in rats 24 h after a single amikacin injection in November *(left)* or in April *(right)*. (Redrawn from Cambar et al. 1987)

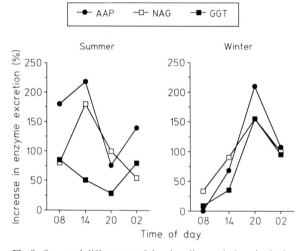

**Fig.8.** Seasonal differences of the circadian variations in the increase of the excretion of three tubular enzymes (gamma glutamyl transferase GGT, alanine aminopeptidase AAP and N-acetyl-beta-D-glucosaminidase NAG) in rats, 24 h after a single gentamicin injection in winter *(right)* or in summer *(left)*. (Redrawn from Pariat et al. 1990)

mum at the end of the animals rest period at 2000 of about 1.5 to 2 times above the control values. In spring (March–April) opposite results were obtained with a maximal increase of brush border enzymes after administration of the drug at 0800. In summer (June–July) the maximum of the enzyme increase was at 1400 with a trough at 2000. Finally, in autumn (October–November) the chronograms were similar to those obtained during the winter with the maximum noted again at 2000 (Fig.8).

In conclusion, all these studies show clearly that the severity of the nephrotoxicity produced by the administration of heavy metals, antibiotics, or anticancerous agents and assessed by enzymuria tests is dependent not only on the circadian time, but also on the season. How can we explain such circannual changes in toxicity or tolerance of living organisms to drugs? Two recent reviews which carefully investigated the problem have brought no answer (Lévi et al. 1987; Cal et al. 1989a). It seems possible that the circadian acrophases of the structural and functional parameters of the cells responsible for the susceptibility-resistance cycle change along the yearly scale as a result of an endogenous circannual rhythm with a period of about 12 months.

Moreover, the circadian rhythms of the different organ target systems may differ along the scale of the year due to the influence of the circannual changes in synthesis and secretion of several hormones known to be involved in organ function. Numerous seasonal changes have been described in enzyme activity, membrane proteins and lipids, or hormonal patterns. An exhaustive recent review also reported in young and in elderly, healthy and diseased human subjects numerous circannual changes in a large number of clinical blood and tissue parameters (Haus et al. 1988).

## Conclusion

Chronotoxicology can provide new complementary approaches to toxicology. Indeed, one can be very surprised by the often important discrepancies in experimental results reported in the literature. Among the numerous causes of variability in experimental biology, one should always take into account age, breed, sex, housing conditions, feeding, and synchronization. In addition, the exact circadian time and season of the year must be considered. Numerous convincing examples have shown the dramatic temporal (circadian and circannual) changes in susceptibility or tolerance of living organisms to toxic substances or drugs. Although the cost of such experiments can be high, the use of drugs presenting severe side effects requires a chronotoxicological approach.

It is regrettable that there are so few experimental and clinical studies on the chronotoxicology of certain nephrotoxic antibiotics, for example aminosides. The development of sophisticated peristaltic programmable computerized infusion systems, such as are used for anticancerous or immunosuppressive agents, must encourage the toxicologist and the pharmacologist as well as the pharmaceutical industry to apply and sponsor chronobiology as one of the most promising means to optimize the administration of dangerous drugs with a poor therapeutic index. We are encouraged by the recent interest of some scientific committees sponsoring lectures about this topic in cancerology, pharmacology, and immunology, or presentations at certain human pathology meetings.

The circannual changes in the circadian variations in susceptibility or tolerance to drugs require caution in any attempt to recommend a definitive absolute optimal time of day or in the year for treatment with potentially toxic agents. Only experimental and clinical investigations will contribute, little by little, to this purpose and will reveal the "optimal time" for the "optimal use" of a drug.

# References

Allison TB, Trenton B (1987) AM vs PM cyclosporin blood levels. Clin Pharmacol Ther 41: 237 A

Belanger PM (1987) Chronobiological variation in the hepatic elimination of drugs and toxic chemical agents. Annu Rev Chronopharmacol 4: 1–46

Boughattas NA (1989) Rythmes de la pharmacocinétique et de la pharmacodynamie de trois agents anticancéreux (cisplatine, oxaliplatine et carboplatine) chez la souris: approche de leurs régulations. Thesis, University of Paris

Bowers LD, Rabatin JT, Wick M, Canafax D, Hrushesky W, Benson E (1986) Circadian pharmacodynamics of cyclosporin in rats and man. Annu Rev Chronopharmacol 3: 219–222

Bruckner JV, Luthra R, Lakatua D, Sackett-Lunden L (1984) Influence of time of exposure to carbon tetrachloride on toxic liver injury. Annu Rev Chronopharmacol 1: 373–376

Bruguerolle B (1984) Circadian chronotoxicity of procainamide. IRCS Med Sci 12: 579–582

Bruguerolle B, Prat M, Douylliez C, Dorfman P (1988) Are there circadian and circannual variations in acute toxicity of phenobarbital in mice? Fundam Clin Pharmacol 2: 301–304

Cal JC (1983) Approche chronobiologique de l'intoxication aigüe par le chlorure mercurique chez les rongeurs. DEA de Nutrition, University of Bordeaux

Cal JC, Desmouliere A, Cambar J (1984) Variations circadiennes et circannuelles de la mortalité induite par le chlorure mercurique chez la souris. In C. R. 109 ème Congrès National des Sociétés Savantes, Dijon, pp 7–18

Cal JC, Dorian C, Cambar J (1985) Circadian and circannual changes in nephrotoxic effects of heavy metals and antibiotics. Annu Rev Chronopharmacol 2: 143–176

Cal JC, Larue F, Guillemain J, Cambar J (1986) Chronobiological approach of protective effect of mercurius corrosivus

against mercury induced nephrotoxicity. Annu Rev Chronopharmacol 3: 99–102

Cal JC, Dorian C, Catroux P, Cambar J (1989a) Nephrotoxicity of heavy metals and antibiotics. In: Lemmer B (ed) Chronopharmacology: cellular and biochemical interaction. Dekker, New York, pp 655–681

Cal JC, Dorian C, Catroux P, Pariat C, Cambar J (1989b) Les facteurs saisonniers comme source d'erreur en toxicologie expérimentale. Sci Tech Anim Lab 14: 121–127

Cambar J, Cal JC (1982) Etude des variations circadiennes de la dose léthale 50 du chlorure mercurique chez la souris. C R Acad Sci [III] 294: 149–152

Cambar J, Cal JC (1985) Rythmes biologiques et toxicité des agents physiques, chimiques et médicamenteux. Pharm Biol 158: 259–269

Cambar J, Cal JC, Desmouliere A, Guillemain J (1983) Etude des variations circadiennes de la mortalité de la souris vis-à-vis du sulfate de cadmium. C R Acad Sci [III] 296: 949–952

Canafax DM, Cipolle RJ, Min DI, Hrushesky WJM, Rabatin JT, Graves NM, Sutherland DER, Bowers LD (1988) Increased evening exposure to cyclosporin and metabolites. Annu Rev Chronopharmacol 5: 5–8

Cardoso SS, Scheving LE, Halberg F (1970) Mortality of mice as influenced by the hour of the day of the drug (araC) administration. Pharmacologist 12: 302 A

Carlson A, Serin F (1950) Time of day as a factor influencing the toxicity of nikethamide. Acta Pharmacol Toxicol (Copenh) 6: 181–186

Caussanel JP (1989) Chronothérapie des cancers par les complexes du platine: étude phase I de l'oxaliplatine. Medical dissertation, University of Paris

Cavallini M, Magnus G, Halberg F, Tao L (1983) Benefit from circadian timing of cyclosporin revealed by delay of rejection of murine heart allograft. Transplant Proc 15: 2960–2965

Cipolle RJ, Canafax DM, Bowers LD, Rabatin JT, Sutherland DER, Hrushesky WJM (1988) Two dose chronopharmacokinetic optimization of cyclosporin in pancreas transplant patients. Annu Rev Chronopharmacol 5: 13–16

Desgagne M, Boutet M, Belanger PM (1988) The mechanism of the chronohepatotoxicity of chloroform in rat: correlation between binding to hepatic subcellular fractions and histologic changes. Annu Rev Chronopharmacol 5: 235–238

Dorian C, Bordenave C, Cambar J (1986) Circadian and seasonal variations in amikacin-induced acute renal failure evaluated by gamma glutamyl transferase excretion changes. Annu Rev Chronopharmacol 3: 111–114

Dorian C, Catroux P, Cambar J (1988) Chronobiological approach to aminoglycosides. Arch Toxicol 12: 151–157

Feuers RJ, Scheving LE (1988) Chronobiology of hepatic enzymes. Annu Rev Chronopharmacol 4: 209–256

Fish J, Yonovitz A, Smolensky M (1984) Effects of circadian rhythm on kanamycin-induced hearing loss. Annu Rev Chronopharmacol 1: 385–388

Fochem K, Michalica W, Picha E (1967) Über die pharmakologische Beeinflussung der tagesrhythmischen Unterschiede in der Strahlenwirkung und zur Frage der jahreszeitlich bedingten Unterschiede der Strahlensensibilität bei Ratten und Mäusen. Strahlentherapie 133: 256–261

Frudden L, Wellso SG (1968) Daily susceptibility of house flies to malathion. J Econ Entomol 61: 1692–1694

Fujimura A, Ebihara A (1987) Chronotoxicity of beta adrenoreceptor blocking agent in spontaneously hypertensive rats. Clin Exp Pharmacol Physiol 14: 805–809

Gruber SA, Cipolle RJ, Canafax DM, Rabatin JT, Erdmann

GR, Hynes PE, Ritz JA, Gould FH, Hrushesky WJM (1988) Circadian-shaped intrarenal cyclosporin delivery. Annu Rev Chronopharmacol 5: 29 A

Halberg F, Stephens AN (1959) Susceptibility to ouabain and physiologic circadian periodicity. Proc Minn Acad Sci 27: 139–143

Halberg F, Bittner JJ, Gully RJ, Albrecht PG, Brackney EL (1955a) Hour periodicity and audiogenic convulsions in mice of various ages. Proc Soc Exp Biol Med 88: 169–173

Halberg F, Spink WW, Albrecht P, Gully RJ (1955b) The influence of *Brucella* somatic antigen upon the temperature rhythm of intact mice. J Clin Endocrinol Metab 15: 887

Halberg F, Johnson EA, Brown BW, Bittner JJ (1980) Susceptibility rhythms of *Escherichia coli* endotoxin and bioassay. Proc Soc Exp Biol Med 103: 142–144

Harris RN, Anders MW (1980) Effect of fasting, diethyl maleate and alcohols on carbon tetrachloride induced hepatotoxicity. Toxicol Appl Pharmacol 56: 191–198

Haus E, Halberg F (1959) 24-hour rhythm in susceptibility of C mice to a toxic dose of ethanol. J Appl Physiol 14: 878–880

Haus E, Nicolau GY, Lakatua D, Sackett-Lundeen L (1987) Reference values for chronopharmacology. Annu Rev Chronopharmacol 4: 333–424

Hayes DK, Morgan NO (1988) Insects as animal models for chronopharmacological research. Annu Rev Chronopharmacol 5: 243–246

Hrushesky W (1985) Circadian timing of cancer chemotherapy. Science 228: 73–75

Jaeger RJ, Conolly RB, Murphy SD (1973) Diurnal variation of hepatic glutathione concentration and its correlation with 1-1,dichloroethylene inhalation toxicity in rats. Res Commun Chem Pathol Pharmacol 6: 465–471

Kabbaj K (1989) Chronopharmacocinétique et chronotoxicité de la cyclosporine chez le rat. Medical dissertation, University of Toulouse

Kuhl JFW, Grace TB, Halberg F, Rosene G, Scheving LE, Haus E (1973) Ellen effect: tolerance of adriamycin by Bagg Albino mice and Fisher rats depends on circadian timing of injection. Int J Chronobiol 1: 335–343

Lavigne JG, Belanger PM, Dore F, Labrecque G (1986) Temporal variations in chloroform induced hepatotoxicity in rats. Toxicology 26: 267–273

Lemmer B, Simrock R, Hellenbrecht D, Smolensky MH (1980) Chronopharmacological studies with propranolol in rodents: implications for the management of CODP patients with cardiovascular disease. In: Smolensky MH, Reinberg A (eds) Recent advances in chronobiology of allergy and immunology. Pergamon, New York, pp 195–208

Lévi F (1982) Chronopharmacologie de trois agents doués d'activité anticancéreuse chez le rat et chez la souris. Chronoefficacité et chronotolérance. Thesis, University of Paris

Lévi F, Hrushesky WJ, Halberg F, Langevin E, Haus E, Kennedy BJ (1982a) Lethal nephrotoxicity and hematologic toxicity of *cis* diaminedichlorplatinum ameliorated by optimal circadian timing and hydration. Eur J Cancer Clin Oncol 18: 471–477

Lévi F, Hrushesky WJ, Blomquist CH, Lakatua DJ, Haus E, Halberg F (1982b) Reduction of cisplatin nephrotoxicity by optimal drug timing. Cancer Res 42: 950–955

Lévi F, Boughattas NA, Blazsek I (1987) Comparative murine chronotoxicity of anticancer agents and related mechanisms. Annu Rev Chronopharmacol 4: 283–331

Liu T, Cavallini M, Halberg F, Cornelissen G, Field J, Sutherland DER (1986) More on the need for circadian circaseptan

and circannual optimization of cyclosporine therapy. Experientia 42: 20–22

Luke DR, Vadiei K, Brunner LJ (1988) Influence of circadian changes in triglyceride concentrations on the pharmacokinetics and experimental toxicity of cyclosporin. Annu Rev Chronopharmacol 5: 31–34

Lutsch EF, Morris RW (1967) Circadian periodicity in susceptibility to lidocaine hydrochloride. Science 156: 100–102

Malmary MF, Kabbaj K, Oustrin J (1988) Circadian dosing-stage dependence in metabolic effects of cyclosporin in the rat. Annu Rev Chronopharmacol 5: 35–38

MacGeachy SM, Leduc G (1988) The influence of season and exercise on the lethal toxicity of cyanide to rainbow trout. Arch Environ Contam Toxicol 17: 313–318

MacLeay DJ, Munro JR (1979) Photoperiodic acclimatization and circadian variations in tolerance of juvenile rainbow trout to zinc. Bull Environ Contam Toxicol 23: 552–557

Moore-Ede MC (1973) Circadian rhythms of drug effectiveness and toxicity. Clin Pharmacol Ther 14: 925–935

Mormont MC, Boughattas N, Lévi F (1989) Mechanisms of circadian rhythms in the toxicity and efficacy of anticancer drugs: relevance for the development of new analogs. In: Lemmer B (ed) Chronopharmacology: cellular and biochemical interactions. Dekker, New York, pp 395–437

Nakano S, Ogawa N (1982) Chronotoxicity of gentamycin in mice. IRCS Med Sci 10: 592–593

Pariat C (1986) Etude expérimentale de la chrono-susceptibilité rénale de trois aminoglycosides. Thesis, University of Poitiers

Pariat C, Courtois P, Cambar J, Piriou A, Bouquet S (1988) Circadian variations in the renal toxicity of gentamicin in rats. Toxicol Lett 40: 175–182

Pariat C, Ingrand P, Cambar J, de Lemos E, Piriou A, Courtois P (1990) Seasonal effects on the daily variations of gentamicin induced nephrotoxicity. Arch Toxicol 64: 205–209

Pati A, Florentin I, Lemaigre G, Mechkouri M, Lévi F (1988) Chronopharmacologic optimization of oral cyclosporin A in mice: a search for a compromise between least renal toxicity and highest immunosuppressive effects. Annu Rev Chronopharmacol 5: 43 A

Pericin C (1981) Effect of season on the acute toxicity of cyclophosphamide in the Chinese hamster and the mouse under laboratory conditions. Experientia 37: 401–402

Prat M (1990) Variations circadiennes de la toxicité aigüe et de la pharmacocinétique plasmatique, cardiaque et cérébrale de trois anesthésiques locaux (bupivacaïne, étidocaïne et mépivacaïne) chez la souris; approche de leur mécanisme. Thesis, University of Aix-Marseille

Prat M, Bruguerolle B (1988) Chronotoxicity and chronokinetics of two local anesthetic agents, bupivacaine and mepivacaine, in mice. Annu Rev Chronopharmacol 5: 263

Ramon M, Morel D, Penornie F, Grellet J, Potaux L, Saux MC, Brachet-Liermain A (1989) Variations mycthemérales de la cyclosporine administrée par voie orale à des transplantés renaux. Thérapie 44: 371–374

Reinberg A, Halberg F (1971) Circadian chronopharmacology. Annu Rev Pharmacol 2: 455–492

Roberts DR, Smolensky MH, Hsi BP, Scanlon JE (1974) Circadian pattern in susceptibility of *Aedes aegypti* (L.) larvae to Dursban. In: Scheving LE, Halberg F, Pauly J (eds) Chronobiology. Igaku-Shoin, Tokyo, pp 612–616

Ross FHN, Sermons AL, Owasoyo JO, Walker CA (1981) Circadian variation of diazepam acute toxicity in mice. Experientia 37: 72–73

Scheving LE, Vedral DF (1966) Circadian variation in susceptibility of the rat to several different pharmacological agents.
Anat Rec 154: 417

Scheving LE, Vedral D, Pauly J (1968) A circadian susceptibility rhythm in rats to pentobarbital sodium. Anat Rec 160:
741–750

Scheving LE, Cardoso SS, Pauly JE, Halberg F, Haus E (1974a)
Variations in susceptibility of mice to the carcinostatic agent
arabinosylcytosine. In: Scheving LE, Halberg F, Pauly J (eds)
Chronobiology. Igaku-Shoin, Tokyo, pp 213–217

Scheving L, von Mayersbach H, Pauly JE (1974b) An overview
of chronopharmacology. J Eur Toxicol 7: 203–227

Schnell CR, Bozigian HP, Davies MH, Merrick BA, Johnson
KL (1983) Circadian rhythm in acetaminophen toxicity: role
of nonprotein sulfhydryls. Toxicol Appl Pharmacol 71: 353–
358

Thompson HH, Walker CH, Hardy AR (1988) Avian esterases
as indicators of exposure to insecticides. The factor of diurnal
variations. Bull Environ Contam Toxicol 41: 4–12

Tsai TH, Burns ER, Scheving LE (1974) Circadian differences
in anti-cancer immunity induced by Bacillus-Calmette-
Guerin (BCG). Anat Rec 178: 478

von Mayersbach H (1976) Time. A key in experimental and
practical medicine. Arch Toxicol 36: 185–216

Wachsmuth ED (1982) Quantification of acute cephaloridine
nephrotoxicity in rats: correlation of serum and 24 hour urine
analyses with proximal tubular injuries. Toxicol Appl Pharmacol 63: 429–445

# Rhythms in Drug-Induced Teratogenesis

I. Sauerbier

## Introduction

Despite the lessons to be learned from the thalido-mide disaster some 20 years ago, recent studies indicate that pregnant women are still consuming therapeutic agents in alarming quantities. Moreover the developmental risks imposed by the maternal use of substances of abuse are, perhaps more than many other risk factors, preventable. The etiology of human malformations includes both (1) genetic factors and (2) drugs and environmental agents (for established teratogens see Brendel et al. 1985).

The influence a teratogenic drug has on the conceptus depends mainly on: (1) physical properties and metabolic patterns of the compounds. The situation is especially complex in the prenatal stage since maternal and fetal pharmacokinetic factors contribute to fetotoxic actions. Thus developmental toxicity involves the following parameters: maternal pharmacokinetics (including the rates of absorption, distribution, metabolic conversion and elimination), placental transfer, distribution within the embryonic or fetal organism, prenatal drug metabolism in the fetus, final concentration at the target cell within the fetal compartment. Thus agents may by direct or indirect mechanisms influence fetal development. (2) Developmental stage. Stage sensitivity indicates that susceptibility to teratogenesis varies during gestation. The most critical period in the development or growth of a particular tissue or organ is during the time of most rapid cell division. During the period of organogenesis (from day 18 through about day 60 of gestation in the human conceptus which corresponds to gestational days 7–13 in mice, e.g.) mammalian embryonic tissue is an especially sensitive target for attack by teratogenic insults. (3) Dose of compound. Dose-response relationships refer to the phenomenon that, as the exposure or dosage increases, frequency and severity of the teratogenic effect increase, as well. Also threshold effects are known. (4)

Background genotype. (5) Physiological and pathological status of the mother.

## Chronobiological Aspects in Teratology

Circadian influences on many biological systems have been well documented. Included among the phenomena which exhibit circadian rhythms in susceptibility are responses to drugs in adult organisms (Reinberg and Smolensky 1983). These rhythmic variations in drug response have been predominantly attributed to variations in pharmacokinetics relative to treatment scheduling.

There are only a few interesting and significant observations regarding the circadian phase influence of drug exposure on subsequent dysgenesis of the fetus. This has been reported for cortisone (Isaacson 1959), dexamethasone (Sauerbier 1986a), hydroxyurea (Clayton et al. 1975), 5-fluorouracil (Sauerbier 1986b), cyclophosphamide (Schmidt 1978; Sauerbier 1981, 1983), cytosine arabinoside (Endo et al. 1987) and ethanol (Sauerbier 1987, 1988).

While performing these teratological studies it will be worthwhile limiting the mating regimen of the animals to a few hours during a 24-h span to give an approximation of the embryonic age. Otherwise the variation in response may result from different developmental stages. This is of particular importance for animals with a short gestational period such as mice, e.g. Thus it can be clearly shown that the circadian phase of drug administration has a bearing on teratogenicity. Also the success rate of breeding is significantly dependent on the circadian stage, being greatest in the morning in mice (Sauerbier 1982).

In the following a short survey is given of those experimental studies that apply to the circadian phase dependent in utero exposure of potential teratogens with emphasion ethanol chronotoxicity.

## Glucocorticoids

Glucocorticoids such as cortisone and dexamethasone are known to produce left palates in rodents at the time of palatal shelf closure (e. g., days 11–15 of gestation for mice). Significantly more cleft palates have been observed during the dark phase, particularly near the onset of the light phase as compared to the other circadian stages (Isaacson 1959; Sauerbier 1986a). In general the frequency distribution of cleft palates is negatively correlated to the circadian fluctuations in serum corticosteroid levels (Sauerbier 1986a), which in turn may be mediated by changes in the hypothalamic-pituitary-adrenal system as having a regulatory influence. Probably circadian-based hormonal effects that modify functional activity through membrane-receptor concentrations may influence teratogenesis (Sauerbier 1989).

## Anticancer Drugs

### Hydroxyurea

Clayton et al. (1975) have examined the teratogenic effects of 750 mg/kg hydroxyurea in rat fetuses following maternal treatment on day 12 of gestation. The incidence of deformities (most common malformations, digital and limb defects) is greatest during the light phase with the maximal amplitude at the dark to light transition. The temporal pattern of teratogenesis has been found to be correlated with motor activity, mitotic rates and DNA synthesis.

### 5-Fluorouracil

Circadian variability of fetal tissues to 5-fluorouracil toxicity has been observed in mice (Sauerbier 1986b, 1989). One hundred percent of the embryos were malformed (mainly digital defects and kinky tail) by the 1400 hours injection of 30 mg/kg on day 11 of gestation whereas 78.2% are so affected when the drug is given at 0100 hours (fetal loss, 32.2% vs. 16.7%).

### Cyclophosphamide

Cyclophosphamide is widely used as a cancer chemotherapeutic agent and as an immunosuppressant. Its use is associated, however, with certain undesirable or toxic effects including fetal malformations. The developmental toxicity of cyclophosphamide has been studied in a variety of mammalian species (Chernoff et al. 1989), which is manifested as lethality, weight reduction and/or a spectrum of external malformations including digital defects, encephalocele, cleft palate, open eyelid, abnormal vertebral

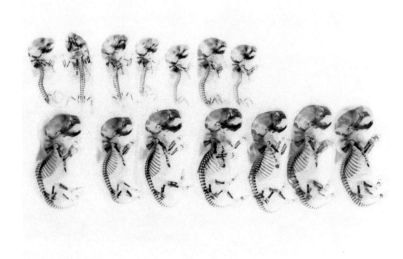

**Fig. 1.** Alizarin red-stained skeletons of day 18 mouse fetuses after maternal exposure to a single dose of 20 mg/kg cyclophosphamide on day 12 at 0700 hours, the time of maximal susceptibility *(upper line)* and at 0100 hours, the time of minimal susceptibility *(lower line)*. Within a given litter all fetuses are similarly affected

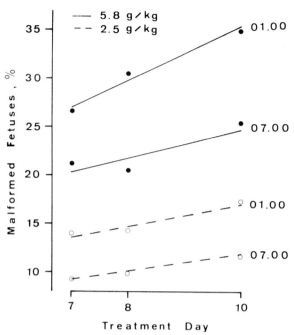

**Fig. 3.** Effect of a single ethanol exposure in mice on teratogenicity as a function of the gestational day of treatment and the circadian phase ($N = 12$ litters/group)

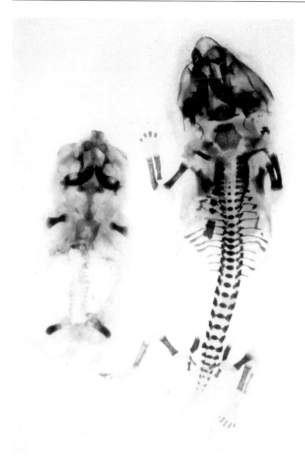

**Fig. 2.** Alizarin red-stained skeletons of day 18 mouse fetuses after a single injection of 20 mg/kg cyclophosphamide on day 12 at 0700 hours *(left)* and at 0100 hours *(right)*. Note the difference in size and ossification as well as possible seasonal differences (see Fig. 1)

column and missing, supernumerary or wavy ribs. This is dependent on the developmental stage during exposure and exhibits a dose-response relationship as well as a strong circadian phase dependence (Sauerbier 1989). Moreover the circadian periodic effect on teratogenesis has been shown for consecutive days of gestation (days 11–13) in mice. The highest incidence of fetal malformations always occurs at the transition from dark to light (0700 hours) while the severity and number of external malformations are significantly reduced at 0100 hours (Figs. 1, 2). For example, the data obtained from fetal examination near term (day 18) demonstrate that administration of 20 mg/kg cyclophosphamide on one specific day of gestation (days 11–13) causes a malformation rate ranging from 82.2% to 65.3% at 0700 hours and from 32.6% to 27.8% at 0100 hours, with a peak of teratogenic sensitivity on day 12. This

has been found to be highly correlated with fetal weight and fetal death.

The potent cytotoxicity of cyclophosphamide has been well documented, suggesting a relationship between cell cycle perturbation and cell death in cyclophosphamide teratogenesis (Chernoff et al. 1989). This has also been demonstrated for excessive cell death in murine limb buds and subsequent limb malformations. Moreover, the location of embryonal bud cell death corresponds to the location of the highest DNA synthesis rates in the limb bud (Chernoff et al. 1989). Consequently, an interaction between circadian variation in DNA synthesis and teratogenic outcome is suggested. However, since cyclophosphamide requires extraembryonic metabolism in maternal liver for biological activity to be teratogenic (Hales and Slott 1987), it becomes quite obvious that maternal rhythms may also play a role while interfering with the circadian variability of fetal tissues to cyclophosphamide.

Cytosine Arabinoside

In contrast to all the previous reports on chronoteratology, Endo et al. (1987) have found no difference in circadian susceptibility to cytosine arabinoside in

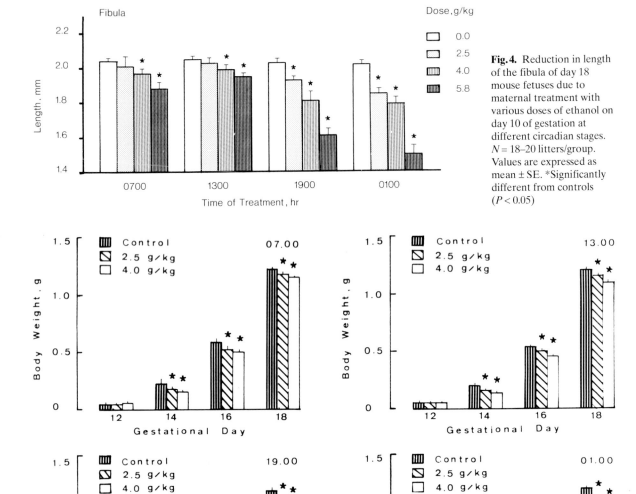

**Fig. 4.** Reduction in length of the fibula of day 18 mouse fetuses due to maternal treatment with various doses of ethanol on day 10 of gestation at different circadian stages. $N = 18$–20 litters/group. Values are expressed as mean ± SE. *Significantly different from controls ($P < 0.05$)

**Fig. 5.** Effect of a single ethanol exposure on fetal body weight after maternal administration on day 10 at four selected time points as gestation progressed. $N = 10$–14 litters/group. Values are expressed as mean ± SE. *Significantly different from controls ($P < 0.05$)

mice when studying the embryotoxic and teratogenic effects at 1000 hours and 1600 hours on day 11 of gestation during different lighting regimens. Several explanations may account for this including species and strain differences as well as differences in the experimental design.

## Alcohol

There is now no longer any doubt as a result of clinical, epidemiological and experimental research that ethanol is a teratogen. Excessive ingestion of ethanol by either pregnant women or experimental animals is known to cause a characteristic pattern of developmental defects known as the fetal alcohol syndrome (Jones and Smith 1973). Along with characteristic craniofacial dysmorphogenesis, CNS dysfunction and various congenital anomalies, especially cardiac and skeletal defects, pre- and postnatal growth retardation is one of the most common deficits observed in the fetal alcohol syndrome (Abel 1984, 1985). Both chronic and acute ingestion of ethanol can exert profound pharmacological as well as pathological defects on developmental processes which can be related to the developmental stage and to the dosage.

Although numerous reports have confirmed the circadian variation in ethanol metabolism in human and adult laboratory animal species (for review see Sauerbier 1987), recently animal studies have also documented the circadian variability of fetal tissues to ethanol toxicity (Sturtevant and Garber 1985; Sauerbier 1987, 1988). Chronovariation in the sensitivity of developing tissues has been described for cerebellar neurogenesis and several embryonic parameters such as weight, size, gross malformations and fetal death. Consistent with all findings the deficit is significantly greater in those embryos, fetuses or pups with maternal ethanol exposure during the early dark hours until the mid-dark hours. The highest incidence of external abnormalities, in utero absorption and reduction in fetal body weight is seen in those groups treated at 0100 hours compared with the other circadian phases of administration, particularly at 0700 hours when examining the fetuses on day 18 after a single ethanol exposure on day 7, 8 or 10 of gestation. Moreover, the severity of damage is strongly related to the dose (Fig. 3).

Since fetal insult is positively correlated with maternal blood ethanol concentrations at the different circadian stages, a direct effect of ethanol on embryonic/fetal tissues can be assumed (Sauerbier (1987). Moreover, the circadian variation in fetotoxicity may be attributed to changes in the rate of etha-

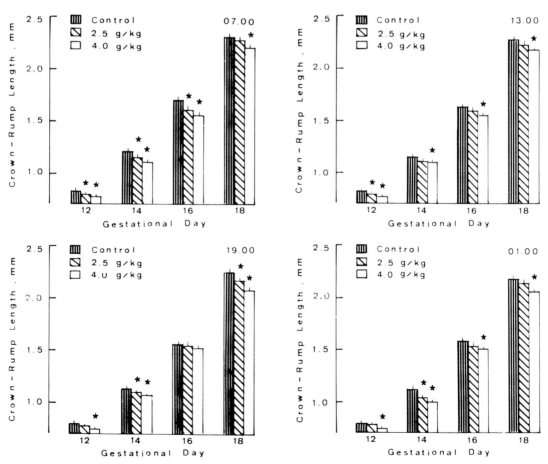

**Fig. 6.** Effect of a single ethanol exposure (2.5 g/kg, 4.0 g/kg) on crown-rump length following treatment of mice on day 10 at different circadian stages. $N = 10$–14 litters/group. Values are expressed as mean ± SE. *Significantly different from controls ($P < 0.05$)

nol metabolism. For example, circadian rhythms of the alcohol dehydrogenase (ADH) and the microsomal ethanol-oxidizing system (MEOS) have been reported (Sturtevant and Garber 1985, 1986). Sessa et al. (1987) have pointed out that in early stages of fetal growth ethanol does not seem to be oxidized because the activity of fetal liver ADH is either absent or very low.

Intrauterine growth retardation is one of the principal features of the fetal alcohol syndrome (Abel 1985), manifested by an overall reduction in somites, fetal crown-rump length and fetal weight, e.g. (Figs. 4–6). All these growth parameters vary in a circadian manner which can be documented for different gestational days following ethanol exposure until term. Although there is a general developmental stage related increase in weight and size for all groups as gestation progresses ethanol-treated animals remain more or less retarded. There is evidence that tissue repair after ethanol-induced damage may only occur at lower doses (1.5 and 2.0 g/kg) throughout the latter prenatal period (unpublished data). Also the total protein content is significantly decreased 48 h posttreatment (Sauerbier 1988).

It has widely been discussed whether fetal growth retardation represents a direct effect of ethanol and/or one of its metabolites, or whether it is due to indirect mechanisms such as maternal restricted carolic intake, e.g. Both, ethanol and acetaldehyde have been implicated as causative agents in the fetal alcohol syndrome. Both substances cross the placental barrier and enter the fetal circulation, producing adverse effects in the fetus (Guerri and Sanchis 1985).

Ethanol has been determined to cause metabolic acidosis and fetal hypoxia due to either a change in placental functions resulting from ethanol damage or to contraction of umbilical vessels caused directly by ethanol (Jones et al. 1981; Altura et al. 1982; Abel 1985). Moreover, Savoy-Moore et al. (1989) assume a similar constrictive effect on placental vessels, thereby increasing umbilicoplacental resistance. It is known there is a characteristic pattern in uterine activity and in umbilical blood flow (Walker et al. 1977; Walsh et al. 1984), which additionally influence the nutrient transfer. The placentotoxic effects include blood diminution, altered placental growth, impaired placental protein synthesis, reduced placental folic receptor activity, inhibition of placental zinc transport and reduction of placental transfer of glucose and amino acids, the latter probably inhibiting the enzyme Na, K-

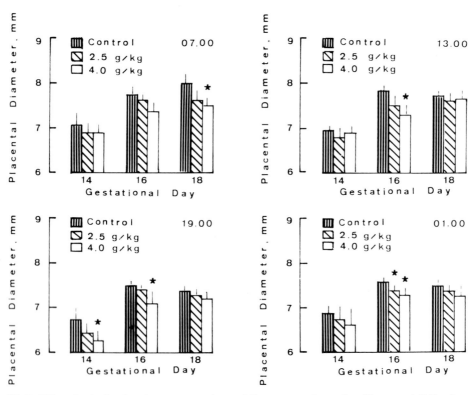

**Fig. 7.** Effect of a single ethanol exposure on placental diameter after administration of 2.5 or 4.0 g/kg ethanol to mice on day 10 of gestation. The mean ± SE has been calculated from 10–14 litters/group. *Significantly different from controls ($P < 0.05$)

ATPase (Abel 1985; Aufrere and Le Bourhis 1987). Previous study have indicated a distinct circadian rhythm in placental protein content 48 h posttreatment, showing a significant correlation between dose level and protein content (Sauerbier 1988). Also the placental diameter is significantly affected by maternal ethanol exposure (Fig. 7), indicating severe intoxication. Interestingly placental sensitivity seems to precede embryonic sensitivity by approximately 6 h, which possibly relates in turn to effects of hormone levels such as prolactin and cortisol, e. g., by affecting the clearance of ethanol. It may generally be based on a complex process regulated by circadian hormonal interactions between the mother and the fetus.

## Conclusion

Intrauterine drug exposure has become a major cause of perinatal morbidity and mortality. Since one of the main goals of developmental toxicology is to evaluate and to exclude potential hazards induced pre- and perinatally in humans, it is necessary that data obtained in experimental animals can be extrapolated to the situation in men. A better understanding of the mode of embryotoxic drugs and of the reasons for the special susceptibility of embryonic tissues to certain toxic events also includes additional research on the influence of circadian susceptibility rhythms of embryonic tissues in mediating the teratogenic outcome as illustrated above. It will increase our understanding of the mechanism(s) involved and make it possible to protect the embryo from hazardous reactions.

## References

Abel El (1984) Fetal alcohol syndrome. Fetal alcohol effects. Plenum, New York

Abel El (1985) Prenatal effects of alcohol on growth: a brief overview. Fed Proc 14: 2318–2322

Altura BM, Altura BT, Carella A, Chatterjee M, Halevy S, Tejani N (1982) Alcohol produces spasms of human umbilical blood vessels: relationship to fetal alcohol syndrome (FAS). Eur J Pharmacol 86: 311–312

Aufrere G, Le Bourhis B (1987) Effect of alcohol intoxication during pregnancy on foetal and placental weight: experimental studies. Alcohol Alcohol 22: 401–407

Brendel K, Duhamel RC, Shepard TH (1985) Embryotoxic drugs. Biol Res Pregnancy 6: 1–54

Chernoff N, Rogers JM, Alles AJ, Zucker RM, Elstein KH, Massero EJ, Sulik KK (1989) Cell cycle alterations and cell death in cyclophosphamide teratogenesis. Teratogenesis Carcinog Mutagen 9: 199–209

Clayton DL, Mullen AW, Barnett CC (1975) Circadian modification of drug-induced teratogenesis in rat fetuses. Chronobiologia 2: 210–217

Endo A, Sakai N, Ohwada K (1987) Analysis of diurnal differences in teratogen (Ara-C) susceptibility in mouse embryos by a progressive phase-shift method. Teratogenesis Carcinog Mutagen 7: 475–482

Guerri C, Sanchis R (1985) Acetaldehyde and alcohol levels in pregnant rats and their fetuses. Alcohol 2: 267–270

Hales BF, Slott VL (1987) The role of reactive metabolites in drug-induced teratogenesis. Prog Clin Biol Res 253: 181–191

Isaacson RJ (1959) Investigation of some of the factors involved in the closure of the secondary palate. Thesis, Minnesota University, Minneapolis

Jones JH, Leichter J, Lee M (1981) Placental blood in rats fed alcohol before and during gestation. Life Sci 29: 1153–1159

Jones KL, Smith DW (1973) Recognition of the fetal alcohol syndrome in early infancy. Lancet 2: 999–1001

Reinberg A, Smolensky MH (1983) Biological rhythms and medicine. Springer, Berlin Heidelberg New York

Sauerbier I (1981) Circadian system and teratogenicity of cytostatic drugs. Prog Clin Res 59 C: 143–149

Sauerbier I (1982) Unterschiedliche Konzeptionsfähigkeit der weiblichen Han: NMRI Maus. Einfluß der Verpaarung am Abend und am Morgen. Z Versuchstierkd 23: 161–164

Sauerbier I (1983) Embryotoxische Wirkung von Zytostatika in Abhängigkeit von der Tageszeit der Applikation bei Mäusen. Verh Anat Ges 77: 147–149

Sauerbier I (1986a) Circadian variation in teratogenic response to dexamethasone in mice. Drug Chem Toxicol 9: 25–31

Sauerbier I (1986b) Circadian modification of 5-fluorouracil-induced teratogenesis in mice. Chronobiol Int 3: 161–164

Sauerbier I (1987) Circadian modification of ethanol damage in utero in mice. Am J Anat 178: 170–174

Sauerbier I (1988) Circadian influence on ethanol-induced intrauterine growth retardation in mice. Chronobiol Int 5: 211–216

Sauerbier I (1989) Embryotoxicity of drugs: possible mechanisms of action. In: Lemmer B (ed) Chronopharmacology. Cellular and biochemical interactions. Dekker, New York, pp 683–697 (Cellular clocks, vol 3)

Savoy-Moore RT, Dombrowski MP, Cheng A, Abel EA, Sokol RJ (1989) Low dose alcohol contracts the human umbilical artery in vitro. Alcoholism (NY) 13: 40–42

Schmidt R (1978) Zur zirkadianen Modifikation der pränataltoxischen Wirkung von Cyclophosphamid. Biol Rundsch 16: 243–248

Sessa A, Desiderio MA, Perin A (1987) Ethanol and polyamine metabolism in adult and fetal tissues: possible implication in fetus damage. Adv Alcohol Subst Abuse 6: 73–85

Sturtevant RP, Garber SL (1985) Circadian exposure to ethanol affects the severity of cerebellar cell dysgenesis. Anat Rec 211: 187

Sturtevant RP, Garber SL (1986) Chronopharmacology of ethanol: acute and chronic administration in the rat. Annu Rev Chronopharmacol 4: 47–76

Walker AM, Oakes GK, McLanghlin MK, Ehrenkranz RA, Chez RA, Alling DW (1977) 24-hr rhythms in uterine and umbilical blood flows of conscious pregnant sheep. Gynecol Invest 8: 288–298

Walsh SW, Ducsay CA, Novy MJ (1984) Circadian hormonal interactions among the mother, fetus and amniotic fluid. Am J Obstet Gynecol 150: 745–753

# Chronobiology of Pregnancy and the Perinatal Time Span

P. J. Meis

## Introduction

Most biological rhythms detected in nonpregnant women persist in the pregnant state. Many of these rhythms are altered during the process of physiological adaptation to pregnancy. A discussion of these rhythms and changes is outside the limited scope of this chapter, which will focus on chronobiological events unique to pregnancy, the fetus, and the newborn. We shall discuss results obtained from the study of humans with some additional data obtained from study of nonhuman primates.

## Labor and Birth

Biological rhythms related to labor and birth have been recognized since at least 1829 (Buek, cited by Smolensky et al. 1972). Since that time, the circadian periodicity of time of birth has been rediscovered many times by many authors. Important descriptions and summaries were published by Malek et al. (1962), Kaiser and Halberg (1962), and Longo and Yellon (1988). Almost all authors reported an increased rate of births occurring during the dark period, usually peaking in early morning hours.

Kaiser and Halberg (1962), in their review of this subject, discuss some of the pitfalls and artifacts introduced in some published series. Births resulting from induced labor or delivered by cesarean section are influenced by the work schedule of birth attendants and are likely to occur in the daytime between 0800 and 1700 hours. King (1960) studied birth in different hospitals in Richmond, Virginia, and found phases of birth frequency approximately 180° apart in different hospitals, explained by one of the hospitals including a relatively high rate of births resulting from induced labors. Other artifacts may include a

possibility of birth attendants lacking some degree of accuracy in reporting the time of birth. Evidence for this possibility includes series with peaks of birth around midnight and a greater tendency for births at even clock hours compared with odd-numbered hours.

The circadian rhythm of the onset of labor has been examined by many authors, with early systematic studies by Malek et al. (1962) and Kaiser and Halberg (1962). The results of these later studies present an apparent paradox. The timing of the peak time for onset of labor was found to be 0100 hours and the peak time of delivery between 0300 and 0400 hours. This small phase difference seems inconsistent with the usual reports of length of labor for primigravidas of 9–10 h and for multigravidas of 6 h. Malek et al. (1962) attempted to resolve this inconsistency by examining the duration of labor by time of onset, and discovered that labors beginning during the night hours were shorter than those beginning during the daytime. Labors whose initial event was rupture of the fetal membranes showed a similar rhythm and phase timing to labors whose onset was the occurrence of uterine contractions.

Smolensky et al. (1972) have provided the largest collection of reports concerning rhythms of labor and birth and have presented elegant data analysis by the use of group mean cosinor analysis. They collected and analyzed more than 2 000 000 cases of spontaneous labors for time of birth and more than 200 000 cases of spontaneous labor for time of onset of labor. Smaller numbers of cases of induced labor and stillbirth were analyzed by time of birth. Results of their analysis are shown in Fig. 1. They found highly statistically significant circadian rhythms for the following: spontaneous onset of labor showed acrophase at 0100 hours and an amplitude of 36% of the mesor. Birth after spontaneous onset of labor showed an acrophase at 0416 hours and a smaller amplitude of 13%. The peak timing of births after induced labor was 1336 hours and of stillbirth was

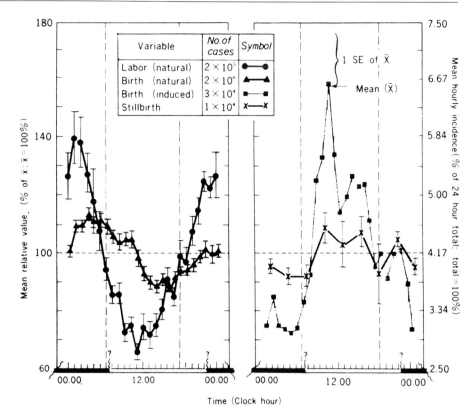

**Fig. 1.** Data collected from many reports showing circadian rhythms of spontaneous labor and birth *on the left*, and birth following induced labor and stillbirth *on the right*. (From Smolensky et al. 1972)

1708 hours. The considerably smaller amplitude of the rhythms of time of birth compared with that of onset of labor suggests greater variability in the length of labor. Thus, while there is a strong tendency for labors to begin around 0100 hours, the time of termination in birth is more variable. This variability along with the previously demonstrated shorter length of night-time labors explains the apparent paradox of the acrophases of labor and birth differing by such a small amount of time.

Relatively little has been written concerning the circadian rhythm of premature labor and birth. Malek et al. (1962) mention that premature births showed a similar timing to term births. Cooperstock (1987a, b) examined births registered in the Collaborative Perinatal Project. The time of onset of labors beginning with contractions was similar in term and preterm deliveries, with peak times ranging from 0030 to 0330 hours. Birth associated with histological signs of inflammation of the fetal membranes showed a different phase of onset of labor with a peak at 1945 hours. Labors beginning with rupture of the fetal membranes showed a peak time of rupture of 0200–0500 hours for both term and preterm labors. Those labors whose time of rupture was near 0200 hours had a shorter latent time (time elapsing

until onset of contractions). Those labors beginning with rupture of the membranes and associated with inflammation of the membranes showed no circadian rhythm of time of rupture.

The timing of birth in other mammalian species has been examined by many authors. Extensive reviews have been made by Jolly (1972) and by Longo and Yellon (1988). In a wide variety of mammalian species, birth tends to occur during the dark or night phase for diurnally active animals including most primate species and during the day or light phase for nocturnally active animals such as rodents or prosimian primates.

An interesting examination of the periodicity of length of gestation was recently performed by Brown (1988). In a comparison of gestational lengths of 213 types of terrestrial placental mammals, Brown found a tendency for gestation length to cluster around the period of 30 days and its multiples. For example, human pregnancy length of 266 days closely approximates 9 lunar cycles. Brown conjectures that moonlight may be a possible zeitgeber for determining the length of gestation.

## Uterine Activity

The description of circadian rhythms of uterine activity is of recent origin (Harbert et al. 1970), and almost all published reports to date have studied nonhuman primate species. Although Harbert et al. (1970) described circadian rhythms of uterine activity in nonpregnant rhesus monkeys, the first description of these rhythms in pregnant animals was made by Harbert in 1977. Harbert measured intraamniotic pressure and uterine blood flow in rhesus *Macaca* and used Fourier analysis to demonstrate statistically significant circadian rhythms for these functions in individual animals. The acrophase of intraamniotic pressure occurred near the onset of the dark phase (1600 hours in his studies) with a range from 1458 to 1833 hours. These changes in intraamniotic pressure were related to contractions of the uterine smooth muscle. The amplitude of the intraamniotic pressure rhythms derived from hourly means ranged from 9 % to 31 % of the 24-h mean values for each animal. Uterine blood flow also showed consistent circadian rhythms whose peaks tended to be in antiphase to the intraamniotic pressure rhythms. This observations is not surprising since contraction of uterine smooth muscle is known to impede blood flow through the organ. In two animals Harbert demonstrated a phase shift in uterine blood flow and intraamniotic pressure rhythms after shifting the L/D cycle by 12 h. Harbert and Spisso (1980) extended these observations to the last part of gestation and monitored 17 animals through labor and delivery. Once again, Fourier analysis showed significant rhythms of intraamniotic pressure and uterine blood flow which tended to be in antiphase to one another. The timings of the rhythms in individual animals were not consistent with one another, and the time of delivery was not consistent. When delivery occurred during the dark phase, intraamniotic pressure rhythms phase shifted 24–36 h prior to delivery resulting in peak activity during the dark phase. Ducsay et al. (1983) also studied intraamniotic pressure in rhesus monkeys. A group of five animals showed highly consistent and statistically significant rhythms of uterine activity with peak hourly activity (circadian amplitude) 2.5 times that of 24-h mean contraction activity. The peak time (acrophase) occurred at 2400 hours (2 h after lights off). The uterine activity rhythm was abolished by the infusion of dexamethasone to the pregnant monkeys. More recently, using similar methodology, Ducsay et al. (1990) demonstrated that rhythms of intraamniotic pressure and maternal plasma concentrations of progesterone, cortisol, estrone and estradiol could be shifted by alteration of the light/dark cycle.

Although assessment of intraamniotic pressure provides useful information, it is an indirect measurement of uterine activity. The use of electromyographic (EMG) recording of uterine muscle activity provides an alternative technique which yields a direct measurement of the activity of uterine smooth muscle. Utilizing this methodology, Tayler et al. (1983) examined circadian rhythms of uterine activity in groups of monkeys who received either EMG implantation alone or EMG plus fetal catheters. Analysis of variance was used to examine rhythmicity. Two different types of EMG patterns were identified. Type I patterns (short bursts of activity lasting 0.5–1.2 min) exhibited prominent circadian rhythms with peaks around 2400 hours (5 h after lights off). The amplitude of these rhythms increased prior to delivery and also increased when the fetus was stressed by inserting catheters in fetal vessels. In some preparations, the fetus died, and this event resulted in diminution of EMG activity and ablation of circadian rhythmicity. Taylor concluded that circadian rhythms of pregnant uterine activity were regulated by the fetus.

Recently, Figueroa et al. (1990) using EMG-instrumented pregnant rhesus monkeys found that uterine activity switches from contractures (type II pattern of Taylor) to contractions (type I of Taylor) shortly after lights out. Mean time of peak contraction activity was 2.6 h after lights off. Shifting the L/D cycle by 6 h shifted peak contraction activity by 4 h.

Little has been published concernig circadian rhythms of human uterine activity. TambyRaja and Hobel (1983) reported studies of pregnant women using external tocodynamometry. They found no measurable uterine activity prior to 30 weeks gestation, irregular contractions at 0600–1900 hours between 30 and 36 weeks, and the appearance of contractions at 2000–2400 hours only after 36 weeks. Several groups of investigators are currently attempting to measure rhythms in pregnant human uterine activity using more sophisticated techniques of ambulatory monitoring.

### Practical Applications

Much effort is made in contemporary practice of obstetrics in attempting to control the timing of birth. These efforts include the use of tocolytic drugs or other therapies to stop premature labor and prolong pregnancy. Equally important are efforts made to induce labor in problem pregnancies to minimize risk to the fetus and mother. The circadian rhythmicity of birth and the rhythms of uterine activity in humans which are the likely mechanism of these birth rhythms suggest that approaches to chronotherapy may be useful. These strategies might employ increases in tocolytic drugs during the dark phase when uterine activity is likely to be high and labor and birth most commonly occur. Further information about circadian rhythms of uterine activity in healthy pregnant women and women in preterm labgor is urgently needed. Studies of the chronopharmacology of drugs employed to stop uterine activity or to enhance uterine activity are also urgently needed. It seems likely that pharmacologic or other therapies to induce labor may be more effective during the dark phase, when spontaneous labor is most likely to begin.

## Fetal Rhythms

Until the advent of noninvasive tools and techniques to measure fetal behavior, the only way of measuring human fetal activity was by maternal perception of fetal movement. Although clinicians and investigators recognized that the fetus responded to stimuli such as sound and vibration (Sontag 1941), no relation to fetal activity to time of day was appreciated (Schmiedler 1941). This lack of detection was due, in part, to the inability to monitor the fetus during maternal sleep. Sterman (1967) used a system of external electrodes on the maternal abdomen to measure fetal activity over 24 h and found a circadian rhythm of fetal movement with marked increases of fetal activity during maternal sleep and, especially, during REM sleep.

The availability of the technique of ultrasound to detect fetal motion brought a new realm of possibility to the study of human fetal anatomy, physiology and chronobiology. The first use of these methods to examine circadian rhythms was published by Boddy and Dawes (1974), who used A-mode ultrasound to study fetal breathing and found it was most frequent in the early evening. Subsequent investigations have used real-time ultrasound to measure fetal breathing and fetal movement and ultrasound transducers to measure fetal heart rate characteristics. A large body of the work published concerning human fetal circadian and ultradian rhythms was performed by the late

John Patrick and his coworkers at the University of Western Ontario, London, Ontario. Dr. Patrick's recent untimely death is a real loss of the field of investigations of fetal physiology and chronobiology.

## Fetal Movement

The use of continuous visualization of the fetus permits 24-h monitoring of fetal activity. When fetuses are studied longitudinally over the course of gestation, one finds the development of circadian rhythms of fetal movement over time. DeVries et al. (1985) found no circadian pattern of movement prior to 30 weeks gestation, with increasingly prominent peaks of movement at 2100–0100 hours after this time. Nasello-Paterson et al. (1988), using similar methods, demonstrated circadian rhythms of fetal movement at 24–28 weeks with peak movement between 2300 and 0800 hours.

Ultradian patterns of fetal movement have also been described. Patrick et al. (1978) described the episodic occurrence of these movement patterns and with Richardson et al. (1979) qualitatively assessed a cycle length of about 90 min. Campbell (1980) applied more sophisticated statistical analysis with spectral power analysis and Box-Jenkins time domain techniques. Significant rhythms were found in the range of 60 to 500 min.

## Fetal Breathing

Recent interest in and appreciation of human fetal rhythmic behavior was perhaps first excited by the report by Boddy and Dawes (1974). They extrapolated work from the fetal sheep to the human fetus and used ultrasound to detect fetal breathing activity. In this early report, Boddy describes two characteristics of human fetal breathing that have received much subsequent study: human fetal breathing increases after maternal meals or glucose intake, and time spent in fetal breathing appears to show a circadian rhythm.

Much of the work describing patterns of human fetal breathing was performed by Patrick and his group and summarized by Patrick and Challis (1980). In human fetuses of 30–40 weeks gestation monitored over 24 h, an increase in percentage time spent breathing occurs 1–2 h after meals (or glucose ingestion) and a prominent increase also occurs between 0200 and 0700 hours during maternal sleep. Campbell (1980) applied time domain techniques to Patrick's data and found statistically significant ultradian components, with cycle lengths of 100–500 min.

## Fetal Heart Rate

Noninvasive tools for monitoring human fetal heart rate and fetal heart rate variability are widely employed in clinical obstetrics. Surprisingly little has been published concerning circadian or ultradian rhythmicity in these characteristics. Although Dalton et al. (1977) reported circadian and ultradian rhythms in the fetal sheep, descriptions of these rhythms in the human fetus were first described by Patrick et al. (1982) and Visser et al. (1982a). Patrick et al. (1982) found a prominent circadian rhythm in fetal heart rate with a trough between 0200 and 0600 hours. A similar rhythm was seen in maternal heart rate with a trough between 2000 and 0700 hours.

When beat-to-beat heart rates are examined, one can observe and quantitate the variation in length between one heart beat interval to the next or to a mean of heart beat intervals. Visser et al. (1982a) also found a circadian rhythm in heart rate variability with the peak phase occurring between 2100 and 0300 hours.

Ultradian rhythms in both fetal heart rate and heart rate variability have been found by several investigators. Happenbrowers et al. (1978) used techniques of autospectral analysis and complex demodulation to identify periodicities in mean heart rate with cycle lengths of about 60 min and 3–4 h. Dalton et al. (1986) focused the same techniques on shorter time intervals and found cycles of fetal heart rate of between 10 and 90 s. Visser et al. (1982b) found cycles of heart rate variability of 1–8 cycles/min.

## Fetal State

As the above-cited studies have described, fetal breathing, fetal movement and fetal heart rate variability are shown to display ultradian rhythmicity. Further, these fetal behavior characteristics are interrelated. Visser et al. (1982a, b) and Campbell (1980) have shown that these patterns are related to fetal cortical state or fetal "sleep" patterns and that these cycles in late pregnancy fetuses are similar to those seen in newborn infants. At least three states are recognized in human fetuses, an awake state, an active sleep state and a quiet sleep state. Fetal movement, fetal breathing and heart rate variability are greatly diminished in quiet sleep. Fetal movement and fetal heart rate variability are enhanced in awake states and fetal breathing enhanced during active or REM sleep.

## Fetal Endocrine Rhythms

Since human fetal blood can be sampled prior to birth only infrequently and only by invasive methods, information about fetal endocrine rhythms has mostly been obtained either from nonhuman primate species, using chronic fetal preparations, or by examining fetal hormones or metabolites of fetal hormones found in the blood of pregnant women. Chief among these metabolites is the hormone estriol, which is unique to human pregnancy, produced by the placenta, and 90% of whose precursors are of fetal adrenal (DHEA-S) origin. Thus, maternal plasma estriol concentrations are a reflection of fetal adrenal function. Circadian rhythms of plasma estriol were identified as early as 1973 by Townsley et al. Reviews of these and other endocrine rhythms of pregnancy have been made by Meis et al. (1983) and Honnebier et al. (1989). After Townsley et al. (1973), other investigators found that plasma estriol concentrations tended to be higher around mignight. Patrick et al. (1979) further demonstrated that in women of 34–35 weeks gestation plasma estriol and plasm cortisol showed prominent circadian rhythms that were inversely related to each other. Since maternal cortisol is known to cross the placenta to the fetus, Patrick theorized that maternal adrenal cortical rhythms were a synchronizing influence on fetal adrenal rhythms with negative feedback on the fetal pituitary-adrenal axis. Patrick et al. (1980) later showed that in late pregnancy, as mean plasma estriol concentrations rose, circadian amplitude diminished, suggesting escape of fetal adrenal function from maternal influence.

Studies of nonhuman primate fetuses have tended to corroborate these insights into human fetal endocrine rhythms. Challis et al. (1980), with a limited number of sampling times in 24 h, showed circadian variations in fetal DHEA-S, progesterone and maternal cortisol, but not in fetal cortisol. Walsh et al. (1984) in a more sophisticated study design was able to sample fetal blood at 3-h intervals over a 48-h time span. He demonstrated circadian rhythmicity in fetal cortisol, DHEA-S, progesterone and estrone, and rhythms in maternal estradiol, estrone, cortisol and progesterone (estriol is not produced in large amounts by the rhesus placenta). Walsh et al. (1984) found that fetal adrenal hormones peaked at 0300 hours and maternal cortisol at 0600–1200 hours.

## Fetal Zeitgebers

The demonstration of circadian and other rhythms of fetal behavior raises the question of what environmental signals operate to synchronize these rhythms to the light/dark or other cycles.

As shown above, rhythms of maternal adrenal cortical hormones strongly influence the fetal pituitary-adrenal axis. Interventions which abolish or suppress maternal adrenal cortical rhythms have a dramatic effect on fetal adrenal hormones or products of these hormones such as estriol. Challis et al. (1981) found that treatment of pregnant women with prednisone and dexamethasone abolished rhythms in estriol as well as suppressing maternal cortisol and estriol. Patrick et al. (1981) examined these same patients and found that circadian rhythms of fetal breathing were abolished. Arduini et al. (1986) studied women treated in late pregnancy with dexamethasone and found suppression of maternal cortisol, ablated rhythms of ACTH and estriol, and an ablation of circadian rhythm of fetal heart rate variability. Arduini et al. (1987) also studied a pregnant woman who had had adrenalectomy performed prior to the pregnancy and received replacement of adrenal cortical steroid hormone. Low levels of cortisol and estriol were seen and no rhythm of fetal heart rate variability. Thus, evidence suggests that abolition of maternal adrenocortical rhythms inhibits circadian rhythmicity of fetal behavior as well as fetal adrenal hormones. The exposure to the fetus of maternal adrenal-cortical rhythms may be a major synchronizer or zeitgeber for circadian rhythmicity.

## Practical Applications

Contemporary methods for assessing the well-being of the fetus rely to a large extent on fetal activities that are influenced by fetal cortical state (i.e., fetal movement, fetal breathing, fetal heart rate and heart rate variability). These functions follow both ultradian and circadian periodicity. Although clinicians recognize that the absence of these functions may represent a temporary condition of fetal sleep, more sophisticated analyses may be possible. The expression of these ultradian rhythms is likely to be in itself a sign of fetal well-being. Thus, evaluation of the cyclicity of these variables may be as valuable as the detection of the presence of these behaviors. Further studies of fetal cortical state and rhythms of fetal behavior in both healthy and stressed fetuses are needed.

## Neonatal Rhythms

Rhythmicity in body functions or behavior in newborns and infants has been studied since the early 1900s. Until recent studies, investigators have focused on circadian rhythms. Reviews of significant work in this area were presented by Hellbrugge (1960). He found no circadian rhythms present at birth in any measurable function such as heart rate, sleep/wake cycle, temperature, electrical skin resistance, and renal function. Further descriptions included: "1. Different physiologic functions develop a certain circadian rhythm independently from each other. 2. The circadian rhythm of the different functions becomes apparent at different times after birth. 3. During the development of circadian periodicity an increase in the range of oscillation occurs in all physiologic functions."

Kleitman and Engelmann (1953) described the development and maturation of sleep/activity cycles in infants. Like Hellbrugge (1960), their focus was also on the increasing conformity of infant behavior to a circadian pattern or rhythm. Kleitman and Engelmann (1953) present an interesting graphic plot on the sleep/wake patterns of an infant raised on a "permissive" schedule (feeding on demand). Inspection of this infant's behavior in the first 6 months of life shows a change from ultradian patterns to a free-running cycle of about 25 h at about 8 weeks of age to a synchronized 24-h pattern at about 16 weeks of age.

This concept of the absence of rhythmicity at birth and the gradual development of circadian rhythms was consistent with contemporary studies of the human fetus, which was also thought to possess no rhythmicity in functions or behavior. While these classic studies of neonates gave valuable information, more recent investigations have modified their conclusions. More sophisticated data analysis has permitted the description in both the fetus and neonate of rhythms of much shorter duration than 24 h. In addition, recent investigators have demonstrated circadian rhythmicity in cardiovascular and pulmonary function in newborn infants.

Robertson (1987) performed an interesting study of body movement of 41 individuals studied longitudinally in the fetal state in the last half of pregnancy and, after birth, in the newborn period. Using Fourier and spectral analysis, Robertson (1987) described a highly statistically significant ultradian rhythm of movement with a period of about 2 min. This frequency was consistently present both in the fetus and in the newborn regardless of the behavioral state of the newborn (active sleep, quiet sleep, crying, etc.).

Earlier studies reported by Hellbrugge (1960) suggested that prematurely born infants have a delay in development of circadian rhythms compared with infants born at term. Recent work presented by Updike et al. (1985) studied six 34- to 37-week-old prematurely born infants and measured transcutaneous blood $PO_2$, heart rate, skin temperature and respiration. Cosinor analysis showed significant circadian rhythms of skin temperature in five of six infants, transcutaneous $PO_2$ in three of six infants, respiratory pauses in two of six infants, respiratory peak frequency in three of six infants and heart rate in three of six infants. Thus, more sophisticated statistical methodology can demonstrate significant rhythmicity in the newborn when gross inspection of the data is uninformative.

Recent studies of cardiovascular parameters in term newborn infants have excited interest. Infants were studied using automatically recording blood pressure apparatus for 48 h after birth. Results of this work by Cagnoni et al. (1987) have been described also by Halberg et al. (1989). Cosinor analysis showed frequently but not universally significant circadian rhythms in systolic blood pressure, diastolic blood pressure and heart rate. The group of infants was segregated into those who possessed a family history of hypertension and those who did not have such a history. The former group had a higher group mean systolic and diastolic blood pressure and lower heart rate than the infants born to normotensive families. The most sharply differing measurement on comparing the two groups was the amplitude of the blood pressure rhythms. Thus in this group of infants a rhythmometric variable provides the best measure of discrimination between presumably healthy individuals and individuals at possible risk for disease in future decades.

The work by Swaab et al. (1985) provides insight into the context of the development of circadian patterns over the first months of life. Swaab studied the human suprachiasmatic nucleus (SCN) by immunocytochemical staining with antibodies against arginine vasopressin (AVP), which appears to be a good marker for this nucleus in the human brain. At the time of birth the number of AVP-containing cells in the SCN is very small but over the neonatal period the number of these cells increases substantially. This increase occurs during the time span when circadian rhythmicity becomes obvious for most physiological functions and for infant behavior. Swaab et al. (1985)

argue that this description regarding the development of the human SCN is "in agreement with observations that in the human, spontaneous diurnal rhythms only develop in the neonatal period, and this reinforces the idea that the fetal diurnal rhythms are to a large extent driven by the mother."

## Practical Applications

The utilization of measurement of cardiovascular rhythms in the newborn as an index of risk for health outcomes in the mature individual is a new and dramatic development. Studies are currently underway in several different geographic locales to replicate and extend these findings. The ultimate demonstration of the utility and consistency of these studies may not be available until the individuals studied attain maturity.

In a broader sense, the concept of utilizing insights gained by chronobiological study to uncover genetic predisposition for health and disease is of potential great promise. However, studies to confirm and explore this area must be performed with careful experimental design and with rigorous attention to detail in order that the results of such studies will be convincing to a scientific and medical community skeptical of the value of chronobiological study.

## References

Arduini D, Rizzo G, Parlati E, Giorlandino C, Valensise H, dell'Acqua S, Romanini C (1986) Modification of ultradian and circadian rhythms of fetal heart rate after fetal-maternal adrenal gland suppression: a double blind study. Prenat Diagn 6: 409–417

Arduini D, Rizzo G, Parlati E, dell'Acqua S, Romanini C, Mancuso S (1987) Loss of circadian rhythms of fetal behavior in a totally adrenalectomized pregnant women. Gynecol Obstet Invest 23: 226–229

Boddy K, Dawes GS (1974) Fetal breathing. J Physiol 243: 599–603

Brown FM (1988) Common 30-day multiple in gestation time of terrestrial placentals. Chronobiol Int 5: 195–210

Campbell K (1980) Ultradian rhythms in the human fetus during the last ten weeks of gestation: a review. Semin Perinatol 4: 301–309

Cagnoni M, Tarquini B, Halberg F, Marz W, Cornelissen G, Mainardi G, Panero C, Shinoda M, Scarpelli P, Romano S, Bingham C, Hellbrugge T (1987) Circadian variability of blood pressure and heart rate in newborns and cardiovascular chronorisk. Adv Chronobiol [B]: 145–151

Challis JRG, Socol M, Murata Y, Mannig FA, Martin CB Jr

(1980) Diurnal variations in maternal and fetal steroids in pregnancy rhesus monkeys. Endocrinology 106: 1283–1288

Challis JRG, Patrick J, Richardson B, Tevaarwerk G (1981) Loss of diurnal rhythm in plasma estrone, estriol, and estriol in women treated with synthetic glucocorticoids at 34 to 345 weeks' gestation. Am J Obstet Gynecol 139: 338–343

Cooperstock M, Wolfe RA (1986) Seasonality of preterm birth in the collaborative perinatal project: demographic factors. Am J Epidemiol 124: 234–241

Cooperstock M, England JE, Wolfe RA (1987a) Circadian incidence of premature rupture of the membranes in term and preterm births. Obstet Gynecol 69: 936–941

Cooperstock M, England JE, Wolfe RA (1987b) Circadian incidence of labor onset hour in preterm birth and chorioamnionitis. Obstet Gynecol 70: 852–855

Dalton KJ, Dawes GS, Patrick JE (1977) Diurnal, respiratory, and other rhythms of fetal heart rate in lambs. Am J Obstet Gynecol 127: 414–424

Dalton KJ, Denman DW, Dawson AJ, Hoffman HJ (1986) Ultradian rhythms in human fetal heart rate: a computerized time series analysis. Int J Biomed Comput 18: 45–60

DeVries JIP, Visser GHA, Prechtl HFR (1985) The emergence of fetal behavior. II. Quantitative aspects. Early Hum Dev 12: 99–120

Ducsay CA, Cook MJ, Walsh SW, Novy MJ (1983) Circadian patterns and dexamethasone-induced changes in uterine activity in pregnant rhesus monkeys. Am J Obstet Gynecol 145: 389

Ducsay CA, Yellon SM, Hess DL, Harvey LM, McNutt CM (1990) Regulation of myometrial and endocrine rhythms by photoperiod in the pregnant rhesus macaque (Abstr 136). 37th Annual Meeting of the Society for Gynecologic Investigation, March 21–24, St Louis

Figueroa JP, Barbera M, Honnebier OM, Jenkins S, Nathanielsz PW (1990) Alteration of 24 hour rhythms in myometrial activity in the chronically catheterized pregnant rhesus monkey following a 6 hour shift in the light-dark cycle (Abstr 137). 37th Annual Meeting of the Society for Gynecologic Investigation, March 21–24, St Louis

Halberg F, Bakken E, Cornelissen G, Halberg J, Halberg E, Delmore P (1989) Blood pressure assessment in a broad chronobiologic perspective. In: Refsum H, Sulg JA, Rasmussen K (eds) Heart and brain, brain and heart. Springer, Berlin Heidelberg New York

Happenbrowers T, Ugartechca JC, Combs D, Hodgman JE, Harpe RM, Sterman MB (1978) Studies of maternal-fetal interaction during the last trimester of pregnancy: ontogenesis of the basic rest-activity cycle. Exp Neurol 61: 136–153

Harbert GM Jr (1977) Biorhythms of the pregnant uterus (Macaca mulatta). Am J Obstet Gynecol 129: 401–408

Harbert GM Jr, Spisso KR (1980) Biorhythms of the primate uterus (Macaca mulatta) during labor and delivery. Am J Obstet Gynecol 138: 686–696

Harbert GM, Cornell GW, Thornton WN (1970) Diurnal variation of spontaneous uterine activity in non-pregnant primates (Macaca mulatta). Science 170: 82–89

Honnebier MBOM, Swaab DF, Mirmiran M (1989) Diurnal rhythmicity during early human development. In: Reppert SM (ed) Development of circadian rhythmicity and photoperiodism in mammals. Perinatology, Ithaca, pp 221–244 (Research in perinatal medicine, vol 9)

Hellbrugge T (1960) The development of circadian rhythms in infants. Cold Spring Harb Symp Quant Biol 25: 311–323

Jolly A (1972) Hour of births in primates and man. Folia Primatol (Basel) 18: 108–121

Kaiser IH, Hallberg F (1962) Circadian periodic aspects of birth. Ann NY Acad Sci 98: 1056–1068

King PD (1960) Distortion of the birth frequency curve. Am J Obstet Gynecol 79: 399–400

Kleitman N, Engelmann TG (1953) Sleep characteristics of infants. J Appl Physiol 6: 269–282

Longo LD, Yellon SM (1988) Biological timekeeping during pregnancy and the role of circadian rhythms in parturition. In: Kunzel W, Jensen A (eds) The endocrine control of the fetus. Springer, Berlin Heidelberg New York, pp 173–192

Malek J, Gleich J, Maly V (1962) Characteristics of the daily rhythm of menstruation and labor. Ann NY Acad Sci 98: 1042–1055

Meis PJ, Buster JE, Kundu N, Magyar D, Marshall JR, Halberg F (1983) Individualized cosinor assessment of circadian hormonal variation in third trimester human pregnancy. Chronobiologia 10: 1–11

Nasello-Paterson C, Natale R, Connors G (1988) Ultrasonic evaluation of fetal body movements over twenty-four hours in the human fetus at twenty-four to twenty-eight weeks' gestation. Am J Obstet Gynecol 158: 312–316

Patrick J, Challis J (1980) Measurement of human fetal breathing movements in healthy pregnancies using a real-time scanner. Semin Perinatol 4(4): 275–286

Patrick J, Fetherston W, Vick H, Voegelin R (1978) Human fetal breathing movements and gross fetal body movements at weeks 34 to 35 at gestation. Am J Obstet Gynecol 130: 693–699

Patrick J, Challis J, Natale R, Richardson B (1979) Circadian rhythms in maternal plasma cortisol, estrone, estradiol, and estriol at 34 to 35 weeks' gestation. Am J Obstet Gynecol 135: 791–798

Patrick J, Challis J, Campbell K, Carmichael L, Natale R, Richardson B (1980) Circadian rhythms in maternal plasma cortisol and estriol concentrations at 30 to 31, 34 to 35, and 38 to 39 weeks' gestational age. Am J Obstet Gynecol 136: 325–334

Patrick J, Challis J, Campbell K, Carmichael L, Richardson B, Tevaarwerk G (1981) Effects of synthetic glucocorticoid administration on human fetal breathing movements at 34 to 35 weeks' gestational age. Am J Obstet Gynecol 139: 324–328

Patrick J, Campbell K, Carmichael L, Probert C (1982) Influence of maternal heart rate and gross fetal body movements on the daily pattern of fetal heart rate near term. Am J Obstet Gynecol 144: 533–538

Richardson B, Natale R, Patrick J (1979) Human fetal breathing activity during effectively induced labor at term. Am J Obstet Gynecol 133: 247–255

Robertson SS (1987) Human cyclic motility: fetal-newborn continuities and newborn state differences. Dev Psychobiol 20: 425–442

Schmiedler GR (1941) The relation of fetal activity to the activity of the mother. Child Dev 12: 63–68

Smolensky M, Halberg F, Sargent F (1972) Chronobiology of the life sequence. In: Itoh S, Ogatak K, Yoshimura H (eds) Advances in climatic physiology. Igaku-Shoin, Tokyo, pp 282–318

Sontag LW (1941) The significance of fetal environmental differences. Am J Obstet Gynecol 41: 996–1003

Sterman MB (1967) Relationship of intrauterine fetal activity to maternal sleep stage. Exp Neurol [Suppl] 19: 98–106

Swaab DF, Fliers E, Portiman TS (1985) The suprachiasmatic

nucleus of the human brain in relation to sex, age, and dementia. Brain Res 342: 37–44

TambyRaja RL, Hobel CJ (1983) Characterization of 24 hour uterine activity (UA) in the second half of the human pregnancy (Abstr 576). 30th Annual Meeting of the Society for Gynecologic Investigation, March 17–20, Washington

Taylor NF, Martin MC, Nathanielsz PW, Seron-Ferre M (1983) The fetus determines circadian oscillation of myometrial electromyographic activity in the pregnant rhesus monkey. Am J Obstet Gynecol 146: 557–567

Townsley JD, Dubin IVH, Grannis GF, Gortman J, Crystle CD (1973) Circadian rhythms of serum and urinary estrogens in pregnancy. J Clin Endocrinol 36: 289–295

Updike PA, Accurso FJ, Jones RH (1985) Physiologic circadian rhythmicity in preterm infants. Nurs Res 34: 160–163

Visser GHA, Goodman JDS, Levine, Dawes GS (1982 a) Diurnal and other cyclic variations in human fetal heart rate near term. Am J Obstet Gynecol 142: 535–544

Visser GHA, Carse EA, Goodman JDS, Johnson P (1982 b) A comparison of episodic heart-rate patterns in the fetus and newborn. Br J Obstet Gynecol 89: 50–55

Walsh SW, Ducsay CA, Novy MJ (1984) Circadian hormonal interactions among the mother, fetus, and amniotic fluid. Am J Obstet Gynecol 150: 745–753

# Patterns of Growth Hormone Release During Childhood and Adolescent Development in the Human

A. D. Rogol

Growth hormone (GH) is released in an episodic, burst-like (pulsatile) manner throughout the day, but especially following the onset of slow-wave sleep (stages 3 and 4). Its secretion is controlled by two hypothalamic peptide hormones – growth hormone-releasing hormone (GHRH) and growth hormone release inhibiting hormone (somatostatin). Additional loci of control include brain neutrotransmitters and neuropeptides, insulin-like growth factor-I [IGF-I (long-loop feedback)], GH itself (short-loop feedback) and GH-RH and somatostatin (ultrashort-loop feedback). In addition, metabolic substrates, e.g., glucose and fatty acids, take part in the regulation of growth hormone secretion. It is the quantity and pattern of circulating GH that ultimately permits the animal to grow, although genetic, disease and especially nutritional factors may override any straightforward relationship between GH secretion and linear growth.

## Control of Growth Hormone Release

(for detailed review, see Müller and Nisticò 1989)

### Central Neurotransmitters

Virtually all of the known (classical) neurotransmitters including the catecholamines, the cholinergic, serotoninergic, opiatergic and γ-aminobutyric acid systems have been implicated in the control of GH synthesis and release. Few act directly at the pituitary to effect GH synthesis or release, but generally act indirectly especially at the hypothalamus to modulate GHRH and somatostatin synthesis or release.

### Growth Hormone Releasing Hormones, Somatostatin and Growth Hormone Releasing Peptide

Growth hormone releasing hormone, long postulated but only recently discovered, has been isolated, purified and sequenced. Amino acid analysis revealed three homologous peptides of 37, 40 and 44 amino acids (Rivier et al. 1982). Synthesis of these peptides showed them to be virtually equipotent and that peptides as small as 29 amino acid residues have complete biological activity. The majority of GHRH-containing neurons are located in the ventral medial and arcuate nuclei of the medial basal hypothalamus. Fibers from these nuclei project to the hypophysiotropic area of the median eminence.

Somatostatin is a 14 amino acid cyclic peptide that inhibits TSH as well as GH secretion from the pituitary and has a great number of other effects on the brain, the endocrine and exocrine pancreas, and, especially, the rest of the gastrointestinal tract. Its major site of synthesis (for control of the anterior pituitary) is the periventricular nucleus, although it is widely distributed in other brain areas. These periventricular neurons of the anterior hypothalamus send projections to the external layer of the median eminence and terminate on the fenestrated capillaries that form the inlet for the hypothalamic-hypophysial portal system. Growth hormone enhances the expression of the structural gene for somatostatin and thus controls the synthesis as well as the release of this neurohormone.

The episodic secretion of GH depends upon the specific rhythms of GHRH and somatostatin. The release of these neurohormones is virtually 180° out of phase. A pulse of GH is produced when somatostatin tone is low while GHRH is released (Plotsky and Vale 1985). Such reciprocal changes are responsible for the ultradian pattern of GH secretion. The data are consistent with the hypothesis that somatostatin secretion sets the timing (frequency and duration) and GHRH the magnitude of GH release.

## Pituitary

Growth hormone participates in its own feedback control of release. An autofeedback mechanism had been proposed for many years, but it is only recently that some of the details of this mechanism have been amenable to testing (Berelowitz et al. 1981; Rosenthal et al. 1986). Growth hormone administration to man and animals has been followed within several hours (but not less) by blunted serum GH responses to provocative stimuli, diminished pulsatile GH release or diminished pituitary GH content. Since circulating levels of IGF-I were also increased, the precise mechanism, long-loop versus short-loop feedback, could not be determined. The mode of delivery of GH may also be a critical parameter. Isgaard et al. (1988) found that the pulsatile intravenous administration of GH was more efficacious than continuous administration of the same quantity to raise IGF-I mRNA in skeletal tissues, e.g., skeletal muscle and rib growth plates. However, the liver reacted differently in that both continuous and pulsatile GH infusion elevated the amount of IGF-I mRNA in an indistinguishable manner. The pituitary expresses the IGF-I gene, suggesting that IGF-I may regulate GH secretion by an autocrine/paracrine mechanism as well as in the usual endocrine manner. Many metabolic substrates are also altered by GH administration. Short- and longer-term studies have suggested different mechanisms. At present it appears that in the longer term (i.e., more than 12 h) GH inhibits its own synthesis and release by augmenting circulating IGF-I concentrations acting to decrease GHRH secretion and increase somatostatin release. Shorter-term inhibition is probably mediated by such metabolic substrates as glucose and free fatty acids. However, alterations in adenylate cyclase activity (GHRH activates adenylate cyclase via $G_S$ and somatostatin inhibits this enzyme via $G_i$), calcium fluxes, and phosphatidyl inositol turnover interact strongly with these substrates.

## Peripheral Feedback Signals

Insulin-like growth factor I substantially inhibits GH synthesis and release by the long-loop feedback mechanism noted above. IGF-I specifically inhibits GH mRNA synthesis and GH release both in the basal state and after stimulation with GHRH. In animals antibodies against IGF-I receptors can block the growth-promoting actions of GH on its target cells – strongly suggesting an indirect role for GH in growth.

IGF-I circulates bound to two or more carrier proteins of 150 and 35 kD which prolong the half-life of this growth factor. The predominant 150-kD form is GH dependent, but levels of the 35-kD protein are more rapidly altered and may potentiate IGF-I action. The physiologic role of these various proteins has not been fully clarified: they may be relatively unimportant because of extensive local growth factor synthesis or they may represent a storage depot or a clearance pathway.

Free fatty acids and glucose interact with other modulators to augment or inhibit GH secretion. Administration of free fatty acids or the increase in free fatty acids following GH administration reduce the subsequent GH responses to GHRH, catecholamines, physical exercise and other stimuli.

Other nutritional factors are doubtlessly involved. Amino acids serve not only as building blocks for muscle protein (anabolic action of GH), but also as precursors for the synthesis of some of the neurotransmitters that control the synthesis and release of GHRH and somatostatin. In adult volunteers, fasting quickly causes an increase in GH concentrations and a decrease in IGF-I levels (Isley et al. 1983). Refeeding is associated with a return to normal of both GH and IGF-I levels. The nitrogen and energy contents of the refeeding diet both influence the return to physiologic circulating GH and IGF-I levels. Physical and psychological stress elevate GH levels in humans, but the detailed mechanisms have not been described, especially for the latter. Insulin-like growth factor I levels rise two- to threefold with the more advanced pubertal stages in boys and girls before diminishing again to adult levels. The pubertal rise is well correlated with the sex-hormone-mediated augmentation of GH secretion and accelerated growth in the later stages of puberty. Individuals above the age of 50 years have declining levels of IGF-I. There is, however, the intriguing possibility of interplay between the endocrine actions of circulating GH and IGF-I and the paracrine/autocrine actions consistent with the dual effector hypothesis of GH action. The local effects are probably more important in those tissues with high local IGF-I gene expression (mainly mesenchymal). These nutritional, stress and sex-hormone mediated effects are probably linked at multiple hierarchical levels permitting interactions among them in addition to their individual effects.

## Reliability of Physiologic Monitoring of Growth Hormone Release

This axis can be evaluated physiologically (daily secretory pattern, exercise) or pharmacologically (GH response to neurotransmitters, certain amino acids, insulin-induced hypoglycemia, etc.). For the present purpose I shall focus on the daily secretory pattern and its altered forms in various pathophysiologic conditions (following sections). The method of sampling (continuous versus intermittent), its frequency, methods of assay and especially analysis of the profiles are critical (Evans et al. 1987), but will not be discussed here. The interested reader is referred to a comprehensive review (Urban et al. 1988). The assessment of hormone secretory profiles has been used to determine alterations within the physiologic range or as evidence for the effects of pharmacologic intervention. However, the comparison of individual hormone concentration profiles is based upon the assumption of strict reliability. For GH we have assessed the reliability of the concentration versus time profiles in children (two more 12- or 24-h profiles in the same state of sexual development) and adult women (two 12- or 24-h profiles during the early follicular phase of the menstrual cycle). For the 12-h (21 pairs) or 24-h (18 pairs) in the women no significant mean differences were observed between studies for any parameter of pulsatile release (unpublished observations); however, the reliability as indicated by the correlation coefficients for specific parameters of pulsatile release varied form $r = 0.25$ (nadir concentration) and 0.36 for peak frequency to $r = 0.66$ (integrated concentration) and $r = 0.71$ (incremental peak amplitude). These data suggest a large degree of biologic variability at least over short (24-h) sampling intervals for some of the parameters of pulsatile GH release, e.g., number of peaks, but moderate to good stability in others, e.g., incremental peak increase and integrated concentration. The amount of GH released per day is more stable than the mode by which it is released. Thus, this natural variation must be taken into account before one can point to a pathologic condition or pharmacologic intervention as etiologic for the alteration in a specific parameter of pulsatile release (for example, GH frequency, see below).

## Growth at Adolescence: Physical Changes and Hormonal Mechanisms

Characteristics of linear growth have been reviewed extensively by Tanner et al. (1966). Rapidly decelerating growth for the first 2–3 years is followed by a relatively constant growth rate during childhood. Adolescent development heralds the onset of rapidly accelerating growth in girls and boys. The growth spurt occurs both earlier in time and earlier within the developmental changes of puberty in girls compared to boys. The pattern, however, is the same – the rapidly accelerating phase ceases at approximately mid-puberty and decelerates to a zero rate as bony epiphyseal fusion occurs. Accompanying the marked alterations in statural growth rate are the onset and progression of primary and secondary sexual characteristics (usually defined operationally by stages of genital, pubic hair and breast development, Marshall and Tanner 1970).

We have extensively investigated the alterations in GH secretory dynamics that accompany these explosive pubertal changes. A review of GH secretory profiles at various stages of adolescent development follows.

### Prepubertal Normally Growing Boys

As part of a longitudinal study of the hormonal alterations during pubertal development, we have had the opportunity to evaluate the circulating GH profiles a number of times in individual prepubertal boys and compare the results to those obtained by Martha et al. (1989) for a group of normally growing prepubertal boys each evaluated once. The latter investigators found that the prepubertal boys had similar mean 24-h GH concentrations (µg/l), sum of GH pulse areas [area under the GH versus time curve that is contained *within* pulses (µg/l × min)], and mean GH pulse increment [pulse increment above baseline (µg/l)] when compared to all other groups (early pubertal, postpubertal and adult) except late pubertal who had augmented values. Similar conclusions have been drawn from preliminary evaluation of our longitudinal study of puberty in boys except that the variance for each parameter of GH release is much less, for example approximately 7% versus more than 30% for 24-h integrated area (area under the concentration versus time curve).

## Prepubertal Short Slowly Growing Boys

Several investigators have found differeces in sponta-neous circulating GH levels in some short children when compared to children of normal stature (Bie-rich 1983; Spiliotis et al. 1984; Zadik et al. 1985; Sa-nayama et al. 1987). However, these studies may have been biased by either a failure to properly control for stage of pubertal maturation or the inability to pre-cisely characterize the pulsatile pattern of GH re-lease. Therefore, we investigated the alterations of spontaneous GH release in healthy, short prepubertal boys by employing a closely matched group of healthy prepubertal boys of normal stature and utiliz-ing an objective, statistically based pulse detection algorithm (Kerrigan et al. 1990). A group of GH-defi-cient children was studied to provide a more compre-hensive spectrum of *possible* alterations in the pul-satile properties of circulating GH concentrations in prepubertal boys of normal and short stature. Forty prepubertal boys were divided into three separate groups: 11 healthy boys with height and weight be-tween the 5th and 95th percentiles for *chronologic* age and skeletal maturity within 2 standard devia-tions for *chronologic* age (group 1); 20 healthy boys with height less than or equal to the 3rd percentile for chronologic age and skeletal age normal or delayed for chronologic age (group 2). All had normal GH levels following provocative stimuli and none had evidence of systemic illness, intrauterine growth retardation, dysmorphic features or previous CNS irradiation. Group 3 comprised nine GH-deficient children (four males, four females and a phenotypic female with an XY karyotype). All were less than the third percentile for height (for chronologic age) and the skeletal maturity was markedly delayed in each.

Maximal GH concentrations following two provoca-tive stimuli were $\leq 6.0\ \mu g/l$ and growth velocity was $\leq 4.6$ cm/year.

Spontaneous GH concentration profiles were as-sessed from serum samples obtained every 20 min from 2000 to 0800 hours via an intravenous heparin-lock needle. The pulsatile characteristics of GH concentration profiles were evaluated using the CLUSTER pulse detection algorithm (Veldhuis and Johnson 1986) as previously described (Martha et al. 1988). Comparisons were made between four groups consisting of normal, GH-deficient, short boys with significant delay of skeletal maturation ($n = 7$), and those short boys with normal skeletal age ($n = 13$).

The pulsatile characteristics of GH release for the three major groups as well those for the subset of seven short boys with significantly delayed bone age are presented in Table 1. As a group the short boys had characteristics of pulsatile GH release that were *indistinguishable* from those of normals; however, those of the GH-deficient group were significantly less than those of both other groups. The subset of short boys with delayed bone age had diminished GH pulse area and sum of GH pulse amplitudes when compared to the normal and GH-deficient groups. These seven boys were distinguishable from the 13 short boys without delayed bone age based only on a lower mean GH pulse amplitude ($10.7 \pm 2.3$ versus $16.8 \pm 1.7\ \mu g/l, p = 0.04$). Characteristic sponta-neous GH concentration profiles for each of the study groups are shown in Fig. 1.

Using data from the 20 short boys we detected a significant correlation between bone age delay and the sum of GH pulse amplitudes ($r = 0.48, p = 0.03$). No significant correlations were detected between growth velocity standard deviation score (SDS) or

**Table 1.** Clinical characteristics of the study groups

| Study group | Chronologic age (years) | Bone age (years) | Bone age delay (years) | Height SDS | Growth velocity, SDS | IGF-I U/ml |
|---|---|---|---|---|---|---|
| Normal ($n = 11$) | $9.0 \pm 0.3$ (7.2 – 10.0) | $8.2 \pm 0.4$ (6.0 – 10.5) | $-0.8 \pm 0.2$ ($-2.1 – 0.5$) | $0.1 \pm 0.3$ ($-1.0 – 1.2$) | $0.6 \pm 0.8$ ($-2.8 – 2.5$) | $0.90 \pm 0.13$ (0.34 – 1.77) |
| Short ($n = 20$) | $9.8 \pm 0.5$ (5.6 – 13.9) | $8.6 \pm 0.5$ (3.5 – 12.1) | $-1.2 \pm 0.3$ ($-3.9 – 1.5$) | $-2.8 \pm 0.6^a$ ($-3.9 – 1.9$) | $-0.9 \pm 0.4$ ($-3.6 – 2.1$) | $0.60 \pm 0.06^a$ (0.22 – 1.18) |
| DBA[c] ($n = 7$) | $10.4 \pm 1.0$ (5.6 – 13.9) | $7.6 \pm 0.8$ (3.5 – 10.0) | $-2.8 \pm 0.2^a$ ($-3.9 – 2.0$) | $-3.1 \pm 0.8^a$ (3.9 – 2.0) | $-1.2 \pm 0.8$ ($-3.6 – 2.1$) | $0.65 \pm 0.12^a$ (0.22 – 1.18) |
| GHD ($n = 9$) | $10.5 \pm 1.3$ (4.8 – 15.3) | $8.2 \pm 1.4$ (2.2 – 14.3) | $-2.3 \pm 0.4^a$ ($-4.8 – 1.0$) | $-3.2 \pm 0.4^a$ ($-4.5 – 2.0$) | $-3.8 \pm 0.7^a$ ($-6.8 – 0.0$) | $0.21 \pm 0.02^b$ (0.13 – 0.29) |

Values are means $\pm$ SEM; numbers in parentheses are ranges. Any value followed by a superscript differs significantly ($p < 0.05$) from all other values within a column not followed by the same superscript
[c] Subset of the short group
DBA, delayed bone age

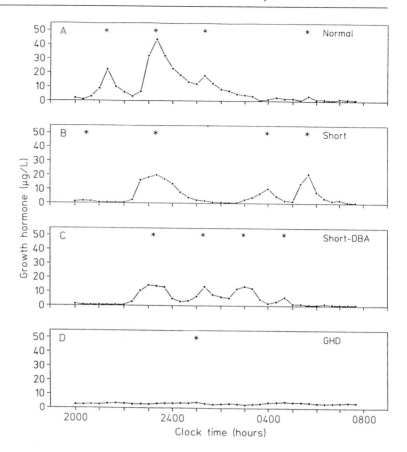

**Fig. 1 A–D.** Spontaneous GH secretory profiles of short boys. Typical spontaneous GH secretory profiles are shown. Each profile is that obtained from a single subject whose GH pulsatile characteristics approximated the respective group mean value. **A** is typical of the normals, **B** is representative of the short boys, **C** is characteristic of the subset of short boys with delayed bone ages and **D** is typical of growth hormone deficient children. The *horizontal axes* are represented as clock time (hours) while the *vertical axes* are respresented as GH concentration (µg/l). An *asterisk* above a data point indicates a GH pulse as determined by the Cluster algorithm (see "Methods" for details of analysis). (From Kerrigan et al. 1990)

height SDS and any of the characteristics of pulsatile GH release. When all groups were combined, significant relationships were detected between height and the sum of GH peak ares ($r = 0.46$, $p = < 0.02$), height SDS and the sum of GH peak areas ($r = 0.38$, $p = 0.02$) and growth velocity SDS and the logarithm of the sum of GH pulse amplitudes ($r = 0.45$, $p < 0.01$).

The observation by several investigators (Hindmarsh et al. 1987; Albertsson-Wikland and Rosberg 1988) that the physiologic pattern of GH release in children is represented by a continuum, rather than as a bimodal distribution, may provide a partial explanation for the difficulty encountered in attempting to distinguish patterns of spontaneous GH release among groups of short children. Hence, the GH concentration profiles of the group of short boys studied were not distinguishable from those of normally growing boys. However, the patterns of pulsatile release of GH from a subset of short boys with significant delay of bone age and for the GH-deficient group differed distinctly when compared to those of normally growing boys. In contrast to the findings of others (Spiliotis et al. 1984), the observed differences

for the subset of short boys did not involve GH pulse frequency (GH-neurosecretory defect) or mean GH concentration, but were manifest as a subnormal sum of pulse areas and sum of pulse amplitudes (see below for GH pulse amplitude modulation with progressive pubertal development). The use of a closely matched control group, different duration of sampling and an objective pulse detection algorithm may partially explain differences between our results and those of previous investigators.

## Prepubertal Short, Slowly Growing Girls with Turner's Syndrome

To investigate the actions of acute and chronic low doses of ethinyl estradiol (EE) on the pulsatile characteristics of GH release, seven girls with Turner's syndrome were evaluated at baseline and after 5 weeks of receiving EE, 100 ng/kg daily, orally (Mauras et al. 1989). After 5 weeks all girls had increased mean GH concentrations (from $7.0 \pm 1.1$ to $13.4 \pm 2.2$ µg/l, $p = 0.008$), mean area within the GH pulses (from $602 \pm 52$ to $1350 \pm 260$ µg/l × min,

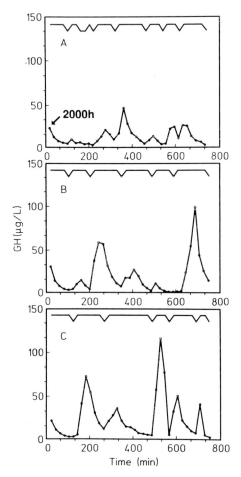

**Fig.2 A–C.** Spontaneous GH secretory profiles in short girls with Turner's syndrome. Twelve-hour profile of serum GH concentrations in a patient (No.3) before (**A**) and after 1 week (**B**) and 5 weeks (**C**) of low-dose EE therapy. Blood was drawn every 20 min for 12 h. Deflections *on top* represent pulses detected by Cluster analysis (From Mauras et al. 1989)

increase in secretory burst frequency (from $5.3 \pm 0.6/12$ h to $7.9 \pm 0.5/12$ h, $p < 0.05$), but no alteration in the maximal rate of GH release (from $1.4 \pm 0.1$ to $1.8 \pm 0.3$ µg/l × min, $p > 0.05$) or GH mass/burst (from $39 \pm 4.4$ to $51 \pm 5.8$ µg/l, $p > 0.05$).

Thus, the marked augmentation of GH concentrations found following ethinyl estradiol administration results from an almost 50% increase in GH secretory burst frequency without change in the mass of GH per burst or the *secretory* burst amplitude. The calculated half-life of endogenous GH did not change (from $19.1 \pm 1.6$ to $18 \pm 1.2$ min, $p = NS$), ensuring that the entire effect depended only upon secretory bursts. Although the precise mechanism for the increased GH production rate following EE therapy cannot be determined from studies of circulating GH concentrations, the marked augmentation of GH secretory burst frequency suggests a significant enhancement of somatotrope sensitivity to GHRH action. The latter may be secondary to a direct facilitative action of the estrogen on the somatotrope, to decreased hypothalamic somatostatin tone, and/or to alterations in GHRH pulse generation.

## Adolescents

### Normally Growing

Martha et al. (1989) have meticulously evaluated GH concentration profiles in normally growing boys evaluated at precisely defined degrees of pubertal development. To investigate the mechanisms subserving physiologic alterations in circulating GH concentrations during puberty, we assessed the GH pulse characteristics of 55 24-h serum GH profiles obtained from healthy male volunteers of normal stature (age 7–27 years) whose physical development spanned the entire pubertal range. Subjects were divided into five study groups based on the degree of pubertal maturation:

| | |
|---|---|
| Prepubertal (Pre) | Tanner stage I physical development in all aspects; testicular size less than 2.5 cm in greatest diameter; prepubertal gonadotropin pattern; 0600 hours serum testosterone concentration < 0.87 µmol/l |
| Early pubertal (Early) | Physical examination evidence for onset of puberty, but pubic hair no greater than Tanner stage III development |

$p = 0.026$), and mean GH pulse amplitude (from $14.0 \pm 2.2$ to $32.8 \pm 6.0$ µg/l, $p = 0.018$). However, there was no detectable alteration in pulse frequency (from $5.3 \pm 0.6$ per 12 h to $5.3 \pm 0.4$, $p = NS$). Despite these significant changes in discrete GH pulse parameters (Fig.2), there was no increase in plasma IGF-I concentration or change in urinary cytologic maturation index, suggesting exquisite sensitivity of the somatotrope to small quantities of biologically active estrogens. Mauras et al. (1990) extended these findings in girls with Turner's syndrome by employing deconvolution analysis (see below), which can distinguish production and metabolic clearance rates. The augmented quantities of GH were accounted for by significant increases in GH production rate (from $194 \pm 22$ to $412 \pm 66$ µg/l × 12 h, $p < 0.05$), with an

| Late pubertal (Late) | Pubic hair development Tanner stage III–IV, but hand epiphyses clearly open |
| --- | --- |
| Postpubertal (Post) | Chronologic age 18 years or less, but hand epiphyses fully fused |
| Adult (Adult) | Healthy young men above age 18 years |

Twenty-four-hour mean concentrations of GH were greater in the late pubertal boys than in all other groups ($p < 0.001$, Fig. 3 A). No differences were detected in the mean 24-h GH concentrations among the Pre, Early, Post and Adult groups. Mean interpulse GH concentrations ($p = 0.08$) and total area under the GH concentrations *versus* time curve during the interpulse periods ($p = 0.17$) were indistinguishable among all groups. In contrast, the amplitude of serum GH pulses was clearly greater in the late pubertal group whether assessed as the mean sum of the GH pulse area ($p \le 0.001$), the mean GH area ($p = 0.004$), or the mean GH pulse amplitude ($p = 0.001$), confirming our previous pilot study (Mauras et al. 1987). As distinguished from most other GH pulse characteristics, the mean number of detectable GH pulses in 24-h GH profiles of the Late boys did not differ statistically from the values for any other study group (Fig. 3 C). Detectable GH pulse frequency in the Early, Post and Adult groups was slightly lower than that for the Pre group ($p = 0.02$). Figure 3 illustrates the values for the average GH pulse area (B) and average number of pulses per 24 h (C) for the five study groups. Note the similarity and dissimilarity of Fig. 2 B and 3 C, respectively, to the profile of the 24-h mean values illustrated in Fig. 3 A.

To assess the influence of regularly occurring rhythms within the GH concentration-time series, we determined the single most dominant rhythm within the data sets. All had a single dominant rhythm which was circadian and invariant across puberty. The best-fit cosine functions for the circadian rhythm for each group are shown in Fig. 4.

Since we did not have height velocity data for our subjects, we compared our 24-h mean GH concentrations to the 50th percentile values for whole-year height velocity for North American boys (Tanner and Davies 1985). The patterns (Fig. 5) are remarkably similar and suggest a strong relationship between these two variables. Our preliminary data from an ongoing longitudinal study of hormonal alterations at puberty confirm a very strong relationship among certain characteristics of GH release (noted above), growth velocity and serum testosterone and IGF-I

levels (on the ascending limb of all three curves, unpublished data). An increase in peripheral GH concentrations during puberty in boys has been suggested by investigators evaluating physiologic, precocious and pharmacologically induced puberty (see below). Thus amplitude modulation of peripheral GH pulses has been implicated as a primary underlying mechanism mediating alterations in linear growth under physiologic and pathophysiologic conditions (Mauras et al. 1987; Albertsson-Wikland and Rosberg 1988; Martha et al. 1989; Mauras et al. 1989; Hindmarsh et al. 1987). Available studies strongly

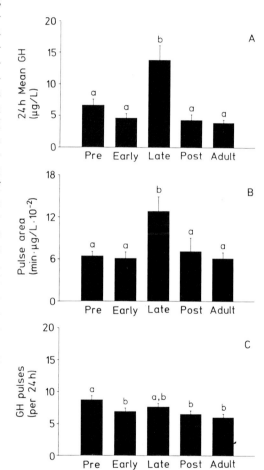

**Fig. 3 A–C.** Growth hormone pulse characteristics of normally growing boys. **A** the mean ($\pm$ SE) 24-h concentrations of GH for the five study groups are illustrated. **B** the mean ($\pm$ SE) area under the GH concentration versus time curve for individual GH pulses, as identified by the Cluster pulse detection algorithm, is presented. **C** the number of GH pulses (mean $\pm$ SE), as detected by the Cluster algorithm, in the 24-h GH concentration profiles for subjects in the five study groups are graphed. In each panel, any *two vertical bars* not identified by the same letter represent statistically different values ($p < 0.05$); *bars* sharing a common letter represent statistically indistinguishable values ($p > 0.05$). (From Martha et al. 1989)

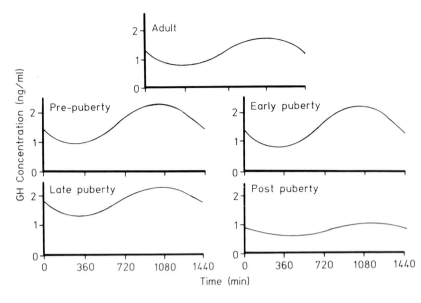

**Fig. 4.** Circadian GH rhythms in normally growing boys. *Subpanels* denote the individual best-fit cosine functions for the group serum GH concentrations sampled at 20-min intervals over 24 h. The data from each group of subjects were fit simultaneously. (From Martha et al. 1989)

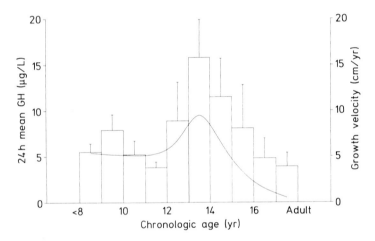

**Fig. 5.** Mean GH concentration for chronologic age in normally growing boys. *Bars* represent values for the 24-h mean (±SE) concentration of GH *(left ordinate)* from all 60 24-h GH profiles subdivided according to chronologic age. An idealized growth velocity curve reproduced from the 50th percentile values for whole-year height velocity of North American boys (Tanner and Davies 1985) is superimposed *(right ordinate)*. (From Martha et al. 1989)

suggest a central role for augmented gonadal steroid hormone levels, particularly circulating testosterone concentrations, in mediating the GH pulse amplitude changes (Mauras et al. 1987; Liu et al. 1987; Ulloa-Aguirre et al. 1990; Tanner et al. 1976). Our data do not permit an examination of the possible role of changes in circulating levels of estradiol 17-$\beta$. Ho et al. (1987) found that only the concentration of (free or total) estradiol, but not that of testosterone, correlated with the 24-h integrated concentration of GH and GH pulse amplitude in normal men and women. It is conceivable that changes in low levels of circulating estradiol, resulting from the aromatization of testosterone, are important in mediating the augmentation of GH pulse amplitude. Direct effects of estrogen (ethinyl estradiol) can alter circulating GH profiles in girls with Turner's syndrome (see above).

## Short Slowly Growing

Exogenous androgen therapy can also cause similar alterations in GH profiles. Mauras et al. (1987) showed that prepubertal boys treated with testosterone enanthate (100 mg every 4 weeks) for 10 weeks had significantly higher mean serum GH pulse amplitudes after compared to prior to therapy ($15.4 \pm 2.4$ vs. $6.8 \pm 1.6$ µg/l, $p = 0.04$), although the GH pulse frequency was not altered. Similar results were obtained with testosterone therapy in an adolescent boy who was agonadal following radiation therapy to the testes for relapse of acute lymphocytic leukemia (Mauras et al. 1989). Marked augmentation of growth velocity and GH pulse amplitude followed therapy with testosterone enanthate only to return to virtual pretreatment values (consistent with GH defi-

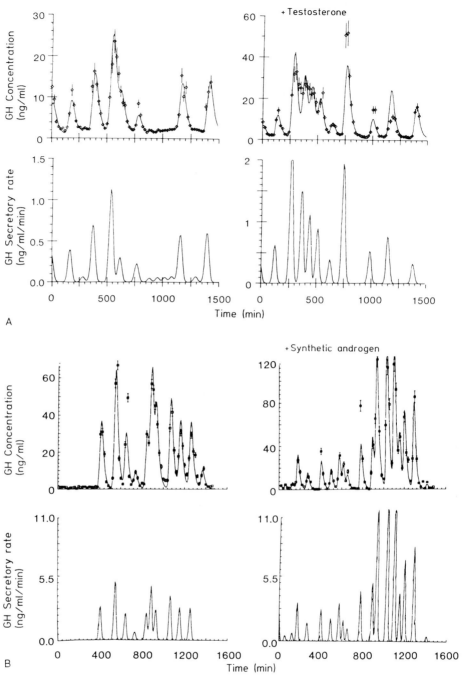

**Fig. 6 A, B.** GH secretory profiles and resolved GH secretory bursts in control and androgen-treated conditions. Illustrative 24-h profiles of serial serum GH concentrations and deconvolution-resolved GH secretory bursts in control and androgen-treated study sessions in four prepubertal boys. Boys underwent blood sampling at 20-min intervals for 24 h under basal (control) and androgen-treated conditions [testosterone (**A**) or oxandrolone (**B**, synthetic androgen)]. The serum GH concentrations were measured by IRMA or RIA, and are depicted in the *upper subpanels*. The continuous curves through the observed serum GH concentration data were calculated by the convolution model. The *lower subpanels* give the computer-resolved GH secretory rates as a function of time. Note the presence of distinct and delimited GH secretory bursts, the amplitudes of which are augmented androgen treatment (ng/ml = µg/l). (From Ulloa-Aguirre et al. 1990)

ciency) when therapy was interrupted for 3 months. We conclude that the augmentation of GH release that occurs during spontaneous pubertal development or following exogenous anabolic steroid therapy is an amplitude-modulated phenomenon, relatively independently of changes in macroscopic GH pulse frequency. Such an effect may be secondary to the action of sex steroid hormones modulating either the responsivity of the somatotrope to endogenous GH-releasing hormone, the amount of GH-releasing hormone, or the tonic inhibitory tone of somatostatin.

In this report I have purposely avoided discussion of GH secretion per se except for a brief note in the section on girls with Turner's syndrome. The fluctuations in circulating GH concentrations result from a dynamic process of simultaneously acting secretion and clearance mechanisms. Although the properties of these fluctuations are no doubt important to the target tissue (this issue of hormone action is beyond the scope of the chapter), information about the *secretory events* from which they arose is not easily extracted. In fact, recent advances in the biophysical description of endocrine pulse detection suggest that the observed peripheral hormonal pulses may be created by *discrete, random* secretory bursts within the gland of origin. These are of short duration and separated by prolonged periods of relative quiescence. The development and employment of new statistically based models that simultaneously consider hormone secretion and clearance [(deconvolution analysis) Veldhuis et al. 1987] provide a new investigative tool to enable precise and accurate estimations of the secretion and clearance rates of GH and other hormones to be made in different physiologic and pathophysiologic states. Ulloa-Aguirre et al. (1990), have studied the effects of androgens (testosterone and the nonaromatizable androgen, oxandrolone) on the mass and rate of growth hormone secretion and on the circulating half-life in pre- and peripubertal boys receiving these anabolic steroids. The latter androgen is not susceptible to aromatization to an estrogen and must therefore act through the androgen receptor (see below). The androgens alter some properties of GH secretion in boys with constitutional delay of growth and adolescence. As noted previously (Link et al. 1986; Mauras et al. 1987), both androgens significantly increased mean 24-h GH concentrations. The increased GH production resulted directly from a greater mass of GH secreted per burst and a greater maximal rate of GH secretion within each burst (Fig. 6 A, B) and androgens amplified the magnitude of the nyctohemeral rhythm in the mass (but not fre-

quency) of the GH secretory episodes. Despite these alterations the number and duration of the GH secretory bursts and the subject-specific circulating GH half-life were unchanged. Thus, androgens acting other than as a substrate for the aromatase enzyme can augment GH secretion via distinct neuroendocrine mechanisms.

Rapidly Growing

Markedly augmented growth hormone secretion leads to acromegaly in adults and gigantism (and acromegaly) in children and adolescents. Few studies have been performed in children and adolescents with growth hormone secreting tumors. We have studied one such patient, an adolescent with the McCune-Albright syndrome [café-au-lait skin lesions, sexual precocity and polyostotic fibrous dysplasia (Benedict 1962)], who in addition had hyperthyroidism and pituitary mammaosomatotrope hyperplasia causing excessive secretion of GH and prolactin (Kovacs et al. 1984). Her GH profile (pretherapy) is shown in Fig. 7. Note the very high baseline levels (no samples < 1 µg/l) and large-amplitude peaks. Partial restoration of the GH profile was accomplished with bromocriptine therapy (data not shown). Complete hypophysectomy was performed during craniofacial surgery for marked hyperostotic changes. All GH levels were < 1.0 µg/l postoperatively. Perhaps a more complete picture of GH secretory dynamics can be obtained from examination of the secretory profiles of an acromegalic patient prior to and following successful therapy. In Fig. 8 are

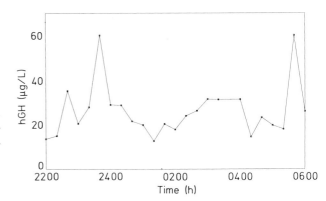

**Fig. 7.** Spontaneous GH secretory profile in an 11-year-old girl with giantism. Growth hormone secretory profile in an 11-year-old girl with McCune-Albright syndrome and giantism. She had excessive secretion of growth hormone from pituitary mammosomatotropes (mammasomatotrophe hyperplasia, see Kovacs et al. 1984)

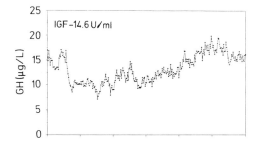

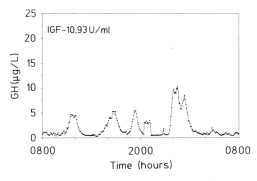

**Fig. 8.** Spontaneous GH secretory profile in an acromegalic woman. Growth hormone secretory profile in a 49-year-old woman. The *top panel* shows results preoperatively and the *bottom panel* after successful transsphenoidal surgery. The nadir in GH following oral glucose loading was 11 μg/l preoperatively and 0.8 μg/l postoperatively. The serum IGF-I levels are shown in the *upper portion of each panel*. (Courtesy of Dr. M. L. Hartman, University of Virginia)

shown the 24-h serum GH profiles (sampling every 5 min) of a 49-year-old woman with acromegaly before and after transsphenoidal removal of an adenoma that induced biochemical and clinical remission. One should note that the basal concentration is lowered and that many of the samples contain < 1 μg/l GH. The postoperative pattern is indistinguishable from normal (Hartman et al. 1990). In general in acromegaly, as can also be surmised for the rare patient with pituitary giantism, the pulsatile pattern of GH release is characterized by augmented basal GH concentrations, increased GH pulse frequency, an attenuation of the normal nyctohemeral rhythm, but markedly augmented nonpulsatile (constitutive?) release of GH. The latter pattern of release may have profound effects on the transduction of the GH signal to IGF-I release by hepatic cells (Hartmann et al. 1990).

## Conclusion

The studies reviewed here have employed objective, validated algorithms to define altered parameters of GH release in the physiologic conditions of growth and pubertal development, in pathologic states of sub- and supranormal growth and following therapy with sex-steroid hormones. The latter have discrete actions on specific parameters of circulating hormonal profiles and on the biophysical parameters that describe secretory burst activity and metabolic clearance. It is clear that the intensive study of hormonal profiles in physiologic states, in pathologic conditions and following therapeutic intervention will permit additional insights into the underlying mechanism of normal and disordered hormone secretion. At present these biophysical studies may not have direct clinical applicability, but the concepts derived from them will permit new and highly innovative studies for optimal drug delivery in various pathologic conditions.

*Acknowledgements.* I am indebted to Drs. Robert M. Blizzard and Johannes D. Veldhuis for continuing support and advice. I acknowledge a group of former and present pediatric endocrine fellows for carrying out many of the original studies: Drs. Kathleen Link, Nelly Mauras, Paul Martha Jr., C. Michele Christie, Jay McDonald, and James Kerrigan.

## References

Albertsson-Wikland K, Rosberg S (1988) Analyses of 24-hour growth hormone profiles in children; relation to growth. J Clin Endocrinol Metab 67: 493–500

Benedict PH (1962) Endocrine features in Albright's syndrome (fibrous dysplasia of bone). Metabolism 11: 30–45

Berelowitz M, Szabo M, Frohman LA (1981) Somatomedin-C mediates GH negative feedback by effects on both the hypothalamus and the pituitary. Science 212: 1279–1281

Bierich JR (1983) Treatment of constitutional delay of growth and adolescence with human growth hormone. Klin Pediatr 195: 309–316

Evans WS, Faria ACS, Christiansen E, Ho KY, Weiss J, Rogol AD, Johnson ML, Blizzard RM, Veldhuis JD, Thorner MO (1987) Impact of intensive venous sampling on characterization of pulsatile GH release. Am J Physiol 252: E549–E556

Hartman ML, Veldhuis JD, Vance ML, Faria ACS, Furlanetto RW, Thorner MO (1990) Somatotropin pulse frequency and basal concentrations are increased in acromegaly and are reduced by successful therapy. J Clin Endocrinol Metab 70: 1375–1384

178    A.D. Rogol

Hindmarsh P, Smith PJ, Brook CGD, Matthews DR (1987) The relationship between height velocity and growth hormone secretion in short prepubertal children. Clin Endocrinol (Oxf) 27: 581–591

Ho KY, Evans WS, Blizzard RM, Veldhuis JD, Merriam GR, Samojlik E, Furlanetto RW, Rogol AD, Kaiser DL, Thorner MO (1987) Effects of sex and age on the 24-hr secretory profile of GH secretion in man: importance of endogenous estradiol concentrations. J Clin Endocrinol Metab 64: 51–58

Isgaard J, Møller C, Isaksson OG, Nilsson A, Mathews LS, Norstedt G (1988) Regulation of insulin-like growth factor messenger ribonucleic acid in rat growth plate by growth hormone. Endocrinology 122: 1515–1520

Isley WL, Underwood LE, Clemmons DR (1983) Dietary components that regulate serum somatomedin C concentrations in humans. J Clin Invest 71: 175–182

Kerrigan JR, Martha PM Jr, Blizzard RM, Christie CM, Rogol AD (1990) Variations of pulsatile growth hormone release in healthy short prepubertal boys. Pediatr Res 28: 11–14

Kovacs K, Horvath E, Thorner MO, Rogol AD (1984) Mammosomatotroph hyperplasia associated with acromegaly and hyperprolactinemia in a patient with the McCune-Albright syndrome. Virchows Arch A 403: 77–86

Link K, Blizzard RM, Evans WS, Kaiser DL, Parker MW, Rogol AD (1986) The effects of androgens on the pulsatile release and the twenty-four-hour mean concentration of growth hormone in peripubertal males. J Clin Endocrinol Metab 62: 159–164

Liu L, Merriam GR, Sherins RJ (1987) Chronic sex steroid exposure increases mean plasma growth hormone concentration and pulse amplitude in men with isolated hypogonadatropic hypogonadism. J Clin Endocrinol Metab 64: 651–656

Marshall WA, Tanner JM (1970) Variation in the pattern of pubertal changes in boys. Arch Dis Child 45: 13–23

Martha PM Jr, Blizzard RM, Rogol AD (1988) Atenolol enhances growth hormone release to exogenous growth hormone-releasing hormone but fails to alter spontaneous nocturnal growth hormone secretion in boys with constitutional delay of growth. Pediatr Res 23: 393–397

Martha PM Jr, Rogol AD, Veldhuis JD, Kerrigan JR, Goodman DW, Blizzard RM (1989) Alterations in the pulsatile properties of circulating growth hormone concentrations during puberty in boys. J Clin Endocrinol Metab 69: 563–570

Mauras N, Blizzard RM, Link K, Johnson ML, Rogol AD, Veldhuis JD (1987) Augmentation of growth hormone secretion during puberty: evidence for a pulse amplitude-modulated phenomenon. J Clin Endocrinol Metab 64: 596–601

Mauras N, Blizzard RM, Rogol AD (1989) Androgen-dependent somatotroph function in a hypogonadal adolescent male: evidence for control of exogenous androgens on growth hormone release. Metabolism 38: 286–289

Mauras N, Rogol AD, Veldhuis JD (1989) Specific, time-dependent actions of low-dose ethinylestradiol administration on the episodic release of growth hormone, follicle stimulating hormone and luteinizing hormone in prepubertal girls with Turner's syndrome. J Clin Endocrinol Metab 69: 1053–1058

Mauras N, Rogol AD, Veldhuis JD (1990) Increased hGH production rate after low dose estrogen therapy in pre-pubertal girls with Turner's syndrome. Pediatr Res 28: 626–630

Müller EE, Nisticó G (1989) Brain messengers and the pituitary. Academic, New York

Plotsky PM, Vale W (1985) Patterns of growth hormone-releasing factor and somatostatin secretion into the hypophysial-portal circulation of the rat. Science 230: 461–463

Rivier J, Spiess J, Thorner M, Vale W (1982) Characterization of a growth hormone releasing factor from a human pancreatic islet tumour. Nature 300: 276–278

Rosenthal SM, Hulse JA, Kaplan SL, Grumbach MM (1986) Exogenous growth hormone inhibits growth hormone-releasing factor-induced growth hormone secretion in normal men. J Clin Invest 77: 176–180

Sanayama K, Noda H, Konda S, Ikeda F, Murata A, Sasaki N, Niimi H, Nakajima H (1987) Spontaneous growth hormone secretion and plasma somatomedin-C in children of short stature. Endocrinol Jpn 34: 627–633

Spiliotis BE, August GP, Hung W, Sonis W, Mendelson W, Bercu BB (1984) Growth hormone neurosecretory dysfunction; a treatable cause of short stature: JAMA 251: 2223–2230

Tanner JM, Davies PW (1985) Clinical longitudinal standards for height and height velocity for Northern American children. J Pediatr 107: 317–329

Tanner JM, Whitehouse RH, Takaishi M (1966) Standards from birth to maturity for height, weight, height velocity, and weight velocity: British children 1965. II. Arch Dis Child 41: 613–635

Tanner JM, Whitehouse RH, Hughes PCR, Carter BS (1976) Relative importance of growth hormone and sex steroids for the growth at puberty of trunk length, limb length, and muscle width in growth hormone-deficient children. J Pediatr 89: 1000–1008

Ulloa-Aguirre A, Ghristie CM, Carcia-Rubi E, Rogol AD, Link K, Blizzard RM, Johnson ML, Veldhuis JD (1990) Testosterone and oxandrolone a non-aromatizable androgen specifically amplify the mass and rate of growth hormone (GH) release secreted per burst without altering GH secretory burst duration or frequency or the half-life. J Clin Endocrinol Metab 71: 846–854

Urban RJ, Rogol AD, Evans JW, Johnson ML, Veldhuis JD (1988) Contemporary aspects of endocrine signal analysis. I. The paradigm of the luteinizing hormone pulse signal in men. Endocr Rev 9: 3–37

Veldhuis JD, Johnson ML (1986) Cluster analysis: a simple, versatile and robust algorithm for endocrine pulse detection. Am J Physiol 250: E486–E493

Veldhuis JD, Carlson ML, Johnson ML (1987) The pituitary gland secretes in bursts: appraising the nature of glandular secretory impulses by simultaneous multiple-parameter deconvolution of plasma hormone concentrations. Proc Natl Acad Sci USA 84: 7686–7690

Zadik Z, Chalew SA, Raiti S, Kowarski AA (1985) Do short children secrete insufficient growth hormone? Pediatrics 76: 355–360

# Rhythms in Biochemical Variables Related to Bone and Their Regulating Hormones

M. E. Markowitz

## Introduction

The presence of circadian rhythms in the blood concentrations of calcium and phosphate (Pi) has been well established by numerous investigators. Both ionized and total calcium (Cai, Cat) measurements show such rhythmicity. Calcium regulates numerous intracellular events. The rhythmic variations in the concentrations of this tightly controlled mineral may thus have significant physiological effects. It is the intent of this chapter to describe the nature and to review several of the possible regulators of these rhythms.

## The Rhythms

The precise patterns of the fluctuations in blood Cai and serum phosphate and Cat concentrations vary somewhat between studies. This is dependent, in part, on the specifics of the research protocols employed. The frequency of sampling, the activities of the subjects, the sleep-wake cycle of the subjects, and the diet all may affect the patterns (Markowitz and Rosen 1990).

## Ionized Calcium

Markowitz et al. (1981) first described the circadian rhythms in Cai concentrations in humans. A simple third-order polynomial equation sufficed as a representation of the mean pattern derived from six healthy adult males. Morning peak levels were followed by a steady decline to a trough occurring in the late afternoon. The amplitude of the peak-trough excursion was about 0.075 mmol or 7% of the 24-h mean value. Considerable variation from the model was evidenced in any single subject, though none of the original subjects was rejected as an outlier after $r$ to $z$ transformation of the correlations of the individual's data set to the putative model. Extension of the data to include a larger number of subjects studied under the same experimental conditions confirmed the validity of the model (Fig. 1). Subsequently, other investigators have found similar pattern shapes in Cai concentrations (Ishida et al. 1983; Portale et al. 1987).

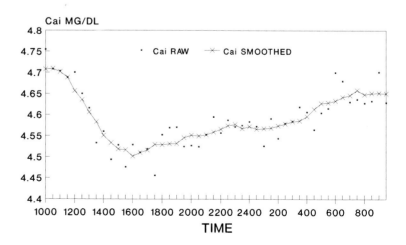

**Fig. 1.** Ionized calcium: raw and smoothed mean data from nine healthy men [8]

However, differences with respect to the timing of events were noted by Perry et al. (1986) in a study of postmenopausal women and also observed in a predominantly female hyperparathyroid group of patients by Lobaugh et al. (1989).

## Total Calcium

Serum Cat concentrations comprise both ionized and bound fractions. Normally about 40% of calcium in serum circulates in the free or unbound state. The remainder is bound chemically to circulating proteins as well as small molecules such as phosphate and citrate. The 24-h pattern of Cat levels is different than that of Cai (Markowitz et al. 1981). Though a late morning peak is again observed, a nocturnal trough occurs (Fig. 2). A shallower late afternoon trough is also observed. Thus the overall shape is of a lopsided "w." The nocturnal decline in Cat concentration has been documented by others as well (Jubiz et al. 1972;

Halloran et al. 1985). No correlation between the ionized and Cat patterns was found (Markowitz et al. 1981). The amplitude of the Cat fluctuations during the 24-h period was 0.125 mmol or about 6% of the 24-h mean (Markowitz et al. 1981; Touitou et al. 1989b).

## Serum Phosphate

In contrast to either of the calcium rhythms, phosphate levels peak at night and trough in the late morning (Fig. 3). In general, an "m"-shaped curve describes the fluctuations (Markowitz et al. 1981; Portale et al. 1987; Logue et al. 1989). A model derived from an eighth-order polynomial fits the mean data from a series of healthy male subjects with a correlation of 0.95. Individual subjects show the least variance from the phosphate model as compared with either of the calcium models. This pattern appears to be the inverse of the Cat rhythm and the correlation

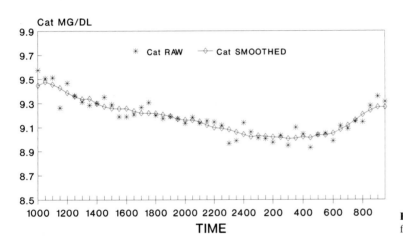

**Fig. 2.** Total calcium: raw and smoothed data from nine healthy men [8]

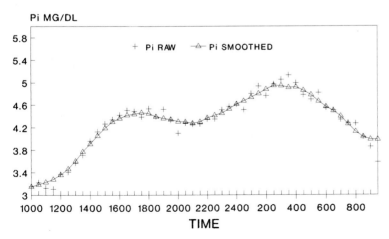

**Fig. 3.** Phosphate: raw and smoothed mean data from nine healthy men [8]

between the two phosphate and Cat models achieved statistical significance (Markowitz et al. 1981). No correlation with the Cai pattern was evident.

## Total Magnesium

The serum concentration of magnesium is kept within a narrow range by the action of a number of neural and humoral control mechanisms. Within this range, however, rhythmic oscillations in several frequency ranges can be identified. A circadian rhythm in total serum magnesium concentration was identified in children, young adults and elderly subjects (Touitou et al. 1978, 1989a; Nicolau et al. 1985) which, although statistically significant, showed a low amplitude, especially in the younger age groups. The circadian acrophase was reported as occurring during daytime by Touitou et al. (1978) in subjects studied in France but was reported during the night time in children and in elderly subjects studied in Romania (Nicolau et al. 1985; Haus et al. 1988). The circadian rhythm in serum magnesium in these studies was out of phase with the circadian rhythm in plasma protein and hematocrit, making a relation to protein binding and hydration unlikely.

A circadian difference in the effect of injected magnesium pyrrolidone carboxylate upon the blood glucose concentration was observed with a glucose elevation after administration at 0800 hours but not at 1400 or 2000 hours (Ferment and Touitou 1986).

In contrast to the small-amplitude circadian rhythm in the serum concentration the circadian rhythm in urinary excretion of magnesium is quite prominent and in its timing appears to coincide with that of calcium during the morning hours (Min et al. 1966; Doe et al. 1960; Heaton and Hodgkinson 1963) or in a group of elderly subjects during the night hours (Nicolau et al. 1985; Haus et al. 1988).

Seasonal variations in urinary excretion were reported with a maximum in spring or summer in some groups of subjects (Halberg et al. 1983; Haus et al. 1988; Touitou et al. 1989a) and with a maximum in fall in others (Nicolau et al. 1985; Haus et al. 1988).

## Ontogeny

Developmental changes in the circadian mineral patterns as a function of aging have been examined only to a limited extent. Neither Cai or Cat show changes in the shape or amplitude of their circadian rhythms

from adolescence through adulthood in males (Markowitz et al. 1984). Menstruating women also show the same patterns. However, Perry et al. (1986) have described a phase-shifted Cai pattern in postmenopausal women, the etiology of which is unclear. The pattern of calcium fluctuations in neonates, if any, has not been examined.

In contrast, phosphate patterns in adolescents appear to be an exaggerated version of the adult rhythm (Markowitz et al. 1984). The peak-trough difference of 1.2 mg/dl observed in adults may be more than doubled in healthy teenagers. The shape and timing of the fluctuations remain the same. These findings are consistent with the higher average phosphate levels found when obtained during random sampling in children undergoing periods of rapid growth (Kruse 1989).

## Regulation of the Mineral Rhythms

### Sources of Blood Mineral

Two major sources may contribute to the changes in the blood concentrations of the minerals observed over time: newly absorbed mineral from the intestine and redistribution from endogenous stores. Both are dependent on local factors, such as a functioning intestinal tract or responsive bone, and on distant influences such as dietary intake of calcium and phosphate as well as the calcium-regulating hormones.

Contribution of Intestinal Absorption

Studies in three mammalian species have examined the effects of food intake on circadian mineral rhythms: rat, dog, human.

Talmage et al. (1975), Staub et al. (1979) and others measured blood calcium periodically in rats kept on a 12-h light/dark cycle. They found patterns similar to those described in humans for Cat; a daytime peak and a nocturnal trough were observed. Staub et al. (1979) found a high correlation between Cai and Cat levels in their studies. Restricting calcium but not caloric intake during the normal nocturnal feeding time of these rodents resulted in a progressive downward displacement of the 24-h Cat pattern, an increase in amplitude swings, but retention of the sine wave shape (Staub et al. 1979). Fasting blunted the diurnal rise in Cat. Hirsch and Hagaman (1982) found that fasting blunted the nocturnal decline in Cat in rats.

The timing of offered food has also been examined. Hirsch and Hagaman (1982) saw no effects of continuous availability of food on the Cat pattern. Shinoda and Seto (1985) confirmed this finding but also demonstrated that 90 % of food consumption occurred during the dark phase regardless of availability. Restricting food intake to only the light part of the cycle for 10 days reversed the timing of the Cat pattern. By 20 days no Cat pattern persisted.

The fall in serum Cat concentrations coincident with calcium ingestion is difficult to reconcile with the findings of Wrobel (1981). Using the everted gut sac technique, maximal intestinal calcium transport occurred nocturnally in rats raised on a 12-h light/dark cycle (Wrobel 1981). Reversing the day/light pattern with food available ad libitum reversed the calcium transport cycle. Restricting food to the daytime portion of the cycle caused asynchronous calcium absorption in immature but not adult rats. Thus an ontogeny in the development of an intestinal calcium transport cycle was suggested.

In dogs prelabeled with $^{45}$Ca, constant levels of both Cat and Cai were described by Wong and Klein (1984). Calcium levels remained constant even if the dogs were fasted or fed low-calcium diets. However the $^{45}$Ca levels did vary rhythmically in the diurnally fed animals with peak levels observed at about 0800 hours and a trough 12 h later. The intestinal absorption of calcium must therefore have contributed to the maintenance of the steady plasma calcium levels in an inverse pattern, i.e., a peak at 2000 hours and a trough at 0800 hours. This contrasts with the decline in calcium levels during the feeding period seen in rats. An interspecies difference is highlighted by the parallel decline and rise in Cat and $^{45}$Ca values throughout the 24-h period seen only in the rodents.

In humans, fasted healthy subjects had no loss in the normal Cat pattern (Jubiz et al. 1972). The effect of fasting or calcium restriction on the Cai pattern has not been reported. However, the effect of dietary phosphate manipulation was examined by Portale et al. (1987). In a group of healthy young men, 10 days on a constant calcium diet with phosphate supplementation downshifted the Cai curve without changes in pattern shape or amplitude. Phosphate restriction shifted the Cai curve upward, also without a change in pattern shape or amplitude.

Shift of the normal nocturnal feeding time of rats to daytime hours by restricting food availability was associated with a phase reversal in the Pi pattern by 10 days (Shinoda and Seto 1985). By limiting Pi intake to less than 500 mg/day in healthy men, Portale et al. (1987) were able to reduce 24-h mean Pi levels by 40 % within the same period. A downshift in the entire Pi curve occurred within 1 day. By 10 days of this diet the afternoon peak in Pi levels had disappeared and the nocturnal peak was delayed by 3 h. Complete fasting flattened the Pi curve (Jubiz et al. 1972). Thus the Pi rhythm appears primarily dependent on the timing and quantity of Pi intake and in turn may influence the Cai pattern.

Contribution of Bone

The studies referred to in the preceding section, which employed the method of preloading animals with $^{45}$Ca, confirm the importance of the contribution of bone to blood calcium concentrations. Both in the dog and rat a circadian pattern of $^{45}$Ca levels in blood is observed (Wong and Klein 1984; Talmage et al. 1975). However, species differences in response to dietary calcium restriction were evident. In the dog, serum calcium levels were maintained at the expense of bone as reflected by a rise in $^{45}$Ca and constant Cat levels (Wong and Klein 1984). In the rat, $^{45}$Ca levels did not rise and Cat concentrations declined (Talmage et al. 1975). No analogous studies have been reported in humans.

Indirect evidence for a circadian rhythm in bone mineral activity in humans can be gleaned from the sequential measurement of serum osteocalcin (Oc) (also called bone gla protein or BGP) concentrations. This protein is produced by osteoblasts and is the second most common protein found in bone (Lian and Friedman 1978). A very small amount of newly synthesized Oc is released into the circulation, where it is measurable by RIA (Gallop et al. 1980). Randomly obtained samples for Oc determination correlate well with parameters of bone turnover and formation obtained by bone biopsy (Slovik et al. 1984). Maximal physiological levels are found during periods of rapid growth (Johansen et al. 1988). Though its precise function in bone is unknown, it is a chemotactic factor for osteoclast precursors and possibly for osteoblasts (Malone et al. 1982; Mundy and Poser 1983). A circadian rhythm in Oc levels occurs in humans (Fig. 4); trough levels are seen diurnally and highest levels occur nocturnally (Gundberg et al. 1985; Saggese et al. 1986). Chronic growth hormone administration to children results in an upward shift in the entire Oc curve (Markowitz et al. 1989). This hormone also increases serum Pi concentrations (in randomly obtained samples). The administration of calcitriol

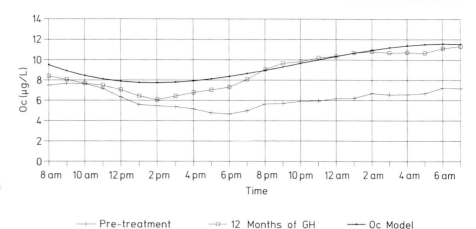

**Fig. 4.** Osteocalcin: effects of growth hormone in children with idiopathic short stature [9]

—+— Pre-treatment        —□— 12 Months of GH        —•— Oc Model

(1,25-dihydroxyvitamin D), a hormone that enhances calcium and Pi absorption, aborts the diurnal fall in Oc levels (Markowitz et al. 1987); corticosteroid treatment blunts the nocturnal rise in Oc (Neilsen et al. 1988). If Oc levels primarily reflect bone mineralization, then maximal mineralization occurs when serum Cat levels are at a nadir. Unfortunately, this elegant dovetailing of aspects of calcium metabolism does not account for the slowly rising Cai concentration observed nocturnally. Whether the circadian changes in Oc are associated directly with changes in the rate of bone mineral accretion and release and thereby do contribute to the observed fluctuations in blood mineral concentrations is presently speculative.

More consistent data are found in studies of rats. Metaphyseal and cartilage mineral deposition occurs with circadian rhythmicity and peaks nocturnally (Simmons et al. 1983; Russell et al. 1984). This coincides with trough Cat and Cai and $^{45}$Ca blood concentrations (Staub et al. 1979).

## Summary of the Contributions of Intestine and Bone

Phosphate rhythms appear strongly dependent on Pi intake. This effect of diet does not appear to be species specific for the three species discussed. In contrast, both bone mineralization and dietary intake of calcium contribute to the calcium rhythms but the proportional effects do appear to be species dependent.

## Calcium-Regulating Hormones

The contribution of the intestine and bone to the blood mineral rhythms may be modulated by endo-crine and paracrine factors. This discussion will be limited to the endocrine effects.

## Parathyroid Hormone

Acute perturbations in blood calcium levels result in inverse responses of the parathyroid glands (Brent et al. 1988). Release of parathyroid hormone (PTH) in turn affects calcium concentrations. This is mediated through PTH effects on bone mineral resorption and renal tubular calcium reabsorption, and indirectly by PTH stimulation of calcitriol (1,25-dihydroxyvitamin D) production in the kidney.

The temporal interrelationship between PTH and calcium concentrations under steady-state conditions has also been investigated. When measured sequentially, nocturnal elevations of PTH levels are consistently noted, regardless of the particular assay employed (Fig. 5) (Jubiz et al. 1972; Logue et al. 1989; Sinha et al. 1975; Radjaipour et al. 1986; LoCascio et al. 1982; Markowitz et al. 1988; Kitamura et al. 1990). In general, PTH fluctuations coincide inversely with Cat concentrations and directly with Pi levels (LoCascio et al. 1982; Markowitz et al. 1988; Kitamura et al. 1990). Analyses of urinary cyclic AMP content led Logue et al. (1989) to conclude that the changes in PTH were rapidly inductive of second messenger production. Presumably, this should result in measurable changes in Cai concentrations. Indeed, a moderately strong inverse relation between PTH and Cai time series matched at concurrent time points has been described with a correlation *(r)* of −0.5 (LoCascio et al. 1982; Markowitz et al. 1988; Kitamura et al. 1990). Further statistical analyses employing the methods of cross-correlation and cross-spectral analyses were performed by Markowitz et al. (1988), in

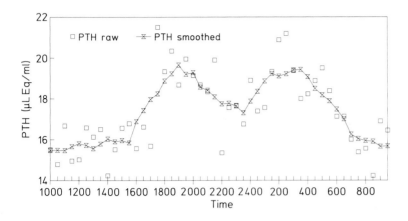

**Fig. 5.** Parathyroid hormone: raw and smoothed mean data from nine healthy men [8]

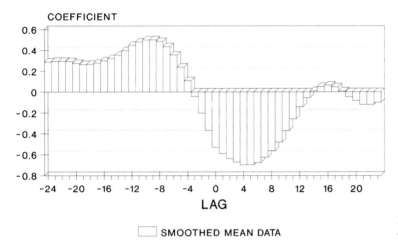

SMOOTHED MEAN DATA

**Fig. 6.** Ionized calcium versus PTH: cross-correlation matrix [8]

which evidence was found for a bidirectional interaction (Fig. 6). The strongest correlation between PTH and Cai occurred when the PTH series was lagged by 2 h, reaching an $r = -0.70$. In addition, PTH changes in turn preceded a similar directional change in Cai levels by 4 h ($r = 0.5$) (Fig. 6). Thus falling Cai levels were following 2 h later by a maximal increase in circulating PTH. This in turn was followed 4 h later by a peak increase in Cai concentrations. The lag between the series and the bidirectional nature is consistent with the known physiology of PTH/Cai interactions (Markowitz et al. 1982; Hochberg et al. 1984). No bidirectionality was found between PTH and either Cat or Pi. Kitamura et al. (1990) did not entirely reproduce these findings; peak inverse correlations were found when all the series were matched at concurrent time points. These authors postulated that differences between the studies may have been predicated by the use of different assays (RIA using an antibody that recognized both intact and c-terminal end species of PTH versus IRMA specific for intact

PTH only) or the frequency of sampling (30 vs 12 min). To determine which study is more reflective of the actual physiological events requires further research.

Parathyroidectomy results in downshift of the intact Cat pattern in rats and downshift and distortion of the Cat pattern in dogs (Staub et al. 1979; Wong and Klein 1984). Hyperparathyroidism in humans causes an upshift in undistorted Cai and Cat patterns (Lobaugh et al. 1989; Jubiz et al. 1972; Sinha et al. 1975; LoCascio et al. 1982; Logue et al. 1990). Thus, absolute concentrations of mineral are affected by abnormalities in PTH with little or no effect on the timing of events in the mineral patterns.

It is therefore doubtful that the circadian mineral rhythms are entirely dependent on PTH fluctuations. At best, PTH correlational studies can account for less than 50% of the variance in the mineral patterns. In conditions of abnormal parathyroid function, mineral rhythms may be maintained in terms of pattern shape and timing though amplitude may vary

(Lobaugh et al. 1989; Markowitz et al. 1982). Thus PTH acts as a modulator of the mineral rhythms.

### Calcitriol

The renal production of hormonal vitamin D, calcitriol, is induced by PTH or by low Pi concentrations. Target organs for this hormone include bone and gut. In the presence of calcitriol there is enhanced absorption of calcium and phosphate from the intestine.

Small diurnal fluctuations in serum calcitriol concentrations in humans were described by Halloran et al. (1985) and Markowitz et al. (1985a). No correlation with Cat, Cai or Pi was found. Cross-correlation or cross-spectral analyses were not performed. The administration of a single near physiological dose of calcitriol to healthy adult men was associated with improved synchronization between mineral patterns and a small increase in the 24-h mean Cai and Pi levels. However, the latter was no greater than the year-to-year variability in 24-h mean mineral levels measured in the same group of subjects under the same experimental conditions (Markowitz et al. 1985b). The amount of dietary phosphate ingested inversely affects the production rate of calcitriol (Portale et al. 1987). As discussed earlier, a downward shift in the Pi pattern accompanied by an upward shift in the Cai rhythm follows 10 days of Pi intake restriction. However, whether the latter is an effect of the increase in calcitriol production or a reduction in serum Pi concentrations is unclear. Anecdotal evidence that calcitriol affects mineral rhythms other than to alter absolute mineral concentrations is provided from case studies in hypoparathyroid children. Normal-appearing Cai patterns were achieved only in children receiving split daily doses of calcitriol, whereas normal range Cai levels without apparent rhythmicity were observed in the children receiving single daily doses of medicine (Markowitz et al. 1982). It appears that calcitriol, like PTH, contributes to mineral levels and may modulate 24-h patterns.

### Calcitonin

Calcitonin is a peptide hormone which participates in the regulation of calcium, magnesium and phosphate metabolism. In rats the thyroid content of immunassayable calcitonin and calcitonin specific RNA showed a high-amplitude circadian rhythm with a peak 4 h before the onset of the daily dark span. This rhythm appeared to be in phase with a low-amplitude circadian variation in plasma calcitonin but apparently not with plasma calcium and phosphate (Lansson et al. 1989).

In clinically healthy human subjects, Hillyard et al. (1977) described a circadian rhythm in immunoreactive calcitonin in plasma with a peak at midday after lunch. Apparently postprandial elevations of plasma calcitonin were reported in patients with medullary thyroid carcinoma (Miyauchi et al. 1980) while Pedrazzoni et al. (1989) in clinically healthy subjects could not find a similar rise 120 min after a standard mixed meal. Robinson et al. (1982) and Emmertsen et al. (1983) could find neither a circadian variation nor postprandial changes of plasma immunoreactive calcitonin in healthy human subjects. Stimulation of calcitonin release by ingestion of magnesium pyrrolidone carboxylate but not by magnesium sulfate showed a circadian variation in healthy subjects with a maximum at 0800 hours (Ferment 1987). The difference between these investigations may be due to the nature of the antibodies used. Although the findings in the rat thyroid tissue are of interest and no comparable data in human subjects are available, the question of a circadian variation in calcitonin production, serum concentrations and action cannot be answered with certainty at the present time.

## Conclusion

This chapter has reviewed recent evidence supporting the presence of circadian rhythms of blood mineral concentrations in mammals. The relative contribution of bone and intestine as sources for the fluctuations in mineral levels appears species specific. Mineral-regulating hormones control absolute levels of calcium and phosphate but do not entirely regulate the shape and phase relationships of the patterns. The physiological significance of these rhythms as they may affect other metabolic processes is unknown. Nevertheless, knowledge of these mineral rhythms is clinically useful. Interpretation of randomly obtained samples for calcium and phosphate must be done in the context of the time of day of sampling and the patient's age, sex and diet.

# References

Brent GA, Leboff MS, Seely EW, Conlin PR, Brown EM (1988) Relationship between the concentration and rate of change of calcium and serum intact parathyroid hormone levels in normal humans. J Clin Endocrinol Metab 67: 944–950

Doe RP, Vennes JA, Flink EB (1960) Diurnal variation of 17-hydroxycorticosteroids, sodium, potassium, magnesium and creatinine in normal subjects and in cases of treated adrenal insufficiency and Cushing's syndrome. J Clin Endocrinol 20: 253–265

Emmertsen K, Christensen SE, Schmitz O, Marqversen J (1983) Diurnal patterns of serum immunoreactive calcitonin in healthy males and in insulin dependent diabetics. Clin Endocrinol (Oxf) 19: 533–537

Ferment O (1987) Rétrocontrôle du magnésium sur la sécrétion de l'hormone parathyroïdienne et de la calcitonine chez l'homme sain. Etude comparée avec le calcium et aspects chronopharmacologiques. Thesis, University of Paris VI

Ferment O, Touitou Y (1986) Time-dependent effect of magnesium on blood glucose in man. Annu Rev Chronopharmacol 3: 395–397

Gallop PM, Lian JB, Hauschka PV (1980) Carboxylated calcium-binding proteins and vitamin K. N Engl J Med 302: 1460–1466

Gundberg CM, Markowitz ME, Mizruchi M, Rosen JF (1985) Osteocalcin in human serum: a circadian rhythm. J Clin Endocrinol Metab 60: 736–739

Halberg F, Lagoguey M, Reinberg A (1983) Human circannual rhythms over a broad spectrum of physiological processes. Int J Chronobiol 8: 225–268

Halloran BP, Portale AA, Castro M, Morris RC, Goldsmith RS (1985) Serum concentration of 1,25-dihydroxyvitamin D in the human: diurnal variation. J Clin Endocrinol Metab 60: 1104–1110

Haus E et al. (1988) Reference values for chronopharmacology. Annu Rev Chronopharmacol 4: 333–424

Heaton FW, Hodgkinson A (1963) External factors affecting diurnal variation in electrolyte excretion with particular reference to calcium and magnesium. Clin Chim Acta 8: 246–254

Hillyard CS, Cooke TJC, Coombes RC, Evans IMA, Macintyre I (1977) Normal plasma calcitonin: circadian variation and response to stimuli. Clin Endocrinol (Oxf) 6: 291–298

Hirsch PF, Hagaman JR (1982) Feeding regimen, dietary calcium, and diurnal rhythms of serum calcium and calcitonin in the rat. Endocrinology 110: 961–968

Hochberg Z, Moses AM, Richman RA (1984) Parathyroid hormone infusion test in children and adolescents. Miner Electrolyte Metab 10: 113–116

Ishida M, Seino Y, Yanaoka K, Tanaka Y, Satomura K, Kurose Y, Yabunchi H (1983) The circadian rhythms of blood ionized calcium in humans. Scand J Clin Lab Invest [Suppl 165] 43: 83–86

Johansen JS, Giwercman A, Hartwell D, Nielsen CT, Price PA, Christiansen C, Skakkebaek NE (1988) Serum bone gla-protein as a marker of bone growth in children and adolescents: correlation with age, height, serum insulin-like growth factor I, and serum testosterone. J Clin Endocrinol Metab 67: 273–278

Jubiz W, Canterbury JM, Reiss E, Tyler F (1972) Circadian rhythm in serum parathyroid hormone concentration in human subjects: correlation with serum calcium, phosphate, albumin and growth hormone levels. J Clin Invest 52: 2040–2046

Kitamura N, Shigeno C, Shiomi K, Lee K, Ohta S, Sone T, Katsushima S, Tadamura E, Kousaka T, Yamamoto I, Dokoh S, Konishi J (1990) Episodic fluctuation in serum intact parathyroid hormone concentration in men. J Clin Endocrinol Metab 70: 252–263

Kruse K (1989) Endocrine control of calcium and bone metabolism. In: Brook GD (ed) Clinical paediatric endocrinology, 2nd edn. Blackwell, Oxford, p 487

Lansson S, Segond N, Milhaud G (1989) Circadian rhythms of calcitonin gene expression in the rat. J Endocrinol 122: 527–534

Lian JB, Friedman PA (1978) The vitamin K-dependent synthesis of carboxyglutamic acid by bone microsomes. J Biol Chem 253: 6623–6626

Lobaugh B, Neelon FA, Oyama H, Buckley N, Smith S, Christy M, Leight GS (1989) Circadian rhythms for calcium, inorganic phosphorus, and parathyroid hormone in primary hyperparathyroidism: functional and practical considerations. Surgery 106: 1009–1017

LoCascio V, Cominacini L, Adami S, Galvanini G, Davoli A, Scuro LA (1982) Relationship of total and ionized serum calcium circadian variations in normal and hyperparathyroid subjects. Horm Metab Res 14: 443

Logue FC, Fraser WD, O'Reilly DSJ, Beastall GH (1989) The circadian rhythm of intact parathyroid hormone (1–84) and nephrogenous cyclic adenosine monophosphate in normal men. J Endocrinol 121: R1–R3

Logue FC, Fraser WD, Gallacher SJ, Cameron DA, O'Reilly DSJ, Beastall GH, Patel U, Boyle IT (1990) The loss of circadian rhythm for intact parathyroid hormone and nephrogenous cyclic AMP in patients with primary hyperparathyroidism. Clin Endocrinol (Oxf) 32: 475–483

Malone JD, Teitelbaum SL, Griffin GL, Senior RM, Kahn AJ (1982) Recruitment of osteoclast precursors by purified bone matrix constituents. J Cell Biol 92: 227–230

Markowitz ME, Rosen JF (1990) The circadian rhythms of calcium, phosphate, and calcium-regulating hormones in human blood. In: Castells S, Finberg L (eds) Metabolic bone disease in children. Dekker, New York, pp 71–83

Markowitz ME, Rotkin L, Rosen JF (1981) Circadian rhythms of blood minerals in humans. Science 213: 672–674

Markowitz ME, Rosen JF, Smith C, DeLuca HF (1982) 1,25-Dihydroxyvitamin $D_3$-treated hypoparathyroidism: 35 patient years in 10 children. J Clin Endocrinol Metab 55: 727–733

Markowitz ME, Rosen JF, Laximinarayan S, Mizruchi M (1984) Circadian rhythms of blood minerals during adolescence. Pediatr Res 18: 456–462

Markowitz ME, Rosen JF, Hannifan MF, Endres DB (1985a) Time-related variations in serum 1,25-$(OH)_2$ vitamin D concentrations in humans. 6th Workshop on Vitamin D, Merano

Markowitz ME, Rosen JF, Mizruchi M (1985b) Effects of 1,25-dihydroxyvitamin $D_3$ administration on circadian mineral rhythms in humans. Calcif Tissue Int 37: 3551–3561

Markowitz ME, Gundberg CM, Rosen JF (1987) The circadian rhythm of serum osteocalcin concentrations: effects of 1,25-dihydroxyvitamin D administration. Calcif Tissue Int 40: 179–183

Markowitz ME, Arnaud S, Rosen JF, Thorpy M, Laximinarian S (1988) Temporal interrelationships between the circadian rhythms of serum parathyroid hormone and calcium concentrations. J Clin Endocrinol Metab 67: 1068–1073

Markowitz ME, Dimartino-Nardi J, Gasparini F, Fishman K, Rosen JF, Saenger P (1989) Effects of growth hormone therapy on circadian osteocalcin rhythms in idiopathic short stature. J Clin Endocrinol Metab 69: 420–425

Min HK, Jones JE, Flink EB (1966) Circadian variations in renal excretion of magnesium, calcium, phosphorus, sodium and potassium during frequent feeding and fasting. Fed Proc 25: 917–921

Miyauchi A, Morimoto S, Onishi T, Takai S-I, Okada Y, Himeno S, Kosaki G, Kumahara Y (1980) Meal-related changes in plasma calcitonin levels in patients with medullary thyroid carcinoma. Endocrinol Jpn 27: 153–156

Mundy G, Poser JW (1983) Chemotactic activity of the γ-carboxyglutamic acid containing protein in bone. Calcif Tiss Int 35: 164–168

Neilsen HK, Charles P, Mosekilde L (1988) The effect of single oral doses of prednisone on the circadian rhythm of serum osteocalcin in normal subjects. J Clin Endocrinol Metab 67: 1025–1030

Nicolau GY, Haus E, Lakatua D, Bogdan C, Petrescu E, Robu E, Sackett-Lunden I, Swoyer J (1985) Chronobiologic observations of calcium and magnesium in the elderly. Rev Roum Med Endocrinol 23: 39–53

Pedrazzoni M, Ciotti G, Davoli L, Pioli G, Girasole G, Palummeri A, Passeri M (1989) Meal-stimulated gastrin release and calcitonin secretion. J Endocrinol Invest 12: 409–412

Perry HM, Province MA, Droke DM, Kim GS, Shaheb S, Avioli LV (1986) Diurnal variation of serum calcium and phosphorus in postmenopausal women. Calcif Tissue Int 38: 115–118

Portale AA, Halloran BP, Morris RC (1987) Dietary intake of phosphorus modulates the circadian rhythm in serum concentration of phosphorus. J Clin Invest 80: 1147–1154

Radjaipour M, Kindtner E, Roesler H, Eggstein M (1986) Circadiane Rhythmik der Konzentrationen von Parathyrin und Calcitonin in Serum. J Clin Chem Clin Biochem 24: 175–178

Robinson MF, Body JJ, Offord KP, Heath H (1982) Variation of plasma immunoreactive parathyroid hormone and calcitonin in normal and hyperparathyroid man during daylight hours. J Clin Endocrinol Metab 55: 538–544

Russell JE, Grazman B, Simmons DJ (1984) Mineralization in rat metaphyseal bone exhibits a circadian stage dependency. Proc Soc Exp Biol Med 176: 342–345

Saggese G, Bertelloni S, Baroncelli GI, Ghirri P (1986) Ritmo circadiano dell'osteocalcina nel bambino. Minerva Pediatr 38: 1035–1037

Shinoda H, Seto H (1985) Diurnal rhythms in calcium and phosphate metabolism in rodents and their relations to lighting and feeding schedules. Miner Electrolyte Metab 11: 158–166

Simmons DJ, Arsenis C, Whitson SW, Kahn SE, Boskey AL, Gollub N (1983) Mineralization of rat epiphyseal cartilage: a circadian rhythm. Miner Electrolyte Metab 9: 28–37

Sinha TK, Miller S, Fleming J, Khairi R, Edmondson J, Johnston CC, Bell NH (1975) Demonstration of a diurnal variation in serum parathyroid hormone in primary and secondary hyperparathyroidism. J Clin Endocrinol Metab 41: 1009–1013

Slovik DM, Gundberg CM, Neer RM, Lian JB (1984) Clinical evaluation of bone turnover by serum osteocalcin measurements in a hospital setting. J Clin Endocrinol Metab 59: 228–230

Staub JF, Perrault-Staub AM, Milhaud G (1979) Endogenous nature of circadian rhythms in calcium metabolism. Am J Physiol 237: R311–R317

Talmage RV, Roycroft JH, Anderson JJB (1975) Daily fluctuations in plasma calcium, phosphate, and their radionuclide concentrations in the rat. Calcif Tissue Res 17: 91–102

Touitou Y, Touitou C, Bogdan A, Beck H, Reinberg A (1978) Serum magnesium circadian rhythm in human adults with respect to age, sex and mental status. Clin Chim Acta 87: 35–41

Touitou Y, Touitou C, Bogdan A, Godard JP (1989a) Physiopathological changes of magnesium metabolism with aging. In: Itokawa Y, Durlach J (eds) Magnesium in health and disease. Libbey, London, pp 103–110

Touitou Y, Touitou C, Bogdan A, Reinberg A, Motohashi Y, Auzeby A, Beck H (1989b) Circadian and seasonal variations of electrolytes in aging humans. Clin Chim Acta 180: 245–254

Wong KM, Klein L (1984) Circadian variations in contributions of bone and intestine to plasma calcium in dogs. Am J Physiol 246: R688–R692

Wrobel J (1981) Development of circadian variations in intestinal calcium transport during maturation of the rat. Acta Physiol Pol 32: 407–417

# Biological Rhythms and Aging

Y. Touitou and E. Haus

## Introduction

Aging in the living organism is a process which starts at birth and ends with the cessation of life. It is, however, not a uniform trend progressing at a uniform rate, but rather shows superimposed periods of rapid growth and development, maturation, and finally times of regression. Superimposed on these changes are biologic rhythms of different frequencies, which at all levels of biologic organization participate in the processes of growth and cell proliferation. Rhythm alterations have been presumed to participate in the processes of cell and organ damage, and finally death of the organism. Since rhythmic changes are found at all levels of organization, they are expected to participate in the aging process at the biochemical and cellular level, and be either causative by being involved or altered, as a consequence of aging changes in organ function. Thus the approach to aging of the biologic time structure has to be extremely broad and involves biochemical cell and organ systems as well as the organism as a whole.

The importance of the problem of aging for our society is accentuated by the epidemiologic fact that an increasing number of subjects in developed countries (and in the future presumably in the developing countries) are found in the population group of 65 years and older. The average life expectancy of individuals having reached the age of 80 years is on the average 7 years. However, it has to be emphasized that the longevity of the individual has not increased substantially during the last 20 years, but that the number of subjects reaching advanced age has increased, thanks to the progress in prevention of disease and the improved medical diagnoses and treatment.

In applying chronobiologic concepts to aging, we have to keep in mind many of the classical biologic approaches and their role in the development, maintenance and finally breakdown and disintegration of the biologic time structure.

The molecular and cellular approach to aging assumes genetic programming of an organism's life span inscribed in the genetic code, i.e., in the DNA of the cell. The replication of DNA in cell growth is a highly periodic process in several frequency ranges, and disturbances at this level can be expected to lead to cell and tissue damage, which may be hypothesized to participate in the aging process. Apart from genetic programming, the "theory of errors" (Orgel 1963) assumes that senescence is due to malfunctioning of the cell brought about through an accumulation of noxious events inflicted from the surroundings, including ionizing radiation and intoxications. This would lead at a given moment to a catastrophic accumulation of such errors, which may be amplified when the damaged cell reproduces itself, and then may lead to a catastrophe for the developing altered cell line. The effects of ionizing radiation and of numerous noxious substances, like carcinogens or many groups of chemotherapeutic agents at the level of the cellular DNA, are highly periodic, ranging in some experimental designs from minimal damage to cell death as a function of the time of exposure (Haus et al. 1974, 1983).

The limited potential for cell divisions in embryonal fibroblasts and the decrease in number of cell divisions of which a fibroblast is capable with increasing age of the subject from which it is derived ($50 \pm 10$ divisions in embryonal cells and $20 \pm 10$ divisions in cells from an individual of around 80 years of age) provide an interesting model to study the interaction of the periodicity of cell proliferation and aging. A similar line of research shows that the number of cell divisions seen in cultured skin fibroblasts of patients with progeria is very low.

Also several of the molecular mechanisms of aging proposed involving DNA and its capacity for repair, alterations of cell membranes and others involve processes known to show periodic variations in different frequencies (see chapter by Edmunds).

# Chronobiologic Approach to Aging

A chronobiologic approach to aging at the cellular and molecular level is essential to obtain meaningful results in the often high amplitude rhythmically functioning systems at this level but raises a number of complex problems some of which are in a different dimension equally pertinent for the physiology and pathology of human aging.

Aging is frequently thought to be related to a loss of the time structure, whereby it remains unclear if this loss is a cause or rather a consequence of the aging process. Aging is accompanied by a decrease, if not by a loss, of the capability for adaptation of different functions of the organism. Numerous rhythms are adaptive in nature: they allow the organism to adapt in phase to the periodic changes of the environment (e. g., of light and darkness, the seasons, daily changes in noise and silence). Rhythms are, therefore, an essential element of the plasticity of a subject in relation to environmental changes.

In general, the living organism (from the single cell to the human body) represents conceptually an oscillating system. This system will function considerably better when it is in resonance with its surroundings, especially when the periodic changes in the environment are identical or similar to the frequency (or frequencies) exhibited by the organism. The growth of tomato plants is optimal when the light-dark cycle occurs with a 24-h frequency and is considerably less favorable when it is different from 24 h. Pittendrigh and Minis (1972) showed that *Drosophila* raised in an environment in which the light-dark cycle covered 24 h (LD 12:12) showed a decrease in their life span when they were transplanted on the 1st day of their adult life into an environment with a different light cycle (e. g., LD 10.5:10.5 or 13.5:13.5), providing a 21- or 27-h day. In principle, aging is synonymous with life: certain cells begin to age from birth on, and from an age of 20–30 years on there is a decrease in our capabilities for adaptation to our environment and to noxious stimuli and aggressions.

The chronobiologic approach to aging attempts to verify the widely held hypothesis that aging corresponds to or is accompanied by an alteration and/or a partial or total loss of our time structure. The chronobiologic study of aging, therefore, aims to explore and to quantify the plasticity of the organism faced with the daily and yearly changes of its environment and tries to interpret the nature of changes detected. The chronobiologic study of aging, furthermore, tries to apply the results of such investigations to the improvement of the conditions and the quality of life of elderly subjects.

# Methodologic Considerations

## Inclusion and Exclusion Factors

For critical interpretation of data in a study of biologic rhythms in the aged, a large number of parameters have to be controlled. These include the synchronization of the subjects participating in the study in regard to times of activity and sleep, meal times, etc. It is necessary to select groups of subjects which are as far as possible homogeneous in regard to age, sex, body weight, physical and intellectual activities, psychologic condition (depression, feeling of abandonment and withdrawal of the elderly, etc.) as well as in regard to their physical condition or impairment (deafness, visual acuity, dementia, etc.).

A strict observation of these principles poses numerous problems in the selection of a group of elderly subjects for study. In addition, certain other facts which are peculiar to elderly subjects have to be considered like polypathology, the functional condition of the kidney (one has to be aware of the decrease in the number of functional nephrons in the aged) and, finally, the polymedication and self-medication prevalent in elderly subjects. Every potential participant in a study involving elderly subjects will have to be examinded clinically and biologically in order to eliminate (depending on the purpose of the study) those subjects which do not correspond to the predetermined criteria. Two other factors are also important to consider because they may influence the results: studies are often performed in an institutionalized environment (hospice or home for the aged) which facilitates the formation of groups of subjects which are synchronized and homogeneous in relation to the inclusion factors chosen. The question has to be raised about the transposition of results obtained under these conditions to the aged person who is not institutionalized and lives in his or her usual familiar surroundings. Another important aspect is the often poor quality of sleep in the aged which can influence a certain number of rhythmic functions (Touitou 1982).

## Transverse and Longitudinal Studies

Longitudinal studies (of the same subjects followed regularly over a time span of many years) are most apt to answer questions on the relations between aging and biologic rhythms since they allow the study of effects of aging in the same organism. Such studies are possible in laboratory animals with a short life span (for example, rats), but are for obvious reasons much more difficult to arrange in human subjects. However, in some instances, the interpretation of aging effects in animals with a short life span may also be difficult due to the existence of circannual rhythms, the effects of which are superimposed on those of the aging process. In this situation one has to differentiate between the effects of aging and circannual rhythms.

Transverse studies are easier to carry out in human subjects in the circadian frequency domain (by comparison of studies over 24 h of groups of subjects of different age, sex, and other chosen parameters) or in the infradian frequency (comparison of groups of subjects sampled several times during the year) but they do not allow the study of aging in the strictest sense (i.e., the kinetics of a potential disturbance of adaptation reactions in human subjects). Transverse studies allow the comparison of subjects from different age groups and allow conclusions to be drawn concerning the effect of age (in a static aspect) but do not allow a study of the dynamic aspect of aging.

## Interpretation of Results

The plasma concentrations of different variables in the blood of clinically healthy elderly subjects have most often been found to be comparable to those in young adults although the urinary output of many of these substances is decreased. It is important to be aware of and to avoid as far as feasible the often encountered difficulty of blood drawing in the aged, who often show poor veins requiring prolonged tourniquet use, which can lead to changes in the analyzed variables. Similarly, the collection of complete urine samples in elderly subjects is difficult, since there is often a significant amount of residual urine which may lead to errors. Finally, we have to consider the very large interindividual variability in aged populations, which results from the differences within these populations in regard to biologic aging, due to the different genetic potential and environmental factors which may lead to marked differences between biologic and chronologic age.

## Hypothalamic-Pituitary-Adrenal Axis

The human adrenal normally produces three main categories of steroid hormone: glucocorticoids (mainly cortisol), mineralocorticoids (mainly aldosterone) and adrenal androgens (mainly dehydroepiandrosterone, DHEA, and its sulfate, DHEA-S). The secretion of cortisol and adrenal androgens is primarily regulated by ACTH, which is in turn regulated by a specific neurohormone, the corticotropin-releasing hormone (CRH) released from the hypothalamus. ACTH and adrenal cortisol production show no clear-cut changes during aging since the responsiveness of cortisol production to stimulation by ACTH is unimpaired as is the ACTH response to infusion of CRH (Pavlov et al. 1986).

The circadian rhythm of plasma cortisol displays the same general pattern in elderly subjects as in young adults, with a peak between 0600 and 0800 hours (Fig. 1), a progressive decrease during the day, with a minimum at night around 0000 hours and a progressive increase until 0800 hours in the morning (Silverberg et al. 1968; Serio et al. 1970; Jensen and Blichert-Toft 1971; Dean and Felton 1979; Nelson et al. 1980). The same profile was also observed in patients with senile dementia of the Alzheimer type (SDAT) by Touitou et al. (1982). The active free fraction of plasma cortisol was found increased in the elderly, which could be related to a decrease in the concentration of binding proteins or/and to a decrease in binding capacity of these proteins with aging (Touitou et al. 1982, 1983c). A tendency toward phase advance in plasma total and free cortisol and plasma 18-hydroxy-11-deoxycorticosterone (18-OHDOC) in the elderly was described by Touitou et al. (1982), Milcu et al. (1978) and Sherman et al. (1985), but this trend was not found during all seasons (Touitou et al. 1983c). Haus et al. (1989) also showed in elderly subjects a phase advance in plasma DHEA-S. Plasma DHEA and DHEA-S decrease progressively in both sexes during aging and also lose their ability to respond to ACTH (Parker et al. 1981). This is one of the most striking changes in steroid production with age probably related to an enzymatic alteration in the biosynthetic pathway. The findings of a phase advance of cortisol and of other steroids in the elderly point to some similarities between aging

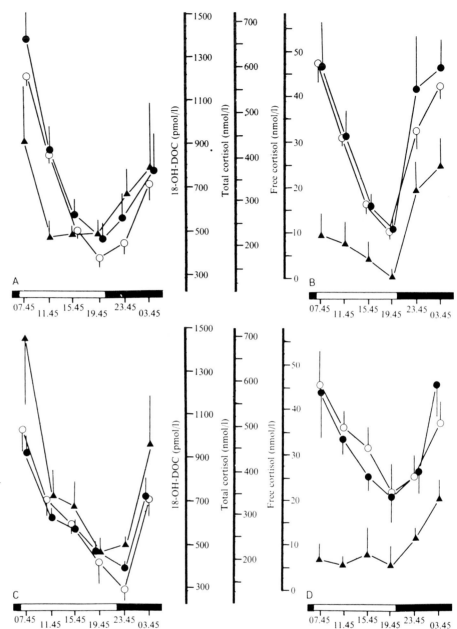

**Fig. 1 A–D.** Circadian variations of plasma concentrations of 18-hydroxy-11-deoxycortico-sterone (*18-OH Doc;* ▲), total cortisol ( ○ ) and free cortisol ( ● ) in (**A**) six elderly men, (**B**) six elderly women, (**C**) seven young men and (**D**) six elderly men and women suffering from senile dementia of the Alzheimer type. Nocturnal rest is shown by a *solid bar.* (From Touitou et al. 1982, 1983c)

and depression, at least in regards to the changes in the rhythmicity of adrenocortical function. However, these preliminary findings need further research since a temporal coincidence does not necessarily imply any causal relationship. By contrast to the persistence of the circadian periodicity of cortisol, no evidence of seasonal rhythmicity of cortisol was found in the elderly (Touitou et al. 1983c; Haus et al. 1988). The absence of a seasonal variation in plasma cortisol concentration is difficult to interpret since

such a variation was found in young adult men (Touitou et al. 1983c) but not in women and children (Haus et al. 1988).

Aldosterone secretion is greatly diminished in the elderly, at least in part due to a decreased secretion of renin which causes the generation of angiotensin, a direct stimulus to aldosterone secretion. A decrease in the circadian rhythm amplitude of plasma aldosterone and renin activity was described by Cugini et al. (1982) and Haus et al. (1989).

## Hypothalamic-Pituitary-Gonadal Axis

Conflicting results have been published in the literature on age-related effects of testosterone production in men. A number of studies carried out in the morning have shown that plasma testosterone concentrations decreased in the aged (Vermeulen et al. 1972; Stearns et al. 1974; Baker et al. 1976) whereas others found normal values in samples obtained during the early afternoon (Harman and Tsitouras 1980; Sparrow et al. 1980). A possible explanation of these contradictory results could be a decrease with aging in the circadian amplitude of plasma testosterone leading to an age difference which in plasma testosterone concentrations may not be detectable at all sampling times. Along these lines, Bremmer et al. (1983) showed that the ability to find an age-related decrease in plasma testosterone concentrations in aged men depends in part upon the time of day during which blood sampling is performed. A circadian rhythm in serum testosterone, with a peak at 0800 hours and a trough between 1900 and 2100 hours, was found in healthy young men but not in clinically healthy elderly men. Thus, elderly men showed significantly lower concentrations of plasma testosterone at certain times of day (0200–1300 hours) but not at other times (1400–2100 hours). Similar results have been found by Marrama et al. (1982) in elderly men and by Simpkins et al. (1981) and Miller and Riegle (1982) in aging rats.

Plasma estradiol and progesterone decrease in women after menopause. Circadian variations of estrone, estradiol, androstenedione and testosterone were not found by Noel et al. (1981) and Lonning et al. (1989), but have recently been documented as statistical phenomena in larger populations (Haus et al. 1988).

The decreasing gonadal activity in the elderly leads through a feedback mechanism to an increased secretion of the gonadotropins, LH and FSH. Both hormones are secreted predominantly in a pulsatile fashion (see the chapter by Veldhuis) and it is difficult to validate a circadian rhythm of plasma LH and FSH in the elderly (Touitou et al. 1981; Haus et al. 1989). In elderly men but not in women a large-amplitude circannual rhythm of LH, with higher levels in April-May, has been described (Touitou et al. 1983b; Haus et al. 1988). Thus, high levels of plasma LH due to weak gonadal activity are compatible with the persistence of a circannual rhythmicity of the hormone, which suggests that this bioperiodicity of the pituitary gland is relatively independent of the decreased

gonadal secretions in the elderly subjects. A presumably similar phenomenon in another area of endocrinology apart from aging is the persistence of the circadian rhythm of plasma ACTH with both a high circadian mean concentration and large amplitude in cortisol-deficient patients (Graber et al. 1965; Oliver et al. 1971). Also the pulsatile patterns of release of the gonadotropin persist in subjects without functional ovaries (Yen et al. 1972; Wallach et al. 1973) or in patients with gonadal dysgenesis (Boyar et al. 1973; Kelch et al. 1973). Thus, the absence of gonadal hormones or their low levels do not affect the qualitative pattern of gonadotropin secretion although plasma concentrations of gonadotropins are increased in such patients.

## Hypothalamic-Pituitary-Thyroid Axis

The plasma thyroxine concentration is not modified (or only slightly decreased) with age whereas thyroidal radioiodide accumulation rates and plasma clearance decrease, which results in a slowing of the thyroxine disposal rate probably related to a decreased activity of enzymes inactivating thyroxine. The therapeutic implication is that the replacement dose of thyroxine should be less in the elderly patients with thyroid insufficiency (Gregerman 1986).

Neither the free hormone level of thyroxine nor the thyroxine-binding-globulin (TBG), the main protein carrier of plasma in man, are significantly affected during aging in man. Plasma values of triiodothyronine ($T_3$) are most often slightly diminished. These normal (or slightly decreased) levels of thyroid hormones are accompanied by normal or, especially at the time of the circadian trough, by slightly increased concentrations of thyroid-stimulating hormone, TSH. The circadian periodicity of the hypothalamic-pituitary-thyroid axis is maintained in elderly human subjects with a peak around 0200 hours for TSH and in the afternoon for total $T_3$ and $T_4$ (Haus et al. 1989).

## Vitamin D, Parathyroid Hormone, Calcitonin and Electrolyte-Related Metabolism

Before vitamin D can exert its biologic effects, two hydroxylation reactions are required. It is first metabolized to 25-hydroxyvitamin D (25-OH-D) in the liver and subsequently in the kidney to $1\alpha$-25-dihydroxyvitamin D ($1,25$-$(OH)_2D$), which is the biologically active metabolite. $1,25$-$(OH)_2D$ plays a major role in the regulation of the mineral metabolism by stimulating the active intestinal absorption of calcium (Ca) and phosphorus (P) and regulating their flux into and out of bone. The renal synthesis and the circulating concentrations of $1,25$-$(OH)_2D$ are, in turn, controlled by the serum levels of Ca and inorganic phosphate (Pi) either directly or, in the case of Ca, through parathyroid hormone (PTH).

The serum concentration of $1,25$-$(OH)_2D$ does not undergo large variations along the 24-h scale but is rather maintained within relatively narrow limits throughout the day (Adams et al. 1979; Prince et al. 1983) though some young men were found to exhibit a circadian rhythm with a small amplitude of about 10% or less (Halloran et al. 1985). Circulating levels of calcitonin are also circadian periodic in normal subjects, with a peak around midday (Hillyard et al. 1977).

The vitamin $D_3$ metabolite concentration would be expected to vary according to climate and sunlight exposure. A seasonal variation (with higher levels during the summer) has been described for 25-OH-D (Stamp and Round 1974; Lester et al. 1977), for 24, 25 $(OH)_2D$ (Kano et al. 1980) and for $1,25$-$(OH)_2D$ (Juttman et al. 1981) in young and elderly subjects though seasonal variations were not found by some investigators for this latter metabolite (Chesney et al. 1981; Lips et al. 1983).

In recent years, osteomalacia has been named as a possible contributory factor in femoral neck fractures, and lower values of vitamin D metabolites are often encountered in the elderly (Lester et al. 1977). Meller et al. (1986) have observed in geriatric patients with long bone fractures that the serum concentrations of vitamin D metabolites correlate positively while that of parathyroid hormone correlates negatively with day length and global solar radiation. The high prevalence of fractures in young and elderly people during the winter months can be related, at least in part, to these biologic variations.

Seasonal variations in serum concentrations of PTH were described in elderly subjects with higher levels in winter when concentrations of 25-OH-D and 24,25-$(OH)_2D$ were lowest, which suggests a possible functional relationship between these hormones (Lips et al. 1983; Meller et al. 1986).

Plasma total Ca and Mg undergo circadian and seasonal variations in the elderly with a small rhythm amplitude (Touitou et al. 1978, 1989; Nicolau et al. 1985). The circadian variations of both cations have been thought to reflect predominantly the diurnal changes in plasma proteins (Jubiz et al. 1972; Markowitz et al. 1981). However, this does not fit well with the large differences in the circadian amplitudes of the plasma total proteins as compared to the much smaller amplitudes of Ca and Mg as found by Touitou et al. (1986b) and the phase difference between plasma Ca and plasma proteins as reported by Nicolau et al. (1985). Markowitz et al. (1988) have described the temporal relationship between the circadian rhythms of serum PTH and Ca concentrations and have found, at least in healthy young men, that the changes in ionized calcium (Cai) concentrations precede inverse changes in PTH levels by 2 h whereas changes in PTH precede similar directional alterations in Cai by about 4 h.

The variations in plasma inorganic phosphorus (Pi) are large with a peak located at night in both young and elderly subjects (Jubiz et al. 1972; Touitou et al. 1989). In both diurnal and nocturnal animals the peak values are observed at the end of the resting period.

## Growth Hormone and Prolactin

Growth hormone (GH) is a pituitary hormone whose secretion is largely pulsatile and in its timing related to the sleep-wake cycle with about three-fourths of the pulses occurring during the first few hours after sleep onset (stages 3 and 4). In old people the night time pulses greatly decrease in both frequency and amplitude (Finkelstein et al. 1972; Blichert-Toft 1975; Murri et al. 1980) whereas basal or daytime concentrations of plasma GH are not modified with age. Prinz et al. (1983) stated that this decrease of GH occurred exclusively in the first 3 h after sleep onset. The biologic effects of GH are believed to be mainly mediated via the stimulation of the hepatic production of somatomedins, especially somatomedin C. Levels of somatomedin C decrease with age, an important physiologic event probably related to decreased levels of GH (Florini et al. 1985). The night time reduction of plasma GH concentrations does not seem to occur secondary to a modification of the sleep structure in the elderly since stages 3 and 4 of sleep, which correspond to the time of major release of GH, are not significantly reduced.

Prolactin is secreted episodically in normal man with a circadian pattern characterized by a nocturnal peak which appears to be in phase with sleep (Parker et al. 1973; Sassin et al. 1972, 1973). Daytime naps are associated with prolactin rises (Parker et al. 1973) and reversal of the sleep-wake cycle results in a peak of plasma prolactin during daytime sleep (Sassin et al. 1973). However, using protocols involving a 7-h advance or a 7-h delay of the sleep-wake cycle in subjects having undergone rapid displacement over several time zones, Désir et al. (1982) also showed the existence of an endogenous component, i.e., an intrinsic circadian rhythmic component in prolactin secretion in addition to the sleep dependency of the hormone. Continuation of circadian rhythmicity of prolactin in the elderly has been reported by some investigators (Touitou et al. 1981, 1983a; Lakatua et al. 1984). However, both unchanged and increased serum concentrations of prolactin were reported, as well as an unchanged or decreased rhythm amplitude (Touitou et al. 1983a; Haus et al. 1989). The reason for the inconsistent findings is not obvious and it is possible that biologic variability and demographic and/or ethnic factors may account for the differences. A circannual rhythm in plasma prolactin has been reported consistently in young and elderly women (Tarquini et al. 1979; Touitou et al. 1983a; Haus et al. 1980, 1989; Nicolau et al. 1984), which could not be found in young and elderly men (Gala et al. 1977; Reinberg et al. 1978; Djursing et al. 1981; Touitou et al. 1983a).

## Catecholamines

Plasma catecholamines (epinephrine, norepinephrine, dopamine) display a circadian rhythmicity with higher daytime levels (Thurton and Deegan 1974; Prinz et al. 1979). Similar variations of norepinephrine have been described in cerebrospinal fluid of man and monkey as well as in brain regions rich in noradrenergic nervous endings (Ziegler et al. 1976). Nocturnal species like rodents are in phase opposition when compared to man, with higher values at night, which strongly suggests that the circadian rhythms of catecholamines are related to the sleep-wake cycle. Daytime and nocturnal plasma levels of norepinephrine are elevated by about 30% and 75% in the elderly compared to young adults (Prinz et al. 1979).

## Pineal Gland

Melatonin (N-acetyl-5-methoxytryptamine) is a hormone mainly secreted by the pineal gland. Its circadian rhythm is similar in humans and in experimental animals of various diurnal and nocturnal species tested, with three to ten times higher nighttime than daytime concentrations (review by Klein 1979). Plasma melatonin declines with age in humans. Touitou et al. (1981) have reported a marked decrease (1.7–2.0 times) in the plasma concentrations of melatonin in elderly subjects in their eighties (Fig. 2). This decrease in the elderly was apparent when compared to young adult subjects both in the circadian and seasonal mean levels of the hormone as well as the mean melatonin concentrations of any clock hour of sampling (Touitou et al. 1984). These data were confirmed in various reports dealing with age effects on melatonin secretion in humans (plasma melatonin: Iguchi et al. 1982; Nair et al. 1986; urinary 6-hydroxymelatonin: Sack et al. 1986) and are in good agreement with previous data by Brown et al. (1979) on the decline of melatonin in human cerebrospinal fluid and by Reiter et al. (1980) on the decreased pineal content in aged Syrian hamsters. In a broad study dealing with 757 unselected elderly subjects in their eighties, Touitou et al. (1985) have found that 53% of the subjects had low (0.17 nmol/l or lower) daytime (08.00–09.30 hours) melatonin concentrations and that 75% of these subjects with low levels had values lower than 0.13 nmol/l.

Several factors could contribute to this decrease. Calcification of the pineal gland is common with aging and rodent species like Syrian hamsters, which characteristically develop heavy calcifications of the pineal, also show a large decrease in the pineal content of melatonin (Reiter et al. 1980). However, studies in humans suggest that calcification of the pineal does not alter the histology of the pinealocyte (Tapp and Huxley 1972) or the pineal activity of hydroxyindole-O-methyltransferase, an enzyme responsible for melatonin synthesis (Wurtman et al. 1964). The mechanism by which melatonin is reduced in elderly subjects is thus poorly understood. $\beta$-Adrenergic receptors in the pinealocyte membranes may decrease in number or may become unresponsive to norepinephrine as described in aged rodents (Greenberg and Weiss 1978). Increased clearance could also account for the decline of plasma melatonin with age but the finding by Sack et al. (1986) of a reduced urinary excretion of 6-hydroxymelatonin in aged human subjects makes this explanation less

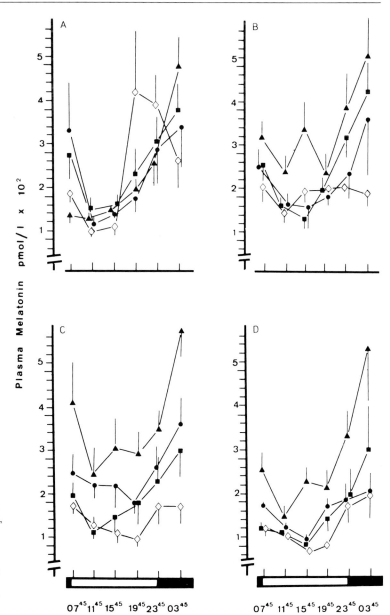

**Fig. 2 A–D.** Plasma melatonin
concentration as a function of time in seven
young healty men ( ▲ ), six elderly men ( ■ ),
six elderly women ( ● ) and six elderly men
and women suffering from senile dementia
of the Alzheimer type ( ◇ ). Blood samples
were taken over 24 h in January (**A**), March
(**B**), June (**C**) and October (**D**). Nocturnal
rest is shown by a *solid bar.* (From Touitou
et al. 1981, 1984)

likely. In addition, the positive correlation between
plasma melatonin and urinary 6-hydroxymelatonin
excretion found by Markey et al. (1985) underlines
the fact that, in subjects with normal liver functions,
low plasma levels of melatonin are not primarily the
result of more rapid metabolism and clearance, but
are the result of lowered pineal production. Modifications of the pineal production of melatonin thus appear to be the main factor to explain the hormonal
decrease with age.

Since melatonin secretion depends on photoperiod, a larger quantity of the hormone is expected
to be produced during the winter months, i.e., the
months corresponding to a short photoperiod. Based
on one or two samplings a day at monthly intervals
(Arendt et al. 1977, 1979), a significant seasonal variation in the plasma melatonin concentration has been
reported in men and women, with peak values in
winter and summer and lower values in spring and
autumn. Touitou et al. (1984) found seasonal variations of the hormone with differences between
young and elderly subjects: plasma melatonin levels
were statistically significantly lower in January in
young men whereas in elderly subjects they were sig-

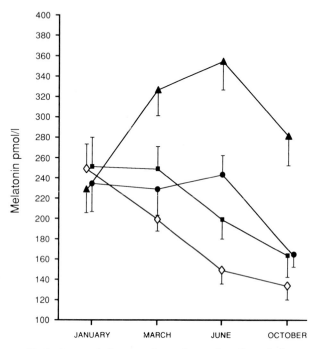

**Fig. 3.** Seasonal changes of nyctohemeral median of plasma melatonin in seven young healthy men (▲), six elderly men (■), six elderly women (●) and six elderly men and women suffering from senile dementia of the Alzheimer type (◇). (From Touitou et al. 1984)

## Body Temperature

Elderly subjects maintain their circadian rhythmicity of body temperature irrespective of sex or mental condition with a peak located in mid or late afternoon (Fig. 4). However, a decrease in the circadian amplitude was reported in healthy elderly subjects (Weitzman et al. 1982; Touitou et al. 1986a; Vitiello et al. 1986) but not in patients with senile dementia of the Alzheimer type (Touitou et al. 1986a). Whether this decrease reflects an internal desynchronization in the elderly is not yet answered. In addition to the circadian periodicity, a seasonal rhythm (Fig. 5) was also described in both healthy and senile demented elderly subjects with a peak during winter and a trough in June (Touitou et al. 1986a).

nificantly lower in October (Fig. 3). The circadian pattern of melatonin secretion was roughly similar in all groups of subjects during all seasons. The remarkable stability throughout the year of the daily peak time of melatonin emphasizes the fact that the seasonal variations of the hormone in man are not related to a shift in the acrophase but rather to seasonal changes in the synthesis, release and/or metabolism of the hormone (Touitou et al. 1984). The highest levels of melatonin are not observed consistently during the winter months as would be expected due to the shortest photoperiod occurring at this time. While Beck-Friis et al. (1984) and Sack et al. (1986) found no seasonal variations in plasma melatonin. Illnerova et al. (1985) reported as the only difference a phase advance (about 1.5 h) in winter without any difference in the peak concentrations or duration. Griffiths et al. (1986) found no seasonal change in the excretion of 6-hydroxymelatonin sulfate in volunteers living in Antarctica although there was some evidence for a phase delay.

## Nonendocrine Frequently Used Laboratory Tests

### Plasma proteins and the problem of volemia

Plasma proteins play an important role in physiology and pharmacology, especially by binding various molecules including a large variety of hormones and drugs. Circadian and seasonal variations of their plasma concentration are therefore doubly important: first, because they can affect the interpretation of the concentrations of the bound and unbound fraction of a physiologic or pharmacologic agent; second, because they must be taken into account in designing a therapeutic protocol that optimizes the tolerance and expected effects of drugs and diminishes their side effects. The concentration of plasma proteins changes predictably in elderly subjects (7%-13%) depending on the hour of sampling and the season. Concentrations decrease noticeably around 04.00 hours, then peak shortly after waking, around 0800 hours with lower levels in the elderly (Fig. 6) (Touitou et al. 1986b). The circadian rhythm of blood volume is not the only component accounting for the rhythms of plasma protein concentrations since the circadian periodicity of plasma proteins differs from that of hematocrit, hemoglobin and red blood cell counts (Figs. 7–9), which are often considered as indexes of blood volume (Touitou et al. 1986b). Circadian fluctuations have also been described in the elderly for $\alpha_1$ acid glycoprotein, $\alpha_1$ antitrypsin and C-reactive protein (Bruguerolle et al. 1989) and for immunoglobulins

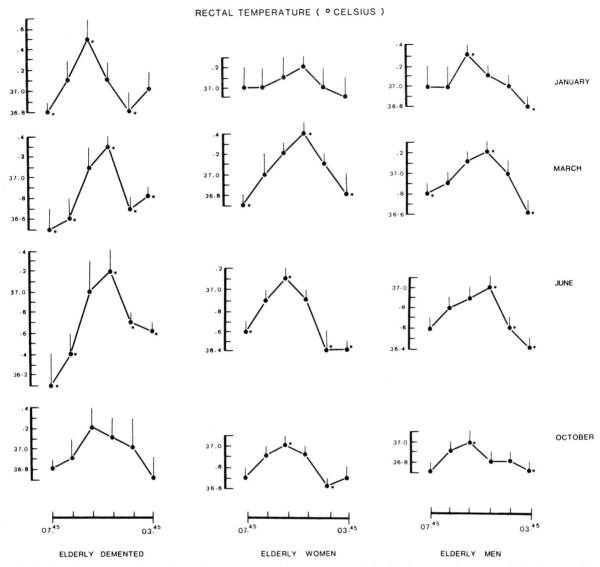

**Fig. 4.** Circadian patterns of rectal temperature measured in January, March, June and October in six elderly men, six elderly women and six elderly men and women suffering from senile dementia of the Alzheimer type. (From Touitou et al. 1986a)

(IgA, IgG, IgM) by Casale et al. (1983). Seasonal variations of plasma proteins (Fig. 10) were also documented and their amplitudes were strikingly higher in the elderly (~8 g/l difference between the peak and the trough) than in young adults (2.5 g/l difference). The seasonal peak occurred in October and the trough in June (Touitou et al. 1986b). These data on plasma proteins can explain, at least in part, that the highest values of plasma-free cortisol occur in June at the time of the seasonal trough of proteins. The circadian and seasonal fluctuations of plasma proteins can produce significant variations in the transport and binding of drugs especially in the aged.

## Electrolytes and Other Frequently Used Laboratory Variables

Serum sodium and chloride concentrations and the circadian rhythms of these parameters are not different in elderly human subjects when compared to young adults. Both of these ions present with a small circadian amplitude, i.e., about 3% total variability for chloride and sodium and 5%–10% for serum potassium (Touitou et al. 1979, 1989; Nicolau et al. 1983). Hypokalemia is common in the elderly and an age-related decrease in total exchangeable sodium and potassium has been reported (Burr et al. 1975;

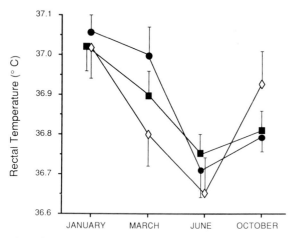

**Fig. 5.** Seasonal pattern of rectal temperature in six elderly men ( ■ ), six elderly women ( ● ) and six elderly men and women suffering from senile dementia of the Alzheimer type ( ◇ ). (From Touitou et al. 1986a)

Touitou 1987). An inadequate dietary intake of the mineral was invoked by some as possibly responsible although it has been shown that neither potassium supplementation nor variation in activity, posture or dietary intakes affected plasma potassium or total exchangeable potassium (Moore Ede et al. 1975). Daily and seasonal variations of calcium, magnesium and phosphorus are present in the aged and have been described in the chapter dealing with vitamin D and parathyroid hormone status in the elderly.

The daily variations of serum urate, urea and creatinine are of relatively small amplitude (6%–12%) whereas the seasonal variations of urea and urate showed a slightly higher amplitude (10%–20%) (Touitou et al. 1989). Casale et al. (1981) found a high-amplitude circadian rhythmicity of plasma iron (peak in the morning) and total iron binding capacity in aged arteriosclerotic patients. This was also found by Nicolau et al. (1987) in a broad study comparing elderly men and women and young subjects. The mechanism responsible for these variations is not fully understood.

## Hematology

The circadian rhythms in the number of circulating red blood cells, neutrophils, lymphocytes, eosinophil leukocytes and platelets in the peripheral blood persist in the aged. Some rhythm parameters, however, show certain differences (Haus et al. 1983; Swoyer et al. 1989). Comparing 23 clinically healthy elderly American subjects (71 ± 5 years of age) with 150 young adults and adults, Swoyer et al. (1989) found the circadian rhythm adjusted mean (mesor) in circulating mature neutrophils in the elderly to be higher than in the young subjects studied at the same geographic location. The circadian amplitude was unchanged. The circadian mean in the "young" neutrophils (band forms) in contrast was decreased in the elderly and a circadian rhythm of these cells was not statistically significant as a group phenomenon. This decrease in mesor may suggest a lower rate of cell replacement. However, any speculation concerning the disappearance of a rhythmic release of these cells from the marrow has to be qualified since the absence of a circadian rhythm detectable as a group phenomenon may also be due to a desynchronization within the group.

An increase in the circadian mesor in the number of circulating monocytes is of interest since no such change with aging has been found with non-time-qualified sampling at single time points (Dybkaer et al. 1981; Munan and Kelly 1979; Nielsen et al. 1984). The circadian acrophase of the number of circulating neutrophils and lymphocytes does occur in the elderly earlier in the day (phase advance) than in the young adult and adult subjects, in spite of a comparable time of rising and retiring. Earlier acrophases in the elderly have been reported for plasma cortisol, PBI, and for some pituitary hormones.

Neither mesor nor amplitude of the circulating lymphocytes showed a statistically significant change in the elderly. Data on circadian rhythms of lymphocyte subtypes in the elderly do not seem to be available at this time.

Casale et al. (1982) studied in Italy a group of 16 institutionalized subjects 76.3 ± 1.32 years of age and also found the circadian rhythms to persist in most variables. The acrophases of the elderly Italian subjects seem to be similar to those found in the Americans. Also Touitou et al. (1986b) observed a circadian rhythm of hematocrit, hemoglobin and red blood cell count in elderly people. The acrophases in this group of subjects, however, appeared to be different.

Other findings of interest in the elderly are a decreased mesor in the number of red cells, and in the mean corpuscular hemoglobin concentration (MCHC), and an increased red cell volume (MCV), and an increased amplitude in the circadian variation of mean corpuscular hemoglobin (MCH) and MCHC (Swoyer et al. 1989).

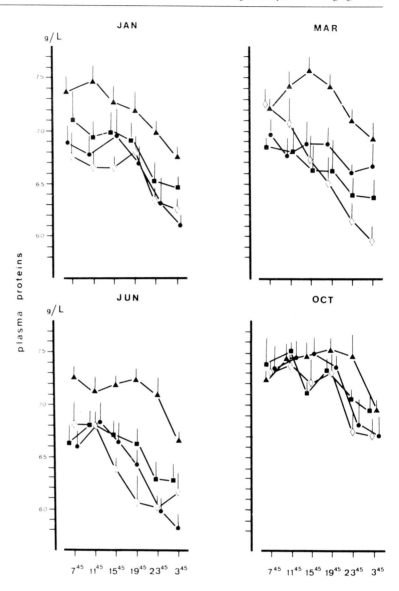

**Fig. 6.** Circadian variations of plasma total proteins in seven young healthy men (▲), six elderly men (■), six elderly women (●), and six elderly men and women suffering from senile dementia of the Alzheimer type (◇). (From Touitou et al. 1986b)

The changes in the mesor of these functions recorded in studies including the entire 24-h span extend earlier, and non-time-qualified investigations in which a slightly lower number of red cells and a relative macrocytosis have been described in the aged.

As these findings indicate, the circadian periodicity in the number of circulating formed elements in the peripheral blood appears to be well maintained in clinically healthy elderly subjects. Among the differences of potential clinical interest are the differences in the circadian acrophase of the neutrophils and lymphocytes. Whether the absence of a circadian rhythm (demonstrable as group phenomenon) and a lowered circadian mean of the neutrophil band forms represent any indication of an alteration in bone mar-row neutrophil cell proliferation or release will have to be explored in more direct studies of the marrow.

## Age-Related Changes in Pharmacokinetics

A rhythm in the therapeutic or toxic action of a drug may be the result of a rhythmic change in one or more of the contributing factors, namely rhythms in absorption, in plasma concentration, in tissue distribution, in drug metabolism, in excretion, in drug concentration at the receptor level, in receptor sensitivity and finally in pharmacologic response. Some of these

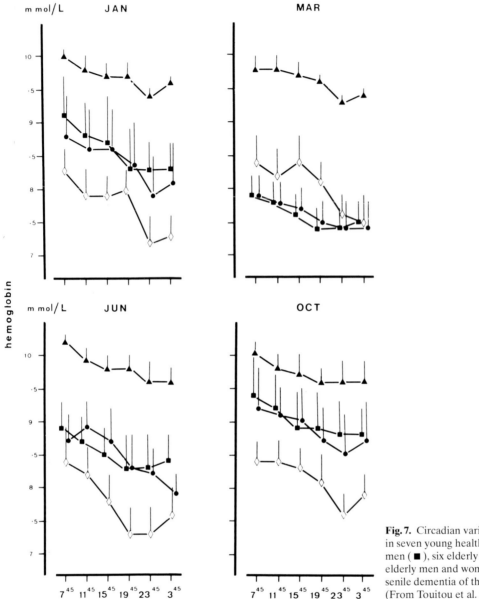

**Fig. 7.** Circadian variations of hemoglobin in seven young healthy men ( ▲ ), six elderly men ( ■ ), six elderly women ( ● ) and six elderly men and women suffering from senile dementia of the Alzheimer type ( ◇ ). (From Touitou et al. 1986b)

parameters interact and these interactions may also be rhythmic. Since all the parameters cited (absorption, distribution, metabolism and excretion) undergo changes with aging, it appears likely that the rhythmic properties of these parameters also change with aging.

Changes in absorption with age may be due to a decrease in intestinal motility or to a decrease in both the secretion of the digesting enzymes and in the blood supply of the gastrointestinal tract, all of which would lead to a decreased absorption. However, it is commonly assumed that the process of absorption is of minor importance in interpreting aging changes in drug effects. In contrast, the distribution volume increases with age for liphophilic drugs, which represent the majority of drugs, and decreases for hydrophilic drugs. The distribution volume is a mathematical concept, defined as the volume of body water necessary to store the amount of drug in the body (if the drug is uniformly present in the same concentration as in the plasma). It indicates the degree to which a drug is concentrated outside the plasma in those fluids and/or tissues which are in equilibrium with it. Changes in distribution with age can be related to a

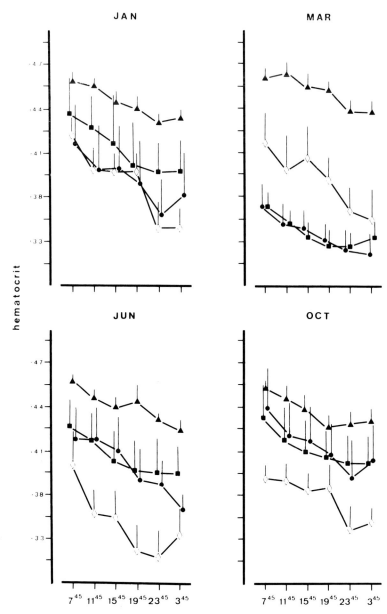

**Fig. 8.** Circadian variations of hematocrit in seven young healthy men ( ▲ ), six elderly men ( ■ ), six elderly women ( ● ) and six elderly men and women suffering from senile dementia of the Alzheimer type ( ◇ ). (From Touitou et al. 1986b)

decrease by about 50% in blood supply to the tissues, to changes in body composition, e.g., a decrease of the total amount of body fluid and of the intracellular water content and an increase in the fat compartment. A decrease in plasma protein concentration with age results in decreased drug binding, which is of special importance for drugs which are to a large extent ( > 90% ) bound to plasma proteins (Touitou et al. 1986b). A small change in binding capacity results in a great increase in the free concentration, which in turn results in a substantial change in volume of distribution, since only the free concentration of the

drug is in equilibrium with the other body compartments. Liver metabolism undergoes change with aging. The circadian variation in the hepatic elimination of drugs and toxic chemical agents which has been described for a large number of molecules (see the chapter by Belanger) can also be modified with aging. However, the data on the effects of aging on the rhythmic variations in drug pharmacokinetics are very scarce. For example, the circadian variations of indomethacin were found to be different in the aged when compared to young adults (Bruguerolle et al. 1986). Because drug consumption is widespread in el-

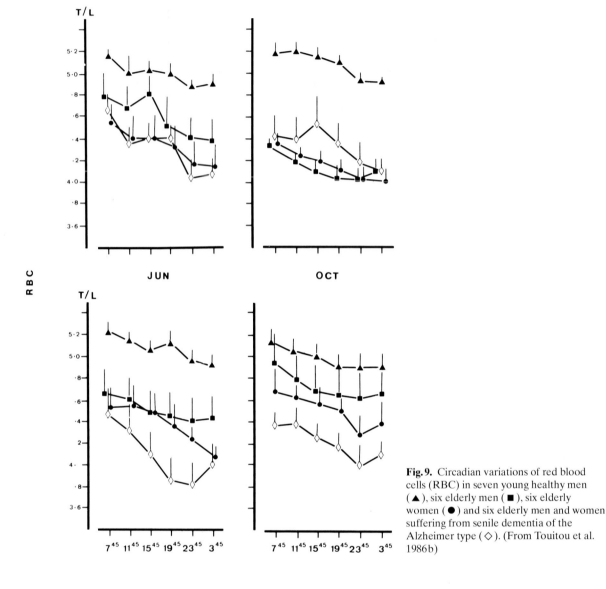

**Fig. 9.** Circadian variations of red blood cells (RBC) in seven young healthy men ( ▲ ), six elderly men ( ■ ), six elderly women ( ● ) and six elderly men and women suffering from senile dementia of the Alzheimer type ( ◇ ). (From Touitou et al. 1986b)

derly subjects, the incidence of drug side effects is also frequent in this population. Modifications of both the pharmacokinetics and biologic rhythms of numerous functions must be considered in the aged to improve drug tolerance and therapeutic effects.

## Conclusions

The data presented here on the influence of aging on biologic rhythms make possible the conclusion that modifications in rhythmicity are different from one function to another. The "aging" of biologic rhythms is not uniform. The rhythms of some functions are modified whereas those of many others are not. Some of the rhythmic functions which are altered in the process of aging are modified earlier and others later. The most frequent modifications when comparing young and elderly human subjects are the change in amplitude, which can be decreased (e.g., melatonin, aldosterone, DHEA-S, body temperature), or which in contrast may be increased (e.g., insulin, GH, calcium, magnesium, NEFA). Also frequent is a change in the 24-h mean (circadian mesor) of the rhythmic variables (e.g., melatonin, free cortisol) and only sometimes a change in the peak time location, i.e.,

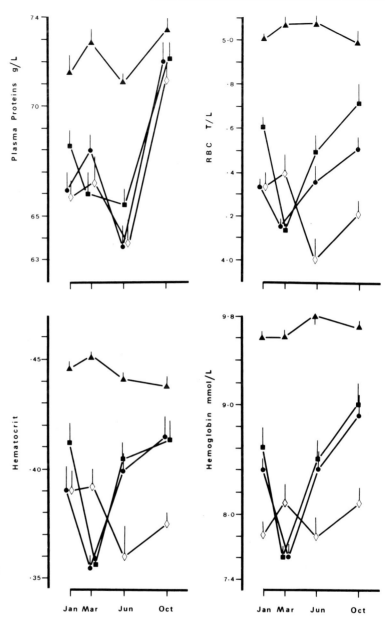

**Fig. 10.** Seasonal variations of plasma total proteins, hemoglobin, hematocrit and red blood cells (RBC) in seven young healthy men (▲), six elderly men (■), six elderly women (●) and six elderly men and women suffering from senile dementia of the Alzheimer type (◇). (From Touitou et al. 1986 b)

the tendency toward a phase shift in relation to clock hour and to environmental synchronizer schedules (e. g., a phase advance of cortisol, 18-hydroxy-11-deo-xycorticosterone, DHEA-S). Evidence is accumulating that differences in phase relationships and therefore the incidence of internal desynchronization of a number of rhythms increase with age. This suggests that the temporal organization is changed since a number of pacemakers which were initially mutually coupled become uncoupled. However, the question is not answered of whether this change is contributing to the decline of physiologic functions of the elderly.

Using short-lived insects and rodents, internal desynchronization of circadian rhythms was induced by maintaining the animals in constant light (Pittendrigh and Minis 1972; Albers et al. 1981) or by repeatedly phase shifting the light-dark cycle to simulate jet lag (Aschoff et al. 1971). These experimental conditions leading to a disruption of the circadian organization (forced internal desynchronization) resulted in most but not in all instances in a reduced life span of the insects and rodents.

In conclusion, aging is a complex phenomenon that obviously cannot be understood by studying a single cause or mechanism. The existence of biologic rhythms, their alterations or their modifications in the aged must be taken into account as useful elements for the understanding of biologic age. Similar to the other biologic approaches to aging, chronobiology alone is not able to provide a response to the fundamental question of the cause of aging. However, it is essential to integrate the concept of a biologic time structure with rhythmic functions of several fequencies in any study dealing with aging.

# References

Adams ND, Gray RW, Lemann J (1979) The effects of oral CaCO$_3$ loading and dietary calcium deprivation on plasma 1,25-dihydroxyvitamin D concentrations in healthy adults. J Clin Endocrinol Metab 48: 1008

Albers HE, Gerall AA, Axelson JF (1981) Circadian rhythm dissociation in the rat: effects of long-term constant illumination. Neurosci Lett 25: 89–94

Arendt J, Wirz-Justice A, Bradtke J (1977) Annual rhythm of serum melatonin in man. Neurosci Lett 7: 327–330

Arendt J, Wirz-Justice A, Bradtke J, Kornemark M (1979) Long term studies on immunoreactive human melatonin. Ann Clin Biochem 16: 307–312

Aschoff J, Saint-Paul U, Wever R (1971) Die Lebensdauer von Fliegen unter dem Einfluß von Zeit-Verschiebungen. Naturwissenschaften 58: 574

Baker HWG, Burger HG, de Kretser B, O'Connor S, Wang C, Mirovis A, Court J, Dunlop M, Rennie GC (1976) Changes in the pituitary-testicular system with age. Clin Endocrinol (Oxf) 5: 349–372

Beck-Friis J, von Rosen D, Kjellman BF, Ljunggren JG, Wetterberg L (1984) Melatonin in relation to body measures, sex, age, season and the use of drugs in patients with major affective disorders and healthy subjects. Psychoneuroendocrinology 9: 261–277

Blichert-Toft M (1975) Secretion of corticotrophin and somatotrophin by the senescent adenohypophysis in man. Acta Endocrinol (Copenh) 78: 1–157

Boyar PM, Finkelstein JW, Roffwarg H, Kappen S, Weitzman E, Hellman L (1973) Twenty four hour luteinizing hormone and follicle stimulating hormone secretory patterns in gonadal dysgenesis. J Clin Endocrinol Metab 37: 521–525

Bremmer WJ, Vitiello MV, Prinz PN (1983) Loss of circadian rhythmicity in blood testosterone levels with aging in normal men. J Clin Endocrinol Metab 56: 1278–1281

Brown GM, Young SN, Gauthier S, Tsui H, Grota LJ (1979) Melatonin in human cerebrospinal fluid in daytime: its origin and variation with age. Life Sci 25: 929–936

Bruguerolle B, Barbeau G, Bélanger PM, Labreque G (1986) Chronokinetics of indomethacin in elderly subjects. Annu Rev Chronopharmacol 3: 425–428

Bruguerolle B, Arnaud C, Levi F, Focan C, Touitou Y, Bouvenot G (1989) Physiopathological alterations of alpha-1-acid glycoprotein temporal variations: implications for chronopharmacology. In: Alpha 1-acid glycoprotein: genetics, biochemistry, physiological functions and pharmacology. Liss, New York, pp 199–214

Burr ML, Leger ASS, Westlake CA, Davies HEF (1975) Dietary potassium deficiency in the elderly: a controlled trial. Age Ageing 4: 148–151

Casale G, Migliavacca A, Bonora C, Zurita IE, de Nicola P (1981) Circadian rhythm of plasma iron, total iron binding capacity and serum ferritin in arteriosclerotic aged patients. Age Ageing 10: 115–118

Casale G, Emiliani S, de Nicola P (1982) Circadian rhythm of circulating blood cells in elderly persons. Hematologica 67: 837–844

Casale G, Marinoni GL, d'Angelo R, de Nicola P (1983) Circadian rhythm of immunoglobulins in aged persons. Age Ageing 12: 81–85

Chesney RW, Rosen JF, Hamstra AJ, Smith O, Mahafrey K, de Luca HF (1981) Absence of seasonal variation in serum concentration of 1,25 (OH)$_2$ vitamin D despite a rise in 25 (OH) vitamin D in summer. J Clin Endocrinol Metab 53: 139–142

Cugini P, Scavo D, Halberg F, Schramm A, Push AJ, Franke H (1982) Methodologically critical interaction of circadian rhythms: sex and aging characterize serum aldosterone of the female adrenopause. J Gerontol 37: 403–411

Dean S, Felton SP (1979) Circadian rhythm in the elderly, a study using a cortisol specific radioimmunoassay. Age Ageing 8: 243–245

Désir D, van Cauter E, L'Hermite M, Refetoff S, Jadot C, Caufriez A, Copinschi G, Robyn C (1982) Effects of jet lag on hormonal patterns. III. Demonstration of an instrinsic circadian rhythmicity in plasma prolactin. J Clin Endocrinol Metab 55: 849–857

Djursing H, Hagen C, Møller J, Christiansen C (1981) Short- and long-term fluctuations in plasma prolactin concentration in normal subjects. Acta Endocrinol (Copenh) 97: 1–6

Dybkaer R, Lauritzen M, Krakauer R (1981) Relative reference values for clinical, chemical and hematological quantities in "healthy" elderly people. Acta Med Scand 209: 1–9

Finkelstein J, Roffwarg H, Boyar R, Kream J, Hellman L (1972) Age-related changes in the twenty-four hour spontaneous secretion of growth hormone. J Clin Endocrinol Metab 35: 665–670

Florini JR, Prinz PN, Vitiello MV, Hintz RL (1985) Somatomedin C levels in healthy young and old men: relationship to peak and 24-hour integrated levels of growth hormone. J Gerontol 40: 2–7

Gala RR, van de Walle C, Hoffman WH, Lawson DM, Piefer DR, Smith SW, Subramanina MG (1977) Lack of a circannual cycle of daytime serum prolactin in man and monkey. Acta Endocrinol (Copenh) 86: 257–262

Graber AL, Givens JR, Nicholson WE, Island DP, Liddle LW (1965) Persistence of diurnal rhythmicity in plasma ACTH concentration in cortisol deficient patient. J Clin Endocrinol Metab 25: 804–807

Greenberg LH, Weiss B (1978) β-adrenergic receptors in aged rat brain: reduced number and capacity of pineal to develop supersensitivity. Science 201: 61–63

Gregerman RI (1986) Mechanisms of age-related alterations of hormone secretion and action. An overview of 3 years of progress. Exp Gerontol 21: 345–365

Griffiths PA, Folkard S, Bojkowski C, English, Arendt J (1986) Persistent 24 h variations of urinary 6-hydroxymelatonin sulphate and cortisol in Antarctica. Experientia 15: 430–432

Halloran BP, Portale AA, Castro M, Morris RC, Goldsmith RS

(1985) Serum concentration of 1,25-dihydroxyvitamin D in the human: diurnal variation. J Clin Endocrinol Metab 60: 1104–1110

Harman SM, Tsitouras PD (1980) Reproductive hormones in aging men. I. Measurements of sex steroids, basal luteinizing hormone and Leydig cell response to human chorionic gonadotrophin. J Clin Endocrinol Metab 51: 35–40

Haus E, Halberg F, Kuhl JFW, Lakatua DJ (1974) Chronopharmacology in animals. Chronobiologia [Suppl 1] 1: 122–156

Haus E, Lakatua DJ, Halberg F, Halberg E, Cornelissen G, Sackett L, Berg HG, Kawasaki T, Ueno M, Uezono K, Matsuoka M, Omae T (1980) Chronobiological studies of plasma prolactin in women in Kyushu, Japan and Minnesota, USA. J Clin Endocrinol Metab 51: 632–640

Haus E, Lakatua DJ, Swoyer J, Sackett-Lundeen L (1983) Chronobiology in hematology and immunology. Am J Anat 168: 467–517

Haus E, Nicolau GY, Lakatua DJ, Sackett-Lundeen L (1988) Reference values for chronopharmacology. Annu Rev Chronopharmacol 4: 333–424

Haus E, Nicolau G, Lakatua DJ, Sackett-Lundeen L, Petrescu E (1989) Circadian rhythm parameters of endocrine functions in elderly subjects during the seventh to the ninth decade of life. Chronobiologia 16: 331–352

Hillyard CJ, Cooke TJC, Coombes RC, Evans IMA, MacIntyre I (1977) Normal plasma calcitonin: circadian variation and response to stimuli. Clin Endocrinol (Oxf) 6: 291–298

Iguchi H, Kato KI, Ibayashi H (1982) Age-dependent reduction in serum melatonin concentrations in healthy human subjects. J Clin Endocrinol Metab 55: 27–29

Illnerova H, Zvolsky P, Vanecek J (1985) The circadian rhythm in plasma melatonin concentration of the urbanized man: the effect of summer and wintertime. Brain Res 328: 186–189

Jensen HK, Blichert Toft M (1971) Serum corticotrophin, plasma cortisol and urinary secretion of 17-ketogenic steroids in the elderly (age group: 66–94 years). Acta Endocrinol (Copenh) 66: 25–34

Jubiz W, Canterbury JM, Reiss E, Taylor FH (1972) Circadian rhythms in parathyroid hormone concentration in human subjects: correlation with serum calcium, phosphate, albumin and growth hormone levels. J Clin Invest 51: 2040–2046

Juttman JR, Visser TJ, Buurman C, de Kam E, Birkenhager JC (1981) Seasonal fluctuations in serum concentrations of vitamin D metabolites in normal subjects. Br Med J 282: 1349–1352

Kano K, Yoshida H, Yata J, Suda T (1980) Age and seasonal variations in the serum levels of 25-OH vitamin D and 24,25 (OH)$_2$ vitamin D in normal humans. Endocrinol Jpn 27: 215–221

Kelch RP, Conte FA, Kaplan SL, Grumbach MM (1973) Episodic secretion of luteinizing hormone (LH) in adolescent patients with the syndrome of gonadal dysgenesis. J Clin Endocrinol Metab 36: 424–427

Klein DC (1979) Circadian rhythms in the pineal gland. In: Krieger DT (ed) Endocrine rhythms. Raven, New York, pp 203–223

Lakatua DJ, Nicolau GY, Bogdan C, Petrescu E, Sackett-Lundeen L, Irvine PW, Haus E (1984) Circadian endocrine time structure in humans above 80 years of age. J Gerontol 39: 654–684

Lester E, Skinner RK, Wills MR (1977) Seasonal variation in serum 25-OH vitamin D in the elderly in Britain. Lancet 1: 979–980

Lips P, Hackeng WHL, Jongen MJM, van Ginkel FC, Netelen-

berg JC (1983) Seasonal variation in serum concentrations of parathyroid hormone in elderly people. J Clin Endocrinol Metab 57: 204–206

Lonning PE, Dowsett M, Jacobs S, Sehem B, Hardy J, Powles TJ (1989) Lack of diurnal variation in plasma levels of androstenedione, testosterone, estrone and estradiol in post menopausal women. J Steroid Biochem 34: 551–553

Markey SP, Higa S, Shih S, Danforth DN, Tamarkin L (1985) The correlation between plasma melatonin levels and urinary 6-hydroxymelatonin excretion. Clin Chim Acta 150: 221–225

Markowitz M, Rotkin L, Rosen JF (1981) Circadian rhythms of blood minerals in humans. Science 213: 672–674

Markowitz ME, Arnaud S, Rosen JF, Thorpy M, Laximinarayan S (1988) Temporal interrelationships between the circadian rhythms of serum parathyroid hormone and calcium concentrations. J Clin Endocrinol Metab 67: 1068–1073

Marrama P, Carani C, Baraghini GF, Volpe A, Zini D, Celani MF, Montanini V (1982) Circadian rhythm of testosterone and prolactin in the ageing. Maturitas 4: 131–138

Meller Y, Kestenbaum RS, Gralinsky D, Shany S (1986) Seasonal variation in serum levels of vitamin D metabolites and parathormone in geriatric patients with fractures in southern Israel. Isr J Med Sci 22: 8–11

Milcu AE, Bogdan C, Nicolau GY, Cristea A (1978) Cortisol circadian rhythm in 70–100 year old subjects. Rev Roum Med Endocrinol 16: 29–39

Miller AE, Riegle G (1982) Temporal patterns of serum luteinizing hormone and testosterone and endocrine response to luteinizing hormone releasing hormone in ageing male rats. J Gerontol 37: 522–528

Moore Ede MC, Brennan MF, Ball MR (1975) Circadian variation of intercompartmental potassium fluxes in man. J Appl Physiol 38: 163–170

Munan L, Kelly A (1979) Age dependent changes in blood monocyte populations in man. Clin Exp Immunol 35: 161–162

Murri L, Barreca T, Cerone G, Massetani R, Galamini A, Baldassarre M (1980) The 24 h pattern of human prolactin and growth hormone in healthy elderly subjects. Chronobiologia 7: 87–92

Nair NPV, Hariharasubramanian N, Pilapil N, Isaac I, Thavundayil JX (1986) Plasma melatonin, an index of brain aging in humans? Biol Psychiatry 21: 141–150

Nelson W, Bingham C, Haus E, Lakatua DJ, Kawasaki T, Halberg F (1980) Rhythm-adjusted age effects in a concomitant study of twelve hormones in blood plasma of women. J Gerontol 35: 512–519

Nicolau GY, Haus E, Lakatua DJ, Bogdan C, Popescu M, Petrescu E, Sackett-Lundeen L, Swoyer J, Adderley J (1983) Circadian periodicity of the results of frequently used laboratory tests in elderly subjects. Rev Roum Med Endocrinol 21: 3–21

Nicolau GY, Lakatua D, Sackett-Lundeen L, Haus E (1984) Circadian and circannual rhythms of hormonal variables in elderly men and women. Chronobiol Int 1: 301–319

Nicolau GY, Haus E, Lakatua DJ, Bogdan C, Petrescu E, Robu E, Sackett-Lundeen L, Swoyer J (1985) Chronobiologic observations of calcium and magnesium in the elderly. Rev Roum Med Endocrinol 23: 39–53

Nicolau GY, Haus E, Lakatua DJ, Bogdan C, Plinga L, Irvine P, Popescu M, Petrescu E, Sackett-Lundeen L, Swoyer J, Robu E (1987) Chronobiology of serum iron concentration in subjects of different ages at different geographic location. Rev Roum Med Endocrinol 25: 63–82

Nielsen H, Blom J, Larsen SO (1984) Human blood monocyte function in relation to age. Acta Pathol Microbiol Immunol Scand [C] 92 (1): 5–10

Noel CT, Reed MJ, Jacobs HS, James VHT (1981) The plasma concentration of oestrone sulphate in postmenopausal women: lack of diurnal variation, effect of ovariectomy, age and weight. J Steroid Biochem 14: 1101–1105

Oliver C, Vague P, Vague J (1971) L'ACTH plasmatique dans les états d'hypocorticisme. Ann Endocrinol (Paris) 32: 868–883

Orgel LE (1963) The maintenance of the accuracy of protein synthesis and the relevance to aging. Proc Natl Acad Sci USA 49: 517–521

Parker DC, Rossman LG, Vanderlaan EF (1973) Sleep-related, nyctohemeral and briefly episodic variation in human plasma prolactin concentration. J Clin Endocrinol Metab 36: 1119–1124

Parker L, Gral T, Perrigo V, Skowsky R (1981) Decreased adrenal androgen sensitivity to ACTH during aging. Metabolism 30: 601–604

Pavlov EP, Harman SM, Chrousos GP, Loriaux DL, Blackman MR (1986) Responses of plasma adrenocorticotropin, cortisol and dehydroepiandrosterone to ovine corticotropin-releasing factor in healthy aging men. J Clin Endocrinol Metab 62: 767–772

Pittendrigh CS, Minis DH (1972) Circadian systems: longevity as a function of circadian resonance in *Drosophila melanogaster*. Proc Natl Acad Sci USA 69: 1537–1539

Prince RL, Wark JD, Omond S, Opie JM, Eagle MR, Eisman JA (1983) A test of 1,25-dihydroxyvitamin D secretory capacity in normal subjects for application in metabolic bone diseases. Clin Endocrinol (Oxf) 18: 127–133

Prinz PN, Halter J, Benedetti C, Raskind M (1979) Circadian variation of plasma catecholamines in young and old men: relation to rapid eye movement and slow wave sleep. J Clin Endocrinol Metab 49: 300–304

Prinz PN, Weitzman ED, Cunningham GR, Caracan J (1983) Plasma growth hormone during sleep in young and old men. J Gerontol 38: 519–524

Reinberg A, Lagoguey A, Cesselin F, Touitou Y, Legrand JC, Delassalle A, Antreassian J, Lagoguey A (1978) Cicadian and circannual rhythms in plasma hormones and other variables of five healthy young human males. Acta Endocrinol (Copenh) 88: 417–427

Reiter RJ, Richardson BA, Johnson L, Ferguson BN, Dinh DT (1980) Pineal melatonin rhythm: reduction in aging Syrian hamsters. Science 210: 1372–1373

Sack RL, Lewy AJ, Erb DE, Vollmer WM, Singer CM (1986) Human melatonin production decreases with age. J Pineal Res 3: 379–388

Sassin JF, Frantz AG, Weitzmann ED, Kapen S (1972) Human prolactin: 24-hour pattern with increased release during sleep. Science 177: 1205–1207

Sassin J, Frantz A, Kapen S, Weitzman E (1973) The nocturnal rise of human prolactin is dependent on sleep. J Clin Endocrinol Metab 37: 436–440

Serio M, Piolanti P, Romano S, de Magistris L, Giustri G (1970) The circadian rhythm of plasma cortisol in subjects over 70 years of age. J Gerontol 25: 95–97

Sherman B, Wysham C, Pfohl B (1985) Age-related changes in the circadian rhythm of plasma cortisol in man. J Clin Endocrinol Metab 61: 439–443

Silverberg A, Rizzo F, Krieger DT (1968) Nyctohemeral periodicity of plasma 17-OH-CS levels in elderly subjects. J Clin Endocrinol Metab 28: 1661–1666

Simpkins JW, Kalra PS, Kalra SP (1981) Alterations in daily rhythms of testosterone and progesterone in old male rats. Exp Aging Res 7: 25–32

Sparrow D, Bone R, Rowe JW (1980) The influence of age, alcohol consumption and body build on gonadal function in men. J Clin Endocrinol Metab 51: 508–512

Stamp TCB, Round JM (1974) Seasonal changes in human plasma levels of 25-hydroxyvitamin D. Nature 247: 563–565

Stearns EL, MacDonnell JA, Kaufman BJ, Padua R, Lucman TS, Winter JSD, Faiman C (1974) Declining testicular function with age, hormonal and clinical correlates. Am J Med 57: 761–766

Swoyer J, Irvin P, Sackett-Lundeen L, Conin L, Lakatua DJ, Haus E (1989) Circadian hematologic time structure in the elderly. Chronobiol Int 6(2): 131–137

Tapp E, Huxley M (1972) The histological appearance of the human pineal gland from puberty to old age. J Pathol 108: 137–144

Tarquini B, Gheri R, Romano S, Costa A, Cagnoni M, Lee JK, Halberg F (1979) Circadian mesor hyperprolactinemia in fibrocystic mastopathy. Am J Med 66: 229–237

Thurton MB, Deegan T (1974) Circadian variations of plasma catecholamine, cortisol and immunoreactive insulin concentrations in supine subjects. Clin Chim Acta 55: 389–397

Touitou Y (1982) Some aspects of the circadian time structure in the elderly. Gerontology 28: 53–67

Touitou Y (1987) Le vieillissement des rythmes biologiques chez l'homme. Pathol Biol (Paris) 35: 1005–1012

Touitou Y, Touitou C, Bogdan A, Beck H, Reinberg A (1978) Serum magnesium circadian rhythm in human adults with respect to age, sex and mental status. Clin Chim Acta 83: 35–41

Touitou Y, Touitou C, Bogdan A, Chasselut J, Beck H, Reinberg A (1979) Circadian rhythm in blood variables of elderly subjects. In: Reinberg A, Halberg F (eds) Chronopharmacology. Pergamon, Oxford, pp 283–290

Touitou Y, Fevre M, Lagoguey M, Carayon A, Bogdan A, Reinberg A, Beck H, Cesselin F, Touitou C (1981) Age and mental health-related circadian rhythm of plasma levels of melatonin, prolactin, luteinizing hormone and follicule-stimulating hormone in man. J Endocrinol 91: 467–475

Touitou Y, Sulon J, Bogdan A, Touitou C, Reinberg A, Beck H, Sodoyez JC, van Cauwenberge H (1982) Adrenal circadian system in young and elderly human subjects: a comparative study. J Endocrinol 93: 201–210

Touitou Y, Carayon A, Reinberg A, Bogdan A, Beck H (1983a) Differences in the seasonal rhythmicity of plasma prolactin in elderly human subjects. Detection in women but not in men. J Endocrinol 96: 65–71

Touitou Y, Lagoguey M, Bogdan A, Reinberg A, Beck H (1983b) Seasonal rhythms of plasma gonadotrophins: their persistence in elderly men and women. J Endocrinol 96: 15–20

Touitou Y, Sulon J, Bogdan A, Reinberg A, Sodoyez JC, Demey-Ponsart E (1983c) Adrenocortical hormones, ageing and mental condition: seasonal and circadian rhythms of plasma 18-hydroxy-11-deoxycorticosterone, total and free cortisol and urinary corticosteroids. J Endocrinol 96: 53–64

Touitou Y, Fevre M, Bogdan A, Reinberg A, de Prins J, Beck H, Touitou C (1984) Patterns of plasma melatonin with ageing and mental condition: stability of nyctohemeral rhythms and differences in seasonal variation. Acta Endocrinol 106: 145–151

Touitou Y, Fèvre-Montagne M, Proust J, Klinger E, Nakache JP (1985) Age- and sex-associated modification of plasma melatonin concentration in man. Relationship to pathology malignant or not, and autopsy finding. Acta Endocrinol (Copenh) 108: 135–144

Touitou Y, Reinberg A, Bogdan A, Auzéby A, Beck H, Touitou C (1986a) Age-related changes in both circadian and seasonal rhythms of rectal temperature with special reference to senile dementia of Alzheimer type. Gerontology 32: 110–118

Touitou Y, Touitou C, Bogdan A, Reinberg A, Auzéby A, Beck H, Guillet P (1986b) Differences between young and elderly subjects in seasonal and circadian variations of total plasma proteins and blood volume as reflected by hemoglobin, hematocrit and erythrocyte counts. Clin Chem 32: 801–804

Touitou Y, Touitou C, Bogdan A, Reinberg A, Motohashi Y, Auzéby A, Beck H (1989) Circadian and seasonal variations of electrolytes in aging humans. Clin Chim Acta 180: 245–254

Vermeulen A, Rubens R, Verdonck L (1972) Testosterone secretion and metabolism in male senescence. J Clin Endocrinol Metab 34: 730–735

Vitiello MV, Smallwood RG, Avery DH, Pascualy RA, Martin DC, Prinz PN (1986) Circadian temperature rhythms in young adult and aged men. Neurobiol Aging 7: 97–100

Wallach EE, Decherney AH, Russ A, Duckett G, Gracia CR, Root AW (1973) Episodic secretion of LH and FSH after ovariectomy. Secretory patterns response to estrogen and progesterone. Obstet Gynecol 41: 227–233

Weitzman ED, Moline ML, Czeisler CA, Zimmerman JC (1982) Chronobiology of aging: temperature, sleep-wake rhythms and entrainment. Neurobiol Aging 3: 299–302

Wurtman R, Axelrod J, Barchas JD (1964) Age and enzyme activity in the human pineal. J Clin Endocrinol Metab 24: 299–301

Yen SSC, Tsai CC, Naftolin F, Vandenberg G, Ajabor L (1972) Pulsatile patterns of gonadotropin release in subjects with and without ovarian function. J Clin Endocrinol Metab 34: 671–675

Ziegler MG, Lake CR, Wood JH, Ebert MH (1976) Circadian rhythm in cerebrospinal fluid noradrenaline of man and monkey. Nature 264: 656

# Chronobiology of Mental Performance

T. H. Monk

## Introduction

With so much of the body and brain's physiology and chemistry changing in a rhythmic circadian manner, it is hardly surprising to note that there are equivalent changes in a person's mood, subjective activation and performance efficiency. Thus, an individual's mental performance abilities are very different from one time of the day to another, and these changes over time can be categorized and studied using similar circadian techniques to those developed for the physiological measures (e.g., body temperature, blood pressure) more often studied by the chronobiologist. There are, however, several major differences that must be recognized if mental performance rhythms are to be studied properly. This chapter will start with a dicussion of these differences, then move on to discuss intertask differences in circadian mental performance rhythms, and the oscillatory changes underlying them.

## Research Methods

Acquiring a *physiological* data point from a subject may be as easy for him as spitting into a paper cup or as irksome as having a syringe of blood taken, but, in either case, once permission has been granted, the attitude of the subject has no bearing on the accuracy or otherwise of the data point obtained. A blood sample grudgingly given is just as useful and accurate to the experimenter as one happily given. The same is not true for mental performance (or, indeed, any psychological measure). An alienated, unmotivated or poorly trained subject can give the experimenter data which are totally unreliable and misleading; worse, indeed, than missing data.

In studying the chronobiology of mental perfor-

mance it is thus vital that considerable effort is put in to ensuring that the subject knows what to do, knows why he or she is doing it, and understands its importance. All too often the obviously medical aspects of the study are emphasized by the experiments (and/or the technician actually running the study), while the psychological measures such as mental performance are relegated to the position of a second-class "side show." To do that is a waste of time for all concerned.

Training (and/or experimental design) must also take into account "practice" or "learning curve" effects. These effects are dramatic, and reflect progressive improvements in performance ability as subjects get more experienced at doing the task. Depending upon what is required of them, subjects' performance may still be improving after dozens (or even hundreds) of test sessions. There are two major approaches to removing practice effects. The first is simply to fit a straight line (or ascending curve) to the time series, and then transform each data point to its deviation from the fitted line. An alternative technique is to balance "session number" against "time of day" using several subjects, each doing the test at six times of day (for example), but starting the first session at different times of day. This has been referred to as the "cyclic latin square" design, and has been used in several different time of day studies (Folkard 1975; Folkard and Monk 1980; Monk and Leng 1982).

## Circadian Performance Rhythms

There are several misconceptions which can impede our understanding of circadian performance rhythms. First many people are under the erroneous impression that circadian performance differences are either trivial in magnitude or simply comprise a square wave function with poor performance at night contrasting with good performance during the day. This implies that there is little change within the day or within the night. Neither is true: even over the normal waking day (0800–2400 hours) there are gradual fluctuations comprising differences as great as those associated with consuming the legal limit of alcohol, for example (Folkard et al. 1977). Indeed in one aspect of mental performance (immediate memory for prose material) waking day (0800–2300 hours) time of day effects can account for changes of greater than 20% above and below the daily mean (Laird 1925; Folkard and Monk 1980). Circadian performance rhythms comprise gradual but substantial changes over the entire 24 h.

A further misconception is that interindividual variability is too great to make any definitive statement about *population* mental performance rhythms. Interindividual differences do exist (primarily in the form of morningness-eveningness (Horne and Ostberg 1976), but these effects are usually only seen in population extremes. As many different studies have shown (Laird 1925; Blake 1967; Hockey and Colquhoun 1972; Folkard and Monk 1985), one *can* make generalizations about the population as a whole, plotting the time of day effect in performance ability for a particular task or test.

## Intertask Differences

A particularly resilient misconception regarding circadian performance rhythms is that they all simply echo the circadian body temperature rhythm, showing a trough in the early hours of the morning and a peak in the mid evening (Colquhoun 1971). This misconception has its roots in the writing of Kleitman (1963), who was struck with the parallelism that he found between simple repetitive tasks such as card sorting, and the body temperature rhythm. Indeed, he even went as far as to assert that measures of oral temperature could replace time-consuming performance testing in assessing the efficiency of shift workers at different times during the 24-h day (Kleitman and Jackson 1950). Kleitman believed the relationship between body temperature and performance to be a *causal* one, with faster thought processes resulting from a warmer physiological milieu.

Colquhoun and his colleagues in Cambridge retained the notion of a parallelism when they took up the study of time of day effects in performance during the 1960s (Blake 1967; Colquhoun 1971). However,

they were careful to avoid ascribing a *causal* relationship, instead postulating that there was a diurnal variation in basal arousal (broadly equivalent to the inverse of sleepiness) which was parallel to body temperature because both were driven by the same underlying oscillator (Hockey and Colquhoun 1972). Performance changes were then held to be mediated by changes in basal arousal. This comprised the "arousal model" of circadian performance rhythms, which was only later specified in detail and quantified (Monk 1982).

A major advantage of the arousal model was that it could be used to explain time of day effects in performance measures which did not happen to show the same evening peak as one typically gets in body temperature. Although often forgotten, early time of day studies by educational psychologists (see Gates 1916; review by Lavie 1980) had found *morning* performance to be superior for many cognitive tasks such as those involving mental arithmetic and short-term memory. In his 1967 paper, Blake showed that, although most of his tasks showed the approximate

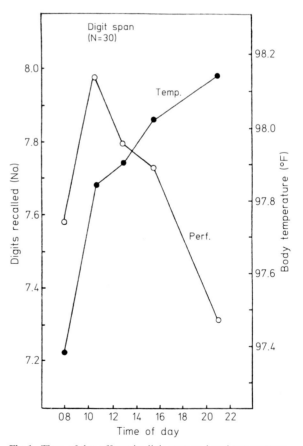

**Fig. 1.** Time of day effects in digit span and oral temperature. (After Blake 1971)

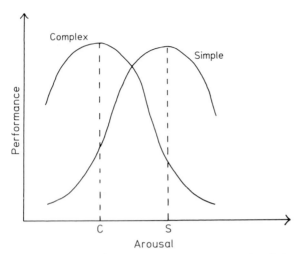

**Fig. 2.** Schematic diagram showing inverted "U" functions linking arousal and performance with a lower optimal arousal for complex task (c) than for simple ones (s).

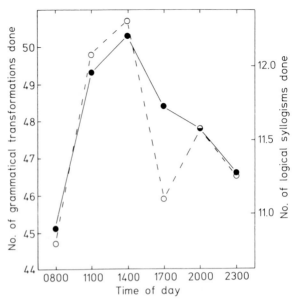

**Fig. 3.** Time of day effects in logical reasoning. (After Folkard 1975)

2100 hours peak found for body temperature, digit span was unique in showing superior performance at 1030 hours (Fig. 1). By invoking the arousal model, Colquhoun and his colleagues were able to explain this apparent anomaly by postulating that evening arousal levels were superoptimal for this task, i.e., on the falling, rather than rising, arm of the "inverted U" linking arousal and performance (see Fig. 2). Thus for cognitively more complex tasks, optimal arousal levels were reached sooner in the day (i.e., at a lower arousal level) than was the case for the more simple repetitive tasks, whose optimal arousal lay beyond the normal diurnal range. Colquhoun and his colleagues then performed a series of experiments manipulating arousal by other means (e.g., knowledge of results, white noise), and determining how that modified the observed time of day function (Blake 1971). The "superoptimal arousal" explanation also held for the morning superiority in short-term memory which had been found by Baddeley et al. (1970) and Hockey et al. (1972), using word lists.

Notwithstanding the ability of the arousal model to account for these different effects, the dominant view through the early 1970s was one of homogeneity in circadian performance rhythms, with the memory tasks being generally regarded as anomalous exceptions. This changed, however, in the mid 1970s with the publication of Folkard's (1975) study of time of day effects in logical reasoning. Using Baddeley's (1968) verbal reasoning tests and a logical syllogisms test, Folkard showed a mid-day peak in performance (Fig. 3) which was different to both the morning peak found in short-term memory and the evening peak found in simple repetitive tasks (and body temperature).

This finding "opened up" the study of time of day effects in human performance, changing the emphasis of the area to one in which intertask differences were investigated, and more attention given to the *mechanisms* underlying circadian performance rhythms. Rather than being concerned solely with the question of at which times of day performance was "better," attention turned to the question of how performance *differed* from one time of day to another. Thus, interest turned to how processing strategies might change as a function of time of day.

## Memory and Time of Day

As described earlier in this chapter, studies in the early 1970s had shown short-term memory performance to be better in the morning than in the afternoon. In a series of studies, Folkard proceeded to investigate this effect, bringing to bear such manipulations as enforced vocalization, articulatory suppression, and acoustic/semantic confusability. In a standard free-recall paradigm using visual presentation of 15-word lists, Folkard and Monk (1980) were able to demonstrate that the normal morning superiority could be eliminated by suppressing subvocal rehearsal of the items as they were presented (accomplished by requiring subjects to repeatedly count aloud from one to ten). This finding, together with others in which elimination of subvocal rehearsal attenuated time of day differences (see Folkard and Monk 1985), led to the hypothesis that there was a *qualitative* difference between the way in which verbal material was processed between morning and afternoon sessions. Folkard hypothesized that in the morning attention was focused on the *sound* of the words (syntactic processing), while in the afternoon it was focused on their *meaning* (semantic processing). The validity of this hypothesis was nicely confirmed in a study in which word lists were confusable either semantically ("big, large, huge, enormous, . . .") or phonemically ("bath, ball, bat, bass, . . ."). When short-term memory was measured, phonemic confusablity had its greatest effect in the morning, semantic confusability in the evening (Folkard 1979).

## Simple Repetitive Performance and Time of Day

In a similar vein, attention was directed towards the simple repetitive tasks whose performance had always been considered to parallel the circadian temperature rhythms, showing an *evening* superiority. In a meta-analysis of several different studies, Monk and Leng (1982) showed that the time of day functions so obtained were far from identical, with some showing a decline over the waking day, and others an improvement. More importantly, they showed an inverse relationship (Fig. 4) between the speed and the accuracy with which a serial search task was performed. Thus, although performance was *fastest* in the evening, it was also least accurate. Using other studies, Monk and Leng concluded that performance changes over the day could be explained by hypothesizing that subjects adopted an increasingly fast but inaccurate strategy as the day wore on.

The ramifications of this finding are that researchers should no longer assert that performance is "better" at one time of day compared to another. Whether performance is "better" or "worse" depends upon the value one places on speed as opposed to accuracy, and the two measures may have very different circadian properties.

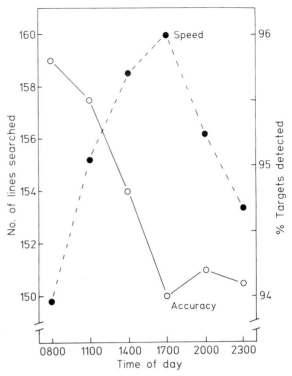

**Fig. 4.** Time of day effects in the speed and accuracy with which a serial search task was performed ($N = 36$). (After Monk and Leng 1982)

## Underlying Oscillatory Control

Once attention had switched to the heterogeneity of circadian performance rhythms, the question naturally arose as to whether they were indeed under the sole control of the circadian oscillator controlling body temperature (the "strong" oscillator), as the arousal theory assumes, or whether other circadian processes were coming into play. Evidence for the latter explanation came from phase shift studies (Hughes and Folkard 1976; Monk et al. 1978) in which the memory load of a task appeared to determine not only the timing of peak and trough (phase) of the performance rhythm, but also the *rate* at which that phase reentrained to an abrupt change in routine.

Experiments to examine the oscillatory control underlying circadian performance rhythms have involved subjects living in temporal isolation. Three types of protocols have been used and all three have produced evidence that intertask differences are also reflected in differences in oscillatory control, although the exact form that such differences take is not always the same.

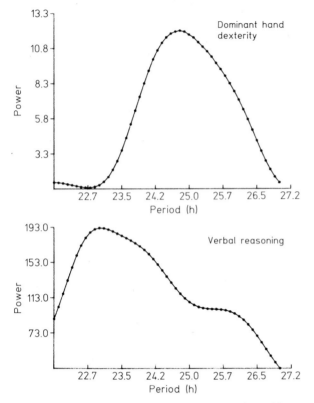

**Fig. 5.** Power frequency spectrum of a free-running subject's performance rhythms for manual dexterity and verbal reasoning tasks. (After Monk et al. 1984)

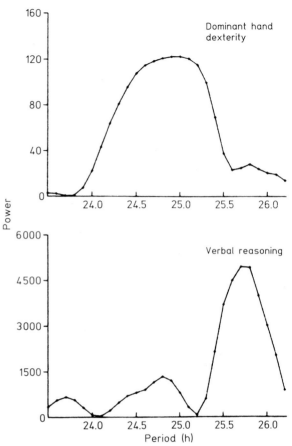

**Fig. 6.** Power of frequency spectrum of performance rhythms for manual dexterity and verbal reasoning tasks in a subject entrained to a 25.8-h day during temporal isolation. (After Monk et al. 1983)

First, there are the standard "free-running" experiments in which subjects in temporal isolation are free to select their own timings of sleep, wakefulness, and meals. A battery of performance tests is given several times throughout each waking day, the experiment lasting several weeks (Monk et al. 1984). By determining the dominant periodicities in the different performance measures, one can plot the equivalent of a power spectrum for each measure. Intertask difference in these plots then suggests possible differences in underlying oscillatory control. An example is illustrated in Fig. 5, which shows results of a free-running subject whose circadian rhythm in the Baddeley (1968) verbal reasoning test had a shorter period than it did in a manual dexterity task (Purdue peg board). Although not all subjects show such dramatic differences, it would appear that there is a tendency for the free-running rhythms of cognitive tasks to have shorter periods than those of simple repetitive tasks.

In the second and third types of temporal isolation protocol, artificially long or short days are imposed upon the subject. In "fractional desynchronization" experiments, day lengths are progressively shortened or lengthened, with each day length different from the one before it. Differences in underlying oscillatory control then become apparent as different circadian rhythms reach the limit of their particular range of entrainment and break away from the progressively shortening or lengthening sleep-wake cycle at different periods (Wever 1979). When a battery of performance tests is given using this technique, cognitive tasks are found to break off at a shorter period than the "break off" periods of simple repetitive tasks (Folkard et al. 1983).

In "forced desynchronization" experiments, one artificially long or short day length, outside the range of entrainment of the strong oscillator, is enforced for many weeks. This oscillator then runs at its "natural" period (typically close to 25 h), while the sleep-wake cycle remains driven at the imposed day length. As in the free-running studies, the equivalent of power spectra can be plotted for different performance measures, and intertask differences in these plots used to suggest differences in underlying oscillatory control. An example is shown in Fig. 6, where the imposed sleep-wake cycles was of a 25.8-h period, and the temperature oscillator ran at a period of 24.8 h. When simple manual dexterity was again compared with the Baddeley (1968) verbal reasoning test, it was found that the rhythmic behavior of dexterity tended to be dominated by the temperature oscillator, while that of verbal reasoning tendend to be dominated more by the cycle of sleep and wakefulness (and/or the rhythmic processes underlying it) (Monk et al. 1983). Thus, this experiment again showed intertask differences related to cognitive complexity, although these were not directly equivalent to those observed in the free-running and fractional desynchronization experiments.

## Conclusion

There is not one single performance rhythm, but many. Different tasks are associated with different time-of-day functions, adjusting at different rates to a change in schedule, and under different oscillatory control. The cognitive complexity or working memory load of a task appears to be an important index of its circadian behavior, although even within the area of memory there are dramatic differences between short- and long-term retention. The mechanisms underlying these differences in both memory and nonmemory tasks appear to be based primarily upon changes in strategy rather than solely upon changes in capacity. The oscillatory control underlying these changes appears to depend upon both the temperature oscillator and sleep-wake cycle, with the particular pattern again dependent upon various aspects of the task.

## References

Baddeley AD (1968) A three-minute reasoning test based on grammatical transformation. Psychom Sci 10: 341–341

Baddeley AD, Hatter JE, Scott D, Snashall A (1970) Memory and time of day. Q J Exp Psychol 22: 605–609

Blake MJF (1967) Time of day effects on performance in a range of tasks. Psychom Sci 9: 349–350

Blake MJF (1971) Temperament and time of day. In: Colquhoun WP (ed) Biological rhythms and human performance. Academic, London, p 109

Colquhoun WP (1971) Circadian variations in mental efficiency. In: Colquhoun WP (ed) Biological rhythms and human performance. Academic, London, p 39

Folkard S (1975) Diurnal variation in logical reasoning. Br J Psychol 66: 1–8

Folkard S (1979) Time of day and level of processing. Mem Cognit 7: 247–252

Folkard S, Monk TH (1980) Circadian rhythms in human memory. Br J Psychol 71: 295–307

Folkard S, Monk TH (1985) Circadian performance rhythms. In: Folkard S, Monk TH (eds) Hours of work – temporal factors in work scheduling. Wiley, New York, p 37

Folkard S, Monk TH, Bradbury R, Rosenthall J (1977) Time of day effects in school children's immediate and delayed recall of meaningful material. Br J Psychol 68: 45–50

Folkard S, Wever RA, Wildgruber CM (1983) Multioscillatory control of circadian rhythms in human performance. Nature 305: 223–226

Gates AI (1916) Variations in efficiency during the day, together with practice effects, sex differences, and correlations. Univ Calif Publ Psychol 1: 1–156

Hockey GRJ, Colquhoun WP (1972) Diurnal variation in human performance: a review. In: Colquhoun WP (ed) Aspects of human efficiency: diurnal rhythm and loss of sleep. English Universities Press, London, p 39

Hockey GRJ, Davies S, Gray MM (1972) Forgetting as a function of sleep at different times of day. Q J Exp Psychol 24: 389–393

Horne JA, Ostberg O (1976) A self-assessment questionnaire to determine morningness/eveningness in human circadian rhythms. Int J Chronobiol 4: 97–110

Hughes DG, Folkard S (1976) Adaptation to an 8 h shift in living routine by members of a socially isolated community. Nature 264: 432–434

Kleitman N (1963) Sleep and wakefulness. University of Chicago Press, Chicago

Kleitman N, Jackson DP (1950) Body temperature and performance under different routines. J Appl Physiol 3: 309–328

Laird DA (1925) Relative performance of college students as conditioned by time of day and day of week. J Exp Psychol 8: 50–63

Lavie P (1980) The search for cycles in mental performance from Lombard to Kleitman. Chronobiologia 7: 247–256

Monk TH (1982) The arousal model of time of day effects in human performance efficiency. Chronobiologia 9: 49–54

Monk TH, Leng VC (1982) Time of day effects in simple repetitive tasks: Some possible mechanisms. Acta Psychol (Amst) 51: 207–221

Monk TH, Knauth P, Folkard S, Rutenfranz J (1978) Memory based performance measures in studies of shiftwork. Ergonomics 21: 819–826

Monk TH, Weitzman ED, Fookson JE, Moline ML, Kronauer RE, Gander PH (1983) Task variables determine which biological clock controls circadian rhythms in human performance. Nature 304: 543–545

Monk TH, Weitzman ED, Fookson JE, Moline ML (1984) Circadian rhythms in human performance efficiency under free-running conditions. Chronobiologia 11: 343–354

Wever R (1979) The circadian system of man: results of experiments under temporal isolation. Springer, Berlin Heidelberg New York

# Biological, Behavioral and Intellectual Activity Rhythms of the Child During Its Development in Different Educational Environments

H. Montagner, G. de Roquefeuil, and M. Djakovic

## Introduction

Hellbrügge and Rutenfranz and their coworkers were the first to do systematic research on the characteristics and ontogeny of biological rhythms in children (Hellbrügge 1960, 1968, 1977; Hellbrügge et al. 1959, 1967; Rutenfranz and Hellbrügge 1957; Rutenfranz and Colquhoun 1979). Their research is one of the basic references for all investigators who study the circadian rhythms of children of all ages. In particular, they showed that circadian rhythmicity only develops during about the 3rd or 4th week after birth for a certain number of physiological and psychophysiological variables (alternation of wake and sleep; pulse; potassium and sodium urine elimination; body temperature; etc.).

Kleitman and Engelman (1953), Parmelee (1961), Kleitman (1963) and Dreyfus-Brisach (1967) were also pioneers whose studies allowed a better understanding of the development and characteristics of the sleep-wakefulness rhythm in infancy and childhood.

More recently, extensive research using electrophysiological methods has been conducted on the ontogeny and regulations of the temporal distribution of sleep states in the child's first days, weeks or months (Anders et al. 1971; Anders and Keener 1987; Harper et al. 1976, 1981; Hoppenbrouwers et al. 1978, 1982; Benoit 1981, 1992; Anders 1982; Navelet 1984, 1989; Navelet et al. 1982; Guilleminault 1987). But, as stated by Anders and Keener (1985), there are still few longitudinal studies "describing the ontogenesis of sleep-wake patterns in normal infants." The picture is the same for older children and especially when they are attending a daycare center, a kindergarten, a primary school or a high school. However, the daily and weekly observations of children in the playroom and playground of the daycare center and the classroom of a kindergarten or a primary school show that a majority of children are not ready to perform any tasks at just any time of the day, whatever the personality and skills of the teacher. Some periods appear to be more favorable than others to mobilize the child's capacities of attention, comprehension, memory, etc., whereas other periods are particularly unfavorable and the teaching act is thus more uncertain. At the same time, it can be seen that these phenomena are more or less marked in different children in the same class, either regularly from one day to the next or on certain days. These current observations are confirmed by studies on the daily fluctuations of behaviors and biological rhythms in children attending kindergarten and primary schools (Montagner 1978, 1983; Koch et al. 1984, 1987; Soussignan et al. 1985, 1988), and daily fluctuations of memory and intellectual activity (Blake 1967, 1971; Folkard 1975, 1980; Folkard et al. 1977; Testu 1982, 1986, 1989).

There is some controversy on the temporal and functional relationship which could exist between the circadian rhythm and daily peaks of alertness (or arousal) and psychological functions including mood and performance efficiency, and those of the core body temperature rhythm (which is often considered to be governed by an endogenous oscillator of the human circadian system) and the daily cycle of sleep and wakefulness (which is considered to be governed by a second oscillator of the human circadian system) (see Wever 1975, 1979; Kronauer et al. 1982). Since there are no data available on the simultaneous recording of these rhythms in children we will not refer to this theoretical assumption, whatever the importance of the work which is done in this field in adolescents and adults (Monk 1987; Monk et al. 1983; Monk and Leng 1986). This paper will only deal with studies concerning the child's biological rhythms during schooltime and the behavioral and physiological fluctuations which appear to be associated with these rhythms, and especially the sleep-wakefulness rhythm. It will also report recent data on the daily fluctuations of intellectual activity in children at school.

## Temporal Evolution of the Sleep-Wakefulness Rhythm of Children from 2 to 5 Years of Age Attending a Kindergarten

Since there was no study available on the sleep-wakefulness rhythm of children attending a kindergarten from one day to the next and throughout the schoolyear and only few data on their adaptation processes to the new requirements and demands of this institution when they begin to attend, i.e., in France from 2 to 3 years of age, our research group collected extensive data on the sleep-wakefulness of children aged 2–5 years (Koch et al. 1984; Soussignan et al. 1985, 1988).

This study was made on 250 children aged from 2 to 5 years in eight classes in seven kindergartens. However, only the information for the 107 children who attended school regularly was retained. Each day the teachers and the parents measured the duration of the afternoon nap and the length of nocturnal sleep from September to June. The previous year the teachers had met with the research workers three times and were thus able to harmonize the filling in of the forms concerning the afternoon nap. The teachers also prepared the forms with the information they received each morning from the parents concerning the time when the child fell asleep and its waking time and the events that occurred during the previous night's sleep of each child.

This method provided data for 10000 periods of afternoon naps and 11000 night sleeps which were computed with a Commodore microcomputer.

The normality was checked and the use of parametric tests was justified. The average and standard deviation of the duration of the nap, nocturnal and total sleep per day were taken into consideration. The frequency of the naps was also considered, i.e., the relationship between the number of times when the children took a nap and the total number of nap situations that occurred each day (including the time when the children did not sleep).

The temporal evolution of the sleep-wakefulness rhythm between 2 and 5 years appears to be related to two processes: the progressive disappearance of the last period of day sleep (nap) without any or only a slight modification of its duration when there is a modification, and the slow decrease of the night sleep during the year (Fig. 1) (Koch et al. 1984; Soussignan et al. 1985).

When kindergarten children only have one period of daytime sleep there is a negative correlation between the duration of night sleep and the duration of the nap during the following day and between the duration of the nap and that of the night sleep which follows (Koch et al. 1984; Soussignan et al. 1985). Thus, a deficit in night sleep can be compensated for, at least partially, by an extension of the duration of the nap on the following day. These data are consistent with those obtained by Klackenberg (1971) with Swedish children and with those of Basler et al. (1980) with Swiss children. These converging results

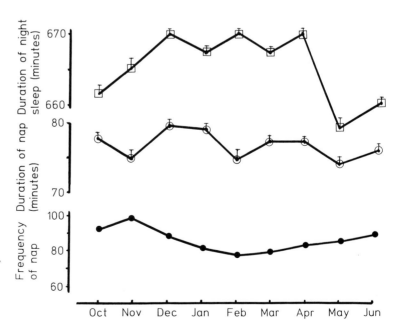

**Fig. 1.** Evolution of the duration of night and nap sleep in minutes and the frequency of the naps in 107 children aged from $2^1/_2$ to 5 years in eight kindergarten classes throughout the schoolyear (October to June). (See Koch et al. 1984)

reinforce the potential value of a free subject-time arrangement at the beginning of the school afternoon: children with a night sleep deficit could then compensate, at least partially, for their lack of sleep and the consequent fatigue. However, it has also been observed that the correlation between the duration of night sleep and the duration of the nap can be zero or positive, either occasionally or regularly, according to the individual child. Thus, children who could be characterized as long or short night sleepers can also be those who have long or short naps.

The duration of the last period of daytime sleep tends to vary little in a given child when this period of sleep tends to disappear, but varies considerably from one child to another. For example, at 4 years of age the duration of the nap in some children is usually 5–15 min, whereas that of other children is regularly 90–120 min or more. It can be asked whether the differences in frequency and duration that are observed at the beginning of the afternoon in vigilance (or drowsiness) in older children, adolescents or adults are not linked to the characteristics of the last period of daytime sleep in early childhood. This way the interindividual differences in the vigilance and attention of individuals at the beginning of the afternoon could be better understood and taken into account.

## Modifications Observed in the Behavior and Biological Rhythms of Children Who Go From the Last Year of Kindergarten (5–6 Years of Age) to the First Year of Primary School (6–7 Years of Age)

Twenty-one children aged 5–6 years (mean age, 71.2 months; SD ± 2.7 months; 11 boys and 10 girls) belonging to a very wide range of socioeconomic groups were studied in four different classes in four different kindergartens in very different suburbs of the town of Besançon located in the northeast of France.

### Behavioral Data

The children were filmed on video, from 0900 to 1130 hours and from 1400 to 1630 hours, for 1 week in each class during the third school term (May–June). The camera was placed in the corner of the room near the teacher's desk so that all the children being studied were in viewing range. The same children were again filmed 6 months later in four 1st-year primary school classes in four different schools from the same suburbs. The children were filmed continuously from 0830 to 1130 hours and from 1400 to 1700 hours for a full week.

Video recordings were analyzed in the laboratory with a microcomputer used as an "ethological keyboard." Each key corresponds to a behavior which was coded with a letter on the computer screen. The ANAFILM program (Koch et al. 1987) enabled the worker to record the frequency and the actual duration of all the behaviors which were coded using the focal child sampling method (Altmann 1974), where a particular child is the focus of observation during a sample period.

The result of the Kruskal-Wallis one-way analysis of variance by ranks showed that there were no significant differences for each of the observational measures in either the four kindergartens ($H < 4.3$; d.f. = 3; $P > 0.05$) or four primary school classes ($H < 7.3$; d.f. = 3; $P > 0.05$). The distribution of behavioral durations can thus be considered to be similar between the four kindergarten classes and between the four primary school classes.

Three categories of behaviors were measured:
1. *Off-task behaviors:* yawning; manipulation of the mouth (biting, sucking a finger or another object); automanipulation of other parts of the body; manipulating objects or carrying out a manual activity not connected with the school task; rhythmical stereotypies; changing position; posturing; moving around the classroom without the permission of the teacher when this movement was not related to the school task.
2. *Communications with peers.*
3. *Task behaviors.*

Yawning was selected among "off-task behaviors" for this paper since it indicates a low behavioral arousal state.

### Cardiovascular Variables

The cardiovascular variables were investigated in 17 of the 21 children in the kindergarten and then in the primary school: heart rate as measured by the pulses and the Holter technique; systolic and diastolic blood pressure were measured with a sphygmomanometer of the Riva-Rocci electronic type (UEDA, I, Health digital 8000).

These variables were studied for a week first during the third term (May and June) in the kindergarten and then 5–6 months later during the first term (October and November) in the primary school.

In kindergarten and primary school classes the readings were taken every day of each week at the same time: 0900, 1000 and 1100 hours and at 1400, 1500 and 1600 hours on Monday, Tuesday, Thursday and Friday (in France Wednesday is a vacation day). On Saturday they were only taken at 0900, 1000 and 1100 hours since schools are closed in the afternoon. Each reading was made in both sections after each child had moved from his/her usual desk to a desk located in the back of the classroom and had sat down in a rest position for 3 min. All the children were trained to this experimental procedure for several days during the previous weeks. We collected 918 items of data for heart rate and 918 items of data for blood pressure.

## Anthropometric Variables

Due to the close relationship between cardiovascular and anthropometric variables (Voors et al. 1976), the weight and the height of the children were recorded in both kindergarten and primary school.

There was no significant difference between the four kindergarten classes and the four primary school classes in Quetelet's anthropometric indices $\frac{W \times 100}{h^2}$, where $W$ is the weight and $h$ the height of the children (Billewicz et al. 1962; Subash Babu and Chuttani 1979): $F(1.86) = 1.12$; $= \bar{X} = 0.1576$ (kindergarten classes) and $F(1.80) = 0.20$; $\bar{X} = 0.1589$ (primary classes). We found no significant difference in Quetelet's anthropometric indices of the children in the kindergartens and the primary schools: $F(1.15) = 2.41$.

## Beginning of the Afternoon

As Hellbrügge, Rutenfranz and others (Hellbrügge 1960, 1968, 1977; Hellbrügge et al. 1959, 1967; Rutenfranz and Colquhoun 1979) demonstrated in their studies of the biological rhythms of children belonging to different age categories, the beginning of the afternoon is characterized by a high proportion of children being drowsy or at least not really vigilant. When children are able to organize their own activities during the day, more than 80% of the 3- to 6-year-olds appeared in these studies to be asleep or

drowsy at 1400 hours. The percentage was approximately 80% for the 6–11-year-olds and 40% for the 12- to 16-year-olds. Our data are in keeping with these observations. As we were able to follow up the same groups of children in the kindergarten (5- to 6-year-olds) and then in the first class of the primary school (6 to 7 years of age), we observed that 68% of the children yawned between 1430 hours and 1500 hours in the first class of primary school (the same percentage was observed between 0900 and 0930 hours: Fig.2 B) and that the yawning frequency peaked at this time and between 0900 and 0930 hours when all children were considered together (Koch et al. 1987). The Cochran test for matching series showed a significant difference between the different 30-min observation periods, whether we considered the percentage of yawning children ($Q = 32.3$, 6 ddl, $P < 0.001$) or the frequency of yawnings (mean $= 7.4$; $\delta = 2.9$; $Q = 358$, 6 ddl, $P < 0.00001$). This study thus confirms that at the beginning of the afternoon there is in most children a decrease (or a sudden decline for some children) in vigilance or an increase in drowsiness (the psychophysiological mechanisms have yet to be defined). This has been observed not only in children but also in adults.

The simultaneous recording of the electrocardiogram of the same children using the Holter system (Koch et al. 1987; Soussignan et al. 1988) has shown that during the 1st year of primary school the heart rate is much higher at 1400 hours (the second half of the school day starts at 1345 hours) than at the previous (1100 hours) and subsequent (1600 hours) readings (Scheffé test, $P < 0.05$), and the readings at the same time several months previously in the same children at the kindergarten (however, the Scheffé test is oriented toward describing a nonsignificant difference) (Fig.3). Moreover, except for pathological cases, the heart rate decreases with age, which is observed in the morning between 0900 and 1100 hours and at 1500 and 1600 hours, when children in kindergarten are compared to themselves 5–6 months later as they are attending the first class of the primary school. We also found that the amplitude of the variations observed in the heart rate throughout the schoolday is significantly higher in the primary school than in the kindergarten, especially at 1400 hours (Fig.4): an analysis of variance gave $F(1.310) = 3.25$ ($p < 0.01$) for the kindergarten and $F(1.300) = 27.5$ ($p < 0.001$) for the primary school.

In accordance with the classical studies of human chronobiology done mainly with adults, but also with children (Hellbrügge 1960; Mills 1975; Wever 1979), the highest values of the heart rate in children attend-

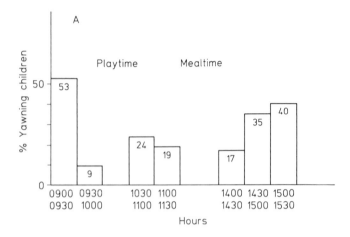

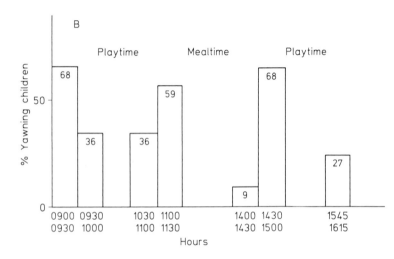

Fig. 2 A, B. Percentage of children in kindergarten (**A**) (mean age, 71.2 months) and 5–6 months later in the first class of the primary school (**B**) (mean age, 76.2 months) who yawn throughout the schoolday (N = 21 children). For more details see Koch et al. (1987)

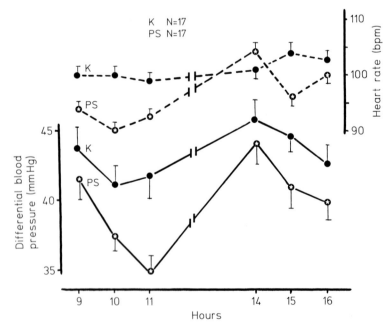

Fig. 3. Variations in heart rate in beats per minute (bpm) and differential blood pressure (mmHg) (systolic-diastolic blood pressure) throughout the schoolday in 17 children attending four different kindergartens (K) in May and June and the 1st year of primary school (PS) 5–6 months later. (October and November)

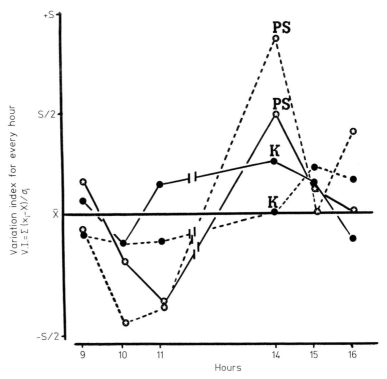

**Fig. 4.** Variation index of heart rate *(broken lines)* and differential blood pressure *(unbroken lines)* in relation to the mean X̄ from 1 h to the next throughout the schoolday in the 17 children under study in Fig. 3, in May and June in the kindergarten (*K*) and 5–6 months later (October and November in the 1st year of primary school (*PS*). The variation index for each hour is $\sum \frac{xi - \bar{X}}{\delta i}$, where xi is the mean of measures at a given hour, $\delta i$ is the standard error and X̄ the mean of all measures at all hours.

ing the first class of primary school and the acrophase of heart rate in children attending the kindergarten (Koch et al. 1987) are at the beginning of the afternoon.

It must be noted that according to the current theory (Lacey 1967; Graham and Clifton 1966) an increase in heart rate appears to be associated with reactions of defense and refusal of information (in comparison a decrease in heart rate is accompanied by reactions of orientation and acceptance of information).

Thus, the beginning of the afternoon can be characterized for the majority of children in the 1st year of primary school by a decrease in the attention capacity or an increase in drowsiness and by an increase in the probability of having reactions of defense and information refusal. The picture was very different with the same children when they attended kindergarten 5–6 months earlier: the proportion of yawning children was only 35% between 1430 hours and 1500 hours (Fig. 2 A); the frequency of yawnings appeared to be approximately five times lower than in the 1st year of the primary school; the chi-square test indicated a significant difference in the frequency of yawnings between the 30-min periods throughout the schoolday (mean = 1.6; $\delta$ = 1.5; chi-square = 19.6, 6 ddl, $p < 0.01$); however, there was no significant dif-

ference in the proportion of yawning children (chi-square = 7.8, 6 ddl, NS).

As no obvious changes were observed either in the Quetelet's anthropometric index $\frac{W \times 100}{h^2}$ or in the children's diet and in the parent's life rhythms, it can be thought that the observed differences at the beginning of the afternoon between the kindergarten and the first class of the primary school were related to an increase in school constraints in the primary school, and especially between 1345 and 1500 hours. In the last year of kindergarten, children were allowed to choose from 1345 to 1500 hours their games, activities and rest periods, whereas a few months later when they were attending the 1st year of primary school they were not. They must increase from 1345 to 1500 hours the duration, frequency and efficiency of their attention processes to cope with the demanding tasks of reading, writing and arithmetic which are imposed on them by the teacher.

## Beginning of the Schoolday

The recording of behavior, heart rate and blood pressure also showed that the proportion of yawning children, the frequency of yawning for all children

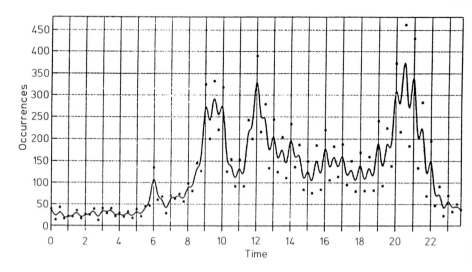

**Fig. 5.** Mean occurrences every 15 min *(each dot)* of falling asleep in 12 children throughout the day (0–24 h) when they were followed up from the 4th to the 15th month of age. Number of days, 2845. Number of sleep episodes and falling asleep times, 11 394. Falling asleep peaks from 0900 to 1000 hours, at about noon and from 2000 to 2100 hours

considered together, and the values for heart rate and blood pressure were higher between 0900 and 0930 hours (for the yawnings) and at 0900 hours (for heart rate and blood pressure) than at any other times in the morning school period, either in the kindergarten or in the primary school (Figs. 2, 3; Koch et al. 1987).

Two hypotheses were put forward to explain these results:

1. They could be related to the shorter duration of night sleep on school days than on vacation days, i.e., Wednesdays and Sundays (see below), due to an earlier awakening on school days (see below). It has yet to be explained why they were observed from 2 to 3 h after the awakening time. Is it one of the lowest values of an ultradian rhythm or the bathyphase of a circadian rhythm of vigilance and attention processes?

2. They could be related to ontogenetic and development processes in infancy. From a recent longitudinal study carrried out with 12 infants from the 4th month to the 15th month it appears that the frequency of falling asleep peaks between 0900 and 1000 hours, at about noon and between 2000 and 2100 hours (Fig. 5) (de Roquefeuil et al., in preparation). Thus, it may be that the yawning peak and relatively high values of the heart rate and blood pressure around 0900 and 0930 hours in 5- to 6-year-old kindergarten children and 6- to 7-year-old primary schoolchildren are related to the falling asleep peak which is observed in 4- to 15-month-old infants.

## High Variability of Biological Rhythms from one Child to Another

A great variability was observed in the circadian heart rate and blood pressure rhythms as well as in the sleep-arousal or sleep-wakefulness rhythms from one child to another in our comparative studies between children in the last year of kindergarten and those in the 1st year of primary school (Koch et al. 1984, 1987; Soussignan et al. 1985).

A close examination of other studies in the field of chronobiology also shows, when all the data are reviewed, the great variability of the rhythms from one child to another at the same age.

For example, between 2 and 3 years of age (children attending daycare centers) and between 2 and 6 years of age (children attending kindergartens) some children tend to have circadian urinary corticosteroid curves (17-OHCS and cortisol) which are stable or relatively stable from one day to the next (Montagner et al. 1978, 1982). Even if these curves can fluctuate on Mondays, as is the case with other children, they tend to be synchronized with the life rhythm or routine that is imposed upon these children during the week (they are taken to the daycare center or the kindergarten at a particular time in accordance with the working hours of their parents and the institutions opening and closing times) to such an extent that the curves for Tuesday and Friday are often identical. When the behavior of these children is followed from one day to the next it is found that they have the most stable relational system; their appeasing behavior is frequent and, at the same time, their aggression and isolation are infrequent and do not last long. However, there are other children who have curves that fluctuate on Mondays and who are characterized by curves which vary from one day to the next and with generally high 17-OHCS levels at several times during the day. The curves of these children generally appear to be out of synchrony with the routine of the center or the kindergarten. Their aggressive behavior and the alternation of these behaviors with isolation are significantly more frequent than those of children with stable curves. These characteristics of behavioral and biological rhythms appear to be closely related to life and work rhythms that are constraining for both parents and particularly for the mother, and the relational systems with the child in which threatening and aggression are significantly more frequent than in families with children who have stable curves. The sleep-arousal rhythm of the children with the most fluctuating curves and relational behavior is also more variable and more often disturbed than that of other children.

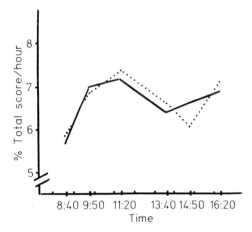

**Fig. 6.** Percentage of the total score per hour in two psychotechnic tests by 6- to 7-year-old children attending the first class of a primary school. Striking out of numbers, *unbroken line;* striking out of figures, *broken line.* (Testu 1989)

## Modifications Observed in the Intellectual Performance of Children Attending Primary School and High School

The research carried out on the variations of the child's intellectual activity (Blake 1971; Testu 1982, 1986, 1989) led to results in keeping with our data.

Thus, Testu (1982, 1986, 1989) showed that, from the first class of the primary school to University, the level of intellectual performances is relatively low at the beginning of the school day (between 0800 and 0900 hours), becomes higher until about 1100–1200 hours, drops after lunch (at about 1400 hours) and finally becomes higher (more or less, according to age) in the middle of the afternoon (1500–1600 hours; Fig. 6). These variations in the intellectual activity not only find their expression in psychotechnic test performances (striking out of figures and numbers; spatial structuralization tests) and performance in school activities (conjugation, dictations, operations, etc.) but also in the strategies chosen to resolve mathematical problems. However, Testu (1982, 1986, 1989) also showed that these variations are more or less marked depending on the children, and may even be absent in some children. For instance, the performance of children attending special education classes *(Sections d'Education Spécialisées),* whose education level is low, vary markedly according to the daily fluctuations shown in Fig. 6, whereas those of good pupils of the same age attending current classes in high-school remain relatively steady throughout the school-day. Other variables operate as well, particularly at the learning stage. Thus, the variations in the performance of pupils attending the last year of the primary school (10- to 11-year-olds) having to cope with mathematical problems are observed only at the beginning of the learning process. The nature and difficulty of the tasks also have an influence on the daily fluctuations of children's intellectual activity. And perhaps, as it was found in adults by Monk and Leng (1986) "... while the circadian performance rhythm of a simple repetitive task (manual dexterity) ran at the period of the deep temperature oscillator, that of a logical reasoning task exhibited significant periodicities at both temperature oscillator and sleep-wake cycle periodicities, with the latter being the more predominant."

## Evolution of the Duration of Night Sleep During Childhood and the Self-regulation Phenomena It Reveals

Two studies have thrown light on the temporal evolution of the duration of night sleep and on the regulations which can be observed in accordance with the individual development and environmental factors. The first study was carried out with children aged from $2^{1}/_{2}$ to $5^{1}/_{2}$ years attending eight kindergartens (Koch et al. 1984; Soussignan et al. 1985). The second study was carried out with children from five classes in the 4th year of primary school (CM1: *cours moyen 1ère année,* 9–10 years of age) and five classes in the

5th year of primary school (CM2: *cours moyen 2ème année*, 10–11 years of age) in five different primary schools,[1] and six classes from the 1st year (11–12 years of age), 2nd year (12–13 years of age), 3rd year (13–14 years of age) and 4th year (14–15 years of age) high-school students from three different high schools and eight classes from the 5th year (15–16 years of age), 6th year (16–17 years of age) and 7th year students (17–18 years of age) in two different high schools.

It was under the guidance of our research group that the schoolteachers from the ten classes of CM1 and CM2 organized the collection of data from the children.

The schoolteachers, all of whom were volunteers, were brought together before the beginning of the study in order to make the aims of the study clear to them, to familiarize them with the documents to be distributed among all the teachers involved in the study, the parents and the youngsters themselves, and also to give them a detailed account of the grids' contents to be completed with and by the children. Every morning and on 2 consecutive weeks per month, each child from each class in the sample filled in the line corresponding to the day of the week. He or she thus recorded: the estimated time of his/her falling asleep; the time of his/her waking up while specifying whether the latter was provoked or spontaneous; periods of drowsiness that might have occurred during the previous day; possible nightmares and night wakings during the previous night, and so on.

The children's weight and height were measured every 2 weeks, either by the teacher or by the school nurse.

The method used in the high schools was approximately the same, the data collection being organized in both cases by Natural Science or Physical Education teachers. Weight and height were measured by the school nurse. In each school, meetings were organized including the Principal, the Headmaster or Headmistress, the teachers involved in the study (and sometimes all the teachers working in the school), the parent's representatives and all the students from the classes concerned.

During each meeting the main aims of the study were made clear, namely,
1. To acquire a better knowledge of the evolution of the sleep duration according to age and puberty changes.

2. To obtain more information on the fluctuations of the sleep duration according to the day of the week, the month and changes in climatic conditions.
3. To increase our knowledge of the influence of the falling asleep time and waking up time on the sleep duration.
4. To undertake an investigation into the possible correlations between the sleep duration and the family features, and particularly the working day arrangement of both parents.
5. To carry out an analysis of the evolution of the sleep duration for each student from the month of October to the month of June.

On the whole, 14 261 sleep durations were processed with regard to 14 children in each section. The introduction after 2300 data had been processed of a new set of data yielded by the study of other students did not alter the results already obtained from the factorial analysis of correspondence, variance analyses, or correlation tests.

From studies carried out in some holiday camps, it emerged that, with regard to the falling asleep time of youngsters aged 10–14 and 14–17 years, there is never more than a 10-min variation between the time reported by an adult observer and that estimated by the youngsters themselves. It can thus be considered that, with regard to our study, there may be a 10-min variation between the effective falling asleep time and that reported by the students.

The following conclusions were drawn:
1. There is relatively little difference between the duration of night sleep of 10- to 11-year-olds (last year of primary school; Table 1) and that of children from 2 to 5 years of age in the kindergarten (Koch et al. 1984). The children studied came from the same region (the suburbs of Besançon, a town in the east of France), the climatic conditions were the same and the families belonged to the same culture (the sample population only included children of European ethnic origin) and to the same socioeconomic categories. For kindergarten children the mean duration of night sleep was 667 min (Koch et al. 1984; Fig. 1) and for children in their last year of primary school it was 619 min (Table 1). Thus, the decrease in night sleep did not exceed 10 min/year between the mean age at kindergarten (3–4 years) and the mean age of the last year at primary school (approximately 10.5 years). The major difference between the two populations resided in the existence of a nap with a mean duration of 83 min at the beginning of the afternoon in kinder-

[1] In France there are 5 years of primary school and 7 years of high school.

**Table 1.** Statistical data showing the duration of night sleep (in minutes) of 59 children in the 5th year of primary school (10–11 years of age) (Note: in France there are only 5 years of primary school)

|  | No. of observations | Mean (min) | Standard deviation | Standard error | Median (min) | Maximum values (min) | Minimum values (min) |
|---|---|---|---|---|---|---|---|
| **Days of the week** | | | | | | | |
| Sunday | 279 | 614.2 | 60.48 | 3.62 | 610.8 | 780 | 225 |
| Monday | 279 | 609 | 63.16 | 3.78 | 604.8 | 900 | 270 |
| Tuesday | 279 | 630.1 | 75.40 | 4.51 | 631.2 | 840 | 325 |
| Wednesday | 279 | 618.4 | 49.41 | 2.95 | 619.9 | 780 | 440 |
| Thursday | 270 | 612.9 | 59 | 3.53 | 612.8 | 810 | 390 |
| Friday | 279 | 610 | 70.11 | 4.19 | 607.5 | 780 | 375 |
| Saturday | 279 | 636.3 | 88.20 | 5.28 | 635.2 | 860 | 295 |
| **Months of the year** | | | | | | | |
| October | 413 | 616.9 | 63.33 | 3.11 | 617.5 | 840 | 225 |
| November | 266 | 626.3 | 72.73 | 4.45 | 626.6 | 860 | 325 |
| December | 371 | 625 | 63.49 | 3.29 | 625.4 | 840 | 320 |
| January | 378 | 627.8 | 73.46 | 3.77 | 628.5 | 900 | 270 |
| February | 273 | 605.2 | 61.94 | 3.74 | 603.8 | 840 | 420 |
| March | – | – | – | – | – | – | – |
| April | 168 | 596.3 | 65.17 | 5.02 | 596.4 | 810 | 420 |
| May | 84 | 623.4 | 80.34 | 8.76 | 624 | 811 | 295 |
| **Type of arousal** | | | | | | | |
| Spontaneous arousal | 1105 | 621.6 | 73.22 | 2.20 | 619.6 | 900 | 225 |
| Provoked arousal | 848 | 614.9 | 60.76 | 2.08 | 615.3 | 860 | 375 |
| Total | 1953 | 618.7 | 68.15 | 1.54 | 617.5 | 900 | 225 |

**Table 2.** Statistical data showing the duration of night sleep (in minutes) of 68 children in the 1st year of high school in three different high schools (11–12 years of age)

|  | No. of observations | Mean (min) | Standard deviation | Standard error | Median (min) | Maximum values (min) | Minimum values (min) |
|---|---|---|---|---|---|---|---|
| **Days of the week** | | | | | | | |
| Sunday | 188 | 611.3 | 55.77 | 4.06 | 620.2 | 780 | 450 |
| Monday | 188 | 584.3 | 59.54 | 4.34 | 591.9 | 765 | 420 |
| Tuesday | 188 | 646.2 | 87.73 | 6.39 | 658 | 900 | 375 |
| Wednesday | 188 | 605.2 | 68.38 | 4.98 | 604.3 | 795 | 235 |
| Thursday | 188 | 597.7 | 58.84 | 4.29 | 599.6 | 855 | 375 |
| Friday | 188 | 584.1 | 85.18 | 6.21 | 576.1 | 851 | 210 |
| Saturday | 187 | 653.9 | 102.3 | 7.47 | 648.2 | 920 | 210 |
| **Months of the year** | | | | | | | |
| October | 476 | 592.1 | 68.06 | 3.11 | 594.9 | 920 | 210 |
| November | 363 | 621.7 | 84 | 4.40 | 622.2 | 895 | 235 |
| December | 112 | 627.8 | 70.42 | 6.65 | 624.9 | 840 | 440 |
| January | 119 | 617.9 | 78.77 | 7.22 | 613.1 | 840 | 375 |
| February | 112 | 644.9 | 101.9 | 9.63 | 609 | 900 | 210 |
| March | 28 | 644.2 | 63.72 | 12.04 | 644.3 | 853 | 540 |
| April | 63 | 601.1 | 87.06 | 10.89 | 606.7 | 805 | 415 |
| May | 42 | 595.4 | 69.79 | 10.77 | 588.1 | 755 | 420 |
| **Type of arousal** | | | | | | | |
| Spontaneous arousal | 847 | 623.1 | 81.86 | 2.81 | 623.5 | 900 | 210 |
| Provoked arousal | 468 | 591.3 | 71.95 | 3.32 | 602.8 | 920 | 270 |
| Total | 1315 | 611.8 | 79.92 | 2.20 | 613.3 | 920 | 210 |

**Table 3.** Statistical data showing the duration of night sleep (in minutes) of 60 children in the 2nd year of high school in three different high schools (12–13 years of age)

| | No. of observations | Mean (min) | Standard deviation | Standard error | Median (min) | Maximum values (min) | Minimum values (min) |
|---|---|---|---|---|---|---|---|
| **Days of the week** | | | | | | | |
| Sunday | 240 | 603.8 | 59.98 | 3.87 | 601.6 | 810 | 450 |
| Monday | 240 | 553.9 | 57.61 | 3.71 | 563.8 | 750 | 420 |
| Tuesday | 240 | 636.9 | 70.16 | 4.52 | 635.2 | 840 | 475 |
| Wednesday | 240 | 570.8 | 49.37 | 3.18 | 575.4 | 750 | 405 |
| Thursday | 240 | 574.6 | 53.65 | 3.46 | 576.3 | 720 | 380 |
| Friday | 240 | 553.3 | 79.18 | 5.11 | 558.8 | 740 | 315 |
| Saturday | 238 | 637.7 | 89.94 | 5.83 | 640.8 | 890 | 340 |
| **Months of the year** | | | | | | | |
| October | 420 | 585 | 71.57 | 3.49 | 583.8 | 840 | 340 |
| November | 301 | 598.3 | 82.57 | 4.75 | 598.6 | 890 | 315 |
| December | 182 | 597.4 | 77.73 | 5.76 | 601.9 | 810 | 380 |
| January | 168 | 586.1 | 67.67 | 5.22 | 588.5 | 840 | 405 |
| February | 168 | 597.9 | 65.37 | 5.04 | 600 | 760 | 435 |
| March | 180 | 591.3 | 74.63 | 5.56 | 595.6 | 840 | 340 |
| April | 161 | 578.1 | 68.56 | 5.40 | 580.1 | 780 | 375 |
| May | 98 | 584 | 91.89 | 9.28 | 596.5 | 840 | 390 |
| **Type of arousal** | | | | | | | |
| Spontaneous arousal | 728 | 611.1 | 79.72 | 2.95 | 602.6 | 890 | 315 |
| Provoked arousal | 950 | 574 | 66.72 | 2.16 | 583.6 | 840 | 340 |
| Total | 1678 | 590.1 | 74.92 | 1.82 | 593.1 | 890 | 315 |

garten (Koch et al. 1984; Soussignan et al. 1985; Fig. 1).

2. There was no significant difference in the overall average of the duration of night sleep between the children in the last year of primary school (10–11 years), whose mean duration of night sleep was 619 min, and the children in the 1st year of high school, whose mean duration of night sleep was 612 min, despite the changes in routine and school rhythms that the passage from primary school to high school brings about. It can be noted, however (Tables 1, 2), that the duration of night sleep of 1st-year high school students is shorter, on the average, than that of primary school children and that it is so 5 days out of 7, but it is longer during the nights from Tuesday to Wednesday and from Saturday to Sunday, i.e., the nights when the children do not have school the next day (French children do not go to school on Wednesdays: see below). However, there was a significant decrease in the duration of night sleep from the 1st to the 2nd year of high school (Tables 2, 3; Fig. 7), and from the 2nd to the 3rd year of high school (Fig. 7). The mean duration of night sleep went from approximately 612 min in the 1st year to 590 min in the 2nd year and 567 min in the 3rd year of high school. This represents a decrease of 45 min in 2 years, i.e., much more than in the previous years.

As no other particular modifications were observed in the daily routine and work rhythms of the parents of the children or in the timetable, space or school activities in the three high schools, from the 1st to the 4th year of high school, the major cause of the sudden

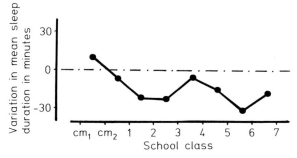

**Fig. 7.** Variation in the mean duration of nightsleep in minutes in children and adolescents attending the last two classes of primary school [the 4th year CM1 (*Cours moyen 1ère année*): 9- to 10-year-olds; the 5th year CM2 (*Cours moyen 2ème année*): 10- to 11-year-olds (in France, there are 5 years of primary school and 7 years of high school)], the first 4 years of high school (the 1st year: 11- to 12-year-olds; the 2nd year: 12- to 13-year-olds; the 3rd year: 13- to 14-year-olds; the 4th year: 14- to 15-year-olds) and the last 3 years of high school (the 5th year: 15- to 16-year-olds; the 6th year: 16- to 17-year-olds and the 7th or final year: 17- to 18-year-olds). For example, there is a decrease of 22 min in the mean duration of night sleep from the 1st to the 2nd year and 23 min from the 2nd to the 3rd year of high school

decrease in the duration of night sleep must be looked for in the modifications in the child's individual development which are specific to the age categories in the 2nd (12–13 years) and 3rd (13–14 years) years. This is the time when a majority of children reach the age of puberty. Thus, it can be suggested that the onset of puberty is characterized, in particular, by a sudden decrease in the duration of night sleep. The data for each adolescent, from October to June, and information concerning the onset of puberty obtained directly and confidentially from each adolescent confirmed this hypothesis.

From the 5th to the 6th year of high school there was another large decrease, i.e., 32 min in the duration of night sleep. As 5th- and 6th-year students, unlike during the previous years, have school on Wednesday mornings, it can be suggested that this decrease is related to the new routine and requirements of the 6th year which has an increased work load and prepares the students for their matriculation or university entrance examination at the end of the final or 7th year at high school.

3. High school students have a self-regulating capacity in relation to the duration of their night sleep provided that their home and the school environments enable them to do so. When students accumulate a deficit in the duration of their sleep most of them tend to go to sleep earlier and to wake up later if their home routine allows them to do so and their school timetable does not systematically require them to go to school early. Thus, when high school students know that on the next day they do not have to go to school until 0900 or 1000 hours and not at 0800 or 0900 hours as is usually the case in France, most of them tend to go to sleep earlier and wake up later, thus increasing the duration of their sleep sometimes by 2–3 h. These data would thus suggest that the self-regulating sleep capacities are directly dependent on the plans that each student makes to arrange his or her timetable for the following day, the conditions of his home life, the attitudes of each family in relation to sleep and the capacities of the school system, all of which have to be favorable to make these self-regulations possible.

It can be assumed that the differences observed from one day to the next during the week mirror

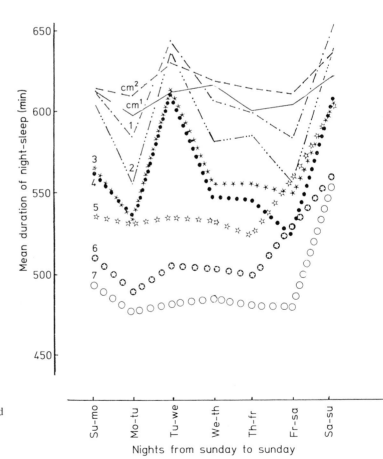

**Fig. 8.** Variation of the mean duration of night sleep from Sunday *(Su–Mo)* to Saturday *(Sa-Su)* throughout the week in children and adolescents attending the last two classes in primary school (4th CM1 and 5th CM2 classes) and the seven classes of high school: 1st, 2nd, 3rd, 4th, 5th, 6th and 7th years

these self-regulating capacities (Fig. 8). For example, it is observed that when Wednesday is a holiday (up until the 4th year of high school), there is a significant increase in the duration of sleep during the night from Tuesday to Wednesday. Moreover, a comparison between Table 1 (last year of primary school) and Table 2 (1st year of high school) shows that high school students clearly increase the duration of their sleep on Tuesday nights more than the primary school pupils do, as if they were spontaneously compensating for the sum of the higher deficits during the other nights.

The increase in the duration of sleep on Tuesday night appears to occur more generally from the last year of primary school to the 4th year of high school in relation to a sudden decrease in the duration of sleep during the previous night. Thus, if we wish to encourage the capacities of children to self-regulate the duration of their sleep, especially if they sleep a lot (see above), it would appear logical to maintain Wednesday as a holiday, at least while the amount of time spent on school activities during the day is not decreased and the arrangement of the daily timetable is not changed.

The longest night is certainly Saturday night for classes from the second last year of primary school to the 7th year of high school even if its duration can often be compared with that of Tuesday night when Wednesday is a school holiday. This is the time when, in almost all the classes, the standard deviation and the maximum values are the highest. But it is also at this time that the lowest values, or those that are among the lowest in relation to the other nights of the week, are observed (lowest minimum values for the second last year of primary school, 240 min; 1st year high school, 210 min; 4th year high school, 210 min; 5th year high school, 180 min; minimum values among the lowest in the last year of primary school, 295 min; 2nd year high school, 340 min; 3rd year high school, 270 min). These values can be compared with the routine of some families and/or some students on Saturday night, i.e., to go to sleep late. However, as has been emphasized, Saturday night enables most of the students from the same schools to compensate, at least partially, for the deficits accumulated during the previous nights, i.e., in the students from the second last class of primary school to the 7th year of high school. Here again, as for Tuesday night for the students in the second last class in the primary school to the 4th year at high school, the increase in the duration of sleep during Saturday night appears to be proportional to the decrease in the duration of sleep during the previous night.

The shortest nights are Monday and Friday. Monday could represent the breaking of the home routine in relation to the passage from the home environment on Sunday to the school environment on Monday (Montagner 1978, 1983; Montagner et al. 1978, 1982), and for Friday night the effect of the fatigue accumulated during the school week and in particular for a school day in France which is among the longest in the world (6 h of classwork in the primary school; 7 h in the first 4 years of high school; 7–8 h in the last 3 years of high school).

4. All our experiments confirm the existence of enormous individual differences in children and adolescents in relation to the duration of night sleep in all school classes and in all age categories (Tables 1–3).

For example, with children in the last year of primary school, the maximum duration of sleep is 900 min when arousal is spontaneous and 860 min when it is provoked, while the shortest durations are 225 and 375 min, respectively. The mean duration of sleep is 618.7 min and more than half of the children have a mean duration that is lower than 600 min; a majority of children sleep 13–15 h on Tuesday and Saturday nights.

## Conclusions

We are beginning to perceive when and how the sleep-wakefulness rhythm, the behavioral and biological daily fluctuations and the variations of the child's intellectual activity are structured during the schooldays in relation to other days, and how they evolve according to age, individual characteristics and environmental factors.

1. A majority of children have a high yawning frequency between 0900 and 0930 hours either when they are attending the kindergarten or 5–6 months later in the first class of the primary school. This constitutes a sign of a relatively low level of vigilance and attention processes in relation to the outside world. Blake (1967) and Testu (1982, 1986, 1989) have demonstrated that at this time of the schoolday most of the intellectual performances which have been measured in children attending primary schools are very often significantly lower than those at 1000 or 1100 hours. The data obtained would suggest that there is a relationship between firstly nonvigilance (or drowsiness), and attention capacity and the low

performances observed at the beginning of the schoolday, and secondly the accumulated deficit of sleep in children. Another relationship could exist between those phenomena and a periodical decrease of arousal due to an ultradian rhythm of wake-arousal related to that of falling asleep in infants (Fig.5). Thus, it would appear essential to take into account the sleep-wakefulness or sleep-arousal rhythms of pupils (and also the sleep-wakefulness and work rhythms of parents) when attempting to re-arrange the schoolday. However, the temporal organization of primary school activities seems to influence also the beginning of school time as 68% of the children yawn at 0900 hours as compared to 53% in the last year of kindergarten.

2. A majority of children also yawn between 1430 hours and 1500 hours in the 1st year of primary school when, without a transition period after several years of attending a kindergarten, the children are faced with the strict requirements of the school curriculum. They must focus their attention and their cognitive processes to master the three basic skills of reading, writing and arithmetic often in continuation with the morning learning tasks. It is also at around 1400 hours that the heart rate of all children in the first class of primary school shows a mean increase of more than 10%. Besides the fact that yawning is not an indication of increased vigilance, the increase in heart rate is often considered to be a defense reaction and a rejection of information (Lacey 1967; Graham and Clifton 1966). At the same time, a decrease in heart rate would be considered as related to an orientation reaction and an acquisition of information. Also, Hellbrügge et al. (1959) observed that the beginning of the afternoon is characterized by a large increase in the percentage of children who are asleep or sleepy in the three age categories: 2–6 years (about 80%), 6–11 years (about 80%) and 11–15 years (about 40%). Many chronobiological data are consistent with these findings and show that the beginning of the afternoon is a period of reduced vigilance and/or increased sleepiness in all age categories including that of adults. They are in keeping with Blake's (1971) and Testu's (1982, 1986, 1989) works on children attending the primary school (and the highschool for Testu): the school performances which have been tested, decrease at the beginning of the afternoon (1400 hours) in relation to 1100 and 1600 hours. Thus, it seems questionable to maintain at the beginning of the afternoon demanding school tasks such as paying attention and the so-called cognitive tasks.

A comparison with data obtained in the last year of kindergarten with the same children reinforces this question. At the beginning of the afternoon the proportion of yawning children is only 35% from 1430 to 1500 hours (as compared to 68% in primary school) and as the heart rate peaks around 1500 hours this increase is smaller than 5% in relation to 1400 and 1100 hours. However, at this time of the day the kindergarten children can lie down or do relaxing activities which are spontaneously selected or accepted without hesitation.

Associated with these findings is the fact that short-term memory appears to be at a favorable level in the middle of the first half of the schoolday in the morning, and that the best time for long-term memory is the middle of the second half of the schoolday in the afternoon (Folkard 1975, 1980).

It was also found that behavioral instability (children who stand up and then sit down and who turn their eyes away from the teacher) increases significantly between 1030 and 1100 hours and between 1545 and 1645 hours for children in the 1st year of primary school, i.e., 2–2$^1/_2$ h after the beginning of each half schoolday (Soussignan et al. 1988).

All these data would suggest reducing the duration of each schoolday and concentrating the school activities into two periods of 2 h each when the vigilance and attention capacities and intellectual performance appear to be at their highest levels; one in the morning between 0900–0930 and 1100–1130 hours and the other between 1500 and 1700–1730 hours. The period from 1100–1130 to 1500 hours could be a free subject time for the children and take place in different environments.

3. The recording of information on the sleep-wakefulness rhythm of children belonging to different age groups and attending the different classes of primary schools and high schools reveals self-regulating processes in the duration of night sleep from one day to the next. This kind of process could help children and adolescents to compensate, at least partially, for their deficits of sleep and cope more efficiently with the demands and temporal organization of tasks within the schoolday throughout the week.

However, as there are important differences from one child to another in the sleep-wakefulness rhythm, biological rhythms and variations in intellectual tasks from one hour to the next, it would be important to have more data available on the correlations and influences which exist between these variables and how self-regulating processes as well as environmental factors interact to allow the different children to adapt to school activities.

Therefore, it would be useful to keep in mind that many developmental processes are slow, especially in young children, and that important changes which occur in growth, physiology and psychology at certain periods of the individual development can have marked influences on attention capacities, efficiency in intellectual activity and school tasks and more generally on adaptation processes of children and adolescents.

However, our knowledge of the child's and adolescent's circadian and ultradian rhythms must be increased from the kindergarten to university levels to try and define even more precisely at what age, under what influences and how the intellectual, psychological and physiological variables fluctuate at every stage of individual development. Methodological progress has now made it possible to continuously record behavioral variables and in particular those which precede, accompany or follow the different types of school activities and the physiological variables in most different children. These studies can be carried out by pluridisciplinary teams of researchers (ethologists, physiologists, psychologists, pediatricians, child psychiatrists, sociologists, linguists, etc.) working with the parents, teachers and other co-educators. This type of research would lead to a more detailed knowledge of the alternations in attention and inattention, intellectual and relational availability and nonavailability, memory capacity and difficulties, relaxation activities and school performances throughout the day according to the individual and family characteristics of each child.

*Acknowledgement.* We would like to thank Franck Dugoul for his help with the translation of this paper.

# References

Altmann J (1974) Observational study of behavior: sampling methods. Behaviour 49: 227–267

Anders TF (1982) Annotation. Neurophysiological studies of sleep in infants and children. J Child Psychol Psychiatry 23: 75–83

Anders TF, Keener M (1985) The developmental course of nighttime sleep-wake patterns in full-term and premature infants during the first year of life. Sleep 8: 173–192

Anders TF, Keener M (1987) Developmental course of nighttime sleep-wake patterns in full-term and premature infants during the first year of life. I. In: Guilleminault C (ed) Sleep and its disorders in children. Raven, New York, pp 173–192

Anders TF, Emde R, Parmelee A (1971) A manual of standardized terminology, techniques and criteria for scoring states of sleep and wakefulness in newborn infants. Brain Research Institute, Los Angeles (UCLA Brain Information Service)

Basler K, Largo RH, Molinari L (1980) Die Entwicklung des Schlafverhaltens in den ersten fünf Lebensjahren. Helv Paediatr Acta 35: 211–223

Benoit O (1981) Le rythme veille-sommeil chez l'enfant. I. Physiologie. Arch Fr Pediatr 38: 619–626

Benoit O (ed) (1992) Le sommeil humain, Masson, Paris

Billewicz WZ, Kemsley WF, Thompson AM (1962) Indices of adiposity. J Prev Soc Med 16: 183–188

Blake MJF (1967) Time of day effects on performance in a range of tasks. Psychon Sci 9: 349–350

Blake MJF (1971) Temperament and time of day. In: Colquhoun WP (ed) Biological rhythms and human performance. Academic, London, pp 109–148

Dreyfus-Brisach C (1967) Ontogenèse du sommeil chez le prématuré humain: études polygraphiques à partir de 24 semaines d'âge conceptionnel. In: Minkowski A (ed) Regional maturation of the brain in early life. Blackwell, Oxford, pp 437–457

Folkard S (1975) Diurnal variation in logical reasoning. Br J Psychol 66: 1–8

Folkard S (1980) A note on "Time of day effects in school children's immediate and delayed recall of meaningful material" – the influence of the importance of the information tested. Br J Psychol 71: 95–97

Folkard S, Monk TH, Bradbury R, Rosenthall J (1977) Time of the day effects in school children's immediate and delayed recall of meaningful material. Br J Psychol 68: 45–50

Graham FK, Clifton RK (1966) Heart-rate change as a component of the orienting response. Psychol Bull 65: 305–320

Guilleminault C (1987) Sleep and its disorders in children. Raven, New York

Harper RM, Hoppenbrouwers T, Sterman MB, McGinty DJ, Hodgman J (1976) Polygraphic studies of normal infants during the first six months of life. I. Heart rate and variability as a function of state. Pediatr Res 10: 945–951

Harper RM, Leake B, Miyahara L, Mason J, Hoppenbrouwers T, Sterman MB, Hodgman J (1981) Temporal sequencing in sleep and waking states during the first 6 months of life. Exp Neurol 72: 294–307

Hellbrügge T (1960) The development of circadian rhythms in infants. Cold Spring Harbor Symp Quant Biol 25: 311–323

Hellbrügge T (1968) Ontogenèse des rythmes circadiens chez l'enfant. In: Cycles biologiques et psychiatrie. Symposium Bel Air, vol 3. Masson, Paris, pp 159–183

Hellbrügge T (1977) Physiologische Zeitgestalten in der kindlichen Entwicklung. Nova Acta Leopold 46: 365–387

Hellbrügge T, Lange J, Rutenfranz J (1959) Schlafen und Wachen in der kindlichen Entwicklung. Untersuchungen über die zeitlichen und tageszeitlichen Verschiebungen des Schlafes. Enke, Stuttgart (Beihefte zum Archiv für Kinderheilkunde, vol 39)

Hellbrügge T, Pechstein J, Ullner R, Reindl K (1967) Zum Verständnis der Periodik-Analyse in der Medizin. Fortschr Med 85: 289–295

Hoppenbrouwers T, Harper RM, Hodgman JE, Sterman MB, McGinty DJ (1978) Polygraphic studies of normal infants during the first six months of life. II. Respiratory rate and variability as a function of state. Pediatr Res 12: 120–125

Hoppenbrouwers T, Hodgman J, Harper RM, Sterman MB

(1982) Temporal distribution of sleep states, somatic and autonomic activity during the first half year of life. Sleep 5: 131–144

Klackenberg G (1971) The development of children in a Swedish urban community. A prospective longitudinal study. VI. The sleep behaviour of children up to three years of age. Acta Paediatr Scand [Suppl] 187: 105–121

Kleitman N (1963) Sleep and wakefulness. University of Chicago Press, Chicago

Kleitman N, Engelman TG (1953) Sleep characteristics in infants. J Appl Physiol 6: 269–282

Koch P, Soussignan R, Montagner H (1984) New data on the wake-sleep rhythm of children aged from $2^1/_2$ to $4^1/_2$ years. Acta Paediatr Scand 73: 667–673

Koch P, Montagner H, Soussignan R (1987) Variation of behavioural and physiological variables in children attending kindergarten and primary school. Chronobiol Int 4: 525–535

Kronauer RE, Czeisler CA, Pilato SF, Moore-Ede MC, Weitzman ED (1982) Mathematical model of the human circadian system with two interacting oscillators. Am J Physiol 242: R3–R17

Lacey JI (1967) Somatic response patterning and stress: some revisions of activation theory. In: Appley MH, Trumbull R (eds) Psychological stress. Appleton-Century-Crofts, New York, pp 14–42

Mills JN (1975) Development of circadian rhythms in infancy. Chronobiologia 2: 363–371

Monk TH (1987) Subjective ratings of sleepiness – the underlying circadian mechanisms. Sleep 10: 343–353

Monk TH, Leng VC (1986) Interactions between interindividual and inter-task differences in the diurnal variation of human performance. Chronobiol Int 3: 171–177

Monk TH, Leng VC, Folkard S, Weitzman ED (1983) Circadian rhythms in subjective alertness and core body temperature. Chronobiologia 10: 49–55

Montagner H (1978) L'enfant et la communication. Stock, Paris

Montagner H (1983) Les rythmes de l'enfant et de l'adolescent. Stock, Paris

Montagner H, Henry JC, Lombardot M, Benedini M, Burnod J, Nicolas RM (1978) Behavioural profiles and corticosteroid excretion rhythms in young children. II. Circadian and weekly rhythms in corticosteroid excretion levels of children as indicators of adaptation to social contexts. In: Reynolds V, Blurton Jones NG (eds) Human behaviour and adaptation. Taylor and Francis, London, pp 229–265

Montagner H, Restoin A, Henry JC (1982) Biological defense rhythms, stress and communication in children. In: Hartup WW (ed) Review of child development research vol 6. Chicago University Press, Chicago, pp 291–319

Navelet Y (1984) Développement du rythme veille-sommeil chez l'enfant. In: Benoit O (ed) Physiologie du sommeil. Masson, Paris, pp 127–141

Navelet Y (1989) L'enfant insomniaque. Rev Prat 39: 26–30

Navelet Y, Benoit O, Bouard G (1982) Nocturnal sleep organization during the first months of life. Electroencephalogr Clin Neurophysiol 54: 71–78

Parmelee A (1961) Sleep patterns in infancy: a study of one infant from birth to eight months of age. Acta Paediatr 50: 160–170

Rutenfranz J, Colquhoun WP (1979) Circadian rhythms in human performance. Scand J Work Environ Health 5: 167–177

Rutenfranz J, Hellbrügge T (1957) Über Tagesschwankungen der Rechengeschwindigkeit bei 11-jährigen Kindern. Z Kinderheilkd 80: 65–82

Soussignan R, Koch P, Montagner H (1985) Contribution à l'étude de l'évolution temporelle du sommeil de nuit et du sommeil de sieste chez de jeunes enfants. C R Acad Sci 300: 359–362

Soussignan R, Koch P, Montagner H (1988) Behavioural and cardiovascular changes in children moving from kindergarten to primary school. J Child Psychol Psychiatry 29: 321–333

Subash Babu D, Chuttani CS (1979) Anthropometric indices independent of age for nutritional assessment in school children. J Epidemiol Community Health 33: 177–179

Testu F (1982) Les variations journalières et hebdomadaires de l'activité intellectuelle de l'élève. Monographics Françaises de Psychologie Vol 59. CNRS, Paris

Testu F (1986) Diurnal variations of performances and information processing. Chronobiologia 13: 319–328

Testu F (1989) Chronopsychologie et rythmes scolaires. Etude expérimentale des variations journalières et hebdomadaires de l'activité intellectuelle de l'élève. Masson, Paris

Voors AW, Foster TA, Frerichs RR, Webber LS, Berenson GS (1976) Studies of blood pressures in children, age 5–14 years, in a total biracial community. Circulation 54: 319–327

Wever R (1975) The circadian multi-oscillator system of man. Int J Chronobiol 3: 19–55

Wever R (1979) The circadian systems of man. Springer, Berlin, Heidelberg, New York

# Chronobiology of Physical Performance and Sports Medicine

C. M. Winget, M. R. I. Soliman, D. C. Holley, and J. S. Meylor

## Introduction

If a soccer coach wants his team to execute several new and exacting plays for next week's championship game, the science of chronobiology suggests that the best time to introduce the plays is 3 in the afternoon. In fact, almost everyone can expect a better athletic performance if workouts and qualifying events are held between 12 noon and 2100 hours (Winget et al. 1985). This expectation is based on the fact that many of the physiological determinants of physical performance, from the activity of metabolic pathways to the creation of cognitive thought, vary in the course of 24 h (see Table 1). The quantitative study of biological phenomena that fluctuate over time (i. e., chronobiology) has helped to define and identify times when the sum of these physiological mechanisms should be functioning optimally. Recommendations based on these findings (and discussed in this chapter) hold wide implications to space flight, agriculture, military training, teaching, business, medicine, psychiatry and other fields besides competitive athletics which require "peak" mental and physical performance. This chapter will also review the impact of abnormal or desynchronized biological rhythms on physical performance and outline countermeasures to deal with possible performance decrements.

The earliest recognition of daily repeated cycles in physiological function and their effect on physical performance may have antedated recorded history; the first scientific reports concerning biological oscillations appeared in the literature over 300 years ago. Daily changes in body weight and dimension were reported by Sanctorius in 1647 (Lusk 1933). He also reported a 30-day cycle in urine turbidity. By the latter half of the last century, it was recognized by some physicians that there were characteristic daily variations in a number of physiological parameters; one well-known phenomenon was the rhythmic rise and fall of body temperature each day. A study in the late

1800s (Masso 1887) reported that such a rhythm persisted on its usual time schedule for at least several days despite reversal of activity patterns from working in the day and sleeping at night to sleeping during the day and working at night.

Circadian (Latin for "about a day") rhythms or daily rhythms are terms given for the regularly occurring daily biological rhythms, some times called biological clocks. These terms should not be confused with the popular "biorhythm" craze that postulates a 23-day physical, 28-day emotional and 33-day intellectual cycle beginning with a person's date of birth (Thommen 1973). The biorhythm model is oversim-

**Table 1.** Circadian rhythms and athletic performance (comprehensive review in Winget et al. 1985)

Circadian rhythmic components of athletic performance
    Sensorimotor: simple reaction time
    Psychomotor: hand/eye coordination
    Sensory perceptual: pain threshold
    Cognitive: information processing
    Neuromuscular: strength
    Psychological
        Affective: mood
        Psychophysiological: arousal
    Cardiovascular: heart rate
    Metabolic: body temperature, resting $O_2$ consumption
    Aerobic capacity: $VO_2max$

Factors which can influence circadian variation in athletic performance
    Physical workload
    Psychological "stressors"
    Motivation

Circadian chronotype (morningness/eveningness) and personality
    Social interaction
    Light and temperature
    Performance proficiency and training
    Sleep
    Postlunch "dip" in performance
    Altitude
    Dietary constituents and meal timing
    Age and gender
    Altered sleep/wake schedules and transmeridian flight

plified and does not accommodate unique fluctuations in the myriad physiological systems that contribute to overall physical performance nor does it incorporate the concept of rhythm rephasal. The time taken for a particular physiological function to oscillate through a single cycle is known as its period. The spectrum of biological periodicity ranges from ultradian rhythms with periods of less than 20 h to infradian rhythms with periods greater than 28 h. Most physiological functions are known to exhibit circadian rhythmicity and a growing body of literature indicates that performance ability is correlated with these physiological cycles. It is for this reason that a careful use of chronobiological principles is viewed by many athletic directors as the ultimate weapon to give their aspiring gold medalists that "winning edge".

Most athletes or dancers, who perform magnificently in the afternoon or evening, would not welcome the prospect of having to play at 0300 hours. Instead, many athletic performers (e.g., marksmen, archers) like to postpone fine and exacting tasks until afternoon when muscular coordination and strength, not to mention timing and acuity, are optimal (Winget et al. 1985).

Evidence of biological rhythmicity has been observed in all eukaryotes, from unicellular organisms to humans. Postulated causes of these rhythms are: (1) oscillations dependent on rhythmic impulses from the environment (exogenous) or (2) self-sustained internal oscillations (endogenous). The daily rise and fall of mammalian body temperature is a classic example of the latter type. Conversely, the daily rhythms that govern an organism's physiology and performance seem to be cued by periodic alterations in the environment called "zeitgebers" (German for "time giver"). Circadian rhythms for humans are primarily synchronized by the light/dark cycle, as well as by everyday social contacts and interaction among people (Wever 1979; Holley et al. 1981). Research expanding on this concept indicates that the biological clock in humans can now be manipulated, moved forward or backward, by properly using broad-spectrum high-intensity white light (Czeisler et al. 1986). Theoretically, an athlete can train in one country for a future sporting event any place in the world by manipulation of the light/dark cycle for the training period.

A poor competitive performance may result when an athlete does not take into consideration his circadian performance profile. An athletic performance that occurs several hours before or after the circadian peak "window" is potentially subject to less than op-

timal performance efficiency. Also, as will be discussed later in this chapter, the sensitivity of the body to drugs such as ethanol, caffeine or medications is quite different at one time of day versus another, and this may then influence subsequent performance (Walker et al. 1981).

Taking circadian rhythms into consideration can produce major benefits in tasks involving endurance, mental function, physical strength and others. Selecting the best circadian time for a performance can mean as much as a 10% difference in how well an athlete or others may perform. A 10% decrement in peak performance can be compared to trying to do one's best with less than 3 h of sleep (Folkard and Monk 1983), drinking the legal limit of alcohol (0.09% in blood alcohol) (Folkard and Monk 1983) or taking 500 mg hexobarbital (Klein et al. 1967). For example, long-term memory recall, in which data must be retained for 1 week or longer, is 8% higher when the material is presented at 1500 h than when it is presented at 0900 h (Folkard et al. 1977). The implications here, for example, are for the timing of coaching instructions and strategy, since the 8% difference in memory retention is similar to the performance decrement induced when sleep is restricted to 3 h.

Another important factor involving circadian rhythm is the personality "chronotype": whether one is a morning person ("lark") who gets up early and goes to bed early, or an evening person ("owl"), who wakes up late and retires late. While there is a difference of about 65 min in the body temperature rhythm peaks between morning and evening types, morning types secrete significantly more epinephrine in the morning than evening types do. Furthermore, the timing of mood and activity rhythms differs by several hours between distinct morning and evening types (Winget et al. 1985). However, Hill et al. (1988) found that the diurnal variations in most responses to exercise are the same for both morning types and evening types.

Two aspects of chronobiology that are important to physical performance will be discussed further. The first is the normal cyclic variation observed in the physiological mechanisms that contribute to overall athletic performance. The other is the effect of abnormal rhythmicity or desynchronization on athletic performance and the countermeasures to improve performance which would normally be degraded by this.

## Circadian Components of Physiology and Athletic Performance

Most physiological functions exhibit circadian rhythmicity: maximum and minimum function occur at specific times of day. In humans, circadian rhythms are expressed as oscillations in physiological processes (e.g., body temperature, heart rate, hormone levels) which are responsive either to internal (e.g., neurotransmitters, electrolytes, or metabolic substrates) or external (e.g., environmental factors, drugs, food, or stressors) stimuli.

A number of excellent general reviews document the myriad studies of the rhythmical nature of our physiology and behavior (Aschoff et al. 1982; Brown and Graeber 1982; Conroy and Mills 1970; Minors and Waterhouse 1981 b; Moore-Ede et al. 1982). It is now widely accepted that most, if not all, parameters when examined with high sample frequency will show rhythmicity. In 1985 we compiled an extensive listing of work dealing with human circadian rhythms and athletic performance (Winget et al. 1985). Table 1 summarizes the many factors that must be considered when evaluating this rather complex subject.

### Circadian variability of Physiology and Behavior

Diurnal variation in athletic performance has been reported for numerous sports. For example, better evening performance has been reported for swimming, running, shot-putters and rowing crews (see Table 2). We have compiled a graphic summary of studies which show the circadian nature of several physiological and behavioral components that contribute to athletic performance. Figure 1 is a phase map of these relationships. The circadian rhythm indices presented in Fig. 1 represent a review of data from several studies (Winget et al. 1985). These circadian indices represent either the peak time (the clock time at which the circadian rhythm reaches maximum value) or acrophase (estimated peak time from harmonic analysis) depending on the method employed in each study.

### Performance (Nonathletic)

We have identified a number of performances that can be measured in the laboratory which, we feel, contribute to athletic performance. These include short-term memory, neuromuscular coordination and

**Table 2.** Cyclic variation in athletic performance and in selected physiological variables that contribute to athletic performance; peak time estimates are given for each of the variables (comprehensive review in Winget et al. 1985)

| Variable | Peak time estimate (hours) |
|---|---|
| Performance | |
| Vigilance | 1900 |
| Cognition | 1800 |
| Manual dexterity | 1600 |
| Neuromuscular coordination | 1500 |
| Athletic performance | |
| Swimming | 2200 |
| Running | 1900 |
| Shot put | 1700 |
| Rowing | 1700 |
| Physiological | |
| Cortisol (plasma) | 0700 |
| Epinephrine (plasma) | 1300 |
| Heart rate | 1800 |
| Body temperature | 1600 |
| Sweat rate | 1300 |
| Respiration rate | 1400 |
| Heat production | 1500 |

manual dexterity, reaction time, vigilance or monitoring performance, and cognitive ability. All of these show circadian rhythmicity and peak around 1400–2000 hours with lows occurring between 0300 and 0600 hours (see Fig. 1). It is interesting to note that many performances are correlated to the core body temperature circadian rhythm. It may be possible for an individual to ascertain subjective peak performance time by establishing his/her own temperature profile using a simple oral thermometer and plotting hourly (during awake time) oral temperature versus time over several days. Theoretical peak performance should correlate with core temperature ±3 h.

Daily variations in the efficiency of performance have long been recognized. Klein et al. (1976) reported circadian cycles of performance in psychomotor, symbol cancellation, reaction time, and digit summation tasks in human subjects. These variations tended to correspond with rhythmic changes in body temperature. Fort et al. (1973) found this relationship was present whether temperature changes were part of the natural circadian rhythm or were artificially induced. Behavioral performance rhythms can be modified by a number of factors such as disposition, practice, motivation, personality (introverts versus extroverts), sleep deprivation and work shifts (Klein et al. 1976).

Folkard (1975) found that speed in performing two logical reasoning tasks improved markedly from 0800 to 1400 hours and fell off fairly rapidly there-

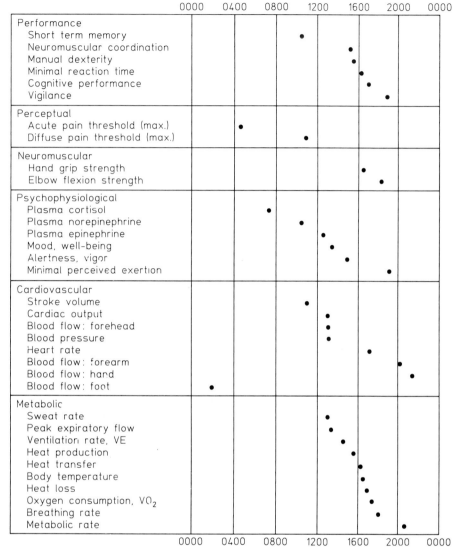

Fig. 1. Circadian rhythm acrophases or peak times in several components contributing to athletic performance at rest. Peak time estimates are shown as *dots*. In studies in which peak time was expressed as a time range, the midpoint was selected as an approximation of mean peak time. In cases in which data from multiple studies were available for a given variable, acrophases or peak times were averaged to provide a single peak time estimate

after. The accuracy decreased linearly over the day. It was concluded that previous work relating performance to time of day actually found functions resulting in differences in task demands rather than from individual differences. The suggestion was made that the higher the memory or articulatory component of the task, the earlier in the day the performance on such a task peaks. It was assumed that the logical tasks employed involved both "short-term" memory and immediate processing.

Buck (1977) conducted a study utilizing a step-input pursuit tracking task which confirmed Folkard's results. Performance varied with the time of day in a manner and to an extent dependent on the choice of index so that circadian rhythms for speed scores were in inverse phase with those for accuracy score. Increased short-term memory demands disrupted the movement-time rhythm, indicating that psychomotor and short-term memory functions vary in inverse phase with the time of day, with short-term memory peaking earlier in the day.

Rutenfranz and Colquhoun (1979) suggested that performance rhythms are closely connected to the cycle of "sleep need." Blake (1967) found circadian performance variance in vigilance, card sorting, reaction time, serial reaction, letter cancellation, and calculation tests. All exhibited the "late peaking" phenomenon which may be due to a final surge of motivation or an interaction between practice effect and time of day.

Rutenfranz and Colquhuon (1979) also purpose that there are two types of tasks, immediate information processing tasks and storage-of-information tasks. The former are roughly in phase with body temperature rhythm (perhaps coincidentally) and the latter are out of phase with this rhythm. A coach may take advantage of the time of day effect in relation to long-term memory and information processing by giving complex coaching instructions in the afternoon rather than in the early morning (Winget et al. 1985).

## Neuromuscular

Stockton et al. (cited in Reilly 1990) reported that of 17 motor performance measures related to actual "sports" performance, 12 showed typical circadian variability. These included hand grip strength, vertical jump performance, measures of muscular coordination, and running. Hand grip strength and elbow flexion strength demonstrate circadian rhythmicity with time of peak between 1400 and 1830 h (see Fig. 1). Furthermore, there appears to be separate distinct circadian rhythms for the dominant and nondominant hand, each driven by a different oscillator (Reinberg et al. 1988). Subjects performing the standing broad jump also have been found to demonstrate time of day proficiency (Reilly and Down 1986).

## Psychophysiological

We include rhythmic hormonal changes, mood/affective variability, arousal/depression cycles, and sensory/perceptual parameters (pain threshold) in the psychophysiological category (see Fig. 1). It should be noted that one or more central circadian oscillators have been postulated to compose the "biological clock" mechanism in man. A currently popular theory holds that two central oscillators prevail, one which is correlated with the body temperature circadian rhythm and another that is correlated with the daily sleep/wake cycle (Moore-Ede and Czeisler 1984). The latter also has an ultradian component with a period of about 90 min. When considering the various types of performances or parameters that fall within the psychophysiological category one must realize that the rhythmic aspect of each may be influenced by one or more of these central mechanisms.

Several reports indicate that subjective "perception" of effort varies over the period of a day with evening being the time of least effort or strain to perform a given task (Faria and Drummond 1982; Reilly and Hales 1988; Wilby et al. 1987; Carton and Rhodes 1985). This may hold significance for scheduling training sessions or team practices requiring maximal efforts (Hill et al. 1989). Also, pain perception is cyclic with the greatest tolerance being later in the day (Procacci et al. 1974; Strempel 1977).

## Cardiovascular (Resting)

Cardiovascular parameters displaying circadian rhythmicity include: stroke volume, heart rate, cardiac output, systemic blood pressure, capillary resistance, and blood flow to the various tissues (see Fig. 1). In a 1972 review (Beljan et al. 1972) of circadian variability of cardiovascular parameters we found numerous published reports including those on circadian rhythmicity of: pulse rate, ECG, blood pressure, blood flow, erythrocytes, leukocytes, clotting components, hematocrit, plasma volume, specific gravity, viscosity, and pH.

## Metabolic

The body temperature rhythm is one of the most reliable indicators of the phase of the internal biological clock mechanism. The correlations between oxygen consumption, metabolic rate, body temperature and biochemistry form the foundation for the study of bioenergetics. Circadian variability exists in parameters identified with these topics including: respiratory functions, i.e., rate, minute ventilation, respiratory quotient, peak expiratory flow, oxygen consumption and metabolic rate; whole body heat transfer and core-periphery heat exchange; sweating and cutaneous blood flow (Stephenson et al. 1984; Winget et al. 1985).

## Circadian Variation in Athletic Performance

### Response to Submaximal Ergonomic Demand

Several studies have examined the circadian aspects of submaximal exercise (Winget et al. 1985). Time of peak of a number of pertinent variables have been documented including: minute ventilation (VE), oxygen consumption ($VO_2$) and $O_2$ cost/unit work,

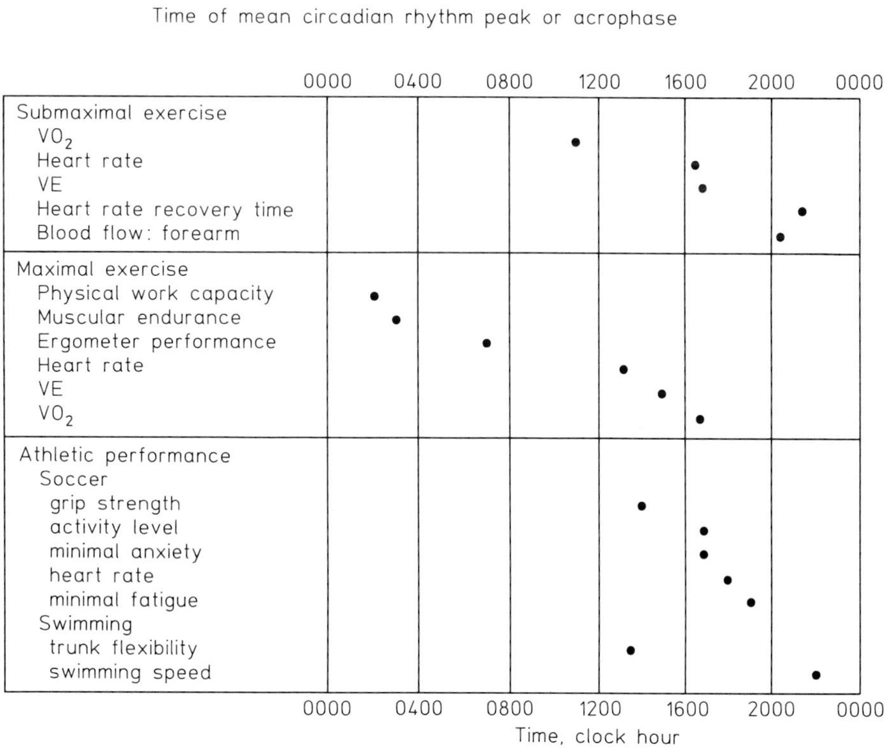

Time of mean circadian rhythm peak or acrophase

| | 0000 | 0400 | 0800 | 1200 | 1600 | 2000 | 0000 |
|---|---|---|---|---|---|---|---|
| **Submaximal exercise** | | | | | | | |
| VO₂ | | | | • | | | |
| Heart rate | | | | | • | | |
| VE | | | | | • | | |
| Heart rate recovery time | | | | | | • | |
| Blood flow: forearm | | | | | | • | |
| **Maximal exercise** | | | | | | | |
| Physical work capacity | • | | | | | | |
| Muscular endurance | | • | | | | | |
| Ergometer performance | | | • | | | | |
| Heart rate | | | | • | | | |
| VE | | | | • | | | |
| VO₂ | | | | | • | | |
| **Athletic performance** | | | | | | | |
| Soccer | | | | | | | |
|   grip strength | | | | • | | | |
|   activity level | | | | | • | | |
|   minimal anxiety | | | | | • | | |
|   heart rate | | | | | • | | |
|   minimal fatigue | | | | | • | | |
| Swimming | | | | | | | |
|   trunk flexibility | | | | • | | | |
|   swimming speed | | | | | | • | |

| | 0000 | 0400 | 0800 | 1200 | 1600 | 2000 | 0000 |

Time, clock hour

**Fig. 2.** Circadian rhythm acrophases or peak times in several components of athletic performance during exercise and in variables measured during athletic performance. Peak time estimates are shown as *dots.* In studies in which peak time was expressed as a time range, the midpoint was selected as an approximation of mean peak time. In cases in which data from multiple studies were available for a given variable, acrophases or peak times were averaged to provide a single peak time estimate

heart rate and recovery heart rate, muscle blood flow, forearm blood flow, forearm minimum vasodilation threshold, minimum sweating threshold, and rectal temperature. As with many of the physiological parameters listed above, these peak in the late afternoon (see Fig. 2).

Response to Maximal Ergonomic Demand

Because of its intuitive appeal as an indicator of overall athletic performance the VO₂ at maximal exercise has been studied by many investigators hoping to define optimal times for training and event scheduling. Unfortunately, these studies are conflicting for a variety of reasons (Winget et al. 1985). Other parameters showing circadian variability following maximal exercise include: heart rate and heart rate recovery time, work capacity, and muscle (finger) endurance (Winget et al. 1985).

Reilly (1990) has recently reviewed the investigations assessing VO₂max and makes several interesting observations. Among these is the estimation that, for a well-trained individual with a body weight of about 70 kg, the normal amplitude of the VO₂ rhythm at rest would be less than 0.5 % of the mean maximal value. He states that it would be difficult to

detect this against the background of biological variation and inherent measurement error associated with assessing VO₂max. In his review Reilly (1990) also deals with an interesting hypothesis that rhythms in vigorous exercise performance may result from a combination of motivation and psychological drive that may be reflected in anaerobic power output. Again, lack of tests with sufficient sensitivity have hampered efforts to establish a causal relationship.

Effects of Jet Lag on Athletic Performance, Including Rhythmic Aspects of Athletic Events

Rapid deployment in jet aircraft across time zones exposes the traveler to environmental cues at the destination which are shifted in time relative to the person's internal "biologic clock" which corresponds to home time. The biological clock and overt physiological and psychological rhythms subsequently rephase to the new environmental conditions. This transient period of rephasal results in circadian rhythm desynchrony (rhythms are temporarily out of phase), a condition referred to as desynchronosis. Often associated with rhythm desynchrony, many travelers experience a syndrome which is known popularly as "jet lag." The symptoms may include one or more of the following: irritability, disorientation or confusion, distorted estimation of time and/or distance, various aches and pains, digestive disorders including constipation, sleep disturbances (hypersomnia, insomnia, or multiple awakening), and decrements in mental and physical performance (Holley et al. 1981; Winget et al. 1984, 1985). Although a primary consequence of the syndrome is fatigue, research has been plagued by problems of definition, measurement, and interpretation of the meaning of "fatigue." It had been suggested that the term be replaced by more explicit physiological criteria, such as sleep loss or neural inhibition, by performance criteria such as performance decrement, or by psychological criteria such as impaired mood.

Wiley Post was the first to recognize the adverse effects of time zone displacement. In his historic circumnavigation of the globe (1931), he reportedly shifted his sleep-wake cycle prior to the trip so that he would be at maximum awake efficiency when he was on the long critical leg over the Soviet Union after many sleepless hours. Not until 1952 did Strughold publish the first scientific investigation dealing with jet lag. Transmeridian flight results in a phase shift between the circadian rhythms of the body, which are synchronized to departure time and the zeitgebers (external environmental synchronizers) at the destination. These flights result ultimately in a differential rephasing of the many oscillating body systems. Since many system components rephase at different rates, during that transient rephasal period internal rhythm desynchrony can exist and this is associated with the jet lag symptomology (see Winget et al. 1984).

Among flight crews, sleep disturbances were found in 59%–78% of the personnel on the first night following transmeridian flight, falling to 25%–30% with sleep disturbances on the third night postflight (Lavernhe et al. 1968; Raboutet et al. 1958). Up to 41% of the crew members experienced gastrointestinal problems. Female flight attendants had a significantly higher incidence of irregular menstrual cycles (Preston et al. 1973) and exacerbated dysmenorrhea (Schmidova 1966). Recently Suvanto et al. (1990), in a study of 342 male and female Finnish flight attendents flying transmeridian routes, found decreased sleep quality after eastward flights of ten time zones. Preston (1973) and co-workers (Preston et al. 1973) studied the sleep patterns of an aircrew operating on world wide operations with sleep logs and subjective assessment of fatigue. They reported that the aircrew consistently experienced acute sleep loss amounting to as much as 5–6 h per night flight. Cumulative sleep loss during one extended flight operation totaled 10–28 h over a 15-day period in five subjects. Physiological rhythms affected by external desynchronization include, for example, the sleep-wake cycle; body temperature; heart rate; respiration; arterial pressure; diuresis and excretion of potassium, sodium 17-hydroxycorticosteroid, and 17-ketosteroids in urine.

Most flight experiments reported in the literature indicate that reentrainment and recovery usually occur at a faster rate following westerly flights, which involve a phase delay, than after eastward flights (phase advance). For example, following a westbound flight from the Federal Republic of Germany to the United States, 3 days were required to reach 95% resynchronization of the psychomotor performance rhythm, whereas 8 days were required following the return flight in the easterly direction (Klein and Wegmann 1980). If the measures from seven flight studies (all involving flights in both directions) are averaged, then mean reentrainment shift rate for all variables is 92 min/day after a westbound flight, but only 57 min/day after an eastbound flight (Aschoff et al. 1975; Klein and Wegmann 1974). The actual rate of shift varies during the period of reentrainment. It is most rapid during the first 24 h and decreases thereafter exponentially. This phenomenon in which that time necessary for reentrainment differs for advance versus delay shifts is termed the "asymmetry effect."

Following transmeridian flight, shift-work scheduling, or unusual work-rest scheduling, there is considerable variation in the rate of readaptation between individuals for a given rhythm and between rhythms in a given individual. For example, studies have shown that 25%–30% of transmeridian travelers have little or no difficulty adjusting to the temporary circadian desynchronization and are asymptomatic; however, about an equal percentage do not adjust well at all. The individual factors associated with widely differing rates of rhythmic readaptation are: directional asymmetry (advance versus delay shift); zeitgeber strength (strong versus weak); rhythm stability (stable versus labile); rhythm amplitude (low versus high); behavioral traits (extroversion/introversion, neuroticism); rhythm chronotype (morning person versus evening person); age (young versus old); motivation (low versus high); sleep habits (rigid versus flexible); and performance task type (simple versus complex, high memory load versus low memory load). Several researchers suggest that a small rhythm amplitude for certain variables may serve as an index of an individual's ability to phase shift and adjust easily to new rest-activity schedules.

Subjects with low amplitudes in the body temperature rhythm show less resistance to phase shift; the rate of rephasal is more rapid in these individuals than in those with large circadian amplitudes (see Winget et al. 1984).

Athletes are intuitively aware of fluctuations in their muscle tone, endurance, mood, and various other factors that make practicing at one time of day preferable to other times. The time of day that athletes are at their "peak" is highly individualized and is dependent upon numerous physiological and psychological factors (as listed above). The fluctuation in mood, alertness/arousal, and energy levels reflect the dynamically changing internal environment of the body.

In the realm of international and transcontinental athletic competition, teams send their elite athletes to competitions at various sites around the world. When traveling to the 1988 Olympics in Seoul, Korea, the United States athletes crossed 8 to 11 time zones, depending on their departure point. For most performance rhythms that have been studied, the peaks occur somewhere between 1500 and 1800 hours (Winget et al. 1985). If competition was set for 1400 hours in Seoul and the United States athletes had not adapted to the new time zone (i.e., their systems were still functioning on home time), it would be equivalent to competing at about 0600 h home time. The athlete would be expected to turn in a world class performance at a time of his/her physiological low point. An important concern relative to athletic performance, chronobiology, and rephasal in the new time zone is: what happens to athletic performance during the transient period of internal rhythm desynchronization when the athlete is adapting to the new time zone? Is athletic performance compromised during this time of concurrent jet lag symptomology?

Jet lag may affect many aspects of performance both mental and physical (athletic). Several groups of investigators have used pilots and passengers as subjects, with tests designed for occupational realism (e.g., cockpit simulators). Following eastward flights of six to nine time zones, significant decreases in efficiency lasting from 1 to 5 days postflight were found in flight simulator performance ($-3\%$ to $-9\%$), psychomotor performance ($-1.8\%$ to $-3.5\%$), and hand-eye coordination ($-8\%$). Performance decrements in both simple tasks such as reaction time and the complex sensorimotor skills required to operate flight simulators were more severe following easterly flights than westerly flights. Simpler tasks recovered to baseline after a few days, but decrements in complex tasks persisted for up to 5 days postflight (see Winget et al. 1984).

A variety of performance tasks and indices of psychophysiological complaint were evaluated in a study of military personnel following an eastward transatlantic flight through six time zones (Wright et al. 1983). Following the flight 50% of the personnel reported fatigue and sleep difficulty and 40% reported subjective feelings of weakness. These symptoms decreased significantly by the 5th day postflight. Postflight performance deterioration was found in dynamic arm strength ($-6.1\%$ to $-10.8\%$), elbow flexor strength ($-13.3\%$), sprint times ($-8.4\%$ to $-12\%$), a lift and carry task ($-9.5\%$), logical reasoning ($-15\%$), and encoding/decoding performance.

There are two major aspects of chronobiology that are important to the athlete. The first is the cyclic variation in many physiological variables that contribute to athletic performance (see above). Diurnal variation in athletic performance had been reported for numerous sports. For example, better evening performances have been reported for swimming, running, shot-putters, and rowing crews (Winget et al. 1985). The other important chronobiological aspect for the athlete is the effect that disruption of circadian rhythms (i.e., circadian desynchrony) has on athletic performance. This disruption commonly occurs following a transmeridian flight to a distant site for competition, and there is evidence that decrement in athletic performance may be yet another symptom of the jet lag syndrome.

In 1980, Sasaki conducted a series of surveys on the scores of international games held in Japan and related them to the length of stay (or the period of reentrainment) of the visiting team. The analysis revealed a systematic trend of recovery from jet lag. As the visiting team rephased to the new time zone, their level of performance increased as was evident by the outcome of their matches. Yaroslavtsev (1968) reported longer recovery periods after cardiovascular loading in athletes for the first few days following a 6-h time zone change, with complete recovery requiring 7–14 days. A time zone change of as little as 3 h has been shown to adversely affect athletes (Gingst 1970). On the 1 day in the new time zone, the majority of athletes complained of fatigue, apathy, sluggishness, reduced appetite, poor or interrupted sleep, and headaches. In addition, Gingst (1970) noted disturbances in cardiovascular and respiratory systems and shifts in metabolism and body temperature rhythms. In a separate study on elite athletes, rephasal after crossing seven time zones required 11–

15 days for the subject's subjective status, 21–22 days for average daily heart rate, and 28 days for the daily arterial pressure rhythm (Yezhov 1979). Antal (1975) reported that 28 sharpshooters who traversed 12 time zones (Britain to New Zealand) averaged about 8 days to achieve pretravel form, with some never fully recovering during the 2 weeks at the destination. From work with saber fencers Reinberg et al. (1985) suggest that athletes with internally synchronized rhythms may have a better chance of winning than athletes with internally desynchronized rhythms.

O'Connor and Morgan (1990), however, have recently reviewed the topic of athletic performance following rapid traversal of time zones and they evaluate critically the separate studies by Sasaki (1980), Antal (1975) and Wright et al. (1983). They argue that limited data exist dealing specifically with trained athletes and time zone shifts. It is their view that no compelling evidence exists demonstrating that air travel adversely influences "athletic" performance. Possible detrimental effects of time zone travel noted in some studies using untrained subjects should not be generalized to include "athletes." Krombholtz (1990) studied two groups of ten runners performing a 24-h relay race and found a significant circadian rhythm in performance of only one of the teams. He speculates that the difference may be related to the different levels of fitness of the two teams.

## Therapy for Desynchronosis

A number of treatments have been tested or suggested to alleviate the deleterious symptomatology associated with circadian rhythm desynchronosis. These treatments include exercise, pre-adaptation by altering sleep/wake cycles, altered meal timing and constituents, drug administration, relaxation techniques, electrosleep therapy and even acupuncture (Holley et al. 1981). Recent effort in the search for effective chronotherapeutic treatment of circadian desynchronosis has largely concentrated on the administration of drugs or manipulation of meal timing and constituents, or use of bright light.

It is possible that exercise can accelerate adaptation to new time zones. Although conclusive data in support of this are lacking in athletes, physical training intervention has been found to reduce work-dependent fatigue and musculoskeletal symptoms in shift workers (Harma et al. 1988a). It may also aid in adaptation to shift work (Harma et al. 1988b).

Altering bedtime for a few days before departure may lessen the disruptive effects of traveling across time zones. The changes in sleep schedules should correspond with the direction of the intended travel (advance or delay), so that the change of circadian rhythms is smooth. Since only the behavior and sleep-wake cycles are adjusted, the alterations in phase of the rhythms occur at a slower rate than in time zone transition (Reilly and Maskell 1988). Sleep, rather than meal times or social activity, is the main synchronizer of the rhythms; prolonged naps at the new location should be avoided as these would operate against adaptation by anchoring the rhythm at its previous phase (Minors and Waterhouse 1981a). Anchoring circadian rhythms is a tactic used by non-sports personnel regularly crossing time zones for short spells abroad; its use by athletes would be practical when they fly in for single contests. To "anchor" or keep one's circadian rhythms locked to home time, at least 4 h of sleep should be taken within the "window" of normal sleep based on the person's home time zone. This 4-h minimum is referred to as "anchor sleep."

Timed exposure to light also appears to offer a promising and exciting alternative for chronotherapeutic treatment of circadian rhythm desynchronosis. Czeisler et al. (1986) induced a 6-h phase shift in temperature rhythm of a human by exposure to 4 h of bright light (700–12000lux, equivalent to ambient outdoor light at dawn) at the appropriate circadian phase. This type of therapy, as do others, would require some knowledge of an individual's circadian phase at the time of exposure and whether rephasal would be speeded by phase advance or phase delay. It is well known in chronobiology that an intervention that has an effect on the central clock mechanism can do so by either advancing the clock or delaying the clock. This is known as the phase response characteristic or phase response curve (Moore-Ede et al. 1982).

Controlled or restricted access to food in mice can be made to oppose or augment the effects of a lighting regimen and is claimed to act as a true rhythm synchronizer (Nelson et al. 1975). In humans, the upset meal schedules resulting from transmeridian flight may be reflected in disturbances of the cycles of waste elimination, blood amino acids and other visceral activities (Beljan et al. 1972). However, meal timing in humans does not alter rhythms for simple task performance (Graeber et al. 1978).

Certain food constituents are claimed to have rhythm synchronizer characteristics (Ehret et al. 1978, 1980). Ehret and Scanlon (1983) proposed a dietary regimen which would accelerate adaptation following rhythm desynchronosis in shift workers and transmeridian travelers. This plan includes pre-flight alteration of light and heavy meals, alteration of the protein/carbohydrate constituent ratio in the diet and programmed use of tea (theophylline) and coffee (caffeine) to speed rhythm readaptation. Military personnel following a similar diet experienced fewer sleep disturbances and reported less subjective fatigue on the first two postflight days than did control group subjects (Graeber 1982). These investigators indicated that high-protein/low-carbohydrate breakfast and high-carbohydrate/low-protein dinner facilitate rephasal following advance phase shifts. A high-carbohydrate/low-protein meal increases insulin secretion, which facilitates the uptake of the amino acid tryptophan into the brain, since insulin release leads to large reduction in the blood levels of other large neutral amino acids that compete with tryptophan for brain uptake. Conversion of tryptophan to the neurotransmitter serotonin may then induce drowsiness and sleep. The high-protein/low-carbohydrate diet results in increased uptake of tyrosine, which is converted to the neurotransmitter norepinephrine, which may increase arousal levels (Ehret et al. 1980; Wurtman 1982). Although appropriate timing of food constituents may hasten the adaptation of circadian rhythms to the timing of environmental synchronizers at the athlete's destination, it may be difficult to implement due to limitations set by commercial airlines and/or restrictions of the training diet. Additionally, Moline et al. (1990) recently reported that this diet failed to produce significant changes in the rate of core temperature reintrainment or in mood or performance of 15 subjects exposed to a 6-h phase advance.

The administration of chronobiotic drugs has also been considered as a means to facilitate readaptation following phase shifts. Chronobiotic drugs are drugs which specifically affect some aspect of biological time structure. Chronobiotic drugs influence the duration of readaptation following phase shifts by acting as synchronizers forcing desynchronized physiological components into phase (Richter 1965) or by intensifying the effects of other synchronizers. Evidence has been presented to implicate the role of brain neurotransmitters in mediating the effects of chronobiotic drugs on circadian rhythms (Soliman and Walker 1986). To facilitate adaptation by shifting the phases of rhythms in the direction of new time zones, chronobiotic drugs must be administered at the right time of the day.

A numer of putative chronobiotic drugs have been evaluated over the last decade. The reports of trials have indicated varying degrees of efficacy. For example, barbiturates have been commonly prescribed to drive the sleep rhythm onto a new schedule following transmeridian flight (Simpson 1980). Recently, it has been suggested that short-acting benzodiazepines such as triazolam and midazolam could reset the circadian clock (Wee and Turek 1989). However, the use of short-acting benzodiazepines or hypnotics for promoting adaptation to jet lag is inadvisable, since the possible side effects on performance from such treatment have not been determined (De looy et al. 1988). Additionally there are reports that use of triazolam has been accomanied by bouts of anterograde amnesia (Morris and Estes 1987; Penetar et al. 1989). Drugs such as alcohol and caffeine are commonly used by aircrews in a nonchronobiotic fashion (not regulated by timed administration) to facilitate sleep or reduce symptoms of fatigue and increase alertness. Quiadon, a serotonindepleting tranquilizer, and corticosteroids were tested for chronobiotic suitability in a human phase shift study (Simpson et al. 1973) and transmeridian flight studies (Christie and Moore-Robinson 1970), but results are inconsistent and inconclusive.

Melatonin, a pineal gland hormone, has shown potential for alleviating jet lag. Experimental treatment with melatonin was found to accelerate resynchronization of the circadian rhythm in cortisol and melatonin after an 8-h eastward flight (Arendt et al. 1987). Jet lag symptoms were alleviated following melatonin administration, although no important differences were noted in oral temperature or performance data (Arendt et al. 1987). However, Petrie et al. (1989) indicate that melatonin can successfully reduce the symptoms of jet lag and speed up recovery time with no adverse side effects.

In general, although the administration of certain chronobiotic drugs may be useful in facilitating adjustment to time zone changes, they may not be appropriate for the competing athlete due to drug restrictions and the possibility of undesirable side effects. Further research is needed on the use of bright light, meal scheduling and constituents, development of safer chronobiotic drugs as well as proper dosage and timing of administration in order to find the most effective and safe way to facilitate adaptation following phase shifts.

## Conclusion

The quality of any physical or mental performance will ultimately depend upon the effective coordination of many physiological systems; most of these systems display circadian rhythmicity (e.g., metabolic rate, mood, hormone levels). The central circadian mechanism which dictates this biological periodicity may actually be a combination of several mechanisms, one correlating with the sleep/wake cycle and the other with body temperature cycles. The actual oscillations defined by these mechanisms can be altered by changes in cues derived from both internal and external stimuli. The quality of a particular physical or mental performance, therefore, will be dependent on the time of day and the degree to which relevant physiological rhythms are kept in phase. For instance, better performances can be expected in the morning for activities dependent on short-term memory, while afternoon is optimal for tasks involving long-term memory. Deviations from the optimal circadian "window" can lead to a significant decrement in peak performance. Proper consideration of pertinent circadian rhythms is important for all activities dependent on optimum performance.

Time shifts in environmental cues (e.g., light/dark cycle) due to transmeridian air travel can cause the periodicity of biological rhythms to shift out of phase. This desynchrony of circadian rhythms will inhibit the ability to achieve optimal physical and mental performance. Allowing time for reentrainment of circadian variables through systematic application of internal or external stimuli is recommended when optimal performance is critical. Reentrainment can be brought about by a number of treatments including exercise, sleep regimes, meal timing, and drug administration. The success of the recovery depends on a thorough knowledge of the types of treatments and the potential physiological impacts each can impart.

Much more work remains to be done to improve our understanding of how chronobiology relates to physical performance. Clearly, mechanisms for the circadian oscillators must be elucidated. In addition, improved definitions are required for "fitness" and "fatigue" so we can quantify how indices describing these concepts relate to the myriad physiological components displaying circadian rhythmicity.

*Acknowledgements.* We thank Dr. Paul X. Callahan for critical review of the manuscript and Ms. Jill Erickson, Ms. Sheli Jones, and Ms. Joann Stevenson for their technical assistance.

## References

Antal LC (1975) The effects of the changes of the circadian body rhythm on the sports shooter. Br J Sports Med 9: 9

Arendt J, Aldhous M, English J, Marks V, Arendt JH, Marks M, Folkard S (1987) Some effects of jet lag and their alleviation by melatonin. Ergonomics 30: 1379–1393

Aschoff J, Hoffman K, Pohl H, Wever R (1975) Re-entrainment of circadian rhythms after phase-shifts to the zeitgeber. Chronobiologia 2: 23–78

Aschoff J, Daan S, Groos GA (eds) (1982) Vertebrate circadian systems: structure and physiology. Springer, Berlin Heidelberg New York

Beljan JR, Rosenblatt LS, Hetherington NW, Lyman JL, Flaim ST, Dale GT, Holley DC (1972) Human performance in aviation environment. NASA Rep NAS2-6657, part I-A

Blake MJF (1967) Time of day effects on performance in a range of tastes. Psychon Sci 9: 349–350

Brown F, Graeber R (eds) (1982) Rhythmic aspects of behavior. Erlbaum, Hillsdale

Buck L (1977) Circadian rhythms in step-input pursuit tracking. Ergonomics 20: 19–31

Carton RL, Rhodes EC (1985) A critical review of the literature on rating scales for perceived exertion. Sports Med 2 (3): 198–222

Christie GA, Moore-Robinson M (1970) Project Pegasus: circadian rhythms and new aspects of corticosteroids. Clin Trials J 7: 7–135

Conroy RTWL, Mills JN (1970) Human circadian rhythms. Churchill, London

Czeisler CA, Allan JS, Strogatz SH, Ronda JM, Sanchez R, Riox CD, Fritag WO, Richardson GS, Kronauer RE (1986) Bright light resets the human circadian pacemaker independent of the timing of the sleep-wake cycle. Science 233: 667–671

De looy A, Minors D, Waterhouse J, Reilly T, Tunstall Pedoe D (1988) The coach's guide to competing abroad. National Coaching Foundation, Leeds

Ehret CF, Scanlon LW (1983) Overcoming jet lag. Berkeley, New York

Ehret CF, Groh KR, Meinert JC (1978) Circadian dyschronism and chronotypic ecophilia as factors in aging and longevity. Adv Exp Med Biol 108: 185–213

Ehret CF, Groh KR, Meinert JC (1980) Considerations of diet in alleviating jet lag. In: Scheving LE, Halberg F (eds) Chronobiology principles and applications to shifts in schedules. Sijthoff and Noordhoff, Rockville, pp 393–402

Faria IE, Drummond BJ (1982) Circadian changes in resting heart rate and body temperature, maximal oxygen consumption and perceived exertion. Ergonomics 25 (5): 381–386

Folkard S (1975) Diurnal variation in logical reasoning. Br J Psychol 66: 1–8

Folkard S, Monk TH (1983) Chronopsychology: circadian rhythms and human performance. In: Gale A, Edwards JA (eds) Attention and performance. Academic, New York, pp 57–78 (Physiological correlates of human behavior, vol 2)

Folkard S, Monk TH, Bradbury R, Rosenthal J (1977) Time of day effects in school children's immediate and delayed recall of meaningful material. Br J Psychol 68: 45–50

Fort A, Harrison MT, Mills JN (1973) Psychomotor performance in circadian rhythms and effect of raising body temperature. J Physiol 231: 114P–115P

Gingst V (1970) Functional displacement in the human organism during passage in place from a 3-hour displacement (in Russian). Teor Prakt Fiz Kul't 3: 39–40

Graeber RC (1982) Alterations in performance following rapid transmeridian flight. In: Brown FM, Graeber RC (eds) Rhythmic aspects of behavior. Erlbaum, Hillsdale, pp 173–212

Graeber RC, Gatty R, Halberg F, Levine H (1978) Human eating behavior: preferences, consumption patterns and biorhythms. US Army Natick Research and Development Command, Natick (Food sciences laboratory report, technical report TR-78-022)

Harma MI, Ilmarinen J, Knauth P, Rutenfranz J, Hanninen O (1988a) Physical training intervention in female shift workers. I. The effects of intervention on fitness, fatigue, sleep and psychomotor symptoms. Ergonomics 31: 39–50

Harma MI, Ilmarmen J, Knauth P, Rutenfranz J, Hanninen O (1988b) Physical training intervention in female shift workers. II. The effects of intervention on the circadian rhythms of alertness, short term memory, and body temperature. Ergonomics 31: 51–63

Hill DW, Cureton KJ, Collins MA, Grisham SC (1988) Diurnal variations in response to exercise of "morning types" and "evening types". J Sports Med Phys Fitness 28: 213–219

Hill DW, Cureton KJ, Collins MA (1989) Circadian specificity in exercise training. Ergonomics 32: 79–82

Holley DC, Winget CM, DeRoshia CW, Heinold MP, Edgar DM, Kinney NE, Langston SE, Markley CL, Anthony JA (1981) Effects of circadian rhythms phase alteration on physiological and psychological variables: implications to pilot performance. Ames Research Center, Moffett Field (NASA Technical Memorandum TM-81277)

Klein K, Wegmann H (1974) The resynchronization of human circadian rhythms after transmeridian flights as a result of flight direction and mode of activity. In: Scheving LE, et al. (eds) Chronobiology. Igaku-Shoin, Tokyo, pp 564–570

Klein K, Wegmann H (1980) Significance of circadian rhythms in aerospace operations. NATO, Neuilly-sur-Seine (Advisory Group for Aerospace Research and Development Report AGARD-AG-247, AGARD)

Klein KE, Bruner H, Wegmann HM, Bouszat P (1967) Die Veränderung der psychomotorischen Leistungsbereitschaft als Folge pharmakodynamischer Einwirkung verschiedener Substanzen mit potentiell sedierendem Effekt. Arzneimittelforschung 17: 1048–1051

Klein K, Wegmann H, Athanassenas G, Hohloweck H, Kuklinski P (1976) Air operations and circadian performance rhythms. Aviat Space Environ Med 47: 221–230

Krombholz H (1990) Circadian rhythm and performance during a 24-hr relay race. Percept Motor Skills 70: 603–607

Lavernhe J, LaFontaine E, Pasquet J (1968) Les réactions subjectives et objectives aux ruptures des rythmes circadiens lors des vols commerciaux longs courriers est-ouest et vice-versa. Rev Med Aeronaut Spat 7: 121–123

Lusk G (1933) Clio medica. Hoeber, New York, p 45

Masso V (1887) Recherches sur l'inversion des oscillations diurnes de la température chez l'homme normal. Arch Ital Biol 8: 177–185

Minors DS, Waterhouse JM (1981a) Anchor sleep as a synchronizer of abnormal routine. Int J Chronobiol 7: 165–188

Minors DS, Waterhouse JM (1981b) Circadian rhythms and the human. Wright, Bristol

Moline ML, Pollack CP, Wagner DR, Zendell S, Lester LS, Salter CA, Hirsch E (1990) Effects of the "jet lag" diet on the adjustment to phase advance. 2nd Meeting of the Society for Research on Biological Rhythms, May 9–13, Jacksonville

Moore-Ede MC, Czeisler CA (eds) (1984) Mathematical models of the circadian sleep-wake cycle. Raven, New York

Moore-Ede M, Sulzman F, Fuller C (1982) The clocks that time us: physiology of the circadian timing system. Harvard University Press, Cambridge

Morris HH, Estes ML (1987) Traveler's amnesia; transient global amnesia secondary to Triazolam. JAMA 258: 945–946

Nelson W, Scheving L, Halberg F (1975) Circadian rhythms in mice fed a single daily meal at different stages of lighting regimen. J Nutr 105: 171–184

O'Connor PJ, Morgan WP (1990) Athletic performance following rapid traversal of multiple time zones: a review. Sports Med 10: 20–30

Penetar DM, Belenky G, Garrigan JJ, Redmond DP (1989) Triazolam impairs learning and fails to improve sleep in a long-range aerial deployment. Aviat Space Environ Med 60: 498–594

Petrie K, Conaglen JV, Thompson L, Chamberlain K (1989) Effect of melatonin on jet lag after long haul flights. Br Med J 298: 705–707

Preston FS (1973) Further sleep problems in airline pilots on worldwide schedules. Aerospace Med 44: 775–782

Preston FS, Bateman SC, Short RV, Wilkinson RT (1973) Effects of flying and of time changes on menstrual cycle length and performance in airline stewardess. Aerospace Med 44: 438–443

Procacci P, della Corte M, Zoppi M, Maresca M (1974) Rhythmic changes of the cutaneous pain threshold in man. A general overview. Chronobiologia 1: 77–96

Raboutet J, Bousquet B, Granotier E, Angiboust R (1958) Trouble du sommeil et du rythme de vie chez le personnel navigant effectuant des vol à longue distance. Med Aeronaut 13: 311–322

Reilly T (1990) Human circadian rhythms and exercise. Critical Reviews in Biomedical Engineering 18: 165–180

Reilly T, Down A (1986) Circadian variation in the standing broad jump. Percept Motor Skills 62: 830

Reilly T, Hales AJ (1988) Effects of partial sleep deprivation on performance measures in females. In: Mega ED (ed) Contemporary ergonomics 1988. Taylor and Francis, London, pp 509–514

Reilly T, Maskell P (1988) Effects of altering the sleep-wake cycle in human circadian rhythms and motor performance. 1rst IOC Congress on Sport Sciences, Colorado Springs

Reinberg A, Proux S, Bartal JP, Levi F, Bicakova-Rocher A (1985) Circadian rhythms in competitive sabre fencers: internal desynchronization and performance. Chronobiol Int 2: 195–201

Reinberg A, Motohashi Y, Bourdeleau P, Andlauer P, Levi F, Bicakova-Rocher A (1988) Alteration of period and amplitude of circadian rhythms in shift workers. With special reference to temperature, right and left hand grip strength. Eur J Appl Physiol 57: 15–25

Richter CP (1965) Biological clocks in medicine and psychiatry. Thomas, Springfield

Rutenfranz J, Colquhuon P (1979) Circadian rhythms in performance. Scand J Work Environ Health 5: 167–177

Sasaki T (1980) Effect of jet lag on sports performance. In: Scheving LE, Halberg F (eds) Chronobiology: principles and applications to shifts in schedules. Sijthoff and Noordhoff, Rockville, pp 417–434

Shmidova VF (1966) The effect of high-altitude and high-speed flights on the functioning of air hostesses obstetric organs (in Russian). Gig Tr Prof Zabol 10: 55–57

Simpson HW (1980) Chronobiotics: selected agents of potential value in jet lag and other dyschronisms. In: Scheving LE, Halberg F (eds) Chronobiology: principles and applications to shifts and schedules. Sijthoff and Noordhoff, Rockville, pp 433–446

Simpson HW, Bellamy N, Halberg F (1973) Double blind trial of a possible chronobiotic (quiadon): field studies in N.W. Greenland. Int J Chronobiol 1: 287–311

Soliman MRI, Walker CA (1986) Effects of chronobiotic drugs on the cholinergic enzyme system of various rat brain regions. Annu Rev Chronopharmacol 3: 39

Stephenson LA, Wenger CB, O'Donovan BH, Nadel ER (1984) Circadian rhythm in sweating and cutaneous blood flow. Am J Physiol 246: R321–R324

Strempel H (1977) Circadian cycles of epicritic and protopathic pain threshold. J Interdiscip Cycle Res 8: 276–280

Suvanto S, Prtinen M, Harma M, Ilmarinen J (1990) Flight attendent's desynchronosis after rapid time zone changes. Aviat Space Environ Med 61: 543–547

Thommen GS (1973) Is this your day? How biorhythm helps you determine your life cycles. Crown, New York

Walker CA, Winget CM, Soliman KRA (eds) (1981) Chronopharmachology and chronotherapeutics. Florida A&M University Press, Tallahassee

Wee BE, Turek FW (1989) Midazolam, a short acting benzodiazepine, resets the circadian clock of the hamster. Pharmacol Biochem Behav 32: 901–906

Wever R (1979) The circadian system of man. Springer, Berlin Heidelberg New York

Wilby J, Linge K, Reilly T, Troup JDG (1987) Spinal shrinkage in females: circadian variation and the effect of circuit weight training. Ergonomics 30: 47–54

Winget CM, DeRoshia CW, Markley CL, Holley DC (1984) A review of human physiology and performance changes associated with desynchronosis of biological rhythms. Aviat Space Environ Med 55: 1085–1096

Winget CM, DeRoshia CW, Holley DC (1985) Circadian rhythms and athletic performance. Med Sci Sports Exerc 17: 498–516

Wright J, Vogel A, Sampson J, Knapik J, Patton J, Daniels W (1983) Effects of travel across time zones (jet lag) on exercise capacity and performance. Aviat Space Environ Med 54: 132–137

Wurtman RJ (1982) Nutrients that modify brain function. Sci Am 246: 50–59

Yaroslavtsev VL (1968) Diurnal cycle of physiological functions of man in conditions of permanent residence during long trips (in Russian). Eastern Siberia, Irkutsk

Yezhov SN (1979) Effect of long latitudinal flights on the functional state of athletes. State Medical Institute, RSFSR Ministry of Public Health, Tyumen

# Night and Shift Work and Transmeridian and Space Flights

A. E. Reinberg and M. H. Smolensky

## Introduction

Shift workers, including night workers in industry, hospitals, communication systems, surface transportation (railroad, bus and truck drivers), air transportation (pilots, flight attendants and passengers of transmeridian flights) are exposed to shift(s) of those environmental factors which synchronize human biologic rhythms. In this chapter we shall refer exclusively to a phase shift ($\Delta\Phi$) of synchronizers (or zeitgebers) equal to or larger than 5 h, whether they result from a transmeridian flight across five time zones (e.g., Paris to New York) or a shift from a daytime to a night time work schedule. This qualification is related to two important facts: (a) synchronizing effects of environmental clue and cue are identical whether or not the geographical location has changed (Halberg and Reinberg 1967; Reinberg and Smolensky 1983). In other words, chronobiologic problems are similar for both the air hostess with a transmeridian flight and the nurse shifting from a day to night schedule. (b) A single $\Delta\Phi \leqslant 5$ h of synchronizers has either minor or more frequently no detectable chronobiologic effects (Halberg and Reinberg 1967; Klein and Wegmann 1979; Gundel and Wegmann 1989). A 1-h $\Delta\Phi$ resulting, for example, transition to or from daylight saving time has no objectively demonstrable effect on either adults or schoolchildren (Reinberg et al. 1989a). Each organism exposed to a $\Delta\Phi \geqslant 5$ h of environmental factors has to resynchronize its biologic time structure, which in practical terms implies that all its biologic clocks have to be reset to the "new" local time or adjusted to the "new" timing of work hours. We call this process "circadian adjustment of rhythms."

Throughout history man has been confronted with problems of night work (Scherrer 1978). Modern industry with its technologic imperatives to maintain "continuous fire" (casts and steel industries, paper mills, oil refineries, power plants, etc.) is at the beginning of an increasing availability of shift work positions. Presently, in the developed countries, about 10% of the active workforce is involved in night and shift work. It is commonly postulated, without experimental and clinical evidence, that: (a) man is able to work and to rest at any time along the 24-h scale and (b) all human subjects react in the same manner when exposed to night work or to a transmeridian flight. However, from chronobiologic observations made to date it is no longer possible to state that anyone is able to perform any task at any time of day or night.

Therefore, several problems need to be considered and solved (biologic rhythm properties, significance and pertinence to shift work). We shall focus here on problems of synchronization. Other problems relate to ergonomy and occupational medicine including interindividual as well as group differences in tolerance to shift work.

Despite the fact that many scientific meetings with published proceedings (Colquhoun et al. 1975; Reinberg et al. 1981; Johnson et al. 1981; Haider et al. 1985), review papers (Aschoff et al. 1975; Aschoff 1978; Smolensky et al. 1985) and books (Reinberg 1979; Folkard and Monk 1985) have been devoted to the effects of night and shift work and transmeridian flights, decision makers seldom take into account biologic aspects when timing and schedules of work-rest cycles along the 24-h scale are discussed. In fact, employees, employers, unions, and management of industries and governments are making their decisions based mainly upon economic, social and/or political aspects rather than chronobiologic facts even though they are available and most pertinent.

Figure 1 summarizes what should be the multifactorial approach to tolerance to shift work. Persons adhering to shift work must adjust to a greater variety of circumstances than those adhering to habitual daytime work only. Several major factors affecting one's ability to comply with and tolerate shift work have to be considered. The factors have been designated for

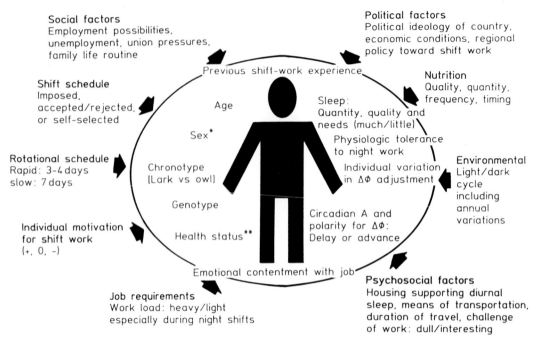

**Fig.1.** Major factors underlying tolerance to shift work. The *ring* delineates the interface between exogenous (external) and endogenous (internal) factors (Smolensky et al. 1985). In certain countries such as France, night and shift work is forbidden for women by law, except for nurses, airline stewardesses and aircraft personnel. * A difference between males and females for tolerance to shift work has not been documented experimentally. ** Personal and/or family history for certain diseases such as peptic ulcer, epilepsy, depressive illness, and diabetes

the purpose of discussion as either external (exogenous) or internal (endogenous) in nature (Fig.1). Only biologic and chronobiologic factors will be reviewed in this chapter.

## Synchronizers or Zeitgebers (Time Givers)

These are signals resulting from periodic changes of environmental factors. Most plant and animal species use the day/night alteration with a period $\tau = 24$ h to synchronize their circadian rhythms, pertinent signals being dawn and/or dusk. In a laboratory setting the artificial light/dark cycle is used to mimic day/night alteration. In human subjects, the prominent synchronizing effects relate to social factors. Our diurnal activity and nocturnal rest are calibrated by a set of imperative hours around which our societal life is built, e.g., work time. Without consciously knowning it we use a large variety of interhuman relationships and their timing to reset our internal clocks. The influence of social factors has been demonstrated by Apfelbaum et al. (1969) and Aschoff et al. (1974) in

experiments during which groups of subjects were isolated from environmental time cues and clues; their biologic rhythms desynchronized from a period of 24 h but all members of the same group exhibited the same circadian periodicity (e.g., with $\tau = 24.8$ h) resulting from interindividual interactions involving visual, auditive as well as physical signals.

However, light may have a synchronizing effect for human subjects particularly when its intensity reaches the rather high threshold level of 2500 lux at 1 m from the source. In comparison, 0.5 lux light intensity suffices to reset circadian oscillators of a hamster. The mechanism commonly accepted (Moore-Ede et al. 1982; Turek 1983; Arendt 1989) is that the light signal goes from the retina to the suprachiasmatic nuclei (SCN) via a retinohypothalamic pathway, then goes to the pineal through the cervical superior ganglia. The circadian rhythm of melatonin secretion by the pineal is under the control of the SCN acting as a primary oscillator. Melatonin secretion occurs only at night starting with lights-off and ending with lights-on. Therefore, a high plasma melatonin level might be the endocrine message corresponding to night time, at least for many mammalian species. In healthy human subjects a minimum of

2500 lux in light intensity is required to induce a dramatic fall in the nocturnal secretion of melatonin. Scheduled times of exposure to bright light (Daan and Lewy 1984) as well as the dosing time of melatonin pills (Arendt et al. 1986a,b) have been proposed to alleviate jet lag effects.

## Alterations of the Organism's Biologic Time Structure Resulting from Synchronizer Manipulations

Biologic rhythms persist when organisms are kept in a constant environment without known zeitgebers. In this free-running condition the period ($\tau$) of circadian rhythm differs from 24 h. Moreover, when isolation conditions are maintained for several weeks, the biologic circadian time structure of certain subjects can split into several components that free run with different frequencies, a state that is called *internal desynchronization* (Aschoff and Wever 1981). For example, the body core temperature rhythm may free run with a $\tau = 25$ h while the sleep-wake rhythm may free run with a $\tau = 33$ h. Obviously, in such a case the acrophase ($\varnothing$) of these endogenous rhythms will drift along the 24-h scale instead of stabilizing in one place.

To a certain extent, transmeridian and space flights mimic some aspects of isolation experiments and therefore favor the physiologic tendency of circadian rhythms to free run with 24 h. However, the major effect of shift work and transmeridian flights concerns the phase shift ($\Delta$) of the circadian acrophase ($\varnothing$) resulting from the phase shift ($\Delta$) of synchronized periodic signals ($\Phi$). For example, the phase delay (lengthened wakefulness) resulting from a Paris to New York flight with $\Delta\Phi = 5$ h is followed by a circadian acrophase delay with $\Delta\varnothing = 5$ h when adjustment is completed. Basically a phase shift $\Delta\Phi$ of zeitgebers is followed by a phase shift $\Delta\varnothing$ of the circadian acrophase of each oscillator, $\Delta\Phi$ and $\Delta\varnothing$s having the same magnitude (number of hours) and the same direction (phase delay or phase advance) when the subject has adjusted his or her time structure. The speed (or the duration) of an adjustment (span of time needed to get $\Delta\varnothing$s equal to $\Delta\Phi$) varies (1) from variable to variable in a given individual; (2) from one individual to another for a given variable; and (3) with the direction of the shift, advance or delay (Fig. 2):

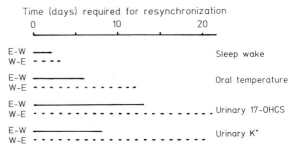

**Fig. 2.** Time (days) required for resynchronization ($\Delta\varnothing = 6$ h) after a transmeridian flight ($\Delta\Phi = 6$ h) from Paris to New York (phase delay) and New York to Paris (phase advance). It varies from function to function for a given subject (here a 43-year-old healthy human male) with regard to sleep wake cycle, oral temperature, urinary excretion of 17-OHCS, and K. It varies with the direction of flight. There is a slow adjustment after an eastbound flight (W-E; phase advance) and a rapid adjustment after a westbound flight (E-W; phase delay). (From Reinberg, unpublished data)

### Physiologic Variables

Some variables adjust rapidly, within 2–5 days, like the sleep-wake rhythms; others require more time like 4–7 days for body temperature, and 5–15 days and even more for adrenocortical activities and related circadian rhythms (plasma cortisol, plasma and urinary potassium, etc.) (Haus et al. 1968; Reinberg 1979) (Fig. 2).

### Subjects

With regard to the same variable, e.g., plasma cortisol, some subjects will adjust rapidly and others more slowly (Reinberg 1979).

### Direction of Shift

A phase delay (delayed ending of activity) resulting for instance from a Paris to New York flight is associated with a faster adjustment than a phase advance (New York to Paris flight). This is not a general rule since about 15%–20% of subjects adjust better and faster after a phase advance than a phase delay. Obviously, the effect of direction of $\Delta\Phi$ applies for shift workers as well.

Between the end of the zeitgeber $\Delta\Phi$ and full adjustment of the organism the time structure is altered with $\varnothing$s drifting at paces different from their former normal position to a new one. One has to deal then with a *transient desynchronization* of biologic rhythms. It does not persist as compared with internal desynchronization, a phenomenon which may persist for longer time spans irrespective of phase shifts.

## Clinical Symptoms of Intolerance to Shift Work

Clinical intolerance to shift work was defined (Andlauer et al. 1979; Reinberg et al. 1986) by the existence and intensity of a set of medical complaints.

1. *Sleep alterations* like poor sleep quality, difficulty in falling asleep (delayed sleep onset > 30 min when retiring), frequent awakenings, etc., documented by subjective self-rating and by objective means of evaluation (subjects must keep a sleep log for at least 1 week).
2. *Persisting fatigue* which does not disappear after sleep, weekends, days off and vacations, thus differing from physiologic fatigue due to physical and/or mental effort.
3. *Changes in behavior* consisting of unusual irritability, tantrums, malaise and feeling of being inadequate to perform.
4. *Digestive troubles* ranging from dyspepsia to epigastric pain and peptic ulcer objectivated by endoscopy. In fact, digestive troubles seem to be less frequent than they were 20 years ago.
5. *The regular use of sleeping pills* (barbiturates, benzodiazepines, phenothiazines, tranquillizers, etc.) is almost a pathognomonic indicator of intolerance to shift work which cannot be controlled or even reduced by these medications (Reinberg et al. 1988).

Symptoms 1, 2 and 5 are present in any intolerant subject. The two others (3 and 4) are less frequent. In addition, one symptom may predominate over the others. Practically, this means that both intensity and number of symptoms vary from subject to subject.

The occurrence of intolerance to shift work is not related to age, duration of shift work, type of industry or type of rotation (Andlauer et al. 1979; Reinberg et al. 1978b, 1986, 1988a). For example, a 25-year-old subject became intolerant after 1 year of shift work while a 56-year-old subject did so after having 30 years of shift work without any complaints or medical problems (Reinberg et al. 1988a).

The diagnosis of intolerance may and often should lead to a recommendation for the subject to be discharged from night shifts. This decision is usually made by occupational health physicians knowing the person, general working conditions and workplace situations and has to be made with his or her agreement but must not be postponed when symptoms are severe. In such a case full and rather rapid (within 1 year) recovery can be expected after the return to day work only without shifts. If the decision to transfer an intolerant shift worker is delayed, the subject may continue to suffer from sleep disturbances and persisting fatigue.

## Night Work as an Accident Risk Factor

For many years various risks associated with night work were considered to be mainly, if not exclusively, a personal problem. Special attention was given to tolerance of shift work, risk of accidents, risk of developing cardiovascular or gastrointestinal diseases and sleep disorders. In addition, however, since 1978 (Ehret 1981; Price and Holley 1981) catastrophies which occurred during night work (Three Mile Island, Chernobyl, plane and bus accidents) have stressed the security aspects for the night worker and his/her surroundings. Therefore, prevention of individual accidents and their possible consequences for the community have to be emphasized in chronobiologic research on shift work.

Circadian rhythms of psychophysiologic variables have been demonstrated in healthy adolescents and adults of both sexes. Results of tests performed around the clock such as reaction time, eye-hand skills, random addition, self-rated fatigue and drowsiness (visual analog scales), tracking, logical reasoning, letter cancellation, and performance of multiple tasks have shown that human subjects synchronized with diurnal activity and nocturnal rest are more efficient during daytime than at night (Reinberg 1979; Folkard and Monk 1985). Large-amplitude differences between diurnal peaks and nocturnal troughs in performance are observed. Moreover, these variations in performance are true circadian rhythms able to free run under constant environmental conditions and to be phase shifted after a $\Delta\Phi$ of synchronizers (Reinberg et al. 1978a; Monk et al. 1984). It is presently well documented that accident risk has a circadian rhythmicity with a prominent nocturnal peak around 0300 hours and one or two diurnal troughs with, in the latter case, a secondary diurnal peak around noon. This is the case for truck drivers (Harris 1977), locomotive engineers (Hildebrandt et al. 1974), bus drivers (Blom and Pokorni 1985), air fighter pilots (Rybak et al. 1983), etc. Sleep deprivation worsens the nocturnal decrease in performance. Åkerstedt and Fröberg (1976) demonstrated in healthy volunteers that circadian rhythm of body temperature, plasma catecholamines and performance (a) persisted during sleep deprivation but (b) performance oscillated around a rapidly decreasing slope while other physiologic variables exhibited no change in amplitude and 24-h adjusted mean. One has to keep in mind that day sleep corresponds to a sleep deprivation. Day sleep is always altered to some extent with regard to night sleep (Benoît 1984).

## Chronobiologic Optimization for Rotation Schedules of Shift Work

From a chronobiologic point of view a number of major items have to be taken into account to optimize shift work. They are (1) the speed of rotation, e. g., weekly rotation of shifts, rapid rotation (every 3–4 days), slow rotation (every 2–4 weeks); (2) the time at which night shift ends and morning shift starts; (3) the duration of shift (conventionally three shifts of 8 h each/24 h); and (4) phase delay and phase advance distribution according to night shift and days off.

### Speed of Rotation

In terms of tolerance to shift work subjective well-being, psychologic advantages as well as reduced alteration of the time structure, a rapid rotation system appears to be a better choice than weekly or slow rotation systems (Reinberg et al. 1979; Rutenfranz et al. 1976). In fact, a rapid rotation system does not allow sufficient time to adjust circadian rhythms during a night shift, a condition which facilitates the return to normal day work. There is no doubt that a rapid rotation aims to increase individual tolerance to night work. However, since it does not increase the nocturnal performance at the work place – because the subject does not phase shift his or her rhythms – this type of rotation is of no help in decreasing the nocturnal risk of accident(s). To achieve the latter it has been proposed (Ehret 1981) to use permanently rhythm-inverted night workers with diurnal sleep time. Despite its theoretical interest, such a proposal has been rejected by many employees (since both social and

family life are dramatically disturbed) as well as by many employers since it is difficult to prevent the temptation of a second job during day time which contributes to a decrease in performance during the night shift.

### Starting Time of Morning Shift

There is an optimal time to retire for sleep. It is around midnight with interindividual differences in our electrically illuminated society (Weitzman et al. 1979; Benoît 1992). The longer the sleep time is delayed, the poorer the quality of sleep with regard to both subjective quality and alteration of EEG sleep patterns. In addition, too-early awakening may have, for many subjects, adverse effects similar to those of sleep deprivation. In one of our studies (Reinberg 1979) involving oil refinery operators, end of night shift and start of morning shift was 0500 hours. This timing favored those workers ending the night shift but penalized, in terms of sleep deprivation, those starting the morning shift since they had to set their alarm clock at 0330 hours to reach the refinery on time. Our proposal to start the morning shift at 0600 hours (instead of 0500 hours) was accepted and after several months was considered to give a critical improvement in shift work conditions (Reinberg 1979).

### Duration of Shift

Traditionally the 24-h span has been divided into three shifts of 8 h duration. The 8-h shift is still the most common and fits with regulations of work duration with regard to both daily and weekly schedules. A rather odd system called $12 \times 12$ (12 h of work alternating with 12 h of rest for 2–4 days) has recently been introduced (Johnson et al. 1981). Despite its short rotation system, this schedule does not fit with the chronobiologic requirements including prevention of both individual and collective risk of accidents. In fact, one of the most interesting schedules (moreover, a time-honored one) is that of the navy ward in western European countries. The night from 2000 to 0800 hours is split into three sections of 4 h each. For a given individual the 4 h shift moves from day to day (e. g., 2000–0000, 0000–0400 and 0400–0800 hours). Sleep deprivation as well as $\Delta \varnothing$s is thus limited. In addition, being allowed a night's sleep sailors reduce their $\Delta \varnothing$ as demonstrated by Minors and Waterhouse (1981). A few hours of night sleep contribute »anchor« ∅s of biologic rhythms and prevent desynchronization. Unfortunately, this rather good system (which is nevertheless far from perfect) can be used only on a ship with a confined population; for city-dwelling workers this system is unbearable since it disturbs both family and social life.

## Distribution of Shifts During Rotations

The rationale of optimizing shift work schedules is to use procedures that minimize circadian rhythm alteration involving direction of shift and position of off-days. Proposals have been formulated (Aschoff 1978; Rutenfranz et al. 1976) to favor phase delay with a certain order of consecutive shifts and to position off-days after certain phase advance and/or night shifts. The availability of such programs is of major interest.

## Comments

Chronobiologic requirements have to be integrated with other aspects summarized in Fig. 1.

In order to get rid of emotional reactions associated with the adoption of shift work schedules in industries, Knauth and Rutenfranz (1982) proposed the following criteria to help make a decision in choosing the design of a shiftwork system: (1) good physiologic tolerance to shiftwork (this aspect will be discussed in a separate section); (2) determinations of hours of best and worst performance; (3) peak and trough times at which individual and group accidents occur; (4) comparative satisfaction of workers who experienced several systems of rotation; and (5) state of health and social problems with regard to a rotation system. From the point of view of this last criterion a rapid rotation is to be preferred whenever possible.

The type of work must be taken into account with specified particulars. This is not as obvious as it would appear. Let us take as an example the night work of nurses. Work load, mental pressure, and physical constraints depend upon the hospital department (intensive care units or head and neck surgery cannot be compared with rather more quiet departments) as well as the physical exercise involved, e. g., due to the architecture of the building. This means that a standard schedule for night- and shift-working nurses is not feasible since it does take into account interindividual differences, type of responsibility or local constraints.

Besides shift work allowing time programming since a certain regularity is predictable (e. g., in most industries: post, police, hospitals, air control) some shift workers such as pilots, flight attendants, and bus and locomotive drivers are faced with irregular night work schedules. This poses additional safety problems since irregularity is added to a rapidly changing shift system with work hours occurring often at times of a minimum in performance.

In this context night and shift work regulations should take into account day to night changes and nocturnal hour to hour differences in the working capability of a given person.

The issue to optimize rotations and shift work schedules has not been reached. Only a transfactorial approach, including chronobiologic aspects, can be of some help.

## Tolerance to Shift Work

### Subject Type

Interindividual differences with regard to shift work problems (Reinberg 1979; Kerkhof 1985) must be taken into account especially with regard to the problem of tolerance. Indices have to be identified which can be used (a) to predict whether or not a given subject will develop an intolerance and (b), when intolerance is present, to estimate its severity since both medical and social decisions need to be made.

One of the first attempts to find a predictive index was to consider subject typology. Horne and Östberg (1976) proposed a test questionnaire to detect morning types (MTs) versus evening types (ETs) of subjects. Roughly, MTs are early-rising larks (also early retiring) while ETs are late-retiring owls (also late risers). These typologic variants in behavior are associated with chronobiologic differences; for example, the body temperature acrophase ($\emptyset$) occurs earlier in MTs than in ETs. It is also assumed that MTs will better tolerate shift work than ETs. This may be valid but has a limited practical significance since "true" MTs and/or ETs represent only 5%–10% of randomly investigated subjects. The fact that neither type predominates widely minimizes the actual predictive value of the MT vs. ET typology. Similar conclusions can be drawn when considering potential predictive values of introvert types (lonesome and silent subjects not prone to communicate and socialize) versus extrovert types. Even if true introverts may tolerate shift work better than true extroverts, they also represent only a very small fraction of investigated subjects. According to Folkard (personal communication), another typology should be considered: subjects having a certain flexibility with regard to the time at which they retire to sleep versus rigid subjects who need to go to bed at a fixed clock hour. "Flexible" subjects would tolerate shift work better than "rigid" ones. The interest of this hypothesis is that such a typology involves larger groups of subjects than the others (e. g., MT vs. ET); however, it remains to be tested on a large number of shift workers.

## Alteration of Period and Amplitude of Circadian Rhythm

During isolation experiments, Aschoff (1978), Wever (1979) and Forêt et al. (1981) observed that a rapid shift of body temperature circadian acrophase (∅) can be associated with a small circadian amplitude *(A). The larger the Δ∅, the smaller the A.* On the other hand, Andlauer et al. (1977, 1979) observed that *individuals poorly tolerant to shift work have a lower circadian amplitude of body temperature than subjects with no complaints.* This has been confirmed by Leonard (1981). Putting together these two findings, we hypothesized that subjects with good tolerance have large-amplitude circadian rhythms (e. g., in body temperature) associated with small Δ∅; in contrast, *poorly tolerant individuals should have small As associated with large Δ∅s; in other terms they should be more prone to desynchronizing their circadian rhythms than subjects with good tolerance.*

In a set of experiments, Reinberg et al. (1978a,b, 1980, 1981, 1988a) found that *tolerant persons exhibited larger amplitude circadian rhythms of oral temperature, right- and left-hand grip strength and heart rate* (but not of peak expiratory flow, PEF) *associated with smaller shifts of the circadian acrophases than subjects with poor tolerance* whatever the age, speed of rotation, schedules, duration of shift work and type of industry (Fig. 3).

The exception of PEF (bronchial diameter) can be explained by the fact that the *A* of this circadian rhythm is partially age related (Reinberg et al. 1978). The nocturnal dip in bronchial patency which increases with age presumably interferes with amplitude changes related to or revealed by zeitgeber manipulations.

All physiologic variables were self-measured four to five times a day for 2 (at least)–4 weeks including day and night shifts and days off. Amplitude end points were estimated on an individual basis using at least two methods: actual peak to trough differences and cosinor. The reason we stress this methodologic aspect of data gathering is that day-to-day variability may obscure the phenomenon when time series are too short. Both Costa and Gafferi (1983) and Härmä et al. (1990) failed to find a tolerance-related difference in the circadian amplitude of body temperature. However, their experiments lasted only 1 or 2 days, a methodology which does not favor the detection of such a change in *As*.

The problem of changes in the circadian period length $\tau$ of some variables will now be considered. As mentioned earlier, differences of $\tau$s in a set of circadian rhythms lead to a state of internal desynchronization. Is there a relationship between the latter state and tolerance to shift work? In fact, some symptoms of intolerance (e. g., persistent fatigue, sleep disturbances, alteration of mood) may also be found in af-

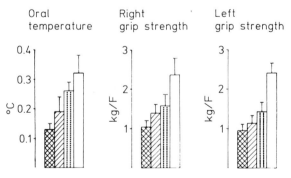

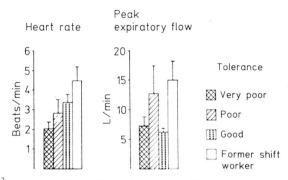

**Fig.3.** Mean circadian amplitude of five variables (± SEM) with regard to the four subgroups and their tolerance to shift work. ANOVA validated the relationship between tolerance to shift work and circadian rhythm amplitude with regard to oral temperature ($P < 0.02$) right ($P < 0.02$) and left ($P < 0.01$)-hand grip strength, heart rate ($P < 0.02$) and even PEF ($P < 0.01$). The $\chi^2$ test was used to verify that tolerance to shift work was associated with large-amplitude rhythms (and poor tolerance with small *As*). This was the case for oral temperature ($\chi^2 = 7.86$; $P < 0.01$), left-hand grip ($\chi^2 = 7.63$; $P < 0.01$), right-hand grip ($\chi^2 = 4.43$; $P < 0.05$), and heart rate ($\chi^2 = 5.27$; $P < 0.05$). (From Reinberg et al. 1988a)

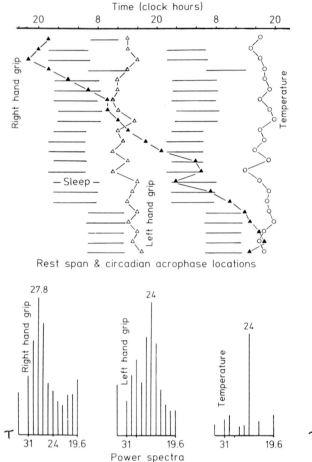

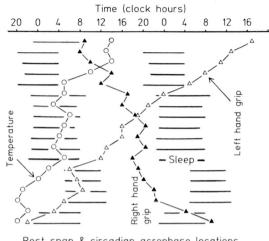

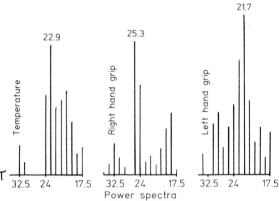

Rest span & circadian acrophase locations

Power spectra

**Fig. 4.** Rest span and circadian acrophase locations of three variables *(top)* power spectra of the same variables *(bottom)*. Double plot of hours of rest span (from lights-out to lights-on), *horizontal bar.* Acrophase location of oral temperature, ○; right ▲ and left △ -hand grip strength. The *tallest of the lines* of any spectrum and the *figure at the top* corresponds to the prominent period of that variable. Subject: oil refinery operator; left handed; age, 39 years; shift working for 14 years; good tolerance to shift work. (From Reinberg et al. 1988a)

**Fig. 5.** Same as Fig. 4. Subject: oil refinery operator; right handed; age, 31 years; shift working for 3 years; very poor tolerance to shift work. (From Reinberg et al. 1988a)

fective disorders. Many studies (Kripke 1981; Wehr and Goodwin 1983; Pflug et al. 1983; Halaris 1987; Bicakova-Rocher et al. 1989) have shown that depressive states are frequently associated with some forms of time structure alteration such as phase advance, phase delay, phase instability as well as internal desynchronization of some variables such as body temperature, cortisol, and TSH. Despite the fact that affective disorders and intolerance to shift work differ with regard to their respective clinical features, evolution, treatment and prognosis, it could well be that the role played by circadian rhythm alterations has some similarities in both these diseases. More-

over, studying the pathologic consequences of shift work in retired workers, Michel-Briand et al. (1981) reported there were more cases of affective disorders in retired shift workers than in retired day workers.

Keeping these reasons in mind, Reinberg et al. (1984a, b, 1988a) and Motohashi et al. (1987) documented circadian rhythms in sleeping and working, oral temperature, grip strength of both hands, peak expiratory flow and heart rate in both tolerant and nontolerant shift workers.

These studies confirmed for oral temperature and extended to other variables (grip strength of both hands, heart rate) that intolerance to shift work is frequently associated with both internal desynchronization and small circadian amplitude.

Illustrative examples of internal desynchronization involving four variables are given in Fig. 4 and 5. Figure 4 summarizes results of cosinor (∅) and power spectrum ($\tau$) analyses of a 22-day-long time series of a 29-year-old left-handed oil refinery operator who had been shift working for 14 years with good tolerance. Both acrophase (day-to-day) locations and power spectra show that sleep-wake, oral temperature and left and grip strength rhythms had a $\tau = 24$ h while right (non-dominant)-hand grip had a circadian period of 27.8 h. In this case only one variable was desynchronized from the others. Figure 5 provides time series analyses (18 days) of another subject (oil refinery operator; right handed; 31 years of age; shift working for 3 years) who developed a very poor tolerance to shift work. The entrained period of sleep/wake rhythm was 24 h as a mean, while $\tau$s of the other variables differed from 24 h as well as from each other. $\tau$s were 22.9 h for oral temperature, 25.3 h for right- and 21.7 h for left-hand grip strength. It can be seen that the occurrence of intolerance to shift work coincided with an internal desynchronization among the circadian rhythms of the investigated variables.

Differences between individual subject's $\tau$s and their relation to tolerance of shift work in these subjects is shown in Fig. 6. The prominent circadian $\tau$s in hours were plotted with regard to oral temperature and right- and left-hand grip strength in relation to tolerance to shift work. Most shift workers with good tolerance and former shift workers had a prominent circadian $\tau$ equal to 24 h in body temperature and left-hand grip strength whereas most shift workers with poor and very poor tolerance and $\tau$s differing from 24 h. The $\chi^2$ test was used to validate the hypo-

thesis that $\tau$s of some variables may differ from 24 h in nontolerant shift workers. For only two variables, oral temperature ($\chi^2 = 30.2$; $P < 0.001$) and left-hand grip strength ($\chi^2 = 14.4$; $P < 0.002$), was good tolerance associated with a prominent circadian $\tau = 24$ h. No statistically significant relationship was obtained between tolerance to shift work and prominent circadian period for right-hand grip strength, PEF or heart rate rhythm.

Neither the kind of work nor the speed of shift had any effect upon the circadian $\tau$s. The only age-related difference was found for left-hand grip strength. Only the $\tau$s of oral temperature and right-hand grip strength were correlated ($P < 0.05$).

Changes in circadian $A$s were also related to desynchronization at least for oral temperature, left-hand grip strength and heart rate, but not for right-hand grip strength and PEF rhythms. Similar findings have been obtained using other methods (e.g., continuous recording of temperature; Reinberg et al. 1989b) and in other populations (e.g., 22 young Japanese of both sexes; Motohashi 1989).

Despite the occurrence of an internal desynchronization and a reduced circadian amplitude of certain rhythms (more frequently in subjects nontolerant of shift work), we cannot recommend considering these rhythm alterations as a specific and reliable chronobiologic index of shift work tolerance since internal desynchronization and low circadian amplitudes can be observed without clinical complaint in apparently healthy subjects who have under-

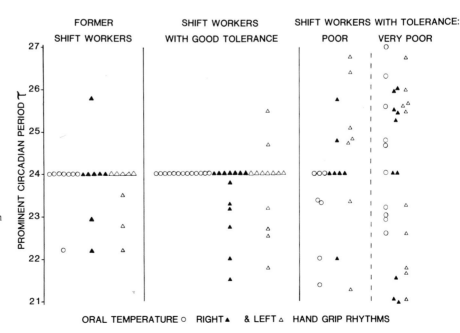

**Fig. 6.** Prominent circadian period resulting from power spectrum analyses for all of the variable and subjects were plotted with regard to each of the four groups and their tolerance to shift work. (From Reinberg et al. 1988)

gone no manipulation of zeitgebers (Bicakova-Rocher et al. 1981, 1984; Reinberg et al. 1984a, b). Some healthy subjects may present with desynchronized circadian rhythms of one or several variables (e.g., body temperature, grip strength, heart rate) without any decrease in performance (Reinberg 1984c, d), any symptoms of shift work intolerance or any affective disorder (Monk et al. 1984; Reinberg 1984).

To conclude:
1. Internal desynchronization and low-amplitude circadian rhythms appear to be frequently associated with symptoms of intolerance to shift work as well as with affective disorders.
2. However, internal desynchronization can also be demonstrated in apparently healthy subjects regularly synchronized in many of their functions with a diurnal activity and a nocturnal rest pattern ($\tau = 24$ h).
3. Interindividual differences seem to exist in the likelihood of a subject developing an internal desynchronization and/or suffering clinical symptoms from the latter.
4. Nontolerant shift workers, jet lag sufferers, and patients with affective disorders might be sensitive to internal desynchronization (or to transient cycles during phase shift and related phenomena).

## Are There Any Medications and/or Methods To Alleviate Night Work and Jet LAG Symptoms?

### Sleeping Pills

As stated earlier, the permanent use of sleeping pills or tranquillizers is a pathognomonic sign of intolerance to shift work as well as to transmeridian flights, i.e., in pilots and flight attendants. In fact, *there is no medication able to phase shift and resynchronize the time structure of a subject altered by a $\Delta\Phi > 5$ h.* The hunting of a wonder pill (or chronobiotic?) for resetting all biologic clocks continues (Reinberg et al. 1988b). Despite the fact that theophylline in the rat (Ehret et al. 1975) and the benzodiazepine triazolam in the hamster (Turek and Losa-Olson 1986) were able to phase shift some rhythms, there is no clinical evidence that such substances (or many others such as pentobarbital, caffeine, ethanol) are able to reset biologic clocks of human subjects.

However, acute and transient use of a sleeping pill (e.g., a benzodiazepine) may be of some help in obtaining daytime sleep after a night's work or a transmeridian flight (Seidel et al. 1984; Walsh et al. 1984). As stated in a previous section, sleep deprivation as well as poor quality of daytime sleep results in a decrement of performance. But it must be emphasized that drug-induced sleep at a certain time is not equivalent to resynchronization. The exceptional use of a sleeping pill which can be of some help for a passenger after a transmeridian flight should not be the rule for crew members or shift workers.

### Melatonin

The case of melatonin as an anti-jet lag medication deserves special attention. Endogenously produced, melatonin exhibits circadian rhythmicity with high levels from lights off to lights on. In pharmacologic but well-tolerated doses, melatonin produces a mild hypnotic effect and is able to phase shift the circadian rhythms of the sleep-wake cycle and fatigue (Arendt et al. 1986b).

A double-blind, placebo-controlled study was conducted by Arendt et al. (1986a, b) to test the potency of melatonin to alleviate jet lag resulting from air travel across eight time zones when flying east, from San Francisco to London. For 3 days before the flight different groups of subjects took either a daily 5-mg dose of melatonin or a placebo at 1800 hours San Francisco time (which is equivalent to 0000 hours in London). On arrival in Britain, the subjects continued to take the same preparation between 2200 and 0000 hours (local time in London) for an additional 4 days. On day 7 after their return to London the subjects were asked to rate their jet lag using a 10-cm visual analog scale varying from 0 (insignificant) to 100 (very bad). Jet lag was deliberately not defined as its nature and severity vary from person to person, but it was considered to be present when scores were 50 or higher. Six of the nine subjects who took placebo rated their jet lag greater than 50 on the visual analog scale. None of the eight subjects who took melatonin rated their jet lag at more than 17. Fisher's exact test for small size samples indicated that jet lag was significantly less severe among those treated with melatonin ($P = 0.009$).

This very interesting study used a specified melatonin dosing-time with emphasis being given to alleviation of jet lag symptoms rather than resynchronization of biologic rhythms. As is the case with benzodiazepines, melatonin produced an apparent phase shift of sleep and mood rhythms. However, further studies are needed to better understand the role of melatonin rhythm alterations with regard to certain forms of depression (e. g., seasonal affective disorders with low nocturnal plasma levels in winter) as well as the effect of certain medications (e. g., circadian rhythm of melatonin is obliterated by the $\beta$-blocking agent propranolol without patients' complaining). Therapeutic possibilities of melatonin deserve to be well defined.

## Bright Light
( > 2500 Lux at 1 m from the Light Source)

This physical agent acts as a synchronizer of human circadian rhythm and has therapeutic properties to control seasonal affective disorders (Kripke 1984; Lewy et al. 1984). Again, the intriguing role of melatonin, in relation to exposure to bright light and its manipulation, remains to be elucidated as well as plasma melatonin changes in shift workers (Touitou et al. 1990). There is no doubt that bright light exposure (with specified duration – e.g., 1 h – and timing) may help to phase advance or phase delay some human circadian rhythms (Czeisler et al. 1986, 1990). One of the major questions to be answered is to whom and under what circumstances should the method be recommended. In our opinion, potential candidates are passengers suffering from jet lag after transmeridian flights. The acceleration of their circadian rhythm adjustment may help to alleviate jet lag and reduce its duration. However, it seems that a rapid adjustment of shift workers, pilots and crew members does not necessarily have to be favored under all circumstances. Let us remember that one of the advantages of a rapid rotation system is precisely that this schedule does not allow the subject to adjust. In other words, the pertinence of applying bright light has to be discussed with regard to each work situation and presumably for each individual.

## Meal Timing

In animals such as rats, mice and rabbits the timing of a restricted access to food (and water) has a synchronizing effect. For example, when food is presented only during the light phase and for a limited number of hours, these nocturnal animals will phase shift their rhythm and become active during the day (Reinberg 1983). Based on animal experiments, Ehret et al. (1980) proposed a diet associated with a

specific timing in order to alleviate jet lag. Despite lack of clinical evidence that such a program is effective, it has been adopted and advertised as a premium for first-class passengers of certain airlines.

In fact, a large dinner as compared with a large breakfast will phase shift the circadian rhythms of plasma insulin and iron while other variables will maintain their acrophase in the same physiologic location (Goetz et al. 1976). Moreover, in shift workers, despite the fact that meals (breakfast, lunch, dinner, snacks) are for social reasons taken at fixed clock hours, all documented variables were phase shifted during night shifts. Therefore, if meal time shifts may phase shift some variables (e. g., insulin) it does not seem to be a practical method to alleviate jet lag.

To conclude, melatonin and bright light deserve to be investigated in depth to learn for whom and under what circumstances they may help to alleviate jet lag of passengers after transmeridian flights. Obviously, it is premature to recommend these methods to shift workers. Sleeping pills, as well as manipulation of meal times, are of no help to the latter.

*Acknowledgement.* This work was supported by the Laurette Veza grant in aid of research in chronobiology and chronotherapy.

# References

Åkerstedt T, Fröberg JE (1970) Individual differences in circadian patterns of catecholamine excretion body temperature, performance and subjective arousal. Biol Psychol 4: 277–292

Andlauer P, Carpentier J, Cazamian P (1977) Ergonomie du travail de nuit et des horaires alternants. Education permanente, Université de Paris I. Cujas, Paris

Andlauer P, Reinberg A, Fourré L, Battle W, Duverneuil G (1979) Amplitude of the oral temperature circadian rhythm and the tolerance of shift-work. J Physiol (Paris) 75: 507–512

Apfelbaum M, Reinberg A, Nillus P, Halberg F (1969) Rythmes circadiens de l'alternance veille-sommeil pendant l'isolement souterrain de 7 jeunes femmes. Presse Med 77: 879–882

Arendt J (1989) Melatonin and the pineal gland. In: Arendt J, Minors DS, Waterhouse JM (eds) Biological rhythms in clinical practice. Wright, London, pp 184–206

Arendt J, Aldhous M, Marks V (1986a) Alleviation of jet lag by melatonin: preliminary results of controlled double blind trial. Br Med J 292: 1170

Arendt J, Aldhous M, Marks V (1986b) Alleviation of jet lag by melatonin. Annu Rev Chronopharmacol 3: 49–51

Aschoff J (1978) Features of circadian rhythms relevant for the design of shift schedules. Ergonomics 39: 739–754

Aschoff J, Wever R (1981) The circadian system of man. In: Aschoff J (ed) Handbook of behavioural neurobiology. Plenum, London, pp 311–331

Aschoff J, Fatranska M, Gerecke U, Giedke H (1974) Twenty-four hour rhythms of rectal temperature in humans: effects of sleep-interruptions and of test-sessions. Pflugers Arch 346: 215–222

Aschoff J, Hoffmann K, Pohl H, Wever R (1975) Reentrainment of circadian rhythms after phase-shifts of the Zeitgeber. Chronobiologia 2: 23–78

Benoît O (1992) Le sommeil humain. Masson, Paris

Bicakova-Rocher A, Gorceix A, Reinberg A (1981) Possible circadian rhythm alterations in certain healthy human adults (effects of placebo). In: Reinberg A, Vieux N, Andlauer P (eds) Night and shift work. Biological and social aspects. Pergamon, Oxford, pp 297–310

Bicakova-Rocher A, Gorceix A, Nicolaï A (1984) Circadian and ultradian rhythm period alteration in apparently healthy subjects with and without placebo. Annu Rev Chronopharmacol 1: 169–171

Bicakova-Rocher A, Reinberg A, Gorceix A, Nouguier J, Nouguier-Soulé J (1989) Rythmes de la température axillaire: prédominance d'une période ultradienne lors de troubles affectifs majeurs. CR Acad Sci [III] 309: 331–335

Blom DHJ, Pokorny MLI (1985) Accidents of bus drivers. An epidemiological approach. Instituut voor Praeventieve Gezondheidszorg TNO, Leiden (Publ no 85023)

Colquhoun P, Folkard S, Knauth P, Rutenfranz J (1975) Experimental studies of shift-work. Westdeutscher Verlag, Opladen

Costa G, Gaffuri E (1983) Circadian rhythms, behavior characteristics and tolerance to shift work. Chronobiologia 10: 395

Czeisler CA, Allan JA, Strogatz SH, Ronda JM, Sanchez R, Rios CD, Freitag WO, Richardson GS, Kronauer RE (1986) Bright light resets the human circadian pacemaker independent of the timing of the sleep-wake cycle. Science 233: 667–671

Czeisler CA, Johnson MP, Duffy JF, Brown EN, Ronda JM, Kronauer RE (1990) Exposure to bright light and darkness to treat physiologic maladaptation to night work. N Engl J Med 322: 1253–1259

Daan S, Lewy AJ (1984) Scheduled exposure to day-light. A potential strategy to reduce jet lag following transmeridian flight. Psychopharmacol Bull 20: 566–568

Ehret CF (1981) New approaches to chronohygiene for the shift worker in the nuclear power industry. In: Reinberg A, Vieux N, Andlauer P (eds) Night and shift work. Biological and social aspects. Pergamon, Oxford, pp 26–30–270

Ehret CF, Potter VR, Dobra KW (1975) Chronotypic action of theophylline and of pentobarbital as circadian Zeitgebers in the rat. Science 188: 1212–1215

Ehret CF, Groh KH, Meinert JC (1980) Consideration of diet on alleviating jet lag. In: Scheving LE, Halberg F (eds) Chronobiology. Principles and applications to shifts in schedules. Sijthoff and Nordhoff, Amsterdam

Folkard S, Monk TA (1985) Hours of work: temporal factors in work scheduling. Wiley, Chichester

Forêt J, Benoît O, Merle B (1981) Circadian profile of long and short sleepers. In: Johnson LC, Tepas DI, Colquhoun WP, Colligan MJ (eds) Biological rhythms, sleep and shift work. Spectrum, New York, pp 499–511

Fröberg J, Karlsson CG, Lévi L, Lidberg L (1972) Circadian variations in performance, psychological ratings, catecholamine excretion and diuresis during prolonged sleep deprivation. Int J Physiol 2: 23–36

Goetz F, Bishop J, Halberg F, Sothern RB, Brunning R, Senske B, Greenberg B, Minors D, Stoney P, Smith FD, Rosen GD,

Haus E, Apfelbaum M (1976) Timing of single daily meal influence relations among human circadian rhythms in urinary cyclic AMP and hemic glucagon, insulin and iron. Experientia 32: 1081–1084

Gundel A, Wegmann HM (1989) Transition between advance and delay response to eastbound transmeridian flights. Chronobiol Int 6: 147–156

Haider M, Koller M, Cervinka R (1985) Night and shift-work: long term effects and their prevention. Lang, Frankfort

Halaris A (ed) (1987) Chronobiology and psychiatric disorders. Elsevier, Amsterdam

Halberg F, Reinberg A (1967) Rythmes circadiens et rythmes de basse fréquence en physiologie humaine. J Physiol (Paris) 59: 117–200

Härmä M, Knauth P, Ilmarinen J, Ollila H (1990) The relation of age to the adjustment of the circadian rhythm of oral temperature and sleepiness to shiftwork. Chronobiol Int 7: 227–233

Harris W (1977) Fatigue, circadian rhythm, and truck accidents. In: Mackie R (ed) Vigilance, theory, operational performance, and physiological correlates. Plenum, New York, pp 1033–1046

Haus E, Halberg F, Nelson W, Hillman (1968) Shifts and drifts in phase of human circadian system following intercontinental flights and in isolation. Fed Proc 27: 224

Hildebrandt G, Rohmert W, Rutenfranz J (1974) 12 and 24 h rhythms in error frequency of locomotive drivers and the influence of tiredness. Int J Chronobiol 2: 175–180

Horne JA, Östberg O (1976) A self-assessment questionnaire to determine morningness-eveningness in human circadian rhythms. Int J Chronobiol 4: 97–110

Johnson LC, Tepas DI, Colquhoun P, Colligan MJ (1981) Biological rhythms, sleep and shift-work. Adv Sleep Res 7

Kerkhof GA (1985) Interindividual differences in the human circadian system, a review. Biol Psychol 20: 83–112

Klein KE, Wegmann HM (1980) Significance of circadian rhythms in aerospace operations. NATO-AGARD, Neuilly-sur-Seine (AGARDograph no 247)

Knauth P, Rutenfranz J (1982) Development of criteria for the design of shiftwork systems. J Hum Ergol [Suppl] 11: 337–367

Kripke DF (1981) Phase advance theories for affective illnesses. In: Wehr TA, Goodwin FK (eds) Biological rhythms and psychiatry. Boxwood, Pacific Grove, pp 41–70

Kripke DF (1984) Critical interval hypotheses for depression. Chronobiol Int 1: 73–80

Leonard R (1981) Amplitude of the temperature circadian rhythm and tolerance to shift-work. In: Reinberg A, Vieux N, Andlauer A (eds) Night and shift work. Biological and social aspects. Pergamon, Oxford, pp 323–329

Lewy AJ, Sack RL, Singer CM (1984) Assessment and treatment of chronobiologic disorders using plasma melatonin levels and bright light exposure: the clock-gate model and the phase response curve. Psychopharmacol Bull 20: 561–565

Michel-Briand C, Chopard JL, Guiot A, Paulmier M, Studer G (1981) The pathological consequences of shift work in retired workers. In: Reinberg A, Vieux N, Andlauer P (eds) Night and shift work. Biological and social aspects. Pergamon, Oxford, pp 399–407

Minors DS, Waterhouse JM (1981) Anchorsleep as a synchronizer of rhythms on abnormal routines. In: Johnson LC, Tepas DI, Colquhoun WP, Colligan MJ (eds) Biological rhythms, sleep and shift work. Spectrum, New York, pp 399–414 (Advances on sleep researchs)

Minors DS, Waterhouse JM (1989) Circadian rhythms in

general practice and occupational health. In: Arendt J, Minors DS, Waterhouse JM (eds) Biological rhythms in clinical practice. Wright, London, pp 207–224

Monk T, Weitzman ED, Fookson JE, Moline ML (1984) Circadian rhythms in human performance efficiency under free-running conditions. Chronobiologia 11: 343–354

Moore-Ede M, Sulzman FM, Fuller CA (1982) The clocks that time us. Harvard University Press, Cambridge

Motohashi Y (1989) Desynchronization of oral temperature and grip strength circadian rhythms in healthy subjects with irregular sleep wake-behavior. Chronobiologia 16: 162–163

Motohashi Y, Reinberg A, Lévi F, Nouguier J, Benoît O, Forêt J (1987) Axillary temperature: a circadian marker rhythm for shift workers. Ergonomics 30: 1235–1247

Pflug B, Johnsson A, Martin W (1983) Alterations in the circadian temperature rhythms in depressed patients. In: Wehr TA, Goodwin FK (eds) Biological rhythms and psychiatry. Boxwood, Pacific Grove, pp 71–76

Price WJ, Holley DC (1981) The last minutes of flight 2860. In: Reinberg A, Vieux N, Andlauer P (eds) Night and shift work. Biological and social aspects. Pergamon, Oxford, pp 287–294

Reinberg A (1979) Chronobiological field studies of oil refinery shift workers. Chronobiologia [Suppl] 1

Reinberg A (1983) Biological rhythms and nutrition. In: Biological rhythms and medicine. Reinberg A, Smolensky M (eds). Springer Berlin Heidelberg New York

Reinberg A, Smolensky M (eds) (1983) Biological rhythms and medicine. Springer, Berlin Heidelberg New York

Reinberg A, Vieux N, Ghata J, Chaumont A, Laporte A (1978a) Is the rhythm amplitude related to the ability to phase-shift circadian rhythms of shift workers? J Physiol (Paris) 74: 405–409

Reinberg A, Vieux N, Ghata J, Chaumont AJ, Laporte A (1978b) Circadian rhythm amplitude and individual ability to adjust to shift work. Ergonomics 21: 763–766

Reinberg A, Migraine C, Apfelbaum M, Brigant L, Ghata J, Vieux N, Laporte E, Nicolaï A (1979) Circadian and ultradian rhythms in the feeding behaviour and nutrient intakes of oil refinery operators with shift-work every 3–4 days. Diabete Metab 5: 33–42

Reinberg A, Andlauer P, Guillet P, Nicolaï A (1980) Oral temperature, circadian rhythm amplitude, ageing and tolerance to shift work. Ergonomics 23: 55–64

Reinberg A, Vieux N, Andlauer P (eds) (1981) Night and shift work. Biological and social aspects. Pergamon, Oxford

Reinberg A, Andlauer P, Bourdeleau P, Lévi F, Bicakova-Rocher A (1984a) Rythme circadien de la force des mains droites et gauches: désynchronisation chez certains travailleurs postés. C R Acad Sci [III] 299: 633–636

Reinberg A, Andlauer P, de Prins J, Malbecq W, Vieux N, Bourdeleau P (1984b) Desynchronization of the oral temperature circadian rhythms and intolerance to shift work. Nature 308: 272–274

Reinberg A, Brossard T, André MF, Joly D, Malaurie J, Lévi F, Nicolaï A (1984c) Interindividual differences in a set of biological rhythms documented during the high arctic summer (79°N) in three healthy subjects. Chronobiol Int 1: 127–138

Reinberg A, Lévi F, Bicakova-Rocher A, Blum JP, Ouechni MM, Nicolaï A (1984d) Biologic time related changes in antihistamine and other effects of chronic administration of me-

quitazine in healthy adults. Annu Rev Chronopharmacol 1: 61–64

Reinberg A, Andlauer P, Lévi F (1986) Chronobiologie et travail posté. Encycl Med Chir (Paris) 16785: 1–5

Reinberg A, Motohashi Y, Bourdeleau P, Andlauer P, Lévi F, Bicakova-Rocher A (1988a) Alteration of period and amplitude of circadian rhythms in shift workers. Eur J Appl Physiol 57: 15–25

Reinberg A, Smolensky M, Labrecque G (1988b) The hunting of a wonder pill for resetting all biological clocks. Annu Rev Chronopharmacol 4: 171–208

Reinberg A, Di Costanzo G, Guérin N, Boulenguiez S, Guran P (1989a) Heure d'été, heure d'hiver: nos horloges biologiques supportent bien. Recherche 20: 1396–1397

Reinberg A, Motohashi Y, Bourdeleau P, Touitou Y, Nouguier J, Nouguier J, Lévi F, Nicolaï A (1989b) Internal desynchronization of circadian rhythms and tolerance of shift work. Chronobiologia 16: 21–34

Rybak J, Ashkenazi IE, Klepfish A, Avgar D, Tall J, Kallner B, Noyman Y (1983) Diurnal rhythmicity and air force flight accidents due to pilot error. Aviat Space Environ Med 54: 1096–1099

Rutenfranz J, Knauth P, Colquhoun WP (1976) Hours of work and shiftwork Ergonomics 19: 331–340

Scherrer J (1981) Man's work and circadian rhythm through the age. In: Reinberg A, Vieux N, Andlauer P (eds) Night and shift work. Biological and social aspects. Pergamon, Oxford, pp 1–10

Seidel WF, Roth T, Roehrs T, Zorik F, Dement WC (1984) Treatment of a 12-hour shift of sleep schedule with benzodiazepine. Science 224: 1262–1274

Smolensky MH, Paustenbach DT, Scheving LE (1985) Biological rhythms, shift-work and occupational health. In: Cralley L, Cralley L (eds) Biological responses. Wiley, New York, pp 175–312 (Industrial hygiene and toxicology, vol 3B, 2nd edn)

Touitou Y, Motohashi Y, Reinberg A, Touitou C, Bourdeleau P, Bogdan A, Auzéby A (1990) Effect of shift work on the night time secretory patterns of melatonin, prolactin, cortisol and testosterone. Eur J Appl Physiol 60: 288–292

Turek FW (1983) Neurobiology of circadian rhythms in mammals. Bioscience 33: 439–444

Turek FW, Losee-Olson S (1986) A benzodiazepine used in the treatment of insomnia phase shifts the mammalian circadian clock. Nature 321: 167–168

Walsh JK, Muehlbach MJ, Sweitzer PK (1984) Acute administration of triazolam for the daytime sleep of rotating shift workers. Sleep 7: 223–229

Wehr TA, Goodwin FK (eds) (1983) Circadian rhythms in psychiatry. Boxwood, Pacific Grove

Weitzman ED, Czeisler CA, Moore-Ede M (1979) Sleep-wake, neuroendocrine and body temperature circadian rhythm under entrained and non-entrained (free-running) conditions in man. In: Suda M, Hayaishi P, Nakagawa H (eds) Biological rhythms and their central mechanism. Elsevier/North-Holland, Amsterdam, pp 199–227

Wever R (1979) The circadian system of man. Results of experiments under temporal isolation. Springer, Berlin Heidelberg New York

# Chronobiology of Human Sleep and Sleep Disorders

B. Hayes and C. A. Czeisler

## Sleep and the Circadian Clock

### Introduction: Sleep-Wake, the Most Conspicuous Circadian Rhythm

Across cultures and through the millenia humans have not only adapted to but organized their existence around periodicities present in their environment and within themselves. "While the earth remaineth," the author of Genesis writes, "seedtime and harvest, and cold and heat, and summer and winter, and day and night shall not cease." But while numerous schemes from sundials to calendars have been devised for parsing time, the unit upon which they all agree as a starting point for multiplication and division is the period of the earth's rotation about its axis, the day. And it is not only the drama of the alternation of environmental darkness and light, but the experience of activity and repose, of consciousness and unconsciousness that has invested the day with its human meaning and those biological rhythms of "about a day" with their compelling scientific significance. Simply put, the sleep-wake cycle is our most conspicuous circadian rhythm. As such it has been one of the most accessible to scientific study – and the most sensitive to environmental disturbance and intrinsic pathology. We have, however, only recently begun to explore and understand the physiological processes that control this important periodicity.

### Endogenous Core Body Temperature Rhythm

Beginning in the mid-19th century the existence of another prominent circadian rhythm, that of body temperature, began to attract scientific interest (Kleitman 1963). It became clear that the tempera-

Supported in part by research and training grants (NIA-1-RO1-AG06072, NIMH-1-RO1-MH45130) from the National Institutes of Health, and by the Brigham and Women's Hospital.

ture rhythm was normally coupled to the sleep-wake cycle, with the daily minimum of body temperature occurring during sleep. Experimental manipulations showed that the rise and fall of temperature was not entirely attributable to environmental factors such as changes in ambient temperature, food intake, or activity; it persisted in situations that held these nearly constant (Kleitman 1963). Other physiological parameters from heart rate to levels of hormone secretion also showed measurable 24-h variations.

The advent of the industrial revolution brought with it night work and rotating shifts. Thomas Edison's invention of electric light stretched the boundaries of the day. And the development of railroad, steamship, and air travel brought the possibility of rapid travel across time zones, which made it apparent that there were some biological limitations inherent in irregular or nonconventional sleep-wake schedules. But it was not until 1938 when, serving as both investigators and subjects in an experiment designed to determine whether humans could adapt to non-24-h sleep-wake schedules, Nathaniel Kleitman and a colleague descended into a Kentucky cavern (where they could live in an environment nearly free of external time cues) that an important discovery was made about the chronobiology of sleep (Kleitman 1963). While one of the subjects was able to easily adapt both his sleep-wake cycle and his temperature rhythm to a 28-h day (19 h wake/9 h sleep), the other showed a persistent near-24-h temperature cycle while on an enforced sleep-wake schedule of 28 h. This was the first time that it was demonstrated that a robust physiological rhythm, namely the body temperature cycle, could oscillate with a period independent of that of the sleep-wake schedule. Kleitman also noted that it was easy to sleep well when his scheduled sleep time bore the normal relationship to his body temperature rhythm, but that he slept poorly when he tried to sleep at other times – despite being in bed in complete silence and darkness after having been awake for the previous 19 h.

Subsequent cave studies in Germany, France, and England demonstrated that human circadian rhythms continue to oscillate in the absence of time cues (Aschoff and Wever 1962; Mills 1964; Siffre 1965). These studies also demonstrated that under such free-running conditions the subjects' self-selected bedtimes and wake times were not always synchronized with their other physiological rhythms, including the rhythm of sleep tendency itself (Aschoff et al. 1967). Even when uncoupled from the sleep-wake cycle, however, core body temperature rhythm continued to oscillate with a near-24-h period.

## Phenomenology of the Sleep-Wake Rhythm

### Sleep in Time Isolation

Studies in time isolation have revealed that in the absence of external time cues an approximately 2:1 ratio of time awake to time asleep is maintained over an extended period. In addition, sleep generally remains consolidated into one primary episode per cycle. However, there are narrower limits to the ability of many other circadian rhythms to accommodate non-24-h schedules. Free-running subjects generally adopt a somewhat longer than 24-h period of the behavioral rest-activity cycle, although periods as short as 16 h or as long as 50 h are sometimes observed. Experiments have shown that it is difficult to entrain the physiological rhythms to day lengths of less than 23 h or more than 27 h. Under normal environmental conditons, the clock of most people is reset daily in order to counter an inherent tendency of the daily sleep episode to drift to later hours.

### Sleep Tendency, REM Tendency

Despite the fact that the majority of sleep time is consolidated into a single daily bout of around 8 h, there is a second peak of sleepiness and sleep tendency that occurs midway through the waking day. The time of maximal alertness and performance emerges shortly thereafter, in the early evening. Laboratory data are consistent with statistical analyses of traffic accidents (Lavie et al. 1986) caused by sleepiness, which showed peak occurrences at the times when research subjects are most sleepy in the lab. Even in the absence of additional sleep, alertness will increase after each of the two peaks in sleep tendency, forming mid-morning and early evening "wake maintenance zones" (Strogatz et al. 1987). When sleep does occur,

there is a propensity for Rapid Eye Movement (REM) sleep to peak, just after the minimum of the body temperature cycle (Fig. 1). Just as there are REM-non-REM cycles during sleep of about 90 min (see "Sleep Architecture" below), there is some evi-

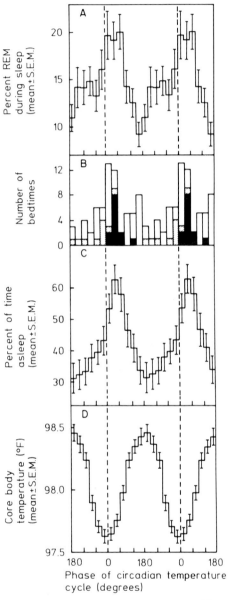

**Fig. 1 A–D.** Variation of the occurrence and internal organization of sleep with circadian phase. Sleep and temperature data from four subjects on self-selected light-dark cycles (94 days total) that were not in synchrony with their circadian temperature rhythm (Czeisler 1980 b). The data are double-plotted for a single day. There is a peak in the number of times subjects who chose to go to bed (**B**) or were already asleep (**C**) corresponding to the ascending phase of the temperature curve. In addition, there was a daily rhythm of REM propensity within sleep

dence of a 90-min rhythm of alertness similar to the basic rest activity cycle (BRAC) first proposed by Kleitman (1963) persisting throughout the day.

## Sleep Length

The length of a sleep bout varies between individuals and from day to day in the same individual. The interindividual variation has been associated with temperament and personality factors. The night to night variation may within certain limits be related to prior sleep or lack of it, but, more interestingly, sleep bout length within an individual is related to circadian phase at sleep onset. The spontaneous duration of a sleep bout initiated at the trough of the temperature curve is shorter than one initiated at the crest (Czeisler et al. 1980a).

## Sleep Physiology

### Sleep Stages

The development of the electroencephalograph (EEG) and other electrophysiological instruments has made it possible to understand sleep not simply as the absence of wakefulness, but as an active process and a system with its own internal dynamics. Subjective and behavioral sleep has been correlated with physiological measurements. There are distinct changes in electroencephalographic activity, eye movements, and muscle tone that can be routinely measured in the sleep laboratory. By convention these parameters are used to divide sleep into two main categories, rapid eye movement (REM) sleep and non-rapid eye movement (NREM) sleep (Rechtschaffen and Kales 1968). REM sleep is associated with wake-like EEG activity, flaccid paralysis of skeletal muscles, irregularity in heart rate and breathing, and dreaming. Non-REM sleep is associated with relaxed musculature, more synchronous EEG activity, as well as some characteristic EEG features. There is little mental activity in NREM sleep, and what there is lacks the vivid, bizarre, and narrative character of REM dreams.

Within NREM sleep there is a division into stages (1–4) of increasing depth. NREM sleep is also known as slow-wave sleep because the frequency of the EEG is much slower than that seen in waking or REM. Amplitude also increases; the deepest stages of sleep, 3 and 4 (also known as delta sleep), are associated with the highest amplitude (greater than 75 MV) and slowest waves (less than 3 cps). Light sleep, stage 1, is a transition between sleep and wakefulness. Stage 2 sleep is associated with theta frequencies (3–7 cps) of moderate amplitude and has characteristic EEG features known as sleep spindles (bursts of 12–14 cps activity from the sensorimotor cortex) and k-complexes (isolated delta waves of a characteristic morphology).

Selective sleep deprivation of individual stages yields a variety of effects. REM deprivation tends to affect memory and induce temporary hyperarousal. Slow-wave sleep deprivation, especially of stages 3 and 4, has been associated with muscular aches and feelings of malaise.

### Sleep Architecture

A normal night's sleep progresses from sleep onset (within a few minutes of getting into bed and trying to sleep) through progressively deeper stages of sleep. Approximately 90 min later the first REM episode occurs, and then the cycle is repeated. The episodes of REM become progressively longer during the night and the episodes of NREM, especially stages 3 and 4, become shorter.

### Neurophysiology of Sleep and the Sleep-Wake Rhythm

While lesion and stimulation studies in animals have provided much information about the neurophysiology of sleep, a single, precisely located "sleep center" has not been found. Even so, it is clear that the activity of a number of loci and pathways extending from the brain stem to the basal forebrain play an important role in the maintenance of wakefulness, and the generation of NREM sleep. The generation of REM sleep appears to be somewhat more restricted to the pons, with different areas being responsible for the REM-related muscle atonia and the other features of REM sleep.

In contrast to the *generation* of sleep and wakefulness, their *timing* is associated with a well-localized brain region. The persistence of circadian rhythms under constant conditions and in environments free of time cues first suggested that there was an endogenous pacemaker driving these rhythms, which is synchronized to the 24-h day by the environmental light-dark cycle in nearly all animals, including humans (Czeisler et al. 1981, 1989; Czeisler 1986). In 1972 lesion studies indicated that the suprachiasmatic nuclei (SCN) of the hypothalamus serve as the central neural pacemaker of the mammalian circadian timing system (Stephan and Zucker 1972; Moore and Eichler 1972). A monosynaptic pathway, the retinohypothalamic tract (RHT), serves as a direct conduit of photic information from the retina to the SCN (Moore and Card 1985). Unlike other tracts from the retina, which are involved in visual perception and project to the visual cortex, tracts from the retina to the SCN modulate the unconscious perception of ambient light levels, which is the mechanism through which the geophysical environment entrains the endogenous circadian pacemaker.

## Circadian Rhythm Sleep Disorders

### Nosology of Sleep Disorders

Among the properties of an intrinsically driven circadian rhythm are period (cycle length), phase (the temporal location of the periodic events), amplitude (the difference between the mean and extreme values), and resetting capacity (ability to be reentrained to new environmental stimuli). While other sleep disorders are classified by the International Classification of Sleep Disorders of the American Sleep Disorders Association (ASDA 1990) as intrinsic dyssomnias, extrinsic dyssomnias, parasomnias, or medical/psychiatric sleep disorders, the circadian rhythm sleep disorders have been categorized separately, in part to acknowledge that in most cases the etiology of circadian disorders is a mixture of both internal and environmental factors, or a temporal mismatch between the two.

### Diagnostic Tests and Techniques

History-Taking and Sleep-Wake Logs

In the differential diagnosis of any sleep disorder there is no substitute for a careful and detailed history. The presenting symptoms are usually either insomnia (i.e., difficulty falling asleep or remaining asleep or subjective experience of poor sleep quality), or hypersomnia, manifested as excessive sleep at night or sleepiness at inappropriate times during the day.

While these symptoms are common in patients with circadian rhythm sleep disorders, there are many other causes to which they may be attributed,

and which must be definitively excluded before the diagnosis of a circadian rhythm sleep disorder can be made with confidence. The practice of sleep disorder medicine is now a well-developed subspecialty, and a full discussion of it is beyond the scope of this chapter (Kryger et al. 1989). But certain symptoms in an individual complaining of a sleep disorder should alert the practitioner to the possibility that there may be a noncircadian etiology. Uncontrollable sleep attacks accompanied by vivid dreams or hallucinations and muscular paralysis should prompt a full diagnostic workup for narcolepsy. A bed partner's report of loud snoring or lapses in breathing during sleep indicate the possibility of sleep apnea, while a report of repetitive leg jerks suggest periodic limb movement during sleep (PLMS, also known as nocturnal myoclonus). It is also important to remember that some circadian disorders may coexist with these disorders and persist after they are treated.

In the patient with no other current sleep pathology a daily log of activities, meals, exercise, naps, and bedtimes is an essential tool in evaluating circadian rhythm sleep disorders. These logs should be kept for 2 weeks or longer, since a disturbance due to shiftwork or travel across time zones can have effects on sleep and daytime alertness weeks after the fact.

Actigraphy and Thermography

Patients who are poor record keepers or whose diaries prove uninformative may be better studied with portable activity and body temperature monitors. These devices are small microprocessors with sensors for movement (usually placed on the wrist) and memory sufficient to monitor a subject's habits for days or weeks. Other devices can also monitor core body temperature (usually a flexible rectal sensor), although ambulatory temperature records are of limited clinical utility due to the marking effects of activity on temperature. In addition to their diagnostic advantages as objective measures of the subject's activity, these devices can be very useful in assessing the efficacy of treatment strategies by comparing pre- and postintervention recordings.

Polysomnography, MSLT, and Ambulatory EEG

Overnight polysomnographic sleep recordings during which a dozen or more physiological variables are recorded in a laboratory setting are a useful, and often essential, tool for ruling out the non-circadian sleep disorders mentioned above, but they are usually not very informative in patients with circadian problems. Polysomnography is an impractical way to study patients for days or weeks at a time, as is often required for observation of circadian pathology. The multiple sleep latency test (MSLT), a series of polygraphically recorded naps at 2-h intervals during the day following polysomnography, makes a sleep lab evaluation more informative for the circadian patient by providing an objective measure of the amount and the timing of sleepiness during the day. Portable recorders developed for ambulatory EEG have made it possible to collect sleep-wake data for continuous periods of 24 h or longer in the patient's natural environment, and the development of computerized systems for evaluating the massive amounts of data collected thereby have made this a useful tool for studying these patients.

## Constant Routine

A great obstacle to the assessment of the period, amplitude, and phase of the human circadian system is that the "noise" imposed by the effects of exercise, food intake, postural changes, ambient temperature, environmental light, and even sleep itself on core temperature, for example, are on the same order as the daily oscillation (approximately 2°F). In an attempt to control for these factors a technique known as the constant routine was devised (Mills et al. 1978; Czeisler 1986; Czeisler et al. 1989). It involves keeping a subject semirecumbent and awake in bed for about 40 h in constant room-level light. The subject is fed isocaloric, nutritionally balanced snacks every hour. Core temperature is monitored continuously, urine is collected every 3 h, and blood is sampled every 20 min through an indwelling venous catheter. Curve-fitting routines are used to determine the minimum of the body temperature curve, which is the best available indicator of the phase of the endogenous circadian pacemaker. This technique is not yet widely available clinically, although it may in some cases be the only way to confidently establish a diagnosis.

## Circadian Rhythm Sleep Disorders

## Time Zone Change (Jet Lag) Syndrome

Transmeridian (across time zones) travel has become commonplace in the modern world. Time zone change (jet lag) syndrome is the name applied to the cluster of symptoms that often accompany such travel. Complaints include insomnia, daytime sleepiness, gastrointestinal upset, and general malaise. These symptoms are usually transient, lasting from 2 days to 2 weeks depending on the number of time zones crossed. The severity of the symptoms also varies with the magnitude of the time difference and increases with age, although there is a great deal of unexplained interindividual difference; not everyone who travels across time zones experiences jet lag. Eastward travel (which, in effect, temporarily shortens one's day) is said to be more difficult than westward travel (which temporarily lengthens it); this is consistent with the slightly longer than 24-h period of the human circadian pacemaker.

As a disorder, jet lag is easily explained in terms of a mismatch between internal clock time and the artificially imposed schedule of the new time zone. If the new time zone is 9 h different from the home environment, for example, one is sleeping or attempting to sleep at a time when one's natural propensity, as evidenced by a persistence of the habitual body temperature rhythm, is to be awake. The underlying oscillator adjusts readily to a change of about 1 h/day under ordinary conditions (which explains why slower travel across time zones by ship never posed the same problems that emerged with travel by air). In contrast, a time zone shift of 9 h might require more than a week to adjust. People who stay in hotel rooms seem to take longer to adjust than those who go outdoors (Klein and Wegmann 1974), presumably because of the resetting effects of sunlight. We have shown in our laboratory that daily bright light exposure administered at the right time of day can reset the circadian pacemaker to any number of hours time difference in a matter of 2 or 3 days (Czeisler et al. 1989).

## Shift Work Sleep Disorder

More than 7 million Americans work on either permanent night shifts or rapidly rotating shifts involving days, evenings, and nights (Mellor 1986). This practice leads many of these workers to adopt sleep-wake schedules that are permanently misaligned with their internal clocks. The symptoms are similar to those of jet lag: insomnia, sleepiness, GI distress, and malaise. But while most people suffering from jet lag experience these symptoms only transiently, people suffering from shift work sleep disorder often have a chronic mismatch between their internal clocks and their external schedules. It is perhaps easiest to understand why workers on a rapidly rotating shift fail to adjust to their schedules. It would be analogous to shuttling back and forth across time zones, staying just long enough to become adjusted before shifting back again.

The shift worker has it worse than the jet traveler, however. The jet traveler is attempting to adapt to a schedule being kept by the majority of the people in the new time zone; the shift worker is not only at odds with his or her internal clock, but with society's schedule as well. Because of environmental disturbances and social pressures, coupled with biological factors, there are inherent difficulties in sleeping during the day. Exposure to ambient daylight occurs at a time that subverts attempts at entrainment to a night schedule. Moreover, many night shift workers attempt to revert to a regular schedule on days off, making it even more difficult for them to adjust to working nights. However, as was first recognized at the turn of the century, despite years on a permanent night work schedule, even in the absence of such days off, the circadian system may fail to adjust (Benedict 1904). We have recently reported that scheduled exposure to bright (7000–12000 lux) artificial light during the night and restricted exposure to daylight during the day (sleeping in completely darkened rooms from 0900 to 1700 hours each day) not only synchronized the rhythms of the workers to their work schedules,

but produced significant improvement in alertness and performance during the nighttime hours (Czeisler et al. 1990).

## Irregular Sleep-Wake Pattern

This sleep disorder, characterized by a disorganized and fragmented pattern of waking and sleeping is rare among the general population, but occurs more frequently among bedridden individuals and among institutionalized patients with diffuse brain dysfunction who have no regular daily routine. Although the total sleep time in any 24-h period remains normal, there are usually frequent naps, with the major sleep time broken into three or more episodes per day, and these sleep episodes show no regular circadian or ultradian organization. The cause of this disorder is unclear. It has been suggested that in patients with diffuse brain dysfunction there may be an involvement of the systems that govern sleep and/or the timing of sleep, especially since this pattern closely resembles that of the infant. It may be, however, that an initial period of irregular sleep has become a self-perpetuating condition in which napping leads to insomnia and vice versa. This disorder is often chronic, although there is a need for controlled studies of treatments with bright light and strict enforcement of

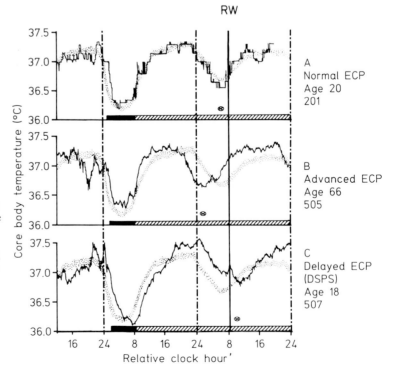

**Fig. 2 A–C.** Unmasking of endogenous circadian phase *(ECP)* with the constant routine. Three individuals are shown who were recorded continuously in the laboratory for 60 h (see text). During day 1 temperature minima occur during sleep. On day 2, during sleep deprivation, temperature minima are somewhat higher due to removal of the evoked effects of sleep on temperature. The subject in **A** has a normal phase relationship between his temperature rhythm and his sleep behavior. The older woman shown in **B** (phase advance) has a temperature minimum that occurs at an earlier relative clock hour compared to normal *(stippled line)*. The DSPS patient in **C** has a temperature minimum that occurs at a much later clock hour than normal

sleep hygiene, which would seem to hold some promise, especially with patients who have otherwise normal brain function.

Delayed Sleep Phase Syndrome

A concept of particular relevance to this disorder is phase delayed (Fig. 2). One's sleep phase is delayed if it occurs at a clock time that is generally stable from day to day, but is consistently later than desired. Individuals with this disorder find it impossible to advance their daily sleep episodes to the desired clock hour by going to bed earlier at night or awakening earlier in the morning, even when oversleeping or morning sleepiness causes severe life disruption and decrements in school or work performance. It can be postulated that such individuals have at some point followed the natural tendency to phase delay caused by the longer than 24-h period of the human circadian pacemaker but are for some reason unable to phase advance to correct for that tendency, either because of some intrinsic dysfunction of the neural pacemaker or a relative insensitivity to the stimuli, either social or environmental, that normally synchronize the circadian pacemaker to the 24-h day. The disorder seems to occur more often in adolescents or young adults, and care must be taken to distinguish it from chronic sleep deprivation, which is also common in that group. The patient must indeed have a chronic inability to fall asleep at an earlier hour, not merely an unwillingness to go to bed on a regular schedule. This condition must also be distinguished from transient difficulties falling asleep and getting up in the morning due to previous episodes of sleep deprivation and subsequent late recovery sleep,

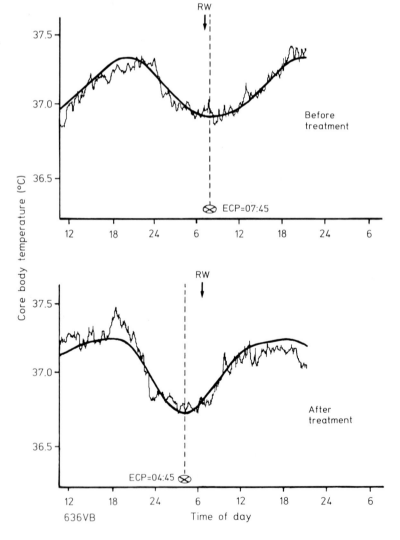

**Fig. 3.** A patient with delayed sleep phase syndrome (DSPS) successfully treated with morning exposure to bright light. Before treatment the patient fell asleep and awoke at a relatively late clock hour and had a relatively late (0745 hours) temperature minimum. After three successive mornings of bright artificial light exposure, sleep onset, the time of awakening, and the endogenous circadian temperature phase were concurrently shifted to an earlier clock hour

which may be self-reinforcing. Patients with DSPS are generally not helped by hypnotic medication, although the condition may encourage a dependency on drugs or alcohol. DSPS patients should be differentiated from those with avoidant personality disorders or other adjustment or psychological problems. A history of being able to awaken on vacations or in environments in which the patient derives no secondary gain from getting up late makes true DSPS a less likely diagnosis.

One treatment that has been used on some patients with this disorder is chronotherapy, which involves delaying the daily sleep episode by several hours each day for several successive days until it reaches the desired clock time. While this has been successful for some patients, great care must be taken not to allow any deviation from the new time of awakening lest it again become stubbornly delayed and require a repeat treatment.

Once we demonstrated the phase-shifting effects of bright light on human circadian rhythms it became clear that individuals with this disorder might be helped with properly timed light exposures. A recent controlled study of 20 patients with DSPS (Rosenthal et al. 1990) has confirmed this finding and demonstrated that morning exposure to bright light followed by restricted exposure to light in the evening can effectively phase advance both the sleep-wake and ambulatory temperature rhythms of some patients to an earlier clock hour. Figure 3 is an example of a patient with DSPS who was successfully treated with morning bright light phototherapy. It is not clear whether or not these patients will relapse without frequent repeat treatments under these conditions, but since the treatments are otherwise benign and the apparatus for the light exposure has become readily available commercially (due to its use in the treatment of seasonal affective disorder), this may not prove to be a major problem in managing such patients.

### Advanced Sleep Phase Syndrome

Advanced sleep phase syndrome is similar to delayed sleep phase syndrome, except that the symptoms are a consistent inability to stay awake in the evening and spontaneous awakening at a clock hour much earlier than desired. This condition more often affects the elderly, and may be related to a shortening of the intrinsic period of the circadian pacemaker with advancing age. Studies also suggest that the elderly are consistently exposed to less bright environmental light than are younger people. Evening exposure to

bright artificial light shows promise in the treatment of this condition.

### Non-24-h Sleep-Wake Disorder

Patients with non-24-hour sleep-wake disorder typically have difficulty falling asleep at night and waking up in the morning, but, unlike the DSPS patient, the awakening time is not stable but is delayed to later and later hours. Rather than having an inability to make a time correction of a few hours in a particular direction, the non-24-h patient is unable to entrain to a 24-h day and his or her pacemaker behaves as though it were free running. Some individuals give up trying to adapt to a 24-h schedule and follow a longer than 24-h day (putting them at odds with the rest of society), but most attempt to live on a regular schedule and find themselves periodically unable to sleep and periodically hypersomnolent. On the occasions when their sleep episodes coincide with their desired clock hours they are able to sleep normally at night and remain alert during the day.

This condition is frequently found, not surprisingly, in the blind. Presumably these individuals have a pathology that prevents the stimulus of environmental light from reaching the SCN.

## Chronobiological Aspects of Other Sleep Disorders

There are many other sleep disorders in the International Classification that have not been categorized as circadian rhythm sleep disorders per se, but which may have some aspect that makes them either conceptually or phenomenologically related to circadian processes. In some cases the links are clear, in others they are more speculative, but further research into these circadian influences seems indicated in nearly every dyssomnia and sleep-related disorder.

The sleep disturbances associated with mood disorders are of interest here partly because of the periodic appearance of the symptoms, but also because the phenomenon themselves have to do with the timing of sleep. These sleep disturbances are also associated with amplitude changes and changes in phase relationships among other rhythms, although the hope that a cause for mood disorders could be quickly elucidated by studying their relationship to circadian systems has not been fulfilled.

One of the hallmarks of endogenous depression is an insomnia characterized by early morning awakenings. A patient who is both anxious and depressed may also have trouble falling asleep, which makes this disorder difficult to differentiate from delayed sleep phase syndrome. In this case polysomnography might be useful, since another of the characteristics of sleep in depression is a change in the timing of REM sleep. In normal individuals and those who have a pure phase delay or advance, the first REM period usually occurs about 90 min after sleep onset, but in depressed patients it occurs considerably earlier. Depressed patients also have a greater amount of REM sleep (at the expense of delta sleep, stages 3 and 4), and this REM sleep tends to have a greater number of eye movements than that of normals.

While sleep apnea has generally been considered to be a sleep-related breathing disorder, recent studies suggest that there is a nocturnal circadian decrement in chemosensitivity (Raschke and Moller 1989) independent of sleep that may reduce respiratory drive or blunt the response to hypoxia or hypercapnia, worsening the condition. There is also a well-known sleep disruption due to nocturnal asthma. Since the disorder is associated with no particular stage of sleep and occurs late in the sleep episode rather than at sleep onset, it seems likely that the etiology of this disorder is more circadian than sleep related. There are a number of other medical conditions that occur during sleep and cause sleep disturbance, such as gastroesophageal reflux and nocturnal cardiac ischemia. Further investigation of these and other sleep disorders seems likely to find links to physiological changes elicited by the circadian pacemaker.

# References

American Sleep Disorders Association (1990) The international classification of sleep disorders. Diagnostic and coding manual. Allen, Lawrence

Aschoff J, Wever R (1962) Spontanperiodik des Menschen bei Ausschluß aller Zeitgeber. Naturwissenschaften 49: 337–342

Aschoff J, Gerecke U, Wever R (1967) Desynchronization of human circadian rhythms. Jpn J Physiol 17: 450–457

Benedict FG (1904) Studies in body-temperature. I. Influence of the inversion of the daily routine; the temperature of night-workers. Am J Physiol 11: 145–169

Czeisler CA (1986) A clinical method to assess the status of the human circadian pacemaker: applications to aging and shift-work. Symposium on Human Chronobiology, Annual AAAS Meeting, Philadelphia

Czeisler CA, Weitzman ED, Moore-Ede MC, Kronauer RE, Zimmerman JC, Campbell C (1980a) Human sleep: Its duration and structure depend on the interaction of two separate circadian oscillators. Sleep Res 9: 270

Czeisler CA, Zimmerman JC, Ronda JM, Moore-Ede MC, Weitzman ED (1980b) Timing of REM sleep is coupled to the circadian rhythm of body temperature in man. Sleep 2: 329–346

Czeisler CA, Richardson GS, Zimmerman JC, Moore-Ede MC, Weitzman ED (1981) Entrainment of human circadian rhythms by light-dark cycles: a reassessment. Photochem Photobiol 34: 239–247

Czeisler CA, Allan JS, Strogatz SH, Ronda JM, Sanchez R, Rios CD, Freitag WO, Richardson GS, Kronauer RE (1986) Bright light resets the human circadian pacemaker independent of the timing of the sleep-wake cycle. Science 233: 667–671

Czeisler CA, Kronauer RE, Allan JS, Duffy JF, Jewett ME, Brown EN, Ronda JM (1989) Bright light induction of strong (Type 0) resetting of the human circadian pacemaker. Science 244: 1328–1333

Czeisler CA, Johnson MP, Duffy JF, Brown EN, Ronda JM, Kronauer RE (1990) Exposure to bright light and darkness to treat physiologic maladaptation to night work. N Engl J Med 322: 1253–1259

Klein KE, Wegmann HM (1974) The resynchronization of human circadian rhythms after transmeridian flights as a result of flight direction and mode of activity. In: Scheving LE, Halberg F, Pauly JE (eds) Chronobiology. Igaku-Shoin, Tokyo, pp 564–570

Kleitman N (1963) Sleep and wakefulness. University of Chicago Press, Chicago

Kryger MH, et al. (eds) (1989) Principles and practice of sleep medicine. Saunders, Philadelphia

Lavie P, Wollman M, Pollack I (1986) Frequency of sleep related traffic accidents and hour of day. Sleep Res 15: 275

Mellor EF (1986) Shift work and flexitime: how prevalent are they? Monthly Lab Rev 11: 14–21

Mills J, Minors D, Waterhouse J (1978) The effect of sleep upon human circadian rhythms. Chronobiologia 5: 14–27

Mills JN (1964) Circadian rhythms during and after three months in solitude underground. J Physiol (Lond) 174: 217–231

Moore RY, Card JP (1985) Visual pathways and the entrainment of circadian rhythms. Ann NY Acad Sci 453: 123–133

Moore RY, Eichler VB (1972) Loss of a circadian adrenal corticosterone rhythm following suprachiasmatic lesions in the rat. Brain Res 42: 201–206

Raschke F, Moller KH (1989) The diurnal rhythm of chemosensitivity and its contribution to nocturnal disorders of respiratory control. Pneumologie 43: 568–571

Rechtschaffen A, Kales A (eds) (1968) A manual of standardized terminology, techniques and scoring system of sleep stages of human subjects. United States Public Health Service, US Government Printing Office, Washington

Rosenthal NE, Joseph-Vanderpool JR, Levendosky AA, Johnston SC, Allen R, Kelly KA, Souetre E, Schultz PM, Starz KE (1990) Phase-shifting effects of bright morning light as treatment for delayed sleep phase syndrome. Sleep 13: 354–361

Siffre M (1965) Beyond time. Chattle and Windus, London

Stephan FK, Zucker I (1972) Circadian rhythms in drinking behavior and locomotor activity of rats are eliminated by hypothalamic lesions. Proc Natl Acad Sci USA 69: 1583–1586

Strogatz SH, Kronauer RE, Czeisler CA (1987) Circadian pacemaker interferes with sleep onset at specific times each day: role in insomnia. Am J Physiol 253: R172–R178

# The Complex Circadian Pacemaker in Affective Disorders

D. F. Kripke, M. D. Drennan, and J. A. Elliott

## Introduction

Affective disorders are remarkably prevalent severe mental illnesses [58]. Most suicides – a major mechanism of mortality associated with mental illnesses – are associated with depressions. In terms of years of working life lost, suicide ranks with heart disease and cancer among the most economically significant health impairments of society [41]. In addition, affective disorders cause crippling and economically devastating disabilities [140].

Progress has been made in the nosology of affective disorders and in their treatment, yet the underlying etiologic mechanisms remain largely mysterious. A considerable component is known to be genetic. Hope that the defect had been localized on chromosome 11 has been set back by new data indicating that the genetics are either more complicated or quite different from what had been thought [57]. More genetic work is needed. Numerous theories involving neurotransmitters, receptors, membrane defects, hormones, and metallic ions compete with theories on the cognitive and social levels of organization. So far, a satisfactory synthesis and validation of theories has not been achieved to explain what must be a multifactorial process. There is today enough evidence to believe that chronobiologic or photoperiodic mechanisms must be involved.

Decades ago, Halberg [40] proposed that periodic depressions might result from free-running circadian rhythms. A variety of theories of how phototherapy works have been proposed, including phase-advance and phase-delay models, correction of low circadian amplitude, effects on a low melatonin syndrome, effects on a critical interval, effects on total daily photon exposure, etc. [20, 61, 67, 113, 135, 137]. None of these models has yet seemed fully satisfactory, we

Supported by MH00117 and the Department of Veterans Affairs.

suspect, because they do not recognize the complexity of the circadian pacemaker in humans. In this discussion, we shall see how studies of animal photoperiodism and analyses of the complex circadian pacemaker in lower species lead to a more complex theory of how circadian pacemakers might be disordered in human affective disorders.

## Photoperiodism

Many aspects of animal photoperiodism have recently been elegantly explained. Seasonal responses in hamsters and other species are under photoperiodic control mediated by the circadian system [35]. It is not merely the duration of daylight – but rather the time relationship of photic exposures to the circadian system - which controls the seasonal neuroendocrine response. Even very brief pulses of light can induce long-day responses in experimental rodents when they are timed correctly. Such responses include such gross endocrine responses as testicular growth or atrophy, as well as complex behaviors including reproductive behavior. Perhaps depression arises from such seasonal response mechanisms gone awry.

An interesting model was described by Eskes and Zucker [37], who showed that deuterium could facilitate gonadal growth the delaying the circadian system such that the critical photosensitive interval of hamsters was partly exposed to light in short days. This demonstration that a phase-delaying drug could alter photoperiodic responsiveness was a particularly attractive model for a possible mechanism of antidepressant drugs.

Mechanisms of photoperiodism have been further elucidated. Substantial evidence suggests that day length controls seasonal responses through regulation of the *duration* of nocturnal melatonin secretion [17, 149]. Melatonin synthesis and secretion rise remarkably at night in both nocturnal animals such as rodents and in diurnal animals such as man. Light exposure promptly suppresses melatonin in nocturnal rodents [25, 99]. Lewy et al. [74] showed that diurnal humans require light much brighter than ordinary artificial illumination to suppress melatonin. Evidently, melatonin acts on the hypothalamus to alter neuroendocrine functions controlling metabolism, reproduction, behavior, etc.

## Multiple-Oscillator System

Vertebrate circadian systems are comprised of multiple circadian oscillators and multiple driven rhythms [86, 92, 93]. Normal physiologic function requires that these rhythms maintain appropriate phase relationships to one another. This temporal organization is achieved through entrainment of one or more central pacemakers by the 24-h periodicity of the environment (principally light).

There is much evidence that circadian rhythms of activity and melatonin are controlled by a complex circadian pacemaker consisting of at least two mutually coupled oscillators [47, 51, 94]. The evening oscillator, E, controls activity onset in nocturnal rodents and the onset of melatonin secretion. The morning oscillator, M, controls activity offset in nocturnal rodents and the offset of melatonin secretion. The phase-angle difference between E and M ($psi_{E-M}$) determines properties of the complex pacemaker (e. g., period length of the coupled system, and responses to light) and of the rhythms it drives (e. g., duration of activity time and duration of pineal melatonin secretion). This two-oscillator model of the complex circadian pacemaker was first described by Pittendrigh and Daan [94] based on observations of running-wheel activity, and responses to light and dark in several rodent species. Later, Illnerova, Elliott and colleagues presented evidence that these same two components control the pineal NAT rhythm in rats (N-acetyltransferase is the key enzyme in circadian control of pineal melatonin synthesis) and pineal melatonin rhythms in hamsters [25, 38, 47, 50, 51]. Normally, under entrainment to a constant photoperiod, the evening and morning oscillators are *coupled* together in a stable phase relationship; however, the phase-angle difference, $psi_{E-M}$, varies as a function of photoperiod, during resetting transients, and during rhythm splitting in constant light [33, 34, 51, 94].

## Melatonin

The pineal hormone melatonin plays an important role in vertebrate circadian systems [132]. In lower vertebrates (e. g., some birds and lizards), pinealectomy results in circadian period changes, dissociation of circadian rhythms, or even abolition of circadian rhythmicity. In lizards, pinealectomy alters the PRC to light pulses, and melatonin injections entrain activity rhythms through a circadian phase-dependent resetting mechanism (PRC to melatonin) [131]. Recent evidence in rats, hamsters and humans suggests that melatonin may also be a weak entraining agent of mammalian rhythms [2, 27]. In Syrian hamsters and mice, continuous melatonin administration via silastic capsule implants blocked changes in activity time (alpha) produced by entrainment to non-24-h LD cycles, suggesting an effect on E-M coupling (Elliott, unpublished). However, the physiologic importance of melatonin as a hormonal coupling agent and synchronizer of mammalian rhythms remains somewhat controversial. To date the most impressive function of melatonin in mammals is its crucial role in regulating seasonal change [6, 38, 132].

Long (e. g., winter) nights seem to produce longer intervals of nocturnal melatonin secretion, which, in turn, mediate winter hormonal responses. In rats, melatonin synthesis and secretion are controlled largely by NAT (N-acetyltransferase). Illnerova and colleagues have reported that in rats the duration of nocturnal melatonin secretion is controlled by a complex circadian pacemaker with evening oscillator and morning oscillator components [46, 49]. The phase-response curves of the evening NAT onset and morning NAT offset seem only partially coupled. It appears possible to so compress the phase-angle between the two oscillator components that, for several nights, NAT does not rise at all. This may be somewhat analogous to Czeisler's concept of light pushing the human circadian oscillator through a phase singularity [21]. Thus, photoperiod seems to work by regulating the coupling of the evening and morning components of the complex pacemaker, which in turn regulate the duration of melatonin secretion.

Studies employing timed daily square-wave infusions of melatonin to pinealectomized hamsters and sheep have indicated that the duration of the nocturnal peak in melatonin is of primary importance in transmitting the photoperiodic message [6, 36]. The interval of endogenous melatonin secretion is longer in the long nights and short days of winter than in the short nights and long days of summer. It is this interval of melatonin secretion (i.e., peak width modulation) which appears to transmit the seasonal (photoperiodic) message to the neuroendocrine system.

The question of how the duration of the melatonin signal is read by the neuroendocrine system(s) which mediate seasonally appropriate responses to melatonin has proved difficult to answer [36, 44]. The response to melatonin peak duration appears independent of photoperiod duration and of circadian timing of infusions [6, 27, 36]. Nonetheless, melatonin pulses of appropriate duration must be infused at approximately circadian intervals to effectively transmit a photoperiodic message, and the duration of the melatonin-free interval is also important [36, 44]. Results with melatonin injections in pineal intact and pinealectomized hamsters have suggested the possibility that a rhythm of sensitivity to melatonin may play some role in regulating the hormone's effects [36, 126].

Thus, the influence of the circadian timing of the melatonin signal has not been ruled out. It is possible that some effects of melatonin on seasonal responses are circadian-phase-dependent, or that melatonin is involved as an internal synchronizer which influences the coupling between circadian oscillators. Indeed, the available data are consistent with the possibility that the hormone acts by entraining oscillators which may be only weakly or indirectly entrainable by light. In the broad comparative view, all circadian and photoperiodic functions of melatonin may derive from this hormone's actions as an internal synchronizer. In this context, we feel it is important to further evaluate the influence of melatonin as a synchronizer of mammalian circadian rhythms and to test the hypothesis that melatonin may play a role in mediating the effects of photoperiod on $psi_{E-M}$ and on the amplitude of the circadian pacemaker's response to light.

Important discoveries have been made of the concrete neuronal systems underlying circadian phenomena. Richter's initial report that the circadian clock was in the anterior hpyothalamus has been clarified to show that circadian rhythms in rodents are controlled by the suprachiasmatic nucleus. Photic stimuli of the retinae are conveyed by the retinohypothalamic tract to the suprachiasmatic nuclei, which control pineal conversion of serotonin to melatonin. The SCN may contain the complex pacemaker which regulates melatonin [85]. Besides serotonin, the SCN-pineal pathway involves cholinergic and noradrenergic neuronal relays, thus combining the main neurotransmitters which have been implicated in depression [73].

More progress has been made in understanding the structure and connections of the suprachiasmatic nucleus. The SCN seems to play the governing role in many circadian rhythms, especially in endocrine functions [110]. Substantial evidence points to the SCN as the predominant circadian pacemaker in mammals (thus, probably in humans). Several groups have shown that SCN transplants restore circadian rhythms to arrhythmic SCN-lesioned animals [28, 29]. Menaker's group found a hamster clock mutant and showed that transplant of fetal SCNs can transplant the donor's genetic circadian clock frequency [96].

Though the complex pacemaker model seems strongly supported as the controller of mammalian photoperiodism (and the complex hypothalamic responses it entails), the mechanism by which melatonin secretion duration acts on the hypothalamus remains to be elucidated. Recent evidence has tended to localize melatonin receptors both in the infundibular area (where melatonin may regulate the secretion of hypothalamic-releasing peptides into the portal circulation) and in the SCN [100, 133]. Thus, the SCN may regulate the secretion of melatonin which feeds back upon itself. It is tempting to speculate that a similar mechanism may play a role in development of human depression.

## Melatonin-Pacemaker Interactions

Melatonin's regulation by light and photoperiod is complex. It cannot be adequately explained either by acute suppression by light or by simple circadian entrainment. Rather, the photoperiod regulates the interval of melatonin synthesis through regulation of the complex circadian pacemaker [26, 36, 47, 52]. More specifically, by regulating the phase relationship, $psi_{E-M}$, the time interval between evening and morning components of the complex pacemaker, photoperiod regulates the duration of melatonin secretion, and thus seasonal changes in physiology (reproduction, metabolism, etc.). This has been termed an internal-coincidence model of photoperiodism. Photoperiod also regulates the resetting properties of the pacemaker – the range and amplitude of its phase-response curve (or PRC) [33, 38]. By inference, by regulating $psi_{E-M}$ and thus the strength of coupling of E and M, photoperiod affects not only the range and amplitude of the PRC but, in addition, the relative strength and amplitude of the complex circadian pacemaker. Theoretically, changes in $psi_{E-M}$ of the pacemaker, changes in complex pacemaker amplitude, and changes in PRC amplitude are all likely to affect the internal phase relationship among circadian rhythms driven by the pacemaker. We hypothesize that analogous changes in pacemaker state may underlie both seasonal and nonseasonal forms of depression.

## Light Intensity

When it was first reported that relatively high intensity artificial light ( > 2000 lux) is required to entrain human circadian rhythms or inhibit nocturnal melatonin secretion, it appeared that man was quantitatively very different from other mammals with respect to these photic responses [74, 140, 145]. However, it was later reported that a much greater light intensity is required for inhibition of melatonin synthesis in wild-caught as compared to laboratory-reared ground squirrels [98]. This raises the possibility that previous exposure to very high intensities of natural sunlight may result in an increased response threshold. Species differences, including differences between diurnal and nocturnal mammals, clearly exist in the response to light intensity [97]. However, the influence of prior photic history has been largely neglected. Thus, exposure to very bright artificial light or natural illumination during all or part of the day, as compared to exposure to only dim artificial light, may influence circadian photic responses. Such considerations may be particularly important for humans, in whom bright versus dim light has such evident consequences for circadian entrainment and because of the growing clinical promise of bright-light phototherapy [70, 145].

## Circadian Pacemakers and Depression

The human circadian system has been conceptualized as consisting of a "strong" oscillator driving core temperature, cortisol, and, possibly, melatonin secretion, and a "weak" oscillator driving the rest/activity cycle [68, 69]. Theoretically, depression might be caused by any of the following disturbances of the circadian pacemaker:
1. There may be an abnormal phase angle either internally, between oscillators, or between the oscil-

lators and the environment. This may, for example, be a phase advance in the "strong" oscillator relative to the "weak" [60, 139].
2. There may be a disturbance of circadian *amplitude*, perhaps due to insufficient or poorly timed light exposure [11, 13, 20]. Alternatively, an excessive constitutional sensitivity to light could cause excessive suppression of melatonin at "room light" intensities.
3. A more elaborate model might conceptualize the "strong oscillator" as a complex circadian pacemaker with coupled morning and evening oscillator components [94] and process S (or the sleep-wake alternation) as a third process or qualitatively distinct oscillatory component [24]. Abnormal coupling of the components of the complex circadian pacemaker might contribute to sleep/wake disturbances, just as such coupling alters patterns of activity and rest in rodent models. An abnormally acute phase angle between morning and evening oscillators could cause low melatonin via a "compression" mechanism [46]. Abnormalities of melatonin secretion suggest that just such pacemaker disorders may exist in patients with major depressive disorders.

Over a period of years, Wetterberg, Beck-Friis, and colleagues have developed evidence for melatonin abnormalities in patients with endogenous depressions, particularly a "low melatonin syndrome" [8–10, 141, 142]. Confirmation of decreased melatonin secretion in patients with major depressions has come from several laboratories [14, 15, 18, 84, 119]. Melatonin also decreases consistently with aging [87, 112]. In concordance with these findings, several studies have shown that antidepressant drugs can increase melatonin secretion [42, 116, 121, 127], which might suggest correction of "low melatonin syndrome" as a mode of antidepressant action. Nevertheless, one of the most careful contrasts of depressed patients with controls uncovered no abnormality at all in the amplitude of nocturnal melatonin secretion, though there was some hint of decreased duration of secretion in the depressed group [128]. Meticulous studies of depressed outpatients by the Pittsburgh group (Jarrett, personal communication) also have failed to reveal melatonin abnormalities. Possibly, in some previous studies, hospitalized patients and controls did not experience similar lighting environments in the days immediately before the studies.

Somewhat similar discrepancies have appeared regarding melatonin sensitivity to light suppression. Initially, Lewy et al. [76] reported that patients with bipolar disorders were abnormally sensitive to suppression of nocturnal melatonin by intermediate illumination intensities, e. g., 500 lux. Nurnberger et al. [88] found that increased sensitivity to suppression of nocturnal melatonin was a trait marker in families affected by affective disorders. Recently, some groups have reported that nocturnal melatonin can be suppressed by illumination intensities as low as 200–300 lux, which Lewy et al. had previously reported did not suppress melatonin in normal adults [12, 81–83]. In San Diego, in two studies, [19, 70] we have been unable to demonstrate abnormal sensitivity of melatonin to light suppression among either bipolar or unipolar depressed drug-free patients contrasted to controls.

Several factors should be considered in trying to explain discrepancies in reports from different groups. First, affective disorders are undoubtedly a somewhat heterogeneous group of illnesses. Second, there is considerable evidence that biological features of depression such as dexamethasone resistance may vary geographically, especially with latitude. Indeed, environmental illumination and temperatures may be factors in geographic differences in the patterns of depression. Finally, depressed patients may not be well matched with controls for such factors as daytime illumination exposure.

A crucial experimental variable is the timing of bright light used to suppress melatonin. Lewy's initial studies tested bright light from 0200 to 0400 hours, whereas our San Diego studies tended to start an hour or two earlier. Because neither group obtained all-night baseline data before attempting melatonin suppression, it is possible that some apparent melatonin suppression in bipolars was due to a phase-timing advance in the spontaneous morning offset of melatonin secretion (similar to that demonstrated by Dr. Parry among patients with PMS) [91]. Our San Diego patients and controls also tended to be older, which could be a factor.

Considering the potential importance of melatonin abnormalities among patients with affective disorders and, furthermore, the discrepancies among reports, it is apparent that additional studies must be obtained with more careful attention to subject groups of adequate size, careful control for age, sex, demography, and environmental factors including prior illumination history, and better standardization of subjects including collection of baseline melatonin profiles.

## Circadian Phase and Depression

Several studies support the notion that, as a population, patients with winter depression have phase-delayed (i. e., slower) circadian rhythms when depressed compared with normal control subjects. Studies not supporting this idea have come from the NIMH group. Levondovsky et al. showed no temperature phase abnormalities, and no changes with light treatment relative to normal controls [108]. Likewise, Rosenthal found no abnormalities in nocturnal melatonin secretory onset or offset in SAD [107]. The studies suffered from a lack of controlled sleep timing, which is known to mask the temperature rhythm [143] and lack of control over daytime light exposure, which could influence nocturnal melatonin secretory phase. When sleep timing and light exposure have been controlled, delayed circadian phase has been observed in most SAD patients. Thus, the core temperature minimum appeared delayed under normal [30] and "constant routine" conditions [4, 5]. Avery also demonstrated a delay in TSH and cortisol rhythms in the same experiment. Lewy found that, when evening light exposure is enhanced, morning light is avoided, and sleep phase held constant, seasonal depressives had a greater delay in dim light melatonin onset (DLMO) [78, 79]. Likewise, when sleep and light exposure were simply controlled, winter depressives had a later DLMO (by 1–1/2 h) than normal controls [5, 77]. Some studies have found no overall effect of clinically significant light treatment on phase. This may be due to design problems, causing inability to detect abnormal phase relative to normal controls [108] or use of a relatively insensitive 4-h urinary melatonin collection [147]. Caution must be used in drawing the conclusion that bright light works in SAD *only* via phase advance. A substantial proportion of patients may respond to either morning or evening light [124]. Evening light would be expected to exacerbate phase-delay abnormalities. Furthermore, midday phototherapy is effective in SAD [53], though this finding could be explained by the phase-advance induced by midday light in primates [45] and humans [22].

The UCSD group has pursued studies of circadian rhythm phases in nonseasonal depression since 1973. Although UCSD and other groups have described a tendency for phase advance (or even short tau) among manic-depressives, the largest UCSD study suggested increased circadian phase heterogeneity among affective patients without consistent mean phase abnormalities [62]. A recent study in Japan [130] reached a similar conclusion. Elsewhere, several authors have described phase advances among patients with major depressive disorders [39, 80, 134]; however, there have also been careful studies which failed to demonstrate consistent abnormalities of mean phase within a group of depressed patients [3, 109, 118]. Parry et al. [90] have found that women with premenstrual depression have earlier rectal temperature minima than controls at all menstrual phases. Likewise, the *offset* (but not the onset) of nocturnal melatonin secretion is advanced in women with PMS as compared to controls [91]. This begins to suggest that the morning oscillator (but perhaps not the evening oscillator) may be phase advanced among women with premenstrual depression.

Perhaps inconsistencies in this area derive from efforts to measure the global phase of the circadian system, without analyzing the separate phase angles of the morning and evening components of a complex circadian pacemaker. In conclusion, diverse findings regarding circadian phase and melatonin abnormalities in depression need to be clarified with more systematic and meticulous studies.

## Phototherapy

Inspired by the basic science of photoperiodism, in 1981, the UCSD group began what may have been the first *controlled* trial of phototherapy for treatment of an affective disorder [59]. Our focus has been on hospitalized veterans with nonseasonal major depressive disorders. At about the same time, the NIMH group dramatically demonstrated the benefit of phototherapy in one patient with winter depression, initiating the contemporary interest in seasonal affective disorders and their response to phototherapy [75]. Wehr and Goodwin [134] have shown that uncontrolled use of sunlight to treat depression can be traced to antiquity. The use of artificial light treatments was popular around the beginning of this century [56] but was not supported by controlled studies for treatment of depression. Controlled light trials were the achievement of the 1980s. By 1990, preliminary studies had described antidepressant phototherapy effects in both seasonal and nonseasonal depressions, and in premenstrual depression. Phototherapy may relieve depressive and sleep disturbances in shift workers [23, 32].

Chronobiologic and phototherapeutic techniques have rapidly attracted interest as a new approach to affective disorders. The potential impact of photo-

therapy can be guaged from the explosive growth of the new Society For Light Treatment and Biological Rhythms, which in its 2 years has grown to about 400 members interested in phototherapy. Rough epidemiologic estimates suggest that, in some parts of the United States, 10%–20% of the population or more might benefit from phototherapy [54, 55]. Very encouraging controlled phototherapeutic results have been obtained with winter depressions [102], with premenstrual depressions [89], and with episodic nonseasonal depressions [67], but the mechanisms of action are quite controversial and require further explanation. Careful long-term trials of phototherapy, determination of optimal regimens, and controlled contrasts with standard antidepressant therapies have scarcely begun.

An excellent book on the subject of phototherapy for seasonal affective disorder has been previously published [102]. Lewy [75] described phototherapy in a seasonally depressed man. A full description of the syndrome "seasonal affective disorder" with the first randomized trial of phototherapy appeared later [105]. Such patients become depressed in the fall/winter with the depression resolving in the spring/summer. Depressive symptoms are usually described as "atypical" with prominent hypersomnia (85% of cases), hyperphagia (70%), weight gain (75%), carbohydrate craving (75%), low activity or psychomotor retardation (94%) and low energy [126]. Up to 30% may present with typical endogenous symptomatology (insomnia, weight loss, etc.) when depressed [104, 105]. Nearly 50% of patients report onset of symptoms in childhood or adolescence [104, 105]. Epidemiologically, women appear to be affected more frequently than men, by 3–4:1 in self-referred populations. Seasonal patients may represent up to 16.1% of specialty clinic recurrent depressives [126].

Phototherapy might become the treatment of choice for seasonal affective disorder. Though open trials have described successful long-term treatment [16, 105], critical treatment parameters have been explored using trials of only 1–2 weeks duration, using self-referred patients recruited chiefly by advertisement [106]. The original articles describing benefits of phototherapy used a combination of bright morning and evening light to elicit effects. However, bright early morning light, i.e., (0600–0800 hours) alone may provide optimal efficacy in deficiency for most SAD patients [77, 113, 124, 125]. Meta-analysis revealed that, in crossover designs, 59% responded preferentially to morning light, but few (10%) responded preferentially to evening light. Alternatively, 53% of SAD patients respond with normalization of symptoms to either morning or a combination of morning and evening light, whereas less than or equal to 38% of patients normalize using evening light. These differential responses disappeared in the most severely affected patients [125].

Critical duration of exposure has been tested primarily using 2500 lux fluorescent light, which is sufficient to suppress melatonin secretion [74]. Most studies support increasing efficacy over the 30-min to 2-h range [124, 146]. Thus, 2 h of 2500 lux lighting would appear to be an optimal duration. One study, however, demonstrates that 30 min of 10 000 lux light may be an effective replacement for 2 h of 2500 lux light [119].

Few studies have addressed the question of lighting spectrum. Lam et al. [72] concluded that ultraviolet spectrum lighting augments the antidepressant response using a UV-filtering mask paradigm. However, others have disagreed with this finding [77, 106]. It appears that fluorescent lighting excluding the ultraviolet spectrum is fully effective. Use of such lighting is desirable in view of the known risks to skin from UV radiation.

The side effects of phototherapy include eye redness and watering, headaches, and irritability which may resemble hypomania.

Unlike phototherapy for seasonal depression, there is no currently established circadian intervention for nonseasonal depression which can be recommended for routine use. Proposed treatments fall into two categories: phototherapy and the manipulation of sleep phase.

The UCSD group has now completed four studies showing significant benefits of bright light treatment for hospitalized veterans with nonseasonal major depressive disorders (three significant contrasts with dim light placebo, one significant contrast with baseline) [63–67]. In our most recent study, the patients treated with bright white light in the evening experienced reduced Hamilton and Beck depression ratings, even though subjects *expected more benefit from placebo red light*. It should be emphasized that these results are only for 7 days of treatment, and phototherapy has not yet been contrasted with standard antidepressant therapies. Nevertheless, because the magnitude of benefit from *1 week* of bright light was of comparable magnitude to benefits of *4 weeks* of antidepressant drugs, further study of this therapeutic modality is certainly indicated. Our data do *not* demonstrate whether morning or evening bright light treatment is more promising for nonseasonal depression.

A number of other groups have likewise reported promising results with phototherapy of nonseasonal depressions [31, 43, 129]. Particularly interesting are the data of Prasko et al. [95], who found that the addition of 6 days phototherapy to antidepressant drugs could double the rate of recovery. In this study, morning phototherapy produced more benefit than afternoon or evening phototherapy; however, the morning phototherapy also entailed sleep deprivation (which was not balanced), leaving the question of optimal timing still unresolved.

Our colleague, Dr. Barbara Parry, has been studying women who suffer from premenstrual depressions of disabling intensity in the late luteal phase of their menstrual cycles. There is evidence in a few patients that evening phototherapy (which might delay the complex circadian pacemaker) may be antidepressant [89].

An excellent review of sleep deprivation therapy for major depression has been recently published [148] and is beyond the scope of this text. In studies including over 1700 total patients, 59% of patients with major depression responded dramatically to one night's total sleep deprivation, but relapsed with recovery sleep [148]. This finding is of interest to circadian physiologists because it may be partly explicable via a "phase advance hypothesis." Sleep deprivation may be thought of as advancing (causing to occur at an earlier time) wakefulness onset such that it temporarily corrects a hypothesized phase angle disturbance between a temperature rhythm which "runs fast" or "early" in depression and a sleep/wake rhythm which is scheduled "normally" to external synchronizers. Alternatively, a "sleep sensitive interval" may become phase-advanced into the normal sleeping period, causing depression [139], and sleep deprivation during this critical interval would, therefore, be antidepressant. Support for both hypotheses comes from the discovery that early morning partial sleep deprivation is antidepressant, whereas deprivation in the first half of the night is not [114].

Furthermore, a 6-h phase advance in sleep timing without actual sleep deprivation appeared antidepressant [138] and useful as adjunctive therapy in refractory, medicated patients [111]. One inconclusive study noted that two out of ten normal subjects experienced significant depression after a forced acute 6-h phase delay in the sleep/wake cycle, but there were no overall significant effects on mood [122].

Sleep deprivation may be practically useful, since pretreatment with mood-active drugs appears to prolong benefit [7]. However, circadian explanations for its antidepressant efficacy are only one of several potential hypotheses (reviewed in [148]).

## Isolation and Depression

If the general hypothesis that circadian rhythm disruptions can trigger affective symptoms is correct, it would be predictable that certain temporal isolation regimes might trigger depression. Indeed, in our opinion, there is considerable evidence that depressions have, in fact, been triggered by temporal isolation studies. For example, one subject committed suicide soon after experiencing internal circadian desynchronization [101]. A women's record-holder for isolation in a cave committed suicide about 1 year after that experience [1], and a psychiatrist who interviewed her feels she was depressed when she left the cave. The young Italian woman who broke that prior cave record became so depressed she suffered a 20 % weight loss underground [115, 117]. Monk [84] has reported systematic data showing steadily lowering mood in isolation from two subjects, but unfortunately much of the systematic mood data collected from hundreds of previous isolation subjects have never been reported in print.

Several possible mechanisms by which isolation might cause depression must be considered. First, since temporal isolation can lead to internal circadian desynchronization, an unfavorable set of internal phase relationships between circadian systems might occur. Second, because of the loss of circadian synchronizers, the amplitude of circadian rhythms might be dampened. Third, most of the isolation experiments have been conducted in very low light conditions, which might affect melatonin regulation. Finally, there are the social aspects of isolation, which may include certain stressful or coercive aspects [117], as well as loneliness and interruption of habitual social supports.

From a theoretical viewpoint, the depressions which temporal isolation may precipitate are of great interest in exploring the etiology of affective disorders. From a practical viewpoint, this potential for life-threatening depression demands deliberate caution in future isolation designs.

## Conclusions

Circadian rhythm theories addressing the etiology of affective disorders have both exploited and moved beyond Halberg's original hypothesis. Phototherapy appears useful in seasonal affective disorder (SAD) and may be useful in nonseasonal and premenstrual depressions. Simple phase-related hypotheses may partially explain the symptoms of SAD, and its response to phototherapy, but some patients do not fit the model. Considerable evidence has accumulated against a simple disturbance of phase in other affective disorders, such as nonseasonal and premenstrual depressions. Advances in the animal literature describing a complex circadian oscillator may provide a basis for explaining many of the abnormalities in both phase *and* amplitude found in affective disorders. Further investigation of this new hypothesis and more empirical testing of phototherapy is required.

## References

1. Anonymous (1990) Woman in isolation test is found dead. L A Times, January 19: A28
2. Armstrong SM (1989) Melatonin and circadian control in mammals. Experientia 45: 932–938
3. Avery DH (1987) REM sleep and temperature regulation in affective disorder. In: Halaris A (ed) Chronobiology and psychiatric disorders. Elsevier, New York, pp 75–101
4. Avery D, Dahl K, Savage M, Brengelmann G (1989) Phase-typing of seasonal affective disorder using a constant routine. Annual Meeting of the Society for Light Treatment and Biological Rhythms, 14 (abstract)
5. Avery D, Dahl K, Savage M, Brengelmann G (1990) Rectal temperature and TSH in winter. Sleep Res 19: 91
6. Bartness TJ, Goldman BD (1989) Mammalian pineal melatonin: a clock for all seasons. Experientia 45: 939–945
7. Baxter LR Jr, Liston EH, Schwartz JM, et al. (1986) Prolongation of the antidepressant response to partial sleep deprivation by lithium. Psychiatry Res 19: 17–23
8. Beck-Friis J, Kjellman BF, Aperia B, Unden D, von Rosen D, Ljunggren JG, Wetterberg L (1985) Serum melatonin in relation to clinical variables in patients with major depressive disorder and a hypothesis of low melatonin syndrome. Acta Psychiatr Scand 71: 319–330
9. Beck-Friis J, Kjellman BF, Ljunggren J-G, Wetterberg L (1984) The pineal gland and melatonin in affective disorders. In: Brown GM, Wainwright SD (eds) The pineal gland endocrine aspects. Pergamon, New York, pp 313–325
10. Beck-Friis J, von Rosen D, Kjellman BF, et al. (1984) Melatonin in relation to body measures, sex, age, season and the use of drugs in patients with major affective disorders and healthy subjects. Psychoneuroendocrinology 9: 261–277
11. Beck-Friis J, Borg G, Wetterberg L (1985) Rebound increase of nocturnal serum melatonin levels following evening suppression by bright light exposure in healthy men: relation to cortisol levels and morning exposure. In: Wurtman RJ, Baum MJ, Potts JT Jr (eds) The medical and biological effects of light. New York Academy of Sciences, New York, pp 371–375
12. Bojkowski CJ, Aldhous ME, English J, et al. (1987) Suppression of nocturnal plasma melatonin and 6-sulphato-

xymelatonin by bright and dim light in man. Horm Metab Res 19: 437–440

13. Borg G, Beck-Friis J, Kjellman BF, Thalen BE, Unden F, Wetterberg L (1987) Nocturnal serum melatonin levels in response to different intensities of evening bright light exposure in healthy man. In: Eur Pineal Study Group Newslett [Suppl] 7: 124

14. Brown RP, Kocsis JH, Caroff S, et al. (1985) Differences in nocturnal melatonin secretion between melancholic depressed patients and control subjects. Am J Psychiatry 142: 811–816

15. Brown RP, Kocsis JH, Caroff S, et al. (1987) Depressed mood and reality disturbance correlate with decreased nocturnal melatonin in depressed patients. Acta Psychiatr Scand 76: 272–275

16. Byerley WF, Brown J, Lebegue B (1987) Treatment of seasonal affective disorder with morning light. J Clin Psychiatry 48: 477–478

17. Carter DS, Goldman BD (1983) Antigonadal effects of timed melatonin infusion in pinealectomized male Djungarian hamsters: duration is the critical parameter. Endocrinology 113: 1261–1267

18. Claustrat B, Chazot G, Brun J, Jordan D, Sassolas GA (1984) Chronobiological study of melatonin and cortisol secretion in depressed subjects: plasma melatonin, a biochemical marker in major depression. Biol Psychiatry 19: 1215–1228

19. Cummings MA, Berga SL, Cummings KL, et al. (1989) Light suppression of melatonin in unipolar depressed patients. Psychiatry Res 27: 351–355

20. Czeisler CA, Kronauer RE, Mooney JJ, Anderson JL, Allan JS (1987) Biologic rhythm disorders, depression, and phototherapy. Psychiatr Clin North Am 10: 687–709

21. Czeisler CA, Kronauer RE, Allan JS, et al. (1989) Bright light induction of strong (type 0) resetting of the human circadian pacemaker. Science 244: 1328–1332

22. Czeisler CA, Kronauer RE, Johnson MP, Allan JS, Johnson TS, Dumont M (1989) Action of light on the human circadian pacemaker: treatment of patients with circadian rhythm sleep disorders. In: Horne J (ed) Sleep '88. Fischer, Stuttgart pp 42–47

23. Czeisler CA, Johnson MP, Duffy JF, Brown EN, Ronda JM, Kronauer RE (1990) Exposure to bright light and darkness to treat physiologic maladaptation to night work. N Engl J Med 322: 1253–1259

24. Daan S, Beersma GM, Borbely A (1984) Timing of human sleep: recovery process grated by a circadian pacemaker. Am J Physiol 15: R161–R183

25. Darrow JM, Goldman BD (1985) Circadian regulation of pineal melatonin and reproduction in the Djungarian hamster. J Biol Rhythms 1: 39–54

26. Darrow JM, Goldman BD (1986) Circadian regulation of pineal melatonin and reproduction in the Djungarian hamster. J Biol Rhythms 1: 39–53

27. Darrow JM, Goldman BD (1986) Effect of pinealectomy and timed melatonin infusions on wheel-running rhythms in Djungarian hamsters. Neurosci Abstr 12: 843

28. DeCoursey PJ, Buggy J (1988) Restoration of circadian locomotor activity in arrhythmic hamster by fetal SCN transplants. Comp Endocrinol 7: 49–54

29. DeCoursey PJ, Buggy J (1989) Circadian rhythmicity after neural transplant to hamster third ventricle: specificity of suprachiasmatic nuclei. Brain Res 500: 263–275

30. Depue R (1989) Effects of light on the biology of seasonal affective disorder. Annual Meeting of the Society for Light Treatment and Biological Rhythms, 16

31. Dietzel M, Saletu B, Lesch OM, Sieghart W, Schjerve M (1986) Light treatment in depressive illness: polysomnographic, psychometric and neuroendocrinological findings. Eur Neurol 25: 93–103

32. Eastman CI (1987) Bright light in work-sleep schedules for shift workers: application of circadian rhythm principles. In: Rensing L, van der Heiden U, Mackey MC (eds) Temporal disorder in human oscillatory systems. Springer, Berlin Heidelberg New York, pp 176–185

33. Elliott JA, Pittendrigh CS (1990) Photoperiodic regulation circadian phase response curves in Syrian hamsters. I. Time course of resetting following single light pulses.

34. Elliott JA, Pittendrigh CS (1990) Photoperiodic regulation of circadian phase response curves in Syrian hamsters. II. After-effects of entrainment to light cycles.

35. Elliott JA, Stetson MH, Menaker M (1972) Regulation of testis function in golden hamsters: a circadian clock measures photoperiodic time. Science 178: 771–773

36. Elliott JA, Bartness TJ, Goldman BD (1989) Effect of melatonin infusion duration and frequency on gonad, lipid and body mass in pinealectomized male Siberian hamsters. J Biol Rhythms 4: 439–455

37. Eskes GA, Zucker I (1978) Photoperiodic regulation of the hamster testis: dependence on circadian rhythms. Proc Natl Acad Sci USA 75: 1034–1038

38. Goldman BD, Elliott JA (1988) Photoperiodism and seasonality in hamsters: role of the pineal gland. In: Stetson MH (ed) Processing of environmental information in vertebrates. Springer, Berlin Heidelberg New York, pp 203–218

39. Gwirtsman HE, Halaris AE, Wolf AW, DeMet E, Piletz JE, Marler M (1989) Apparent phase advance in diurnal MHPG rhythm in depression. Am J Psychiatry 146: 1427–1433

40. Halberg F (1968) Physiologic considerations underlying rhythmometry, with special reference to emotional illness. Symp Bel-Air III, Masson, Paris, pp 73–126

41. Hamburg DA, Elliott GR, Parron DL (1982) In: Health and behavior: frontiers of research in the behavioral sciences. National Academy Press, Washington

42. Hariharasubramanian N, Nair NPV, Pilapil C, Isaac I, Quirion R (1986) Effect of imipramine on he circadian rhythm of plasma melatonin in unipolar depression. Chronobiol Int 3: 65–69

43. Heim M (1988) Zur Effizienz der Bright-Light-Therapie bei zyklothymen Achsensyndromen – eine cross-over-Studie gegenüber partiellem Schlafentzug. Psychiatr Neurol Med Psychol (Leipz) 40: 269–277

44. Herbert J (1989) Neural systems underlying photoperiodic time measurement: a blueprint. Experientia 45: 965–972

45. Hoban TM, Sulzman FM (1985) Light effects on circadian timing system of a diurnal primate, the squirrel monkey. Am J Phsyiol 249: R274–R280

46. Illnerova H, Humlova M (1990) The rat pineal N-acetyltransferase rhythm persists after a five-hour, but disappears temporarily after a seven-hour advance of the light-dark cycle: a six-hour shift may be a turning point. Neurosci Lett 110: 77–81

47. Illnerova H, Vanecek J (1982) Two-oscillator structure of the pacemaker controlling the circadian rhythm of N-acetyltransferase in the rat pineal gland. J Comp Physiol 145: 539–548

48. Illnerova H, Vanecek J (1982) Two-oscillator structure of the pacemaker controlling the circadian rhythm of *N*-acetyltransferase in the rat pineal gland. J Comp Physiol 145: 539–548

49. Illnerova H, Vanecek J (1987) Entrainment of the circadian rhythm in the rat pineal *N*-acetyltransferase activity by prolonged periods of light. J Comp Physiol 161: 495–510

50. Illnerova H, Vanecek J (1987) Dynamics of discrete entrainment of the circadian rhythm in the rat pineal *N*-acetyltransferase activity during transient cycles. J Biol Rhythms 2: 95–108

51. Illnerova H, Vanecek J (1989) Complex control of the circadian rhythm in pineal melatonin production. In: Mess B, Ruzsas C, Tima L, Pevet P (eds) The pineal gland: current state of pineal research, Elsevier, New York, pp 137–153

52. Illnerova H, Vanecek J, Hoffmann K (1989) Different mechanisms of phase delays and phase advances of the circadian rhythm in rat pineal *N*-acetyltransferase activity. J Biol Rhythms 4: 187–200

53. Jacobsen FM, Wehr TA, Skwerer RA, Sack DA, Rosenthal NE (1987) Morning versus midday phototherapy of seasonal affective disorder. Am J Psychiatry 144: 1301–1305

54. Kasper S, Rogers SLB, Yancey A, Schulz PM, Skwerer RG, Rosenthal NE (1989) Phototherapy in individuals with and without subsyndromal seasonal affective disorder. Arch Gen Psychiatry 46: 837–844

55. Kasper S, Wehr TA, Bartko JJ, Geist PA, Rosenthal NE (1989) Epidemiological findings of seasonal changes in mood and behavior. Arch Gen Psychiatry 46: 823–833

56. Kellogg JH (1910) In: Light therapeutics: a practical manual of phototherapy for the student and practitioner. Good Health, Battle Creek

57. Kelsoe JR, Ginns EI, Egeland JA, et al. (1989) Re-evaluation of the linkage relationship between chromosome 11p loci and the gene for bipolar affective disorder in the Old Order Amish. Nature 342: 238–243

58. Klerman GL, Weissman MM (1989) Increasing rates of depression. JAMA 261: 2229–2235

59. Kripke DF (1981) Photoperiodic mechanisms for depression and its treatment. In: Perris C, Struwe G, Jansson B (eds) Biological psychiatry. Elsevier/North-Holland, Amsterdam, pp 1249–1252

60. Kripke DF (1983) Phase advance theories for affective illnesses. In: Wehr T, Goodwin F (eds) Circadian rhythms in psychiatry: basic and clinical studies. Boxwood, Pacific Grove, pp 41–69

61. Kripke DF (1984) Critical interval hypotheses for depression. Chronobiol Int 1: 73–80

62. Kripke DF, Mullaney DJ, Atkinson MS, Huey LY, Hubbard B (1979) Circadian rhythm phases in affective illnesses. Chronobiologia 6: 365–375

63. Kripke DF, Risch SC, Janowsky D (1983) Bright white light alleviates depression. Psychiatry Res 10: 105–112

64. Kripke DF, Risch SC, Janowsky DS (1983) Lighting up depression. Psychopharmacol Bull 19: 525–530

65. Kripke DF, Gillin JC, Mullaney DJ, Risch SC, Janowsky DS (1987) Treatment of major depressive disorders by bright white light for 5 days. In: Halaris A (ed) Chronobiology and psychiatric disorders. Elsevier, New York, pp 207–218

66. Kripke DF, Gillin JC, Mullaney DJ (1989) Bright light benefit unrelated to REM latency. In: APA (ed) APA 142nd Annual Meeting, new research program and abstracts. American Psychiatric Association, Los Angeles, p 137

67. Kripke DF, Mullaney DJ, Savides TJ, Gillin JC (1989) Phototherapy for nonseasonal major depressive disorders. In: Rosenthal NE, Blehar NC (eds) Seasonal affective disorders and phototherapy. Guilford, New York, pp 342–356

68. Kronauer RE (1987) A model for the effect of light on the human "deep" circadian pacemaker. Sleep Res 16: 621

69. Kronauer RE, Czeisler CA, Pilato SF, Moore-Ede MC, Weitzman ED (1982) Mathematical model of the human circadian system with two interacting oscillators. Am J Physiol 242: R3–R17

70. Lam RW, Kripke DF, Gillin JC (1989) Phototherapy for depressive disorders: a review. Can J Psychiatry 34: 140–147

71. Lam RW, Berkowitz AL, Berga SL, Clark CM, Kripke DF, Gillin JC (1990) Melatonin suppression in bipolar and unipolar mood disorders. Psychiatry Res 33: 129–134

72. Lam RW, Buchanan A, Clark C, Remick RA (1989) UV vs non-UV light therapy for SAD. Annual Meeting of the Society for Light Treatment and Biological Rhythms 12

73. Lewy AJ (1983) Biochemistry and regulation of mammalian melatonin production. In: Relkin R (ed) The pineal gland. Elsevier, New York, pp 77–128

74. Lewy AJ, Wehr TA, Goodwin FK, Newsome DA, Markey SP (1980) Light suppresses melatonin secretion in humans. Science 210: 1267–1269

75. Lewy AJ, Kern HA, Rosenthal NE, Wehr TA (1982) Bright artificial light treatment of a manic-depressive patient with a seasonal cycle. Am J Psychiatry 139: 1496–1498

76. Lewy AJ, Nurnberger JI, Wehr TA, et al. (1985) Supersensitivity to light: possible trait marker for manic-depressive illness. Am J Psychiatry 142: 725–728

77. Lewy AJ, Sack RL, Miller LS, Hoban TM (1987) Antidepressant and circadian phase-shifting effects of light. Science 235: 352–354

78. Lewy AJ, Sack RL, Singer CM (1985) Bright light, melatonin, and biological rhythms: implications for the affective disorders. Psychopharmacol Bull 21: 368–372

79. Lewy AJ, Sack RL, Singer CM (1988) Winter depressives may have phase delayed phase response curves. Soc Bio Psychiatry 43: 309–310

80. Linkowski P, Mendlewicz J, Kerkhofs M, et al. (1987) 24-hour profiles of adrenocorticotropin, cortisol, and growth hormone in major depressive illness: effect of antidepressant treatment. J Clin Endocrinol Metab 65: 141–52

81. McIntyre IM, Norman TR, Burrows GD, Armstrong SM (1989) Human melatonin suppression by light is intensity dependent. J Pineal Res 6: 149–156

82. McIntyre IM, Norman TR, Burrows GD (1990) Melatonin supersensitivity to dim light in seasonal affective disorder. Lancet 1: 488

83. Mendlewicz J, Branchey L, Weinberg U, Branchey M, Linkowski P, Weitzman ED (1980) The 24 hour pattern of plasma melatonin in depressed patients before and after treatment. Commun Psychopharmacol 4: 49–55

84. Monk TH (1989) A visual analogue scale technique to measure global vigor and affect. Psychiatry Res 27: 89–99

85. Moore RY (1983) Organization and function of a central nervous system circadian oscillator: the suprachiasmatic hypothalamic nucleus. Fed Proc 42: 2783–9

86. Moore-Eda MC, Sulzman FM (1981) Internal Temporal Order. In: Aschoff J (ed) Biological rhythms. Plenum, New York, pp 215–241 (Handbook of behavioral neurobiology, vol 4)

87. Nair NPV, Hariharasubramanian N, Pilapil C, Isaac I, Thavundayil JX (1986) Plasma melatonin – an index of brain aging in humans? Biol Psychiatry 21: 141–150
88. Nurnberger JI Jr, Berrettini W, Tamarkin L, Hamovit J, Norton J, Gershon E (1988) Supersensitivity to melatonin suppression by light in young people at high risk for affective disorder: a preliminary report. Neuropsychopharmacology 1: 217–223
89. Parry BL, Berga SL, Mostofi N, Sependa PA, Kripke DF, Gillin JC (1989) Morning versus evening bright light treatment of late luteal phase dysphoric disorder. Am J Psychiatry 146: 1215–1217
90. Parry BL, Mendelson WB, Duncan WC, et al. (1989) Longitudinal sleep EEG, temperature and activity measurements across the menstrual cycle in patients with premenstrual depression and in age-matched controls. Psychiatry Res 30: 285–303
91. Parry BL, Berga SL, Laughlin GA, et al. (1990) Altered waveform of plasma nocturnal melatonin secretion in premenstrual depression. Arch Gen Psychiatry 47: 1139–1146
92. Pittendrigh CS (1981) Circadian systems: general perspective. In: Aschoff J (ed) Biological rhythms. Plenum, New York, pp 57–80 (Handbook of behavioral neurobiology, vol 4)
93. Pittendrigh CS (1981) Circadian systems: entrainment. In: Aschoff J (ed) Biological rhythms. Plenum, New York, pp 95–124 (Handbook of behavioral neurobiology, vol 4)
94. Pittendrigh CS, Daan S (1976) A functional analysis of circadian pacemakers in nocturnal rodents. V. Pacemaker structure: a clock for all seasons. J Comp Physiol 106: 333–355
95. Prasko J, Foldmann P, Praskova H, Zindr V (1988) Hastened onset of the effect of antidepressive drugs when using three types of timing of intensive white light. Cs Psychiatry 84: 373–383
96. Ralph MR, Foster RG, Davis FC, Menaker M (1990) Transplanted suprachiasmatic nucleus determines circadian period. Science 247: 975–978
97. Reiter RJ (1985) Action spectra, dose response relationships, and temporal aspects of light's effects on the pineal gland. Ann NY Acad Sci 453: 215–230
98. Reiter RJ, Hurlbut EC, Richardson BA, King TS, Wang LCH (1982) Studies on the regulation of pineal melatonin production in the Richardson's ground squirrel (Spermophilus richardsonii). In: Reiter RJ (ed) The pineal and its hormones. Liss, New York, pp 57–65
99. Reiter RJ, Richardson BA, King TS (1983) The pineal gland and its indole products: their importance in the control of reproduction in mammals. In: Relkin R (ed) The pineal gland. Elsevier, New York, pp 151–199
100. Reppert SM, Weaver DR, Rivkees SA, Stopa EG (1988) Putative melatonin receptors in a human biological clock. Science 242: 78–81
101. Rockwell DA, Winget CM, Rosenblatt LS, Higgins EA, Hetherington NW (1978) Biological aspects of suicide: circadian disorganization. J Nerv Mental Dis 166: 851–858
102. Rosenthal NE, Blehar M (eds) (1989) Seasonal affective disorders and phototherapy. Guilford, New York
103. Deleted
104. Rosenthal NE, Wehr TA (1987) Seasonal affective disorders. Psychiatr Ann 17: 670–674
105. Rosenthal NE, Sack DA, Gillin JC, et al. (1984) Seasonal affective disorder: a description of the syndrome and preliminary findings with light therapy. Arch Gen Psychiatry 41: 72–80
106. Rosenthal NE, Sack DA, Skwerer RG, Jacobsen FM, Wehr TA (1989) Phototherapy for seasonal affective disorder. In: Rosenthal NE, Bleher M (eds) Seasonal affective disorders and phototherapy. Guilford, New York, pp 273–294
107. Rosenthal NE, Sack DA, Jacobsen FM, et al. (1985) The role of melatonin in seasonal affective disorder (SAD) and phototherapy. In: Wurtman RJ, Waldhauser F (eds) Melatonin in humans. Center for Brain Sciences and Metabolism Charitable Trust, Cambridge, MA, pp 233–241
108. Rosenthal NE, Lavendosky AA, Skwerer RG, et al. (1990) Effects of light treatment on core body temperature in seasonal affective disorder. Biol Psychiatry 27: 39–50
109. Rubin RT, Poland RE, Tower BB, Hart PA, Blodgatt ALN, Forster B (1981) Hypothalamo-pituitary-gonadal function in primary endogenously depressed men: preliminary findings. In: Fuxe K, Gustafsson JA, Wetterberg L (eds) Steroid hormone regulation of the brain. Pergamon, Oxford, pp 387–396
110. Rusak B (1989) The mammalian circadian system: models and physiology. J Biol Rhythms 4: 121–134
111. Sack DA, Nurnberger J, Rosenthal NE, Ashburn E, Wehr TA (1985) Potentiation of antidepressant medications by phase advance of the sleep-wake cycle. Am J Psychiatry 142: 5
112. Sack RL, Lewy AJ, Erb DL, Vollmer WM, Singer CM (1986) Human melatonin production decreases with age. J Pineal Res 3: 379–388
113. Sack RL, Lewy AJ, White DM, Singer CM, Fireman MJ, Vandiver R (1990) Morning vs evening light treatment for winter depression. Arch Gen Psychiatry 47: 343–351
114. Schilgen B, Tolle R (1980) Partial sleep deprivation as therapy for depression. Arch Gen Psychiatry 37: 267–271
115. Siffre M (1975) Six months alone in a cave. Nat Geogr 147: 426
116. Skene DJ (1987) The effect of chronic antidepressant administration on rat pineal function. Eur Pineal Study Group Newslett 17: 46–48
117. Sobel D (1980) In the sleep lab: losing touch. New York Times. July 1, 8, 15, 22: Sect C1
118. Souetre E, Salvati E, Savelli M, et al. (1988) Rythmes endocriniens en période de dépression et de rémission. Psychiatr Psychobiol 3: 19–27
119. Steiner M, Brown GM (1984) Melatonin/cortisol ratio: a biological marker? In: APA (ed) APA Annual Meeting, new research abstracts. American Psychiatric Association, Los Angeles
120. Stetson MH, Watson-Whitmyre M (1986) Effects of exogenous and endogenous melatonin on gonadal function in hamsters. J Neural Transm [Suppl] 21: 55–80
121. Stewart JW, Halbreich U (1989) Plasma melatonin levels in depressed patients before and after treatment with antidepressant medication. Biol Psychiatry 25: 33–38
122. Surridge-David M, McLean AW, Coulfer M (1987) Mood change following acute delay of sleep. Psychiatry Res 22: 149–158
123. Terman JS, Terman M, Schalger D, et al. (1990) Efficacy of brief, intense light exposure for treatment of winter depression. Psychopharmacol Bull 26: 3–11
124. Terman M, Quitkin FM, Terman JS, Stewart JW, McGrath PJ (1987) The timing of phototherapy: effects on clinical response and the melatonin cycle. Psychopharmacol Bull 23: 354–357

125. Terman M, Terman JS, Quitkin FM, McGrath PJ, Stewart JW, Rafferty B (1989) Light therapy for seasonal affective disorder: a review of efficacy. Neuropsychopharmacology 2: 1–22

126. Thase M (1989) Comparison between seasonal affective disorder and other forms of recurrent depression. In: Rosenthal NE, Bleher M (eds) Seasonal affective disorder and phototherapy. Guildferd, New York, pp 64–78

127. Thompson C, Checkley SA, Arendt J (1987) Circadian rhythms of melatonin and cortisol during treatment with desipramine in depressed and normal subjects. In: Halaris A (ed) Chronobiology and psychiatric disorders. Elsevier, New York, pp 103–116

128. Thompson C, Franey C, Arendt J, Checkley SA (1988) A comparison of melatonin secretion in depressed patients and normal subjects. Br J Psychiatry 152: 260–265

129. Tsujimoto T, Hanada K, Shioiri T, Daimon K, Kitamura T, Takahashi S (1990) Effect of phototherapy on mood disorders. 2nd Meeting of the Society for Research on Biological Rhythms, May 9–13, New York.

130. Tsujimoto T, Yamada N, Shimoda K, Hanada K, Takahashi S (1990) Circadian rhythms in depression. II. Circadian rhythms in inpatients with various mental disorders. J Affect Dis 18: 199–210

131. Underwood H (1990) The pineal and melatonin: regulators of circadian function in lower vertebrates. Experientia 46: 120–128

132. Underwood H, Goldman BD (1987) Vertebrate circadian and photoperiodic systems: role of the pineal gland and melatonin. J Biol Rhythms 2: 279–315

133. Vanecek J, Pavlik A, Illnerova H (1987) Hypothalamic melatonin receptor sites revealed by autoradiography. Brain Res 435: 359–362

134. Wehr TA, Goodwin FK (1981) Biological rhythms and psychiatry. In: Arieti S (ed) American handbook of psychiatry. Basic, New York, pp 46–74

135. Wehr TA, Jacobsen FM, Sack DA, Arendt J, Tamarkin L, Rosenthal NE (1986) Phototherapy of seasonal affective disorder. Arch Gen Psychiatry 43: 870–875

136. Wehr TA, Rosenthal NE (1989) Seasonality and affective illness. Am J Psychiatry 146: 829–840

137. Wehr TA, Wirz-Justice A (1982) Circadian rhythm mechanisms in affective illness and in antidepressant drug action. Phamacopsychiatry 15: 31–39

138. Wehr TA, Wirz-Justice A, Goodwin FK, Duncan W, Gillin JC (1979) Phase advance of the circadian sleep-wake cycle as an antidepressant. Science 206: 710–713

139. Wehr TA, Lewy AJ, Wirz-Justice A, Craig C, Tamarkin L (1982) Antidepressants and a circadian rhythm phase-advance hypothesis of depression. In: Gollu R, et al. (eds) Brain peptides and hormones. Raven, New York, pp 263–276

140. Wells KB, Stewart A, Haya RD, et al. (1989) The functioning and well-being of depressed patients: results from the medical outcomes study. JAMA 262: 914–920

141. Wetterberg L, Aperia B, Beck-Friis J, et al. (1981) Pineal-hypothalamic-pituitary function in patients with depressive illness. In: Fuxe K, et al. (eds) Steroid hormone regulation of the brain. Pergamon, Oxford, pp 397–403

142. Wetterberg L, Beck-Friis J, Kjellman BF, Ljunggren JG (1984) Circadian rhythms in melatonin and cortisol secretion in depression. Ann Hum Biol 197–205

143. Wever RA (1979) The circadian system of man: results of experiments under temporal isolation. Springer, Berlin Heidelberg New York

144. Wever RA (1985) Use of light to treat jet lag: differential effects of normal and bright artificial light on human circadian rhythms. Ann NY Acad Sci 453: 282–304

145. Wever RA (1989) Light effects on human circadian rhythms: a review of recent Andechs experiments. J Biol Rhythms 4: 161–185

146. Wirz-Justice A, Schmid AC, Graw P, et al. (1987) Dose relationships of morning bright white light in seasonal affective disorders (SAD). Experientia 43: 574–575

147. Wirz-Justice A, Graw P, Krauchi K, et al. (1989) Most SAD patients are phase-delayed in winter but respond equally well to morning or evening light. Annual Meeting of the Society for Light Treatment and Biological Rhythms, 19

148. Wu JC, Bunney WE (1990) The biological basis of an antidepressant response to sleep deprivation and relapse: review and hypothesis. Am J Psychiatry 147: 14–21

149. Yellon SM, Bittman EL, Lehman MN, Olster DH, Robinson JE, Karsch FJ (1985) Importance of duration of nocturnal melatonin secretion in determining the reproductive response to inductive photoperiod in the ewe. Biol Reprod 32: 523–529

# Rhythmic and Nonrhythmic Modes of Anterior Pituitary Hormone Release in Man

J. D. Veldhuis, M. L. Johnson, A. Iranmanesh, and G. Lizarralde

## Introduction

Variations in plasma hormone concentrations have been studied extensively in various conditions of health and disease in the human and experimental animals. A wide range of environmental, metabolic, nutrient, and hormonal stimuli are known to modulate normal patterns of anterior pituitary hormone release and/or alter the metabolic clearance of the endocrine effector substance. The ability to evaluate quantitatively not only variations in plasma hormone *concentrations* but also regulated features of hormone *secretory* events has been acquired recently through the use of so-called deconvolution techniques [20, 26, 32, 36, 41, 48, 54, 56, 60, 62, 72, 71, 74]. *Deconvolution* represents a procedure in which available plasma hormone concentration measurements are interpreted mathematically as the specific consequence of definable secretory events and hormone-specific metabolic clearance. The examination of calculated in vivo secretory events is of particular importance pathophysiologically, when a clinician and investigator wish to assess regulation of the actual secretory behavior of the endocrine gland. In contrast, evaluation of plasma hormone concentrations in the conventional manner discloses information about the hormonal milieu to which the target gland is exposed. Although the majority of available clinical investigative work has focused on variations in plasma hormone concentrations (and hence, the signal made available to the target tissue), recent advances in analytical tools have allowed detailed studies of the secretory rhythms inherent in the regulated output of the endocrine gland. Accordingly, here we will emphasize not rhythms in pituitary hormone concentrations in plasma, but rhythms in the *secretory behavior* of the anterior pituitary gland under a range of normal and pathological conditions in man.

There are special problems and challenges in the clinical evaluation of rhythms in anterior pituitary hormone release in man, and to a lesser extent in experimental animals. For example, in clinical investigative endocrinology, serial hormone measurements cannot always be obtained at frequent intervals over extended periods, which therefore leaves the investigator with short noisy data series that are particularly difficult to analyze [60, 71, 74]. Moreover, as discussed further below, the vast majority of neuroendocrine events that involve the anterior pituitary gland and its hormones are episodic or random, rather than explicitly rhythmic or periodic. As importantly, the exact waveform of the underlying secretory event is typically not known to the clinician or investigator, thus requiring the development and implementation of waveform-independent models to evaluate the secretory behavior of the endocrine gland in vivo [71, 73, 74]. Finally, there are confounding effects of sometimes large interindividual differences in metabolic clearance rates, dose-dependent experimental uncertainty introduced by the measurement system [e. g., radioimmunoassay (RIA) immunoradiometric assay (IRMA), bioassay], and variability caused by the in vivo experimental paradigm (e. g., nonuniformities in hormone admixture in plasma, sample withdrawal, centrifugation, freezing, thawing, defibrination). Accordingly, our thesis is that the clinical and experimental evaluation of human anterior pituitary function in vivo requires specialized tools predicated upon valid biophysical models and supported by adequate statistical treatments. In the absence of such tools, short noisy hormone time series containing undefined secretory waveforms confounded by subject-specific metabolic clearance are essentially uninformative. In addition, we believe that such analytical tools must be submitted to independent validation under defined conditions of known biological behavior so as to provide not only mathematically relevant but also biologically validated instruments for the study of episodic hormone release in various pathophysiological states. Accordingly, given the critical nature of valid quantitative neuroendocrine observations in different clinical states, where appropriate we will emphasize the limitations and strengths of available analyses applied to any particular hormone or condition.

## Circadian Rhythms in Plasma Hormone Concentrations

Among the best-studied phenomena associated with the release of anterior pituitary hormones in vivo are the circadian or more broadly the nyctohemeral rhythms in plasma hormone concentrations. However, rhythms in plasma hormone *concentrations* must be distinguished from rhythms in glandular *secretory events*. Secretory events combined with subject- and hormone-specific metabolic clearance (as well as hormone distribution and interconversion) together specify the plasma hormone concentration profiles. Techniques are available to the clinician and investigator to estimate in vivo secretory and clearance functions that jointly give rise to the measured plasma hormone concentration pattern. Thus, secretory information is of a derived nature. This will be reviewed later. Here, we will first summarize data on 24-h rhythms in plasma hormone concentrations per se.

As shown in Table 1, in the normal man, significant 24-h rhythms in the plasma concentrations of luteinizing hormone (LH) follicle-stimulating hormone) (FSH), thyroid-stimulating hormone (TSH), prolactin, growth hormone (GH), adrenocorticotrophic hormone (ACTH), and β-endorphin can be demonstrated by appropriate statistically based methods of cosinor analysis. Cosinor analysis is a nonlinear curve-fitting procedure, which allows one to specify the best-fit cosine function (sinusoidal rhythm) that characterizes the observed data (e.g., plasma hormone concentrations over time) associated with some given periodicity. Here, for the sake of simplicity, we will consider a rhythm having a 24-h periodicity as "nyctohemeral" rather than "circadian", because the term "circadian" carries additional implications in chronobiology. For example, circadian rhythms are synchronized to environmental cues, are generated internally, are temperature dependent, and exhibit free-running or sustained oscillations during temporary withdrawal from the cuing environment [38]. Such characteristics typically apply to cortisol and ACTH and apparently β-endorphin rhythms [1, 2, 7, 9, 11, 13, 15, 17, 23, 63, 64, 76, 78, 80, 82–85, 88], but to our knowledge have not been so rigorously satisfied for other adenohypophyseal hormones. In some circumstances, a nyctohemeral pattern of hormone release occurs in part because of food or sleep-entrained release mechanisms. Such mechanisms can be defined by prolonged fasting and/or acute sleep-wake cycle reversal studies; GH is released especially during stage III and IV of sleep [1, 18, 83, 85]; and prolactin and TSH with pre-sleep and sleep-stage dependence [12, 14, 49, 52, 65]. In contrast, food and sleep deprivation, temporal isolation, and sleep-wake reversal studies reveal that ACTH and cortisol release exhibit intrinsic 24-h rhythmicity that is not exclusively sleep-entrained, but has free-running characteristics. Less is known about the nyctohemeral and sleep-specified release mechanisms for LH, FSH, TSH, and β-endorphin [5, 8, 12, 25, 70, 77], although recent studies indicate that β-endorphin release is very closely coupled to that of ACTH diur-

**Table 1.** Twenty-four-hour rhythms in plasma concentrations of anterior pituitary hormones

| Hormone (units) | Amplitude (concentration) | Acrophase (clock time) | Mesor (concentration) | Ratio of amplitude/mesor (%) |
|---|---|---|---|---|
| LH (mIU/ml) | 0.60 ± 0.10 | 2400 ( ± 114 min) | 7.6  ± 0.53 | 8 |
| FSH (mIU/ml) | 0.38 ± 0.08 | 2253 ( ± 170 min) | 6.8  ± 0.08 | 6 |
| Prolactin (ng/ml) | 1.4  ± 0.39 | 0151 ( ±  96 min) | 4.7  ± 0.57 | 30 |
| TSH (μU/ml) | 0.51 ± 0.17 | 0233 ( ±  51 min) | 1.8  ± 0.38 | 28 |
| GH (ng/ml) | 0.41 ± 0.16 | 0032 ( ±  64 min) | 0.50 ± 0.18 | 82 |
| ACTH (pg/ml) | 5.9  ± 1.1 | 0915 ( ±  86 min) | 13    ± 1.8 | 45 |
| β-Endorphin (pg/ml) | 5.2  ± 0.81 | 0626 ( ± 116 min) | 32    ± 3.9 | 16 |

Data are mean estimates ($N$ = six to eight normal, middle-aged men studied for each hormone). Plasma hormone concentrations were determined by radioimmunoassay (LH, FSH) or radioimmunometric assay (others) in blood collected at 10-min intervals over 24 h.
Amplitude, one-half the difference between the zenith and nadir of the 24-h rhythm
Acrophase, time when the maximal value of the 24-h rhythm occurs
Mesor, mean value about which the 24-h rhythm oscillates

nally at least in the unperturbed healthy individual maintained in an unmanipulated environment [23]. Finally, some authors have emphasized that 24-h plasma hormone concentration rhythms may originate from the combined influence of an endogenous circadian oscillator and sleep-specified activation or suppression of hormone secretion [10, 38, 63]. For example, this construct seems relevant to the gonadotropic hormone, LH (and its testicular product, testosterone), as well as TSH, since sleep-wake reversal is accompanied by a release of these hormones at the new time of sleep as well as to some degree at the former time of sleep [4, 27, 43, 44, 45, 46, 64, 66].

If the amplitude of the 24-h rhythm in anterior pituitary hormone concentration is expressed as a percentage ratio of the mesor (cosinor mean 24-h plasma hormone concentration), then the respective percentage variations over 24 h are 8% for LH, 6% for FSH, 28% for TSH, 30% for prolactin, 82% for GH, 45% for ACTH, and 16% for $\beta$-endorphin. As discussed further below, exactly how these 24-h variations in plasma hormone concentrations are generated mechanistically cannot be deduced facilely from mere inspection of the plasma hormone concentration profiles. Additional analysis is required to determine the presence or absence of underlying secretory bursts, evaluate their amplitude and frequency distribution over 24 h, and assess whether any interpulse tonic (basal or constitutive) mode of hormone secretion exists that is also regulated with a 24-h periodicity.

## Mode of Secretion of Adenohypophyseal Hormones

As alluded to above, mere visual inspection of plasma hormone concentration profiles reveals episodic variations in the data. Such variations when characterized by an abrupt increase in the plasma hormone concentration followed by a more gradual decrease have been termed "pulses" or "peaks" [60]. Moreover, direct catheterization studies in experimental animals have revealed that the pituitary gland releases hormones in a burst-like manner characterized by the rapid onset of increased hormone secretion with a subsequent waning of the secretion rate. This is illustrated for LH and FSH in Fig. 1. Similarly, in vitro studies utilizing intact, sectioned, or dispersed anterior pituitary tissue reveal typically minimal levels of basal hormone release in the absence of

added secretagogue, whereas bursts of hormone secretion can be triggered by the administration of relevant agonists [81]. In many circumstances, such as in vitro perifusion studies of corticotrope cells, release episodes under perifusion conditions appear to be exclusively dependent upon the acute availability of relevant agonist, but occasional reports also exist of apparently random "mini-pulses" of hormone release that occur in ostensibly undisturbed perifusion systems in the absence of administered secretagogue [17, 50, 53, 55]. Whether such in vitro variations in effluent concentrations of pituitary hormones result from one or more artifacts and/or disturbances inherent in perifusion technology, or whether such release episodes are truly autonomous and independent of hypothalamic regulation cannot be inferred categorically from available data. Moreover, in vivo catheterization procedures that collect blood samples and/or tissue fluids from the hypothalamopituitary neuroendocrine unit do not yet rule definitely on this question of low levels of basal secretion independently of

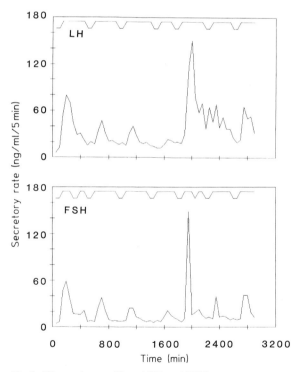

**Fig. 1.** Illustrative profiles of LH and FSH secretory rates assessed *in vivo* in the horse by repetitively sampling blood from the petrosal vein. Blood was sampled at 5-min intervals from pituitary venous effluent and submitted to assay to determine the content of LH and FSH (top and bottom panels respectively). The profiles of LH and FSH secretory rates indicate that a burst-like mode of hormone release occurs from this endocrine gland. (Data provided by S. Alexander and C. Irvine, Christchurch, New Zealand)

hypothalamic signaling. Nonetheless, a general inference is that major episodes of pituitary hormone secretion can be achieved in vivo and in vitro by the administration of brief impulses of relevant hypothalamically derived releasing factors with or without the simultaneous withdrawal of hypothalamic inhibitory substances [47]. The latter may be more important in some species (and for some hormones) than others.

Specific agonists that trigger anterior pituitary hormone secretion in vivo and in vitro are numerous, but are believed to include principally in the case of LH, gonadotrophin-releasing hormone (GnRH); for FSH, putatively GnRH and possibly one or more members of the inhibin superfamily such as activin (stimulate), inhibin, and/or follistatin (the latter two compounds inhibit FSH release at least in in vitro studies); for ACTH, corticotrophin-releasing hormone (CRH) and arginine vasopressin (AVP); for TSH and prolactin, TRH (stimulatory) and dopamine (inhibitory); for GH, GHRH (stimulatory) and somatostatin (inhibitory); various aminergic compounds such as norepinephrine and epinephrine; serotonin; and a host of potentially significant neuropeptides, including but not limited to, neuropeptide Y, bombesin, substance P, acetylcholine, and various opioid peptides, e.g. β-endorphin, met-enkephalin [40, 55]. In general, the acute exposure of anterior pituitary cells in vivo and in vitro to relevant secretagogue(s) with concomitant withdrawal of inhibitors results in a sudden burst-like release of hormone. Consequently, a burst-like model of anterior pituitary hormone secretion has emerged as a plausible general inference, whereas the presence and/or degree of associated interburst basal (tonic or constitu-

**Table 2.** Apparent modes of anterior pituitary hormone secretion in normal men

| Hormone | Discrete bursts only | Purely tonic secretion | Combined burst and tonic secretion |
|---|---|---|---|
| LH | + | − | − |
| FSH | + | − | − |
| Prolactin | − | − | + |
| TSH | − | − | + |
| GH | + | − | − |
| ACTH | + | − | − |
| β-Endorphin | + | − | . − |

" + " denotes a significant ($P < 0.05$) and " − " a nonsignificant pattern of endogenous anterior-pituitary hormone secretion as estimated by a waveform-independent deconvolution technique (see text) in six to eight healthy men. Hormones were assayed by immunological methods (see Table 1)

tive) hormone release is less clearly established in health 60, 69, 72, 74, 86, 87] (Table 2). Of considerable interest, the timing of most pituitary hormone secretory bursts appears to be random, i.e., serial interpulse interval values are statistically independent numbers [6, 51, 59, 69–71, 74–78]. An exception is LH interpulse intervals in the luteal phase [51], and GH interpulse intervals in normal men [18]. In both of these exceptions, serial interpulse interval values are negatively autocorrelated, so that long and short intervals (between peaks) tend to alternate. Such alternations suggest intraaxis feedback regulation.

If a burst-like model of pituitary hormone release obtains, one must then ask the question: *"How can a burst-like mode of apparently randomly occurring secretory events give rise to a circadian and/or ultradian rhythm in plasma hormone concentrations?"* This interesting and fundamental question has been addressed recently using deconvolution techniques, which are capable of defining in vivo hormone secretory patterns quantitatively from available plasma hormone concentration measurements. Such techniques have permitted the demonstration that alterations in hormone secretory burst frequency and/or amplitude and/or interpeak basal secretory rates can all serve as physiological mechanisms by which the 24-h rhythms in plasma hormone concentrations are generated in normal individuals (vide infra). Moreover, deconvolution procedures provide a far more refined fit of the plasma concentration data than can be accomplished typically by cosinor analysis per se. This concept is illustrated for three 24-h serum hormone profiles (ACTH, prolactin, and GH) in a normal man) (Fig. 2). Note that overall 24-h variations are relatively minimal compared to more striking minute-to-minute fluctuations in blood hormone concentrations. The basis for such rapid episodic changes is discussed further below, as disclosed by the neuroendocrine tool of deconvolution analysis.

In order to present the results of deconvolution analyses of anterior pituitary gland secretory behavior, a short review of deconvolution techniques follows. More detailed reviews are available in the neuroendocrinology literature [36, 37, 41, 42, 48, 56, 60, 62, 69, 71, 72, 74]. In brief, deconvolution refers to a mathematical procedure for resolving or dissecting the values of one or more underlying rate-dependent processes that operate simultaneously to specify some observed overall outcome. For example, hormone secretion and clearance operate simultaneously to specify the net observed plasma hormone concentration at any given instant. Deconvolution is concerned with determining the values for the secre-

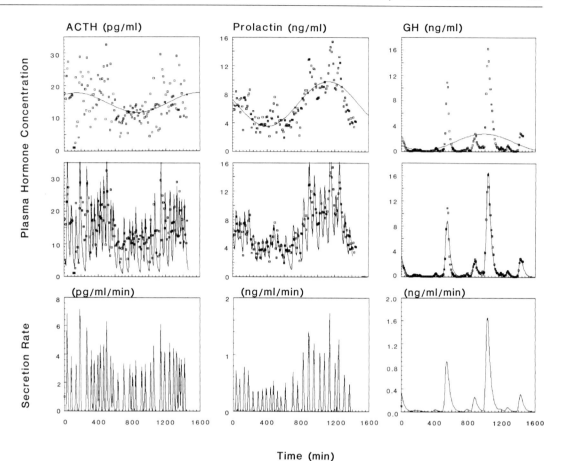

**Fig. 2.** Simple cosinor fit of 24-h plasma hormone concentration profiles compared to convolution-model fitting. The 3 *upper* subpanels depict (from left to right) 24-hour plasma or serum concentrations of ACTH, prolactin and GH in blood withdrawn at 10-min intervals in individual men. The sinusoidal curve that is plotted represents the best-fit 24-h cosine rhythm. The *middle* subpanels show the multiple-parameter deconvolution fits of the same data. The *lower* subpanels give the deconvolution-specified ACTH, prolactin, and GH secretory bursts. (Veldhuis et al. unpublished)

tion or the clearance function (or both) given knowledge of the plasma hormone concentrations over time. Thus, deconvolution can provide a powerful and informative tool by which to investigate the secretory conduct of the neuroendocrine ensemble.

In general, two classes of deconvolution techniques are available to aid in the evaluation of neuroendocrine secretory behavior: waveform-specific and waveform-independent techniques. These two types of deconvolution approach are illustrated schematically in Fig. 3.

Waveform-specific deconvolution procedures assume some algebraically definable waveform for the secretory event, and then solve for specific features (e.g., duration and magnitude) of the secretory event. For example, if the secretion burst is assumed to be approximated by a Gaussian function, then any given secretory burst can be described quantitatively by the

standard deviation of that secretory Gaussian, its mean (or location in time), and its amplitude [69, 71, 72]. The integral of the Gaussian secretory burst (its area) equals the apparent mass of hormone secreted within that burst. Its amplitude represents the maximal rate of hormone secretion attained within the burst, whereas the secretory burst SD or half-duration (duration at half-maximal amplitude) provides information about the duration in time of the secretion event. Using such a waveform-specific model, one is able not only to solve for specific features of secretion (e. g., secretory burst number, location in time, amplitude, and duration), but also simultaneously to calculate hormone and subject-specific half-life [69, 71, 72]. Accordingly, the waveform-specific model has considerable utility when a presumptive waveform can be postulated reasonably, and information is desired about both secretion and clearance.

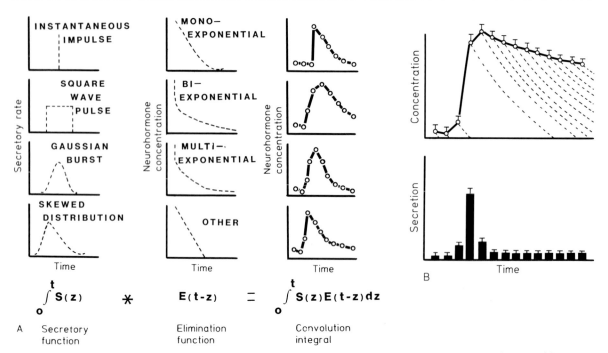

$$\int_{o}^{t} S(z) \quad * \quad E(t-z) \quad = \quad \int_{o}^{t} S(z)E(t-z)\,dz$$

A    Secretory          Elimination          Convolution
     function           function             integral

**Fig. 3. A** Schema of waveform-specific deconvolution models, in which secretory bursts of algebraically explicit form are assumed to be generated at random or fixed intervals and the resultant secreted molecules removed by relevant hormone elimination kinetics that are subject and condition specific. A variety of waveforms *(leftmost subpanels)* could be chosen for the secretory event, and an array of elimination functions *(middle panels)* can be considered to describe the dissipation of the molecules from the sampling compartment. **B** Waveform-inde-pendent deconvolution procedure. In order to avoid assumptions about the secretion waveform, the apparent underlying secretion rate can be estimated in each sample by model-independent techniques, which assume a priori knowledge of the kinetics of hormone distribution and disposal. The summation of the combined effects of all relevant sample secretory rates *(lower panel)* and appropriate elimination kinetics *(dotted curves)* yields the observed plasma hormone concentration profile *(upper panel)*

Given definitive information about hormone clearance in any given pathophysiological setting (i.e., the half-life is known for the hormone of interest under the conditions of study), one or more general waveform-independent deconvolution techniques can be employed [18, 36, 41, 56, 60, 62, 71]. Such approaches provide a means to calculate hormone secretory rates associated with each of the sample hormone concentrations. Estimates of secretion do not assume any particular secretory waveform or the presence or absence of tonic interburst basal secretion [71]. Thus, as shown schematically in Fig. 4, the fundamental mode(s) of hormone secretion can be inferred quantitatively: viz., a burst-like pattern, a tonic (constitutively or sustained) release profile, or a combined burst and tonic mode. In short, a waveform-independent procedure is of particular relevance when the hormone half-life is well defined a priori, but it cannot be used to evaluate both sample secretory rate and hormone half-life simultaneously [to calculate both secretion and clearance rates simultaneously without any assumed

waveform or special properties of the secretory event yields a nearly infinite number of possible mathematical solutions to the set of convolution integrals, simply because any particular estimate of secretion will yield a highly dependent (correlated) estimate of clearance for any given series of hormone concentrations] [71].

Recently, versions of both the waveform-specific and the waveform-independent models have been developed with appropriate statistical methodology, so that the statistical confidence limits for secretion and clearance estimates are dependent upon the degree of dose-dependent experimental uncertainty in the data as well as the statistical error inherent in the half-life prediction (e.g., the standard deviation associated with the independently estimated half-life value) [24, 71, 72]. Thus, the two general categories of deconvolution method can yield useful complementary information, whereas one particular approach may be more desirable under certain experimental conditions. As discussed in the next section, the application of such deconvolution technology has per-

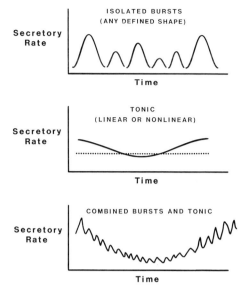

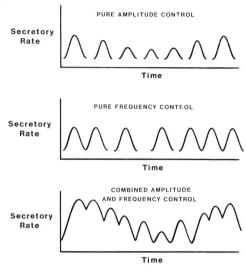

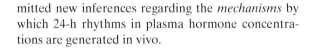

**Fig. 4.** Schematic modes of possible in vivo hormone secretion involving either isolated bursts of any defined shape *(upper panel)*, tonic (either linear or non-linear) secretory rates over 24 h *(middle panel)*, or a pattern of combined bursts and tonic secretion that varies over 24 h *(lower panel)*. The presence of an in vivo tonic mode of hormone release must be evaluated by a statistically valid method that distinguishes between low rates of interburst secretion and random experimental variance

**Fig. 5.** Schematic model of secretory control mechanisms for generating circadian plasma hormone concentration rhythms. The *upper subpanel* denotes pure amplitude control as a means by which 24-h variations in circulating hormone concentrations could be attained. The *middle panel* illustrates the possible occurrence of secretory bursts of uniform amplitude but of diurnally varying frequency. The *lower subpanel* depicts a possible model in which 24-h plasma hormone concentration rhythms are built up by combined variations in the amplitude and frequency of underlying secretory events

mitted new inferences regarding the *mechanisms* by which 24-h rhythms in plasma hormone concentrations are generated in vivo.

## Circadian and Ultradian Rhythms in Anterior-Pituitary Hormone Secretory Events

In principle, a 24-h rhythm in plasma hormone concentrations could arise as a result of circadian variations in one or more of the following dynamic variables: (a) the amplitude of randomly dispersed hormone secretory bursts; (b) the frequency of secretory events; (c) the rate of basal interburst hormone secretion (tonic or constitutive hormone release upon which bursts may be superimposed); and/or (d) the hormone half-life. These concepts are illustrated in Fig. 5. To our knowledge, major variations in the metabolic clearance rates of anterior pituitary hormones have not been documented as the proximate basis for circadian rhythms in the corresponding plasma hormone concentration. On the

other hand, recent evidence from deconvolution studies indicates that individual hormones exhibit 24-h rhythms in their plasma concentrations in response to nyctohemeral modulation of either secretory burst amplitude and/or number and/or tonic interburst secretion rates. Thus, we will discuss the various anterior pituitary hormones in relation to their individual mechanistic categories, as defined by one particular technique of waveform-independent deconvolution analysis. This methodology assumes a uniform hormone half-life over the 24-h period of observation (a time-invariant metabolic clearance rate) and allows for model-free statements about possible circadian variations in secretory burst-frequency and/or amplitude and/or tonic secretion rates. The bases for such analyses are illustrated in Fig. 6. In brief, cosinor analysis is applied not to the plasma hormone concentrations but to maximal peak secretory rates, secretory burst areas, interpulse (basal, tonic) secretory rates, or interburst intervals (time elapsing between consecutive peak maxima). Thus, one asks the question: "*Is there a significant rhythm* (e. g., of some given cosine periodicity) *in the hormone secretory feature of interest* (e. g., secretory burst amplitude, frequency, interpeak basal secretion rate)?"

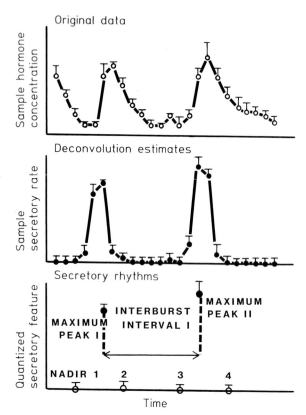

**Fig. 6.** Schema of the procedure used to evaluate whether the originally observed serum hormone concentration data that exhibit circadian rhythmicity are generated by 24-h rhythmic regulation of secretory burst amplitude, frequency, and/or interburst nadir secretory rates. The *upper subpanel* denotes the form of the original data, which typically consist of serial sample hormone concentrations and the associated measurement variability (*vertical marks* associated with each data point) for samples distributed at uniform intervals over time. The *middle subpanel* denotes the calculated sample secretory rates and their statistical confidence limits (*vertical marks* denote standard deviations of the estimated sample secretory rate). In the *lower subpanel,* one or more features of the secretory signal are depicted schematically. Note that the value of each secretory feature is given in relation to its time of occurrence. For example, the individual nadir secretory rates (nadirs numbered *1–4*) are shown with their standard deviations at the times of their occurrences, as are the maximal secretory rates of peak one and peak two, and the first interburst interval (whose center is located in time). Cosinor analysis can then be applied to determine whether any particular, relevant secretory feature (secretory rate nadir, maximum, or interburst interval) is distributed in a significantly periodic manner (see text)

## Frequency Control of Hormone Secretory Bursts

Among the anterior pituitary hormones LH, FSH, TSH, prolactin, GH, ACTH, and β-endorphin, none exhibits a 24-h rhythm in *secretory burst frequency* only. However, several hormones exhibit varia-

**Table 3.** Twenty-four-hour rhythms in specific secretory features of anterior pituitary hormones in normal men.

| Hormone | Secretory burst amplitude | Interburst ("basal") secretory rate | Secretory burst frequency |
|---|---|---|---|
| LH | + | – | – |
| FSH | – | ( + ) | – |
| Prolactin | + | + | – |
| TSH | + | – | + |
| GH | + | – | + |
| ACTH | + | – | – |
| β-Endorphin | + | – | + |

Parentheses denote that the absolute value of this basal rate was not distinguishable from zero

tions in secretory burst frequency and amplitude (Table 3).

## Amplitude Modulation of Pituitary Secretory Bursts

Adrenocorticotrophic hormone secretory events exhibit pure amplitude modulation over 24 h; i.e., there is a strong 24-h rhythm in the amplitude of ACTH secretory bursts in normal men, but there is no 24-h variation in ACTH secretory burst frequency or in the very low levels of interburst (basal) ACTH secretory rates (Fig. 7). Thus, the 24-h plasma ACTH *concentration* rhythm derives from the 2.2-fold nyctohemeral variation in ACTH *secretory* burst amplitude (maximal rate of ACTH secretion attained per release episode). Since the mass of hormone secreted per burst is directly proportional to the amplitude of the secretion burst, there is also a 2.2-fold 24-h variation in ACTH secretory burst mass. This observation corresponds well to the approximately similar quantitative variation in plasma ACTH concentrations. The mechanisms that generate a periodic variation in ACTH secretory rate and mass in vivo are not known in detail, but presumably involve the concerted effects of relevant in vivo secretagogues of ACTH (CRH, AVP, etc.) with or without corresponding withdrawal of inhibitors of ACTH release (e. g., somatostatin, feedback effects of glucocorticoids) [40]. Like that of ACTH, the 24-h rhythm in LH release is explicitly *amplitude* specified in normal men (in women, both amplitude and frequency modulation of LH secretory events can be recognized at certain stages of the menstrual cycle [51]).

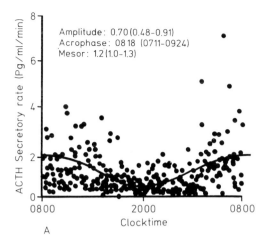

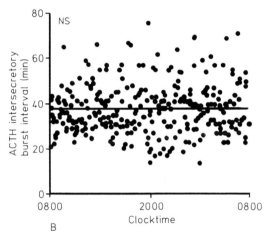

**Fig.7A, B.** Twenty-four-hour rhythms in ACTH secretory burst amplitudes (**A**), but not ACTH intersecretory burst intervals (**B**). The 24-h rhythms in measures of ACTH *secretion* were described by plotting the individual secretory burst amplitudes (maximal ACTH secretory rates attained within a release episode) against the corresponding times of the burst centers. Data are from 8 adult male subjects (320 resolved secretory events in total). Cosinor analysis was then used to determine the presence and magnitude of any significant 24-h rhythm in ACTH seretory measures. The amplitude, acrophase, and mesor of the 24-h rhythm in ACTH secretory-burst amplitudes were all significant (**A**). In contrast, as shown in **B**, there was no significant 24-h rhythm in ACTH intersecretory burst intervals, plotted as the duration of the interburst interval (time in minutes separating consecutive ACTH secretory burst centers) against the time of the mid-point of the interval. [From 78]

## Tonic Interburst (Basal or Constitutive) Hormone Secretion

Only FSH *secretion* profiles in normal men exhibited a 24-h variation in solely interburst basal (tonic) release. However, the basal rates of FSH secretion that are estimated between apparently randomly dispersed FSH secretory events are extremely low, approaching zero. In fact, the majority of FSH interburst secretory rates have standard deviations that overlap zero, i.e., do not achieve statistical significance for the half-life of FSH used in this analysis. Thus, at least in normal men, the 24-h variation in interburst basal FSH secretory rates presumably has minimal physiological significance, since the majority of these interpeak secretory rates approach zero. However, observations in healthy older men and postmenopausal women who have substantially increased basal serum FSH concentrations and higher interburst secretion rates suggest that circadian variations in "tonic" FSH secretion may contribute more importantly to the definition of 24-h variations in plasma FSH concentrations in older individuals (Veldhuis and Johnson, unpublished).

## Frequency and Amplitude Modulation of Secretory Bursts

Several hormones exhibit combined modulation of secretory burst amplitude and frequency: namely, GH, TSH, and β-endorphin. In this regard, biophysical modeling of secretory behavior indicates that combined regulation by both frequency and amplitude control yields an effective means for specifying large variations in plasma hormone concentrations [67, 69].

## Secretory Burst Amplitude and Interburst Nadir (Basal) Secretory Rates

Only prolactin secretory events manifest a 24-h rhythm in amplitude accompanied by a significant 24-h variation in interburst basal hormone secretory rates. As illustrated in Fig. 8, this mechanism is utilized effectively to achieve significant 24-h rhythms in circulating levels of prolactin in normal men, since the combined changes in basal secretory rate and secretory burst amplitude will yield additive changes in plasma hormone concentrations [67]. The low (but significant) level of tonic (basal) interpeak prolactin secretion may reflect the operation of incompletely inhibitory dopamine pathways in vivo in normal men. Exactly how such basal and/or the associated bursts of prolactin secretion are coordinately or differentially regulated over 24 h in various pathophysiological settings is not known at present.

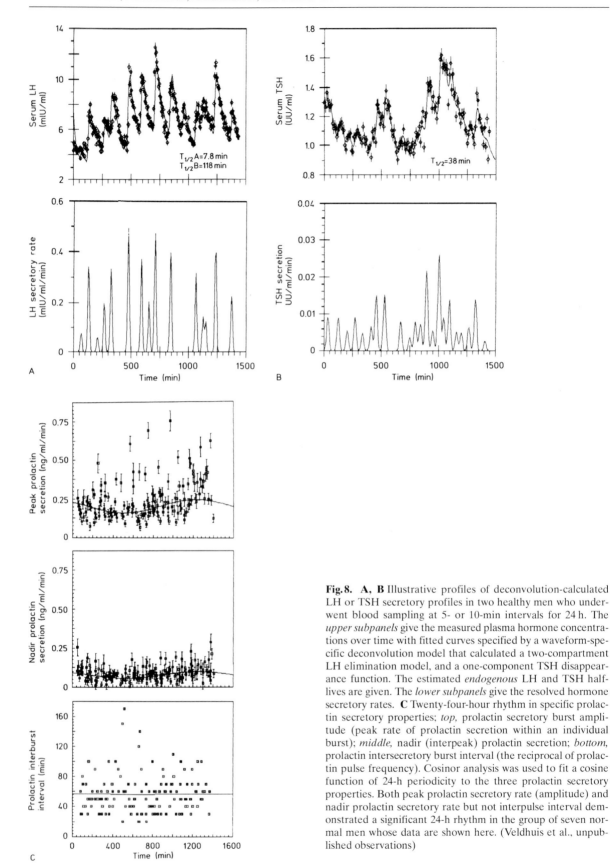

**Fig. 8. A, B** Illustrative profiles of deconvolution-calculated LH or TSH secretory profiles in two healthy men who underwent blood sampling at 5- or 10-min intervals for 24 h. The *upper subpanels* give the measured plasma hormone concentrations over time with fitted curves specified by a waveform-specific deconvolution model that calculated a two-compartment LH elimination model, and a one-component TSH disappearance function. The estimated *endogenous* LH and TSH half-lives are given. The *lower subpanels* give the resolved hormone secretory rates. **C** Twenty-four-hour rhythm in specific prolactin secretory properties; *top*, prolactin secretory burst amplitude (peak rate of prolactin secretion within an individual burst); *middle*, nadir (interpeak) prolactin secretion; *bottom*, prolactin intersecretory burst interval (the reciprocal of prolactin pulse frequency). Cosinor analysis was used to fit a cosine function of 24-h periodicity to the three prolactin secretory properties. Both peak prolactin secretory rate (amplitude) and nadir prolactin secretory rate but not interpulse interval demonstrated a significant 24-h rhythm in the group of seven normal men whose data are shown here. (Veldhuis et al., unpublished observations)

## Secretory Burst Amplitude and Frequency As Well As Interburst Basal Secretory Rates

No hormone secretory profile as resolved by wave-form-independent deconvolution analysis exhibited a threefold modulatory mechanism, in which significant 24-h variations occur in secretory burst amplitude, frequency, and interburst basal secretory rates. For example, since interpeak basal GH secretory rates typically approach zero in normal, fed individuals [18], the major contributions to the 24-h rhythm in plasma GH concentrations arise by way of combined frequency and amplitude modulation of GH secretory bursts. This finding does not exclude a quantitatively minor circadian rhythm in very low levels of currently undetectable interpeak GH secretion. Alternatively, the latter may in fact represent low-amplitude GH secretory bursts, which cannot be resolved by most present GH assays [74]. However, based upon the limits of detection of the GH immunoradiometric assay and the mean plasma GH concentrations observed, we can estimate that any rhythm in GH secretion below the detection limit would contribute less than 10% to the total GH secretory (production) rate. These qualifications apply only to GH, since the levels of LH, FSH, ACTH, prolactin, TSH, and $\beta$-endorphin are uniformly detectable over a full 24 h in current immunoassays.

## Secretory Burst Frequency and Interburst Basal Hormone Secretory Rates

Studies have not yet identified any anterior pituitary hormone whose secretory profile over 24 h can be accounted for by combined regulation of secretory burst frequency and interburst basal secretory rates. Such a pattern might occur in one or more pathophysiological conditions associated with altered pituitary secretion.

Table 3 summarizes the apparent mechanisms responsible for generating 24-h rhythms of various anterior pituitary hormones.

## Regulation of the Burst-Like Mode of Anterior Pituitary Hormone Secretion in Health and Disease

Although plasma concentrations of various anterior pituitary hormones are known to vary widely under different pathophysiological conditions, far less is known about the specific regulation and pathological alteration in pituitary hormone *secretory* events in man or experimental animals. Here, we will note selected pathophysiological settings in which significant changes in anterior pituitary hormone *secretory* patterns have been reported:
1. Nutrient and metabolic signals (fasting, obesity, diabetes mellitus)
2. Hypo- and hyperthyroidism
3. Normal puberty
4. The human menstrual cycle
5. Healthy aging
6. Pathological hypersecretory states (acromegaly)

In brief, one or more metabolic cues (the precise nature of which has not yet been defined) elicit marked alterations in the amplitude and detectable frequency of GH *secretory* bursts, e.g. (a) obesity is associated with suppressed GH secretory burst amplitude, as well as decreased endogenous GH half-life [18, 79]; and, conversely (b) insulin-dependent diabetes mellitus and fasting evoke significantly increased GH secretory burst frequency and amplitude with no evident alteration in endogenous GH half-life [3, 18, 21, 74]. Such findings are consistent with various hypotheses of metabolic regulators of GH secretory dynamics, e.g., that free fatty acid, insulin, or insulin-like growth factor type I (IGF-I) suppresses episodic GH secretion [74]. Thyroid disorders have been associated with decreased serum GH and increased serum LH concentrations (hypothyroidism), or increased serum GH and LH concentrations (hyperthyroidism). These variations in mean circulating (glyco)-protein hormone concentrations in turn appear to be due to a correspondingly altered amplitude and/or frequency of episodic GH and LH secretion unassociated with any change in hormone metabolic clearance (hyperthyroidism) or associated with decreased LH metabolic clearance (hypothyroidism) (unpublished) [22].

Major changes in plasma concentrations of various anterior pituitary hormones occur in puberty: see Chapter of Rogol. The very limited deconvolution analyses of GH *secretion* in this physiological context have revealed prominent *amplitude*-dependent variations across different pubertal stages [33, 57]. An associated but less striking finding is both frequency and amplitude control of episodic GH release, which can also be induced by exogenous sex-steroid hormones (either androgens or estrogen [34, 35, 57]). Estrogen also amplifies the amplitude of prolactin secretory bursts in the human [68].

Episodic LH *secretion* (assessed by a waveform-specific deconvolution technique) varies as a function of stage of the human menstrual cycle [51]. Specifically, although the estimated half-life of endogenous LH and the daily (endogenous) LH secretory rate are stage-of-cycle invariant, the amplitude, frequency, *and* duration of LH secretory bursts all exhibit significant regulation across the menstrual cycle [51]. The exact physiological impact of such regulated variations in the mode of LH release on the ovary is not known at present.

Healthy aging is accompanied by various endocrine disturbances [58], but few data are available regarding specific alterations in anterior pituitary-hormone secretion and/or clearance. Our preliminary observations indicate a significant decline in the reserve capacity of bioactive LH secretion in healthy older men [61], and in detectable GH secretory burst frequency and amplitude in healthy older men (Veldhuis and Iranmanesh, unpublished). Much more study of age-related changes in pituitary hormone secretion and clearance is needed.

There is virtually no published information of which we are aware regarding deconvolution-specified alterations in secretory and/or clearance properties of most anterior pituitary hormones in pathological hypersecretory states, e.g., prolactin, GH, FSH, TSH, or $\alpha$-subunit secreting pituitary adenomas. Limited data in depression regarding plasma ACTH pulsatility indicate increased peak frequency [39], whereas studies of ACTH secretion in Cushing's disease presumptively associated with pituitary adenomas have suggested *amplitude* enhancement of ACTH secretory bursts with attenuation of 24-h rhythms in plasma ACTH concentrations (reviewed in [63]). Our preliminary findings in acromegaly indicate a marked increase in tonic (nonpulsatile) GH secretion in this condition of GH excess [19]. Considerable additional investigations are required to delineate the particular mechanisms, as well as their specificity and their reversibility, that subserve increased plasma adenohypophyseal concentrations in pathological hypersecretory states. Where possible such studies should: (a) evaluate both secretion and clearance and (b) determine the extent to which augmented amplitude and/or frequency of under-lying secretory events contributes to the pathological increase in circulating pituitary hormone concentrations.

## Selected Unresolved Issues in the Pathophysiology of Anterior Pituitary Hormone Secretory Rhythms

At least the following issues should be considered in further investigations of rhythmic and nonrhythmic secretion of anterior pituitary hormones in man:

1. The physiological and biochemical mechanisms that stipulate 24-h variations in secretory burst frequency and/or amplitude and/or basal secretion rates for each adenohypophyseal hormone
2. The *nature of* alterations in rhythmic and nonrhythmic pituitary hormone secretion in changing physiological states (e.g., puberty, menopause, aging) and various pathological states (e.g., hormone-secreting anterior pituitary tumors)
3. The *impact* of pathophysiologically altered pituitary-hormone secretory signals on the target gland (e.g., effects of increased GH secretory burst amplitude and frequency in diabetes mellitus on gluconeogenesis, lipolysis, IGF-I synthesis)
4. The importance to the subject of adaptive responses of circadian and ultradian regulation of pituitary-hormone secretory burst number and/or amplitude
5. The extent to which distinct hormone secretory profiles (e.g., GH and LH) are coordinately or independently regulated, and the implications of nonrandom associations between different pituitary-hormone (secretory) pulse trains
6. The relevance to pathophysiology of interburst "tonic" (basal, constitutive) hormone release (e.g., interpeak GH secretion in acromegaly).

We conclude that further investigations are required of rhythmic and nonrhythmic features of adenohypophyseal hormone secretion in both normal physiological states and pathological conditions. Such studies are likely to aid in clarifying the clinical pathophysiology of disrupted pituitary function and to help ultimately in delineating the mechanisms that control target tissue responses to pituitary hormone signaling in health and disease.

*Acknowledgments.* We thank Patsy Craig for her skillful preparation of the manuscript; Paula P. Azimi for the artwork; and Sandra Jackson and the expert nursing staff at the University of Virginia Clinical Research Center for conduct of the research protocols. This work was supported in part by NIH grant # RR00847 to the University of Virginia Clinical Research Center, RCDA # 1 KO4 HD 00634 (JDV), NIH Grant No. AM-30302 and GM-28928 (MLJ), Veterans Administration Medical Research Funds (AI), Diabetes Endocrine Research Center Grant DK-38942, and NIH-supported Clinfo Data Reduction Systems, as well as funds from the Pratt Foundation and Biodynamics Institute.

# References

1. Alford FP, Baker HWG, Burger HG, deKretser DM, Hudson B, Johns MW, Masterton JP, Patel YC, Rennie GC (1973) Temporal patterns of integrated plasma hormone levels during sleep and wakefulness. I. Thyroid-stimulating hormone, growth hormone and cortisol. J Clin Endocrinol Metab 37: 841–847

2. Antoni FA (1986) Hypothalamic control of adrenocorticotropin secretion: advances since the discovery of 41-residue corticotropin-releasing factor. Endocr Rev 7: 351–378

3. Asplin CM, Faria ACS, Carlsen EC, Vaccaro VA , Barr RE, Iranmanesh A, Lee MM, Veldhuis JD, Evans WS (1989) Alterations in the pulsatile mode of growth hormone release in men and women with insulin-dependent diabetes mellitus. J Clin Endocrinol Metab 69: 239–245

4. Boyar R, Finkelstein J, Roffwarg H, Kapen S, Weitzman E, Hellman L (1972) Synchronization of augmented luteinizing hormone secretion with sleep during puberty. N Engl J Med 287: 582–586

5. Brabant G, Brabant A, Ranft U, Ocran K, Kohrle J, Hesch RD, von zur Muhlen A (1987) Circadian and pulsatile thyrotropin secretion in euthyroid man under the influence of thyroid hormone and glucocorticoid administration. J Clin Endocrinol Metab 65: 83–88

6. Butler JP, Spratt DI, O'Dea LS, Crowley WF Jr (1986) Interpulse interval sequence of LH in normal men essentially constitutes a renewal process. Am J Physiol 250: E338–E340

7. Carnes M, Kalin NH, Lent SJ, Barksdale CM, Brownfield MS (1988) Pulsatile ACTH secretion: variation with time of day and relationship to cortisol. Peptides 9: 325–331

8. Chan V, Jones A, Liendo-Ch P, McNeilly A, Landon J, Besser GM (1978) The relationship between circadian variations in circulating thyrotrophin, thyroid hormones and prolactin. Clin Endocrinol 9: 337–349

9. Clayton GW, Librik L, Gardner RL, Guillemin R (1973) Studies on the circadian rhythm of pituitary adrenocorticotropin release in man. J Clin Endocrinol Metab 23: 975–981

10. Czeisler CA, Weitzman ED, Moore-Ede MC, Zimmerman JC, Kronauer RS (1980) Human sleep, its duration and organization depend on its circadian phase. Science 210: 1264–1267

11. Desir D, van Cauter E, Goldstein J, Fang VS, Leclercq R, Refetoff S, Copinschi G (1980) Circadian and ultradian variations of ACTH and cortisol secretion. Horm Res 13: 302–316

12. Desir D, van Cauter E, L'Hermite M, Retetoff S, Jadot C, Caufriez A, Copinschi G, Robyn C (1982) Effects of "jet lag" on hormonal patterns. III. Demonstration of an intrinsic circadian rhythmicity in plasma prolactin. J Clin Endocrinol Metab 55: 849–856

13. Follenius M, Brandenberger G, Simeoni M, Reinhardt B (1982) Diurnal cortisol peaks and their relationships to meals. J Clin Endocrinol Metab 55: 757–761

14. Frantz AG (1979) Rhythms in prolactin secretion. In: Krieger DT (ed) Endocrine rhythms. Raven, New York, p 175

15. Gallagher TF, Yoshida K, Roffwarg HD, Fukushida DK, Weitzman ED, Hellman L (1973) ACTH and cortisol secretory patterns in man. J Clin Endocrinol Metab 36: 1058–1073

16. Gambacciani M, Liu JH, Swartz WH, Tueros VS, Rasmussen DD, Yen SSC (1987) Intrinsic pulsatility of ACTH release from the human pituitary in vitro. Clin Endocrinol 26: 557–563

17. Graber AL, Givens JR, Nicholson WE, Island DP, Liddle GW (1965) Persistence of diurnal rhythmicity in plasma ACTH concentrations in cortisol-deficient patients. J Clin Endocrinol Metab 25: 804–807

18. Hartman ML, Faria ACS, Vance ML, Johnson ML, Thorner MO, Veldhuis JD (1991) Temporal structure of in vivo growth hormone secretory events in man. Am J Physiol 260: E101–E110

19. Hartman ML, Veldhuis JD, Vance ML, Faria ACS, Furlanetto RW, Thorner MO (1990) Somatotropin pulse frequency and basal growth hormone concentrations are increased in acromegaly. J Clin Endocrinol Metab 70: 1375–1384

20. Henery RJ, Turnbull BA, Kirkl M, McArthur JW, Gilbert I, Besser GM, Rees LH, Pedoe DST (1989) The detection of peaks in luteinizing hormone secretion. Chronobiol Int 6: 259–265

21. Ho KY, Veldhuis JD, Johnson ML, Furlanetto R, Evans WS, Alberti KGMM, Thorner MO (1988) Fasting enhances growth hormone secretion and amplifies the complex rhythms of growth hormone secretion in man. J Clin Invest 81: 968–975

22. Iranmanesh A, Lizarralde G (1989) Primary hypothyroidism alters the pulsatile mode of LH and FSH release by amplitude modulation (Abstr 113 A). 71 Annual Meeting of the Endocrine Society, Seattle

23. Iranmanesh A, Lizarralde G, Johnson ML, Veldhuis JD (1989) Circadian, ultradian and episodic release of beta-endorphin in men, and its temporal coupling with cortisol. J Clin Endocrinol Metab 78: 1019–1026

24. Johnson ML, Lassiter AE, Veldhuis JD (1990) A waveform-independent deconvolution technique to analyze in vivo hormone secretion (Abstr 905). 72nd Annual Meeting of the Endocrine Society, Atlanta

25. Judd HL (1979) Biorhythms of gonadotropins and testicular hormone secretion. In: Krieger DT (ed) Endocrine rhythms. Raven, New York, p 299

26. Jusko WJ, Slaunwhite WR, Aceto T (1975) Partial pharmacodynamic model for the circadian-episodic secretion of cortisol in man. J Clin Endocrinol Metab 40: 278–289

27. Kapen S, Boyar RM, Finkelstein JW, Hellman L, Weitzman

ED (1974) Effect of sleep-wake cycle reversal on luteinizing hormone secretory pattern in puberty. J Clin Endocrinol Metab 39: 293–299

28. Keller-Wood ME, Dallman MF (1984) Corticosteroid inhibition of ACTH secretion. Endocr Rev 5: 1–24

29. Krieger DT (1979) Rhythms in CRF, ACTH and corticosteroids. In: Krieger DT (ed) Endocrine rhythms. Raven, New York, pp 123–142

30. Krieger DT, Aschoff J (1979) Endocrine and other biological rhythms. Endocrinology 11: 2079

31. Liddle GW, Island D, Meador CK (1962) Normal and abnormal regulation of corticotropin secretion in man. Recent Prog Horm Res 18: 125–166

32. Linkowski P, Mendlewicz J, Leclercq R, Brasseur M, Hubain P, Golstein J, Copinschi G, van Cauter E (1985) The 24-hour profile of adrenocorticotropin and cortisol in major depressive illness. J Clin Endocrinol Metab 61: 429–438

33. Martha PM, Rogol AD, Veldhuis JD, Kerrigan JR, Goodman DW, Blizzard RM (1989) Alterations in the pulsatile properties of circulating growth hormone concentrations during puberty in boys. J Clin Endocrinol Metab 69: 563–570

34. Mauras N, Blizzard RM, Link K, Johnson ML, Rogol AD, Veldhuis JD (1987) Augmentation of growth hormone secretion during puberty: evidence for a pulse amplitude-modulated phenomenon. J Clin Endocrinol Metab 64: 596–601

35. Mauras N, Rogol AD, Veldhuis JD (1989) Specific, time dependent actions of low-dose estradiol administration on the episodic release of GH, FSH, and LH in prepubertal girls with Turner's syndrome. J Clin Endocrinol Metab 69: 1053–1058

36. McIntosh RP, McIntosh JEA (1985) Amplitude of episodic release of LH as a measure of pituitary function analysed from the time-course of hormone levels in the blood: comparison of four menstrual cycles in an individual. J Endocrinol 107: 231–239

37. McIntosh RP, McIntosh JEA, Lazarus L (1988) A comparison of the dynamics of secretion of human growth hormone and LH pulses. J Endocrinol 118: 339–345

38. Moore-Ede MC, Czeisler CA, Richardson GS (1983) Circadian timekeeping in health and disease. N Engl J Med 309: 469, 530

39. Mortola JF, Liu JH, Gillen JC, Rasmussen DD, Yen SSC (1987) Pulsatile rhythms of adrenocorticotropin (ACTH) and cortisol in women with endogenous depression: evidence for increased ACTH pulse frequency. J Clin Endocrinol Metab 65: 962–968

40. Negro-Vilar A, Johnston CA, Spinedi E, Valenca M, Lopez F (1987) Physiological role of peptides and amines in the regulation of ACTH secretion. Ann NY Acad Sci 512: 218

41. Oerter KE, Guardabasso V, Rodbard D (1986) Detection and characterization of peaks and estimation of instantaneous secretory rate for episodic pulsatile hormone secretion. Comput Biomed Res 19: 170–191

42. O'Sulivan F, O'Sullivan J (1988) Deconvolution of episodic hormone data: an analysis of the role of season on the onset of puberty in cows. Biometrics 44: 339–353

43. Parker DC, Rossman LG, Vanderlaan EF (1973) Sleep-related nyctohemeral and briefly episodic variation in human plasma prolactin concentrations. J Clin Endocrinol Metab 36: 1119–1124

44. Parker DC, Pekary AE, Hershman JM (1976) Effect of normal and reversed sleep-wake cycles upon nyctohemeral rhythmicity of plasma thyrotropin. Evidence suggestive of an inhibitory influence in sleep. J Clin Endocrinol Metab 43: 318–329

45. Parker DC, Rossman LG, Pekary AE, Hershman JM (1987) Effect of 64-hour sleep deprivation on the circadian waveform of thyrotropin (TSH): further evidence of sleep-related inhibition of TSH release. J Clin Endocrinol Metab 64: 157–161

46. Patel YC, Alford FP, Burger HG (1972) The 24-hour plasma thyrotrophin profile. Clin Sci 43: 71–77

47. Plotsky PM, Vale W (1986) Patterns of growth hormone-releasing factor and somatostatin secretion into the hypophysial-portal circulation of the rat. Science 230: 461–465

48. Rebar R, Perlman D, Naftolin F, Yen SSC (1973) The estimation of pituitary luteinizing hormone secretion. J Clin Endocrinol Metab 37: 917–927

49. Sassin JF, Frantz AG, Kapen S, Weitzman ED (1973) The nocturnal rise of human prolactin is dependent on sleep. J Clin Endocrinol Metab 37: 436–440

50. Shin SH, Reifel CW (1981) Adenohypophysis has an inherent property for pulsatile prolactin secretion. Neuroendocrinology 32: 139–144

51. Sollenberger ML, Carlson EC, Veldhuis JD, Evans WS (1990) Nature of LH secretory events throughout the human menstrual cycle. J Neuroendocrinol 2: 845–852

52. Sowers JR, Catania RA, Hershman JM (1982) Evidence for dopaminergic control of circadian variations in thyrotropin secretion. J Clin Endocrinol Metab 54: 673–675

53. Stewart JK, Clifton DK, Koerker DJ, Rogol AD, Jaffe T, Goodner CJ (1985) Pulsatile release of growth hormone and prolactin from the primate pituitary in vitro. Endocrinology 116: 1–5

54. Swartz CM, Wahby VS, Vacha R (1986) Characterization of the pituitary response in the TRH test by kinetic modeling. Acta Endocrinol (Copenh) 112: 43–48

55. Thorner MO (1992) The adenohypophysis. In: Williams RH (ed) Textbook of endocrinology. Saunders, Philadelphia

56. Toutain PL, Laurentie M, Autefage A, Alvinerie M (1988) Hydrocortisone secretion: production rate and pulse characterization by numerical deconvolution. Am J Physiol 255: E688–E695

57. Ulloa-Aguirre A, Blizzard RM, Garcia-Rubi E, Rogol AD, Link K, Christie CM, Johnson ML, Veldhuis JD (1990) Testosterone and oxandrolone, a non-aromatizable androgen, specifically amplify the mass and rate of growth hormone (GH) secreted per burst without altering GH secretory burst duration or frequency or the GH half-life. J Clin Endocrinol Metab 71: 846–854

58. Urban RJ, Veldhuis JD (1988) Hypothalamo-pituitary concomitants of aging. In: Sowers JR, Felicetta JV (eds) The endocrinology of aging. Raven, New York, pp 41–74

59. Urban RJ, Veldhuis JD (1991) Modulating effects of dihydrotestosterone and estradiol on the dynamics of FSH secretion and clearance in young men. J Androl 12: 27–35

60. Urban RJ, Evans WS, Rogol AD, Johnson ML, Veldhuis JD (1988) Contemporary aspects of discrete peak detection algorithms. I. The paradigm of the luteinizing hormone pulse signal in men. Endocr Rev 9: 3–37

61. Urban RJ, Veldhuis JD, Blizzard RM, Dufau ML (1988) Attenuated release of biologically active luteinizing hormone in healthy aging men. J Clin Invest 81: 1020–1029

62. Van Cauter E (1979) Method for characterization of 24-hour temporal variation of blood components. Am J Physiol 237: E255–E264

63. Van Cauter E, Refetoff S (1985) Multifactorial control of the 24-hour secretory profiles of pituitary hormones. J Endocrinol Invest 8: 381–391

64. Van Cauter E, Leclercq R, Vanhaelst L, Golstein J (1974) Simultaneous study of cortisol and TSH daily variations in normal subjects and patients with hyperadrenalcorticism. J Clin Endocrinol Metab 39: 645–652

65. Van Cauter E, L'Hermite M, Copinschi G, Refetoff S, Desir D, Robyn C (1981) Quantitative analysis of spontaneous variations of plasma prolactin in normal man. Am J Physiol 241: E355–E363

66. Vanhaelst L, van Cauter E, Degaute JP, Golstein J (1972) Circadian variations of serum thyrotropin levels in man. J Clin Endocrinol Metab 35: 479–482

67. Veldhuis JD (1989) Mean plasma hormone concentration is controlled in a linear manner by secretory impulse frequency. Am J Physiol 257: E299–E300

68. Veldhuis JD, Evans WS (1989) Mechanisms subserving estradiol's induction of increaed prolactin concentrations in man: evidence for amplitude modulation of spontaneous prolactin secretory bursts. Am J Obstet Gynecol 161: 1149–1158

69. Veldhuis JD, Johnson ML (1988) A novel general biophysical model for simulating episodic endocrine gland signaling. Am J Physiol 255 (18): E749–E759

70. Veldhuis JD, Johnson ML (1988) In vivo dynamics of luteinizing hormone secretion and clearance in man: assessment by deconvolution mechanics. J Clin Endocrinol Metab 66: 1291–1300

71. Veldhuis JD, Johnson ML (1990) New methodological aspects of evaluating episodic neuroendocrine signals. In: Yen SSC, Vale W (eds) Advances in neuroendocrine regulation of reproduction. Plenum, New York, p 123–139

72. Veldhuis JD, Carlson ML, Johnson ML (1987) The pituitary gland secretes in bursts: appraising the nature of glandular secretory impulses by simultaneous multiple-parameter deconvolution of plasma hormone concentrations. Proc Natl Acad Sci USA 84: 7686–7690

73. Veldhuis JD, Clifton D, Crowley WF, Cutle GB, Filicori M, Johnson ML, Maciel RJ, Merriam GR, Santoro MF, Steiner RA, Santen RJ (1988) Preferred attributes of objective pulse analysis methods. In: Crowley WF, Hofler JG (eds) The episodic secretion of hormones. Wiley, New York, pp 111–118

74. Veldhuis JD, Faira A, Vance ML, Evans WS, Thorner MO, Johnson ML (1988) Contemporary tools for the analysis of episodic growth hormone secretion and clearance in vivo. Acta Paediatr Scand 347: 63–82

75. Veldhuis JD, Iranmanesh A, Clarke I, Kaiser DL, Johnson ML (1989) Random and non-random coincidence between LH peaks and FSH, alpha subunit, prolactin, and GnRH pulsations. J Neuroendocrinol 1: 1–10

76. Veldhuis JD, Iranmanesh A, Lizarralde G, Johnson ML (1989) Amplitude modulation of a burst-like mode of cortisol secretion gives rise to the nyctohemeral glucocorticoid rhythm in man. Am J Physiol 257: E6–E14

77. Veldhuis JD, Johnson ML, Dufau ML (1989) Physiological attributes of endogenous bioactive luteinizing hormone secretory bursts in man: assessment by deconvolution analysis and in vitro bioassay of LH. Am J Physiol 256: E199–E207

78. Veldhuis JD, Iranmanesh A, Johnson ML, Lizarralde G (1990) Amplitude, but not frequency, modulation of ACTH secretory bursts gives rise to the nyctohemeral rhythm of the corticotropic axis in man. J Clin Endocrinol Metab 71: 452–463

79. Veldhuis JD, Iranmanesh A, Ho KKY, Lizarralde G, Waters MJ, Johnson ML (1991) Dual defects in pulsatile growth hormone secretion and clearance subserve the hyposomatotropism of obesity in man. J Clin Endocrinol Metab 72: 51–59

80. Veldhuis JD, Johnson ML, Iranmanesh A, Lizarralde G (1990) Temporal structure of in vivo adrenal secretory activity estimated by deconvolution analysis. J Biol Rhythms 5: 247–255

81. Watanabe T, Orth DN (1987) Detailed kinetic analysis of adrenocorticotropin secretion by dispersed anterior pituitary cells in a microperifusion system: effects of ovine corticotropin-releasing factor and arginine vasopressin. Endocrinology 121: 1133–1145

82. Weeke J, Gundersen HJG (1978) Circadian and 30 minute variations in serum TSH and thyroid hormones in normal subjects. Acta Endocrinol (Copenh) 89: 659–672

83. Weitzman ED (1976) Circadian rhythms and episodic hormone secretion in man. Annu Rev Med 27: 225–243

84. Weitzman ED, Fukushima D, Nogeire C, Roffwarg H, Gallagher TF, Hellman L (1971) Twenty-four hour pattern of the episodic secretion of cortisol in normal subjects. J Clin Endocrinol Metab 33: 14–22

85. Weitzman ED, Nogeire C, Perlow M, Fukushima D, Sassin J, McGregor P, Gallagher TF, Hellman L (1974) Effects of a prolonged 3-hour sleep-wake cycle on sleep stages, plasma cortisol, growth hormone and body temperature in man. J Clin Endocrinol Metab 38: 1018–1030

86. Winters SJ (1990) Inhibin is released together with testosterone by the human testis. J Clin Endocrinol Metab 70: 548–550

87. Winters SJ, Troen PE (1986) Testosterone and estradiol are co-secreted episodically by the human testis. J Clin Invest 78: 870–872

88. Yates FE, Maran JW (1980) Stimulation and inhibition of adrenocorticotropin release. In: Knobil E, Sawyer CW (eds) The pituitary gland and its neuroendocrine control, part 2. American Physiological Society, Washington, pp 367–404 (Handbook of physiology, vol 4)

# Chronobiology of the Hypothalamic-Pituitary-Adrenal and Renin-Angiotensin-Aldosterone Systems

A. Angeli, G. Gatti, and R. Masera

## Introduction

Intercellular communication is an essential part of the functioning of multicellular organisms. The last 20 years have seen the recognition and characterization of many signaling molecules produced by specialized cells. It has become clear that classical hormones are inserted in multifaceted systems of information together with other substances that carry signals for activity, growth and differentiation of a wide range of cells. As discussed elsewhere in this book, cells with specific functions such as immunity and phagocytosis are capable of producing soluble signaling molecules that profoundly influence the nervous and the endocrine system. In this conceptual framework, where all attempts to rigidly classify biological systems fail as a new knowledge is acquired and where, e.g., considerable functional overlap is apparent between hormones, neurotransmitters, neuromodulators, cytokines and growth factors, one could simply state that the so-called hypothalamic-pituitary-adrenal (HPA) system, which is responsible for the release into the bloodstream of glucocorticoids, and the renin-angiotensin-aldosterone (RAA) system, which is responsible for the release into the bloodstream of the major mineralocorticoid hormone, are exceedingly complex and that some classical views on their functioning are no longer tenable. Once the relations between the components of a given system are liable to be continuously changed by influences of other signaling substances, classical feedbacks and axes are too simplistic terms and it is more appropriate to think at each anatomical or functional station of the system as a node of a network controlled in a multifactorial fashion.

Bearing these considerations in mind, it is held that rhythmic fluctuations characterize almost all steps of the signaling programs that involve both systems. They have been studied extensively for the HPA system because of the assumed importance of glucocorticoids in allowing the body to resist stress. More recently, it has become apparent that also mineralocorticoid hormone secretion is organized in a complex temporal fashion. Application to clinical medicine of what has been learned about the rhythmic organization of both systems has provided new insights into pathophysiology and offered new strategies to the treatment of many disease conditions.

## Chronoendocrinology of the Hypothalamic-Pituitary-Adrenal System

It is well established that in mammals glucocorticoid hormones are secreted rhythmically according to multifrequency endogenous programs [124]. A prominent feature of this temporal organization refers to the circadian cycle.

The human adrenals secrete cortisol in intermittent secretory bursts interspersed with intervals of quiescence [126, 257]; episodic secretion, however, does not override a fundamental circadian program, which in diurnally active subjects emerges as a clustering of intense secretory episodes during the early morning hours and is easily documented with methods of time series analysis [155, 256]. The concentration profile in the plasma consists of a series of pulses the size and frequency of which allow, in most circumstances, application of the cosinor analysis, with the angular frequency (period) assumed as 24 h [157, 168]. In the face of the remarkable advances in understanding of mechanisms that are involved in the control of the circadian rhythm of plasma glucocorticoid levels, much less is known about the significance of this high-amplitude oscillation for the economy of the organism. Evidence has been presented that the circadian periodicity of cortisol secretion represents an important signal to gain synchronization of the human temporal structure at different levels, i.e., to place other biological rhythmicities along an appropriate time scale in order to meet better the vicissitudes and emergencies of daily life. Accordingly, the term "endogenous synchronizer" has been proposed [13, 101]. Independently of the distinct circadian surge, typified by elevated systemic concentrations of cortisol coincident with the beginning of the organism's daily activity cycle, in a certain number of individuals another low-amplitude circadian peak of glucocorticoid secretion is apparent. It follows the midday meal and is located in the early afternoon hours [78, 187].

Present knowledge indicates that the rhythmic secretion of any endocrine structure is basically an expression of the genetic chrono-organization of the hormone-producing cells. Classical studies have demonstrated that rhythmic steroidogenesis occurs at the adrenal level also in the absence of cyclic adrenocorticotrophic hormone (ACTH) inputs [11, 153, 247]. Specific nervous and hormonal signals, however, act as organizers of the genetically determined oscillators in order to have appropriate and complementary rhythmicities and to gain compliance with respect to environmental changes [19, 103]. The so-called ACTH-dependent steroidogenesis is to a great extent dependent on hierarchical information, i.e., on the recognition of specific signals and proper response at subsequent levels. The circadian pattern of plasma glucocorticoids shows close correlations with those of ACTH and other pro-opiomelanocortin (POMC)-related hormones like $\beta$-lipotropin and $\beta$-endorphin [79, 85, 126, 172]. This supports the view that the timing of specific POMC synthesis or processing serves to modulate in terms of amplitude and acrophase the adrenocortical program by driving endogenous and environmental inputs in the most effective way. A 41 amino acid peptide now known as corticotrophin-releasing factor (CRF)-41 or corticotrophin-releasing hormone (CRH), originally identified in ovine hypothalamic tissue, is generally accepted as the major factor responsible for the release of ACTH [248].

The present consensus, however, is that the hypophysiotropic message specifically conveyed from the brain to the POMC-processing pituicytes is multifactorial in composition [21, 82]. A schematic diagram of the information flow through the HPA system, with emphasis on the hypothalamic station, is presented in Fig. 1. Environmental synchronizers such as the light/dark alternation, memorized social cues such as the rest/activity routine, and actual or perceived challenges (stressors) are first encoded into neurochemical messages and then relayed to specialized neurosecretory cells in the paraventricular nuclei (PVN) of

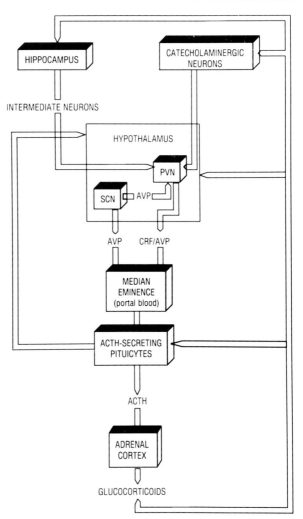

**Fig. 1.** Schematic view of bidirectional influences between the hypothalamopituitary-adrenal (HPA) and the nervous systems. *PVN*, paraventricular nuclei; *SCN*, suprachiasmatic nuclei; *AVP*, arginine vasopressin; *CRF*, corticotropin-releasing factor; *ACTH*, adrenocorticotropic hormone

the hypothalamus via multiple, distributed pathways [185]. Probably, several neuropeptides involved in the control of ACTH release are secreted by different cells in a stimulus-specific fashion. The provisional list includes CRF-41, arginine vasopressin (AVP), oxytocin and angiotensin II; these substances are thought to be released from nerve terminals in the zona externa of the median eminence into the primary portal plexus [86, 184]. This area, which is characterized by a dense fenestrated capillary plexus, lacks a blood-brain barrier and is thus subject to modulatory influences from systemically circulating substances [184]. Antibody sequencing studies on hypothalamic extracts have determined that CRF bioactivity of rat stalk median eminence extracts is not due solely to CRF-41 and AVP, the best-known ACTH secretagogues, but also to other substances which may interact synergistically [149]. Independently of compounds that may reach the ACTH-secreting cells from the peripheral circulation or as a consequence of hypothalamic secretion into the portal vessels of the median eminence, there are several neurotransmitter mechanisms that are influential on PVN activity and hence on ACTH control [24, 86, 149, 184, 236]. Recently, another interesting molecule has been added to the cohort of modulatory substances that deal with ACTH secretion. Interleukin-1 (IL-1) is a relatively small protein of about 17.5 kD that is most commonly secreted by macrophages but is also produced in a variety of tissues, including the brain. Administered systemically or intracerebrally, it has a potent activating effect on the HPA system, increasing plasma concentrations of ACTH and glucocorticoids. The site of this action is still controversial; evidence that it works through the secretion of CRF-41 by PVN is compelling, but a direct action on the pituitary under certain circumstances has not been excluded [65]. In summary, although CRF-41 has to be considered an essential component of the hypothalamic control of ACTH secretion, there are a host of modulatory substances that can act singly or in concert at different levels. Additionally, paracrine and/or autocrine mechanisms arguably operate upon the ACTH-secreting cells. This is an emerging area of research that could be of relevance to the circadian program of POMC synthesis and processing. Opioid peptides and fragments of POMC cleavage have been found to affect the ACTH response to stressors and to exogenous CRF-41 at least under certain circumstances [8, 97]. The role of molecules produced by the intermediate stellate cells of the adenohypophysis, such as interleukin-6 (IL-6), is unclear too, but preliminary evidence suggests that they have to do with the responsiveness of pituicytes to secretagogues and, more specifically, with ACTH secretion and action [150, 165, 208]. In conclusion, the search for a single clock mechanism, a brain pacemaker able to drive the consequential cascade CRH→ACTH→adrenal cortex in a circadian fashion by enhancing PVN activity at a certain time in the early morning hours, is possibly naive in the light of such a complex network of signals that concern ACTH secretion. Rather, it seems that a genetic program develops in accord with the recognition of exogenous synchronizers in order to gather during a relatively short span of the 24-h cycle those ACTH secretagogues that interact synergistically in the most efficient way and then those

signals that turn off or effectively dampen ACTH secretion, thus preventing the HPA system from overshooting. Long-lasting activation of glucocorticoid secretion, in fact, may be detrimental to the organism's economy; for instance, glucocorticoids exert inhibitory effects on substances that share the characteristic of being important elements of the body's normal defense mechanism [167].

The concept of a general program for the circadian chrono-organization of HPA system involving several complementary mechanisms (in space and time) fits well with the observations that such a program extends beyond the rhythmic release of hormones into the peripheral blood and involves the regulation of the hormonal action at the target level. In the 1980s, chronophysiology of the corticosteroid-recognizing proteins, including intracellular receptors, became an expanding area of study. The presence of a circadian rhythm of the corticosteroid-binding globulin (CBG, transcortin) binding capacity in the peripheral plasma of different species was documented [17, 27, 39, 176]. Available evidence suggests that, similar to CBG, the glucocorticoid receptors of circulating cells are also in vivo modulated by glucocorticoids themselves according to a circadian program [41, 215]. Traffic and function of diverse lymphocyte subpopulations are admittedly controlled by endocrine signals; circadian variations in human T- and B-lymphocyte activities, and in natural killer (NK) activity as well, have been repeatedly reported [41, 143]. It is now clear that POMC processing, eventually leading to hormonal molecules such as ACTH and β-endorphin, occurs in lymphocytes [30]. As a consequence, an immunomodulatory circuit has been hypothesized to operate between lymphocytes and the HPA system, such circuits being susceptible to environmental changes and to rhythmic modulation (Gatti and Angeli, this book).

In contrast to other pituitary hormones, first luteinizing hormone (LH) and occasionally prolactin (PRL), there is no evidence for ACTH that the intermittent secretory episodes are the expression of a predictable ultradian periodicity amenable to the "basic rest-activity cycle" (BRAC) of Kleitman [15, 118, 190]. In studies in which blood samples were drawn as often as 10–20 min, the episodic release of ACTH into the bloodstream, and the corresponding release of cortisol, as inferred from the measurement of the hormonal concentrations in plasma, could not meet the criteria for definition as recurrent phenomena (algorithmically validated). It cannot be excluded, however, that during the second part of sleep, concomitantly with the emergence of multiple brain ultradian rhythms related to BRAC, e.g., REM-non-REM alternation during sleep, there may be particular neurotransmitter conditions allowing the expression of such basic oscillation also for ACTH. In the remaining part of the 24-h cycle, this basic pattern would be obscured, at least in individual adults. Whether in infants it is recognizable and postnatal maturation of the brain accomplishes the superimposition of the circadian program of HPA activity to an almost complete degree is still matter of speculation [105, 108].

With regard to infradian rhythmicities, most available data on plasma cortisol levels do not suggest the occurrence of significant changes throughout the human menstrual cycle [17]. In a careful study performed in a homogeneous group of 12 normally cycling volunteers, Carandente et al. [42] have recently documented by cosinor analysis that the circadian midline estimating statistics of rhythm (MESOR) of plasma cortisol albeit to a minor degree tends indeed to rise in the luteal part of the cycle.

A circannual, low-amplitude, rhythm of glucocorticoid secretion has been suggested to occur in adult humans on the basis of statistical validation of transversal and longitudinal time series of data on plasma cortisol and urinary metabolites [1, 192]. A roughly similar pattern has been described for radioimmunoassayable plasma ACTH [48]. The issue is still open, since alternative explanations of the results, i.e., changes in the metabolic clearance rate or in the plasma protein-binding capacity, have not been ruled out.

## Sleep, Brain Clocks, Neurotransmitters and the Circadian Rhythm of the HPA System

The link between sleep and body temperature was known from the work of German physiologists in the mid-nineteenth century. They established the clinically important fact that humans normally experience daily fluctuations in core body temperature that have their minimum at night during sleep. They also documented that this rhythmic change is not a reflex response to the daily fluctuations of ambient light and temperature. Rather, it is an expression of the general rhythmic program of the organism that places sleep strategically on the descending limb of the body temperature curve, when heat generation from food consumption decreases, muscles become inactive and an energy conservative state is achieved.

The last 30 years have witnessed a remarkable growth in understanding of sleep. It is now agreed that the mammalian sleep cycle is nested within the intricate brain system controlling circadian rhythms. As stated before, the view that a single circadian clock, housed in a well-defined region of the brain and entrained by the light/dark cycle to beat at a frequency of once every 24 h, accounts for the organization of all circadian-rhythmic functions relevant to the rest-activity cycle is now outdated. There are structures, containing very specialized cell groups, that act as organizers, or phase-resetters, of the numerous clocks genetically programmed to oscillate in a circadian fashion; their proper functioning prevents clocks from free-running. The suprachiasmatic nuclei of the hypothalamus (SCN) do certainly belong to these structures. After the demonstration of the retinohypothalamic tract, which provides a pathway for the entraining effect of light upon SCN, and that of the circadian firing pattern of SCN neurons removed and kept alive in organ culture [204, 219, 245], it has become clear that output signals from SCN are of primary importance in synchronizing the activity of the rest of the hypothalamus, and in a broader sense the activity of the rest of the brain, in such a way to bring them to a higher level of harmony with the environmental cycle of light/dark [108, 161, 246]. We do not know at present the exact nature of the links between SCN and the lower brain stem stations that control the sleep cycle, on the one hand, and between SCN and PVN that produce the major ACTH secretagogues on the other. Efferent projections have been found to leave the SCN via ventrocaudal, lateral, dorsal and rostral pathways, with a number of these projections terminating in other hypothalamic nuclei, as well as in the midbrain [161, 229, 234]. There is evidence that PVN receives monosynaptic projections [234]. It has been suggested that SCN exert their influence on the sleep-wake cycle by conveying specific messages to the non-REM-REM oscillator located in the pons [108]. While REM sleep matures early in intrauterine life, non-REM sleep matures only after birth and presupposes the development of a specific oscillator related to BRAC that allows the proportion of total sleep devoted to non-REM sleep to increase progressively during the first 2 years of life [230, 255]. An apparent parallelism exists between the postnatal maturation of the non-REM-REM oscillator and that of the circadian organization of HPA system. Not surprisingly, also the complete maturation of the retinohypothalamic tract is achieved postnatally and presumably within this time [57, 72, 107]. It appears likely, therefore, that sequential complementary mechanisms subserve during the nocturnal hours the activation of the non-REM-REM oscillator and the up-resetting of ACTH secretion, as a consequence of coordinating signals that at a certain time superimpose the circadian program upon oscillations of higher frequency. These signals also convey information by exogenous synchronizers; with regard to the light/dark cycle, the SCN play a central, yet probably not exclusive, role [162, 246]. The complexity and the individual variability of the circadian/ultradian interplay could account for discrepancies in the literature on ACTH-cortisol surges and REM sleep [7, 152, 227, 249]. Notwithstanding these limitations, available data suggest that the highest probability of first detecting episodes of increased ACTH and cortisol secretion is normally between the first and the second REM stages [34, 129]. In any event, the conceptual framework of multiple clocks sharing common mechanisms of circadian control that develop postnatally fits well with the observation that in early months after birth not only sleep but also rest/activity alternation, feeding sequence, alertness, and consciousness of the parental presence need to be progressively organized in a circadian way. The maternal role to synchronize "socially" the baby is primarily important; it may be teleologically sound that hormonal signals to the developing brain are indeed present in the milk with circadian fluctuations. With regard to HPA maturation, we know that the human milk contains cortisol in appreciable amounts, that the hormone is rapidly available since CBG concentrations are very low in comparison with the plasma counterpart, and that milk cortisol is circadian rhythmic, roughly paralleling the plasma pattern [2, 179, 214]. Accordingly, in the case of breast feeding, the infant's brain receives more cortisol in the morning than in the evening. One relevant point is that the newborn adrenal gland is very much different from the adult one. The so-called permanent cortex, which is responsible for the glucocorticoid secretion, is only a small portion of the total gland; the large amount of the primitive or fetal cortex atrophies soon after birth, whereas the permanent cortex proliferates. Complete replacement of the fetal cortex and the adult zonal configuration of the gland are accomplished only by the 1st year of life [138, 225]. It is obvious that a harmonic development of the HPA system presupposes complex signaling between brain and periphery; in this light, the "circadian training" by the mother, in terms of psychoperceptive and biochemical inputs as well (timing and composition of food), may be crucial for the complete expression of the genetic program during adult life

[145]. In experimental animals there is evidence that pharmacological manipulation of brain transmitters or maternal deprivation when applied immediately after birth eventually lead to macroscopic abnormalities of the circadian glucocorticoid rhythm in adulthood [122, 123, 199, 264]. In the rat, the rhythm appears much more closely tied to the meal pattern than in the human [121, 236, 237]. This is conceivable because, in the rat, the activity phase that begins with the onset of darkness corresponds rigidly to the feeding pattern due to the type of muscular motion that necessitates energy substrate availability. This is not the case in the adult human, especially in our social environment; in fact, changing meal time and composition did not lead to apparent modifications of the plasma cortisol rhythm, at least in acute experiments [94]. The issue of the relationships between feeding pattern and HPA rhythms in the human, however, is far from being clarified [94]; notwithstanding episodes of ACTH and glucocorticoid release during the 1st year of life that have been linked with the feeding time [7], the mechanisms accounting for the postprandial surge of cortisol could still share common neurotransmitter control mechanisms with the major circadian surge [7]. That the rest/activity alternation, sleep and feeding have complex relationships is obvious from simple observations; e.g., we are not hungry when asleep, taste is absent while dreaming, physical exercise has a tonic effect on sleep, and sleep deprivation or night-eating have deleterious effects on work proficiency. Disorders of eating behavior such as anorexia nervosa are often characterized by abnormalities of the HPA system, with increased ACTH and cortisol secretion due to increased secretion of hypothalamic CRH and, most interestingly, with alterations of the circadian rhythm [74, 83, 93, 113, 254]. In turn, the two major ACTH secretagogues, CRF-41 and AVP, have been demonstrated to affect profoundly the control of food intake. Gastric emptying and other variables related to the feeding pattern have been associated with the ultradian BRAC together with the non-REM-REM oscillators [140, 190] and rhythms in human performance [117]. Once again, one key point of the human chronophysiology is the method of interaction of the two genetic basic programs, the circadian (related to the diurnal activity of the human species and expressed by the activity of the HPA system) and the ultradian BRAC. The interaction is such that the acrophase of the former coincides with a powerful damping of amplitude of the latter. In other words, the circadian cycle has a potent masking effect upon the ultradian cycle. The interaction develops postnatally and the rhythmic

profile of plasma glucocorticoid levels is the most accessible marker.

Turning back to the coordinating signals linked to the activity of SCN, one candidate could be melatonin (or other biologically active substance(s) produced by the pineal) since the SCN-pineal circuit is well established in all mammalian species [147, 186]. Melatonin (MT) is secreted in a circadian manner entrained by the light/dark cycle; serum levels display a prominent peak at night and are persistently low during daytime [197]. Studies on the relationships between MT and the HPA system have been prompted by the observations of Wetterberg [259] and Wetterberg et al. [260], who first proposed a mutual relationship between MT and cortisol rhythms, on the basis of the coincidence of high cortisol and low serum MT in depressive patients. Present knowledge, however, indicates that in the human MT and cortisol rhythms are uncoupled. Moreover, the presence of each rhythm does not depend on the presence of the other [253]; also the exercise-induced rises in serum MT and cortisol were found to be dissociated [233]; finally, no clear modification of the circadian cortisol rhythm was recorded after administration of pharmacological doses of MT, both over a short period [151] and over a long period [240]. In patients with Cushing's syndrome the sustained endogenous hypercortisolism without apparent circadian rhythmicity does not seem to lead to appreciable changes of the circadian profile of MT secretion [183]. In conclusion, available evidence speaks against a primary role of the pineal hormone MT as a circadian coordinating signal for the HPA system.

Vasopressinergic neurons are numerous in SCN, and AVP may be suggested as a candidate for circadian signaling. There is a clear-cut circadian rhythm in the amount of AVP in the SCN, at least in the rat [173]; a close relation has been suggested between the circadian pattern of AVP concentrations and electrical activity recorded from SCN neurons in vitro [91]. AVP levels in the cerebrospinal fluid (CSF) are also circadian-rhythmic and appear to be entrained by the light/dark cycle [180, 198, 217]. Interestingly enough, the AVP rhythm in CSF shows roughly coincidental acrophases across the mammalian species and has an effective insulation from the osmotic regulation of the circulating AVP levels [47, 223]. A growing body of evidence indicates that an independent neuroanatomical vasopressinergic system participates in intracerebral communication; the SCN are an important station of the system that operates separately from the classical AVP-containing neurons identified in the hypothalamic neurohypophyseal tract [216, 222].

The AVP history is becoming increasingly complex. It is known that the hormone derives from a larger precursor molecule, the propressophysin, that during axonal transport to nerve endings the prohormone is cleaved in neurosecretory vescicles to give rise to AVP, the respective carrier protein neurophysin and a 39 amino acid C-terminal glycoprotein (CPP) [139]. No precise functions have yet been attributed to neurophysin or CPP. Heterogeneity of AVP receptors has also been demonstrated [243]. As has been found for POMC, processing of propressophysin could be regulated in a different way not only among species but also among AVP-containing neurons of various brain structures. Recently, it was reported that AVP concentrations in CSF have to do with the circadian amplitude of sleep rhythmicities [37]. Whether the AVP-synthesizing neurons of PVN, which project to the median eminence and are involved in the control of ACTH secretion, are different from the neurons which project to the posterior lobe of the pituitary and are coupled with the "intracerebral" vasopressinergic system is yet unknown. AVP and CRF-41 coexist in nerve terminals at the level of the median eminence and in cells of PVN, suggesting that the two peptides are cosecreted [141, 201, 212, 250]. Potentiation of CRF-41-induced ACTH release by exogenously administered AVP or AVP-agonists is a well-known phenomenon [26, 58, 92]. Since the amplitude of the circadian rhythm of CRF-41 secretion is most probably much lower than the corresponding amplitude of ACTH secretion [209], it is tempting to think that the activation of vasopressinergic neurons at a critical time in the night potentiates the secretagogue effect upon the ACTH-secreting pituicytes. The same concept could apply to oxytocinergic neurons, still unsettled with regard to their role in the control of ACTH secretion. Conflicting results have suggested that oxytocin could have both inhibitory and stimulatory effects depending on the experimental circumstances [22, 87, 181, 209]. Also oxytocin is circadian rhythmic in CSF [198]. In conclusion, it appears that, in contrast to melatonin, vasopressin may represent an important circadian regulator of the HPA system, although the relationships between the vasopressinergic neurons of SCN and PVN remain unclear at present.

In addition to the regulation of pituitary ACTH secretion CRFs are involved in the physiological regulation of central and peripheral nervous activities, which in turn results in metabolic and cardiovascular effects [36, 66, 77, 164]. It is thought that these neuroendocrine-induced effects in metabolic substrate availability and cardiovascular function coordinate a generalized response to stressful events that is mediated at the hypothalamic level. In doing so, CRFs interplay with brain neurotransmitters; information is bidirectional and it is not surprising that catecholamines, acetylcholine and serotonin, which are mostly implicated with autonomic efferent pathways and the central control of food intake, have been repeatedly suggested as modulatory signals of the circadian rhythm of the HPA system [12, 90, 120, 244]. Pertinently, sleep research is focusing on the same transmitters as markers of the REM-on and REM-off neuronal populations [108, 136, 231]; interestingly enough, catecholaminergic activation has been suggested to suppress REM sleep and to switch off the non-REM-REM oscillator. Is it merely coincidental that catecholaminergic activation has an enhancing effect on CRF secretion and action? Catecholaminergic innervation of PVN has been well characterized; noradrenergic inputs, for example, innervate all parvicellular parts of PVN rich in CRF-41-neurosecretory cells and those parts of the magnocellular portions in which AVP-containing cells are numerous [235]. The mature pattern of distribution is reached postnatally [23, 115]. Systemic administration of $\alpha$- or $\beta$-adrenergic agonists which do not penetrate the blood-brain barrier is generally regarded as ineffective on ACTH secretion in the human [4–6, 69]; however, infusion of $\alpha$-adrenergic agonists which do cross the blood-brain barrier resulted in elevated ACTH and cortisol secretion [4]. These observations suggest that the human ACTH-secreting cells are relatively insensitive to systemic catecholamines, whereas a facilitatory role of the hypothalamic catecholaminergic milieu upon CRF secretion is an attractive hypothesis, corroborated by many studies performed in rodents [185]. As extensively reviewed by Plotsky et al. [185], current evidence suggests that catecholamines facilitate CRF-41 secretion and in addition both food-related surges and the circadian enhancement of ACTH secretion in humans may be at least partly dependent upon $\alpha_1$-adrenergic activation. There are many more questions than answers concerning the issue of catecholaminergic modulation of HPA function, and they are the logical consequence of current experimental inadequacies. The situation in man concerning acetylcholine and serotonin is more confused. It has been suggested that both transmitters participate in the control of the circadian ACTH surge [88, 89, 226], but the data should be interpreted with care in view of the multiple pharmacological actions of most of the agents that have been used. In particular, the enhancement of serotoninergic activity was suggested to activate ACTH secretion rhythmically in the second part of the night [45, 213], but subsequent studies have yielded confusing results. In summary, complex neurotransmitter changes at the hypothalamic level certainly contribute to the circadian activation of the HPA system; in this light, more intense catecholaminergic inputs to CRF-secreting neurons, possibly complemented by consensual changes in serotoninergic and opposite changes in the cholinergic inputs, and by $\alpha_1$-adrenergic receptor upregulation [4, 38, 82, 134], are at present the most likely events. Advances in our understanding will come only from experiments with far more sophisticated protocols and methods than those applied in the last 2 decades.

## Circadian Chronosusceptibility of the HPA System to Endogenous and Exogenous Modifiers

As mentioned before, current knowledge indicates that the HPA system is circadian rhythmic in its function as an expression of a fundamental genetic program that organizes nervous and metabolic activities for diurnal activity and nocturnal sleep. A coordinating system that develops postnatally serves to reset the phase of a number of cellular clocks and to synchronize hierachically CRF inputs from the hypothalamus to the anterior pituitary, ACTH stimulation of the adrenal cortex, and glucocorticoid action at the periphery. Multifactorial regulation accounts for the physiological rhythmic parameters (acrophase, amplitude, MESOR). This regulation is precisely tuned by servo-mechanisms, which include classical feedback circuits and a network of feed-sideward signals (neuromodulators, paracrine products, peptide fragments, metabolites) capable of modifying target cell responsiveness to specific informative inputs. The robustness of the circadian program, which has a masking effect upon a second fundamental rhythmic program (the ultradian BRAC), is documented by the substantial consistency of its parameters in different groups of healthy individuals [102, 104, 125, 258], in different biological fluids, such as venous blood, capillary blood and saliva [242], in the same subjects over several consecutive days and under different conditions of daily activity [125, 155]. There are no appreciable sex differences. Aging too was found to have minor effects *per se* [104, 137, 171, 218, 241]. Transmeridian flying or shift working intrude in a relatively "difficult" way into the genetic program, and physiologically the signaling network responsible for the rhythmic parameters is able to buffer effectively rapid changes of exogenous synchronizers [61, 100, 101, 125, 242]. Whether this occurs for a definite time, over months or years, or indefinitely, and in genetically different individuals as well (e. g., in the so-called "morning" or "evening" type), and in subjects with clinical or subclinical neurotransmitter dysfunction such as those experienced in anxiety or depressive disorders, and whether endogenous adjustment of the circadian program following changes of environmental synchronizers leads to modified conditions of interplay with the ultradian BRAC (with alterations, e. g., in sleeping and feeding schedule) are but a few of the clinically important questions concerning not only chrononeuroendocrinology or chronopharmacology, but also social and preventive medicine [104, 109, 114, 146]. Experiments designed to answer these questions have built up a vast body of literature (see chapters in this book) and are still in progress, but will require time before conclusions due to the complexity of the issue and to obvious methodological difficulties. It is generally held that postnatal development of servomechanisms within the regulatory network in adulthood is a prerequisite for both robustness and flexibility of the circadian program. The most investigated servomechanism is certainly the feedback inhibition by glucocorticoids of ACTH release and subsequent steroid production from the adrenal. For a long time, it has been agreed that the negative feedback effect of glucocorticoids is a complicated mechanism involving both the brain and the pituitary [13, 56, 124, 135, 211]. It was also established that the circadian profile of circulating ACTH concentrations is not brought about by rhythmic changes in the feedback control, since it is still present after adrenalectomy or in patients with Addison's disease [96, 174]. In these cases, however, ACTH levels are markedly higher than those seen in normal subjects; adrenocortical steroids, therefore, do not cause but modulate the oscillatory pattern of ACTH secretion. Thus, it could be said that they are an important component of the multifactorial regulation of the circadian parameters.

It is recognized that the effects of exogenous administered glucocorticoids upon ACTH secretion are profoundly different as a function of dose, time and mode of administration (pulse(s) or infusion). It is also recognized that such effects cannot be extrapolated to those of endogenous glucocorticoid levels, which are an expression of a secretory program built up in a complex way to cope with the vicissitudes of daily life and to provide a circadian synchronizer for many metabolic, nervous and immune functions [13, 18, 29, 43, 99, 220]. Thus, the data obtained in experiments dealing with pharmacological inhibition need to be interpreted with great care. Notwithstanding these limitations, current opinion is that there are three types of feedback in terms of the timing of the inhibitory effect: fast, intermediate and delayed or slow [82, 135]. The fast feedback that operates within minutes depends on the rate of corticosteroid increase and represents the dynamic, differential mechanism of regulation. The rapidity of the effect suggests that protein synthesis is not involved and that the feedback inhibition is preferentially on CRF or ACTH secretion into the bloodstream [82]. Intermediate feedback effects appear after relatively short durations of glucocorticoid exposure or after repeated discontinuous increases in circulating glucocorticoids, and are related to the absolute concentration of steroid reached in plasma. The delayed feedback that operates within hours in the rat and perhaps within days in the human [135] depends on the dose and on the duration of steroid exposure and represents the integral mechanism of regulation. It is logical to think that the slow feedback acts to reduce messenger ribonucleic acid (mRNA) for precursor protein of CRF or ACTH [64]. The relative role and integration of the different types of feedback are still uncertain. The fast and intermediate feedback effects would be expected to occur under physiological conditions in order to prevent overstimulation by the circadian drive and by stressful events. The slow feedback would be expected to occur only in pathological conditions (Cushing's syndrome) or after pharmacological treatment with corticosteroids for several days. What is clear is that glucocorticoid feedback inhibition is not equally effective on the different inputs to the HPA system and is not equally effective throughout the 24-h day. The circadian activation appears much more sensitive than abrupt challenges caused by neurogenic or systemic stressors [82]. Furthermore, evidence obtained from different studies strongly suggests that in the human a peculiar chrono-organization results in effectiveness of the servomechanism during a limited time span with respect to the remaining hours [13]. In experiments using single-dose metyrapone, the magnitude of ACTH release was clearly much higher in the hours following midnight than during daytime [13, 16] and in experiments using low-dose infusion of glucocorticoids, the magnitude of ACTH inhibition pointed in the same direction [13, 43, 44]. A clinically sound concept is that the inhibition of ACTH is greater when corticoid administration takes place during or just prior to the circadian rise, i.e., at midnight or shortly before [18, 43, 170]. Accordingly, the time of administration of glucocorticoid compounds to patients is often adjusted in accordance with the action required: if suppression of endogenous production is to be avoided, then administration in the morning and/or early afternoon is more appropriate; if suppression is required, then late evening medication is preferable [13, 18, 160]. This dichotomy has received unanimous support, yet represents an oversimplification of the chronopharmacological background, since it takes into account only one effect (in most indications, the undesired one). Dosing should also be considered. Classical studies have documented that chronosusceptibility of the HPA system to exogenous glucocorticoids is dose dependent [44]; at least acutely, with higher doses rapid and complete suppression can be achieved at any time of the 24-h cycle. These observations lend indirect support to the concept that temporal organization of the negative glucocorticoid feedback concerns primarily the endogenous signal and to a lesser degree the pharmacological inhibition. In other words, the physiological feedback loop has an intrinsic chronoeffectiveness and is programmed to operate during the hours corresponding to the circadian rise of ACTH secretion. It appears to be an economical servomechanism that allows the ACTH secretion to be quantitatively programmed for the entire 24-h day, on the basis of well-timed information from the periphery and the overall HPA system, to refrain from overactivity and to comply with diurnal stressors [13]. Whether the exquisite susceptibility to the glucocorticoid inhibition during a critical span of the circadian cycle depends on a type of central supersensitivity with the prevailing intervention of the fast feedback mechanism or on the presence at that time of specific neurotransmitter or neurosecretory activities is an unanswered question. What appears to emerge from recent investigations is that the site of action of corticosteroids in inhibiting the circadian ACTH secretion (and the adrenalectomy-induced secretion as well) is on the activity of central neural components of the HPA system, rather than at the corticotroph level on POMC gene transcription [56, 103]. In an elegant study using adrenalectomized rats, the brain was demonstrated to be more sensitive than the pituitary to the physiological feedback of endogenous glucocorticoids [144]. Recently, attention has been focused on hippocampal neurons that are an important target of the central glucocorticoid action and appear to be involved in the feedback loop. Their role will be discussed later. Here it is pertinent to say that glucocorticoids are able to depress AVP syn-

thesis in AVP-containing neurons and specifically to affect the staining intensity and the number of AVP immunoreactive nerve fibers in the median eminence [82, 116, 263]. In the rat, following perturbation of the HPA system by adrenalectomy, levels of transcript homologous to vasopressin mRNA increased in CRF-immunoreactive neurons in PVN, and immunocytochemical studies have established the presence of vasopressin in parvocellular CRF-containing neurons [116, 263]. Not surprisingly, results of clinical studies in man, reviewed by Dallman et al. [55, 56], indicate that the magnitude of ACTH responses to an injection of AVP administered either at the nadir or at the peak (acrophase) of the endogenous circadian glucocorticoid activity are consistently different. AVP did not appreciably stimulate ACTH release when given at the time of nadir, but stimulated a marked ACTH response when given at the time of the circadian peak. While the rhythm of changes in POMC gene transcription at the corticotroph level is compatible with the subsequent synchronization of those of POMC mRNA concentrations and POMC-derived peptide secretion [154], that of immunoreactive CRF-41 in CSF does not seem to be in accordance with the program of circadian pituitary activation [84]. Explanation of the CRF pattern in CSF may well involve peptide synthesis from neurons that do not project to the median eminence, but could also be interpreted as indirect evidence that the rhythmic up-resetting of POMC and POMC-derived peptides does require cooperative inputs and/or synergism of more than one ACTH secretagogue. As a general comment, the bulk of these results can be viewed as fitting well with the hypothesis of a specifically timed co-exposure of ACTH-secreting pituicytes to CRF-41 and AVP and with an endogenous feedback mechanism programmed to turn off AVP.

In experimental rodents, the chronofeedback mechanisms are complemented by chronofeedforward mechanisms compatible with the evidence of a genetic program to amplify and reset the adrenal response to ACTH rhythmic inputs. It has been shown that the glucocorticoid response to injections or infusions of natural and synthetic ACTH preparations differs as a function of the circadian stage in which the stimulus is applied, the peak time being roughly coincidental with that of the endogenous ACTH secretion [54, 110, 175]. These observations suggest that at least in certain species multiple rhythmic changes not only in the signal production but also in the signaling action contribute to the synchronization and the stability of the HPA activity. Also in humans, the chronosusceptibility of the adrenal cells to exogenous

ACTH has been repeatedly documented [14, 73, 130, 163, 194, 195, 239], but the extent of the morning to evening changes is apparently minor and is not enough to recommend testing for diagnostic purposes at a particular clock hour [239]. Interestingly enough, indication of the circadian timing seems to be of clinical value for testing HPA activity with ovine CRF-41. Both plasma immunoreactive ACTH and cortisol responses were found to vary inversely with the baseline plasma cortisol concentrations. But while the time of administration does seem to have only a minor influence on the plasma ACTH response, the cortisol response appears much greater in the evening than in the morning [59]. With regard to the underlying mechanisms, one explanation could be sought in circadian changes of the transduction step between the formation of the ACTH-receptor complex on the cell membrane and the activation of adenylate cyclase [194, 239]. Alternatively, temporal variations in adrenal blood flow, thereby in the ACTH delivery rate, could play a role [110]. In any event, the human adrenal cortex and in a broader sense the HPA system tend to respond to provocative challenges in a relatively more intense way when the endogenous rhythmic activity is at its lower levels, i.e., in the evening hours. This has been documented for a variety of stimuli [131, 238] and, as far as the exposure to exogenous ACTH is concerned, may be viewed to be at variance with respect to laboratory rodents. It is unknown whether the circadian drive in the early morning hours plays a facilitatory role for subsequent responses of the HPA system to other stimuli, as it has been postulated to occur for some neurogenic stressors [82].

## Clinical Aspects of HPA Rhythmicities

The introduction of chronobiological concepts into the clinical practice and the recognition of the circadian rhythm of the HPA system as an endocrine entity in its own right (a signal to be received by target cells), preparatory to daily activity and expression of a fundamental genetic program, have led in the past 2 decades to a rapidly expanding literature on the pathological conditions associated with disordered glucocorticoid profiles and on the use of drugs potentially interfering with the circadian program, i.e., corticosteroids. It was Pincus [182] who 50 years ago gave the first description of the nyctohemeral changes in the urinary excretion of 17-ketosteroids. Since then, a host of clinical studies have dealt with the diagnostic significance of evaluating the circadian pattern of urinary and, more recently, plasma corticosteroid concentrations in disorders of the HPA system. The results of the most classical studies have been repeatedly reviewed [12, 124, 155]. Textbooks of medicine report worldwide that profound disturbances of the circadian rhythm of plasma cortisol occur regularly in Cushing's syndrome and that the most suitable screening test for evaluating individuals suspected of hypercortisolism is the overnight single-dose dexamethasone suppression test, which represents a typical chronobiological approach. In adulthood, probands ingest 1 mg dexamethasone at 2300 hours, and a plasma cortisol value is obtained at 0800 hours the next morning. This test can be performed on an outpatient basis and accurately separates the normal population from those with pathology. It is a cost-effective application of the chronosusceptibility of ACTH secretion to the feedback inhibition by glucocorticoids. Another widely used screening approach in the exploration of adrenal function is sampling morning or evening plasma cortisol concentrations, since the consistency of the circadian rhythm in the healthy population makes it normal to find high values in the morning and low levels in the evening. Thus, adrenal insufficiency is suspected when morning cortisol levels are low and, conversely, adrenal hyperfunction is suspected when evening cortisol levels are high. In any event, clinicians must be aware that episodic variations in hormone secretions may occur at any time throughout the 24-h day and lead to rapid changes in the plasma concentrations, thus masking the values to be expected at the different stages of the circadian cycle [104].

Independently of the aforementioned implications for diagnostic endocrinology, the chronobiology of the HPA system has offered to clinicians a number of opportunities for better diagnosis and monitoring of patients with endocrine and nonendocrine disease. The recognition of rhythm parameters as valuable endpoints in describing normalcy and the availability of computer programs to quantify rhythms satisfactorily have led to meaningful results, e.g., in psychiatric disorders. In a spectrum of clinical conditions, including anorexia nervosa, anxiety and depression, abnormalities of the circadian parameters of glucocorticoid secretion are frequently encountered [10, 103, 128, 133, 189, 191, 206]. Attention has been centered on the so-called depression-associated hypercortisolism, which is thought to be related to abnormalities of neuroendocrine regulatory processes of CRF secretion involving also a relative refractoriness to cortisol feedback servo-mechanisms [10]. Indeed, patients with major depression show significantly elevated CSF concentrations of CRF-41-like immunoreactivity [169, 202]. It is beyond the scope of the present chapter to review current knowledge about the issue whether psychiatric disturbances, i.e., endogenous depression, are related to pathological interplay between the fundamental circadian and ultradian programs leading to abnormal internal phase relationships between sleep and HPA activation. Several clinical features of depression have led investigators to hypothesize that alterations in the timing of the circadian program are involved in the pathophysiology of psychic symptoms and of neuroendocrine abnormalities as well. In the 1980s, this hypothesis was called the phase-advance hypothesis of depression [127, 132]. It is pertinent, however, to point out the current opinion that in a certain number of depressive patients hypersecretion of glucocorticoids is secondary to a defective feedback inhibitory mechanism which leads to inadequately restrained CRF-ACTH activity [10, 71, 210, 211]. This feedback abnormality is thought to exist only during depression, because recovered patients respond to the overnight single-dose dexamethasone testing with normal suppressibility of plasma ACTH [10]. It remains unclear if the defect represents a typical example of chronopathology in that it involves the endogenous glucocorticoid signal during a critical span of the 24-h cycle (reduced nocturnal effectiveness) or if the feedback otherwise normally effective is insufficient to counteract increased CRF secretion due to activating the hypothalamic milieu. Noteworthy is the fact that a similar refractoriness to feedback inhibition with hypercortisolism and subtle abnormalities of the circadian parameters may be found in the elderly, especially if mentally deteriorated such as in Alzheimer's disease and in vascular (multi-infarct) dementia [75, 178, 210]. A major advance in understanding the mechanism underlying hypercortisolism in these conditions, and in depression as well, has been the appreciation of the hippocampus as an important glucocorticoid negative feedback site [20, 71]. Both in the rat and in primates the hippocampus is a glucocorticoid target tissue, given its high concentrations of glucocorticoid receptors; ACTH secretion increases after hippocampal lesions and the magnitude of HPA system activation parallels the degree of hippocampal damage; the inhibitory role of the hippocampus depends on glucocorticoid action on this structure. As a logical assumption from accumulating evidence, it has been proposed that a discrete population of hippocampal neurons operate as mediators of the endogenous feedback loop [71]. Would this be true, one

crucial question could be raised: does the hippocampal-hypothalamic inhibitory circuit operate in a circadian way, within the network of coordinating signals that allow the nocturnal-early morning surge of the HPA system to be definite in terms both of intensity and of timing? Available data suggest that in the rat the influence of the hippocampal neurons upon CRF-ACTH secretion has indeed circadian fluctuations [76, 158], but there is no information about the effects of environmental synchronizers and of lesions of the SCN upon this mechanism. Yet the mechanisms by which glucocorticoid-dependent hippocampal activation exert this role are unknown. Except for the septum, the limbic structures do not project to the PVN directly [177]. Interestingly enough, it has been suggested that the hippocampus regulates AVP secretion preferentially with respect to the major secretagogue CRF-41 [71]. Perhaps, the pieces of a puzzle depicting AVP as key molecule in the circadian organization of HPA system are becoming less scarce, yet still remain scattered. In any event, there is no doubt that during the past few years clinical research has been increasingly oriented to the study of the HPA system during senescence, with the aim of finding neuroendocrine markers of aging brain and related disorders. In the aging rat and the aging human as well (yet to a lesser degree), there is hippocampal neuron loss and/or hippocampal glucocorticoid receptor loss that has been suggested to account for the refractoriness to single-dose dexamethasone inhibition and relevant hyperadrenocorticism [20, 25, 46, 53, 71, 211]. Chronobiologically speaking, it is noteworthy that in the aging rat higher levels of plasma corticosterone may be found only in the morning, i.e., at the physiological circadian nadir of the HPA system, and in the aging human higher levels of plasma cortisol may be found only in the late evening or at midnight, i.e., at the corresponding circadian nadir [20, 178]. It is obviously premature to affirm that midnight plasma cortisol is a chronobiological marker of aging brain; longitudinal studies are needed. Its candidate role, however, cannot be ruled out by previous studies based on single morning samples or even on accurate evaluation of the daily secretion. The senile deterioration of brain, in fact, could theoretically be expressed by a specific vulnerability of the endogenous glucocorticoid feedback loop, which has a peculiar circadian chronoeffectiveness. Anatomically (neuronal loss or damage) or functional impairment of the brain circuit(s) involved in this chronoeffectiveness could lead to alterations of the circadian program of the HPA system and eventually to hypercortisolism in different clinical conditions such as affective disorders, Alzheimer's disease and other dementia states, and in the elderly themselves. At the beginning of the 1990s this is a fascinating area of investigation.

Turning back to the second point of clinical relevance of the circadian rhythm of the HPA system, i.e., the use of drugs that are potentially interfering and hormonally active (ACTH and glucocorticoid-agonists or antagonists), a body of findings of clinico-pharmacological studies performed mostly in the 1970s have indicated that for many clinical indications the corticosteroid administration has to be concentrated in the morning and/or early afternoon. Injection of 40 mg methylprednisolone in asthmatic children was more effective at 0800 and 1500 hours than at 0300 and 1900 hours [193]. In a double-blind, crossover, placebo-controlled study, Reinberg et al. [195] have found that chronic administration of corticoids at 0800 and 1500 hours was more effective in controlling asthma and enhancing peak expiratory flow (PEF) values than the same agents and dose given at 1500 and 2000 hours. Considering that Soutar et al. [224] were unable to reduce the nocturnal dip in PEF despite giving sufficient doses as infusion to maintain constant and rather high plasma steroid levels, one could state that glucocorticoids are more effective when administered in the morning and/or early afternoon hours as far as the control of allergic asthma and bronchial patency are concerned. This particular chronoeffectiveness could be partially explained by the circadian synchronizing effect of glucocorticoids on $\beta_2$-adrenoreceptor density [98, 142, 196]. Available evidence also points to an improvement of the immunosuppressive effect of glucocorticoids when the major part of the steroid dose is given in the morning with the remaining part administered in the afternoon [62, 119]. Using this schedule, patients given cadaveric renal allografts were maintained under satisfactory immunological control by low-dose prednisone regimens. Whether these results depend upon an intrinsic chronoeffectiveness of the corticoid treatment (gain in the intensity of desired effects in a part of the 24-h cycle) or upon a lesser degree of pituitary-adrenal suppression is still a matter of debate. Possibly both mechanisms concur. In any event, there is now agreement that in a large number of patients more than a single morning dose of corticoids (even on alternate days) does not seem to be necessary for control of an inflammatory process.

It has also been shown that Addison's patients on conventional replacement therapy distributed in two unequal parts (two-thirds on awakening and one-third in the afternoon) may be doing well and experience no particular discomfort yet have extremely low plasma cortisol levels during a large part of the 24-h day [13]. Insofar as the findings with exogenously administered corticoids can suggest, human target tissues and cells are prepared to respond to glucocorticoid information along a programmed temporal scale that favors several effects during the first half of the daytime hours, lagging behind by several hours the expected crest-time (acrophase) of the plasma cortisol rhythm. This fits well with the current concepts on glucocorticoid action at the cellular level and on the role of the HPA system as a circadian synchronizer of the human temporal structure. In nocturnally active rodents the same condition applies with a phase shift of approximately 12 h [67, 221]. It is attractive to think that physiologically glucocorticoid action does not need a continuous supply of hormone molecules to target cells but appropriate waves complying with the peripheral susceptibility. These considerations have prompted in the 1980s clinical research on the possible "chronization" of the human circadian system with the use of agonists of the HPA system hormones. Chronizer was defined as an exogenous molecule the administration of which at a proper dose and timing is able to reset some correlated biorhythmicities, i.e., resynchronizes them with a new schedule, possibly more physiological or more useful, for instance, to attain therapeutic aims [19]. A clinically valuable feature of the chronizer action is conceivably the capability of restoring or enhancing the rhythmic organization of some key functions, thus opposing endogenous or exogenous events that may disrupt the sophisticated signaling system controlling these functions. As it has been stated before, subtle chronoabnormalities of the adrenocortical function have been correlated with diverse brain conditions, including affective disorders and senile deterioration. The same can be said for diverse chronic disease conditions, arterial hypertension, malnutrition, etc. An ACTH-agonist peptide, the heptadecapeptide alsactide, was proven effective both in experimental animals and in humans in ameliorating psychophysical variables and the tolerance of antimitotic drugs when administered at low doses early in the morning, i.e., just before the peak of endogenous cortisol would reach its maximum. These doses were found to be able to raise plasma cortisol level for a relatively short time, thus enhancing intensity and duration of the endogenous wave by no more than 50% [252]. Whether alterations in timing of periodic variables correlated or, better, synchronized by adrenocortical function can be at least partially corrected by HPA chronization deserves to be investigated extensively. Another application of this chronization strategy may be indicated in the jet-lag syndrome, especially in subjects in whom subjective feeling suggests slow adaptation to time-zone transition [43]. In conclusion, recent knowledge from different clinical research areas has not only pointed out the high complexity of the informational network that controls the circadian program of the HPA system, but has also indicated novel chrononeuroendocrinological approaches to preventive and curative medicine.

## Chronoendocrinology of the Renin-Angiotensin-Aldosterone System

The zona fasciculata anatomically is the larger of the adrenal cortex, and its major secretory product is cortisol in man and corticosterone in the rat. It is the glucocorticoid-secreting station of the HPA system. Its activity is typically rhythmic, with a very prominent circadian component which, as previously mentioned, is an expression of a genetically determined circadian program controlled by a complex signaling network primarily operating in the brain. The zona glomerulosa, the outermost zone of the adrenal cortex, is very thin and discontinuous in some areas where the zona fasciculata extends to the glandular capsule. In the human, this zone is still less evident than in other species. It specifically produces aldosterone, the most important mineralocorticoid hormone. Its activity is also rhythmic but in a different way with respect to the inner zone. In the circadian domain, a clear-cut prevailing monocomponent rhythm is not apparent; rather, at least two circadian aldosterone surges are distinguishable in normal diurnally active, nocturnally resting individuals, and secretory episoded relatively often mask during daytime the hormonal values to be expected on the basis of the circadian variations. As a consequence, the circadian rhythm parameters as computed by cosinor analysis of serial plasma aldosterone measurements throughout the 24-h cycle are characterized by reduced amplitude and delayed acrophase in comparison with the cortisol counterpart [49, 51]. In summary, in the face of the high-amplitude robust rhythm of glucocorticoid secretion, which represents a synchronizing signal for a number of cellular clocks, the circadian changes of aldosterone secretion are less apparent and are not likely to be an important chronobiological message for target cells. The latter assumption, indeed, is indirectly supported by current knowledge on mineralocorticoid physiology. Mineralocorticoid can be classified an adrenal corticosteroid which stimulates active transepithelial sodium transport at physiological concentrations by interacting with specific cellular receptors [232]. Transport of sodium in one direction is frequently accompanied by passage of potassium and hydrogen ion in the opposite direction. Aldosterone is the most potent mineralocorticoid, but other mineralocorticoids are secreted by the zona fasciculata of the adrenal cortex, in other words by the peripheral station of the HPA system. The principal mineralocorticoids produced by this zone are cortisol and deoxycorticosterone (DOC) [33]; the zona fasciculata also maintains significant peripheral levels of 18-hydroxydeoxycorticosterone (18-OH-DOC), corticosterone, 11-deoxycortisol, steroid molecules able to interact with specific mineralocorticoid receptors. Aldosterone is perhaps 500- to 1000-fold more active than cortisol in experimental models of mineralocorticoid action, but its secretion rate is 100-fold lower than that of cortisol and is equal or less than that of DOC, a steroid with a circadian pattern parallel to cortisol and with a potent mineralocorticoid activity [232]. Thus, it is logical to think that if the mineralocorticoid target cells need to be temporally arranged along a circadian scale by the hormonal information, this can be accomplished by the "ACTH-dependent" mineralocorticoids, which show a high-amplitude and basic circadian program. The foregoing considerations also offer an explanation as to why there is a dearth of information on the underlying mechanisms and the clinical aspects of the circadian changes of aldosterone secretion. These changes are currently thought to reflect the multifactorial regulation of the cells of the zona glomerulosa; in particular, the two aforementioned circadian components are attributed to ACTH and angiotensin II stimulation, respectively [28, 33].

It is held that, whereas the steroidogenic activity of the inner part of the adrenal cortex is almost exclusively controlled by the anterior pituitary, the production of aldosterone by the zona glomerulosa is adapted to the sodium and potassium concentrations and to the volume of extracellular fluid by a complex mainly extrapituitary control system [9, 188, 232]. The most important hormonal factor in this system is the octapeptide angiotensin II. However, a considerable number of known or poorly defined stimulators and inhibitors also contribute to the regulation of aldosterone biosynthesis, i.e., $K^+$, $Na^+$, the atrial natriuretic peptide (ANP), dopamine, prostaglandins, pituitary products such as a glycoprotein termed aldosterone-stimulating factor (ASF), and POMC-derived peptides other than ACTH [188]. The role of

ACTH in this array of modifiers or modulators is still uncertain. Due to its action on different steps of steroidogenesis, it seems obvious that the pituitary hormone participates in a stimulatory way in the control of aldosterone biosynthesis; acute injection of ACTH or agonist peptides is followed by a rapid rise of plasma aldosterone concentrations [32, 33, 63]. However, the physiological significance of endogenous ACTH stimulation is generally regarded as not important and, in a sense, permissive or facilitatory of the major aldosterone secretagogue, angiotensin II. ACTH, indeed, conveys hierarchically to the adrenal cortex a fundamental chronobiological message that allows the steroid-secreting cells to be synchronized properly according to the circadian program. Aldosterone-secreting cells too receive the message. Thus, ACTH is probably weakly effective as a stimulator of aldosterone production, at least at physiological concentrations, but is likewise probably highly effective as a chrono-organizer, as it increasingly recognized [28, 49, 156, 188]. In fact, in normal individuals assuming upright posture from 0800 to 2200 hours and then remaining supine until the following morning, the pattern of plasma aldosterone concentrations is clearly affected by postural influences, but displays an overt circadian component which persists in subjects recumbent for the entire 24-h cycle and peaks in the early morning hours roughly at the time of awakening [9, 49, 50, 112]. By sampling blood as frequently as every 10–20 min, it was documented that also the aldosterone release into the bloodstream is arranged in episodes with interspersed pauses, that these episodes are often but not always concomitant with those of cortisol release, and that in diurnally active subjects there is a second apparent circadian component of the aldosterone-producing activity, which peaks later than the first one, at about midday [156, 205, 261]. This component is generally related to the temporal organization of renin release, as inferred by the measurement of plasma renin activity (PRA) [95, 156, 159]. Renin release from the juxtaglomerular cells of the kidney is the primary determinant of the activity of the RAA system. In brief, the proteolytic action of renin splits off the inactive decapeptide angiotensin I from angiotensinogen, a glycoprotein produced in the liver. Angiotensin-converting enzyme (ACE) turns angiotensin I into the active octapeptide angiotensin II. Receptors for angiotensin II are numerous on the surface of the cells of the adrenal zona glomerulosa and the peptide has been demonstrated to directly stimulate aldosterone biosynthesis at different stages [188]. During the past decade, much has been learned about the complex neuroendocrine system that governs renin release and about the mechanisms accounting for the temporal variations of this fundamental signal. In normally active subjects the circadian profile of PRA is roughly as follows: PRA levels begin to rise prior to awakening in the early morning hours, then a reflex increase in renin release produced by assuming the upright posture accentuates and prolongs the ascending wave until approximately midday; in the afternoon PRA levels decrease until the nadir in the cycle in late evening [50, 112]. The endogenous component of PRA rhythmicity is a circadian entity in its own right; it is apparent in subjects maintained resting in bed throughout the 24-h day and it has complex relationships with the ultradian BRAC and related non-REM-REM alteration during sleep [35, 166, 203]. There are three major pathways that control renin secretion, the so-called bareceptor pathway, the macula densa pathway and the neuroendocrine pathway [80]. They operate in an integrated fashion, but it is generally believed that the renal nerves and catecholamines are primarily responsible for the postural increase with the mediation of beta-adrenergic receptors [68]. Interestingly enough, the increment of PRA as a response to upright posture is larger in the morning than in the afternoon [9]. In a number of models, glucocorticoids are able to increase the number and/or the functional affinity of beta-adrenoreceptors [196]. It is reasonable to think that the circadian signal of plasma glucocorticoid levels operates as an endogenous synchronizer also at the level of the juxtaglomerular cells, up-resetting their reactivity to postural modification just after awakening. In turn, the postural generation of angiotensin II would account for a second secretory wave of aldosterone secretion and the particular shape form of the plasma aldosterone profile. Evidence that a circadian component is yet detectable in plasma aldosterone levels after suppression of the postural PRA rise by administration of an adrenergic beta-blocker [49] and after suppression of the HPA circadian rhythm by administration of exogenous glucocorticoids [32] indicates that both ACTH and renin-angiotensin are physiologically influential. This has also been confirmed by studies on laboratory animals [106] and on experimental renovascular hypertension in which dissociated behaviors between PRA and aldosterone rhythmicities have been documented [60, 81]. The same applies to cases of primary hyperaldosteronism, in which plasma aldosterone levels display significant circadian variations in spite of suppressed PRA [31, 265]. Although other components of the multifactorial system that control the activity of the adrenal zona glomerulosa normally display low-amplitude circadian changes (e. g., plasma ACE activity or ANP concentrations), there is no doubt that the most important pathophysiological and clinical implications of chronoendocrinology of RAA should come out from the definition of the relationships between rhythms of aldosterone, PRA and cortisol in different conditions. Unfortunately, this is a relatively neglected area of study, in the face of obvious interest to clinicians. In a most recent review on stimulation and suppression of the mineralocorticoid hormones in normal subjects and adrenocortical disorders [111], there are only few generic words on the circadian pattern of plasma aldosterone levels under different clinical conditions. A Diagnosis of dexamethasone-suppressible hyperaldosteronism (DSH), for example, is strongly supported by the documentation of a circadian rhythm presenting with an amplitude higher than normal and an acrophase synchronous with that of cortisol [70, 265]. A diagnosis of idiopathic primary hyperaldosteronism (IHA), due to bilateral hyperplasia of the zona glomerulosa, on the other hand, is supported by the lack of appreciable circadian variations in the recumbent position, complemented by an appreciable postural rise when the patient assumes the upright position. The mechanism underlying this pattern can be sought in the hypersensitivity of the glomerulosa cells to very low plasma angiotensin II concentrations, since renin release is basically suppressed as in other forms of primary hyperaldosteronism. The differential diagnosis of these cases from those due to an aldosteronoma (aldosterone-producing adenoma, APA) is indeed corroborated by the chronobiological approach in that adenomas are autonomous in their dependency on angiotensin II stimulation but often maintain some degree of ACTH dependency, as seen by the significant circadian rhythm with acrophase early in the morning [31]. More generally, it could be said that the relative degree of influence of the two major circadian components of the aldosterone secretion changes as a function of the volume of the extracellular fluid and sodium balance. In conditions characterized by hypovolemia and negative sodium balance (high PRA with secondary hyperaldosteronism) the levels of circulating aldosterone appear to be more closely tied to the renin-angiotensin rhythm; when body sodium is replete and PRA is low, ACTH rhythmicity becomes more important [156]. However, on the chronobiology of mineralocorticoid steroids much remains to be investigated, e. g., in experimental and clinical hypertension and as a response to pharmacological treatments that interfere with the

RAA system. An important challenge would be to determine how the adrenocortical rhythmicities interface with those of other vasoactive hormones such as catecholamines [40, 148, 228], atrial natriuretic peptide(s) [200, 207, 262] and angiotensin II [52, 251] in the control of blood pressure. It is perhaps only through elucidation of this issue that an integrated picture of different forms of what is now called essential hypertension can be achieved.

## Conclusions

It is evident from the material presented in this chapter that the HPA system plays a fundamental role in the human circadian system. We discussed the signaling network that controls the expression of multiple circadian clocks hierarchically organized within the system. The recent appreciation that multiple secretagogues can act in concert to regulate the ACTH secretion is raising interest on the role of AVP besides that of CRF-41 as a key molecule in this network. Postnatal maturation of the circadian program for diurnal activity and feeding and nocturnal resting and sleeping leads the HPA system to acquire robust rhythm parameters and complex relationships with another basic ultradian program (BRAC). Target cells for glucocorticoids too are programmed to respond to the hormonal input differently as a function of the circadian stage. Rhythms in the susceptibility to glucocorticoid action may well be phase shifted among different targets. Available evidence suggests that chronoeffectiveness also concerns the negative feedback loop which is an important control mechanism of the rhythmic parameters. Anatomical or functional impairment of the brain circuit(s) involved in such chronoeffectiveness could be of clinical relevance. In spite of the vast body of applications to medicine of the chronobiological concepts relevant to HPA function, much still has to be done for the RAA system. Many questions remain about the relationships between the ACTH-dependent corticosteroids and aldosterone, and about the physiological significance of the circadian variations of renin release. Recent developments on the multifactorial control of aldosterone secretion may open new questions. It is hoped that chronobiological research will soon provide information useful to clinicians for better diagnosis and management of hypertension and disorders of water and electrolyte balance.

# References

1. Agrimonti F, Angeli A, Frairia R, Fazzari AM, Tamagnone C, Fornaro D, Ceresa F (1982) Circannual rhythmicities of cortisol levels in the peripheral plasma of healthy adult subjects. Chronobiologia 9: 107–114
2. Agrimonti F, Frairia R, Fornaro D, Torta M, Borretta G, Trapani G, Bertino E, Angeli A (1982) Circadian and circaseptan rhythmicities in corticosteroid-binding globulin (CBG) binding capacity of human milk. Chronobiologia 9: 281–290
3. Alford FP, Baker HW, Burger HB, Dekretzer DM, Hudson B, Johns MW, Masterton YP, Patel GC, Rennie GC (1973) Temporal patterns of integrated plasma hormone levels during sleep and wakefulness. I. Thyroid-stimulating hormone, growth hormone and cortisol. J Clin Endocrinol Metab 37: 841–847
4. Al-Damluji S (1988) Adrenergic mechanisms in the control of corticotrophin secretion. J Endocrinol 113: 5–14
5. Al-Damluji S, Cunnah D, Perry L, Grossman A, Besser G (1987) The effect of alpha adrenergic manipulation on the 24 hours pattern of cortisol secretion in man. Clin Endocrinol (Oxf) 26: 61–66
6. Al-Damluji S, Cunnah D, Grossman A, Besser G (1987) Effect of adrenaline on basal and corticotropin-releasing factor stimulating ACTH secretion in man. J Endocrinol 112: 145–150
7. Al-Damluji S, Iveson T, Thomas JM (1987) Food induced cortisol secretion is mediated by central $\alpha$-adrenoreceptor modulation of pituitary ACTH release. Clin Endocrinol (Oxf) 26: 629–636
8. Allolio B, Schulte HM, Deuss U (1987) Effect of oral morphine and naloxone on pituitary-adrenal response in man induced by corticotropin-releasing hormone. Acta Endocrinol (Copenh) 114: 509–514
9. Ambruster H, Vetter W, Beckeroff R, Nussberger J, Vetter H, Siegenthaler W (1975) Diurnal variation of plasma aldosterone in supine man. Relationship to plasma renin activity and plasma cortisol. Acta Endocrinol (Copenh) 80: 95–103
10. Amsterdam JD, Maislin G, Gold P, Winokur A (1989) The assessment of abnormalities in hormonal responsiveness at multiple levels of the hypothalamic-pituitary-adrenocortical axis in depressive illness. Psychoneuroendocrinology 14: 43–62
11. Andrews RW (1968) Temporal secretory response of cultured hamster adrenals. Comp Biochem Physiol 26: 179–193
12. Angeli A (1982) Rhythmicity of pituitary hormone secretion. In: Motta M, Zanisi M, Piva F (eds) Pituitary hormones and related peptides. Academic, New York, pp 319–336
13. Angeli A (1983) Glucocorticoid secretion: a circadian synchronizer of the human temporal structure. J Steroid Biochem 19: 545–554
14. Angeli A (1984) Peptide hormone analogues and novel clinical applications. Ric Clin Lab 14: 123–135
15. Angeli A, Carandente F (1988) An update on clinical chronoendocrinology. Adv Biosci 73: 319–333
16. Angeli A, Fonzo D, Bertello P, Gaidano G, Ceresa F (1975) Circadian rhythm of urinary 17-hydroxycorticosteroids during metyrapone-induced ACTH release in normal subjects. Chronobiologia 2: 133–144
17. Angeli A, Frairia R, Dogliotti L, Crosazzo C, Rigoli F, Ceresa F (1978) Differences between temporal patterns of plasma cortisol and corticosteroid binding globulin binding capacity throughout the 24 hours day and the menstrual cycle. J Endocrinol Invest 1: 31–38
18. Angeli A, Carandente F, Halberg F (1983) Temporal aspects of glucocorticoid action and clinical implications. Ric Clin Lab 13: 203–217
19. Angeli A, Carandente F, Dammacco F, Halberg F, Martini L (1987) Alsactide: ACTH-agonist for use in microdoses in brain-adrenal and other feedsidewards. Chronobiologia 14: 99–143
20. Angelucci L, Valeri P, Grossi E (1980) Involvement of hippocampal corticosterone receptors in behavioural phenomena. In: Brambilla G, Racagni G, de Wied G (eds) Progress in psychoneuroendocrinology. Elsevier, Amsterdam, pp 186–199
21. Antoni FA (1986) Hypothalamic control of adrenocorticotropic secretion: advances since the discovery of 41-residue corticotropine-releasing factor. Endocr Rev 7: 351–373
22. Antoni FA, Holmes MC, Jones MT (1983) Oxytocin as well as vasopressin potentiate ovine CRF in vitro. Peptides 4: 411–415
23. Asano Y (1971) The maturation of the circadian rhythm of brain norepinephrine. Life Sci 10: 883–894
24. Assenmacher I. Szafarczyk A, Alonso G, Ixart G, Barbanel G (1987) Physiology of neural pathways affecting CRH secretion. Ann NY Acad Sci 512: 149–161
25. Ball M (1987) Neuronal loss, neurofibrillary tangles and granulovacuolar degeneration in the hippocampus with aging and dementia. Acta Neuropathol (Berl) 37: 111–119
26. Bardeleben U, Holboer F, Stalla G, Muller O (1985) Combined administration of human corticotropin-releasing factor and lysine vasopressin induces cortisol escape from dexamethasone suppression in healthy subjects. Life Sci 37: 1613–1618
27. Barnett JL, Winfield CG, Cronin GM, Makin AW (1981) Effect of photoperiod and feeding on plasma corticosteroid concentration and maximum corticosteroid binding capacity in the pig. Aust J Biol Sci 34: 557–585
28. Bartter FC, Chan JCM, Simpson HW (1979) Chronobiological aspect of plasma renin activity, plasma aldosterone and urinary electrolytes. In: Krieger DT (ed) Endocrine rhythms. Raven, New York, pp 49–132
29. Berczi I (1986) The influence of pituitary-adrenal axis on the immune system. In: Berczi I (ed) Pituitary function and immunity. CRC, Boca Raton, pp 49–132
30. Blalock JE, Smith EM, Meyer WJ (1985) The pituitary-adrenocortical axis and the immune system. Clin Endocrinol Metab 14: 1021–1038
31. Biglieri EG, Irony I (1990) Primary aldosteronism. In: Biglieri EG, Melby JC (eds) Endocrine hypertension. Raven, New York, pp 71–85
32. Biglieri EG, Shambelan M, Slaton PE (1969) Effect of adrenocorticotropin on deoxycorticosterone corticosterone and aldosterone secretion. J Clin Endocrinol Metab 29: 1090–1101
33. Biglieri EG, Arteaga E, Kater CE (1988) Effect of ACTH on aldosterone and other mineralocorticoid hormones. Ann NY Acad Sci 512: 426–437
34. Born J, Kern W, Bieber K, Fehm-Wolfsdorf G, Schiebe M, Fehm HL (1986) Night-time plasma cortisol secretion is associated with specific sleep stages. Psych 21: 1415–1424

35. Branderberger G, Follenius M, Muzet A, Erhart J, Schieber JP (1985) Ultradian oscillations in plasma renin activity: their relationship to meals and sleep stages. J Clin Endocrinol Metab 61: 280–284

36. Britton KT, Koob GF (1988) Behavioral effects of corticotropin-releasing factor. In: Schatzberg A, Nemeroff CE (eds) The hypothalamic-pituitary-adrenal axis. Physiology, pathophysiology and psychiatric implications. Raven, New York, pp 55–66

37. Brown MH, Nunez AA (1989) Vasopressin-deficient rats show a reduced amplitude of the circadian sleep rhythms. Physiol Behav 46: 759–762

38. Calogero AE, Bernardini R, Margioris AN, Bagdy G, Gallucci WT, Munson PJ, Tamarkin I, Tomai TP, Brady L, Gold PW, Chrolisos GP (1989) Effects of serotoninergic agonists and antagonists on corticotropin-releasing hormone secretion by explanted rat hypothalami. Peptides 10: 189–200

39. Calvano SE, Reynolds RW (1984) Circadian fluctuations in plasma corticosterone, corticosterone-binding activity and total protein in male rats: possible disruption by serial blood sampling. Endocrinol Res 10: 11–27

40. Cameron OG, Curtis GC, Zelink T, McCann D, Roth T, Guire K, Huber-Smith M (1987) Circadian fluctuation of plasma epinephrine in supine humans. Psychoneuroendocrinology 12: 41–51

41. Carandente F, Angeli A, de Vecchi A, Dammacco F, Halberg F (1988) Multifrequency rhythms of immunological functions. Chronobiologia 15: 7–23

42. Carandente F, Angeli A, Candiani GB, Crosignani PG, Dammacco F, de Cecco L, Marrama P, Massobrio M, Martini L (1990) Rhythms in the ovulatory cycle. 3rd cortisol and dehydroepiandrosterone sulphate (DHEA-S). Chronobiologia 17: 209–217

43. Ceresa F, Angeli A (1977) Chronotherapie corticoide. In: Mirouze J (ed) 14ième Congrès International de Thérapeutique. Rapports généraux. Expansion Scientifique Francaise, Paris, pp 211–223

44. Ceresa F, Angeli A, Boccuzzi G, Molino G (1969) Once-a-day neurally stimulated and basal ACTH secretion phased in man and their response to corticoid inhibition. J Clin Endocrinol Metab 29: 1074–1082

45. Chihara K, Kato Y, Maeda K, Matsukura S, Imura H (1976) Suppression by cyproheptadine of human growth hormone and cortisol secretion during sleep. J Clin Invest 57: 1393–1402

46. Coleman P, Flood D (1987) Neuron numbers and dendritic extent in normal aging and Alzheimer's disease. Neurobiol Aging 8: 521–540

47. Coleman RJ, Reppert SM (1985) The cerebrospinal fluid vasopressin is effectively insulated from osmotic regulation of plasma vasopressin. Am J Physiol 248: E346–E352

48. Copinschi G, Leclercq R, Goldstein J, Robin C, Delsir D, Delaet MH, Virasoro E, Vanhaelst L, L'Hermite M, van Cauter E (1977) Seasonal modifications of circadian and ultradian variations of ACTH, cortisol, βMSH and TSH in normal man. Chronobiologia 4: 106

49. Cornelissen G, Halberg F, Haus E, Smith D, Harrison JR (1982) Circadian acrophase lead of human circulating aldosterone and ACTH with respect to cortisol. Chronobiologia 9: 346

50. Cugini P, Manconi E, Serdoz N, Mancini A, Meucci R, Scavo D (1980) Rhythm characteristics of plasma renin, aldosterone and cortisol in five types of mesor-hypertension. J Endocrinol Invest 3: 143–149

51. Cugini P, Salandi E, Lisanu M, Lucia P, Tomassini R, Scavo D (1982) Age-related chronopathology in circadian and circannual rhythm of plasma renin and aldosterone in hypertensives. Eur Rev Med Pharmacol Sci 4: 15–20

52. Cugini P, Letizia C, Scavo D (1988) The circadian rhythmicity of serum angiotensin converting enzyme: its phasic relation with the circadian cycle of plasma renin and aldosterone. Chronobiologia 15: 229–232

53. Dallman MF (1984) Viewing the ventral hypothalamus from the adrenal gland. Am J Physiol 246: R1–R3

54. Dallman MF, Engeland WC, Rose JC, Wilkinson CW, Shinsako J, Siedenburg F (1978) Nycthemeral rhythm in adrenal responsiveness to ACTH. Am J Physiol 235: R210–R218

55. Dallman MF, Akana SF, Cascio CS, Darlington DN, Jacobson L, Levin N (1987) Regulation of ACTH secretion. Variation on a theme of B. Recent Prog Horm Res 43: 113–167

56. Dallman MF, Akana SF, Jacobson L, Levin N, Cascio CS, Shinsako J (1987) Characterization of corticosterone feedback regulation of ACTH secretion. Ann NY Acad Sci 512: 402–414

57. Davis FC (1981) Ontogeny of circadian rhythms. In: Aschoff J (ed) Biological rhythms. Plenum, New York, pp 257–274 (Handbook of behavioral neurobiology, vol 4)

58. DeBold CR, Sheldon WR, Decherney GS, Jackson RV, Alexander AN, Vale W, Rivier J, Orth DN (1984) Arginine vasopressin potentiates ACTH release induced by ovine corticotropin releasing factor. J Clin Invest 73: 533–538

59. Decherney GS, de Bold CR, Jackson RV, Sheldon WR, Island DP, Orth DN (1985) Diurnal variation in the response of plasma adrenocorticotropin and cortisol to intravenous ovine corticotropin releasing hormone. J Clin Endocrinol Metab 61: 273–279

60. De Forrest JM, Davis JO, Freeman RH (1979) Circadian changes in plasma renin activity and plasma aldosterone concentration in two-kidney hypertensive rats. Hypertension 1: 142–149

61. Desir D, van Cauter E, Fang V, Martino E, Tadot C, Spire JP, Noel P, Refetoff S, Copinschi G, Goldstein J (1981) Effect of "jet-lag" on hormonal patterns. I. Procedures, variations in total plasma proteins and disruption of adrenocorticotropin-cortisol periodicity. J Clin Endocrinol Metab 52: 628–641

62. De Vecchi A, Ponticelli G (1982) A survey of current steroid regimens for transplanted patients. Heart Transplant 2: 64–70

63. Dolan LM, Carey RM (1989) Adrenal cortical and medullary function: diagnostic tests. In: Vaughan ED, Carey RM (eds) Adrenal disorders. Thieme, Stuttgart, pp 81–145

64. Drouin J, Sun YL, Nemer M (1990) Regulatory elements of the pro-opiomelanocortin gene. Pituitary specificity and glucocorticoid repression. Trends Endocrinol Metab 1: 219–225

65. Dunn AJ (1990) Interleukin-1 as a stimulator of hormone secretion. PNEI 3: 26–34

66. Emeric-Sauval E (1986) Corticotropic-releasing factor (CRF). A review. Psychoneuroendocrinology 11: 277–294

67. Eratalay YK, Simmons DJ, El-Mofty SK, Rosenberg GD, Nelson W, Haus E, Halberg F (1981) Methylprednisolone chronopharmacology: bone growth in the rat mandible following every-day or alternate day schedules. Arch Oral Biol 26: 769–777

68. Espiner EA (1987) The effects of stress on salt and water balance. Ballieres Clin Endocrinol Metab 1: 375–390

69. Evans PJ, Dieguez C, Rees LH, Hall R, Scanlon MR (1986) The effect of cholinergic blockade on the ACTH, beta-endorphin and cortisol responses to insulin-induced hypoglycemia. Clin Endocrinol (Oxf) 24: 687–691

70. Fallo F, Mantero F (1990) Dexamethasone-suppressible hyperaldosteronism. In: Biglieri EG, Melby JC (eds) Endocrine hypertension. Raven, New York, pp 87–97

71. Feldman S, Conforti N (1980) Participation of the dorsal hyppocampus in the glucocorticoid feed-back effect on adrenocortical activity. Neuroendocrinology 30: 52–56

72. Felong M (1976) Development of the retinohypothalamic projection in the rat. Anat Rec 184: 400–401

73. Ferrari E, Bossolo PA, Schianca GPC, Solerte SB, Fioravanti M, Nascimbene M (1982) Adrenocortical responsiveness to the synthetic ACTH 1-17 analogue given at different circadian times. Chronobiologia 9: 133–141

74. Ferrari E, Fraschini F, Brambilla F (1990) Hormonal circadian rhythms in eating disorders. Biol Psychiatry 27: 1007–1020

75. Ferrier N, Pascual J, Charlton BG, Wrigth C, Leake A, Griffiths HW, Fairbairn AF, Edwardson JA (1988) Cortisol, ACTH and dexamethasone concentrations in a psychogeriatric population. Biol Psychiatry 23: 252–260

76. Fischette C, Komisurak B, Ediner H (1980) Differential fornix ablations and the circadian rhythmicity of adrenal corticosterone secretion. Brain Res 195: 373–382

77. Fisher LA, Brown MR (1983) Corticotropin-releasing factor: central nervous system effects on the sympathetic nervous system and cardiovascular regulation. In: Ganten G, Pfaff D (eds) Central cardiovascular control, basic and clinical aspects. Springer, Berlin Heidelberg New York, pp 87–101

78. Follenius M, Brandenberger G, Hietter B (1982) Diurnal cortisol peaks and their relation to meals. J Clin Endocrinol Metab 55: 757–761

79. Follenius M, Simon C, Brandenberger G, Lenzi P (1987) Ultradian plasma corticotropin and cortisol rhythms: time-series analyses. J Endocrinol Invest 10: 261–266

80. Fray JCS, Park CS, Valentine AN (1987) Calcium and the control of renin secretion. Endocr Rev 8: 53–93

81. Freeman RH, Davis JO, Williams GM, Seymour AA (1982) Circadian changes in plasma renin activity and plasma aldosterone concentration in one-kidney hypertensive rats. Proc Soc Exp Biol Med 169: 86–89

82. Gaillard RC, Al-Damluji S (1987) Stress and the pituitary-adrenal axis. Ballieres Clin Endocrinol Metab 1: 319–354

83. Garfinkel PE (1984) Anorexia nervosa: an overview of hypothalamic-pituitary function. In: Brown GM, Koslow SH, Reichlin S (eds) Neuroendocrinology and psychiatric disorders. Raven, New York, pp 301–314

84. Garrick N, Hill J, Szele T, Tomai T, Gold P, Murphy D (1987) Corticotropin-releasing factor: a marked circadian rhythm in primate cerebrospinal fluid peaks in the evening and is inversely related to the cortisol circadian rhythm. Endocrinology 121: 1329–1338

85. Gennazzani AR, Petraglia F, Nappi C, Martignoni E, de Leo M, Facchinetti F (1983) Endorphins in peripheral plasma: origin and influencing factors. In: Muller EE, Gennazzani AR (eds) Central and peripheral endorphins: basic and clinical aspects. Raven, New York, pp 89–97

86. Gibbs DM (1985) Measurement of hypothalamic corticotropin releasing factors in hypophyseal portal blood. Fed Proc 44: 203–209

87. Gibbs DM (1986) Stress specific modulation of ACTH secretion by oxytocin. Neuroendocrinology 42: 456–458

88. Gibbs DM, Vale W (1983) Effect of the serotonin up-take inhibitor fluoxetine on corticotropin-releasing factor and vasopressin secretion into hypophyseal portal blood. Brain Res 280: 176–179

89. Gilad GM (1987) The stress-induced response of the septo-hyppocampal cholinergic system. A vectorial outcome of psychoneuroendocrinological interactions. Psychoneuroendocrinology 12: 167–184

90. Gilbert F, Dourish CT, Brazell C, McClue S, Stahl SM (1988) Relationship of increased food intake and plasma ACTH levels to 5-HT receptor activation in rats. Psychoneuroendocrinology 13: 471–478

91. Gillette MU, Reppert SM (1987) The hypothalamic suprachiasmatic nuclei: circadian patterns of vasopressin secretion and neuronal activity in vitro. Brain Res Bull 19: 135–139

92. Gillies GE, Linton EA, Lowry PJ (1982) Corticotropin releasing activity of the new CRF is potentiated several times by vasopressin. Nature 299: 355–357

93. Gold PW, Gwirtsman H, Avgerinos PC, Nieman LK, Gallucci WT, Chrousos GP (1986) Abnormal hypothalamic-pituitary-adrenal function in anorexia nervosa. N Engl J Med 314: 1355–1362

94. Goldman J, Wajchenberg BL, Liberman B, Nery M, Achando S, Gernek OA (1975) Contrast analysis for the evaluation of the circadian rhythms of plasma cortisol, androstenedione and testosterone in normal men and the possible influence of meals. J Clin Endocrinol Metab 60: 164–171

95. Gordon RD, Wolfe LK, Island DP, Liddle GW (1966) A diurnal rhythm in plasma renin activity in man. J Clin Invest 45: 1587–1592

96. Graber AL, Givens J, Nicholson W, Island DP, Liddle GW (1965) Persistence of diurnal rhythmicity in plasma ACTH concentrations in cortisol deficient patients. J Clin Endocrinol Metab 25: 804–807

97. Grossman A, Clement-Jones VC, Besser GM (1985) Clinical implications of endogenous opioid peptides. In: Muller EE, MacLeod RM, Frohman LA (eds) Neuroendocrine perspectives. Elsevier, Amsterdam, pp 243–294

98. Haen E (1987) The peripheral lymphocyte as clinical model for receptor disturbances. Asthmatic diseases. Bull Eur Physiopathol Respir 22: 539–541

99. Halberg F (1969) Chronobiology. Annu Rev Physiol 31: 675–725

100. Halberg F (1982) Physiologic 24 hours rhythms: a determinant of response to environmental agents. In: Schaefer KE (ed) Man's dependence on the earthly atmosphere. MacMillan, New York, pp 49–98

101. Halberg F, Visscher MB, Flink EB, Berge K, Bock F (1951) Diurnal rhythmic changes in blood eosinophil levels in health and in certain diseases. Lancet 71: 312–319

102. Halberg F, Reinhard J, Baratter FC (1969) Agreement in endpoints from circadian rhythmometry on healthy human beings living on different continents. Experientia 25: 107–112

103. Halberg F, Sanchez de La Pena S, Cornelissen G (1985) Circadian adrenocortical cycle and the central nervous system. In: Redfen PH, Campbell LC, Davies JA, Martin KE

(eds) Circadian rhythms in the central nervous system. VCH Weinheim, pp 57–79

104. Haus E, Lakatua DJ, Sackett-Lundeen LL, Swoyer J (1986) Chronobiology in laboratory medicine. In: Rietveld WJ (ed) Clinical aspects of chronobiology. Hoechst Medication Service, Amsterdam, pp 13–83

105. Hellbrugge T (1974) The development of circadian and ultradian rhythms of premature and full-term infants. In: Scheving LE, Halberg F, Pauly JE (eds) Chronobiology. Igaku-Shoin, Tokyo, pp 339–351

106. Hilfenhalis M, Hertig T (1979) Effect of inverting the light dark cycle on the circadian rhythm of urinary secretion of aldosterone, corticosterone and electrolytes in the rat. In: Reinberg A, Halberg F (eds) Chronopharmacology. Pergamon, Oxford, pp 49–55

107. Hiroshige T, Sato T (1970) Postnatal development of circadian rhythm of corticotropin releasing activity in the rat hypothalamus. Endocrinol Jpn 17: 1–6

108. Hobson JA (1989) Sleep. Scientific American Library, New York

109. Horne JA, Ostberg O (1976) A self-assessment questionnaire to determine morningness-eveningness in human circadian rhythms. Int J Chronobiol 4: 97–110

110. Kaneko M, Kaneko K, Shinsako J, Dallman MF (1981) Adrenal sensitivity to adrenocorticotropin varies diurnally. Endocrinology 109: 70–75

111. Kater CE, Biglieri EG, Brust N, Chang B, Hirai J, Irony I (1989) Stimulation and suppression of the mineralcorticoid hormones in normal subjects and adrenocortical disorders. Endocr Rev 10: 149–164

112. Katz F, Romfh P, Smith JA (1975) Diurnal variation of plasma aldosterone, cortisol and renin activity in supine man. J Clin Endocrinol Metab 40: 125–134

113. Kaye WF, Gwirtsman HE, George DT, Ebert MH, Jimesson DC, Tomai TP, Chrousos GP, Gold PW (1987) Elevated cerebrospinal fluid levels of immunoreactive corticotropin-releasing hormone in anorexia nervosa: relation to state of nutrition, adrenal function and intensity of depression. J Clin Endocrinol Metab 64: 203–211

114. Kerkhof GA (1985) Interindividual differences in human circadian rhythms. A review. Biol Psychol 20: 83–112

115. Khachaturian H, Sladek J (1980) Simultaneous monoamine histofluorescence and neuropeptide immunocytochemistry. III. Ontogeny of catecholamine varicosities and neurophysin neurons in the rat supraoptic and paraventricular nuclei. Peptides 1: 77–81

116. Kiss J, Mezey E, Skirboll L (1984) Corticotropin-releasing factor immunoreactive neurons in the paraventricular nucleus become vasopressin positive after adrenalectomy. Proc Natl Acad Sci USA 81: 1854–1858

117. Klein R, Armitage R (1979) Rhythms in human performance: $1\frac{1}{2}$ hours oscillations in cognitive style. Science 204: 1236–1237

118. Kleitman N (1982) Basic rest-activity cycle 22 years later. Sleep 5: 311–317

119. Knapp MS, Pownall R (1980) Chronobiology, pharmacology and the immune system. Int J Immunopharmacol 2: 91–93

120. Krieger DT (1973) Neurotransmitter regulation of ACTH release. M Sinai J Med (NY) 40: 302–314

121. Krieger DT (1974) Food and water restriction shifts corticosterone, temperature, activity and brain amine periodicity. Endocrinology 95: 1195–1201

122. Krieger DT (1974) Effect of neonatal hydrocortisone on corticosteroid circadian periodicity, responsiveness to ACTH and stress in prepuberal and adult rats. Neuroendocrinology 16: 355–363

123. Krieger DT (1975) Effect of intraventricular neonatal 6-OH dopamine and 5,6-dihydroxytryptamine administration on the circadian periodicity of plasma corticosteroid levels in the rat. Neuroendocrinology 17: 63–74

124. Krieger DT (1979) Rhythms in CRF, ACTH and corticosteroids. In: Krieger DT (ed) Endocrine rhythms. Raven, New York, pp 123–142

125. Krieger DT, Aschoff J (1979) Endocrine and other biological rhythms, vol 3. In: de Groot LJ (ed) Endocrinology. Grune and Stratton, New York, pp 2079–2109

126. Krieger DT, Allen W, Rizzo F (1971) Characterization of the normal pattern of plasma corticosteroid levels. J Clin Endocrinol Metab 32: 266–284

127. Kripke DF (1983) Phase-advance theories for affective illness. In: Wehr TA, Goodwin FK (eds) Circadian rhythms in psychiatry. Boxwood, Pacific Grove, pp 41–69

128. Krishnan KRR, Ritchie JC, Manepalli AN, Venkataramam S, France RD, Nemeroff CB, Barroll BJ (1988) What is the relationship between plasma ACTH and plasma cortisol in normal humans and depressed patients? In: Schatzberg AF, Nemeroff CB (eds) The hypothalamic-pituitary adrenal axis: physiology, pathophysiology and psychiatric implications. Raven, New York, pp 115–131

129. Kupfer DJ, Bulik CM, Jarrett DB (1983) Nighttime plasma cortisol secretion and EEG sleep: are they associated? Psychiatry Res 10: 191–199

130. Iannotta F, Magnoli L, Visconti G, Rampini A, Facchinetti A, Giuliani P (1987) Differences in cortisol, aldosterone and testosterone responses to ACTH 1-17 administered at two different times of the day. Chronobiologia 14: 39–46

131. Ichikawa Y, Nishikai M, Kawagoe M, Yoshida K, Homma M (1972) Plasma corticotropin, cortisol and growth hormone responses to hypoglycemia in the morning and evening. J Clin Endocrinol Metab 34: 859–898

132. Jarrett DB, Coble PA, Kupfer DJ (1983) Reduced cortisol latency in depressive illness. Arch Gen Psychiatry 40: 506–511

133. Jarrett DB, Coble P, Kupfer DJ (1985) Cortisol secretion during sleep in patients with a severe depressive illness. Psychiatr Med 3: 101–110

134. Joanny P, Steinberg J, Zamora A, Conte-Devolx B, Millet Y, Oliver C (1989) Corticotropin-releasing factor release from in vitro superfused and incubated rat hypothalamus. Effect of potassium, norepinephrine and dopamine. Peptides 10: 903–911

135. Jones MT, Gillham B, Greenstein BD, Beckford U, Holmes MC (1982) Feedback actions of adrenal steroid hormones. In: Ganten D, Pfaff D (eds) Adrenal action on brain. Springer, Berlin Heidelberg New York, pp 45–68 (Current topics in neuroendocrinology, vol 2)

136. Jouvet M (1984) Mécanismes des états du sommeil. In: Benoit O (ed) Physiologie du sommeil. Son exploration fonctionnelle. Masson, Paris, pp 1–18

137. Lakatua DJ, Nicolaug Y, Bogdan C, Petrescu E, Sackett-Lundeen LL (1984) Circadian endocrine time structure in humans above 80 years of life. J Gerontol 39: 648–654

138. Lanaman JT (1953) Fetal zone of the adrenal gland. Medicine (Baltimore) 32: 389–396

139. Land H, Schutz G, Schmale H, Richter D (1982) Nucleotide sequence of cloned cDNA encoding bovine arginine-vasopressin-neurophysin II precursor. Nature 295: 299–303

140. Lavie P, Kripke DF, Hiatt JF, Harrison J (1978) Gastric rhythms during sleep. Behav Biol 23: 526–530
141. Lechan RM (1978) Neuroendocrinology of pituitary hormone regulation. Clin Endocrinol Metab 16: 475–501
142. Lemmer B, Lang PH (1986) Daily variation in the beta-adrenoreceptor-adenylatecyclase-cAMP-phosphodiesterase system. In: Middeke M, Hotzgreve H (eds) New aspects in hypertension: adrenoreceptors. Springer, Berlin Heidelberg New York, pp 155–165
143. Levi F, Canon C, Blum JP, Mechkouri M, Reinberg A, Mathe G (1985) Circadian and circahemidian rhythms in mice lymphocyte-related variables from peripheral blood of healthy subjects. J Immunol 134: 217–222
144. Levin N, Shinsako J, Dallman MF (1988) Corticosterone acts on the brain to inhibit adrenalectomy-induced adrenocorticotropin secretion. Endocrinology 122: 694–700
145. Levin R, Fitzpatrick KM, Levine S (1976) Maternal influences on the ontongeny of basal levels of plasma corticosterone in the rat. Horm Behav 7: 41:48
146. Levy D (1987) Circadian rhythms and personality characteristics: an empirical investigation. Rass Psicol 4: 47–56
147. Lewy AJ (1983) Effects of light on melatonin secretion and the circadian system of man. In: Wehr TA, Goodwin FK (eds) Circadian rhythms in psychiatry. Boxwood, Pacific Grove, pp 203–233
148. Linsell CR, Lightman SL, Mullen PE, Brown MJ, Causon RC (1985) Circadian rhythms of epinephrine and norepinephrine in man. J Clin Endocrinol Metab 60: 1210–1215
149. Linton EA, Gillies GE, Lowry PJ (1983) Ovine corticotropin-releasing factor and vasopressin: antibody quenching studies on hypothalamic extracts of normal and Brattleboro rats. Endocrinology 113: 1878–1883
150. MacLeod RM, Spangelo BL (1990) Newly identified relationships among immunopeptides and neuroendocrine hormones (Abstr). 1st International Congress of the International Society for Neuroimmunomodulation (ISNIM), May 23–26, Florence
151. Mallo C, Zaidan R, Faure A, Brun J, Chazot G, Claustrat B (1988) Effects of a four-day nocturnal melatonin treatment on the 24 h plasma melatonin, cortisol and prolactin profiles in humans. Acta Endocrinol (Copenh) 119: 474–480
152. Mandell MP, Mandell AJ, Rubin RT, Brill P, Rodnick J, Sheff R, Chaffey B (1966) Activation of the pituitary adrenal axis during rapid eye movement sleep in man. Life Sci 5: 583–587
153. Meier AM (1976) Daily variation in concentration of plasma corticosteroids in hypophysectomized rats. Endocrinology 93: 1475–1479
154. Millington WR, Blum M, Knight R, Mueller GP, Roberts JL, O'Donohue TL (1986) A diurnal rhythm in proopiomelanocortin messenger ribonucleic acid that varies concomitantly with the content and the secretion of $\beta$-endorphin in the intermediate lobe of the rat pituitary. Endocrinology 116: 829–834
155. Minors DS, Waterhouse JM (1981) The endocrine system. In: Circadian rhythms and the human. Wright, London, pp 140–165
156. Minors DS, Waterhouse JM (1981) The kidney and hormones affecting it. In: Circadian rhythms in the human. Wright, London, pp 68–94
157. Minors DS, Waterhouse JM (1988) Mathematical and statistical analysis of circadian rhythms. Psychoneuroendocrinology 13: 443–464
158. Moberg G, Scapagnini U, de Groot J (1971) Effect of sectoring the fornix on diurnal fluctuation in plasma corticosterone levels in the rat. Neuroendocrinology 7: 11–18
159. Modunger RS, Sharif-Zadek K, Ertel NH, Gutkin M (1976) The circadian rhythm of renin. J Clin Endocrinol Metab 43: 1276–1282
160. Moeller H (1985) Chronopharmacology of hydrocortisone and $9\alpha$-fluorohydrocortisone in the treatment for congenital adrenal hyperplasia. Eur J Pediatr 144: 370–373
161. Moore RJ (1980) Suprachiasmatic nucleus, secondary synchronicity stimuli and the central neural control of circadian rhythms. Brain Res 183: 13–28
162. Moore-Ede MC (1983) The circadian timing system in mammals: two pacemakers preside over many secondary oscillators. Fed Proc 42: 2802–2808
163. Morgano A, Puppo F, Criscuolo D, Lotti G, Indiveri F (1987) Evening administration of alpha-interferon-relationship with the circadian rhythm of cortisol. Med Sci Res 15: 615–616
164. Morley JE (1987) Neuropeptide regulation of appetite and weight. Endocr Rev 8: 256–287
165. Morris CS, Hitchcock E (1985) Immunocytochemistry of folliculo-stellate cells of normal and neoplastic pituitary gland. J Clin Pathol 38: 481–488
166. Mullen PE, James VHT, Lightman SL, Linsell C, Peart WS (1980) A relationship between plasma renin activity and the rapid eye movement phase of sleep in man. J Clin Endocrinol Metab 50: 466–469
167. Munck A, Gutre PM, Molbrook NI (1984) Physiological functions of glucocorticoids in stress and their relationship to pharmacological actions. Endocr Rev 5: 25–44
168. Nelson W, Tong YL, Lee J-K, Halberg F (1979) Methods for cosinor-rhythmometric. Chronobiologia 6: 305–323
169. Nemeroff C, Widerlov E, Bisette G, Walleus H, Karlsson I, Eklund K, Kilts C, Loosen P, Vale W (1984) Elevated concentrations of CSF corticotropin releasing faktor-like immunoreactivity in depressed patients. Science 226: 1342–1345
170. Nichols T, Nugent CA, Tyler GH (1965) Diurnal variation in suppression of adrenal function by glucocorticoids. J Clin Endocrinol Metab 25: 343–349
171. Nicolau GY, Haus E (1989) Chronobiology of the endocrine system. Rev Roum Med Endocrinol 27: 153–183
172. Nicolau GY, Haus E, Lakatua D, Popa M, Marinescu I, Ionescu B, Bogdan C, Popescu M, Sackett-Lundeen L, Robu E, Petrescu E (1986) Circadian rhythm of beta-endorphin in the plasma of clinically healthy subjects and in patients with adrenocortical disorders. Rev Roum Med Endocrinol 24: 185–195
173. Noto T, Hashimoto H, Doi Y, Nawajima T, Kato N (1983) Biorhythms of arginine-vasopressin in the paraventricular supraoptic and suprachiasmatic nuclei of rats. Peptides 4: 875–878
174. Oliver C, Vague P, Vague J (1971) L'ACTH plasmatique dans les etats d'hypocorticisme. Ann Endocrinol (Paris) 32: 868–883
175. Ottenweller JE, Meier AH, Ferrell BR, Horseman MD, Procton A (1978) Extrapituitary regulation of the circadian rhythm of plasma corticosteroid concentrations in rats. Endocrinology 103: 1875–1879
176. Ottenweller JE, Miller AH, Russo AC, Frenzke ME (1979) Circadian rhythms of plasma corticosterone binding activity in the rat and the mouse. Acta Endocrinol (Copenh) 91: 150–157

177. Palkovits M (1987) Anatomy of neural pathways affecting CRH secretion. Ann NY Acad Sci 512: 139–148

178. Parnetti L, Mecocci P, Neri C, Palazzetti D, Fiacconi M, Santucci A, Santucci C, Ballatori E, Reboldi GP, Caputo N, Signorini E, Senin U (1990) Neuroendocrine markers in aging brain: clinical and neurobiological significance of dexamethasone suppression test. Aging 2: 173–179

179. Payne DW, Peng LH, Dearlman WH, Talbert LM (1976) Corticosteroid-binding proteins in human colostrum and milk, and rat milk. J Biol Chem 251: 5272–5276

180. Perlow MJ, Reppert SM, Hartman HA, Fischer DA, Self SM, Robinson AG (1982) Oxytocin, vasopressin and estrogen-stimulated neurophysin: daily patterns of concentration in cerebrospinal fluid. Science 216: 1416–1418

181. Petraglia F, Facchinetti F, d'Ambrogio G, Volpe A, Genazzani AR (1986) Somatostatin and oxytocin infusion inhibits the rise of plasma $\beta$-endorphin, lipotrophin and cortisol induced by insulin hyperglycemia. Clin Endocrinol (Oxf) 24: 609–616

182. Pincus G (1943) A diurnal rhythm in the secretion of urinary 17-ketosteroids in young men. J Clin Endocrinol 3: 195–199

183. Piovesan A, Terzolo M, Borretta G, Torta M, Bunvia T, Osella G, Paccotti P, Angeli A (1990) Circadian profile of serum melatonin in Cushing's disease and acromegaly. Chronobiol Int 7: 259–261

184. Plotsky PM (1987) Regulation of hypophysiotropic factors mediating ACTH secretion. Ann NY Acad Sci 512: 205–217

185. Plotsky PM, Cunningham ET, Widmaier R (1989) Catecholaminergic modulation of corticotropin-releasing factor and adrenocortical secretion. Endocr Rev 10: 437–458

186. Preslock JP (1984) The pineal gland: basic implications and clinical correlations. Endocr Rev 5: 282–308

187. Quigley ME, Yen SSC (1979) A mid-day surge in cortisol levels. J Clin Endocrinol Metab 49: 945–947

188. Quinns SJ, Williams GH (1988) Regulation of aldosterone secretion. Annu Rev Physiol 50: 409–426

189. Raskind M, Peskind E, Rivard MF, Veith R, Barnes R (1982) Dexamethasone suppression test and cortisol circadian rhythm in primary degenerative dementia. Am J Psychiatry 139: 1468–1471

190. Rasmussen DD (1986) Physiological interaction of the basic rest-activity in the brain: pulsatile luteinizing hormone secretion as a model. Psychoneuroendocrinology 11: 389–405

191. Ratge P, Kroll E, Diever U, Hadjimas A, Wisser H (1982) Circadian rhythm of catecholamines, cortisol and prolactin is altered in patients with apallic syndrome in comparison with normal volunteers. Acta Endocrinol (Copenh) 101: 428–435

192. Reinberg A, Lagoguey M (1978) Annual endocrine rhythms in healthy young adult men: their implications in human biology and medicine. In: Assenmacher I, Farner DS (eds) Environmental endocrinology. Springer, Berlin Heidelberg New York, pp 113–121

193. Reinberg A, Halberg F, Falliers G (1974) Circadian timing of methylprednisolone effects in asthmatic boys. Chronobiologia 1: 333–347

194. Reinberg A, Guillemant S, Ghata NJ, Guillemant J, Touitou Y, Dupont W, Lagoguey M, Bourgeois P, Briere L, Fraboulet G, Guilett P (1980) Clinical chronopharmacology of ACTH 1-17. I. Effects on plasma cortisol and urinary 17-hydroxycorticosteroids. Chronobiologia 7: 513–523

195. Reinberg A, Gervais P, Chaussade M, Fraboulet G, Duburque B (1983) Circadian changes in effectiveness of corticosteroids in light patients with allergic asthma. J Allergy Clin Immunol 71: 425–433

196. Reinhardt D, Becker B, Nagel-Hemke M, Schiffer R, Zehmisch T (1983) Influence of beta-receptor-agonists and glucocorticoids on alpha-and-beta-adrenoreceptors of isolated blood cells from asthmatic children. Pediatr Pharmacol 3: 293–402

197. Reiter RJ (1982) Neuroendocrine effects of the pineal gland and of melatonin. In: Ganong WF, Martini L (eds) Frontiers in neuroendocrinology, vol 7. Raven, New York

198. Reppert SM, Schwartz WJ, Uhl GR (1987) Arginine vasopressin: a novel peptide rhythm in cerebrospinal fluid. Trends Neurosci 10: 76–80

199. Reppert SM, Weaver DR, Rivlees SA (1988) Maternal communications of circadian phase to the developing mammal. Psychoneuroendocrinology 13: 63–78

200. Richards AM, Tonolo G, Fraser R, Morton JJ, Leckie BJ, Ball SG, Robertson JIS (1987) Diurnal change in plasma atrial natriuretic peptide concentrations. Clin Sci 73: 489–495

201. Roth KA, Weber E, Barchas GD (1982) Immunoreactive corticotropin-releasing factor (CRF) and vasopressin are colocalized in a subpopulation of the immunoreactive vasopressin cells in the paraventricular nucleus of the hypothalamus. Life Sci 31: 1875–1860

202. Roy A, Pickar D, Paul S, Doran A, Chrousos G, Gold PW (1987) CSF corticotropin-releasing hormone in depressed patients and normal control subjects. Am J Psychiatry 144: 641–645

203. Rubin RT, Poland RE, Gouin PR, Tower BB (1978) Secretion of hormones influencing water and prolactin during sleep in normal adult men. Psychosom Med 40: 44–59

204. Rusak B, Boulos Z (1981) Pathways for photic entrainment of mammalian circadian rhythms. Photochemistry 34: 267–273

205. Schanbelan M, Brust NL, Chung BCF, Slater KL, Biglieri EG (1976) Circadian rhythm and the effect of posture on plasma aldosterone concentration in primary aldosteronism. J Clin Endocrinol Metab 43: 115–131

206. Sachar EJ, Hellman L, Roffwarg H, Halpern E, Fukushima D, Gallagher T (1973) Disrupted 24-hour patterns of cortisol secretion in psychotic depression. Arch Gen Psychiatry 28: 19–24

207. Sakurai H, Naruse M, Naruse K, Obana K, Higashida T, Kurimoto F, Demura H, Inagani T, Schizume K (1987) Postural suppression of plasma atrial natriuretic polypeptide concentrations in man. Clin Endocrinol (Oxf) 26: 173–178

208. Salas MA, Evans SW, Levell MJ, Whicher JT (1990) Interleukin-6 and ACTH act synergistically to stimulate the release of corticosterone from adrenal gland cells. Clin Exp Immunol 79: 470–473

209. Sapolsky RM, Plotsky PM (1990) Hypercortisolism and its possible neural bases. Biol Psychiatry 27: 937–952

210. Sapolsky RM, Krey LC, McEwen BS (1986) The neuroendocrinology of stress and aging: the glucocorticoid cascade hypothesis. Endocr Rev 7: 284–301

211. Sapolsky RM, Armanini M, Packan D, Tombaugh G (1987) Stress and glucocorticoids in aging. Clin Endocrinol Metab 16: 965–994

212. Sawchenko PE, Swanson LW, Vale WW (1984) Co-expression of corticotropin-releasing factor and vasopressin

immunoreactivity in parvocellular neurosecretory neurons. Proc Natl Acad Sci USA: 81: 1883–1887

213. Scapagnini U, Moberg GP, van Loon GR, DeGroot J, Ganong WF (1971) Relations of brain 5-hydroxytryptamine content to the diurnal variations in plasma corticosterone in the rat. Neuroendocrinology 7: 90–96

214. Schams D, Karg H (1986) Hormones in milk. Ann NY Acad Sci 464: 75–86

215. Schlechte JA, Ginsberg BH, Sherman BM (1982) Regulation of the glucocorticoid receptor in human lymphocytes. J Steroid Biochem 16: 69–74

216. Schwartz WJ, Repperts M (1985) Neural regulation of the circadian vasopressin rhythm in cerebrospinal fluid: a prominent role for the suprachiasmatic nuclei. J Neurosci 5: 2771–2778

217. Schwartz WJ, Coleman RJ, Reppert SM (1983) A daily vasopressin rhythm in rat cerebrospinal fluid. Brain Res 263: 105–112

218. Serio M, Piolanti P, Romano S, de Magistris L, Giusti G (1970) The circadian rhythm of plasma cortisol in subjects over 70 years of age. J Gerontol 25: 95–97

219. Shibata S, Liou SY, Ueki S, Oomura Y (1984) Influence of environmental light-dark cycle and enucleation on activity of suprachiasmatic neurons in slice preparations. Brain Res 302: 75–81

220. Signore A, Cugini P, Letizia C, Lucia P, Murano G, Pozzilli P (1985) Study of the diurnal variation of human lymphocyte subsets. J Clin Lab Immunol 17: 25–28

221. Smolensky MH, Halberg F, Pitts G, Nelson W (1981) The chronopharmacology of methylprednisolone: clinical implications of animal studies with special emphasis on the moderation of growth inhibition by timing to circadian rhythms. In: Smolensky MH, Reinberg A, MacGovern JP (eds) Recent advances in the chronobiology of allergy and immunology. Pergamon, Oxford, pp 137–171

222. Sofroniew MA, Weindl A (1980) Identification of paracellular vasopressin and neurophysin neurons in the suprachiasmatic nucleus of a variety of mammals including primates. J Comp Neurol 193: 659–665

223. Sorensen PS, Hammer M (1985) Vasopressin in plasma and ventricular cerebral spinal fluid during dehydration, postural changes and nausea. Am J Physiol 248: R78–R83

224. Soutar CA, Costello J, Ijaduoca O, Turner-Warwick M (1975) Nocturnal and early morning asthma. Relationship to plasma corticosteroid and response to cortical infusion. Thorax 30: 436–444

225. Sperling A (1980) Newborn adaptation: adrenocortical hormones and ACTH. In: Tulchinsky D, Ryan KJ (eds) Maternal-fetal endocrinology. Sanders, Philadelphia, pp 387–408

226. Spinedi E, Negro-Vilar A (1983) Serotonin and adrenocorticotrophin (ACTH) release: direct effects at the anterior pituitary level and potentiation of arginin-vasopressin induced ACTH release. Endocrinology 112: 1217–1223

227. Steiger A, Herth T, Holsboer F (1987) Sleep-electroencephalography and the secretion of cortisol and growth hormone in normal controls. Acta Endocrinol (Copenh) 116: 36–42

228. Stene M, Panagiotis N, Tuck ML, Sowers JR, Mayers D, Berg G (1980) Plasma norepinephrine levels are influenced by sodium intake, glucocorticoid administration and circadian changes in normal man. J Clin Endocrinol Metab 51: 1340–1345

229. Stephan FK, Berkely KJ, Moss KL (1981) Efferent connections of the rat suprachiasmatic nucleus. Neuroscience 6: 2625–2641

230. Sterman MB (1972) The basic rest-activity cycle and sleep: developmental consideration in man and cats. In: Clemente CD, Purpura DP, Mayer FE (eds) Sleep and the nervous system. Academic, New York, pp 175–197

231. Stock G (1982) Neurobiology of REM sleep. A possible role for dopamine. In: Ganten D, Pfaff D (eds) Sleep. Clinical and experimental aspects. Springer, Berlin Heidelberg New York, pp 1–36

232. Stockigt JR (1976) Mineralcorticoid hormones. Adv Steroid Biochem Pharmacol 3: 161–238

233. Strassman RJ, Appenzeller O, Lewy AJ, Qualls CR, Peake GT (1989) Increase in plasma melatonin, $\beta$-endorphin and cortisol after a 28.5-mile mountain race: relationship to performance and lack of effect of naltrexone. J Clin Endocrinol Metab 69: 540–545

234. Swanson LW, Cowan WM (1975) The efferent connections of the suprachiasmatic nucleus of the hypothalamus. J Comp Neurol 160: 1–12

235. Swanson LW, Sawchenko P, Berod A, Harman B, Helle K, Vandoren D (1981) An immunohistochemical study of the organization of catecholaminergic cells and terminal fields in the paraventricular and supraoptic nuclei of the hypothalamus. J Comp Neurol 196: 271–276

236. Szafarczyk A, Malaval F, Laurent A, Gibaud R, Assenmacher I (1987) Further evidence for a central stimulatory action of catecholamines on adrenocorticotropin release in the rat. Endocrinology 121: 883–892

237. Takahashi K, Inoui K, Takahashi Y (1977) Parallel shift in circadian rhythms of adrenocortical activity and food intake in blinded and intact rats exposed to continuous illumination. Endocrinology 100: 1097–1107

238. Tabeke K, Setaishi C, Hirama M, Yamamoto M, Horiuchi Y (1966) Effects of a bacterial pyrogen on the pituitary-adrenal axis at various times in 24 hours. J Clin Endocrinol Metab 26: 437–442

239. Terzolo M, Piovesan A, Osella G, Puligheddu B, Torta M, Paccotti P, Angeli A (1990) Morning to evening changes of human pituitary and adrenal responses to specific stimuli. J Endocrinol Invest 13: 181–185

240. Terzolo M, Piovesan A, Panarelli M, Torta M, Osella G, Paccotti P, Angeli A (1990) Effects of long-term, low-dose, time-specified melatonin administration on endocrine and cardiovascular variables in adult men. J Pineal Res 9: 113–124

241. Touitou Y (1982) Some aspects of the circadian time structure in the elderly. Gerontology 28: 53–67

242. Touitou Y, Motohashi Y, Patti A, Levi F, Reinberg A, Ferment O (1986) Comparison of cortisol circadian rhythms documented in samples of saliva, capillary (finger tips) and venous blood from healthy subjects. Annu Rev Chronopharmacol 3: 297–299

243. Tribollet E, Barberis C, Jard S, Dubois-Dauphin M, Dreifuss JJ (1988) Localization and pharmacological characterization of high affinity binding sites for vasopressin and oxytocin in the rat brain by light microscopy autoradiography. Brain Res 442: 105–118

244. Tuomisto J, Mannisto P (1985) Neurotransmitter regulation of pituitary hormones. Pharmacol Rev 37: 249–332

245. Turek FW (1985) Circadian neural rhythms in mammals. Ann Rev Physiol 47: 49–64

246. Turek FW, van Cauter E (1988) Rhythms in reproduction. In: Knobil E, Neilly J (eds) The physiology of reproduction. Raven, New York, pp 1789–1830

247. Ungar F, Halberg F (1962) Circadian rhythm of the in vitro response of mouse adrenal to adrenocorticotropic hormone. Science 137: 1058–1060

248. Vale W, Spiess J, Rivier C, Rivier J (1981) Characterization of a 41-residue ovine hypothalamic peptide that stimulates secretion of corticotropin an $\beta$-endorphin. Science 213: 1394–1397

249. Van Cauter E (1984) Rythmes hormonaux et sommeil. In: Benoit O (ed) Physiologie du sommeil. Son exploration fonctionnelle. Masson, Paris, pp 85–98

250. Vandesande F, Diericks K, Demey J (1977) The origin of the vasopressinergic and oxytoninergic fibers of the external region of the median eminence of the rat hypophysis. Cell Tissue Res 180: 443–452

251. Veglio F, Pietrandrea R, Ossola M, Vignani A, Angeli A (1987) Circadian rhythm of the angiotensin converting enzyme (ACE) activity in serum of healthy adult subjects. Chronobiologia 14: 21–25

252. Veglio F, Padoan M, Gambino M, Paccotti P, Terzolo M, Angeli A (1988) Plasma steroid responses to circadian-stage-specified injection of different doses of the ACTH analogue alsactide (ACTH 1-17) in healthy adult man. Ric Clin Lab 18: 95–104

253. Waldhauser F, Frisk H, Kratgasser-Gasparotti A, Schober E, Wieglmaier C (1986) Serum melatonin is not affected by glucocorticoid replacement in congenital adrenal hyperplasia. Acta Endocrinol (Copenh) 111: 355–361

254. Walsh BT, Roose SP, Katw JL, Dyrenfurth I, Wright L, Vandewiele R, Glassman AH (1987) Hypothalamic-pituitary-adrenal-cortical activity in anorexia nervosa and bulimia. Psychoneuroendocrinology 12: 131–140

255. Webb W (1974) The rhythms of sleep and waking. In: Scheving LE, Halberg F, Pauly JE (eds) Chronobiology. Igaku-Shoin, Tokyo, pp 482–486

256. Weitzman ED (1976) Circadian rhythms and episodic hormone secretion in man. Annu Rev Med 27: 225–243

257. Weitzman ED, Fukushima D, Nogeire C (1971) Twenty-four hour pattern of episodic secretion of cortisol in normal subjects. J Clin Endocrinol Metab 33: 14–22

258. Weitzman ED, Nogeire C, Perlow M (1974) Effects of a prolonged 3-hours sleep-wake cycle on sleep stages, plasma cortisol, growth hormone and body temperature in man. J Clin Endocrinol Metab 38: 1018–1030

259. Wetterberg L (1978) Melatonin in humans: physiological and clinical studies. J Neural Transm [Suppl] 13: 289–310

260. Wetterberg L, Beck-Frus J, Kjellman BF, Ljunggren JG (1984) Circadian secretion in depression. In: Usdin E (ed) Frontiers in biochemical and pharmacological research in depressives. Raven, New York, pp 197–207

261. Williams GH, Cain JP, Dluhy RG, Underwood RH (1972) Studies on the control of plasma aldosterone concentrations in normal man. I. Response to posture, acute and chronic volume depletion and sodium loading. J Clin Invest 51: 1950–1957

262. Winters CJ, Sallman AL, Vesely DL (1988) Circadian rhythm of prohormone atrial natriuretic peptides 1–30, 31–67 and 99–126 in man. Chronobiol Int 5: 403–409

263. Wolfson B, Manning R, Davis L, Arentzen R, Baldino F (1985) Co-localization of corticotropin-releasing factor and vasopressin in RNA in neurones after adrenalectomy. Nature 315: 59–61

264. Yamazaki J, Takahashi K (1983) Effect of change of mothers and lighting conditions on the development of the circadian adrenocortical rhythm in blinded rat pups. Psychoneuroendocrinology 8: 237–244

265. Young WF, Klee GC (1988) Primary aldosteronism. Diagnostic evaluation. Clin Endocrinol Metab 17: 367–395

# Chronobiology of the Hypothalamic-Pituitary-Gonadal Axis in Men and Women

C. H. Blomquist and J. P. Holt, Jr.

## Introduction

Gonadal secretion of steroid hormones is episodic in men and women and is subject to neuroendocrine control by the hypothalamus and pituitary. In turn, pulsatile secretion of gonadotropin-releasing hormone (GnRH) by the hypothalamus and luteinizing hormone (LH) and follicle-stimulating hormone (FSH) by the pituitary are modulated by steroids secreted by the gonads. Current data also suggest GnRH secretion is regulated by hypothalamic catecholamines and endogenous opioids (Knobil 1980; Yen 1986; Marshall and Kelch 1986). The availability of highly sensitive and specific methods for quantitating hormone levels in small amounts of serum has allowed for sampling at short time intervals over extended periods. This and the ongoing development of new statistical methods for data analysis (Merriam and Wachter 1982; Veldhuis et al. 1986a; Urban et al. 1988, see Chap. of Veldhuis et al.) and the detection of low-amplitude rhythms (Filicori et al. 1984; Haus et al. 1984) continue to reveal the chronobiologic complexity of the hypothalamopituitary-gonadal axes. The documentation of changes in rhythms during prepubertal development, as well as new knowledge of variations in secretion patterns associated with gonadal dysfunction and infertility in adults, has led to a better understanding of reproductive pathophysiology. In this chapter we will review the chronobiology of normal gonadal function in men and women, prepubertal and pubertal development of these patterns and changes associated with some aspects of reproductive pathophysiology.

## Development and Maturation of the Hypothalamopituitary-Gonadal Axes

Development and maturation of the hypothalamopituitary-gonadal axes in the human occur in four stages: (1) a developmental stage which encompasses fetal development and early infancy, (2) a stage which lasts through childhood and during which gonadotropin and steroid hormone levels are low, (3) a period of pubertal development during which gonadotropin and steroid secretion patterns change, and (4) sexual maturation and the appearance of characteristic adult functional patterns (Marshall and Kelch 1986). A fifth stage, associated with aging in men and the postmenopausal period in females, is also of importance.

Accumulated data from a number of laboratories suggest pubertal development involves modulation of a preexisting episodic pattern of gonadotropin secretion which is highly active in the neonatal period and infancy and then largely inhibited during childhood. Pubertal maturation is associated with changes in GnRH and gonadotropin pulse amplitude and frequency, the appearance of gonadal steroid-dependent regulatory mechanisms, and changes in gonadotropin structural heterogeneity and biopotency (Marshall and Kelch 1986). Low-amplitude, high-frequency pulses of LH with an interpulse interval of $12.7 \pm 1.7$ min have been observed in experiments in which pituitary tissue from 21- to 23-week fetuses was perifused in vitro (Gambacciani et al. 1987). These observations are suggestive of intrinsic, intrapituitary pulse-generating mechanisms at this stage of development.

Pulsatile secretion of LH has been detected in newborn girls at 7 days of age (Danon et al. 1982). Waldhauser et al. (1981) sampled three male and three female infants at 6–12 weeks of age every 30 min over an 8-h period. In the male infants, LH values ranged from 3.6 to 34.7 mIU/ml with a mean

value of $11.3 \pm 6.08$mIU/ml and FSH from 1.8 to
4.6 mIU/ml with a mean value of $3.3 \pm 0.64$ mIU/ml.
In the female infants, LH values ranged from undetectable to 4.7 mIU/ml with a mean value of $1.8 \pm 0.7$
mIU/ml; FSH levels were higher with a mean value of
$11.2 \pm 2.5$ mIU/ml and a range of 6.5 to 22.7 mIU/ml.
Secretion was pulsatile in both sexes with an interpulse interval of approximately 90 min.

During the prepubertal stage of childhood, mean
daily plasma estrogen, androgen and gonadotropin
levels are low (Jenner et al. 1972; Bidlingmaier et al.
1973). FSH to LH ratios are relatively high (Jakacki
et al. 1982). In females, plasma LH concentrations
increase 100-fold from 7 years of age (0.01–0.06
mIU/ml) to adulthood (2.8–7.4 mIU/ml, cycle days
4–7) and FSH/LH ratios decrease from approximately 20 in prepubertal girls to less than 1.0 after
menarche (Apter et al. 1989). In males, LH levels
range from less than 0.1 mIU/ml in prepubertal boys
(Wennink et al. 1988; Dunkel et al. 1990) to approximately 5.0 mIU/ml in adults (Urban et al. 1988).

Early studies of gonadotropin secretion in
children during the peripubertal period led to the
suggestion that the appearance of nocturnal, sleep-
entrained episodic LH secretion heralded the onset
of puberty (Boyar et al. 1972). Subsequent studies
have supported the concept that changes observed at
the onset of puberty are modulations of preexisting
episodic secretion patterns.

Pulsatile LH secretion has been documented in
prepubertal children before physical evidence of sexual maturation is apparent (Parker et al. 1975; Penny
et al. 1977; Jakacki et al. 1982). Sleep-entrained increases in plasma LH and FSH levels occur in both

prepubertal and pubertal children (Kulin et al. 1976;
Beck and Wuttke 1980; Jakacki et al. 1982). Observation of nocturnal increases in testosterone levels and
LH pulse frequency and amplitude followed, as
morning approaches, by decreases in pulse frequency
and then pulse amplitude (Parker et al. 1975; Corley
et al. 1981; Ross et al. 1983; Kelch et al. 1985) has led
to the hypothesis (Marshall and Kelch 1986) that in
both girls and boys sleep-related changes in both the
amplitude and frequency of GnRH pulses play a fundamental role in the regulation of LH and FSH secretion during sexual maturation.

More recent analyses have confirmed nocturnal
changes in pulsatile secretion patterns and have led
to the suggestion that, in boys, nocturnal increases in
LH and GnRH pulse frequency occur 1–2 years before the clinical onset of puberty concurrent with an
increase in pituitary responsiveness to GnRH (Fig.1).
Though apparent in the peripubertal period, the
greatest changes in LH pulse amplitude occur after
the clinical onset of puberty (Wu et al. 1990).

Because of the extremely low levels of plasma LH
in prepubertal girls, the presence or absence of pulsatile secretion and its properties have been difficult
to establish. Current data are consistent with sleep-
related increases in GnRH pulse amplitude and frequency resulting in increased LH and FSH secretion
and incomplete follicle development during the early
peripubertal period. After mid-puberty, a continuous
circhoral GnRH and LH pulse frequency is achieved
resulting in the circatrigintan pattern of hormone formation and ovulation characteristic of the menstrual
cycle (Marshall and Kelch 1986).

Reiter et al. (1987) detected pulsatile secretion of

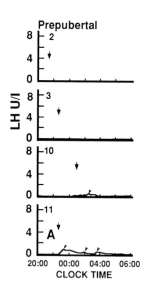
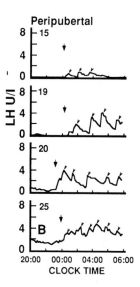
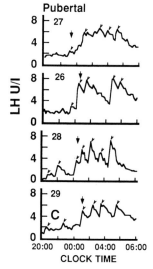

**Fig. 1.** Profiles of plasma LH in prepubertal, peripubertal and pubertal boys. The *arrows* indicate the onset of sleep, the *arrowheads* indicate statistically significant LH pulses. (From Wu et al. 1990)

bioactive LH (1 pulse/150 min) in five of seven and immunoactive LH (1 pulse/212 min) in six of seven prepubertal girls at Tanner breast stage I. Only two of six pubertal girls (one at Tanner stage II and the other at stage IV) exhibited pulsatile LH secretion. In these latter subjects, bioactive LH pulses occurred at a frequency of 1 pulse/120 min while pulses of immunoactive LH were detected once every 6 h. These data indicate episodic secretion patterns of bioactive and immunoactive LH differ among prepubertal and pubertal girls and suggest variations in gonadotropin structure as expressed in biopotency and immunoreactivity may be of fundamental importance in the regulation of sexual maturation.

Circadian rhythms in LH, FSH and testosterone have been detected in children at 11 years of age (Haus et al. 1988). Peak levels of LH occurred at approximately 0300 hours for both boys and girls. FSH maxima were at 0300 hours for boys and 0500 hours for girls. A more significant difference in timing was observed for testosterone, with peak levels occurring at approximately 0330 hours in boys and 1000 hours in girls.

Although diurnal, ultradian and pulsatile patterns of secretion during sexual maturation are becoming better defined, less is known of circannual or seasonal rhythms not only in sexual maturation but in reproductive function and sexual activity in humans, as well. Some years ago, data which indicated menarcheal onset occurred most frequently during December and January were reviewed (Valsik 1965). More recent discussions have emphasized the difficulty in clarifying the relative roles of exogenous factors such as photoperiod and endogenous physiologic phenomena such as melatonin secretion in establishing seasonal rhythms in sexual function and behavior (Smolensky et al. 1981; Surbey et al. 1986).

## Developmental Pathophysiology

Developmental pathophysiology of the hypothalamopituitary-gonadal axis in both males and females is expressed as: (1) abnormal gonadal development or function, (2) incomplete or abnormal physical sexual maturation, (3) abnormal plasma gonadotropin, estrogen and androgen levels, and (4) changes in episodic secretory patterns of GnRH, gonadotropins and steroid hormones. Precocious pubertal development can be divided into the diagnostic subgroups central precocious puberty, peripheral precocious

puberty or combined central and peripheral precocious puberty (Pescovitz et al. 1986). Central precocious puberty in both boys and girls results from early activation of the hypothalamopituitary-gonadal axes. This may be associated with hypothalamic tumors or other CNS lesions, or may be idiopathic. Children with peripheral precocious puberty have peripheral sex steroid production in the absence of maturation of the hypothalamopituitary-gonadal axes. In patients with central precocious puberty, basal levels of LH and FSH are elevated as are peak responses to GnRH. Episodic nocturnal LH secretion has also been observed (Crowley et al. 1981).

Gonadotropin-releasing hormone given daily in a nonpulsatile manner is effective in reducing gonadotropin levels and responses to GnRH in central precocious puberty but is not efficacious in peripheral precocious puberty (Pescovitz et al. 1986).

In girls with premature thelarche but lacking other signs of precocious puberty, pulsatile LH secretion and sleep-related, episodic LH release have been reported (Beck and Stubbe 1984). The major response to GnRH is an increase in serum FSH (Pescovitz et al. 1988; Wang et al. 1990) while basal LH and peak LH response are less than in precocious puberty. This has led to the suggestion that premature thelarche and central precocious puberty represent different stages of GnRH neuron maturation within the hypothalamus (Pescovitz et al. 1988).

A delay in the onset of puberty may reflect a primary defect in hypothalamic or pituitary function or a delay in maturation of the hypothalamopituitary-gonadal axes. It is often difficult to differentiate between the two situations clinically because both are characterized by low serum levels of gonadotropins. Clinical studies in a number of laboratories have demonstrated that prepubertal boys with either condition will respond to GnRH. The presence of nocturnal, sleep-related episodic LH secretion and an enhanced response to GnRH appear to differentiate delayed puberty from gonadotropin deficiency (Wagner et al. 1986; Ehrmann et al. 1989). Evidence has been presented (Delemarre-van de Waal 1985) that pulsatile administration of GnRH at 90-min intervals can change the pituitary response from a pubertal into an adult pattern in both girls and boys. It is of interest that in men pulsatile gonadotropin levels and secretory patterns achieved by pulsatile administration of GnRH are more effective in stimulating testicular growth, but not necessarily sperm output, than stable gonadotropin concentrations achieved by bolus administration of human chorionic gonadotropin and human menopausal gonadotropin (Liu et al. 1988).

# Chronobiologic Characteristics of the Menstrual Cycle

The menstrual cycle is an example of a genetically fixed, periodic phenomenon (Haus et al. 1984). Characteristic episodic patterns of gonadotropin secretion by the anterior pituitary and steroid hormone secretion by the ovaries are associated with follicular growth and development, oocyte maturation and ovulation. This pattern exists within a time structure domain in which these parameters exhibit a variety of circannual or seasonal, circadian, ultradian and pulsatile rhythms. In women during the reproductive years, repetitive menstrual cycles exhibit a circatrigintan rhythm (median 28 days, range 26–35 days) characterized by a midcycle surge in LH secretion by the pituitary. This peak is often used as the reference time point for characterizing the time course of ovarian secretion of steroids, and LH and FSH secretion by the pituitary. In this case, for a time frame of reference, the day of the LH peak is taken as day 0. Alternatively, the 1st day of menses is considered the 1st day of the cycle. In a normal cyle, though follicle recruitment and development actually begin during the late luteal phase of the preceding cycle, the follicular phase of the cycle encompasses days 1 through 14 or the period from day $-14$ or $-15$ to day 0 and the luteal phase from day 15 through 28 or day 0 through day $+14$ or $+15$ (Yen 1986).

## Episodic Secretion of Ovarian Steroids

Steroid formation by the ovary is compartmentalized and involves a variety of cell types – thecal, granulosa and stromal cells during the follicular phase and luteinized granulosa and thecal cells of the corpus luteum during the luteal phase. Normal steroidogenesis during both phases of the menstrual cycle is absolutely dependent on FSH and LH (Mais et al. 1986; McLachlan et al. 1989).

In normal premenopausal women the ovaries synthesize over 95% of the estradiol produced during a normal menstrual cycle (Abraham 1974; Baird and Fraser 1974). The mean daily secretion rate and serum level are low during the early follicular phase. Estradiol formation increases rapidly after day $-5$, reaching a peak at approximately day $-2$, and then decreases sharply coincident with the LH surge. This is followed by a second increase beginning at day $+1$ or $+2$ of the luteal phase and reaching peak levels at approximately day $+9$ or $+10$. The production rate decreases rapidly after day $+10$, reaching early follicular phase levels by day $+13$ to $+15$. Peak increases are 8- to 10-fold in the peripheral circulation and 200- to 250-fold in ovarian venous blood during the late follicular phase. Peak values are somewhat lower during the luteal phase (Baird and Fraser 1974; Soules et al. 1989a) The ovary containing the active follicle or corpus luteum following ovulation accounts for the increased estradiol formation (Baird and Fraser 1974; Aedo et al. 1980a, b).

As with estrogens, the ovary is the principal site of progesterone formation during the menstrual cycle. Progesterone production is low during the early follicular phase, increases approximately fourfold in the preovulatory period and reaches a peak during the midluteal phase. The peak production rate is approximately 25-fold greater than that in the early follicular phase (Abraham et al. 1971; de Jong et al. 1974). The increase during the luteal phase is due to progesterone synthesis by the corpus luteum (Aedo et al. 1980b).

In early studies, subjects were sampled on a daily basis or individual subjects were sampled once on a particular day of the cycle (Abraham et al. 1971; de Jong et al. 1974). With the development of radioimmunassay methods for the quantitation of steroids in small volumes of serum, it became possible to sample at more frequent intervals. This approach was taken, in particular, to clarify the time relationships between the beginning and peak of LH secretion and steroid secretion (Korenman and Sherman 1973; Thorneycroft et al. 1974; Landgren et al. 1977; Hoff et al. 1983). Backstrom et al. (1982) presented evidence of episodic secretion of estradiol. Maximum frequency was observed in the mid-follicular phase with a decline in the luteal phase. They concluded that the increased frequency was associated with an increase in LH pulse frequency.

Veldhuis et al. (1988a) have shown that progesterone is secreted in a pulsatile fashion with an interpulse interval of approximately 118 min in the midluteal phase. Both progesterone and LH showed multiple ultradian rhythms with periodicities of 48–241 min. In all cases there was a significant cross-correlation between progesterone and LH secretory patterns (Fig. 2). They also detected circadian rhythms in LH and progesterone secretion with peak values at 1944 hours and 1802 hours, respectively. A more recent study by Rossmanith et al. (1990) has demonstrated a close coupling of estradiol, progesterone and LH pulses in the midluteal phase (Fig. 3).

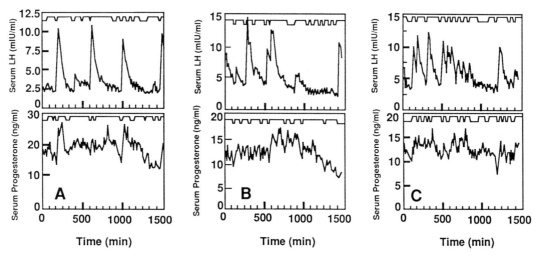

**Fig. 2 A–C.** Luteinizing hormone and progesterone secretion patterns in three women sampled in the midluteal phase of their menstrual cycles. (Adapted from Veldhuis et al. 1988a)

**Fig. 3.** Twenty-four-hour pulse profiles of LH, estradiol and progesterone in two women in the midluteal phase of their menstrual cycles. (Adapted from Rossmanith et al. 1990)

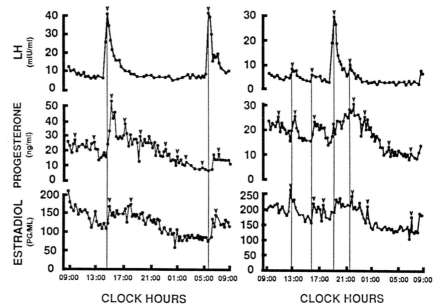

They also made the interesting observation of discrete pulses of estradiol and progesterone not preceded by an LH pulse, suggestive of autonomous pulsatile steroid secretion by the corpus luteum.

A recent population-mean cosinor analyses (Carandente et al. 1989) has shown the presence of a circadian rhythm in estradiol in the late luteal phase (peak at 1400 hours). In the same study, rhythms in progesterone secretion were detected in all phases of the menstrual cycle, with peak values being reached at approximately 1000, 0900, 0100 and 1700 hours in the early follicular, late follicular, early luteal and late luteal phases, respectively. The authors suggest the

adrenal glands may account for a significant portion of rhythmic progesterone secretion in the follicular phase.

## Neuroendocrine Regulation of Ovarian Steroidogenesis

Luteinizing hormone and FSH are secreted by the pituitary in a pulsatile manner. Patterns of gonadotropin release vary across the menstrual cycle and re-

**Table 1.** Representative estimates (mean ± SD) of LH secretion parameters during the follicular phase of the menstrual cycle

| Cycle day | 1 – 4 | 5 – 9 | 10 – 14 |
|---|---|---|---|
| Concentration[a] | 18 ± 2 | 23 ± 1 | 35 ± 16 |
| Frequency[b] | 6.8 ± 0.6 | 8.2 ± 0.6 | 8.8 ± 0.4 |
| Amplitude[c] | 21 ± 2 | 28 ± 2 | 25 ± 16 |
| Interval[d] | 36.2 ± 3.9 | 50.0 ± 6.2 | 46.2 ± 2.7 |
| (From Backstrom et al. 1982) | | | |
| Cycle day | 3 – 5 | – | 10 – 12 |
| Concentration | 6.7 ± 1.0 | – | 9.0 ± 1.4 |
| Frequency | 11.8 ± 0.6 | – | 14.3 ± 1.0 |
| Amplitude | 4.0 ± 0.7 | – | 3.3 ± 0.6 |
| Interval | – | – | – |
| (From Reame et al. 1984) | | | |
| Cycle day | 2 – 4 | 5 – 9 | 10 – 14 |
| Concentration | 8.0 ± 0.4 | – | 15.7 ± 2.5 |
| Frequency | – | – | – |
| Amplitude | 6.5 ± 0.4 | 5.1 ± 0.8 | 7.2 ± 1.2 |
| Interval | 94 ± 4 | 67 ± 3 | 71 ± 4 |
| (From Filicori et al. 1986) | | | |
| Cycle day | – | 4 – 9 | – |
| Concentration | – | 6.5 ± 0.5 | – |
| Frequency | – | 11.0 ± 0.8 | – |
| Amplitude | – | 2.9 ± 0.4 | – |
| Interval | – | – | – |
| (From Nippoldt et al. 1989) | | | |

Units are: [a] concentration = mIU/ml; [b] frequency = pulses/12 h; [c] amplitude = mIU/ml; [d] interpulse interval = minutes

**Table 2.** Representative estimates (mean ± SD) of LH secretion parameters during the luteal phase of the menstrual cycle

| Cycle day | 15 – 18 | 19 – 23 | 24 – 28 |
|---|---|---|---|
| Concentration[a] | 21.5 ± 5.2 | 6.9 ± 1.9 | 6.8 ± 1.2 |
| Frequency[b] | – | – | – |
| Amplitude[c] | 12.3 ± 2.2 | 10.7 ± 4.6 | 8.6 ± 3.4 |
| Interval[d] | 99 ± 20 | 162 ± 33 | 173 ± 20 |
| (From Filicori et al. 1984) | | | |
| Cycle day | – | 18 – 20 | 24 – 26 |
| Concentration | – | 7.7 ± 2.0 | 5.3 ± 0.9 |
| Frequency | – | 8.0 ± 2.0 | 7.8 ± 1.0 |
| Amplitude | – | 10.3 ± 4.4 | 6.3 ± 2.4 |
| Interval | – | – | – |
| (From Reame et al. 1984) | | | |
| Cycle day | 15 – 18 | 19 – 24 | 25 – 28 |
| Concentration | 21.4 ± 2.2 | – | 5.8 ± 0.7 |
| Frequency | – | – | – |
| Amplitude | 14.9 ± 1.7 | 12.2 ± 2.0 | 7.6 ± 1.1 |
| Interval | 103 ± 8 | 206 ± 51 | 216 ± 39 |
| (From Filicori et al. 1986) | | | |
| Cycle day | – | 20 – 22 | – |
| Concentration | – | 5.3 ± 0.5 | – |
| Frequency | – | 4.2 ± 0.6 | – |
| Amplitude | – | 8.2 ± 4.5 | – |
| Interval | – | 159 ± 15 | – |
| (From Veldhuis et al. 1988a) | | | |
| Cycle day | – | 22 – 23 | – |
| Concentration | – | 4.7 ± 0.4 | – |
| Frequency | – | 4.2 ± 0.5 | – |
| Amplitude | – | 5.0 ± 0.9 | – |
| Interval | – | – | – |
| (From Nippoldt et al. 1989) | | | |
| Cycle day | – | 20 – 22 | – |
| Concentration | – | 7.9 ± 0.7 | – |
| Frequency | – | 3.3 ± 0.3 | – |
| Amplitude | – | 10.3 ± 1.6 | – |
| Interval | – | 228 ± 30 | – |
| (From Rossmanith et al. 1990) | | | |

Units are: [a] concentration = mIU/ml; [b] frequency = pulses/12 h; [c] amplitude = mIU/ml; [d] interpulse interval = minutes

flect the frequency and amplitude of the pulsatile release of GnRH from the hypothalamus (Knobil 1980). GnRH release and the response of pituitary gonadotrophs to it are modulated by estradiol and progesterone (Marshall and Kelch 1986). Though the episodic release of gonadotropins and steroids has been known for many years, only recently have the statistical methods available been powerful enough to definitively characterize rhythms and temporal relationships between gonadotropin and steroid pulses.

Sampling frequency and the time interval of sampling markedly affect estimates of pulse frequency and amplitude (Reame et al. 1984). A detailed analysis of the effects of sampling frequency indicates sampling at 2- to 3-min intervals would be needed to identify 90% of LH pulses in both women and men. Sampling at 10- to 20-min intervals under some conditions will recover as little as 50% of the total pulses. The accuracy of estimates of pulse frequency based on the longer sampling intervals is greatly improved by increasing the sampling period, e.g., to 24 h. Pulse amplitude is less sensitive to sampling frequency and the length of the period of sampling (Veldhuis et al. 1986b).

Recent studies also indicate there may be significant quantitative differences in measurements of LH levels and pulsatile behavior depending on whether bioactive or immunoactive LH is measured (Veldhuis et al. 1984, 1989). Results from a number of analyses of LH secretion during normal menstrual cycles are summarized in Tables 1 and 2. Mean LH concentration and pulse frequency increase and the interpulse interval decreases during the follicular phase. In contrast, pulse amplitude is relatively constant (Fig. 4).

During early luteal phase, the concentration of LH exceeds that in the follicular phase but decreases such that by late luteal phase it is less than that of early follucular phase. Pulse amplitude is highest during early luteal phase and decreases with time, concomitant with increases in interpulse interval. The detection of twofold to threefold pulsatile variations in LH and

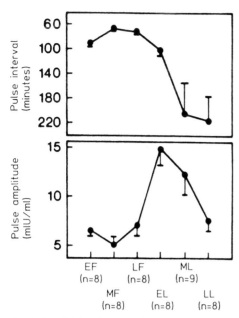

**Fig. 4.** Luteinizing hormone interpulse interval and amplitude in the different stages of the menstrual cycle. (Adapted from Filicori et al. 1986)

progesterone levels during the luteal phase has led to the suggestion that single progesterone determinations may not be adequate to characterize corpus luteum secretory activity (Filicori et al. 1984).

Gonadotropin dynamics during midcycle at the time of the LH surge and at the luteal-follicular transition at menses are less well characterized. Filicori et al. (1986) observed one woman on day 0, the time of the LH peak. With this individual the LH pulse interval was identical to that of the late follicular phase while pulse amplitude increased approximately fivefold. With a second women in whom menses began at the end of the 24-h sampling period, LH pulse frequency increased to a value characteristic of the early follicular phase while pulse amplitude was characteristic of late luteal phase.

In the majority of studies, radioimmunoassay has been used to quantitate immunoactive LH. In one study, Veldhuis et al. (1984) used both a bioassay and an immunoassay to quantitate LH in samples from six women, each sampled at three stages of the menstrual cycle. They detected statistically significant differences between bioactive and immunoactive LH mean concentrations and pulse amplitudes in early follicular and late follicular phase samples. The bioactive to immunoactive ratios were approximately two in samples from both phases. Bioactive and immunoactive LH values were not discrepant in the luteal phase samples.

Wolfram et al. (1989) examined the pulsatile pattern of LH secretion in women undergoing gonadotropin-induced follicle development and oocyte maturation for in vitro fertilization. When compared with women with normal cycles, no differences were noted in the number, amplitude, interval and area of LH pulses during the follicular phase. However, the decrease in LH pulse frequency characteristic of the normal luteal phase was not seen in the hyperstimulated patients. Interestingly, the length of the luteal phase in the treated women was the same as that in the control population.

There are sleep-related changes in the pulsatility of LH secretion. LH pulse frequency decreases during sleep (Fig. 5) (Soules et al. 1985; Filicori et al. 1986). The interpulse interval is maximal at night or early morning, increasing from a mean value of 115 min in the 0800–1400 hours interval to a peak value of 185 min between 2000 and 0200 hours (Veldhuis et al. 1988a). This slowing in LH pulsatility has been observed during both the follicular (Soules et al. 1986; Filicori et al. 1986) and the luteal phase (Filicori et al. 1986; Veldhuis et al. 1988a).

Results pertaining to circadian variations in gonadotropin secretion have been conflicting. Kapen et al. (1973) reported a sleep-related decrease in LH secretion during the early follicular phase. Soules et al. (1986) were unable to confirm that observation. A circadian rhythm in LH in the luteal phase (days +6 to +8) with maximal levels being reached at 1940 ± 1.7 hours was reported by Veldhuis et al. (1988a). Using cosinor analysis, Carandente et al. (1989) were unable to detect circadian periodicity in LH in any of the menstrual cycle stages when all 15 cases in their study were analyzed as a group. However, in some individuals, a rhythm was present in the late follicular and early luteal phases.

Evidence of circannual or seasonal rhythms in the timing of the LH surge has been presented. Testart et al. (1982) reported seasonal changes in the timing of the onset of the surge in a series of patients with normal menstrual cycles but manifesting sterility of tubal origin. When analyzed on the basis of 3-h time intervals, the highest surge frequency (28.6%) during spring cycles occurred at 1500 ± 1.5 hours and at 0300 ± 1.5 hours (42.5%) during the rest of the year. They also estimated that ovulation occurred primarily in the evening during autumn and winter and primarily in the morning during spring. Seibel at al. (1982) studied a group of periovulatory women with normal menstrual cycles. Their results indicated the LH surge occurred most often in the morning between 0500 hours and 0900 hours. They did not de-

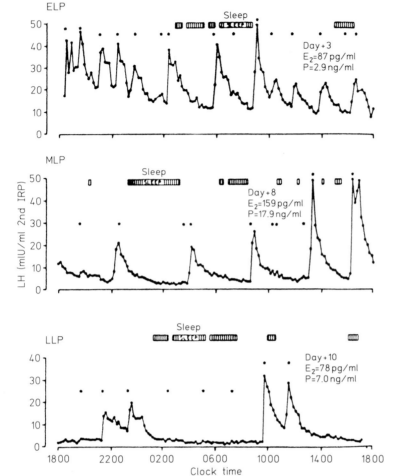

**Fig. 5.** Patterns of episodic LH secretion during the luteal phase of the menstrual cycle. The mean estradiol (E2) and progesterone (P) levels for the 24-h period are given in the figure. Periods of sleep are indicated by *hatched bars*. (Adapted from Filicori et al. 1986)

tect a seasonal variation. Djahanbakhch et al. (1984) measured serum LH and estradiol during the periovulatory period in a series of five women. They observed that the LH surge started between 2400 and 0800 hours in all cases.

Mean daily levels of plasma FSH increase during the early follicular phase of a normal menstrual cycle (Cargille et al. 1969). During late follicular phase, levels decrease reaching a preovulatory nadir at approximately day – 2. This is followed by a rapid increase with a peak coincident with the LH peak. Following this, levels decrease during luteal phase until approximately day + 10, after which an increase is observed. This increase, which continues into the early follicular phase of the next cycle, appears to be necessary for normal folliculogenesis (DiZerega and Hodgen 1981) and may be due in part to the reduced GnRH pulse frequency characteristic of the luteal phase.

Estimates of the pattern of episodic FSH secretion have differed significantly. Backstrom et al. (1982) re-

ported that FSH pulses increased from $2.3 \pm 0.3$ pulses/6 h to $3.8 \pm 0.3$ pulses/6 h in late follicular phase. Following ovulation, plasma FSH levels dropped as did pulse frequency to $0.8 \pm 0.2$ pulses/6 h. Pulse amplitude also decreased. The authors concluded there was a close correlation between the timing of FSH and LH pulses.

In studies by Reame et al. (1984), pulsatile FSH release was evident only in some subjects in early follicular or late luteal phase when estradiol levels were low. Filicori et al. (1986) concluded that the episodic pattern of FSH could not be quantitated reliably because of low pulse amplitude. In their study mean daily plasma FSH levels decreased from $10.0 \pm 0.3$ mIU/ml in early follicular phase to a low value of $7.2 \pm 0.5$ mIU/ml in late follicular phase. Following ovulation, levels declined during luteal phase to a nadir of $4.7 \pm 0.7$ mIU/ml just prior to menstruation. They too concluded that FSH secretion was significantly correlated with LH.

In summary, the pattern of daily plasma FSH differs from that of LH. FSH levels are highest during late luteal phase and the subsequent early follicular phase. Pulsatile secretion is evident but of low amplitude in comparison with that of LH. This may reflect in part the relatively long half-life of FSH compared to LH (Yen et al. 1968, 1970a).

Circadian rhythm in FSH secretion has been detected in early luteal phase (peak at 1500 hours) and late luteal phase (peak at 2300 hours) (Carandente et al. 1989).

## Pathophysiology of the Menstrual Cycle

### Hypothalamic Amenorrhea

Impaired secretion of GnRH is the underlying cause of a form of amenorrhea in which patients exhibit normal or diminished plasma gonadotropin levels with an absence of normal cyclical changes characteristic of the menstrual cycle but in whom the response to administered GnRH is normal (Lachelin and Yen 1978). Current findings suggest reduced pulse frequency of GnRH secretion is the mechanism of ovulatory dysfunction (Reame et al. 1985, Berga et al. 1989). In patients with hypothalamic amenorrhea, LH and GnRH pulse frequency in the early follicular phase resembles that of the late luteal phase in normal women (Reame et al. 1985). In addition, pulse amplitude is highly variable and interpulse intervals are irregular, also characteristic of the late luteal phase in normal women. A low LH pulse frequency during the day is a relatively consistent finding. However, secretion patterns are more variable at night. They may approach normal follicular phase frequencies in some patients (Khoury et al. 1987). In general, these patients respond well to pulsatile administration of GnRH at a frequency of 1.0 pulse every 90 min (Reid et al. 1981; Hurley et al. 1984; Armar et al. 1987).

### Luteal Phase Deficiency

The term luteal phase deficiency (LPD) describes a clinical situation characterized by decreased luteal phase progesterone formation by the corpus luteum in association with infertility or habitual abortion (McNeely et al. 1988). The mechanism underlying the defect in progesterone production is unknown. Since

events in the follicular phase are known to influence progesterone formation in the luteal phase, attention has focused on characterizing possible changes in episodic secretion of gonadotropins throughout the cycle as well as abnormalities in luteal phase progesterone secretion patterns.

Soules et al. (1984, 1989a, b) have characterized episodic hormone secretion patterns in LPD patients. In their studies of patients with a complaint of either infertility or recurrent abortion and whose luteal phase biopsies satisfied criteria of LPD, integrated luteal phase levels of estradiol and progesterone were decreased in comparison with normal women. The midcycle LH surge was less, and both bioactive and immunoreactive LH levels during the luteal phase were decreased. LH pulse frequency in the early follicular phase was higher ($12.8 \pm 1.4/12$ h) and pulse amplitude reduced in patients with LPD than in normal women ($8.2 \pm 0.7/12$ h) (Fig. 6). In the luteal phase, progesterone pulse amplitude and integrated

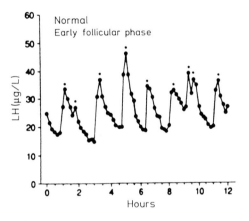

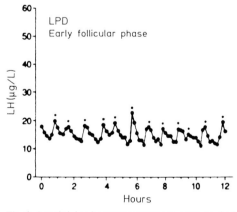

**Fig. 6.** Luteinizing hormone secretory patterns during the early follicular phase in a normal woman and in a patient with luteal phase deficiency *(LPD)*. The *black dots* indicate statistically significant peaks. (Adapted from Soules et al. 1989b)

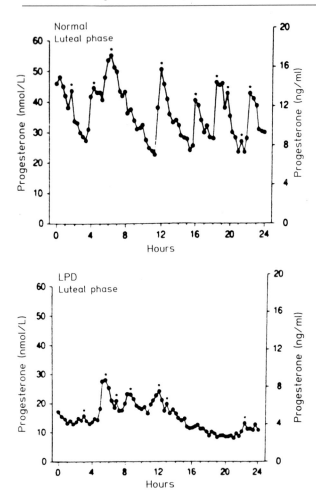

**Fig. 7.** Episodic pattern of progesterone secretion during the luteal phase in a normal woman and in an LPD patient. Statistically significant peaks are indicated by *black dots.* (Adapted from Soules et al. 1989b)

mean progesterone levels were reduced, while progesterone pulse frequency was the same as that of normal women (Fig. 7).

In a separate study by this same group (Soules et al. 1987) it was demonstrated that LPD could be induced by the imposition of a rapid gonadotropin secretion pattern with frequent intermittent doses of GnRH. A GnRH pump was used to administer a dose of peptide every 30 min beginning on day 2,3 or 4 of the follicular phase and continuing until ovulation. The total daily dose was within the normal range used for ovulation induction. However, the frequency of administration was significantly greater than the physiological frequency of every 60–90 min. These observations have led to the proposal (Soules et al. 1989b) that increased LH pulse frequency during the follicular phase may result in inadequate luteal phase

progesterone production and manifest itself in luteal phase deficiency.

## Polycystic Ovarian Disease

Polycystic ovarian disease (PCOD) is a major cause of infertility and subfertility in women. The syndrome characterizes women with chronic anovulation. LH levels may be normal or elevated relative to FSH, giving rise to an increased LH/FSH ratio (Yen et al. 1970b; Rebar et al. 1976). These patients often have enlarged polycystic ovaries in association with hirsutism and obesity. The cause of PCOD is unknown (McKenna 1988). The degree to which increases in LH pulse frequency and amplitude account for the increases in plasma LH levels remains a matter of some dispute. Current results suggest increases in pulse frequency and amplitude, possibly related to changes in serum estradiol levels, and deviations in circadian periodicity characterize LH secretion in PCOD. Burger et al. (1985) compared PCOD patients with patients expressing secondary amenorrhea not related to PCOD and with normally cycling women in the follicular phase of the cycle. Luteinizing hormone pulse amplitude was higher in PCOD patients and lower in non-PCOD patients compared to normal women.

Kazer et al. (1987) compared LH secretion patterns in PCOD patients and normal women in early follicular (day 2–3) and mid follicular (day 6–10) phases of their cycles. LH frequencies were indistinguishable. LH pulse amplitude ($12.2 \pm 2.7$ mIU/ml) in the PCOD patients was higher than normal women in either early ($6.2 \pm 0.8$ mIU/ml) or mid ($6.4 \pm 0.6$ mIU/ml) follicular phase.

Venturoli et al. (1988) compared PCOD patients with normal women on day 5–6 of the follicular phase. PCOD patients had higher mean LH levels and LH pulse amplitude was higher than controls ($11.6 \pm 3.7$ mIU/ml versus $5.2 \pm 1.8$ mIU/ml). They did not detect any differences in frequency or interpulse interval. When LH secretion was studied over a 24-h period, the PCOD group showed a consistent circadian rhythm in plasma LH with highest values at 1720 hours unrelated to sleep and different from the control population. They concluded that LH secretion in their group of PCOD patients deviated from normal with increased pulse amplitude and abnormal circadian periodicity.

Waldstreicher, et al. (1988), in a study of 12 PCOD patients sampled at 10-min intervals for periods of 12–24 h, found that mean serum LH concentration and pulse amplitude were increased compared to normal women at early, mid and late follicular stages of the cycle. In addition, LH pulse frequency was faster in women with PCOD ($24.8 \pm 0.9$ pulses/24 h) than that in early ($15.6 \pm 0.7$), mid ($22.2 \pm 1.1$) and late ($20.8 \pm 1.2$) follicular phase. The increased pulse freuquency correlated with serum estradiol levels leading the authors to suggest a possible etiological role for the steroid in increasing the frequency of GnRH release.

In contrast to hypothalamic amenorrhea patients, pulsatile GnRH administration is often ineffective in PCOD patients. This ineffectiveness appears to be related to an abnormal pituitary sensitivity to GnRH, with excessive ovarian androgen production and obesity as complicating factors. Recent clinical findings (Filicori et al. 1988, 1989; Surrey et al. 1989) suggest pulsatile GnRH treatment can be effective if it is preceded by treatment with a GnRH analog which renders the patient hypogonadotropic.

## Chronobiologic Characteristics of Testosterone and Gonadotropin Secretion in Men

In normal men, testosterone secretion by the testis and gonadotropin secretion by the anterior pituitary are episodic. The end of puberty in boys is characterized by the appearance of a pulsatile LH and FSH secretion pattern throughout the daily 24-h period in contrast to the augmented nocturnal pattern characteristic of pubertal stages.

The Leydig cells of the adult testis are the source of testosterone and estradiol. A pulsatile pattern of secretion of both steroids is apparent in testicular vein blood. Winters and Troen (1986), in a study of six men with varicocele-assoicated infertility, observed that testicular vein testosterone levels ranged from 1 to 1540 ng/ml. Mean pulse frequency was $4.0 \pm 0.3$ pulses/4 h and pulse amplitude was $176 \pm 42$ ng/ml. Estradiol was also secreted episodically in close correlation with testosterone.

Veldhuis et al. (1987) measured testosterone in peripheral blood samples drawn from five normal men at 15-min intervals over a 36-h period. The median interpulse interval was 106 min. Median peak duration was 96 min with a median incremental amplitude of 2.6 ng/ml. A significant circadian periodicity was also detected. The mean amplitude of the 24-h rhythm was 1.85 ng/ml with the peak value occurring at 0630 hours.

That LH secretion in men is episodic has been known for many years (Santen and Bardin 1973). However, only recently have chronobiologic parameters and correlations been rigorously characterized (Urban et al. 1988). LH and FSH both exhibit pulsatile secretory patterns closely correlated with testosterone secretion (Fig. 8). In a recent study of eight normal men sampled at 5-min intervals over a 24-h period, Veldhuis et al. (1988b) estimated median values of 55 min for LH interpulse interval, 40 min for peak duration with a pulse amplitude of 37% or 1.8 mIU/ml. FSH secretion is also episodic and closely coupled temporally to LH secretion (Veldhuis et al. 1987). For FSH, the median interpulse interval was 70 min, median peak duration 50 min, median peak height and incremental amplitude 7.2 mIU/ml and 1.2 mIU/ml, respectively. Both LH and FSH also exhibited significant circadian periodicities when analyzed at 15-min intervals over a

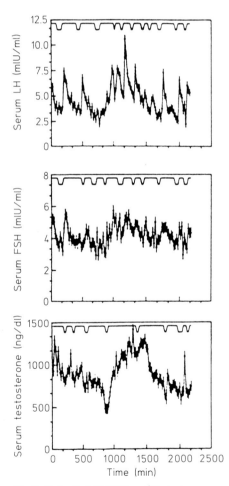

**Fig. 8.** Episodic LH, FSH and Testosterone secretion in a normal man. The *solid line* in each figure indicates statistically significant peak areas. (Adapted from Veldhuis et al. 1987)

24-h period (Veldhuis et al. 1987), with peak levels being reached at 0736 and 0708 hours, respectively.

Findings from a number of laboratories indicate GnRH and gonadotropin secretion in the male are regulated, in part, by testosterone and estradiol. In hypogonadal men, testosterone levels are low. Mean basal LH levels and pulse amplitude and frequency are increased compared to normal values (Winters and Troen 1983). Administration of testosterone to primary hypogonadal men decreases LH and FSH levels and pulse frequency and amplitude, consistent with a negative feedback effect of the androgen (Matsumoto and Bremner 1984).

Results from a number of laboratories also strongly implicate estradiol in the regulation of LH secretion in men. Hyperestrogenism due to the presence of an estrogen-producing adrenal tumor has been shown to be associated with marked suppression of LH levels as well as pulse amplitude and frequency. These parameters increased following surgical removal of the tumor (Veldhuis et al. 1985). Steady-state intravenous infusion of estradiol into normal men has also been shown to suppress LH concentration and pulse amplitude and frequency. That these latter effects were due to the action of estradiol itself on the hypothalamopituitary axis was further supported by the observation that oral administration of the antiestrogen tamoxifen enhanced LH pulse frequency and amplitude (Veldhuis and Dufau 1987).

In summary, the chronobiology of the hypothalamopituitary-gonadal axis is complex in both men and women. The application of new analytical and statistical techniques has uncovered statistically significant rhythms in various time domains in both sexes. It is to be expected that continued application of these methods will lead to a better understanding of human reproduction and the pathophysiology of developmental disorders and infertility.

# References

Abraham GE (1974) Ovarian and adrenal contribution to peripheral androgens during the menstrual cycle. J Clin Endocrinol Metab 39: 340

Abraham GE, Odell WD, Swerdloff RS, Hopper K (1971) Simultaneous radioimmunoassay of plasma FSH, LH, progesterone, 17-hydroxyprogesterone and estradiol-17beta during the menstrual cycle. J Clin Endocrinol Metab 34: 312–318

Aedo A-R, Pederson PH, Pedersen SC, Diczfalusy E (1980a)

Ovarian steroid secretion in normally menstruating women. I. The contribution of the developing follicle. Acta Endocrinol (Copenh) 95: 212–221

Aedo A-R, Pedersen PH; Pedersen SC, Diczfalusy E (1980b) Ovarian steroid secretion in normally menstruating women. II. The contribution of the corpus luteum. Acta Endocrinol (Copenh) 95: 222–231

Apter D, Cacciatore B, Alfthan H, Stenman U-H (1989) Serum luteinizing hormone concentrations increase 100-fold in females from 7 years of age to adulthood, as measured by time-resolved immunofluorometric assay. J Clin Endocrinol Metab 68: 53–57

Armar NA, Tan SL, Eshel A, Jacobs HS, Adams J, Sutherland IA (1987) Practical aspects of pulsatile LHRH therapy. Br J Hosp Med 37: 429–436

Backstrom CT, McNeilly AS, Leask RM, Baird DT (1982) Pulsatile secretion of LH, FSH, prolactin, oestradiol and progesterone during the human menstrual cycle. Clin Endocrinol (Oxf) 17: 29–42

Baird DT, Fraser IS (1974) Blood production and ovarian secretion rates of estradiol – $17\beta$ and estrone in women throughout the menstrual cycle. J Clin Endocrinol Metabol 38: 1009–1017

Beck W, Stubbe P (1984) Pulsatile secretion of luteinizing hormone and sleep-related gonadotropin rhythms in girls with premature thelarche. Eur J Pediatr 141: 168–170

Beck W, Wuttke W (1980) Diurnal variations of plasma luteinizing hormone, follicle-stimulating hormone, and prolactin in boys and girls from birth to puberty. J Clin Endocrinol Metab 50: 635–639

Berga SL, Mortola JF, Girton L, Suh B, Laughlin G, Pham P, Yen SSC (1989) Neuroendocrine aberrations in women with functional hypothalamic amenorrhea. J Clin Endocrinol Metab 68: 301–308

Bidlingmaier F, Wagner-Barnack M, Butenandt O, Knorr D (1973) Plasma estrogens in childhood and puberty under physiologic and pathologic conditions. Pediatr Res 7: 901–907

Boyar RM, Finkelstein J, Roffwarg H, Kapen S, Weitzman E, Hellman L (1972) Synchronization of augmented luteinizing hormone secretion with sleep during puberty. N Engl J Med 287: 582

Burger CW, Korsen T, van Kessel H, van Dop PA, Caron FJM, Schormaker J (1985) Pulsatile luteinizing hormone patterns in the follicular phase of the menstrual cycle, polycystic ovarian disease (PCOD) and non-PCOD amenorrhea. J Clin Endocrinol Metab 61: 1126–1132

Carandente F, Angeli A, Candiani GB, Crosignani PG, Dammacco F, de Cecco L, Marrama P, Massobrio M, Martini L (1989) Rhythms in the ovulatory cycle. $2^{nd}$: LH, FSH, estradiol and progesterone Chronobiologia 16: 353–363

Cargille CM, Ross GT, Yoshimi T (1969) Daily variations in plasma follicle stimulating hormone, luteinizing hormone and progesterone in the normal menstrual cycle. J Clin Endocrinol Metab 29: 12–19

Corley KP, Valk TW, Kelch RP, Marshall JC (1981) Estimation of GnRH pulse amplitude during pubertal development. Pediatr Res 15: 157–162

Crowley WF Jr, Comite F, Vale W, Rivier J, Loriaux DL, Cutler GB Jr (1981) Therapeutic use of pituitary desensitization with a long-acting LHRH agonist: a potential new treatment for idiopathic precocious puberty. J Clin Endocrinol Metab 52: 370–372

Danon M, Beitins IZ, Velez O, Ostrea T, Crawford JD (1982)

Dynamics of bioactive LH during the first seven days of life. Pediatr Res 16: 137A

De Jong FH, Baird DT, van der Molen HJ (1974) Ovarian secretion rates of oestrogens, androgens and progesterone in normal women and in women with persistent ovarian follicles. Acta Endocrinol (Copenh) 77: 575–587

Delemarre – van de Waal HA, van den Brande JL, Schoemaker J (1985) Prolonged pulsatile administration of luteinizing-releasing hormone in prepubertal children: Diagnostic and physiologic aspects. J Clin Endocrinol Metab 61: 859–867

DiZerega GS, Hodgen GD (1981) Folliculogenesis in the primate ovarian cycle. Endocr Rev 2: 27–45

Djahanbakhch O, Warner P, McNeilly AS, Baird DT (1984) Pulsatile release of LH and oestradiol during the periovulatory period in women. Clin Endocrinol (Oxf) 20: 579–589

Dunkel L, Alfthan H, Stenman U-H, Perheentupa J (1990) Gonadal control of pulsatile secretion of luteinizing hormone and follicle-stimulating hormone in prepubertal boys evaluated by ultrasensitive time-resolved immunofluorometric assays. J Clin Endocrinol Metab 70: 107–114

Ehrmann DA, Rosenfield RL, Cuttler L, Burstein S, Cara JF, Levitsky L (1989) A new test of combined pituitary-testicular function using the gonadotropin-releasing hormone agonist Nafarelin in the differentiation of gonadotropin deficiency from delayed puberty: pilot studies. J Clin Endocrinol Metab 69: 963–967

Filicori M, Butler JP, Crowley WF Jr (1984) Neuroendocrine regulation of the corpus luteum in the human. J Clin Invest 73: 1638–1647

Filicori M, Santoro N, Merriam GR, Crowley WF Jr (1986) Characterization of the physiological pattern of episodic gonadotropin secretion throughout the human menstrual cycle. J Clin Endocrinol Metab 62: 1136–1144

Filicori M, Campaniello E, Michelacci L, Pareschi A, Ferrari P, Bolelli G, Flamigni C (1988) Gonadotropin-releasing hormone (GnRH) analog suppression renders polycystic ovarian desease patients more susceptible to ovulation induction with pulsatile GnRH. J Clin Endocrinol Metab 66: 327–333

Filicori M, Flamigni C, Campaniello E, Valdiserri A, Ferrari P, Merriggiola MC, Michealcci L, Pareschi A (1989) The abnormal response of polycystic ovarian disease patients to exogenous pulsatile gonadotropin-releasing hormone: characterization and management. J Clin Endocrinol Metab 69: 825–831

Gambacciani M, Lin JH, Swartz WH, Tueros VS, Yen SSC, Rasmussen DD (1987) Intrinsic pulsatility of luteinizing hormone release from the human pituitary in vitro. Neuroendocrinology 45: 402–406

Haus E, Lakatua DJ, Sackett-Lundeen LL, Swoyer J (1984) Chronobiology in laboratory medicine. In: Rietveld WJ (ed) Clinical aspects of chronobiology. Bakker, Baarn, pp 13–82

Haus E, Nicolau GY, Lakatua D, Sackett-Lundeen L (1988) Reference values for chronopharmacology. Annu Rev Chronopharmacal 4: 333–424

Hoff JD, Quigley ME, Yen SSC (1983) Hormonal dynamics at midcycle: a reevaluation. J Clin Endocrinol Metab 57: 792–796

Hurley DM, Brian R, Outch K, Stockdale J, Fry A, Hackman C, Clarke I, Burger HG (1984) Induction of ovulation and fertility in amenorrheic women by pulsatile low-dose gonadotropin-releasing hormone. N Engl J Med 310: 1069–1074

Jakacki RI, Kelch RP, Sauder SE, Lloyd JS, Hopwood NJ, Marshall JC (1982) Pulsatile secretion of luteinizing hormone in children. J Clin Endocrinol Metab 55: 453–458

Jenner MR, Kelch RP, Kaplan SL, Grumbach MM, (1972) Hormonal changes in puberty. IV. Plasma estradiol, LH, and FSH in prepubertal children, pubertal females, and in precocious puberty, premature thelarche, hypogonadism, and in a child with a feminizing ovarian tumor. J Clin Endocrinol Metab 34: 521–530

Kapen S, Boyar R, Perlow M, Hellman L, Weitzman ED (1973) Luteinizing hormone: changes in secretory patterns during sleep in adult women. Life Sci 13: 693–701

Kazer RR, Kessel B, Yen SSC (1987) Circulating luteinizing hormone pulse frequency in women with polycystic ovary syndrome. J Clin Endocrinol Metab 65: 233–236

Kelch RP, Hopwood NJ, Sauder S, Marshall JC (1985) Evidence for decreased secretion of gonadotropin-releasing hormone in pubertal boys during short-term testosterone treatment. Pediatr Res 19: 112–117

Khoury SA, Reame NE, Kelch RP, Marshall JC (1987) Diurnal patterns of pulsatile luteinizing hormone secretion in hypothalamic amenorrhea: reproducibility and responses to opiate blockade and an alpha2-adrenergic agonist. J Clin Endocrinol Metab 64: 755–762

Knobil E (1980) Neuroendocrine control of the menstrual cycle. Recent Prog Horm Res 36: 53–88

Korenman SG, Sherman BM (1973) Further studies of gonadotropin and estradiol secretion during the prevulatory phase of the human menstrual cycle. J Clin Endocrinol Metab 36: 1205–1209

Kulin HE, Moore RG Jr, Sautner SJ (1976) Circadian rhythms in gonadotropin excretion in prepubertal and pubertal children. J Clin Endocrinol Metab 42: 770–773

Lachelin GCL, Yen SSC (1978) Hypothalamic chronic anovulation. Am J Obstet Gynecol 130: 825–831

Landgren BM, Aedo AR, Nunez M, Cekan SZ, Diczfalusy E (1977) Studies on the pattern of circulating steroids in the menstrual cycle. Acta Endocrinol (Copenh) 84: 620–632

Liu L, Chandari N, Corle D, Sherins RJ (1988) Comparison of pulsatile subcutaneous gonadotropin-releasing hormone and exogenous gonadotropins in the treatment of men with isolated hypogonadotropin hypogonadism. Fertil Steril 49: 302–308

Mais V, Kazer RR, Cetel NS, Rivier J, Vale W, Yen SSC (1986) The dependency of folliculogenesis and corpus luteum function on pulsatile gonadotropin secretion in cycling women using a gonadotropin-releasing hormone antagonist as a probe. J Clin Endocrinol Metab 62: 1250–1255

Marshall JC, Kelch RP (1986) Gonadotropin-releasing hormone: role of pulsatile secretion in the regulation of reproduction. N Engl J Med 315: 1459–1468

Matsumoto AM, Bremner WJ (1984) Modulation of pulsatile gonadotropin secretion by testosterone in man. J Clin Endocrinol Metab 58: 609–614

McKenna TJ (1988) Pathogenesis and treatment of polycystic ovary syndrome. N Engl J Med 318: 558–562

McLachlan RI, Cohen NL, Vale WW, Rivier JE, Burger HG, Bremner WJ, Soules MR (1989) The importance of luteinizing hormone in the control of inhibin and progesterone secretion by the human corpus luteum. J Clin Endocrinol Metab 68: 1078–1085

McNeely MJ, Soules MR (1988) The diagnosis of luteal phase deficiency: a critical review. Fertil Steril 50: 1–15

Merriam GR, Wachter KW (1982) Algorithms for the study of episodic hormone secretion. Am J Physiol 243: E310–E31

Nippoldt TB, Reame NE, Kelch RP, Marshall JC (1989) The roles of estradiol and progesterone in decreasing luteinizing hormone pulse frequency in the luteal phase of the menstrual cycle. J Clin Endocrinol Metab 69: 67–76

Parker DC, Judd HL, Rossman LG, Yen SSC (1975) Pubertal sleep-wake patterns of episodic LH, FSH and testosterone release in twin boys. J Clin Endocrinol Metab 40: 1099–1109

Penny R, Olambiwonnu NO, Frasier SD (1977) Episodic fluctuations of serum gonadotropins in pre- and post-pubertal girls and boys. J Clin Endocrinol Metab 45: 307–311

Pescovitz OH, Comite F, Hench K, Barnes K, McNemar A, Foster C, Kenigsberg D, Loriaux L, Cutler GB (1986) The NIH experience with precocious puberty: diagnostic subgroups and response to short-term luteinizing releasing hormone analogue therapy. J Pediatr 108: 47–54

Pescovitz OH, Hench KD, Barnes KM, Loriaux DL, Cutler GB Jr (1988) Premature thelarche and central precocious puberty: the relationship between clinical presentation and the gonadotropin response to luteinizing hormone-releasing hormone. J Clin Endocrinol Metab 67: 474–479

Reame NE, Sauder SE, Kelch RP, Marshall JC (1984) Pulsatile gonadotropin secretion during the human menstrual cycle: evidence for altered frequency of gonadotropin-releasing hormone secretion. J Clin Endocrinol Metab 59: 328–337

Reame NE, Sauder SE, Case GD, Kelch RP, Marshall JC (1985) Pulsatile gonadotropin secretion in women with hypothalamic amenorrhea: evidence that reduced frequency of gonadotropin-releasing hormone secretion is the mechanism of persistent anovulation. J Clin Endocrinol Metab 61: 851–858

Rebar R, Judd HL, Yen SSC, Rakoff J, Vandenberg G, Naftolin F (1976) Characterization of the inappropriate gonadotropin secretion in polycystic ovary syndrome. J Clin Invest 57: 1320–1329

Reid RL, Leopold GR, Yen SSC (1981) Induction of ovulation and pregnancy with pulsatile luteinizing hormone releasing factor: dosage and mode of delivery. Fertil Steril 36: 553–559

Reiter EO, Biggs DE, Veldhuis JD, Beitins IZ (1987) Pulsatile release of bioactive luteinizing hormone in prepubertal girls: discordance with immunoreactive luteinizing hormone pulses. Pediatr Res 21: 409–413

Ross JL, Loriaux DL, Cutler Jr GB (1983) Developmental changes in neuroendocrine regulation of gonadotropin secretion in gonadal dysgenesis. J Clin Endocrinol Metab 57: 288–293

Rossmanith WC, Laughlin GA, Mortola JF, Johnson ML, Veldhuis JD, Yen SSC (1990) Pulsatile cosecretion of estradiol and progesterone by the midluteal phase corpus luteum: temporal link to luteinizing hormone pulses. J Clin Endocrinol Metab 70: 990–995

Santen RJ, Bardin CW (1973) Episodic luteinizing hormone secretion in man: pulse analysis, clinical interpretation, physiologic mechanisms. J Clin Invest 52: 2617–2628

Seibel MM, Shine W, Smith DM, Taymor ML (1982) Biological rhythm of the luteinizing hormone in women. Fertil Steril 37: 709–711

Soules MR, Steiner RA, Clifton DK, Bremner WJ (1985) Abnormal patterns of pulsatile luteinizing hormone in women with luteal phase deficiency. Obstet Gynecol 63: 626–629

Soules MR, Steiner RA, Cohen NL, Bremner WJ, Clifton DK (1985) Nocturnal slowing of pulsatile luteinizig hormone secretion in women during the follicular phase of the menstrual cycle. J Clin Endocrinol Metab 61: 43–49

Soules MR, Clifton DK, Bremner WJ, Steiner RA (1987) Corpus luteum insufficiency induced by a rapid gonadotropin-releasing hormone-induced gonadotropin secretion

pattern in the follicular phase. J Clin Endocrinol Metab 65: 457–464

Soules MR, McLachlan RI, Ek M, Dahl KD, Cohen NL, Bremner WJ (1989a) Luteal phase deficiency: characterization of reproductive hormones over the menstrual cycle. J Clin Endocrinol Metab 69: 804–812

Soules MR, Clifton DK, Cohen NL, Bremner WJ, Steiner RA (1989b) Luteal phase deficiency: abnormal gonadotropin and progesterone secretion patterns. J Clin Endocrinol Metab 69: 813–820

Smolensky MIt, Reinberg A, Bicakove-Rocher A, Sanford J (1981) Chronoepidemiological search for circannual changes in the sexual activity of human males. Chronobiologia 8: 217–230

Surbey MK, de Catanzero D, Smith MS (1986) Seasonality of conception in Hutterite colonies in Europe (1758–1881) and North America (1858–1944). J Biosoc Sci 18: 337–345

Surrey ES, de Ziegler D, Lu JKH, Chang RJ, Judd HL (1989) Effects of gonadotropin-releasing hormone (GnRH) agonist on pituitary and ovarian responses to pulsatile GnRH therapy in polycystic ovarian disease. Fertil Steril 52: 547–552

Testart J, Frydman R, Roger M (1982) Seasonal influence of diurnal rhythms in the onset of the plasma luteinizing hormone surge in women. J Clin Endocrinol Metab 55: 374–377

Thorneycroft IH, Sribyatta B, Tom WK, Nakamura RM, Mishell DRJr (1974) Measurement of serum LH, FSH, progesterone, 17-hydroxyprogesterone and estradiol-17beta at 4-hour intervals during the periovulatory phase of the menstrual cycle. J Clin Endocrinol Metab 39: 754–758

Urban RJ, Evans WS, Rogol AD, Kaiser DL, Johnson ML, Veldhuis JD (1988) Contemporary aspects of discrete peak-detection algorithms. I. The paradigm of the luteinizing hormone pulse signal in men. Endocr Rev 9: 3–37.

Valsik JA (1965) The seasonal rhythm of menarche: a review. Hum Biol 37: 79–90

Veldhuis JD, Dufau ML (1987) Estradiol modulates the pulsatile secretion of biologically active luteinizing hormone in man. J Clin Invest 80: 631–638

Veldhuis JD, Beitins IZ, Johnson ML, Serabian MA, Dufau ML (1984) Biologically active luteinizing hormone is secreted in episodic pulsations that vary in relation to stage of the menstrual cycle. J Clin Endocrinol Metab 58: 1050–1058

Veldhuis JD, Sowers JR, Rogol AD, Klein FA, Miller N, Dufau M (1985) Pathophysiology of male hypogonadism associated with endogenous hyperestrogenism. N Engl J Med 312: 1371–1375

Veldhuis JD, Weiss J, Mauras N, Rogol AD, Evans WS, Johnson ML (1986a) Appraising endocrine pulse signals at low circulating hormone concentrations: use of regional coefficients of variation in the experimental series to analyze pulsatile luteinizing hormone release. Pediatr Res 20: 632–637

Veldhuis JD, Evans WS, Johnson ML, Wills MR, Rogol AD (1986b) Physiological properties of the luteinizing hormone pulse signal: impact of intensive and extended venous sampling paradigms on its characterization in healthy men and women. J Clin Endocrinol Metab 62: 881–891

Veldhuis JD, King JC, Urban RJ, Rogol AD, Evans WS, Kolp LA, Johnson ML (1987) Operating characteristics of the male hypothalamo-pituitary-gonadal axis: pulsatile release of testosterone and follicle-stimulating hormone and their temporal coupling with luteinizing hormone. J Clin Endocrinol Metab 65: 929–941

Veldhuis JD, Christiansen E, Evans WS, Kolp LA, Rogol AD,

Johnson ML (1988a) Physiological profiles of episodic progesterone release during the midluteal phase of the human menstrual cycle: analysis of circadian and ultradian rhythms, discrete pulse properties, and correlations with simultaneous luteinizing hormone release. J Clin Endocrinol Metab 66: 414–421

Veldhuis JD, Evans WS, Urban RJ, Rogol AD, Johnson ML (1988b) Physiological attributes of the luteinizing hormone pulse signal in the human: cross-validation studies in men. J Androl 9: 69–74

Veldhuis JD, Urban RJ, Beitins IZ, Blizzard RM, Johnson ML, Dufau ML (1989) Pathophysiological features of the pulsatile secretion of biologically active luteinizing hormone in man. J Steroid Biochem 33: 739–749

Venturoli S, Porcu E, Fabbri R, Magrini O, Gammi L, Paradisi R, Forcacci M, Bolzani R, Flamigni C (1988) Episodic pulsatile secretion of FSH, LH, prolactin, oestradiol, oestrone, and LH circadian variations in polycystic ovary syndrome. Clin Endocrinol (Oxf) 28: 93–107

Wagner TOF, Brabant G, Warsch F, Hesch RD, von zur Muhlen A (1986) Pulsatile gonadotropin-releasing hormone treatment in idiopathic delayed puberty. J Clin Endocrinol Metab 62: 95–101

Waldhauser F, Frisch H, Pollak A, Weissenbacher G (1981) Pulsatile secretion of gonadotropins in early infancy. Eur J Pediatr 137: 71-74

Waldstreicher J, Santoro NF, Hall JE, Filicori M Crowley MF Jr (1988) Hyperfunction of the hypothalamic-pituitary axis in women with polycystic ovarian disease: indirect evidence for partial gonadotroph desensitization. J Clin Endocrinol Metab 66: 165–172

Wang C, Zhong CQ, Leung A, Low LCK (1990) Serum Bioactive follicle-stimulating hormone levels in girls with precocious sexual development. J Clin Endocrinol Metab 70: 615–619

Wennink JMB, Delemarre-van-de Waal HA, van Kessel H, Mulder GH, Foster JP, Schoemaker J (1988) Luteinizing hormone secretion patterns in boys at the onset of puberty measured using a highly sensitive immunoradiometric assay. J Clin Endocrinol Metab 67: 924–928

Winters SJ, Troen P (1986) Testosterone and estradiol are co-secreted episodically by the human testis. J Clin Invest 78: 870–873

Wolfram J, Siegberg R, Apter D, Alfthan H, Stenman U, Laatikainen T (1989) Pulsatility of serum-luteinizing hormone during hyperstimulation with clomiphene citrate and human menopausal gonadotropin for in vitro fertilization. Fertil Steril 52: 817–820

Wu FCW, Butler GE, Kelnar CJH, Sellar RE (1990) Patterns of pulsatile luteinizing hormone secretion before and during the onset of puberty in boys: a study using an immunoradiometric assay. J Clin Endocrinol Metab 70: 629–637

Yen SSC (1986) The human menstrual cycle. In: Yen SSC, Jaffe BR (eds) Reproductive endocrinology. Saunders, Philadelphia, pp 200–236

Yen SSC, Llerena LA, Little B, Pearson OH (1968) Disappearance rates of endogenous luteinizing hormone and chorionic gonadotropin in man. J Clin Endocrinol Metab 28: 1763–1767

Yen SSC, Llerena LA, Pearson OH, Littell AS (1970a) Disappearance rates of endogenous follicle-stimulating hormone in serum following surgical hypohysectomy in man. J Clin Endocrinol Metab 30: 325–329

Yen SSC, Vela P, Rankin J (1970b) Inappropriate secretion of follicle stimulating hormone and luteinizing hormone in polycystic ovarian disease. J Clin Endocrinol Metab 30: 435–442

# Chronobiology of the Hypothalamic-Pituitary-Thyroid Axis

G. Y. Nicolau and E. Haus

## Introduction

The hypothalamic-pituitary-thyroid (HPT) axis is part of a complex web of neuroendocrine functions. It shows an intricate time structure with rhythmic variations of multiple frequencies found at all levels of the system from hypothalamic neurons to the cells of the peripheral target tissues. The frequencies observed range from rapid neuronal discharges to ultradian rhythms, and/or pulsatile secretions to circadian and circannual rhythms. The rhythmic variations are superimposed upon aging trends. The time-dependent rhythmic (and nonrhythmic) variations of the HPT system interact with, modulate and are modulated by similar time-dependent variations of other neuroendocrine, metabolic and immune functions.

## Thyrotropin-Releasing Hormone

The hypothalamic regulation of the HPT axis centers around the neurotransmitter named after its best-explored action *thyrotropin-releasing hormone (TRH)*. TRH is a tripeptide (pyroglutamyl-histidyl-prolineamide) which, in addition to its capacity to stimulate the release of thyroid-stimulating hormone (TSH) from the anterior pituitary, also stimulates prolactin and shows a wide distribution, not only in different regions of the hypothalamus, but also in extra hypothalamic brain regions and beyond the confines of the central nervous system (for reviews see Jackson 1982; Prasad 1985; Metcalf and Jackson 1989). Outside the central nervous system TRH biosynthesis occurs in many peripheral tissues including the spleen (Simard et al. 1989).

Hypothalamic TRH is transported to the anterior pituitary through the portal vessels and acts on the thyrotrope as a classic hormone. At different extrapituitary locations, TRH plays a role in neurotransmission (neurocrine function) and in cell to cell regulation (paracrine function). TRH stimulates TSH release after attachment to high-affinity receptors on the thyrotrope membrane, with activation of adenylatecyclase and subsequent generation of cyclic AMP as secondary messenger.

## Circadian Rhythm in Hypothalamic Content of TRH

The content of TRH in the hypothalamus of rats exposed to a standard light/dark cycle regimen shows a circadian variation (Collu et al. 1977; Mannisto et al. 1978; Koivusalo and Leppaluoto 1979; Kerdelhue et at. 1981; Brammer et al. 1979), the timing of which varies between different studies and investigators. Martino et al. (1985) and Kerdelhue et al. (1981) found in rats the highest values of TRH in the middle of the daily light span (the resting span of the animals) and the lowest values in the middle of the daily dark span (activity span). Covarrubias et al. (1988) measured hypothalamic TRH and TRH messenger RNA (TRHmRNA) during the circadian cycle. In adult animals, the hypothalamic content in TRH and TRHmRNA was similar in its timing with the highest levels observed at the onset of the daily dark span. The appearance of the TRH peak, just before the activity period of the rats, did correlate with the TSH circadian rhythm in the same animals. The well-defined rhythm in the whole hypothalamus measured by these investigators suggested that most of the TRH-ergic cycling nuclei are synchronized. In contrast, Mannisto et al. (1978) reported a phase difference in TRH concentration between anterior hypothalamus and the medial basal hypothalamus, the latter of which showed higher TRH concentrations during the dark span and lower concentrations during light.

## Ontogenetic Development of Circadian Rhythm in Hypothalamic TRH

Studying the HPT axis during the development of rats between 5 and 30 days of life, Covarrubias et al. (1988) reported the gradual establishment of a light/dark synchronized circadian rhythm in TRH and TRHmRNA in the hypothalamus. The immature animals showed different patterns and different timing at different stages of development. In spite of these differences within the age groups, the TRH and TRHmRNA peaks did occur mostly at the same time. During maturation the circadian rhythm of these agents apparently developed in the entire group of animals at about the same age and was recognizable as a group phenomenon. Throughout this study, the TRHmRNA rhythm showed a higher amplitude than that of TRH. The establishment of the circadian rhythm during maturation of the animals appears to be a multifactorial event involving the cellular maturation of the TRH-ergic neurons and the establishment of negative feedback regulations by the thyroid hormones and possibily the development of other control mechanisms (Segerson et al. 1987; Covarrubias et al. 1987).

## Thyrotropin-Releasing Hormone and Neuronal High Frequency Rhythm

Acting as a neuropeptide, TRH shows a potent stimulatory effect on respiration (Dekin et al. 1985). In in vitro preparations TRH induces rhythmic bursting in neurons in the respiratory division of the nucleus tractus solitarius (NTS) apparently by modulating the membrane excitability of NTS neurons and allowing them to express endogenous bursting activity. TRN alters the activity in some NTS neurons from a nonrhythmic to a rhythmic pattern in a frequency range of seconds. This action of TRH supporrts the hypothesis that this neuropeptide takes part in the control of rhythmic breathing in mammals.

## Thyrotropin-Releasing Hormone in Extraneural Distribution

In extraneural distribution, TRH has been reported at numerous sites and in structures which are known to show rhythmic variation in several frequencies. These include the gastrointestinal tract and pancreas (i.e., the islets of Langerhans) of the rat. The occurrence of TRH at these locations suggests that TRH is part of a diffuse neuroendocrine system and outside the nervous system is localized in the "neuroendocrine program" cells, which migrate to the endoderm during embryonic development (Pearse and Takor 1979). In other locations, TRH-like material has been reported in the reproductive system of male rats, including the prostate, testes, epididymis, and seminal vesicles (Pekary et al. 1980), in the spleen (Simard et al. 1989), and in the human placenta (Shambaugh et al. 1979). Although several of these organs are known to show rhythms of several frequencies, time-related actions of TRH outside the HPT system are at this time not documented.

## Thyrotropin-Releasing Hormone Determinations in Body Fluids

Although radioimmunoassays for TRH have become available, the wide anatomic distribution of the tripeptide and the dilution of the hypophyseal portal vessel blood in the systemic circulation make it unlikely that the measurement of TRH levels in body fluids can be used to draw any conclusion concerning hypothalamic TRH activity. The material measured as immunoreactive TRH in the peripheral circulation and in urine cannot be assumed with certainty to be derived from the hypothalamus (Jackson and Reichlin 1979). A direct assay of pituitary portal vessel blood is not feasible in human subjects.

Thyrotropin-releasing hormone-like material has been detected in the cerebrospinal fluid of normal subjects and is increased in patients with major depression (Kirkegaard et al. 1979). In the spinal fluid of rhesus monkeys, Berelowitz et al. (1981) showed a circadian variation of immunoreactive TRH and somatostatin. Also here, however, the origin of the TRH and its significance need further investigation.

## Manipulation of Lighting Regimen and Circadian Rhythm of TRH

In rats exposed to continuous light or dark the mean concentration of hypothalamic TRH is comparable (Martino et al. 1985; Brammer et al. 1979). However, the circadian variation of the hypothalamic TRH content did disappear as a group phenomenon. This does not necessarily imply a disappearance of circadian periodicity as such but may rather be due to a desynchronization of the circadian rhythm within the group. It seems unlikely that the pineal gland may be involved in this phenomenon, since the circadian variation of hypothalamic TRH persists in rats after pinealectomy (Brammer et al. 1979). In the rat retina the immunoreactive TRH was found to be directly related to light exposure (Schaeffer et al. 1977).

## Thyrotropin-Releasing Hormone Effects on Multiple Systems

Thyrotropin-releasing hormone affects several physiological and behavioral processes independently from its action on the pituitary. After central nervous system administration of TRH to conscious animals, changes have been reported in body temperature, respiration, blood pressure, electroencephalogram, and motor activity. TRH reverses the hypothermia and CNS depression induced by several CNS-acting compounds. The naturally induced state of CNS seasonal depression (hibernation) in squirrels can be reversed by microinjection of TRH into the dorsal hippocampus. These results suggest that TRH participates in the control of body temperature and the modulation of the level of arousal in the CNS, two characteristically circadian periodic functions. The effects of TRH on hippocampal neurons are circadian rhythm stage dependent. When administered to awake and behaviorally activated squirrels, TRH produces a dose-dependent decrease in body temperature accompanied by behavioral quieting and reduction in metabolic rate and electromyographic activity. In contrast, when the peptide is administered during slow-wave sleep it produced increased thermogenesis and an increase in electromyographic activity and in the amount of electroencephalographic desynchronization (Stanton et al. 1981).

## Interaction of TRH with Other Neurotransmitters

Thyrotropin-releasing hormone characteristically interacts with other neurotransmitters, e.g., 5-hydroxytryptamine (5-HT), dopamine and norepinephrine. TRH immunoreactivity has been found to occur at some locations in neurons containing other chemical neurotransmitters and/or peptides which are known to show rhythmic variations in and/or outside the CNS. These include 5-HT, substance P like immunoreactivity, and human growth hormone like immunoreactivity (Hökfelt et al. 1989). TRH interacts with catecholamines, i.e., dopamine and norepinephrine, in brain and spinal cord (Bennett et al. 1989). The interactions between TRH and these neurotransmitters indicate that TRH may also act indirectly

upon cycling structures in the central nervous system altering the release and action of other neurotransmitters in neurons and interneurons at different locations, e.g., TRH stimulates the release of norepinephrine in both brain and spinal cord (Bennett et al. 1989).

These interactions are of special interest since the effects upon the central adrenergic, dopaminergic and serotoninergic systems in the CNS may be related to motor and behavioral functions and disorders some of which are accompanied by circadian and other rhythm alterations. Also, the effects of central nervous system active drugs like the antidepressants may be mediated or modulated by such interactions (Marsden et al. 1989; Bennett et al. 1989).

## Pituitary Actions of TRH, TRH Receptors

The action of TRH upon the pituitary leads to release and *de novo* synthesis of thyroid-stimulating hormone (TSH). TRH acts as an equally potent stimulator of prolactin synthesis and secretion; however, its function in normal prolactin physiology remains unclear. The effects of TRH on thyrotrophs and mammotrophs are initiated by the interaction with specific membrane receptors (Hinkle 1989). The TRH receptor on the pituitary cell membrane is only slowly degraded (Hinkle and Tashjian 1975) and thus does not undergo rapid down regulation. A short pulse of exposure to TRH, therefore, can lead to a prolonged secretion of TSH from the pituitary cell (of about 25 min after a 0.4-min exposure). TRH dissociates from its receptor intact and quite rapidly. Thus, the thyrotropic pituitary cells can respond to repeated challenges with TRH with bursts of hormone secretion with relatively little diminution of response beyond that expected from the depletion of the intracellular hormone stores (Aizawa and Hinkle 1985).

## Regulation of TRH Receptors and Hormonal Interactions at the Receptor Level

The concentration of TRH receptors is regulated both homologously and heterologously. Occupancy of TRH receptors by TRH leads to a slow loss of TRH-binding sites (homologous down regulation), requiring about 24 h to reach a maximum (Gershengorn 1978; Hinkle and Tashjian 1975). The density of TRH receptors can be modulated by other hormones in what appears to be a physiologically significant heterologous receptor regulation. Thyroid hormones exert powerful negative feedback control over the pituitary response to TRH (Morley 1981; Jackson 1982), part of which, at least, occurs at the level of the TRH receptor (Gershengorn 1978; Hinkle et al. 1979). This component of thyroid hormone feedback control is slow, requiring 24–48 h and is fully reversible.

At the TRH receptor level occurs an interaction with gonadal and with adrenal steroids. Estrogens can increase the TRH receptor levels *in vitro* and *in vivo* (DeLean et al. 1977; Gershengorn et al. 1979). Also glucocorticoids lead to an increase in TRH receptor density (Tashjian et al. 1977), an effect which opposes the effect of thyroid hormones to deplete TRH receptor density. The heterologous modulation of TRH receptor density by gonadal and adrenal steroids suggests the possibility of rhythmic interactions between the thyrotropic, gonadotropic and adrenotropic axes. If this should be the case it would affect predominantly the lower frequencies since the changes in receptor density are slow, consistent with effects on the rate of receptor synthesis.

## Regulation of TSH Secretion

### Regulation by TRH

Thyroid-stimulating hormone (TSH) is released from the thyrotrope cells of the pituitary in response to impulse of TRH, which also initiates synthesis of the hormone in a more protracted response. TRH is the main stimulator of TSH release and production and is thought to be the most important factor in the regulation of the pulsatile secretion and the circadian rhythm of TSH.

### Other Factors Acting Upon TSH Secretion, Catecholamines

Many of the factors acting upon the synthesis and release of TRH act indirectly on the production and secretion of TSH. It can tentatively be concluded that the noradrenergic system is stimulatory (Chan et al. 1979; Vijayan and McCann 1978) and the serotoninergic system inhibitory (Krulich et al. 1979; Mess et al. 1986). Only alpha-adrenergic receptors appear to be involved in the regulation of TSH release (Greenspan et al. 1986).

### Dopaminergic Effects

Although there is some circumstantial evidence implicating endogenous dopamine as exerting an inhibitory control on the circadian rhythm of TSH (Scanlon et al. 1980; Sowers et al. 1982), dopamine agonist and antagonist studies have proved contradictory. Scanlon et al. (1980) found that the TSH response to dopamine receptor blockage with metoclopramide was significantly greater at 2300 hours than at 1100 hours. Treatment with the dopamine agonist bromocriptine eliminated the early nocturnal TSH surge, suggesting that dopamine might modulate the circadian rhythm of TSH (Sowers et al. 1982). In a patient with mild hyperthyroidism due to inappropriate thyrotropin secretion and a normal circadian variation of TSH, the dopamine agonist bromocriptine failed to suppress serum TSH and the circadian rhythm was not altered by this treatment (Magee et al. 1986). This observation suggests that dopamine does not control the circadian variation of TSH in nontumoral TSH-mediated hyperthyroidism. There was in this patient a strongly exaggerated TSH response to the dopamine antagonist domperidone which was similar in both morning and evening, contrary to previous investigations in normal subjects which had shown that a greater response occurred at 2300 hours (Scanlon et al. 1980; Rodriguez-Arnao et al. 1982). It appears that in nontumoral TSH hypersecretion the diurnal variation of TSH persists despite an exogenously induced increased dopaminergic tone which is similar in the morning and in the evening (Magee et al. 1986).

Supraphysiological levels of dopamine were required to abolish the circadian variation in serum TSH (Kerr et al. 1987). Thus dopamine does not appear to be the prime physiologic modulator of the TSH circadian rhythm at the pituitary level. It is probable that prolactin rhythmicity in contrast to TSH periodicity is controlled by alterations in dopamine tone, i.e., since no close temporal association was found between the prolactin and TSH rhythms (Kerr et al. 1987).

### Glucocorticoid Effects

Both natural and synthetic glucocorticoids inhibit thyroid function in man. This effect has been shown to be due to inhibition of TSH secretion and is exerted both at the hypothalamic and pituitary levels depending upon the dose and duration of steroid exposure (Brabant et al. 1987).

## Gamma-aminobutyric Acid

Gamma-aminobutyric acid (GABA) functions as a neurotransmitter in the mammalian central nervous system in particular in the hypothalamus. In rats GABA has an inhibitory effect on central thyrotropin control via an inhibition of TRH release from the hypothalamus and may be partly responsible for the low trough levels of serum TSH observed during its circadian rhythm (Vijayan and McCann 1978; Jordan et al. 1981, 1983). Chronopharmacologic manipulation of the synthesis of GABA or the blockage of the GABA-ergic active sites can lead in rats to the appearance of an inappropriate nocturnal peak during the daily dark span (Jordan et al. 1981). Human data on GABA effects on TSH secretion are not available in the literature.

## Opioid Peptides

The role of opioid peptides on TSH secretion is still controversial. It appears that opioid peptides inhibit the tonic hormone release by the HPT axis under resting conditions and suppress the reactive changes in the system following specific stimulations (e.g., goiterogens, thyroidectomy). Opioids appear to act mainly at the hypothalamic level suppressing the release of TRH. However, in addition to this hypothalamic inhibitory mechanism endogenous opioids may also have a direct action on the pituitary influencing TSH release (Mess et al. 1986).

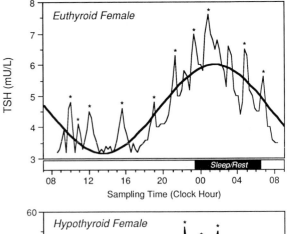

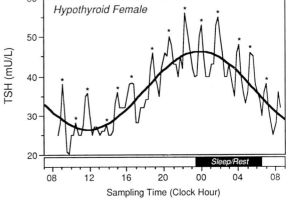

**Fig. 1.** Plasma concentrations of TSH sampled at 15-min intervals in a euthyroid 35-year-old woman during the early follicular phase ($T_4 = 88.8$ nmol/l) and a 48-year-old (moderately) hypothyroid woman ($T_4 = 45.0$ nmol/l) amenorrheic for 3 years. Peaks identified by cluster program of Veldhuis and Johnson (1986). Episodic secretions and circadian rhythm (characterized by best-fitting cosine curve) are seen in both patients

## Multifrequency Rhythms in TSH Secretion

Thyrotropin-stimulating hormone is secreted in a rhythmic fashion with multiple frequencies including short-term variations of less than 1 h (Weeke and Gundersen 1978), episodic or pulsatile secretions with a more or less regular frequency of hours, a circadian rhythm and seasonal variations.

## Pulsatile Secretion of TSH

Thyroid-stimulating hormone is normally secreted from the pituitary gland in a series of discrete pulses (Fig. 1). The amplitude and according to some investigators also the frequency of these pulses increase at night in normal individuals leading to the circadian

elevation of TSH concentration at this time (Goldstein et al. 1981; Sowers et al. 1982; Caron et al. 1986; Greenspan et al. 1986; Rossmanith et al. 1988; Brabant et al. 1990; Samuels et al. 1989, 1990).

The thyrotrophs in the pituitary seem to respond better to intermittent than to continuous TRH stimulation (Snyder and Utiger 1972; Spencer et al. 1980). Normal TSH pulse patterns may be necessary for intact thyroid gland function. Studies with sampling at 10- or 15-min intervals for 24 h have shown an average TSH pulse frequency of 9 pulses/24 h in normal men and women (range 7–12) with a slight but in some studies significant increase in the number of episodes at night. The TSH pulse amplitudes average over the 24-h span 2.9 mU/l in normal individuals (range, 1.1–7.0). However, the mean pulse amplitude increases significantly from 2.2 mU/l during the day to 3.6 mU/l at night (Samuels 1988; Samuels et al.

1990). The relatively high peak values of individual TSH pulses during the evening hours have to be kept in mind since the values reached during this time may be slightly above the usually accepted normal range. The TSH pulses were found to be statistically significantly concordant with the pulses of LH, FSH, and the alpha-subunit of the glycoprotein hormones, suggesting that an underlying unified signal coordinates the pulsatile hormone secretion from both gonadotrophs and thyrotrophs (Samuels et al. 1990).

*Changes in Pulse Patterns by Fasting and Drugs.* Prolonged fasting leads to a lowering of the 24-h mean plasma TSH concentration with decrease in mean TSH pulse amplitude. The TSH pulse frequency during fasting remained unchanged (Romijn et al. 1990). The impact of fasting is most impressive during the time of the nocturnal TSH surge with decrease also in the circadian TSH amplitude (Spencer et al. 1983). The decline in TSH secretion induced by starvation is probably not caused by an alteration in endogenous TRH release as infusions of TRH are unable to prevent the reduction of plasma TSH concentrations during fasting (Spencer et al. 1983). Starvation, metoclopramide infusion (Rossmanith et al. 1988) and alterations of sleep patterns (Brabant et al. 1990) all cause significant changes in TSH pulse amplitude but no change in pulse frequency. Also many medications which suppress plasma TSH levels like glucocorticoids, dopamine and somatostatin lower the TSH pulse amplitude with little effect on pulse frequency. It thus appears that the intrinsic TSH pacemaker rhythm represents a rather persistent biologic phenomenon. Patients taking these medications do not usually develop overt thyroid dysfunctions and continue to have measurable TSH levels although they may be as low as 0.1 mU/l especially with high doses of glucocorticoids. Abnormalities in TSH pulsatile release can also be found in extrathyroidal disease states, e.g., chronic renal failure (Wheatley et al. 1989).

*Pulse Patterns in Primary and Secondary Hypothyroidism.* Patients with primary hypothyroidism (thyroid gland failure) and elevated baseline serum TSH levels have a normal number of TSH pulses over 24 h but the TSH pulse amplitude is increased (Samuels 1988, Samuels et al. 1990). A night-time increase in TSH pulse amplitude may be preserved when TSH levels are moderately high (e.g., 10–60 mU/l) but is markedly decreased or lost when TSH levels reach or surpass 80 mU/l. Patients with moderate primary hypothyroidism may in the course of their episodic vari-

ations have spontaneous changes in serum TSH concentrations of 20 mU/l or more. A change in serum TSH concentration of this extent may, therefore, not necessarily indicate a change in the patient's clinical condition.

Patients with disorders of the pituitary or hypothalamus who develop hypothyroidism as a result of TSH deficiency may show inappropriately low or normal circadian mean plasma TSH concentrations. Some of these patients may have a normal average 24-h TSH pulse frequency and amplitude, but fail to have a night-time increase in pulse amplitude. It thus appears that a loss of the usual nocturnal variations in TSH amplitude may be sufficient to cause clinical hypothyroidism (Samuels 1988).

*Pulse Patterns in Hyperthyroidism.* In hyperthyroidism secondary to intrinsic thyroid disease, a pulsatile secretory pattern could not be recognized by older techniques of TSH determination due to suppression and extremely low serum concentrations of TSH. More recently using a highly sensitive immunochemoluminometric assay, Evans et al. (1986) observed a circadian rhythm in plasma TSH values also in patients with hyperthyroidism, suggesting a continuation of the pulsatile release with higher amplitudes during evening and/or night also under the effect of an increased thyroid hormone feedback.

In patients with TSH-induced hyperthyroidism resulting from a TSH-secreting pituitary tumor or from pituitary resistance to thyroid hormone feedback, TSH pulses showed a fairly normal frequency but increased amplitudes (Samuels et al. 1989).

## Circadian Periodicity of TSH

In the study of the circadian periodicity of TSH, longer sampling intervals (e.g., 3 or 4 h) are frequently used which do not allow the characterization of individual TSH pulses. This can lead to a substantial degree of aliasing if single subjects or small groups are studied over a single or over only a few 24-h spans. There is no way to know if a spotcheck as obtained by a blood-drawing taken at a certain time was obtained at the peak, the slope or the trough of a TSH pulse. It can be predicted only with statistical probability that it is more likely to find at certain times higher values than at others due to the differences in pulse amplitude and frequency. Studies of circadian periodicity with sampling intervals which do not allow the recognition of the pulsatile pattern of plasma TSH concentration require a long enough

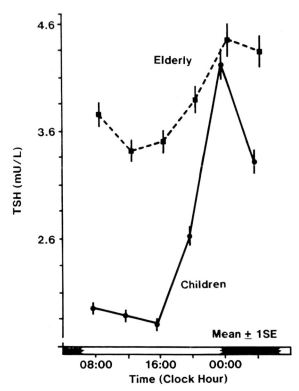

**Fig. 2.** Plasma concentrations of TSH in 193 euthyroid ($T_4 = 105$ nmol/l) children ($11 \pm 1.5$ years of age) and in 277 clinically euthyroid ($T_4 = 88.2$ nmol/l) elderly men and women ($74 \pm 11$ years of age) sampled at 4-h intervals over a 24-h span. The TSH concentrations in the elderly are markedly higher at the circadian trough (i.e., during the daytime) but are similar at the time of the circadian peak (around 0000 hours)

sampling span (numerous periods) in a single individual or a large enough group size to allow a statistically valid quantitative description of a circadian rhythm. With inadequate sample size even apparently rhythmometrically "described" rhythms may be spurious, not reproducible and misleading. Whenever the circadian frequency component in the time structure of TSH production and release is discussed it has to be realized that this rhythm is determined by the sum of the pulsatile episodes which are superimposed upon the underlying circadian oscillation which in turn modulates their amplitude and frequency.

The circadian frequency component of plasma immunoreactive TSH shows peak values during the night and early morning hours with low values during daytime (Fig. 2). The TSH concentrations close to the acrophase are in children more than twice the concentration found at the trough of the circadian rhythm. In elderly subjects, the TSH amplitude was found to be decreased due to an increase of the

trough values during daytime with little difference in the peak values during evening and night.

The circadian TSH acrophase together with that of parameters of thyroid function is shown in Fig. 3 as determined by population mean cosinor in subjects of both sexes ranging in age from childhood to senescence, studied on three continents with comparable methodology. The extent of the circadian variations in plasma TSH concentration, thyroid hormones, and related parameters is shown in Fig. 4, for the same populations as in Fig. 3.

*Sex and Sex Hormones.* Sex differences in the timing of the daily maximum of circulating TSH concentrations were described by some investigators (Golstein et al. 1981). In adult women this maximum was reported to occur in the early morning hours between 0300 and 0600 hours (Vanhaelst et al. 1972) while in men the peak TSH levels have been found by several investigators between 2100 and 0100 hours before the onset of sleep (Parker et al. 1976, 1987; Azukizawa et al. 1976; Lucke et al. 1977; Chan et al. 1979; Sowers et al. 1982; Caron et al. 1986). Others did not find such a sex difference in groups of subjects of different ages studied simultaneously under comparable conditions (Custro and Scaglione 1980; Nicolau et al. 1982, 1984c, 1987; Haus et al. 1988; Nicolau and Haus 1989) or if present were found to be rather minor (Haus et al. 1988). The administration of oral contraceptives does not result in significant changes of the TSH secretory pattern (Van Cauter and Refetoff 1985).

*Age.* The circadian rhythm in TSH in children 5 years of age or older is comparable to that found in adults (Nicolau et al. 1987; Haus et al. 1988; Nicolau and Haus 1989; Rose and Nisula 1989). No data in newborn and children younger than 5 years of age have been found in the available literature. In the elderly, there is a tendency for elevation of plasma TSH concentrations especially during trough time of the circadian rhythm (Fig. 2).

*Sleep.* The circadian rhythm in TSH appears to be independent of sleep. The circadian peak concentrations have been found by some investigators prior to the onset of sleep (Parker et al. 1976, 1987) but by others were reported to occur during the sleep time of the subjects (Weeke 1973; Van Cauter et al. 1974; Nicolau et al. 1982, 1984a, b, c, 1987; Haus et al. 1988; Nicolau and Haus 1989). When subjects were maintained on a 21-h sleep-wake cycle for 12 calendar days the evening elevation remained 24-h periodic, suggesting an endogenous circadian rhythm that per-

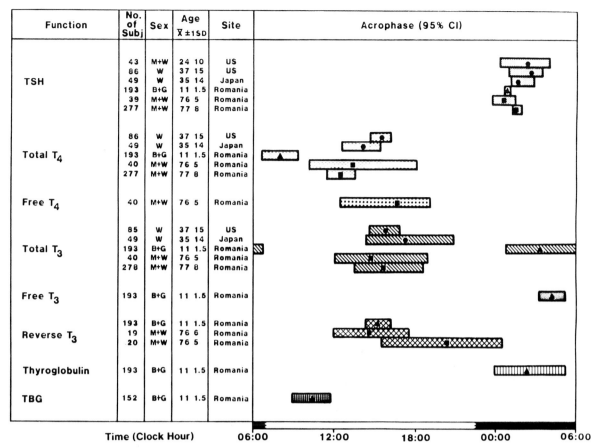

| Function | No. of Subj | Sex | Age X̄ ±1SD | Site | Acrophase (95% CI) |
|----------|-------------|-----|-------------|------|---------------------|
| TSH | 43 | M+W | 24 10 | US | |
| | 86 | W | 37 15 | US | |
| | 49 | W | 35 14 | Japan | |
| | 193 | B+G | 11 1.5 | Romania | |
| | 39 | M+W | 76 5 | Romania | |
| | 277 | M+W | 77 8 | Romania | |
| Total T₄ | 86 | W | 37 15 | US | |
| | 49 | W | 35 14 | Japan | |
| | 193 | B+G | 11 1.5 | Romania | |
| | 40 | M+W | 76 5 | Romania | |
| | 277 | M+W | 77 8 | Romania | |
| Free T₄ | 40 | M+W | 76 5 | Romania | |
| Total T₃ | 85 | W | 37 15 | US | |
| | 49 | W | 35 14 | Japan | |
| | 193 | B+G | 11 1.5 | Romania | |
| | 40 | M+W | 76 5 | Romania | |
| | 278 | M+W | 77 8 | Romania | |
| Free T₃ | 193 | B+G | 11 1.5 | Romania | |
| Reverse T₃ | 193 | B+G | 11 1.5 | Romania | |
| | 19 | M+W | 76 6 | Romania | |
| | 20 | M+W | 76 5 | Romania | |
| Thyroglobulin | 193 | B+G | 11 1.5 | Romania | |
| TBG | 152 | B+G | 11 1.5 | Romania | |

Time (Clock Hour)   06:00   12:00   18:00   00:00   06:00

**Fig. 3.** Circadian acrophase chart of TSH, thyroid hormones and related parameters. The patients were studied at different geographic locations, and all chemical determinations were done in the same laboratory (Haus et al. 1988)

sisted despite the imposed noncircadian schedule (Rossman et al. 1981).

*Thyroid Hormone Feedback.* Although in clinically healthy subjects thyroid hormones do not seem to be responsible for the circadian variation in TSH concentration (Weeke and Gundersen 1978) their contribution is essential for the presence of the circadian rhythm of TSH, which although still present in patients with mild hypothyroidism (Fig. 1), is absent in patients with severe hypothyroidism but can be restored by replacement therapy with thyroxin (Weeke and Laurberg 1976).

*Cortisol, Dexamethazone, Hypercorticism, Somatostatin.* Studies with simultaneous measurements of cortisol and TSH have suggested that the circadian variations of TSH also are not caused by a negative feedback regulation by cortisol (Van Cauter et al. 1974). In spite of the suppressive effect of cortisol on TSH there is no close relationship between the circa-

dian profiles of the two hormones (Van Cauter and Refetoff 1985; Salvador et al. 1985, 1988). Dexamethazone administration does not alter the circadian TSH profile (Chan et al. 1979). However, the circadian rhythm of TSH was found to be abolished in some patients with hypercortisolism (Van Cauter et al. 1974), presumably reflecting the inhibitory action of chronic supraphysiological concentrations of cortisol on TSH secretion.

Somatostatin infusion obliterates the early nocturnal TSH surge without affecting daytime levels (Weeke et al. 1980). Since somatostatin inhibits TRH-stimulated TSH secretion, this observation supports the assumption that the nocturnal TSH peak is most likely caused by a surge in hypothalamic TRH. This assumption is also supported by the observation that the extent of the rise in TSH after intravenous administration of TRH is strongly positively correlated with the height of the nocturnal circadian peak in the same subjects.

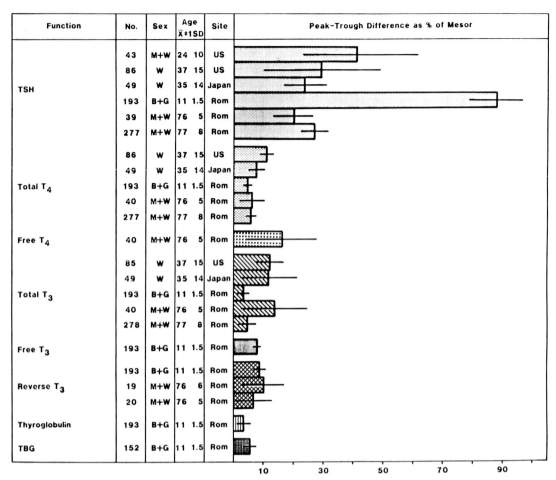

**Fig.4.** Extent of the circadian variation in TSH, thyroid hormones and related parameters in same subjects as shown in Fig.3. The extent of the circadian variation is expressed as the peak-trough difference (of the best-fitting cosine curve in the cosinor analysis = the double amplitude) in percent of the rhythm-adjusted circadian mean (the MESOR)

## Circadian Variation in the Response of the Pituitary to TRH

In clinically healthy men and women, the peak TSH levels as well as the increases in TSH after TRH stimulation were found to be higher at 2300 than at 1100 hours (Scanlon et al. 1980; Rodriguez-Arnao et al. 1982). Also Caroff and Winokur (1984) and Sensi et al. (1988, 1990) reported a significantly higher increase of TSH values after administration of TSH at night. In contrast Tatar et al. (1983) reported after testing at 0800 and 2000 hours in healthy males, who had been fasting for 12 h prior to the study, a more marked increase of TSH after TRH administration in the morning and Nathan et al. (1982) injecting at 0900 and 1800 hours found no difference in response. The higher TSH response reported by Caroff and Winokur (1984) in the evening disappeared after acute inversion of the rest-activity schedule. The question of a circadian cycle in TSH response to TRH will require some additional work with consideration of the season and with sampling at more than two time points and in relation to the sleep-wakefulness/activity-rest schedule rather than clock hour.

## Circadian Chronopathology: TSH secretion

Alterations of the circadian rhythm of plasma TSH have been investigated in a limited number of pathological states. A normal pattern is preserved in patients with mild or moderate primary hypothyroidism as well as in hypothyroid patients treated with thyroxine (Weeke and Laurberg 1976). However, the circadian variation appeared to be abolished in untreated patients with severe primary hypothyroidism

(Caron et al. 1986; Weeke and Laurberg 1976). The evening rise in TSH was found to be absent in central hypothyroidism (Caron et al. 1986; Rose and Nisula 1987). Children with clinically and biochemically euthyroid endemic goiter in the goiter region of Tirgoviste, Romania (Nicolau et al. 1987, 1989), and elderly subjects with type II (non-insulin-dependent) diabetes mellitus (Nicolau et al. 1984a, b) showed a normal circadian rhythm in TSH. Suppression of the nocturnal increase in plasma TSH has been observed during development of ketosis in diabetic patients studied during insulin withdrawal (Schmitz et al. 1981) and in patients with chronic renal failure (Wheatley et al. 1989).

Abnormal circadian variations of TSH have been reported in male and female patients with endogenous depression (Schmitz et al. 1981; Weeke and Weeke 1978) and in this condition may be related to abnormalities in TRH production in the central nervous system (Kirkegaard et al. 1979) with elevated TRH concentrations in the spinal fluid and down-regulation at the pituitary level.

$T_3$ ($fT_3$) during the late night and early morning hours (Nicolau et al. 1987; Haus et al. 1988) similar to a group of subjects reported by Weeke and Gundersen (1978). The elderly subjects in our series showed no statistically significant circadian rhythm in $fT_3$. The acrophase of reverse $T_3$ ($rT_3$) in children was found during the afternoon and in elderly subjects during late afternoon and evening. A circadian rhythm in thyroglobulin (TG) showed in a study of 194 Romanian children the acrophase during the night hours (Nicolau et al. 1987; Haus et al. 1988), while Guagnano et al. (1986) reported an acrophase of TG in four young adult males during the afternoon. Thyroxin-binding globulin (TBG) showed a circadian rhythm in children with a peak during the forenoon (Nicolau et al. 1987; Haus et al. 1988). A comparison of the circadian timing of the HPT axis in children and in elderly subjects showed a slight phase delay in the circadian acrophase of TSH, in contrast to a much more marked phase delay in $TT_3$, $TT_4$, and $rT_3$ in the elderly. The extent of the circadian rhythm in thyroid and thyroid-related functions is shown in Fig. 4.

## Circadian Periodicity of Thyroid Hormones

Low amplitude circadian rhythms of total thyroxine ($TT_4$) and triiodothyronine ($TT_3$) were found, the timing of which varies to some extent between different studies by different investigators and in our laboratory in subjects of different age groups and geographic locations (Fig. 3) (Haus et al. 1988). In most studies the acrophase of $TT_4$ was found during the forenoon or noon hours with lower values during the night (Halberg et al. 1981a; Weeke et al. 1980; Nicolau et al. 1982, 1984c; Haus et al. 1988; Nicolau and Haus 1989). DeCostre et al. (1971) suggested that these low-amplitude variations may be caused by the posture-dependent daily variations of total plasma protein concentrations with higher protein levels during ambulation than during recumbency. However, the acrophase differences in $TT_4$ and $TT_3$ between subjects of different age groups reported by Haus et al. (1988), which were independent of the serum protein variations in the same subjects, make such an interpretation unlikely. The circadian rhythm in free $T_4$ ($fT_4$) showed in elderly subjects a peak during daytime (Fig. 3). The circadian rhythm in $TT_3$ showed in elderly subjects an acrophase in the late afternoon in contrast to a group of children $11 \pm 1.5$ years of age, which showed the acrophase of $TT_3$ and also of free

## Seasonal Variations and/or Circannual Rhythms of the HPT Axis

### Adaptation of the HPT Axis to the Environmental Temperature

The HPT axis shows marked seasonal variations which are to a large degree but probably not exclusively a response to changes in environmental temperature (Lungu et al. 1966; Ogata et al. 1966). The seasonal variation in thyroid hormones, protein-bound iodine (PBI) and related parameters continues in experimental animals when all known environmental factors (i.e., feeding and lighting) except environmental temperature are kept as far as feasible constant (Lungu et al. 1967; Nicolau and Teodoru 1975, 1976). However, also some other environmental factors seem to contribute to the variable circannual amplitude under otherwise comparable environmental conditions (Lungu and Nicolau 1973; Moraru and Nicolau 1977).

Stimulation of TRH secretion by the hypothalamus and of TSH release in response to cold has been reported in animals (Jobin et al. 1975) and in human subjects (Wilber and Baum 1970; Tuomisto et al. 1977; Reed et al. 1986) including newborn infants (Fisher and Odell 1969). Cold exposure has been reported to provoke an increase in thyroid hormone production (Eastman et al. 1974). The changes after acute cold exposure, however, are complex and may show differences in the response of $TT_3$ and $TT_4$ concentrations and of their free fractions. The adaptive response to cold exposure may include changes in thyroid hormone binding capacity and in the equilibrium between extracellular and intracellular $fT_3$ and $fT_4$ (Solter et al. 1989). In a study of chronic cold exposure during and after prolonged residence in Antarctica, Reed et al. (1986) studying their subjects first in the northern hemisphere and then during and after a 42-week stay in Antarctica in the southern hemisphere found an increase in the integrated TSH response to TRH administration by 50% over the warm climate response levels but found no statistically significant changes in the plasma concentrations of $fT_4$, $TT_4$, $fT_3$ and $TT_3$. This study was done without regard to the possibility of an endogenous circannual rhythm in HPT function. Among the factors leading to the discrepancy between this and other studies on cold adaptation may be the interaction between an endogenous circannual rhythm and the cold stimulus examined by the investigators.

A sex difference in the response of TSH during heat acclimatization from a cool-day to a hot-day climate was reported by Buguet et al. (1988). The men (Caucasians as well as Africans) showed during heat acclimatization a decrease in the circadian mesor of TSH while the women did not show such a change.

## Seasonal Variation in Serum TSH Concentrations

Seasonal variations in TSH in euthyroid subjects were reported by numerous investigators (Nicolau et al. 1984c, 1987; Halberg et al. 1981a, 1983; Hugues et

al. 1983; Lagoguey and Reinberg 1981; Haus et al. 1988). The seasonal peak in plasma TSH concentration, however, was quite variable in different studies. It was found by Halberg et al. (1981b), in adult women studied in Minnesota, United States, and in Kyushu, Japan, during winter; by Lagoguey and Reinberg (1981), in young adult men in France during spring and by Nicolau et al. (1984c, 1987) and by Haus et al. (1988) in elderly women and in children with and without endemic goiter in Romania during summer and fall. In the Romanian children the circannual acrophase in TSH did precede the circannual acrophase of the thyroid hormones, suggesting that the seasonal variation of the latter is not a direct result of an increase in plasma TSH concentrations (Fig. 5). The circannual TSH peak found in France in spring (Lagoguey and Reinberg 1981; Hugues et al. 1983) and in Romania during the early fall (Nicolau et al. 1984c, 1987; Haus et al. 1988) makes a temperature dependence of the circannual rhythm of TSH unlikely.

## Seasonal Variation in Response of TSH to TRH

In experimental animals, seasonal differences in response to TRH injection were reported for serum TSH and thyroid hormone concentrations (Khurana and Madan 1986). Seasonal differences to melatonin injection were found in rats in the response of $T_3$, $T_4$, TSH, and in $I^{131}$ thyroid uptake (Rom-Bugoslavskaya and Shcherbakova 1986).

During treatment with constant doses of L-$T_4$, patients with primary hypothyroidism show a seasonal variation in plasma TSH concentrations with the peak during winter (Konno and Morikawa 1982) and

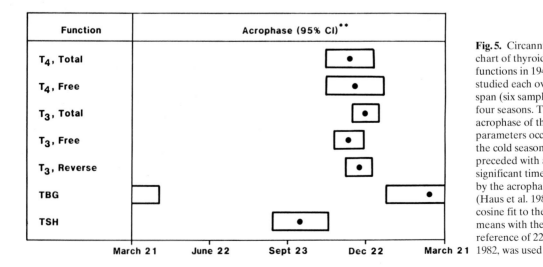

**Fig. 5.** Circannual acrophase chart of thyroid and related functions in 194 children studied each over one 24-h span (six samples) during all four seasons. The circannual acrophase of the thyroid parameters occurs during the cold season but is preceded with a statistically significant time difference by the acrophase of TSH (Haus et al. 1988). A single cosine fit to the circadian means with the phase reference of 22 December, 1982, was used

the 95% confidence region of the acrophase extending from December to February. Also, the TSH responses to TRH were in hypothyroid patients higher in winter. However, the seasonal variation in TSH response found in the hypothyroid patients was not recognizable in clinically healthy subjects (Nagata et al. 1976; Konno 1978, 1980; Hedstrand and Wide 1973). Konno and Morikawa (1982) ascribed the seasonal variation in TRH response in the treated hypothyroid patients to an alteration in thryoid hormone feedback. While basal TSH concentrations were within the usual range in summer they increased significantly during winter with five of the ten patients having supranormal values. In view of the constant dose of L-$T_4$ given it seems possible that $T_4$ utilization may be stimulated in the cold (Golstein-Golaire et al. 1970; Ingbar et al. 1954) leading to a drop in $T_4$ and a resultant increase in TSH production. This mechanism would not be operational in the same way in intact human subjects in whom the plasma thyroid hormone concentrations rise to higher levels and peak during the winter.

## Seasonal Variation in Thyroid Hormone Concentrations

In human populations living under climatic conditions showing seasonal variations in temperature (irrespective of the yearly temperature average) most investigators report a seasonal variation in serum or plasma $TT_3$ and $TT_4$ concentrations inversely related with the seasonally changing environmental temperature (Smals et al. 1977; Halberg et al. 1981a; Nicolau et al. 1984c; Haus et al. 1988; Perez et al. 1980; Oddie et al. 1979). Seasonal variations of other indices of thyroid function with a high during the cold season have been reported for basal metabolic rate (Osiba 1957), serum protein-bound iodine concentrations (Osiba 1957; DuRuisseau 1965; Lungu and Nicolau 1973; Nicolau and Teodoru 1975), radioiodine uptake (Lewitus et al. 1964) and urinary $T_3$ and $T_4$ excretion (Rastogi and Sawhney 1976).

While in children (Fig.5) (Nicolau et al. 1987), the circannual variations in thyroid parameters in moderate climates are highly reproducible, elderly subjects showed by ANOVA no seasonal variation in $TT_4$ but did show a circannual rhythm in TSH with the highest values found during summer and fall (Nicolau et al. 1984c; Haus et al. 1988; Nicolau and Haus 1989). $TT_3$ showed in the elderly a seasonal variation with a peak during winter (Haus et al. 1988). The apparent absence of a group-synchronized seasonal variation in

thyroid hormones raises the question of a defect in cold adaptation in the elderly, but phenomena like desynchronization of a circannual rhythm within the group or a free-running circannual rhythm cannot be excluded.

Studies in a subtropical climate with little or no seasonal temperature variations failed to demonstrate a seasonal variation in circulating $TT_3$ and $TT_4$ concentrations (Rastogi and Sawhney 1976; Halberg et al. 1983).

It is of interest for the question of a fetal-maternal synchronization of circannual periodicity that higher values of serum $T_4$ and TSH were found in newborn infants born in winter as compared with those born during summer (Rogowski et al. 1974). Similar results were reported by Oddie et al. (1979) in umbilical cord serum.

## Seasonal Variation in Serum Thyroglobulin Concentration

Guagnano et al. (1986) reported in clinically healthy adults in Italy that thyroglobulin shows a circannual variation with acrophase in mid-May with a 95% confidence interval ranging from April to July. In Romanian children no such variation was found (Nicolau et al. 1987; Haus et al. 1988).

## Seasonal Chronopathology of the Thyroid

### Seasonal Variation in Thyroid Volume and Endemic Goiter Incidence

The seasonal variations and/or circannual rhythms in HPT axis function are accompanied by seasonal changes in thyroid morphology and thyroid pathology. The thyroid volume measured ultrasonically in clinically healthy men changes by approximately 23% with minimum values during summer and maximum values in winter (Hegedüs et al. 1987). Similarly the incidence of endemic goiter among schoolchildren in southern Tasmania, Australia, was lower during the (local) winter with an increase in later spring despite constant and apparently adequate iodine prophylaxis (Gibson et al. 1960).

## Seasonal Variation in Iodine Uptake and/or Metabolism

Jonckheer et al. (1982) found by X-ray fluorescence in patients with nontoxic endemic goiter in Belgium a seasonal variation in stable intrathyroidal iodine with the acrophase during April/May and the lowest values in September/October. Romanian children with and without endemic goiter showed a comparable circannual rhythm in $TT_4$, $fT_4$, $TT_3$ $fT_3$, $rT_3$ and thyroxin-binding globulin (TBG) with high values in late fall and low values in spring and with low serum TSH concentrations in spring and peak TSH values during early fall (Nicolau et al. 1987). The same children showed a seasonal variation in iodine excretion, with higher values in late December and March and a low in late June and late September, when the values indicated borderline iodine deficiency (Nicolau et al. 1989). Neither in the Belgian nor in the Romanian subjects were there known dietary factors which could have been responsible for this seasonal variation although such factors can at this time not be excluded.

In the United Kingdom, where a similar seasonal variation in iodine excretion was found with lower values in summer as compared to winter (Broadhead et al. 1965), this variation was ascribed to a seasonal variation in the iodine content of cow's milk (Broadhead et al. 1965; Wenlock et al. 1982), which was thought to be due almost entirely to the high iodine content of food supplements given to the animals during the winter months (Broadhead et al. 1965). Farther down in the food chain this leads to a substantially increased human iodine consumption during this time of the year (Nelson and Phillips 1985).

## Seasonal Variation in Incidence of Thyrotoxicosis

In the endemic goiter area of Tasmania, Australia (Connolly et al. 1970; Connolly 1971), but also in New Zealand (Ford 1988) and in the United Kingdom (Phillips et al. 1985), an increased incidence of thyrotoxicosis was reported during (local) spring and summer and a decreased incidence during (local) winter. This seasonal variation in the incidence of thyrotoxicosis was observed in Tasmania after iodinization of the bread, which was constant throughout the year. The spring-summer increase in the clinical diagnosis of thyrotoxicosis in England and Wales in contrast was ascribed to the seasonal variation in iodine uptake, the peak of which was reported to precede the clinical manifestations by several months (Phillips et al. 1985). The elevated incidence in this report, however, extended in a sustained plateau throughout the second and third quarter of the year with a decrease from a high value in August to low values in November. These observations which are supported by laboratory data confirm earlier clinical observations in Germany (Hirsch 1930; Jacobowitz 1932) and Denmark (Iversen 1948) all of which had reported a spring-summer excess in onset and presentation of thyrotoxicosis without information on possible seasonal variations in iodine intake.

## Other Seasonal HPT Chronopathology (Subacute Thyroiditis, TSH in Prostatic Cancer)

Subacute thyroiditis, a self-limiting inflammatory disease of the thyroid, presumably of viral origin, shows a seasonal variation with high prevalence in summer and a peak in July-August corresponding to that of established infections due to some enteroviruses (Martino et al. 1987). In patients with prostatic cancer, the seasonal variations in TSH were absent in contrast to concomitantly studied clinically healthy men of the same age (Halberg et al. 1981b).

## Conclusions

The HPT axis is characterized by multifrequency rhythmic variations at all levels. The hypothalamic oscillators serving as pacemakers for the HPT axis are subject to two-way or multiple-way interactions with most known neurotransmitter systems. The neurotransmitter most centrally involved in the regulation of the HPT axis, TRH, apparently has a broad spectrum of functions within and without the central nervous system.

Thyrotropin-releasing hormone and in consequence TSH are secreted in a pulsatile fashion with changes in pulse amplitude and frequency leading to or being an expression of the circadian rhythm component of the system. Pulsatile secretion of TSH seems to be necessary for maintaining "normal" thyroid function. The circadian rhythm in TSH is of sufficiently high amplitude to have clinical implications. In its measurement by a few spotchecks, however, the superimposed pulsatile variations can lead in individual patients to a substantial degree of aliasing of the circadian pattern.

The circadian rhythms of $TT_3$ and $TT_4$ are of low amplitude and do not pose diagnostic problems. The circadian rhythms in $fT_3$ and $fT_4$ show a relatively higher amplitude and have shown more variability in their rhythm parameters.

The pulsatile and the circadian oscillations are endogenous in nature and presumably genetically fixed. Although the pulse amplitude and frequency of TSH increases during the evening and/or night hours no more precise synchronization of the pulsatile secretions with environmental factors or other frequencies is known. The circadian rhythm of the HPT system on the other hand is synchronized by the activity-rest schedule with our periodic surroundings.

The seasonal variations observed in different parameters of the HPT axis raise the question of their environmentally induced or endogenous nature. The available information suggests that both important exogenous factors and an endogenous circannual rhythm may be involved. Of the environmental factors temperature seems to be of primary importance since the seasonal variation continues under a constant diet and lighting regimen throughout the year. Also the decrease in amplitude or lack of a demonstrable seasonal variation in subtropical climates speaks for a primarily exogenously induced reactive periodicity. However, in analogy to other circannual rhythms, e.g., that of the adrenal or of cell proliferation in numerous organs, the exogenously induced seasonal variation may serve as a synchronizer for an endogenous circannual rhythm. Seasonal variations in iodine intake and/or metabolism may also play a role and may be related to certain forms of thyroid pathology, i.e., endemic goiter and thyrotoxicosis.

# References

Aizawa T, Hinkle PM (1985) Differential effects of thyrotropin releasing hormone, vasoactive intestinal peptide, phorbol ester, and depolarization in GH4C1 rat pituitary cells. Endocrinology 116: 909–919

Azukizawa M, Pekary AE, Hershman JM, Parker DC (1976) Plasma thryotropin, thyroxine, and triiodothyronine relationships in man. J Clin Endocrinol Metab 43: 533–542

Bennett GW, Marsden CA, Fone KCF, Johnson JV, Heal DJ (1989) TRH-catecholamine interaction in brain and spinal cord. Ann NY Acad Sci 533: 106–120

Berelowitz M, Perlow MJ, Hoffman HJ, Frohman LA (1981) The diurnal variation of immunoreactive thyrotropin-releasing hormone and somatostatin in the cerebrospinal fluid of rhesus monkey. Endocrinology 109: 2102–2109

Brabant G, Brabant A, Ranft U, Ocran K, Kohrle J, Hesch RD, von zur Muhlen A (1987) Circadian and pulsatile thyrotropin secretion in euthyroid man under the influence of thyroid hormone and glucocorticoid administration. J Clin Endocrinol Metab 65: 83–88

Brabant G, Prank K, Ranft U, Schuermeyer T, Wagner TOF, Hauser H, Kummer B, Feistner H, Hesch RD, von zur Muhlen A (1990) Physiological regulation of circadian and pulsatile thyrotropin secretion in normal man and woman. J Clin Endocrinol Metab 70: 403–409

Brammer GL, Morley JE, Geller E, Yuwiler A, Hershman JH (1979) Hypothalamus-pituitary-thyroid axis interactions with pineal gland in the rat. Am J Physiol 236: E416–420

Broadhead GC, Pearson IB, Wilson GM (1965) Seasonal changes in iodine metabolism. I. Iodine content of cows' milk. Br Med J 5431: 343–348

Buguet A, Gati R, Souboni JP, Hanniquet AM, Livecchi-Gonnot G, Bittel J (1988) Seasonal changes in circadian rhythms of body temperatures in humans living in a dry tropical climate. Eur J Appl Physiol 58: 334–339

Caroff S, Winokur A (1984) Hormonal response to thyrotropin-releasing hormone following rest-activity reversal in normal men. Biol Psychiatry 19: 1015–1025

Caron PJ, Nieman LK, Rose SR, Nisula BC (1986) Deficient nocturnal surge of thyrotropin in central hypothyroidism. J Clin Endocrinol Metab 62: 960–964

Chan V, Jones A, Liendoch P, McNeilly A, Landon J, Besser GM (1979) The relationship between circadian variations in circulating thyrotropin, thyroid hormones and prolactin. Clin Endocrinol (Oxf) 9: 337–349

Collu R, DuRuisseau P, Tache Y, Durcharme JR (1977) Thyrotropin-releasing hormone in rat brain: nyctohemeral variations. Endocrinology 100: 1391–1393

Connolly RJ (1971) Seasonal variation in thyroid function. Med J Aust 1: 633–636

Connolly RJ, Vidor GI, Stewart JC (1970) Increase in thyrotoxicosis in endemic goiter area after iodation of bread. Lancet 1: 500–502

Covarrubias L, Uribe RM, Mendes M, Charli JL, Joseph-Bravo P (1987) Hypothalamic TRH mRNA regulation under different physiological conditions. Ann NY Acad Sci 553: 476–478

Covarrubias L, Uribe RM, Mendes M, Charli JL, Joseph-Bravo P (1988) Neuronal TRH synthesis: developmental and circadian TRH mRNA levels. Biochem Biophys Res Commun 151: 615–622

Custro N, Scaglione R (1980) Circadian rhythm of TSH in adult men and women. Acta Endocrinol (Copenh) 95(4): 405–471

DeCostre P, Buhler U, DeGroot LJ, Refetoff S (1971) Diurnal rhythm in total serum thyroxine levels. Metabolism 20: 782–791

Dekin MS, Richerson GB, Gatting PA (1985) Thyrotropin-releasing hormone induces rhythmic bursting in neurons of the nucleus tractus solitarius. Science 229: 67–69

DeLean A, Ferland L, Drouin J, Kelly PA, Labrie F (1977) Modulation of pituitary thyrotropin-releasing hormone receptor levels by estrogens and thyroid hormones. Endocrinology 100: 1496–1504

DuRuisseau JP (1965) Seasonal variation of PBI in healthy Montrealers. J Clin Endocrinol Metab 25: 1513–1515

Eastman CJ, Ekins RP, Leith IM, Williams ES (1974) Thyroid hormone response to prolonged cold exposure in man. J Physiol 24: 175–181

Evans PJ, Weeks I, Jones MK, Woodhead JS, Scanlon MF (1986) The circadian variation of thyrotropin in patients with primary thyroidal disease. Clin Endocrinol (Oxf) 24: 393–398

Fisher DA, Odell WD (1969) Acute release of thyrotropin in the newborn. J Clin Invest 48: 1670–1677

Ford HC (1988) Seasonality of thyrotoxicosis in Wellington. N Z Med J 101: 72–73

Gershengorn MC (1978) Biohormonal regulation of the thyrotropin-releasing hormone receptor in mouse pituitary thyrotropic tumor cells in culture. J Clin Invest 62: 937–943

Gershengorn MC, Marcus-Samuels BE, Geras E (1979) Estrogens increase the number of thyrotropin-releasing hormone receptors on mammotropic cells in culture. Endocrinology 105: 171–176

Gibson HB, Howeler JF, Clements FW (1960) Seasonal epidemics of endemic goiter in Tasmania. Med J Aust 47: 875–880

Golstein J, Vanhaelst L, van Cauter E (1981) Physiological factors influencing nyctohemeral pattern of TSH. In: van Cauter E, Copinschi G (eds) Human pituitary hormones: circadian and episodic variations. Nijhoff, The Hague, pp 118–131

Golstein-Golaire J, Vanhaelst L, Bruno OD, Leclercq R, Copinschi G (1970) Acute effects of cold on blood levels of growth hormone, cortisol, and thyrotropin in man. J Appl Physiol 29: 622–626

Greenspan SL, Kilbanski A, Schoenfeld D, Ridgway EC (1986) Pulsatile secretion of thyrotropin in man. J Clin Endocrinol Metab 63: 661–668

Guagnano MT, Angelucci F, Cervone L, Menduni P, Sensi S (1986) Oscillazioni circadiane e circannuali della tireoglobulina nell uomo adulto sano. Boll Soc Ital Biol 62: 315–320

Halberg F, Cornelissen G, Sothern RB, Wallach LA, Halberg E, Ahlgren A, Kuzel M, Radke A, Barbosa J, Goetz F, Buckley J, Mandel J, Shuman L, Haus E, Lakatua D, Sackett L, Berg H, Kawasaki T, Ueno M, Uezono K, Matsuoka M, Omae T, Tarquini B, Cagnoni M, Garcia Sainz M, Perez VE, Griffiths K, Wilson D, Donati L, Tatti P, Vasta M, Locatelli I, Camagna A, Lauro R, Tritsch G, Wetterberg L, Wendt HW (1981 a) International geographic studies of oncological interest on chronobiologic variables. In: Kaiser HE (ed) Neoplasms – comparative pathology of growth in animals, plants, and man. Williams and Wilkins, Baltimore, pp 553–596

Halberg F, Tarquini B, Lakatua D, Halberg E, Seal U, Haus E, Cagnoni M (1981 b) Circadian and circannual plasma TSH rhythm, human mammary and prostatic cancer and step toward chrono-oncoprevention. Lab J Res Lab Med 8: 251–257

Halberg F, Lagoguey M, Reinberg A (1983) Human circannual rhythms over a broad spectrum of physiological processes. Int J Chronobiol 8: 225–268

Haus E, Nicolau GY, Lakatua D, Sackett-Lundeen L (1988) Reference values for chronopharmacology. Annu Rev Chronopharmacol 4: 333–424

Hedstrand H, Wide L (1973) Serum thyrotropin and lipid levels in summer and winter (Letter). Br Med J 4: 420

Hegedüs L, Rasmussen N, Knudsen N (1987) Seasonal variation in thyroid size in healthy males. Horm Metab Res 19: 391–392

Hinkle PM (1989) Pituitary TRH receptors. Ann NY Acad Sci 553: 176–187

Hinkle PM, Tashjian AH (1975) Thyrotropin releasing hormone regulates the number of its own receptors in the GH3 strain of pituitary cells in culture. Biochemistry 14: 3845–3851

Hinkle PM, Perrone MH, Greer TL (1979) Thyroid hormone action in pituitary cells. Differences in the regulation of thyrotropin-releasing hormone receptors and growth hormone synthesis. J Biol Chem 254: 3907–3911

Hirsch D (1930) Beitrag zum Basedowproblem. Dsch Arch Klin Med 168: 331–346

Hökfelt T, Tsuruo Y, Ulfhake B, Cullheim S, Arvidson V, Foster GA, Schultzberg M, Schalling M, Arborelius L, Freedman J, Post C, Visser T (1989) Distribution of TRH-like immunosensitivity with special reference to coexistence with other neuroactive compounds. Ann NY Acad Sci 553: 76–105

Hugues JN, Reinberg A, Lagoguey M, Modigliani E, Sebaoun J (1983) Les rythmes biologiques de la secretion thyréotrope. Ann Med Interne (Paris) 134: 83–94

Ingbar SH, Kleeman CR, Quinn M, Bass DE (1954) The effect of prolonged exposure to cold on thyroid function in man. Clin Res Proc 2: 86

Iversen K (1948) Temporary rise in the frequency of thyrotoxicosis in Denmark, 1941–1945. Roseekilde and Bagger, Copenhagen, p 68

Jackson IMD (1982) Throtropin releasing hormone. N Engl J Med 306: 145–155

Jackson IMD, Reichlin S (1979) Distribution and biosyntheses of TRH in the nervous system. In: Collu R, Barbeau A, Ducharme JR, Rochefort J-G (eds) Central nervous system effects of hypothalamic hormones and other peptides. Raven, New York, pp 3–54

Jacobowitz H (1932) Basedow und Jahreszeit. Z Klin Med 122: 307–315

Jobin M, Ferland L, Cote J, Labrie F (1975) Effect of exposure to cold on hypothalamic TRH activity and plasma levels of TSH and prolactin in the rat. Neurendocrinology 18: 204–212

Jonckheer M, Coomans D, Broeckaert I, van Paepegem R, Deconinck F (1982) Seasonal variation of stable intrathyroidal iodine in nontoxic goiter disclosed by x-ray fluorescence. J Endocrinol Invest 5: 27–31

Jordan D, Veisserire M, Mornex R (1981) Controle aminergique du rythme circadien de la TSH serique chez le rat. Ann Endocrinol (Paris) 42: 21

Jordan D, Poncet C, Veisseire M, Mornex R (1983) Role of GABA in the control of thyrotropin secretion in the rat. Brain Res 268: 105–110

Kerdelhue B, Palkovits M, Karteszi M, Reinberg A (1981) Circadian variations in substance P, luliberin (LH-RH) and thyroliberin (TRH) contents in hypothalamic and extrahypothalamic brain nuclei of adult male rats. Brain Res 206: 405–413

Kerr DJ, Singh VK, McConway MG, Beastall GH, Connell JMC, Alexander WD, Davies DL (1987) Circadian variation of thyrotropin, determined by ultrasensitive immunoradiometric assay, and the effect of low dose nocturnal dopamine infusion. Clin Sci 72: 737–741

Khurana ML, Madan ML (1986) Seasonal influence on thyroidal response to thyrotropin-releasing hormone in cattle and buffalo. J Endocrinol 108: 57–61

Kirkegaard C, Faber J, Hummer L, Rogowski P (1979) Increased levels of TRH in cerebrospinal fluid from patients with endogenous depression. Psychoneuroendocrinology 4: 227–235

Koivusalo F, Leppaluoto J (1979) Brain TRF immunoreactivity during various physiological and stress conditions in the rat. Neuroendocrinology 29: 231–236

Konno N (1978) Comparison between the thyrotropin response to thyrotropin releasing hormone in summer and that in winter in normal subjects. Endocrinol Jpn 25: 635–639

Konno N (1980) Reciprocal changes in serum concentrations of triiodothyronine and reverse triiodothyroinine between summer and winter in normal adult men. Endocrinol Jpn 27: 471–476

Konno N, Morikawa K (1982) Seasonal variation of serum thyrotropin concentration and thyrotropin response to thyrotropin-releasing hormone in patients with primary hypothyroidism on constant replacement dosage of thyroxine. J Clin Endocrinol Metab 54: 1118–1124

Krulich L, Vijayan E, Coppings RL (1979) On the role of the central serotoninergic system in the regulation of the secretion of TSH and PRL: TSH inhibiting and prolactin releasing effects of 5-hydroxytryptamine and quipazine in the male rat. Endocrinology 105: 276–283

Lagoguey M, Reinberg A (1981) Circadian and circannual changes of pituitary and other hormones in healthy human males: their relationship with gonadal activity. In: van Cauter E, Copinschi G (eds) Human pituitary hormones: circadian and episodic variations. Nijhoff, The Hague, pp 261–278

Lewitus Z, Hasenfratz J, Toor M, Massry S, Rabinowitch E (1964) [131]I uptake studies under hot climatic conditions. J Clin Endocrinol Metab 24: 1084–1086

Lucke C, Herhmann R, von Mayersbach H, von zur Muhlen A (1977) Studies on circadian variations of plasma TSH, thyroxine and triiodothyronine in man. Acta Endocrinol (Copenh) 86: 81–88

Lungu A, Nicolau GY (1973) Circannual rhythm of thyroid function in rat. Rev Roum Endocrinol 10: 365–372

Lungu A, Nicolau GY, Cocu F, Teodoru V, Dinu I (1966) Protein-bound iodine variations and spontaneous atmospheric temperature oscillations. Rev Roum Endocrinol 3: 279–282

Lungu A, Cocu F, Mitrache L, Teodoru V, Nicolau GY, Dinu I (1967) Influenta oscilatiilor temperaturii ambiante asupra colesterolului seric si a iodului hormonal din singe. Culegere de Lucrari ale Institutului Meteorologic pe Anul 1965, Bucuresti, pp 503–510

Magee B, Sheridan B, Scanlon MF, Atkinson AB (1986) Inappropriate thyrotrophin secretion, increased dopaminergic tone and preservation of the diurnal rhythm in serum TSH. Clin Endocrinol (Oxf) 24: 209–215

Mannisto PT, Pakkanen J, Ranta T, Koivusalo F, Leppaluoto J (1978) Diurnal variations of medial basal and anterior hypothalamic thyroliberin (TRH) and serum thyrotropin (TSH) concentrations in male rats. Life Sci 23: 1343–1350

Marsden CA, Bennett GW, Fone KCF, Johnson JV (1989) Functional interactions between TRH and 5-hydroxytryptamine (5-HT) and prolactin in rat brain and spinal cord. Ann NY Acad Sci 553: 121–134

Martino E, Bambini G, Vaudagna G, Breccia M, Baschieri L (1985) Effects of continuous light and dark exposure on hypothalamic thyrotropin-releasing hormone in rats. J Endocrinol Invest 8: 31–33

Martino E, Buratti L, Bartalena L, Mariotti S, Cupini C, Aghini-Lombardi F, Pinchera A (1987) High prevalence of subacute thyroiditis during summer season in Italy. J Endocrinol Invest 10 (3): 321–323

Mess B, Ruzsas C, Rekasi Z (1986) Central monoaminergic and opioidergic regulation of thyroid function and its ontogenic differentiation. Monogr Neural Sci 12: 117–127

Metcalf G, Jackson IMD (eds) (1989) Thyrotropin releasing hormone: biomedical significance. Ann NY Acad Sci 553: 1–631

Moraru S, Nicolau GY (1977) Relationships between the small ions in the atmosphere and the circadian and circannual biorhythm of the rabbit and rat thyroid. Rev Roum Med Endocrinol 15: 189–194

Morley JE (1981) Neuroendocrine control of thyrotropin secretion. Endocr Rev 2: 396–436

Nagata H, Izumiyama T, Kamata K, Kono S, Yukimura Y, Tawata M, Aizawa T, Yamada T (1976) An increase of plasma triiodothyronine concentration in man in a cold environment. J Clin Endocrinol Metabol 43: 1153–1156

Nathan RS, Sachar EJ, Tabrizi MA, Asnis GM, Halbreich U, Halpern FS (1982) Diurnal hormonal responses to thyrotropin-releasing hormone in normal man. Psychoneuroendocrinology 7: 235–238

Nelson M, Phillips DI (1985) Seasonal variations in dietary iodine intake and thyrotoxicosis. Hum Nutr Appl Nutr 39: 213–216

Nicolau GY, Haus E (1989) Chronobiology of the endocrine system. Rev Roum Med Endocrinol 27: 153–183

Nicolau GY, Teodoru V (1975) Les variations saisonnieres de la fonction thyroidienne. Rev Roum Biol Biol Anim 20: 141–145

Nicolau GY, Teodoru V (1976) Zone critique de la température atmosphérique pour l'adaptation thyroïdienne chez les bovins. Rev Roum Biol Biol Anim 21: 45–48

Nicolau GY, Haus E, Lakatua D, Bogdan C, Petrescu E, Sackett-Lundeen L, Berg H, Ioanitu D, Popescu M, Chioran C (1982) Endocrine circadian time structure in the aged. Rev Roum Med Endocrinol 20: 165–176

Nicolau GY, Haus E, Lakatua D, Bogdan C, Petrescu E, Robu E, Sackett-Lundeen L, Swoyer J, Adderley J (1984a) Circadian time structure of endocrine and biochemical parameters in adult onset (type II) diabetic patients. Rev Roum Med Endocrinol 22: 227–243

Nicolau GY, Haus E, Lakatua D, Dumitriu L, Bogdan C, Sackett-Lundeen L, Stelea P, Stelea S, Petrescu E (1984b) Circadian rhythm of TSH in adult onset non-insulin dependent (type II) diabetics with altered thyroid state. Rev Roum Med Endocrinol 22: 117–124

Nicolau GY, Lakatua DJ, Sackett-Lundeen L, Haus E (1984c) Circadian and circannual rhythms of hormonal variables in clinically healthy elderly men and women. Chronobiol Int 1: 301–319

Nicolau GY, Dumitriu L, Plinga L, Petrescu E, Sackett-Lundeen L, Lakatua D, Haus E (1987) Circadian and circannual variations of thyroid function in children 11 ± 1.5 years of age with and without endemic goiter. Prog Clin Biol Res 227B: 229–247

Nicolau GY, Haus E, Dumitriu L, Plinga L, Lakatua D, Ehresman D, Adderley J, Sackett-Lundeen L, Petrescu E (1989) Circadian and seasonal variations in iodine excretion in children with and without endemic goiter. Rev Roum Med Endocrinol 27: 73–86

Oddie TH, Klein AH, Foley TP, Fisher DA (1979) Variation in values for iodothyroinine hormones, thyrotropin and thyroxine-binding globulin in normal umbilical cord serum with season and duration of storage. Clin Chem 25: 1251–1253

Ogata K, Sasaki T, Murakami N (1966) Central nervous and metabolic aspects of body temperature regulation. Bull Inst Const Med Kumamoto Univ [Suppl] 16: 1–67

Osiba S (1957) The seasonal variation of basal metabolism and activity of thyroid gland in man. Jpn J Physiol 7: 335–365

Parker DC, Pekary AE, Hershman JM (1976) Effect of normal and reversed sleep-wake cycles upon nyctohemeral rhythmicity of plasma thyrotropins: evidence suggestive of an inhibitory influence in sleep. J Clin Endocrinol Metab 43: 318–329

Parker DC, Rossman LG, Pekary AE, Hershman JM (1987) Effect of 64-hour sleep deprivation on the circadian waveform of thyrotropin (TSH): further evidence of sleep-related inhibition of TSH release. J Clin Endocrinol Metab 64: 157–161

Pearse AG, Takor TT (1979) Embryology of the diffuse neuroendocrine system and its relationship to the common peptides. Fed Proc 38: 2288–2294

Pekary AE, Meyer NV, Vaillant C, Hershman JM (1980) Thyrotropin-releasing hormone and a homologous peptide in the male rat reproductive system. Biochem Biophys Res Commun 95: 993–1000

Perez PR, Lopez JG, Mateos IP, Escribano AD, Sanchez MLS (1980) Seasonal variations in thyroid hormones in plasma. Rev Clin Esp 156: 245–247

Phillips DI, Barker DJ, Morris JA (1985) Seasonality of thyrotoxicosis. J Epidemiol Community Health 39: 72–74

Prasad C (1985) Thyrotropin releasing hormone. In: Lajtha A (ed) Handbook of neurochemistry, vol 8. Plenum New York, pp 175–200

Rastogi RK, Sawhney RC (1976) Thyroid function in changing weather in a subtropical region. Metabolism 25: 903–908

Reed HL, Burman KD, Shakir KMM, O'Brian JT (1986) Alterations in the hypothalamic-pituitary-thyroid axis after prolonged residence in Antarctica. Clin Endocrinol (Oxf) 25: 55–65

Rodriguez-Arnao MD, Weithman DR, Hall R, Scalon MF, Camporro JM, Gomez-Pan A (1982) Reduced dopaminergic inhibition of thyrotropin release in states of physiological hyperprolactinemia. Clin Endocrinol (Oxf) 17: 15–19

Rogowski P, Siersbaek-Nielsen K, Hansen M (1974) Seasonal variation in neonatal thyroid function. J Clin Endocrinol Metab 39: 919–922

Rom-Bugoslavskaya ES, Shcherbakova VS (1986) Seasonal changes in melatonin influence on thyroid function. Bull Eksp Biol Med 101: 268–269

Romijn JA, Adriannse R, Brabant G, Prank K, Endert E, Wiersinger WM (1990) Pulsatile secretion of thyrotropin during fasting: a decrease of thyrotropin pulse amplitude. J Clin Endocrinol Metab 70: 1631–1636

Rose SR, Nisula BC (1987) Deficient nocturnal thyrotropin surge in children with central hypothyroidism (Abstr 291). 69th Meeting of the Endocrine Society

Rose SR, Nisula BC (1989) Circadian variation of thyrotropin in childhood. J Clin Endocrinol Metab 68: 1086–1090

Rossman LG, Parker DC, Pekary AE, Hershman JM (1981) Effect of an imposed 21 hour sleep-wake cycle upon the rhythmicity of human plasma thyrotropin. In: van Cauter E, Copinschi G (eds) Human pituitary hormones: circadian and episodic variations. Nijhoff, The Hague, pp 96–113

Rossmanith WG, Mortola JF, Laughlin GA, Yen SS (1988) Dopaminergic control of circadian and pulsatile pituitary thyrotropin release in women. J Clin Endocrinol Metab 67: 560–564

Salvador J, Wilson DW, Harris PE, Peters JR, Edwards C, Foord SM, Dieguez C, Hall R, Scanlon MF (1985) Relationships between the circadian rhythms of TSH, prolactin and cortisol in surgically treated microprolactinoma patients. Clin Endocrinol (Oxf) 22: 265–272

Salvador J, Dieguez C, Scanlon MF (1988) The circadian rhythms of thyrotropin and prolactin secretion. Chronobiol Int 5: 85–93

Samuels M (1988) Neuroendocrine control of TSH secretion. Proceedings of the American Thyroid Association National Meeting, Montreal

Samuels M, Wood WM, Gordon DF, Kleinschmidt-Demasters

BK, Lillehei K, Ridgway EC (1989) Clinical and molecular studies of a thyrotropin secreting pituitary adenoma. J Clin Endocrinol Metab 68: 1211–1215

Samuels MH, Lillehei K, Kleinschmidt-Demasters BK, Stears J, Ridgway EC (1990) Patterns of pulsatile pituitary glycoprotein secretion in central hypothyroidism and hypogonadism. J Clin Endocrinol Metab 70: 391–395

Scanlon MF, Weetman AP, Lewis M, Pourmand M, Rodriguez-Arnao MD, Weightman DR, Hall R (1980) Dopaminergic modulation of circadian thyrotropin rhythms and thyroid hormone levels in euthyroid subjects. J Clin Endocrinol Metab 51: 1251–1256

Schaeffer JM, Brownstein MJ, Axelrod J (1977) Thyrotropin-releasing hormone-like material in the rat retina: changes due to environmental lighting. Proc Natl Acad Sci USA 74: 3579–3581

Schmitz O, Alberti KG, Hreidarsson AB, Laurberg P, Weeke G, Orskov H (1981) Suppression of the night increase in serum TSH during development of ketosis in diabetic patients. J Endocrinol Invest 4: 403–407

Segerson TP, Kauer J, Wolfe HC, Mobtaker H, Wu P, Jackson IMD, Lechan RM (1987) Thyroid hormone regulates TRH biosynthesis in the paraventricular nucleus of the rat hypothalamus. Science 238: 78–80

Sensi S, Capani F, DeRemigis PL, Guagnan MT, Casulini G, d'Emilio A, Rugge L, Nepu A, Bucciarelli R, Menduni P, Sensi SL, Vitullo R (1988) Circadian time structure of pituitary and adrenal responsiveness to CRF, TRH and LHRH (Abstr). Chronobiologia 15: 253

Sensi S, Capani F, DeRemigis PL, Guagnan MT (1990) Circadian time structure of pituitary and adrenal responsiveness to CRF, TRH and LHRH (Abstr). 2nd World Conference on Clinical Chronobiology, Monte Carlo

Shambaugh G III, Kubek M, Wilber JF (1979) Thyrotropin-releasing hormone activity in the human placenta. J Clin Endocrinol Metab 48: 483–486

Simard M, Pekary AE, Smith VP, Hershman JM (1989) Thyroid hormones modulate thyrotropin-releasing hormone biosynthesis in tissues outside the hypothalamic-pituitary axis of male rats. Endocrinology 125: 529–531

Smals AGH, Ross HA, Kloppenborg PWC (1977) Seasonal variation in serum $T_3$ and $T_4$ levels in man. J Clin Endocrinol Metab 44: 998–1001

Snyder PY, Utiger RD (1972) Response to thyrotropin releasing hormone (TRH) in normal man. J Clin Endocrinol Metab 34: 380–385

Solter M, Brkic K, Petek M, Posavec L, Sekso M (1989) Thyroid hormone economy in response to extreme cold exposure in healthy factory workers. J Clin Endocrinol Metab 68: 168–172

Sowers JR, Catania RA, Hershman JM (1982) Evidence for dopaminergic control of circadian variations in thyrotropin secretion. J Clin Endocrinol Metab 54: 673–675

Spencer CA, Greenstadt MA, Wheeler WS, Kletzky OA, Nicoloff JT (1980) The influence of a long-term low dose thyrotropin-releasing hormone infusions of serum thyrotropin and prolactin concentrations in man. J Clin Endocrinol Metab 51: 771–775

Spencer CA, Lum SM, Wilber JF, Kaptein EM, Nicoloff JT (1983) Dynamics of serum thyrotropin and thyroid hormone changes in fasting. J Clin Endocrinol Metab 56: 883–888

Stanton TL, Beckman AL, Winokur A (1981) Thyrotropin-releasing hormone effects in the central nervous system: dependence on arousal state. Science 214: 678–681

Tashjian AH Jr, Osborne R, Maina D, Knaian A (1977) Hydrocortisone increases the number of receptors for thyrotropin-releasing hormone on pituitary cells in culture. Biochem Biophys Res Commun 79: 333–340

Tatar P, Strbak MR, Vigas M (1983) Different response of TSH to TRH in the morning and evening. Horm Metab Res 15: 461

Tuomisto J, Mannisto P, Lamberg BA, Linnoila M (1977) Effect of cold exposure on serum thyrotropin levels in man. Acta Endocrinol (Copenh) 83: 522–527

Van Cauter E, Refetoff S (1985) Multifactorial control of the 24-hour secretory profiles of pituitary hormones. J Endocrinol Invest 8: 381–389

Van Cauter E, Leclercq R, Vanhaelst L, Golstein J (1974) Simultaneous study of cortisol and TSH daily variations in normal subjects and patients with hyperadrenalcorticism. J Clin Endocrinol Metab 39: 645–652

Vanhaelst L, van Cauter E, Deguate JP, Golstein J (1972) Circadian variations of serum thyrotropin levels in man. J Clin Endocrinol Metab 35: 479–482

Veldhuis JD, Johnson ML (1986) Cluster analysis: a simple, versatile and robust algorithm for endocrine pulse detection. Am J Physiol 250: E486–493

Vijayan E, McCann SM (1978) Effects of intraventricular gamma aminobutyric acid (GABA) on plasma growth hormone and thyrotropin in conscious ovariectomized rats. Endocrinology 103: 1888–1893

Weeke J (1973) Circadian variation of serum thyrotropin levels in normal subjects. Scand J Clin Lab Invest 31: 337–342

Weeke J, Gundersen HJG (1978) Circadian and 30 minute variations in serum TSH and thyroid hormones in normal subjects. Acta Endocrinol (Copenh) 89: 659–672

Weeke J, Laurberg P (1976) Diurnal TSH variations in hypothyroidism. J Clin Endocrinol Metab 43: 32–37

Weeke A, Weeke J (1978) Disturbed circadian variation of serum thyrotropin in patients with endogenous depression. Acta Psychiatr Scand 57: 281–289

Weeke J, Christensen SE, Hansen AP, Laurberg P, Lundbaek K (1980) Somatostatin and the 24-h levels of serum TSH, $T_3$, $T_4$ and reverse $T_3$ in normals, diabetics and patients treated for myxoedema. Acta Endocrinol (Copenh) 94: 30–37

Wenlock RW, Buss DH, Moxon RE, Bunton NG (1982) Trace nutrients. IV. Iodine in British food. Br J Nutr 47: 381–390

Wheatley T, Clark PMS, Clark JDA, Holder R, Raggatt PR, Evans DB (1989) Abnormalities of thyrotropin (TSH) evening rise and pulsatile release in haemodialysis patients: evidence for hypothalamic-pituitary changes in chronic renal failure. Clin Endocrinol (Oxf) 31: 39–50

Wilber JF, Baum D (1970) Elevation of plasma TSH during surgical hypothermia. J Clin Endocrinol Metab 31: 372–375

# The Pineal

J. Arendt

## Introduction

The inclusion of the pineal gland as a separate chapter, rather than a passing reference, in a book devoted to matters of clinical interest is witness to the enormous interest in the function of this organ in recent years. The investigation of the pineal gland in animals and humans has led not only to an understanding of its physiology but also to major advances in our understanding of biological rhythms and their importance in humans. Moreover, current research points to therapeutic approaches based on the fundamental science of pineal function, of importance particularly in psychiatry and occupational health.

The human pineal, like that of many mammals, tends to calcify during development. Most adults show evidence of more or less calcification [1]. In the past, this phenomenon led clinicians to asume that the gland was functionless and merely useful as a radiological marker. We now know that calcification does not relate to decreased secretory activity during adult life [2]. The gland remains active throughout the human life span albeit with declining function in very old age. A clear physiological role for the human pineal is not yet established with certainty. We may nevertheless make informed judgements as to its likely function based on a substantial body of knowledge derived largely from animal work. Most importantly, perhaps, we may use the characteristics of pineal secretion of melatonin (5-methoxy-$N$-acetyl tryptamine) to assess pathological processes accompanied by disturbed rhythmic function.

## Possible Functions of the Human Pineal

In order to assess the role of the human pineal, it is necessary to refer to its known function in animals. The gland is essentially concerned with the transmission of environmental information to body physiology [3, 4]. It is thereby intimately concerned with the control of rhythmic functions, be they seasonal or circadian. Although there is some evidence for pineal transmission of information about temperature variations [5], by far the most important input to the pineal is that of light-dark transitions which, in nonequatorial regions, indicate day length or photoperiod. Many seasonal species use photoperiod as the primary environmental input to, reproduction, puberty, coat growth, behaviour and other time-dependent events [4, 6, 7]. Removal or denervation of the pineal destroys this ability to perceive changes in day length [6, 7]. With a sufficiently long life span, seasonal events continue but are desynchronised from the yearly periodicity of the environment [8]. Thus the pineal appears to be a zeitgeber (synchroniser) of an underlying endogenous annual rhythm of seasonality. There is no doubt that pineal secretion of melatonin conveys the environmental message. Melatonin is secreted during the dark phase of the day in a normal environment [9]. Removal or denervation of the pineal abolishes rhythmic production [10, 11]. The rhythm is endogenous [12] and both entrained [13] and suppressed [14, 15] by light, i. e. with a sufficiently long day the duration of nighttime secretion is shortened by light either in the morning or the evening or possibly, in some cases, both morning and evening. Essentially, therefore, melatonin production correlates positively to the length of the night. It is generally accepted that the duration of melatonin secretion encodes the photoperiodic message and suitable administration of melatonin is, in all respects, able to mimic the effects of a given photoperiod [3, 4].

With respect to circadian rhythms, the pineal is of very considerable importance in some birds and lower vertebrates, acting in some cases as a master rhythm generator and in others as a coupling agent. In mammals the role of the pineal is more subtle. Pinealectomy in rodents enhances the rate of the circadian rest-activity rhythm adaptation to forced phase-shift of the light-dark cycle ([16] and references therein): suitably timed melatonin administration likewise either enhances the rate of adaptation and/or changes the direction of re-entrainment [16]. This apparent paradox may be resolved if one considers that rate of adaptation is positively and specifically related to the rate of change of the melatonin rhythm. Superposition of exogenous melatonin may in effect counteract the retarding effect of the endogenous rhythm and indeed there is evidence for more rapid entrainment of the endogenous rhythm in the presence of carefully timed exogenous administration. To date, only phase advances induced by exogenous melatonin have been described in rats [17], and these only by administration during a narrow window in the late subjective day. There is no doubt that in rodents the maternal pineal transmits photic entraining information to the developing circadian system ([18] cited in [118]).

Considering, therefore, these known characteristics of pineal function in animals, it is possible to propose similar functions in humans. Evidently the importance of the pineal in humans depends on the extent to which we are responsive to changes in light-dark cycles. Any physiological or pathological features which are demonstrably photoperiod-dependent in humans are likely to involve the pineal gland in a coordinating role. There is no doubt that bright light is of considerable importance as a circadian zeitgeber in humans [19]. Seasonality in humans is well-described [20]; however, to what extent human seasonal rhythms are photoperiod-dependent is largely unknown. The best example is perhaps that of seasonal affective disorder (S.A.D. or winter depression) in which appropriate treatment with bright light is a spectacularly successful therapy [21]. Nonetheless, even here there is considerable controversy as to the mode of action of bright light and this may not involve 'photoperiodic' mechanisms at all.

Classical techniques used to demonstrate the function of endocrine glands (e.g. ablation) are evidently impossible in normal humans, although some surgically pinealectomised patients are currently under study in a number of centres. Moreover, unequivocal demonstration of pineal-dependent effects is a very long-term commitment insofar as seasonal rhythms are concerned. Accidents of nature, such as the occurrence of pineal tumours, are always informative and in this regard pineal tumours are frequently associated with abnormal pubertal development [22]. It may be that we shall have to infer rather than to demonstrate the role of the pineal in humans. As a general but subtle coordinator of mammalian rhythm

physiology the pineal should be implicated in the timing of a multitude of functions, of which the most important are probably sleep, reproductive function, mood and the immune system.

## Melatonin Production and Metabolism in Humans

Pathways involved in the production of melatonin have primarily been investigated in rodents. Tryptophan, an essential dietary amino acid, is hydroxylated in the 5 position to 5-hydroxytryptophan, which is subsequently decarboxylated to serotonin. N-acetylation of serotonin followed by O-methylation of $N$-acetyl serotonin completes the synthesis of melatonin (Fig. 1) [23]. Serotonin $N$-acetyl transferase (NAT) is usually considered to be the rate-limiting enzyme in melatonin production in view of its massive increase in activity at night. This may not necessarily be true in all circumstances and certainly mass action effects of inreasing serotonin availability have been described [24].

The pineal is innervated by postganglionic fibres from the superior cervical ganglion (SCG); there is additional evidence for central innervation, however, sympathetic input governs the synthesis and secretion of melatonin [1]. The transmitter appears to be noradrenaline (NA) acting via $\beta_1$-receptors with some $\alpha$-receptor participation [25]. In humans there is good evidence for most of these assertions. $\beta_1$-receptor antagonists will suppress human melatonin production [26, 27]. The NA uptake inhibitor desmethylimipramine stimulates early onset of melatonin secretion when given in the later afternoon [28] and the serotonin uptake inhibitor fluvoxamine greatly increases the amplitude of melatonin secretion [24]. An $\alpha$-receptor antagonist, prazosin, slightly decreases melatonin secretion [29]. Most monoamine oxidase inhibitors appear to increase melatonin production, although it is possible that specifically monoamine oxidase B inhibitors are effective (see [30] for references). $\gamma$-Amino butyric acid (GABA) receptors have been found in rodent pineals as has vasoactive intestinal peptide (VIP) innervation [31, 32]. The benzodiazepine alprazolam will with inhibit human melatonin production and VIP is stimulatory [33, 34].

In those animal species studied to date, the rhythm of melatonin production is generated in the suprachiasmatic nucleus (SCN) [12]. Entrainment by the light-dark cycle occurs through the retino-hypotha-

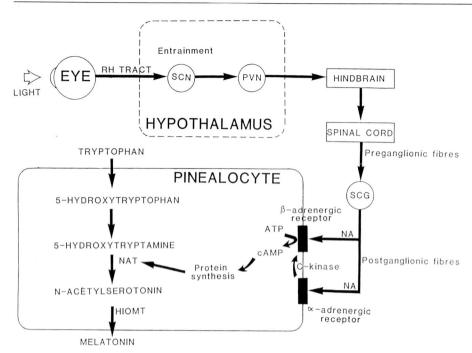

**Fig. 1.** Neural and biochemical pathways controlling melatonin production. HIOMT, Hydroxyindole-O-methyl-transferase; for other abbreviations see text

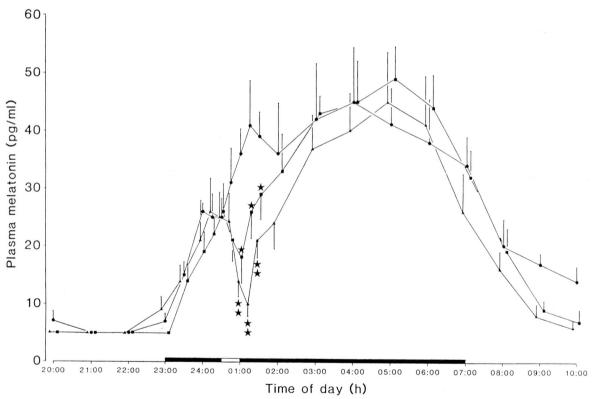

**Fig. 2.** Circadian rhythm of melatonin in humans and suppression of melatonin production by light. The *black bar* indicates the dark phase when subjects (*n* = 5) were kept in light intensity of < 0.1 lux (●—●) or received 30 min light between 24.30 and 0100 hours of 300 lux (■—■) or 2500 lux (▲—▲); 300 lux significantly suppresses melatonin but 2500 lux is more effective. (From [41] with permission)

lamic (RH) projection to the SCN (Fig.1). The paraventricular nucleus (PVN) acts as a relay station [35] before neural pathways convey the signal to the SCG. Although there is no direct evidence for SCN-PVN involvement in human melatonin production, other aspects of these pathways in humans are consistent with animal work. Some blind people may have free-running, phase delayed or phase advanced melatonin rhythms [36, 37]. Most information relates to free-running rhythms in the blind, thus underlining the importance of light-dark cycles in human circadian physiology. In pathological conditions with sympathetic degeneration, e.g. diabetic autonomic neuropathy, Shy-Drager syndrome and cervical cord lesion, the rhythmic production of melatonin is lost, although continuous low-level secretion probably occurs (see [38] for references).

The effects of light on the control of human melatonin production have been of very particular interest. Early attempts to suppress melatonin production with light were unsuccessful [39]; however, Lewy et al., using a light intensity ( > 2500 lux) higher than in previous experiments, found that such bright light (akin to natural light which ranges up to 100000 lux) was indeed able fully to suppress melatonin in humans, although partial suppression can be found with lower intensities (Fig. 2) [40, 41]. This observation has led to a re-evaluation of the role of light in human physiology, in particular as a circadian zeitgeber and to new treatments for depression.

Wever and coworkers [19] have shown very clearly that this intensity of light, considerably higher than in all previous human rhythm experiments, is a strong zeitgeber, both in extending the range of entrainment and in speeding up the rate of adaption to a forced phase-shift of a number of 'strong' oscillator variables [42].

More recently, our group, that of Lewy and others have demonstrated the ability of suitably timed bright light treatment to phase-shift the melatonin rhythm. It is not yet known, however, what the limits of light intensity and duration are for entrainment of human melatonin. Certainly 500 lux (approximately domestic intensity) is insufficient to modify the rhythm in the presence of conflicting alternative zeitgebers, such as sleep, meal times and social cues [43].

Light as a seasonal zeitgeber in humans has only been explored in one study [43]. Broadway et al. [43] investigated the effect of a 'skeleton' spring photoperiod, 1 h of 2500 lux at 0700–0800 hours and 1830–1930 hours, on the melatonin rhythm during the Antarctic winter. An advance phase-shift was induced comparable to that seen in summer (Fig.3). Thus a

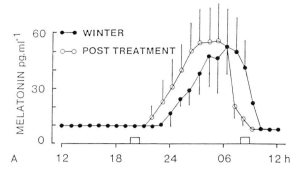

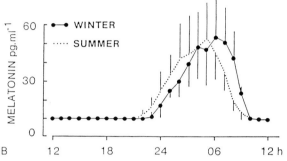

**Fig.3 A, B.** Human plasma melatonin response to bright ( > 2500 lux) light treatment for 2 h daily during the Antarctic winter. Two sessions of 1 h light were given daily for 6 weeks at the times shown by *open blocks*. The response to light treatment is comparable to the phase advance found in summer compared to winter. **A** 24 h mean plasma melatonin (*n* = 5) measured at the winter solstice (●) and after 6 weeks of light treatment following the winter solstice (○); **B** measured at the winter solstice (●) and at the summer solstice (...). Standard errors are shown. A clear phase advance of approximately 2 h is seen (*p* < 0.0001) following light treatment, similar to the phase advance seen in summer. No significant effect of dim ( < 500 lux) light was found. (From [43] with permission)

natural seasonal change can be induced by photoperiodic manipulation.

## Physiological Variations of Melatonin Secretion

In humans, as in virtually all other species studied, melatonin production occurs at night in a normal environment. Daytime levels tend to be around the limit of detection of most assays (below 10 pg/ml). The evening rise is initiated around 2100 hours; maximum values occur usually from 0100 to 0500 hours, dropping to daytime levels by 1000 hours. Obviously these statements apply to average profiles. One of the most important facets of melatonin secretion is the extremely reproducible pattern observed from day to day in the same individual in contrast to the very large inter-individual variation [38]. In our experience individual variation of maximum melatonin levels varies from < 10 pg/ml to 250 pg/ml in apparently healthy subjects. Thus it is necessary, when comparing melatonin levels between groups, to use very large numbers of subjects.

The major melatonin metabolite 6-sulphatoxymelatonin (aMT6s) is an excellent reflection, both quantitative and qualitative, of melatonin production in plasma and in urine [44]. As with melatonin there is remarkable intra individual consistency and inter-individual variation (Fig.4). Urine measurement of this compound is particularly useful, in that long-term noninvasive studies can be performed in populations otherwise difficult to sample, for example children, and in field studies. It is simple to use individuals as their own controls in these circumstances and thus to reduce the number of subjects required to make observations. For most purposes, in healthy individuals aMT6s can be used interchangeably with melatonin as an index of pineal function. In addition to the daily rhythm of melatonin production, there is a seasonal phase-shift with a phase advance in summer (Fig.3) in both temperate [45, 46] and polar [43] latitudes. According to some authors [47] there is a small increase in duration of secretion in winter and a bimodal distribution of daytime values with peaks in summer and winter, albeit of very low amplitude [46]. It is possible that the phase delay in winter is due to the weakened zeitgeber input of the short photoperiod, leading to weakened coupling with the light-dark cycle. With the natural free-running period of most human rhythms being > 24 h this would lead to a phase delay. Menstrual influence on melatonin is controversial. Several reports suggest low values prior to ovulation and phase-shifts during the cycle. Others report no change (see [38] for references).

There is certainly evidence from animal work that low melatonin precedes ovulation [48] and that the pineal is involved in the circadian timing mechanisms controlling the LH surge [49, 50]. In view of the strong relationship between the pineal and seasonality it is of interest that the LH surge in humans has a circadian timing component with a seasonal phase-shift [51]. Melatonin certainly has the potential to modifiy ovarian and testicular function at many levels of the hypothalamic-pituitary-gonadal axis. Increased amplitude and duration of melatonin secretion are seen in hypothalamic amenorrhea, but any causal relationships remain to be established [52].

During very early life the mesor and amplitude of the plasma melatonin rhythm is very low, rises from 0 to 3 years and slowly declines thereafter to reach adult levels in the late teens [53, 54]. This decline has been related to pubertal development, assuming an anti-gonadotrophic function of the pineal. Although undoubtedly a negative correlation exists between melatonin production and sexual development in humans, a causal relationship has never been demonstrated. In rats, however, injection of melatonin during one of two 'windows' during the subjective day will retard development of the hypothalamus-pituitary-gonadal axis [55]. This effect may well relate to the inhibitory effects of short photoperiods in rats. There is some evidence for abnormal pubertal development in blind people and here again we may invoke a possible pineal influence [20].

In fact the decline in plasma melatonin during de-

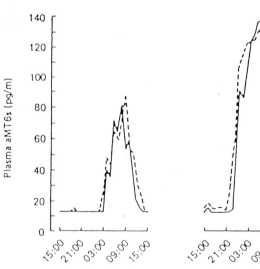
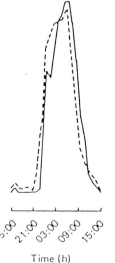
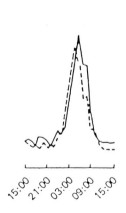

**Fig.4.** Reproducibility of the aMT6s rhythm in human plasma. Three male subjects were sampled at hourly intervals for 24 h on two occasions, autumn and winter, in Antarctica. Nighttime samples (2300–0800 hours) were taken in dim red light. The *solid line* represents the first sampling and the *dashed line* the second sampling for each subject. These individual profiles are taken from the mean autumn and winter profiles in [43]

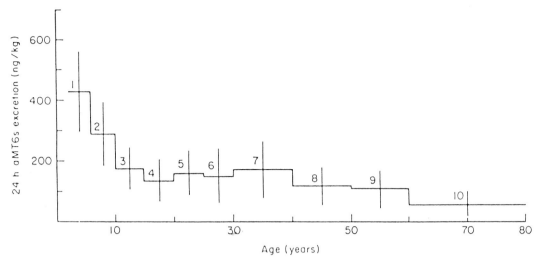

**Fig. 5.** Variation of 24 h aMT6s excretion with age in ten age groups expressed as a function of body weight (means ± SD, ng aMT6s/kg). Group 1: 2–5.9 ($n = 10$); group 2: 6–9.9 ($n = 9$); group 3: 10–14.9 ($n = 12$); group 4: 15–19.9 ($n = 9$); group 5: 20–24.9 ($n = 31$); group 6: 25–29.9 ($n = 15$); group 7: 30–39.9 ($n = 16$); group 8: 40–49.9 ($n = 9$); group 9: 50–59.9 ($n = 8$); group 10: 60–80 ($n = 11$). Differences between groups were as- sessed by analysis of variance and further statistical significance was determined by Duncan's multiple-range test. Significant differences were found between group 1 and all other groups ($p < 0.001$); group 2 and groups 3–10 ($p < 0.01$–$p < 0.001$); group 10 and groups 1–3 and 5–7 ($p < 0.05$–$p < 0.001$). (From [3] with permission)

velopment is only reflected in declining aMT6s con- centrations in urine if the latter are expressed per unit body weight (Fig. 5) [2, 56]. Thus from its early maxi- mum, pineal output probably varies little in early years and variations are accounted for by other factors such as growth. There may also be some steroid suppression of melatonin production as the gonads develop.

During the bulk of adult life melatonin in plasma [54] and aMT6s in urine [2] remain rather constant. In old age, however, there is a highly significant de- cline in the amplitude of production [2, 57, 58]. This is probably one of the most consistent observations in pineal research (Fig. 5). If melatonin does indeed function as a coordinator of biological rhythms, then disruption of the circadian system in old age may well relate to this decline in amplitude.

The possible pulsatile nature of melatonin secre- tion has been the subject of a number of reports. It seems likely that secretion is pulsatile, particularly in view of animal work to this effect involving sampling from the confluens sinus [59]; however, pulses are dif- ficult to detect in the peripheral circulation.

### Pathological Changes in Melatonin Secretion

A multitude of clinical studies purport to describe modifications of melatonin production in conditions as varied as psoriasis and diabetes. Given the inter- esting relationships between the pineal and the im- mune system, this is perhaps not surprising. It would be difficult to review the entire field in the space available. Rather, selected aspects will be treated which, in the opinion of the author, are most impor- tant and/or interesting, and reference to comprehen- sive reviews will be given.

### Psychiatry

Without a doubt, most of the clinical attention to the pineal occurs in psychiatry. The mechanisms control- ling melatonin production involve indoleamines and catecholamines, adrenergic receptors, cyclic adeno- sine monophosphate (cAMP), methylation, light sen- sitivity, and central rhythm-generating systems, all of which are of considerable interest to psychiatrists, particularly those concerned with affective disorders.

Abnormalities in circadian rhythms have been postulated [60]. Specifically, disruption of sleep, espe- cially early waking, early-onset episodes of rapid eye movement (REM) and possible earlier timing of the cortisol and temperature rhythms are associated with endogenous depression. Diurnal variation in mood may also come into this category.

Periodic episodes of depression and mania, particularly the rapid cycling variety, may relate to disorders of the timekeeping systems. An 'internal desynchronisation' hypothesis has been advanced whereby two central rhythm-generating systems with slightly different periods might generate a 'beat' phenomenon manifested as periodic clinical symptoms [61, 62].

The oscillatory structure of the circadian system is a matter for considerable debate; however, there is some evidence that the melatonin rhythm results from a strongly endogenous oscillator(s) also concerned with the generation of cortisol and temperature rhythms [63]. Thus we might expect to find differently timed melatonin rhythms in unipolar and bipolar disorders if any of these ideas are correct. In general, this does not appear to be the case, with the possible exception of S.A.D [64]. There is undoubtedly some evidence for a decline in the amplitude of the melatonin rhythm in depression [65, 66, 67]. This may reflect abnormal coupling of the circadian system, but other factors may be involved (see below) and not all studies are consistent [68].

A non-uniform seasonal distribution of affective disorders has been known for a considerable time. It has been shown in large populations that the incidence of depression peaks in spring, with a secondary peak in autumn [69]. One theory proposed to explain this phenomenon is related to the rapid change in length of day at these times. As the light-dark cycle is a major synchroniser of circadian rhythms it is possible that coupling of the circadian system is less robust at the equinoxes [70] and most particularly after the forced phase-shift of **artificial summer time.**

If such is the case, then abnormal phase relationships within the circadian system should be demonstrable, for example, melatonin in relation to the **sleep-wake cycle.** No consistent observations have been made; however, it is conceivable that very small changes are present and are important but extremely difficult to detect. The melatonin rhythm itself is highly reproducible within the same individual in the normal population [38]. Careful investigation of this reproducibility and its relation to other rhythms within individual patients would be worthwhile, if difficult to accomplish. A further category of depression with a seasonal incidence is S.A.D. In fact these patients may be **bipolar** in view of reported mild spring mania following winter depressive episodes [62]. The treatment originally proposed for this disorder was the creation of an artificial summer day length using 3 h of bright full-spectrum light (Vitalite, 2500 lux) morning and evening [71, 21]. The **'melatonin hypothesis'** predicted that such light treatment would shorten the duration of melatonin secretion, thus

generating a summer day length signal, by analogy with animal work. The light treatment appears to be efficient, even though the very existence of this syndrome is questioned by some, and a major placebo effect has also been proposed. Nonetheless, it does not appear necessarily to work through melatonin. For example, light treatment in the middle of the day had no effect on the melatonin rhythm but induced remission of symptoms [72]. Checkley et al. [62] administered light treatments of different duration (3 h and 1 h) morning and evening in such a way that an identical effect (delayed onset of secretion) was seen on melatonin, but only the longer duration of light was an effective treatment.

Rosenthal et al. [73] used a $\beta$-blocker (atenolol) to suppress melatonin secretion in S.A.D. patients but did not find a significant improvement in mood. However, atenolol has many effects other than suppression of melatonin, and the precise profile of secretion was unfortunately not assessed.

Another series of observations in a group of S.A.D. patients suggested that the melatonin rhythm itself was delayed compared with that in normal subjects [64]. Bright light in the morning in this group of patients shifted the delayed melatonin rhythm backwards to a more normal phase position and was more efficient than evening light in treating this group. The inference was made that the shift in melatonin may relate to the treatment efficacy. Indeed, these same authors [74] have proposed phase typing the melatonin rhythm of patients with possible rhythm disorder and using this information to design appropriate rhythm-shifting treatments (especially bright light). This approach is of considerable theoretical interest but remains an isolated observation. Moreover a number of other observations should be taken into account. Normal volunteers have a phase delayed melatonin rhythm in winter compared with that in summer [43, 45, 46]. Bright light will also shift the melatonin rhythm in healthy men but has little effect on mood [43].

Pharmacological doses of melatonin at different times of day given to normal volunteers have not been shown to modify **mood** in a number of reports. One study, however, found that giving gigantic doses of melatonin during the daytime exacerbated depression in a small group of subjects [75]. In contrast, when used as a possible treatment for **jet lag** travelling eastward, evening melatonin administration (phase advanced) tended to improve mood, at the same time shifting the endogenous melatonin rhythm to an earlier phase position [76] (see below).

Patients with S.A.D. may have delayed, advanced

or normal melatonin rhythms such that when grouped they appear to compare well with the general population. However, we are still short of information in this regard. So far there is little information as to the relationship of the melatonin rhythm to other circadian variables such as cortisol and temperature in the same individual with S. A. D., and the maintained overall structure of the circadian system may be of much greater importance than observations of a single rhythm such as melatonin. It is possible that light functions as a synchroniser to strengthen the coupling of the system at all times of the day. It remains debatable, however, whether good internal synchronisation of rhythms is important for well-being or not.

Most pharmacological antidepressant treatments will increase the availability of catecholamines and/or serotonin by acting through uptake mechanisms, increased precursor formation or inhibition of catabolism. A suggested functional deficiency in biogenic amine function has proved to be a highly productive hypothesis. Moreover, observations in depressed patients fairly consistently support a state of serotonin deficiency [77]. Functional deficiency of NA, serotonin or both has the potential to lower pineal melatonin production. There are conflicting observations on this subject [68], but the weight of paper probably favours decreased melatonin in depression [65, 66, 67]. One report also suggests state-dependent higher levels in mania [78].

Of greater interest, perhaps, is the response of melatonin to different antidepressant treatments. Most monoamine oxidase inhibitors appear to increase production, as do several amine reuptake inhibitors and precursors (see [30] for references). The data are not yet consistent enough to suggest a link between an increase in melatonin production and efficacy of treatment but this possibility certainly merits exploration. The melatonin response may also serve to differentiate between the modes of action of different agents. For example, a single dose of DMI (desmethylimipramine, 100 mg dose at 1600 hours), primarily a NA uptake inhibitor, advances the onset of melatonin secretion in the evening in normal people, leading to a longer duration of secretion [28]. A single doese of fluvoxamine, a serotonin uptake inhibitor, does not alter the timing of onset, but increases the amplitude and duration of melatonin secretion in normal volunteers [24]. It may be that the onset of secretion is dependent on availability of NA and the amplitude on the availability of serotonin in healthy volunteers.

One area of promise using melatonin as a psychiatric probe is concerned with the light sensitivity of bipolar patients. The complete suppression of melatonin production in healthy volunteers at night requires rather bright light (2500 lux) [40], although partial suppression may be achieved with 300–500 lux (Fig. 2) [41]. Early reports suggest that in bipolar patients complete suppression may be achieved with 500 lux [79]. These observations require further confirmation. They may prove to be of both diagnostic and therapeutic importance. Familial sensitivity is currently being investigated as a possible genetic marker [80]. Lithium treatment appears to reduce light sensitivity in bipolar patients [81], hence this phenomenon may be strongly related to the manifestation of the disease. An involvement of biological rhythms is again suggested – possibly over-responsiveness of the circadian system to the unpredictable exposure to artificial light in domestic environments. In this area we must, however, tread with care; light sensitivity is determined by a number of factors and especially previous light exposure [14]. We will have most particularly to take into account people's lifestyles – outdoor workers see up to 100000 lux on a sunny day, while office workers in winter rarely see more than 1000 lux [82].

## Oncology

The pineal and in particular melatonin has an inhibitory effect on the growth of a number of different tumour types, including dimethylbenzanthracene (DMBA)-induced mammary tumours, prostatic tumours and induced prolactin-secreting pituitary tumours in rats, although the effects are by no means always consistent and some authors have reported opposite effects (see [83] for reviews). Most observations are complicated by the known time- and photoperiod-dependent nature of the effects of melatonin. In many cases these factors have not been properly addressed. Some of the most convincing work, by Tamarkin and colleagues [84] and Blask and co-workers [85], reports clear inhibitory effects of melatonin on DMBA-induced mammary cancer in rats and on human ovarian cancer cell lines in vitro, respectively. Recently the inhibitory effects of melatonin and a number of its analogues have been confirmed on several human cancer cell lines (ovary, bladder, breast); the doses used were very large, however [86]. The in vivo effects of melatonin may well be indirect via the neuroendocrine and immune systems [83].

A relationship between the pineal gland and tumour development has been known for a very long time (for recent reviews see [83]). There is a considerable body of evidence that a number of pineal related products, such as synthetic indoleamines, pteridines and $\beta$-carbolines, suppress the growth of human melanoma cells in vitro. However, crude pineal extracts may be more effective suggesting synergy of action and/or unknown suppressive factors [87].

The most recent work has concentrated on the hormone-dependent cancers, as there would appear to be a potential therapeutic role for melatonin here. Photoperiod is reported to influence the growth of a number of different tumours [88, 89, 90]. Clearly the role of the pineal in transmitting photoperiodic information implies a direct influence on growth in this case. In some [91] but not all [92] studies, the production of melatonin itself in patients with malignant growths is reported to be abnormal. Nothing clear or consistent is evident from the accumulated literature. Although the weight of evidence suggests reduced melatonin production in cancer of the reproductive organs, it must be stated that extreme care is necessary in matching patients and controls, particularly with respect to age [92].

## Immune System

There is no doubt that rhythmicity is a major characteristic of the immune system. Pinealectomy is reported to suppress various aspects of the immune system and, in many studies, enhances tumour growth. Thus we may assume that the pineal does indeed influence the immune system. Likewise, suppression of melatonin production depresses the immune response and melatonin administration is reported to reverse this effect [93]. Thus melatonin is the most likely candidate for this particular pineal function.

In humans, circadian rhythms are present in most, if not all, immune functions. In some cases (e. g. circulating lymphocytes) the peak level (acrophase) is situated close to that of melatonin in a normal (24 h) environment [94]. Evidently it would not be surprising to find correlations between melatonin production and any circadian phenomenon, providing that adjustments are made for phase differences. Such correlations are not necessarily meaningful. For melatonin to be implicated in the control of immune system rhythmicity, carefully controlled experiments are indicated using pinealectomy, timed administration of physiological profiles of melatonin and classical chronobiological techniques of forced phase-shift and conditions of free-running rhythms. It would be of considerable interest to investigate whether daily variations in immune responses are truly endogenous, i. e. persist in a time-free environment. If such is the case, then their possible dependence on the SCN as a central pacemaker is clearly desirable to investigate. There is very little information to date on the entrainment of rhythms in the immune system even to the light-dark cycle and other potential zeitgebers. The little evidence available does not suggest a major effect of melatonin on the spontaneous temporal variations of the immune system.

The effects of different photoperiods on humoral and cell-mediated immunity have been the subject of a few reports. Photoperiod-induced changes in splenic weight and lymphocyte and macrophage counts suggest stimulation by short photoperiod [95]. This would not be consistent with circannual variations in lymphocytes, which, in general, appear to peak in long day lengths. Melatonin shows clear changes in duration of secretion, positively related to the night length in animals. In humans, duration change is not a major feature: rather a shift in acrophase, delayed in winter and advanced in summer, is evident together with a minor 6 month rhythm in daytime levels with nadirs in spring and autumn. Data on circannual variations in the immune system are not sufficiently detailed to attempt meaningful correlations with melatonin. Melatonin is, however, the major photoperiod transducer with regard to seasonal rhythms. In so far as photoperiod governs the immune system the effector molecule is likely to be melatonin. According to some reports, circaseptan variations exist in both immune response [94] and melatonin [96]. Little is known of the origin of either of these periodicities.

One of the most interesting observations concerning particularly DMBA-induced mammary cancer in rats is that blinding and underfeeding halts the development of breast adenocarcinomas and that this effect is pineal-dependent [97]. Underfeeding is associated with high amplitude of the melatonin rhythm and blinding leads to free-running melatonin. In theory, therefore, we have a strong signal operating at the natural endogenous periodicity of the animal and hence conceivably, greatly strengthened coupling of at least parts of the circadian system. Such reinforcement may well optimise physiological defense mechanisms.

## General Endocrine Relationships in Humans

Although the pineal-reproductive axis is beginning to be understood, other observations related to the endocrine system remain difficult to interpret and are often inconsistent. A review by Vaughan [98] covers much of the earlier literature. The most frequent observations concern cortisol secretion. For example [99] and references therein), an inverse relationship between cortisol and melatonin levels in depressed patients has been proposed and the administration of melatonin in both animals and humans may hasten the resynchronisation of cortisol rhythms following an abrupt phase-shift [76]. It is difficult to support a close physiological relationship between these two hormones, however, in the light of other observations. For example, the resynchronisation of the melatonin rhythm following time zone change is more rapid than that of cortisol [100].

Prolactin, a light- and melatonin-dependent hormone in animals, is undoubtedly influenced by exogenous melatonin in humans (e.g. [101]) but reports concerning melatonin levels in hyperprolactinaemia are inconsistent. Neither chronic nor acute milligram doses of melatonin were effective in influencing a large number of hormones (growth hormone, luteinizing hormone, follicle-stimulating hormone, thyroxine, testosterone, cortisol) in two recent studies [101, 102]. However, thyroid hormones are photoperiod-dependent in animals and thus may be pineal related in humans. The acute modification of prolactin and growth hormone by sleep and the lack of such an effect on melatonin [103] does not argue

for a close relationship. If melatonin influences central rhythm-generating systems in humans as it does in animals there is nevertheless considerable scope for modulatory effects on endocrine rhythms. Any hormone with a circadian rhythm may well show correlations with melatonin production, but this does not imply a causal relationship.

One curious observations is that a small percentage of apparently normal men have no detectable melatonin rhythm [9] implying that the pineal has no major functional importance in adult men. It is difficult, in the light of this phenomenon, to ascribe any major importance to findings of low melatonin in various clinical conditions.

## Effects of Melatonin on Rhythmic Function in Humans

Another approach to the study of the human pineal is via the administration of melatonin. In suitable doses melatonin has mild hypnotic properties, is sedative and anti-convulsant and has low toxicity in animals [104, 105]. It has been used experimentally in humans to induce short-term sleep and fatigue and to shift (advance) the timing of fatigue together with its own endogenous rhythm [106, 107]. Chronic, daily, low dose administration in the afternoon has little effect on anterior pituitary hormones [102]. The latter observations, coupled with the phase-shifting properties in rats, prompted an investigation into its use in the treatment of jet lag over eight time zones eastwards.

**Fig. 6.** Difference in jet lag ratings (placebo minus melatonin on a visual analogue score) for each subject (*n* = 52). Positive values show improvement with melatonin.
□ melatonin taken on westward flight;
■ melatonin taken on eastward flight. (From [109] with permission)

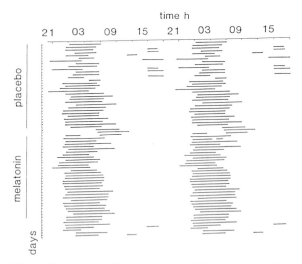

**Fig. 7.** Sleep record (daily log) of a blind man treated at 2330 hours daily with placebo for 30 days and melatonin for 31 days as shown, with a 3-day wash-out period interposed. *Horizontal bars* represent sleep times. Data from each day are plotted twice for clarity. No daytime naps were recorded during melatonin treatment. (From [112] with permission)

A small number of subjects reported insignificant jet lag compared with a placebo-treated group, using a carefully timed protocol of melatonin administration [108]. The main efects were to improve sleep and alertness and to readapt rapidly its own circadian rhythm [76]. Thus, it is possible that melatonin is able to increase the speed of resynchronisation of some circadian rhythms in humans. Recent reports on larger numbers of subjects (Fig. 6) suggest that it is effective both eastward and westward [109], and during simulated phase-shift it will speed up the resynchronisation of a number of circadian rhythms including cortisol [110]. In situations where disturbed rhythms contribute to lack of well-being such as jet lag, shiftwork, insomnia, blindness and old age, melatonin has considerable therapeutic potential. When used to entrain the free-running sleep-wake cycle of a blind man (Fig. 7) [111] it was not, however, able also to entrain cortisol and temperature [112], although a number of studies [76, 102, 113] suggest an ability to phase-shift melatonin's endogenous rhythm. These pharmacological studies do not pinpoint a physiological role for melatonin.

A function for the human pineal remains elusive although a coordinating role in rhythm physiology is a strong possibility. An important approach to this problem would be the administration of physiological levels of melatonin to human subjects pinealectomised for therapeutic reasons. This should reveal short-term events associated with physiological changes in melatonin secretion. Whether such an approach can clarify the possible season-related effects of melatonin in humans remains to be seen.

## Sites of Action of Melatonin

The mechanism of action of melatonin is currently the subject of intense investigation. Although in vitro it is known to affect a multitude of systems, only recently has convincing evidence been obtained for sites of action within the brain. The photoperiod-related effects of melatonin are primarily induced through the hypothalamus, as shown by the elegant experiments of Glass and Lynch [114]. Convincing high-affinity specific binding sites for melatonin are found in the SCN, the pars tuberalis, the retina and usually, to a lesser extent, some other brain areas: the area postrema, the preoptic area and the hippocampus, inter alia [115–120]. Some species differences are evident although binding in the pars tuberalis is seen in all species examined. The pars tuberalis has no known physiological function as yet but evidently a site of action for melatonin within the SCN is coherent with its function in the regulation of a number of biological rhythms.

## Conclusion

The production of melatonin by the pineal gland is an endogenously generated circadian rhythm, entrained and suppressed by light. It is highly reproducible within individuals and provides an excellent 'marker' rhythm for circadian studies, particularly in psychiatry. It conveys photoperiodic information in animals and has important functions in the control of circadian rhythms particularly in lower vertebrates and birds. It may well have similar functions in humans. Certainly its administration suffices to entrain sleep-wake cycles. Therapeutic possibilities include the treatment of rhythm abnormalities (jet lag, shiftwork, insomnia, blindness, old age) and anti-tumour activity.

*Acknowledgements.* This review was written during tenure of grants from the Wellcome Trust and the SERC. I would like to thank them and also particularly my coworkers M. Aldhous, C. Bojkowski, J. Broadway, S. Deveson (for Fig. 1), J. English and D. J. Skene.

# References

1. Vollrath L (1981) The pineal organ. Springer Berlin Heidelberg New York

2. Bojkowski CJ, Arendt J (1990) Factors influencing urinary 6-sulphatoxymelatonin, a major melatonin metabolite, in normal human subjects. Clin Endocrinol 33: 435–444

3. Tamarkin L, Baird CJ, Almeida OFX (1985) Melatonin: a co-ordinating signal for mammalian reproduction. Science 227: 714–720

4. Arendt J (1986) Role of the pineal gland and melatonin in seasonal reproductive function in mammals. Oxf Rev Reprod Biol 8: 266–320

5. Vivien-Roels B, Arendt J (1981) Environmental control of pineal and gonadal function: preliminary results on the relative roles of photoperiod and temperature. In: Ortavant R, Pelletier J, Ravault JP (eds) Photoperiodism and reproduction. INRA Paris, pp 272–278

6. Lincoln GA, Short RV (1980) Seasonal breeding: nature's contraceptive. Recent Prog Horm Res 36: 1–52

7. Reiter RJ (1980) The pineal and its hormones in the control of reproduction in mammals. Endocr Rev 1: 109–131

8. Herbert J (1981) The pineal gland and photoperiodic control of the ferret's reproductive cycle. In: Follett BK, Follett DE (eds) Biological clocks in seasonal reproductive cycles. Bristol, Wright, pp 261–276

9. Arendt J (1985) Mammalian pineal rhythms. Pineal Res Rev 3: 161–213

10. Arendt J, Brown WB, Forbes JM, Marston A (1980) Effect of pinealectomy on immunoassayable melatonin in sheep. J Endocrinol 85: 1–2P

11. Neuwelt EA, Lewy AJ (1983) Disappearance of plasma melatonin after removal of a neoplastic pineal gland. N Engl J Med 19: 1132–1135

12. Moore, RY, Klein DC (1974) Visual pathways and the central neural control of a circadian rhythm in pineal serotonin N-acetyltransferase. Brain Res 71: 17–33

13. Lincoln GA, Ebling FJP, Almeida OFX (1985) Genration of melatonin rhythms. In: Evered D, Clark S (eds) Photoperiodism, melatonin and the pineal. Pitman, London, pp 129–148 (CIBA foundation symposium 117)

14. Reiter RJ (1985) Action spectra, dose-response relationships and temporal aspects of light's effects on the pineal gland. In: Wurtman RJ, Baum MJ, Potts JT Jr (eds) The medical and biological effects of light. Ann N Y Acad Sci 453: 215–230

15. Illnerova H, Backström M, Säärf J, Wetterberg L, Vangbo B (1978) Melatonin in rat pineal gland and serum: rapid parallel decline after light exposure at night. Neurosci Let 9: 189–193

16. Armstrong SM (1989) Melatonin and circadian control in mammals. Experientia 45: 932–939

17. Armstrong SM, Thomas EMV, Chesworth MJ (1989) Melatonin-induced phase-shifts of rat circadian rhythms. In: Reiter RJ, Pang SF (eds) Advances in pineal research 3. Libbey, London, Paris, pp 265–270

18. Davis DC, Mannion J (1986) Society for Neuroscience Abstracts 12: 212

19. Wever RA, Polásek J, Wildgruber CM (1983) Bright light affects human circadian rhythms. Pflugers Arch 396: 85–87

20. Parkes AS (1976) Patterns of sexuality and reproduction. Oxford University Press, Oxford

21. Rosenthal NE, Sack DA, Gillin JC, Lewy AJ, Goodwin FK et al. (1984) Seasonal affective disorder. A description of the syndrome and preliminary findings with light therapy. Arch Gen Psychiatr 41: 72–79

22. Kitay JI, Altschule MD (1954) The pineal gland. Harvard University Press, Cambridge, pp 79–95

23. Klein DC (1985) Photoneural regulation of the mammalian pineal gland. In: Evered D, Clark S (eds) Photoperiodism, melatonin and the pineal. Pitman, London, pp 38–56 (Ciba foundation symposium 117)

24. Demisch K, Demisch L, Bochnik JH et al. (1986) Melatonin and cortisol increase after fluvoxamine. Br J Clin Pharmacol 22: 620–622

25. Sugden D, Weller JL, Klein DC, Kirk KL, Creveling CR (1984) $\alpha$-adrenergic potentiation of $\beta$-adrenergic stimulation of rat pineal N-acetyltransferase: studies using citazoline and fluorine analogs of norepinephrine. Biochem Pharmacol 33: 3947–3950

26. Hanssen T, Heyden T, Sundberg I, Wetterberg L (1977) Effect of propanolol on serum-melatonin. Lancet II: 309

27. Cowen PJ, Fraser S, Sammons R, Green AR (1983) Atenolol reduces plasma melatonin concentrations in man. Br J Clin Pharmacol 15: 579–581

28. Franey C, Aldhous M, Burton S, Checkley S, Arendt J (1986) Acute treatment with desipramine stimulates melatonin and 6-sulphatoxymelatonin in man. Br J Clin Pharmacol 22: 73–79

29. Checkley SA, Palazidou E (1988) Melatonin and anti-depressant drugs: clinical pharmacology. In: Miles A, Philbrick DRS, Thompson C (eds) Melatonin: clinical perspectives. Oxford University Press, Oxford, pp 190–204

30. Arendt J (1989) Melatonin – a new probe in psychiatric investigation. Br J Psychiatr 155: 585–590

31. Ebadi M, Govitrapong P (1986) Neural pathways and neurotransmitters affecting melatonin synthesis. J Neural Transm (Suppl) 21: 125–158

32. Moller M, Mikkelsen JD (1989) Vasoactive intestinal peptide (VIP) and peptide histidine isoleucine (PHI) in the mammalian pineal gland. In: Reiter RJ, Pang SF (eds) Advances in pineal research, vol 3. Libbey, London, Paris, pp 1–10

33. Yuwiler A (1983) Vasoactive intestinal peptide stimulation of pineal serotonin-N-acetyltransferase activity: general characteristics. J Neurochem 41: 141–156

34. Yuwiler A (1983) Light and agonists alter pineal N-acetyltransferase induction by vasoactive intestinal poly-peptide. Science 220: 1082–1083

35. Klein DC, Smoot R, Weller JL, Higa S, Markey SP, Creed GJ, Jacobowitz SM (1983) Lesions of the paraventricular nucleus area of the hypothalamus disrupt the suprachiasmatic spinal cord circuit in the melatonin rhythm generating system. Brain Res Bull 10: 647–652

36. Smith JA, O'Hara J, Schiff AA (1981) Altered diurnal serum melatonin rhythm in a blind man. Lancet II: 933

37. Lewy AJ, Newsome DA (1983) Different types of melatonin circadian secretory in some blind subjects. J Clin Endocrinol Metabol 56: 1103–1107

38. Arendt J (1988) Melatonin. Clin Endocrinol 29: 205–229

39. Arendt J (1978) Melatonin assays in body fluids. J Neural Transm (Suppl) 13: 265–278

40. Lewy AJ, Wehr TA, Goodwin FK, Newsome DA, Markey SP (1980) Light suppresses melatonin secretion in humans. Science 210: 1267–1269

41. Bojkowski C, Aldhous M, English J et al. (1987) Sup-

pression of nocturnal plasma melatonin and 6-sulphatox-
ymelatonin by bright and dim light in man. Horm Metab
Res 19: 437–440

42. Wever RA (1985) Use of light to treat jet-lag: differential
effects of normal and bright artificial light on human circadian rhythms. In: Wurtman RJ, Baum MJ, Potts JR Jr (eds)
The medical and biological effects of light. Ann N Y Acad
Sci 453: 282–304

43. Broadway J, Arendt J, Folkard S (1987) Bright light phase
shifts the human melatonin rhythm during the Antarctic
winter. Neurosci Let 79: 185–189

44. Bojkowski C, Arendt J, Shih M, Markey SP (1987) Assessment of melatonin secretion in man by measurement of its
metabolite: 6-sulfatoxymelatonin. Clin Chem 33: 1343–
1348

45. Illnerova H, Zvolsky P, Vanecek J (1985) The circadian
rhythm in plasma melatonin concentration of the ur-
banised man: the effect of summer and winter time. Brain
Res 328: 186–189

46. Bojkowski C, Arendt J (1988) Annual changes in 6-sulpha-
toxymelatonin excretion in man. Acta Endocrinol 117:
470–476

47. Beck-Friis J, von Rosen D, Kjellman BF, Ljungen JG, Wetterberg L (1984) Melatonin in relation to body measures,
sex, age, season and the use of drugs in patients with major
affective disorders and healthy subjects. Psychoneuroendocrinology 9: 261–277

48. Ozaki Y, Wurtman RJ, Alonso R, Lynch HJ (1978) Melatonin secretion decreases during the proestrous stage
in the rat estrous cycle. Proc Nat Acad Sci USA 75: 531–
534

49. Morello H, Caligaris L, Haymal B, Taleisuik S (1989) The
pineal gland mediates the inhibition of proestrous luteinising hormone surge and ovulation in rats resulting from
stimulation of the medial raphe nucleus or injection of
5HT into the third ventricle. J Neuroendocrinol 1: 195–197

50. Brzezinski A, Lynch HJ, Wurtman RJ, Siebel MM (1987)
Possible contribution of melatonin to the timing of the luteinising hormone surge. N Engl J Med 316: 1550–1551

51. Testart J, Frydman R, Roger M (1982) Seasonal influence
of diurnal rhythms in the onset of the plasma luteinizing
hormone surge in women. J Clin Endocrinol Metabol 55:
374–377

52. Berga SL, Mortola JF, Yen SSC (1988) Amplification of
nocturnal melatonin secretion in women with functional
hypothalamic amenorrhea. J Clin Endocrinol Metab 66:
242–244

53. Das Gupta D, Riedel L, Frick JH, Attanasio A, Ranke MB
(1983) Circulating melatonin in children: in relation to
puberty, endocrine disorders, functional tests and racial
origin. Neuroendocrinol Let 5: 63–78

54. Waldhauser F, Steger H, Vorkapic P (1987) Melatonin secretion in man and the influence of exogenous melatonin
on some physiological and behavioural variables. Adv Pineal Res 2: 207–223

55. Sizonenko PC, Lang V, Rivest RW, Aubert ML (1985) The
pineal and pubertal development. In: Evered D, Clark S
(eds) Photoperiodism, melatonin and the pineal. Pitman,
London, pp 208–230 (Ciba foundation symposium 117)

56. Young IM, Francis PL, Leone AM, Stovell P, Silman RE
(1986) Night/day urinary 6-hydroxymelatonin production
as a function of age, body mass and urinary creatinine levels: a population study in 110 subjects aged 3–80. J Endocrinol 111 (Suppl) abstr 32

57. Iguchi H, Kato KI, Ibayashi H (1982) Melatonin serum levels and metabolic clearance rate in patients with liver cirrhosis. J Clin Endocrinol Metab 54: 1025–1027

58. Touitou Y, Fèvre M, Lagoguey M, Carayon A, Bogdan A,
Reinberg A, Beck H, Cesselin F, Touitou C (1981) Age-
and mental health-related circadian rhythms of plasma
levels of melatonin, prolactin, luteinizing hormone and
follicle-stimulating hormone in man. J Endocrinol 91: 467–
475

59. Cozzi B, Ravault JP, Ferraudi B, Reiter RJ (1988) Melatonin concentration in the cerebral vascular sinuses of
sheep and evidence for its episodic release. J Pineal Res 5:
535–544

60. Wehr TA, Goodwin FK (1981) Biological rhythms and psychiatry. In: Arieti S, Brodie HKL (eds) American handbook of psychiatry vol 7, 2nd edn. Basic Books, New York,
pp 46–74

61. Halberg F (1968) Physiological considerations underlying
rhythmometry with special reference to emotional illness.
In: Ajuriagnerra J (ed) Cycles biologiques et psychiatrie.
Georg, Geneve and Masson, Paris

62. Checkley S (1989) The relationship between biological
rhythms and the affective disorders. In: Arendt J, Minors
DS, Waterhouse J (eds) Biological rhythms in clinical practice. Butterworth, London

63. Wever RA (1986) Characteristics of circadian rhythms in
human functions. J Neural Transm. (Suppl) 2: 323–374

64. Lewy AJ, Sack RL, Miller LS, Hoban TM (1987) Antidepressant and circadian phase-shifting effects of light.
Science 235: 352–354

65. Wetterberg L, Aperia B, Beck-Friis J, Kjellman BF et al.
(1982) Melatonin and cortisol levels in psychiatric illness.
Lancet II: 100

66. Claustrat B, Chazot G, Brun J (1984) A chronobiological
study of melatonin and cortisol secretion in depressed subjects: plasma melatonin, a biochemical marker in major depression. Biol Psychiatr 19: 1215–1228

67. Frazer A, Brown R, Kocsis J (1986) Patterns of melatonin
rhythms in depression. J Neural Transmission 21: 269–290

68. Thompson C, Franey C, Arendt J,Checkley SA (1988) A
comparison of melatonin secretion in depressed patients
and normal subjects. Br J Psychiatr 152: 260–265

69. Eastwood MR, Peacocke J (1976) Seasonal patterns of
suicide depression and electroconvulsive therapy. Br J Psychiatr 129: 472–475

70. Papousek M (1975) Chronobiologische Aspekte der
Zyklothymie. Fortschr Neurol Psychiatr 43: 381–440

71. Lewy JJ, Kern HE, Rosenthal NE, Wehr TA (1982) Bright
artificial light treatment of a manic-depressive patient with
a seasonal mood cycle. Am J Psychiatr 139: 1496–1498

72. Wehr TA, Jacobsen FM, Sack DA, Arendt J, Tamarkin L,
Rosenthal NE (1986) The efficacy of phototherapy in seasonal affective disorder appears not to depend on its timing
or its effect on melatonin secretion. Arch Gen Psychiatr 43:
870–875

73. Rosenthal NE, Jacobsen FM, Sack DA, Arendt J, James
SP, Parry BL (1988) Atenolol in seasonal affective disorder: a test of the melatonin hypothesis. Am J Psychiatr 145:
52–56

74. Lewy AJ, Sack RL, Singer CM (1985) Treating phase-
typed chronobiologic sleep and mood disorders using appropriately timed bright artificial light. Psychopharmacol
Bull 21: 368–372

75. Carman JS, Post RM, Buswell R, Goodwin FK (1976) Ne-

gative effects of melatonin on depression. Am J Psychiatr 133: 1181–1186

76. Arendt J, Aldhous M, English J, Marks V, Arendt JH, Marks M, Folkard S (1987) Some effects of jet lag and their alleviation by melatonin. Ergonomics 30: 1379–1393

77. Heninger GR, Charney DS, Sternberg DE (1984) Sertonergic function in depression. Arch Gen Psychiatr 41: 398–402

78. Lewy AJ, Wehr TA, Gold PW, Goodwin FK (1978) Plasma melatonin in manic-depressives illness. In: Usdin E, Kopin IJ, Barchas J (eds) Catecholamines: basic and clinical frontiers, vol 2. Pergamon, Oxford, pp 1173–1175

79. Lewy AJ, Wehr TA, Goodwin FK et al. (1981) Manic depressive patients may be supersensitive to light. Lancet I: 383–384

80. Nurnberger JI, Berrettini W, Tamarkin L, Hamovit J, Norton J, Gershon E (1988) Supersensitivity to melatonin suppression by light in young people at high risk for affective disorder. Neuropsychopharmacology 1: 217–223

81. Carney PA, Seggie J, Vojtechovsky M, Parker J, Grof E, Grof P Bipolar patients taking lithium have increased dark adaptation threshold compared with controls. Pharmacopsychiatry (in press)

82. Okudaira N, Kripke DF, Webster JB (1983) Naturalistic studies of human light exposure. Am J of Physiol 245: R613–615

83. Gupta D (1988) Neuroendocrine signals in cancer. In: Gupta D, Attanasio A, Reiter RJ (eds) The pineal gland and cancer. Müller und Bass, Tubingen, pp 9–28

84. Tamarkin L, Cohen M, Roselle D, Reichert C, Lippman M, Chabner B (1981) Melatonin inhibition and pinealectomy enhancement of 7,12-dimethylbenz-(a) anthracene-induced mammary tumours in the rat. Cancer Res 41: 4432–4436

85. Blask DE (1984) The pineal: an oncostatic gland? In: Reiter RJ (ed) The pineal gland. Raven, New York, pp 253–284

86. Hill SM, Blask DE (1988) Effects of the pineal hormone melatonin on the proliferation and morphological characteristics of human breast cancer cells (MCF-7) in culture. Cancer Res 48: 6121–6125

87. Lapin V, Ebels I (1981) The role of the pineal gland in neuroendocrine control mechanisms of neoplastic growth. J Neural Transm 50: 257–264

88. Stanberry R, Das Gupta TK, Beattie CW (1983) Photoperiodic control of melanoma growth in hamsters: influence of pinealectomy and melatonin. Endocrinology 113: 469–475

89. Bartsch H, Bartsch C (1981) Effect of melatonin on experimental tumours under different photoperiods and times of administration. J Neural Transm 52: 269–279

90. Aubert C, Janiaud P, Lecalves J (1980) Effect of pinealectomy and melatonin on mammary tumour growth in Sprague-Dawley rats under different conditions of lighting. J Neural Transm 47: 121–130

91. Tamarkin L, Danforth D, Lichter A, de Moss E, Cohen M, Chabner B, Lippman M (1982) Decreased nocturnal plasma melatonin peak in patients. Science 216: 1003–1005

92. Skene DJ, Bojkowski CJ, Currie JE, Wright J, Boulter PS, Arendt J (1990) 6-sulphatoxymelatonin production in breast cancer patients. J Pineal Res 8: 269–276

93. Maestroni GJ, Conti A, Pierpaoli W (1988) Role of the pineal gland in immunity. III Melatonin antagonises the im-

munosuppressive effect of acute stress via an opiatergic mechanism. Immunology 63: 465–469

94. Levi F, Reinberg A, Canon C (1989) Clinical immunology and allergy. In: Arendt J, Minors DSM, Waterhouse JM (eds) Biological rhythms in clinical practice. Wrigth, London, pp 99–135

95. Brainard G, Knobler RL, Podolin PL, Lavasa M, Lublin FD (1987) Neuroimmunology: modulation of the hamster immune system by photoperiod. Life Sci 40: 1319–1326

96. Vollrath L, Welker HA (1988) Atypical 24-h rhythm of serotonin N-acetyltransferase activity in the rat pineal gland. Chronobiol Int 5: 115–120

97. Blask DE, Hill SM, Pelletier DB (1988) Oncostatic signalling by the pineal gland and melatonin in the control of breast cancer. In: Gupta D, Attanasio A, Reiter RJ (eds) The pineal gland and cancer. Müller und Bass, Tübingen, W Germany, pp 195–206

98. Vaughan GM (1984) Melatonin in humans. Pineal Res Rev 2: 141–201

99. Lang V, Sizonenko PC (1988) Melatonin and human adreno-cortical function. In: Miles A, Philbrick DRS, Thompson C (eds) Melatonin, clinical perspectives. Oxford University Press, pp 62–78

100. Fevre-Montagne M, van Canter E, Refetoff S, Desir D, Tourniaire J, Copinschi G (1981) Effects of "jet-lag" on hormonal patterns. II Adaptation of melatonin circadian periodicity. J Clin Endocrinol Metab 52: 642–649

101. Waldhauser F, Steger H, Vorkapic P (1987) Melatonin secretion in man and the influence of exogenous melatonin on some physiological and behavioural variables. Adv Pineal Res 2: 207–223

102. Wrigth J, Aldhous M, Franey C, English J, Arendt J (1986) The effects of exogenous melatonin on endocrine function in man. Clin Endocrinol 24: 375–382

103. Akerstedt T, Gillberg M, Wetterberg L (1982) The circadian covariation of fatigue and urinary melatonin. Biol Psychiatr 17: 547–554

104. Cramer H (1978) Melatonin and sleep. Fourth european congress on sleep research. Tirgu-Mures. Karger, Basel, pp 204–210

105. Sugden D (1983) Psychopharmacological effects of melatonin in mouse and rat. J Pharmacol Exp Ther 227: 587–591

106. Vollrath L, Semur P, Gammel G (1981) Sleep induction by intranasal application of melatonin. Adv Biosci 29: 327–329

107. Arendt J, Bojkowski C, Folkard S, Franey C, Minors DS, Waterhouse JM, Wever RA, Wildgruber C, Wright J (1985) Some effects of melatonin and the control of its secretion in man. In: Evered D, Clark S (eds) Photoperiodism, melatonin and the pineal. Pitman, London, pp 266–283 (Ciba foundation symposium 117)

108. Arendt J, Aldhous M, Marks V (1986) Alleviation of jetlag by melatonin: preliminary results of controlled double-blind trial. Br Med J 292: 1170

109. Skene DJ, Aldhous M, Arendt J (1989) Melatonin, jet-lag and the sleep-wake cycle. In: Horne J (ed) Sleep '88, proceduring of european sleep congress, Jerusalem. Karger, Basel

110. Samel A, Maas H, Vejroda M, Wegman HM (1989) Influence of melatonin treatment on human circadian rhythmicity. Aviat Space Environ Med abstr 485, p 52

111. Arendt J (1988) Synchronisation of a disturbed sleep-wake cycle in a blind man by melatonin treatment. Lancet I: 772–773

112. Folkard S, Arendt J, Aldhous M, Kennett H (1990) Mela-

tonin stabilises sleep onset time in a blind man without entrainment of cortisol or temperature rhythms. Neurosci Let 113: 193–198

113. Sack RL, Lewy AJ, Hoban TM (1987) Free-running melatonin rhythms in blind people: phase shifts with melatonin and triazolam administration. In: Rensing L, an der Heiden V, Mackey MC (eds) Temporal disorder in human oscillatory systems. Springer, Berlin Heidelberg New York, pp 219–224

114. Glass JD, Lynch GR (1982) Evidence for a brain site of melatonin action in the white-footed mouse, peromyscius leucopus. Neuroendocrinology 34: 1–6

115. Vanecek J, Pavlik A, Illnerova H (1987) Hypothalamic melatonin receptor sites revealed by autoradiography. Brain Res 435: 359–362

116. Morgan PJ, Williams LM, Davidson G, Lawson W, Howell E (1989) Melatonin receptors on ovine pars tuberalis: characterisation and autoradiographical localisation: J Neuroendocrinol 1: 1–4

117. De Reviers MM, Tillet Y, Pelletier J (1991) Melatonin binding sites in the brain of sheep exposed to light or pinealectomised. Neurosci Lett 121: 17–20

118. Reppert SM, Weaver DR, Rivkees SA, Stopa EG (1988) Putative melatonin receptors in a human biological clock. Science 242: 78–81

119. Dubocovich ML (1985) Characterisation of a retinal melatonin receptor. J Pharmacol Exp Ther 234: 395–401

120. Laudon M, Zisapel N (1986) Characterization of central melatonin receptors using $I_{125}$-melatonin. FEBS 197: 9–12

# Chronobiology of Endocrine-Immune Interactions

G. Gatti, A. Angeli, and R. Carignola

## Introduction

Accumulating evidence suggests the existence of modulatory interactions and bidirectional communications between the neuroendocrine and immune systems [2, 34, 60, 108]. Accordingly, it is postulated that alterations in either of these systems may be functionally felt in the other. Thus, psychosocial and neuroendocrine influences affect proneness to many diseases with a major immunologic component including infections, cancer, and autoimmune disorders [39, 59, 93, 94, 110]. Conversely, immunologic cell-derived hormones and soluble factors may elicit, in the course of an immune response, endocrine and brain functional changes [12, 18].

Biological rhythms are a ubiquitous feature of physiology. The existence of a temporal order in the organism allows both coordination of vital processes with recurring (and thus predictable) changes in the environment and adjustment to emergencies and nonpredictable changes. In this regard, hormones participate at multiple levels in the integrative and controlling mechanisms that provide a complex hierarchical framework for the outcome of the genetically encoded program of rhythms. In human beings, throughout life virtually all neuroendocrine variables and hormone actions are organized according to a multifrequency spectrum of rhythms. Furthermore, hormones are messengers and oscillating systems are among their targets.

Accordingly, circadian and infradian variations in human T and B lymphocyte activities and in various immune responses have been convincingly documented (reviewed in this book and [53]). The existence of a temporal order of immunoregulation implies that the regulatory outcome of immune functions will be the result of endogenous rhythmic neuroendocrine signals superimposed on autoregulatory immunologic signals elicited in the course of antigen challenge. The secretion by the immune effectors of soluble messengers (e.g., hormones, lymphokines) can evoke information in the central nervous system about the type of immune response, which in turn is transformed into neuroendocrine-dependent regulatory mechanisms that optimize the efficiency and functional flexibility of the immune response. In this regard it has been stated that the maintenance of endocrine rhythmic signals that influence and probably coordinate rhythmic immune functions may be of value in the improved treatment of a number of immunologic diseases [53], thus suggesting a more integrative view of immunoregulation and neuroendocrine-immune connections.

Due to the complexity of the interactions, it is obviously difficult to extrapolate from time-unspecified experiments, or even from single measurements, the role of different modulators. The hierarchies of the neuroendocrine-immune network may well change during different stages of the endogenous rhythms involved and conceivably in health vs disease. Furthermore, studies of rhythmic variations in the biological events related to endocrine-immune interactions are relatively anecdotal and the vast immunologic literature continues to be based largely on the belief that all body functions are regulated to maintain constant internal conditions.

In this chapter, we attempt to review current knowledge concerning interactions between the neuroendocrine and the immune systems, taking into account chronobiologically related events. In addition, some chronoimmunotherapeutic implications will be discussed.

## Multiplicity of Extracellular Signals

No cell in the organism lives in isolation. An elaborate cell-to-cell communication system with thousands of chemical signals operates in metazoans. The current classification of these extracellular signals is based on the distance over which the signal must act. In *endocrine* signaling, hormones are released in the blood and act on distant target cells. In *paracrine* signaling, the target cell is close to the signaling cell. In *autocrine* signaling, cells respond to substances that they themselves release in the extracellular space. The same molecule sometimes acts in two or even three types of signaling. Epinephrine and certain peptides, for example, function both as neurotransmitters or neuromodulators (paracrine) and as systemic hormones (endocrine). Moreover, communication by extracellular signals usually involves several steps: synthesis and release of the substance by the signaling cell; transport to the target; detection of the signal by a specific receptor; transduction of the signal from the receptor to other molecules; and the biochemical response. Some signals induce a modification in the activity of one or more enzymes already present in the target cell; other signals primarily alter the pattern of gene expression. Finally, the signaling molecules reciprocally regulate their synthesis, release, degradation, transport, and action.

It is clear that the signaling system of the organism exceeds in complexity any model of physics. In this conceptual framework feedback axes are an incomplete view, since the relationships between two signaling entities are liable to be continuously modified by influences of other entities. Control mechanisms that were held as single and deterministic need to be considered multifactorial and flexible [7, 51].

The identification of neuropeptide and steroid hormone receptors on cells of the immune system has been critical in this regard (Table 1). Endocrine signals modulate immune function through their interaction with specific receptors which appear to be structurally and functionally identical to the receptors found on neuroendocrine tissues [17, 23, 90, 91, 92, 126]. Thus the two systems rely on common signaling and recognition in order to "crosstalk" and coordinate responses which may initially be perceived by one system exclusively.

Rhythmic changes characterize virtually all steps of cell-to-cell signaling; temporal changes in the susceptibility to hormonal action have been generally found, often phase-shifted among different targets. Moreover, temporal variations exist in the binding-activated process associated with hormone-receptor interaction and/or in the induced transcriptional and translational events. Present concepts of molecular chronobiology suggest that rhythmic patterns may indeed involve all the above mentioned steps.

The temporal organization of hormone receptor expression on cells of the immune system remains to be explored. However, some reports point to the existence of a circadian rhythmic pattern of receptor expression on lymphocytes. For instance, a 24-h oscillation of the binding activity of adrenergic receptors on human immunocompetent cells has been demonstrated [85]. Data obtained with the use of anti-glucocorticoid molecules (the norsteroid RU 486) and a monoclonal antibody, anti-corticosteroid-binding globulin (CBG), are compatible with a role of both high-affinity glucocorticoid receptors and CBG-like molecules in mediating some aspect of the inhibitory action of cortisol on immune effectors [5, 41]. Interestingly enough, it has also been suggested that the binding capacity of CBG-like molecules and specific glucocorticoid receptors in human lymphocytes is modulated in vivo by glucocorticoids themselves and displays a circadian rhythmicity [4, 101], phase-shifted with respect to plasma cortisol levels. Rhythmic changes may reportedly occur also for other hormone-receptors interactions, conceivably a general feature which characterizes these systems. The binding of hormones to specific recognizing molecules in immune cells may thus be a key mechanism to further cell target responsiveness at a proper circadian stage. Although there are as yet few answers to the many questions surrounding this issue, recognition of the rhythmic events that coordinate receptor expression and activity may prove crucial in the field of endocrine-immune connections.

**Table 1.** Specific receptors for hormones and neurotransmitters on lymphocytes

Steroid hormones: glucocorticoids, sex hormones

Peptide hormones: CRH, ACTH, opiates, TSH, PRL, GH, AVP, oxitocin, somatostatin, substance P, VIP, insulin

Neurotransmitters: catecholamines, acetylcholine, histamine 5-hydroxytryptamine

Arachidonic acid-derived molecules: leukotriene $B_4$, $PGE_2$

CRH, corticotropin releasing hormone; ACTH, adrenocorticotropic hormone; TSH, thyroid stimulating hormone; PRL, prolactin; GH, growth hormone; AVP, arginine vasopressin; VIP, vasoactive intestinal peptide; PG, prostaglandin

# Reciprocal Interactions Between the Neuroendocrine and Immune Systems

## The Chronoimmunomodulatory Role of Hormones

Evidence shows that cells of the immune system are deeply influenced by hormones and neurotransmitters. Obviously, levels of control concern general cellular metabolism and cell division, although several reports point to the fact that hormones can operate at more refined levels on specific cell functions. For instance, the intracellular levels of cyclic nucleotides participating in the processes of activation or suppression of immunocompetent cells can be affected by hormones [47, 50]. Further more, a variety of endocrine signals operate at the immune effector level in a manner similar to that of cytokines/lymphokines, through receptors coupling to the same G-proteins [63]. Thus, common messenger pathways are shared by endocrine, paracrine, and probably autocrine peptide molecules, which influence immune functions and are conventionally classified as hormones, cytokines/lymphokines, and growth factors.

Glucocorticoids are the final products of the hypothalamic-pituitary-adrenal (HPA) system and are crucial mediators of endocrine-immune interactions. Generally speaking, they exert inhibitory effects on immune functions, but it should be emphasized that they display some stimulatory effects too. The matter has been extensively reviewed [9, 25, 30, 82, 117] and relevant data on the actions of glucocorticoids are summarized in Table 2.

One key aspect of the immunomodulatory properties of glucocorticoids refers to the putative control by these hormones of traffic and compartementalization of lymphocyte subsets. It is generally held that this control accounts for the circadian bioperiodicity of *immune* cell circulation [36, 37]. In fact, in diurnally active, nocturnally resting individuals, the lowest cell counts of circulating lymphocytes are usually found to follow, by approximately 3–4 h, the peak of cortisol concentrations. Although some studies failed to validate statistically correlations between T lymphocyte subsets and plasma cortisol [69], it was logical to think that plasma glucocorticoid levels operate as an endogenous synchronizer, able to drive other rhythmicities along an appropriate time scale [1, 27, 64, 80, 114]. Glucocorticoid action at the cellular level indeed requires a certain lag time before effects can be detected. It has also been found that blood lymphocyte subpopulations are not syn-

**Table 2.** Glucocorticoids and immune functions

Classic (dexamethasone binding; type I and type II) glucocorticoid receptors in immune cells. Transcortin (CBG)-like binding in some lymphocytes.

Regulation of traffic and distribution of circulating effectors

Synchronization of circadian rhythms

Inhibition of T lymphocyte functions: mitogen-induced proliferation, IL-1- and IL-2-dependent activation, mRNA synthesis for IL-2 receptor, lymphokine production

Modulation of B lymphocyte functions: dose-dependent activation or inhibition of antibody production

Inhibition of monocyte/macrophage functions: monocyte differentiation, antigen presentation, IL-1 production

Inhibition of NK cell functions: precursor recruitment and NK differentiation, cytotoxic activity

chronous in their circadian variations; selective compartmentalization effects and/or differential sensitivities to soluble messengers (glucocorticoids and others) could underlie phase-shifting. Confirmation of the role played by glucocorticoids has come from the recent demonstration of a glucocorticoid dependency in the expression of lymphocyte "homing" receptors [24]. Specific ligand/receptor interactions between "homing" receptors and "vascular addressins" expressed by endothelial cells may, in fact, represent a crucial event in the regulation of immune responses and lymphocyte compartmentalization.

In recent years, the demonstration of specific receptors for pro-opiomelanocortin (POMC)-derived peptides on cells of the immune system has raised the question of their functional significance. Available data are often conflicting and not comparable due to differences in the experimental protocols (Table 3). Virtually all studies did not take into account the timing, i.e., the circadian stage at which a particular experimental procedure was performed. Just to focus on one example, beta-endorphin has been suggested to have facilitatory effects in vitro on some immune functions and inhibitory effects on others and to enhance the lymphocyte proliferative response to mitogens in rats but not in humans [46, 77]. Investigators are often troubled with the variability of results within the same set of experiments [22, 61, 62, 67, 75] and attempt to explain the bidirectionality of the effects observed for several molecules [124]. The in vitro concentrations are an important variable, since some effects are clearly dependent on a particular range of concentrations. As mentioned before the timing has been practically ignored in most of these studies. The molecular conformation of signaling

**Table 3.** Pro-opiomelanocortin-derived peptide effects on immune effectors

ACTH
  Inhibits antibody production
  Enhances B lymphocyte growth and differentiation
  Inhibits IFN-$\gamma$ production
  Blocks IFN-$\gamma$ activation of macrophages
  Enhances NK cell activity

ENDOGENOUS OPIATES
$\beta$ Endorphin
  Inhomogeneous effects on B lymphocytes
  Inhibition of T lymphocyte response to mitogens
  Enhancement of IFN production by NK cells
  Enhancement of NK cell activity
$\alpha$ Endorphin
  Reduction of plaque forming cells in the spleen
  Inhomogeneous effects on T lymphocytes
Leu- and met-enkephalins
  Reduction of plaque forming cells in the spleen
  Enhancement of NK cell activity

peptides has attracted more attention. In the case of endorphins it has been suggested that those with unmodified N terminals, which bind opiate receptors, act in one direction, whereas sequences with modified N terminals, which virtually bind only to nonopiate receptors, act in the opposite direction [124]. Turning back to POMC-derived molecules, ACTH and closely related peptides are also capable of acting directly on immune cells. Once again no clear statement can be made on the effects. ACTH has been found to enhance both growth and differentiation of B cells and immunoglobulin synthesis [3], whereas it appears to be inhibitory on interferon production and T cell-dependent functions [22, 55, 62, 75, 96, 118].

Aside from those for glucocorticoids and POMC related peptides, specific binding sites for a number of hormones have been found in lymphocytes. The literature is very fragmented and it is a very difficult task to evaluate functionally each hormone. Most notable are the immunomodulatory properties of sex steroids [48, 49], prolactin [10, 28, 56, 57], thyroid stimulating hormone (TSH) and thyroid hormones [18, 65, 86], corticotropin releasing hormone (CRH) [88, 113], substance P, somatostatin, and vasoactive intestinal peptide (VIP) [84, 89, 91]. There is also evidence for a role of the pineal in immunity. The pineal has been demonstrated to be an important structure of the mammalian circadian system. In experimental rodents, pinealectomy or pharmacological blockade of pineal function impairs immune reactivity, as determined by the primary antibody response and mixed lymphocyte reactions [71–73]. Interestingly

enough, a circadian stage dependency of the effects of melatonin, the most important pineal hormone, has emerged from studies indicating that immunoaugmenting effects are expected when melatonin administration occurs during evening or night, i.e., shortly prior or at the time when hormone synthesis is high. Furthermore, at least in mice, melatonin is able to interact with corticosteroids by counteracting some of their immunosuppressive effects. Data in animals and humans suggest that the opioid system and melatonin have important relationships; an opiatergic mechanism has indeed been suggested as a mediator of the immunoaugmenting properties of the pineal hormone [72–74].

In summary, circuits involving hormone signals that are clearly circadian rhythmic, such as those generated by glucocorticoids, melatonin, or beta-endorphin, have been suggested to influence a variety of immune related phenomena. However, as already pointed out, little if any information is available on the chrononeuroendocrine-immune network that coordinates the temporal aspects of immunomodulation. We do not know whether there are significant changes of phase and amplitude in endocrine signaling related to physiological or pathological immune responses or whether there is specific information from environmental synchronizers which is conveyed to the immune signaling system. However, it seems logical to postulate that biochemical interactions are genetically programmed to link neurologic endocrine, and immune activities.

## Release of Hormones by Lymphocytes and Autocrine/Paracrine Functional Regulation

Recently the immune system has been shown to produce a variety of hormones. Lymphocytes can secrete immunoreactive ACTH and endorphin-like peptides in response to immunostimulants such as Newcastle disease virus (NDV), or following lipopolysaccharide-induced proliferation, or exposure to tumor cells [52, 104, 106]. Moreover, a direct stimulation by monocytes of cortisol production from cultured human adrenocortical cells has been convincingly documented [123].

Interestingly enough, the regulation of POMC related peptide release by lymphocytes is similar to that of pituitary cells; thus, lymphocytes may secrete POMC-derived peptides in response to CRH and the release into the extracellular space is suppressible by dexamethasone [107]. In other words, similar control mechanisms operate between glucocorticoids and the hypothalamic-pituitary unit and between glucocorticoids and immunocompetent cells.

Equally as important as demonstrating lymphocyte POMC-derived hormone production was to demonstrate that ACTH from lymphocytes had steroidogenic activity. As would be expected from the identical physicochemical properties of lymphocyte ACTH and its pituitary-derived counterpart (including amino acid sequence), lymphocyte-derived ACTH is also able to stimulate steroidogenesis both in vitro and in vivo [52, 78, 105, 107]. Studies in hypophysectomized mice have demonstrated that ACTH produced by lymphocytes during immune stimulation can result in steroidogenesis in the absence of the pituitary-derived hormone and that animals could undergo an appreciable glucocorticoid response following NDV infection [105]. In the human, the amounts of POMC related products released into the extracellular space by lymphocytes are extremely low and inadequate to effectively act as systemic hormones. These products conceivably act as local endocrine/paracrine signals among cells of the immune system.

The temporal organization of lymphocyte POMC synthesis and release is not known. At least in rat pituicytes, circadian changes occur not only at the level of POMC mRNA but also in the rate of POMC gene transcription; significant changes in the rate of POMC gene transcription were detected some hours before the corresponding changes in mRNA synthesis and the subsequent secretion of POMC-derived peptides [79]. We do not know whether a similar circadian organization applies to lymphocyte POMC gene transcription and POMC release. It is tempting to think that intrinsic circadian variations of POMC gene expression occur as a function of a genetically determined cellular clock, possibly modulated by signaling control mechanisms. Evidence exists for CRH production by lymphocytes themselves [112] and for glucocorticoid-dependent regulation of lymphocyte POMC gene expression and transcription [15, 98]. Thus, mutual interaction and cooperation seem to link a multiplicity of oscillatory signals involved in the regulation of a cellular rhythmic program.

An understanding of the time-dependent hierarchies of the neuroendocrine-immune network is extremely difficult to achieve. One has to consider, for example, that hormone production by the immune system is not limited to POMC-derived peptides, but now includes hypothalamic and pituitary hormones including TSH, somatostatin, prolactin, growth hormone, and gonadotropins [18], whose temporal organization could be a prominent feature.

## Immune-Derived Lymphokines and Immune-Neuroendocrine Circuits

The demonstration of neuroendocrine changes during the immune response is a less controversial argument in favor of the theory of bidirectional endocrine-immune cooperation. The existence of a family of lymphokines, which can act as immunohormones and bring about these neuroendocrine responses, has been convincingly documented. Prominent features refer to modulation of the HPA system (Table 4). From a chronobiological point of view, it seems clear that such influences may intrude into the temporal organization of the adrenocortical cycle, hence interfering with the synchronizing action of glucocorticoids on the human circadian structure. Thus it is not surprising to find alterations of the temporal order of hormone release and metabolite production in the course of pharmacological manipulation of immunity [29] or during pathological conditions affecting the immune system, such as HIV infections and related immunodeficiency syndromes [121].

In this context, evidence points to interleukin-1 (IL-1) as the most important trigger of the HPA system when an immunologic challenge is presented [12, 13]. IL-1 is a monokine, produced by antigenically challenged macrophages, that induces the production of interleukin-2 (IL-2), a T cell derived lymphokine that stimulates proliferation of T cells [33]. In addition, IL-1 is involved in modulating many biological processes including pyrogenesis, hematopoiesis, neutrophil activation, and tumor cell destruction. Two distinct IL-1 subtypes have been identified in humans, IL-1$\alpha$ and IL-1$\beta$. An interesting aspect of the

**Table 4.** Lymphomonokines and the hypothalamic-pituitary-adrenal (HPA) system

| Lymphokine or monokine | Effect on HPA system |
|---|---|
| $\alpha$-Interferon | Stimulation of steroid secretion by direct action on adrenal cells |
| $\beta$-Interferon | Reduction of CRH-induced release of ACTH by pituitary cells |
| Interleukin-1 | Stimulation of ACTH release from pituitary cells |
| Interleukin-2 | Possible induction of ACTH secretion by pituitary cells |
| Interleukin-6 | Induction of ACTH release from pituitary cells |
| Tumor necrosis factor-$\alpha$ | Induction of ACTH release from pituitary cells |

action of IL-1 in activating the HPA system is that only IL-1$\beta$ appears to be effective [120]. Besedovsky et al. showed that supernatants from concanavalin A stimulated lymphocytes increased plasma corticosterone levels when administered to rats and that this effect was indeed due to the production of IL-1 by activated monocytes [11]; as little as 0,5 mg IL-1 elicited an increase in plasma ACTH and corticosterone, which peaked 2 h after injection and returned to baseline values 4 h after injection [11]. Several studies have been performed to identify the site of action of IL-1 and most evidence now points to the hypothalamus. Although an IL-1-dependent stimulation of ACTH release from cultured pituitary cells has been described [125], there is plenty of data that indicate an action on CRH secreting neurons [14, 97]. The IL-1 induced increase in ACTH is highly specific, in that blood levels of prolactin, growth hormone, and melanocyte stimulating hormone (MSH-related peptides) are not affected [14] and can be counteracted by anti-CRH antibodies [97].

It is well accepted that IL-1 is produced by several cell types other than monocytes. Interestingly enough, IL-1 has been shown to be synthesized by glial cells [38] and, surprisingly, by neurons of a number of brain regions, including the hypothalamus [21]. Preliminary reports point to a circadian rhythmic pattern of IL-1 production and action [81]; moreover certain phase relationships to sleep stages and/or light-dark alternation and pineal activity, and hence to pleiotropic neuroendocrine interactions with environmental synchronizers, have been suggested [19, 66]. Since the inhibition by glucocorticoids of IL-1 gene expression is well established [68], these observations are consistent with the hypothesis of regulatory circuits among immunocytes, the HPA axis, and specific brain regions.

Cells of the immune system produce a variety of molecules that, in addition to IL-1, can influence the HPA system. Recent studies have demonstrated the presence of IL-2 receptors on pituitary At T-20 cells which can be up-regulated by CRH [20]. This observation is especially important since IL-2 can stimulate POMC expression in pituitary cells, and a circadian stage dependency of the biological response to IL-2 has been reported [100], being higher at the beginning of the activity cycle, i.e., coincidental with the circadian peak of POMC production. Also IL-6 and tumor necrosis factor (TNF)-$\alpha$ have been recently reported to induce ACTH release from pituitary cells [109, 122], whereas interferon-$\alpha$ (IFN-$\alpha$) has been shown to stimulate steroid secretion by a direct effect on adrenal cells [16, 95]. There is also evidence for the existence of immune system-derived inhibitors of the HPA axis, including IFN-$\gamma$, which was shown to reduce CRH stimulated ACTH release from cultured rat pituitary cells [119], or the neurophil-derived human peptide HP-4, which possesses corticostatic activity by inhibiting ACTH induced steroidogenesis [103].

Taken together, these findings strongly suggest the existence of an immune-HPA loop, in which the immune system serves as a sensory organ for noncognitive stimuli, such as bacteria, virus, or tumor cells, and is able to convey information to the neuroendocrine system via soluble mediators; conversely, in response to cognitive stimuli, including changes of environmental synchronizers, cells of the neuroendocrine system can produce hormones potentially affecting immune responses and their temporal ordering. In the case of the HPA system, long and short feedback mechanisms may be operating (Fig. 1). Long feedback is initiated with the stimulation by immune-derived lymphokines of pituitary ACTH, resulting in

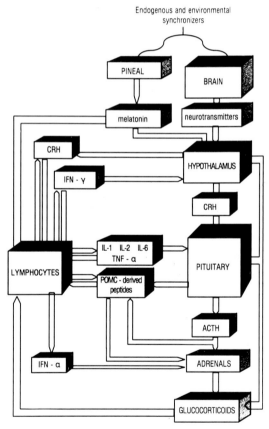

**Fig. 1.** Bidirectional interactions between lymphocytes and the hypothalamic-pituitary-adrenal (HPA) system. CRH, corticotropin releasing hormone; IL, interleukin; POMC, pro-opiomelanocortin; TNF, tumor necrosis factor; ACTH, adrenocorticotropic hormone; IFN, interferon

adrenal steroidogenesis. The loop is completed by the negative feedback inhibition of glucocorticoids on lymphokine production. In addition, POMC related peptides from the pituitary may act as modulators of immune cell responsiveness, participating in feedback loops with the role of avoiding, in some instances, overshooting of steroid-dependent immunosuppression. Short feedback includes the production of POMC derived hormones by cells of the immune system, with potential effects upon glucocorticoid production. Also, under these circumstances secretion of soluble factors with corticostatic activity could exert a compensatory "brake" effect on adrenal activation. Finally, endocrine-mediated effects on immune function are also expected to be influenced by the action of peptides of immune origin, which have regulatory effects in an autocrine/paracrine fashion in the context of ultra-short feedback mechanisms. Most of these points are obviously speculative and their significance with regard to the temporal ordering of endocrine-immune interactions remains a focus for future investigation.

## Natural Killer Cells:
## A Model for Studying Chronobiological Aspects of Endocrine-Immune Interactions

Natural killer (NK) cells are a heterogeneous lymphocyte subset that mediates nonmajor histocompatibility complex (MHC)-restricted and antibody-dependent cytotoxicities [116]. They appear to play a role in a variety of human diseases, including cancer, viral infections, and autoimmune disorders. A considerable number of regulatory factors, both soluble and cellular, have been shown to influence NK cell function. Among the NK activating lymphokines, both IFN-$\gamma$ and IL-2 appear more closely associated with this function, supporting the reactivity of pre-NK cells and their differentiation to active effectors. Hormones are also known to be important regulators of NK cell activity; in particular, the HPA system seems to provide multifunctional control. The final products of the system, i.e., glucocorticoids, are potent in vitro and in vivo inhibitors of NK cell activity [41, 42, 47, 58, 87]; Our own data are consistent with the view that products of POMC cleavage may effectively counterbalance the inhibitory effect of glucocorticoids [45, 76]. In other words, the HPA system could provide bidirectional modulation of NK cell activity, notably in the course of events activating this

system. Under such circumstances, e. g., during stress, it appears likely that the release of ACTH and other peptides from endocrine (and/or immune cells) can prevent an overshooting effect of glucocorticoids on this particular immune function.

On this basis, NK cells represent a reliable model for studying chronobiological implications of endocrine-immune interactions. A prominent feature of the temporal organization of NK cells refers to circadian rhythms [40, 43]. Investigating the circadian profile of NK cell activity of peripheral blood mononuclear (PBM) cell preparations, obtained serially at 4 h intervals throughout the 24 h cycle in 25 healthy adult subjects of both sexes, circadian variations were apparent. NK cell activity was at its maximum at the end of the night or in the early morning, somewhat later in women than in men, and then declined during the afternoon. The 24 h rhythmic pattern was validated in its statistical significance by cosinor analysis. Maximum enhancement of the activity resulting from exposure of PBM cell preparations to IFN-$\gamma$ was also observable in the second part of the night or in the early morning, i. e., in phase with the peak of the spontaneous activity. A significant circadian rhythm of the percent increase above control levels was validated by cosinor analysis. When PBM cell preparations were exposed to inhibitory concentrations of glucocorticoids, the maximum effect occurred earlier in the night, apparently phase-shifted with respect to spontaneous activity. A significant circadian rhythm of the percent reduction below control levels was validated by cosinor analysis. Interestingly, the circadian chronosusceptibility of NK cells and ACTH secreting cells to specific cortisol-dependent inhibition were roughly comparable in time. In terms of molecular endocrinology, intracellular $Ca^{2+}$-dependent pathways have been suggested to mediate both effects [42].

The circadian variations of NK cell activity are conceivably connected with circadian stage-dependent changes, which characterize both the secretion and the response of immune effectors to soluble regulatory factors. As a first insight into this area of research, we focused our attention on melatonin, both because an increasing body of experimental results speaks for an immunomodulatory role of the pineal hormone and because the circadian acrophase of the spontaneous and lymphokine induced NK cytotoxicities were located at the end of the night, just following peak serum melatonin concentrations. We have investigated the effects in vitro and in vivo of melatonin on spontaneous human NK cell activity and responsiveness to biological modifiers [8, 44]. In vitro melatonin was totally ineffective over a vast range of biologically active concentrations. In vivo melatonin was effective under specific experimental conditions. In a double-blind, placebo-controlled study we administered a single oral dose of 1 mg or 100 mg melatonin to male adult volunteers in the morning at 0800 or in the evening at 1800. Blood samples were taken before and then serially until 14 h after administration. Most of the data on NK cell activity after melatonin were not significantly different from the corresponding ones after placebo, but a clear-cut, dose-dependent augmentation of NK cell responsiveness to IFN-$\gamma$ was apparent when melatonin was given at 1800. Differences in the effectiveness vs the corresponding doses given at 0800 were striking. In male adult volunteers, we then administered, as a single oral dose, 2 mg melatonin at 1800 daily for 60 days. Chronic administration of the hormone was able to augment significantly not only responsiveness to IFN-$\gamma$ (the dynamics of this action were complex throughout the 2 months of exposure to exogenous melatonin and call for further studies), but also spontaneous NK cell activity. During melatonin treatment, a highly significant correlation was also apparent between the number of phenotypically identifiable NK effectors (HNK-1$^+$) and the level of spontaneous activity (lytic units). Since such a correlation hardly comes out in PBM preparations obtained with blood sampling under baseline conditions, a facilitory role has been attributed to melatonin in the recruitment of NK cells from precursors (pre-NK), a process admittedly up-modulated by paracrine/autocrine IFN-$\gamma$.

Taken together, these results support the view that melatonin is indeed a component of the chrononeuroendocrine network of signals that coordinate human NK cell activity and that it has modulatory effects upon lymphokine recognition and/or action during the nocturnal, but not during the diurnal, part of the 24 h cycle.

It seems obvious that a host of soluble factors other than melatonin provide neuroendocrine information to NK cells and other immune cells in temporally programmed scales. The physiological significance of our results awaits scrutiny from a clinical viewpoint. Chronobiology could offer novel approaches to improving immunotherapeutic strategies using lymphokine activated immune effectors.

## From General View of The Chronoendocrine-Immune Network to a Clinical Application

An alteration of the temporal structure, notably of the coordination of circadian bioperiodicities, has frequently been reported in disease. It may be viewed as resulting from an endogenous alteration at one or several nodes in the network of rhythms or as the action of exogenous agents, including drugs. Alterations of the neuroendocrine time structure have been noted in diverse conditions, aging and psychiatric illness among others, that have immunopathologic correlates [32, 35, 70, 83, 99]. Immunodepression after exposure to stressful events may be a feature which suggests a link between stress related disruption of rhythmic neuroendocrine signaling and immune abnormalities [2, 34, 102, 111].

Clinical investigations in humans are still scarce. A prominent question remains, whether in these and other conditions resetting or restoring of the periodicities related to endocrine functions could bring about a reconstitution of the efficiency of the immune system. The clinical application of a short chain synthetic, COOH-terminally amidated, ACTH agonist analogue (alsactide) has allowed some advances in this field [7]. An important feature of this molecule is its flexibility for use as a stimulator of adrenocortical activity. The term "chronizer" seems appropriate for alsactide in that it has the documented capability of resetting or enhancing the rhythmic organization of some functions, especially of those which are more dependent on glucocorticoid modulation. The results of studies in experimental animals and humans have emphasized the possible use of alsactide for gaining compliance of severely ill patients with regard to chronic disease and/or treatment affecting immunosurveillance and for gaining tolerance to a number of potentially damaging xenobiotics [6, 7, 26, 31, 54, 115].

The chronosusceptibility of immune effectors to endogenous modulators corroborates the view that pharmacologically administered modifiers could exert different effects on the endocrine-immune network as a function of time. Appropriate analytical and computational techniques will help to understand the relationships between chronopharmacology and the neuroendocrine-immune network and to validate time-qualified immunosuppressive and immunostimulating therapies.

# References

1. Abo T, Kawate T, Itoh K, Kumagai K (1981) Studies on the bioperiodicity of the immune response. I.Circadian rhythms of human T, B and K cell traffic in the peripheral blood. J Immunol 126: 1360–1363

2. Ader R (ed) (1981) Psychoneuroimmunology. Academic Press, New York

3. Alvarez Mon M, Kehrl JH, Fauci AS (1985) A potential role for adrenocorticotropin in regulating human B lymphocyte function. J Immunol 135: 3823–3826

4. Angeli A (1983) Glucocorticoid secretion: a circadian synchronizer of the human temporal structure. J Steroid Biochem 19: 545–554

5. Angeli A, Carandente F (1985) Clinical implications of circadian changes of steroid-protein interactions. In: Chronopharmacology and chronotherapeutics Institute Scientifique Roussel, Paris, pp 38–40 (Table Ronde Roussel, vol 54)

6. Angeli A, Carandente F, Dammacco F, Martini L (1985) Progress in chronophysiology of peptide hormones with emphasis on clinical application of analogues. Chronobiologia 12: 293–304

7. Angeli A, Carandente F, Dammacco F, Halberg F, Martini L (1986) Alsactide: ACTH-agonist for use in microdoses in brain-adrenal and other feedsidewards. Chronobiologia 14: 99–143

8. Angeli A, Gatti G, Sartori ML, Delponte D, Carignola R (1988) Effects of exogenous melatonin on human natural killer (NK) cell activity. An approach to the immunomodulatory role of the pineal gland. In: Gupta D, Attanasio A, Reiter R (eds) The Pineal gland and cancer. Brain Research Promotion, London-Tübingen, pp 145–156

9. Bateman A, Singh A, Kral T, Solomon S (1989) The immune-hypothalamic-pituitary-adrenal system. Endocr Rev 10: 92–112

10. Bernton EW (1989) Prolactin and immune host defenses. Progr Neuroendocrinoimmunol 2: 21–29

11. Besedowsky HO, Del Rey A, Sorkin E (1981) Lymphokine-containing supernatants from Con-A stimulated cells increase corticosterone blood levels. J Immunol 126: 385–387

12. Besedowsky HO, Del Ray A, Sorkin E (1985) Immunoneuroendocrine interactions. J Immunol 135 (Suppl): 750–754

13. Besedowsky HO, Del Rey A, Sorkin E, Dinarello CA (1986) Immunoregulatory feedback between interleukin-1 and glucocorticoid hormones. Science 233: 652–654

14. Berkenbosch F, Del Rey A, Van Oers JWAM, Tilders FJH, Besedowsky HO (1989) Feedback circuit involving the immune hormone interleukin-1 and the hypothalamic-pituitary-adrenal system. In: Van Laon GR, Kvetnansky R, Mc Carty R, Axelrod J (eds) Stress: neurochemical and humoral mechanisms. Gordon and Breach, New York, pp 523–535

15. Birnberg NC, Lissitzky JC, Hinman M, Herbert E (1983) Glucocorticoids regulate pro-opiomelamocortin gene expression in vitro at the levels of transcription and secretion. Proc Natl Acad Sci USA 80: 6982–6986

16. Blalock JE, Harp C (1981) Interferon and ACTH induction of steroidogenesis, melanogenesis and antiviral activity. Arch Virol 67: 45–49

17. Blalock JE, Bost KL, Smith EM (1985) Neuroendocrine peptide hormones and their receptors in the immune system. Production, processing and action. J Neuroimmunol 10: 31–40

18. Blalock JE, Harbour-Mc Menonin D, Smith EM (1985) Peptide hormones shared by the neuroendocrine and immunologic system. J Immunol 135 (Suppl): 750–754

19. Blatteis CM, Llanos QJ, Awmed MS (1988) Interactions between interleukin-1 and opioid binding sites in guinea pig brain. In: Spectar NH (ed) Neuroimmunomodulation, Gordon and Breach, New York, pp 103–105

20. Bost KL, Blalock JE (1989) Activation of the pituitary-adrenal axis by the immune system. In: Van Laon GR, Kvetnansky R, Mc Carty R, Axelrod J (eds) Stress: neurochemical and humoral mechanisms. Gordon and Breach, New York, pp 513–522

21. Breder CH, Dinarello CA, Saper CB (1988) Interleukin-1 immunoreactive innervation of the human hypothalamus. Science 240: 321–324

22. Brown SL, Van Epps DE (1986) Opioid peptides modulate production of interferon-gamma by human mononuclear cells. Cell Immunol 103: 19–26

23. Carr DJJ (1988) Receptors for neuroendocrine peptides on cells of the immune system. Progr Neuroendocrinoimmunol 1: 20–21

24. Chiappelli F, Nguyen L, Esmail I, Fahey J (1990) Regulation of the expression of the homing receptor Leu 8 by glucocorticoids in vivo. Abstract Book of the 4th international workshop on neuroimmunomodulation, Florence, May 23–26, abstract 345

25. Claman HN (1972) Corticosteroids and lymphoid cells. New Engl J Med 287: 388–397

26. Cornelissen G, Sanchez De La Pena S, Halberg F (1985) From drug synergism and antagonism to endocrine chrono-immunomodulators. J Interdiscipl Cycle Res 16: 239–242

27. Cove-Smith JR, Kabler P, Pownall R, Knapp MS (1978) Circadian variations in an immune response in man. Brit Med J ii: 253–254

28. Cross RJ, Roszman TL (1989) Neuroendocrine modulation of immune function: the role of prolactin. Progr Neuroendocrinoimmunol 2: 17–20

29. Cugini P, Letizia C, Scavo D, Di Palma L, Sepe M, Battisti P, Pozzilli P, Cassisi A, Cioli AR, Scibilia G, Macchiarelli AG, Marino B, Cavallini M (1990) Circadian rhythms of T lymphocyte subsets, cortisol and cyclosporine in heart transplanted subjects. J Immunol Res 2: 59–63

30. Cupps TR, Fauci AS (1982) Corticosteroid-mediated immunoregulation in man. Immunol Rev 65: 133–155

31. Dammacco F, Campobasso N, Altomare E, Iodice G (1984) Analogues in immunology. La Ricerca Clin Lab 14: 137–147

32. Darko DF, Gillin JC, Risch CS, Bulloch K, Shahrokh G, Tasevska Z, Hamburger RN (1988) Immune cells and the hypothalamic-pituitary axis in major depression. Psychiatry Res 25: 173–179

33. Dinarello CA (1988) Biology of interleukin-1. Faseb J 2: 108–115

34. Dunn AJ (1989) Psychoneuroimmunology for the psychoneuroendocrinologist: a review of animal studies of nervous system-immune system interactions. Psychoneuroimmunology 14: 251–274

35. Fabris N, Moccheggiani E, Muzzioli M, Provinciali M (1988) Immune-neuroendocrine interactions during aging. Progr Neuroendocrinoimmunol 1: 4–9

36. Fauci AS (1975) Mechanism of corticosteroid action on lymphocyte subpopulations. I. Redistribution of circulating T and B lymphocytes to the bone marrow. Immunology 28: 669–680

37. Fauci AS, Dale DC (1975) The effects of hydrocortisone on the kinetics of normal human lymphocytes. Blood 46: 235–243

38. Fontana A, Weber E, Dayer JM (1984) Synthesis of interleukin-1/endogenous pyrogen in the brain of endotoxin-treated mice: a step in fever induction? J Immunol 133: 1696–1698

39. Fox BH (1981) Psychosocial factors and the immune system in human cancer. In: Ader R (ed) Psychoneuroimmunology. Academic Press, New York, pp 103–157

40. Gatti G, Cavallo R, Delponte D, Sartori ML, Masera R, Carignola R, Carandente F, Angeli A (1987) Circadian changes of human natural killer (NK) cells and their in vitro susceptibility to cortisol inhibition. Annu Rev Chronopharmacol 3: 75–78

41. Gatti G, Cavallo R, Sartori ML, Delponte D, Masera R, Salvadori A, Carignola R, Angeli A (1987) Inhibition by cortisol of human natural killer (NK) cell activity. J Steroid Biochem 26: 49–58

42. Gatti G, Masera R, Cavallo R, Sartori ML, Delponte D, Carignola R, Salvadori A, Angeli A (1987) Studies of the mechanism of cortisol inhibition of human natural killer cell activity: effects of calcium entry blockers and calmodulin antagonists. Steroids 49: 601–616

43. Gatti G, Cavallo R, Sartori ML, Carignola R, Masera R, Delponte D, Salvadori A, Angeli A (1988) Circadian variations of interferon-induced enhancement of human natural killer (NK) cell activity. Cancer Detect Prevent 12: 431–438

44. Gatti G, Carignola R, Masera R, Sartori ML, Salvadori A, Magro E, Angeli A (1989) Circadian-stage-specified effects of melatonin on human natural killer (NK) cell activity: in vitro and in vivo studies. Annu Rev Chronopharmacol 5: 25–28

45. Gatti G, Masera R, Sartori ML, Magro E, Carignola R, Parvis G, Angeli A (1989) Intermodulation between POMC-derived peptides and glucocorticoids in the control of human natural killer (NK) cell activity. Abstract book of the ENA 12th Annual Meeting/EBBS 21 st Annual Meeting, Turin 3–7 September, Supplement 2 Europ J Neurosci, p 199

46. Gilman SC, Schwartz JM, Milner RJ, Bloom FE, Feldman JD (1982) Beta-endorphin enhances lymphocyte proliferative response. Proc Natl Acad Sci USA 79: 4226–4230

47. Goldstein D, Dawson J, Laszlo J (1988) Suppression of natural killer cell activity by hydrocortisone. J Biol Regul Homeost Agents 2: 25–30

48. Grossman CJ (1984) Regulation of the immune system by sex steroids. Endocrine Rev 5: 435–455

49. Grossman CJ (1985) Interactions between the gonadal steroids and the immune system. Science 227: 257–261

50. Hadden JW (1983) Cyclic nucleotides and related mechanism in immune regulation: a minireview. In: Fabris N, Geraci E, Hadden J, Mitchison NA (eds) Immunoregulation, Plenum, New York, pp 201–219

51. Halberg F, Sanchez De La Pena S, Cornelissen G (1985) Circadian adrenocortical cycle and the central nervous system. In: Redfarm PH, Campbale IC, Davies JA, Martin KF (eds) Circadian rhythms in the central nervous system. VCH Weinheim (FRG); pp 57–59

52. Harbour-McMenamin D, Smith EM, Blalock JE (1985) Bacterial lipopolysaccharide induction of leukocyte-derived corticotropin and endorphins. Infection and Immunity 48: 813–817

53. Haus E, Lakatua DJ, Swoyer J, Sackett-Lundeen L (1983) Chronobiology in hematology and immunology. Am J Anat 168: 467–517

54. Haus E, Nicolau G, Lakatua DJ, Sackett-Lundeen L, Petrescu E (1989) Circadian rhythm parameters of endocrine functions in elderly subjects during the seventh to the ninth decade of life. Chronobiologia 16: 331–352

55. Heijnen CJ, Zijlstra J, Kaverlaars A, Croiset G, Ballieux RE (1987) Modulation of the immune response by POMC-derived peptides. I. Influence on proliferation of human lymphocytes. Brain Behav Immun 1: 284–291

56. Hie Stand PC, Mekler P, Nordmann R, Grieder A, Permmongkol G (1986) Prolactin as a modulator of lymphocyte responsiveness provides a possible mechanism of action for cyclosporin. Proc Natl Acad Sci USA 83: 2599–2603

57. Holaday JW, Bryant HU, Kenner JR, Bernton EW (1988) Pharmacologic manipulation of the endocrine-immune axis. Progr Neuroendocrinoimmunol 1: 6–8

58. Holbrook NJ, Cox WI, Horner HC (1983) Direct suppression of natural killer activity in human peripheral blood leukocyte cultures by glucocorticoids and its modulation by interferon. Cancer Res 43: 4019–4025

59. Irwin J, Livnat S (1987) Behavioral influences on the immune system: stress and conditioning. Progr Neuropsychopharmacol Biol Psychiat 11: 137–143

60. Jankovic BD (1989) Neuroimmunomodulation: facts and dilemmas. Immunol Lett 21: 101–118

61. Johnson HM, Smith EM, Torres BA, Blalock JE (1982) Neuroendocrine hormone regulation of an in vitro antibody production. Proc Natl Acad Sci USA 79: 4171–4175

62. Johnson HM, Torres BA, Smith EM, Dion LD, Blalock JE (1984) Regulation of lymphokine (gamma-interferon) production by corticotropin. J Immunol 132:246–250

63. Johnson HM, Russel JK, Torres BA (1988) Second messenger pathways of the immune system. Progr Neuroendocrinoimmunol 1: 24–25

64. Kawate T, Abo T, Hinna S, Kumagai K (1981) Studies on the bioperiodicity of the immune response. II. Covariations of murine T and B cells and a role for corticosteroid. J Immunol 126: 1364–1367

65. Keast D, Taylor K (1982) The effect of tri-iodothyronine on the phytohaemoagglutinin respons of T lymphocytes. Clin Exp Immunol 47: 212–220

66. Krueger JM (1988) Interleukin-1: a promotor of slow-wave sleep. In: Spector NH (ed) Neuroimmunomodulation Gordon and Breach, New York, pp 103–105

67. Kusnecov AW, Husband AJ, King MG, Pang, Smith R (1987) In vivo effects of beta-endorphin on lymphocyte proliferation and interleukin-2 production. Brain Behav Immun 1: 88–97

68. Lee SW, Tso AP, Chan H, Thomas J, Petrie K, Eugui EM, Allison AC (1988) Glucocorticoids selectively inhibit the transcription of the interleukin-1-beta gene and decrease the stability of interleukin-1-beta mRNA. Proc Natl Acad Sci USA 85: 1204–1208

69. Levi FA, Canon C, Touitou Y, Sulon J, Mechkouri M, Demey Ponsart E, Touboul JP, Vannetzel JM, Mowzowicz I, Reinberg A, Mathè G (1988) Circadian rhythms in a circulating T-lymphocyte subtypes and plasma testosterone,

total and free cortisol in five healthy men. Clin Exp Immunol 71: 329–335

70. Levy SM, Herberman RB, Simons A, Whiteside T, Lee J, Mc Donald R, Beable M (1989) Persistently low natural killer cell activity in normal adults. Immunological hormonal and mood correlates. Natl Immunol Cell Growth Regul 8: 173–186

71. Maestroni GJM, Conti A, Pierpaoli W (1986) Role of the pineal gland in immunity. Circadian synthesis and release of melatonin modulates the antibody response and antagonizes the immunosuppressive effect of corticosterone. J Immunol 13: 19–30

72. Maestroni GJM, Conti A, Pierpaoli W (1986) Melatonin regulates immunity via an opiatergic mechanism. Clin Neuropharmacol 9 (Supplement 4): 479–481

73. Maestroni GJM; Conti A, Pierpaoli W (1987) Role of the pineal gland in immunity. II. Melatonin enhances the antibody response via an opiatergic mechanism. Clin Exp Immunol 68: 384–391

74. Maestroni GJM, Conti A (1990) The pineal-endogenous opioid system-immune network. Mechanism and significance. Abstract Book 4th international workshop on neuroimmunomodulator, Florence, May 23–26. abstract 347

75. Mandler RN, Biddison WE, Handler R, Serrate S (1986) Beta-endorphin augments the cytolytic activity and interferon production of natural killer cells. J Immunol 136: 934–939

76. Masera R, Gatti G, Sartori ML, Carignola R, Magro E, Angeli A (1989) In vitro effects of beta-endorphin and cortisol on human natural killer (NK) cell activity. J Endocrinol Invest 12 (Suppl 4): 67–70

77. Mc Cain HW, Lamster IB, Bozzone JM, Grbic JT (1982) Beta-endorphin modulates human immune activity via non-opiate receptor mechanisms. Life Sci 31: 1619–1624

78. Meyer WJ, Smith EM, Richards GE, Cavallo A, Morril AC, Blalock JE (1987) In vivo immunoreactive ACTH production by human mononuclear leukocytes from normal and ACTH-deficient individuals. J Clin Endocrinol Metab 64: 98–104

79. Millington WR, Blum M, Knight R, Mueller GP, Roberts JL, O'Donohue TL (1986) A diurnal rhythm in proopiomelanocortin messenger ribonucleic acid that varies concomitantly with the content and secretion of beta-endorphin in the intermediate lobe of the rat pituitary. Endocrinology 118: 829–834

80. Miyawaki T, Taga K, Nagaoki T, Seki H, Suzuki Y, Taniguchi N (1983) Circadian changes of lymphocyte subset in human peripheral blood. Clin Exp Immunol 55: 618–622

81. Moldofsky H, Lue FA, Eisen J, Keystone E, Gorczynski RM (1986) The relationship of interleukin-1 and immune functions to sleep in humans. Psychosomat Med 48: 309–318

82. Munck A, Guyre PM, Holbrook NJ (1984) Physiological functions of glucocorticoids in stress and their relation to pharmacological actions. Endocr Rev 5: 25–44

83. Nerozzi D, Santoni A, Bersani G, Magnani A, Bressan A, Pasini A, Antonozzi I, Frajese G (1989) Reduced natural killer cell activity in major depression: neuroendocrine implications. Psychoneuroendocrinology 14: 295–301

84. O'Dorisio MS, Wood CL, O'Dorisio TM (1985) Vasoactive intestinal peptide and neuropeptide modulation of the immune response. J Immunol 135 (Suppl): 792–796

85. Pangerl A, Remien J, Haen E (1986) The number of beta-adrenoceptor sites on intact human lymphocytes depends on time of day, on season and on sex. In: Reinberg A, Smolensky M, Labrecque G (eds) Annual review of chronopharmacology, vol 3. Pergamon, Oxford, pp 331–334

86. Papic M, Stein-Streilen J, Zakarija M, Mc Kenzie JM, Guffee J, Fletcher MA (1987) Suppression of peripheral blood natural killer cell activity by excess thyroid hormone. J Clin Invest 79: 404–408

87. Parrillo JE, Fauci AS (1978) Comparison of the effector cells in human spontaneous cellular cytotoxicity and antibody-dependent cellular cytotoxicity: differential sensitivity of effectors cells to in vivo and in vitro corticosteroids. Scand J Immunol 8: 99–107

88. Pawlikowski M, Zelazowski P, Dohler K, Stepien H (1988) Effects of two neuropeptides, somatoliberin (GRF) and corticoliberin (CRF) on human lymphocyte natural killer activity. Brain Behav Immun 2: 50–56

89. Payan DG, Goetzl EJ (1985) Modulation of lymphocyte function by sensory neuropeptides. J Immunol 135 (Suppl): 783–786

90. Payan DG, Mc Gillis JP, Goetzl EJ (1987) Neuropeptide modulation of lymphocyte function. Ann NY Acad Sci 496: 182–191

91. Pert CB, Ruff MR, Weber RJ, Herkenham M (1985) Neuropeptides and their receptors: a psychosomatic network. J Immunol 135 (Suppl): 820–826

93. Plaut SM, Friedman CR (1981) Psychosocial factors in infectious disease. In: Ader R (ed) Psychoneuroimmunology. Academic Press, New York, pp 3–30

92. Plaut SM (1987) Lymphocyte hormone receptors. Ann Rev Immunol 5: 621–669

94. Riley V (1981) Psychoneuroendocrine influences on immunocompetence and neoplasia. Science 212: 1100–1109

95. Roosth J, Pollard RB, Brown SL, Meyer WJ (1986) Cortisol stimulation by recombinant interferon-alfa 2. J Neuroimmunol 12: 311–316

96. Ruff MR, Pert CB (1986) Neuropeptides are chemoattractants for human monocytes and tumor cells: a basis for mind-body communication. In: Plotnikoff NP, Faith RE, Murgo AJ, Good RA (eds) Enkephalins and endorphins: stress and the immune system. Plenum, New York, pp 387–398

97. Sapolsky R, River C, Yamamoto C, Plotsky G, Vale V (1987) Interleukin-1 stimulates the secretion of hypothalamic corticotropin-releasing factor. Science 238: 522–524

98. Schachter BS, Johnson LK, Baxter JD, Roberts JL (1982) Differential regulation of proopiomelanocortin mRNA levels in the anterior and intermediate lobes of rat pituitary. Endocrinology 110: 1442–1444

99. Schattner A, Steinbock M, Tepper R, Schonfeld A, Vaisman N, Hahn T (1990) Tumor necrosis factor production and cell-mediated immunity in anorexia nervosa. Clin Exp Immunol 79: 62–66

100. Scheving LE, Tsai TH, Pauly JE, Scheving LA, Fevers RJ, Kannabrocki EL, Lucas EA (1989) Temporal organization. Some reasons why it should not be ignored by biological researchers as well as by practicing oncologists. Chronobiologia 16: 307–329

101. Schlechte JA, Ginsberg BH, Sherman BM (1982) Regulation of the glucocorticoid recetpor in human lymphocytes. J Steroid Biochem 16: 69–74

102. Shavit Y, Terman GW, Martin FC, Lewis JW, Liebeskind JC, Gale RP (1985) Stress, opioid peptides, the immune system and cancer. J Immunol 135 (Suppl): 834–837

103. Singh A, Bateman A, Zhu Q, Shimasaki S, Esch F, Solo-

mon S (1987) Structure of a novel human granulocytic peptide with anti-ACTH activity. Biochem Biophys Res Comm 155: 524–529

104. Smith EM, Blalock JE (1981) Human lymphocytes production of corticotropin and endorphin-like substances: association with leukocyte interferon. Proc Natl Acad Sci USA 78: 7530–7534

105. Smith EM, Meyer WJ, Blalock JE (1982) Virus-induced corticosterone in hypophysectomized mice: a possible lymphoid adrenal axis. Science 218: 1311–1312

106. Smith EM, Harbour-Mc Menamin D, Blalock JE (1985) Lymphocyte production of endorphins and endorphin-mediated immunoregulatory activity. J Immunol 135 (Suppl): 779–782

107. Smith EM, Morril AC, Meyer WJ, Blalock JE (1986) Corticotropin releasing factor induction of leukocyte-derived immunoreactive ACTH and endorphins. Nature 322: 881–882

108. Spector NH (ed) (1988) Neuroimmunomodulation. Gordon and Breach, New York

109. Spangelo BL, Judd AM, Isakson PC, Mac Leod RM (1989) Interleukin-6 stimulates anterior pituitary hormone release in vitro. Endocrinology 125: 575–577

110. Stein M (1985) Bereavement, stress and immunity. In: Guillemin R, Cohn M, Melnechuk T (eds) Neural modulation of immunity. Raven, New York, pp 29–44

111. Stein M, Keller SE, Schleifer SJ (1985) Stress and immunomodulation: the role of depression and neuroendocrine function. J Immunol 135 (Suppl): 827–833

112. Stephanou A, Jessop DS, Knight RA, Lightman SL (1990) Corticotrophin-releasing-factor-like immunoreactivity and mRNA in human leukocytes. Brain Behav Immun 4: 67–73

113. Stepien H, Zelazowski P, Pawlikowski M, Dohler D (1987) Corticotropin releasing factor (CRF) suppression of human peripheral blood leukocyte chemotaxis. Neuroendocrinol Lett 9: 225–230

114. Tavadia HB, Fleming KA, Hume PD, Simpson HW (1975) Circadian rhythmicity of human plasma cortisol and PHA induced lymphocyte transformation. Clin Exp Immunol 22: 190–193

115. Touitou Y, Touitou C, Bogdan A, Chasselut J, Beck H, Reinberg A (1979) Circadian rhythm in blood variables in elderly subjects. In: Reinberg A (ed) Chronopharmacology, Pergamon, Oxford, pp 283–290

116. Trinchieri G, Perussia B (1984) Human natural killer cells: biological and pathological aspects. Lab Invest 50: 489–512

117. Tsokos GC, Balow JE (1986) Regulation of human cellular immune response by glucocorticosteroids. In: Plotnikoff NP, Faith RE, Murgo AJ, Good RA (eds) Eukephalins and endorphins: stress and the immune system. Plenum, New York, pp 159–171

118. Van Epps DE, Saland L (1984) Beta-endorphin and met enkephalin stimulate human peripheral blood mononuclear cell chemotaxis. J Immunol 132: 3046–3053

119. Vankelecom H, Carmellet H, Heremans H, Van Damme J, Dijkmans R, Billiau A, Denef C (1990) Interferon-gamma inhibits stimulated adrenocorticotropin, prolactin and growth hormone secretion in normal rat anterior pituitary cell culture. Endocrinology 126: 2919–2926

120. Vehara A, Gottschall PE, Dahl RR, Arimura A, (1987) Stimulation of ACTH release by human interleukin-1-beta, but not by interleukin-1-alfa, in conscious, freely moving rats. Biochem Biophys Res Comm 146: 1286–1290

121. Villette JM, Bourin P, Doinel C, Mansour I, Fiet J, Boudou P, Dreux C, Rove R, Debord M, Levi F (1990) Circadian variations in plasma levels of hypophyseal, adrenocortical and testicular hormones in men infected with human immunodeficiency virus. J Clin Endocrinol Metab 70: 572–577

122. Walton PE, Cronin MJ (1989) Tumor necrosis factor-alfa inhibits growth hormone secretion from cultured human adrenocortical cells. Endocrinology 125: 925–929

123. Whitcomb RW, Linehan WM, Wahl LM, Knazek RA (1988) Monocytes stimulate cortisol production by cultured human adrenocortical cells. J Clin Endocrinol Metab 66: 33–38

124. Williamson SA, Knight RA, Lightman SL, Hobbs JR (1987) Differential effects of beta-endorphin fragments on human natural killing. Brain Behav Immun 1: 329–335

125. Wolosky BRNMJ, Smith EM, Meyer WJ, Fuller GM, Blalock JE (1985) Corticotropin-releasing activity of monokines. Science 230: 1035–1037

126. Wybran J (1986) Enkephalins as molecules of lymphocyte activation and modifiers of the biological response. In: Plotnikoff NP, Faith RE, Murgo AJ, Good RA (eds) Enkephalins and endorphins: stress and the immune system. Plenum, New York, pp 253–262

# Chronobiology, Nutrition, and Diabetes Mellitus

L. Méjean, M. Kolopp, and P. Drouin

With the collaboration of C. Michaud and N. Musse

## Introduction

Blood glucose levels are commonly used as a marker of carbohydrate metabolism, and subsequent oscillations are related to various phenomena. Blood glucose levels are controlled by several hormones, i.e., insulin and the counterregulatory hormones. The circulating levels and the metabolic action of these hormones undergo different cyclic variations (Aparicio et al. 1974; Hautecouverture et al. 1975; Holst et al. 1983; Lefebvre et al. 1987; Waldhausl 1989). These cyclic oscillations may influence the hormonal action on glucose consumption by peripheral tissues, glycogen degradation, and gluconeogenesis. The exogenous supply of glucose depends on food intake, produced by a food behavior which is regulated by pre- and peri-ingestive factors which may stimulate or inhibit any motivation states such as hunger, satiety, satiation, or appetite (Nicolaïdis and Burlet 1988).

Moreover, these states are controlled by peripheral or central systems, among which the hypothalamus-hypophyseal axis is implicated (Leibowitz 1988). Any secreted peptides, may increase, i.e., neuropeptide Y (Levine and Morley 1984), or decrease, i.e., corticotropin releasing factor (CRF), neurotensin, vasopressin, oxytocin, serotonin (Fantino 1986), food intake. These peptides are located in the suprachiasmatic and ventromedian nucleis, which may be considered as a pacemaker of biological rhythms (Block and Page 1978).

It appears therefore that interrelationships between the chronobiology of nutrition and the treatment of diabetes mellitus, which were suspected for a long time (Mollerstrom 1928, 1953), are complex. Previous studies on the relationships between chronobiology and nutrition (Debry et al. 1977) or on the chronobiology of glucose tolerance (Hautecouverture et al. 1975) have been published.

In the present synthesis we will briefly review the recent developments on the chronobiology of blood glucose levels, insulin levels, and food intake and their potential consequences on the treatment of type I (insulin-dependent) and type II (non-insulin-dependent) diabetes mellitus.

## Chronobiology of Blood Glucose Levels and Glucose Tolerance

Blood glucose levels are influenced by multiple metabolic phenomena. In the study of biorhythmicity, blood sugar levels and glucose tolerance must be separated.

Circadian rhythms of glycemia have been frequently discussed: Freinkel et al. (1968) and Floyd et al. (1974) found such variations while Faiman and Moorhouse (1967), Hautecouverture et al. (1974), Malherbe et al. (1969), Schlierf and Raetzer (1972) and Deschamps et al. (1969) were not able to demonstrate a consistent circadian variation. These studies used a number of different experimental designs. More recently, Swoyer et al. (1984) found a circadian rhythm (amplitude: $4.1\% \pm 0.6\%$ of the MESOR, acrophase: $6.32 \pm 1.50$ h) on a sample of 47 healthy men and women, who were strictly synchronized in regard to their diurnal activity and sleep. The differences between the published results may be explained by the low amplitude of this rhythm.

Mejean et al. (1988) demonstrated by continuous blood glucose monitoring that concentrations of blood glucose throughout the day are not constant. The results of continuous blood glucose monitoring performed on five healthy volunteers were analyzed using both cosinor (Nelson et al. 1979) and spectral analysis (De Prins and Malbecq 1983) (Tables 1 and 2). A circadian rhythm of blood glucose was not statistically significant. However, an ultradian rhythm (period 6 h (amplitude 0.44 mmol/l, MESOR $4.71 \pm 0.14$ mmol/l, acrophase 2.56 h) was identified and confirmed using spectral analysis. Moreover a "subject" effect was found to be a main factor, as shown by the prominent periods of each individual. This ultradian rhythm with a high amplitude (between 10% and 16% of the MESOR values) was detected in each of the five subjects.

A conventional display of pooled raw data (Fig. 1 A) shows large swings with four peaks during 24 h. The first three peaks are related to the three main meals. The fourth (at the end of the night) may reflect the so-called dawn phenomenon. It appears that the ultradian rhythmicity observed in healthy volunteers is not exclusively related to carbohydrate intake but suggests a circadian variation of glucose tolerance. Several experimental results favor this hypothesis.

**Table 1.** Cosinor analysis of bioperiodicities of blood glucose (BG) and immunoreactive insulin (IRI) calculated from a sample of five healthy volunteers[a]

|  | Time period (h) | Rhythm detection | M (mmol/l + 1 SE) | Amplitude (mmol/l) | Acrophase (h:min)[b] |
|---|---|---|---|---|---|
| BG | 6 | $p < 0.005$ | $4.71 \pm 0.14$ | 0.44 | 2:56 |
|  | 24 | $0.10 > p > 0.05$ | $4.74 \pm 0.14$ | – | 16:05 |
| IRI | 6 | $0.10 > p > 0.05$ | $34.75 \pm 4.05$ | – | 3:46 |
|  | 24 | $p < 0.05$ | $33.15 \pm 3.35$ | 13.45 | 14:22 |

M, rhythm adjusted mean for the considered time periods shown in column 1
[a] For prominent periods of BG and IRI see Table 2 footnotes
[b] Using midnight as the reference point

**Table 2.** Spectral analysis of bioperiodicities of blood glucose (BG) and immunoreactive insulin (IRI) calculated from a sample of five healthy volunteers[a]

|  | Trial period (h:min)[b] | | | | | | | | |
|---|---|---|---|---|---|---|---|---|---|
|  | 1:00 | 1:30 | 3:00 | 6:00 | 8:00 | 12:00 | 24:00 | 1:30 | 6:00 |
| BG | 0.028 | 0.089 | 0.155 | 0.461 | 0.033 | 0.061 | 0.30 | 0.078 | 0.344 |
| IRI | 0.85 | 5.25 | 5.90 | 9.50 | 4.20 | 0.40 | 15.60 | 4.55 | 6.45 |

Prominent periods of BG and IRI (in hours) resulting from spectral analysis: subject 1: BG 12, IRI 12; subject 2: BG 6.9 > 8 > 24; IRI 12 > 4; subject 3: BG 8 > 6, IRI 24; subject 4: BG 5.3 > 4.8, IRI 24; subject 5: BG 4.8 > 5.3 = 24, IRI 24
[a] Without 24 and 12 hour component
[b] As measured by the amplitude (mmol/l)

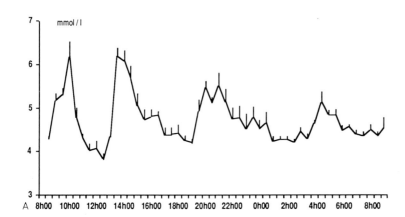

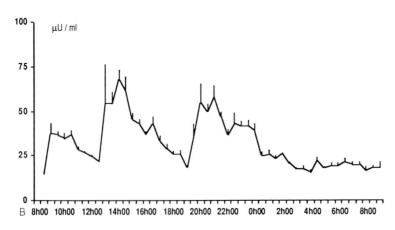

**Fig. 1 A, B.** Variations of blood glucose (**A**) and plasma insulin (**B**) levels measured in five healthy volunteers during a 24 h span

The blood glucose responses during oral glucose tolerance tests performed in the afternoon are higher than the results of the same tests performed at 0800 hours. Several studies show that this phenomenon is not modified by the age or the sex of the subjects (Jarrett and Keen 1969, 1970; Zimmer et al. 1974), the duration of fasting before the test (Bowen and Reeves 1967; Jarrett and Keen 1969), the composition of the last meal (Caroll and Nestel 1973), or the administered glucose load (Ben Dyke 1971; Jarrett et al. 1972). This phenomenon is also independent of digestive absorption since the same results are found when intravenous glucose tolerance tests (ivGTT) are performed (Abrams et al. 1968; Nemeth et al. 1970; Whichelow et al. 1974). Similarly, the hypoglycemic effect of i. v. tolbutamide is lower when the test is performed during the afternoon (Baker and Jarrett 1972), whereas discrepancies are noted when hypoglycemia is induced via i. v. insulin injections (Gibson and Jarrett 1972).

Circadian variations of glucose tolerance were reported to be absent in patients with impaired glucose tolerance such as in obese subjects (Marliss et al. 1970) or diabetic patients (Caroll and Nestel 1973; Jarrett 1974). However, Ghata and Reinberg (1979) did find a rhythm of blood and urinary glucose in type I diabetic patients.

These earlier studies were greatly expanded when the dawn phenomenon became the focus of clinical interest. This phenomenon, consisting of a rise of blood glucose concentrations in the early morning hours, was found by numerous investigators in diabetic patients (Bolli and Gerich 1984; Campbell et al. 1986; Gale et al. 1979; Gale 1985; Koivisto et al. 1986; Schmidt et al. 1979, 1981, 1984; Skor et al. 1983) and in healthy nondiabetic subjects (Schmidt et al. 1984; De Feo et al. 1986; Bolli et al. 1984; Van Cauter et al. 1989). In contrast, Atiea et al. (1988), Menneilly et al. (1986), Marin et al. (1988), and Simon et al. (1988) could not find comparable blood glucose changes in their subjects. The pathogenesis of the dawn phenomenon was first thought to be linked to the synthesis of degradation of insulin (Bolli et al. 1984; Campbell et al. 1985; Kerner et al. 1984; Widmer et al. 1988). It is now widely regarded to be caused by the nocturnal secretion of growth hormone (Arias et al. 1985; Beaufrere et al. 1988; Orskov et al. 1985; Skor et al. 1984) and is considered to be independent of blood glucose control in diabetic patients (Vaughan et al. 1986).

## Ultradian and Circadian Rhythms of Plasma Insulin Levels

In similar studies the findings on time-dependent variations of plasma insulin appear to be more homogeneous. The studies by Freinkel et al. (1968), Floyd et al. (1974), Malherbe et al. (1969), and Hautecouverture et al. (1974), despite their discrepancies with regard to blood glucose rhythmicity, all agree in regard to a circadian variation of plasma insulin. However, interindividual differences must be taken into account for both insulin and blood glucose. In an earlier study (Mejean et al. 1988), we measured plasma insulin levels in venous blood samples (every 5 min during meals and every 30 min at other times) (Table 1). Cosinor and spectral analysis were performed on the pooled raw data and showed a significant circadian rhythm (MESOR $= 33 \pm 3$ mU/ml; amplitude $= 14$ mU/ml; acrophase $= 14.22$ h). These results confirmed the work of Swoyer et al. (1984). The spectral analysis showed not only the prominence of the circadian bioperiodicity, but also the presence of an ultradian rhythm (period 6 h) which does not appear using the cosinor method. When analyzed on an individual basis, the results showed that two subjects had a prominent rhythm with a 12 h period, while the other three had only a circadian bioperiodicity. It is worth noting that the meal related insulin responses were higher after lunch than after breakfast or dinner (Fig. 1 B). The same holds true when insulin secretion was measured after various stimuli such as the oral glucose tolerance test (oGTT), ivGTT, or i. v. tolbutamide test.

The chronobiological variation of insulin secretion may be linked to either an endogenous rhythm of $\beta$ cell function or a rhythmicity of insulin receptors and insulin binding.

Several lines of evidence tend to justify the first hypothesis, namely, the demonstration of a circadian rhythm of the nuclear volume of the $\beta$ cell (Hellmann and Hellestrom 1959), the circadian variations of the in vivo pulsatility of pancreatic islet peptides (Jaspan et al. 1986), the absence of detectable rhythmicity of insulin release in patients with islet cells tumors (Villaume et al. 1984), and finally by the work of Nicolau et al. (1983) which confirms a circadian rhythm of C-peptide in healthy subjects. However, the second hypothesis is supported by several investigations indicating a circadian variation of insulin binding to the receptor of erythrocytes or monocytes (Beck-Nielsen and Pedersen 1978; Hung et al. 1986; Pedersen et al. 1982; Schultz et al. 1983 a,b; Wu et al. 1986).

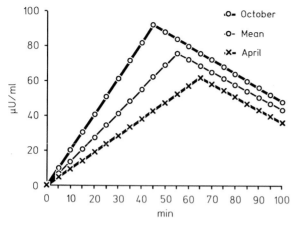

**Fig. 2.** Seasonal variations of poststimulative insulin secretion during oGTT in healthy subjects

Circannual rhythmicities of glucose tolerance and insulin-secretion were also reported. Campbell et al. (1975 a, b) showed a significant increase in the glucose/insulin ratio during autumn (March) when compared to the same ratio measured in spring (September) in 12 volunteer subjects who were living in the Antarctic. In the same way, our group compared, over a 2 year period on a monthly basis, the insulin responses during oGTT performed on a sample of healthy subjects. The insulin responses (Fig. 2) were lower and delayed in spring (April) than in autumn (October) (Mejean et al. 1977). More recently, Nicolau et al. (1983) (meal tests in elderly subjects), Haus et al. (1983) (meal tests in a group of young women), and Del Ponte et al. (1984) (tolbutamide tests in healthy subjects) detected the same pattern despite the differences of the methodological approach.

## Chronopharmacology of Insulin

### Chronobiology of Insulin Requirement in Diabetes Mellitus

In diabetic patients, the aim of insulin therapy is to restore a near normal blood glucose control and to prevent both hyper- and hypoglycemic episodes.

As mentioned in the first part of this review, the spontaneous rhythms of blood glucose are not found in diabetic patients. However, other biological rhythms are still present in these patients, which may interfere with the pharmacodynamic effects of the hypoglycemic drugs. Ruegemer et al. (1990) have re-

cently shown, in patients with type I diabetes, that the hypoglycemic risk induced by physical exercice is higher in the evening than in the morning. Similarly, Calles Escando et al. (1989) described a circadian variation of the insulin secretion patterns induced by meal tests in type II diabetic patients.

From the clinical point of view, few studies have been published showing the interrelationships of chronobiology and diabetes, either on a daily basis or for a longer period. The observations of Mirouze and Collard (1973) and Mirouze et al. (1977a) have clearly demonstrated that the injected insulin efficiency increases in the afternoon as compared to the morning. The insulin requirement decreases step by step from the morning to the night. Also, Debry et al. (1977) showed that insulin requirements are higher during the morning than during the afternoon. These observations were corroborated by Bruns et al. (1981) and Castillo et al. (1983), who demonstrated the reality of circadian variations in insulin requirements of type I diabetic patients.

Recently, we studied the ultradian, circadian, and circannual variations of blood glucose and injected insulin in type I diabetic patients (Kolopp et al. 1986). Six type I C-peptide negative diabetic patients treated by intensified conventional insulin therapy were included in the study for more than 1 year (12–27 months). On a monthly basis, using cosinor and spectral analysis, an ultradian rhythm of blood glucose (period 6 h) was found almost in all instances, whereas a circadian rhythm was not. However, a circadian rhythm was detected more frequently in patients with good or fair blood glucose control, suggesting that a normalization of blood glucose levels may contribute to the reappearance of the rhythmicity noted in healthy controls.

Emphasis has to be put on the fact that in patients with good metabolic control, blood glucose and insulin requirement were relatively independent. These results, obtained on a small number of patients, have to be confirmed, but they suggest that rhythmic changes in blood glucose and injected insulin should be recognized when designing a realistic strategy for timing and dosing of insulin therapy. However, one must keep in mind the large inter- and intraindividual variations and the rhythmicity of the pharmacokinetics of insulin (Mirouze et al. 1977b).

## Chronobiological Variations of Food Behavior

On a daily basis the existence of rhythms in food intake may be illustrated by the alimentary pattern of the newborn. Similarly, Collier et al. (1972, 1973) in rats and Migraine and Reinberg (1974) in humans have shown the existence of a circadian rhythm of food intake. Identical results were obtained by Debry et al. (1975), who demonstrated a circadian rhythm of spontaneous food intake for 4-year-old children as a group phenomenon.

The bioperiodicity of food intake does not only show daily but also weekly and annual rhythms. To illustrate this point, we may use the results of a study that our group performed on a cohort including 51 young women (21 ± 1 years old) who were students in Nancy. For 12 months, one week out of four, they reported their alimentary consumption (types and quantity of food, times of meals) and all the factual events which may have induced any modifications of their food behavior (stress, illness, tiredness, menstruation, etc.). The raw data werde analyzed using a personal software (Musse et al. 1989). We will discribe here only energy, protein, lipid, and carbohydrate intake expressed in grams and in the percentage of ingested energy and the ratio of energy consumed during each meal and between meals.

The analysis of the weekly variations in food behavior shows some differences at two particular times of the week: one on Saturday and Sunday with a high intake of protein and fats, the other on Wednesday and Thursday linked to a high intake of carbohydrates and sugars. No other study, performed under these experimental conditions, has been published to our knowledge. In a comparison with other studies reported in the literature, the methodological designs and the subjects studied differed substantially.

Our results are in agreement with those of Debry et al. (1975) on 4- to 5-year-old children: these authors found the acrophase of a circaseptan rhythm of spontaneous food consumption on Saturday and Sunday. However, no difference was found during the middle of the week. With the regard to the circannual variations of food intake, in this group of women, Fig. 3 and 4 illustrate the observed variations. As can be seen, the analysis of variance shows a progressive increase of the caloric intake, from spring to winter.

Protein intake increased parallel to the total caloric intake. The highest fat intake (expressed in g/day or as % of the total caloric intake) is observed during summer, whereas carbohydrate consumption increased during autumn and winter.

These results may be compared to those of three similar and previously published studies. In adults, Sargent (1954) reported that the food intake is higher in autumn and winter than in spring and summer. Debry et al. (1975) observed a maximum of caloric and carbohydrate consumption in summer (present study in autumn), a high fat intake in summer (present study also in summer) and the maximum of protein intake in spring and autumn (present study in summer and autumn). The methodologies used by our group and by Debry et al. (1975) are different. In the present study, the same young women were included and studied every week while Debry et al. (1975) compared the food behavior of several different samples of 4- to 5-year-old children which were different at each study period. Sargent (1954) showed a circannual rhythm of food behavior in a sample of 6–10-month-old newborns. He observed an increase in body weight at the end of summer and the beginning of autumn. The subjects ate a more lipid rich diet in spring and summer than in autumn and winter. Carbohydrate consumption was highest in summer and autumn but individual variations were noted. In our study, more than 60% of the subjects showed an increase in body weight before winter as compared to summer.

Whatever the period considered, the synchronizers of these rhythms must be discussed. An endogenous component of circadian food behavior has been demonstrated in rats by Collier et al. (1972, 1973), who established that the food behavior of these animals showed large genetic components which reflected their place in the alimentary cycle and their situation in the ecological group.

The environmental events which may influence usual food intake are characterized by the following four parameters: times, quantity, frequency, and choices of food at mealtime. If one of them is modified, the other three are changed to obtain a new balance, but the fundamental rhythm which characterizes the species is not altered. Migraine and Reinberg (1974) showed a fundamental circadian rhythm of food intake which persists when the subject lives underground without outside time cues.

However, the influence of exogenous environmental factors was discussed by Reinberg et al. (1984) and Hellbrugge (1967). For instance, Hellbrugge showed in children a circadian periodicity which appears after the first weeks of life and that environmental factors such as sleep, care, bathing periods, etc., greatly influence this phenomenon and may be con-

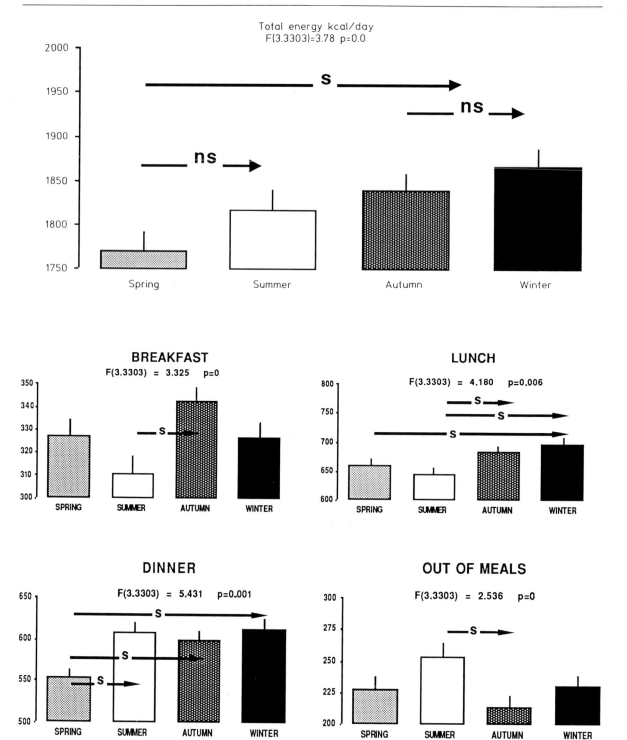

**Fig. 3.** Seasonal variations of food intake measured in 51 healthy students

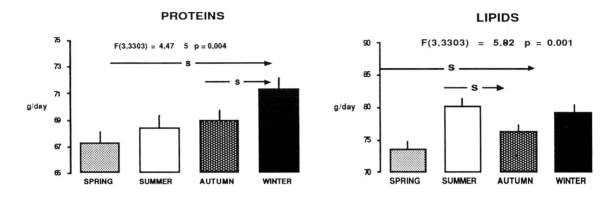

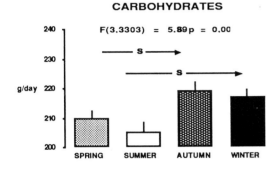

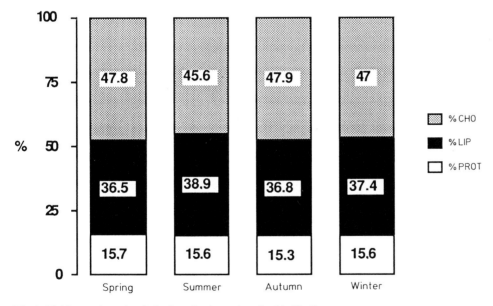

**Fig. 4.** Lipid, protein and carbohydrate intake as described in Fig. 3

sidered as synchronizers. In particular the alternation between waking and sleeping appears to be an important synchronizer of these rhythms (Minors et al. 1989). Moreover, De Castro (1987, 1988) has clearly shown that the motivation states, including hunger and satiety, are primarily modified by the effects of external oscillators.

With regard to the analysis of weekly variations in food behavior, the results observed by Debry et al. (1975) and by our group are unlikely to be related to an endogenous phenomenon. In fact, Marcocchi et al. (1987) observed a high caloric and fat intake in healthy adults during Saturday and Sunday; this abnormality seems related to the weekend phenomenon. Similarly, Debry et al. (1968) observed the same increase of energy and fat intake during days off in a group of shiftworkers.

The annual rhythmicity of food behavior appears to be biphasic, as reported by Sargent (1954) who suggests the following concepts: The *first phase* (from April to October) corresponds to a period of activity ending in preliminary steps to inactivity or inanition. The *second phase* (from October to April) is just the opposite. During the first phase, the tissues use mainly sugar as energy and stock lipids which will be used during the second phase. The seasonal variations of human food intake would thus be similar to those described in hibernating animals and would be a function of nutritional requirements. Annual changes in the photoperiod were thought to represent the adequate informative stimulus.

According to Sargent (1954), the seasonal bioperiodicities of human physiological and metabolic functions are due to an "old mechanism the function of which mimics that of hibernation in lower animals". Indeed, Haberey et al. (1967) studying the Lerot *(Eliomys quercinus)* have shown the existence of a circannual rhythm of the pentose phosphate pathway. This metabolical pathway appears to be less important during the winter months than in June or August when compared to the Embden-Meyerhoff metabolic pathway. Under these conditions, the hibernant would possess a potential NADPH reductor tank more important in summer than in winter. This would also explain that a significant increase of lipid synthesis leads to an increase in body weight.

When assessed in a population living in economically developed countries, the bioperiodicity of food intake may be greatly altered by environmental and sociocultural factors. Although energy homeostasis and requirements are still important factors in the food cycle, the permanent search for food does not appear to be the first preoccupation of such populations. The new and varying organization of the 24 h day, the week, and the year, resulting from socioeconomic influences and including weekends, holidays, and shiftwork, greatly modify food behavior. Under these conditions metabolic signals, although they are still present, lose much of their importance as compared to neurobiological functions. However, according to Reinberg et al. (1984) periodic variations of lifestyle may be synchronizers which do not induce some rhythms but may modify them.

## Conclusion

This review demonstrates the importance of chronobiological phenomena in the treatment of diabetes mellitus. The efficiency of ingested drugs or injected insulin and the effects of food intake vary, depending on when they are administered to the subjects. Under these circumstances, the development of a dietetic and pharmacological chronotherapy is justified to improve the quality of the treatment of diabetes mellitus.

*Acknowledgments.* We wish to thank C. Michaud and N. Musse for their contribution to this work.

## References

Abrams RL, Cherchio GM, Graver AI (1968) Circadian variation of intravenous glucose tolerance in man. Diabetes 17: 314

Aparicio NJ, Puchulu FE, Gagliardino JJ, Ruiz R, Llorens JM, Ruiz J, Lamas A, Miguel R (1974) Circadian variation of the blood glucose plasma insulin and human growth hormone levels in response to an oral glucose load in normal subjects. Diabetes 23: 132–137

Arias P, Kerner W, Pfeiffer EF (1985) Suppression of the dawn phenomenon by somatostatin. Diabetologia 27: 252 A

Atiea JA, Vora JP, Owens DR, Luzio S, Read GF, Walker RF, Hayes TM (1988) Non-insulin-dependent diabetic patients (NIDDMs) do not demonstrate the dawn phenomenon at presentation. Diabetes Res Clin Pract 5: 37–44

Baker G, Jarrett RJ (1972) Diurnal variation in the blood sugar and plasma insulin response to tolbutamide. Lancet 2: 945–947

Beaufrere B, Beylot M, Metz C, Ruitton A, Francois R, Riou JP, Mornex R (1988) Dawn phenomenon in type I (insulin-dependent) diabetic adolescents: influence of nocturnal growth hormone secretion. Diabetologia 31: 607–611

Beck-Nielsen H, Pedersen O (1978) Diurnal variation in insulin binding to human monocytes. J Clin Endocrinol Metab 47: 385–390

Ben Dyke R (1971) Diurnal variation of oral glucose tolerance in volunteers and laboratory animals. Diabetologia 7: 156–159

Block GD, Page TL (1978) Circadian pacemakers in the nervous system. Annu Rev Neuroscience 1: 19–34

Bolli GB, Gerich JE (1984) The "dawn phenomenon", a common occurrence in both non-insulin-dependent and insulin-dependent diabetes mellitus. N Engl J Med 310: 746–750

Bolli GB, De Seo P, De Cosmo S, Perriello G, Ventura MM, Calcinaro P, Loll C, Campbell P, Brunetti P, Gerich JE (1984) Demonstration of a dawn phenomenon in normal human volunteers. Diabetes 33: 1150–1153

Bowen AJ, Reeves RL (1967) Diurnal variation in glucose tolerance. Arch Intern Med 119: 261–264

Bruns W, Jutzi E, Fischer U, Bombor H, Woltanski KP, Jahr D, Wodrig W, Albrecht G (1981) Circadian variations in insulin concentrations and blood glucose in non-diabetics as well as insulin requirements in insulin-dependent diabetics. Z Gesamte Inn Med 36: 258–260

Calles-Escandon NJ, Jaspan J, Robbins DC (1989) Postprandial oscillatory patterns of blood glucose and insulin in NIDDM: abnormal diurnal secretion patterns and glucose homeostasis independent of obesity. Diabetes Care 12: 709–714

Campbell IT, Jarrett RJ, Keen H (1975a) Diurnal and seasonal variation in oral glucose tolerance: studies in the Antarctic. Diabetologia 11: 139–146

Campbell IT, Jarrett RJ, Rutland P, Stimmler L (1975b) The plasma insulin and growth hormone response to oral glucose: diurnal and seasonal observations in the Antarctic. Diabetologia 11: 147–150

Campbell PJ, Bolli GB, Cryer PE, Gerich JE (1985) Pathogenesis of the dawn phenomenon in patients with insulin-dependent diabetes mellitus. N Engl J Med 312: 1473–1479

Campbell PJ, Cryer P, Gerich JE (1986) Occurrence of the dawn phenomenon without a change in insulin clearance in patients with insulin-dependent diabetes mellitus. Diabetes 35: 749–752

Caroll KF, Nestel PJ (1973) Diurnal variation in glucose tolerance and in insulin secretion in man. Diabetes 22: 333–348

Castillo M, Nemery A, Verdin E, Lefebvre PJ, Luyckx AS (1983) Circadian profiles of blood glucose and plasma free insulin during treatment with semisynthetic and biosynthetic human insulin, and comparison with conventional monocomponent preparations. Eur J Clin Pharmacol 25: 767–771

Collier G, Hirsch E, Hamkin P (1972) The ecological determinants of reinforcement in the rat. Physiol Behav 9: 705–726

Collier G, Hirsch E, Kanarek R, Marwin A (1973) Environmental determinants of feeding patterns. Int J Chronobiol 1: 322 (abstract)

Debry G, Barbier JM, Girault P, Lefort JM (1968) Les habitudes alimentaires des travailleurs à feux continus. Cah Nutr Diet 3: 14–19

Debry G, Bleyer R, Reinberg A (1975) Circadian, circannual and other rhythms in spontaneous nutrient and caloric intake of healthy four year olds. Diabetes Metab 1: 91–99

Debry G, Mejean L, Villaume C, Drouin P, Martin JM, Pointel JP, Gay G (1977) Chronobiologie et nutrition humaine. XIVème Congrès International de Thérapeutique. L'Expansion Scientifique Française, Paris, pp 225–245

De Castro JM (1987) Circadian rhythms of the spontaneous meal patterns, macronutrient intake, and mood of humans. Physiol Behav 40: 437–446

De Castro JM (1988) Physiological, environmental and subjective determinants of food intake in humans: a meal pattern analysis. Physiol Behav 44: 651–659

De Feo P, Perriello G, Ventura MM, Calcinaro F, Basta G, Lolli C, Cruciani C, Dell'olio A, Santeusanio F, Brunetti P (1986) Studies on overnight insulin requirements and metabolic clearance rate of insulin in normal and diabetic man: relevance to the pathogenesis of the dawn phenomenon. Diabetologia 29: 475–480

Del Ponte A, Guagnano MT, Sensi S (1984) Circannual rhythm of insulin release and hypoglycemic effect of tolbutamide stimulus in healthy man. Chronobiol Int 1: 225–228

De Prins J, Malbecq W (1983) Analyse spectrale pour données non équidistantes. Bull Classe Sci Acad Royale de Belgique 64: 287–294

Deschamps I, Heilbronner J, Canivet J, Lestradet H (1969) Les variations spontanées de l'insuline au cours des vingt-quatre heures chez des sujets normaux. Presse Méd 77: 1815–1817

Faiman C, Moorhouse JA (1967) Diurnal variation in the levels of glucose and related substances in healthy and diabetic subjects during starvation. Clin Sci 31: 111–126

Fantino M (1986) Opiacés endogènes et prise alimentaire. Cah Nutr Diet 10: 347–358

Floyd JC, Pek S, Schteingart DE, Fajans SS (1974) Diurnal changes of plasma glucose (G), insulin (IRI) and massively obese subjects during fasting. Diabetes 33: 371

Freinkel M. Mager M, Vinnick L (1968) Cyclicity in the interrelationships between plasma insulin and glucose during starvation in normal young men. J Lab Clin Med 71: 171–178

Gale EAM (1985) The dawn phenomenon, fact or artefact. Neth J Med 28: 50

Gale EAM, Kurtz AB, Tattersall RB (1979) In search of the Somogyi effect. Lancet II: 279–282

Ghata J, Reinberg A (1979) Circadian rhythms of blood glucose and urinary potassium in normal adult males, and in similarly aged insulin dependent diabetic subjects. In: Reinberg A, Halberg F (eds) Chronopharmacology. Pergamon, Oxford, pp 315–332

Gibson T, Jarrett RJ (1972) Diurnal variation in insulin sensibility. Lancet II: 947–948

Haberey P, Dantlo C, Kaiser C (1967) Evolution des voies métaboliques du glucose et le choix alimentaire spontané chez un hibernant: le lérot. Arch Sci Physiol 21: 59–66

Haus E, Nicolau G, Halberg F, Lakatua D, Sackett-Lunden L (1983) Circannual variations in plasma insulin and C-peptide in clinically healthy subjects. Chronobiologia 10: 132

Hautecouverture M, Slama G, Assan R, Tchobroustsky G (1974) Sex diurnal variations in venous blood and plasma insulin levels. Effect of estrogens in men. Diabetologia 10: 725–730

Hautecouverture M, Slama G, Tchobroutsky G (1975) Cycle nychthéméral de la glycémie et de l'insulino-sécrétion. In: Journées annuelles de diabétologie de l'Hotel Dieu. Flammarion, Paris, pp 79–83

Hellbrugge T (1967) Ontogénèse des rythmes circadiens chez l'enfant. In: De Ajuriaguerra (ed) Cycles biologiques et psychiatrie. Masson, Paris, pp 159–181

Hellman B, Hellestrom C (1959) Diurnal changes in the function of the pancreatic islets of rats as indicated by nuclear size in the islet cells. Acta Endocrinol 31: 267–281

Holst JJ, Schwartz TW, Lovgren NA, Pedersen O, Beck-Nielsen H (1983) Diurnal profile of pancreatic polypeptide, pancreatic glucagon, gut glucagon and insulin in human morbid obesity. Int J Obes 7: 529–538

Hung CT, Beyer J, Schulz G (1986) Fasting and feeding variations of insulin requirements and insulin binding to erythrocytes at different times of the day in insulin-dependent diabetics assessed under the condition of glucose-controlled insulin infusion. Horm Metab Res 18: 466–469

Jarrett RJ (1974) Diurnal variation in glucose tolerance: associated changes in plasma insulin, growth hormone and non-esterified fatty acids and insulin sensitivity. In: Aschoff J, Ceresa F, Halberg F (eds) Chronobiological aspects of endocrinology. Schattauer Stuttgart, pp 229–238 (Symposia Medica Hoechst, vol 9)

Jarrett RJ, Keen H (1969) Diurnal variation of oral glucose tolerance; a possible pointer to the evolution of diabetes mellitus. Brit Med J 2: 341–344

Jarrett RJ, Keen H (1970) Further observations on the diurnal variation in oral glucose tolerance. Brit Med J 4: 334–337

Jarrett RJ, Baker IA, Keen H, Oakley NW (1972) Diurnal variation in oral glucose tolerance blood sugar and plasma insulin levels morning, afternoon and evening. Brit Med J 1: 199–201

Jaspan JB, Lever E, Polonsky KS, Van Cauter E (1986) In vivo pulsatility of pancreatic islet peptides. Am J Physiol 251: E 215–E 226

Kerner W, Navascues I, Torres AA, Pfeiffer EF (1984) Studies on the pathogenesis of the dawn phenomenon in insulin-dependent diabetic patients. Metabolism 33: 458–464

Koivisto VA, Yki-Jarvinen H, Helve E, Karonen SL, Pelkonen R (1986) Pathologenesis and prevention of the dawn phenomenon in diabetic patients treated with SCII. Diabetes 35: 78–82

Kolopp M, Bicakova-Rocher A, Reinberg A, Drouin P, Mejean L, Levi F, Debry G (1986) Ultradian, circadian and circannual rhythms of blood glucose and injected insulins documented in six self-controlled adult diabetics. Chronobiol Intern 3: 265–280

Lefebvre PJ, Paolisso G, Scheen AJ, Henquin JC (1987) Pulsatility of insulin and glucagon release: physiological significance and pharmacological implications. Diabetologia 30: 443–452

Leibowitz SF (1988) Neurotransmetteurs centraux et contrôle des appétits spécifiques pour les macronutriments. Ann Endocrinol 49: 133–140

Levine AS, Morley JE (1984) Neuropeptide Y: a potent inducer of consummatory behavior in rats. Peptides 5: 1025–1029

Malherbe C, De Gasparo M, De Hertogh R, Hoett JJ (1969) Circadian variations of blood sugar and plasma insulin levels in man. Diabetologia 5: 397–404

Marcocchi N, Musse JP, Kohler F, Michaud C, Michel F, Schwertz A, Drouin P, Mejean L (1987) Une expérience d'enquête nutritionnelle informatisée à la foire de Nancy. Cah Nutr Diet 22: 185–195

Marin G, Rose SR, Kibarian M, Barnes K, Cassorla F (1988) Absence of dawn phenomenon in normal children and adolescents. Diabetes Care 11: 393–396

Marliss EB, Aoki TT, Unger RM, Soeldner J, Gahill GF (1970) Glucagon levels and metabolic effects in fasting man. J Clin Invest 49: 2256–2270

Mejean L, Reinberg A, Gay G, Debry G (1977) Circannual changes of the plasma insulin response to glucose tolerance test of healthy young human males. Proceedings of the XXVII th international congress physiol sciences, p 498

Mejean L, Bicakova-Rocher A, Kolopp M, Villaume C, Levi F, Debry G, Reinberg A, Drouin P (1988) Circadian and ultradian rhythms in blood glucose and plasma insulin of healthy adults. Chronobiol Int 5: 227–236

Menneilly GS, Elahi D, Minaker KL, Rowe JW (1986) The

dawn phenomenon does not occur elderly subjects. J Clin Endocrinol Metab 63: 292–296

Migraine C, Reinberg A (1974) Persistance des rythmes circadiens de l'alternance veille sommeil et du comportement alimentaire d'un homme de 20 ans pendant son isolement souterrain et sans montre. C R Acad Sci 279: 331–334

Minors DS, Rabbitt PM, Worthington H, Waterhouse JM (1989) Variation in meals and sleep-activity patterns in aged subjects; its relevance to circadian rhythm studies. Chronobiol Int 6: 139–146

Mirouze J, Collard F (1973) Continuous blood glucose monitoring in brittle diabetes. Proceeding of the 8th Congress of the Diabetes Federation, pp 532–545

Mirouze J, Selam JL, Pham TC, Cavadoce D (1977a) Evaluation of exogenous insulin homeostasis by the artificial pancreas in insulin dependent diabetes. Diabetologia 13: 273–278

Mirouze J, Selam JL, Pham TC, Orsetti A (1977b) Le pancréas artificiel extracorporel, nouvelle orientation du traitement insulinique. In: Mirouze J (ed) XIVème congrès international de thérapeutique. Expansion Scientifique, Paris, pp 79–91

Mollerstrom J (1928) Om dygnsvariationer; blod-och urinsocker-kurvan hos diabetiker. Hygieia (Stockholm) 91: 23

Mollerstrom J (1953) Rhythmus, Diabetes und Behandlung. Verhandlungen der dritten Konferenz der Internationalen Gesellschaft für biologische Rhythmusforschung (Hamburg 1949). Acta Med Scand (Suppl) 278: 1949

Musse N, Michaud C, Musse JP, Mejean L (1989) Gestion informatisée de l'enquête alimentaire. In: Nutrition et santé: le défi scientifique du XXI ème siècle, Montreal. Abstract 4, p 53

Nelson W, Tong YK, Jueng-Kuen L, Halberg F (1979) Methods for Cosinor rhythmometry. Chronobiologia 6: 305–323

Nemeth S, Vigas M, Macho L, Stukovsky R (1970) Diurnal variations of the disappearance rate glucose from the blood of healthy subjects is revealed by iv GTT. Diabetologia 6: 641

Nicolaidis S, Burlet C (1988) Les régulations de la prise alimentaire. Ann Endocrinol 49: 87–88

Nicolau GY, Haus E, Lakatua DJ, Bogdan C, Popescu M, Petrescu E, Sackett-Lundeen L, Stelea P, Stelea S (1983) Circadian and circannual variations in plasma immunoreactive insulin (IRI) and C-peptide concentrations in elderly subjects. Endocrinologie 21: 243–255

Orskov H, Schmitz O, Christiansen J (1985) More evidence of growth hormone's role in the dawn phenomenon. Diabetes Res Clin Pract (Suppl 1): 426

Pedersen O, Hjollund E, Lindsky HO, Beck-Nielson H, Jensen J (1982) Circadian profiles of insulin receptors in insulin-dependent diabetics in usual and poor metabolic control. Am J Physiol 242: E 127–E 136

Reinberg A, Debry G, Levi F (1984) Chronobiologie et Nutrition. Encyclopédie Médicochirurgicale, Paris, 10390A10, pp 1–30

Ruegemer JJ, Squires RW, Marsh HM, Haymond MW, Cryer PE, Rizza RA, Miles JM (1990) Differences between prebreakfast and late afternoon glycemic response to exercise in IDDM patients. Diabetes Care 13: 104–110

Sargent F (1954) Season and the metabolism of fat and carbohydrate, a study of vestigial physiology. Meteorol Monogr (USA) 2: 68–80

Schlierf G, Raetzer H (1972) Diurnal patterns of blood sugar, plasma insulin free fatty acid and triglyceride levels in normal subjects and patients with type IV hyperlipoproteinemia and the effect of meal frequency. Nutr Metab 17: 113–126

Schmidt MI, Hadji-Georgopoulos A, Rendell M, Margolis S,

Kowarsky D, Kowarsky A (1979) Fasting hyperglycemia and associated free insulin and cortisol changes in "Somogyi-like" patients. Diabetes Care 2: 457–464

Schmidt MI, Hadji-Georgopoulos A, Rendell M, Kowarsky A, Margolis S, Kowarsky D (1981) The dawn phenomenon, an early morning glucose rise implications for diabetic intraday blood glucose variation. Diabetes Care 4: 579–585

Schmidt MI, Lin QX, Gwynne JT, Jacobs S (1984) Fasting early morning rise in peripheral insulin: evidence of the dawn phenomenon in nondiabetes. Diabetes Care 7: 32–35

Schulz B, Greenfield M, Reaven GM (1983a) Diurnal variation in specific insulin binding to erythrocytes. Exp Clin Endocrinol 81: 273–279

Schulz B, Ratzmann KP, Albrecht G, Bibergeil H (1983b) Diurnal rhythm of insulin sensitivity in subjects with normal and impaired glucose tolerance. Exp Clin Endocrinol 81: 263–272

Simon C, Brandenberger G, Follenius M (1988) Absence of the dawn phenomenon in normal subjects. J Clin Endocrinol Metab 67: 203–205

Skor DA, White HN, Thomas L, Shah SD, Santiago JV (1983) Examination of the role of the pituitary-adrenocortical axis, counterregulatory hormones and insulin clearance in variable nocturnal insulin requirements in insulin-dependent diabetics. Diabetes 32: 403–407

Skor DA, White NH, Thomas L, Santiago JV (1984) Relative roles of insulin clearance and insulin sensitivity in the pre-breakfasting increase in insulin requirements in insulin-dependent diabetic patients. Diabetes 33: 60–63

Swoyer J, Haus E, Lakatua D, Sackett-Lundeen L, Thompson M (1984) Chronobiology in the clinical laboratory. In: Haus E, Kabat J (eds) Proceedings of the XVth International Conference of the International Society of Chronobiology. Basel, Karger, p 533–543

Van Cauter E, Desir D, Decoster C, Fery F, Balasse EO (1989) Nocturnal decrease in glucose tolerance during constant glucose infusion. J Clin Endocrinol Metab 6: 604–611

Vaughan NJ, Rao RH, Kurtz AB, Buckell HM, Spathis GS (1986) Strict nocturnal diabetic control diminishes subsequent glycemic escape during acute insulin withdrawal. Metabolism 35: 136–142

Villaume C, Beck B, Dollet JM, Pointel JP, Drouin P, Debry G (1984) 28-hour profiles of blood glucose (BG), plasma immunoreactive insulin (IRI) and IRI/BG ratio in four insulinomas. Ann Endocrinol (Paris) 45: 155–160

Waldhausl W (1989) Circadian rhythms of insulin needs and actions. Diabetes Res Clin Pract 6: 17–24

Whichelow MJ, Sturge RA, Keen H, Jarrett RJ, Stimmler L, Grainger S (1974) Diurnal variation in response to intravenous glucose. Brit Med J 1: 488–491

Widmer A, Keller U, Pasquel M, Berger W (1988) Alterations in insulin clearance and hepatic blood flow during the night do not contribute to the "dawn phenomenon" in type I diabetes. Horm Res 29: 197–201

Wu MS, Ho LT, Jap TS, Chen JJ, Kwok CF (1986) Diurnal variation of insulin clearance and sensitivity in normal man. Proc Natl Sci Counc Repub China 10: 64–69

Zimmer PZ, Wall JR, Rome R, Stimmler L, Jarrett RJ (1974) Diurnal variation in glucose tolerance: asociated changes insulin growth hormone and non esterified fatty acids. Brit Med J 1: 485–491

Dawn phenomena in diabetes. Editorial. (1984) Lancet 1: 1333–1334

# Endocrine and Neuroendocrine Axis Testing in Clinical Medicine: Evidence for Time Dependency

Y. Touitou and A. Bogdan

## Introduction

Numerous tests are used in clinical practice to assess the integrity of different parts of the endocrine and neuroendocrine systems. In these procedures the relationship between the inhibitory or activating effect of the diagnostic agent and the plasma concentration of the responding hormone are most often considered, without any regard to temporal aspects. As will be shown in this review, however, the time of testing represents a variable of great importance in the interpretation of the data. Furthermore, the time-related differences in the effects of the drugs used as diagnostic but also as therapeutic agents may be critical in the interpretation of both stimulation and suppression tests of endocrine and neuroendocrine functions (Touitou 1989).

## Hypothalamo-Pituitary-Adrenal Axis

The glucocorticoid hormones (mainly cortisol and corticosterone), the adrenal androgens (mainly DHEA, DHEA-S) and some 18-hydroxylated steroids (18-hydroxy-corticosterone, 18-hydroxy-11-deoxycorticosterone) are under the control of the secretion of ACTH from the anterior pituitary. ACTH release is, in turn, regulated by a specific neurohormone, the corticotropin-releasing hormone (CRH), secreted by the hypothalamus (Rivier and Plotsky 1986). The release of CRH by the hypothalamus is governed by a complex control mechanism including: (1) an endogenous oscillator presumably related to the suprachiasmatic nucleus of the hypothalamus determining circadian (and presumably other) rhythmic function; (2) superimposed stimulatory effects from the central nervous system; and (3) a negative feedback through endogenous and exogenous steroid hormones. The hypothalamus, pituitary, and adrenal cortex thus form a neuroendocrine axis. CRH released from the hypothalamus into the hypophyseal portal vessels activates the release of ACTH from the anterior pituitary and this latter hormone, in turn, activates the secretion of cortisol from the adrenal cortex. The secretion of ACTH (and probably CRH) is episodic in nature, and the frequency, duration, and temporal distribution of these episodes result in parallel circadian rhythms in secretion and plasma concentration of ACTH-dependent steroids, e.g., cortisol.

Stimulation tests of ACTH secretion involve the use of CRH, metyrapone, insulin, pyrogen, and vasopressin. Suppression tests of ACTH secretion use synthetic glucocorticosteroids, mainly dexamethasone. The stimulation test of adrenocortical function is based on the use of ACTH.

## Stimulation Tests of ACTH Secretion

The hypothalamic-pituitary-adrenal axis exhibits in mammals a well-known circadian rhythm. In humans, non-human primates, and other diurnal species, the peak concentrations of plasma cortisol and ACTH occur before or at the time of awakening, i.e., around the onset of daylight in the early morning hours, whatever the age and sex of the individual (Krieger 1979, Touitou et al. 1982). Data obtained from nocturnally active rodents indicate that the peak of plasma corticosterone concentrations occurs around the onset of darkness, i.e., at the time preceding their period of activity.

Many different types of physical and psychological stress cause ACTH release. Since the ACTH response is thought to be mediated by CRH and inhibited by natural and synthetic glucocorticoids, any interference with the synthesis of cortisol, and its negative feedback on ACTH release (e.g., Addison's disease, congenital adrenal hyperplasia, and drug-induced enzyme blockade), will result in increased secretion of ACTH.

## Corticotropin-Releasing Hormone

CRH, a major physiological regulator of pituitary ACTH secretion, is secreted by the hypothalamus, then distributed via the hypophyseal portal venous system. Hypothalamic and median eminence CRH bioactivity and CRH-like immunoreactivity exhibit a complex pattern. Some data indicate that CRH synthesis and release occur in the hypothalamus before darkness, although some other types of circadian variations were also reported (Hiroshige and Sakakura 1971; Szafarczyk et al. 1980; David-Nelson and Brodish 1969; Moldow and Fischman 1984). As CRH concentrations are difficult to determine in plasma, continuous sampling of cerebrospinal fluid (CSF) over 24 h was performed in rhesus monkeys and showed peak values in the evening (at 1930 hours) and trough values in the morning (0745 hours). Simultaneously sampled cortisol peaked around 0900 hours with a nadir around 2230 hours (Garrick et al. 1987). CRH and CRH receptors are also widely distributed in other brain regions of primates and rodents. The unexpected difference in the timing of the circadian rhythms of CRH and cortisol suggests that CRH in CSF could reflect or mediate some nonhypophysiotropic function of this peptide in the brain (Garrick et al. 1987).

CRH has been isolated from ovine hypothalami and its primary structure has been determined (Vale et al. 1983; Spiess et al. 1981). Synthetic ovine type CRH was found to stimulate ACTH secretion from the human pituitary (Vale et al. 1981; Chan et al. 1982; Grossman et al. 1982; Suda et al. 1983; Lamberts et al. 1984). It became therefore possible to document the response of the pituitary to the administration of CRH (Tsukada et al. 1983; Schulte et al. 1985; Watabe et al. 1985).

Plasma ACTH and cortisol responses to CRH were determined in the morning (0900 hours) and evening (2200 hours) in seven healthy men by Tsukada et al. (1983). Synthetic ovine CRH (100 μg) or saline was given intravenously and blood samples were collected before and 15, 30, 45, 60, 90, and 120 min thereafter. Plasma ACTH and cortisol concentrations were significantly higher after CRH injection in the morning than in the evening. However, neither the maximum increments in plasma ACTH

and cortisol above the control values nor the increments at each time point following CRH administration in the morning differed significantly from those after CRH administration in the evening. Also, increments in the area under the concentration curves of ACTH and cortisol following CRH injection in the morning did not differ significantly from those in the evening. Since it is unlikely that the metabolic clearance rate of ACTH and cortisol is different between morning and evening, these data suggest that the responsiveness of the pituitary to CRH in the morning and in the evening does not differ significantly (Tsukada et al. 1983). The same conclusions were reached with the administration of 1 μg/kg of synthetic ovine CRH to six healthy men at 0900 or 1700 hours, using the same protocol (Watabe et al. 1985). In contrast to these data, some investigators have reported a significantly larger response of ACTH and cortisol at 0900 than at 2200 hours when administering 1 μg/kg of CRH (Schulte et al. 1985), whereas others have reported a greater plasma cortisol (but not ACTH) response to CRH when this agent is given in the evening (Copinschi et al. 1983; De Cherney et al. 1985). These differences cannot be attributed to the dose of CRH administered because no significant difference in the response was found with doses ranging from 1 μg/kg to 100 μg/subject (Orth et al. 1983; Schürmeyer et al. 1984). The reason for this discrepancy is therefore poorly understood. The limitation in the study of circadian rhythms at two time points only and possible differences in the synchronization of the subjects may have to be considered. Also, since the activity of the hypothalamo-pituitary-adrenal axis is thought to depend upon the sleep-wake schedule of each individual and is greatest at the time of (or shortly after) awakening and lowest at around the time of retiring (Orth et al. 1967; Tanaka et al. 1978), it is possible that a clearer difference could be observed if the responses to CRH were compared at times closer to awaking and retiring. An another possibility is that CRH response also depends upon the months of the year. Unfortunately, in none of the above quoted papers was the month of the experiments specified, although seasonal or circannual rhythms have been demonstrated in the plasma concentration of a number of hormones (Touitou et al. 1983a,b,c, 1984; Haus et al. 1989) and in the response of hormones to a number of drugs (Reinberg et al. 1983).

Continuous and/or closely repetitive stimulation with gonadotropin-releasing hormone (GnRH) or growth hormone-releasing hormone (GH-RH) leads to blunted Gn (Knobil 1980; Crowley et al. 1981) or

GH (Losa et al. 1984; Webb et al. 1984; Vance et al. 1985) responses in human subjects due to pituitary desensitization. In contrast, the increases in plasma ACTH during repetitive human CRH administration (100 µg, i. v.) to normal subjects were within the same range, independent of the intervals (60, 90, 180 min) between the CRH pulses. When CRH was infused continuously (100 µg/h for 3 h) after an initial CRH bolus (100 µg i. v.), ACTH, and cortisol remained elevated at a nearly constant level during the infusion. These data suggest that there is no desensitization of the CRH receptor or depletion of a readily releasable ACTH pool, as is observed with other pituitary hormones after releasing hormone stimulation (Schopohl et al. 1986).

A pulsatile administration of human CRH in patients with secondary adrenal insufficiency is able to restore both cortisol secretion and a normal plasma cortisol circadian pattern when the pulses are timed similar to the seven to nine spontaneous endogenous plasma cortisol episodes (Avgerinos et al. 1986).

### Insulin, Pyrogen, and Vasopressin

Circadian variations of ACTH and cortisol secretion in response to several stimuli such as insulin-induced hypoglycemia (Takebe et al. 1969; Ichikawa et al. 1972), pyrogen (Takebe et al. 1966, 1968), or vasopressin (Clayton et al. 1963) have been reported.

*Insulin-induced hypoglycemia* is based upon the observation that a fall in blood glucose to less than 0.40 g/l (2.2 mmol/l) results in a strong promotion of ACTH release and cortisol production. Plasma ACTH, GH, and cortisol responses to an i. v. injection of insulin (0.1 U/kg) were documented in healthy subjects at 0900 and 2100 hours with blood sampling at 0, 15, 30, 45, 60, 90, and 120 min after injection. No significant differences were found between the absolute values of plasma ACTH, GH, or cortisol concentrations observed following insulin hypoglycemia induced in the morning and in the evening. In contrast, the increment of plasma cortisol concentration at 45, 60, and 90 min after insulin injection was significantly greater in the experiment performed at 2100 than in that performed at 0900 hours. The apparent discrepancy observed between the increment of cortisol concentration and ACTH concentration in plasma may be explained by response characteristics of the adrenal cortex to ACTH. It was suggested that the amount of ACTH necessary for a specific increment in plasma cortisol concentration from a low basal level is less than that necessary for the same increment in cortisol concentration from a higher basal level (Ichikawa et al. 1972). This finding may also be the expression of the circadian cycles of sensitivity of the adrenal to ACTH described in experimental animals (Haus et al. 1974).

No significant difference according to the time of administration could be observed in the peak levels of cortisol attained by *insulin* (Takebe et al. 1969), *pyrogen* (Takebe et al. 1966, 1968), or *vasopressin* (Clayton et al. 1963). Also, the urinary excretion of steroids 6 or 24 h after injection of pyrogen at 2200 was not different from the value obtained after an injection at 1000 hours (Takebe et al. 1968). In contrast, Takebe et al. (1966) reported in three healthy subjects that plasma bioassayable ACTH after pyrogen injection at 2300 was elevated, whereas no response was observed after the injection at 0900.

### Metyrapone Test

Metyrapone (2-methyl-1,2-di-3-pyridyl-1-propanone; SU 4885; metopirone) is a drug used to test the ability of the pituitary to respond to a decreased concentration of plasma cortisol. Its administration represents therefore an indirect stimulation test of the pituitary. This drug is currently used to inhibit adrenal steroidogenesis by suppressing the excess of steroids, e.g., in adrenal cortical carcinomas. Metyrapone inhibits, among other enzymes, adrenal 11 $\beta$-hydroxylase activity, which results in a fall in the production of cortisol associated with a rise of 11-deoxycortisol (S), its precursor. Due to the reduction of cortisol secretion, a compensatory increase in ACTH release follows and the secretion of 11-deoxycortisol and its urinary metabolite (tetrahydro-11-deoxycortisol, THS) become elevated. Production of other 11-hydroxylated compounds such as corticosterone and aldosterone is also inhibited, with a concomitant increase in secretion of the corresponding 11-deoxy compounds.

The in vivo effects of metyrapone on adrenal steroid synthesis have been studied mainly to find out which is the best way to induce a steroid response. Changes in the effects of metyrapone on adrenal steroidogenesis as a function of the timing of its administration have been reported by several authors (Martin and Hellman 1964; Sprunt et al. 1967; Jubiz et al. 1970; Kohlberg et al. 1972; Stahl et al. 1972; Takebe 1973; Perlow et al. 1974; Tucci 1975). Timing of drug administration has been studied either with a single dose per 24 h at different clock hours or with repeated doses given at different time intervals. Touitou et al. (1976, 1977) showed a circadian variation of both S and urinary THS after a 36 h sustained administration of metyrapone (0.75 g every 4 h; total dose 7.5 g) to healthy human subjects. Thus, the partial replacement of a physiologic secretion (cortisol) by the drug-induced secretion of a cortisol precursor (11-de-

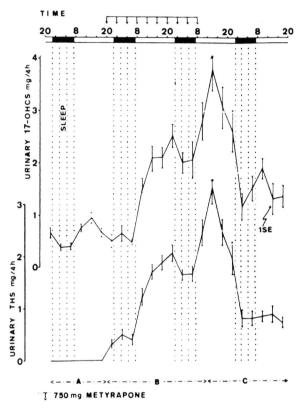

**Fig. 1.** Circadian rhythms of urinary 17-hydroxycorticosteroids *(17-OHCS)* and tetrahydro-11-deoxycortisol *(THS)* in response to a 36 h, 4 hourly, sustained administration of metyrapone to young healthy men. A rhythmic response is observed despite sustained administration of the drug. (From Touitou et al. 1976, 1977) *A,* before administration of metyrapone; *B,* during administration of metyrapone; *C,* after administration of metyrapone

oxycortisol) did not alter the circadian rhythm of the involved secretions including the timing of their peaks (Figs. 1 and 2). The circadian variations in the secretion of S, the precursor of cortisol, are similar to and slightly precede those of cortisol (Touitou 1976). These studies also showed that, even with a sustained administration of the drug (fixed dose at regular time intervals), a circadian rhythm of both plasma S and urinary THS occurred. Two major hypotheses can be considered to explain the observed changes: (1) the circadian rhythms of plasma S and urinary THS reflect exclusively the rhythm of the pituitary adrenal system, 11 $\beta$-hydroxylase inhibition being constant, or (2) the rhythms of plasma S and urinary THS are also related to a circadian variation in the inhibitory effect of metyrapone (Touitou et al. 1976). In these studies, the estimated portion of urinary 11-hydroxylated tetrahydrocorticocosteroids (corresponding to the fraction of the secretion of cortisol not inhibited) does not show a significant rhythm during metyrapone administration. Thus, these data favor the hypothesis of a circadian change in the response of the pituitary adrenal system to metyrapone. Data by Angeli et al. (1977) agree with these findings although the experimental protocol was different. A single (750 mg) dose of metyrapone was administered orally to healthy men in separate experiments, in 1 week intervals, at different fixed times (0000, 0400, 0800, 1200, 1600, and 2000 hours).

Blood samples were drawn before ingestion and then at 1 h intervals for five consecutive hours. The inhibition of 11 $\beta$-hydroxylase, expressed as the percentage of S among the glucocorticoids, i. e., $S/S + F$, varied along the 24 h scale reaching its maximum at 2000 and its minimum at 0800 hours. The most pronounced release of ACTH after metyrapone administration, indirectly evaluated by the increase of plasma total glucocorticoids, was observed when the drug was administered at 0000 and 0400 hours and

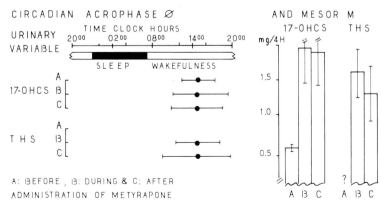

**Fig. 2.** Urinary 17-hydroxycorticosteroids *(17-OHCS)* and tetrahydro-11-deoxycortisol *(THS)* before *(A)*, during *(B)*, and after *(C)* a sustained administration of metyrapone in healthy man. A statistically significant rise of 17-OHCS and THS is detected during *(B)* and after *(C)* with reference to before *(A)*. The acrophases of detected rhythms are located around 1400. (From Touitou et al. 1976, 1977)

very little or not at all at any of the other times studied. All these data indicate that: (1) the inhibitory effect of metyrapone on 11 $\beta$-hydroxylase in humans shows a circadian rhythm with a peak at 2000 hours corresponding to the highest level of S and (2) when removing cortisol feedback by low doses of metyrapone, ACTH release is increased only during a portion of the 24 h scale, namely between 0000 and 0400 hours, independent of the circadian variations in 11 $\beta$-hydroxylase inhibition by metyrapone. This suggests strongly that the pituitary is able to understand the feedback signal coming from the adrenals, i.e., the plasma cortisol levels, only during the early morning hours (0000–0400 hours) at the beginning of the expected elevation of cortisol secretion. In addition to these data, toxic effects of metyrapone (300–400 mg/kg) in the mouse were found to be circadian stage-dependent, with a peak significantly higher at 1600 hours, in animals standardized under a light-dark regimen with light from 0600 to 1800 hours (Ertel et al. 1964). Recent data indicate that the in vitro inhibition of 11 $\beta$-hydroxylase activity by metyrapone in mouse adrenals is circadian stage-dependent (Y. Touitou unpublished data).

## Stimulation Tests of Adrenocortical Function

### Corticotropin (ACTH)

In clinical practice, ACTH testing is mainly used to screen for adrenal insufficiency, to document adrenal recovery after withdrawal of glucocorticoid therapy, and to screen for the nonclassical or late-onset forms of congenital adrenal hyperplasia. At the standard 250 μg dose of ACTH 1–12, an i.v. bolus injection stimulates adrenal secretion as effectively as a 2 h continuous ACTH infusion (Munabi et al. 1986).

Circadian changes in the effects of ACTH as a function of its time of administration have been reported both in vitro and in vivo. Using mouse adrenals, Ungar and Halberg (1962) found that the in vitro response of corticosterone to ACTH was highest when plasma corticosterone levels were lowest. In contrast, Dallman et al. (1978) showed that in rats the corticosterone response to ACTH was in phase with the circadian rhythm of plasma corticosterone.

With the analogue ACTH 1-17 ($\beta$-1-alanyl-17-lysyl-corticotropin-1-17-heptadecapeptide-4-amino-$N$-butylamide), Reinberg et al. (1980) reported that in

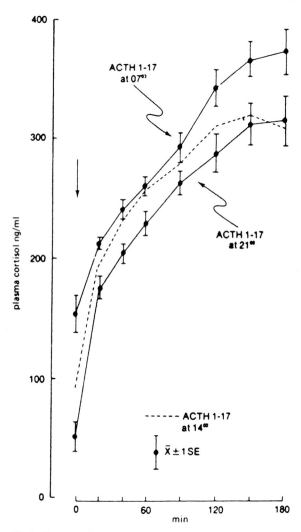

**Fig. 3.** Compared response curves of plasma cortisol after injection of ACTH 1-17 at three different time points (From Reinberg et al. 1980)

healthy men studied during winter, the largest response, gauged both by plasma cortisol and urinary 17-hydroxycorticosteroids (17 OHCS) was obtained when the peptide was injected at 0700 rather that at 1400 or 2100 hours (Figs. 3 and 4). However, different data were obtained with the same peptide at the same dose (100 μg i. m.) by Günther et al. (1980) in arthritic male and female patients studied in August and in healthy females studied in May by Ferrari et al. (1982). To test the hypothesis that these discrepancies could be related to changes in the effects of ACTH as a function of season, Reinberg et al. (1983) studied the same subjects who had been studied previously during winter (Reinberg et al. 1980) again with a comparable protocol during summer. The subjects

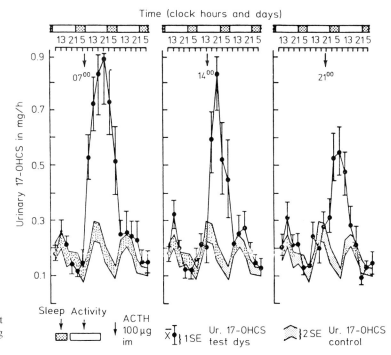

**Fig. 4.** Compared responses of urinary excretion of 17-hydroxycorticosteroids *(17-OHCS)* after injection of ACTH 1-17 at three different time points. (From Reinberg et al. 1980)

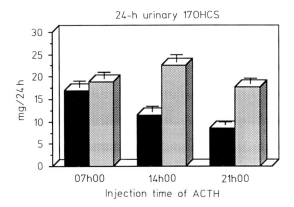

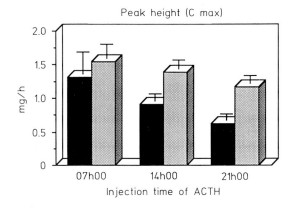

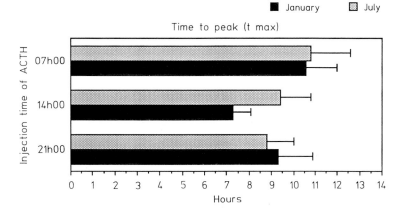

**Fig. 5.** Seasonal changes in the effect of ACTH on adrenocortical function in humans: role of the time of year (January vs July) on the 24 h urinary output of the 17-hydroxycorticosteroids *(17-OHCS)* and the peak height *(C max)* and the time to peak *(t max)* after administration of ACTH 1-17 at three different time points in January and July. (From Reinberg et al. 1983)

were sampled before and for 3 h (seven samples) after an i.m. injection of physiologic saline (control) or 100 µg ACTH 1-17 either at 0700, 1400, or 2100 hours. Each subject was his own control. A period of 1 week separated each injection. The study lasted 6 weeks for each experiment. The 24 h urinary 17-OHCS excretion was largest for ACTH injected at 0700 in winter and 1400 hours in summer and the minimum occurred after ACTH given at 2100 hours (Fig. 5). The highest peak of urinary 17-OHCS was found after ACTH at 0700 both in winter and summer (Fig. 5). In addition, both in winter and summer, the injection of ACTH at 0700 was followed by the greatest decrease in self-rated fatigue (24 h mean) and the largest increase (24 h mean) both in grip strength and peak expiratory flow rate (bronchial patency) as compared to the injections at 1400 and 2100 hours (Reinberg et al. 1983). The strongest stimulation of glucocorticoid secretion thus occurs when ACTH is administered around the beginning of the activity span, when the activity of the adrenals is close to its circadian peak.

Beside its action on steroid synthesis, ACTH is known to increase GH secretion in humans (Köbberling et al. 1976), an effect that has been related either to a direct action on pituitary somatotrophic cells,

with changes in the neurotransmitters controlling GH secretion, or to an effect on GH-RH secretion by the hypothalamus. ACTH 1-17, a potent ACTH analogue, was also found to increase GH secretion, with the highest GH response 40 min after the administration of ACTH at 1400 or 2100 and the lowest response after the 0700 hours administration (Fig. 6). This effect was found not to be mediated by a concomitant increase in GH-RH the main physiologic stimulant of ACTH release (Touitou et al. 1988, 1990). However the effect of ACTH in stimulating GH secretion should be reexamined taking into account possible time-dependent variations in the density and/or sensitivity of GH receptors.

## Suppression Tests of ACTH Secretion

### Synthetic Glucocorticoids

Both *cortisol and synthetic glucocorticoids* inhibit the secretion of ACTH by the pituitary through a negative feedback regulatory mechanism between ACTH release and plasma glucocorticoid concentration (Sayers and Sayers 1947; Yates et al. 1961; Dallman and Yates 1969). Many studies have demonstrated in humans and rats that the suppressive effect of ACTH by glucocorticoids is dependent on the time of their administration. When administered just prior to the circadian rise of ACTH, the inhibition of the pituitary is greater than at any other time of administration whether in humans (Di Raimondo and Forsham 1956; Nichols et al. 1965; Krieger et al. 1971, 1974) or rats (Nicholson et al. 1985; Akana et al. 1986). Various corticoids including dexamethasone (Di Raimondo and Forsham 1956; Nichols et al. 1965; Nugent et al. 1965), triamcinolone (Grant et al. 1965), methylprednisolone (Ceresa et al. 1969), and cortisol (Perlow et al. 1974) have been shown to suppress ACTH secretion in a time-dependent fashion. Ceresa et al. (1969) have demonstrated that administration of 6-methyl prednisolone (660 µg per hour) intravenously to healthy subjects resulted in a strong inhibition of cortisol secretion when infused between 0000 and 0400, a weaker inhibition when infused between 0400–0800, 0400–1600, and 1600–2000, and no effect at all when administered between 0800 and 1600 hours (Fig. 7). Therefore, the time of administration of glucocorticoids to humans must be adjusted in accord with the effect required: if suppression of endogenous secretion of cortisol is required, the admin-

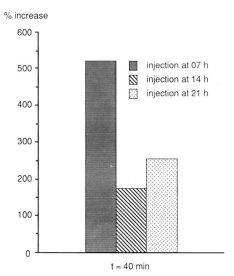

**Fig. 6.** Increment of plasma GH concentration after ACTH 1-17 injection at three different circadian steps. Each column represents the percent increase of GH in comparison with saline at the same time. (From Touitou et al. 1990)

|  | GH concentrations (ng/ml) | | |
|---|---|---|---|
|  | 07 h 40 | 14 h 40 | 21 h 40 |
| ACTH | 6.41 | 12.95 | 11.84 |
| Saline | 1.04 | 4.73 | 3.36 |

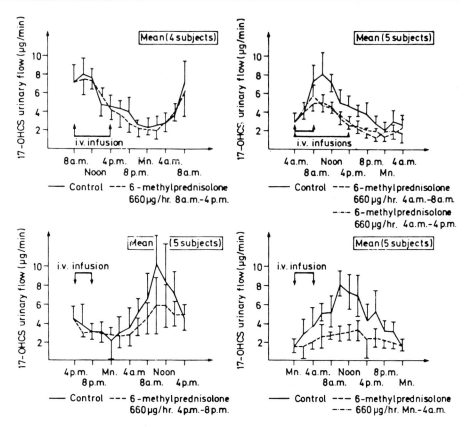

**Fig.7.** Changes in the urinary profiles of 17-hydroxycorticosteroids *(17-OHCS)* after i.v. administration of 6-methylprednisolone (660 µg/h) at different time points in the 24 h scale. (From Ceresa et al. 1969)

istration must take place in the evening; if this suppression is not required, then the administration must take place in the morning or the early afternoon. These data also suggest that the circadian rhythmicity of the hypothalamo-pituitary-adrenal axis involves two different oscillating functions: a "basal" circadian secretory rhythm extending throughout the entire day and superimposed, an "impulsive" rhythm with a tendency for increased number and/or height of "episodic" secretions and apparently also increased sensitivity to feedback effects. This impulsive phase is found during a limited time span, from 0000 to 0700 hours. Corticosteroid inhibition of ACTH and metyrapone stimulation are more effective during the highly secretory "impulsive" phase of ACTH secretion and less effective during the basal phase.

*RU 486*, a synthetic steroid molecule with antagonist activity to both glucocorticosteroids and progesterone in various animal models, was given to healthy subjects at 0200 or 1400 hours and plasma ACTH and cortisol levels were measured for 4 h. No ACTH or cortisol response to RU 486 was observed after administration at 1400, whereas a strong response was observed after administration at 0200 hours (Bertagna et al. 1984). This study was con-

firmed and extended with a different timing when RU 486 was administered twice, at a 1 week interval, to the same subjects at either 0000 or 1000 hours; the antiglucocorticoid effect was only apparent during the morning hours after midnight administration. These data demonstrate the possibility of optimizing the antiprogestational effect of this drug and its potential use for human fertility control by modifying the dose and time of administration (Gaillard et al. 1984).

The most widely used diagnostic application of these findings is the *dexamethasone suppression test (DST)* introduced by Liddle in 1960 for the diagnosis of Cushing's disease. The drug is given (1 mg per os) at 2300 and plasma cortisol is assayed at 0800 and compared to the 0800 hours concentration on the day before administration of dexamethasone (Dex). Besides its use in endocrinology, DST has been used extensively in recent years as a test of hypothalamic-pituitary-adrenal axis function in the diagnosis of endogenous depression. Nonsuppression of cortisol has been reported to occur in about 50 % of patients with endogenous depression (Carroll et al. 1981). However, despite the large number of studies, the clinical utility of the DST in psychiatric patients is not yet clearly established, mainly due to a large number of variables involved, e.g., differences in the dose of Dex used, the sampling times, cutoff criteria, the validity of the diagnosis, concomitant pathologies and/or drug administration, and the age of the patients. In addition, the bioavailability of Dex is a critical factor in determining DST results (in Lowy and Meltzer 1987). In psychiatric patients, in normal controls, and/or in patients with Cushing's syndrome, Dex plasma levels vary markedly 9 and 16 h after a 1 mg dose of Dex, and this variability contributes in determining the outcome of the DST status. Therefore, simultaneous measurements of both serum cortisol and Dex levels are necessary for an appropriate evaluation of DST results.

Circadian oscillators in major depressive illness may be phase advanced by several hours, e. g., prolactin, MHPG (3-methoxy-4-hydroxyphenylglycol), the hypothalamo-pituitary-adrenal axis, REM, the activity cycle, body temperature, sleep (see the chapter by Kripke et al.). In this case, a 2300 hours dose of Dex (as usually given) might fail to influence ACTH secretion, resulting in an apparent rather than genuine resistance to DST. Pepper et al. (1983) tested this hypothesis by administering Dex to major depressive patients at 1900 and 2300 hours on separate evenings. DST results were unaffected by altering the clock time of administration of the drug, which suggests that resistance to feedback suppression in depressive illness is likely to be of primary CNS origin rather than secondary to the timing of Dex administration in relation to a phase-shift and the hypothalamo-pituitary-adrenal axis in the depressed patient.

## Hypothalamo-Pituitary-Gonadal Axis

The gonads secrete their relevant steroid hormones (i.e., testosterone, estrogen) in response to a trophic signal from the pituitary, i.e., the peptide hormones LH and FSH. The secretion of gonadotrophins by the pituitary is stimulated by releasing factors, i.e., peptides secreted by the hypothalamus (e.g., GnRH). The gonadal steroids, in turn, act through feedback mechanisms to modulate the hypothalamic and pituitary secretion of (neuro)hormones.

The secretion of LH by the pituitary is pulsatile, probably reflecting a pulsatility of GnRH secretion (Clarke et al. 1982; Gross et al. 1985). The characteristics of the pulsatility (frequency, temporal distribution, width and amplitude of pulses) seem to be

a major factor in HPG axis regulation. The availability of synthetic GnRH allowed to evoke a pulse-like release of LH through a bolus intravenous injection of a low dose of this hypothalamic releasing factor.

Besides the direct gonadal stimulation by human chorionic gonadotropin (HCG) infusion, the gonadotrophins response to GnRH (and/or clomiphene) administration has become a widely used method to assess HPG axis integrity. Among the factors potentially influencing the test are the age and treatment of the subjects, the time of drug administration, and the technical design of the study (bio- or radioimmunoassay).

The determination of plasma gonadotrophins with a bioassay or a radioimmunoassay can lead to a different appreciation of the studied phenomenon. Winters et al. (1982) studied LH and FSH secretion with radioimmunoassays in young ($n = 10$; 20–29 years) and elderly men ($n = 12$; 65–80 years). Whereas the mean morning concentrations of immunoreactive LH and FSH in the aged were 2- to 3-fold higher than in young men, they found no age-related difference in the frequency or amplitude of the pulses of LH or in the areas under the response curve of LH and FSH after GnRH administration. The major age-related modifications observed were a delayed peak time of the LH response and a slower decrease of LH levels following GnRH stimulation. Similar data were reported by Celani et al. (1984). These investigations studied both bioactive and immunoreactive LH in six aged men ($\approx 72$ years) and in nine young men ($\approx 27$ years). They showed that both bioactive (B) and immunoreactive (I) LH levels were higher in the aged men, but that their B/I ratio was smaller than in the young men. Stimulation by GnRH revealed interesting differences. Whereas mean immunoreactive LH increments were not significantly different between the age groups, bioreactive LH increments were significantly smaller in the aged although no difference could be seen in the B/I ratios within each age group. The bioactive LH mean maximum response was smaller in the aged, whereas no difference was observed in the mean response areas under the curve. Celani et al. (1984) found some difference in the mean peak time of response to GnRH, i.e., 45 min in the aged vs 30 min in young men, with a slower LH decrease in elderly. This latter finding might be related to differences of metabolic clearance (MCR) (Veldhuis et al. 1986). These investigators showed that following a bolus injection of highly purified human LH, the bioactive LH MCR was $24.1 \pm 4.7$ ml/min (mean $\pm$ SD), whereas the immunoreactive LH MCR was 56.2 ± 12 ml/min. Bioactive and immunoreactive LH can react differently to other parameters. It has been shown in 21 men (21–29 years) that estradiol i.v. infusion resulted in a larger decrease of bioactive LH than of immunoreactive LH and therefore in a decrease of the B/I LH ratio (Veldhuis et al. 1987). In contrast, antiestrogen (tamoxifen) administration resulted in an increased B/I LH ratio and increased bioactive LH pulse frequency and amplitude. Also, low dose pulsed infusion of LH-RH increases the B/I LH ratio (Veldhuis et al. 1987).

Tenover et al. (1987) could not find age-related differences in bioactive FSH levels between 23 young men (23–35 years) and 16 aged men (65–84 years), whereas immunoreactive FSH was significantly increased in the aged, resulting in smaller B/I FSH ratios. They observed no difference in B/I FSH ratios according to the time of day. These findings add to those of Perez-Lopez et al. (1983) who reported that the repetitive (0,90, and 180 min) combined i.v. administration of TRH and GnRH to eight young men (20–29 years) resulted in similar serum levels of LH, FSH, and prolactin whether performed at 0200 or 0900 hours.

Lagoguey and Reinberg (1981) studied the testicular response to HCG stimulation in four young men at 0700, 1400, and 2000 hours in May–June and again in October–November. They showed that the increase in serum testosterone was significant only for the stimulation at 2000 hours and that this effect was stronger in spring than in autumn.

## Hypothalamo-Pituitary-Thyroid Axis

The thyroid gland, which secretes the triiodothyronine (T3) and thyroxine (T4), is under the control of the pituitary gland via the thyroid stimulating hormone (TSH). In turn, T3 and T4 exert a negative feedback control over TSH formation. The release of TSH by the pituitary is, at least in part, initiated by the binding of a hypothalamic hormone, the thyrotropin releasing hormone (TRH), to the receptors of the membrane of the thyrotrope cells. In the healthy adult plasma TSH has a circadian rhythm with a peak around midnight. Whether TRH is the major factor generating the rhythm of TSH is not yet clearly established since it has been shown that continuous infusion of TRH in humans failed to suppress the TSH rhythm (Spencer et al. 1980).

TSH response to exogenous TRH stimulation is the test most commonly performed to assess HPT axis regulation. Discordant results have been reported according to the time at which the test is performed. Spencer et al. (1980) found in healthy adult men a more pronounced response after a 500 µg TRH injection at 2300 than to an injection of the same dose at 1100. This was confirmed by Perez-Lopez et al. (1981) who found in healthy adult women a more marked response to 200 µg TRH at 0200 than at 1000 hours. In contrast, Nathan et al. (1982) found in young men no difference in the response of TSH to a 500 µg TRH injection at either 0900 or 1800 hours. These data are also different from those of Tatar et al. (1983) who observed a lesser response to the injection of 500 µg TRH 15 and 30 min after an injection at 2000 than to the injection of the same dose at 0800 hours. In contrast, the TSH concentration 45 min after injection remained higher after the injection at 2000 than at 0800 hours. Aging apparently does not influence the responsiveness of the HPT axis since Blicher-Toft et al. (1975) found no difference either in basal morning TSH levels or in TSH response to TRH between young and elderly subjects.

Among the factors interfering in this test of pituitary-thyroid function, major depressive disorders were shown to be responsible for a significantly lesser response than found in healthy subjects. In the same study the authors also observed a more markedly decreased response in patients during the acute phase than in those in complete remission (Unden et al. 1986).

## Parathyroid Hormone and Calcitonin

Parathyroid hormone (PTH), calcitonin, and 1,25-dihydroxy vitamin D are the principal hormones implicated in calcium (Ca) and magnesium (Mg) metabolism. In a longitudinal study in young healthy men, Ferment et al. (1987) explored the effects of a relatively moderate dose of Ca (4.25 mmol) and Mg (7.08 mmol) on plasma immunoreactive PTH and calcitonin. Two magnesium salts were used, magnesium sulphate ($MgSO_4$) and magnesium pyrrolidone carboxylate (MgPC, a magnesium salt of 5-oxoproline) and their effects were compared to Ca gluconate and NaCl. Each salt was administered intravenously to each subject at 0800, 1400, and 2000 hours. When administered at 0800 both $MgSO_4$ and Ca gluconate produced a statistically significant 30% decrease in plasma PTH levels 45 min after the injection. This effect was more sustained with Ca (2 h) than with Mg (45 min) which suggests that, on a molar basis, Mg was less potent than Ca in regulating PTH secretion in humans (Ferment et al. 1987). The i.v. administration of the salts at 1400 or 2400 hours produced a less consistent effect on PTH secretion. The plasma calcitonin increase in response to Ca gluconate or to $MgSO_4$ was not found to be circadian stage dependent (O. Ferment and Y. Touitou, unpublished data). These findings strongly suggest that the hormonal regulation of Ca and Mg are not constant throughout the 24 h scale.

## Insulin and Glucose Tolerance Test

The endocrine activity of the pancreas acts upon the metabolism of glucose through the secretion of insulin and glucagon, both polypeptidic hormones secreted by the $\beta$- and $\alpha$-cells of the Langerhans' islets, respectively. The relations between the secretion of insulin and plasma glucose concentrations are still not completely clear. Lang et al. (1979), using 1 min sampling intervals and autocorrelation techniques of rhythm evaluation, showed that in humans the basal plasma insulin concentrations cycled regularly with a mean period of 13 min, whereas glucose cycled 2 min in advance of insulin.

The oral glucose tolerance test is most commonly practiced to assess insulin secretion by the pancreas. Large differences in the effects of glucose intake are evident according to the hour of ingestion. The afternoon intake of glucose results in higher plasma glucose concentrations and a larger area under the curve than the morning intake (Jarrett 1972; Jarrett and Keen 1970, 1972). This time-related difference was larger in elderly than in young subjects. The intravenous test revealed the same time-related difference (Wichelow et al. 1974), excluding differences in glucose resorption as a determining factor. A partial explanation for this difference in response to glucose loads lies in the response of immunoreactive insulin (Capani et al. 1972; Zimmer et al. 1974).

Circadian rhythms of the effects of insulin have been reported (Gibson et al. 1975), e.g., intravenous injection of insulin in the morning produced a greater decrease of plasma glucose concentration (48%) than injection in the afternoon (30%). The same phenomenon was observed on plasma free fatty acid con-

centration, whereas an opposite effect was demonstrated on plasma GH. The morning injection was less efficient in increasing plasma GH concentrations than the afternoon injection (Gibson et al. 1975).

A circannual rhythm of the insulin response to the oral glucose tolerance test has been described. The peak of plasma insulin concentration obtained in response to a 50 g oral intake of glucose was greater and occurred faster in September than in April (Méjean et al. 1977). It has to be emphasized that the basal values of plasma insulin concentration were the same in September and in April.

Many external factors play a role in glucose-insulin interaction. The effects of meals (time and size) are a matter of controversy. Some have reported a higher blood glucose concentration and larger insulin requirements after breakfast (Göschke et al. 1974; Hepp et al. 1977; Mirouze et al. 1977; Schulz et al. 1982; Rewers et al. 1985), whereas other investigators reported that postprandial plasma glucose and insulin concentrations were directly related to meal size but were not affected by time of day (Touitou 1989).

Physical exercise also modifies plasma glucose and insulin concentrations. Physical training has been shown to accelerate free fatty acid utilization and generate a glycogen sparing effect (Issekutz et al. 1964; Mole et al. 1971). Physical exercise also produces changes at the insulin receptor level, since Leblanc et al. (1979) showed that exercice increased insulin binding in human monocytes. These elements may, at least in part, explain why athletes have reduced insulin requirements in the presence of elevated plasma glucose concentrations produced by glucose intake (Leblanc et al. 1979; Björntorp et al. 1970).

## Prolactin Testing

Prolactin (PRL), a polypeptide hormone secreted by the pituitary, is released episodically over the 24 h span and the frequency and magnitude of the episodes are, in part, under the control of central nervous system rhythms (time of day, sleep-wake cycle). The circadian pattern is characterized by a minimum around noon and a nighttime increase with a maximum around midsleep. Whereas the part played by PRL in conditions such as lactation and amenorrhea is well-known, its role in the genesis of mammary cancer in human beings is still controversial (Haus et al. 1980).

Pituitary lactotroph function has been assessed by, among others, the insulin tolerance test. Seven young

**Table 1.** Drug effects on melatonin secretion in humans

| Increase | Decrease | Inconclusive | Phase-shift | No effect |
|----------|----------|--------------|-------------|-----------|
| Chlorpromazine | Alprenolol | Ritodrine | Desipramine | Carbidopa |
| Clorgyline | Atenolol | Sprinting | Light | L-Dopa |
| Tranylcypromine | Metoprolol | | | Dexamethasone |
| Destyrosine-$\gamma$-endorphin | Propranolol | | | Fludrocortisone |
| Fluvoxamine | Benzodiazepine (450191-s) | | | Isoproterenol |
| Lithium | Clonidine | | | Orciprenaline |
| Psoralens | Flunitrazepam | | | LH-RH |
| Tetrahydrocannabinol | | | | TRH |
| | | | | LH-RH + TRH |
| | | | | Deamino-D-Arg-Vasopressin |
| | | | | Flupenthixol |
| | | | | Fluphenazine |
| | | | | Deprenyl |
| | | | | Maprotiline |
| | | | | Insulin-induced hypoglycemia |
| | | | | Pneumoencephalography |
| | | | | Electroshock therapy |
| | | | | Sleep deprivation |
| | | | | Psychosociological stress |

( $\approx$ 23 years) healthy men received a 0.1 U/kg i.v. injection of insulin at either 0900 or 1830 hours. Subsequent frequent blood sampling showed that although basal PRL levels were equal, there was a significantly greater response (i.e., PRL increase) after the 1830 injection. This difference was attributed to a possible circadian variation in hypothalamic serotoninergic activity in humans (Nathan et al. 1979).

The study of Poleri et al. (1981) showed that long-term bromocriptine administration to elderly men (70–85 years) resulted in decreased PRL values throughout the 24 h span and led to a substitution of the circadian rhythmicity present before the treatment by episodes with a 12 h period. These data differ to some extent from those of Sowers et al. (1983) who found that bromocriptine treatment in four men (42–61 years), besides markedly decreasing PRL levels and eliminating its circadian rhythmicity, did not induce any ultradian frequencies.

Stimulation of PRL release with constant TRH infusion (0.4 $\mu$g/min) in aging men (50–69 years and 70–99 years) revealed that the oldest men had an earlier and greater maximum response (Blackman et al. 1986). It appears that this would not be continuous process over a wide age range since both Blackman et al. (1986) and Arnetz et al. (1986) could not find an increase of PRL concentrations between young ( $\approx$ 30–40 years) and middle aged ( $\approx$ 60–65 years) men. The PRL response to TRH stimulation, however, was in both studies less in the middle-aged men.

Differences in the PRL response to TRH stimulation according to the time of TRH infusion have been reported. A bolus injection of 500 $\mu$g TRH to five young men ( $\approx$ 28 years), at either 0900 or 1800, showed a significantly greater response ($p < 0.025$) in the evening (Nathan et al. 1982). Since no difference could be seen in basal PRL levels at 0900 and 1800 hours, this would suggest a diurnal variation in the pituitary lactotroph responsiveness to TRH. Concordant data were obtained using insulin-induced hypoglycemia (Nathan et al. 1979) or using haloperidol as the PRL releasing stimulus (Nathan et al. 1981). Others failed finding such time-related differences in pituitary responsiveness to TRH stimulation (Spencer et al. 1980); their protocol, however, differed in its timing (infusion at 1100 and 2300 hours).

Depressive disorders have been shown to impair TSH response to TRH stimulation but no such an effect was seen on PRL. The bolus injection of 200 $\mu$g TRH and 100 $\mu$g LH-RH at 0900 hours to 26 patients with major depressive disorders (either in acute phase or in complete remission) and to 23 controls showed no difference in the PRL response between the three groups of subjects (Unden et al. 1987).

## Pineal Gland Testing

Melatonin, a hormone mainly secreted by the pineal gland, is significantly higher at night (i.e., during the daily dark span irrespective of diurnal or nocturnal activity patterns) in the gland and in biological fluids of all species tested, including humans, whatever the age, sex, and season (Touitou et al. 1981, 1984; Iguchi et al. 1982). This nighttime rise is one of the main methodological problems when documenting melatonin secretion, since a proper investigation of this hormone requires multiple nocturnal blood samplings. Therefore, the stimulation (or suppression) of melatonin secretion by a diagnostic agent (a drug), as commonly performed for other endocrine glands, would probably allow an easier and more accurate investigation of pineal testing. Unfortunately, it appears that up to now no drug can be properly used to test the pineal function. However, the possible effects of stress through, e. g., a prolonged physical activity, and of psoralens in increasing melatonin secretion in humans are worth ascertaining in further research (review in Touitou et al. 1987). Suppression of the nocturnally elevated plasma melatonin concentration by bright light has been widely studied and may be of some diagnostic importance in some psychiatric disorders, e. g., depression (Lewy et al. 1981, 1984).

## Conclusion

The chronobiological approach is useful for practical implications in testing endocrine and neuroendocrine functions. Numerous tests used in clinical practice to assess the endocrine or neuroendocrine systems have been shown to respond differently according to the time of testing. Some of these results may find a practical therapeutic application, e. g., the suppressive effects of exogenous corticosteroids vary along the day according to the variation in sensitivity of the ACTH secreting system to feedback inhibition. From this background, recent data, obtained in patients receiving prednisolone for up to 10 years, have shown that the pituitary-adrenal axis is preserved when the drug is given in the morning, thus at a distance from the impulsive phase of ACTH secretion (Reinberg et al. 1988).

## References

Akana SF, Cascio CS, Du JZ, Levin N, Dallman MF (1986) Reset of feedback in the adrenocortical system: an apparent shift in sensitivity of adrenocorticotropin to inhibition by corticosterone between morning and evening. Endocrinology 119: 2325–2332

Angeli A, Frajria R, Fonzo D, Bertello P, Gaidano G, Ceresa F (1977) Proc. XII intern conference int. soc. for chronobiology. Il Ponte, Milan, pp 189–196

Arnetz BB, Lahnborg G, Eneroth (1986) Age related differences in the pituitary prolactin response to thyrotropin-releasing hormone. Life Sci 39: 135–139

Avgerinos PC, Schürmeyer TH, Gold RW, Tomai TP, Loriaux DL, Sherins RJ, Cutler GA Jr, Chrousos GP (1986) Pulsatile administration of human corticotropin-releasing hormone in patients with secondary adrenal insufficiency: restoration of the normal cortisol secretory pattern. J Clin Endocr Metab 62: 816–821

Bertagna X, Bertagna C, Luton JP, Husson JM, Girard F (1984) The new steroid analog RU 486 inhibits glucocorticoid action in man. J Clin Endocr Metab 59: 25–28

Björntorp P, Holm G, Jacobson B (1970) Effect of physical training on insulin production in obesity. Metabolism 19: 631–638

Blackman MR, Kowatch MA, Wehman RE, Harman SM (1986) Basal serum prolactin levels and prolactin responses to constant infusions of thyrotropin-releasing hormone in healthy aging men. J Gerontol 41: 699–705

Blicher-Toft M, Hummer L, Dige-Petersen H (1975) Human serum thyrotrophin level and response to thyrotrophin-releasing hormone in the aged. Gerontol Clin 17: 191–203

Capani F, Caradonna P, Carotenuto M, Camilli G, Sensi S (1972) Circadian rhythm of immunoreactive insulin under glycemic stimulus. Biochem Biol Sper 10: 115–124

Carroll BJ, Feinberg M, Greden JF, Tarika J, Albaala AA, Haskett RJ, James NM, Kronfol Z, Lohr N, Steiner M, de Vigne JP, Young E (1981) A specific laboratory test for the diagnosis of melancholia. Arch Gen Psychiat 38: 15–22

Celani MF, Montanini V, Baraghini GF, Carani C, Marrama P (1984) Effects of acute stimulation with gonadotropin releasing hormone (GnRH) on biologically active serum luteinizing hormone (LH) in elderly men. J Endocrinol Invest 7: 589–595

Ceresa F, Angeli A, Boccuzzi P, Molino G (1969) Once a day neurally stimulated and basal ACTH secretion phases in man and their response to corticoid inhibition. J Clin Endocr Metab 29: 1074–1082

Chan JSD, Lu CL, Seidah NG, Chretien M (1982) Corticotropin releasing factor (CRF): effects on the release of proopiomelanocortin (POMC)-related peptides by human anterior pituitary cells in vitro. Endocrinology 111: 1388–1390

Clarke IJ, Cummins JT (1982) The temporal relationship between gonadotropin releasing hormone (GnRH) and luteinizing hormone (LH) secretion in ovariectomized ewes. Endocrinology 111: 1737–1739

Clayton GW, Librik L, Gardner RL, Guillemin R (1963) Studies on the circadian rhythm of pituitary adrenocorticotropic release in man. J Clin Endocr Metab 23: 975–980

Copinschi G, Beyloos M, Bosson D, Desir D, Golstein J, Robyn C, Linkowski P, Mendlewicz J, Marcel Franckson JA (1983) Immediate and delayed alterations of adrenocorticotropin and cortisol nyctohemeral profiles after corticotropin releasing factor in normal man. J Clin Endocr Metab 57: 1287–1291

Crowley WF Jr, Vale W, Rivier J, Mac Arthur JW (1981) LH-RH in hypogonadotropic hypogonadism. In: Zatuchini GI, Schelten JD, Sciarra JJ (eds) LH-RH Peptides as female and male contraceptives. Harper, Philadelphia, pp 321–333

Dallman MF, Yates FE (1969) Dynamic asymetrics in the corticosteroid feedback path and distribution-metabolism-binding elements of the adrenocortical system. Ann NY Acad Sci 156: 696–721

Dallman MF, Engeland WC, Rose JC, Wilkinson CW, Shinsako J, Siedenburg F (1978) Nycthemeral rhythm in adrenal responsiveness to ACTH. Am J Physiol 235: R210–R218

David-Nelson MA, Brodish A (1969) Evidence for a diurnal rhythm of corticotropin-releasing factor (CRF) in the hypothalamus. Endocrinology 85: 861–866

De Cherney GS, De Bold CR, Jackson RV, Sheldon WR Jr, Island DP, Orth DN (1985) Diurnal variation in the response of plasma adrenocorticotropin and cortisol to intravenous ovine corticotropin-releasing hormone. J Clin Endocr Metab 61: 273–279

Di Raimondo VC, Forsham PH (1956) Some clinical implications of spontaneous diurnal variation in adrenal cortical secretion activity. Am J Med 21: 321–323

Ertel RJ, Halberg F, Ungar F (1964) Circadian system phase-dependent toxicity and other effects of methopyrapone (SU-4885) in the mouse. J Pharmacol Exp Ther 146: 395–399

Ferment O, Garnier PE, Touitou Y (1987) Comparison of the feedback effect of magnesium and calcium on parathyroid hormone secretion in man. J Endocr 113: 117–122

Ferrari E, Bossolo PA, Mandelli BM, Carnevale Schianca GP, Fioravanti M (1982) Adrenocortical responsiveness to the synthetic analogue ACTH 1-17 given at different circadian stages. Chronobiologia 9: 133–141

Fischman AJ, Moldow RL (1984) In vivo potentiation of corticotropin releasing factor activity by vasopressin analogues. Life Sci 35: 1311–1319

Gaillard RC, Riondel A, Muller AF, Herrman W, Baulieu EE (1984) RU 486: a steroid with antiglucocorticosteroid activity that only disinhibits the human pituitary-adrenal system at a specific time of day. Proc Natl Acad Sci USA 81: 3879–3882

Garrick NA, Hill JL, Szele FG, Tomai TP, Gold PW, Murphy DL (1987) Corticotropin-releasing factor: a marked circadian rhythm in primate cerebrospinal fluid peaks in the evening and is inversely related to the cortisol circadian rhythm. Endocrinology 121: 1329–1334

Gibson T, Stimler L, Jarrett RJ, Ruthland P, Shiu M (1975) Diurnal variation in the effects of insulin in blood glucose, plasma non-esterified fatty acids and growth hormone. Diabetologia 11: 83–88

Göschke H, Denes A, Girard J, Collard F, Berger W (1974) Circadian variations of carbohydrate tolerance in maturity onset diabetics treated with sulfonylureas. Horm Metab Res 6: 386–391

Grant SD, Forsham PH, Di Raimondo VC (1965) Suppression of 17-hydroxycorticosteroids in plasma and urine: single and divided doses of triamcinolone. N Engl J Med 273: 1115–1118

Gross KM, Matsumoto AM, Southworth MB, Bremmer WJ (1985) Evidence for decreased luteinizing hormone-releasing hormone pulse frequency in man with selective elevation of follicle-stimulating hormone. J Clin Endocr Metab 60: 197–202

Grossman A, Kruseman ACN, Perry L, Tomlin S, Schally AV, Coy DH, Rees LH, Comaru-Schally AM, Besser GM (1982) New hypothalamic hormone, corticotropin-releasing factor, specifically stimulates the release of adrenocorticotropic hormone and cortisol in man. Lancet I: 921–922

Günther R, Herold M, Halberg E, Halberg F (1980) Circadian placebo and ACTH effects on urinary cortisol in arthritics. Peptides 1: 387–390

Haus E, Halberg F, Kuhl JFW, Lakatua DJ (1974) Chronopharmacology in animals. In: Aschoff J, Ceresa F, Halberg F (eds) Chronobiological aspects of endocrinology. Schottauer, Stuttgart, pp 269–304

Haus E, Lakatua DJ, Halberg F, Halberg E, Cornelissen G, Sacket LL, Berg HG, Kawazaki T, Ueno M, Uezono K, Matsuoka M, Omae T (1980) Chronobiological studies of plasma prolactin in women in Kyushu, Japan and Minnesota, USA. J Clin Endocrinol Metab 51: 632–640

Haus E, Nicolau G, Lakatua DJ, Sackett-Lundeen L, Petrescu E (1989) Circadian rhythm parameters of endocrine functions in elderly subjects during the seventh to the ninth decade of life. Chronobiologia 16: 331–352

Hepp KD, Renner R, von Funckne HJ, Mehnert H, Haerten WR, Kresse H (1977) Glucose homeostasis under continuous intravenous insulin therapy in diabetics. Horm Metab Res 7: 72–76

Hiroshige T, Sakakura M (1971) Circadian rhythm of corticotropin-releasing activity in the hypothalamus of normal and adrenalectomized rats. Neuroendocrinology 7: 25–36

Ichikawa Y, Nishikai M, Kawagoe M, Yoshiba K, Homma M (1972) Plasma corticotropin, cortisol and growth hormone responses to hypoglycemia in the morning and evening. J Clin Endocr Metab 34: 895–898

Iguchi H, Kato I, Ibayashi H (1982) Age-dependent reduction in serum melatonin concentrations in healthy human subjects. J Clin Endocr Metab 55: 27–29

Issekutz B Jr, Miller HI, Paul P, Radahl K (1964) Aerobic work capacity and plasma FFA turnover. J Clin Invest 43: 904–910

Jarrett RJ (1972) Circadian variations in blood glucose levels, in glucose tolerance and in plasma immunoreactive insulin. Acta Diabetol Lat 9: 263–275

Jarrett RJ, Keen H (1970) Further observations on the diurnal variation in oral glucose tolerance. Br Med J 4: 334–337

Jarrett RJ, Keen H (1972) Diurnal variations of oral glucose tolerance: a possible pointer of the evolution of diabetes mellitus. Br Med J 2: 341–345

Jubiz W, Matsukura S, Meikle AW, Harada G, West CD, Tyler FH (1970) Plasma metyrapone, adrenocorticotrophic hormone, cortisol and deoxycortisol levels. Sequential changes during oral and intravenous metyrapone administration. Arch Int Med 125: 468–471

Knobil E (1980) Neuroendocrine control of the menstrual cycle. Recent Prog Horm Res 36: 53–88

Köbberling J, Jüpper H, Hesch RD (1976) The stimulation of growth hormone release by ACTH and its inhibition by somatostatin. Acta Endocr (Copenh) 81: 263–269

Kohlberg IJ, Doret AM, Paunier L, Sizonenko PC (1972) Assessment of the pituitary adrenal cortex axis in children by a single dose metyrapone test. Helv Paediat Acta 27: 437–447

Krieger DT (1979) Rhythms in CRF, ACTH and corticosteroids. In: Krieger DT (ed) Endocrine rhythms. Raven, New York, pp 123–142

Krieger DT, Gewitz GP (1974) The nature of the circadian periodicity and suppressibility of immunoreactive ACTH in Addison's disease. J Clin Endocr Metab 39: 46–56

Krieger DT, Allen W, Rizzo F, Krieger HP (1971) Characterization of the normal temporal pattern of plasma corticosteroid levels. J Clin Endocr Metab 32: 266–284

Lagoguey M, Reinberg A (1981) Circadian and circannual changes of pituitary and other hormones in healthy human males: their relationship with gonadal activity. In: Van Cauter E, Copinschi G (eds) Human pituitary hormones. Circadian and episodic variations. Nijhoff, The Hague, pp 261–278

Lamberts SWJ, Verleun T, Oosterom R, Jong F, Hackeng WH (1984) Corticotropin-releasing factor (ovine) and vasopressin exert a synergistic effect on adreno-corticotropin release in man. J Clin Endocr Metab 58: 298–303

Lang DA, Matthews DR, Phil D, Peto J, Turner RC (1979) Cyclic oscillations of basal plasma glucose and insulin concentrations in human being. N Engl J Med 301: 1023–1027

Leblanc J, Nadeau A, Boulay M, Rousseau-Migneron S (1979) Effects of physical training and adiposity on glucose metabolism and $^{125}$I-insulin binding. J Appl Physiol 46: 235–239

Lewy AJ, Wehr TA, Goodwin FK, Newsome DA, Rosenthal NE (1981) Manic depressive patients may be supersensitive to light. Lancet I: 383–384

Lewy AJ, Sack RL, Singer CM (1984) Assessment and treatment of chronobiologic disorders using plasma melatonin levels and bright light exposure, the clock-gate model and the phase response curve. Psychopharmacol Bull 20: 561–565

Liddle A (1960) Test of pituitary-adrenal suppressibility in the diagnosis of Cushing's syndrome. J Clin Endocr Metab 20: 1539–1560

Losa M, Bock J, Schopohl J, Stalla GK, Muller OA, Von Werder K (1984) Growth hormone releasing factor infusion does not sustain elevated GH-levels in normal subjects. Acta Endocr (Copenh) 107: 462–470

Lowy MT, Meltzer HY (1987) Dexamethasone bioavailability implications for DST research. Biol Psychiat 22: 373–385

Martin MM, Hellman DE (1964) Temporal variations in SU-4885 responsiveness in man: evidence in support of circadian variations in ACTH secretion. J Clin Endocr 24: 253–260

Méjean L, Reinberg A, Guy G, Debry G (1977) Circannual changes of the plasma insulin response to glucose tolerance test in healthy young human males. In Proc XXVIIth International Congress Physiological Sciences, 18–23 July, 1475: 498

Mirouze J, Selam JL, Pham TC, Calvadore E (1977) Evaluation of exogenous insulin homeostasis by the artificial pancreas in insulin-dependent diabetes. Diabetologia 13: 273–278

Moldow RL, Fischman AJ (1984) Circadian rhythm of corticotropin releasing factor-like immunoreactivity in rat hypothalamus. Peptides 5: 1213–1215

Mole PA, Oscai LB, Holloszy JO (1971) Adaptation of muscle to exercise: increase in levels of palmityl CoA synthetase, carnitine transferase and palmityl CoA dehydrogenase and the capacity to oxydize fatty acids. J Clin Invest 90: 779–778

Munabi AK, Feuillan P, Staton RC, Rodbard D, Chrousos GP, Anderson RE, Strober MD, Loriaux DL, Cutler GB Jr (1986) Adrenal steroid responses to continuous intravenous adrenocorticotropin infusion compared to bolus injection in normal volunteers. J Clin Endocr Metab 63: 1036–1040

Nathan RS, Sachar EJ, Langer G, Tabrizi MA, Halper FS (1979) Diurnal variations in the response of plasma prolactin, cortisol and growth hormone to insulin-induced hypoglycemia in normal men. J Clin Endocr Metab 49: 231–235

Nathan RS, Perel JM, McCarthy T, Jarrett DB (1981) Diurnal pattern in the prolactin responses to haloperidol in man. Neuroendocrinol Lett 5: 324

Nathan RS, Sachar EJ, Tabrizi MA, Asnis GM, Halbreich U, Halpern FS (1982) Diurnal hormonal responses to thyrotropin-releasing hormone in normal men. Psychoneuroendocrinology 7: 235–238

Nichols T, Nugent CA, Tyler FH (1965) Diurnal variation in suppression of adrenal function by glucocorticoids. J Clin Endocr Metab 25: 343–349

Nicholson S, Lin JH, Mahmoud S, Campbell E, Gillham B, Jones M (1985) Diurnal variations in responsiveness of the hypothalamo-pituitary adrenocortical axis of the rat. Neuroendocrinology 40: 217–224

Nugent CA, Nichols T, Tyler FH (1965) Diagnosis of Cushing's syndrome: single dose dexamethasone suppression test. Arch Intern Med 116: 172–176

Orth DN, Island DP, Liddle GW (1967) Experimental alteration of the circadian rhythm in plasma cortisol (17 OHCS) concentration in man. J Clin Endocr Metab 27: 549–555

Orth DN, Jackson RV, De Cherney GS, Debold CR, Alexander AN, Island DP, Rivier J (1983) Effect of synthetic ovine corticotropin-releasing factor. J Clin Invest 71: 587–595

Pepper GM, Davis KL, Davis BM, Krieger DT (1983) DST in depression is unaffected by altering the clock time of its administration. Psychiatry Res 8: 105–109

Perez-Lopez FR, Gomez Agudo G, Abos MD (1981) Serum prolactin and thyrotropin responses to thyrotropin-releasing hormone at different times of day in normal women. Acta Endocr (Copenh) 97: 7–11

Perez-Lopez FR, Legido A, Abos MD, Lafarga L (1983) Human pituitary chronoendocrinology: repetitive stimulation with LRH/TRH at different times of the day. Acta Endocr (Copenh) 102: 327–331

Perlow M, Weitzman ED, Hellman L (1974) Effect of cortisol infusions on endogenous cortisol secretion in man. Endocrinology 95: 790–795

Poleri A, Masturzo P, Murialdo G, Agnoli A (1981) Circadian rhythmicity of prolactin secretion in elderly subjects: changes during bromocriptine treatment. J Endocrinol Invest 4: 317–321

Reinberg A, Guillemant S, Ghata N, Guillemant J, Touitou Y, Dupont W, Lagoguey M, Bourgeois P, Brière L, Fraboulet G, Guillet P (1980) Clinical chronopharmacology of ACTH 1-17. Effects on plasma and urinary 17-hydroxycorticosteroids. Chronobiologia 7: 513–523

Reinberg A, Touitou Y, Lévi F, Nicolai A (1983) Circadian and seasonal changes in ACTH-induced effects in healthy young men. Eur J Clin Pharmacol 25: 657–665

Reinberg A, Touitou Y, Botbol M, Gervais P, Chaouat D, Lévi F, Bicakova-Rocher A (1988) Oral morning dosing corticosteroids in long-term treated cortico-dependent asthmatics: increased tolerance and preservation of adrenocortical function. Ann Rev Chronopharmacol 5: 209–212

Rewers M, Dmochowski K, Walczak M (1985) Diurnal variation in carbohydrate tolerance to mixed meal in insulin-dependent diabetic adolescents during continuous intravenous insulin effusion (CIVII). Endokrynol Pol 36: 1–7

Rivier CL, Plotsky PM (1986) Mediation by corticotropin releasing factor (CRF) of adenohypophysial hormone secretion. Annu Rev Physiol 48: 475–494

Sayers G, Sayers MA (1947) Regulation of pituitary adrenocorticotrophic activity during the response of the rat to acute stress. Endocrinology 40: 265–273

Schopohl J, Hauer A, Kaliebe T, Stalla GK, Von Werder K, Müller OA (1986) Repetitive and continuous administration of human corticotropin releasing factor to human subjects. Acta Endocr (Copenh) 112: 157–165

Schürmeyer TH, Avgerinos PC, Gold PW, Gallucci WT, Tomai TP, Cutler GB, Loriaux DL, Chrousos GP (1984) Human corticotropin-releasing factor in man: pharmacokinetic properties and dose-response of plasma adrenocorticotropin and cortisol secretion. J Clin Endocr Metab 59: 1103–1108

Schulte HM, Chrousos GP, Oldfield EH, Gold PM, Cutler GB, Loriaux DL (1985) Ovine corticotropin-releasing factor administration in normal men. Horm Res 21: 69–74

Schulz G, Beyer J, Happ J, Gordes U, Küstner E, Hassinger N (1982) Variation of basal carbohydrate dependent insulin consumption during a 24 hour period of control. Horm Metab Res 12: 221–224

Sowers JR, Viosca SP, Windsor C, Korenman SG (1983) Influence of dopaminergic mechanisms on 24-hour secretory patterns of prolactin, luteinizing hormone and testosterone in recumbent men. J Endocrinol Invest 6: 9–15

Spencer CA, Greenstadt MA, Wheeler WS, Kletzsky OA, Nicoloff JT (1980) The influence of long-term low dose thyrotropin releasing hormone infusions on serum thyrotropin and prolactin concentration in man. J Clin Endocr Metab 51: 771–775

Spiess J, Rivier J, Rivier C, Vale W (1981) Primary structure of corticotropin-releasing factor from ovine hypothalamus. Proc Natl Acad Sci USA 78: 6517–6521

Sprunt JG, Brownie MCK, Hannach DH (1967) Some aspects of the pharmacology of metyrapone. Proc R Soc Med 60: 908–909

Stahl M, Kapp J, Zachman M, Girard J (1972) Effect of a single oral dose of metyrapone on secretion of growth hormone and urinary tetrahydro-11-deoxycortisol and tetrahydro-11-deoxycorticosterone excretion in children. Helv Paed Acta 27: 147–153

Suda T, Tozawa F, Mouri T, Sasaki A, Shibasaki T, Demura H, Shizume K (1983) Effects of cyproheptadine, reserpin, and synthetic corticotropin-releasing factor in pituitary glands from patients with Cushing's disease. J Clin Endocr Metab 56: 1094–1099

Szafarczyk A, Hery M, Laplante E, Ixart G, Assenmacher I, Kordon C (1980) Temporal relationships between the circadian rhythmicity in plasma levels of pituitary hormones and in hypothalamic concentrations of releasing factors. Neuroendocrinology 30: 369–376

Takebe K (1973) Temporal rhythm in the response of urinary 17-OHCS to metopirone in Cushing's syndrome due to bilateral adrenal hyperplasia. J Clin Endocr 36: 433–438

Takebe K, Setaishi C, Hirama M, Yamamoto M, Horiuchi Y (1966) Effects of a bacterial pyrogen on the pituitary-adrenal axis at various times in the 24 hours. J Clin Endocr Metab 26: 437–442

Takebe K, Setaishi C, Yamamoto M, Horiuchi Y (1968) Pattern of urinary 17-hydroxycorticoid excretion following the injection of pyrogen in normal subjects. J Clin Endocr Metab 28: 924–926

Takebe K, Kunita H, Sawano S, Horiuchi Y, Mashimo K (1969) Circadian rhythms of plasma growth hormone and cortisol after insulin. J Clin Endocr Metab 29: 1630–1633

Tanaka K, Nicholson WE, Orth DN (1978) Diurnal rhythm and disappearance half-time of endogenous plasma immunoreactive β-MSH (LPH) and ACTH in man. J Clin Endocr Metab 46: 883–890

Tatar P, Strbak V, Vigas M (1983) Different response of TSH to TRH in the morning and evening. Horm Metab Res 15: 461

Tenover JS, Dahl KD, Hsueh AJW, Lim P, Matsumoto AM, Bremmer WJ (1987) Serum bioactive and immunoreactive follicle-stimulating hormone levels and the response to clomiphene in healthy young and elderly men. J Clin Endocr Metab 64: 1103–1108

Touitou Y (1989) Time-dependent effects of drugs used as diagnostic agents on the hypothalamo-pituitary-adrenal axis. In: Lemmer B (ed) Chronopharmacology. Cellular and biochemical interactions. Dekker, New York, pp 597–614

Touitou Y, Limal JM, Bogdan A, Reinberg A (1976) Circadian rhythms in adrenocortical activity during and after a 36 hour 4-hourly sustained administration of metyrapone in humans. J Steroid Biochem 7: 517–520

Touitou Y, Bogdan A, Limal JM, Touitou C, Reinberg A (1977) Circadian rhythms in urinary steroids in response to a 36-hour sustained metyrapone administration in eight young men. Horm Metab Res 9: 314–321

Touitou Y, Fèvre M, Lagoguey M. Carayon A, Bogdan A, Reinberg A, Beck H, Cesselin F, Touitou C (1981) Age- and mental health-related circadian rhythms of plasma levels of melatonin, prolactin, luteinizing hormone and follicle-stimulating hormone in man. J Endocr 91: 467–475

Touitou Y, Sulon J, Bogdan A, Touitou C, Reinberg A, Beck H, Sodoyez JC, Demey-Ponsart E, Van Cauwenberge H (1982) Adrenal circadian system in young and elderly human subjects: a comparative study. J Endocr 93: 201–210

Touitou Y, Carayon A, Reinberg A, Bogdan A, Beck H (1983a) Differences in the seasonal rhythmicity of plasma prolactin in elderly human subjects: detection in women but not in men. J Endocr 96: 65–71

Touitou Y, Lagoguey M, Bogdan A, Reinberg A, Beck H (1983b) Seasonal rhythms of plasma gonadotrophins: their persistence in elderly men and women. J Endocr 96: 15–21

Touitou Y, Sulon J, Bogdan A, Reinberg A, Sodoyez JC, Demey-Ponsart E (1983c) Adrenocortical hormones, aging and mental condition: seasonal and circadian rhythms of plasma 18-hydroxy-11-deoxycorticosterone, total and free cortisol and urinary corticosteroids. J Endocr 96: 53–63

Touitou Y, Fèvre M, Bogdan A, Reinberg A, De Prins J, Beck H, Touitou C (1984) Patterns of plasma melatonin with ageing and mental condition: stability of nyctohemeral rhythms and differences in seasonal variation. Acta Endocr (Copenh) 106: 145–151

Touitou Y, Bogdan A, Claustrat B, Touitou C (1987) Drugs affecting melatonin secretion in man. In: Trentini GP, De Gaetani C, Pevet P (eds) Fundamentals and clinics in pineal research. Raven, New York, pp 349–356

Touitou Y, Castagno L, Reinberg A, Garnier P, Donnadieu M, Motohashi Y, Evain-Brion D, Bogdan A (1988) The corticotropin-induced GH increase is not mediated by GH-RH. Neuroendocrinol Lett 10: 33–38

Touitou Y, Garnier P, Reinberg A, Castagno L, Donnadieu M, Bogdan A, Motohashi Y (1990) Growth hormone (GH) and GH-releasing hormone response to ACTH 1-17 at different times of day. Evidence of a circadian stage dependence. Eur J Clin Pharmacol 38: 149–152

Tsukada T, Nakai Y, Koh T, Tsujii S, Imura H (1983) Plasma adrenocorticotropin and cortisol responses to intravenous injection of corticotropin releasing factor in the morning and evening. J Clin Endocr Metab 57: 869–871

Tucci JR (1975) Metyrapone test in Cushing's disease. J Clin Endocr 40: 521–523

Unden F, Ljunggren JG, Kjellman BF, Beck-Friis J, Wetterberg L (1986) Twenty-four hour serum levels of T4 and T3 in relation to decreased TSH serum levels and decreased response to TRH in affective disorders. Acta Psychiat Scand 73: 358–365

Unden F, Ljunggren JG, Kjellman BF, Beck-Friss J, Wetterberg L (1987) Unaltered 24 h serum PRL levels and PRL response to TRH in contrast to decreased 24 h serum TSH levels and TSH response to TRH in major depressive disorder. Acta Psychiatr Scand 75: 131–138

Ungar F, Halberg F (1962) Circadian rhythm in the in vitro response of mouse adrenal to adrenocorticotropic hormone. Science 137: 1058–1060

Vale W, Spiess J, Rivier C, Rivier J (1981) Characterization of a 41-residue ovine hypothalamic peptide that stimulates secretion of corticotropin and $\beta$-endorphin. Science 213: 1394–1397

Vale W, Rivier C, Brown MR, Spiess J, Koob G, Swanson L, Bilezikjian L, Bloom F, Rivier J (1983) Chemical and biological characterization of corticotropin releasing factor. Recent Progr Horm Res 39: 245–270

Vance ML, Kaiser D, Evans S, Thorner MO, Furlanetto R, Rivier J, Vale W, Perisutti G, Frohman LA (1985) Evidence for a limited growth hormone (GH)-releasing hormone (GH-RH)-releasable quantity of GH: effects of 6-hour infusions of GH-RH on GH secretion in normal man. J Clin Endocr Metab 60: 370–375

Veldhuis JD, Dufau ML (1987) Estradiol modulates the pulsatile secretion of biologically active luteinizing hormone in man. J Clin Invest 80: 631–638

Veldhuis JD, Fraioli F, Rogol AD, Dufau ML (1986) Metabolic clearance of biologically active luteinizing hormone in man. J Clin Invest 77: 1122–1128

Watabe T, Tanaka K, Hasegawa M, Miyabe S, Shimizu N (1985) Responses of plasma adrenocorticotropin and cortisol to in-travenous injection of synthetic ovine corticotropin releasing factor in the morning and early evening in normal human subjects. Endocrinol Jpn 32: 771–779

Webb CB, Vance ML, Thorner MO, Perisutti G, Thominet J, Rivier J, Vale W, Frohman LA (1984) Plasma growth hormone response to constant infusion of human pancreatic growth hormone releasing factor. J Clin Invest 74: 96–103

Wichelow MJ, Sturge RA, Keen H, Jarrett RJ, Stimmler L, Grainger S (1974) Diurnal variation in response to intravenous glucose. Br Med J 2: 488–491

Winters SJ, Troen P (1982) Episodic luteinizing hormone (LH) secretion and the response of LH and follicle-stimulating hormone to LH-releasing hormone in aged men: evidence for coexistent primary testicular insufficiency and an impairment of gonadotropin secretion. J Clin Endocr Metab 55: 560–565

Yates FE, Leeman SE, Glenister DW, Dallman MF (1961) Interaction between plasma corticosterone concentration and adrenocorticotrophin-releasing stimuli in the rat: evidence for the reset of an endocrine feedback control. Endocrinology 69: 67–80

Zimmet PZ, Wall JR, Rome R, Stimmler L, Jarrett RJ (1974) Diurnal variation in glucose tolerance: associated changes in plasma insulin, growth hormone and non-esterified fatty acids. Br Med J 2: 485–488

# Biological Rhythms in Hepatic Drug Metabolism and Biliary Systems

P. M. Bélanger and G. Labrecque

## Introduction

The liver is the largest organ in mammals and has many complex metabolic functions. These include the metabolism and storage of carbohydrates, lipids, minerals, proteins, hormones, and vitamins, the formation of bile, and the synthesis of blood coagulation factors and plasma proteins. One liver function that is very important in pharmacology is the biotransformation of drugs and toxic agents into metabolites that are more readily excreted into the bile or urine. In some cases, the liver metabolizes exogenous compounds into pharmacologically active products (Drayer 1976, 1982; Garattini 1985) or into chemical intermediates which could react with biological macromolecules to cause cellular necrosis (Connors 1976; Drayer 1976, 1982; Boyd et al. 1983; Anders 1985).

The literature reviews of Reinberg and Smolensky (1982) and Bruguerolle (1987) indicated that circadian and other temporal variations have been reported in the pharmacokinetics of many drugs. Therefore, it becomes relevant to study the chronobiological variations in the hepatic drug metabolizing system in order to better understand the changes observed in the kinetics and toxicity of many compounds. The aim of this report is to review these chronobiological changes in the processes involved in the hepatic metabolism of drugs. The chronobiology of biliary excretion will also be briefly considered.

## Biological Rhythms in Hepatic Drug Metabolism

The liver is the major site of drug metabolism. The reactions involved in drug metabolism and the localization of the enzymatic systems that catalyze them within the hepatocyte are presented in Table 1. Circa-

**Table 1.** General Pathways of Hepatic Drug Metabolism[a]

| Reactions | Cellular Localization of Enzyme |
|---|---|
| **Oxidations** | |
| Aliphatic hydroxylation | Microsomes[b] |
| Aromatic hydroxylation | Microsomes |
| N-, O- and S-dealkylation | Microsomes |
| N- and S-oxidation | Microsomes |
| Deamination | Microsomes, mitochondria |
| Epoxidation | Microsomes |
| **Reductions** | |
| Ketone reduction | Microsomes, cytosol[c] |
| Nitroreduction | Microsomes |
| **Hydrolyses** | |
| Deesterification | Microsomes, cytosol, lysosomes |
| Deamidation | Microsomes, cytosol |
| **Conjugations** | |
| Acetylation | Cytosol |
| Glucuronidation | Microsomes |
| Glutathion conjugation | Cytosol |
| Glycine conjugation | Mitochondria |
| Sulfate conjugation | Cytosol |

[a] Adapted from Gillette (1966) and Jakoby (1980)
[b] Refers to particles of endoplasmic reticulum isolated by ultracentrifugation of the 9,000 or 10,000 xg supernatant of liver homogenates
[c] Refers to the soluble fraction of liver homogenate

dian variations in the activity of different enzyme systems have been identified and they explain some of the time-dependent changes in drug kinetics.

## The Hepatic P-450 Monooxygenase System

Radzialowski and Bousquet (1967, 1968) were the first to report a circadian variation in hepatic drug metabolism. In the liver of mice synchronized under an alternating light-dark cycle (light on: 0630–2200 hours), they found that p-nitroanisole-o-demethylase activity of rat liver was highest at 0200 and lowest at 1400 hours, respectively (Fig. 1). These investiga-

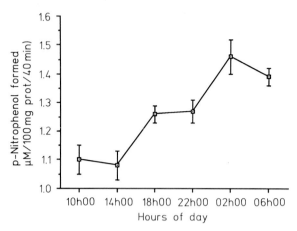

**Fig. 1.** Circadian rhythm in the *p*-nitroanisole-*O*-demethylase activity of rat liver. Rats synchronized under an alternating light-dark cycle (light: 0630–2200 hours) were killed at different hours of the day. Livers were dissected out and the activity of *p*-nitroanisole-*O*-demethylase was determined in vitro. (Redrawn and adapted from Radzialowski and Bousquet 1967)

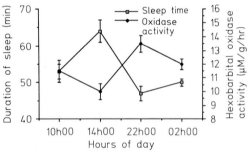

**Fig. 2.** Time-dependent variations in the hexobarbital sleeping time and in the hexobarbital oxidase activity of rat liver. Rats synchronized on a 12 h light-dark cyle (light on: 0600 hours) received an intraperitoneal doses of 150 mg/kg hexobarbital four times a day. Liver hexobarbital oxidase activity was determined in whole liver homogenate. (Redrawn and modified from Nair and Casper, 1969)

tors obtained similar temporal variations in aminopyrine demethylase and hexobarbital oxidase activities in both male and female rats and in male mice. They reported also that the temporal variation in demethylase activity was due essentially to a modification of the microsomal component and was not related to any component in the soluble fraction.

Similar temporal variations have been detected in various oxidative reactions catalyzed by microsomal monooxygenase systems. These variations were found with different substances such as aniline, benzphetamine, benzpyrene, biphenyl, imipramine, steroids, aminopyrine, hexobarbital, and *p*-nitroanisole. Bélanger (1988) recently reviewed the literature in

this area. The work of Nair and Casper (1969) is of special interest to pharmacologists because these investigators reported an inverse relationship between the hexobarbital oxidase activity of rat liver and the sleeping time produced by administration of the barbiturate to mice at different times of day. These data are of interest because hexobarbital is hypnotic drug which is eliminated from the body by hepatic microsomal oxidations to inactive products. As shown in Fig. 2, maximal hexobarbital oxidase activity and minimal sleeping time were obtained at 2200, whereas the lowest enzymatic activity and the longest sleeping time were found at 1400 hours. This was the first time a temporal correlation could be identified between the metabolism and the activity of a drug.

Although many studies have reported temporal variability in the hepatic P-450 monooxygenase system, very few attempts have been made to determine the mechanism for these rhythmic variations. More recently, Bélanger et al. (1986) determined the activity of aniline hydroxylase and *p*-nitroanisole demethylase and the concentration of cytochrome P-450 in purified microsomes isolated from the same rat livers. As shown in Fig. 3, the temporal variation in aniline hydroxylase activity was similar to that of the microsomal concentration of cytochrome P-450, with peak and trough values obtained at 2200 and 1700 hours, respectively. The excellent correlation coefficient between hydroxylase activity and P-450 concentration at the different times of the day suggests the dependence of the temporal variation in oxidase activity on that of the concentration of the enzyme. However, this was not exactly the case for *p*-nitroanisole demethylase activity, which was determined in the same hepatic homogenate fraction under similar conditions. In contrast to the times of the peaks and troughs of hydroxylase activity and of the P-450 concentration, the temporal variability in demethylase activity was maximal and minimal at 0900 and 1700 hours, respectively. Considering that these reactions are essentially catalyzed by the same enzyme system, the rather poor correlation coefficient (0.598) between demethylase activity and P-450 concentration at different hours of the day indicates that the temporal variations in hepatic monooxygenase activity cannot be explained solely on the basis of similar time-dependent variations in the total concentration of P-450. Similar data were found by Tredger and Chhabra (1977) in their studies of benzphetamine-*N*-demethylase, the hydroxylation of biphenyl at two different sites of the molecule, the rate of P-450 reductase, and the microsomal concentration of P-450. These data indicate that other factors must be in-

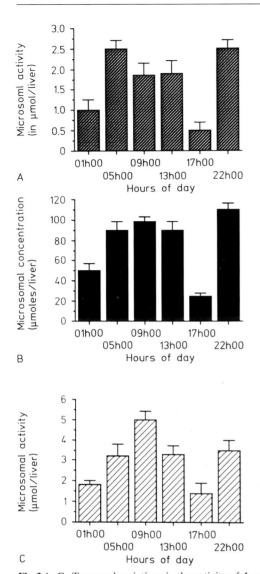

**Fig.3 A–C.** Temporal variations in the activity of **A** aniline hydroxylase and **B** in the concentration of cytochrome *P*-450 and **C** in the activity of p-nitroanisole-*O*-demethylase in rat liver. Livers were dissected from rats synchronized on a 12 h light-dark cycle (light on: 0700–1900 hours). The activities of aniline hydroxylase **(A)** and *p*-nitroanisoledemethylase **(B)** were quantified in vitro and the *P*-450 concentrations **(C)** were determined in the same animals. Each point represents the mean value ± SE of three animals. (Reproduced from Bélanger et al. 1986)

volved in the different temporal relationships in the catalytic activity of the monooxygenase system.

It may be suggested that a temporal variation in the microsomal concentration of the various cytochrome P-450 isoenzymes may explain the different time-dependent variations in the catalytic activity of the monooxygenase system. This hypothesis is sup-

ported by the work of Miyasaki et al. (1990), who recently identified time-dependent changes in the activity of P-450 isoenzymes. Using testosterone hydroxylation as an experimental model, these investigators reported two types of temporal variations in these hydroxylation reactions: (1) a circadian change was found in the activity of P-450 UT-2, as the peak values for testosterone 2-alpha and 16-alpha hydroxylations were both observed at 0800 hours and (2) an ultradian rhythm was detected in the P-450 PB-1 activity because peaks in the 2-beta and 6-beta-hydroxylations were found at 1200 and 2400 hours. No temporal change was observed in the activity of P-450 UT-2, which is known to catalyze the 16-beta hydroxylation of testosterone. Further research is needed in this area.

## The Conjugation Reactions

The time-dependent variations in nonoxidative pathways of drug metabolism (i.e., reduction, hydrolysis, and conjugation) have not been investigated as thoroughly as the monooxygenase system. Of interest is the work on the chronobiology of the hepatic transferases involved in glucuronide and sulfate conjugation that was carried out in freely fed rats synchronized on a 12 h light-dark cycle (Bélanger et al. 1985). The enzyme-substrate characteristics of glucuronosyltransferase and sulfotransferase presented in Table 2 indicate that the maximal rate of the glucuronidation of *p*-nitrophenol was significantly higher at 2100 (i.e., during the activity period of the animal) than at 0900 (i.e., during the resting span). By contrast, the sulfation of phenol was twice as great at 0900 than at 2100 hours. There was no difference in the apparent affinity of the glucuronosyltransferase for its substrate at either time of day, as estimated by the $K_m$ values, but the apparent affinity of the sulfotransferase for its phenolic substrate was increased by fourfold at 2100 hours. It is interesting to note that these time-dependent variations of both transferases were not present in fasting rats.

The temporal variation in hepatic glucuronidation and sulfation may explain the chronopharmacokinetics of acetaminophen (paracetamol), which is known to be eliminated mainly by these two pathways in humans and in rats (Shively and Vesell 1975; Bélanger et al. 1987). In their human studies, Shively and Vesell (1975) reported that the plasma $t_{1/2}$ of this analgesic was 15 % longer at 0600 than at 1400 hours, and the mean ratio of the glucuronide conjugate over unchanged acetaminophen, excreted in the first 3.5 h

**Table 2.** Enzyme-Substrate Characteristics of the Transferases Involved in Glucuronide and Sulfate Conjugation of Rat Liver at 09h00 and 21h00

| Enzyme (Cellular Homogenate Used) | Substrate | Kinetic Parameter[a] | Time of Day | |
|---|---|---|---|---|
| | | | 0900 | 2100 |
| Glucuronosyl-Transferase (Microsomes) | p-Nitrophenol | Vmax (nmol/g 15 min) | 5.6 ± 1.2 | 8.8 ± 0.8[b] |
| | | km (mM) | 0.20 ± 0.03 | 0.15 ± 0.05 |
| Sulfotransferase (Soluble Fraction) | Phenol | Vmax (μmol/g/5 min) | 3.5 ± 0.1 | 1.6 ± 0.2[b] |
| | | km (mM) | 0.18 ± 0.01 | 0.04 ± 0.02[b] |

[a] Vmax, maximal rate of an enzymic reaction; km, Michaelis constant representing the substrate concentration at half the maximal rate of the reaction
[b] $p < 0.05$

urine sample, varied between 5.2 at 0600 and 7.8 at 1400 hours. In rats receiving a single dose of acetaminophen at 0900 or 2100 hours, we found that the total clearance and the total metabolism of the drug, determined as the extraction ratio following administration of a single 40 mg/kg dose to rats, were both greater in the evening than in the morning (Bélanger et al. 1987). The results obtained in human and in rats are in good agreement if one considers that the rat is a nocturnal animal and that its biological rhythms are inversely synchronized to those of humans: acetaminophen metabolism was significantly greater during the activity period in both species.

The conjugation reaction is a pathway which is particularly important in toxicology. Reduced glutathione (GSH) has two main cellular functions: (1) it removes peroxides and other oxygen radicals produced by various enzyme systems and (2) it forms an adduct with electrophilic compounds or with free-radical species or other reactive intermediates of toxic drugs which could be produced during the metabolism of these xenobiotics by the cytochrome P-450 monooxygenase system. As we reported recently (Bélanger 1988), many investigators have found temporal variations in the hepatic concentrations of GSH. One recent example was obtained in rats synchronized under a 12 h light-dark cycle (light on: 0700–1900 hours). Figure 4 illustrates that maximal and minimal hepatic levels of GSH were found at 0900 and 2100 hours, respectively (Bélanger et al. 1988).

This time-dependent variation in hepatic GSH levels has an important implication for hepatotoxic agents whose necrotic effects have been shown to be dependent on the covalent binding of reactive intermediates to proteins, nucleic acids, or other macromolecules of the cell. A higher concentration of GSH in the liver leads to a greater formation of GSH conjugate and to a decrease in the alkylation of cellular components. Inversely, lower hepatic GSH levels increased the covalent binding to macromolecules which often caused cellular dysfunction and necrosis. Thus, the circadian variation in hepatic GSH levels appears to be the major determinant of the chronohepatotoxicity of agents such as 1,1,-dichloroethylene (Jaeger et al. 1973), allyl alcohol (Hanson and Anders 1978), chloroform (Lavigne et al. 1983; Bélanger et al. 1988), and styrene (Desgagné and Bélanger 1986).

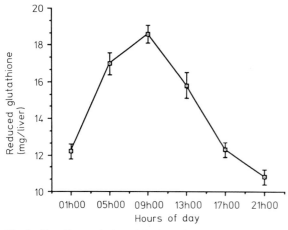

**Fig. 4.** Circadian variation of the hepatic concentration of reduced glutathione (GSH) in rats. The concentration of reduced GSH was determined in the livers of rats synchronized on a 12 h light-dark cycle (light on: 0700–1900 hours). The results are the mean values obtained in three rats/hour of day and they are expressed as mg GSH/liver. (Reproduced from Bélanger et al. 1988)

## Factors Modifying Time-Dependent Changes in Drug Metabolism

The time-dependent variations in the metabolism of drugs summarized above could be caused by rhythmic changes in endogenous factors, such as the rate of blood flow to the liver and the binding of drugs to plasma proteins, and in the endocrine system.

The extent of protein binding could explain some circadian variations in drug metabolism because hepatic enzymes metabolize only drugs which are not bound to plasma proteins. This factor will not be important for drugs with a very high rate of metabolism (i.e., drugs with a high extraction ratio). However, for drugs such as diazepam, phenytoin, theophylline, and warfarin, which are metabolized mainly by the liver but at a much slower rate (i.e., drugs with a low hepatic ratio), the elimination rate will be dependent upon both the rate of metabolism and the extent of protein binding. Further research is needed before a clear correlation can be established between circadian variations in protein binding and those in the metabolism of drugs.

The circadian variations in hepatic blood flow (Fig. 5) could also account for some time-dependent changes in drug metabolism (Labrecque et al. 1988). This factor will be very important for drugs with a high hepatic extraction ratio ($E > 0.7$) but its contribution will be minimal with drugs whose extraction ratio is low ($E < 0.3$).

Circadian rhythms in the endocrine system could also influence the rate of drug metabolism. Indeed, in rodents, the temporal variation in the activity of some hepatic enzymes appeared to be inversely related to the plasma corticosterone level. Maintenance of constant plasma levels of this hormone abolished the rhythmicity of aminopyrine-$N$-demethylase and $p$-nitroanisole-$O$-demethylase, but it had no effect on the activity of 4-dimethylaminoazobenzene reductase (Radzialowski and Bousquet 1967). The in vivo

importance of this factor is still unclear because adrenalectomy decreased the clearance of antipyrine in rats without any effect on the day-night variation of this kinetic parameter.

Finally, the effect of food on the temporal variations in drug metabolism is still unclear. While Radzialowski and Bousquet (1967, 1968) reported that fasting had no effect on the time-dependent variations in enzyme activity, we showed that fasting the rats overnight abolished the morning-evening differences in the enzyme characteristics of the hepatic glucuronosyltransferase and sulfotransferase activities (Bélanger et al. 1985). It must be pointed out also that there are many reports indicating that fasting abolished the rhythmicity in the hepatic concentration of GSH.

Thus, investigations carried out in the last 20 years have provided insights into the mechanisms of the time-dependent variations in the enzymatic systems involved in hepatic drug metabolism. However, further research is needed to identify clearly these mechanisms.

## Biological Rhythms in the Biliary System

Although biliary excretion appears to be more important in laboratory animals than in humans, this process is an important hepatic elimination route for many xenobiotics (Smith 1973; Garrett 1978; Levine 1983). The mechanisms involved in biliary excretion are less well known than those of urinary excretion because our knowledge of biliary physiology is relatively incomplete. The current concept suggests that bile flow and composition are important determinants of both the rate and extent of biliary excretion of endogenous compounds such as bile acids, phospholipids, and cholesterol (Smith 1973; Vonk et al. 1978b; Levine 1983; Okolicsanyl et al. 1986).

Vonk et al. (1978a) appear to be the only investigators to report a temporal variation in the biliary excretion of a foreign compound. They studied the day-night variation in the biliary concentration and excretion rate of dibromosulphthalein (DSP), which is a dye exclusively cleared by the bile and not metabolized by the liver. The data indicated that the biliary excretion of DSP was 25 % higher at midnight than at noon. A parallel increase in bile flow and modifications of the bile composition at different times of the day were correlated with these results (Bortz and Steele 1973; Mitropoulous et al. 1973; Ho and Drummond 1975; Vonk et al. 1978b).

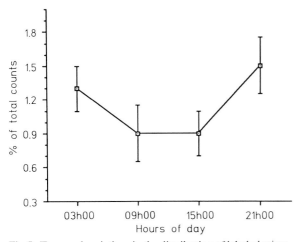

**Fig. 5.** Temporal variations in the distribution of labeled microspheres to the liver of rats. Four groups of rats received [85]strontium-labeled microspheres four times a day. The radioactivity in the liver was quantified 5 min after intravenous injection of the microspheres. (Reproduced from Labrecque et al. 1988)

# Conclusions

Investigations carried out in the last 20 years have indicated clearly that the activities of hepatic enzymes involved in the metabolism of drugs and other xenobiotics are not constant during a 24 h period. Circadian variations were detected mainly in the monooxygenase system, conjugation reactions, and, to a much lesser extent, in biliary excretion. Further research is needed to better characterize the mechanisms of time-dependent changes in the activity of hepatic enzymes. As indicated earlier by Bélanger et al. (1984), it has now become possible to set guidelines to predict the metabolic and pharmacokinetic behavior of a chemical substance.

From our point of view, the time-dependent variation in drug metabolism is a very important cause of around-the-clock changes in pharmacokinetics. Circadian changes in hepatic metabolism could account, in part, for the fact that the intensity of desired and undesired drug effects fluctuates during the active or resting periods of the organism.

# References

Anders MW (1981) Bioactivation of foreign compounds. Academic, New York

Bélanger PM (1988) Chronobiologic variation in the hepatic elimination of drugs and toxic agents. Annu Rev Chronopharmacol 4: 1–46

Bélanger PM, Labrecque G, Doré F (1984) Rate limiting steps in the temporal variations of the pharmacokinetics of some selected drugs. In: Haus E, Kabat H (eds) Chronobiology 1982–1983. Karger, Basel, p 359

Bélanger PM, Lalande M, Labrecque G, Doré F (1985) Diurnal variations in the transferases and the hydrolases involved in glucuronide and sulfate conjugation of rat liver. Drug Metab Disp 13: 386–389

Bélanger PM, Bruguerolle B, Desgagné M (1986) Temporal variation in the inductive effect of 3,4-benzpyrene in rat. Annu Rev Chronopharmacol 3: 361–364

Bélanger PM, Lalande M, Doré F, Labrecque G (1987) Time-dependent variations in the organ extraction ratios of acetaminophen in rat. J Pharmacokinet Biopharm 15: 133–143

Bélanger PM, Desgagné M, Boutet M (1988) The mechanism of the chronohepatoxicity of chloroform in rat: correlation between covalent binding to hepatic subcellular fractions and histologic changes. Annu Rev Chronopharmacol 5: 235–238

Bortz WM, Steele LA (1973) Synchronization of hepatic cholesterol synthesis, cholesterol and bile acid content, fatty acid synthesis and plasma free fatty acid levels in fed and fasted rat. Biochem Biophys Acta 306: 85–94

Boyd SC, Grygiel JJ, Michim RF (1983) Metabolic activation as a basis for organ-selective toxicity. Clin Exp Pharmacol Physiol 10: 87–99

Bruguerolle B (1987) Chronopharmacokinetics. Pathol Biol 35: 925–934

Connors TA (1976) Bioactivaction and cytotoxicity. In: Bridge JW, Chasseaud LF (eds) Progress in drug metabolism, vol 1. Wiley, London, p 41

Desgagné M, Bélanger PMB (1986) Chronohepatotoxicity of styrene in rat. Annu Rev Chronopharmacol 3: 103–106

Drayer DE (1976) Pharmacologically active drug metabolites: therapeutic and toxic activities, plasma and urine data in man, accumulation in renal failure. Clin Pharmacokin 1: 426–443

Drayer DE (1982) Pharmacologically active metabolites of drugs and other foreign compounds: clinical, pharmacological, therapeutic and toxicological considerations. Drugs 24: 519–542

Garratini S (1985) Active drug metabolites: an overview of their relevance in clinical pharmacokinetics. Clin Pharmacokin 10: 216–227

Garrett ER (1978) Pharmacokinetics and clearances related to renal processes. Int J Clin Pharmacol 16: 155–172

Gillette JR (1966) Mechanisms of oxidation and reduction by enzymes in hepatic endoplastic reticulum. Adv Pharmacol 4: 219–261

Ho KJ, Drummond JL (1975) Circadian rhythm of biliary excretion and its control mechanisms in rats with chronic biliary drainage. Am J Physiol 229: 1427–1437

Jacoby WB (1980) Enzymatic basis of detoxication, vols 1, 2. Academic, New York

Jaeger RJ, Conoly RB, Murphy ID (1973) Diurnal variation of hepatic glutathione concentration and its correlation with 1,1, dichloroethylene inhalation toxicity in rats. Res Commun Chem Pathol Pharmacol 6: 465–471

Hanson SK, Anders MW (1978) The effect of diethyl maleate treatment, fasting and time of administration of allyl alcohol hepatoxicity. Toxicol Lett 1: 301–305

Labrecque G, Bélanger PMB, Doré F, Lalande M (1988) 24-hour variations in the distribution of labelled microspheres to the intestine, liver and kidneys. Annu Rev Chronopharmacol 5: 445–447

Lavigne JG, Bélanger PM, Doré F, Labrecque G (1983) Temporal variations in chloroform-induced hepatoxicity in rats. Toxicology 26: 267–273

Levine WG (1983) Excretion mechanisms. In: Caldwell J, Jacoby WB (eds) Biological basis of detoxification. Academic, New York, p 251

Mitropoulos KA, Balasubramaniam S, Murant NB (1973) The effect of interruption of the enterohepatic circulation of bile acids and of cholesterol feeding on cholesterol 7a-hydroxylase in relation to the diurnal rhythm in its activity. Biochem Biophys Acta 326: 428–438

Miyasaki Y, Yatagai M, Imaoka S, Funae Y, Motohashi Y, Kobayashi Y (1990) Temporal variations in hepatic cytochrome P-450 isoenzymes in rats. Annu Rev Chronopharmacol 7: 149–152

Nair V, Casper R (1969) The influence of light on daily rhythm in hepatic drug metabolizing enzymes in rat. Life Sci 8: 1291–1298

Okolicsanyi L, Lirussi F, Strazzabosco M, Jemmolo RM, Orlando G, Nassuato G, Muraca M, Grepaldi G (1986) The effects of drugs on bile flow and composition. An overview. Drugs 31: 430–448

Radzialowski FM, Bousquet WF (1967) Circadian rhythm in

hepatic drug metabolizing activity of the rat. Life Sci 6: 2545–2548

Radzialowski FM, Bousquet WF (1968) Daily rhythmic variation in the hepatic drug metabolism in the rat and mouse. J Pharmacol Exp Ther 163: 229–238

Reinberg A, Smolensky MH (1982) Circadian changes of drug disposition in man. Clin Pharmacokin 7: 401–420

Shively CA, Vesell ES (1975) Temporal variations in acetaminophen and phenacetin half-life in man. Clin Pharmacol Ther 18: 413–424

Smith RL (1973) The excretory function of the bile. Chapman and Hall, London

Tredger JM, Chhabra RS (1977) Circadian variations in microsomal drug metabolizing enzyme activities in rat and rabbit tissues. Xenobiotica 7: 481–489

Vonk RJ, Scholtens E, Strubbe JB (1978a) Bile excretion of dibromosulphthalein in the freely moving unanesthetized rat: circadian variation and effects of deprivation of food and pentobarbital anesthesia. Clin Sci Molec Med 55: 399–406

Vonk RJ, Van Doorn ABC, Strubbe JB (1978b) Bile secretion and bile composition in the freely mooving unanesthetized rat with a permament biliary drainage. Influence of food intake on bile flow. Clin Sci Molec Med 55: 235–359

# Chronobiology of the Gastrointestinal System

J. G. Moore

Rhythmicity characterizes many functions of the human organism, including those of the gastrointestinal (GI) system. Important and significant time-dependent changes have been documented in GI motility patterns [1, 2], drug intestinal absorption rates [3, 4], small bowel mucosal enzyme activities [5], mucosal DNA synthesis rates [6], and gastric acid secretion [7], among others [8]. The literature on chronobiologically oriented investigations in humans and animals, at both the organ and cellular level, is vast and beyond the scope of this discussion. This review will focus only on large amplitude rhythms of human GI motor, absorptive, and secretory function with real or potential influence on disease expression and/or treatment. Circadian rhythms will be emphasized although important ultradian rhythms of GI function will also be mentioned.

## Ultradian and Circadian GI Motility Patterns

Motor activity of the human GI tract is characterized by peristaltic contractions and relaxations with resultant cephalocaudal flow of intestinal contents. Evidence of this activity is observed in both fed and fasted subjects. In fed subjects gastric motor activity is dominated by phase 3 peristalic contractions. These waves, occurring in humans at the frequency of 3/min, originate as pacesetter potentials from high on the gastric greater curvature. In the presence of food or antral distention (e. g., by air), the electrical pacesetter potential converts to an action potential and muscular contraction. Figure 1 displays the normal pacesetter potential frequency which, if converted to action potentials, generates peristaltic contractions [9]. These contractions are responsible for the grinding of digestible solid particles to a size (less than 1.0 mm) that permits passage from the antrum into the duodenum. Disruption of this motor pattern may result in abnormal retention of gastric contents. Figure 1 also illustrates a rare example of a dysrhythmic disorder in phase 3 motor activity. In this example, from a 5 month old infant, the pacesetter potential is seen to originate from an antral site, rather than high on the greater curvature, at increased frequency (tachygastria). The abnormality resulted in propagation of foodstuffs in the retrograde direction and persistent symptoms of retention and vomiting. The symptoms were unrelieved by pyloromyotomy or by gastrojejunostomy but, at 1 year of age, finally responded to resection of the distal three-fourths of the stomach.

In the fasted GI tract, motor activity is dominated by the migrating motor complex (MMC) that originates in the stomach and travels caudally through the small bowel at 90–120 min intervals in humans. These powerful "housekeeping" waves are responsible for gastric emptying of indigestible solid particles (greater than 1.0 mm) and maintaining cephalocaudal flow of intestinal contents during fasting. Their appearance is stimulated by fasting and the hormone motilin; they are promptly inhibited by feeding [10–13]. Figure 2 displays a 24 h small intestinal motility pattern in a fasted healthy subject [10]. Note the peristaltic aboral direction and regularity in timing of the MMC. Alterations in MMC patterns have been documented in bacterial overgrowth and infection, pregnancy, myotonic dystrophy, systemic sclerosis, partial intestinal obstruction, and other disorders [11, 13]. Figure 3 illustrates disruption of MMC activity in a patient with intestinal pseudo-obstruction [13]. Thus, cyclical rhythmic motor activity in the ultradian time domain is characteristic of the healthy human GI tract and disruption of this activity is observed in a variety of disorders. However, the motor abnormalities described have not been defined in enough detail to be of diagnostic aid nor has any cause and effect relationship to the development of any particular symptom been established.

Circadian rhythmic motor patterns, superimposed

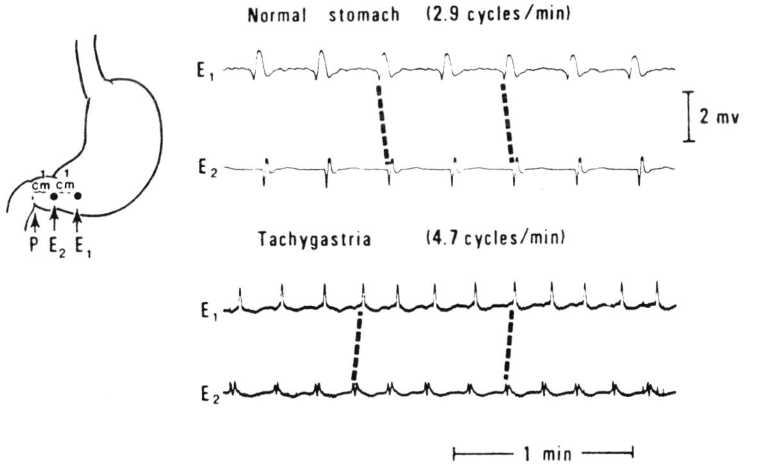

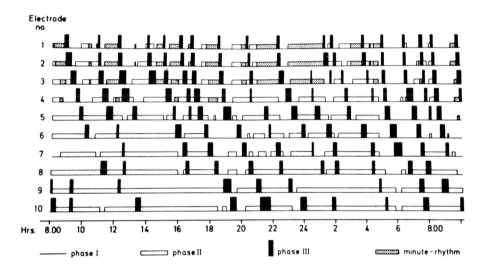

**Fig. 1.** Operative recordings of gastric electrical activity taken from antral region in healthy human stomach *(top panel)* and 5 month old patient with gastric retention *(bottom panel). Dotted lines* indicate that propagation appears to be in an oral direction in the patient, in contrast to the usual aboral direction. In addition, the patient recording exhibits more rapid electrical activity (tachygastria). (From [9])

**Fig. 2.** Representative MMCs from 10 sites (duodenum to ileum) in a fasting healthy volunteer studied for 24 h. Phase 3 waves correspond to MMCs. Note the aboral migration of most complexes. In the fasting state, MMCs occur at 90–120 min intervals. (From [10])

on ultradian patterns, are also characteristic of the healthy human GI tract. Figure 4 displays circadian rhythmicity in the speed of MMC propagation in the small bowel of a group of healthy subjects and a group of patients with irritable bowel symptoms intubated with long intestinal tubes equipped with pressure sensors [1]. The daytime velocity of MMC propagation (cm/s) was more than double the nocturnal value in both groups. Gastric emptying rates for meals also varied over the 24 h period in the same direction [2]. In a two time point study (Fig. 5), gastric emptying rates for meals administered at 2000 h were over 50% slower than emptying rates for the same meal administered to the same subjects at 0800 hours. Thus, gastric and small bowel motor activity is slower at night than in the daytime. However, esophageal motor activity does not appear to exhibit circadian variation, suggesting that nocturnal slowing of GI

motility is not a generalized characteristic of the entire GI tract [14].

The major control for gut motor activity is believed to reside in the enteric nervous system, or "gut brain", since all of the fasted and fed motor patterns are observed in denervated gut preparations [1, 11, 12, 13]. However, circadian modulation of these motility patterns may well be under central nervous system control, perhaps through vagus nerve innervation. The propagating velocity of the ovine MMC is reduced after vagotomy, and vagotomy also abolishes the circadian rhythm of acid secretion in humans [15, 16].

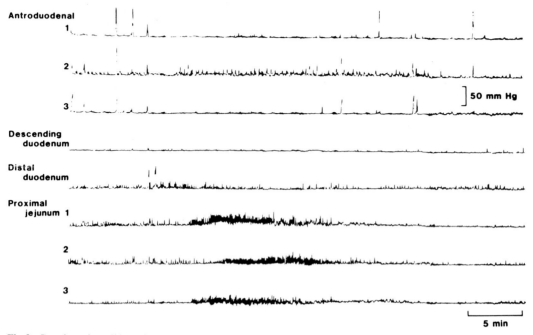

**Fig.3.** Gastric and small bowel motor activity in a patient with chronic idiopathic intestinal pseudo-obstruction. Note lack of propagation of MMC activity at the level of the proximal intestine. (From [13])

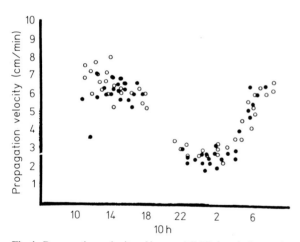

**Fig.4.** Propagation velocity of human MMCs in relation to the 24 h clock. ● patients; ○ normal subjects. Gap at 1800 hours represents feeding time. In companion study in pigs, feeding did not influence circadian pattern. (From [1])

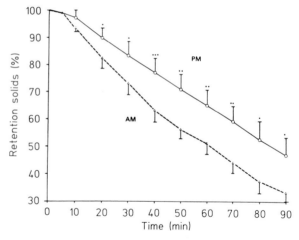

**Fig.5.** Solid-phase morning (AM) and evening (PM) gastric emptying curves. Morning values were significantly faster at all dotted timing intervals. Sixteen healthy young subjects ($\bar{x}$ age = 32 years) were studied. Each was fed a radiolabeled, standardized, 300 g 206 Kcal meal on separate study days. (From [2])

## Circadian Drug Absorption Patterns

The findings of reduced gastric and small bowel motility during the nightime have important implications in the pharmacokinetics of orally administered drugs. Oral aspirin (ASA), theophylline, ethanol, indomethacin, and ketoprofen all exhibit either higher plasma concentrations ($C_{max}$), shorter times to peak plasma concentrations ($T_{max}$), or faster drug disappearance rates when administered during the morning compared to the evening [3, 17–20]. These pharmacokinetic observations may be explained, at least in part, by circadian differences in gastric emptying

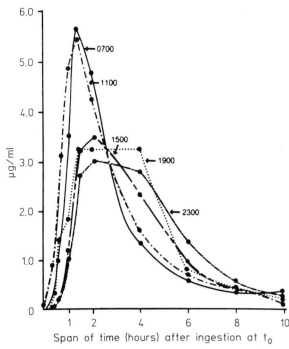

**Fig. 6.** Mean plasma concentrations of indomethacin in nine healthy subjects each given a single 100 mg dose at different times. $T_0$ was 0700, 1100, 1500, and 2300 h scheduled at weekly intervals. Samples missing for 2 subjects at $T_0 = 0700$ and for 4 subjects at $T_0 = 1100$. (From [3])

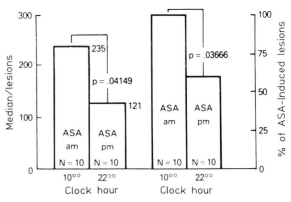

**Fig. 7.** Comparison of median number of morning and evening aspirin-induced mucosal lesions *(left)* and as a percentage of the total number of lesions produced by morning administration of aspirin ( = 100%) *(right)*. For the group ($n = 10$, healthy male volunteers) evening administration produced 37% fewer lesions. (From [25])

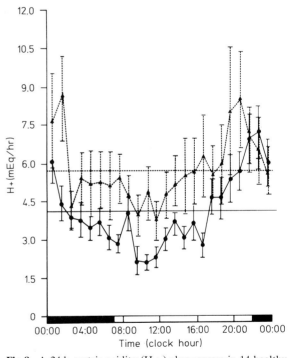

**Fig. 8.** A 24 h gastric acidity (H + ) chronogram in 14 healthy volunteers ( ● ) and 21 patients with active peptic ulcer disease ( ▲ ). Each point represents mean ± SEM values. Note low morning and high evening rates of secretion in both groups. *Dashed line* represents mean 24 h secretory rate for the ulcer group ($5.76 \pm 0.98$ mEq H + /h); *solid line* represents mean rate for the healthy group ($4.12 \pm 0.40$ mEq H + /h). (From [28])

rates. For example, the time-to-peak plasma concentrations after oral theophylline were increased 67% with evening administrations (1900 hours) compared with morning administration (0700 hours) [18]. A 54% increase in evening meal emptying time, together with a decreased nocturnal MMC propagating velocity, would contribute to the observed circadian variation in theophylline absorption. The absorption of indomethacin, a commonly prescribed anti-inflammatory drug, is similarly influenced by circadian factors (Fig. 6) [3]. As is apparent, $T_{max}$ and $C_{max}$ values were significantly shorter and higher, respectively, when the dose (100 mg) was given at 0700 hours rather than at 1900 hours.

## Circadian Drug Tolerance and GI Toxicity

ASA and other nonsteroidal anti-inflammatory drugs (NSAIDs) produce predictable dose-related acute damage to the human gastric mucosa [21, 22]. The lesions range from punctate hemorrhages to deep ulcerations. Huskisson, in 1976, reported that five of

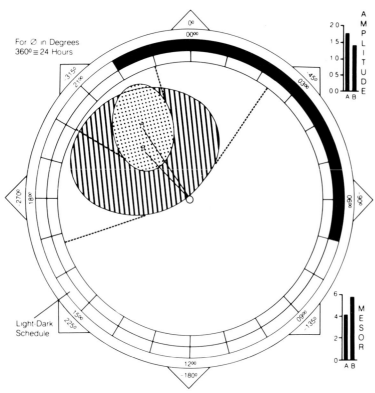

**Fig.9.** Population mean cosinor polar plot of data shown in Fig.8 for healthy *(dotted ellipse)* and ulcer *(hatched ellipse)* groups. Comparison of results revealed significant circadian rhythms of basal gastric acid secretion with similar amplitudes and acrophases. (From [28])

| Key To Ellipses | P | PR | No. Obs. | Mesor | SE | Amplitude (95% CL) | | Acrophase (∅) (95% CL) | | |
|---|---|---|---|---|---|---|---|---|---|---|
| A. Health | <0.001 | 38 | 14 | 4.12 | 0.40 | 1.78 | (0.97  2.61) | −326° (−300 | −345) | |
| B. Ulcer | 0.034 | 33 | 21 | 5.76 | 0.98 | 1.38 | (0.09  2.72) | −316° (−251 | −35) | |

P = Probability of hypothesis amplitude = 0; No. Obs. = Number of observations; PR = Percent rhythm (percentage of variability accounted for by cosine curve); 95% CL = Conservative 95% confidence limits derived from cosinor ellipse

ten patients experienced central nervous or GI side effects after morning ingestion of indomethacin (100 mg), while only one of these patients had side effects if the same dose was taken during the evening hours [23]. In a larger study, including 517 osteoarthritic patients, orally administered sustained release indomethacin produced a significantly ($p < 0.001$) greater number of undesirable side effects when delivered at 0800 hours (30% of patients) than at 2000 hours (9% of patients) [24]. Of the therapeutic withdrawals encountered during the study, 29 (66%) were associated with morning ingestion. The side effects reported in these studies appear to correlate with the higher peak plasma indomethacin concentrations and the more abrupt rise to these high levels following the morning dose. In an endoscopic study on ten healthy male volunteers administered 1300 mg of ASA at 0800 or 2000 hours on separate study days, the evening dose produced 37% fewer hemorrhagic lesions, as counted during upper GI endoscopy (Fig.7) [25]. Thus, a small but consistent series of reports indicate that nightime administration of NSAIDs is better tolerated than morning administration. The mechanism underlying increased morningtime NSAID acute damage is unknown, but it does not appear to be due to either delayed gastric emptying (Fig.6) or to increased basal gastric acidity (Fig.8) at those hours, factors that theorectically should lead to increased acute NSAID damage. In the rat model, dissociation of acid secretion and mucosal damage by ASA and butyric acid over the circadian period has been described, suggesting that the level of gastric acid alone is not the sole determinant of mucosal damage produced by these agents [26, 27].

## Circadian Gastric Acid Secretory Patterns

A circadian rhythm in basal gastric acid secretion has been reported in healthy men and in men with active duodenal ulcer disease [7, 28]. In these studies, quantitative gastric acid collections were obtained in the fasting state by nasogastric tube suction techniques; hourly collections for 24 h were measured for volume, pH, and total titratable acidity. The 24 h chronogram and cosinor polar plots for total titratable acidity are displayed in Fig. 8 and 9, respectively. Note that in both groups acid output is highest in the evening and lowest in the morning hours. The mean hourly acid output is 30% higher in the ulcer patients, an expected observation. The mechanism (s) underlying circadian rhythmicity in basal gastric is not fully understood. The rhythmic changes in acidity are not accompanied by rhythmic changes in serum gastrin, an endogenous hormone known to stimulate acid secretion [29]. However, vagal nerve integrity appears to be important in maintaining circadian rhythmicity because rhythmicity is lost in postvagotomy patients with measurable levels of basal gastric acidity [16].

## Circadian Gastric Acid Secretory Patterns Under H$_2$-Receptor Antagonist Blockade

Does circadian rhythmicity in fasting gastric acidity influence H$_2$-receptor antagonist therapy? Figure 10 depicts the 24 h intragastric (IG) pH pattern in a group ($n = 15$) of ambulating patients with healed duodenal ulcer disease studied in the fasted state [30]. The measurements were obtained from tethered intragastrically placed pH sensors. The technique provides continuous IG pH readings (q 4 s), may be left in place for several days, and has greatly simplified the monitoring of intragastric acid measurements, especially in the study of antisecretory compounds [31]. In the study depicted in Fig. 10, subjects received either intravenous (IV) saline (placebo control) or IV ranitidine, 6.25 mg/h/24 h, or IV ranitidine, 10 mg/h/ 24 h, during one of three separate study days. During the placebo control period a low IG pH was maintained throughout the 24 h (24 h median hourly IG pH = 1.19). Ranitidine significantly ($p < 0.001$) elevated IG pH when compared to placebo in a dose-dependent manner (6.25 mg/h/24 h median hourly pH = 5.98; 10.0 mg/h/24 h median hourly IG pH =

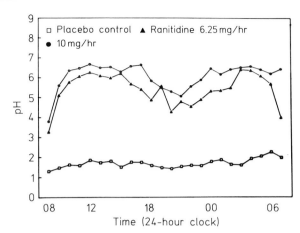

**Fig. 10.** Hourly 24 h intragastric pH values in 15 fasted, healed, duodenal ulcer patients. In a randomized design, patients received either IV placebo (□—□), IV ranitidine 6.5 mg/h (▲—▲), or IV ranitidine, 10 mg/h (●—●) for 24 h. Note evening fall in pH values, corresponding in time to increased acid output observed in ulcer and healthy groups displayed in Fig. 9. (From [30])

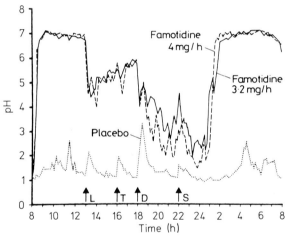

**Fig. 11.** Median 24 h intragastric (IG) pH profiles in 12 fed duodenal ulcer patients. *Dotted line* represents control (placebo) 24 h study; *solid line* represents IV continuous infusion (CI) of famotidine, 3.2 mg/h × 24 h; *dashed line* represents IVCI famotidine, 4.0 mg/h × 24 h. Meals are shown at bottom by *arrows: L* (lunch), *T* (tea), *D* (dinner), and *S* (snack). Note pronounced fall in IG pH during the late afternoon and evening hours in these subjects. (With permission [33])

6.77). However, during both ranitidine infusion schedules, IG pH was observed to decline during the late afternoon and evening hours and then rose again to pre-decline levels for the remaining study hours. Plasma ranitidine concentrations did not decline during the evening time frame. Thus, "breakthrough" in IG pH control by ranitidine occurred during the time frame corresponding to peak acid secretory rates observed in Fig. 8. This data suggests that the $H_2$-receptor antagonist dose/acid inhibiting response is not fixed over the circadian period. Hill equation-generated dose/ response curves for the morning and evening resulted in an $IC_{50}$ value (the dose of ranitidine required to produce 50% inhibition of acid) of 75 ng/ml for the morning and 150 ng/ml for the evening time frame. In another study, employing the same IG pH measurement techniques in a group ($n = 12$) of patients with healed duodenal ulcer disease, "breakthrough" of IG pH control by ranitidine was again demonstrated during the evening time frame [32].

In the fed state, compared to fasting, "breakthrough" of $H_2$-receptor blockade is even more pronounced. Merki et al. performed 24 h IG pH studies in a group of ambulating, healed, duodenal ulcer patients allowed regularly scheduled meals [33]. $H_2$-receptor blockade was provided by IV continuous infusion of famotidine in two dosing schedules (3.2 mg and 4.0 mg/h/24 h). Figure 11 shows the results demonstrating "breakthrough" in IG pH control by famotidine during the afternoon and evening time frame. Intragastric pH values fell below 2.0 in these fed patients, compared to falls in IG pH to levels of 4 in the above mentioned fasting studies. Thus, time of day and feeding markedly influence the effectiveness of $H_2$ blockade treatment, and either more $H_2$ blocker or another non-histamine receptor sensitive, acid suppressing drug may be required to maintain elevated IG pH values during the evening time frame.

## Conclusion

Chronobiological factors in experimental and clinical gastroenterology are receiving increased attention and study. Ultradian ($< 20$ h period), circadian (about a 24 h period), and infradian (greater than 28 h period) GI rhythms have been described in a variety of animal species, including humans. This review has focused on only a few large amplitude rhythms that have been adequately documented in the human GI tract. The relevance of these rhythms to the development of disease or their relationship to the production of GI symptoms as they are reported around the ultradian, circadian, and infradian clocks remains an intriguing but speculative endeavor. Elucidation of the molecular and physiologic basis underlying rhythmic phenomenon will likely lead to a greater understanding of their role in health and dis-

ease. In the meantime, it is now clear that chronobiological factors, including the chronopharmacokinetic and chronopharmacodynamic action of drugs, must be taken into account in the treatment of specific diseases. That these rhythmic processes and time-dependent drug effects are clinically relevant is best exemplified by the use of theophylline in the treatment of asthma [34]. Chronotherapy (i.e., the optimization of drug effects and/or minimization of toxicity by timing of administration of drugs in consonance with naturally occurring biological rhythms) should lead to improved drug dosage schedules for a variety of disease processes, including those of the GI tract.

## References

1. Kumar D, Wingate D, Ruckebush Y (1986) Circadian variation in the propagating velocity of the migrating motor complex. Gastroenterology 91: 926–930
2. Goo RH, Moore JG, Greenberg E, Alazraki NP (1987) Circadian variation in gastric emptying of meals in man. Gastroenterology 93: 515–518
3. Clench J, Reinberg A, Dziewanowski Z, Ghata J, Smolensky MH (1981) Circadian changes in the bioavailability and effects of indomethacin in healthy subjects. Eur J Clin Pharmacol 20: 359–369
4. Reinberg A, Smolensky MH (1982) Circadian changes of drug disposition in man. Clin Pharmacokin 7: 401–420
5. Markiewicz A, Karminski M, Tarquini B, Halberg F (1981) Circadian rhythm of enzymatic activity in human jejunum. Int J Chronobiol 7: 282
6. Buchi KN, Rubin NH, Moore JG (1988) Circadian cellular proliferation in human rectal mucosa. Annu Rev Chronopharm 5: 355
7. Moore JG, Englert E (1970) Circadian rhythm of gastric acid secretion in man. Nature 226: 1261–1262
8. Vener K, Moore JG (1988) Chronobiologic properties of the alimentary canal affecting xenobiotic absorption. Annu Rev Chronopharm 4: 257–281
9. Telander RL, Morgan KG, Krevlen DL, Schmalz PF, Kelly K, Szurszewski JH (1978) Human gastric atony with tachygastria and gastric retention. Gastroenterology 75: 497–501
10. Fleckenstein P, Oigaard A (1978) Electrical spike activity in the human small intestine. Am J Dig Dis 23: 776–780
11. Weisbrodt NW (1987) Motility of the small intestine. In: Johnson LR (ed) physiology of the gastrointestinal tract. 2 edn. Raven, New York
12. Sarna S (1985) Cyclic motor activity: migrating motor complex. Gastroenterology 89: 894–913
13. Malagaleda JR, Camilleri M (1985) Disorders of motility of the stomach. In: Berk JE (ed) Bockus gastroenterology, 4 edn. Saunders, Philadelphia, chap 76
14. Avots-Avotins AE, Ashworth WD, Stafford BD, Moore JG (1990) Day and night esophageal motor function. Am J Gastroenterol 85 (6): 683–685
15. Gregory PC, Rayner DV, Wenham G (1984) Initiation of migrating myoelectric complex in sheep by duodenal acidi-

fication and hyperosmolarity: role of vagus nerves. J Physiol 355: 509–521

16. Moore JG (1973) High gastric acid after vagotomy and pyloroplasty in man. Evidence for non-vagal mediation. Am J Dig Dis 18: 661–669

17. Markiewicz A, Semenowicz K (1979) Time dependent changes in the pharmacokinetics of aspirin. Int J Clin Pharmacol Biopharm 17: 409–411

18. Kyle GM, Smolensky MH, Thorne LG, Hsi BP, Robison A, McGovern JP (1986) Circadian rhythm in the pharmacokinetics of orally administered theophylline. In: Smolensky MH, Reinberg A, McGovern JP (eds) Recent advances in the chronobiology of allergy and immunology. Pergamon, Oxford

19. Reinberg A, Clench J, Aymard N, Gaillot M, Bourdon R, Gervais P, Abulker C, Dupont J (1975) Variation circadiennes des effects de l'ethanolemie chez l'homme adulte sain (etude chronopharmacologique). J Physiol (Paris) 70: 1–22

20. Reinberg A, Levi F, Toviton Y, Leliboux A, Simon A, Frydman A, Bickora-Rocher A, Bruguerolle B (1986) Clinical chronokinetic changes in a sustained release preparation of ketoprofen (SRK). Annu Rev Chronopharmacol 3: 317–320

21. Graham DY, Smith JL (1986) Aspirin and the stomach. Ann Intern Med 104: 390–398

22. Katz KA, Sunshine AG, Cohen S (1987) The effect of nonsteroidal antiinflammatory drugs on upper gastrointestinal symptoms and mucosal integrity. J Clin Gastroenterol 92 (2): 142–148

23. Huskisson EC (1976) Chronopharmacology of anti-rheumatic drugs with special reference to indomethacin. In: Huskisson EC, Velo GP (eds) Inflammatory arthropathies. Excerpta Medica, Amsterdam

24. Levi F, LeLouarn C, Reinberg A (1984) Chronotherapy of osteo arthritic patients: optimization of indomethacin sustained release (ISR). Annu Rev Chronopharmacol 1: 345–348

25. Moore JG, Goo RH (1987) Day and night aspirin-induced gastric mucosal damage and protection by ranitidine in man. Chronobiol Int 4 (Ulcerogenesis suppl): 111–116

26. Ventura U, Carandente F, Montini E, Ceriani T (1987) Circadian rhythmicity of acid secretion and electrical function in intact and injured rat mucosa – the relation of timing to ulcerogenesis. Chronobiol Int 4 (Ulcerogenesis suppl): 43–52

27. Szabo S, Pfeiffer DC, Oishi T (1986) The pharmacology of drug induced gastric and duodenal ulcers. Annu Rev Chronopharmacol 3: 383–384

28. Moore JG, Halberg F (1986) Circadian rhythm of gastric acid secretion in men with active duodenal ulcer. Dig Dis Sci 31 (11): 1185–1191

29. Moore JG, Wolfe ME (1974) Circadian plasma gastrin levels in feeding and fasting man. Digestion 11: 226–232

30. Sanders SW, Moore JG, Buchi KN, Bishop AL (1988) Circadian variation in the pharmacodynamic effect of intravenous ranitidine. Annu Rev Chronopharm 5: 335–338

31. Fimmel CJ, Etienne A, Cillufo T, Ritter C, Gasser T, Rey JP, Caradonna-Moscatelli P, Sabbitini F, Pace F, Buhler HW, Bauerfind P, Blum AL (1985) Long-term ambulatory gastric pH monitoring: validation of a new method and effect of H2-antagonists. Gastroenterology 88 (6): 1842–1851

32. Ballesteros MA, Hogan DL, Koss MA, Isenberg JI (1990) Bolus or intravenous infusion of ranitidine: effects on gastric pH and acid secretion: a comparison of cost and relative efficacy. Ann Intern Med 112: 334–339

33. Merki HS, Witzel L, Kaufman D, Kempf M, Neuman J, Rohmel J, Walt RP (1988) Continuous infusion of famotidine maintains high intragastric pH in duodenal ulcer. Gut 29: 453–457

34. Smolensky MH, McGovern JP, Scott PH, Reinberg A (1987) Chronobiology and asthma. II. Body-time-dependent differences in the kinetics and effects of bronchodilator medication. J Asthma 24 (2): 91–134

# Cardiovascular Chronobiology and Chronopharmacology

B. Lemmer

Circadian rhythms in the functions of the cardiovascular system are now well established (for reviews see [1–5]). Biological rhythms in heart rate and blood pressure were already described at the end of the eighteenth and during the nineteenth century by Falconer, Autenrieth, Knox, Wilhelm, Zadek, and others (for references see [4]). Following these early reports, numerous and more sophisticated studies have provided convincing evidence for circadian rhythms in heart rate and in blood pressure in both in healthy subjects and in patients suffering from cardiovascular diseases. The development and use of automatic, 24 h, blood pressure monitoring devices has greatly contributed to our present knowledge about circadian rhythms in the cardiovascular system.

While the rhythms in heart rate and blood pressure are the most well known periodic functions in the cardiovascular system, other parameters have been shown to exhibit circadian variations as well. Some of these, pertinent to an understanding of the chronopathology of cardiovascular disease and the chronopharmacokinetics and dynamics of cardiovascular active drugs, will be mentioned below. In Tables 1 and 2, cardiovascular active drugs are compiled for which, in humans, daily variations were reported in the drugs' pharmacokinetic and/or effects [6]. If not otherwise stated, references are compiled in the review articles mentioned above.

## Chronobiology of the Cardiovascular System

In Fig. 1 some basic mechanisms are compiled which are involved in the regulation of the cardiovascular system. All of these were shown to be modified by the circadian stage, thus being of importance for the understanding of chronopharmacological findings with cardiovascular active drugs.

**Table 1.** Chronopharmacodynamics of cardiovascular active drugs in humans. (From [6])

| β-Blockers | Calcium channel blockers | ACE inhibitors | Diuretics | Nitrates | Others |
|---|---|---|---|---|---|
| Acebutolol | Amlodipine | Captopril | Hydrochlorothiazide | Glyceryl trinitrate | Clonidine |
| Atenolol | Nifedipine | Enalapril | Indapamide | Isosorbide dinitrate | Prazosin |
| Bopindolol | Nisoldipine | | Piretanide | Isosorbide 5-mononitrate | Potassium chloride |
| Labetolol | Nitrendipine | | Xipamide | | |
| Mepindolol | Verapamil | | | | |
| Nadolol | | | | | |
| Oxprenol | | | | | |
| Pindolol | | | | | |
| Propranolol | | | | | |
| Sotalol | | | | | |

**Table 2.** Chronopharmacokinetics of cardiovascular active drugs in humans. (From [6])

| β-Blockers | Calcium channel blockers | Nitrates | Glycosides | Others |
|---|---|---|---|---|
| Propranolol | Diltiazem | Isosorbide dinitrate | Digoxin | Dipyridamol |
| | Nifedipine | Isosorbide 5-mononitrate | Metildigoxin | Potassium chloride |
| | Verapamil | (immediate and sustained release) | | |

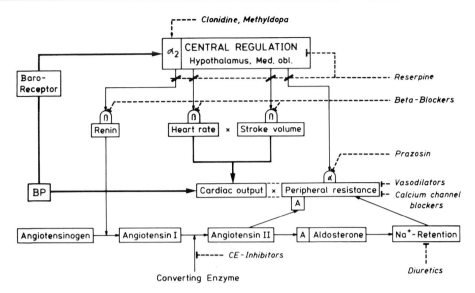

**Fig. 1.** Simplified scheme of main mechanism involved in the regulation of the cardiovascular system and main sites of action of antihypertensive drugs. Drug-induced blockade is indicated by |– – – – and stimulation by ◄ – – – –. (From [5, 6])

Aside from the rhythms in heart rate and in blood pressure which were already mentioned, stroke volume and cardiac output (Fig. 1) are also higher during the activity phase in humans.

Circadian rhythm in blood flow (e.g., forearm, calf) in humans peaks around the early afternoon hours with trough values at night. In rats the hepatic flow is greatest at the early activity period at night.

Capillary resistance has also been shown to vary over 24 h. A circadian rhythm in adrenergic vascular response, monitored by an increase in arterial blood pressure due to the infusion of norepinephrine, has been demonstrated in humans [7]. A dose-dependent blood pressure increase was found throughout a 24 h period. The effect, however, was more marked during the daytime and greatly reduced during the night [7]. An early morning increase in vascular reactivity, around 0300–0400 hours, was prominent. This is of interest because an early morning rise in blood pressure has been described frequently in normotensive and hypertensive patients (see below). However, at present there are no data demonstrating whether or not this pattern is mainly endogenous in nature or due to masking effects.

Blood volume in humans increases during the evening and falls at midnight. A causal explanation for the circadian rhythm in blood volume is not readily apparent. Blood volume is controlled by a balance between the influx and efflux of water, electrolytes, plasma proteins, and red blood cells and by hemodynamic factors. Circadian rhythms are present in renal water and electrolyte excretion, renal plasma flow, and glomerular filtration rate, with peak values during the daytime and trough values during the night.

Total plasma protein, hematocrit, and blood viscosity, for example, also manifest circadian rhythms with trough values late at night (see below). These may also contribute to the rhythm in blood volume. Similar findings have been obtained in rats, in which the cerebral blood volume was highest at the onset of the resting period during the light phase. Various parameters obtained from ECG recordings and cardiac functions in humans may also vary with time of day, such as PQ, QRS, and QT intervals, systolic time interval (STI), corrected preejection period (PEPI), left ventricular ejection time (LVETI), etc. QT intervals, e.g., were larger during sleep than during waking hours and diurnal variations were blunted in transplanted patients. Recently, no diurnal variations were found in LVETI and external isovolumic contraction time (EICT) in five normal subjects. In patients with coronary heart disease, the EICT was, however, significantly prolonged from 0000 to 0400 hours, suggesting a decrease in left ventricular contractility in these patients [8]. Urinary and, more importantly, plasma or serum concentrations of norepinephrine or epinephrine in humans exhibit higher values during early daytime hours than during the night, indicating a daily variation in sympathetic tone. This rhythm continued whether subjects had a normal activity-rest cycle or were recumbent over a 24 h period [9]. In agreement with these findings, plasma levels of the second messenger cyclic adenosine monophosphate (cAMP) in normal adults were higher during daytime activity than during nocturnal sleep (Fig. 2) [10]. Similarly, the turnover of norepinephrine in the rat heart was significantly higher in the dark than in the light span, indicating that sympa-

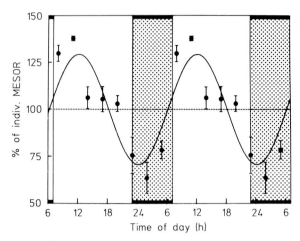

**Fig. 2.** Circadian rhythm in the cAMP content in human plasma, mean values ± SEM of 6 healthy subjects as a percent of the 24 h mean; double plot. Individual cAMP values were in the range of 8–14 pmol/ml plasma. (From [37])

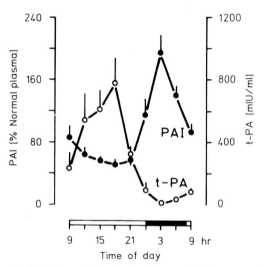

**Fig. 3.** Daily variations in the plasma activities of t-PA (tissue-type plasminogen activator, *filled circles*) and PAI (plasmi-nogen activator inhibitor, *open circles*) over 24 h; shown are mean values ± SEM. (Figure redrawn from [12])

thetic tone is higher in the activity period during darkness than in the resting period during light ([11], see in [5]).

Significant daily rhythms are also present in the plasma concentrations of renin, angiotensin, aldosterone, and atrial natriuretic hormone. The renin-angiotensin-aldosterone system constitutes another regulator of the cardiovascular system (Fig. 1). A circadian rhythm in plasma renin and aldosterone, with peak values at or just before the usual time of waking from night rest, was described in subjects who re-

mained recumbent throughout 24 h and in subjects who were investigated while adhering to a normal activity-rest cycle. In night active rats, pineal angiotensin converting enzyme (ACE) exhibited a pronounced circadian rhythm with a peak at the end of the rest period, during the daytime. If this phase relationship is true for humans, the occurrence of increased ACE activity also would be expected to coincide with peak values in angiotensin II and aldosterone before or at the onset of the human activity cycle (see above). Thus, there is at least a coincidence in the occurrence of circadian peak values of pressor hormones, such as the catecholamines, renin, aldosterone, and angiotensin II, although this coincidence might not represent a causal relationship!

Blood viscosity, hemoglobin, hematocrit, and adrenaline- or ADP-induced platelet aggregation in humans display significant daily variations which are greatest during the morning hours. Red blood cell count and plasma protein concentrations are also circadian phase-dependent. Recently it has been shown that fibrinolytic activity is greatly reduced during early morning hours related to increased plasminogen activator inhibition [12] (Fig. 3).

In rat tissues (heart, brain) significant daily rhythms can be traced down to the level of the β-adrenoceptor-adenylate cyclase-cAMP-phosphodiesterase system ([13–15] see in [5]).

In conclusion, these data clearly point to a pronounced circadian organization of the cardiovascular system.

## Chronopathology of the Cardiovascular System

Aside from the circadian rhythms in physiological functions, various clinical reports clearly demonstrate that the onset of cardiovascular diseases and symptoms exhibits a pronounced temporal dependency (for review see [1–5]). Representative findings concerning symptoms of angina pectoris are compiled in Fig. 4, demonstrating that in patients suffering from angina pectoris, ST segment elevations occur more frequently at night, whereas ST segment depressions are registered more often during daytime hours, indicating differences in etiology. Also, angina attacks do not occur at random during the 24 h of a day (Fig. 4). As mentioned above, it has been shown that fibrinolytic activity is greatly reduced during early morning hours related to increased plasminogen activator in-

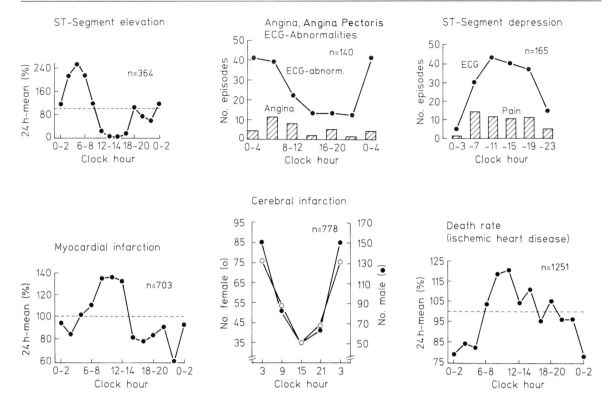

**Fig. 4.** Summary of circadian stage-dependent pathophysiological findings in cardiovascular diseases. Angina pectoris: ST segment elevation in 25 patients with variant angina [38]; angina attacks and ECG abnormalities such as ST segment elevation, T wave pseudonormalization in 13 patients with variant angina [39]; ST segment depression and painful episodes (*n* = 165) during ambulatory monitoring in patients subsequently undergoing coronary angiography [40]. Myocardial infarction: onset evaluated by the MB creatinase method in 703 patients [41]. Cerebral infarction: evaluated in 778 male and female patients [42]. Death rate: evaluated in 1251 patients with ischemic heart disease [43]. (Figures were redrawn according to the references mentioned in [5, 6])

hibiton ([12] Fig. 3). This finding is of great interest in the light of the morning peak in myocardial infarction, which has been reported by more than 12 groups independently (see also [1–5]).

## Hypertension

As already indicated in Fig. 1, both various groups of drugs and individual compounds are used in the treatment of arterial hypertension. It has been convincingly demonstrated in animal experiments and in clinical studies that the effects (desired and undesired) and the pharmacokinetics of drugs used in the treatment of cardiovascular disorders are likely to exhibit circadian temporal dependencies. These reports have been reviewed recently [2–5] and will not be included in the present review.

Nearly all drug classes of antihypertensives were shown to exhibit daily variations in their effects on blood pressure (BP) and heart rate (HR), even after chronic administration. In general, lowering of BP and HR by various $\beta$-blockers and calcium channel blockers was more pronounced during daytime than nighttime hours [2–5]. $\beta$-receptor blocking drugs are still of great therapeutic value in the treatment of various cardiovascular disorders, e.g., coronary heart disease, hypertension, and arrhythmias. The various $\beta$-receptor blocking drugs differ not only in their specific effects (receptor affinity and selectivity, intrinsic sympathomimetic activity) and in their nonspecific effects (related to lipophilicity of the compound), but also in their main routes of elimination (lipophilic drugs, mainly hepatic biotransformation; hydrophilic drugs, renal elimination). Various $\beta$-receptor blocking drugs have been intensively studied in animal experiments [16–18] and in patients [2–4, 19, 20] in relation to circadian time. However, detailed chronopharmacokinetic studies with different $\beta$-blockers have only been performed in rats [16–18]. In humans, only the pharmacokinetics of propranolol were investigated at different times of day [19, 20]. In our own

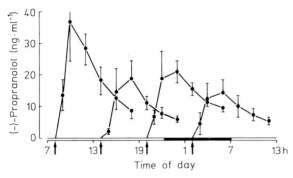

**Fig.5.** Chronopharmacokinetics of oral propranolol in healthy subjects after drug ingestion of 80 mg racemic propranolol at four different times of day, as indicated by the *arrows;* mean values ± SEM. (From [19])

study significant daily variations in peak drug concentration ($C_{max}$), area under the curve (AUC), and elimination half-life were found when racemic propranolol (Dociton®) was administered at four circadian times (0800, 1400, 2000, 0200 h) to healthy subjects (Fig.5, Table 3). Trough values in $C_{max}$ and AUC were found after drug application at 0200 h and peak values in $C_{max}$ and AUC were obtained at 0800 h [19]. As has been shown for other lipophilic drugs, the absorption of propranolol was clearly circadian stage-dependent, being greatest when ingested at 0800 hours (Table 3). At any time of drug intake, the ratio of the plasma concentrations of (−)- to (+)-propranolol was about 1.5 [19]. Thus, the stereospecific metabolism of propranolol did not display a circadian phase dependency. Interestingly, in humans shorter

elimination half-lives of either (−)- or (+)-propranolol were found when the drug was given at 0800 hours rather than at 2000 hours [19]. Table 3 demonstrates that the changes in HR in relation to the respective circadian control values (hemodynamic functions always measured in the sitting position) were markedly different depending on the time of propranolol ingestion: After administration of propranolol at 0200 hours the HR was only slightly affected within the first 6 h after drug intake. However, 2 h later, at the onset of the activity span when sympathetic tone was increasing again (as demonstrated by the rhythm in plasma norepinephrine and cAMP), the HR lowering effect was pronounced again and about equally to that found after drug application at 0800 hours. Thus, peak drug effects coincided with peak drug concentrations only after propranolol intake at 0800 hours and 1400 hours, and were delayed after propranolol dosing at 2000 hours and at 0200 hours (Fig.6), indicating a circadian time dependency in the dose-response relationship of propranolol. Furthermore, these data clearly demonstrate that the chronopharmacokinetics of propranolol cannot mainly be responsible for the daily variations in the drug's hemodynamic effects. The daily rhythm in the level of sympathetic tone seems mainly to determine the degree in $\beta$-adrenoceptor blockade, thus leading to more pronounced effects in the activity period of humans during daytime hours, in which sympathetic tone is high.

Recently, daily variations in the pharmacokinetics of several calcium channel blockers were reported in abstract form, including verapamil [21], diltiazem [22], and nifedipine [23, 24]. In the study of Hla et al. [21] eight volunteers received 80 mg verapamil on six

**Table 3.** Pharmacokinetic and hemodynamic parameters of oral propranolol (80 mg) administered at four different times of day in healthy volunteers. (From [19])

|  | Time of propranolol administration (hours) | | | |
|---|---|---|---|---|
|  | 0800 | 1400 | 2000 | 0200 |
| Pharmaco-kinetics |  |  |  |  |
| $C_{max}$ | 38.6 ± 11.2 | 20.0 ± 6.5 | 26.2 ± 5.3 | 18.4 ± 4.4* |
| $t_{max}$ (h) | 2.5 ± 0.5 | 3.5 ± 0.5 | 3.0 ± 0.6 | 3.5 ± 1.0 |
| AUC (ng/ml × h) | 169 ± 47 | 106 ± 30 | 140 ± 23 | 92 ± 22 |
| $t_{1/2}\,\beta$ (h) | 3.3 ± 0.4 | 4.2 ± 0.5 | 4.9 ± 0.2 | 4.4 ± 0.6** |
| $C_{max}/t_{max}$ (ng/ml/h) | 17.9 ± 6.4 | 7.5 ± 3.9 | 10.6 ± 3.7 | 7.1 ± 2.4* |
| Hemodynamics (heart rate) |  |  |  |  |
| $E_{max}$ (b/min) | 16.0 ± 2.4 (b/min) | 11.7 ± 1.8 | 16.3 ± 1.5 | 15.3 ± 4.6 |
| $T_{max}$ (h) | 2.3 ± 0.6 | 4.5 ± 1.0 | 6.5 ± 1.5 | 7.0 ± 1.0 |

ANOVA: *$p < 0.05$; **$p < 0.01$

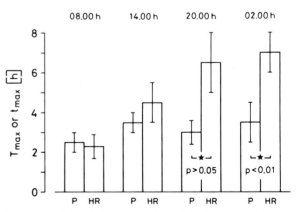

**Fig.6.** Correlation between time-to-peak (−)-propranolol concentrations ($t_{max}$, P) and time-to-peak effect in decreasing heart rate ($T_{max}$, HR) after propranolol ingestion at four different times of day. (From [6, 19])

different occasions (04, 08, 12, 16, 20, 24 hours local time). Daily variations were found in $C_{max}$, $t_{max}$, and AUC, with maximum values in $C_{max}$ and AUC at 8 h and minimum values in $t_{max}$ at 12 hours [21]. In the second study 12 volunteers received 60 mg diltiazem every 6 h (06, 12, 18, 24 hours) for 5 days and plasma concentrations were determined on day 5 immediately before and after each drug administration [22]; $t_{max}$ was longest and $C_{max}$ and AUC were lowest after administration at 24 h [22]. Recently, we were also able to demonstrate significant daily variations in the pharmacokinetics of an immediate release preparation of nifedipine in healthy subjects [23, 24]. In this study nifedipine (Cordicant, Kapsel) was given to healthy subjects at a dose of 10 mg either at 0800 or at 1900 hours. The pharmacokinetic analysis revealed that not only was $t_{max}$ significantly longer after evening than after morning administration (37.5 vs 22.5 min), but also that $C_{max}$ (45.7 vs 82.0 ng/ml), AUC (85 vs 130 ng/ml/h), and drug absorption ($C_{max}/t_{max}$: 1.5 vs 4.5 ng/ml/h) were significantly smaller in the evening than in the morning. Thus, a significant daily variation in the bioavailability was found, with a reduction in bioavailability of about 35% in the evening. Since the ratio in the AUC of the main nitropyridine metabolite to the parent compound was not significantly different at the two time points of drug application, it was assumed that the reduction in nifedipine bioavailability must mainly be due to a reduced absorption and/or an increased presystemic metabolism in the evening [23, 24]. Only minor daily variations were found in the cardiovascular effects in normotensive subjects of this nifedipine preparation [23, 24]. Thus, it turned out that the dose-response relationship of the calcium channel blocker nifedipine was different during daytime vs during the night. In a very recent study we could demonstrate that a retard formulation of nifedipine (Cordicant retard) did not show significant daily variations in its pharmacokinetics when applied twice daily (20 mg at 0800 hours and at 1900 hours) for 7 days [25]. Time-to-peak drug effects on BP decrease, however, was significantly shorter after evening than after morning application whereas the duration in BP decrease was slightly longer after morning drug application ([25] and unpublished results).

The data described, which were obtained in independent studies, clearly give evidence that the pharmacokinetics of various types of calcium channel blockers are circadian phase-dependent. Consistently, lower $C_{max}$ and/or longer $t_{max}$ values and a reduced bioavailability were found after evening dosing compared to morning dosing for all three

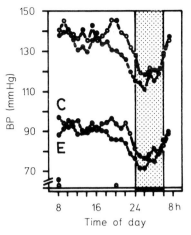

**Fig.7.** Effects of chronic captopril ($C$, 37.5 mg b.i.d.) and enalapril ($E$, 20 mg o.i.d.) on circadian variation in BP of hypertensive patients. (Redrawn from [26])

compounds. In accordance with chronopharmacokinetic findings for the $\beta$-blocker propranolol [19], absorption of the calcium channel blockers was also delayed in the evening [21–25]. In the first nifedipine study [23, 24] the hemodynamic drug-induced effects – though small in the normotensive subjects – were also circadian time-dependent. Recently, clinical studies showed that the antihypertensive effects of ACE inhibitors are also dependent on the time of day. In a randomized crossover study, the two ACE inhibitors captopril and enalapril, differing greatly in their respective elimination half-lives (1.9 h vs 11 h), were studied in ten hypertensive patients after monotherapy with once a day enalapril (20 mg o.i.d) and twice a day captopril (37.5 mg b.i.d.) for 8 weeks [26]. Figure 7 shows that the drugs' efficacies were dissimilar over 24 h, with enalapril lowering BP significantly greater than captopril during daytime and over 24 h. Interstingly, the rapid early morning rise in BP was not affected by the ACE inhibitors (Fig. 7), as already observed for various $\beta$-blockers [2–5]. Similar findings were reported for twice daily captopril (50 mg b.i.d; 25 mg b.i.d; [27, 28]), whereas a once a day dosage of 25 mg captopril seems to be insufficient to lower BP over 24 h [27]. The addition of hydrochlorothiazide (25 mg) to 25 mg captopril once a day seemed to normalize 24 h BP in most patients [26]. An earlier study with the diuretic xipamide in hypertensive patients showed that this drug reduced the elevated BP throughout 24 h when administered either during daytime or nighttime [29]. Quite similar findings were reported after a 6 week treatment with once daily 6 mg piretanide in mild hypertension [30].

In conclusion, recent findings on the chronopharmacology of calcium channel blockers, ACE inhibitors, and the diuretic piretanide further support the notion that the circadian rhythm in BP in hypertensive patients may be differently affected by antihypertensive treatment at different circadian times [2–5]. As already described for oral propranolol [19, 20], the chronopharmacokinetics of the calcium channel blockers may not be responsible for the daily variation in their antihypertensive effects. However, the reduced bioavailability in the evening hours is interesting to note in light of the discussion on bioequivalence of compounds within one class of drugs and thus deserves further explanation. It is becoming more and more evident that the parameters which determine a pharmacokinetic model are bound to circadian variations [2–5]. In addition, the findings with the ACE inhibitors indicate that inhibition of plasma ACE may not be the crucial point in explaining the daily variation in the drugs' antihypertensive effects.

## Congestive Heart Failure

Convincing evidence has recently been presented on the existence of daily variation in the pharmacokinetics of digoxin in six elderly male patients after a single oral dose of 0.5 mg in a randomized crossover study [31]. The $C_{max}$ values were higher after a 7 h than after a 19 h administration; however, the relative bioavailability was not affected by time of day as can be assumed from the nonsignificant difference in AUC after morning and evening dosing [31].

## Coronary Heart Disease

It has already been pointed out that the various forms of angina pectoris can exhibit different circadian patterns, regarding the symptoms of the disease (Fig. 3). Earlier findings on the chronopharmacology of nitroglycerine and propranolol in coronary heart disease are compiled in the reviews mentioned [2–5]. Most interestingly, the circadian variation in ischemic episodes (type of ischemia documented by coronary arteriography) was abolished by a 5 day treatment with once a day, long-acting propranolol (80 mg); the morning peak (0600–1200 hours) in the number of ischemic episodes was reduced by β-adrenoceptor blockade [32]. Similar findings were reported on metoprolol treatment (200 mg b.i.d.) for 1 week concerning the total number and the duration of ischemic episodes in ten patients [33].

Finally, in healthy subjects significant daily vari-

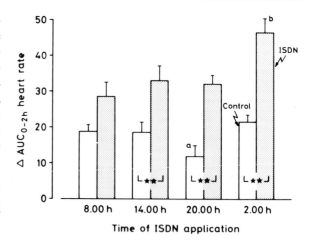

**Fig. 8.** Daily variation in orthostasis under control conditions and under single drug application of isosorbide dinitrate (20 mg) at four different times of day in healthy subjects. Changes in heart rate from supine to 3 min upright (difference in $AUC_{0-2\,h}$). (From [34, 44]

**Table 4.** Pharmacokinetic and hemodynamic parameters of an immediate release preparation of isosorbide-5-mononitrate (60 mg p.o.) administered at two different times of day to eight healthy volunteers. (From [34, 35])

|  | Time of immediate release IS-5-MN administration | |
|---|---|---|
|  | 0630 hours | 1830 hours |
| **Pharmacokinetics** | | |
| $C_{max}$ | 1605 ± 175 | 1588 ± 173 |
| $t_{max}$ (h) | 0.9 ± 0.3 | 2.1 ± 0.4* |
| AUC (ng/ml × h) | 9539 ± 827 | 10959 ± 707 |
| $t_{1/2}\,\beta$ (h) | 4.6 ± 0.4 | 4.2 ± 0.4 |
| **Hemodynamics** | | |
| BP systolic decrease $T_{max}$ (mmHg) | 0.7 ± 0.1 | 1.1 ± 0.1 |
| BP diastolic decrease $T_{max}$ (mmHg) | 0.4 ± 0.1 | 0.6 ± 0.2 |
| HR increase $T_{max}$ (b/min) | 0.8 ± 0.3 | 0.9 ± 0.2 |

Wilcoxon test: * $p < 0.01$

ations were described for the cardiovascular effects and/or the pharmacokinetics of the oral nitrates isosorbide dinitrate and two galenic formulations of isosorbide-5-mononitrate [34–36]. These studies provided the first evidence that independent of whether or not the pharmacokinetics varied with time of day, the decrease in BP and the reflex-induced increase in HR occurred at lower drug concentrations in the evening than in the morning, thus showing that there are daily variations in the dose-response relationship of oral nitrates. Moreover, the study with isosorbide dinitrate (ISDN) clearly indicated that not only daily

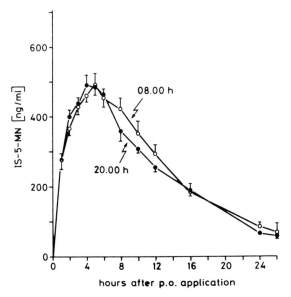

**Fig. 9.** Plasma concentrations-time curves of isosorbide-5-mononitrate (IS-5-MN) after oral intake of 60 mg of a sustained release preparation by 10 healthy subjects at two different times of day; mean ± SEM. (From [36])

**Table 5.** Pharmacokinetic and hemodynamic parameters of sustained release isosorbide 5-mononitrate (60 mg p. o.) at two different times of day. (From [36])

Time of sustained release IS-5-MN administration

| | | |
|---|---|---|
| Pharmacokinetics | | |
| $C_{max}$ | 509 ± 31 | 530 ± 26 |
| $t_{max}$ (h) | 5.2 ± 0.7 | 4.9 ± 0.3 |
| AUC (ng/ml × h) | 6729 ± 375 | 6418 ± 199 |
| $t_{1/2}\beta$ (h) | | |
| Hemodynamics | | |
| BP systolic decrease $T_{max}$ (mmHg) | 5.0 ± 0.6 | 2.8 ± 0.5* |
| BP diastolic decrease $T_{max}$ (mmHg) | 6.0 ± 0.7 | 2.9 ± 0.5** |
| HR increase $T_{max}$ (b/min) | 5.2 ± 1.0 | 3.8 ± 0.6 |

Mean values ± SEM of ten subjects
*$p < 0.05$; **$p < 0.005$

variations in orthostasis are present under control conditions, with peak values at night, but that the degree of orthostasis is greatly enhanced by ISDN mainly during the night (Fig. 8).

Interesting results were obtained with the two isosorbide-5-mononitrate (IS-5-MN) preparations [35, 36]: With the immediate release preparation of IS-5-MN (IS-5-MN Stadapharm), clear-cut daily variations were found in regard to $t_{max}$, being significantly shorter after the morning (0.9 ± 0.3 h) than after the evening (2.1 ± 0.4 h) administration (Table 4). Most

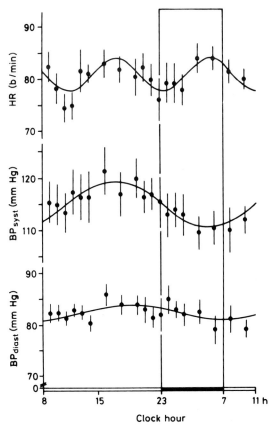

**Fig. 10.** Daily variations in heart rate and systolic and diastolic blood pressure after 3 min of standing upright in 10 male healthy subjects. Cosinor analysis revealed a significant ($p < 0.01$) 12 h rhythm in heart rate and a 24 h rhythm in systolic ($p < 0.00001$) and diastolic ($p < 0.038$) blood pressure. Mean values ± SEM of 10 subjects. (From [36])

interestingly, time-to-peak drug effects in decreasing BP and reflexively increasing HR coincided with the $t_{max}$ pharmacokinetics in the morning, but were in advance by about 1 h in the evening (Table 4). No daily variations were observed in the pharmacokinetics of the sustained release formulation of IS-5-MN (Coleb-Duriles, [42], Fig. 9, Table 5). Figure 10 shows the daily variations in the mean control values in systolic and diastolic BP and in HR after 3 min of standing upright [36]. Rhythm (cosinor) analysis of the group mean data revealed significant rhythmic variations for the 24 h rhythms of systolic ($p < 0.00001$) and diastolic BP ($p < 0.038$) and for the 12 h-rhythm in HR ($p < 0.001$) [36] (Fig. 10). Rhythm adjusted means for systolic and diastolic BP and HR were 115.0 ± 0.4 mm Hg, 82.4 ± 0.4 mm Hg, and 80.8 ± 0.5 beats/min, respectively. Peak values of rhythms (acrophases) occurred at 17.23 ± 0.4 h (systolic BP), at 18.98 ± 1.24 h (diastolic BP) and at 17.58 ± 0.4 and

5.58 ± 0.4 h (HR) (see also Fig. 10). Thus, these hemodynamic data clearly demonstrate that significant
daily variations are present also when hemodynamic
functions are measured in the standing position
throughout the 24 h of a day. Peak drug effects in
lowering BP and increasing HR again coincided with
$t_{max}$ in the drug's pharmacokinetics (around 0500
hours) in the morning, but occurred about 2 h earlier
after drug administration in the evening (Table 5).
This again points to daily variations in the dose-response relationship of drugs, as already mentioned
before for propranolol and nifedipine (see below). In
addition, the data obtained with the two different formulations of IS-5-MN indicate that the kind of drug
formulation may also be of importance, whether or
not chronokinetics can be observed.

## Conclusion

Recent findings in the chronopharmacology of cardiovascular active drugs clearly demonstrate that
compounds such as $\beta$-blockers, calcium channel
blockers, ACE inhibitors, diuretics, digoxin, and oral
nitrates exhibit daily variations in their pharmacokinetics and/or in their hemodynamic effects [4–6,
44]. Thus, the dose-response relationship seems to be
circadian phase-dependent as well. Finally, these data
indicate that galenic formulations may also of be importance, whether or not a circadian phase dependency can be found in a drug's pharmacokinetics.

## References

1. Smolensky MH, Tatar SE, Bergmann SA, Losmann JG, Barnard CN, Dacso CC, Kraft IA (1976) Circadian rhythmic aspects of human cardiovascular function: a review by chronobiologic statistical methods. Chronobiologia 3: 337–371
2. Lemmer B (1987) Cardiovascular medications. In: Kuemmerle HP, Hitzenberger G, Spitzy KH (eds) Klinische Pharmakologie. Landsberg, ecomed 4th edn, pp 1–14
3. Lemmer B (1989) Circadian variations in the effects of cardiovascular active drugs. In: von Arnim T, Maseri A (eds) Predisposing conditions for acute ischemic syndromes. Steinkopff, Darmstadt, pp 1–11
4. Lemmer B (1989) Temporal aspects of the effects of cardiovascular active drugs in humans. In: Lemmer B (ed) Chronopharmacology – Cellular and biochemical interactions. Dekker, New York, pp 525–541
5. Lemmer B (1989) Circadian rhythms in the cardiovascular system. In: Arendt J, Minors DS, Waterhouse JM (eds) Biological rhythms in clinical practice. Butterworth, London pp 51–70
6. Lemmer B, Scheidel B, Behne S (1991) Chronopharmacokinetics and chronopharmacodynamics of cardiovascular active drugs: propranolol, organic nitrates, nifedipine. Ann NY Acad Sci 618: 166–181
7. Hossmann V, Fitzgerald GA, Dollery CT (1980) Circadian rhythm of baroreflex reactivity and adrenergic vascular response. Cardiovasc Res 14: 125–129
8. Aronow WS, Harding PR, DeQuattro V, Isbell M (1973) Diurnal variation of plasma catecholamines and systolic time intervals. Chest 63: 722–726
9. Tuck ML, Stern N, Sowers JR (1985) Enhanced 24 h norepinephrine and renin secretion in young patients with essential hypertension: relation with the circadian pattern of arterial blood pressure. Am J Cardiol 55: 112–115
10. Lemmer B (1988). Die Bedeutung der Chronopharmakologie für die medikamentöse Therapie. Verh Dtsch Ges Inn Med 94: 420–430
11. Lemmer B, Saller R (1974) Influence of light and darkness on the turnover of noradrenaline in the rat heart. Naunyn Schmiedeberg's Arch Pharmacol 282: 75–84
12. Andreotti F, Davies GJ, Hackett DR, Khan MI, De Bart ACW, Aber VR, Maseri A, Kluft C (1988) Major circadian fluctuations in fibrinolytic factors and possible relevance to time of onset of myocardial infarction, sudden cardiac death and stroke. Am J Cardiol 62: 635–637
13. Lemmer B, Bärmeier H, Schmidt S, Lang PH (1987) On the daily variation in the beta-receptor – adenylate cyclase – cAMP – phosphodiesterase – system in rat forebrain. Chronobiol Int 4: 469–475
14. Lemmer B, Witte K (1989) Circadian rhythm of the in vitro stimulation of adenylate cyclase in rat heart tissue. Eur J Pharmacol 159: 311–315
15. Lemmer B, Wald C (1990) Influence of age on the rhythm in basal and forskolin-stimulated adenylate cyclase activity of the rat heart. Chronobiol Int 7: 107–111
16. Lemmer B, Winkler H, Ohm T, Fink M (1985) Chronopharmacokinetics of beta-receptor blocking drugs of different lipophilicity (propranolol, metoprolol, sotalol, atenolol) in plasma and tissues after single and multiple dosing in the rat. Naunyn Schmiedeberg's Arch Pharmacol 330: 42–49
17. Lemmer B, Bathe K, Lang PH, Neumann G, Winkler H (1983) Chronopharmacology of $\beta$-adrenoceptor blocking drugs: pharmacokinetics and pharmacodynamic studies in rats. J Am Coll Toxicol 2: 347–358
18. Lemmer B, Bathe K (1982) Stereospecific and circadian-phase-dependent kinetic behaviour of d,1-, 1- and d-propranolol in plasma, heart and brain of light-dark-synchronized rats. J Cardiovasc Pharmacol 4: 635–644
19. Langner B, Lemmer B (1988) Circadian changes in the pharmacokinetics and cardiovascular effects of oral propranolol in healthy subjects. Eur J Clin Pharmacol 33: 619–624
20. Semenowicz-Siuda K, Markiewicz A, Korczynska-Wardecka J (1984) Circadian bioavailability some effects of propranolol in healthy subjects and liver cirrhosis. Int J Clin Pharmacol Ther Toxicol 22: 653–658
21. Hla KK, Henry JA, Phull R, Volane QN, Latham AN (1988) Chronopharmacology of verapamil. 25th anniversary international symposium calcium antagonists in hypertension, Basel
22. Guillet P, Dubruc C, Bianchetti G, Thenot JP, Thebault JJ

(1988) The pharmacokinetics of diltiazem vary according to its time of administration. Int Conf Pharmaceut Sci Clin Pharmacol, Jerusalem, Abstr p 55

23. Lemmer B, Behne S, Becker HJ (1989) Chronopharmacology of oral nifedipine in healthy subjects. Eur J Clin Pharmacol 36 (Suppl): A 177

24. Behne S, Becker HJ, Liefhold J, Lemmer B (1990) On the chronopharmacokinetics, effects on cardiovascular functions and plasma cAMP of oral nifedipine in healthy subjects. In: Lemmer B, Hüller H (eds) Clinical chronopharmacology, Klinische Pharmakologie/Clinical pharmacology vol 6. Zuckschwerdt, Munich 6: 49–63

25. Lemmer B, Nold G, Behne S, Becker HJ, Liefhold J, Kaiser R (1990) Chronopharmacology of oral nifedipine in healthy subjects and in hypertensive patients. Eur J Clin Pharmacol 183: 521

26. Blankenstijn PJ, Wentig GJ, Schalekamp MADH (1989) 24 hour blood pressure profiles during ACE inhibition. A comparative study of twice daily captopril versus once daily enalapril. Int symp ACE inhibition, London, Abstr F 164

27. Boxho G (1988) Twenty-four-hour blood pressure recording during treatment with captopril twice daily. Curr Ther Res 44: 361–366

28. Meijer JL, Ardesch HG, van Rooijen JC, De Bruijn JHB (1987) Captopril plus hydrochlorothiazide once daily normalize 24 h blood pressure in patients with essential hypertension. Br J Clin Pharmacol 23: 83S–88S

29. Raftery EB, Melville DI, Gould BA, Mann S, Whittington JR (1981) A study of the antihypertensive action of xipamide using ambulatory intra-arterial monitoring. Br J Clin Pharmacol 12: 381–385

30. Palma JL, Isasa D, Senor J (1986) Ambulatory blood pressure monitoring in mild hypertension treated with piretanide. 2nd Int conf diuretics. Cascais, Portugal, Abstr p 55

31. Bruguerolle B, Bouvenot G, Bartolin R, Manolis J (1988) Chronopharmacocinétique de la digoxine chez le sujet de plus de soixante-dix ans. Therapie 43: 251–253

32. Joy M, Pollard CM, Nunan TO (1982) Diurnal variation in exercise in angina pectoris. Br Heart J 48: 156–160

33. Willich SN, Pohjola-Sintonen S, Bhatia SJS, Shook TL, Tofler GH, Muller JE, Curtis DG, Williams GH, Stone PH (1989) Suppression of silent ischemia by metoprolol without alteration of morning increase of platelet aggregability in patients with stable coronary artery disease. Circulation 79: 557–565

34. Lemmer B, Scheidel B, Stenzhorn G, Blume H, Lenhard G, Grether D, Renczes J, Becker HJ (1989) Clinical chronopharmacology of oral nitrates. Z Kardiol 78: 61–63

35. Scheidel B, Lenhard G, Blume H, Becker HJ, Lemmer B (1989) Chronopharmacology of isosorbide-5-mononitrate (immediate release, retard formulation) in healthy subjects. Eur J Clin Pharmacol 36 (Suppl): A 177

36. Lemmer B, Scheidel B, Blume H, Becker HJ (1991) Clinical chronopharmacology of oral sustained-release isosorbide-5-mononitrate in healthy subjects. Eur J Clin Pharmacol 40: 71–75

37. Lemmer B (1990) Chronopharmakologie antiischämisch wirksamer Arzneimittel. In: Rudolph W (ed) Therapie der koronaren Herzerkrankung – Aktuelle Aspekte. Springer, Berlin Heidelberg New York, pp 76–84

38. Waters DD, Miller DD, Bouchard A, Bosch X, Theroux P (1984) Circadian variation in variant angina. Am J Cardiol 54: 61–64

39. Araki H, Koiwaya Y, Nakagaki O, Nakamura M (1983) Diurnal distribution of ST-segment elevation and related arrhythmias in patients with variant angina: a study by ambulatory ECG monitoring. Circulation 67: 995–1000

40. Arnim von T, Höfling B, Schreiber M (1985) Characteristics of episodes of ST elevation or ST depression during ambulatory monitoring in patients subsequently undergoing coronary angiography. Br Heart J 54: 484–488

41. Muller JE, Stone PH, Turin ZG, Rutherford JG, Czeisler CA, Parkers C, Poole WK, Passamani E, Roberts R, Robertson T, Sobel BE, Willerson JT, Braunwald E (1985) The milis study group: circadian variation in the frequency of onset of acute myocardial infarction. N Engl. J Med 313: 1315–1322

42. Marshall J (1977) Diurnal variation in occurrence of strokes. Stroke 8: 230–231

43. Mitler MM, Hajdukovic RM, Shafor R, Hahn PM, Kripke DF (1987) When people die. Cause of death versus time of death. Am J Med 82: 266–274

44. Lemmer B (1990) Recent advances in the chronopharmacology of cardiovascular diseases. In: Lemmer B, Hüller H (eds) Clinical chronopharmacology. Klinische Pharmakologie/Clinical pharmacology, vol 6, Zuckschwerdt, Munich: 45–52

# Chronobiologic Blood Pressure Assessment from Womb to Tomb

G. Cornélissen, E. Haus, and F. Halberg

## Introduction

Like many other physiologic functions, and perhaps even more so, blood pressure (BP) varies greatly. It varies, of course, between the heart's relaxation and its contraction. While the differences between systolic (S) and diastolic (D) BP have long been recognized and utilized, it is often not realized that the DBP at a given time can be higher than the SBP measured in the same person at another time on the same day. Both SBP and DBP vary in adulthood on the average by more than 50 mmHg within each day (Cornélissen 1987; Halberg et al. 1984a). Part of this variability is predictable insofar as it is rhythmic. Among the rhythms that characterize BP, circadians (with a frequency of one cycle in about a day) are prominent during most of the human lifespan. Ultradian rhythms (with a frequency higher than circadian) and infradians (with a frequency lower than circadian), notably with circaseptan (about weekly) and circannual (about yearly) components, also characterize BP, as do trends with growth, development, maturation, and aging. Variance transpositions from one frequency domain to another are observed in association with development and aging (Anderson et al. 1989; Halberg 1963; F. Halberg et al. 1991; Hillman et al. to be published; Ikonomov et al. to be published).

Against this background, the value of casual single measurements of BP may be questioned unless they

___

*Metaanalysis:* Retrospective evaluation and integration of results from previous studies (Mann C (1990) Meta-analysis in the breech. Science 249: 476–480)

*Chronometaanalysis:* Analysis containing additional and/or revised information by a so-called 'microscopic' analysis of previously published data to complement the discussion and/or documentation of a point related to the scope of the original publication(s) and/or a new point emerging independently of the original publication(s) (Halberg F, Cornélissen G, Nelson W (1981) Circadian murine ouabain chronotolerance revisited. Chronobiologia 8: 275–281)

are time specified and evaluated in the light of chronobiologic reference standards that account for rhythmic behavior. Apart from the use of time-qualified single measurements or means derived from a limited number of measurements, the assessment of the very dynamic characteristics of the changes in BP, observed as a function of time, has the merit of adding new sensitive gauges of BP deviation. The rhythm characteristics are indicative not only of physiologic functions in apparent good health and of overt pathology, but also of cardiovascular disease risk. The mapping of rhythms in clinical health care yields an altogether new, positive, and quantitative definition of good health, as opposed to the present negative definition as the nondetection of disease, if not its absence, which is hardly ever ascertained. Assessment of early indications of an elevated risk of cardiovascular disease provides both a harbinger and a rational basis for the prevention of major diseases of our civilization, such as certain heart and kidney diseases and stroke.

## Ambulatory Monitoring of Blood Pressure

With automatic instrumentation for indirect noninvasive BP monitoring, it is possible to follow the time course of BP variation around the clock (or week and seasons) and in large groups of subjects to establish reference norms for the interpretation of single values and of rhythm characteristics. The circadian pattern of BP, with its higher values during the day, intermediate values in the evening, and lower values during the night in most people, is determined by both endogenous and exogenous factors. The variability of human BP along the 24-h scale was recognized over 100 years ago by Zadek (1881), and the wealth of classical early literature has been reviewed (Smolensky et al. 1972; Halberg et al. 1988a; J. Halberg et al. 1984). The existence of an underlying intrinsic rhythmic mechanism is supported by the observation, validated statistically, that BP rises before awakening and that this cannot result only from a change in posture or activity (E. Halberg et al. 1981a; F. Halberg et al. 1988a). The circadian rhythm in blood pressure reflects the activation of the adrenergic system and the adrenal cortex preparing for the start of the day's motor activities (F. Halberg 1953).

The automatic monitoring of physiologic functions in general and of BP in particular, by means of miniaturized ambulatory recorders, has helped diagnostic medicine to become more efficient, perhaps at the cost of being more expensive. This claim may be true when the analysis of the data provided by these recorders remains limited to the computation of a mean and standard deviation and their inspection by the naked eye. Due to this apparent increase in cost, the use of such monitors has at the present remained reserved for solving special problems in a limited number of patients. In contrast, the chronobiologic approach recommends that BP be monitored (manually if not with ambulatory recorders) in every person, irrespective of gender, age, ethnicity, or social status (F. Halberg 1970). Reference norms thus become available and the time structure of BP coordination can be resolved for the individual as a set of new endpoints. Chronobiology offers tools for a new interpretation of single and/or serial BP values and for a refined diagnosis. Prevention of undue BP elevation and of BP-related pathology is a major aim of clinical chronobiology. The automatic monitors become cost effective once they contribute to this goal by means of encouraging self help regarding health care, as should generally be practiced and taught as part of the curriculum in schools and in adult classes. Using an automatic monitor is much easier than driving a car, and interpreting a computer print-out of the data is no more complicated than adhering to traffic regulations.

Self-help involves chronobiology in two major ways: first, in resolving a structure in time involving underlying rhythms with multiple frequencies and trends as a function of age; and second, in taking the appropriate steps for primary prevention in time rather than in response to a catastrophic event. Self-measurements implemented and analyzed chronobiologically encourage patient compliance. Contrary to unsupported worries about producing hypochondriacs, laid to rest decades ago at the Mayo Clinic (Brown 1930; Mueller and Brown 1930), self-measurements also have a desirable placebo effect (Scarpelli et al. 1987). Chronobiology calls for active participation by patients in their health care by lowering risks. To achieve this goal, the patient must be able to interpret the data obtained manually or automatically, and health care personnel can reinforce self-help by education. By delivering "high touch" as much as "high tech", the prognosis of patients with a slight or malignant BP elevation can be improved (Scarpelli et al. 1987).

## Status Quo

That the traditional approach to the study of BP and its deviations is unsatisfactory is suggested by scholarly retrospective analysis (or metaanalysis) (Wilcox et al. 1986) – albeit not a chronometaanalysis (Halberg et al. 1988a) – which raises the question whether clinical practice should be based on the results of clinical trials. While the foregoing authors answer in the negative, others (Pötschke-Langer and Apfelbach 1986) advocate the use of a smorgasbord of drugs on the basis of the results from these same clinical trials. The chronobiologically unacceptable design of the reviewed studies must be considered. The outcome of large-scale clinical trials such as the Australian Therapeutic Trial in Mild Hypertension (1980) reflects the situation best (F. Halberg et al. 1988a). Out of nearly 2000 individuals in this study who received a placebo, almost half reportedly responded to this treatment, most of them within less than a year. It is uncertain as to how many cases represented false positive diagnoses at the start of the study, how many were false negatives at the completion of the study, and how many were placebo responders. What seems certain is that reliance on single or a few casual BP measurements, checked against arbitrarily (consensus-derived) fixed thresholds (e.g., of 140/90 mmHg set by WHO), can lead to a large number of incorrect diagnostic and/or therapeutic decisions. The unwarranted status quo of the conventional approach has led to undesired side effects and a constant financial drain as consequences of, perhaps, unneeded medication. Other (e.g., dietary) interventions which do not present such unwarranted side effects may be indicated, at least for some individuals. Decreases in BP following weight loss, if not following sodium restriction, have been documented for adults with mild elevations of BP (F. Halberg et al. 1988a). Exercise and dietary or other restrictions should also be specified chronobiologically (F. Halberg et al. 1988a). It has been reported that pharma-

## BLOOD PRESSURE AND RISK OF MESOR-HYPERTENSION (RMH)

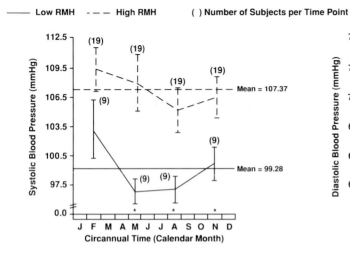

—— Low RMH    - - - High RMH        ( ) Number of Subjects per Time Point        ∗ P < .05 ( t-Test) Not Adjusted for Multiple Testing

|  | Low RMH | | | High RMH | | | | |
|---|---|---|---|---|---|---|---|---|
| Season | Mean | S.E. | N.Obs. | Mean | S.E. | N.Obs. | t | P |
| Winter | 103.34 | 2.99 | 9 | 109.48 | 2.26 | 19 | −1.57 | 0.1285 |
| Spring | 96.87 | 1.29 | 9 | 108.09 | 2.84 | 19 | −2.83 | 0.0141 |
| Summer | 97.15 | 1.39 | 9 | 105.30 | 2.25 | 19 | −2.37 | 0.0253 |
| Fall | 99.76 | 1.72 | 9 | 106.61 | 2.09 | 19 | −2.08 | 0.0467 |

|  | Low RMH | | | High RMH | | | | |
|---|---|---|---|---|---|---|---|---|
| Season | Mean | S.E. | N.Obs. | Mean | S.E. | N.Obs. | t | P |
| Winter | 65.90 | 1.44 | 6 | 69.89 | 2.16 | 5 | −1.46 | 0.1704 |
| Spring | 64.17 | 0.79 | 6 | 67.91 | 3.82 | 5 | −1.04 | 0.3213 |
| Summer | 63.79 | 1.50 | 6 | 67.64 | 2.89 | 5 | −1.24 | 0.2453 |
| Fall | 63.67 | 1.35 | 6 | 71.27 | 2.81 | 5 | −2.56 | 0.0296 |

**Fig. 1.** Statistically significant differences in circadian MESOR of SBP between 99 and 107 mm Hg and of DBP between 64 and 69 mm Hg are observed between clinically healthy women (while recumbent) at low vs. high risk of developing a high BP and/or related cardiovascular disease, as assessed by question-naire (RMH). This result suggests the desirability of entering the physiologic range to seek differences such as those illustrated in this figure with the aim of early risk detection that should lead to timely preventive intervention (Halberg et al. 1981 a)

cological agents can be advantageously replaced by non-drug, e.g., dietary and exercise, treatments in as many as 47% of "mild hypertensives" (Stamler et al. 1985).

Chronobiologic studies relying on 24-h monitoring at 10-min intervals reveal that the 24-h mean of SBP of adult women at high vs. low familial and personal risk of developing high BP, measured in the recumbent position, varies during the seasons between 110 and 105 vs. 103 and 97 mmHg, respectively (F. Halberg et al. 1986a). Differences between the two risk groups are also found for the 24-h mean of DBP, which varies by season between 68 and 71 vs. 64 and 66 mmHg, respectively (Fig. 1). For a group, risk can thus be recognized by means of relatively small differences in BP that are found well within the physiologic range and even within the range of imprecision of single manual measurements. International womb-to-tomb studies of BP and heart rate with a view of outcomes are necessary to establish the minimal data requirements that can be extended from groups to in-dividuals. In this context, however, it has already been documented that groups of women with differ-ent average BP values below 120/75 (SBP/DBP) dur-ing pregnancy have dramatically different outcomes, with the incidence of intrauterine growth retardation (IUGR) being five in one group and zero in another (p < .05), notwithstanding the conventionally aver-age but not chronobiologically acceptable BP values being well below conventional limits (Rigó et al., in press). Only the application of a chronobiologic ap-proach to strategically placed serial data allows a meaningful evaluation and interpretation of such small differences. Early risk assessment gains in im-portance in the light of recent results on the evolution of high BP in an initially healthy population of 946 subjects 18–38 years of age (Julius et al. 1990), which led the authors to conclude that "borderline hyper-tension is neither transient nor innocuous. Its associ-ation with other predictors of atherosclerosis calls for clinical attention."

## Blood Pressure Deviation in the Light of Time-Specified Reference Norms (Chronodesms)

Arbitrary consensus-derived fixed thresholds reflect mostly morbidity and mortality statistics describing previous generations. The chronobiologist prefers to determine reference limits that are rhythm-stage dependent. These are derived from large data bases collected on clinically healthy individuals of each gender in different age groups. Further specification is desirable, when possible, as to ethnicity, socioeconomic status, and preferably a negative personal and familial history of high BP and related cardiovascular diseases. Time-specified reference norms (chronodesms) can be constructed in different ways. They can be model dependent or model independent, and they can be computed as tolerance or prediction intervals (F. Halberg et al. 1978; Nelson et al. 1983). A tolerance interval comprises a given proportion of the reference population with a stated confidence, whereas a prediction interval includes, on the average, a given proportion of the reference population. Reference limits built around a given model rely on the assumption that the model (e.g., a 24-h cosine curve) is adequate and that the random variability around the model follows a normal distribution, the characteristics of which do not depend upon rhythm stage. Model-independent norms can be constructed, for instance, as 90% prediction limits, accounting for both inter- and intraindividual variability within a given interval which is progressively displaced throughout one cycle of the periodicity investigated. Such limits account for changes both in mean and in variance as a function of rhythm stage.

Once reference limits are specified as a function of time on the basis of large data bases of information on clinically healthy, preferably low-disease-risk peers, the evaluation of a BP profile and the recognition of abnormality can be done rigorously and objectively. Whenever an individual's profile strays outside of these limits at a given time, it may be termed an excess (or deficit) as defined for that person with respect to the peer group. If, for example, the individual's profile is above the upper limit of the peer group by 10 mmHg for, say, 2 h, the area between the profile and the limit represents the product of pressure and time excess. To estimate the area accurately, it has to be calculated by numerical integration by computer (F. Halberg et al. 1990a). If and only if the days of monitoring are representative of other days and even if that individual is normotensive, on the average, for

2 h of the day, there will be, on the average, a 2-h × 10 mmHg excess day after day. Over a long time span, such cumulative excess could contribute to increasing the likelihood of morbidity and mortality. Replotting the BP excess on the basis of renewed monitoring in this case after allowing sufficient time for a corrective action (e.g., medication, sodium intake manipulation, etc.) will objectively show the results of the therapy, as compared to the sole reliance upon measurements at office visits. Any excess (or deficit) is computed (in mmHg × h) as the area delineated by the reference limit and the individual's profile when it is outside the chronodesm, integrated over one cycle (24 h). Fractionated excess over consecutive 3-h spans of the 24-h day provides additional information regarding the circadian pattern of BP excess. In principle, the concept of BP excess is similar to that of pack-years of cigarettes or of person-years of exposure to asbestos. Rather than asking whether a BP is too high or too low, new and more pertinent questions can be raised, such as "To what extent does a BP exceed the upper limit of peers at a given time?"; "How long does BP remain above or below a time-varying threshold?"; "What is the total excess or deficit in 24 h and its long-term projection (e.g., over 10 years)?"; and "At what circadian rhythm stages does most of the excess (or deficit) occur?"

## Scenarios of Outcomes with or Without a Chronobiologic Approach to Blood Pressure Assessment

Abnormality may occur, for instance, during rest by night while BP is above or even below the limits set by WHO, but above the time-varying limits of healthy peers. Such 'odd-hour' deviations in BP cannot be diagnosed during regular office hours. In the absence of domiciliary and/or self-help for the early detection of an undesirable condition (and also for rigorously assessing the effects of any intervention), a considerable BP excess may accumulate and lead to target organ damage (Germanò 1988; McDonald et al. 1990; Reeves et al. 1984; Verdecchia et al. 1990), often without obvious symptoms, since high BP is the silent disease par excellence.

For many individuals who may lead a long life, only on rare occasions and only transiently will their cumulative risk from an elevated BP exceed the limits of acceptability (scenario A). Even so, monitoring may be indicated to assess the effects of environmental factors that also contribute to a deviant BP. At the other extreme (scenario B), some individuals may die prematurely, possibly from the consequences of high BP, since cumulative but "silent" excess was never recognized. Their risk may be genetic in nature, stem from an undesirable lifestyle, or come from a combination of factors. The aim of chronobiology is to avoid scenario B and to strive toward scenario A by means of early prevention.

The status quo, unfortunately, consists more often than not of scenario C, in which an elevated BP is detected only after the damage to target organs has led to a catastrophic event such as a stroke or a heart attack. Early detection of a deviant BP is not sufficient, however, if it is not accompanied by appropriate preventive intervention or compliance with treatment and/or monitoring. Without compliance, the person at risk may only delay the occurrence of high BP and its sequelæ (scenario D). The chronobiologic approach (scenario E) is early screening in the context of education and consistent chronotherapy as soon as values exceeding a peer group norm are found. (With refined, unobtrusive instrumentation for longitudinal monitoring from womb to tomb, it becomes feasible to establish personal norms which should eventually replace peer-group reference standards). By compliance with treatment and/or monitoring, the individual at risk may avoid the fate of scenarios B–D in favor of scenario A (E. Halberg et al. 1990).

Where other approaches fail, chronobiologic approaches make it possible recognize the risks of developing high BP, even at birth (Cornélissen et al. 1987, 1990a; F. Halberg et al. 1986c, 1988a, 1990b). Chronobiology thus promotes primary prevention as well, thereby reducing human suffering or the loss of human life, while also curtailing staggering health care costs.

Chronobiology obviously also promotes secondary prevention, that is to forestall the complications of an already established high BP before it converts into catastrophic disease (Güllner et al. 1979; F. Halberg et al. 1966, 1988a). There is, for instance, a growing body of evidence indicating that the process of atherosclerosis is largely preventable and to a substantial degree even reversible (Blankenhorn 1975; Buchwald et al. 1974; Katz et al. 1958; Wissler and Vesselinovitch 1976; Zelis et al. 1970; for review see F. Halberg et al. 1986b).

## Are Changes Within the Usual Range Trival or Do They Provide Critical, Partly Novel Information for the Practicing Physician?

### Endogenous Rhythmic Components Characterize BP and Heart Rate Rather Than Mere Responses to Exogenous Stimulation

It is a widely held misconception that the variability in BP and heart rate relates to a large extent, if not exclusively, to changes in activity, posture, and other external stimuli. An intrinsic rhythmic component in the within-day variations of cardiovascular functions such as heart rate was suggested as early as 1797 by Falconer, and the feature of free-running was presented in support of endogenicity 169 years later for BP (Halberg et al. 1966). In a study using 24-h ambulatory monitoring, Marler et al. (1988) made adjustments for different activities (sleeping, resting supine, sitting, standing, and exercise) to determine, idiometrically, treatment effects on the circadian BP rhythm, as also advocated by us (Cornélissen 1987; Cornélissen et al. 1988; F. Halberg et al. 1984a, 1988a, 1989). Even after accounting for the effect of all these activities on BP, the circadian rhythm remains prominent and statistically significant. Although changes in posture usually contribute to the extent of within-day change, circadian rhythmicity persists with statistical significance under conditions of bedrest (F. Halberg et al. 1988a; Stadick et al. 1987).

Not only is there an endogenous rhythm (apart from activity) with a circadian period, but there is also one with a period of about 7 days, which free-runs in social isolation (Fig. 2) while it is usually 7-day synchronized on those adhering to the routine of a social week (Fig. 3). The circadian and the circaseptan components can both free-run; in some instances, one component may assume a period slightly longer than 24 h while the other assumes a period slightly shorter than 7 days. Moreover, whereas both circadian and circaseptan phases can change following a transmeridian flight, they behave differently: the circadian phase advances as anticipated after a flight from west to east and delays after a flight from east to west; by contrast, the circaseptan may undergo a single unidirectional phase change. Thus, after a transmeridian flight, the internal timing of the two components characterizing a given variable is drastically and chronically altered (Halberg et al. 1988c; Hillman et al. 1988).

Rather than the rhythms being the sole response to daily activities, the rhythm stage critically deter-

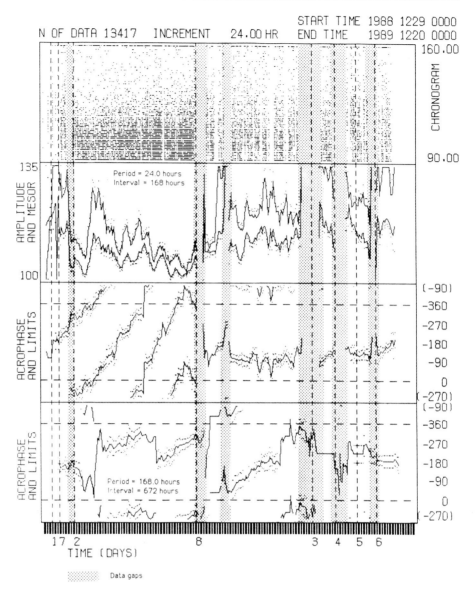

START TIME 1988 1229 0000
END TIME   1989 1220 0000

N OF DATA 13417    INCREMENT    24.00 HR

**Fig. 2.** Chronobiologic serial section of SBP of a 27-year-old woman who spent 4 months in the isolation of a cave in New Mexico (event *lines 7* and *8*). She measured her BP and heart rate at about 30-min intervals during isolation and afterward, using an ambulatory monitor from Colin Medical Instruments (San Antonio, TX 78249). The *top row* illustrates the data as a function of time. *Rows 2* and *3* display the circadian MESOR *(lower curve, row 2),* amplitude (distance between the two curves in *row 2*) and acrophase *(row 3)* of a 24-h cosine curve fitted by least squares to data in a 168-h interval which is progressively displaced by increments of 24 h throughout the time series. The acrophase reference is Dec. 25, 1988. The *lower dashed line* in row 3 represents midnight and the *upper dashed line* represents the following midnight. The drift seen in the acrophase sequence indicates that during isolation, the cir-

cadian period differs from exactly 24 h: the high values occur later and later each day. A 24-h synchronized rhythm is resumed after isolation, as seen by a relatively stable acrophase sequence following event line marker 8. *Row 4* shows the acrophase sequence for the circaseptan (about 7-day) component, assessed over intervals of 672 h progressively displaced by 24 h. In this row, the *lower dashed line* represents midnight from Saturday to Sunday on Dec. 25., 1988, while the *upper dashed line* represents midnight from Saturday to Sunday one week later (January 1, 1989). It can be seen that the circaseptan component free-runs during isolation and afterward, being 7-day resynchronized only after reinduction of the menstrual cycle (by medication; event *line 3*). Event *lines 1–6* indicate menses (Halberg et al. 1990c).

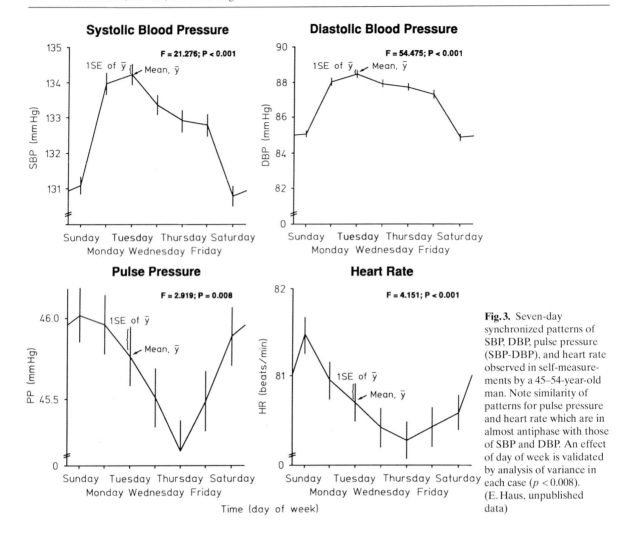

**Fig. 3.** Seven-day synchronized patterns of SBP, DBP, pulse pressure (SBP-DBP), and heart rate observed in self-measurements by a 45–54-year-old man. Note similarity of patterns for pulse pressure and heart rate which are in almost antiphase with those of SBP and DBP. An effect of day of week is validated by analysis of variance in each case ($p < 0.008$). (E. Haus, unpublished data)

mines the responses to activity. Reading aloud and immersing the hand for a minute into ice water have drastically different effects at different circadian stages (Halberg et al. 1986e). A decrease or an increase in BP may be observed immediately following such stimuli, as is the case in response to exercise (Cabri et al. 1988; Halberg et al. 1988a; Levine et al. 1977).

In individuals who follow a 24-h social routine, the circadian period is usually 24-h in length. The innate nature of the circadian rhythm in BP can be documented indirectly by a so-called free-run, that is by the demonstration of periods that deviate consistently from precisely 24 h. Such an internal circadian desynchronization of human BP was first suggested in the case of a child studied at the University of Minnesota Hospital. For several weeks, three shifts of nurses measured several body functions around the clock. All variables were found to be 24-h syn-

chronized, except for SBP (F. Halberg et al. 1988a). Endogenous components of rhythmicity are also documented during long-term human social isolation; not only free-running circadian, but also free-running circasemiseptan (about 3.5-day) and circaseptan (about 7-day) rhythms are found with periods close to those of the socioecologic environment, yet differing in period from a precise environmental cycle length of one day, half a week, or one week (F. Halberg et al. 1990c).

Multifrequency rhythms characterize most if not all vital functions. In human heart rate, the characteristics of these dynamic processes such as the MESOR, amplitude, and acrophase all have inherited features. These results are revealed by intraclass correlation coefficients comparing the among-twin-pair and within-twin-pair variability in each of these rhythm characteristics. Heritability of circadian features other than the MESOR is validated by the fact that the intra-class correlation coefficients computed for each parameter are statistically significantly different from zero (Cornélissen et al. 1985; F. Halberg 1983; Hanson et al. 1984).

Major national and international endeavors are currently concerned with the mapping of the human genome. Initially, metabolic disturbances in selected genetic (e.g., glycogen storage) diseases were in the forefront. Also of concern are presumably polygenic contributions to major multifactorial diseases such as high BP. It is important to realize that in the absence of overt illness, the "best" state of health and states of elevated disease risk also have genetic roots. Health and risk can now be defined positively by the penetration of chronobiology into the normal range to resolve subtle differences in dynamic characteristics of change, if not in rhythm-adjusted means between groups at various degrees of risk of developing a host of civilization's diseases. Genetic mapping will have to focus sooner or later upon the range in which usual function occurs (F. Halberg et al., to be published). This is all the more important, as twin studies reveal that the measurements used to trace inherited characteristics and thus the coefficients of heritability themselves undergo a circadian rhythm. This is documented for SBP and for an electrocardiographic variable, intraventricular cardiac conductivity gauged by the QRS (Zaslavskaya et al. 1990).

## How Can the Chronobiologic Approach be Evaluated in the Absence of Actuarial Morbidity and Mortality Statistics Related to Its Practice?

The presence of rhythms of large amplitude in BP can no longer be disputed. Odd-hour BP elevation has already been associated with overt pathology, as in the case of severe preeclampsia (Miyamoto et al. 1988; Redman et al. 1976) and of increased risk of cardiovascular disease (Germanó 1988; McDonald et al. 1990; Reeves et al. 1984). Moreover, by 14 years of age, correlations are found between the circadian amplitude of DBP and target organ involvement, namely the thickness of the interventricular cardiac septum determined by M-mode echocardiography (Croppi et al. 1986; F. Halberg et al. 1988 a). In clinically healthy Japanese volunteers 20–40 years of age, with an acceptable BP according to WHO criteria, the individuals with a chronobiologic BP excess of at least 25 mmHg × h (over 24 h) exhibited a statistically significantly larger interventricular septal thickness, posterior wall thickness, left ventricular mass (with or without correction for body surface area), and

relative wall thickness than did controls (showing a BP excess of less than 25 mmHg × h) (Saito et al. 1990).

As noted above, pregnant women with "office hypertension" but with a 24-h BP mean remaining on the whole within the putative physiologic range suffer more complications, such as intrauterine growth retardation of the fetus, than "office-normotensive" pregnant women (Rigó et al., to be published). Groups of pregnant women with SBP values all under 140 mmHg (group I), with less than 5% of the values above 140 mmHg (group II), and with more than 5% of the values above 140 mmHg (group III) differ substantially in their rhythm-adjusted mean. Whereas the MESORs are well within currently accepted limits for each group, the differences are between $103.4 \pm 0.7$ ($n = 44$ in group I), $110.4 \pm 1.5$ ($n = 21$ in group II), and $124.23 \pm 2.0$ mmHg ($n = 24$ in group III) (Cornélissen et al. 1989).

Values of SBP above 140 mmHg in around-the-clock profiles collected with ambulatory monitors are found in a larger proportion of pregnant women who are to develop preeclampsia, gestational hypertension, or an otherwise elevated BP than in women with uncomplicated pregnancies ($p < .05$). Larger-scale studies are needed. Practicing physicians and researchers alike have been invited to meet the challenge offered by chronobiology in a comparative study of outcomes (F. Halberg et al. 1990 b). Data on outcomes are presently being collected in the context of an international womb-to-tomb study (Cornélissen 1990; Cornélissen et al. 1990 a; F. Halberg et al. 1990 b).

## Pitfalls to be Avoided in the Evaluation of an Individual's Response to Treatment

*Gradual Rather than Immediate Response.* The response to the initiation (or withdrawal) of treatment is not necessarily immediate. Studies evaluating the effect of the removal and reinstitution of antihypertensive medication on the rhythm-adjusted mean of BP have been conducted longitudinally over spans of weeks or months (F. Halberg et al. 1984 b; Levine and Halberg 1972; Little et al. 1990). In a chronobiologic study, the BP MESOR continued to fall for about 2 months following resumption of hydrochlorothiazide treatment in a patient who had been off medication for 6 weeks (Levine and Halberg 1972). In patients treated with lisinopril, without chronobiologic considerations, a further decrease in BP was still apparent after 6 weeks on treatment (Zachariah et al.

1988). Using around-the-clock automatic monitoring of BP and heart rate of a subject living in complete social isolation, the "washout" of the effects of a beta-blocker was demonstrated to take more than a month (F. Halberg et al. 1984b).

*Amplitude Hypertension.* The assessment of a mean value may not be sufficient to evaluate the efficacy of treatment. Decreases in MESOR following treatment with beta-blockers may be accompanied by increases in circadian amplitude (Cornélissen et al. 1988; F. Halberg et al. 1988a). It is thus possible, at least for some individuals, that BP is not reduced at the most critical time, i.e., when BP reaches its overall highest values during the day. A decrease rather than an increase in the circadian amplitude of BP was reported for patients treated with atenolol in an Asian population (Kumagai et al. 1989)

*Individualization.* The specific treatment and timing should be individualized. Not everybody responds in the same fashion to a given intervention such as sodium restriction. While it has been known that BP is not reduced in all individuals after sodium restriction, it is less well known that some persons may even increase their BP after sodium restriction (F. Halberg et al. 1988a). Parameter tests are available (Bingham et al. 1982) to evaluate the statistical significance for the individual of a response to a given intervention. These tests are more powerful than conventional statistical procedures since, in the presence of a prominent circadian rhythm, the MESOR is both more accurate and more precise than the arithmetic mean (F. Halberg et al. 1990a, d). Chronobiologic tasks also include information provided by the amplitude and acrophase representing the dynamics of BP variability. Cosinor techniques need not be restricted to the analyses of BP data themselves; they are equally applicable to indices of BP excess (or deficit) computed over short intervals (e.g., 3 h, as fractionated hyper- (or hypo-)baric indices. An individualized assessment of the timing of excess thus becomes possible, on the basis of which the most opportune time to administer treatment can be determined.

*Account of Regression Toward the Mean.* Longitudinal monitoring is desirable in order to avoid misinterpretations due to regression toward the mean (F. Halberg et al. 1988a, 1990b). Day-to-day changes in BP can occur spontaneously in the absence of intervention. These changes result in part from infradian components and in part from random fluctuations due to a multitude of factors that cannot all be controlled. In studies involving groups of individuals, it is important to design experiments wherein, in addition to control spans on a placebo and, if possible, untreated control groups, each subject undergoes all or at least most treatments, the sequence of treatments being randomized.

*Need for a Placebo.* Antihypertensive medication is expensive and causes side effects which may occasionally be serious. Treatment should thus be maintained at the lowest dose which assures acceptable BP behavior. When, by rigorous and adequate chronobiologic monitoring, an acceptable BP can be observed with the administration of a placebo or without treatment, the medication should be discontinued. This is suggested by results from the Australian Therapeutic Trial in Mild Hypertension (1980) (see F. Halberg et al. 1988a) and from a number of chronobiologically studied individual cases (Little et al. 1990).

According to current conventional criteria of upper limits of 140/90 mmHg (SBP/DBP), the 48% of the 1943 individuals with "borderline hypertension" who were assigned to the placebo group in the Australian Therapeutic Trial and who, 3 years later, were found to have acceptable BP values should not have been treated at all, being false positives (Halberg et al. 1988a). In view of the recognition of risk by chronobiologic means, however, it seems important to suggest that such individuals may benefit from nondrug intervention if indeed repeated monitoring reveals BP excess with reference to time-specified limits derived for low-risk peers. To return to conventional thinking, office hypertensives or white coat hypertensives must not necessarily be regarded as normotensives, but deserve further scrutiny, as documented separately by the results of long-term tracking by Julius et al. (1990) and also chronobiologically during pregnancy (Rigó et al., to be published; Zaslavskaya et al., to be published).

Variations in BP during treatment, also in cases of chronobiologically established (i.e., mild) hypertension, may be found as a result of lower-frequency rhythms, i.e., circaseptans, circatrigintans, and circannuals (Sothern and Halberg 1986, 1989) or may be the result of changes in lifestyle (e.g., weight loss, Fig. 4, exercise, diet) and/or social loads (Fig. 5) (Haus et al., to be published). For all of the foregoing reasons and for scientific rigor in the practice of medicine, the need for a placebo cannot be overemphasized, as shown in the following two case studies.

*Case 1.* To examine the need for antihypertensive therapy and its timing, a 46-year-old woman with a 10-

**Fig. 4.** A reduction in dietary calorie intake for 2 months is associated with a statistically significant decrease in SBP in 10 out of 11 individuals diagnosed conventionally as borderline hypertensive. Change was gauged individually (by a test of equality of circadian MESOR). In 8 of them, a decrease in DBP is also validated with statistical significance. Subjects 2 and 12 did not contribute data during stage III and are hence not listed. (Halberg et al. 1988a; see also Lee et al. 1982)

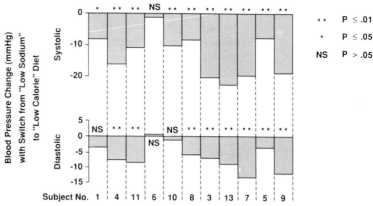

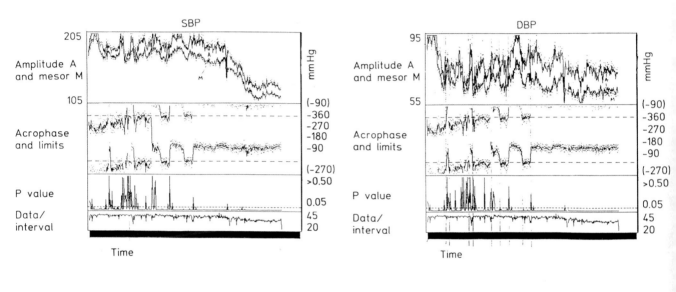

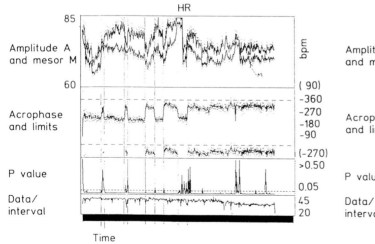

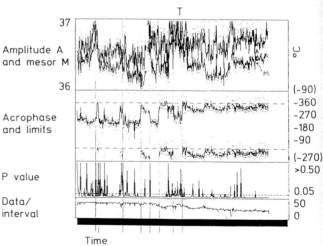

**Fig. 5.** Longitudinal view with 14, 046 sets of self-measurements of the aging human circulation and metabolism, the latter gauged by core temperature, the former by SBP, DBP, and heart rate. (From Haus et al. to be published)

year history of "mild to moderate hypertension", treated for that span with 50 mg of hydrochlorothiazide per day, usually taken before retiring, monitored her BP for several months with an ambulatory instrument from Colin Medical Instruments. During this span, the timing of treatment was changed in a double-blind study and replaced at times by a placebo (Little et al. 1990). While the patient clearly, albeit only gradually, responded to the treatment which appeared to be superior in the morning or at noon than in the evening, her BP was found to be acceptable on placebo. Her BP remained within currently acceptable limits after she was taken off medication for over a year.

The change from a drug to a placebo, of course, must not be done indiscriminately and not without strict and continuous medical supervision. In the North American climate, possibilities for litigation raise their ugly head, but must not forestall action in the best interests of a patient. Clearly, automatic around-the-clock monitoring for 48 h is indicated while the treatment continues. Such monitoring, or at least self-measurements at 1-h intervals during the day with one interruption of sleep at night, or some domiciliary measurements by a family member, are indicated while the patient is being switched to placebo. Since the wash-out of a drug may take days, weeks, or even longer, domiciliary monitoring is mandatory if automatic monitoring is unavailable. Proper analyses can be carried out rapidly on a computer on a daily basis, so that an increase in BP that appears to be medically significant can prompt the resumption of treatment. There is an added problem in cases such as the one discussed above, namely the question whether reduction of BP by medication, within the physiologic range, is a goal to be pursued at the price of eventually incurring side effects. Such studies should be carried out preferably on monozygotic twins who present BP MESORs within the acceptable physiologic range. One twin in each pair would receive the treatment, while the other would receive the placebo. Since the intratwin-pair variability in circadian characteristics can be anticipated to be much smaller than the inter-twin-pair variability in these parameters, differences in outcomes in relation to treatment kind and extent of BP change in response to antihypertensive medication could be assessed on relatively small groups of monozygotic twin pairs.

*Case 2.* By means of five to eight manual BP self-measurements a day, an elderly widowed physician collected 14,046 sets of data between September 24, 1972 (at 78 years of age) and March 5, 1981 (at 87 years of age) (Haus et al., to be published). From 1972 to 1975,

she lived alone in Europe; during the summers of 1972–1974 she visited her family in the USA. In 1975, she moved permanently to the USA, close to her family. She had been diagnosed as having essential hypertension in 1972 and was treated first (11/Feb/72–3/Jan/74) with a combination of reserpine (0.2 mg/day), dihydroergocistine (1.16 mg/day), and clopamide (10 mg/day), and from 1974 to 1978 with a daily combination dose of 0.075 mg 2-(2,6-dichlorphenylamino)-2-imidazoline and 15 mg 1-oxo-3-(3'-sulfamyl 4'-chlorphenyl) isoindoline, a diuretic. For a seizure disorder, she received 400 mg phenytonin per day throughout the entire observation span. After she moved to the USA, her SBP, DBP, and pulse rate began to drop, which led in 1978 to discontinuation of her antihypertensive therapy. In spite of the cessation of treatment, the BP dropped further while the subject was in relatively good health and followed her usual daily activity (Haus et al., to be published).

Figure 5 shows the time course of SBP (top left), DBP (top right), heart rate (bottom left), and temperature (bottom right) in so-called chronobiologic serial sections. In each graph, the lower curve in the top row represents the MESOR; its standard error is indicated by the dots below that curve. The circadian amplitude is shown by the distance between the upper and lower curves in the top row. The dots above the upper curve indicate one standard error of the amplitude. In the second row of each serial section, the acrophase, a measure of the timing of overall high values, is shown; it is expressed in (negative) degrees with 360 ° equated to 24 h and the reference time chosen as local midnight in Austria. The lower dashed line represents midnight on one day and the upper dashed line midnight on the following day. When the acrophases occur around midnight, they are plotted twice. Dots on either side of the acrophase curve represent their 95 % confidence intervals. The bottom two rows illustrate the statistical significance of the circadian rhythm by the $p$-value from the zero amplitude test and the number of observations in each interval.

This figure shows a steady drop in SBP from about 190 mmHg at the start of monitoring to less than 110 mmHg over 8 years later. For both SBP and DBP, a drift in acrophase is seen at the beginning of the study. These two variables free-run with a period longer than 24 h (they slant upward). By contrast, the acrophases of heart rate and temperature show changes in time location over short spans that are associated with round trips between Innsbruck, Austria, and Minneapolis, Minnesota, while the subject had homes in both locations. Eventually, the trips ceased since she established her home in Minnesota.

The entire circadian system is now synchronized on the same frequency, albeit with different circadian acrophases among the different variables. The figure also shows the gradual drop in SBP (and to a lesser extent in DBP), which continues over the years, after the discontinuance of all antihypertensive treatment. Such spontaneous decreases must be kept in mind whatever the underlying factors may be. Indeed, the pressure should be checked under a placebo on a regular basis, so that any spontaneous if gradual changes are not overlooked.

## Mapping the Blood Pressure Chronome[1] from Womb to Tomb

The large extent of spontaneous change within the physiologic range of BP can be resolved and should be exploited. New information is being provided by the resolution of rhythms and by the study of their intermodulations as gauges of health and harbingers of elevated risk. The spectral aspect of the circulation and other systems codetermines the pattern of their responses to the environment, as expressed, for example, in nonrandom morbidity and mortality patterns of outcomes by day, week, and year. Such patterns are found in relation to sudden infant death syndrome (Wu et al., to be published) and myocardial infarctions, and sudden and other cardiac death (F. Halberg 1984; Haus 1989; Master and Jaffee 1952; Muller et al. 1985, 1987; Otto et al. 1982; Reinberg et al. 1973; Smith 1861; Smolensky et al. 1972; Tofler et al. 1987; WHO 1976), (Fig. 6, 7), and have even been shown to differ among various kinds of strokes (Johansson et al. 1990). These patterns in time that make the difference between death and survival invite research into a budding chronoepidemiology (Cornélissen et al. 1990b); they are demonstrable by the time of birth in variables such as BP that are monitored longitudinally or transversely (Cornélissen et al. 1990a; F. Halberg et al. 1990b). In order to detect rhythm alteration early enough to implement preven-

[1] The chronome is defined as the expression of genetic, epigenetic, and broader environmental interactions in the form of rhythms (in an inferentially resolved mathematical multifrequency spectrum), trends (with development, maturation and aging) and the residual variation which holds and as yet hides much added information about the relative contributions of nature and nurture to chronos (our structure in time) and chaos (the residual variability, which may still exhibit some "deterministic" rather than completely "random" features) (F. Halberg et al., to be published b).

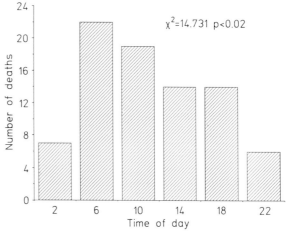

**Fig. 6.** Routine-synchronized circadian (about 24-h) variation in the occurrence of cardiac death in elderly subjects living in a nursing home under apparently uniform environmental conditions. (Haus 1989)

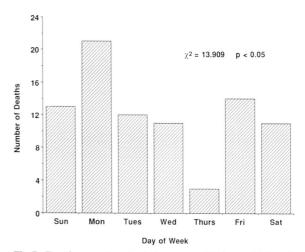

**Fig. 7.** Routine-synchronized circaseptan (about weekly) variation in the occurrence of cardiac death in elderly subjects living in a nursing home under apparently uniform environmental conditions. (Haus 1989)

tive intervention, it is critical to map the human chronome, the expression of genetic and developmental factors in the form of rhythms and trends (F. Halberg et al., to be published b). As a step toward this goal, chronobiologic reference values are critical.

## Results on Pregnant Women

Circadian reference limits for BP and heart rate in clinically healthy pregnant women in each trimester of pregnancy are illustrated in Table 1 (Cornélissen et

**Table 1.** Time-varying 90% prediction limits for systolic (S) and diastolic (D) blood pressure (BP) of clinically healthy Minnesotan pregnant women in the three trimesters (1, 2, 3)

| Time of day | SBP | | | DBP | | |
|---|---|---|---|---|---|---|
| | 1 | 2 | 3 | 1 | 2 | 3 |
| 0030 | 79-121 | 79-118 | 81-125 | 43-70 | 43-68 | 44-74 |
| 0130 | 79-118 | 77-115 | 79-121 | 42-67 | 41-67 | 43-71 |
| 0230 | 75-**117** | 77-113 | 82-121 | 41-**66** | 42-65 | 44-71 |
| 0330 | 74-121 | 76-**111** | 79-**118** | **39**-70 | **40-64** | **42-69** |
| 0430 | 74-120 | 75-112 | 79-121 | 39-69 | 40-65 | 43-72 |
| 0530 | 77-118 | 76-112 | 79-121 | 40-68 | 41-65 | 43-72 |
| 0630 | **73**-130 | **74**-119 | **76**-125 | 39-75 | 40-72 | 43-74 |
| 0730 | 76-131 | 77-123 | 78-131 | 39-79 | 43-74 | 44-79 |
| 0830 | 81-133 | 79-128 | 82-137 | 44-78 | 44-78 | 47-83 |
| 0930 | 84-136 | 86-131 | 86-136 | 46-82 | 49-79 | 47-85 |
| 1030 | 87-136 | 88-133 | 92-136 | 49-83 | 50-80 | 52-85 |
| 1130 | 90-138 | 89-134 | 94-136 | 49-85 | 50-81 | 54-82 |
| 1230 | 92-135 | 91-134 | **95**-134 | 50-82 | 50-80 | 52-83 |
| 1330 | 89-141 | 90-133 | 94-136 | 47-**86** | 52-80 | 55-83 |
| 1430 | 88-**143** | 84-136 | 90-140 | 50-83 | 48-80 | 49-**87** |
| 1530 | 87-139 | 87-135 | 94-139 | 48-84 | 50-81 | 52-84 |
| 1630 | 89-138 | 90-**137** | 94-**142** | 49-81 | 51-**84** | **56**-84 |
| 1730 | 89-137 | 91-136 | 94-139 | 49-81 | 51-81 | 54-84 |
| 1830 | **94**-136 | **93**-134 | 93-139 | 49-81 | **53**-79 | 54-85 |
| 1930 | 94-136 | 90-135 | 93-142 | **52**-80 | 51-81 | 53-86 |
| 2030 | 93-131 | 88-136 | 91-138 | 51-78 | 49-80 | 52-83 |
| 2130 | 88-131 | 88-131 | 89-138 | 48-78 | 48-79 | 52-83 |
| 2230 | 80-132 | 85-127 | 87-138 | 44-78 | 47-75 | 49-83 |
| 2330 | 80-125 | 79-123 | 82-134 | 40-73 | 43-73 | 45-80 |

These time-specified peer-group-based limits of acceptability (chronodesms, in mm Hg) reveal that the lower or upper limit changes within any one day over a range (given by **bold** numbers) is much larger than the change from trimester to trimester. These limits may be used in lieu of a fully arbitrary horizontal line, drawn at 140/90 (systolic/diastolic) or any other fixed limit

al. 1991). A statistically significant circadian rhythm in BP is found for a large majority (85%) of clinically healthy pregnant women monitored mostly at hourly intervals for 48 h with an ambulatory instrument (ABPM-630, manufactured by Colin Medical Instruments, San Antonio, TX 78249). The added consideration of the second harmonic (cosine curve with a 12-h period) improves the approximation of the circadian waveform by 20%–25%. Overall, the consideration of the first six harmonics accounts for nearly 60% of the total variance. The overall range of change within a day averages 58 and 47 mmHg for SBP and DBP, respectively. The highest SBP value exceeds 140 mmHg in about 50% of the cases, and the highest DBP value exceeds 90 mmHg in about 25% of the cases. A positive correlation is found between the BP MESOR and circadian amplitude, on the one hand, and, on the other, a cardiovascular disease risk score (Cornélissen et al. 1989) based on family history and experience at any prior pregnancy (Table 2). A breakdown of the risk score into compo-

**Table 2.** Cardiovascular risk scale

| Person | Risk factors | Degree of risk[a] |
|---|---|---|
| *Pregnant woman* | Previous incidence of eclampsia or toxemia | 2 |
| | Previous incidence of gestational "hypertension" or high blood pressure | 1 |
| *Parents* | Presence of cardiovascular disease | 1 |
| | Presence of high blood pressure | 0.5 |
| *Grandparents* | Presence of cardiovascular disease | 0.5 |
| | Presence of high blood pressure | 0.25 |

[a] Risk listed per individual; it is additive within each generation and across generations; maximal risk per generation is 2, i.e., if both high blood pressure and cardiovascular disease are present, only the higher risk score is considered. Presence of high blood pressure or cardiovascular disease in siblings, aunts, and uncles is not considered here to avoid bias caused by difference in the number of family members and in view of usual lack of information on family members other than parents and grandparents. Risk scale thus spans interval from 0 to 6

nents specific to the maternal or paternal family history and to a history of high BP vs. that of overt cardiovascular disease indicates the paternal history to contribute most to the effect. Changes as a function of gestational age revealed: for BP, a decrease between the 12th and 15th gestational week, reaching minimal values around the 20th week, and an increase between the 30th and 32nd weeks; for heart rate, a steady increase as a function of gestational age is observed (Cornélissen et al. 1989). The circadian amplitude of BP continues to decrease throughout pregnancy. As compared to uncomplicated pregnancies, the decrease in BP MESOR up to the 20th gestational week may not occur in pregnancies that are to terminate with gestational hypertension or an otherwise elevated BP (F. Halberg et al. 1991 c). In contrast, already at the transition between the first to the second trimester, the circadian amplitude of BP decreases more in those pregnancies than in uncomplicated ones. In the case of heart rate, the circadian amplitude increases more in the women at risk than in clinically healthy women (F. Halberg et al. 1991 c).

## Results on Newborns

Several hundred newborns have been monitored for about 2 days during the first week of life. Although the circadian amplitude is still relatively small and a rhythm is demonstrable (at the 5% probability level) in about 35% of the data series, overall, a circadian rhythm comes to the fore as a group phenomenon which is more prominent in full-term than in preterm infants. The BP MESOR of girls is slightly but statistically significantly higher than that of boys (Wu et al., in preparation). The timing of high values (acrophase) occurs early in the morning on the average, on the first day of life, and progressively shifts toward the afternoon. During the first week of life, several spectral components, resolved transversely, already complement the circadian cardiovascular rhythms. Ultradian components are likely associated with the feeding patterns. As a group phenomenon, the circaseptan and circannual components both modulate the circadian system of the newborn (Cornélissen et al. 1990a; F. Halberg et al. 1990b). By referring data to the time of birth and after a normalization of the data to reduce interindividual differences, a statistically significant circaseptan pattern emerges that has an amplitude much larger (by a factor of about 5) than the circadian or circannual components (Cornélissen et al. 1987; Wu et al. 1990). This circaseptan group component is not seen by reference to the day of the week; it is clearly endogenous, synchronized by the process of birth and a feature of single stimulus manifestation, if not induction (since the event of birth itself carries no 7-day information, even if, as a group phenomenon, there is a 7-day synchronized circaseptan in the incidence of birth as such) (Marazzi et al. 1990). That we are not dealing merely with a response to birth per se can be documented by studies on prematures, confirming, longitudinally, the large-scale transverse assessment of circaseptans (Halberg et al. 1991).

The dynamic characteristics of change describing these multifrequency rhythms constitute sensitive gauges of risk. They reveal the effects of the intrauterine exposure to beta-adrenergic agonists. Drugs inhibiting the contractility of uterine smooth muscle, beneficial for the mother and her fetus in that they prevent premature delivery, are associated with a larger circadian amplitude of BP in the newborn, possibly expressing a higher risk of developing an elevated BP later in life (F. Halberg et al. 1988b, 1991; Syutkina et al. 1991; Tarquini et al. 1990). A large circadian amplitude of BP was earlier associated with an increased cardiovascular disease risk (F. Halberg et al. 1986c, 1988a).

## Blood Pressure Monitoring in Children

Autorhythmometry in Minnesotan schools (F. Halberg 1970; F. Halberg et al. 1973, 1974; R. Haus et al. 1987; Rabatin et al. 1981) and elsewhere (LaSalle et al. 1983; Scarpelli et al. 1987; Scheving et al. 1982) showed the feasibility of physiologic monitoring and established that a successful program of self-measurements was possible in high school classes. Chronobiologic self-measurements have been described in technical detail (F. Halberg et al. 1972) and remain a practical step in school screening. Children as young as 9 years of age were taught how to self-measure BP and heart rate and were asked to fill out a questionnaire inquiring about family histories regarding cardiovascular disease risk. Results indicated that in the absence of any difference in overall mean, the circadian amplitude and acrophase of SBP differed in children with a negative or positive family history of high BP. This difference came about primarily because of a larger amplitude in children with a positive vs. negative family history of high BP (Scarpelli et al. 1987). This study and others revealed the importance of at least one and preferably two nightly measurements and the need for a monitoring span of at least 48 h to obtain a reliable estimate of the circadian group amplitude. To apply risk assessment to the individual, longitudinal studies are needed, and the minimal sampling requirements for such studies remain to be established by the decimation of automatically collected longitudinal time series (del Pozo et al. 1991; Halberg et al. 1991e).

Systematic self-measurements several times a day on most days for several years in a few individuals have revealed long-term trends (infradian variation) but not necessarily an increase with age. Actually, the variability in BP is larger within one day than are the expected changes in MESOR as a function of age in adult life in health (F. Halberg et al. 1981a, 1988a; Sothern and Halberg 1989). Monitoring for about 3.5 years before and after menarche in two sisters (F. E. Halberg et al. 1985; F. Halberg et al. 1988a; J. Halberg et al. 1985) indicates, in the older sister, an increase in SBP and a lesser increase in DBP with an increased (upward) dispersion around age 13, about 1.5 years before menarche, with no apparent change just prior to or after menarche. In the younger sister, several months before the start of menarche, an increase is seen in the (upward) dispersion of SBP. At menarche, there is a rise in the overall circadian MESOR. These results show that cardiovascular changes of considerable extent may characterize prepubertal girls and await longitudinal individualized assessment on a larger scale.

Ambulatory monitoring of school children 13–19 years of age (Ferencz et al. 1986) led to the specification of reference standards for circadian parameters, based on data from children with a negative family

**Table 3.** Location and dispersion indices of circadian MESOR and amplitude in 177 clinically healthy boys and girls, 13–19 years of age, with different family history of high blood pressure

| | MESOR | | | Amplitude | | |
|---|---|---|---|---|---|---|
| | Mean | SD | Median (90% PL[a]) | Mean | SD | Median (90% PL[a]) |
| **Negative family history of high BP** | | | | | | |
| **Boys (n = 46)** | | | | | | |
| SBP | 113.2 | 8.2 | 112.5 (99.3, 127.0) | 7.5 | 4.1 | 7.8 (0.64, 14.42) |
| MAP | 80.3 | 6.1 | 79.9 (69.9, 90.7) | 5.0 | 3.0 | 4.2 (0.00, 10.01) |
| DBP | 63.9 | 7.3 | 64.7 (51.5, 76.3) | 4.8 | 3.1 | 4.6 (0.00, 10.02) |
| PP | 49.3 | 9.6 | 49.7 (33.0, 65.5) | 6.2 | 4.0 | 5.2 (0.00, 13.05) |
| HR | 75.4 | 8.9 | 74.7 (60.4, 90.5) | 10.5 | 4.4 | 10.1 (3.09, 17.90) |
| SBP × HR | 85.5 | 13.0 | 83.9 (64.4, 108.5) | 16.8 | 7.8 | 17.3 (3.57, 30.09) |
| **Girls (n = 43)** | | | | | | |
| SBP | 104.1 | 7.3 | 104.0 (91.6, 116.5) | 9.2 | 5.0 | 8.7 (0.69, 17.69) |
| MAP | 76.9 | 5.0 | 76.3 (68.4, 85.4) | 6.6 | 3.9 | 5.5 (0.00, 13.28) |
| DBP | 63.3 | 6.3 | 63.9 (52.6, 73.9) | 6.0 | 3.8 | 5.7 (0.00, 12.50) |
| PP | 40.8 | 9.3 | 40.2 (25.1, 56.6) | 6.4 | 2.9 | 6.4 (1.51, 11.31) |
| HR | 81.9 | 6.5 | 82.5 (70.9, 93.0) | 12.8 | 5.6 | 12.3 (3.36, 22.28) |
| SBP × HR | 86.5 | 10.0 | 86.3 (69.6, 103.5) | 20.4 | 9.2 | 19.8 (4.77, 35.99) |
| **Positive family history of high BP** | | | | | | |
| **Boys (n = 39)** | | | | | | |
| SBP | 115.9 | 10.7 | 115.9 | 11.1 | 6.0 | 11.4 |
| MAP | 81.9 | 5.1 | 82.2 | 7.0 | 3.1 | 6.8 |
| DBP | 64.9 | 5.3 | 64.4 | 6.0 | 3.1 | 6.0 |
| PP | 50.9 | 11.6 | 50.9 | 9.2 | 5.3 | 8.4 |
| HR | 78.3 | 8.3 | 79.0 | 12.1 | 5.3 | 11.8 |
| SBP × HR | 92.2 | 14.2 | 90.2 | 22.7 | 10.0 | 20.6 |
| **Girls (n = 49)** | | | | | | |
| SBP | 106.0 | 9.0 | 105.7 | 9.1 | 3.9 | 9.0 |
| MAP | 79.1 | 5.6 | 79.3 | 7.0 | 3.3 | 6.8 |
| DBP | 65.6 | 5.9 | 65.0 | 6.5 | 3.5 | 6.2 |
| PP | 40.3 | 9.1 | 38.9 | 5.6 | 3.4 | 4.9 |
| HR | 82.5 | 14.0 | 80.0 | 11.4 | 5.2 | 10.9 |
| SBP × HR | 88.0 | 16.6 | 84.2 | 19.2 | 7.9 | 18.2 |

SBP, systolic blood pressure; MAP, mean arterial pressure; DBP, diastolic blood pressure; PP, pulse pressure (SBP-DBP); HR, heart rate; SD, standard deviation. SBP, MAP, DBP, and PP are expressed in mmHg, HR in beats/min, and SBP × HR in mmHg × beats/min ($\times 10^{-2}$). BP and HR were measured every 7.5 min for 24 h by automatic ambulatory monitoring.
[a] 90% PL = 90% prediction limits

**Table 4.** Comparison of rhythm characteristics of systolic blood pressure separately for boys and girls as a function of whether their parents did or did not have a high blood pressure

Population-mean cosinor

| Gender | Parent with high BP | No. of subjects | MESOR, M | Amplitude, A (95% confidence limits) | Acrophase, Φ[a] |
|---|---|---|---|---|---|
| girls | no | 43 | 104.1 (101.9, 106.3) | 7.19 (5.60, 8.79) | −246 (−232, −260) |
| | yes | 49 | 106.0 (103.4, 108.6) | 6.98 (5.56, 8.40) | −244 (−232, −256) |
| boys | no | 46 | 113.2 (110.8, 115.6) | 5.92 (4.62, 7.22) | −246 (−232, −258) |
| | yes | 39 | 115.9 (112.4, 119.3) | 8.43 (6.47, 10.40) | −251 (−234, −267) |

Parameter tests

| | M | | A | | Φ | | (A, Φ) | |
|---|---|---|---|---|---|---|---|---|
| Groups compared[b] | F | (p) | F | (p) | F | (p) | F | (p) |
| NPH vs. PPH (girls) | 1.193 | (0.278) | 0.040 | (0.843) | 0.052 | (0.820) | 0.043 | (0.958) |
| NPH vs. PPH (boys) | 1.752 | (0.189) | 4.925 | (0.029) | 0.283 | (0.596) | 2.409 | (0.093) |

[a] In degrees, with 360° = 24 h; 0° = 0000. M and A in mm Hg
[b] NPH, no parent(s) with high blood pressure; PPH, parent(s) with high blood pressure

**Table 5.** Regression with age, separately for MESOR and circadian amplitude of systolic and diastolic blood pressure of boys and girls 13–14 years of age

| Group | | SBP | | | | DBP | | | |
| | | MESOR | | Amplitude | | MESOR | | Amplitude | |
| | | r | (p) | r | (p) | r | (p) | r | (p) |
|---|---|---|---|---|---|---|---|---|---|
| NPH | girls | 0.091 | 0.567 | 0.033 | 0.826 | −0.066 | 0.676 | 0.467 | 0.002 |
| | boys | 0.429 | 0.003 | 0.020 | 0.888 | 0.180 | 0.231 | 0.033 | 0.824 |
| PPH | girls | 0.135 | 0.641 | 0.025 | 0.861 | −0.042 | 0.774 | 0.329 | 0.020 |
| | boys | 0.434 | 0.006 | 0.072 | 0.668 | 0.272 | 0.090 | 0.062 | 0.710 |

SBP, systolic blood pressure; DBP, diastolic blood pressure; NPH, no parent(s) with high blood pressure; PPH, parent(s) with high blood pressure; r, Pearson product-moment correlation; p, p value from test: r = 0

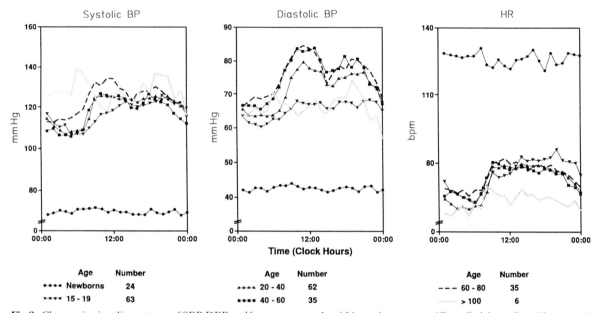

**Fig. 8.** Changes in circadian pattern of SBP, DBP, and heart rate as a function of age from newborns to centenarians (males). Each curve represents hourly means averaged over all individuals within a given group. (Compiled from Cornélissen et al. 1990a; Halberg et al. 1990b; Ikonomov et al., to be published; Scarpelli et al. 1988; Tarquini et al. 1990)

history of high BP (Table 3; F. Halberg et al. 1988a). It can be seen that the extent of predictable variation within a day in BP can be substantial; the double circadian amplitude amounts to 15 and 22 mmHg for SBP in the case of boys with a negative or positive family history of high BP, respectively. Differences can also be seen in circadian parameters such as the amplitude of SBP as a function of cardiovascular disease risk. A comparison of rhythm characteristics as a function of the parental history of high BP again finds a larger circadian amplitude in boys with a positive history (Table 4; F. Halberg et al. 1986d). Differences as a function of gender and age are also found (Table 5; Cornélissen et al. 1986). New endpoints are thus available to the practitioner for a refined diagnosis of any early BP deviation.

## Chronobiology of Blood Pressure Coordination and Age

Figures 8 and 9 show, for each gender, changes in SBP, DBP, and heart rate as a function of circadian stage for different age groups from newborns to centenarians. The relatively flat curves of newborns and centenarians contrast with those of other age groups exhibiting a circadian group rhythm. The apparent lack of circadian rhythmicity in the mean BP curves of newborns and centenarians cannot be interpreted as the absence of circadian periodicity. This artifact results from the averaging over the group of predominantly noncircadian BP changes, ultradian and infradian, that are not necessarily synchronized among the newborns and in the elderly. In the case of

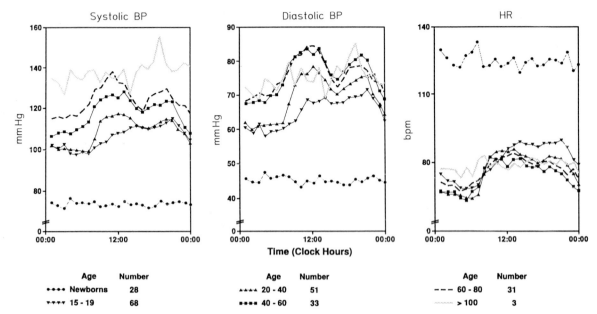

**Fig. 9.** Changes in circadian pattern of SBP, DBP, and heart rate as a function of age from newborns to centenarians (females). Each curve represents hourly means averaged over all individuals within a given group. (Compiled from Cornélissen et al. 1990a; Halberg et al. 1990b; Ikonomov et al., to be published; Scarpelli et al. 1988; Tarquini et al. 1990)

centenarians, there is a variance transposition from the 24-h to the 12-h component and to even higher frequency components for BP, whereas for heart rate the major component remains the circadian one (Ikonomov et al. 1990). There is a dissociation between BP and heart rate. The heart, responding to higher coordinating centers in the hypothalamus and in response to stimuli hitting the brain, maintains a primary circadian behavior. This is not the case for BP which shows ultradian prominence, perhaps as a result of changes in the periphery.

Whether the circadian-to-ultradian variance transposition in the centenarian is beneficial, indifferent or detrimental remains to be established on the basis of outcome studies. That the multifrequency rhythm structure is critical is apparent from the fact that it is associated with the gates of susceptibility for a number of disease conditions that change predictably as a function of about-daily, about-weekly, and about-yearly stage, tipping the scale between death and survival. If a relation can be established between the extent of alteration in the circadian pattern of BP and heart rate on the one hand and longevity and the quality of life of the elderly on the other hand, strategies could be developed to manipulate either the circadian pattern of BP or that of heart rate or both, be it by exercise, by diet, or by other means, to optimize the processes of aging. Such optimization may perhaps be attempted by endocrine manipula-

tion as well should exercise and diet in themselves fail to achieve the goal. Such intervention may have to be implemented before the spectral alteration becomes so drastic that it may become irreversible. A rational prevention will depend upon the manipulation of the coordinating centers in the hypothalamus, which with age undergo changes comparable to those in a supra-chiasmatic lesion, namely changes in amplitude and in timing of a number of rhythms at all levels of organization from temperature to cell division (Halberg 1981; Halberg et al. 1979, 1981; Scheving et al. 1983).

According to some authors, high BP, although an important modulatory factor in young patients, has "limited influence on the cardiovascular regulatory functions in older subjects whose autonomic functions have already been substantially altered by advancing age" (Kawamoto et al. 1989). An increased risk for the development of cardiovascular diseases has been associated with excess pressure (Balsano 1988; Borhani 1988; Stokes et al. 1989). A relation between BP and death from coronary heart disease has, for instance, emerged from the Framingham Study (Stamler et al. 1989). In this study, risks of death were also substantially increased among those "hypertensive" men who had already had end-organ damage, emphasizing the importance of early treatment to prevent such damage (F. Halberg et al. 1986b).

Circadian rhythms in a number of variables such as catecholamine excretion persist in the elderly

**Table 6.** Rhythmometric summary of blood pressure in clinically healthy Minnesotan and Japanese women of various ages monitored mostly in recumbency

| Group | Minnesota | | | | | Japan | | | | |
|---|---|---|---|---|---|---|---|---|---|---|
| | No. | % rhythm ± SE | MESOR (mmHg ± SE) | Double amplitude (mm Hg ± SE) | N (%) series with $p < 0.05$ | No. | % rhyhtm ± SE | MESOR (mmHg ± SE) | Double amplitude (mm Hg ± SE) | N (%) series with $p < 0.05$ |
| **SBP** | | | | | | | | | | |
| Ad | 22 | 14.0 ± 2.8 | 98.4 ± 1.8 | 9.7 ± 1.2 | 16 (73) | 8 | 20.3 ± 4.1 | 105.2 ± 3.2 | 9.1 ± 1.5 | 7 (88) |
| YA | 32 | 23.5 ± 2.8 | 102.1 ± 1.4 | 13.1 ± 1.3 | 32 (100) | 6 | 4.3 ± 2.6 | 102.0 ± 2.2 | 4.3 ± 1.6 | 1 (17) |
| PM | 27 | 22.1 ± 3.3 | 109.2 ± 1.8 | 15.2 ± 1.7 | 25 ( 92) | 8 | 22.6 ± 4.5 | 110.2 ± 4.4 | 14.9 ± 1.7 | 7 (88) |
| **DBP** | | | | | | | | | | |
| Ad | 22 | 10.0 ± 2.5 | 62.4 ± 0.9 | 5.8 ± 0.9 | 13 (59) | 8 | 10.1 ± 2.7 | 63.8 ± 2.7 | 6.9 ± 1.0 | 6 (75) |
| YA | 32 | 21.7 ± 3.0 | 65.5 ± 0.8 | 9.4 ± 1.0 | 29 (91) | 6 | 13.8 ± 5.4 | 66.1 ± 0.8 | 8.6 ± 2.0 | 5 (83) |
| PM | 27 | 24.1 ± 3.8 | 71.7 ± 1.1 | 10.6 ± 1.4 | 21 (78) | 8 | 15.4 ± 4.0 | 75.3 ± 2.8 | 8.6 ± 1.6 | 7 (88) |

AD, adolescent; YA, young adult; PM, postmenopausal woman; SBP, systolic blood pressure; DBP, diastolic blood pressure (in mmHg).
Adapted from F. Halberg et al. (1981a)

(Haus et al. 1988; Lakatua et al. 1987). Circannual rhythms have also been reported in the elderly for serum concentrations of albumin, bilirubin, calcium, chloride, CPK, globulin, glucose, LDH, potassium, sodium, triglycerides, uric acid, and total protein (Haus et al. 1988; Nicolau et al. 1984; Reilly et al. 1987). A positive correlation between the circadian mean of DBP and aldosterone was reported for the elderly; a negative correlation between the circadian means in norepinephrine excretion and SBP and DBP in the elderly contrasts with a positive correlation observed in children between these variables (Thompson et al. 1987).

In animal models, a decrease in the circadian amplitude of body core temperature gauged longitudinally by telemetry as a function of aging was found in one strain of rats, whereas in another strain, this amplitude decrease was accompanied by an acrophase advance (J. Halberg et al. 1981; Yunis et al. 1974). These changes with age simulate the alteration in circadian rhythm characteristics seen after the bilateral removal of the suprachiasmatic nuclei (F. Halberg et al. 1977; Powell et al. 1980). In human beings, a statistically significant phase advance of the circadian SBP rhyhtm with a numerically smaller amplitude is found in seven healthy elderly (mean age 64.0 ± 1.8 years, range 61–66 years) as compared to seven healthy young men (mean age 24.0 ± 3.1 years; range 20–29 years) (Otsuka et al. 1989). Circadian acrophase advance with increasing age in the elderly has also been reported for plasma cortisol (Haus et al. 1988, 1989; Milcu et al. 1978), dehydroepiandrosterone-sulfate (Haus et al. 1989), and protein bond iodine (Nicolau et al. 1979). In some human variables, but not in all, the circadian amplitude decreases with age (Tables 6,

7) (F. Halberg 1981; F. Halberg et al. 1981a; Haus et al. 1989; Nelson et al. 1980). Using spectrometry, Kane et al. (1986) reported "several dominant ultradian rhythms with periods of 47 min, 80 min, and 6–8 hours, the latter rhythms being present only in subjects aged over 60 [years]." In subjects over 90 years of age, Rossi et al. (1986) found a group circadian rhythm for heart rate by population – mean cosinor, but could not detect such a rhythm for premature beats by the same method; instead, a major component of 8 h characterized the premature atrial beats, and one of 4 h the ventricular beats. Thus, on the average, there is a frequency multiplication for ECG abnormality with advancing age.

A cross-sectional study involving the ambulatory monitoring of BP in 40 clinically healthy 20–60-year-old North American men did not reveal any statistically significant changes with age (F. Halberg et al. 1984a). Chronobiologic analyses of a large multicenter study involving the ambulatory monitoring of BP for 24 h (Scarpelli et al. 1988) showed that the relative prominence of ultradians (vs. the circadian component) increases with age for DBP (Anderson et al. 1989). A lesser ultradian increase in prominence, seen for SBP, was not yet detected as statistically significant. An increase in short-term vs. whole-day variability for both SBP and DBP was also reported for "hypertensive" patients (Tochikubo et al. 1987).

In centenarians, a circadian component persists for BP and heart rate (Stoynev et al. 1990). A decrease in the circadian quotient (derived from the percentage of the total variance contributed by the circadian rhythm) gauges primary aging, with a relative ultradian prominence (as compared to the

**Table 7.** Effects of age on circadian MESOR and amplitude of 12 plasma hormones in clinically healthy Minnesotan and Japanese women sampled around the clock in each of four menstrual stages (when applicable) and four seasons

| Hormone (units) | No. | Age[a] I | Age[a] II | Age[a] III | ANOVA[b] F | ANOVA[b] p | Contrast (II vs. III) t | Contrast (II vs. III) P |
|---|---|---|---|---|---|---|---|---|
| **A. MESOR** | | | | | | | | |
| LH (mlU/ml) | 27 | 14.9 | 18.6 | 80.1 | 74.80 | < 0.001 | | |
| Prolactin (ng/ml) | 29 | 22.9 | 19.5 | 12.7 | 10.37 | < 0.001 | | |
| E1 (pg/ml) | 27 | 75.0 | 94.9 | 38.3 | 14.66 | < 0.001 | | |
| E2 (pg/ml) | 26 | 81.3 | 113.2 | 17.0 | 14.52 | < 0.001[c] | 6.46 | 0.001 |
| 17-OH progesterone (pg/ml) | 29 | 602.0 | 716.0 | 158.0 | 22.50 | < 0.001[c] | 7.90 | < 0.001 |
| Cortisol (µg/dl) | 30 | 8.7 | 9.0 | 9.4 | 0.12 | | | |
| Aldosterone (ng/dl) | 25 | 6.1 | 6.6 | 4.9 | 2.16 | | | |
| DHEA-S (ng/ml) | 28 | 3090.0 | 1400.0 | 960.0 | 20.65 | 0.001[c] | 1.93 | |
| TSH (µlU/ml) | 29 | 3.3 | 6.4 | 5.0 | 1.51 | | | |
| T3 (ng/dl) | 28 | 94.7 | 98.7 | 101.7 | 0.35 | | | |
| T4 (µg/dl) | 29 | 6.8 | 7.0 | 7.2 | 0.15 | | | |
| Insulin (µU/ml) | 29 | 26.9 | 19.3 | 22.5 | 3.08 | | | |
| **B. Amplitude** | | | | | | | | |
| LH (mlU/ml) | 27 | 4.0 | 4.8 | 13.4 | 38.99 | < 0.001[c] | 7.13 | < 0.001 |
| Prolactin (ng/ml) | 29 | 12.4 | 16.5 | 11.0 | 4.34 | 0.027 | | |
| E1 (pg/ml) | 27 | 17.6 | 15.9 | 11.4 | 4.30 | 0.03[c] | | |
| E2 (pg/ml) | 26 | 28.0 | 28.4 | 8.1 | 10.00 | < 0.001[c] | 5.30 | 0.001 |
| 17-OH progesterone (pg/ml) | 29 | 181.0 | 196.0 | 128.0 | 10.00 | < 0.001 | | |
| Cortisol (µg/dl) | 30 | 7.1 | 7.5 | 7.0 | 1.23 | | | |
| Aldosterone (ng/dl) | 25 | 4.1 | 2.5 | 1.8 | 11.49 | < 0.001[c] | 2.36 | 0.056 |
| DHEA-S (ng/ml) | 28 | 580.0 | 370.0 | 230.0 | 19.62 | < 0.001 | 3.42 | 0.006 |
| TSH (µlU/ml) | 29 | 0.73 | 1.95 | 1.04 | 2.48 | | | |
| T3 (ng/dl) | 28 | 12.9 | 12.5 | 13.2 | 0.17 | | | |
| T4 (µg/dl) | 29 | 0.58 | 0.59 | 0.67 | 2.01 | | | |
| Insulin (µU/ml) | 29 | 20.5 | 15.6 | 18.0 | 2.85 | | | |

Adapted from F. Halberg et al. (1981) and from W. Nelson et al. (1980).
ANOVA, analysis of variance; F, value taken by F test in ANOVA; P, corresponding $p$ value; LH, luteinizing hormone; E1, estrone; E2, estradiol; DHEA-S, dehydroepiandrosterone sulfate; TSH, thyrotropic hormone; T3, triiodothyronine; T4, thyroxine.
Each subject contributed one value, her average circadian MESOR, to the analysis.
[a]  I, II and III indicate mean values for ages 15–21, 29–36, and 44–59 years (postmenopausal women), respectively
[b]  Only $p$ values < 0.05 are indicated
[c]  Inhomogeneity of variance ($p < 0.05$)

circadian) in both SBP and DBP, but not (yet?) in heart rate (Hillman et al., to be published; Ikonomov et al. 1990, to be published). A change with age is also seen from an overall lead to a lag of heart rate vs. the acrophase of DBP (Hillman et al., to be published). That the DBP "ages" more rapidly than heart rate (if the variance transposition from the circadian to the ultradian spectral domain is accepted as a marker of aging) is also suggested by the fact that the 12-h and/or 8-h component peak in the spectrum of DBP, whereas the 24-h component is the most prominent feature of the heart rate spectrum.

## Conclusion

The naked eye can detect neither histologic cell structure nor the time "microscopic" structure of biologic variability. Computer software can resolve the temporal structure in seemingly irregular data. Systematic data collection as a function of time allows a chronobiologic interpretation usually based on age trends and/or a multifrequency rhythm spectrum. Whereas automatic instrumentation greatly helps the measurement of BP, among other physiologic functions, manual self-measurements are also practical, provided their limitations are understood and the subjects studied by autorhythmometry are evaluated longitudinally. The reference standards are then derived from the subject's own BP history and the age change mapped on a peer group studied by comparable methodology (with only peer-group reference standards being available at the outset of a study).

The major challenge today is to reach the intelligent lay public through education. At little cost for sphygmomanometers and/or other high school science equipment, the early initiation in chronobiology becomes part of self-help for health literacy. The teaching of self-help is practical and a necessity in many fields, as in the case of diabetes, for instance. The concern about BP deviation is another field that could greatly benefit from self-help for health care. Chronobiologic BP screening is recommended for everyone, starting in the family, during the education of the pregnant woman, and in earliest schooling. Reference standards for different age groups for the interpretation of single BP values are already available and have to be updated, improved, and complemented by personalized reference standards as outcomes are assessed. Rhythm charactersitics as new endpoints can be exploited to better recognize early rhythm alteration indicative of a heightened cardiovascular disease risk (Cornélissen et al. 1990c). Once risk is recognized, preventive intervention can be initiated with or without antihypertensive medication. A whole multifrequency spectrum of rhythms, to be quantified at birth (Cornélissen et al. 1990a; F. Halberg et al. 1990b) awaits regular scrutiny as to its merits in assessing as early as possible the individualized risk of developing high BP later in life. Indeed, ongoing research suggests that prevention of high BP rests in the hands of obstetricians and neonatologists (F. Halberg et al. 1990e).

Since in many cases the chronological approach to risk assessment has been validated by questionnaire, one should consider whether the questionnaire may not be simpler and more cost-effective than chronobiologic monitoring and interpretation. The argument that information concerning family history may not be available in certain cases does not hold for a majority of individuals, even if it may concern quite a few in times of war when parents die on the battlefield prior to the overt manifestation of cardiovascular or other diseases. What is critical is the fact that family history cannot be altered whereas a chronobiologic index can constitute a gauge of a desired or undesired effect of drug or nondrug intervention. A case in point is the effect of betamimetics, drugs that are in themselves invaluable in preventing premature labor. That the benefit during pregnancy must be viewed in the context of possible long-term deleterious effects that may be apparent only in late maturity must be rigorously considered. Indeed, a chronobiologic approach does reveal late betamimetic effects on the cardiovascular rhythm of the offspring. It is up to future research with outcomes to find the percentage of false positives and of false negatives in terms of the diagnosis of a high risk state when one relies on questionnaires on the one hand or on the chronobiologic approach on the other hand. This is one of the major purposes of a womb-to-tomb study of the BP and heart rate chronome, with a view of quantifying health as a state of low cardiovascular disease risk (among other disease risks).

# References

Anderson S, Cornélissen G, Halberg F, Scarpelli PT, Cagnoni S, Germanó G, Livi R, Scarpelli L, Cagnoni M, Holte JE (1989) Age effects upon the harmonic structure of human blood pressure in clinical health. Proc 2nd Ann IEEE Symp on Computer-Based Medical Systems, Minneapolis, June 26–27, 1989. Computer Society Press, Washington, DC, pp 238–243

Australian National Blood Pressure Study Management Committee (1980) The Australian therapeutic trial in mild hypertension. Lancet i: 1261–1267

Balsano F (1988) Opening remarks. J Hypertens 6 (Suppl 1): S1

Bingham C, Arbogast B, Cornélissen Guillaume G, Lee JK, Halberg F (1982) Inferential statistical methods for estimating and comparing cosinor parameters. Chronobiologia 9: 397–439

Blankenhorn DH (1975) Evidence for regression/progression of atherosclerosis in man. In: Proc Int Workshop Conference on Atherosclerosis, London, Ont

Borhani NO (1988) Isolated systolic hypertension in the elderly. J Hypertens 6 (Suppl 1) S15–S19

Brown GE (1930) Daily and monthly rhythm in the blood pressure of a man with hypertension: a three-year study. Ann Intern Med 3: 1177–1189

Buchwald H, Moore RB, Varco RL (1974) The partial ileal bypass operation in treatment of hyperlipemias. Adv Exp Med Biol 63: 221–230

Cabri J, De Witte B, Clarys JP, Reilly T, Strass D (1988) Circadian variation in blood pressure responses to muscular exercise. Ergonomics 31: 1559–1565

Cornélissen G (1987) Instrumentation and data analysis methods needed for blood pressure monitoring in chronobiology. In: Scheving LE, Halberg F, Ehret CF (eds) Chronobiotechnology and Chronobiological Engineering. Nijhoff, Dordrecht, pp 241–261

Cornélissen G (1990) Challenge to the statistician interested in the individualized assessment of health in the inner cities and urban areas: a tribute to Franz Halberg. (Health of Inner Cities and Urban Areas, International Conference, Cardiff, Wales, September 4–7, 1989). Statistician 39: 105–109

Cornélissen G, Halberg F, Tuna N, Hanson B, Bouchard T Jr (1985) Heritability of circadian characteristics of heart rate (HR): implications for pacing and medications. XVII Int Conf Int Soc Chronobiol, Little Rock, Ark, Nov 3–6, 1985. Chronobiologia 12: 238

Cornélissen G, Wilson D, Halberg F, Ferencz F (1986) Age and sex effects upon circadian characteristics in children of parents with and without a high blood pressure. Int Symp Chronobiology and VI Conf Indian Soc Chronobiology, Osmania University, Hyderabad, India, Nov 19–21, 1986, pp 25–27

Cornélissen G, Halberg F, Tarquini B, Mainardi G, Panero C, Cariddi A, Sorice V, Cagnoni M (1987) Blood pressure rhythmometry during the first week of human life. In: Tarquini B, Vergassola R (eds) Social diseases and chronobiology: Proc III Int. Symp. social diseases and chronobiology, Florence, Nov 29, 1986. Esculapio, Bologna, pp 113–122

Cornélissen G, Scarpelli PT, Halberg F, Halberg J, Halberg Francine, Halberg E (1988) Cardiovascular rhythms: their implications and applications in medical research and practice. In: Hekkens WTJM, Kerkhof GA, Rietveld WJ (eds) Trends in Chronobiology. Pergamon, Oxford, pp 335–355

Cornélissen G, Kopher R, Brat P, Rigatuso J, Work B, Eggen D, Einzig S, Vernier R, Halberg F (1989) Chronobiologic ambulatory cardiovascular monitoring during pregnancy in group health of Minnesota. Proc 2nd Ann IEEE Symp on computer-based medical systems. Minneapolis, June 26–27, 1989. Computer Society, Washington DC, pp 226–237

Cornélissen G, Sitka U, Tarquini B, Mainardi G, Panero C, Cugini P, Weinert D, Romoli F, Cassanas G, Maggioni C, Vernier R, Work B, Einzig S, Rigatuso J, Schuh J, Kato J, Tamura K, Halberg F (1990a) Chronobiologic approach to blood pressure during pregnancy and early extrauterine life. In: Hayes DK, Pauly JE, Reiter RJ (eds) Chronobiology: its role in clinical medicine, general biology, and agriculture, part A. Wiley-Liss, New York, pp 585–594

Cornélissen G, Halberg F, Montalbetti N, Lakatua DJ, Haus E, Hermida R, Uezono K, Kawasaki T, Omae T, Tarquini B (1990b) Clinical chemistry chronobiometry improves averages as MESORs, transforming confusing variation into new, sensitive, useful parameters. Clin Chem 36: 930–931

Cornélissen G, Bakken E, Delmore P, Orth-Gomér K, Åkerstedt T, Carandente O, Carandente F, Halberg F (1990c) From various kinds of heart rate variability to chronocardiology. Am J Cardiol 66: 863–868

Cornélissen G, Halberg F, Kopher R, Kato J, Maggioni C, Tamura K, Otsuka K, Miyake Y, Ohnishi M, Satoh K, Rigó J Jr, Paulin F, Adam Z, Zaslavskaya RM, Work B, Carandente F (1991) Halting steps in Minnesota toward international blood pressure (BP) rhythm-specified norms (chronodesms) during pregnancy. Chronobiologia 18: 72–73

Croppi E, Livi R, Scarpelli L, Romano S, de Leonardis V., Cagnoni M, Scarpelli PT (1986) Chronobiologically assessed blood pressure (BP) and left ventricular wall thickness in children with and without a family history of high blood pressure. In: Tarquini B, Vergassola R (eds) Social diseases and chronobiology: Proc III Int Symp Social diseases and chronobiology, Florence, Nov 29, 1986. Esculapio, Bologna, pp 55–56

del Pozo F, Rodriguez MJ, Arredondo MT, Otsuka K, Gómez E, Halberg F (1991) Decimation of ambulatory blood pressure (BP) series. Proc Assoc Adv Med Instr Washington, DC, May 11–15, 1991 p 29

Falconer W (1797) Beobachtungen über den Puls. Heinsius, Leipzig

Ferencz C, Brenner JI, Dischinger PC, Wilson PD (1986) Blood pressure variability in adolescents of hypertensive and normotensive parents. JACC 7: 164 A

Germanó G (ed) (1988) Conflicting aspects in the clinical approach to hypertension. Italian Department of Public Health

Project on blood pressure variability, Monte Cassino, Italy, Oct 21–22, 1988

Güllner HG, Bartter FC, Halberg F; Delea C (1979) Circadian temperature and blood pressure rhythms guide timed optimization and gauge antimesorhypertensive prazosin effects. Chronobiologia 6: 105

Halberg E, Halberg F, Shankaraiah K (1981a) Plexo-serial linear-nonlinear rhythmometry of blood pressure, pulse and motor activity by a couple in their sixties. Chronobiologia 8: 351–366

Halberg E, Halberg J, Halberg Francine, Sothern RB, Levine H, Halberg F (1981b) Familial and individualized longitudinal autorhythmometry for 5 to 12 years and human age effects. J Gerontol 36: 31–33

Halberg E, Delmore P, Finch M, Cornélissen G, Halberg F (1990) Chronobiologic assessment of deviant human blood pressure: an invitation for improvements – In: Hayes DK, Pauly JE, Reiter RJ (eds) Chronobiology: its role in clinical medicine, general biology, and agriculture, part A. Wiley-Liss, New York, pp 305–318

Halberg F (1953) Some physiological and clinical aspects of 24-hour periodicity. Lancet (USA) 73: 20–32

Halberg F (1963) Periodicity analysis: a potential tool for biometeorologists. Int J Biometeorol 7: 167–191

Halberg F (1970) Chronobiology and the delivery of health care. In: O'Leary J (ed) A systems approach to the application of chronobiology in family practice. Health Care Research Program, Department of Family Practice and Community Health, University of Minnesota pp 31–96

Halberg F (1981) Biologic rhythms, hormones and aging. In: Vernadakis A, Timiras PS (eds) Hormones in development and aging. Spectrum, New York, pp 451–476

Halberg F (1983) Quo vadis basic and clinical chronobiology: promise for health maintenance. Am J Anat 168: 543–594

Halberg F (1984) Preface. In: Haus E, Kabat H (eds) Chronobiology 1982–83. Karger, Basel, pp V–VIII

Halberg F, Good RA, Levine H (1966) Some aspects of the cardiovascular and renal circadian system. Circulation 34: 715–717

Halberg F, Johnson EA, Nelson W, Runge W, Sothern R (1972) Autorhythmometry – procedures for physiologic self-measurements and their analysis. Physiol Teacher 1: 1–11

Halberg F, Halberg J, Halberg Francine, Halberg E (1973) Reading, 'riting, 'rithmetic – and rhythms: a new 'relevant' 'R' in the educative process. Perspect Biol Med 17: 128–141

Halberg F, Haus E, Ahlgren A, Halberg E, Strobel H, Angellar A, Kühl JFW, Lucas R, Gedgaudas E, Leong J (1974) Blood pressure self-measurement for computer-monitored health assessment and the teaching of chronobiology in high schools. In: Scheving LE, Halberg F, Pauly JE (eds) Chronobiology. Proc Int Soc for the Study of Biological Rhythms, Little Rock, Ark. Thieme Stuttgart pp 372–378

Halberg F, Powell EW, Lubanovic W, Sothern RB, Brockway B, Pasley RN, Scheving LE (1977) The chronopathology and experimental as well as clinical chronotherapy of emotional disorders. In: Halberg F, Carandente F, Cornélissen G, Katinas GS (eds) Glossary of chronobiology. Chronobiologia 4, Suppl 1: 189

Halberg F, Lee JK, Nelson WL (1978) Time-qualified reference intervals – chronodesms. Experientia 34: 713–716

Halberg F, Lubanovic WA, Sothern RB, Brockway B, Powell EW, Pasley JN, Scheving E (1979) Nomifensine chronopharmacology, schedule-shifts and circadian temperature rhythms in di-suprachiasmatically lesioned rats: modeling emotional chronopathology and chronotherapy. Chronobiologia 6: 405–424

Halberg F, Cornélissen G, Sothern RB, Wallach LA, Halberg E, Ahlgren A, Kuzel M, Radke A, Barbosa J, Goetz F, Buckley J, Mandel J, Schuman L, Haus E, Lakatua D, Sackett L, Berg H, Wendt HW, Kawasaki T, Ueno M, Uezono K, Matsuoka M, Omae T, Tarquini B, Cagnoni M, Garcia Sainz M, Perez Vega E, Wilson D, Griffiths K, Donati L, Tatti P, Vasta M, Locatelli I, Camagna A, Lauro R, Tritsch G, Wetterberg L (1981a) International geographic studies of oncological interest on chronobiological variables. In: Kaiser H (ed) Neoplasms. Comparative pathology of growth in animals, plants and man. Williams and Wilkins, Baltimore, pp 553–596

Halberg F, Hayes DK, Powell EW, Scheving LE (1981b) Preface. In: Halberg F, Scheving LE, Powell EW, Hayes DK (eds) Chronobiology, Proc XIII Int Conf Int Soc Chronobiol, Pavia, Italy, September 4–7, 1977. II Ponte, Milan, pp v–x

Halberg F, Drayer JIM, Cornélissen G, Weber MA (1984a) Cardiovascular reference data base for recognizing circadian mesor- and amplitude-hypertension in apparently healthy men. Chronobiologia 11: 275–298

Halberg F, Scheving LE, Lucas E, Cornélissen G, Sothern RB, Halberg E, Halberg J, Halberg Francine, Carter J, Straub KD, Redmond DP (1984b) Chronobiology of human blood pressure in the light of static (room-restricted) automatic monitoring. Chronobiologia 11: 217–247

Halberg F, Halberg E, Carandente F, Cornélissen G, März W, Halberg J, Drayer J, Weber M, Schaffer E, Scarpelli P, Tarquini B, Cagnoni M, Tuna N (1986a) Dynamic indices from blood pressure monitoring for prevention, diagnosis and therapy. In: ISAM 1985, Proc Int Symp Ambulatory Monitoring, Padua, March 29–30, 1985. CLEUP, pp 205–219

Halberg F, Haus E, Halberg E, Cornélissen G, Scarpelli P, Tarquini B, Cagnoni M, Wilson D, Griffiths K, Simpson H, Balestra E, Reale L (1986b) Chronobiologic challenges in social medicine: illustrative tasks in cardiology and oncology. In: Halberg F, Reale L, Tarquini B (eds) Proc 2nd Int Conf Medico-Social Aspects of Chronobiology. Florence, Oct 2, 1984 Istituto Italiano di Medicina Sociale, Rome, pp 13–42

Halberg F, Cornélissen G, Bingham C, Tarquini B, Mainardi G, Cagnoni M, Panero C, Scarpelli P, Romano S, März W, Hellbrügge T, Shinoda M, Kawabata Y (1986c) Neonatal monitoring to assess risk for hypertension. Postgrad Med 79: 44–46

Halberg F, Cornélissen G, Wilson D, Ferencz C (1986d) Circadian systolic amplitude separates boys of parents with or without a high blood pressure. Int Symp Chronobiology and VI Conf Indian Soc Chronobiology, Osmania University, Hyderabad, India, Nov 19–21, 1986, pp 23–24

Halberg F, Kausz E, Winter Y, Wu J, März W, Cornélissen G (1986e) Circadian rhythmic response in cold pressor test. J Minn Acad Sci 51: 14

Halberg F, Cornélissen G, Halberg E, Halberg J, Delmore P, Bakken E, Shinoda M (1988a) Chronobiology of human blood pressure. Medtronic continuing medical education seminars, 4th edn

Halberg F, Maggioni C, Cornélissen G, Cariddi A, Mainardi G, Tarquini B, Panero C, Sorice V, Romuli F, Cagnoni M (1988b) Intrauterine exposure to β2-adrenergic agonist or corticosteroid affects neonatal circadian blood pressure amplitude: diethylstilbestrol revisited? Chronobiologia 15: 265–266

Halberg F, Hillman D, Halberg E (1988c) Circaseptan (about 7-day) cardiovascular adjustment after transmeridian round-trip (west-east-west vs. east-west-east) flights. Chronobiologia 15: 247–249

Halberg F, Bakken E, Cornélissen G, Halberg J, Halberg E, Delmore P (1989) Blood pressure assessment with a cardiovascular summary, the sphygmochron, in broad chronobiologic perspective. In: Refsum H, Sulg JA, Rasmussen K (eds) Heart and Brain, Brain and Heart. Springer, Berlin Heidelberg New York pp 142–162

Halberg F, Bakken E, Cornélissen G, Halberg J, Halberg E, Wu J, Sánchez de la Peña S, Delmore P, Tarquini B (1990a) Chronobiologic blood pressure assessment with a cardiovascular summary, the sphygmochron. In: Meyer-Sabellek W, Anlauf M, Gotzen R, Steinfeld L (eds) Blood pressure measurements. Steinkopff Darmstadt, pp 297–326

Halberg F, Cornélissen G, Bakken E (1990b) Caregiving merged with chronobiologic outcome assessment, research and education in health maintenance organizations (HMOs). In: Hayes DK, Pauly JE, Reiter RJ (eds) Chronobiology: its role in clinical medicine, general biology, and agriculture, part B. Wiley-Liss, New York, pp 491–549

Halberg F, Cornélissen G, Kopher R, Choromanski L, Eggen D, Otsuka K, Bakken E, Tarquini B, Hillman DC, Delmore P, Kawabata Y, Shinoda M, Vernier R, Work B, Cagnoni M, Cugini P, Ferrazzani S, Sitka U, Weinert D, Schuh J, Kato J, Kato K, Tamura K (1990c) Chronobiologic blood pressure and ECG assessment by computer in obstetrics, neonatology, cardiology and family practice. In: Computers and perinatal medicine: proc 2nd World Symp Computers in the Care of the Mother, Fetus and Newborn. Kyoto, Japan, Oct 23–26, 1989. Excerpta Medica, Amsterdam, pp 3–18

Halberg F, Cornélissen G, Halberg J, Bakken E, Delmore P, Wu J, Sánchez de la Peña S, Halberg E (1990d) The sphygmochron for blood pressure and heart rate assessment: a chronobiologic approach. In: Miles LE, Broughton RJ (eds) Medical monitoring in the home and work environment. Raven, New York, pp 85–98

Halberg F, Cornélissen G, Bakken E, Delmore P, Tamura K (1990e) Sphygmochron: chronobiologic assessment of self- or automatically, preferably ambulatorily, measured blood pressure and heart rate. In: Birkenhäger WH, Halberg F, Prikryl P (eds) Proc Int Symp on hypertension, Brno, Czechoslovakia, April 9–10, 1990, Masaryk University, Brno, pp 4–18f.

Halberg F, Wang Z, Cornélissen G, Bingham C, Rigatuso J, Hillman D, Wakasugi K, Kato K, Kato J, Tamura K, Sitka U, Weinert D, Schuh J, Coleman JM, Mammel M, Miyake Y, Ohnishi M, Satoh K, Watanabe Y, Otsuka K, Watanabe H, Johnson D (1991) SIDS and about-weekly patterns in vital signs of premature babies. In: Yoshikawa M, Uono M, Tanabe H, Ishikawa S (eds) New Trends in Autonomic Nervous System Research: Basic and Clinical Interpretations, Selected Proc 20th Int Cong Neurovegetative Research, Tokyo, Sept 10–14, 1990. Excerpta Medica, Amsterdam, pp 581–585

Halberg F, Mikulecky M, Zaslavskaya RM, Suslov MG, Teibloom MM, Hillman DC, Wang Z, Wu J, Hodkova M, Uezono K, Sonkowsky R, Haus E, Bakken E, Cornélissen G (to be published b) Can we map the genome without mapping the chronome? Re: nature vs. nurture in phylogeny, ontogeny, health and the environment. In: Mikulecky M, Horecky J, Cornélissen G, Halberg F (eds) Proc Int Symp biorhythms in clinical medicine. Bratislava, Czecho-Slovakia, Sept 4–5, 1990

Halberg F, Cornélissen G, Kopher R, Kato J, Maggioni C, Tamura K, Otsuka K, Miyake Y, Ohnishi M, Satoh K, Rigó J Jr, Paulin F, Adam Z, Work B, Carandente F, Zaslavskaya RM (1991c) Chronobiology of blood pressure (BP) in uncomplicated pregnancy vs. gestational hypertension (GH) or pree-clampsia (PE). Chronobiologia 18: 73–74

Halberg F, Halberg E, Yatsyk GV, Syutkina EV, Safin SR, Grigoriev AE, Abramian AS (1991) Development of blood pressure and heart rate rhythms and effects of betamimetic drugs. Biull Eksp Biol Med 8: 202–205

Halberg F, Halberg E, Halberg J, Ikonomov O, Otsuka K, Holte J, Tamura K, Saito Y, Hata Y, Uezono K, Wang ZR, Xue ZN, del Pozo F, Hillman DC, Samayoa W, Bakken E, Cornélissen G (1991e) Womb to tomb blood pressure (BP) monitoring: are single or even 24-hour measurements enough? Proc Assn Adv Med Instr, Washington, DC, May 11–15, 1991, p 38

Halberg FE, Halberg E, Halberg J, Southern RB, Halberg F (1985) Inter-individual cardiovascular differences before and around menarche. XVII Int Conf Int Soc Chronobiol, Little Rock, 3–6 Nov 1985. Chronobiologia 12: 2 48

Halberg J, Halberg E, Regal P, Halberg F (1981) Changes with age characterize circadian rhythms in telemetered core temperature of stroke-prone rats. J Gerontol 36: 28–30

Halberg J, Halberg F, Leach CN (1984) Variability of human blood pressure with reference mostly to the non-chronobiologic literature. Chronobiologia 11: 205–216

Halberg J, Halberg E, Halberg Francine, Sothern RB, Halberg F (1985) Systolic but not diastolic blood pressure, heart rate or oral temperature rise at menarche. XVII Int Conf Int Soc Chronobiol, Little Rock, Ark, November 3–6, 1985. Chronobiologia 12: 249

Hanson BR, Halberg F, Tuna N, Bouchard TJ Jr, Lykken DT, Cornélissen G, Heston LL (1984) Rhythmometry reveals heritability of circadian characteristics of heart rate of human twins reared apart. Cardiologia 29: 267–282

Haus E (1989) Study of human biologic rhythms can advance health care. Ramsey Med Bull 3 (3)

Haus E, Nicolau G, Lakatua DJ, Sackett-Lundeen L (1988) Reference values for chronopharmacology. Annu Rev Chronopharm 4: 333–424

Haus E, Nicolau G, Lakatua DJ, Sackett-Lundeen L, Petrescu E (1989) Circadian rhythm parameters of endocrine functions in elderly subjects during the seventh to the ninth decade of life. Chronobiologia 16: 331–352

Haus E, Haus M Sr, Cornélissen G, Wu J, Halberg F (to be published) Circadian desynchronization of elderly blood pressure from heart rate and resynchronization by familial proximity. Proc VIII Biennial Meeting, Ind Soc Chronobiol, Raipur, India, Nov 24–26, 1990. Biochim Clin

Haus R, Halberg E, Haus E, Halberg F, Haus A, Halcomb A, Cornélissen G (1987) Circadian urinary characteristics of adolescents: sensitive dynamic indices complement mean values as new physiologic endpoints. In: Pauly JE, Scheving LE (eds) Advances in chronobiology, Part B, Proc XVII Int Conf Int Soc Chronobiol, Little Rock, Ark, USA, Nov 3–7, 1985. Liss, New York pp 21–30

Hillman D, Halberg E, Halberg F (1988) More on about-weekly (circaseptan) cardiovascular variation after transmeridian (7-h shift of schedule) east-west-east flights. Chronobiologia 15: 249–250

Hillman DC, Halberg E, Ikonomov O. Stoynev A, Madjirova N, Cornélissen G, Halberg F (to be published) The ultradian blood pressure (BP) and heart rate (HR) chronome of centenarians vs. that of young adults. Proc VIII Biennial Meeting, Ind Soc Chronobiol, Raipur, India, Nov 24–26, 1990. Biochim Clin

Ikonomov O, Hillman DC, Stoynev A, Madjirova N, Germanò

G, Scarpelli PT, Cornélissen G, Halberg E, Halberg F (1990) Ultradian and circadian aspects of the centenarian blood pressure (BP) and heart rate (HR) chronome. Abstract, 2nd Int Cong Hypertension in the Elderly: Focus on Risk Factors. Rome, December 5–7, 1990, p 86

Ikonomov O, Stoynev G, Cornélissen G, Stoynev A, Hillman D, Madjirova N, Halberg F (to be published) The chronobiology of blood pressure and heart rate in centenarians.

Johansson BB, Norrving B, Widner H, Wu J, Halberg F (1990) Stroke incidence: circadian and circaseptan (about-weekly) variations in onset. In: Hayes DK, Pauly JE, Reiter RJ (eds) Chronobiology: its role in clinical medicine, general biology, and agriculture, part A. Wiley-Liss, New York, pp 427–436

Julius S, Jamerson K, Mejia A, Krause L, Schork N, Jones K (1990) The association of borderline hypertension with target organ changes and higher coronary risk. Tecumseh blood pressure study. JAMA 264: 354–358

Kane RL, Kiersch ME, Yates FE, Benton L, Solomon DH, Satz P, Beck JC (1986) Dynamic assessment of cognitive and cardiovascular performance in the elderly. Isr J Med Sci 22: 225–230

Katz LN, Stamler J, Pick R (1958) Nutrition and atherosclerosis. Lea and Febiger, Philadelphia

Kawamoto A, Shimada K, Matsubayashi K, Chikamori T, Kuzume O, Ogura H, Ozawa T (1989) Cardiovascular regulatory functions in elderly patients with hypertension. Hypertension 13: 401–407

Kumagai Y, Fujimura A, Sugimoto K, Ebihara A (1989) Effect of a beta-blocker, atenolol, on circadian rhythm of blood pressure. J Jap Soc Clin Pharmacol 20: 415–419

Lakatua DJ, Nicolau GY, Bogdan C, Plinga L, Jachimowitz A, Sackett-Lundeen L, Petrescu E, Ungureanu E, Haus E (1987) Chronobiology of catecholamine excretion in different age groups. In: Pauly JE, Scheving LE (eds). Advances in chronobiology, part B. Liss, New York, pp 31–50

LaSalle D, Sothern RB, Halberg F (1983) Sampling requirements for description of circadian blood pressure (BP) amplitude (A). Chronobiologia 10: 138

Lee JY, Gillum RF, Cornélissen G, Koga Y, Halberg F (1982) Individualized assessment of circadian rhythm characteristics of human blood pressure and pulse after moderate salt and weight restriction. In: Takahashi R, Halberg F, Walker (eds) Toward Chronopharmacology. Pergamon, Oxford

Levine H, Halberg F (1972) Circadian rhythms of the circulatory system. Literature review. Computerized case study of transmeridian flight and medication effects on a mildly hypertensive subject. US Air Force Report SAM-TR-72-3, April 1972

Levine H, Saltzman W, Yankaskas J, Halberg F (1977) Circadian state-dependent effect of exercise upon blood pressure in clinically healthy men. Chronobiologia 4: 129–130

Little J, Sánchez de la Peña S, Cornélissen G, Abramowitz P, Tuna N, Halberg F (1990) Longitudinal chronobiologic blood pressure monitoring for assessing the need and timing of antihypertensive treatment. In: Hayes DK, Pauly JE, Reiter RJ (eds) Chronobiology: its role in clinical medicine, general biology, and agriculture, part B. Wiley-Liss, New York, pp 601–611

Marazzi A, Wang Z, Paccaud F, Hillman DC, Cornélissen G, Halberg F (1990) Circadian, circaseptan and secular variation in human birth and stillbirth. Abstract, 2nd World Conf. on Clinical Chronobiology, Monte Carlo, April 10–13, 1990 Chronobiologia 17: 177

Marler MR, Jacob RG, Lehoczky JP, Shapiro AP (1988) The

statistical analysis of treatment effects in 24-hour ambulatory blood pressure recordings. Stat Med 7: 697–716

Master AM, Jaffee HL (1952) Factors in the onset of coronary occlusion and coronary insufficiency. JAMA 148– 794–798

McDonald K, Sánchez de la Peña S, Cavallini M, Olivari MT, Cohn JN, Halberg F (1990) Chronobiologic blood pressure and heart rate assessment of patients with heart transplants. In: Hayes DK, Pauly JE, Reiter RJ (eds) Chronobiology: its role in clinical medicine, general biology, and agriculture, part B. Wiley-Liss, New York, pp 471–480

Milcu SM, Bogdan CM, Nicolau GY, Cristea AL (1978) Cortisol circadian rhythm in 70-100-year-old subjects. Rev Roum Med Endocrinol 16: 29–39

Miyamoto S, Shimokawa H, Sumioki H, Touno A, Nakano H (1988) Circadian rhythm of plasma atrial natriuretic peptide, aldosterone and blood pressure during the third trimester in normal and preeclamptic pregnancies. Am J Obst Gynecol 158: 393–399

Mueller SC, Brown GE (1930) Hourly rhythms in blood pressure in persons with normal and elevated pressures. Ann Intern Med 3: 1190–1200

Muller JE, Stone PH, Turi SG, Rutherford JD, Czeisler CA, Parker C, Poole WK, Passamani E, Roberts R, Robertson T, Sobel BE, Willerson JT, Braunwald E, MILIS Study Group (1985) Circadian variation in the frequency of onset of acute myocardial infarction. N Engl J Med 313: 1315–1322

Muller JE, Ludmer PL, Willich SN, Tofler JH, Aylmer G, Klangos I, Stone PI (1987) Circadian variation in the frequency of sudden cardiac death. Circulation 75: 131–138

Nelson W, Bingham C, Haus E, Lakatua DJ, Kawasaki T, Halberg F (1980) Rhythm-adjusted age effects in a concomitant study of twelve hormones in blood plasma of women. J Gerontol 35: 512–519

Nelson W, Cornélissen G, Hinkley D, Bingham C, Halberg F (1983) Construction of rhythm-specified reference intervals and regions, with emphasis on 'hybrid' data, illustrated for plasma cortisol. Chronobiologia 10: 179–193

Nicolau GY, Bogdan C, Cristea AL, Milcu SM (1979) La variation circadienne de la concentration de l'iode protéique plasmatique (PBI) chez les vieillards de 70–100 ans. Rev Roum Med Endocrinol 17: 119–126

Nicolau GY, Lakatua D, Sackett-Lundeen L, Haus E (1984) Circadian and circannual rhythms in hormonal variables in elderly men and women. Chronobiol Int 1: 301–319

Otsuka K, Kitazumi T, Matsubayashi K, Kawamoto A, Sadakane N, Chikamori T, Kuzume O, Shimada K, Ogura H, Ozawa T (1989) Age-related alterations in the circadian pattern of blood pressure. Am J Noninvas Cardiol 3: 159–165

Otto W, Hempel WE, Wagner CV, Bert A (1982) Einige periodische und aperiodische Variationen der Herzinfarkt Sterblichkeit in der DDR. Z Gesamte Inn Med 37: 756–763

Pötschke-Langer M, Apfelbach J (1986) Milde Hypertonie. Auf dem Wege zu einem internationalen und nationalen Konsens. Fortschr Med 44: 852–856

Powell EW, Halberg F, Pasley JN, Lubanovic W, Ernsberger P, Scheving LE (1980) Suprachiasmatic nucleus and circadian core temperature rhythm in the rat. J Therm Biol 5: 189–196

Rabatin JS, Sothern RB, Halberg F, Brunning RD, Goetz FC (1981) Circadian rhythms in blood and self-measured variables of ten children, 9 to 14 years of age. In: Halberg F, Scheving LE, Powell EW, Hayes DK (eds) Chronobiology. Proc XIII Int Conf Int Soc Chronobiol, Pavia, Italy, September 4–7, 1977. Il Ponte, Milan, pp 373–385

Redman CWG, Beilin LJ, Bonnar J (1976) Reversed diurnal blood pressure rhythm in hypertensive pregnancies. Clin Sci Mol Med 51: 687s–689s

Reeves RA, Johnson AM, Shapiro AP, Traub YM, Jacob R (1984) Ambulatory blood pressure monitoring: methods to assess severity of hypertension, variability and sleep changes. In: Weber MA, Drayer JIM (eds) Ambulatory blood pressure monitoring. Springer Berlin Heidelberg New York pp 27–34

Reilly C, Nicolau GY, Lakatua DJ, Bogdan C, Sackett-Lundeen L, Petrescu E, Haus E (1987) Circannual rhythms of laboratory measurements in serum of elderly subjects. In: Pauly JE, Scheving LE (eds) Advances in chronobiology, part B: Liss, New York, pp 51–72

Reinberg A, Gervais P, Halberg F, Gaultier M, Poynette N, Abulker C, Dupont J (1973) Mortalité des adultes: rythmes circadiens et circannuels. Nouv Presse Med 2: 289–294

Rigó J Jr, Paulin F, Adám Z, Halberg F (to be published) Unacceptable cardiovascular disease risk, notably in pregnancy, with conventionally acceptable blood pressure. Mikulecky M, Horecky J, Cornélissen G, Halberg F (eds) Proc Int Symp biorhythms in clinical medicine. Bratislava, Czecho-Slovakia, Sept 4–5, 1990

Rossi A, Storza C, Carandente F (1986) Heart rate and extrasystolic arrhythmias in active subjects aged over 90. A chronobiological study. Chronobiologia 13: 309–318

Saito Y, Mukaiyama S, Ishii H, Komaya T, Tamura K (1990) Clinical evaluation of chronobiological blood pressure data validated by echocardiogram. In: Hayes DK, Pauly JE, Reiter RJ (eds). Chronobiology: its role in clinical medicine, general biology, and agriculture, part A. Wiley-Liss, New York, pp 319–324

Scarpelli PT, Romano S, Livi R, Scarpelli L, Cornélissen G, Cagnoni M, Halberg F (1987) Instrumentation for human blood pressure rhythm assessment by self-measurement. In: Scheving LE, Halberg F, Ehret CF (eds). Chronobiotechnology and chronobiological engineering. Nijhoff, Dordrecht, pp 141–188

Scarpelli PT, Livi R, Cagnoni S, Croppi E, Scarpelli L, Pieri A, Cornélissen G, Halberg F, Corsi V, Germanò G (1988) Italian multicentric study on blood pressure variability. Chronobiological reference limits and analysis of deviant subjects. Proc III Meeting on 'Conflicting aspects in the clinical approach to hypertension'. Monte Cassino, Oct 21–22, 1988

Scheving LE, Shankaraiah K, Halberg F, Halberg E, Pauly JE (1982) Self-measurements taught and practiced in public high schools in Little Rock, Arkansas, reveal rhythms and bioergodicity. Chronobiologia 9: 346

Scheving LE, Tsai TS, Powell EW, Pasley JN, Halberg F, Dunn J (1983) Bilateral lesions of suprachiasmatic nuclei affect circadian rhythms in [³H]-thymidine incorporation into deoxyribonucleic acid in mouse intestinal tract, mitotic index of corneal epithelium, and serum corticosterone. Anat Rec 205: 239–249

Smith E (1861) Draft of historic review: from periodic fever to chronobiology. Cited after P. Lavie

Smolensky M, Halberg F, Sargent F II (1972) Chronobiology of the life sequence. In: Itoh S, Ogata K, Yoshimura H (eds) Advances in climatic physiology. Igaku-Shoin, Tokyo, pp 281–318

Sothern RB, Halberg F (1986) Circadian and infradian blood pressure rhythms of a man 20 to 37 years of age. In: Halberg F, Reale L, Tarquini B (eds) Proc 2nd Int Conf Medico-Social Aspects of Chronobiology. Florence, Oct 2, 1984. Istituto Italiano di Medicina Sociale, Rome, pp 395–416

Sothern RB, Halberg F (1989) Longitudinal human multifrequency structure of blood pressure self-measured for over 2 decades. Proc 2nd Ann. IEEE Symp on Computer-Based Medical Systems, Minneapolis, June 26–27, 1989. Computer Society Press, Washington DC, pp 288–294

Stadick A, Bryans R, Halberg E, Halberg F (1987) Circadian cardiovascular rhythms during recumbency – In: Tarquini B, Vergassola R (eds) Social diseases and chronobiology: Proc III Int Symp Social Diseases and Chronobiology, Florence, Nov 29, 1986. Esculapio, Bologna, pp 191–200

Stamler J, Neaton JD, Wentworth DN (1989) Blood pressure (systolic and diastolic) and risk of fatal coronary heart disease. Hypertension 13: 2–12

Stamler R, Stamler J, Grimm R, Dyer A, Gosch FC, Berman R, Elmer P, Fishman J, Van Heel N, Civinelli J, Hoeksema R (1985) Nonpharmacological control of hypertension. Prev Med 14: 336–345

Stokes J III, Kannel WB, Wolf PA, D'Agostino RB, Cupples LA (1989) Blood pressure as a risk factor for cardiovascular disease. The Framingham study: 30 years of follow-up. Hypertension 13 (Suppl I): 13–18

Stoynev A, Ikonomov O, Hillman DC, Halberg F (1990) Circadian rhythms in centenarian blood pressure (BP) and heart rate (HR). Abstract, 2nd World Conference on Clinical Chronobiology. Monte Carlo, April 10–13, 1990. ALM, Florence, p 67

Syutkina EV, Safin SR, Grigoriev AE, Abramian AS, Polyakov YA, Taybloom M, Halberg E, Halberg F (1991) Intrauterine exposure to betamimetics affects adolescent circadian blood pressure (BP) and heart rate (HR) rhythms. Biochim Clin 15: 155–156

Tarquini B, Fernández JR, Wu J, Cornélissen G, Maggioni C, Mainardi G, Panero C, Hermida RC, Cagnoni M, Halberg F (1990) Infradian, notably circannual, cardiovascular variation gauging effect of intrauterine exposure to β-adrenergic agonists. In: Hayes DK, Pauly JE, Reiter RJ (eds) Chronobiology: its role in clinical medicine, general biology, and agriculture, part A. Wiley-Liss, New York, pp 595–604

Thompson ME, Nicolau GY, Lakatua DJ, Sackett-Lundeen L, Plinga L, Bogdan C, Robu E, Ungureanu E, Petrescu E, Haus E (1987) Endocrine factors of blood pressure regulation in different age groups. In: Pauly JE, Scheving LE (eds) Advances in chronobiology, part B. Liss, New York, pp 79–95

Tochikubo O, Miyazaki N, Yamada Y, Fukuoka M, Kaneko Y (1987) Mathematical evaluation of 24-hour blood pressure variability in young, middle-aged and elderly hypertensive patients. Jpn Circ J 51: 1123–1130

Tofler GH, Brezinski D, Schafter AI, Czeisler CA, Rutherford JD, Willich SN, Gleason RE, Williams GH, Muller JE (1987) Concurrent morning increase in platelet aggregability and the risk of myocardial infarction and sudden cardiac death. N Engl J Med 316: 1514–1518

Verdecchia P, Schillaci G, Guerrieri M, Gatteschi C, Benemio G, Boldrini F, Porcellati C (1990) Circadian blood pressure changes and left ventricular hypertrophy in essential hypertension. Circulation 81: 528–536

Wilcox RG, Mitchell JRA, Hampton JR (1986) Treatment of high blood pressure: should clinical practice be based on results of clinical trials? Br Med J 293: 433–437

Wissler WR, Vesselinovitch D (1976) Studies of regression of advanced atherosclerosis in experimental animals and man. Ann NY Acad Sci 275: 363–368

World Health Publication (1976) Myocardial infarction community registers, public health in Europe. WHO, Copenhagen

Wu J, Cornélissen G, Tarquini B, Mainardi G, Cagnoni M, Fernández JR, Hermida RC, Tamura K, Kato J, Kato K, Halberg F (1990) Circaseptan and circannual modulation of circadian rhythms in neonatal blood pressure and heart rate. In: Hayes DK, Pauly JE, Reiter RJ (eds) Chronobiology: its role in clinical medicine, general biology, and agriculture, part A. Wiley-Liss, New York, pp 643–652

Wu J, Cornélissen G, Halberg F (to be published) Multifrequency spectrum of rhythms underlies incidence of human sudden death. In: Mikulecky M, Horecky J, Cornélissen G, Halberg F (eds) Proc Int Symp Biorhythms in Clinical Medicine. Bratislava, Czecho-Slovakia, Sept 4–5, 1990

Wu J, Mainardi G, Tarquini B, Cagnoni M, Cornélissen G, Halberg F (in preparation) Gender differences in neonatal rhythm characteristics of blood pressure and heart rate.

Yunis EJ, Fernandes G, Nelson W, Halberg F (1974) Circadian temperature rhythms and aging in rodents. In: Scheving LE, Halberg F, Pauly JE (eds) Chronobiology. Proc Int Soc for the Study of biological rhythms, Little Rock, Ark, Thieme, Stuttgart pp 358–363

Zachariah PK, Sheps SG, Schwartz GL, Schirger A, Ilstrup DM, Long CR, Carlson CA (1988) Antihypertensive efficacy of lisinopril: ambulatory pressure monitoring. Am J Hypertens 1: 274s–279s

Zadek I (1881) Die Messung des Blutdrucks am Menschen mittels des Basch'schen Apparates. Z Klin Med 2: 509–551

Zaslavskaya RM, Suslov MG, Taybloom MM (1990) Genetic aspects of biorhythmological adaptation of blood circulation to medium and high mountain hypoxy. Abstract, 6th Meeting Eur Soc Chronobiol Barcelona, Spain, July 6–8, 1990. pp 55–56

Zaslavskaya RM, Cornélissen G, Rigó J Jr, Paulin F, Adám Z, Rigo J, Sr, Maggioni C, Mello G, Scarpelli PT, Hermida R, Tarquini B, Cagnoni M, Otsuka K, Miyake Y, Ohnishi M, Satoh K, Watanabe Y, Quadens O, Cugini P, Ahlgren A, Tamura K, Bakken E, Halberg E, Halberg F (to be published) Tracking blood pressure womb-to-tomb with the socially important mapping of the broad human chronome.

Zelis R, Mason DT, Braunwald E, Levi R (1970) Effects of hyperlipoproteinemias and their treatment on the peripheral circulation. J Clin Invest 49: 1007–1015

# Nocturnal Asthma: Mechanisms and Chronotherapy

M. H. Smolensky and G. E. D'Alonzo

## Introduction

Asthma is a chronic obstructive airways disease. The underlying abnormalities are excessive contraction of tracheobronchial smooth muscle and hypersecretion of mucus plus musosal edema in association with airways inflammation (Spector 1982). Historical accounts of the disease describe a worsening of symptoms overnight. Nonetheless, most clinicians question the time dependency of this disease or feel that nocturnal asthma is a special subtype of asthma. In the majority of untreated patients asthma worsens, or occurs only, overnight. The increase in cough,

wheeze, and breathlessness at this time causes substantial problems for patients (Pfeiffer et al. 1989). A number of hypotheses have been proposed to explain why asthma is so common overnight. These include day-night variations in certain environmental factors such as barometric pressure relative humidity, and ambient temperature; proximity and concentration of various offending antigens; accumulative effects of psychological and physiological stresses during the day; and assumption of a supine posture at night. An alternate explanation for the time dependency of this disease stresses the role of endogenous circadian bioperiodicities in relationship to changes in the external environment during each 24 h (Barnes 1984a; Smolensky et al. 1981, 1986a). With regard to chronobiological considerations, successful management of patients entails not only the institution of environmental control methods, but also an understanding of the circadian features of the disease to achieve a chronotherapy of anti-asthma medications (Reinberg et al. 1988a, b; Smolensky et al. 1986b, 1987a).

## Circadian Rhythm of Dyspnea

The nocturnal worsening of asthma is well-recognized by patients, although not always appreciated by practitioners. It is rather surprising that relatively few investigators (Dethlefsen and Repges 1985; Douglas 1985; Mascia 1968; Reinberg et al. 1963; Turner-Warwick 1988) have objectively evaluated the magnitude of the day-night variation in the occurrence of the disease. Dethlefsen and Repges (1985), in an investigation involving adult asthmatics, examined the time of day of 1631 episodes of dyspnea in a large group of untreated patients studied prior to the multicenter evaluation of a once a day, sustained release theophylline formulation. Episodes of dyspnea were mainly restricted to the overnight span, primarily between

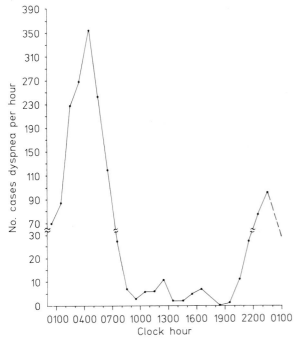

**Fig. 1.** Day-night difference in the hourly occurrence of asthma (1631 episodes of dyspnea) in a large group of untreated patients (3129, mostly asthmatic). Symptoms were most common around 0400–0500 hours (Data from Dethlefsen and Repges 1985)

0200 and 0700 hours, with the peak number manifested between 0400 and 0500 hours. There was more than a 50-fold difference in the number of attacks overnight as compared to the daytime, between 1600 and 1700 hours (Fig. 1). A large survey conducted by Turner-Warwick (1988) on more than 7600 asthma patients found that approximately 90% experienced nocturnal symptoms severe enough to cause awakenings from sleep at least periodically. Almost 40%, regardless of age and gender, awakened nightly due to asthma and over 60% awakened at least three nights per week.

From a clinical perspective, the day-night pattern in the onset and severity of asthma has important ramifications, especially since death from severe exacerbations is possible. Douglas (1985), reviewing four studies pertaining to asthma mortality, found that 93 of 219 deaths occurred between midnight and 2000 hours. In this regard, the presence of large amplitude circadian variation in conjunction with a less than predicted 24 h mean level of airways patency is a recognized risk factor of fatal asthma (Benatar 1986; Hetzel et al. 1977). Today, even with the availability of a variety of potent bronchodilator medications, asthma mortality is on the rise.

## Circadian Rhythm in Bronchial Patency

In a sense, nocturnal asthma is due to an amplification and overexpression of the slight increase in airway tone that normally occurs overnight in diurnally active healthy persons (Smolensky and Halberg 1977). The nocturnal decline in airways caliber is synchronous in both asthmatics and normal individuals adhering to a routine of diurnal activity and nighttime sleep. However, asthmatic patients exhibit markedly greater day-night variation in airflow, amounting to 25% of the 24 h mean level in mild patients and up to 40%–60% in more severe patients (Hetzel and Clark 1980; Lewinsohn et al. 1960; Smolensky et al. 1986a).

Features of the 24 h pattern in bronchial patency of asthmatic patients and healthy persons differ greatly. Cosinor analyses (Halberg et al. 1972; Nelson et al. 1979) have proven useful for objectively detecting and quantifying such bioperiodicities in terms of the mesor, the rhythm-adjusted mean, a middle value around which predictable temporal change occurs; amplitude, a measure of the within-rhythm variability (numerically equivalent to one-half the peak-to-trough difference); and acrophase ($\phi$), the peak time

of the rhythm relative to local midnight. Statistical significance of rhythmicity is tested by determining if the amplitude differs from zero (Nelson et al. 1979). Cosinor analysis of measures of airways patency obtained several times daily over one or more days by asthma patients typically reveals statistically significant, moderate to large amplitude, circadian rhythms (Smolensky et al. 1986a). The large circadian amplitude of bronchial patency is a diagnostic feature of obstructive airways diseases, in particular asthma (Hetzel and Clark 1980; Smolensky et al. 1986a). In general, the more severe the asthma, the greater the circadian amplitude and, concomitantly, the more reduced the 24 h level (mesor) of airways patency (Smolensky and Halberg 1977; Smolensky et al. 1986a). This increased amplitude of airways tone results primarily from the reduction of bronchial caliber overnight. Circadian rhythmicity of bronchial patency in healthy subjects either is undetectable by spirometry, i.e., amplitude equals zero, or is of quite low amplitude, and the 24 h mean is much higher than that of asthmatics. The acrophase, indicative of the clock time of greatest bronchial patency, occurs typically between 1200 and 1700 hours, both in diurnally active asthmatics and healthy subjects (Smolensky and Halberg 1977; Smolensky et al. 1986a).

The nocturnal decline in airflow in asthma patients is often referred to as the "morning" or "nocturnal dip." However, this is somewhat misleading. The reduced nocturnal patency reported in most studies of diurnally active asthmatics commonly represents the comparison of measurements of airways status obtained before bedtime at night and after awakening in the morning. In all but a minority of patients, bronchial patency is not constant throughout the daytime and suddenly falling during the night; rather, it is continually changing, more or less as a sinusoidal function of 24 h duration. Those patients with a medical history of true sudden nocturnal or morning dips in airways patency must be monitored carefully and treated aggressively, since it is these individuals in particular who are at great risk of ventilatory arrest and death due to asthma (Benatar 1986; Hetzel et al. 1977).

## Circadian Changes in Neuroendocrine Processes

Although a clear understanding of the pathogenesis of asthma remains elusive, the staging of several circadian periodicities contributes to the worsening of

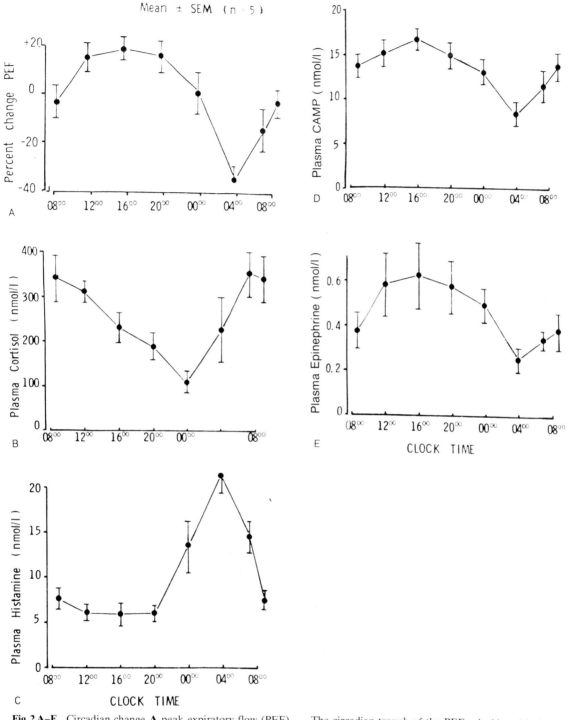

**Fig. 2 A–E.** Circadian change **A** peak expiratory flow (PEF), **B** plasma cortisol, **C** histamine, **D** cyclic adenosine monophosphate (cAMP), and **E** epinephrine. Shown are the mean and S.E.M. values for five asthmatics at each clock hour over 24 h. The circadian trough of the PEF coincides with the circadian minimum of cAMP and epinephrine and the peak of histamine. (Reproduced from Barnes et al. 1980)

asthma overnight. While it is tempting to ascribe a cause and effect relationship between the staging of certain identified circadian rhythms and nocturnal symptoms, causality is difficult to establish since a coincidence in time of phenomena itself does not necessarily imply such.

*Circadian Rhythm in Plasma Cortisol.* In diurnally active subjects, plasma cortisol levels are highest around the time of awakening from sleep and lowest during the middle of the night (Barnes et al. 1980; Reinberg et al. 1988a; Soutar et al. 1975). The peak-trough variation in this circadian rhythm is generally very large (Fig.2). Since cortisol can affect airways function in asthmatic persons, it was initially proposed that differences in bronchial patency between them and nonasthmatic individuals could be due to disparities in cortisol secretion. However, no differences have been detected; both groups exhibit comparable circadian patterns. This does not, however, negate the possibility that the nocturnal decline in plasma cortisol plays a role in the exacerbation of nighttime asthma, but its role alone appears to be limited. Since the airways of asthmatics generally are chronically inflamed, circadian changes in plasma cortisol, which is expected to exert an anti-inflammatory effect, could contribute to the day-night difference in airways patency and the risk of asthma.

*Circadian Rhythm in Adrenergic Function.* Airways tone is strongly regulated by adrenergic mechanisms. In asthmatic individuals, adrenergic factors play a role in the mechanism of nocturnal bronchospasm (Barnes 1985; Postma et al. 1985). Asthmatic patients at times appear to exhibit impaired cardiovascular and metabolic responses to $\beta$-adrenergic agonist medication and reduced density of $\beta$-receptors on circulating leukocytes has been reported (Brooks et al. 1979; Shelhamer et al. 1980) Although such observations could represent the effect of previous adrenergic therapy (Barnes 1984b), comparable findings have been reported also from studies on nontreated asthmatic patients. Thus, reduced $\beta$-adrenergic receptor density and responsiveness could be contributory.

The nocturnal exacerbation of asthma also may result from an increased $\alpha$-adrenergic responsiveness (Barnes 1986; Henderson et al. 1979), particularly at night. Although asthmatic patients may exhibit an exaggerated bronchoconstrictor response to an inhaled $\alpha$-agonist, there is no direct information on whether the effect varies over the 24 h.

The smooth muscle of human airways lacks direct sympathetic intervention which implies that $\beta$-receptors are regulated by circulating plasma catecholamines. In both healthy and asthmatic persons, the plasma levels of epinephrine exhibit a comparable, large amplitude, circadian rhythm. Typically, both the urine and plasma concentrations of epinephrine are highest around mid-afternoon and lowest around 0400 hours (Fig.2), the time when airflow is most reduced (Barnes et al. 1980). In asthmatics, the nocturnal decline in circulating epinephrine may result in bronchoconstriction, directly by a reduction in the endogenous stimulation of $\beta_2$-receptors located in the smooth muscle of the airways or indirectly by affecting other cells in the airways (Barnes 1984c). Epinephrine inhibits mediator secretion from inflammatory cells in the airways of asthma patients; withdrawal of endogenous $\beta$-agonist stimulation overnight is associated with an increased release of bronchoconstrictive mediators, e.g., histamine (Fig. 2) (Barnes et al. 1980). Although the data are limited, epinephrine also may have an inhibitory influence on cholinergic fiber and ganglionic neurotransmission, thereby modulating cholinergic bronchoconstrictor tone (Skoogh and Svedmyr 1984).

The airways themselves exhibit circadian differences in their reaction to $\beta$-agonist administration. The 24 h pattern in the response of $\beta$-receptors to $\beta$-agonist agents by asthmatic patients is comparable to that of nonasthmatics; however, the overall effect in asthmatics is greater since airways patency is lower during the 24 h in the patients than it is in normal persons (FitzGerald et al. 1980; Gaultier et al. 1977, 1988).

*Circadian Rhythm of Plasma Cyclic Adenosine Monophosphate.* Plasma cyclic adenosine monophosphate (cAMP) exhibits a prominent high amplitude, circadian rhythm, with the highest level at 1600 hours and the lowest one at 0400 hours (Fig.2), the time of lowest airways patency (Barnes et al. 1980; Mikuni et al. 1978). The 24 h pattern of plasma cAMP in asthmatics is similar to that of healthy persons, although the 24 h mean tends to be somewhat greater in the former. Changes in plasma cAMP over the day and night probably reflect epinephrine-mediated activation of tissue adenyl cyclase.

*Circadian Rhythm in Plasma Histamine.* Plasma histamine (Fig.2) has a circadian rhythm in asthmatic persons, with the peak level coinciding in time with greatest bronchoconstriction, i.e., at 0400 hours (Barnes et al. 1980). Plasma histamine apparently does not increase nocturnally in nonasthmatics. This

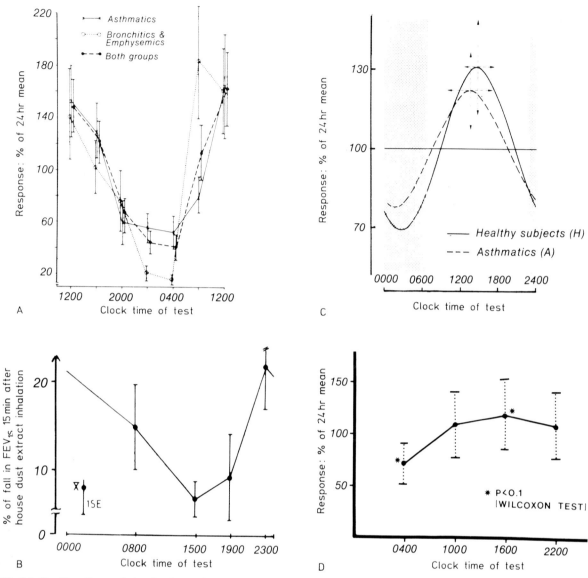

**Fig. 3 A–D.** Circadian variation in airways hyperreactivity to **A** aerosolized histamine, plotted as percentage change during the 24 h in threshold concentration for 20% change in $FEV_{1.0}$ from test time baseline; **B** aerosolized house dust, plotted as clock time-dependent depression in $FEV_{1.0}$ from baseline level; **C** aerosolized acetylcholine, plotted as percentage change during the 24 h in threshold concentration for 20% change in $FEV_{1.0}$ from test time baseline; **D** plotted as percentage variation in the effect of hyperventilation of cold air on $FEV_{1.0}$

implies that, in asthma patients, sensitized mast cells release preformed histamine and perhaps other bronchoconstrictor inflammatory mediators (Kunkel et al. 1988), particularly overnight when plasma epinephrine declines. When epinephrine is infused parenterally into asthmatics at 0400 hours plasma histamine concentration is reduced by 50% and the early morning decrease in airflow is attenuated (Barnes et al. 1980). Thus, the circadian stage-related reduction in plasma epinephrine at night may be permissive to the nonstimulated release of preformed histamine from inflammatory cells and the subsequent alteration of airways status.

*Cholinergic Reflex Mechanisms.* Cholinergic tone increases at night and this also may contribute to the pathogenesis of nocturnal asthma (Gaultier et al. 1977; Postma et al. 1985). The airways are densely innervated by cholinergic fibers and neuronal stimulation under certain conditions can induce bronchoconstriction. The reduction in endogenous $\beta$-receptor stimulation at night leads to increased cholinergic-mediated bronchoconstriction. Studies involving cholinergic blockade with either intravenous atropine (Morrison et al. 1988) or ipratropium bromide (Catterall et al. 1988) demonstrate that enhanced parasympathetic tone contributes to, but does not fully explain, the development of nocturnal asthma. Airways irritant receptor stimulation, for example by inflammatory mediators, airways cooling, or esophageal reflux of gastric acid, likely activates the cholinergic nervous system. It also is possible that the threshold of axon reflex mechanisms may be lowered at night. The role of nonadrenergic, noncholinergic, neuronal activity has yet to be demonstrated (Barnes 1986).

*Circadian Rhythm in Serum IgE.* IgE is a major immunoglobulin of the allergic diseases, including asthma. Gaultier et al. (1987) documented a fivefold greater mean level of IgE in asthmatics than in nonallergic controls. Moreover, a high amplitude circadian rhythm equal to 30% of the 24 h mean was detected in asthmatics although not in controls. In asthmatics, highest serum IgE levels occur around midday; lowest levels occur at night. The day-night variation in serum IgE could reflect temporal differences in tissue or cell-bound IgE. Such a hypothesis is consistent with the observation that experimentally stimulated histamine release from basophils or mast cells obtained from asthma patients during the night (as compared to the daytime) is low when, at the same time, plasma histamine levels are highly elevated (Kunkel and Jusuf 1984).

*Circadian Rhythm in Bronchial Hyperreactivity.* The severity of nocturnal asthma is highly correlated with the degree of bronchial reactivity (Ryan et al. 1982). In general, the greater the decline in airways tone at night, the greater the bronchoconstrictor response to histamine during the day. The nocturnal worsening of asthma may in part represent circadian rhythmicity in airways hyperreactivity (Fig. 3). Asthmatic patients, both adults (De Vries et al. 1962; Tammeling et al. 1977) and children (Van Aalderen et al. 1987), when challenged by the inhalation of histamine or methacholine aerosols at different clock hours of the day and nighttime, exhibit large circadian variations in the threshold dose required to produce either a 15% or 20% alteration in bronchial patency. In diurnally active patients, the airways are more reactive when challenged in the early morning, when asthma patients are typically asleep, than during the day. Simi-

lar results have been found when house dust challenge (Gervais et al. 1977) was performed with house dust sensitive asthmatics. For all provoking agents thus far studied, except for bronchial challenge with cold dry air and eucapneic hyperventilation (Sly and Landau 1986), the hyperreactivity of the airways during the nighttime is far greater than during the daytime. When asthmatic children were challenged with cold dry air and eucapneic hyperventilation, a greater degree of bronchoconstriction resulted from afternoon than from overnight tests (Sly and Landau 1986). Overall, these findings indicate that the large amplitude circadian rhythmicity in bronchial hyperreactivity is a general characteristic of asthma, and the time when asthmatics are exposed to specific triggering agents may be almost as important as the quantitative aspects of the exposures.

*Circadian Rhythm in Airways Inflammation.* Airways inflammation and concomitant mucosal edema is thought to increase nocturnally and seems to be responsible, in part, for the worsening of asthma overnight (Barnes 1988; Barnes et al. 1988; Martin et al. 1988). However, carefully designed studies are yet needed to further investigate this hypothesis. One report suggest that there is a higher number of inflammatory cells recovered when bronchoalveolar lavage is performed on asthmatics at 0400 than at 1600 hours (Martin et al. 1988). Studies conducted by Labrecque and colleagues on laboratory animals showed that the processes underlying inflammation are circadian rhythmic (Labrecque and Bélanger 1989).

## Exogenous Factors Related to Nocturnal Asthma

The role of exogenous environmental factors warrants discussion for a complete understanding of the etiology of asthma. Exposure to specific triggering agents during the daytime not only is capable of provoking acute immediate asthma, but also so-called late-phase asthma in the evening or overnight. In some cases, even daytime provocation of the airways may result in recurrent nocturnal asthma (Cockcroft et al. 1984). The overall mechanism of the late-phase reaction following allergen exposure, whether it be recurrent or not, especially in relation to the circadian determinants of airways status has yet to be defined. Allergen-induced bronchoconstriction, particularly the late-phase reaction, may contribute to nocturnal wheezing in some asthmatics; however, it does not appear to explain the phenomenon of nocturnal asthma by itself, since a similar circadian pattern of bronchoconstriction is found in patients classified as having intrinsic (endogenous) asthma, in whom there are no apparent allergic factors (Connolly 1979). Too, a recent survey showed that nocturnal asthma is just as frequent in intrinsic as in extrinsic (environmentally triggered) asthma (Turner-Warwick 1984). Finally, the dependence or independence of late-phase asthma, especially when it occurs overnight, from circadian rhythmic factors has yet to be evaluated.

*Sleep in the Supine Position.* The relationship between sleep and asthma is not clear. Deprivation of nightly sleep may affect the nocturnal level of bronchial patency in some patients (Ballard et al. 1989; Catterall et al. 1986). However, the circadian rhythm of airways patency persists even in stable asthmatics kept awake overnight or who sleep in the upright position (Hetzel and Clark 1979). Catterall et al. (1986) studied 12 asthmatic patients during two nights; one night they were allowed to sleep, but during the other they were kept awake. Airflow decreased during both nights, but its decline was somewhat attenuated when sleep was foregone. Thus, although the loss in airways patency is often in-phase with sleep, sleep per se does not seem essential for the development of nocturnal asthma. Asthmatics adhering to unusual work-rest schedules experience bronchoconstriction during sleep regardless of the time it is taken (Hetzel and Clark 1979). When shiftworkers with asthma change from day to night work, asthma occurs during the daytime hours of sleep. In one study the shift in the circadian acrophase of the rhythm in airflow obstruction occurred quickly, with the first sleep following shift change. In seasoned shiftworkers, the time required for readjustment of the circadian system may be rather short (Halberg et al. 1977; Reinberg 1979) and thus the finding of asthma during the first daytime sleep of a nightwork shift is not entirely unexpected. Nocturnal asthma also does not seem to be related to postural change from an upright to supine position. Patients who are confined to bed throughout an entire 24 h period still display day-night variation in airways tone (Clark and Hetzel 1977). Additionally, a recent study failed to find evidence of sustained bronchoconstriction after lying down during the usual activity span (Whyte and Douglas 1989).

Several studies have investigated the relationship between bronchoconstriction and stage of sleep, with conflicting results. In animals, there is marked fluctu-ation in airways resistance during rapid eye movement (REM) sleep (Sullivan et al. 1979), but this is not the case in humans (Lopes et al. 1983). No relationship has yet been found between asthma attacks which cause awakening from sleep and the stage of sleep in either adults or children (Kales et al. 1970a, 1970b; Malo et al. 1985), and there is no obvious relationship to REM sleep (Montplasir et al. 1982). However, the findings of these studies are difficult to interpret since the threshold of awakening because of bronchoconstriction may vary with the stage of sleep. The monitoring of oxygen saturation overnight has shown that declines in saturation are more frequent in asthmatics than nonasthmatics, although this might be explained by their lower resting arterial oxygen tension (Catterall et al. 1982). There is a yet no clear relationship between hypoxemic episodes and bronchoconstriction.

*Snoring.* In certain patients, asthma may arise from the vibration of the upper airways during snoring. In some asthmatics who snore, but who do not have the features of sleep apnea syndrome, nocturnal asthma is improved by the application of continuous positive airway pressure (CPAP) (Chan et al. 1988; Guilleminault et al. 1988). Nonetheless, the exact role snoring plays in the etiology of nocturnal bronchospasm, overall, is as yet unknown.

*Timing of Bronchodilator Medication.* The inhalation of traditional (nonsustained effect) sympathomimetic and anticholinergic medications before bedtime is unlikely to confer protection throughout the entire nighttime period in the majority of patients. The duration of activity of these aerosol medications is relatively short, unlike that of sustained release medications, which possess a pharmacokinetic profile suitable for the overnight control of the disease. Thus, nocturnal asthma has been explained from time-to-time as a failure of $\beta$-aerosol bronchodilator therapy overnight. However, since even in untreated patients asthma is 50- to 70-fold more common overnight than in the afternoon, loss of effect of $\beta$-agonist aerosols does not explain the nighttime prevalence of the disease.

*Gastroesophageal Reflux.* Gastroesophageal reflux has been said to be a common phenomenon in patients with asthma. The use of both theophyllines and $\beta_2$-agonists relaxes the lower esophageal sphincter, thereby facilitating acid reflux from the stomach, especially during the night when assuming the supine position for sleep. However, it has not been convincingly demonstrated that spontaneous gastroesophageal reflux results in nocturnal bronchospasm (Nagel et al. 1988; Hughes et al. 1983; Martin et al. 1982). During a trial of an $H_2$-antagonist (which inhibits gastric acid production) in patients with both symptomatic reflux and nocturnal bronchospasm, patients showed only a small, although significant, improvement in asthma symptoms. However, there was no change in morning peak expiratory flow (PEF) (Goodall et al. 1981). While certain patients can be provoked into asthma by an esophageal challenge with dilute acid (Davis et al. 1983; Wilson et al. 1985), it seems doubtful that nocturnal asthma in patients is due to gastroesophageal reflux alone (Spaulding et al. 1982).

*Impaired Mucociliary Clearance.* In asthmatic patients, excessive production and accumulation of mucus worsens breathing distress due to bronchospasm. Mucus accumulation may be particularly troublesome at night during sleep since mucociliary clearance in normal persons is decreased then (Bateman et al. 1978).

*Airway Cooling.* Cooling of the upper airways may result in bronchospasm in certain patients (Deal et al. 1979). The 1 °C decline in core body temperature normally occurring overnight, however, does not seem in itself to constitute an important factor in nocturnal asthma. Although the airways of asthmatics exhibit greater reactivity to histamine, acetylcholine, and antigens during the nighttime than the daytime, they are more sensitive to cool air during the day than the night. Also, Chen and Chi (1982) found that sleep in a warm humid environment attenuated, but did not eliminate, the decline in airflow found when breathing cool dry air.

## Chronotherapy of Nocturnal Asthma

Specific treatment strategies must be implemented if nocturnal asthma is to be managed properly and effectively. Ideal bronchodilator chronotherapy should confer its optimal effect when bronchoconstriction is likely to be worst; for most asthmatic patients, this is in the early morning (Turner-Warwick 1988). Thus, the pharmacokinetics and pharmacodynamics of bronchodilator medications, the temporal pattern of susceptibility-resistance to asthma, and the synchronizer (activity/sleep) schedule of the patient constitute the crucial factors that must be known to effectively devise an individualized (chrono)therapy of the disease. Knowledge of differences in the dose-response of the targeted tissues of the respiratory tract to various bronchodilator and anti-inflammatory medications as a function of their administration time is required as well (D'Alonzo et al. 1990; FitzGerald et al. 1980; Gaultier et al. 1977, 1988; Reinberg et al. 1988 a,b; Smolensky et al. 1987 b, 1987 c; Reinberg and Levi 1990).

Primarily, it has been specific, oral, sustained release theophylline, $\beta$-agonist, and tablet corticosteroid medications which have been formulated and used as chronotherapies for nocturnal asthma. Chronotherapy as used here refers to the design and utilization of galenic formulations to optimize the delivery of needed medications with regard to a specific administration time and in accordance with such criteria as rhythm dependencies in: (a) the pharmacokinetics of the formulation, (b) the susceptibility of the target tissue to the medication and/or to its active metabolites, (c) the need for medication with regard to the temporal pattern in the disease processes, and (d) the individual differences between patients in the need for medication with respect to time during the 24 h (or other time scales, e.g., menstrual cycle) and to dose.

The duration of action of the current generation of $\beta$-adrenergic and anticholinergic bronchodilator aerosol medications is too brief to ensure protection against bronchoconstriction throughout the entire nighttime period (Joad et al. 1987). Although longer duration-in-effect, $\beta$-agonist, aerosol, bronchodilator agents are currently being tested, published reports regarding their effectiveness for nocturnal asthma are still few in number. Too, even though inhalative anti-inflammatory agents are now used as a first line treatment of asthma, documentation of the effectiveness of corticosteroid and cromolyn sodium aerosol therapies specifically for the alleviation of nocturnal asthma is still incomplete.

*Sustained Release Theophylline Therapy.* Unlike conventional aerosol $\beta$-agonist therapy, for which the bronchodilator effect is likely to be waning as nocturnal bronchoconstriction is increasing (Joad et al. 1987), certain sustained release theophylline (SRT) products can be administered in a manner that provides an elevated blood level during the time when bronchodilation is most needed. Accordingly, these specifically formulated SRTs are administered once a day in the evening (Busse and Bush 1985; D'Alonzo et al. 1990; Goldenheim et al. 1987; Neuenkirchen et al. 1985; Wilkens et al. 1987) or twice daily by means of an unequal morning-evening dosing strategy (Bruguerolle et al. 1987; Darrow and Steinijans 1987; Smolensky et al. 1987c). Such theophylline chronotherapies have been successful in ameliorating the deterioration of airways function overnight without compromising it during the daytime (D'Alonzo et al. 1990; Dethlefsen and Repges 1985; Neuenkirchen et al. 1985; Reinberg et al. 1987; Wilkens et al. 1987).

The advantage of once daily evening theophylline chronotherapy for nocturnal asthma in comparison to equal-interval, equal-dose therapy is shown in Fig. 5. A word of caution regarding once a day SRT therapy is necessary since not all theophylline formulations are optimal, or even appropriate, to use as

once daily evening treatment for nocturnal asthma. Candidate once a day SRT products must possess an appropriate kinetic profile to ensure optimal pharmacodynamic efficacy during sleep and the kinetics of the formulation must be predictable and safe, even when taken with meals (Jonkman 1987, 1989; Smolensky et al. 1987c; Steinijans et al. 1987). It appears that the efficacy of SRTs for nighttime asthma is related to their producing a serum theophylline level that is sustained as a plateau during the critical overnight span, particularly between 0000 and 0600 hours (Fig. 5).

It must be recognized that the pharmacokinetics of the various galenic formulations of claimed once a day SRTs differ from one another (Fig. 4). Uniphyl

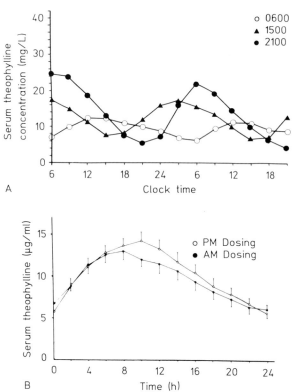

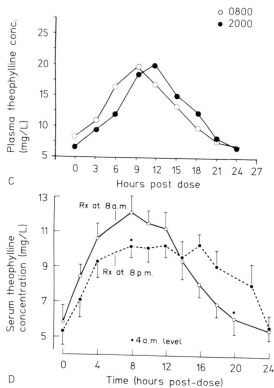

**Fig. 4A–D.** Pharmacokinetic profiles of four different once a day theophyllines. **A** Theo-24 (Searle Pharmaceuticals, Inc., USA); when taken as approved by the Food and Drug Administration (FDA) in the morning (0600) by eight asthmatic children, serum theophylline concentrations (STC) overnight are too low to be as effective as desired for treating nocturnal asthma. When taken at 2100 hours, the bioavailability of the drug is greater resulting in very high STC for a portion of the 24 h dosing interval. **B** Uniphyl (Purdue Frederick, USA) taken in the morning results in therapeutic STC at the time when bronchospasm is not likely to be most severe. When taken in the evening, the STC overnight is maintained in the therapeutic range. **C** Pharmacokinetic profile of STC from TheoDur (Key Pharmaceuticals, USA) when taken by ten asth-

matics once daily. When dosed in the morning, the drug fails to achieve therapeutic STC overnight. Dosed in the evening, TheoDur gives rise to higher STC, primarily during the later portion of the sleep span and during the initial phase of the activity span, times which might be too late to be optimally effective for patients suffering from nocturnal symptoms. **D** Average STC for a group of eight asthmatic adults studied while treated with Armophylline (Armour, France) at 0800 or 2000 hours in separate studies. When taken in the morning, therapeutic STC overnight are not maintained and efficacy is less than desired. When taken in the evening, the therapeutic effect is best. (Data for figure derived or reproduced from Smolensky et al. 1987; Rivington et al. 1985; Frankoff et al. 1987, 1988; Reinberg et al. 1987)

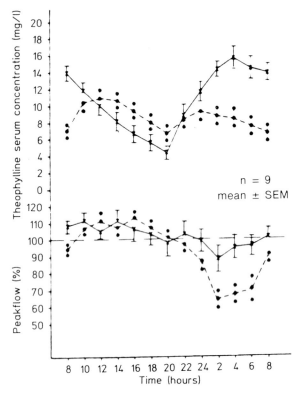

**Fig. 5.** Serum theophylline concentration (mg/l) and peak expiratory flow rate (l/min) for a once daily (●——●) and a twice daily (●---●) administered theophylline formulation in nine patients over a 24 h period. Time of theophylline administration is identified by *arrows* (↑, twice daily administration; ↑↑, once daily administration). (From Wilkens et al. 1987)

(Purdue Frederick, North America; Rivington et al. 1985) and Armophylline (Armour, France; Reinberg et al. 1987) exhibit a kind of plateau in the time-serum theophylline concentration curve. Even though Theo-24 (Searle, USA) exhibits a daytime plateau in the time-serum theophylline concentration profile when taken at 0600 (Smolensky et al. 1987 d), there is an insufficient level of the drug overnight when protection from asthma tends to be most needed. Yet, administration of Theo-24 in the same dose to the same patients at 1700 or 2100 hours results in a different pharmacokinetic profile characterized most notably by an absence of a plateaued drug level overnight and an enhanced (three-to fourfold) bioavailability of theophylline. Finally, Theo-Dur (Schering, USA) fails to exhibit the required well-defined plateau in drug level, desirable for optimal effectiveness, when taken once daily in the evening or morning (Frankoff et al. 1987). A more comprehensive review of the topic of circadian rhythm-adapted theophylline chronotherapeutics for

nocturnal asthma may be found elsewhere (Smolensky et al. 1987 a).

Even when the kinetics of once a day SRTs are proven to be safe and optimally effective when administered in the evening, it must be demonstrated that such formulations are free of adverse effects. Based on initial reports, it appears that theophylline administered in the evening not only improves airways status overnight, but patient-rated sleep integrity and quality as well (D'Alonzo et al. 1990; Reinberg et al. 1987), although minor deterioration of electroencephalogram scored sleep quality has been discussed (Rhind et al. 1984). Overall, the side effect profile for once daily in the evening SRT chronotherapy has been reported to be as good, or even better, than it is for conventionally dosed theophylline (e. g., twice daily at equal intervals and in equal doses) for patients at risk of nocturnal symptoms (D'Alonzo et al. 1990; Wilkens et al. 1987).

In some asthma patients, symptoms may be manifested only during the daytime or they may be unpredictable over the 24 h. In those who suffer daytime difficulties, SRT dosing may be more appropriate once daily in the morning, rather than the evening, as recommended previously (Goldenheim et al. 1987). In those whose symptoms are unpredictable in their occurrence, once a day SRT therapy might well be contraindicated.

*SR β-Adrenergic Agonist Therapy.* Specifically formulated SR β-agonist tablets, like SRTs, can significantly moderate the morning dip in airways patency in asthmatic patients. In this regard, Postma et al. (1984) showed the advantage of an unequal (one-third daily dose at 0800 and the remaining two-thirds at 2000 hours) morning-evening treatment schedule for one SR β-agonist tablet formulation, terbutaline (Bricanyl, Draco). However, more definitive studies documenting the efficacy and side effect profile of evening once a day or an unequal morning-evening dose strategy for β-agonist oral drugs are awaited.

The issue of (chron)optimizing SR β-agonist tablet formulations for nocturnal asthma requires careful evaluation, especially in consideration of the evidence that: (1) the dose-response of the airways is circadian rhythmic (FitzGerald et al. 1980; Gaultier et al. 1988) and (2) the risk of asthma is greater during the night than the day (Turner-Warwick 1988).

During the past few years oral SR β-agonist bronchodilator tablet medications have been marketed in Europe and in North America. Thus far, only a limited number of published investigations have addressed their utility as an evening chronotherapy for alleviating nocturnal asthma. Moore-Gillon (1988) evaluated the efficacy of an 8 mg dose of SR albuterol as Volmax (Glaxo, Europe). In patients with a history of nocturnal symptoms, evening SR albuterol in comparison to placebo treatment resulted in a statistically significant increase in morning PEF and in the number of mornings free from the symptoms of wheeze and shortness of breath. Another study found a once nightly Volmax dose to be as effective as once nightly TheoDur. In comparison to the placebo condition, both drugs were equally effective in reducing the number of sleep-disturbed nights in patients suffering from nocturnal asthma. Furthermore, evening SR albuterol dosing was associated with significantly reduced daytime symptom scores of wheeze, shortness of breath, and mucus production (Creemers 1988).

In spite of the fact that albuterol sulfate as Proventil Repetabs (Schering, USA) is marketed as a candidate treatment for nocturnal asthma, no well-conducted clinical studies substantiating its therapeutic utility for such have been published as yet. Although Proventil Repetabs provide a better sustained blood level of albuterol overnight than do the conventional albuterol tablets (Powell et al. 1987) and appear to be free of meal effect (Bolinger et al. 1989), their effectiveness in ameliorating the nocturnal deterioration of lung function and the symptoms of asthma is not well-documented.

Bambuterol (Draco, Astra; Sweden) also is a candidate once a day evening treatment for nocturnal asthma (Persson 1988). Bambuterol tablets deliver organically bound terbutaline in the form of a prodrug to the lung. In this form, the rate of prodrug metabolism is controlled so that a sustained local concentration of terbutaline is ensured for as long as 24 h. Bambuterol has been studied in asthmatic patients given a single dose of either 10, 20, or 30 mg in the evening (at ~1830, 2000, or 2200 hours, respectively, in different investigations) and compared to other dosing forms and schedules, for example a 5 mg t.i.d. nonsustained release or a 10 mg b.i.d. SR terbutaline regimen (Bricanyl Depot, Draco, Astra; Sweden).

One company sponsored study found that a terbutaline schedule consisting of 5 mg at 0800, 1400, and 2000 hours produced only a slight but statistically significantly better daytime forced expiratory volume $FEV_1$ (by ~2%–3%) than an evening (2000 hours) 20 mg Bambuterol treatment schedule. Of greater interest was the significant anti-asthmatic effect during the night from evening Bambuterol administration, as judged by the decreased reliance upon $\beta$-agonist bronchodilator inhalers to rescue acute asthma flare-

up. Another investigation compared the effect of a single 5, 10, or 20 mg evening (1800–1900 hours) dose of Bambuterol to placebo terbutaline tablets. Bambuterol given in the evening, either in a 10 or 20 mg dose, resulted in a statistically significant increase in morning and evening PEF with reference to the placebo treatment. Administration of both the 10 and the 20 mg dose of Bambuterol in the evening also was associated with a statistically significant decline of $\beta$-agonist aerosol use to relieve acute episodes of airways obstruction. In a published study, Persson (1988) compared the efficacy of the single evening 30 mg Bambuterol dosing schedule with the 10 mg SR terbutaline twice daily (0800 and 2200 hours) treatment regimen. The mean overall PEF was significantly higher during evening Bambuterol dosing and the requirement for $\beta$ aerosols as a rescue for acute asthma was significantly reduced. Overall, the results of studies conducted on Bambuterol suggest that evening administration in a dose of 20–30 mg significantly improves airways patency in asthmatic patients, both during the daytime and overnight. In comparison to terbutaline administered as regular or a SR formulation, Bambuterol dosing has been associated with no greater risk of side effects (tremor, headache, or elevated blood pressure and heart rate).

While the findings of the initial investigations on evening dosed Bambuterol, Volmax, and Repetabs are encouraging as another chronotherapy of nocturnal asthma, further studies are awaited on a larger number of patients. In particular, data are required to ascertain the relative efficacy of the nighttime dosing schedule of these medications for controlling nocturnal asthma and for stabilizing the airways during the day and night. Finally, additional study is required to ensure patient tolerance to the drugs when timed as a single evening administration.

*Anti-Inflammatory Corticosteroid Therapy.* Corticosteroids are frequently used to control asthma. The administration time of these medications, whether ingested, inhaled or infused, may influence the magnitude of both their desired and undesired effects (Reinberg 1989; Reinberg et al. 1974, 1988a). As a historical note, the first significant chronotherapy of asthma marketed by a major pharmaceutical company (Upjohn, USA) was methylprednisolone (Solu-Medrol). When used daily or on alternate days, the morning, coinciding with the commencement of the daily activity span, timing of this medication as tablets proved to be a safe and effective means of managing a variety of steroid-responsive diseases. However, it is still not infrequent, even today, that the daily dose of oral corticosteroids is prescribed for splitting into several small administrations for consumption at breakfast, lunch, dinner, and at bedtime. This (homeostatic) type of treatment schedule is likely to potentiate at least one of the several side effects of oral corticosteroids, adrenal suppression.

The timing of corticosteroid tablet therapy early in the daily activity span lessens the risk of adrenal suppression and potentiates the therapeutic effect on airways patency (Reinberg et al. 1977b, 1983, 1988a; Reinberg 1989). In contrast, the administration of corticosteroid tablets in the same dose at bedtime is less effective in controlling asthma and is more likely to induce adrenal suppression (Reinberg et al. 1977b, 1988a; Reinberg 1989). Recent findings on long-term treatment of patients with synthetic corticosteroid tablets during the initial hours of the daily activity span are consistent with earlier ones demonstrating a better effect of an acute administration of methylprednisolone on the airways patency of asthmatic children when timed at 0700 or 1300 than at 1900 or 0300 (Reinberg et al. 1974).

One corticoid formulation, known as Dutimelan 8-15 (DTM 8-15, Hoechst; Italy), was developed specifically as a chronotherapy. DTM 8-15 consists of two tablets, one possessing 7 mg prednisolone acetate and 4 mg prednisolone alcohol for an oral administration at 0800 and the second having 15 mg cortisone acetate and 3 mg prednisolone alcohol for an oral administration at 1500 hours, daily.

In a double-blind, crossover designed study (Reinberg et al. 1983), eight asthmatic patients during one phase of the investigation received Dutimelan 8-15 as intended, with the major steroid dosing at 0800 and a small one at 1500 plus placebo at 2000 hours (a dosing schedule referred to as DTM 8-15). During another phase, the patients received placebo at 0800 hours, the same small daily corticosteroid dose as received before at 1500, and the major daily dose at 2000 instead of 0800 as before (a dosing schedule referred to as DTM 15-20). During both phases of the investigation, PEF was self-assessed every 4 h during the diurnal span of wakefulness (0700–2300 hours) and also when awakened from sleep due to asthma. When the corticosteroid doses were taken orally at 0800 and 1500 hours, the 24 h mean PEF was statistically significantly greater than when dosing was at 1500 and 2000. Moreover, when the major corticosteroid dose was timed early during the activity span (as DTM 8-15), the nocturnal dip in airways patency, as assessed by PEF, was better controlled than when it was later timed (as DTM 15-20). It is of interest that the DTM 8-15 chronotherapy results in an optimization

of tablet corticosteroid therapy for asthma, with minimal risk of adrenal inhibition (Reinberg 1989; Reinberg et al. 1977b, 1983).

DTM 8-15 (as did morning dosing of Solu-Medrol) demonstrates the practicality of chronocorticotherapy. DMT 8-15, as marketed by Hoechst (Italy) in only two different dosage strengths, however, is limited in its application. Nonetheless, this product demonstrates the utility of chronocorticotherapy. Finally, even when a regular corticosteroid tablet product, such as prednisolone, is prescribed for asthma, the large morning dose schedule is well-tolerated and produces best results. The evidence for this comes from a long-term study of a small number of diurnally active, steroid-dependent, asthmatic patients who regularly take their major daily dose of oral prednisolone at 0800 hours and, in some patients, a small additional dose early in the afternoon. Even after 7 or more years of prednisolone chronotherapy, there is no evidence of an iatrogenic adrenal suppression or alteration of bone density (Reinberg 1989; Reinberg et al. 1988b). Most importantly, each patient's previously unstable asthma condition has been effectively controlled.

Currently, inhalative corticosteroid aerosol medications are popular for managing airways inflammation, which is now considered to be one of the major factors in the pathophysiology of asthma. Although the advantage of a time-dependent dosing schedule for aerosolized corticoids has been explored, investigation of such is just commencing (Toogood 1982); the data thus far reported do not enable a proper evaluation of aerosolized chronocorticotherapy for nocturnal asthma.

## Discussion

Both chronobiological and cyclic environmental factors are implicated in the etiology and pathogenesis of nocturnal asthma. An appreciation of endogenous, interrelated, temporally varying processes in particular is essential for achieving optimization of medications for asthma. Epidemiologic findings clearly demonstrate that the administration of conventional (homeostatically derived) bronchodilator medications in equal doses and at equal intervals does not avert nocturnal episodes of asthma, especially in patients suffering from moderate or severe forms of the disease. Nearly 75% of asthma patients awaken at least once weekly due to breathing distress (Turner-

Warwick 1988) in spite of the fact they are compliant to such treatment schedules. Obviously, new or more aggressive chronotherapies are warranted which take into account day-night differences in the requirement for bronchodilator medications.

Barnes (1989b), in a recent publication, states that the first line of therapy for mild asthma is inhalative $\beta$-agonist medication. However, this is not a reasonable approach when symptoms disturb nightly sleep. When this is the case, reliance upon conventional inhalative $\beta$-agonist aerosols as the principal means of therapy is inappropriate since their duration of effect is too short. Barnes also recommends a stepped therapy for patients with more severe asthma. The first step is aerosolized corticosteroids, since in many patients airways inflammation constitutes the basis for bronchial hyperreactivity. According to Barnes, the second step of therapy consists of adding a $\beta_2$-agonist aerosol medication. Finally, the third step consists of adding theophylline. The merit for such a stepped therapy yet awaits investigative scrutiny, inspite of its recommendation in a prestigious medical journal.

Since airways inflammation must be managed, there is no doubt that inhalative synthetic corticosteroids are appropriate in the therapy of asthma. Aerosolized $\beta_2$-agonist medication also has a prominent role. However, the rationale underlying the particular stepping of pharmacotherapy for asthma, as recommended by Barnes, ist not clear. Inhalation of conventional $\beta_2$-agonist medications before bedtime does not maintain airways patency for a sufficient duration of time to be effective during the entire overnight span (Joad et al. 1987); patients awaken at night with asthma all too commonly. Furthermore, it is unknown whether equal-interval inhalative corticosteroid therapy alone eliminates nocturnal asthma, especially during the initial span of treatment. When bronchodilator medications having appropriate duration of action are not utilized concomitantly with inhalative corticoids, asthma is likely to be manifested overnight, especially during the initial 1–2 months of aerosol corticotherapy.

Although general strategies for the management of nocturnal asthma can be recommended, clinical emphasis must be given to the particular requirements of individual patients. First to be determined is whether or not the patient's symptoms are well-controlled by the prescribed medications according to their dose and scheduling. If the patient complains of persisting bouts of asthma, then specific information pertaining to the time of their occurrence, in relation to the sleep-activity routine of the patient, is required. Depending on the pattern and the severity of

symptoms during the day and nighttime hours, particular chronotherapeutic strategies may be warranted. In those patients whose asthma is especially difficult to manage, use of daily diaries for recording the occurence of symptoms as well as the self-assessment of PEF several times daily is recommended. This approach is indispensable for evaluating the effectiveness of various bronchodilator medications according to their dosage and scheduling in time during the 24 h and for devising an individualized therapy for the patient.

A recent survey of medical journals published in Europe and in North America reveals numerous advertisements claiming the merit of various bronchodilator products for treating nocturnal asthma. Some claims are based on the findings of only daytime-conducted studies which examined the pharmacokinetics and airways response to the given treatments. Rarely have overnight assessments been done to substantiate the supposition that the response to the given bronchodilator therapies is the same or better during the nighttime as it is during the daytime. Claims alone are insufficient. Proof of claims, in terms of efficacy against nighttime symptoms and safety based on data derived from well-designed and executed scientific studies, is an absolute requirement. Physicians and pharmacists must be provided needed information to make informed and wise decisions regarding therapeutic interventions. The reluctance of the pharmaceutical industry to sponsor investigations of candidate bronchodilator medications as potential chronotherapies is unfortunate and may even have resulted in lost opportunities for new drug applications and development.

At this time, only a limited number of bronchodilators have been approved by governmental regulatory agencies for their specific prescription as chronotherapies to control nocturnal symptoms. Primarily, it is the theophyllines which have received the greatest attention and recognition as once daily treatments. However, several $\beta$-agonist tablet preparations are under evaluation as once daily evening chronotherapies for the management of nocturnal asthma. The possible impact of the $\beta$-agonist tablet formulations on nighttime asthma remains to be fully established. While both the once daily theophylline and $\beta$-agonist formulations and specific corticosteroid tablet products represent initial attempts aimed at an effective chronotherapy of asthma, more research and developmental effort are required to achieve appropriate programmed in-time interventions against the disease. This is of major concern today, since the mortality from asthma is reported to be increasing around the world (Benatar 1986; Hetzel et al. 1977; Nicklas 1989).

## Conclusion

The compelling evidence is that nocturnal asthma may be explained by the stage relationships of several circadian rhythms which interact and affect the inflammatory milieu. With regard to those biological processes which undergo circadian rhythmicity, their amplitude can be so great that the patient must be considered as being distinctly different physiologically and biochemically during the daytime than during the nighttime. Therefore, the need for bronchodilator and anti-inflammatory medications to treat asthma may differ markedly according to the time of day that the disease is most likely to express itself. Successful pharmacotherapy of asthma must, as a minimum, succeed in averting the nocturnal decline of airways patency and sleep-disturbing breathing distress. Too, it must result in an overall elevation of the 24 h mean level of bronchial patency. Finally, if the control of nocturnal asthma is the principal goal of therapy, then the quality of sleep and functional performance during the daytime must not be compromised.

## References

Ballard RD, Saathoff MC, Patel DK, Kelly PL, Martin RJ (1989) The effect of sleep on nocturnal bronchoconstriction and ventilatory patterns in asthmatics. J Appl Physiol 67: 243–249

Barnes PJ (1984a) Nocturnal asthma: mechanisms and treatment. Br Med J 288: 1397–1398

Barnes PJ (1984b) Autonomic control of the airways and nocturnal asthma. In: Turner-Warwick M, Levy J (eds) New prospectives in theophylline therapy. International congress and symposium series, no 78. Royal society of medicine, London, pp 5–12

Barnes PJ (1984c) Autonomic control of the airways in nocturnal asthma. In: Barnes PJ, Levy J (eds) Nocturnal asthma. Oxford University Press, Oxford, pp 69–73

Barnes PJ (1985) Circadian variation in airway function. Am J Med 79 (suppl 6 A): 5–9

Barnes PJ (1986) State of the art. Neural control of human airways in health and disease. Am Rev Respir Dis 134: 1289–1314

Barnes PJ (1988) Inflammatory mechanisms and nocturnal asthma. Am J Med 85 (suppl 1 B): 64–70

Barnes PJ (1989a) Autonomic control of the airways and nocturnal asthma as a basis for drug treatment. In: Lemmer B (ed) Chronopharmacology, cellular and biochemical interactions. Cellular clocks series, vol 3. Dekker, New York, pp 53–63

Barnes PJ (1989b) Drug therapy – a new approach to treatment of asthma. N Engl J Med 321: 1517–1527

Barnes P, FitzGerald G, Brown M, Dollery C (1980) Nocturnal asthma and changes in circulating epinephrine, histamine and cortisol. N Engl J Med 303i: 263–267

Barnes PJ, Chung KF, Page CP (1988) Inflammatory mediators and asthma. Pharmacol Rev 40: 49–84

Bateman JRM, Pavia D, Clarke SW (1978) The retention of lung secretions during the night in normal subjects. Clin Sci 55: 523–527

Benatar SR (1986) Fatal asthma. N Engl J Med 314: 423–429

Bolinger AM, Young KYL, Gambertoglio JG, Newth CJL, Zureikat G, Powell M, Leung P, Affrime MB, Symchowicz S, Patrick JE (1989) Influence of food on the absorption of albuterol repetabs. J Allergy Clin Immunol 83: 123–126

Brooks SM, McGowan K, Bernstein IL; Altenau P, Peagler J (1979) Relationship between numbers of beta-adrenergic receptors in lymphocytes and disease severity in asthma. J Allergy Clin Immunol 63: 401–406

Bruguerolle B, Philip-Joet F, Parrel M, Arnaud A (1987) Unequal twice-daily, sustained-release theophylline dosing in chronic obstructive pulmonary disease. Chronobiol Int 4: 381–386

Busse WW, Bush RK (1985) Comparison of morning and evening dosing with a 24 h sustained-release theophylline, Uniphyl, for nocturnal asthma. Am J Med 79 (suppl 6 A): 62–66

Catterall JR, Rhind GB, Stewart IC (1986) Effect of sleep deprivation on overnight bronchial constriction in nocturnal asthma. Thorax 41: 676–680

Catterall JR, Calverley PMA, Brezinova V, Douglas NJ, Brash HM, Shapiro CM, Flenley DC (1982) Irregular breathing and hpyoxaemia during sleep in chronic stable asthma. Lancet 1: 301–304

Catterall JR, Rhind GB, Stewart IC, Whyte KF, Shapiro CM, Douglas NJ (1986) Effect of sleep deprivation on overnight bronchoconstriction in nocturnal asthma. Thorax 41: 676–680

Catterall JR, Rhind GB, Whyte KF, Shapiro CM, Douglas NJ (1988) Is nocturnal asthma caused by changes in airway cholinergic activity? Thorax 43: 720–724

Chan CS, Woolcock AJ, Sullivan CE (1988) Nocturnal asthma: role of snoring and obstruction sleep apnea. Am Rev Respir Dis 137: 1502–1504

Chen WY, Chi H (1982) Airway cooling and nocturnal asthma. Chest 81: 675–680

Clark TJH, Heztel MR (1977) Diurnal variation of asthma. Brit J Dis Chest 71: 87–92

Cockcroft DW, Hoeppner VH, Werner GD (1984) Recurrent nocturnal asthma after bronchoprovocation with western red cedar sawdust: association with acute increase in non-allergic bronchial responsiveness. Clin Allergy 14: 61–68

Connolly CK (1979) Diurnal rhythms in airways obstruction. Br J Dis Chest 73: 357–366

Creemers JD (1988) A multicenter comparative study of salbuterol controlled release (Volmax) and sustained-release theophylline (Theo-Dur) in the control of nocturnal asthma (abstract). Eur Res J 1 (suppl): 333S

D'Alonzo GE, Smolensky MH, Feldman S, Gianotti LA, Emerson MB, Staudinger H, Steinijans VM (1990) Twenty-four-hour lung function in adult patients with asthma: chronoptimized theophylline therapy once daily in the evening versus conventional twice-daily dosing. Am Rev Respir Dis 142: 84–90

Darrow P, Steinijans VW (1987) Therapeutic advantage of unequal dosing of theophylline in patients with nocturnal asthma. Chronobiol Int 4: 349–357

Davis RS, Larsen GL, Grunstein MM (1983) Respiratory response to intraesophageal acid infusion in asthmatic children during sleep. J Allergy Clin Immunol 72: 393–398

Deal EC, McFadden ER, Ingram RH, Strauss RH, Jaeger JJ (1979) Role of respiratory heat exchange in production of exercise-induced asthma. J Appl Physiol 46: 467–475

Dethlefsen U, Repges R (1985) Ein neues Therapieprinzip bei nächtlichem Asthma. Med Klin 80: 40–47

De Vries K, Goei JT, Booy-Noord H, Orie NG (1962) Changes during 24 hours in the lung function and histamine hyperreactivity of the bronchial tree in asthmatic and bronchitic patients. Int Arch Allergy 20: 93–101

Douglas NJ (1985) Asthma at night. Clin Chest Med 6: 663–674

FitzGerald GA, Barnes P, Brown MJ, Dollery CT (1980) The circadian variability of circulating adrenaline and bronchomotor reactivity in asthma. In: Smolensky MH, Reinberg A, McGovern JP (eds) Recent advances in the chronobiology of allergy and immunology. Pergamon, Oxford, pp 89–94

Frankoff HM, Smolensky MH, D'Alonzo GE, Gianotti L, Hsi B, McGovern JP (1987) Comparison of sustained-release theophylline scheduled conventionally (twice-daily, equal interval in equal amount) versus once-daily morning or evenings on circadian pattern of bronchial patency in asthmatics. Chronobiol Int 4: 421–433

Gaultier CL, Reinberg A, Girard F (1977) Circadian rhythms in lung resistance and dynamic lung compliance of healthy children. Effects of two bronchodilators. Respir Physiol 31: 169–182

Gaultier C, De Montis G, Reinberg A, Motohashi Y (1987) Circadian rhythm of serum total immunoglobulin E (IgE) of asthmatic children. Biomed Pharmacother 41: 186–188

Gaultier C, Reinberg A, Motohashi Y (1988) Circadian rhythm in total pulmonary resistance of asthmatic children. Effects of $\beta$-agonist agent. Chronobiol Int 5 (3): 285–290

Gervais P, Reinberg A, Gervais C, Smolensky MH, De France O (1977) Twenty-four-hour rhythm in the bronchial hyperreactivity to house dust in asthmatics. J Allergy Clin Immunol 59: 207–213

Goldenheim PD, Conrad EA, Schein LK (1987) Treatment of asthma by a controlled-release theophylline tablet formulation: a review of the North American experience with nocturnal dosing. Chronobiol Int 4: 397–408

Goodall RJL, Earis JE, Cooper DN, Bernstein A, Temple JG (1981) Relationship between asthma and gastroesophageal reflux. Thorax 36: 116–121

Guilleminault C, Quera-Salva MA, Powell N, Riley R, Romaker A, Partinen M, Baldwin R, Nino-Murcia G (1988) Nocturnal asthma: snoring, small pharynx and nasal CPAP. Eur Respir J 1: 902–907

Halberg F, Reinberg A (1977) Chronobiological serial sections gauge circadian rhythm adjustment following transmeridian flight and life in novel environment. Waking Sleeping 1: 259–279

Halberg F, Johnson EA, Nelson W, Sothern R (1972) Autorhythmometry – procedures for physiologic self-measurements and their analysis. Physiol Teacher 1: 1–11

Henderson WR, Shelhamer JH, Reingold DB, Smith LJ, Evans R III, Kaliner M (1979) Alpha-adrenergic hyperresponsiveness in asthma. N Engl J Med 300: 642–647

Hetzel MR, Clark TJH (1979) Does sleep cause nocturnal asthma? Thorax 34: 749–754

Heztel MR, Clark TJH (1980) Comparison of normal and asthmatic circadian rhythms in peak expiratory flow rate. Thorax 35: 732–738

Hetzel MR, Clark TJH; Branthwaite MA (1977) Asthma: analysis of sudden deaths and ventilatory arrest in hospital. Br Med J 1: 808–811

Hughes DM, Spiers S, Rivlin J, Levinson H (1983) Gastroesophageal reflux during sleep in asthmatic patients. J Pediatr 102: 666–672

Joad JP, Ahrens RC, Lindren SD, Weinberg MM (1987) Relative efficacy of maintenance therapy with theophylline, inhaled albuterol and the combination for chronic asthma. J Allergy Clin Immunol 79: 78–85

Jonkman JHG (1987 a) Food interactions with once-a-day theophylline preparations: a review. Chronobiol Int 4: 449–458

Jonkman JHG (1987 b) Food interactions with sustained release theophylline preparations. A review. Clin Pharmacokin 16: 162–179

Kales A, Beall GN, Bajor GF, Jacobson A, Kales J (1970 a) Sleep studies in asthmatic adults: relationship of attacks to sleep stage and time of night. J Allergy 41: 164–173

Kales A, Kales JD, Sly RM, Sharf MB, Tijiauw LT, Preston TA (1970 b) Sleep patterns in asthmatic children. All night EEG studies. J Allergy 46: 300–308

Kunkel G, Jusuf L (1984) Theoretical and practical aspects of circadian rhythm for theophylline therapy. In: Turner-Warwick M, Levy J (eds) New perspectives in theophylline therapy. International Congress and Symposium Servies, no 78. Royal Society of Medicine, London, pp 149–155

Kunkel G, Nigam S, Herold D, Jusuf L, Albright DL (1988) Arachidonic acid metabolites and their circadian rhythm in patients with allergic bronchial asthma. Chronobiol Int 5: 387–394

Labrecque G, Bélanger PM (1989) The chronopharmacology of the inflammatory process. Annu Rev Chronopharmacol 2: 291–325

Lewinsohn HC, Capel LH, Smart J (1960) Changes in forced expiratory volume throughout the day. Br Med J 1: 462–464

Lopes JM, Tabachnik E, Muller NL, Levison H, Bryan AC (1983) Total airway resistance and respiratory muscle activity during sleep. J Appl Physiol 54: 773–777

Malo JL, Montplaisir J, Monday J, Walsh J (1985) Sleep, breathing and dreams of asthmatic subjects. In: Isles AF, von Wicket P (eds) Sustained release theophylline and nocturnal asthma. Excerpta Medica, Amsterdam, pp 36–43

Martin ME, Grunstein MM, Larson GL (1982) The relationship of gastroesophageal reflux in children with asthma. Ann Allergy 49: 318–322

Martin RJ, Cicutto LC, Ballard RD, Szefler SJ (1988) Airway inflammation in nocturnal asthma (abstract). Am Rev Respir Dis 137: 284

Mascia M (1968) Evaluation of night coughing in asthmatic children. J Asthma Dis 5: 163–169

Mikuni M, Saito Y, Koyama T, Daiguji M, Yamashita I, Yamazaki K, Honma A, Ui M (1978) Circadian variation in plasma. 3':5'-cyclic adenosine monophosphate and 3':5'-cyclic guanosine monophosphate in normal adults. Life Sci 22: 667–671

Montplasir J, Walsh J, Malo JL (1982) Nocturnal asthma: features of attacks, sleep and breathing patterns. Am Rev Respir Dis 125: 18–22

Moore-Gillon J (1988) Volmax (salbuterol CR 8 mg) in the management of nocturnal asthma: a placebo-controlled study (abstract). Eur Res J 1 (suppl 2): 306 S

Morrison JFJ, Pearson SB, Dan HG (1988) Parasympathetic nervous system in nocturnal asthma. Br Med J 296: 1427–1429

Nagel RA, Brown P, Perks WH, Wilson RSE, Kerr JD (1988) Ambulatory pH monitoring of gastroesophageal reflux in "morning dipper" asthmatics. Br Med J 297: 1371–1373

Nelson W, Tong YL, Lee JK, Halberg F (1979) Methods for cosinor rhythmometry. Chronobiologia 6: 305–323

Neuenkirchen H, Wilkens JH, Oellerich M, Sybrecht GW (1985) Nocturnal asthma: effect of a once per evening dose of sustained release theophylline. Eur J Respir Dis 66: 196–204

Nicklas RA (1989) Perspective on asthma mortality–1989. Ann Allergy 63: 578–584

Pfeiffer C, Marsac A, Lockhart A (1989) Chronobiological study of the relationship between dyspnea and airway obstruction in symptomatic asthmatic subjects. Clin Sci 77: 237–244

Persson G, Gnosspelius Y, Anehus S (1988) Comparison between a new once-daily, bronchodilating drug, bambuterol, and terbutaline sustained-release, twice daily. Eur Respir J 1: 223–226

Postma DS, Köeter GH, Meurs H, Keyzer JJ (1984) Slow release terbutaline in nocturnal bronchial obstruction: relation of terbutaline dosage and blood levels with circadian changes in peak flow values. Annu Rev Chronopharmacol 1: 101–104

Postma DS, Keyzer JJ, Löeterm HG, Sluiter HJ, De Vries K (1985) Influence of the parasympathetic and sympathetic nervous system on nocturnal bronchial obstruction. Clin Sci 69: 251–258

Powell ML, Weinberger M, Dowdy Y, Gural R, Symchowicz S, Patrick JE (1987) Comparative steady state bioavailability of conventional and controlled-release formulations of albuterol. Biopharm Drug Disp 8: 461–468

Reinberg A (ed) (1979) Chronobiological field studies of oil refinery shift workers. Chronobiologia 6 (suppl): 1–119

Reinberg A (1989) Chronopharmacology of corticosteroids and ACTH. In: Lemmer B (ed) Chronopharmacology, cellular and biochemical interactions. Dekker, New York, pp 137–167

Reinberg A, Levi F (1990) Dose-response relationships in chronopharmacology. Annu Rev Chronopharmacol 6: 25–46

Reinberg A, Ghata J, Sidi E (1963) Nocturnal asthma attacks; their relationship to the circadian adrenal cycle. J Allergy 34: 323–330

Reinberg A, Gervais P, Morin M, Abulker C (1971) Rythme circadien humain du seuil de la response bronchique a l'acetylcholine. Comptes Rendus Acad Sci (Paris) 272: 1879–1881

Reinberg A, Halberg F, Falliers C (1974) Circadian timing of methylprednisolone effects in asthmatic boys. Chronobiologia 1: 333–347

Reinberg A, Gervais P, Ghata J (1977a) Chronobiologic aspects of allergic asthma. In: McGovern JP, Smolensky MH, Reinberg A (eds) Chronobiology in allergy and immunology. Thomas, Springfield, pp 36–63

Reinberg A, Guillet P, Gervais P, Ghata J, Vignaud D, Abulker C (1977b) One-month chronotherapy (Dutimelan, 8–15 mite). Control of the asthmatic condition without adrenal suppression and circadian rhythm alteration. Chronobiologia 4: 295–312

Reinberg A, Gervais P, Chaussade M, Fraboulet G, Duburque B (1983) Circadian changes in effectiveness of corticosteroids

in eight patients with allergic asthma. J Allergy Clin Immunol 71: 425–433

Reinberg A, Pauchet F, Ruff F, Gervais A, Smolensky MH, Levi F, Gervais P, Chaouat D, Abella M-L, Zidani R (1987) Comparison of once-daily evening versus morning sustained-release theophylline dosing for nocturnal asthma. Chronobiol Int 4: 409–419

Reinberg A, Smolensky MH, D'Alonzo GE, McGovern JP (1988a) Chronobiology and asthma. III. Timing corticotherapy to biological rhythms to optimize treatment goals. J Asthma 25: 219–248

Reinberg A, Touitou Y, Botbol M, Gervais P, Chaouat D, Levi F, Bicakova-Rocher A (1988b) Oral morning dosing of corticosteroids in long term treated cortico-dependent asthmatics: increased tolerance and preservation of the adrenocortical function. Annu Rev Chronopharmacol 5: 209–212

Rhind GB, Connaughton JJ, McFie J, Douglas NJ, Flenley DC (1984) The effect of theophylline on nocturnal wheeze and sleep quality in adults with asthma. Annu Rev Respir Dis 129: A 45

Rivington RN, Calcutt L, Child S, MacLeod PJ, Hodder RV, Stewart JH (1985) Comparison of morning versus evening dosing with a new once-daily oral theophylline formulation. Am J Med 79 (suppl 6 A): 67–72

Ryan G, Latimer KM, Dolovich J, Hargreave FE (1982) Bronchial responsiveness to histamine: relationship to diurnal variation of peak flow rate, improvement after bronchodilator, and airway calibre. Thorax 37: 423–429

Shelhamer JH, Metcalfe DD, Smith LJ, Kaliner M (1980) Abnormal adrenergic responsiveness in allergic subjects: analysis of isoproterenol-induced cardiovascular and plasma cyclic adenosine monophosphate responses. J Allergy Clin Immunol 66: 52–61

Skoogh E-E, Svedmyr N (1984) Beta$_2$-adrenoceptor stimulation inhibits ganglionic transmission in ferret trachea. Am Rev Respir Dis 129: A 232

Sly PD, Landau LI (1986) Diurnal variation in bronchial responsiveness in asthmatic children. Pediatr Pulmonol 2: 344–352

Smolensky MH (1989) Chronopharmacology of theophylline and beta-sympathomimetics. In: Lemmer B (ed) Cellular and biochemical aspects of chronopharmacology. Dekker, New York, pp 65–113

Smolensky MH, Halberg F (1977) Circadian rhythm in airway patency and lung volumes. In: McGovern JP, Smolensky MH, Reinberg A (eds) Chronobiology in allergy and immunology. Thomas, Springfield, pp 117–138

Smolensky MH, Reinberg A, Queng JT (1981) The chronobiology and chronopharmacology of allergy. Ann Allergy 47: 237–252

Smolensky MH, Barnes PJ, Reinberg A McGovern JP (1986a) Chronobiology and asthma. I. Day-night differences in bronchial patency and dyspnea and circadian rhythm dependencies. J Asthma 23: 321–343

Smolensky MH, Scott PH, Barnes PJ, Jonkman JHG (1986b) The chronopharmacology and chronotherapy of asthma. Annu Rev Chronopharmacol 2: 229–273

Smolensky MH, D'Alonzo GE, Kunkel G, Barnes PJ (eds) (1987a) Circadian rhythm-adapted theophylline chronotherapy for nocturnal asthma. Chronobiol Int 4: 301–466

Smolensky MH, D'Alonzo GE, Kunkel G, Barnes PJ (1987b) Day-night patterns in bronchial patency and dyspnea: basis for once-daily and unequally divided twice-daily theophylline dosing schedules. Chronobiol Int 4: 303–318

Smolensky MH, McGovern JP, Scott PH, Reinberg A (1987c) Chronobiology and asthma. II. Body-time differences in the kinetics and effects of bronchodilator medications. J Asthma 24: 91–134

Smolensky MH, Scott PH, Harrist RB, Hiatt PH, Wong TK, Baenzinger JC, Klank BJ, Marbella A, Meltzer A (1987d) Administration-time dependency of the pharmacokinetic behavior and therapeutic effect of a once-a-day theophylline in asthmatic children. Chronobiol Int 4: 435–447

Soutar CA, Costello J, Ijaduola O, Turner-Warwick M (1975) Nocturnal and morning asthma: relationship to plasma corticosteroids and response to cortisol infusion. Thorax 30: 436–440

Spaulding HS, Mansfield LE, Stein MR, Sellner JC, Gremillion DE (1982) Further investigation of the association between gastroesophageal reflux and bronchoconstriction. J Allergy Clin Immunol 69: 516–521

Spector SL (1982) (Chairman, Scientific Assembly on Allergy and Clinical Immunology). ATS News, Fall, no 5

Steinijans VW, Trautmann H, Johnson E, Beier W (1987) Theophylline steady-state pharmacokinetics: recent concepts and their application in chronotherapy of reactive airway diseases. Chronobiol Int 4: 331–347

Sullivan CE, Zamel N, Kozar LF, Murphy E, Phillipson EA (1979) Regulation of airway smooth muscle tone in sleeping dogs. Am Rev Respir Dis 119: 87–99

Tammeling GJ, De Vries K, Kruyt EW (1977) The circadian pattern of the bronchial reactivity to histamine in healthy subjects and patients with obstructive lung disease. In: Smolensky MH, Reinberg A, McGovern JP (eds) Chronobiology in allergy and immunology. Thomas, Springfield, pp 139–150

Toogood JH, Baskerville JC, Jennings B, Lefcoe NM, Johansson SA (1982) Influence of dosing frequency and schedule on the response of chronic asthmatics to the aerosol steroid, budesonide. J Allergy Clin Immunol 70: 28–298

Turner-Warwick M (1984) Definition and recognition of nocturnal asthma. In: Barnes BJ, Levy J (eds) Nocturnal asthma. International Congress and Symposium Series, no 73. Royal Society of Medicine, London, pp 3–5

Turner-Warwick M (1988) Epidemiology of nocturnal asthma. Am J Med 85 (suppl 1 B): 6–8

Van Aalderen WMC, Postma DS, Löeter GH, Gerritsen J, Knol K (1987) Increase in airway hyperreactivity during the night in asthmatic children. Res Dis 135: 460

Whyte KF, Douglas NJ (1989) Posture and nocturnal asthma. Thorax 44: 579–581

Wilkens JH, Wilkens H, Heins M, Kurtin L, Oellerich M, Sybrecht GW (1987) Treatment of nocturnal asthma: the role of sustained-release theophylline and oral $\beta$-2-mimetics. Chronobiol Int 4: 387–396

Wilson NM, Charette L, Thomas A, Silverman M (1984) The acid test for gastroesophageal reflux in children with asthma. Thorax 39: 695–696

Wilson NM, Charette L, Thomson AH, Silverman M (1985) Gastroesophageal reflux in childhood asthma: the acid test. Thorax 40: 592–597

# Renal Excretion: Rhythms in Physiology and Pathology

J. Cambar, J. C. Cal, and J. Tranchot

## Introduction

Biorhythmicity of living organisms, unicellular or multicellular, animal or vegetable, has been empirically observed for many centuries, but were precisely described only recently.

For decades, the most widely described temporal variations were those of blood pressure (Zadek 1881) and urinary excretion (Vogel 1854). Temporal, structural, and functional variations can be shown in all physiological systems. Histological and biochemical changes have been largely reported in many organs including liver, heart, brain, kidney, and intestine.

When comparing the current chronobiological knowledge concerning renal anatomy and physiology with that about the brain, the digestive tract, or the liver, the scarcity of bibliographic data is very surprising. This is most probably due to the heterogeneity of the kidney, one of the main factors accounting for the difficulty in renal research. Nevertheless, chronobiological observations concerning urinary excretions have been widely reported over the last 30 years in humans and, paradoxally, to a lesser extent in experimental animals, especially dogs and rodents.

As urinary excretion can be easily approached noninvasively, it is understandable why numerous studies have reported temporal changes of more than 100 substances eliminated in healthy or pathologic urines.

In spite of the abundance of literature in this field, we still cannot claim today that the chronobiology of the kidney is, at the present time, well-studied and well-understood. Only a few extensive reviews have appeared during the last decade (Minors et Waterhouse 1981; Moore-Ede et al. 1982; Koopman et al. 1985a, b, 1987, 1989a, b; Cambar et al. 1987). All these papers invariabely ask the same questions about the origin of excretion rhythms, the role of external synchronizers such as meal and exercise, and whether existence of endogenous oscillators exist in renal tissue.

We will try, in this chapter, to present the more recent data on renal chronobiology, especially in renal hemodynamics and tubular functions. Indeed, a better knowledge of renal chronobiology seems to be relevant for better clinical use of important drugs such as nephrotoxic agents (antibiotics, antitumor agents, cyclosporine) or substances leading to changes in renal hemodynamics (antihypertensive vasodilators, for example).

## Insights into Renal Structure and Function

In order to understand the temporal aspects of excretory function, the chronobiologist should be familiar with the structure and function of the kidney. Although, a detailed analysis of renal anatomy and physiology cannot be given here, some minimal information, which is required for the comprehension of the chronophysiology and chronopathology of renal excretion, will be presented.

The kidney is surrounded by a fibrous capsule and can be either unilobular (rat, mouse) or multilobular (dog, humans). A transverse section through a kidney permits one to distinguish distinct zones: the outer layer (or cortex) which appears dark and is easily recognized by numerous tiny red dots; and the glomeruli, which are exclusively present in this zone. Internally adjacent to the cortex, the medulla is subdivided into an outer medulla, a richly vascular area, and an inner medulla. The part of the medulla which projects into the pelvis of the ureter is called the papilla. Blood irrigation plays an important role because the kidneys receive nearly 20% of the cardiac output, and this may be in itself an important factor in renal drug effects and toxicity. Blood enters the kidney by the renal artery then proceeds into the interlobular arteries. Each interlobular artery gives rise to a large number of branches, the afferent arterioles, which directly pass to the glomeruli. The intraglomerular circulation is a capillary network, in which the blood leaves through the efferent arterioles.

The zones of the kidney (cortex and medulla) are the result not only of its vascular organization, but also of the arrangement of the nephrons, the basic units of the kidney. Their number varies depending on the species (about 2 millions in humans, 500 000 in the dog, and 40 000 in the rat), but also according to the organism's age. Each nephron consists of a glomerulus and a long tubule with a very heterogeneous structure, which ends by joining a collecting duct common to numerous nephrons. These collecting ducts unite to form the large ducts of Bellini, which run through the papilla.

In order to understand the complexity of renal physiology and morphology, the heterogeneity of the nephron populations and the structural and functional heterogeneity of each nephron must be considered. Indeed, three groups of nephrons can be described:

1. Superficial cortical nephrons, with the glomeruli located just below the renal surface
2. Deep or juxtamedullary nephrons, with the glomeruli located immediatly above the corticomedullary junction
3. The midcortical nephrons, with the glomeruli located between the former two.

The cortical location of the glomeruli of each group of nephrons is well-correlated with the length of the loop of Henle, this being longer in the juxtamedullary nephrons.

In healthy kidneys, most water-soluble solutes can be filtered through the glomeruli with little or hardly any restriction; for solutes with high molecular weights (60 Kda), the glomerular barrier plays a selective filter role to restrict their passage into the ultrafiltrate. Nevertheless, even in health, many milligrams of plasma proteins can be filtered, but more than 99% are reabsorbed at the tubular level. Other macromolecules can be excreted in urine, e. g., molecules secreted by cells along the entire nephron and the urinary tract (enzymes, specific proteins).

Whatever the group of nephrons may be, each has a glomerulus, which contributes to a selective filtration of certain blood components, and a tubule, which modifies the composition of the glomerular ultrafiltrate by large reabsorption and secretion (organic acids and bases, conjugated or free drugs). The glomerulus (or renal corpuscle) consists of a capillary tuft surrounded by the Bowman's capsule. Blood enters the glomerulus through the afferent arteriole and leaves it through the efferent arteriole, the difference between the entering and leaving blood volume being the glomerular ultrafiltrate. The glomerular filtration rate is proportional to the renal blood flow, the glomerular filtration pressure, and the glomerular filter permeability (permselectivity).

Whatever the population of nephrons may be, the most recent studies have reported the existence of numerous different segments, grossly subdivided into proximal tubules (further subdivided into three parts), limb of Henle (subdivided into four parts), distal tubule (recently subdivided into three parts), collecting tubule (subdivided into three parts), and, finally, collecting tubules forming the papillary collecting duct. Such an intra- and internephron anatomical, physiological, and biochemical heterogeneity characterizes the function of the kidney in urine production. All these different segments exhibit a particular role in forming the urine. The differential role of each nephron part is due, to a large extent, to large differences in hormonal control and in passive and active transmembrane flux.

As the fluid moves out of the proximal tubule into the loop of Henle, the classical process of urinary concentration begins, the loop of Henle acting as a countercurrent multiplier. We must consider, at each nephron level, solute and water flux between tubular lumen, tubular cell, and peritubular blood.

The juxtaglomerular apparatus has an important role in the regulation of the renin-angiotensin system. This complex consists of those portions of the afferent and efferent arterioles which lie near the glomerulus together with the macula densa of the distal tubule belonging to the same nephron. This unique structure, the juxtaposing of early distal tubule and glomerular arterioles, suggests complex mechanisms which link tubular functions and renal hemodynamics.

Particular cells, called granular cells, release a proteolytic enzyme, renin. Renin activates a vasoactive substance, angiotensin I, which, with the help of another enzyme, liberates into the blood angiotensin II, the most potent vasoconstricting agent actually known. The renin-angiotensin system can be considered as an important modulator of intrarenal hemodynamics.

This rapid overview of renal structure and function is meant to highlight that urine excretion is controlled by numerous factors (hemodynamic, glomerular, tubular), each showing circadian changes that can explain the circadian variation in urinary excretion.

## Overview of Renal Function Assessment Techniques

For physiological and toxicological studies, we can distinguish two major types of investigations to assess renal function: (1) techniques to assess renal blood (or plasma) flow (RBF) and glomerular filtration rate (GFR) in order to evaluate renal hemodynamics, and (2) techniques to assess tubular functions based on the analysis of the urinary excretion of several substances and, especially, electrolyte solutes, and renal markers. Histological changes of renal parenchyma, especially at the proximal tubular level, can complete this procedure.

Determinations of RBF and GFR are the most useful measurements of renal function. The choice of the technique can be made according to the aim of the experiment and the experimental model. In the case of physiological investigations, the most accurate techniques involve constant infusions of reference substances, such as inulin and para-amino hippuric acid (PAH), and the withdrawal of multiple blood and urine samples. Indeed, inulin and PAH clearance techniques can be considered as the reference for GFR and RBF estimation. Due to their complexity, other easier techniques have been reported, e.g., a single injection clearance technique that needs only a single injection of reference testing substance, multiple blood samples, but no urine collection.

More recently, the renal GFR has been calculated from the analysis of the manner by which kidney takes up radioactive tracer using direct radionuclide imaging studies; this procedure does not require blood sampling.

In patients and in experimental animals in prolonged experiments extending over many days, only noninvasive blood and urine sampling can be used. The GFR can be evaluated by this approach with two commonly used procedures: (1) when only blood samples are collected, the evaluation of blood urea nitrogen (BUN) and creatinine is the most commonly used procedure for monitoring GFR in humans and in animals. (2) when blood and urine are collected, the urea and creatinine clearances can be determined giving an approximation of GFR.

The analysis of urine, considered as a "liquid biopsy", gives a good image of the functional and structural integrity of the kidney. Urine analysis offerts a valuable diagnostic for kidney dysfunction. The nonspecificity of urine analysis also can provide a clinical advantage for other organ pathologies. With the pro-gress of analytical techniques, the number of solutes that can be determined in urine is becoming larger and larger; for example, water, electrolytes, proteinuria, tubular enzymes, casts, hematuria, glucose, and renal cells can be mentioned as the more common ones. An exhaustive review presents the findings on experimental and clinical urinalysis (Piperno 1981).

## Excretion Chronophysiology

Circadian changes in urinary flow in humans have been described during the middle of the last century and present one of the earliest proofs of a temporal structure in humans (Vogel 1854). For over 100 years, and especially since 1950, many clinicans have investigated circadian and circannual variations in many components of human urine (Ghata and Reinberg 1954; Reinberg et al. 1971, 1981; Minors and Waterhouse 1982). In healthy adult subjects, more or less large circadian variations are found for nearly all urinary solutes.

### Electrolytes

Circadian variations of 11 urinary variables were reported in young healthy men with a comparable synchronization. Most circadian peaks in urinary excretion are located in the early afternoon, except for chloride, phosphates, and 17-ketosteroids (Touitou et al. 1978).

A more recent large investigation of healthy adult male subjects reports variations in urinary cations, with the excretion maximum at 1900 hours for sodium and magnesium, 1600 hours for calcium, and 1300 hours for potassium; the excretion minimum was at 0400 hours for all four cations (Kanabrocki et al. 1983).

All authors noticed that the acrophase of urinary solute excretion occurs generally during the daily human activity period, with nevertheless some interindividual changes (Koopman et al. 1989a).

Similar chronobiological studies were reported in experimental animals, such as rats (Cohn et al. 1970; Roelfsema et al. 1980; Cambar et al. 1978), mice (Cambar et al. 1981), dogs (Gordon and Lavie 1985), or monkeys (Moore-Ede and Herd 1977). As in humans, the acrophases of urinary volume and solutes occur during the activity span of the animals. The important role of the photoperiod as synchronizer can

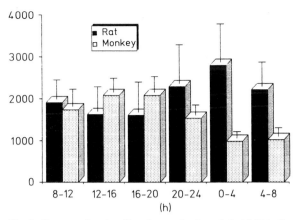

**Fig. 1.** Comparative circadian changes in diuresis (ml/4 h) in the monkey (day active) and in the rat (night active). (Redrawn from Cambar et al. 1981; Moore-Ede et al. 1982) (12/12 DL)

be shown by comparing excretion chronograms in rats (with a nocturnal activity) and in monkeys (with a diurnal activity) (Fig. 1).

The relevance of these observations will be shown later with regard to drug elimination and to the severity of drug nephrotoxicity as a function of the circadian stage of administration or time of intoxication.

Some rhythms, for example those of urinary flow and excretion of sodium, potassium, chloride, and urate, are of considerable amplitude in human subjects and increase during daytime. For example, it can be noted that during the day chloride excretion does not exceed 10% of the total daily excretion and nightly potassium excretion can be four to five times higher than the diurnal one (Bartter et al. 1979).

The phasing of the rhythms of water, sodium, chloride, and potassium are similar but not identical. Thus, chloride and potassium tend to peak earlier than sodium; moreover, the day-by-day variation in the difference in peak times between sodium and chloride is less than that between sodium and potassium. This implies that the mechanisms for sodium and chloride reabsorption are linked more closely than those for sodium and potassium. This is in agreement with what is known about tubular function (Koopman et al. 1989a).

Other noncircadian excretory rhythms have been reported for electrolytes, e.g., ultradian ones in dogs (Gordon et Lavie 1982) and in humans (Lavie et Kripke 1977) with a 90 min period (Lavie et Kripke 1981); or infradian, menstrual cycle, circaseptan (Uezono et al. 1984), or circanual rhythms in humans (Robertson et al. 1977), dogs (L'Azou 1989), and rodents, e.g., hamsters (Haberey et al. 1967) desert rodents (Amirat et al. 1980).

In addition to the electrolyte excretory rhythms, which have been largely described for decades, about ten other solutes also show a rhythmic excretion including hormones, metabolites, biological markers, or renal enzymes.

## Enzymes

Circadian variations in enzyme excretion (enzymuria) have been described in rats (Grotsch et al. 1985; Cal et al. 1987), in children (Feldman et al. 1989), and in adults (Jacey 1968; Maruhn et al. 1977; Lakatua et al. 1982). More recently, an infradian rhythm of the urinary excretion of numerous enzymes has been reported (Burchardt et al. 1988).

When we consider the exclusively tubular origin of the enzymes present in the urine, we can indirectly point out temporal changes in biosynthetic or catabolic activities of the renal tissue, particularly in renal biomembranes. This has been directly shown in renal parenchyma (Van Pilsum and Halberg 1964; Nagai et al. 1975; Franciolini et al. 1979) but not yet at a precise nephron level because of the complexity and heterogeneity of the kidney (Guder and Ross 1984). Compared with the extensive studies on the enzymatic activities in liver, brain, and the digestive tract, similar ones in renal tissue are rather scarce.

In the rat, circadian variations in renal $\gamma$-glutamyl-transferase (GGT) activity have been demonstrated (Hoffman and Hardeland 1981), with a significant increase at the end of the light period. This finding is similar to our own observation on the urinary excretion of this enzyme, which peaks at the beginning of the dark period (Cal et al. 1987) (Fig. 2). Circadian

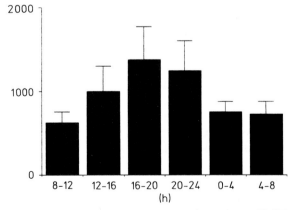

**Fig. 2.** Circadian changes in $\gamma$-glutamyl-transferase (GGT) (mU/4 h) urinary excretion in the rat. (redrawn from Cal et al. 1987) (12/12 DL)

and circannual variations of several enzymes of lysosomal origin have been recently reported in human plasma (Goi et al. 1989).

## Urinary pH and Drug Elimination

Finally, we can report circadian and circanual variations in urinary pH (Robertson et al. 1977). Large pH changes, between 5.0 and 8.0, can be noted during the day, with a minimum during subject sleep and an increase at awakening of the patient, with a maximum about at 1200.

Urinary pH plays an important role in determining the ionized or nonionized state of a molecule and, as a consequence, controls its excretion. The excretion of a drug will increase if it is ionized and the reabsorption of a drug will increase if it is nonionized. There is always an equilibrium between the basic and acidic forms of a drug. So it can be assumed that increased acidity of the nocturnal urine (low pH) will promote the nocturnal excretion of basic drugs and decrease that of acidic ones (Waterhouse and Minors 1989). Excretion of some drugs is particularly affected by urine pH, e.g. hydrochlorothiazide, penicillin, amphetamines, and phenobarbital (Brater and Chennavasin 1984).

Such circadian changes in renal elimination of drugs are an important factor in chronopharmacokinetics, as has been reviewed previously (Reinberg et al. 1984).

## Excretion in Newborn and Aged Subjects

We have presented the urinary excretion rhythms in standard healthy human subjects (often young volunteers such as students or nurses). We can also report similar studies in other age groups, either very young (newborn babies) or in clinically healthy elderly subjects.

In newborn babies, renal immaturity during the first month of life is characterized by low glomerular and tubular functions. Absence or poor amplitude of excretory rhythms in the newborn period are not surprising because of the immaturity of the endogenous circadian system.

During the first week of life, the newborn presents essentially ultradian rhythms and only progressively do the circadian rhythms of the different functions become apparent at different times after birth. Heart rate, urine volume, and sodium and creatinine excretion become circadian periodic between the 16th and 20th week (Hellbrügge 1960; Hellbrügge et al. 1964).

In aged subjects, it is well-established that renal functions decrease dramatically, i.e., GFR decreases because of a decrease in the number of filtering glomeruli. We have to be aware of the practical implications of such a decrease in hemodynamic and tubular function in the medical treatment of elderly subjects, i.e., with potentially toxic pharmacological agents. Circadian rhythm alterations in the elderly have been widely reported in the literature (Minors and Waterhouse 1981; Touitou 1982, Touitou et al. 1989 a, b). A recent paper reports extensively on circadian and seasonal variations of numerous electrolytes and solutes in young adults and aging humans. Temporal changes could be detected in the young adult and in elderly human subjects which, in both age groups, showed relatively small amplitudes (Touitou et al. 1989 b). Moreover, very recent papers about physiopathologic changes in magnesium metabolism during aging show the prevalence of magnesium deficiency in an elderly population and the existence of circadian and seasonal rhythms in plasma magnesium (Touitou et al. 1987, 1989 a). The seasonal variations were even more marked in the elderly subjects than in the young ones (Touitou et al. 1989 a).

## Mechanisms of Excretory Rhythms

Although it is clear that in mammals the source of the circadian rhythm in urinary excretion is endogenous in origin, it is less obvious whether the source is extra- or intrarenal.

The kidney is subject to multiple rhythmic influences, any of which could potentially drive the renal solute excretion rhythms. Most of the major blood components can freely filter at the glomerular level but are almost completely reabsorbed by the tubules. As we have already noted, the amount of substance excreted in the urine results from the balance between what is filtered by the glomeruli and what is reabsorbed or secreted by the tubules.

As has been well-studied by Moore-Ede et al. (1978–1982), there are three possible sources for circadian variations in the excretion of a urinary constituent: exogenous, extrarenal endogenous, and intrarenal endogenous ones.

Very few renal rhythms are directly dependent solely on exogenous sources, such as ingested food or water. Nevertheless, the peak in urinary excretion of water, electrolytes, or nitrogen compounds can occur when food intake is maximal in the rat (Cambar et al. 1981). In contrast, with a constant intravenous or intragastric infusion of sodium chloride or a nutritive solution (without any food or water intake), the circadian excretion rhythms of rats remain unchanged (Roelfsema and Van der Heide 1982, Roelfsema et al. 1982). Thus, the resultant excretion rhythms in human subjects and experimental animals living with ad libitum food intake are a combination of exogenous and endogenous infuences.

Rhythms in the excretion of some urinary constituents, such as cortisol and cortisol metabolites are passive responses to circadian variations in plasma solute concentrations which are under the control of endogenous circadian oscillators. The major determinant of the urinary cortisol (and 17-CS and 17-OH-CS) rhythms is the circadian plasma cortisol concentration, which is driven by a series of endogenous circadian oscillators located in the adrenal cortex and influenced in their timing by other oscillators located in other parts of the body (discussed in another chapter of this book).

The third source of urinary excretion rhythms is intrarenal endogenous oscillators. The already described complexity of the structure and function of the mammalian kidney permits one to understand why it is so difficult to define precisely whether the source of renal rhythmicity is intrarenal or extrarenal.

More detailed explanations on the chronophysiology of urinary potassium excretion were presented in a well-documented review (Moore-Ede et al. 1982). These investigators showed that the circadian

rhythm of plasma cortisol acts as an "internal mediator timing the renal K rhythm". Moreover, the behavior of the potassium rhythm, when cortisol rhythms are manipulated, suggests that there is "an intrarenal circadian oscillator which generates the excretion rhythm."

In a more general consideration (Fig. 3), we will consider the extrarenal sources as systemic hemodynamics (e. g., blood pressure and heart rate), systemic nervous control (e. g., cholinergic or adrenergic), and solute plasma concentration (the filtered solute load depends upon GFR and plasma concentration). We will consider also all the intrarenal factors, such as renal hemodynamics (RBF, GFR, vasoactive hormone regulation) or tubular reabsorption-secretion processes (tubular ultrastructure, metabolism, and hormone rhythms such as ADH, PTH, or aldosterone). As shown in Fig. 3, with all of the rhythmic factors contributing to a rhythmicity in urinary excretion, it is difficult to isolate the respective role of one particular.

The circadian changes in amplitude in plasma concentration of most substances are generally small, except for phosphate or urea. Nevertheless, a circadian rhythm in plasma concentration of a solute may contribute to the circadian rhythm in its filtered load. The relationship between plasma concentration and urine excretion is unclear.

Simultaneous circadian changes in plasma concentration and urine excretion have been described for

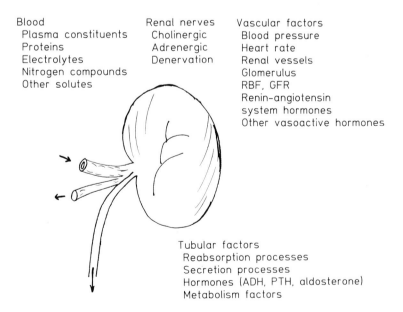

**Fig. 3.** The main factors causing circadian rhythmicity in urine excretion

circulating blood components, e.g., sodium, potassium, or phosphate (Mills 1966). Plasma potassium concentration has a circadian range of about 10% relative to a plasma maximum at the time of the maximum in urinary excretion (Moore-Ede et al. 1975).

Nevertheless, the renal responsiveness, to a change in plasma potassium concentration varies markedly with the time of the day; and the urinary potassium rhythm cannot be considered as a simple passive consequence of the circadian variations in serum potassium concentration (Moore-Ede et al. 1982). It can be concluded that when plasma rhythms influence those of urinary excretion, they cannot be the major determinant.

Among the very numerous factors controlling urinary excretion, systemic and renal hemodynamics can be mentioned first, since the RBF amounts to 20% of the cardiac output.

Blood pressure and heart rate show large circadian changes with a peak between 1600 and 2000 hours and a minimum during the night (Millar-Craig et al. 1978). It is therefore not surprising that the RBF changes during the day. A circadian rhythm in PAH clearance has been reported by some but not by all investigators (Koopman 1989a). The chronogram of the circadian rhythms in PAH clearance seems to be similar (in phase) with that of blood pressure (Brod and Fencl 1957; Wesson 1964; Delea 1979).

Likewise, a GFR circadian variation can be detected in the rat and in humans, with both creatinine or inulin clearance technique. In humans, GFR is generally lowest during sleep at night and presents a maximum in the middle of the day; the amplitude is small (20%–30% as compared with the MESOR) in

comparison with that of electrolyte excretion (Sirota et al. 1950; Wesson and Lauler 1961; Koopman et al. 1989b). Circadian rhythms in GFR have been described also in pathologic conditions in hypertensive subjects (Vagnucci and Wesson 1964) or in heart failure (Wesson 1979).

Likewise, we have shown similar circadian changes in GFR with respect to creatinine clearance in the rat (Cambar et al. 1979) and, very recently, also with inulin clearance (unpublished personal data) (Fig. 4). We noted an increase during the daily dark span. Such circadian changes in local tissue hemodynamics have been found in intestine, liver, and kidney (Labrecque et al. 1988).

Although it has been reported that the phase of GFR rhythm is not exactly identical to those of RBF in humans (Koopman et al. 1989a), we have obtained chronograms in the rat which appeared to be in phase for GFR, RBF, and diuresis (Redon et al. 1989).

In conclusion, it appears that renal hemodynamics (GFR and RBF) present circadian changes with a peak in the middle of the activity period.

The components of the renin-angiotensin system and other circulating vasoactive hormones contribute to regulate renal hemodynamics. All these hormones show circadian changes of their plasma concentration (personal review to be published), including renin, angiotensin, aldosterone, and cortisol or corticosterone in humans (Beilin et al. 1982; Bartter et al. 1979; Richards 1986; Stephenson et al. 1989) and in rats (Hilfenhaus 1976; De Forrest et al. 1979).

Epinephrine and norepinephrine show a peak in humans between 1400 and 1600 (Wisser and Knoll 1975), well-correlated with the acrophase of blood pressure, dopamine, and prostaglandins (Jubiz et al. 1972).

Recently, in our laboratory, closely related circadian changes have been found in normotensive rats in plasma aldosterone, renin, angiotensin II, angiotensin converting enzyme, and atrial natriuretic factor (Laval 1988). However, such a coincidence in the occurrence of the circadian peak values of pressor hormones may not necessarily indicate a causal relationship (Lemmer 1989).

Circadian changes in tubular structure and function can markedly affect the regulation of urinary excretory rhythmicity; but, while the circadian variations in urinary excretion are very well-documented, no temporal variations in tubular anatomy or ultrastructure (for example, the height of the tubular epithelium of the brush border membrane) similar to recent findings in the small intestine (Gardner and Steele 1989), are known. Only one study has detected

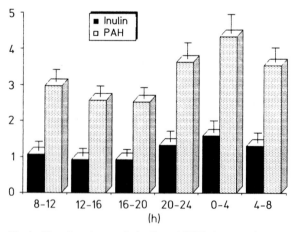

**Fig. 4.** Circadian changes in inulin and PAH clearance (respectively, glomerular filtration rate and renal blood flow) in the rat (Unpublished personal data) (12/12 DL)

circadian variations in autophagic vacuoles in the cytoplasm of rat renal tubular cells, with a maximum during the day (Pfeifer and Scheller 1975).

Circadian changes in hormones known to affect the regulatory mechanisms of renal tubular function (especially reabsorption-secretion processes), e.g., parathyroid hormone, aldosterone, and vasopressin, have already been reported. It has been shown in humans (Jubiz et al. 1972) and in the rat (Simmons et al. 1979) that plasma parathyroid hormone exhibits a circadian rhythm, the maximum level of which coincides with the maximum in phosphate excretion. Likewise, plasma vasopressin exhibits circadian variations in the rat (Greeley et al. 1982) and in humans (Pruzczynski et al. 1984). However, the rhythms in urine flow in well-hydrated men are not correlated with rhythmic secretion of ADH (Lavie et al. 1980). Finally, circadian variations have been reported for aldosterone which correlate with urinary electrolyte excretion in humans (Williams et al. 1972; Bartter et al. 1979; Reinberg et al. 1981; Cugini et al. 1984) and in the rat (De Forrest et al. 1979).

## Excretion Chronopathology

In clinical disorders (e.g., endocrine dysfunction), and particularly in renal pathology (e.g., chronic renal failure, hypertension), renal functions can be dramatically disturbed.

In congestive heart failure, inverted rhythms in GFR, diuresis, and sodium excretion have been reported (Counihan 1979; Wesson 1979; Goldman 1951). Such changes or even a disappearance of excretion rhythms has been reported in other conditions such as hepatic cirrhosis (Goldman 1951; Jones et al. 1952), portal hypertension, orthostatic hypotension (Davidson et al. 1976), or hypoalbuminuric edema with maintenance of a normal potassium excretion rhythm.

Abnormal circadian urinary excretion rhythms have been reported also in diabetic autonomic neuropathy (Bell et al. 1987). In hypertensive patients, comparison of their excretory rhythms with those of healthy subjects has led to conflicting conclusions (Aslanian et al. 1978; Bultasova et al. 1986; Kawasaki et al. 1984; Luft et al. 1980; Natali et al. 1982; Vagnucci et Wesson 1964).

## Endocrine Pathology

Endocrine dysfunction can induce dramatic changes in excretory rhythms. Corticotropic axis dysfunction can change the 24 h plasma cortisol profiles. In Cushing's syndrome, with hypercortisolism (Van Cauter 1989), the circadian variations in plasma cortisol and urinary 17-hydroxy steroids and 17-ketosteroids are invariably absent, with changes in excretory sodium potassium and magnesium rhythms. In contrast, in acutely depressed patients with primary affective disorder, hypercortisolism is accompanied by persistent cortisol circadian rhythmicity.

Likewise, in hypocortisolism, as in Addison's syndrome, in spite of some conflicting results, alterations and even suppression of the circadian rhythm in potassium excretion have been reported (Reinberg et al. 1971). It is assumed that the change in potassium metabolism (plasma and urinary levels) is a result of adrenal dysfunction but presents also a remarkable example of circadian chronopathology (Reinberg 1971) (Fig. 5).

Another example of pathology, enuresis, characterized by excretory disorders, presents as abnormal water and electrolyte rhythms (Lewis et al. 1970; Norgaard et al. 1985; Rittig et al. 1989).

## Renal Pathology

In patients with renal disorders, renal hemodynamics (GFR and RBF) and tubular processes can be decreased. We know the clinical consequences of such a decrease for drug elimination. Concerning this topic, recent and extensive reviews can be consulted (Koopman 1985b; Koopman et al. 1989b). It has been shown that most patients with nephrotic syndrome have a circadian rhythm for albumin, transfer-

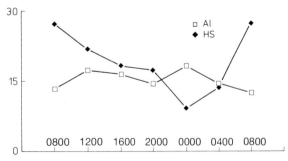

**Fig. 5.** Circadian changes in potassium excretion in healthy subjects *(HS)* and adrenal insufficiency *(AI)* patients. (Redrawn from Reinberg 1971)

rin, and IgG excretion in urine, with a maximum protein excretion at 1600 and a minimum at 0300, even during a regimen of complete bedrest and defined identical meals. Moreover, a circadian rhythm in the selectivity index (IgG clearance/transferrin clearance) was seen in many patients also with a nighttime minimum. More recently, similar circadian variations in urinary $\beta2$-microglobulin excretion were found in a majority of patients (75%) with nephrotic syndrome (Koopman et al. 1987).

### Renal Transplantation

The role of the nervous system in renal physiology is becoming better documented. Renal nerve sympathetic stimulation induces arteriolar vasoconstrictive effects with important renal hemodynamic changes. Renal denervation leads to proximal tubular reabsorption changes, with a significant increase in sodium excretion. An interesting physiopathologic denervated kidney model is represented by renal transplantation. After a transient period during which renal function is disturbed, the patients present a normal renal function without detectable toxic damages due to immunosuppressive agents (corticoids, cyclosporin). We will consider the renal chronotoxicological effects of these agents in another chapter of this book.

Excretion rhythms in renal transplant patients are conflicting, showing either inversion of water and electrolyte rhythms (Berlyne 1968) or rhythms similar to those of healthy subjects (Albertsen 1968; Koene et al. 1973).

In our renal unit, we have investigated the circadian rhythms of electrolyte excretion in transplant patients. We have shown that we must consider the time when the donor kidney was removed and not the time of the remaining kidney of the recipient (Fournier 1983). Circaseptidian rhythm of kidney allograft rejection has been also described (Ratte et al. 1973, 1977; Knapp et al. 1979, 1980; Lévi and Halberg 1981).

### Excretion Chronopharmacology

Consideration of the physiological variations in renal function leads to an understanding of chronopharmacological and chronotoxicological phenomena. Chronopharmacoloy uses information on the organisation in physiological time structure to optimize the clinical effect of a drug by selecting the most favorable time of administration, increasing its desirable pharmacological actions and decreasing its side effects.

Thus far, such approaches have been poorly described for drugs having effects at the renal level, i.e., antihypertensive agents and diuretics. Recent reviews have reported a chronopharmacological approach to antihypertensive agents (Lemmer 1989). Antihypertensive drugs, such as beta-blockers (Raftery et al. 1981), verapamil (Gould et al. 1982), or captopril (Boxho 1988), exhibit circadian variations in their pharmacological effects.

The diuretic, natriuretic, kaliuretic, or aldosteronuric potency of drugs, such as chlorothiazide (Shiotsuka et al. 1973), hydrochlorothiazide (Mills et al. 1977; Simpson 1979), furosemide (Hilfenhaus and Herting 1978), or theophylline (Cambar et al. 1980), is circadian stage-dependent in humans and in the rat.

Chlorothiazide administration to spontaneously hypertensive rats (SHR) does not change plasma sodium and potassium when given at 2200, but markedly decreases them at 1000. In parallel, chlorothiazide-induced potassium urinary excretion reaches its maximum at 0600, which fits well to the maximum of plasma potassium decrease at 1000 (Shiotsuka et al. 1973).

Also in rat, the furosemide-induced potassium and aldosterone loss is circadian stage-dependent, as shown in Fig. 6. When given at 1800, furosemide does not increase potassium and aldosterone excretion, but increases both when given at 0600 (from 1.04 to 1.23 mmol/24 h for kaliuresis and from 12.6 to 26.3 ng/24 h for aldosterone) (Hilfenhaus and Herting 1978) (Fig. 6).

Under the same experimental conditions, we noted a similar chronopharmacological effect with

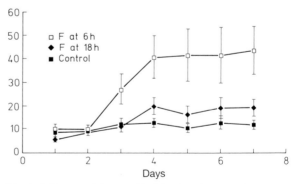

**Fig. 6.** Influence of furosemide (F) administration time on aldosterone excretion (ng/24 h) in the rat (Redrawn from Hilfenhaus and Herting 1978) (12/12 DL)

another diuretic, theophylline, a well-known bronchodilator agent. When given at 1400 hours, theophylline induces a large diuresis and natriuresis increase (respectively, 123% and 223% when compared with controls) but increases diuresis and natriuresis only slightly when given at 0800. Likewise, at 0800, 2000, or 0200 hours, theophylline shows only a slight kaliuretic effect (less than +10%) but dramatically increases it when given at 1400 (+66%) (Cambar et al. 1980).

Similar data have been reported in healthy human subjects with another diuretic, hydrochlorothiazide. Mills et al. (1977) showed that plasma potassium and sodium are lowered more when the drug is given at 1600 and 2000 hours and that natriuresis and diuresis are most markedly increased at 1200. Moreover, headhache and vertigo are most frequently noted at 1600 hours. On the basis of these observations, Mills et al. (1977) suggest that hydrochlorothiazide be given to patients in the morning.

In conclusion, it may be of interest for optimization of the use of diuretics to determine the administration time when water and sodium loss are maximal with a minimum of potassium and aldosterone loss.

The toxicological significance of the physiological temporal variations in renal functions is considered in another section of this book. It seems possible to establish a relation between renal chronophysiology and chrononephrotoxicity. Indeed, the existence of circadian changes in renal hemodynamics with a diurnal maximum can explain why drug elimination during daytime is about 20% higher than during nighttime (Mouveroux 1985). Moreover, the diurnal increase in urine flow in humans leads to a decrease in the drug concentration in the tubular lumen, which diminishes the nephrotoxic side effects of drugs like antibiotics or heavy metals (Cal et al. 1985). For most drugs, especially the water-soluble ones, the main route of elimination is the kidneys. However, only few data exist about the temporal aspects of drug elimination (Waterhouse and Minors 1989).

## Conclusion

The chronobiology of excretion is well-known and has been extensively described, both in humans and in experimental animals, for, e.g., electrolytes, nitrogen compounds, hormones, metabolic components, or trace elements. Excretion chronophysiology is multifactorial and a result of a combination of exogenous and endogenous influences. With such a large number of rhythmic factors contributing to rhythmicity in urinary excretion, it is difficult to isolate the respective role of a particular one. If dietary factors and postural changes are important exogenous influences, these changes are not wholly responsible for renal circadian rhythms, because excretory rhythms remain unchanged when diet and posture are well-controlled. A good knowledge of renal morphophysiology, especially structural and functional nephron heterogeneity, is necessary for a sound renal chronophysiological approach. Thus, a well-established basis of chronophysiology permits to study renal chronopathology, even chronopharmacology, chronotoxicology, and chronotherapeutics, which are very important for practical clinical applications. For such investigations, we must develop in the future new approaches, in vivo and in vitro, for assessing temporal changes in renal ultrastructure and in tubular enzyme activities, especially for ATPases at the brush border membrane. Moreover, for a practical clinical point of view, many mechanistic investigations must be carried out in renal chronopharmacology to explain such temporal changes in the effects of renal drugs such as diuretic, antihypertensive, or vasoactive agents.

## References

Albertsen K (1968) Urinary rhythm after renal transplantation. Lancet 2: 636–637

Amirat Z, Khammar F, Brudieux R (1980) Comparative seasonal variations of adrenal weight in two species of desertic rodents (Psammomys obesus and Gerbillus gerbillus). Mammalia 44: 399–406

Aslanian NL, Assatrian DG, Bagdassarian RA, Kurginian AG, Shukhian UM (1978) Circadian rhythms of electrolyte excretion in hypertensive patients and healthy subjects. Chronobiologia 5: 251–261

Bartter FC, Chan JCM, Simpson HW (1979) Chronobiological aspects of plasma renin activity, plasma aldosterone and urinary electrolytes. In: Krieger DT (ed) Endocrine rhythms. Raven, New York, pp 225–245.

Beilin LJ, Deacon J, Michael C, Vandongen R, Lalor C, Barden A, Davidson L (1982). Circadian rhythms of blood pressure and pressor hormones in normal and hypertensive pregnancy. Clin Exp Pharmacol Physiol 9: 321–326.

Bell GM, Reid W, Ewing DJ, Doig A, Cumming AD, Watson ML, Clarke BF (1987) Abnormal diurnal urinary sodium and water excretion in diabetic automatic neuropathy. Clin Sci 73: 259–265.

Berlyne GM (1968) Abnormal urinary rhythm after renal transplantation in man. Lancet 2: 435–436.

Boxho G (1988) Twenty four hour blood pressure recording

during treatment with captopril twice daily. Curr Ther Res 44: 361–366

Brater DC, Chennavasin P (1984) Effects of renal disease pharmacokinetic considerations. In: Benet LZ, Massoud N, Gambertoglio JG (eds) Physiological basis for drug treatment Raven, New York, pp 119–147

Brod J, Fencl V (1957) Diurnal variations of systemic and renal hemodynamics in normal subjects and in hypertensive disease. Cardiologia 31: 494–497

Bultasova H, Veselkova A, Brodan V, Pinsker P (1986). Circadian rhythms of urinary sodium, potassium and some agents influencing their excretion in young bordline hypertensives. Endocrinol Exp 20: 359–369

Burchardt U, Winkler K, Klagge M (1988) Infradian biorhythms of enzymuria in man? J Clin Chem Clin Biochem 26: 491–496

Cal JC, Dorian C, Cambar J (1985) Circadian and circanual changes in nephrotoxic effects of heavy metals and antibiotics. Annu Rev Chronopharmacol 2: 143–176.

Cal JC, Lemoigne F, Crockett R, Cambar J (1987). Circadian rhythm in gamma glutamyl-transpeptidase and leucine aminopeptidase urinary activity in rats. Chronobiol Int 4: 153–160

Cambar J, Toussaint C, Nguyen BA C (1978) Etude des rythmes circadiens de l'excrétion des électrolytes et des protéines urinaires chez le rat. CR Soc Biol 172: 103–109

Cambar J, Lemoigne F, Toussaint C (1979) Etude des variations nycthémèrales de la filtration glomérulaire chez le rat. Experientia 35: 1607–1609

Cambar J, Toussaint C, Lemoigne F, Canellas J (1980) Circadian rhythm in the diuretic effects of theophylline. In: Smolensky MH (ed) Recent advances in the chronobiology of allergy and immunology. Pergamon, Oxford, pp 126–136

Cambar J, Lemoigne F, Toussaint C, Cales P, Crockett R (1981) Variations circadiennes de l'excrétion des électrolytes et des substances azotées dans l'urine des souris et des rats. J Physiol (Paris) 77: 887–890

Cambar J, Dorian C, Cal JC (1987) Chronobiologie et physiopathologie rénale. Path Biol 35: 977–984.

Cohn C, Webb L, Joseph D (1970) Diurnal rhythm of urinary electrolyte excretion by the rat: influence of feeding habits. Life Sci 9: 803–809

Counihan TB, Dunne A, Ryan MF, Ryan MP (1979) Tissue-related changes in urinary aldosterone and electrolyte excretion during amiloride administration to congestive heart failure patients. Brit J Pharmacol 67: 494 P

Cugini P, Centanni M, Murano G (1984) Toward a chronophysiology of circulating aldosterone. Biochem Med 32: 270–282

Davidson CD, Smith DB, Morgan LM (1976) Diurnal pattern of water and electrolyte excretion and body weight in idiopathic orthostatic hypotension. The effect of three treatments. Am J Med 61: 709–715

De Forrest JM, Davis JO, Freeman RH, Stephens GA, Watkins BE (1979) Circadian changes in plasma renin activity and plasma aldosterone concentration in two-kidneys hypertensive rats. Hypertension 1: 142–149

Delea CS (1979) Chronobiology of blood pressure. Nephron 23: 91–97

Feldmann D, Flandrois C, Jardel A, Phan TM, Aymard P (1989) Circadian variations and reference intervals for some enzymes in urine of healthy children. Clin Chem 35: 864–873

Fournier C (1983) Devenir du rythme nycthéméral de la diurèse chez le transplanté rénal en période post-opératoire. Dissertation, Bordeaux

Franciolini F, Becciolini A, Casati V, Cremoni D, Giache V, Porcian S (1979) Circadian activity of rat kidney enzymes. Experientia 35: 582-583

Gardner MLG, Steele DZ (1989) Is there circadian change in villus height in rat small intestine? Q J Exp Physiol 74: 257

Ghata J, Reinberg A (1954) Variations nycthémèrales, saisonnières et géographiques de l'élimination urinaire du potassium et de l'eau chez l'homme adulte sain. CR Acad Sci 239: 1680–1682

Goi G, Fabi A, Lombardo A, Caimi L, Tetamanti G, Montalbetti N, Cavalleri M Halberg F (1989) Chronobiological study of several enzymes of lysosomal origin in human plasma. Chronobiologia 16: 93–102

Goldman R (1951) Studies on diurnal variation in water and electrolyte excretion: nocturnal diuresis of water and sodium in congestive cardiac failure and cirrhosis of the liver. J Clin Invest 30: 1191–1199

Gordon C, Lavie P (1982) Ultradian rhythms in renal excretion in dogs. Life Sci 31: 2727–2731

Gordon CR, Lavie P (1985) Day-night variations in urine excretion and hormones in dogs: role of autonomic innervation. Physiol Behav 35: 175–181

Gould BA, Mann S, Kieso H, Subramanian VB, Raftery EB (1982) The 24 hour ambulatory blood pressure profile with verapamil. Circulation 65: 22–27

Greeley GH, Morris M, Eldridge JC, Kizer JS (1982) A diurnal plasma vasopressin rhythm in rats. Life Sci 31: 2843–2848

Grotsch H, Hropot M, Klaus E, Malerczyk V, Mattenheimer H (1985) Enzymuria of the rat: biorhythms and sex differences. J Clin Chem Clin Biochem 23: 343–347

Guder WG, Ross BD (1984) Enzyme distribution along the nephron. Kidney Int 26: 101–111

Haberey P, Canguilhem B, Kayser C (1967) Evolution saisonnière de l'élimination urinaire du sodium et du potassium chez le Hamster d'Europe (Cricetus cricetus). CR Soc Biol 161: 2044–2048

Hellbrugge T (1960) The development of circadian rhythms in infants. Biological Clocks. Cold Spring Harbor Symp Quant Biol 25: 311–323

Hellbrugge T, Enrengut-Lange J, Rutenfranz J, Steur K (1964) Circadian toxicity of physiological functions in different stages of infency and childhood. Ann NY Acad Sci 117: 361–373

Hilfenhaus M (1976) Circadian rhythm of the renin-angiotensin aldosterone system in the rat. Arch Toxicol 36: 305–316

Hilfenhaus M, Herting T (1978) Furosemide-induced hyperaldosteronism and potassium loss in rats as modified by the time of drug administration Chronobiologia: 5: 189

Hoffmann J, Hardeland R (1981) Diurnal rhythmicity in rat kidney gamma glutamyltranseptidase activity. Arch Int Physiol Bioch 89: 245–247

Jacey M (1968) Circadian cycles of lectin deshydrogenase in urine and blood plasma: response to high pressure. Aerospace Med 39: 410–412

Jones R, Mc Donald GO, Last JH (1952) Reversal of diurnal variation in renal function in cases of cirrhosis with ascites. J Clin Invest 31: 326–334

Jubiz W, Canterbury JM, Reiss E (1972) Circadian rhythm in serum parathyroid hormone concentration in human subjects: correlation with serum, calcium, phosphate, albumin and growth hormone levels. J Clin Invest 21: 2040–2046

Kanabrocki EL, Scheving LE, Olwin JH, Mařks GE, Mc Cormick JB, Halberg F, Greco J, De Bartolo M, Nemchausky BA, Kaplan E, Sothern R (1983) Circadian variation in the urinary excretion of electrolytes and trace elements in men. Am J Anat 166: 121–148

Kawasaki T, Ueno M, Uezono K, Kawano T, Abe I, Kawazoe N, Eto T, Fukiyama K, Omae T (1984) The renin-angiotensin-aldosterone system and circadian rhythm of urine variables in normotensive and hypertensive subjects. Jpn Circ J 48: 168–176

Knapp MS, Cove-Smith JR, Dugdale R, Mc Kenzie N, Pownall R (1979) Possible effect of tissue on renal allograft rejection. Br Med J 1: 75–77

Knapp MS, Byron NP, Pownall R, Mayor P (1980) Tissue of day taking immunosuppressive agents after renal transplantation: a possible influence on graft survival. Br Med J 280: 1382

Koene R, Van Liebergen F, Wijdeveld P (1973) Normal diurnal rhythm in the excretion of water and electrolytes after renal transplantation. Clin Nephrol 1: 266–270

Koopman MG, Krediet RT, Arisz L (1985a) Circadian rhythms and the kidney. A review. Neth J Med 28: 416–423

Koopman MG, Krediet RT, Zuyderhandt FJM, Demoor EAM, Arisz L (1985b) A circadian rhythm of proteinuria in patients with a nephrotic syndrome. Clin Sci 69: 395–401

Koopman MG, Krediet RT, Zuyderhandt FJM, Demoor EAM, Arisz L (1987) Circadian rhythm of urinary $\beta$2-microglobulin excretion in patients with a nephrotic syndrome. Nephron 45: 140–146

Koopman MG, Minors DS, Waterhouse JM (1989a) Urinary and renal circadian rhythms. In: Arendt J, Minors DS, Waterhouse JM (eds) Biological rhythms in clinical practice. Wright, London, pp 83-98

Koopman MG, Koomen GCM, Krediet RT, Demoor EAM, Hoek FJ, Arisz L (1989b) Circadian rhythm of glomerular filtration rate in normal individuals. Clin Sci 77: 105–111

Labrecque G, Belanger PM, Dore F, Lalande M (1988) 24 hours variations in the distribution of labelled microspheres to the intestine liver and kidneys. Annu Rev Chronopharmacol 5: 445–448

Lakatua DJ, Blomquist CH, Haus E, Sackettlundeen L, Berg H, Swoyer J (1982) Circadian rhythm on urinary N-acetyl-$\beta$-glucosaminidase (NAG) of clinically healthy subjects. Timing and phase relation to other urinary circadian rhythms. Am J Clin Pathol 1: 69–78

Laval M (1988) Etude expérimentale et clinique de la structure temporelle du système rénine-angiotensine. Thèse de doctorat d'état de pharmacie, Université de Bordeaux II

Lavie P, Kripke DF (1977) Ultradian rhythms in urine flow in walking humans. Nature 269: 142–143

Lavie P, Luboshitzky R, Kleinhouse N, Shenorr Z, Barzilai D, Glick SM, Leroith D, Levy J (1980). Rhythms in urine flow are not correlated with rhythmic secretion of ADH in well-hydrated men. Horm Metab Res 12: 66–70

Lavie P, Kripke D (1981) Ultradian circa 1: 1/2 hour rhythms: a multioscillatory system. Life Sci 29: 2445–2450

L'Azou B (1989) Variations circannuelles de l'hémodynamique rénale et de la fonction tubulaire chez le chien éveillé. Dissertation in pharmacy, University of Bordeaux II

Lemmer B (1989) Circadian rhythms in the cardio-vascular system. In: Arendt J, Minors DS, Waterhouse JM (eds) Biological rhythms in clinical practice. Wright, London, pp 51–70

Levi F, Halberg F (1981) Circaseptan (about 7-day) bioperiodicity – spontaneous and reactive – and the search for pacemakers. La Ricerca Clin Lab 12: 323–370

Lewis HE, Lobban MC, Tredre BG (1970) Daily rhythms of renal excretion in a child with nocturnal enuresis. J Physiol 210: 42–43

Luft FC, Weinberger MH, Grim CE, Henry DP, Fineberg NS (1980) Nocturnal urinary electrolyte excretion and its relationships to the renin system and sympathetic activity in normal and hypertensive man. J Lab Clin Med 95: 395–401

Maruhn D, Strozyk K, Gielow L, Bock KD (1977) Diurnal variations of urinary enzyme excretion. Clin Chim Acta 75: 427–433

Millar-Craig MW, Bishop CN, Raftery EB (1978) Circadian variation of blood pressure. Lancet 1: 795–797

Mills JN (1966) Human circadian rhythms Physiol Rev 46: 128–171

Mills JN, Waterhouse JM, Minors DS (1977) Chronopharmacological studies on hydrochlorothiazide. Int J Chronobiol 4: 267–294

Minors DS, Waterhouse JM (1981) The kidney and hormones affecting it. In: Minors DS, Waterhouse JM (eds) Circadian rhythms and the human. Wright, London, pp 68–94

Minors DS, Waterhouse JM (1982) Circadian rhythms of urinary excretion: the relationship between the amount excreted and the circadian changes. J Physiol 327: 39–51

Moore-Ede MC, Herd JA (1977) Renal electrolytes circadian rhythms: independence from feeding and activity patterns. Am J Physiol 232: F 127–F 135

Moore-Ede MC, Brennan MF, Ball MR (1975) Circadian variation of intercompartmental potassium fluxes in man. J Appl Physiol 38: 163–170

Moore-Ede MC, Meguid MM, Fitz-Patrick GF, Boyden CM, Ball MR (1978) Circadian variations in response to potassium infusion. Clin Pharmacol Ther 23: 218–227

Moore-Ede MC, Sulzman FM, Fuller CA (1982) Renal function. In: Moore-Ede MC (ed) The clocks that time us. Harvard University Press, Cambridge, pp 259–277

Mouveroux L (1985) Approche chronobiologique de l'élimination rénale des médicaments. Dissertation in pharmacy, University of Bordeaux II

Nagai K, Suda M, Yamagishi O, Toyama Y, Nakagawa H (1975) Studies on the circadian rhythm of phospho-enol-pyruvate carboxykinase. III. Circadian rhythm in the kidney. J Biochem 77: 1249–1254

Natali G, Acitelli P, Cicogna S, De Lauro G, De Pietro M, Ruggieri M, Silberkuhl G, Spracca G, Trotta A (1982) Electrolyte urinary excretion in hypertensive conditions. Chronobiologia 9: 257

Norgaard JP, Pederson EB, Djurhuus JC (1985) Diurnal anti-diuretic-hormone levels in enuretics. J Urol 134: 1029–1031

Pfeifer U, Scheller H (1975) A morphometric study of cellular autography including diurnal variations in kidney tubules of normal rats. J Cell Biol 64: 608–621

Piperno E (1981) Detection of drug induced nephrotoxicity with urinalysis and enzymuria assessment. In: Book JB (ed) Toxicology of the kidney. Raven, New York, pp 31–55

Pruzczynski W, Caillens H, Drieu L, Moulonguet-Doleris L, Ardaillou R (1984) Renal excretion of antidiuretic hormone in healthy subjects and patients with renal failure. Clin Sci 67: 307–312

Raftery EB, Millar-Craig MW, Mann S, Balasubramanian V (1981) Effects of treatment on circadian rhythms of blood pressure. Biotel Patient Monitg 8: 113–120

Ratte J, Walberg F, Kuhl JFW, Najarian JS (1973) Variations circadiennes du rejet de l'allogreffe rénale chez le rat. Union Med Canada 102: 289–293

Ratte J, Halberg F, Kuhl JFW, Najarian JS (1977) Circadian and circaseptan variations in rat kidney allography rejection. In:

McGovern J (ed) Chronobiology in allergy and immunology. CC Thomas, Springfield, pp 250

Redon P, Laval M, Cambar J (1989) Chronobiological aspects of the renin-angiotensin system: physiology and physiopathology. Annu Rev Chronopharmacol 6: 183–223

Reinberg A (1971) Biological rhythms of potassium metabolism. In: Proceedings 8th Colloquium, Int. Potash Institute, Uppsala, pp 160–180

Reinberg A, Ghata J, Halberg F, Appelbaum M, Gervais P, Boudon P, Abulker C, Dupont J (1971) Distribution temporelle du traitement de l'insuffisance corticosurrénalienne. Essai de chronothérapeutique. Ann Endocrinol Paris 32: 566–573

Reinberg A, Dupont W, Touitou Y, Lagoguey M, Bourgeois P (1981) Clinical chronopharmacology of ACTH 17. Effects on plasma testosterone, plasma aldosterone, plasma and urinary electrolytes (K, Na, Ca, Mg). Chronobiologia 8: 11–31

Reinberg A, Levi F, Smolensky MH (1984) Chronopharmacocinétique clinique. J Pharmacol 15: 95–124

Richards AM, Nicholls MG, Espiner EA, Ikram H, Cullens M, Hinton D (1986) Diurnal patterns of blood pressure heart rate and vasoreactive hormones in normal man. Clin Exp Ther Pract A 8: 153–166

Rittig S, Knudsen UB, Norgaard JP, Pedersen EB, Djurhuus JC (1989) Abnormal diurnal rhythm of plasma vasopressin and urinary output in patients with enuresis. Am J Physiol 256: F664–F671

Robertson WG, Hodgkinson A, Marshall DH (1977) Seasonal variations in the composition of urine from normal subjects: a longitudinal study. Clin Chim Acta 80: 347–353

Roelfsema F, Van der Heide D (1980) Circadian rhythms of urinary electrolytes excretion in freely moving rats. Life Sci 27: 2303–2309

Roelfsema F, Van der Heide D (1982) The influence of intravenous infusion of electrolytes on the diurnal excretory rhythms. Life Sci 30: 771–778

Roelfsema F, Van der Heide D, Smeenk D (1982) The influence of continuous intragastric feeding of liquid food on the diurnal excretion of electrolytes and urea in rats. J Interdiscipl Cycle Res 13: 89–96

Shiotsuka RN, Halberg F, Haus E, Lee JK, Machugh R, Simpson H, Levine H, Ratte J, Najarian J (1973) Results bearing on the chronotherapy of hypertension: saluresis and diuresis without kaliuresis can be produced by properly timing chlorothiazide administration according to circadian rhythms. Int J Chronobiol 1: 358

Simmons DJ, Whiteside LA, Whitson SW (1979) Biorhythmic profiles in the rat skeleton Metab Bone Dis Rel Res 2: 49–64

Simpson HW (1979) Hydrochlorothiazide diuresis in healthy man: review of the circadian mediation. Nephron 23: 98–103

Sirota JH, Baldwin DS, Villareal H (1950) Diurnal variations in renal function in man. J Clin Invest 9: 187–192

Stephenson ZA, Kolka MA, Francesconi R, Gonzalez RR (1989) Circadian variations in plasma renin activity catecholamines and aldosterone during exercise in women. Eur J Appl Physiol Occup Phys 58: 756–764

Touitou Y (1982) Some aspects of the circadian time structure in the elderly. Gerontology 28: 53–67

Touitou Y, Touitou C, Bogdan A, Chasselut J, Beck H, Reinberg A (1978) Circadian rhythm in 11 urinary variables of 7 young healthy men. In: Proceeding IV$^e$ Colloque Biologie Prospective, Masson, Paris 161–165

Touitou Y, Godard JP, Ferment O, Chastang C, Proust J, Bogdan A, Auzeby A, Touitou C (1987) Prevalence of magnesium and potassium deficiencies in the elderly. Clin Chem 33: 518–523

Touitou Y, Touitou C, Bogdan A, Godard JP (1989a) Physiopathological changes of magnesium metabolism with aging. In: Itokawa Y, Durlach J, (eds) Magnesium in health and disease. J Libbey, London, pp 103–110

Touitou Y, Touitou C, Bogdan A, Reinberg A, Motohashi Y, Auzeby A, Beck H (1989b) Circadian and seasonal variations of electrolytes in aging humans. Clin Chim Acta 180: 245–254

Uezono K, Haus E, Swoyer J, Kawasaki T (1984) Circaseptan rhythms in clinically healthy subjects. In: Haus E, Kabat HF (eds) Chronobiology 1982/1983. Karger, Basel, pp 257–262

Vagnucci AI, Wesson LG (1964) Diurnal cycles of renal hemodynamics and excretion of chloride and potassium in hypertensive subjects. J Clin Invest 43: 522–531

Van Cauter E (1989) Endocrine rhythms. In: Arendt J, Minors DS, Waterhouse JU (eds) Biological rhythms in clinical practice. Wright, London, pp 51–70

Van Pilsum JF, Halberg F (1964) Transamidinase activity in mouse kidney: an aspect of circadian periodic enzyme activity. Ann NY Acad Sci 117: 337–353

Vogel J (1854) Klinische Untersuchungen über den Stoffwechsel bei gesunden und kranken Menschen überhaupt und der durch den Urin insbesondere. Arch Vereins Wissensch Heilk 1: 96

Waterhouse JM, Minors DS (1989) Temporal aspects of renal drug elimination. In: Lemmer B (ed) Chronopharmacology cellular and biochemical interactions. Dekker, New York, pp 35–50

Wesson LG (1964) Electrolyte excretion in relation to diurnal cycles of renal function. Plasma electrolyte concentrations and aldosterone secretion before and during salt and water balance changes in normotensive subjects. Medicine 43: 547–592

Wesson LG (1979) Diurnal circadian rhythms of renal function and electrolytes excretion in heart failure. Int J Chronobiol 6: 109–117

Wesson LG, Lauler DP (1961) Diurnal cycle of GFR and sodium and chloride excretion during responses to altered salt and water balance in man. J Clin Invest 40: 1967–1973

Williams GH, Tuck ML, Rose LI (1972) Studies on the control of plasma aldosterone concentration in normal man. Response to sodium chloride infusion. J Clin Invest 51: 2645–2652

Wisser H, Knoll E (1975) Ergebnisse einer Kurzzeitstudie über die Circadiane Rhythmik bei normaler und erhöhter Katecholamineausscheidung im Urin. Clin Chim Acta 59: 1–17

Zadek J (1881) Die Messung des Blutdruks des Menschenmittels des Bach'schen Apparatus. Z Klin Med 2: 509–511

# Inflammatory Reaction and Disease States

G. Labrecque

## Introduction

Clinicians are well aware that the signs and symptoms of rheumatoid arthritis (RA) and osteoarthritis (OA) vary within days and between days. The classic morning stiffness observed in patients with RA is so characteristic that it has become one of the diagnostic criteria of the disease [1]. In the last decade, the circadian and circannual variability in inflammatory and arthritic diseases has been confirmed by clinical studies, and important time-dependent variations have been demonstrated, including the desirable and undesirable side effects of nonsteroidal anti-inflammatory drugs (NSAID). The concepts and principles of chronopharmacology can now be used to optimize the efficacy of NSAID in patients suffering from arthritis, but most physicians must familiarize themselves with this new approach in drug treatment.

The specific objectives of this chapter are: (1) to review the time-reported changes in the clinical signs and symptoms of arthritis, (2) to present the clinical chronpharmacology and chronokinetics of NSAID, (3) to illustrate how the concepts and principles of chronopharmacology can be used to optimize and to individualize drug treatment of arthritis.

## Chronopathology of Inflammatory and Arthritic Diseases

Biological rhythms have been described in experimental inflammation. Investigators have reported that the edema produced by administration of inflammatory agents was larger or appeared faster during the activity period of the animals. Fewer studies attempted to determine the mechanisms of the time-dependent variations in inflammation, and it appears that circadian rhythms in exudation of plasma proteins and in the blood flow to inflamed tissues explain the data. It is not intended to summarize these studies here, because two exhaustive literature reviews have been published recently [2, 3] on the chronobiology, chronopathology, and chronopharmacology of inflammation in laboratory animals. This chapter will review the data obtained in systematic investigations indicating that circadian (~24 h) and circannual (~1 year) variations can be detected in the major symptoms of inflammatory and arthritic states (pain, stiffness, swelling, etc.)

## Arthritic Pain

Temporal patterns in the level of human pain have been described in intractable [4,5] or dental pain [5–8] and in pain induced by heat, cold, or faradic stimuli [6]. Although the hour of appearance of peak or trough pain values differed in experimentally induced pain and in disease-related pain, a circadian rhythm of pain was detected in all studies. In RA patients self-monitoring their symptoms during the waking span while being treated with 100 mg flurbiprofen (Flurbi) administered at 0900 and 2100 hours, Kowanko et al. [9, 10] reported that the intensity of pain varied consistently as a function of the hour of the day. Figure 1 a illustrates the data obtained in one patient and shows that pain was consistently higher after waking in the morning than in the afternoon or evening. A similar pattern was obtained when the data of 15 RA patients were analyzed together [9].

The circadian pattern of pain may be slightly different in patients suffering from other arthritic diseases. For instance, Job-Deslandre et al. [11] studies the self-rated pain in four patients suffering from evolutive OA of the knee or the hip and they found that the peak of self-rated pain occurred at 2100 hours. This difference may be explained by the fact that the osteoarthritic pain in the knee or the hip could be due mainly to mechanical friction within the joints and it

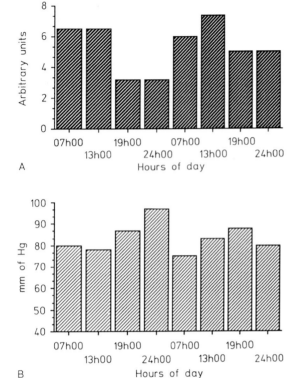

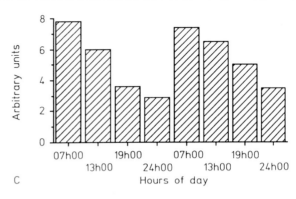

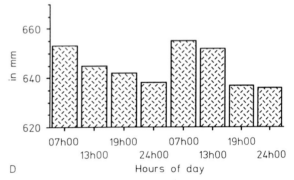

**Fig. 1 A–D.** Circadian variations in the signs of arthritis. Data were obtained from a patient with RA conducting self-assessment of **A** pain, **B** grip strength, **C** stiffness, and **D** joint size several times daily while receiving flurbiprofen at 1300 and 2300 hours. (Modified from [9])

**Table 1.** Interindividual changes in the circadian rhythm of pain in arthritic patients. (From [12] with permission)

| Circadian profiles | Number of patients | Hours of maximal pain |
|---|---|---|
| 1 peak/24 h period | 3 | 0800–1400 hours |
|  | 19 | 1400–2000 hours |
|  | 8 | 2000–0800 hours |
| 2 peaks/24 h period | 23 | 1 peak in the morning 1 peak in the evening |
| No circadian profile detected | 4 | Very little change found over 24 h |

is more likely to occur at the end of the activity period of the patients.

In a large, double-blind, randomized, crossover study involving 68 ambulatory patients with OA of the hip or the knee, Lévi et al. [12] demonstrated an important variation in the 24 h mean self-rated pain ratings. These authors used a visual analogue scale to quantify pain intensity. The arbitrary units used corresponded to the size of the scale and they varied between 0 and 20 mm. The pain ratings of these pa-

tients varied from $1.3 \pm 0.4$ to $15.6 \pm 0.1$ mm with an overall mean of $6.8 \pm 0.6$ mm. These large variations can be explained by important interindividual differences in the circadian pattern of the intensity of self-rated pain. Table 1 shows that 30 of 57 patients reported a single peak of pain occurring either in the morning, afternoon, or at night; 23 patients reported two peaks of pain during the 24 h period, whereas self-rated pain intensity varied little throughout the day in four patients. The circadien rhythms of pain intensity did not seem to be influenced by daily physical activity at the reported times.

Ultradian and circannual periodicities were also observed in 39 patients with spondylarthritis (SA). Alvin et al. [13] evaluated the circadian and circannual changes in the rheumatic symptoms of these patients and detected an ultradian rhythm for pain with peaks between 0600 and 0900 and between 1800 and 2100 hours. A circannual rhythm was also found both for the onset and the relapses of the illness: peak and through pain occurred in winter and in summer, respectively.

## Joint Stiffness, Finger Size, and Grip Strength

In patients suffering from either RA or OA circadian changes could be found in grip strength, joint stiffness, and finger joint size. As shown in Figs 1 b–d, the circadian patterns of joint stiffness and joint size of a RA patient were in phase with the circadian rhythm of pain: peak values of both symptoms were obtained early in the morning [10]. These rhythms differed in phase by approximately 12 h from the circadian changes of the grip strength of the hand: highest grip strength was demonstrated when the subjective ratings of stiffness, pain and joint circumferences were least and vice versa. Harkness et al. [14] obtained similar data in ten patients with classic RA.

As it was for pain, ultradian and circannual variations were found in the vertebral stiffness observed in patients with SA [13]. Vertebral stiffness was highest when waking and getting up (i. e., 0600–0900 hours) with a second but smaller peak occurring in the evening (1800–2100 hours). Stiffness was also more significant in winter than in summer.

## Other Signs and Symptoms of Arthritis

In an interesting but rather incomplete study, Brothers and Hadler [15] obtained data on the chronophysiology of synovial effusions at two times of the day. They performed serial morning and afternoon arthrocenteses in six RA patients who had persistent effusions in the synovial cavity but minimal knee destruction. Consistent diurnal changes were found in the pH, glucose concentration, and protein content of the synovial effusions in five out of six patients. The glucose concentration in the synovial fluid was lower in the afternoon than in the morning samples. No consistent diurnal pattern was observed in the concentration of lactate and in the $pO_2$ values of synovial fluid aliquots.

More recently, Focan et al. [16] determined the time structure of plasma proteins in patients with inflammatory symptoms. The circadian rhythms in total plasma proteins, oromucoid, prealbumin, transferrin, ceruloplasmin and alpha-1-antitrypsin were investigated in ten patients with inflammatory syndrome. In comparison to the data obtained in 14 healthy control subjects, the study suggested that the circadian time structure of these plasma proteins was not greatly affected during inflammation.

The data on the time-dependent changes in the signs and symptoms of arthritis are very interesting from a theorical point of view, because this approach may lead to a better understanding of the basic mechanisms of the arthritic disease itself. The concept of the chronopathology of arthritis is of considerable interest in the day-to-day practice of medicine. Studies have indicated that the moments of maximal and minimal intensity of the signs and symptoms of arthritis varied within a given period of time in the day or the year, with the type of arthritic disease, and with the arthritic patient. Thus, studies must be carried out to determine whether better relief of pain, stiffness, and inflammation can be obtained when anti-arthritic medications are given according to the circadian rhythms of the patients' symptoms rather than at random, as is usually done presently. This was the aim of the chronopharmacological studies carried out with different NSAID in the last 10 years.

## Chronopharmacology of NSAID

## Time-Dependent Changes in the Effects of NSAID

Experiments carried out in animal models of acute and chronic inflammatory diseases have indicated that the anti-inflammatory action of indomethacin (Indo), phenylbutazone (PHZ), or Flurbi was not constant over a 24 h period. As reviewed elsewhere (2, 3), these studies indicated that maximal effectiveness was obtained when the NSAID were administered during the resting period of the animals.

The first studies on the clinical chronopharmacology of NSAID were carried out by Reinberg and his coworkers. As illustrated in Fig. 2, these investigators showed [17] that the 1100 and 1500 hours administration of 100 mg Indo to nine healthy subjects altered significantly the circadian rhythms of oral temperature, random number addition test, eye-hand skill, mood and fatigue self-ratings, grip strength, joint size, urinary pH, 17-hydroxycorticosteroids, and systolic blood pressure. However, the 1900 and 2300 hours doses did not produce any significant change in all circadian rhythms investigated. In another investigation on four patients suffering from OA of the hip or the knee, these investigators found that maximal effectiveness of 100 mg Indo was obtained with the 1200 hours dose, which decreased the mean 24 h self-rated pain and stiffness by 54% and 65%, respectively [11]. Interindividual differences were observed also during this investigation. The 1200 hours administration produced a significant reduction in oral tem-

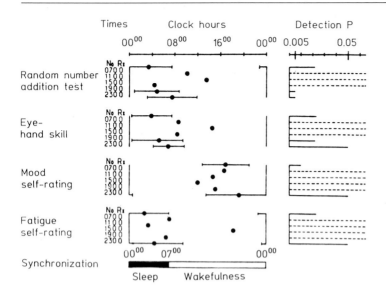

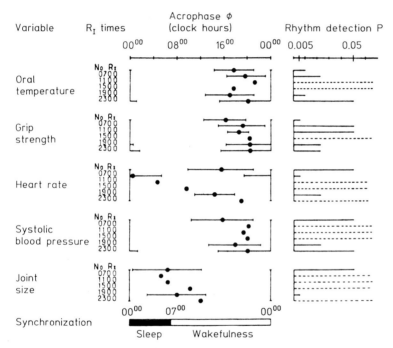

**Fig. 2.** Detection of rhythm and time of acrophase of physiological variables in healthy humans. The *dot* indicates time of peak value for each physiological value during control (no indomethacin, time or after administration of indomethacin (*RI;* 100 mg po) at 5 different hours of the day. Rhythm detection is indicated on the *right side of the graph;* the *dashed line* indicates that no significant rhythm was detected. (Reproduced with permission from [17])

perature in three of the four patients in comparison to two of the four patients after the 0800 hours dose but in only one subject after the evening tablet of Indo. Thus, these studies suggested that the effectiveness of Indo varies according to the hour of administration and that careful selection of time of drug administration could lead to individualization of drug treatment.

The largest, multicenter, chronotherapeutic study on Indo was carried out in France [18–20]. Three

studies involved 497 patients suffering from coxarthrosis (233 patients), gonarthrosis (236 patients), or other arthritic diseases of the joints (28 patients). In these studies, each patient took a 75 mg sustained release Indo capsule at 0800 for 1 week, 1200 hours for another week, and 2000 hours for the last week. Each patient was asked to report side effects and to self-rate pain intensity with a visual analogue scale every 2 h from 0700 to 2300 hours for 1–2 days during the washout period and on the last day of each test week.

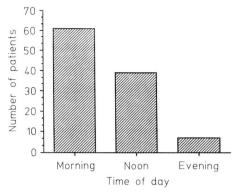

**Fig. 3.** Time-dependent variation in the undesirable effects of indomethacin in arthritic patients. Indo was administered orally at 3 h of the day and patients were asked to report undesirable effects. The graph presents the number of patients who reported undesirable effects at each hour of day. (Modified from [19])

In these studies, the effectiveness of the three schedules of drug administration was found to be very good to excellent by both physicians and patients. The morning ingestion time was considered best by 28% of the patients, whereas 22% and 35% of the patients preferred the noon and evening administration, respectively. Many patients reported side effects consisting of gastrointestinal symptoms or neurosensorial signs. Figure 3 illustrates that 32% patients reported undesirable side effects with the morning ingestion of Indo in comparison to only 7% after the evening dose of the drug. These data on the undesirable effects of Indo in diurnally active patients are supported by the findings of Moore et al. [21], who reported recently that the number of gastric mucosal lesions produced by oral administration of aspirin (ASA) at 1000 hours was twice as larg as those obtained after the 2200 hours dosing. A few months ago, Decousus et al. [22] determined the effect of dosing time on the effect of sustained release ketoprofen (Keto) in 114 patients with OA of the hip and knee. They found that the tolerance of patients for an NSAID was again twice as good for the evening group than for the morning group. It is interesting to note that Reinberg et al. [23] obtained slightly different data in their chronopharmacological study on tenoxicam in five patients with OA of the hip, because the optimal dosing time for this NSAID with a very long half-life (~3 days) was 0800 hours. No explanation is available for these surprising data and further research needs to be carried out with tenoxicam in a larger groups of patients.

Lévi et al. [12, 24] pursued their chronotherapeutic investigation with Indo and studied how the time of drug administration could improve further drug effectiveness. Using the protocol described above for their chronotherapeutic study of a sustained release formulation of Indo, these investigators noted that the time of NSAID administration producing the optimal analgesic effect differed among arthritic patients. This was explained by large interindividual differences in the circadian variation of self-rated pain intensity. Evening administration was most effective in subjects with a predominantly nocturnal pain, while Indo ingested in the morning or at noon was most effective in subjects whose peak pain occurred in the afternoon or evening. When the medication was taken at the time preferred by the subjects, the analgesic effect of the drug was further increased by about 60% over previous values.

Finally, it must be noted that chronopharmacological studies have been carried out also in patients suffering from RA. For instance, in 17 RA patients, Kowanko et al. [9] and Swannel [25] showed that a 200 mg dose of Flurbi administered twice a day was more effective than a 100 mg dose ingested four times daily. Subjective measurements of pain and stiffness indicated that part of the twice daily Flurbi dose must be administered at night to control morning stiffness and pain. Rejholec et al. [26] confirmed these data in a double-blind multicenter study.

In summary, the results obtained in different clinical studies suggest strongly that careful selection of time of administration can improve the effectiveness of NSAID and it can reduce markedly the undesirable effects of these anti-arthritic drugs. The data indicate clearly the need for well-designed studies on time qualification of drug administration in clinical trials involving every NSAID.

## Clinical Chronopharmacokinetics of NSAID

To explain the circadian variations in the effectiveness and side effects of NSAID, it is mandatory to determine whether the kinetics of these drugs vary during a 24 h period. It would be important also to determine whether a correlation can be drawn between the time-dependent changes in the pharmacokinetics and the pharmacodynamics of these drugs. The role of food or meal schedule and time-dependent changes in the susceptibility of the organism must also be examined.

Data in laboratory animals have indicated clearly that the absorption of all NSAID studied so far varied as a function of time of drug administration [2, 3]: significantly higher plasma levels and shorter times to

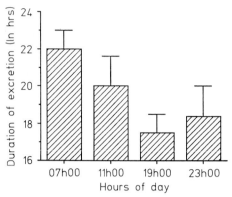

**Fig. 4.** Circadian variation in the duration of excretion of sodium salicylate. Healthy volunteers ingested orally 1 g of sodium salicylate at 4 different h of the day. Urinary excretion was measured until no salicylate was found in the urine. (Modified from [29, 30])

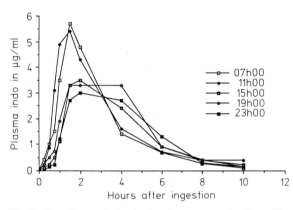

**Fig. 5.** Circadian variation in the plasma concentrations of indomethacin. Nine healthy volunteers ingested orally 100 mg Indo at 5 different h of the day. Plasma levels were determined by a spectroflurometric method. (Reproduced with permission from [17])

peak levels ($t_{max}$) were obtained when the rats received doses of Indo, PHZ, or ASA at the beginning of their activity period (i.e., at 2000 hours). The mechanisms of these time-dependent changes in the absorption of NSAID are not known, but studies carried out in our laboratory revealed that time-dependent changes in the rate of drug absorption is not limited to NSAID only [27]. Indeed, the data obtained after oral administration of antipyrine, acetaminophen, furosemide, hydrochlorothiazide, Indo, and PhZ suggested that the higher evening rate of drug absorption could be found in rats only with lipid-soluble drugs, such as the NSAID, but not with a highly water-soluble drug, such as antipyrine. As blood flow to the gastrointestinal tract has been

shown to be higher in the beginning of the activity period of rats [28], it may very well be that this factor explains the higher rate of absorption observed at this time of day. Further research is needed to prove or disprove this working hypothesis.

Time-dependent changes in the kinetics of NSAID have been described also in humans. The first study was carried out by Reinberg et al. [29, 30], who reported the circadian variation of the urinary excretion of sodium salicylate following the oral administration of 1 g sodium salicylate to six volunteers at four times of the day. Figure 4 illustrates that the duration of urinary excretion of salicylate was shortest when the drug was administered between 1900 and 2300 hours. The kinetics of 1.5 g ASA administered orally to six male volunteers at 0600, 1000, 1800, and 2200 hours also varied as a function of time of administration. Indeed, Markiewick and Semenowicz [31] showed that the highest and lowest plasma salicylate levels and area under the curve (AUC) values were obtained after the 0600 and 2200 hours doses, respectively. Thus, a better oral availability was achieved when ASA was administered at the beginning of the activity period of the subjects.

The first complete chronokinetic study on NSAID was carried out with Indo. In a crossover study, Clench et al. [17] administered 100 mg Indo capsules to nine young adults at five hours of day and determined the time-dependent changes in the pharmacokinetics of the NSAID. Circadian changes were obtained in peak plasma concentration ($C_{max}$) of Indo, $t_{max}$ values, and in the apparent plasma disappearance rate of the drug as determined by the difference in the concentration of the drug at 2 and 4 h after ingestion. Figure 5 shows that administration of Indo at the beginning of the activity period (i.e., at 0700 and 1100 hours) gave the highest $C_{max}$ and shortest $t_{max}$ values. There was no temporal variation in the AUC value.

Guissou et al. [32] studied the pharmacokinetics of the sustained release form of Indo in three different groups of patients with RA. As shown in Fig. 6, they determined the plasma levels of Indo and its demethylated metabolite at different time intervals up to 24 h after the oral administration of 75 mg of the drug at 0800, 1200, or 2000 hours. The highest $C_{max}$ and shortest $t_{max}$ values of Indo were obtained at 1200 and the lowest $C_{max}$ was found at 2000 hours. Figure 6 shows also that the plasma levels of $O$-desmethyl-Indo were higher during the interval of 6–20 h after administration of Indo at 2000 hours. These investigators reported also that the mean AUC value determined after the 2000 hours administration of Indo

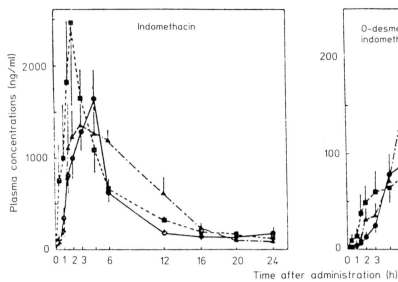

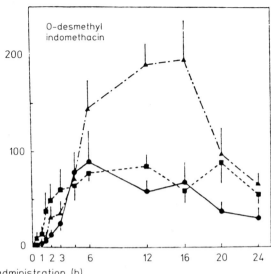

Time after administration (h)

**Fig. 6.** Plasma concentration of indomethacin and o-desmethyl indomethacin in arthritic patients. A single oral dose of 75 mg sustained release Indo was administered at 0900 hours (●—●), 1200 (■---■), and 2000 (▲–·–·–▲) to patients with mild degenerative RA. Each point represents mean ± SE of 5–6 patients. (Reproduced with permission from [32])

and expressed in terms of body weight of patients was nearly half the similar AUC value determined after drug intake at 2000 hours. However, it should be pointed out that the mean body weight of the volunteers who took the drug at 0800 hours was 81 ± 7 kg while that of the patients who participated in the 2000 hours trials was 59 ± 4 kg. If one normalized for the body weight, the circadian changes in AUC becomes much less apparent if of any significance. Finally, Bruguerolle et al. confirmed the time-dependent changes in the kinetics of Indo [33] and showed that the temporal variations in the kinetics of Indo could be modified and even be absent in elderly subjects [34].

The pharmacokinetics of sulindac was investigated in 12 healthy subjects receiving 200 mg of the drug at 0900 and 2100 hours for 7 days [35]. Plasma concentrations of sulindac and its active sulfide and inactive sulfone metabolites were determined on the seventh day of the trial. There was a diurnal variation in the AUC and peak plasma concentration ($C_{max}$) of the active metabolite. Mean AUC and $C_{max}$ values were greater between 0900 and 2100 hours.

Time-dependent changes in the pharmacokinetics of Ketoprofen (Keto) have also been studied [36–38]. When a single oral dose of Keto was administered to human volunteers at 0100, 0700, 1300, and 1900 hours, Queneau et al. [36] found that the $C_{max}$ obtained after the 0700 administration was twice the value obtained at other times of the day; the $t_{max}$ value

was also shorter after the morning administration. Similar time-dependent changes have been reported after the oral administration of a sustained release preparation or 100 mg Keto to healthy subjects [37, 38]. It is interesting to point out that intravenous infusion of Keto at a constant rate did not produce constant plasma levels. In patients receiving a daily dose of 5 mg/kg/24 h Keto perfused at a rate of 2 ml/h, Decousus et al. [39] found a trough of plasma levels of Keto (1.99 ± 0.5 mg/l) when the NSAID was perfused in the morning, whereas peak concentration (3.95 ± 0.5 mg/l) was obtained at 2100 hours.

The studies indicate clearly that the kinetics of NSAID are not constant throughout the day. In agreement with animal studies, the data showed clearly that NSAID absorption was better and probably more complete at the beginning of the activity period of the healthy volunteers or arthritic patients. The exact mechanism of the time-dependent changes in the pharmacokinetics of NSAID has not been elucidated, yet. However, these circadian variations were not dependent upon the pharmaceutical forms of the NSAID because they were reported after the administration of regular capsules or tablets, sustained release preparations, and even after constant infusion of NSAID. They could not be explained either by the presence of food in the stomach because the variations were observed by investigators who studied the chronokinetics of NSAID in fasting subjects or in patients who were taking standardized meals throughout the experiments. Further research is needed to have a better understanding of the mechanisms underlying the circadian variations in the kinetics of NSAID.

## Guidelines for Chronotherapeutic Use of NSAID

The time-dependent changes in the effects of NSAID presented above are obviously relevant for the daily practice of physicians. These chronopharmacological data can be used to prepare guidelines that physicians should use in their daily practice to optimize and to individualize the treatment of arthritic patients with NSAID. These guidelines can be summarized as follows:

1. *The circadian rhythms in the maximal intensity of signs and symptoms of arthritis varies according to the disease.* Chronopathological studies showed that maximal intensity of pain does not occur at the same hours of the day in both OA and RA patients [11, 12, 14, 25]. Thus, physicians must take these differences into consideration before prescribing medications.

2. *Administration of NSAID according to the biological rhythms of arthritis leads to an optimization and individualization of drug treatment of arthritic patients.* Physicians should determine first when the symptoms of arthritis are maximal in each patient and then they must prescribe the medications accordingly. The data on Indo-induced analgesia [12, 24] indicated that the effectiveness of this NSAID was increased by twofold when it was administered a few hours before the moment of the patient's peak pain. Evening doses were more effective in subjects with predominantly nocturnal or early morning pain, whereas morning or noon doses were more effective in patients with greater afternoon or evening pain. Further data is needed with other NSAID (such as tenoxicam), but the work done until now suggests that a chronopharmacological approach to NSAID can be used to optimize and even to individualize the utilization of some NSAID.

3. *When a NSAID treatment is ineffective or is producing too many side effects, a change in the time of drug administration must be tried before prescribing another NSAID.* Studies with Indo [12, 17–20, 24] and Flurbi [9, 25, 26] indicated that the effectiveness

and tolerance of these NSAID vary depending on the time of administration. The chronotolerance of Indo, ASA, or Keto was clearly better after their evening ingestion. Thus if side effects or lack of effectiveness force patients to stop taking their medications, a modified time of administration could restore the efficacy of the drug or reduce its undesired effects.

## Conclusions

This chapter summarized the human chronopharmacology of NSAID. It presented evidence that there were regular, predictable, circadian and circannual variations in the effects and pharmacokinetics of NSAID that could be used in clinical situations to optimize and to individualize drug treatment of arthritic diseases.

Further research is needed before we understand completely the mechanisms of the time-dependent changes in NSAID effectiveness and toxicity. Although the chronokinetic data explain, in part, the high toxicity of NSAID at some hours of the day, they cannot account for the circadian variations in the effectiveness of Indo and related drugs. Future studies must be designed to determine why tissue sensitivity to NSAID varies as a function of the moment of administration with a 24 h or 1 year span. These studies should also investigate the mechanisms of the time-dependent changes in the signs and symptoms of experimental inflammation and arthritis. The chronobiological and chronopathological studies should produce a better understanding of the physiopathology of arthritic diseases. Today's medications are prescribed often for specific clock hours: patients are instructed to take their medicines at meal times or early in the morning and sometimes before retiring at night. It is hoped that this mode of drug administration will increase the patients' compliance or will reduce some side effects of drugs. As shown in the NSAID studies summarized above, more attention must be brought to the administration of drugs at the most appropriate time both for the drug itself and for the patients. The chronotherapeutic studies with NSAID have provided experimental evidence to support the clinical impressions of Huskisson [40], that evening Indo administration to diurnally active arthritic patients was superior to morning ingestions because the drug effectiveness was better at night while CNS side effects were minimal. These data illustrate also that the concepts and findings of chrono-pharmacology are pertinent in clinical situations. They are absolutely required for solving problems of drug optimization and individualization of treatment because they provide a method to enhance the effectiveness of drug and/or to reduce undesired side effects. It is hoped that physicians using the chronopharmacological and chronotherapeutical concepts in their daily practice would be able to prescribe drugs more rationally.

# References

1. Ropes MW, Bennett GA; Cobb S et al. (1958) Revision of diagnostic criteria for rheumatoid arthritis. Bull Rheum Dis 9: 175
2. Labrecque G, Bélanger PM (1986) The chronopharmacology of the inflammatory process. Annu Rev Chronopharmacol 2: 291–325
3. Labrecque G, Reinberg A (1989) Chronopharmacology of nonsteroid anti-inflammatory drugs. In: Lemmer G (ed) Chronopharmacology. Cellular and biochemical interactions, 1st edn. Dekker, New York, p 545
4. Glynn CJ, Lloyd JW, Foldkard S (1975) The diurnal variation in perception of pain. Proc R Soc Med 69: 369–372
5. Pöllman L, Harris PHP (1978) Rhythmic changes in pain sensivity in teeth. Int J Chronobiol 5: 459–464
6. Pöllmann L, Hildebrandt G (1979) Circadian variations of potency of placebos on pain threshold in healthy teeth. Chronobiologia 6: 145
7. Proccacci P, Corte MD, Zoppi M, Maresca M (1974) Rhythmic changes of the cutaneous pain threshold in man. A general review. Chronobiologia 1: 77–96
8. Pöllmann L (1984) Duality of pain demonstrated by the circadian variation in tooth sensitivity. In: Haus E, Kabat H (eds) Chronobiology 1982–1983. Karger, Basel, p 225
9. Kowanko IC, Pownall R, Knapp MS, Swannel AJ, Mahoney PGC (1981) Circadian variations in the signs and symptoms of rheumatoid arthritis and in the therapeutic effectiveness of flurbiprofen at different times of the day. Br J Clin Pharmacol 11: 477–484
10. Kowanko ICR, Knapp MS; Pownall R, Swannel AJ (1982) Domicilliary self-measurement in rheumatoid arthritis and the demonstration of circadian rythmicity. Ann Rheum Diss 41: 453–455
11. Job-Deslandre C, Reinberg A, Delbarre F (1983) Chronoeffectiveness of indomethacin in four patients suffering from an evolutive osteoarthritis of the hip or knee. Chronobiologia 10: 245–254
12. Levi F, LeLouarn C, Reinberg A (1985) Timing optimized sustained indomethacin treatment of osteoarthritis. Clin Pharmacol Ther 37: 77–84
13. Alvin M, Focan-Hensard D, Lévi F, Focan C, Franchimont P (1988) Chronobiological aspects of spondylarthritis. Annu Rev Chronopharmacol 5: 17–20
14. Harkness JAL, Richter MB, Pamayi GS, Van DePete K, Unger K, Pownall R, Geddawi M (1982) Circadian variation in disease activity in rheumatoid arthritis. Br Med J 284: 551–554
15. Brothers GB, Hadler NM (1983) Diurnal effusions in rheumatoid synovial effusions. J Rheumatology 10: 471–474
16. Focan C, Bruguerolle B, Arnaud C, Levi F, Mazy V, Focan-Henrad D, Bouvenot G (1988) Alteration of circadian time-structure of plasma proteins in patients with inflammation. Annu Rev Chronopharmacol 5: 21–24
17. Clench J, Reinberg A, Dziewanowska Z, Ghata J, Smolensky MH (1981) Circadian changes in the bioavailability and effects of indomethacin in healthy subjects. Eur J Clin Pharmacol 20: 359–369
18. Bouchacourt P, LeLouarn C (1982) Etude chronothérapeutique de l'indométacine à effet prolongé dans l'arthrose des membres inférieurs. Trib Méd Suppl 1: 32–35
19. Simon L, Hérisson P, LeLouarn C, Lévi F (1982) Etude hospitalière de chronothérapeutique avec 75 mg d'indométacine à effet prolongé en pathologie rhumatismale dégénérative. Trib Méd Suppl 1: 43–47
20. Lévi F, LeLouarn C, Peltier A (1982) Etude chronopharmacologique en double-aveugle de 75 mg d'indométacine à effet prolongé dans l'arthrose. Trib Méd Suppl 1: 48–53
21. Moore JG, Goo RH (1987) Day and night aspirin-induced gastric mucosal damage and protection by ranitidine in man. Chronobiol Int 4: 11–116
22. Decousus H, Perpoint B, Boissier C, Ollagnier M, Mismetti P, Hocquart J, Queneau P (1990) Timing optimizes sustained release ketoprofen treatment in osteoarthritis. Annu Rev Chronopharmacol 7: 289–292
23. Reinberg A, Kahn MF, Mandredi R, Chaouat D, Chaouat Y, Delcambe B (1990) Tenoxicam chronotherapy of rheumatoid diseases. Annu Rev Chronopharmacol 7: 293–297
24. Lévi F, LeLouarn C, Reinberg A (1984) Chronotherapy of osteoarthritic patients: optimization of indomethacin sustained release (ISR). Annu Rev Chronopharmacol 1: 345–348
25. Swannel AJ (1983) Biological rhythms and their effect in the assessment of disease activity in rheumatoid arthritis. Br J Clin Practice 38 (Suppl 33): 16–19
26. Rejholec V, Vitulova V, Vachtenheim J (1984) Preliminary observations from a double-blind crossover study to evaluate the efficacy of flurbiprofen given at different times of day in the treatment of rheumatoid arthritis. Annu Rev Chronopharmacol 1: 357–360
27. Bélanger PM, Labrecque G, Doré F (1984) Rate-limiting steps in the temporal variations of the pharmacokinetics of selected drugs. In: Haus E, Kabat HF (eds) Chronobiology 1982–1983. Karger, Basel, p 359
28. Labrecque G, Doré F, Bélanger PM, Lalande M (1988) Circadian variation in the blood flow to different organs in the rat. Annu Rev Chronopharmacol 5: 445–448
29. Reinberg A, Zagula-Mally ZW, Ghata J, Halberg F (1967) Circadian rhythm in duration of salicylate excretion referred to phase of excretory rhythms and routine. Proc Soc Exp Biol Med 124: 826–832
30. Reinberg A, Clench J, Ghata J, Albuker F, Dupont F, Zagula-Mally ZW (1975) Rythmes circadiens de l'excretion urinaire du salicylate (chronopharmacocinétique) chez l'adulte sain. CR Acad Sci (Paris) 280: 1697–1700
31. Markiewick A, Semenowick K (1979) Time-dependent changes in the pharmacokinetics of aspirin. J Clin Pharmacol Biopharm 17: 409–411
32. Guissou P, Cuisinaud G, Llorca G, Lejeune E, Sassard J (1983) Chronopharmacologic study of a prolonged release form of indomethacin. Eur J Clin Pharmacol 24: 667–672
33. Bruguerolle B, Desnuelle C, Jadot G, Valli M, Acquaviva

PC (1983) Chronopharmacocinétique de l'indométacine à effet prolongé en pathologie rhumatismale. Rev Int Rhum 13: 263–267

34. Bruguerolle B, Barbeau G, Bélanger PM, Labrecque G (1986) Chronokinetics of indomethacin in elderly subjects. Annu Rev Chronopharmacol 3: 425–428

35. Swanson BN, Boppana VK, Viasses PH, Holmes GI, Monsell K, Ferguson RK (1982) Sulindac disposition when given once or twice daily. Clin Pharmacol Ther 32: 397–403

36. Queneau P, Ollagnier M, Decousus H, Cherrah Y, Perpoint B (1984) Ketoprofen chronokinetics in human volunteers. Annu Rev Chronopharmacol 1: 353–356

37. Reinberg A, Lévi F, Touitou Y, Le Liboux A, Simon J, Frydman A, Bicakova-Rocher A, Bruguerolle B (1986) Clinical chronokinetic changes in a sustained release prep-aration of ketoprofen (SRK). Annu Rev Chronopharmacol 3: 317–320

38. Olagnier M, Decousus H, Cherrah Y, Lévi F, Mechkouri M, Queneau P, Perpoint B (1987) Circadian changes in the pharmacokinetics of oral ketoprofen. Clin Pharmacokin 12: 367–378

39. Decousus H, Olagnier M, Cherrah Y, Perpoint B, Hocquart J, Queneau P (1986) Chronokinetics of ketoprofen infused intravenously at a constant rate. Annu Rev Chronopharmacol 3: 321–322

40. Huskisson, EC (1976) Chronopharmacology of anti-rheumatic drugs with special reference to indomethacin. In: Huskisson EC, Velo GP (eds) Inflammatory arthropathies. Excerpta Medica, Amsterdam, p 99

# Chronobiology of Immune Functions: Cellular and Humoral Aspects

G. Fernandes

## Introduction

Several recent chronoimmunological studies have demonstrated many prominent rhythms in the functional activities of lymphoid cells in both animals and humans (Ford 1975; Fernandes et al. 1976, 1984; Haus et al. 1983; Levi et al. 1985; Caradente et al. 1988b). Indeed, peripheral blood leukocyte (PBL) counts and cells from several other lymphoid tissues undergo large amplitude circadian-dependent changes (Halberg and Visscher 1950; Brown and Dougherty 1956; Fauci 1975; Scheving et al. 1978). The acrophase of the PBL count usually occurs 8–12 h apart from the acrophase of serum corticosterone (Fernandes et al. 1981; Kreiger and Aschoff 1974). This may indicate that endocrine hormonal activity may regulate circadian-dependent movement of lymphoid cells from various lymphoid tissues to peripheral blood. However, in the diseased state, both the acrophase and/or amplitude of many biological rhythms vary considerably (Halberg et al. 1973; McGovern et al. 1976). Descriptions of circadian-dependent changes in cellular and humoral immune responses suggest that both cell loss and changes in cellular antigen density may be under the influence of several endocrine hormones and/or lymphokines. Altered production and synthesis of many biological factors and changes in membrane proteins or their receptor concentrations may significantly modify the immune response during circadian stages and particularly during stress or disease.

## The Immune System

The immune system is fully equipped to recognize and to respond constantly not only to viral and other infectious agents, but also to a large number of invading foreign antigens. Several new sources of environmental antigens are regularly infused into the body by absorption of the daily intake of a variety of common food products and their antigens. Further, insults from carcinogens, food preservatives (e.g., nitrate and nitrite), polycyclic aromatic hydrocarbons from vegetables or fruits, and plant pollens may initiate the activation of immune responses. The generation of an immune response involves a series of interactions between lymphocytes and mononuclear cells which include cell-to-cell communication, generation of immunoreactive molecules, mitotic division, immunoglobulin (Ig) synthesis and secretion, and expression of several cell surface markers not found generally on resting lymphocytes (Roitt 1984). An effective immune response requires balanced functioning of thymic-dependent T lymphocytes, helper (CD4) and cytotoxic lymphocyte subsets (CD8), antibody-producing B lymphocytes (Ig$^+$), macrophages (Mϕ), natural killer (NK) cells, and enhancing or inhibiting soluble factors or cytokines (Table 1). Immunoregulatory processes involving antigenic and functional heterogeneity of T cell subsets and their immunopotentiating molecules give specificity to several immunological responses throughout life. Cell-mediated responses occur that do not involve secretion of specific antibody and involve T lympho-

**Table 1.** Major immune cells and cytokines involved in immune function

| Cells | Cytokines |
|---|---|
| Macrophages | IL-1, TNFs |
| T-Helper (CD4) | IL-2 to IL-8 |
| T Suppressor (CD8) | |
| | IFNs, CSFs |
| B-Cells (Ig$^+$) | |
| NK Cells | |

This work was supported in part by NIH Grant AG03471 and the Klebergs Foundation, San Antonio.

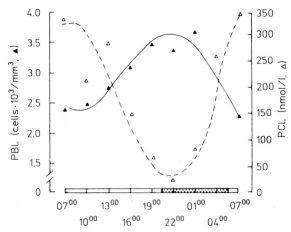

**Fig. 1.** Circadian rhythms ($p < 0.01$) of OKT11$^+$ peripheral blood lymphocytes (PBL) and plasma cortisol levels (PCL) in healthy subjects. (From Carandente et al. 1988b)

cytes as the effector cells. Humoral immune responses (in which the Igs are the effector cells) may be induced resulting in antibody synthesis and secretion. Most antigens that trigger humoral responses are currently divided into two groups on the basis of the additional cells required for the induction of B cells into a state of optimal antibody production. The thymus-independent antigens do not require help from T lymphocytes to induce an antibody response from B cells. In contrast, thymus-dependent antigens require interaction with antigen specific T lymphocytes. Furthermore, accessory cells or Mϕs must collaborate closely in an antigen nonspecific fashion with the antigen specific T lymphocytes to induce specific antibody production by B cells. It appears that circadian-dependent hormones or serum factors can reduce both the number of circulating cells in peripheral blood (Fig. 1) and the functional activity of one or more of the interacting cells, resulting in a reduced immune response or chronic, steady state, induced immune suppression that can impair the immune response to foreign antigens. It is quite apparent that although the circadian rhythm for total lymphocytes may peak during the late hours of the day, the circadian peaks for T and B cells in the peripheral blood of humans are 12 h apart (Table 2). Chronic immunodeficiency may arise by the functional loss of T lymphocyte subsets, especially CD8 cells, resulting in failure to suppress CD4 (TH1 and TH2 subsets) cell activity. An excess of unregulated T cells may regularly result in elevated B cells which may increase oncogene mRNA levels and increase autoantibody production. This dysynchronized B cell activity and reduced circulatory T cell amplitude may eventually lead to the development of a chronic autoimmune disease state, particularly due to overproduction of lymphokines such as interleukin-I (IL-1), IL-4, IL-6, tumor necrosis factor (TNF), etc.

## The Nature and Biological Role of Lymphokines

Lymphokines are glycosylated or nonglycosylated polypeptides. Lymphokines and Mϕ-derived cytokines, which may be produced in larger quantities

**Table 2.** Acrophasogram of some immunological variables in human beings. (From Carandente et al. 1988b)

| variable | rest-activity cycle | | | | PR | $p$ | subjects ($n$) | source |
|---|---|---|---|---|---|---|---|---|
| | $06^{00}$ | $12^{00}$ | $18^{00}$ | $0000$ | | | | |
| neutrophils | | | | | 50 | < 0.001 | 150 | Haus |
| neutrophils | | | | | 18 | < 0.001 | 6 | Dammacco |
| lymphocytes | | | | | 59 | < 0.001 | 150 | Haus |
| lymphocytes | | | | | 34 | < 0.001 | 6 | Dammacco |
| T cells (EA-RFC) | | | | | 68 | < 0.001 | 150 | Haus |
| T cells (EA-RFC) | | | | | | n.s. | 6 | Dammacco |
| $T_S$(CD8$^+$) | | | | | 51 | < 0.005 | 6 | Dammacco |
| $T_H$(CD4$^+$) | | | | | 47 | < 0.05 | 6 | Dammacco |
| B cells | | | | | 76 | < 0.033 | 150 | Haus |
| B cells | | | | | 82 | < 0.05 | 6 | Dammacco |
| NK-cell activity | | | | | | < 0.05 | 5 | Williams |
| NK-cell activity | | | | | 51 | < 0.05 | 6 | Dammacco |
| responsiveness to PHA | | | | | 88 | < 0.001 | 9 | Haus |

**Table 3.** Main features of various lymphokines

| Lympho-kine | Molecular weight | Cell source(s) | Main actions |
|---|---|---|---|
| IFN | 40000–50000 (Dimer) | T cells NK cells | Immunoregulation of lymphocytes, mono-cytes and tissue cells |
| IL-1$\alpha$ IL-1$\beta$ | 33000 (precursor) 17500 | Monocytes Dendritic cells B cells Fibroblasts Epethelial cells Endothelium Astrocytes | Immunoregulation and inflammatory me-diator on T & B cells |
| IL-2 | 15000 | T cells NK cells | Proliferation activa-tion of T and B cells |
| IL-3 | 15000 | T cells | Pan-specific CSF for stem cells |
| IL-4 | 15000 | T cells | Division and differen-tiation of B cells |
| IL-5 | 15000? | T cells | Differentiation of B cells and eosinophils |
| IL-6 | 20000 | T cells Fibroblasts Macrophages | Differentiation of B cells and thymocytes |
| IL-7 | 25000 | T cells | Differentiation of B cells and thymocytes, T cell proliferation |
| IL-8 | 15000? | T cells Macrophages | Stimulate granulocy-tes, fibroblasts, B cells |

under the influence of other lymphokines, exert effects on most, perhaps all, organ systems and should be seen as part of a coordinated response to an immunological challenge. Lymphokines can be secreted by both T cells and B cells, though T cells are assumed to be the major source in cell-mediated responses (Mizel 1989). There is some evidence from work with T cell hybridomas and T cell clones that not all T cells secrete the same range of lymphokines. For instance, it has been sugested that murine T cell clones can be divided into those secreting IL-2, IL-3, IL-4, IL-5, etc., and the lymphokine list already extends from 1 to 8 (Table 3).

Study of the biological role of lymphokines, particularly to establish their action on a circadian basis, has been minimal, mainly because it is extraordinarily difficult. Early work based on crude supernatants of lymphocyte cultures was criticized as phenomenology since the active components were unknown. However, the genes for many of the lymphokines are currently cloned and large quantities of pure substances are available for chronotherapy applications.

Nonetheless, while superficially more promising, in vitro studies can still be biologically misleading, since in vivo no lymphokine would ever operate in isolation. A recent study showed minimum influence of circadian rhythms on IL-2 production in human peripheral blood cells (Smith and Pruett 1989); yet, it is possible that their response to lymphokines varies and carefully controlled experiments are required to reconfirm such a possibility. Both isolation of lymphokines and a well-defined characterization of T lymphocyte subsets, including their interaction with other lymphoid cells, occurred very recently; however, the cytotoxic function of T cells and their active role in rejecting organ transplants have been known for the past 20 years. Thus, attempts to prevent or to delay the rejection of allogenic tissue grafts were immediately employed, using circadian approaches, soon after the role of lymphocytes was defined in rejecting kidney grafts in animals.

### Circadian Cellular Rhythms: Allograft Rejection

Generally, graft rejection displays the two key features of adaptive immunity, namely, memory and specificity. These characteristics can be demonstrated by grafting skin from one animal to another. Memory is demonstrated when a second allogeneic skin graft from one donor is rejected by the recipient more quickly than the first graft from that donor. Only sites in the recipient which are accessible to the immune system are susceptible to the graft rejection phenomenon. There are certain "privileged" sites in the body where allogeneic grafts can survive indefinitely (Cerilli 1988). It is also found that the ability to reject graft can be transferred with previously sensitized lymphocytes. It appears that the mechanisms involved in killing are similar whether cytotoxic T cell, NK, or other lymphoid cells such as phagocytic cells are involved. The binding of effector cells to target cells and the killing mechanism of the former may be enhanced by the availability of lymphokines both during the initial activation of effector cells and during the actual killing of target cells. However, studies are needed to establish whether a circadian rhythmic action can influence both these parameters, and several critical experimental approaches are still required to dissect a safe circadian time for using lymphokines in vivo to decrease and/or enhance cytotoxic events. For instance, not much is known about which lymphocytes can release the maximum amount of lymphokines during the circadian phase and how many lymphokines are active at any one time. Further, it is possible that the effect of a lymphokine depends on the stage of maturation of the Mɸ, and its activation is linked closely to activation of other T cells or NK cells via activation and production of other T cell lymphokines. These and many other observations confirm the pivotal role of the immune system in graft rejection. In view of the unphysiological nature of tissue transplantation, it may seem surprising that the immune system has thrown such a formidable barrier in the path of transplantation surgery. However, it appears that this major side effect of the immune system can be modified to some extent by performing tissue or organ transplants on a circadian basis. For instance, rejection of both allogeneic skin in mice and kidney grafts in rats was delayed significantly when grafted during different circadian phases (Ratte et al. 1973; Halbert et al. 1974).

This finding eventually led to studies on rejection of kidney grafts in humans. In clinical transplantation, allogeneic recognition is often considerably more complicated than recognition in vitro. Hyperacute graft rejection is caused entirely by T cells, NK cells, and cytotoxic antibodies, usually those to different blood group substances or to class I MHC antigens. There is only a very narrow window, perhaps 0–72 h after a kidney transplant, when hyperacute rejection can occur. For instance, in humans, renal allograft rejection was regularly noted at night. In this study cosinor analysis revealed a significant 24 h rhythm in rejection events with the acrophase at 0517 hours (Knapp et al. 1981).

Against the background of circadian work on animals and humans, later transplant studies carried out in humans have resulted in recognition of circaseptan rhythms in rejection of kidney grafts. For instance, the retrospective analysis of data undertaken in 147 transplant patients, with kidneys from cadavers or living relatives, demonstrated a statistically significant circaseptan rhythm in the appearance of the first signs of rejection 1–2 months following transplantation. The concomitant fit of a linear trend and of an 8.1 day cosine curve established the statistical significance of the phenomenon. The 8.1 day fit seems better than a 7 day fit, yet it is pertinent that changes in corticosteroid treatment were carried out at weekly intervals. A larger number of patients, and a proper analysis of the details of treatment are required to clarify any circaseptan rhythm interactions or contributions of weekly changes in therapy to the circaseptan rhythm. Further, it may be necessary to optimize therapy in the light of any endogenous circaseptan and circadian rhythms (Devecchi et al. 1981).

After the discovery of cyclosporine and its valuable use for preventing allograft rejection, studies in the field of chronobiology also led to observations of significant variations in the rejection time of allografts both in rodents and humans. Cyclosporine chronotherapy of pancreas allotransplanted rats revealed, beyond a circadian stage dependence of equal daily doses, a further gain in graft function from doses varying from day-to-day, with about a 7 day periodicity; the first highest dose was given on the third or fifth day after surgery (Liu et al. 1986). In addition, the effects of timing cyclosporine (Cs) adinistration was further tested on nephrectomized dogs bearing an allografted kidney. For instance, 1.5 mg/kg Cs per day was given via an externally programmable, implanted, Medtronic pump at a continuous injection rate or with one of six changing rates, each involving eight different doses, increasing and then decreasing sinusoidally every day. The total dose per day was equal to that in dogs infused at a constant rate. Both the constant and the sinusoidal infusion schedules were better than the others. The assumption that the differences among dogs treated with different sinusoidal schedules varied randomly was rejected by the fit of a 24 h cosine function. This approach demonstrated a statistically significant circadian rhythm in response to sinusoidal Cs treatment. The timing of the best schedule during the dark span is surprising ih view of a similar timing found for the optimal Cs effect in nocturnal rats bearing heart or pancreas allografts. The investigators pointed out the need for further research in this particular area to examine whether this similarity of timing is reproducible, as seems to be the case for melatonin in blood or urine of human beings and nocturnal rodents (Halberg et al. 1983b). The experimental design, including the use of a programmable implantable pump, and the procedures for cosinor analysis, in conjunction with diagnostic regression tests, represent a classical approach to chronotherapy which may be more generally applicable in the future. Although chronopharmacokinetic changes have also been explored by investigators and their rhythmicity documented, these changes alone may not fully account for circadian action; instead other factors must also be considered (Cavallini et al. 1986).

## NK Cells

Many other biological rhythms in physiological processes have been extensively studied and documented by many investigators (Smolensky et al.

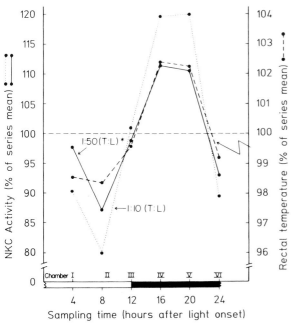

For φ in degrees
360°=24 hours

Light-dark schedule

Single cosinor

**Fig.2.** Natural killer cells activity in spleens of young rats, as determined by ability to lyse tumor cells, is circadian rhythmic; cytotoxicity levels in two target-to-effector cell ratios are found parallel to the rectal temperature rhythm. * Tumor (target): lymphocyte (responder) cell ratio. (From Fernandes et al. 1979)

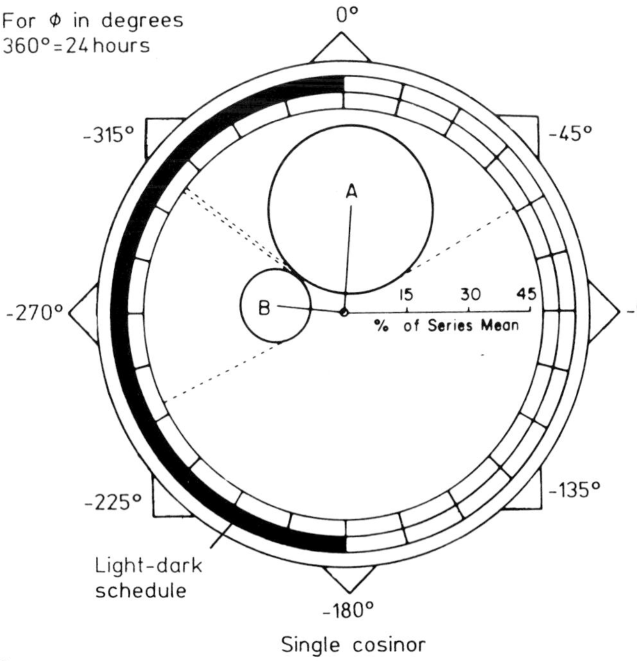

| KEY TO ELLIPSES | P | NO OBS. | PERCENT RHYTHM | AMPLI-TUDE** (95% CL) | ACROPHASE*** (0) (95% CL) |
|---|---|---|---|---|---|
| A MICE | 0.021 | 12 | 57.7 | 24.2 (4.0, 44.0) | − 4°(−307, − 58) |
| B RATS | 0.001 | 12 | 77.5 | 15.9 (7.6, 24.2) | −276°(−244, −309) |

**Fig.3.** Cosinor analysis of circadian rhythm in natural killer cell activity reveals statistically significant change. Acrophase for NKC activity is located during active span of nocturnal rodents. Data transformed as % of series mean (see *). (From Fernandes et al. 1978)

* Timing adjusted to light onset ( ≡ 0°) for male C$_3$H/Umc mice or male Fischer rats, ~2 months of age;
** °Celsius;
*** In angular degrees from time of light onset ( ≡ 0°) with 360° ≡ 24 hrs

1972; Fernandes et al. 1980; Reinberg and Smolensky 1983). In contrast, reports of circadian variations in immunity have been relatively uncommon until comparatively recent (McGovern et al. 1976). The in vivo circadian rhythmicity of a cell-mediated immune process was documented, although phenomena such as cyclic changes in body temperature are associated with the graft vs host reaction (Cornelius et al. 1969). Humorally mediated hypersensitivity reactions in human skin (Reinberg et al. 1969) and lung (Gervais et al. 1977) exhibit circadian variation, and rhythms have been demonstrated for NK cell function both in animals (Fernandes et al. 1979) and humans (Williams et al. 1981). We first described a circadian-dependent NK activity in mice and rats (Figs.2 and 3). We have reported that NK cells in Fischer rats and C3H mice exhibit a significant rhythmic activity in lysing murine lymphoma cells; high activity was associated with higher body temperature (Fernandes et al. 1981). Whether circadian NK cell activity rhythms depend upon hormonal mechanisms is not well understood. In animals, immunological rhythms may be highly reproducible due to the usage of inbred animals and controlled dietary and environmental conditions. Yet, in humans, immunological rhythms may vary from one individual to another due to the dis-

parity of genetic backgrounds and other environmental factors.

If circadian rhythms also exist in cell-mediated immune processes, then some of the diagnostic techniques in clinical immunology and transplantation might need reappraisal to avoid diagnostic and treatment variations. Studies reported by several investigators have suggested that both NK activity (Fig.4) and the toxicity of drugs against tumor growth and survival vary every 24 h span due to changes in hormone levels (Haus et al. 1970; Halberg et al. 1977; Haus et al. 1974a, 1974b; Gatti et al. 1987). Many studies have suggested that it is important to understand the relationship between resistance or susceptibility to tumor growth and the progression and fluc-

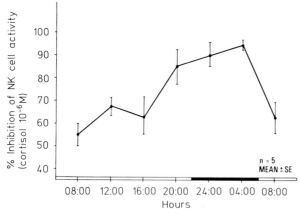

**Fig. 4.** Percent inhibition of NK cell activity by $1 \times 10^{-6}$ cortisol (20 h incubation) throughout the 24 h cycle in healthy women. (From Gatti et al. 1987)

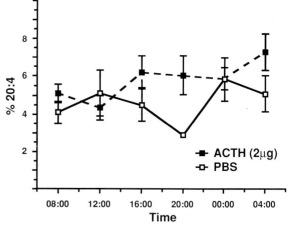

**Fig. 6.** Circadian change in the concentration of arachidonic acid (20:4) in the thymocytes of young C57Bl/6 mice treated with PBS or ACTH (HOE 433) 2μg/kg body weight for six different time periods for 24 h before animals were killed. Fatty acid levels were determined by gas chromotograph using a chloroform-methanol extraction

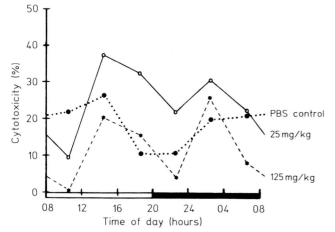

**Fig. 5.** NK activity in splenocytes of old B/6 mice injected with cyclophosphamide. The light and dark periods are indicated as *clear* and *shaded bars* on the abscissa. (From Sandberg and Fernandes 1987)

tuation of immune function, particularly NK activity (Angeli et al. 1988). Results of many studies clearly suggest that rhythms with different frequencies have a role in influencing immune functions in mice and humans (Tavadia et al. 1975; Abo et al. 1981, 1982; Swoyer et al. 1987).

The integrity of both T and B cell immune functions is crucial for survival and resistance to infection with bacteria, fungi, and viruses and for resistance against malignancy. Natural immunity and response to antigens have been established as being rhythmic and predictable. In nocturnally resting and diurnally active human beings, the maximum response of human lymphocytes to PHA occurs in the morning, at 9:00 a.m. (Haus et al. 1983). The relevance of vari-

ations of immunological response to environmental insults and certain diseases and the beneficial aspects of following circadian rhythms of longer duration need to be further investigated, as this will be of considerable importance in maintaining optimum immune reactivity (Levi et al. 1985).

Recently, studies were also carried out to assess circadian-based drug therapy effects on enhancement of NK function. The results revealed that low doses of cyclophosphamide were able to increase NK activity more during the resting period than during periods of high activity (Fig. 5) (Sandberg and Fernandes 1987). It was not clear, however, if the observed increase in NK activity caused by cyclophosphamide in young and old mice was primarily due to decreased activity of immunoregulatory suppressor cells or reduced Mφ effects on lymphokine production.

Indiveri et al. (1985) reported circadian variations in the mixed leukocyte reaction (MLR), with the highest proliferation occurring in cells prepared from blood collected at 0800 hours, and stimulated with autologous PHA-stimulated T cells prepared from blood collected at 2000 hours. The investigators suggested that results of autologous MLRs must be considered carefully, including analyzing abnormalities in these reactions in pathological conditions. Furthermore, stimulation of autologous MLRs have been associated with differences in plasma cortisone concentrations.

The circadian changes in lymphocyte activity may also be closely linked to the changes occurring in plasma lipids and in membrane fatty acid composition. Functional changes in lymphocytes are indeed based on cell-cell interaction, particularly membrane lipid and protein interactions and other membrane receptor functions. Occurrence of circadian changes in fatty acid desaturation in liver and liver microsomes was well-established (Dato et al. 1973). We too recently decided to measure the fatty acid composition of thymocytes obtained from mice killed at six different time points. With and without 24 h prior treatment with ACTH-1-17 (HOE 433), a synthetic short-chain derivative, the results clearly indicate that thymocytes obtained from mice treated with saline undergo a circadian change for several fatty acids, but such a change may be modified by the usage of ACTH (Fig. 6). Thus, future studies are required to establish whether the changes in fatty acids and phospholipids in membranes, which may determine the functional activity of membrane proteins in activating signalling mechanisms in various lymphocyte subsets, indeed modify the response to various ligands on a circadian basis.

## Circadian Humoral Immune Response and Other Longer Rhythms

Several chronoimmunological studies have demonstrated circadian rhythms of different frequencies for humoral immune responses as well (Pownal and Knapp 1980; Kim et al. 1980). Considerable evidence indicates that, as the counts of murine and human lymphocytes (including B cells) and other leukocytes undergo circadian rhythms, it is obvious that humoral immune functions and their response to various T cell-dependent and independent antigens and to lymphokines also may closely follow a circadian rhythmic pattern (Bertough et al. 1983; Besedovsky et al. 1986).

In order to analyze circadian influence on antibody formation, we immunized mice with SRBC, a T cell-dependent antigen, under careful lighting and in temperature controlled rooms and measured serum antibody function (Fernandes et al. 1974). A significant variation within 24 h for antibody titer against SRBC and circadian changes for rectal temperature, hematocrit levels, thymus weight, and spleen weights were also noted. Further, in a separate study, formation of the fewest plaque-forming cells (PFC) in response to SRBC antigen occurred when the latter were injected close to the dark span. We attributed this change to an indirect influence of higher concentrations of steroids or hormones, which may have exerted a suppressive effect on initiation of antibody production (Figs. 7 and 8). In addition, spleen cell analysis revealed that marked differences in the proliferative responses of spleen lymphocytes to mitogens were dependent on circadian rhythmic frequencies including B cell function (Fernades et al. 1976; Calderon and Thomas 1980).

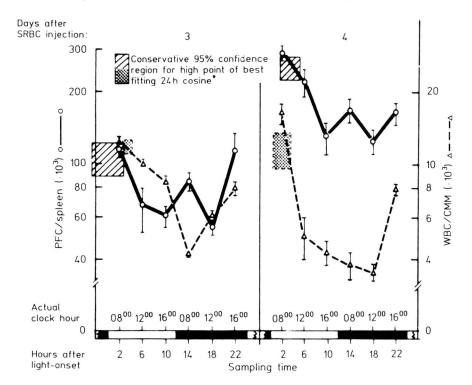

**Fig. 7.** Circadian rhythm in humoral immune response in mice. Circulating WBC and response of PFC to sheep. (From Fernandes et al. 1976)

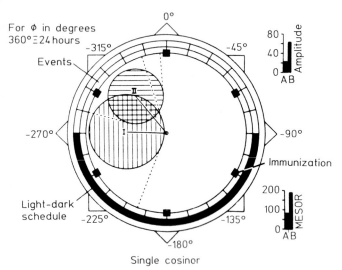

For φ in degrees
360° ≡ 24 hours

Events

Light-dark
schedule

Single cosinor

**Fig. 8.** Cosinor presentation of circadian rhythm in murine immune system: plaque forming responses after injection of sheep RBC (see *). (From Fernandes et al 1976)
* At each of 6 circadian stages, 4 hr apart, a separate group of 4 adult female Balb/C mice was injected i. p. with sheep red blood cells; 72 and 96 hr after injection animals were killed for plaque assay in Exp. I & II respectively

In addition to circadian B cell activity, serum Ig concentration and several other components of serum proteins undergo circadian periodicities (Reinberg et al. 1980; Hayashi and Kikuchi 1982). Changes in lymphocyte subsets and Igs after stressful physical work have also been reported (Hedfors et al. 1983). Further, circulating immune complexes in lupus and rheumatoid arthritis patients (Isenberg et al. 1981) undergo circadian variations which should be closely observed for assessing the severity of the disease and for establishing therapeutic regimen on a circadian basis.

The reaction and also the variation in the immune system to antigenic stimuli may be highly dependent on a network of endogenous circadian, circaseptan, and circannual rhythms. Studies with the goal of modifying biological rhythms by adopting biotechnological approaches for treatment of diseases have been proposed (Kawahara et al. 1980, Scheving et al. 1987). Caradente et al. (1988a) indeed recently treated healthy subjects and multiple myeloma patients with cefodizime at two time points, in the morning or evening, for a week and detected several lymphocyte surface marker rhythms.

Shifrine et al. (1980) studied the effect of seasonal variation in cell-mediated immunity in dogs using whole blood, lectin induced lymphocytes mitogen in a stimulation test over a period of 16 months. The results exhibited seasonal variation in immune re-

sponse with the peak in summer and a poor response in winter. Further, a circannual variation for rubella antibody titer was noted in a large number of patients followed for a period of 7 years, indicating that careful multianalyses are needed to establish the presence of immunity against any infectious diseases (Rosenblatt et al. 1982). Natural body defenses not only follow a physiological rhythmic frequency, but we now have to also consider circannual rhythms for augmenting the optimal immune response, for proper diagnosis, or for effective treatment of human beings in order to increase resistance to various diseases, including preventing damage from exposure to chemical carcinogens.

Common acute lymphoblastic leukemia antigen (CALLA) is currently used to characterize and monitor lymphoid leukemias. CALLA + lymphocytes are not generally present in normal subjects. CALLA is found on the membrane of leukemic lymphoblasts. Cannon et al. (1985) have reported a significant circadian rhythm in the presence of CALLA + lymphocytes. Thus circadian rhythms in lymphocyte subsets may be essential to understand the differentiation processes of immune cells in the bone marrow and thymus in order to similarly adjust the dosage of drugs to treat malignant diseases (Halberg et al. 1937; Hrushesky 1985).

An understanding of the organization of the immune system along the yearly scale may influence the success or failure of immune therapy. Circannual rhythms have been demonstrated in animals kept in standardized environments (Brock 1983). Studies conducted by Levi et al. (1988) have demonstrated seasonal modulation of circadian rhythms of circulating T and NK lymphocyte subsets in healthy volunteers. These studies revealed a rhythm with a period equal to 6 months for circulating T helper cells. Further, not only maximum values for T suppressor and cytotoxic and NK lymphocytes were observed, but the results attributed circadian desynchronization to a spring peak in the incidence of viral infection and Hodgkin's disease in patients (Levi et al. 1988).

Pati et al. (1987) have documented statistically significant circannual rhythms in the proliferative response of lymphocytes to mitogens, a 6 month rhythm in the activation of splenic B lymphocytes, and a circannual rhythm for NK cell activity in mice. These and other studies suggest not only that circadian approaches are important, but also that seasonal modifications of immunotherapy in clinical trials may be equally essential for planning effective immunotherapeutic treatment strategies for various diseases in the future.

## Conclusion

Several physiological variables related to the immune system are closely linked to rhythms of different frequencies. The frequency of chronoimmunological rhythms may range from short duration to much longer duration. Changes in MESOR and amplitude of each rhythm, e.g., ultradian (few hours), circadian (24 h), circaseptan (7 days), circatrigintan (30 days), and rhythms of longer duration, such as seasonal and circannual, are known to play critical roles in modulating various functional aspects of the immune system.

The immune system is highly intricate and includes a delicately regulated network of cells whose lymphokine products are intimately involved in cellular and humoral immune responses. The immune cells include helper and suppressor T cells, B cells, M$\phi$, and NK cells. The primary role of the immune system, particularly in the natural state and/or after activation with antigenic stimuli, is to resist invasive pathogenic microorganism and to discriminate between self and non-self. The ability to react promptly to non-self protects the integrity of the host. Cells of the immune system show a predictable circadian change in their number and function, including proliferative responses to mitogens, NK cell activity, antibody production, phagocytosis, delayed-type hypersensitivity, skin and kidney graft rejection, etc. The risk of rejection of transplanted tissue due to an immune reaction can also be modified. The donor tissue can be placed after inducing a safe circadian time span of host immune unresponsiveness by a suitable, circadian sensitive, immunosuppressive therapy. This would allow the transplant to survive and, most importantly, would reduce the risk of malignancy.

However, the circadian based immune system, which is fully regulated and controlled by other physiological rhythms, including several endocrine hormones, may fail to differentiate self antigens from non-self, mainly because of antigenic changes on the cell surface and over-production of disregulated lymphokines, including interleukins, interferons, prostaglandins, etc. This could initiate an abnormal immune response against self tissue. Such a failure of the immune system occurs generally in autoimmune diseases, such as rheumatoid arthritis, lupus, etc. Attempts are underway to inhibit adverse lymphokine production by adapting drug therapies to a circadian rhythm and by augmenting the cytotoxic function of immune cells by infusing various lymphokines to suppress malignant cell proliferation in vivo.

**Acknowledgements.** I am grateful to Dr. Jaya Venkatraman for her kind help in preparing the manuscript and Cindy Trevino for her excellent typing assistance.

## References

Abo T, Kawata T, Itoh K, Kumagai K (1981) Studies on the bioperiodicity of the immune response. 1. Circadian rhythm of human T, B and cell traffic in the peripheral blood. J Immunol 126: 1360

Abo T, Cooper MD, Balch CM (1982) Characterization of HNK1 + (Leu7) human lymphocytes. I. Two distinct phenotypes of human NK cells with different cytotoxic capability. J Immunol 129: 1752

Angeli A, Gatti G, Sartori ML, Del Ponte D, Carignola R (1988) Effect of exogenous melatonin on human NK cell activity. An approach to the immunomodulatory role of the pineal gland. In: Gupta D, Attanasio A, Reiter R (eds) The Pineal gland an cancer. Brian Research Foundation, London

Bertough JV, Roberts-Thomson PJ, Bradley J (1983) Diurnal variation of lymphocyte subsets identified by monoclonal antibodies. Brit Med J 286: 1171–1172

Besedovsky H, Del Rey A, Sorkin E, Dinarello CF (1986) Immunoregulatory feedback between interleukin-1 and glucocorticoid hormones. Science 233: 652–654

Brock MA (1983) Seasonal rhythmicity in lymphocyte blastogenic response of mice persists in a constant environment. J Immunol 130: 2686–2688

Brown HH, Dougherty JF (1956) The diurnal variation of blood leucocytes in normal and adrenalectomized mice. Endocrinology 58: 365–370

Cannon C, Levi F, Reinberg A, Mathe G (1985) Circulating callapositive lymphocytes exhibit circadian rhythms in man. Leukemia Res 9: 1539–1546

Calderon RA, Thomas DB (1980) In vivo cyclic change in B-lymphocyte susceptibility to T-cell control. Nature (Lond) 285: 662–664

Caradente F, DeVecchi A, Halberg F, Cornelissen G, Dammacco F (1988a) Toward a chronoimmunomodulation by cefodizime in multiple myeloma and chronic uremia. Chronobiologia 15: 61–85

Caradente F, DeVecchi A, Dammacco F, Halberg F (1988b) Multifrequency rhythms of immunological function. Chronobiologia 15: 7–23

Cavallini M, Halberg F, Cornelissen G, Enrichens F, Margarit C (1986) Organ transplantation and broader chronotherapy with implantable pump and computer programs for marker rhythm assessment. J Controlled Release 3: 3–13

Cerilli CJ (1988) Organ transplantation and replacement. JB Lippincott, Maryland Cornelius EA, Yunis EJ, Martinez C (1969) Cyclic phenomena in the graft-versus-host reaction. Proc Soc Exp Biol Med 131: 680–684

Dato SM, Catala A, Brenner RR (1973) Circadian rhythm of felty acid desaturation in mouse liver. Lipids 8: 1–6

Devecchi A, Halberg F, Sothern RB, Cantaluppe A, Ponticelli C (1981) Circaseptan rhythmic aspects of rejection in treated patients with kidney transplant. In: Walker CA, Winget CM, Soliman KFA (eds) Chronopharmacology and chemotherapeutics. Florida A & M University Foundation, Tallahasse, pp 339–353

Fauci AS (1975) Mechanism of corticosteroid action on lymphocyte subpopulations. I. Redistribution of circulating T and B lymphocytes to the bone marrow. Immunology 28: 669–680

Fernandes G, Yunis EJ, Nelson W, Halberg F (1974) Differences in immune response of mice to sheep red blood cells as a function of circadian phase. In: Scheving LE, Halberg F, Pauly JE (eds) Chronobiology. Proceedings of the international Society for the study of biological rhythms. George Thime, Stuttgart, Igaku Shoin, Tokyo, pp 329–338

Fernandes G, Halberg F, Yunis E, Good RA (1976) Circadian rhythmic plaque-forming cell response of spleens from mice immunized by SRBC. J Immunol 117: 962–966

Fernandes G, Carandente F, Halberg E, Halberg F, Good RA (1979) Circadian rhythm in activity of lympholytic natural killer cells from spleens of Fischer rats. J Immunol 123: 622–625

Fernandes G, Halberg F, Good RA (1980) Circadian dependent chronoimmunological responses of T, B and natural killer cells. Allergology 3: 164–170

Fernandes G, Halberg F, Halberg E, Miranda M, Yunis EJ, Good RA (1981) Murine circadian rhythm in natural cell-mediated cytotoxicity against lymphoma cells and chronoimmunology. In: Walker CA, Soliman KFA, Winget CM (ed) Chronopharmacology and chronotherapeutics. A & M University Foundation, Tallahasse, pp 233–245

Fernandes G, Talal N, De Haven J (1984) The effect of circadian rhythm on immune functions and splenic lymphocyte subsets in mice. Annu Rev Chronopharmacol 1: 149–152

Ford WL (1975) Lymphocyte migration and immune responses. Progr Allergy 19: 1–59

Gatti G, Del Ponte D, Cavallo R, Sartori ML, Salvadori A, Carignola, Carandente F, Angeli A (1987) Circadian changes in human natural killer cell activity. In: Pauly JE, Scheving LE (eds) Advances in chronobiology, part A. Liss, New York, pp 399–409

Gervais P, Reinberg A, Gervais C, Smolensky M, DeFrance O (1977) Twenty four hour rhythm in the bronchial hyperactivity to house dust in asthmatics. J Allergy Clin Immunol 59: 207–213

Halberg F, Visscher MB (1950) Regular diurnal physiological variation in eosinophil levels in five stocks of mice. Proc Soc Exp Biol Med 75: 846–847

Halberg F, Haus E, Cardoso SS, Scheving LE, Kuhl JFW, Shiotsuka R, Rosene G, Pauly JE, Runge W, Spalding JF, Lee JK, Good R (1973) Toward a chronotherapy of neoplasia: tolerance of treatment depends upon host rhythms. Experientia (Basel) 29: 909–934

Halberg J, Halberg E, Runge W, Wicks J, Cadotte L, Yunis E, Katinas G, Stutman O, Halberg F (1974) Transplant chronobiology. In: Scheving LE, Halberg F, Pauly JE (eds) Chronobiology. Proceedings of the international Society for the study of biological rhythms. Thieme, Stuttgart, Igaku Shoin, Tokyo, pp 320–328

Halberg F, Duffert D, Von Mayersbach H (1977) Circadian rhythm in serum immunoglobulins of clinically healthy young men. Chronobiologia 4: 114a

Halberg F, Sanchez De La Pena S, Fernandes G (1983a) Immunochropharmacology. In: Hadden J, Chedid L, Dukor P, Spreafico F, Willoughby D (eds) Advances in immunopharmacology. Pergamon, Oxford, pp 173–198

Halberg F, Sanchez De La Pena S, Fernandes G (1983b) Rhythm scrambling and tumorigenesis in CD2F1 mice. In: Tarquini B, Vergassola R (eds) 3rd International symposium of social diseases and chronobiology, Nov 29, 1986, Florence, pp 59–61

Haus E, Halberg F (1970) Circannual rhythm in level and timing of serum corticosteron in standardized inbred mature c-mice. Environ Res 3: 81–106

Haus E, Halberg F, Kuhl JFW, Lakatua DJ (1974a) Chronopharmacology in animals. Chronobiologia 1: (Suppl 1) 122–156

Haus E, Fernandes G, Kuhl JFW, Yunis EJ, Lee JK, Halberg F (1974b) Murine circadian susceptibility rhythm to cyclophosphamide. Chronobiologia 1: 270–277

Haus E, Lakatua DJ, Swoyer J, Sackett-Lundeen L (1983) Chronobiology in hematology and immunology. Am J Anat 168: 467–517

Hayashi O, Kikuchi M (1982) The effects of the light-dark cycle on humoral and cell-mediated immune responses of mice. Chronobiologia 9: 291–300

Hedfors H, Holm G, Ivansen M, Wahren J (1983) Physiological variation of blood lymphocyte reactivity: T-cell subsets, immunoglobulin production, and mixed lymphocyte reactivity. Clin Immunol Immunopathol 27: 9

Hrushesky WJM (1985) Circadian timing of cancer chemotherapy. Science 228: 73–75

Indiveri F, Pierri I, Rogna S, Poggi A, Montaldo P, Romano R, Pende A, Morgano A, Barabino A, Ferrone S (1985) Circadian variations of autologous mixed lymphocyte reactions and endogenous cortisol. J Immunol Methods 82: 17–24

Isenberg DA, Guisp AJ, Morrow WJW, Newham D, Snaith ML (1981) Variations in circulating immune complex levels with diet, exercise and sleep: a comparison between normal controls and patients with systemic lupus erythematosis. Ann Rheumatic Diseases 40: 466–469

Kawahara K, Levi F, Halberg F, Cornelissen G, Sutherland DE, Rynasiewicz J, Gorecki P, Najarian J (1980) Circaseptan bioperiodicity in rejection of heart and pancreas allografts in the rat. Chronobiologia 7: 132

Kim Y, Pallansch M, Carandente F, Reissmann G, Halberg E, Halberg F, Halberg F (1980) Circadian and circannual aspects of the complement cascade – new and old results, differing in specificity? Chronobiologia 7: 189–204

Knapp MS, Pownall R (1980) Chronobiology, pharmacology and the immune system. Int J Immunopharmacol 2: 91–93

Knapp MS, Pownall R, Cove-Smith JR (1981) Circadian variations in cell-mentiated immunity and in the timing of human allograft rejection. In: Walker CA, Winget CM, Soliman KFA (eds) Chronopharmacology and chronotherapeutics. A & M University Foundation, Tallahasse, Florida, pp 329–338

Krieger DT, Aschoff J (1979) Biological Rhythms. In: Degroot L (ed) Endocrinology, vol 2. Grune and Stratton, New York, pp 2079–2109

Levi F, Canon C, Blum JP, Mechkouri M, Reinberg A, Mathe G (1985) Circadian and/or circahemidian rhythms in nine lymphocyte-related variables from peripheral blood of healthy subjects. J Immunol 134: 217–222

Levi FA, Canon C, Touitou Y, Reinberg A, Mathe G (1988) Seasonals modulation of the circadian time structure of circulating T and natural killer lymphocyte subsets from health subjects. J Clin Invest 81: 407–413

Liu T, Cavallini M, Halberg F, Cornelissen G, Field J, Sutherland DER (1986) More of the need for circadian circaseptan and circannual optimization of cyclosporine therapy. Experientia 42: 20–22

McGovern JP, Smolensky M, Reinberg A (eds) (1976) Chronobiology in allergy and immunology. Thomas, Springfield

Mizel SB (1989) The interleukins. FASEBJ 3 (12): 2379–2388

Pati AK, Florentin I, Chung V, DeSousa M, Levi F, Mathe G (1987) Circannual rhythm in natural killer activity and mitogen responsiveness of murine splenocytes. Cellular Immunol 108: 227–234

Pownall R, Knapps MS (1980) Immune responses have rhythms. Are they important? Immunol Today 1: 7–10

Ratte J, Halberg F, Kuhl JFW, Najarian JS (1973) Circadian variation in the rejection of rat kidney allografts. Surgery 73: 102–108

Reinberg A, Smolensky M (1983) Biological rhythms and medicine. Cellular, metabolic, physiopathologic and pharmacologic aspects. Springer, Berlin Heidelberg New York, pp 305

Reinberg A, Zagulla-Mally Z, Ghata J, Halberg F (1969) Circadian reactivity rhythms of human skin to house dust, penicillin and histamine. J Allergy 44: 292–298

Reinberg A, Schuller E, Clench J, Smolensky MH (1980) Circadian and circannual rhythms of leukocytes, proteins and immunoglobulins. In: Smolensky MH, Reinberg A, McGovern JP (eds) Recent advances in the chronobiology of allergy and immunology. Pergamon, Oxford, pp 251–259 (Advances in the biosciences, vol 28)

Roitt IM (1984) Essential immunology, 5th edn. Blackwell, Oxford

Rosenblatt LS, Shifrine M, Hetherington NW, Paglierioni T, Mackenzie MR (1982) A circannual rhythm in rubella antibody titens. J Interdiscipl Cycle Res 13: 81–88

Sandberg LE, Fernandes G (1987) Enhancement of natural killer activity in young and old mice treated with cyclophospha-mide on a circadian basis. In: Pauly JE, Scheving LE (eds) Advances in chronobiology. Liss, New York, pp 411–419

Scheving LE, Burns ER, Pauly JE, Tsai TH (1978) Circadian variation in cell division of the mouse alimentary tract, bone marrow and corneal epithelium. Anat Rec 191: 479–486

Scheving LE, Halberg F, Ehret CF (1987) Chronobiotechnology and chronobiological engineering. Nijhoff, Den Haag

Shifrine M, Taylor N, Rosenblatt LS, Wilson F (1980) Seasonal variation in cell mediated immunity of clinically normal dogs. Exp Hematol 8: 318–326

Smith C, Pruett SB (1989) Circadian variations of human lymphocytes are not responsible for contradictory or variable results in studies of IL-2 production. Immunol Lett 20: 15–20

Smolensky M, Halberg F, Sargent II F (1972) Chronobiology of the life sequence. In: Ito S, Ogata K, Yoshimura H (eds) Advances in climatic physiology. Igaku Shoin, Tokyo, pp 281–318

Swoyer J, Haus E, Sackett-Lundeen L (1987) Circadian reference values for hematologic parameters in several strains of mice. In: Pauly JE, Scheving LE (eds) Advances in chronobiology, part A. Liss, New York, pp 281–296

Tavadia HB, Fleming KA, Hume PD, Simpson HW (1975) Circadian rhythmicity of human plasma cortisol and PHA-induced lymphocyte transformation. Clin Exp Immunol 22: 190–193

Williams RM, Kraus LJ, Inbar M, Dubey DP, Yunis EJ, Halberg F (1981) Circadian bioperiodicity of natural killer cell activity in human blood (individually assessed). In: Walker CA, Winget CM, Soliman KFA (eds) Chronopharmacology and chronotherapeutics. Florida A & M University, Tallahassee, pp 269–273

# Chronobiology of Circulating Blood Cells and Platelets

E. Haus

## Introduction

The great variability in the number of circulating formed elements in the peripheral blood has been noted since techniques for counting these structures became available during the second half of the last century. It was soon recognized that some of these variations do not occur at random, but are the expression of regularly recurring rhythmic events (Japha 1900; Sabin et al. 1927). With improvements in the accuracy and precision of hematologic methods of investigation, it became apparent that some of these periodic variations, especially in the circadian range, are highly reproducible and predictable in their timing and, in some instances, are large enough to be of clinical interest.

In the study of hematologic parameters in the peripheral blood, rhythmic events have been described in the frequency range of a few hours (ultradian) (Sabin et al. 1927), in the prominent circadian range (Halberg et al. 1953; Halberg and Visscher 1950; Brown and Dougherty 1956; Haus 1959; Bartter et al. 1962; Malek et al. 1962), and in the frequency range of about 1 week (circaseptan) (Lévi and Halberg 1982; Derer 1960; Haus et al. 1981, 1983, 1984). These rhythmic variations may be superimposed upon rhythms with periods between 15 and 30 days (Morley 1966), including circavigintan and circatrigintan rhythms and the menstrual frequency range in women, and upon seasonal changes or circannual variations (Kusnetsova et al. 1977; Berger 1980 a, b; Bratescu and Teodorescu 1981; Reinberg et al. 1980; Rocker et al. 1980). Rhythmic events of different frequencies have been found at several levels of organization of the hematopoietic system, e. g., in the proliferation of the hematopoietic elements in the bone marrow (Mauer 1965; Scheving and Pauly 1973; Laerum and Aardal 1981; Laerum et al. 1989; Bartlett et al. 1984; Haus et al. 1983; Smaaland et al. 1987; see the chapter by Smaaland and Laerum) and in the lympoid elements in lymph nodes, thymus, and spleen (Scheving 1981; Haus M et al. 1984). In the bone marrow, periodic events occur in the proliferation and maturation of the red cell and granulocyte precursors within the marrow and in the release of the cells from the marrow. In the peripheral blood, the periodic changes in the number of circulating cells may be the result of: (a) different factors, which may in themselves be rhythmic, e. g., the influx and distribution of some young formed elements; (b) the distribution between the circulating and the marginal cell compartments; and (c) the distribution between different tissues or organs of the body. Also, cell destruction, cell removal, and related reticuloendothelial (Szabo et al. 1977) and immunologic functions (Many and Schwartz 1971; Levi and Halberg 1982; Weigle 1975; Shifrine et al. 1980; see the chapter by Canon and Lévi) show periodicity in several frequency ranges.

## Circadian and Other Rhythms of the Formed Elements in Peripheral Blood in Clinically Healthy Human Subjects

Circadian variations have been described by numerous investigators at different geographic locations for the number of all circulating corpuscular elements in the human blood (e. g., Haus et al. 1983; Halberg et al. 1977; Touitou et al. 1978; Reinberg et al. 1977). The regularity and the amplitudes of these rhythms, however, vary between the different elements and between different populations examined (e. g., Swoyer et al. 1989).

## Red Cell Parameters

The number of circulating red blood cells, hemoglobin, and hematocrit show, in clinically healthy young adults (Touitou et al. 1979, 1986; Haus et al. 1983, 1988) and in elderly subjects (Touitou et al. 1979, 1986; Haus et al. 1988; Swoyer et al. 1989), a highly reproducible and regular but low amplitude circadian rhythm, with acrophase in most studies around 11:00 local time in young adults (Fig. 1) and apparently somewhat earlier in some elderly populations (Haus et al. 1988; Swoyer et al. 1989). The error estimate of the acrophase of this rhythm is relatively narrow, and by cosinor analysis about 50% of the total variability in the number of circulating red cells over a 24 h span can be accounted for by their circadian variation (Haus et al. 1983, 1988). However, the amplitude of the circadian rhythms in red cell parameters is very small and in individual measurements may be close to the imprecision of the method; thus, it is interesting from a physiologic viewpoint but is diagnostically irrelevant. The extent of the circadian

variations in circulating formed elements in the peripheral blood is shown in Fig. 2, with the range of change during a 24 h span expressed as percent of the lowest value observed.

The number of circulating reticulocytes shows a circadian rhythm with an acrophase reported around 0100 hours with a 95% confidence interval (CI) in the cosinor analysis between 1948 and 0428 hours (Haus et al. 1983, 1984). The double amplitude of this rhythm as a group phenomenon was, in a young adult population, only around 10000 cells, the circadian rhythm accounting for about 37% of the total variability encountered over a 24 h span. Due to the irregularity of this rhythm with large intersubject differences, the cosinor amplitude is relatively small. The range of the circadian variation in single subjects, expressed as the highest of each subject's measurements as a percent of the lowest, was around 130%. The circadian rhythm in circulating reticulocytes may indicate a circadian periodic release of these cells from the bone marrow and thus conceivably may serve as a marker rhythm for bone marrow rhyth-

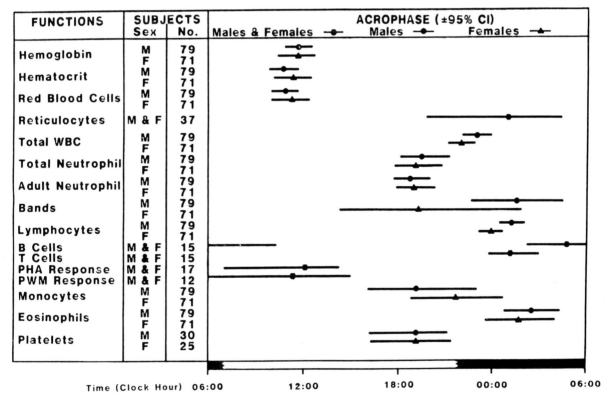

**Fig. 1.** Circadian acrophase chart of hematologic parameters in clinically healthy young adult and adult subjects of both sexes (mean age 24 ± 10 years; range 11–57 years) studied over a 24 h span (6–7 time points at 4 h intervals) at St. Paul-Ramsey Medical Center, St. Paul, MN, USA. All subjects were Caucasians, diurnally active, and on a 3 meal schedule. The acrophase, as determined by population mean cosinor (Nelson et al. 1979), is shown as *full circle* or *triangle* with 95% confidence interval as *horizontal bar* (Haus et al. 1983, 1988; Swoyer et al. 1989)

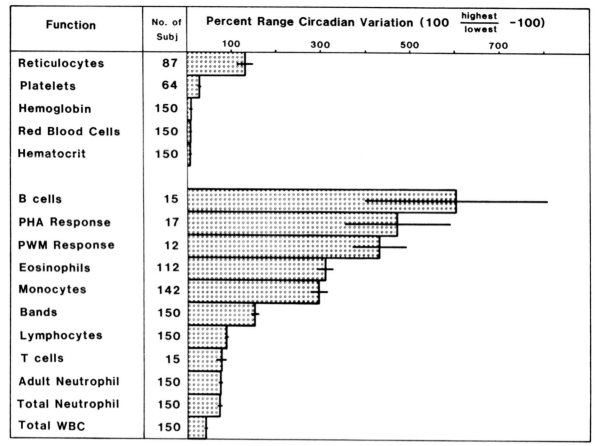

| Function | No. of Subj | Percent Range Circadian Variation ($100 \frac{highest}{lowest} - 100$) |
|---|---|---|
| Reticulocytes | 87 | |
| Platelets | 64 | |
| Hemoglobin | 150 | |
| Red Blood Cells | 150 | |
| Hematocrit | 150 | |
| B cells | 15 | |
| PHA Response | 17 | |
| PWM Response | 12 | |
| Eosinophils | 112 | |
| Monocytes | 142 | |
| Bands | 150 | |
| Lymphocytes | 150 | |
| T cells | 15 | |
| Adult Neutrophil | 150 | |
| Total Neutrophil | 150 | |
| Total WBC | 150 | |

**Fig.2.** Extent of circadian variation in hematologic variables in the peripheral blood, in the same population as shown in Fig. 1 expressed in percent of the lowest value of each subject encountered during the 24-hour span (Group mean ± ISE)

micity. This might be of importance for the timing of therapeutic interventions for which the bone marrow is either the target or the victim of undesirable side effects (e.g., cancer chemotherapy). Additional information will have to be obtained about the sampling requirements to reliably characterize the circadian rhythm in circulating reticulocytes in single subjects.

In regard to the circadian rhythms in red cell parameters, it is of interest that the circadian acrophase in hemoglobin and hematocrit in several strains of mice on a lighting regimen of light-dark (LD) 12:12, with L from 0600 to 1800 hours, also peaks around noon (Haus et al. 1983; Swoyer et al. 1987); in the nocturnal rodent this corresponds to the middle of the daily rest span. Extrapolations in timing from mice to humans can not be based on the assumption of a 12 h time difference between diurnally active humans and nocturnal rodents. This may be of importance if an animal model is used to explore

times of optimal treatment for chemotherapeutic regimens.

A low amplitude circaseptan (about weekly) rhythm in the number of circulating red blood cells and in hematocrit and hemoglobin concentration was found in two of our studies and was reproducible from one year to the other (Haus et al. 1981, 1984). In both studies, which extended from November to February of two consecutive years, with sampling three times per week between 0800 and 0900 hours, the acrophases were found on Monday and Tuesday, respectively.

Seasonal variations in red cell numbers, hemoglobin, and hematocrit were reported by Touitou et al. (1986), the timing of which seemed to vary between different age groups. Human plasma volume increases during the summer by about 9% (Yoshimura 1958; Sjostrand 1962), which may lead to the decrease in hematocrit observed by Rocker et al. (1980) during the same season. Our data were mainly obtained during the months of February and July and do not allow the recognition of a seasonal variation in red cell parameters.

Functional parameters of red cells also show periodic variations. For example, the microviscosity of circulating erythrocytes shows a circadian rhythm, with a peak in the morning (acrophase 0500 with 95% CI from 0250 to 0720 hours). This is out of phase with the rhythm of the erythrocyte membrane-bound methyltransferase I, which peaked during the night hours (acrophase 2200 with 95% from 1720 to 0210 hours) (Lévi et al. 1987).

# White Blood Cell Parameters

## Total White Cell Count

The circadian rhythm in total white blood cell count is highly reproducible, as has been shown in many studies by numerous investigators in different geographic locations, on populations of different ethnic background with different living habits, and during different seasons (Roitman et al. 1975; Halberg et al. 1977; Reinberg et al. 1977). The acrophase is found

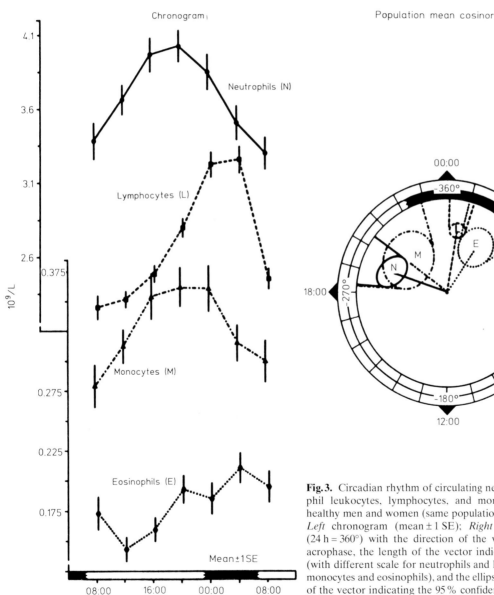

**Fig. 3.** Circadian rhythm of circulating neutrophil and eosinophil leukocytes, lymphocytes, and monocytes in clinically healthy men and women (same population as shown in Fig. 1). *Left* chronogram (mean ± 1 SE); *Right* polar cosinor plot (24 h = 360°) with the direction of the vector indicating the acrophase, the length of the vector indicating the amplitude (with different scale for neutrophils and lymphocytes than for monocytes and eosinophils), and the ellipse surrounding the tip of the vector indicating the 95% confidence region for amplitude and acrophase

during the evening hours (between 2100 and 0000 hours) with a change in total white cell count over the 24 h span in clinically healthy subjects from about 900 to 2000 cells (0.9–2.0 $10^9$/l) (Haus et al. 1983, 1988). The circadian rhythm in total white blood cell count, however, is the composite of the circadian rhythms of the different types of leukocytes (neutrophils, lymphocytes, etc.), some of which show phase differences between each other as shown in Figs. 1 and 3. The circadian rhythm in circulating neutrophil and eosinophil leukocytes, monocytes, and lymphocytes is shown in Fig. 3 in a population of diurnally active young adult subjects. Left, chronogram shown as the mean value measured at each time point with the SEM. *Right,* polar cosinor plot with the *full circle* corresponding to the 24 h period analyzed; the direction of the vector expressing the acrophase; the length of the vector, the amplitude; and the ellipse surrounding the tip of the vector, the 95 % confidence interval of amplitude and acrophase. The non-overlap of the ellipse with the center of the plot indicates rejection of the null hypothesis for amplitude by an F test as part of the cosinor analysis (Nelson et al. 1979).

## Circulating Neutrophils

The circulating neutrophil leukocytes show a circadian rhythm with acrophase during the late afternoon and evening hours (around 1900 with a 95 % CI from 1744 to 2120 hours). The changes in the number of circulating neutrophils observed during the 24 h span are, in clinically healthy subjects, around 600–1000 cells per $cm^3$ (0.6–1.0 $10^9$/l); occasionally, however, values up to 2500 (2.5 $10^9$/l) have been observed.

## Circulating Monocytes

The circadian rhythm in circulating monocytes and eosinophils is shown at the bottom of Fig. 3. Although with the 150 subjects examined, the standard errors of the group means are quite small, the monocytes show a considerable variation in their number, which is expressed in a larger error ellipse in the polar cosinor plot. Due to these very marked individual variations, the circadian reference ranges of the circulating monocytes are so wide as to be of little diagnostic benefit in the application to single samples (Haus et al. 1983).

## Circulating Eosinophils

The circadian rhythm in the number of circulating eosinophils shows its acrophase during the night hours (Fig. 3). Also here, the large variation in the number of these cells in different subjects leads to extremely broad circadian reference ranges. Thus, in spite of the high amplitude of this rhythm in individual subjects, the diagnostic value of a single eosinophil count will not be substantially improved by the time-qualified usual range. Of more importance is the awareness of the physician of the high amplitude circadian periodicity of the eosinophil leukocytes in the evaluation of consecutive samples in the same subject. In its relation to plasma corticosteroid concentrations, the circadian change in the number of circulating eosinophils has in the past been used for the assessment of adrenal cortical function (Halberg 1959; Halberg et al. 1958), and the circadian change (rather than the absolute value) may still be of some interest indicating the peripheral response or responsiveness of some target tissue to corticosteroids. The data on eosinophils shown here were obtained on 200 cell counts of Wright-stained smears, which contributes to the variability due to the small number of cells actually counted. The value of eosinophil counts for circadian rhythm evaluation can be improved if chamber counts of these cells are obtained.

## Total Circulating Lymphocytes

The circadian rhythm in the number of circulating lymphocytes (Fig. 3) is a very regular and highly reproducible phenomenon also in small groups and in individual subjects. The acrophase of the circadian rhythm in circulating lymphocytes occurs during the night hours with highest values found between midnight and 0100 hours. A circadian rhythm in the number of circulating lymphocytes is also evident when individuals are sampled only once, each at different times over a 24 h span. Although the reference ranges for the evaluation of single samples (Haus et al. 1983) are rather broad, the range of change in individual subjects and in serial independently examined subjects sampled only once each shows, between morning and evening samples, a difference of about 800 cells (0.8 $10^9$/l) and between morning and night samples, a difference of over 1000 cells/cmm (1.0 $10^9$/l). In a group of Frenchmen studied in France, Canon and Lévi (see their chapter in this volume) reported a seasonal difference with absence of a demonstrable circadian rhythm in circulating lymphocytes during March. In our studies, such a difference was not obvious. Although we do not have, at this time, a controlled study during March, American subjects studied in Minnesota, USA did show the usual high amplitude circadian variation in the number of circulating lymphocytes during February.

## Circulating Lymphocyte Subtypes

Although the circadian rhythms in the number of circulating lymphocytes are one of the most stable and reproducible periodic functions, circulating lymphocytes are not a homogeneous population but consist of functionally very different subtypes. Several of the subtypes show a circadian rhythm in their number and/or in their relative proportion of the total circulating lymphocyte population.

The demonstration of a circadian rhythm in many of the lymphocyte subsets, its timing, and its extent depend to some degree on the method of identification used (e.g., surface markers identified by different monoclonal antibodies vs E rosette formation, surface immunoglobulin identification, or testing of aspects of cell function). The circadian variation found in ten clinically healthy subjects, studied under conditions comparable to those of the subjects shown in Figs. 1–3, is shown in Fig. 4. Circadian variations in circulating T cells and/or T cell subsets were described (Swoyer et al. 1975; Haus and Halberg 1978; Ritchie et al. 1983; Miyawaki et al. 1984; Haus et al. 1983; Bertouch et al. 1983; Abo et al. 1981; Signore et al. 1985; Canon et al. 1985, 1986; Lévi et al. 1988 a, b) and are discussed more in detail in the chapter by Canon and Lévi. In general, the most consistent variations were found in the CD3 + and CD4 + cells, while the CD8 + cells were found by some to remain consistent over the 24 h span (Ritchie et al. 1983) or to exhibit a 12 h rhythm only (Lévi et al. 1985, 1988 a). The circadian variations in the number of circulating lymphocytes persist under conditions of sleep deprivation (Ritchie et al. 1983). A circadian rhythm in E rosette forming (T) cells was found to persist in vitro in aliquots of a single blood drawing studied over the following 24 h span and was reported to continue still in 4-day-old cell cultures (Gamaleya et al. 1988). These in vitro studies suggest that the rhythmic fluctuations in subpopulations of blood lymphocytes depend on circadian changes in the cells themselves (a cellular oscillator) rather than on environmental factors in the body such as plasma cortisol, which although not causing the circadian rhythm, may act as a synchronizer.

The ratio between CD4 + (T helper-inducer cells or OKT4 + cells) and CD8 + (T suppressor-cytotoxic lymphocytes or OKT8 + cells), which is used widely as an estimate of the balance of immunologic functions in the body, was reported by some investigators (Lévi et al. 1983, 1988 b) to change rhythmically during the 24 h span in individual subjects but not as a group phenomenon (Bertouch et al. 1983), or it was found not to change at all by others (Ritchie et al. 1983).

## Lymphocyte Functions

Not only the number of circulating lymphocytes, but also numerous aspects of lymphocyte function show circadian rhythms. The T cell response to phytohemagglutinin (PHA) has been found to be circadian periodic by several investigators studying this parameter over the entire 24 h span (Haus et al. 1974, 1983; Tavadia et al. 1975). There appears to be some variability in circadian timing, possibly due to seasonal or other factors, which may lead to negative results, i.e., if only two time points along the 24 h scale are examined (Felder et al. 1985).

Mixed lymphocyte reactions (MLR) with PHA activated T cells and also with autologous non-T cell activation showed a statistically significant circadian variation in the proliferative response of T cells in both types of autologous MLR, but with an apparent phase difference between each other. In the MLR with autologous PHA-activated T cells, the proliferation with cells from blood drawn at 1200 was significantly lower than from blood drawn at 0800 and at 2000 hours. In contrast, in MLR with autologous non-T cells, the highest proliferation was obtained with cells drawn at 0800 with a progressive drop at 1200 and 2000 hours. The circadian variations of the autologous MLR seem to reflect changes in the proliferative response of T cells. The difference in timing of the two types of autologous MLR may be due to the differences in the types of cells responding (Damle and Gupta 1982).

No circadian variations were found in the proliferative and stimulating activity of lymphocyte cell populations in the allogeneic MLR (Indiveri et al. 1985).

The predominant B cell activation with pokeweed mitogen (PWM) showed a circadian variation in the hands of some (Haus et al. 1983; Moldofsky et al. 1986) but not of other investigators (Indiveri et al. 1985). It appears that seasonal (Gamaleya et al. 1988; Canon et al. 1986) and perhaps other rhythmic or

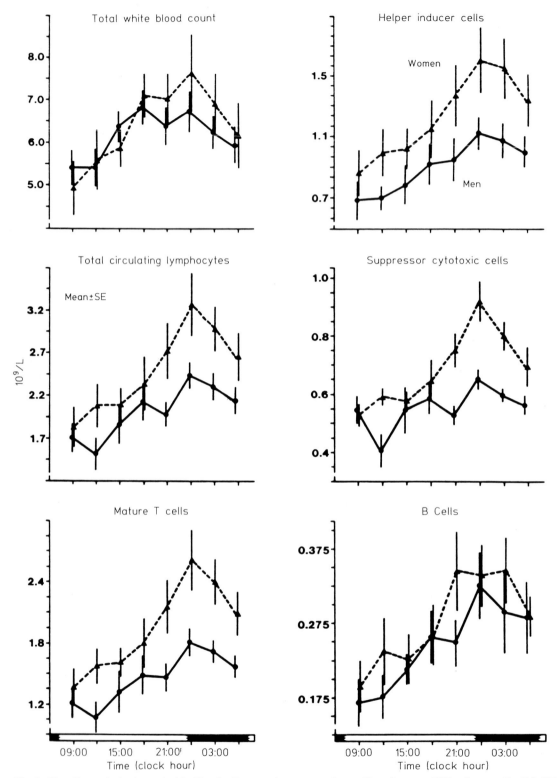

**Fig. 4.** Circadian variation in total white blood cell count, circulating lymphocytes, and some lymphocyte subtypes in ten, clinically healthy, diurnally active, adult subjects. Circadian rhythms of comparable phase were described by population mean cosinor for total lymphocytes, CD3 + (mature T cells), CD4 + (helper-inducer T cells), CD8 + (suppressor-cytotoxic T cells), and B1 + (B cells) subtypes. The CD4 + /CD8 + ratio in this study was 1.7 ± 0.2

nonrhythmic variations may play a role in the modification of lymphocyte functions, which will need further closely controlled studies for clarification.

Sleep deprivation of 40 h led to a phase alteration in the circadian rhythm in the lymphocyte response to PWM, but not to PHA (Moldofsky et al. 1989). A possible relation of the B cell but not of the T cell (and plasma cortisol) response to sleep is of interest.

In a two point study, the frequency of sister chromatid exchange in human blood was significantly less at 0900 than at 2100 hours, suggesting a circadian variation of this parameter (Slozina and Golovachev 1986) which may be of importance for the lymphocyte response to a number of environmental stimuli, including radiation and certain carcinogenic agents.

Circadian rhythms of variables presumably related to lymphocyte function were described for urinary neopterin excretion (Auzeby et al. 1988), which indicates T cell activation (Huber et al. 1984). The peak of urinary neopterin (0630 hours) follows that of the circulating lymphocytes in the same subjects by about 2 h (Lévi et al. 1988 b).

Circadian variations in lymphocyte adrenoreceptor density were reported with peak values around noon and a trough around midnight (Pangerl et al. 1986), and circannual variations were found with high values in spring and summer and low values during winter (Pangerl et al. 1986; Haen et al. 1988). The decrease in circadian MESOR in winter was associated with a relative increase in circadian amplitude. Disturbances in the expression and function of adrenoreceptors appear to play a role in the pathobiology of, e.g., hypertension and in bronchial asthma (Haen 1987), both of which show circadian chronopathology.

A marked circadian variation in the glucocorticoid receptor content was reported, with the values at 2300 being 38% higher than at 0800 hours. The receptor affinity did not change over the 24 h span, and there was no seasonal variation in number and affinity of glucocorticoid receptors (Homo-Delarche 1984).

## Natural Killer Cells

There are some conflicting reports on circadian rhythms in natural killer (NK) cell numbers and activity. A considerable number of regulatory factors, both soluble and cellular, have been shown to influence NK cell functions, including some of the interferons, i.e., interferon (INF)-$\gamma$ and interleukin (IL)-2 (Herberman and Callewaert 1985; Richts-

meier 1985). Some studies using the HNK-1 monoclonal antibody for identification of a lymphocyte population, including NK cells and lymphocytes showing antibody-dependent killer activity, did not show statistically significant circadian variations (Ritchie et al. 1983), presumably due to the lack of NK specificity of this antibody and possible phase differences in the different lymphocyte types targeted by it. In contrast, the NK cell activity in the peripheral blood of clinically healthy adult subjects, studied by biologic end points, shows a reproducible circadian rhythm, with the activity being high in the morning and then declining to a minimum during the night hours (Abo et al. 1981; Williams et al. 1979; Moldofsky et al. 1986, 1989; Gatti et al. 1986, 1988) and thus being out of phase with the numbers of circulating lymphocytes of most other subtypes. Gatti et al. (1988) showed not only a circadian variation in spontaneous NK cell activity (acrophase 0422), but also found a circadian rhythm in the response of NK cell activity to stimulation with INF-$\gamma$, which was synchronous with the spontaneous NK cell activity (acrophase 0403). Investigations by Gatti et al. (1988) also raise the question of a seasonal (circannual) pattern of NK cell activity.

## Other Periods of Lymphocyte Rhythms: Ultradian, Infradian, Circaseptan, and Circannual

Rhythms of different frequencies other than the circadian in lymphocyte numbers and/or functions have been reported. Lévi et al. (1985, 1988 a) described by rhythmometric analysis about a 12 h (circahemidian) rhythm in the number of total circulating lymphocytes, suppressor cytotoxic (OKT8 + ) T cells, and the OKT4 + /OKT8 + ratio. Only a 12 h rhythm was found by these investigators in cytotoxic-suppressor (OKT8 + ) T cells, which is at variance with some other reports (Swoyer et al. 1990). Again questions have to be raised concerning antibody specificity, possible seasonal variations, and masking of a circadian rhythm by environmental stimuli. Circaseptan variations of many immune related functions are well-known (for review see Lévi and Halberg 1982; Haus et al. 1983; Reinberg and Smolensky 1983) and seem to represent a presumably genetically fixed property of immune related structures (including lymphocytes) to respond to an immunologic stimulus with a circaseptan response of both the humoral and the cell-bound immune system. Circaseptan variations in the numbers of circulating T and B cells were observed in the course of the development of an antigenically triggered encephalomyelitis in guinea pigs, with the rhythms of the two cell types being of opposite phase. Relapses in this experimental disease state occurred when the number of circulating T lymphocytes was low (Raine et al. 1978).

Circannual variations in the relative number of circulating B and T lymphocytes, as determined by a bacterial adhesion test, have been described in clinically healthy human subjects by Bratescu and Teodorescu (1981), with a peak of the T cells in late fall and of the B cells during the winter. The circannual variation was apparent in the subtypes only; the total number of lymphocytes showed no statistically significant circannual variation. Lévi et al. (1988a) described circannual changes in the number and circadian rhythm characteristics of circulating total lymphocytes (acrophase in November, zero circadian amplitude in March), pan-T cells (acrophase in March), suppressor-cytotoxic T cells (acrophase in December), and NK lymphocytes (acrophase in October).

Seasonal variations in the response of human (Shifrine et al. 1982a) and canine (Shifrine et al. 1980, 1982b) lymphocytes to lectin-induced transformation suggest a circannual cycle in cell-mediated immunity. Also, certain enzymatic functions in human lymphocytes show circannual variations, which may interfere with a potential diagnostic use of these enzymes (Richter et al. 1978; Paigen et al. 1981).

## Alterations of Circadian Rhythms in Human Lymphocytes

Alterations of circadian periodicity of human lymphocytes have been reported in patients with lymphoid tumors (Swoyer et al. 1975). Infection with the lymphotropic human immunodeficiency virus (HIV) leads to alterations in the circadian rhythm of circulating lymphocytes as an early event in the course of the development of the AIDS related syndrome (ARS) and AIDS (Bourin et al. 1989; Martini et al. 1988b, c; Swoyer et al. 1990). The alterations of circadian periodicity in circulating white blood cells seen in nine subjects infected by HIV is shown in Figs. 5–7 and was discussed in detail by Swoyer et al. (1990). The patients were compared with ten clinically healthy subjects of comparable age studied under comparable conditions. Circadian periodicity was found in the HIV infected patients and in the controls in hemoglobin, hematocrit, and number of circulating red cells, but with significantly lower amplitudes in the patients (Swoyer et al. 1990). In spite of the

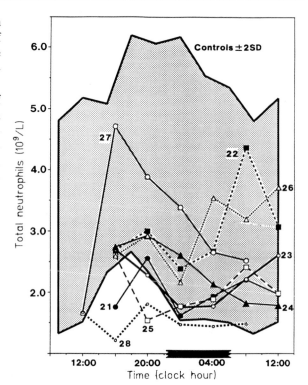

**Fig. 5.** Circadian variation in the number of circulating neutrophil leukocytes in eight HIV positive subjects as compared with the range ( ± 2 SD) derived from ten clinically healthy subjects studied under comparable conditions. Marked rhythm alterations are seen in HIV positive subjects (Swoyer et al. 1990)

marked differences in MESOR and amplitudes, the timing of the circadian rhythms in red cell parameters in the HIV infected patients and in the clinically healthy subjects was comparable. A circadian rhythm was demonstrable as a group phenomenon by cosinor analysis in the controls but not in the patients in the number of circulating neutrophils, monocytes, total lymphocytes, CD3 + , CD8 + and B1 + lymphocytes, and in platelets. In the HIV infected subjects, the CD4 + lymphocytes showed, by cosinor analysis, a circadian variation of similar timing as in the controls but with very markedly decreased circadian MESOR and amplitude. The total number of circulating neutrophils in the HIV infected subjects is shown in Fig. 5 and the number of circulating CD4 + and CD8 + cells in Fig. 6 against the time-qualified reference ranges obtained from the ten control subjects (shown as two standard deviations around the mean). The circadian rhythm alterations of the circulating neutrophils and lymphocyte subtypes were found irrespective of whether these cells, in a given patient, were within, below, or, in the case of CD8 + cells, even above the usual range. In our studies and in

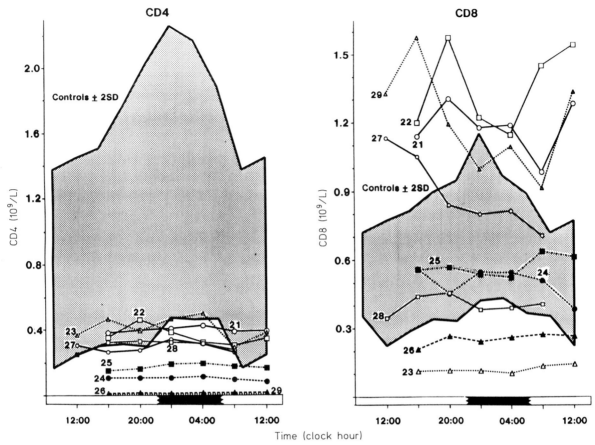

**Fig. 6.** Circadian variations in number of circulating CD4 + (helper-inducer T cells) and CD8 + (suppressor-cytotoxic T cells) in nine HIV positive subjects as compared to the range (± 2 SD) derived from ten clinically healthy subjects studied under comparable conditions. Marked rhythm alterations in CD8 + cells in HIV positive subjects. The minimal circadian variations in CD4 + cells still suggest some circadian rhthmicity as group phenomenon by cosinor analysis (Swoyer et al. 1990)

those of Martini et al. (1988a) and Bourin et al. (1989), circadian rhythm alterations of circulating lymphocytes, and especially CD8 (suppressor-cyto-toxic) lymphocytes, appear to be an early event in the course of HIV infection and were found, by all three investigators, in HIV positive patients who had not yet developed the full clinical symptomatology of AIDS. In these same patients, the circadian rhythm in plasma cortisol was found to be intact (Fig. 7), again indicating a dissociation of the circadian variation in the total number and at least some of the subsets of the circulating lymphocytes from those of plasma cortisol.

The number of CD4 + cells, although markedly depressed in most HIV infected patients, still shows a very low amplitude circadian variation as a group phenomenon with acrophase in the same location as the normal subjects, possibly maintaining their usual phase relation with plasma cortisol. It is not clear whether the phase relation to cortisol of the CD4 + cells represents a causal relationship or not. The CD8 + cells showed high amplitude circadian variations in some subjects, but with marked alterations in timing which were different from patient to patient. This observation is of interest concerning possible mechanism in the regulation of these cell types in the course of a retroviral infection, including the possibility of free-running rhythms.

Martini et al. (1988a) studied 12 patients who had undergone autologous bone marrow transplantation for acute leukemia (10 patients) or non-Hodgkins lymphoma (2 patients) with marrow treated in vitro with cyclophosphamide at the CFUGM-LD 90-95 level (Gorin et al. 1986; Martini and Gorin 1988). Sampling only at 0800 hours and at mignight, Martini et al. (1988d) found rhythm alterations involving the CD4 + cells in 7 of the 12 patients, even if the total CD4 + lymphocyte count was within the usual range.

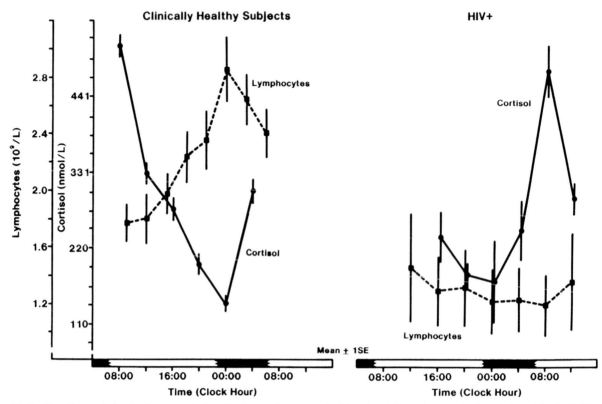

**Fig. 7.** Circadian variation in plasma cortisol and in the number of circulating (total) lymphocytes in clinically healthy subjects and HIV positive patients. Maintenance of rhythm in plasma cortisol but loss of lymphocyte rhythms is found in the HIV positive patients

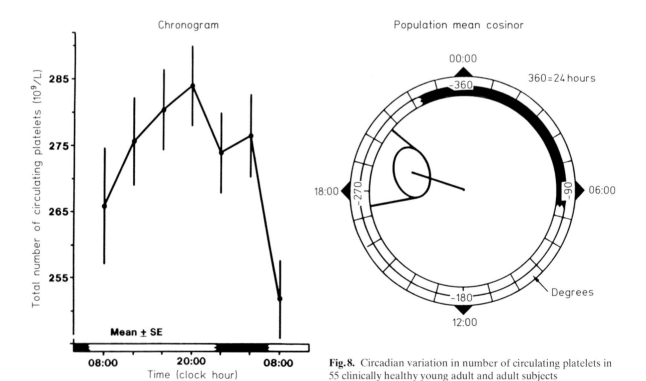

**Fig. 8.** Circadian variation in number of circulating platelets in 55 clinically healthy young adult and adult subjects

Although a characterization of the rhythm alterations with sampling at two time points only is not feasible, the investigators concluded that the evaluation of the CD4 + lymphocyte cycle might be a more sensitive test for an abnormal regulation of this cell line than the absolute CD4 + count itself.

## Chronobiology of Blood Platelets and Platelet Functions

### Number of Circulating Platelets

The number of platelets circulating in the peripheral blood shows a statistically highly significant circadian variation as a group phenomenon (Fig. 8). However, the variance between different individuals in the population is so large and the extent of the circadian variation in platelet numbers is so small that the circadian rhythm-dependent differences in their reference range are clinically irrelevant for the evaluation of single platelet counts.

### Platelet Function Tests (Aggregation, Adhesiveness)

Of more importance are functional changes in the circulating platelets, as expressed by changes in aggregability to stimulation with adenosine diphosphate (ADP) or epinephrine (Petralito et al. 1982; Tofler et al. 1987; Mehta et al. 1989; Haus et al. 1990) or in platelet adhesiveness, as expressed by their retention on glass bead columns (Haus et al. 1990). Figure 9 shows the circadian variation in platelet aggregation in response to stimulation with ADP and epinephrine and in platelet "adhesiveness" as found in the same 10 clinically healthy subjects sampled after at least 30 min recumbency. Platelet aggregation in response to stimulation and platelet "adhesiveness" is both the result of a complex series of reactions with chemical and mechanical factors participating (Kroll and Schafer 1989). The phase difference between the two indicators of platelet function found in our subjects is of interest. Observations on the timing of the circadian peak in platelet aggregation vary between different investigators. Tofler et al. (1987), using the stimulation threshold with ADP and epinephrine as end points, found, in a group of clinically healthy subjects following a "routine series of typical morning activity" including some physical exercise and coffee, a peak in

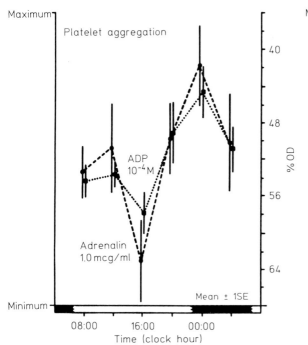

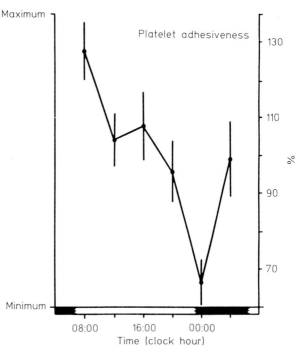

**Fig. 9.** Circadian rhythm in platelet aggregation after stimulation with adenosine diphosphate (ADP) and adrenalin *(left)* and platelet retention (adhesiveness) on a glass bead column (six consecutive 1 ml aliquots of each sample) *(right)* in ten clinically healthy, diurnally active subjects

aggregability at 0900 hours with higher values during daytime than during evening and night hours. The circadian changes in platelet aggregatability seemed to parallel the circadian changes in plasma epinephrine and norepinephrine. The differences in timing of the circadian rhythm in platelet aggregation observed by different investigators may be an expression of the interaction between endogenous circadian rhythmic parameters and exogenous stimulation, which changes with differences in experimental design. The circadian rhythm in platelet function may contribute to the circadian variations in the incidence of sudden cardiac death (Rabkin et al. 1980; Muller et al. 1987), myocardial infarction (Reinberg et al. 1973; Muller et al. 1985), and cerebral infarction (Reinberg et al. 1973; Marshall 1977; Marler et al. 1989), which have been reported to occur most frequently during the early and midmorning hours. The circadian rhythms in procoagulant platelet activities, together with those of other coagulation factors (Haus et al. 1990; Conchonnet et al. 1990; see the chapter by Decousus), may contribute to the transient rhythmic (and thus in their timing predictable) risk states for these conditions and may be of interest for timed treatment and prevention.

The close relationship of platelet physiology and catecholamine (and possibly other neurotransmitters) levels and metabolism is of interest concerning possible mechanisms of rhythm alterations in platelet function and/or of the timing of this rhythm. Mehta et al. (1989), sampling diurnally active clinically healthy subjects during daytime only, found the lowest epinephrine threshold dose for platelet aggregation at 0900 hours. The maximal responses to fixed doses of epinephrine were at 0900 and 2100 hours. There was a circadian rhythm in the $\alpha$-2-adrenoreceptor dissociation constant, which showed an inverse trend with the platelet sensitivity to epinephrine. Mehta et al. (1989) could not find any correlation between catecholamine levels and platelet adrenoreceptor number or affinity.

## Platelet Rhythms Other Than Circadian

Phenolsulfotransferase (PST), the main sulfoconjugating enzyme of catecholamines in human platelets, varies with the menstrual cycle and shows a marked decrease 7–10 days before menstruation (Abenhaim et al. 1981), correlating in time with the increased premenstrual sympathetic activity and excretion of free catecholamines and vanillyl mandelic acid (Kobus et al. 1979).

Circadian and cirannual variations of serotonin uptake in platelets (Wirz-Justice and Richter 1979; Rausch et al. 1982), in benzodiazepine binding sites (Lévi et al. 1987), and in seasonal variations in [$^3$H]imipramine binding to platelets have been reported (Whitaker et al. 1984; Egise et al. 1983), but their reproducibility and biologic significance, i.e., in relation to depressive illness, remain controversial (Arora et al. 1984; Baron et al. 1988). In this context, it has to be emphasized that a certain time relation between rhythms does not imply a causal relationship, although absence of a fixed time relation will make a cause-effect relationship rather unlikely.

## Circadian Variations in the Number of Circulating Stem Cells in the Peripheral Blood

Lasky et al. (1983) found a circadian variation of the noncommitted, circulating, pluripotential, hematopoietic stem cells (CFU-GEMM), with higher values in the samples taken at 1600 than at 0800 hours. The committed stem cell precursors (CFU-GM and BFU-E) apparently showed similar variations which, however, did not reach statistical significance. The timing of the circadian variation in CFU-GM, as suggested in the data of Lasky et al. (1983), is similar to the findings by Verma et al. (1980) and Morra et al. (1984), who both reported higher values in the afternoon, but is at variance with the observation of Ross et al. (1980), who reported a peak in CFU-GM (with, i.e., type I eosinophilic colonies) around 0900 hours (acrophase 0916 with 95% CI from 0640 to 1152). The afternoon circadian rise in CFU-GM paralleled the rise in polymorphonuclear leukocytes (PMN) in the same subjects. The number and proportion of circulating CFU-GM also paralleled the PMN changes after epinephrine injection rather than after cortisol (Morra et al. 1984) and did not follow the decrease in number of circulating lymphocytes after cortisol. In keeping with previous work (Morra et al. 1981a), Morra et al. (1984) suggested that the circadian rise in CFU-GM is due to a change in distribution between compartments rather than to a release from the bone marrow (Morra et al. 1981b, 1984). For further discussion of stem cell and bone marrow rhythms, see the chapter by Smaaland and Laerum.

## Synchronization and Desynchronization of Hematologic Rhythms in Human Subjects

Endogenously determined but environmentally synchronized circadian rhythms follow a shift of their dominant synchronizer not abruptly, but slowly over several and sometimes numerous transient cycles. If several synchronizers act upon a circadian periodic function, this function will tend to follow its dominant synchronizer, although it may be modified in its temporal adjustment by secondary synchronizers acting upon the same function. A phase adaptation of hematologic rhythms has been reported in human subjects after changes in the sleep/wakefulness pattern (Sharp 1960) and the activity/rest cycle as encountered in shift workers or in subjects exposed to transmeridian flights over several time zones. An example of the phase adaptation of the circadian rhythm in circulating lymphocytes in a shift worker is shown in Fig. 10. The subject had spent 3 years on permanent nightshift (5 days of the week) with an almost complete shift in her living habits. She was studied over a 24 h span before and 21 and 42 days after changing to a regular day-shift. After 21 days, neither the circadian rhythm in plasma cortisol nor in circulating lymphocytes (Fig. 11) had adapted fully to her now diurnal activity pattern. Full phase adaptation had occurred

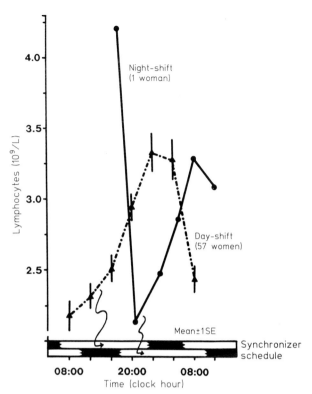

**Fig. 10.** Number of circulating lymphocytes in 57 medical technologists on day-shift (work from 0800 to 1630 hours) and in a 26-year-old technologist on permanent night-shift (work from 0000 to 0830 hours)

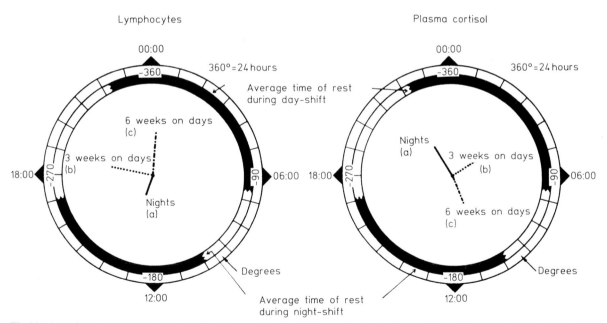

**Fig. 11.** Acrophase of circadian rhythm in plasma cortisol and in number of circulating lymphocytes in a 26-year-old medical technologist *(a)* after 3 years on permanent night-shift (work from 0000 to 0830 hours), *(b)* 21 days after 14 days vacation and change to day-shift (work from 0800 to 1630), *(c)* 42 days after vacation and change to day-shift. Slow phase adaptation of both circadian rhythms reaching target phase (completion) only between 21 and 42 days after end of permanent night-shift

at the time of reexamination after 42 days (Haus et al. 1983, 1984). In nonpermanent shift workers, a full phase adaptation of these functions to unusual shift times probably never occurs. Also, in permanent shift workers, a complete phase adaptation often will not occur if the subjects do not fully shift their living habits during the week and/or if they return to a diurnal activity pattern on weekends.

The timing of food uptake in animals, but to some extent also in human subjects, may influence the completeness of a phase-shift under an altered work schedule and may in itself phase shift certain functions (e.g., neutrophil granulocytes) but not others (e.g., lymphocytes) (Haus et al. 1983).

## Cyclic Hematopoiesis

### Cyclic Hematopoiesis in Clinically Healthy Subjects

Oscillations in the number of circulating granulocytes, with a frequency between 18 and 22 days, have been reported in clinically apparently healthy subjects by some (Morley 1966, 1973) but could not be found by other investigators (Dale et al. 1973; Maughan et al. 1973). Oscillations of the platelet count with the period of 21–35 days were reported in 4 of 11 normal individuals (Morley 1969).

Morley et al. (1970) proposed a model of a control system for erythropoiesis, granulopoiesis, and thrombopoiesis which was based upon the assumption of a double negative-feedback system containing an absolute time delay. The existence of such a rhythmicity in hematopoiesis in clinically healthy subjects is still controversial and will require further investigation. In our laboratory, apparent infradian variations in the number of circulating neutrophils and/or lymphocytes were found in the circavigintan frequency range in clinically healthy subjects in two studies extending from November to February of two consecutive years in groups of nine subjects each, but were not reproducible from one year to the other.

### Cyclic Neutropenia

Human cyclic hematopoiesis may be found in a number of distinctive disease entities (i.e., cyclic neutropenia, cyclic thrombocytopenia), characterized by about 21 day fluctuations of circulating blood neutro-

phils, monocytes, eosinophils, lymphocytes, platelets, and reticulocytes (Guerry et al. 1973; Dale and Hammond 1988). The numbers of monocytes, platelets, and reticulocytes and the concentration of colony-stimulating factor (CSF) frequently oscillate in cyclic neutropenia with the same period as the neutrophil count, but out of phase. Monocytosis tends to occur just as the neutrophil count is beginning to rise and peaks of CSF tend to occur at the nadir of the neutrophil count or at the peak of monocytosis. The symptomatology of the disorder will vary depending upon the cell line primarily involved and the severity of the condition. The recurrent severe neutropenia causes patients to experience periodic symptoms of fever, malaise, mucosal ulcers, and, rarely, life threatening infections. The disease occurs both as a congenital disorder and in an acquired form (Dale and Hammond 1988; Loughran and Hammond 1986; Loughran et al. 1986), with essentially identical phenotypic presentations. Studies of the pathophysiology of cyclic hematopoiesis demonstrate that the abnormality lies in the regulation of cell production and not in peripheral destruction (Guerry et al. 1973; Dale and Hammond 1988). Studies of the bone marrow in such patients show that the hematopoietic progenitor cell numbers fluctuate cyclically (Brandt et al. 1975; Greenberg et al. 1976; Verma et al. 1982). Bone marrow transplantation from a child to her sister with leukemia resulted in the transfer of cyclic neutropenia, suggesting that the basic defect represents a stem cell disorder (Krance et al. 1982). It was suggested that an abnormality in responsiveness of the progenitor cells to granulocyte-macrophage CSF (GM-CSF) may be a characteristic feature in this disease (Wright et al. 1989). Hammond et al. (1989) reported the efficacy of recombinant human granulocyte CSF (G-CSF) in patients with cyclic neutropenia ameliorating the neutropenia and diminishing the frequency and severity of infections. Migliaccio et al. (1990) studied the number and growth factor requirements of committed progenitor cells (CFU-GM and BFU-E) in three patients with cyclic neutropenia (two congenital, one acquired) before and during treatment with recombinant human G-CSF. When the patients with congenital disease were treated with G-CSF, the cycling of blood cells persisted but the cycle length was shortened from 21 days to 14 days and the amplitude of the variations in the cell counts increased in both types of circulating progenitor cells. In the patient with acquired cyclic neutropenia, this effect could not be seen. Erythroid and myeloid bone marrow progenitor cells from untreated patients differed in their growth factor responsiveness, requiring different doses and combinations of GM-CSF plus G-CSF and interleukin-3 (IL-3).

Progenitor cells from cyclic neutropenia patients are five to ten times less responsive to G-CSF than normal progenitors. Their response to IL-3 is normal (Hammond et al. 1989). It appears that long-term treatment of patients with cyclic neutropenia with growth factors leads to an improvement of the condition. In the small number of patients thus far treated over a prolonged time span, no untoward side effects upon the hematopoietic elements have been reported (Migliaccio et al. 1990).

## Cyclic Thrombocytopenia

Cyclic thrombocytopenia is a rare entity in which the platelet count oscillates with a period of 20–40 days. Its sporadic occurrence and onset during adult life suggests that it is an acquired rather than a genetic disorder. Studies in one patient suggest that the platelet's life span was normal (Lewis 1974).

## Cyclic Hematopoiesis
## and Chronic Granulocytic Leukemia

The development and course of chronic granulocytic leukemia (CML) involves a complex sequence of events and the interaction of growth stimulating factors and of growth suppressing or maturation inducing factors (Metcalf 1989). *In vitro,* most myeloid leukemic cells are dependent upon normal growth regulators and some normal regulators are able to suppress some myeloid leukemic cell populations. It appears that the initial chronic phase of CML might, in many instances, be merely a myelodysplastic or "preleukemic" state frequently still responding to growth factor regulation (Metcalf 1989). It appears that CSFs are essential cofactors in the development of myeloid leukemia, although they may act differently upon different target systems and, possibly at different times.

Marked cyclic oscillations in the white blood cell count were reported in a subpopulation of 10%–20% of patients with Phl+ and/or Phl− CML (Morley et al. 1967; Kennedy 1970; Vodopick et al. 1972; Shadduck et al. 1972; Gatti et al. 1973; Chikkappa et al. 1976; Mehta and Agarwal 1980; Iubal et al. 1983; Umemura et al. 1986). Some investigators suggest that a closer longitudinal study might lead to the detection of even larger numbers of patients with more or less pronounced cycling of the leukemic elements.

The periods reported varied between 30 and 120 days (most often around 70 days). In addition, lower frequency oscillations of the leukocyte counts were observed in some patients with peaks occurring at intervals between 9 and 20 months.

In the course of progression of the disease or of lowering of the number of circulating white cells by treatment, the cycling continues following the upward or downward trend of the overall white cell count. Although the period varies between different subjects, it recurs with considerable regularity in each individual. Platelet numbers and hemoglobin concentrations may vary in synchrony with the leukocytes (Umemura et al. 1986) and the CFU-GM in the peripheral blood may, in some instances, do the same.

The cycling of the numbers of circulating granulocytes and their precursors in patients with CML is due to a periodic contraction and expansion of the total blood granulocyte pool (Vodopick et al. 1972) and the periodic influx of progenitor cells from the hematopoietic stem cell pool (Umemura et al. 1986). This appears to be the same phenomenon as in cyclic hematopoiesis, except that the cycle period is longer, possibly due to a longer maturation time of some blood cell precursors in CML (Wheldon et al. 1974). The oscillation seems to arise because of a primary defect in the hematopoietic stem cells and their interaction with hematopoietic feedback control mechanisms and other (cyclic or noncyclic) regulating systems (Chikkappa et al. 1976). The mechanisms involved may be complex and, in addition to feedback mechanisms and endogenous- and/or feedback-induced variations in growth factor production, include a cellular substrate which in itself shows a multifrequency time structure. The rhythmic changes in the cell system involved may determine and/or alter substantially the functioning of feedback mechanisms. Any mathematical models trying to explain the phenomenon of cyclic hematopoiesis and cyclic CML will have to incorporate the concept of nonstationary rhythmically changing cell functions and cell sensitivity to appropriate and abnormal stimulation or inhibition (Schulthess and Mazer 1982; Klein and Valleron 1977; Guiguet et al. 1978).

From the clinical viewpoint, cycling of the leukemic elements in CML can pose problems in the diagnosis and in the monitoring of therapy, in which the spontaneous cyclic variations may be confounded with treatment effects or lack thereof. The cyclicity of many CMLs may provide an opportunity to learn more about the nature of this disorder. At a time when hematopoietic growth factors and/or their antibodies are being identified and becoming available, due to recombinant DNA and other technologies, such knowledge may provide a handle for the clinical management of the disease.

## Rhythmic Changes in Responsiveness, Susceptibility and Resistance – Chronopharmacology and Chronotherapy

In addition to the circadian rhythms in the number of circulating white blood cells, there are circadian periodic variations in their responses to external stimuli, including experimental procedures and medical treatment. Figure 12 shows the circadian variation in the lymphopenic response in adult male and female Balb/C mice, kept on LD 12:12, to handling and to the injection of 1 μCi[³H]thymidine (Haus et al. 1974). The animals were killed 2 h after the [³H]thymidine injection. A control group of animals remained untreated and was disturbed as little as possible prior to being killed.

The undisturbed mice showed the expected circadian variation in the number of circulating lymphocytes, with the lowest numbers observed after onset of the dark span. A marked lymphopenic response was seen in the mice exposed to handling and to [³H]thymidine injection, which, however, was strictly circadian periodic and limited only to certain circadian stages. Before and after the daily low in the number of circulating lymphocytes, the lymphopenia induced by the treatment of the animals was quite marked with a drop in number of over 4000 cells/cmm (4.0 10⁹/l) at the peak time of the circadian rhythm. It was completely absent, however, at the trough time, and at this time no lymphopenic response below the levels of the untreated animals was found. This circadian sensitivity cycle of circulating lymphocytes was not specific for [³H]thymidine but was also observed in separate experiments after handling and saline injection alone, although to a lesser extent. The mechanism involved most likely represents a catecholamine mediated change in distribution between circulating and noncirculating cell compartments.

Of special interest are the sensitivity-resistance

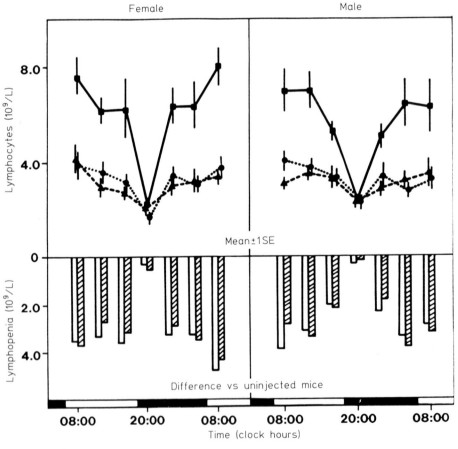

**Fig. 12.** Circadian stage-dependent response in number of circulating lymphocytes in male and female Balb/C mice 2 h after injection of [³H]thymidine (1 mCi/2 ml/20 g body wight i. p.). ■——■ No stimulation; ▲---▲ No stimulation prior to ³H-Thymidine; ●---● Complete Freund's Adjuvant s. c. 4 days prior to ³H-Thymidine

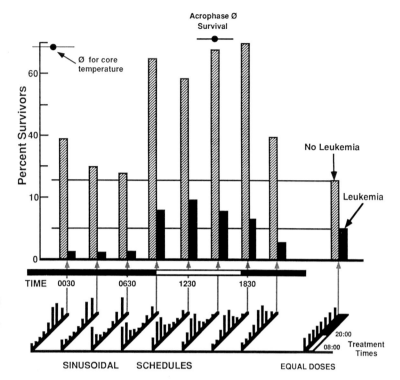

**Fig. 13.** Circadian rhythm in susceptibility of leukemic (L 12: 10 acute lymphocytic leukemia) and nonleukemic BDF₁, mice to arabinoside cytosine (ara-C) (240 mg/kg i. p.) given either in eight equal doses over the 24 h span or in eight sinusoidal treatment schedules. When the highest doses are given between 0930 and 1830 hours (to animals on LD 12:12 with L from 0600 to 1800), there is a markedly improved tolerance of the drug over the eight equal dose treatment and an improved survival of leukemic animals. If the drug is given at the "wrong" time, the leukemic animals show a higher mortality than the controls

cycles of hematopoietic cells to agents used in the treatment of human malignancies, including leukemia. The side effects of these agents on the host, i. e., upon the bone marrow, are very often a limiting factor for therapy. In some forms of treatment, part of the mortality is due to the treatment rather than to the underlying disease. The results of animal experiments suggest that an adaptation of the treatment schedule to the circadian variations in susceptibility of the host to the drug can minimize the undesirable side effects of the treatment. In experimental models, the importance of timing of antileukemic treatment for survival of the animals and curative effect has been widely documented (Haus et al. 1972, 1988; Haus and Halberg 1978, Scheving et al. 1980, 1989). Chronotherapeutic treatment schedules adjusted to the susceptibility-resistance cycles of the host increase the survival of leukemic animals, as compared to the effect of treatment schedules established without regard to the organism's time structure. However, treatment at the "wrong time" will damage and, in some instances, may kill the host (Fig. 13). The implications for scheduling of chemotherapy in human malignancies are obvious but go beyond the scope of this presentation.

In the treatment of leukemic malignancies, the tumor may also show a time structure, with circadian rhythms or rhythms in other frequency ranges. Such rhythms may lead to the identification of cycles of the sensitivity of the leukemic cells to therapeutic agents. Periodicities in hematologic malignancies have been detected in some patients with multiple myeloma (Zinneman et al. 1972, 1974) and some leukemias (Kachergene et al. 1972; Ramot et al. 1976) including CML, as discussed previously.

Although timing of chemotherapy and radiotherapy in chronobiologically determined schedules has been shown to improve host tolerance and therapeutic effects in animal models, there has been little application to human oncology (Hrushesky et al. 1989). In the treatment of acute lymphoblastic leukemia in childhood, Rivard et al. (1985) reported that, in children who had achieved complete remission with a standard induction protocol and had received maintenance therapy with daily 6-mercaptopurine (6-MP) and weekly methotrexate (MTX) (plus monthly vincristine and prednisone), the disease-free survival was better and the risk of relapsing substantially lower when 6-MP and MTX were given in the evening than in the morning hours. This interesting observation will require a follow-up in controlled prospective studies.

## Conclusions

The recognition of a multifrequency time structure in the number and functions of the circulating corpuscular elements in the peripheral blood and in the hematopoietic organs is essential for the scientific and clinical exploration of hematologic parameters.

Circadian acrophase maps of hematologic variables and information on the extent of their circadian variations have become available (Haus et al. 1983, 1988). The high amplitude variations of some parameters, e.g., the number of circulating neutrophils and lymphocytes, may have diagnostic implications in clinical medicine. Time-qualified reference ranges are of importance in functions with high amplitude rhythms. In some of these rhythms, however, e.g., those of the circulating eosinophils, monocytes, and blood platelets, the wide spread of values found in the clinically healthy population makes the time-qualified usual ranges very broad, sometimes to the point that they do not add much diagnostic value to the presently used non-time-qualified reference ranges. Also, in these parameters, the circadian rhythm has to be kept in mind if consecutive samples are obtained in the same subject.

Rhythm alterations may be of pathobiologic and clinical interest, e.g., in viral infections (i.e., with human immunodeficiency virus) and in neoplasia. Low amplitude rhythms usually do not represent diagnostic problems, but their timing may indicate differences in responsiveness of one or the other component of the hematopoietic or the immune system.

The recognition of time-dependent changes in the effects of hematologic regulatory and growth factors and in the susceptibility and resistance to environmental agents, including drugs used in clinical medicine, environmental toxins and chemical carcinogens, is expected to lead to the development of chronopharmacologic treatment schedules and/or temporal adjustment of exposure, e.g., at the workplace, in order to try to take advantage of these predictable changes. Especially in the case of drugs which unavoidably have significant side effects, such as those used in cancer chemotherapy, improvement of host tolerance by timing of treatment appears to be promising. Also, the circadian and circaseptan variations in responsiveness of the immune system to antigen exposure and to challenge of the immunized organism may be of considerable practical importance. The observation of circaseptan rhythms of lymphocyte subtypes in the development of an immune disease is of interest for the study of the mechanisms of immune regulation and for the prediction of times of increased or decreased probability of relapse.

In clinical medicine, chronobiology leads to a redefinition of the usual ranges for certain high amplitude parameters and adds new end points, such as the rhythm parameters of MESOR, amplitude, and acrophase, for the description of normalcy. Alterations in the organism's time structure may be of importance for the early recognition of abnormal function, often before structural disease can be identified. Chronotherapeutic interventions with and without rhythm manipulations are expected to provide a more effective approach to the treatment of hematologic disorders.

## References

Abenhaim L, Romain Y, Kuchel O (1981) Platelet phenolsulfotransferase and catecholamines: physiological or pathological variations in humans. Canadian J Physiol Pharmacol 59: 300–306

Abo T, Kawate T, Itoh K, Kumagai K (1981) Studies on the bioperiodicity of the immune response. I. Circadian rhythms of human T, B, and K cell traffic in the peripheral blood. J Immunol 126: 1360–1363

Arora RC, Kregel L, Meltzer HY (1984) Circadian rhythm in the serotonin uptake in the blood platelets of normal controls. Biol Psychiatry 19: 1579–1584

Auzéby A, Bogdan A, Krosi Z, Touitou Y (1988) Time-dependence of urinary neopterin, a marker of cellular immune activity. Clin Chem 34: 1866–1867

Baron M, Barkai A, Kowalik S, Fieve RR, Quitkin F, Gruen R (1988) Diurnal and circannual variation in platelet $^3$H-Imipramine binding: comparative data on normal and affectively ill subjects. Neuropsychobiology 19: 9–11

Bartlett P, Haus E, Tuason T, Sackett-Lundeen L, Lakatua D (1984) Circadian rhythm in number of erythroid and granulocytic colony forming units in culture (ECFU-C and GCFU-C) in bone marrow of BDF$_1$ male mice. In: Haus E, Kabat H (eds) Chronobiology 1982–1983. Karger, New York, pp 160–164

Bartter FC, Delea CS, Halberg F (1962) A map of blood and urinary changes related to circadian variations in adrenal cortical function in normal subjects. Ann NY Acad Sci 98: 969–983

Berger J (1980a) Circannual rhythms in the blood picture of laboratory rats. Folia Haematol (Leipzig) 107: 54–60

Berger J (1980b) Seasonal influences on circadian rhythms in the blood picture of laboratory mice. Z Versuchstierk 22: 122–134

Bertouch JV, Roberts-Thompson P, Bradley J (1983) Diurnal variation of lymphocyte subsets identified by monoclonal antibodies. Br Med J 286: 1171–1172

Bourin P, Mansour I, Lévi F, Villette JM, Roué R, Fiet J, Rouger P, Doinel C (1989) Perturbations précoces des rhythmes circadiens des lymphocytes T et B au cours de l'infection par le virus de l'immunodeficience humaine (VIH). CR Acad Sci (Paris) 308: 431–436

Brandt L, Forssman O, Mitelman F, Odeberg H, Olofsson T,

Olsson I. Svensson B (1975) Cell production and cell function in human cyclic neutropenia. Scand J Haematol 15: 228–240

Bratescu A, Teodorescu M (1981) Circannual variations in the B cell/T cell ratio in normal human peripheral blood. J Allergy Clin Immunol 68: 273–280

Brown HE, Dougherty TF (1956) The diurnal variation of blood leukocytes in normal and adrenalectomized mice. Endocrinology 58: 365–375

Canon C, Lévi F, Reinberg A, Mathé G (1985) Circulating calla-positive lymphocytes exhibit circadian rhythms in man. Leukemia Res 9: 1539–1546

Canon C, Lévi F, Touitou Y, Sulon J, Demey-Ponsart E, Reinberg A, Mathé G (1986) Variations circadienne et saisonnière du rapport inducteur: suppresseur (OKT4 + :OKT8 + ) dans le sang veineux de l'homme adulte sain. CR Acad Sci Paris 302: 519–524

Chikkappa G, Borner G, Burlington H, Chanana AD, Cronkite EP, Ohl S, Pavelec M, Robertson JS (1976) Periodic oscillation of blood leukocytes, platelets and reticulocytes in a patient with chronic myelocytic leukemia. Blood 47: 1023–1030

Conchonnet Ph, Decousus H, Boissier C, Perpoint B, Raynaud J, Mismetti P, Tardy B, Queneau P (1990) Morning hypercoagulability in man. Annu Rev Chronopharmacol 7: 165–168

Dale DC, Alling DW, Wolff SM (1973) Application of time series analysis to serial blood neutrophil counts in normal individuals and patients receiving cyclophosphamide. Br J Haematol 24: 57–64

Dale DC, Hammond WP IV (1988) Cyclic neutropenia: a clinical review. Blood Rev 2: 178–185

Damle NK, Gupta S (1982) Autologous mixed lymphocyte reaction in man. III. Regulation of autologous MLR by theophylline -resistant and -sensitive human T-lymphocyte subpopulations. Scand J Immunol 15: 493–499

Derer L (1960) Rhythm and proliferation with special reference to the 6-day rhythms of blood leukocyte count. Neoplasma (Brasil) 7: 117–133

Egise D, Desmedt D, Schouteus A, Mendelewicz J (1983) Circannual variations in the density of tritiated imipramine binding sites on blood platelets in man. Neuropsychobiology 10: 101–102

Felder M, Doré CJ, Knight SC, Ansell BM (1985) In vitro stimulation of lymphocytes from patients with rheumatoid arthritis. Clin Immunol Immunopath 37: 253–261

Gamaleya NF,Shisko ED, Cherny AP (1988) Preservation of circadian rhythms by human leukocytes in vitro. Byull Eksper Biol Med 106: 598–600

Gatti RA, Robinson WA, Deinard AS, Nesbit M, McCullough JJ, Ballow M, Good RA (1973) Cyclic leukocytosis in chronic myelogenous leukemia: new perspectives on pathogenesis and therapy. Blood 41: 771–782

Gatti G, Cavallo R, Sartori ML, Marinone C, Angeli A (1986) Cortisol at physiological concentrations and prostaglandin E2 are additive inhibitors of human natural killer cell activity. Immunopharmacology 11: 119–128

Gatti G, Cavallo R, Sartori ML, Carignola R, Masera R, Delponte D, Salvadori A, Angeli A (1988) Circadian variations of interferon-induced enhancement of human natural killer (NK) cell activity. Cancer Detec Prev 12: 431–438

Gorin NC, Donay L, Laporte JP, Lopez M, Mary JY, Najman A, Salmon C, Aegerter P, Stachowiak J, David J, Pene F, Kantor G, Deloux J, Duhamel E, Van der Akker J, Gerota J, Parlier Y, Duhamel G (1986) Autologous bone marrow transplantation using marrow incubated with ASTA Z 7557 in adult acute leukemia. Blood 67: 1367–1376

Greenberg PL, Bax I, Levin J, Andrews TM (1976) Alteration of colony-stimulating factor output, endotoxemia, and granulopoiesis in cyclic neutropenia. Am J Haematol 1: 375–385

Guerry D IV, Dale DC, Omine M, Perry S, Wolff SM (1973) Periodic hematopoiesis in human cyclic neutropenia. J Clin Invest 52: 3220–3230

Guiguet M, Klein B, Valleron AJ (1978) Diurnal variation and the analysis of percent labelled mitoses curves. In: Valleron AJ, Macdonald PD (eds), Biomathmetics and cell kinetics. Biomedical, Elsevier/North Holland, pp 191–198

Haen E (1987) The peripheral lymphocyte as clinical model for receptor disturbances. Bull Europ Physiopath Respir 22: 539–541

Haen E, Langenmayer I, Pangerl A, Liebl B, Remien J (1988) Circannual variation in the expression of $\beta_2$-adrenoceptors on human peripheral mononuclear leukocytes (MNLs). Klin Wschr 66: 579–582

Halberg F (1959) Physiologic 24-hour periodicity: general and procedural considerations with reference to the adrenal cycle. Z Vitamin-, Hormon- and Fermentforsch 10: 225–296

Halberg F, Visscher MB (1950) Regular diurnal physiological variation in eosinophil levels in five stocks of mice. Proc Soc Exp Biol Med 75: 846–847

Halberg F, Visscher MB, Bittner JJ (1953) Eosinophil rhythm in mice: range of occurence; effects of illumination, feeding and adrenalectomy. Am J Physiol 174: 109–122

Halberg F, Barnum CP, Silber R, Bittner JJ (1958) 24-hour rhythms at several levels of integration in mice on different lighting regimens. Proc Soc Exp Biol Med 97: 897–900

Halberg F, Sothern RB, Roitman B, Halberg E, Benson E, Halberg F, Mayersbach von H, Haus E, Scheving LE, Kanabrocki EL, Bartter FC, Delea C, Simpson HW, Tavadia HB, Fleming KA, Hume P, Wilson C (1977) Agreement of circadian characteristics for total leukocyte counts in different geographic locations. Proc XII Int Conf Int Soc of Chronobiology, Il Pointe, Italy, pp 3–17

Hammond WP, Price TH, Souza LM, Dale DC (1989) Treatment of cyclic neutropenia with granulocyte colony-stimulating factor. N Engl J Med 320: 1306–1311

Haus E (1959) Endokrines System and Blut. In: Heilmeyer L, Hittmair A (eds) Handbuch der gesamten Hämatologie 2. Urban und Schwarzenberg, Munich, pp 181–286

Haus E, Halberg F (1978) Cronofarmacologia della neoplasia con speciale riferimento alla leucemia. In: Bertelli A (ed) Farmacologia clinica e terapia. Edizioni, Turin, pp 29–85

Haus E, Halberg F, Scheving LE, Pauly JE, Cardoso S, Kuhl JFW, Sothern RB, Shiotsuka RN, Hwang DS (1972) Increased tolerance of leukemic mice to arabinosyl cytosine with schedule adjusted to circadian system. Science 177: 80–82

Haus E, Halberg F, Kuhl JFW, Lakatua DJ (1974) Chronopharmacology in animals. Chronobiologia 1 (Suppl 1): 122–156

Haus E, Lakatua D, Swoyer J, Sackett-Lundeen L (1983) Chronobiology in hematology and immunology. Am J Anat 168: 467–517

Haus E, Lakatua DJ, Sackett-Lundeen L, Swoyer J (1984) Chronobiology in laboratory medicine. In: Reitveld WT (ed) Clinical aspects of chronobiology. Bakker, Baarn, pp 13–82

Haus E, Sackett LL, Haus M, Babb WK, Bixby EK (1981) Cardiovascular and temperature adaptation to phase shift by intercontinental flights – longitudinal observations. Adv Biosci 30: 375–390

Haus E, Nicolau GY, Lakatua D, Sackett-Lundeen L (1988) Reference values for chronopharmacology. Annu Rev Chronopharmacol 4: 333–424

Haus E, Cusulos M, Sackett-Lundeen L, Swoyer J (1990) Circadian variations in blood coagulation parameters, alpha-antitrypsin antigen and platelet aggregation and retention in clinically healthy subjects. Chronobiol Intern 7: 203–216

Haus M, Sackett-Lundeen L, Lakatua D, Haus E (1984) Circannual variation of $^3$H-thymidine uptake in DNA of lymphatic organs irrespective of relative length of light and dark span. J Minn Acad Sci 49: 19

Herberman RB, Callewaert DH (eds) (1985) Mechanism of cytotoxicity by NK cells. Academic Press, Orlando

Homo-Delarche F (1984) Glucocorticoid receptors and steroid sensitivity in normal and neoplastic human lymphoid tissue: a review. Cancer Res 44: 431–437

Hrushesky WJM, Roemeling v R, Sothern RB (1989) Circadian chronotherapy: from animal experiments to human cancer chemotherapy. In: Lemmer B (ed) Chronopharmacology. Dekker, New York, pp 439–473

Huber C, Batchelor JR, Fuchs D, Hauser A, Lang A, Niederwieser D, Reitnegger G, Swetly P, Troppmair J, Wachter H (1984) Immune response associated production of neopterin. Release from macrophages primarily under control of interferon-gamma. J Exp Med 160: 310–316

Indiveri F, Pierri I, Rogna S, Poggi A, Mantaldo P, Romano R, Pende A, Morgano A, Barabino A, Ferrone S (1985) Circadian variations of autologous mixed lymphocyte reactions and endogenous cortisol. J Immunol Meth 82: 17–24

Iubal A, Aktein E, Barak I, Meytes D, Many A (1983) Cyclic leukocytosis and long survival in chronic myeloid leukemia. Acta Haematol 69: 353–357

Japha A (1900) Die Leukozyten beim gesunden und kranken Säugling. Jahrbuch Kinderheilk 52: 242–270

Kachergene NB, Koshel IV, Nartsissov RP (1972) Circadian rhythm of dehydrogenase activity in blood cells during acute leukemia in childhood. Pediatriia 51: 81–85

Kennedy BJ (1970) Cyclic leukocyte oscillations in chronic myelogenous leukemia during hydroxy-urea therapy. Blood 35: 751–760

Klein B, Valleron AJ (1977) A compartmental model for the study of diurnal rhythms in cell proliferation. J Theor Biol 64: 27–42

Kobus E, Wasilewska E, Bargiel Z (1979) Urinary excretion of catecholamines (CA) and vanilmandelic acid (VMA) during a normal menstrual cycle. Bull Acad Pol Sci 27: 71–74

Krance RA, Spruce WE, Forman SJ, Rosen RB, Hecht T, Hammond WP, Blume KG (1982) Human cyclic neutropenia transferred by allogeneic bone marrow grafting. Blood 60: 1263–1266

Kroll MH, Schafer AI (1989) Biochemical mechanisms of platelet activation. Blood 74: 1181–1195

Kusnetsova SS, Parvdina GM, Yezhova VM (1977) Seasonal variations of some parameters of peripheral blood and haematogenetic organs in mice. Zh Obsliteh Biol 38: 133–140

Laerum OD, Aardal NP (1981) Chronobiological aspects of bone marrow and blood cells. In: Mayersbach von H, Scheving LE, Pauly JE (eds) Biological rhythms in structure and function. Liss, New York, 59C, 87–97

Laerum OD, Smaaland R, Sletvold O (1989) Rhythms in blood and bone marrow: potential therapeutic implications. In: Lemmer B (ed) Chronopharmacology. Dekker, New York, pp 371–393

Lasky LC, Ascensao J, McCullough J, Zanjani ED (1983) Steroid modulation of naturally occurring diurnal variation in circulating pluriopotential haematopoietic stem cells (CFU-GEMM). Br J Haematol 55: 615–622

Lévi F, Halberg F (1982) Circaseptan (about 7-day) bioperiodicity – spontaneous and reactive – and the search for pacemakers. La Ricerca Clin Lab 12: 323–370

Lévi F, Canon C, Blum JP, Reinberg A, Mathé G (1983) Large amplitude circadian rhythm in helper: suppressor ratio of peripheral blood lymphocytes. Lancet II: 462–463

Lévi F, Canon C, Blum JP, Mechkouri M, Reinberg A, Mathé G (1985) Circadian and/or circahemidian rhythms in nine lymphocyte-related variables from peripheral blood of healthy subjects. J Immunol 134: 217–222

Lévi F, Benavides J, Touitou Y, Quarteronet D, Canton T, Uzan A, Auzeby A, Gueremy C, Sulon J, Le Fur G, Reinberg A (1987) Circadian rhythm in the membrane of circulating human blood cells: microviscosity and number of benzodiazepine binding sites, a search for regulation by plasma ions, nucleosides, proteins, or hormones. Chronobiol Int 4: 235–243

Lévi F, Canon C, Touitou Y, Reinberg A, Mathé G (1988a) Seasonal modulation of the circadian time structure of circulating T and natural killer lymphocyte subsets from healthy subjects. J Clin Invest 81: 407–413

Lévi F, Canon C, Touitou Y, Sulon J, Mechkouri M, Ponsart ED, Touboul JP, Vannetzel JM, Mowzowicz I, Reinberg A, Mathé G (1988b) Circadian rhythms in circulating T lymphocyte subtypes and plasma testosterone, total and free cortisol in five healthy men. Clin Exp Immunol 71: 329–335

Lewis ML (1974) Cyclic thrombocytopenia: a thrombopoietin deficiency. J Clin Pathol 27: 242–246

Loughran TP Jr, Hammond WP IV (1986) Adult-onset cyclic neutropenia is a benign neoplasm associated with clonal proliferation of large granular lymphocytes. J Exp Med 164: 2089–2094

Loughran TP Jr, Clark EA, Price TH, Hammond WP (1986) Adult onset cyclic neutropenia is associated with increased large granular lymphocytes. Blood 68: 1082–1087

Malek J, Suk K, Brestak M (1962) Daily rhythm of leukocytes, blood pressure, pulse rate and temperature during pregnancy. Ann NY Acad Sci 98: 1018–1091

Many A, Schwartz RS (1971) Periodicity during recovery of the immune response after cyclophosphamide treatment. Blood 37: 692–695

Marler JR, Price TR, Clark GL, Muller JE, Robertson T, Mohr JP, Hier DB, Wolf PA, Caplan LR, Foulkes MR (1989) Morning increase in onset of ischemic stroke. Stroke 20: 473–476

Marshall J (1977) Diurnal variation in the occurrence of strokes. Stroke 8: 230–231

Martini E, Gorin NC (1988) Lymphocytic populations of the peripheral blood. Applications to the immunological monitoring of autografts of bone marrow. Rev Prat 38: 1997–2004

Martini E, Gorin NC, Gastal C, Doinel C, Roquin H, Najman A, Salmon C (1988a) Disappearance of CD4 lymphocyte circadian cycles in autologous bone marrow transplantation. Biomed Pharmacother 42: 357–359

Martini E, Muller JY, Doinel C, Gastal C, Roquin H, Douay L, Salmon C (1988b) Disappearance of CD4 lymphocyte circadian cycles in HIV infected patients: early even during asymptomatic infection. AIDS 2: 133–134

Martini E, Muller JY, Gastal C, Doinel C, Meyohas MC, Roquin H, Frottier J, Salmon C (1988c) Early anomalies of CD4 and CD20 lymphocyte cycles in human immunodeficiency virus. Presse Med 17: 2167–2168

Martini E, Roquin H, Gastal C, Doinel C (1988d) Reduction of circulating lymphocytes after giving blood. Effects of establishment of reference values for CD4 and CD8 lymphocytes. Ann Biol Clin (Paris) 46: 327–328

Mauer AM (1965) Diurnal variations of proliferative activity in the human bone marrow. Blood 26: 1–7

Maughan WZ, Bishop CR, Pryor TA, Athens JW (1973) The question of cycling of the blood neutrophil concentrations and pitfalls in the statistical analysis of sampled data. Blood 41: 85–91

Mehta BC, Agarwal MB (1980) Cyclic oscillations in leukocyte count in chronic myeloid leukemia. Acta Haematol 63: 68–70

Mehta J, Malloy M, Lawson D, Lopez L (1989) Circadian variation in platelet alpha$_2$-adrenoceptor affinity in normal subjects. Am J Cardiol 63: 1002–1005

Metcalf D (1989) The roles of stem cell self-renewal and autocrine growth factor production in the biology of myeloid leukemia. Cancer Res 49: 2305–2311

Migliaccio AR, Migliaccio G, Dale DC, Hammond WP (1990) Hematopoietic progenitors in cyclic neutropenia: effect of granulocyte colony-stimulating factor *in vitro*. Blood 75: 1951–1959

Miyawaki T, Taga K, Nagaoki T, Seki H, Suzuki Y, Taniguchi N (1984) Circadian changes of T lymphocyte subsets in human peripheral blood. Clin Exp Immunol 55: 618–622

Moldofsky H, Lue FA, Eisen J, Keyston E, Gorczynski RM (1986) The relationship of interleukin-1 and immune functions to sleep in humans. Psychosom Med 48: 309–318

Moldofsky H, Lue FA, Davidson JR, Gorczynski R (1989) Effects of sleep deprivation on human immune functions. FASEB J 3: 1972–1977

Morley AA (1966) A neutrophil cycle in healthy individuals. Lancet II: 1220–1222

Morley AA (1969) A platelet cycle in normal individuals. Aust Ann Med 18: 127–129

Morley AA (1973) Letter to editor. Blood 41: 329

Morley AA, Baikie AG, Galton DAG (1967) Cyclic leukocytosis as evidence for retention of normal homeostatic control in chronic granulocytic leukemia. Lancet II: 1320–1323

Morley A, King-Smith EA, Stohlman F Jr (1970) The oscillatory nature of hemopoiesis. In: Stohlman F Jr (ed) Symposium on hemopoietic cellular proliferation. Grune and Stratton, New York, pp 3–14

Morra L, Ponassi A, Bruzzi P, Parodi GB, Caristo G, Sacchetti C (1981 a) Influence of the spleen on the blood distribution of the colony forming cells (CFU-C) in man. Acta Haematol 66: 81–85

Mora L, Ponassi A, Parodi GB, Caristo G, Bruzzi P, Sacchetti C (1981 b) Mobilization of colony forming cells (CFU-C) into the peripheral blood of man by hydrocortisone. Biomedicine 35: 87–90

Morra L, Ponassi A, Caristo G, Bruzzi P, Bonelli A, Zunino R, Parodi GB, Sacchetti C (1984) Comparison between diurnal changes and changes induced by hydrocortisone and epinephrine in circulating myeloid progenitor cells (CFU-GM) in man. Biomed Pharmacother 38: 167–170

Muller JE, Ludmer PL, Willich N, Tofler GH, Aylmer G, Klangos I, Stone PE (1987) Circadian variation in the frequency of sudden cardiac death. Circulation 75: 131–138

Muller JE, Stone PH, Turi SG, Rutherford JD, Czeisler CA, Parker C, Poole WK, Passamani E, Roberts R, Robertson T, Sobel BE, Willerson JT, Braunwald E (MILIS Study Group) (1985) Circadian variation in the frequency of onset of acute myocardial infarction. N Engl J Med 313: 1315–1322

Nelson WL, Tong YL, Lee JK, Halberg F (1979) Methods for cosinor rhythmometry. Chronobiologia 6: 305–323

Paigen B, Ward E, Reilly A, Houten L, Gurtoo H, Minowada J, Steenland K, Havens MB, Sartori P (1981) Seasonal variation of aryl hydrocarbon hydroxylase activity in human lymphocytes. Cancer Res 41: 2757–2761

Pangerl A, Remien J, Haen E (1986) The number of $\beta$-adrenoceptor sites on intact human lymphocytes depends on time of day, on season and on sex. Annu Rev Chronopharmacol 3: 331–334

Petralito A, Mangiafico RA, Gibiino S, Cuffari MA, Miano MF, Fiore CE (1982) Daily modifications of plasma fibrinogen, platelets aggregation, Howell's time, PTT, TT and antithrombin III in normal subjects and in patients with vascular disease. Chronobiologia 9: 195–201

Rabkin SW, Mathewson FAL, Tate RB (1980) Chronobiology of cardiac sudden death in men. JAMA 244: 1357–1358

Raine CS, Traugott V, Stone SH (1978) Suppression of chronic allergic encephalomyelitis: relevance to multiple sclerosis. Science 201: 445–448

Ramot B, Brok-Simoni F, Chweidan E, Ashkenazi YE (1976) Blood leukocyte enzymes. III. Diurnal rhythm of activity in isolated lymphocytes of normal subjects and chronic lymphatic leukemia patients. Br J Haematol 34: 79–85

Rausch JL, Shoch NS, Burch EA, Donald AG (1982) Platelet serotonin uptake in depressed patients: circadian effect. Biol Chem 17: 121–123

Reinberg A, Smolensky M (1983) Biological rhythms and medicine. Cellular, metabolic, physiopathologic and pharmacologic aspects. Springer, Berlin Heidelberg New York

Reinberg A, Gervais P, Halberg F, Gaultier M, Roynette N, Abulker CH, Dupont J (1973) Mortalité des adultes: rythmes circadiens et circannuels dans un hôpital parisien et en France. Nouv Presse Méd 2: 289–294

Reinberg A, Schuller E, Delasnerie N, Clench J, Helary M (1977) Rhythmes circadiens et circannuels des leucoytes, proteines totales, immunoglobulines A, G et M. Etude chez 9 adultes jeunes et sains. Nouv Presse Méd 6: 3819–3823

Reinberg A, Schuller E, Clench J, Smolensky MH (1980) Circadian and circannual rhythms of leukocytes, proteins and immunoglobulins. In: Smolensky MH (ed) Recent advances in the chronobiology of allergy and immunology. Pergamon, New York, pp 251–259

Richtsmeier WS (1985) Interferon. Present and future prospects. CRC Crit Rev Clin Lab Sci 20: 57–93

Ritchie AWS, Oswald I, Micklem HS, Boyd JE, Elton RA, Jazwinska E, James K (1983) Circadian variation of lymphocyte subpopulations: a study with monoclonal antibodies. Br Med J 286: 1773–1775

Richter A, Kadar D, Liszka-Hagmajer E, Kalow W (1978) Seasonal variaton of aryl hydrocarbon hydroxylase inducibility in human lymphocytes in culture. Res Commun Chem Pathol Pharmacol 19: 453–475

Rivard GE, Infante-Rivard C, Hoyoux C, Champagne J (1985) Maintenance chemotherapy for childhood acute lymphoblastic leukaemia: better in the evening. Lancet II: 1264–1266

Rocker L, Feddersen HM, Hoffmeister H, Junge B (1980) Jahreszeitliche Veränderungen diagnostisch wichtiger Blutbestandteile. Klin Wochenschr 58: 769–778

Roitman B, Sothern RB, Halberg F, Mayersbach von H, Scheving LE, Haus E, Bartter FC, Delea C, Simpson H, Tavadia H, Fleming K, Hume P, Wilson C, Halberg E (1975) Circadian acrophases for total blood leukocytes counted on different continents. Chronobiologia 2 (Suppl 1): 58

Ross DD, Pollak A, Akman SA, Bachur NR (1980) Diurnal variation of circulating human myeloid progenitor cells. Exp Hematol 8: 954–960

Sabin FR, Cunningham RS, Doan CA, Kindwale JA (1927)

526    E. Haus

The normal rhythm of white blood cells. Bull Johns Hopkins Hosp 37: 14–67

Scheving LE (1981) Circadian rhythms in cell proliferation: their importance when investigating the basic mechanism of normal versus abnormal growth. In: Mayersbach von H, Scheving LE, Pauly JE (eds) Biologogic rhythms in structure and function. Prog Clin Biol Res 59C: 39–79

Scheving LE, Pauly JE (1973) Cellular mechanisms involving biorhythms with emphasis on those rhythms associated with the S and M stages of cell cycle. Int J Chronobiol 1: 269–286

Scheving LE, Burns ER, Pauly JE, Halberg F (1980) Circadian bioperiodic response of mice bearing advanced L1210 leukemia to combination therapy with adriamycin and cyclophosphamide. Cancer Res 40: 1511–1515

Scheving LE, Tsai TH, Feuers RJ, Scheving LA (1989) Cellular mechanisms involved in the action of anticancer drugs. In: Lemmer B (ed) Chronopharmacology. Dekker, New York, pp 317–369

Schulthess von GV, Mazer NA (1982) Cyclic neutropenia (CN): a clue to the control of granulopoiesis. Blood 59: 27–37

Shadduck RK, Winkelstein A, Nunna NG (1972) Cyclic leukemia cell production in CML. Cancer 29: 399–401

Sharp GWG (1960) Reversal of diurnal leukocyte variations in man. J Endocrinol 21: 107–114

Shifrine M, Taylor N, Rosenblatt LS, Wilson F (1980) Seasonal variation in cell mediated immunity of clinically normal dogs. Exp Hematol 8: 318–326

Shifrine M, Garsd A, Rosenblatt LS (1982a) Seasonal variation in immunity of humans. J Interdiscipl Cycle Res 13: 157–165

Shifrine M, Rosenblatt LS, Taylor N, Hetherington NW, Matthews VJ, Wilson FD (1982b) Seasonal variations in lectin-induced lymphocyte transformation in Beagle dogs. J Interdiscipl Cycle Res 13: 151–156

Signore A, Cugini P, Letizia C, Lucia P, Murano G, Pozzilli P (1985) Study of the diurnal variation of human lymphocyte subsets. J Clin Lab Immunol 17: 25–28

Sjostrand T (1962) Blood volume. In: Dow P (ed) Handbook of physiology, Sect 2. Circulation, vol 1. Am Physiol Soc, Washington DC, p 51

Slozina NM, Golovachev GD (1986) The frequency of sister chromatin exchanges in human lymphocytes determined at different time within 24 hours. Citologia 28: 127–129

Smaaland R, Sletvold O, Bjerknes R, Lote K, Laerum OD (1987) Circadian variations in cell cycle distribution in human bone marrow. Chronobiologia 14: 239

Swoyer JK, Sackett LL, Haus E, Lakatua DJ, Taddeini L (1975) Circadian lymphocytic rhythms in clinically healthy subjects and in patients with hematologic malignancies. Internat Congr on Rhythmic Functions in Biological Systems. Egerman, Vienna, pp 62–63

Swoyer J, Haus E, Sackett-Lundeen L (1987) Circadian reference values for hematologic parameters in several strains of mice. Prog Clin Biol Res 227: 281–296

Swoyer J, Irvine P, Sackett-Lundeen L, Conlin L, Lakatua DJ, Haus E (1989) Circadian hematologic time structure in the elderly. Chronobiol Int 6: 131–137

Swoyer J, Rhame F, Hrushesky W, Sackett-Lundeen L, Sothern R, Gale H, Haus E (1990) Circadian rhythm alterations in HIV infected patients. In: Hayes D, Pauly J, Reiter R (eds) Chronobiology: its role in clinical medicine, general biology, and agriculture. Wiley, New York, 341A: 437–449

Szabo I, Kovats TG, Halberg F (1977) Circadian rhythm in phagocytic index of CBA mice, replicated in two studies. Chronobiologia 4: 155

Tavadia HB, Fleming KA, Hume PD, Simpson HW (1975) Circadian rhythmicity of human plasma cortisol and PHA-induced lymphocyte transformation. Clin Exp Immunol 22: 190–193

Tofler GH, Brezinski D, Schafer AI, Czeisler CA, Rutherford JD, Willich SN, Gleason RE, Williams GH, Muller JE (1987) Concurrent morning increase in platelet aggregability and the risk of myocardial infarction and sudden cardiac death. N Engl J Med 316: 1514–1518

Touitou Y, Touitou C, Bogdan A, Beck H, Reinberg A (1978) Serum magnesium circadian rhythm in human adults with respect to age, sex, and mental status. Clin Chim Acta 83: 35–41

Touitou Y, Touitou C, Bogdan A, Chasselut J, Beck H, Reinberg A (1979) Circadian rhythms in blood variables in elderly subjects. In: Reinberg A, Halberg F (eds) Chronopharmacology: advances in the biosciences, vol 19. Pergamon, New York, pp 283–290

Touitou Y, Touitou C, Bogdan A, Reinberg A, Auzeby A, Beck H, Guillet P (1986) Differences between young and elderly subjects in seasonal and circadian variations of total plasma proteins and blood volume as reflected by hemoglobin, hematocrit and erythrocyte counts. Clin Chem 32: 801–804

Umemura T, Hirata J, Kaneko S, Nishimura J, Motomura S, Kozuru M, Ibayashi H (1986) Periodical appearance of erythropoietin-independent erythropoiesis in chronic myelogenous leukemia with cyclic oscillation. Acta Haematol 76: 230–234

Verma DS, Fisher R, Spitzer G, Zander AR, McCredie KB, Dicke KA (1980) Diurnal changes in circulating myeloid progenitor cells in man. Am J Hematol 9: 185–192

Verma DS, Spitzer G, Zander AR, Dicke KA, McCredie KB (1982) Cyclic neutropenia and T lymphocyte suppression of granulopoiesis: abrogation of the neutropenic cycles by lithium carbonate. Leukemia Res 6(4): 567–576

Vodopick H, Rupp EM, Edwards CL, Goswitz FA, Beauchamp JJ (1972) Spontaneous cyclic leukocytosis and thrombocytosis in chronic granulocytic leukaemia. N Engl J Med 286: 284–290

Weigle WO (1975) Cyclical production of antibody as a regulatory mechanism in the immune response. Adv Immunol 21: 87–111

Wheldon TE, Kirk J, Finlay HM (1974) Cyclic granulopoiesis in chronic granulocytic leukemia: a simulation study. Blood 43: 379–385

Whitaker PM, Warsh JJ, Stancer HC, Persad E, Vint CK (1984) Seasonal variations in platelet $^3$H-imipramine binding: comparable values in control and depressed populations. Psychiatry Res 11: 127–131

Williams RM, Krause LJ, Dubey DP, Yunis EJ, Halberg F (1979) Circadian bioperiodicity in natural killer cell activity of human blood. Chronobiologia 6: 172

Wirz-Justice A, Richter R (1979) Seasonality in biochemical determinations: a source of variance and a clue to the temporal incidence of affective illness. Psychiatry Res 1: 53–60

Wright DG, LaRussa VF, Salvado AJ, Knight RD (1989) Abnormal responses of myeloid progenitor cells to granulocyte-macrophage colony-stimulating factor in human cyclic neutropenia. J Clin Invest 83: 1414–1418

Yoshimura H (1958) Seasonal variations in human plasma volume. Jpn J Physiol 8: 165–179

Zinneman HH, Thompson M, Halberg F, Kaplan M, Haus E (1972) Circadian rhythms in urinary Bence-Jones protein excretion. Clin Res 20: 798

Zinneman HH, Halberg F, Haus E, Kaplan M (1974) Circadian rhythms in urinary light chains, serum iron and other variables of multiple myeloma patients. Int J Chronobiol 2: 3–16

# Chronobiology of Human Bone Marrow

R. Smaaland and O. D. Laerum

## General Introduction

Bone marrow is the production site for all types of blood cells, which are released into the peripheral blood according to the needs of the body, mediated through different feedback mechanisms. The production occurs as a combination of cell proliferation and gradual maturation, until the end stage is reached with a population of mature cells that can exert their specialized functions but are no longer capable of cell proliferation [1]. It takes approximately 14 days from immature stem cells starting proliferation until mature cells result [2].

In current cancer treatment bone marrow suppression represents the main dose limiting factor and increasingly so due to a more frequent use of combination therapy. Leukopenia or granulocytopenia related to cytotoxic therapy is one of the major concerns due to two related clinical problems which may be encountered in the course of a treatment schedule: (a) low nadir values, which in some cases may be long lasting and lead to serious, potentially life-threatening infections; and (b) a less dramatic fall in the granulocyte count may occur, but the counts may be too low to enable administration of full cytotoxic doses at the scheduled time, leading to dose reduction and/or postponement of treatment courses. Either of these decisions may have serious consequences for the ultimate treatment outcome [1]. The high sensitivity of the bone marrow to cytotoxic treatment is to a large extent related to the high proliferation rate of bone marrow cells [3, 4]. Circadian proliferative rhythms may therefore be especially pertinent for drugs whose toxic-therapeutic ratios are narrow, such as anticancer drugs, especially on the background of increasing evidence that any compromise of dosage or delay in treatment schedule diminish the likelihood of cancer control or cure [5–8].

Little is known about biological rhythms of the different parameters of human bone marrow. It is surprising to see the systematic neglect of the consideration of biological rhythms in the hematological literature. Several large surveys on hematopoiesis have been published without even a mention of time-dependent variations [2]. New information related to proliferation is now emerging, and it has become increasingly clear that hematopoiesis is not a temporally fixed phenomenon. This can be taken advantage of in several ways, both in relation to cytotoxic side effects of cancer therapy and to primary diseases of the bone marrow such as myelodysplastic conditions.

In animal studies it has been shown that protection of host normal tissues, including bone marrow, can be achieved for different cytotoxic agents by determining host toxicity rhythms along the 24-h time scale. Optimal circadian therapeutic schedules can then be devised for the administration of cytotoxic drugs [9–12]. In addition to reduced mortality due to acute toxicity, it has been shown that an increase in effect on tumor or cure rate can be obtained [10, 13–16], or that it is possible to eliminate or reduce drug-induced death due to toxicity, while still using an effective dose [17]. Clinical studies have also demonstrated a circadian dependence of drug cytotoxicity to the bone marrow, resulting in fewer dose reductions, fewer treatment-related complications, and fewer postponements of treatment courses when drugs have been administered at certain times of the day [18, 19], as well as increased long-term survival [20]. Reduced chance of relapse has also been demonstrated when maintenance cytotoxic therapy with 6-mercaptopurine and methotrexate was given in the evening rather than the morning for acute leukemia in children [21]. These studies strongly suggest that an increased dose intensity as well as an increased tumor effect may be feasible by taking circadian rhythms of bone marrow cell proliferation into consideration.

The use of recombinant hematopoietic growth factors to accelerate the regeneration of granulocyte numbers after conventional or high-dose cytotoxic therapy with or without autologous/allogeneic bone marrow transplantation may further take advantage of biological rhythms of the bone marrow by increasing therapeutic effects or reducing side effects. In addition, an optimal use of these growth factors relative to biological rhythms may be potentially beneficial for patients with acquired immunodeficiency syndrome, myelodysplastic syndrome, or aplastic anemia [22].

In recent years it has become increasingly clear that the intracellular nonprotein thiol glutathione (GSH) plays an important role in the cellular defense against cytotoxic damage to different tissues [23–26]. Depleting the local concentration of this peptide enhances the cytotoxicity of several chemotherapeutic agents to tumors. Because the content of the peptide may play an important role relative to bone marrow sensitivity for cytotoxic damage, the GSH content of human bone marrow and its relation to proliferative parameters will also be dealt with in this chapter.

The existence of interindividual differences in circadian time structure implies that monitoring of physiological marker variables would be of benefit to allow individualized chronotherapy. In our studies of bone marrow parameters along the circadian scale, we have sampled blood for monitoring of hematological and hormonal/biochemical parameters. Relations between these peripheral blood parameters and the measured bone marrow parameters will be discussed.

Thus, knowledge about predictable temporal changes of bone marrow proliferation in vitro and in vivo may not only be useful for understanding the regulation of bone marrow proliferation, but also for screening optimal dosing time of cytotoxic agents and hematopoietic growth factors, and for choosing the right time for harvesting bone marrow cells for auto- or allografting. The possibility of treating patients at optimal circadian time, interval, and schedule is today feasible and cost-effective through programmable delivery systems [27].

## Proliferation of Total Human Bone Marrow Cell Population

### Introduction

Since the production and migration of mature granulocytes into peripheral blood is both dependent on the actual needs of the body and on several hormonal and regulatory factors (reactive homeostasis) [28], the proliferative pool in the marrow may vary considerably from time to time [29]. For example, physical exercise as well as cortisone/cortisol are strong mobilizers of granulocytes, as are acute bacterial infections. In addition, there may be strong nonrandom endogenous rhythmic variations in these proliferative and mobilizing processes, which may further complicate the picture, but these are also a part of the organism's homeostasis (predictive homeostasis) [28].

Cell cycle phase-dependent differences in the effects of cytotoxic drugs are well documented [30]. Rapidly proliferating cells are usually more sensitive to cytotoxic drugs than quiescent cells [3, 4]. Several studies in mice have demonstrated circadian variations in total bone marrow cell proliferation, i.e., measurement of DNA synthesis, and/or mitotic index/activity, or duration of mitosis, either by using [³H]TdR labeling, percentage of labeled mitoses, or flow cytometry [31–37]. Studying erythropoiesis, Dörmer et al. reported that DNA synthesis in mice underwent circadian variations, with acrophase during the dark/activity period [38].

Although large differences in proliferation according to circadian time have been found, the results have not been consistent. An explanation of this variation may be obtained by investigations on different species, interstrain differences, animals of different age and sex, and different lighting schedules.

In humans, Mauer found that the activity of [³H]TdR-labeled cells of the myeloid lineage was clearly higher during the day than at midnight in three of four individuals, and with a trend towards lower DNA synthesis in the fourth individual [39]. Bone marrow was sampled four times at different hours along the 24-h span during a 42-h period. He was unable to find any circadian variation of the [³H]TdR labeling index of erythropoiesis. Although this was an in vitro study, as direct exposure of individuals to the isotope was not feasible, the short exposure of 1 h after the sample was obtained should express the DNA synthesis at the time of sampling. The mitotic index was found to be highest at 1800 hours or at midnight and lowest at 0600 hours in five of six individuals. Thus, the time with lowest percentage of cells incorporating the labeled thymidine preceded the time of lowest number of mitotic figures by about 6 h. Killman et al. made corresponding observations in one human volunteer in an earlier study with regard to mitotic indices, demonstrating an increase in this proliferative parameter from early morning to late evening [40].

Recently, we have done serial sampling of bone

marrow from human volunteers to investigate circadian variations of cell cycle distribution of bone marrow cells by using a flow cytometric technique. The rationale for this is that the total activity may be the most relevant parameter when determining periods of low susceptibility to cell cycle-specific cytostatics.

## Materials and Methods

From November 1986 to August 1988 we investigated the cell cycle distribution of bone marrow cells from 16 healthy male volunteers (mean age = 33.7 years; range 19–47 years), five of them undergoing the sampling procedure twice, altogether making twenty-one 24-h periods. All volunteers had given their informed written consent to enter the study, which was performed according to the guidelines of the regional medical ethics committee. Bone marrow was sampled by puncturing the sternum and anterior iliac crests in a randomized sequence every 4 h during a 24-h period, a total of seven times. The start of the experiment was also randomized to either 0800, 1200, or 1600 hours, in order to reduce the possibility that the repeated-puncture procedure itself would give systematic-interference with the results. No premedication was given. The area of the puncture site was infiltrated with a local anesthetic (Lidocaine 20 mg/ml, Astra). Bone marrow (0.3 ml) was aspirated into a 2-cc syringe. One part of the sample was utilized for routine smears, while one droplet was stained directly for DNA flow cytometry. Another droplet was placed onto each of two tilted microscope slides, in order to let the blood slide down and thereby increase the proportion of marrow elements, which thereafter were immediately removed from the slides and stained (indirect staining). Thus, two parallel samples were stained at each time point. Both samples of bone marrow cells were added to 2 ml of ice-cold staining solution consisting of ethidium bromide, detergent, and RNAse according to the method described by Vindeløv [41]. The tubes were sealed and the solution shaken before being placed in an ice bath for at least 10 min. Both single cell suspensions were analyzed thereafter on a Cytofluorograph 50 H (Ortho), interfaced to a Model 2150 computer. Cell cycle distribution was estimated by using the cell cycle analysis program [42]. The mean coefficient of variation (CV) of the DNA histograms was 3.3%. Venous blood was obtained from the same subjects at the same time as bone marrow sampling to determine peripheral blood parameters, including total and differential blood cell counts, in addition to cortisol measurements. The blood was obtained as the initial procedure or immediately after the anesthesia of periost before the bone marrow puncture. In this way an artificially increased level of cortisol as a result of the puncture procedure itself was avoided [43]. All individuals but two had followed a regular diurnal rhythm for at least 3 weeks before the experiment. They continued their usual activities during the study period, apart from the sampling periods. They went to sleep after the 2400-hour sample was taken, but were wakened for 15–30 min for the 0400-hour sample. The analyses of fraction of cells in DNA synthesis were done by taking the mean value of the DNA synthesis-phase (S-phase) of the two differently stained samples at each time point.

## Results

The mean value of the percentage of cells in the S-phase of the two differently stained sampled of bone marrow cells harvested at each time point showed a large variation along the circadian scale for all twenty-one 24-h periods. The range of change from lowest to highest value during the 24-h period for each subject varied between 28.5% and 338.5%. The mean values of the periods with lowest and highest S-phase were 8.9% ± 0.5% (SE) and 17.6% ± 0.6%, respectively, i.e., difference of nearly 100%.

Differences in phasing over the 24-h period were observed between subjects. Thus, between individuals the time of highest and lowest DNA synthesis varied with a mean of 4 h according to the cosinor analyses. An example of two DNA histograms for two time points along the 24-h sampling period is shown in Fig. 1, while four examples of individual circadian stage-dependent variations of the fraction of cells in DNA synthesis are shown in Fig. 2. Lower values in the evening, at midnight, or late at night can be seen. When pooling the data for all subjects for the mean S-phase values, a consistent pattern is seen, with a lower DNA synthesis around midnight than in the day (Fig. 3). The observation period goes over 32 h because the time of sampling started either at 0800, 1200, or 1600 hours. This makes it possible to measure the DNA synthesis for the pooled data for two consecutive day-periods, which corroborates the DNA synthesis values measured at daytime. Due to different phasing between the subjects, the difference between the lowest and highest value is smaller than found for the individual subjects. However, the circadian stage-dependent variation is statistically significant, analyzed both with ANOVA and the cosinor

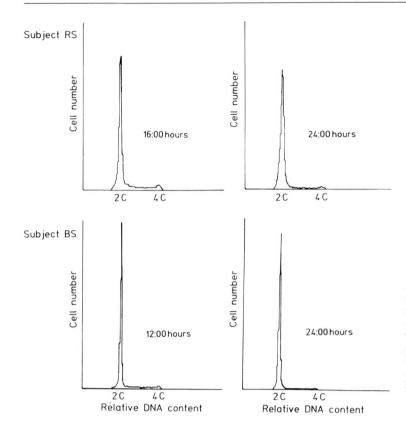

**Fig. 1.** DNA synthesis at two time points on the 24-h time scale (day and midnight). The two peaks (*2C* and *4C*) in each histogram designate the $G_0/G_1$ phase and $G_2/M$ phase. The part of the histogram between the two peaks is the S-phase. The height (i.e., the area) of the S-phase expresses the percentage of cells in DNA synthesis, which is lowest at midnight for both individuals

method, $P = 0.018$ and $P = 0.016$, respectively. The mean S-phase value of the 24-h sampling period varied from 10.9% to 16.6% for the different individuals, i.e., a difference of 52.3%. Due to this interindividual difference, the data were also normalized and expressed as a percentage of the mean value. The $P$ values for ANOVA and cosinor analysis after normalization were 0.006 and 0.012, respectively. The time of highest DNA synthesis (acrophase) estimated by cosinor analysis was 1316 hours, while the time of lowest DNA synthesis (trough) was 0116 hours.

In 19 of the 21 sampling periods a lower fraction of cells in DNA synthesis from 2400 to 0400 hours was measured as compared to DNA synthesis between 0800 hours and 2000 hours ($P = 0.0005$; paired Student's t-test, two-tailed).

Nearly the same pattern of circadian variation was seen for the two methods of preparing the cells, i.e., the direct and the indirect method (Fig. 4). The fraction of cells in DNA synthesis was slightly higher for each time point when the indirect staining method was used. The difference was significant only for two time points, 0800 and 2400 hours, $P < 0.03$ and $P < 0.001$, respectively. However, when comparing the paired data available for all time points ($n = 120$), a highly significant difference was observed between

the two methods, with a larger fraction of cells in DNA synthesis from the sample prepared using the indirect staining method than that with the direct staining method, 14.2% ± 0.03% and 12.7% ± 0.3%, respectively ($P < 0.0001$). In addition, a highly significant correlation was found between the two methods when comparing the two ways of bone marrow sampling ($r = 0.62$; $P < 0.0001$). The fraction of cells in DNA synthesis when using the mean value of the two methods was 13.2% ± 0.3% for the whole material.

No significant difference in S-phase between winter (October–March) and summer (April–September) was observed.

A significantly higher S-phase was measured for samples taken from the sternum ($n = 52$) than from the iliac crests ($n = 89$) (14.6% ± 0.5 vs 12.6% ± 0.3, respectively, $P = 0.001$), but only for the samples obtained at 0800 hours was there a statistically significant difference between the two sites, with marginal statistical significance at 1600 and 2000 hours, $P = 0.06$ and $P = 0.07$, respectively.

To see if the higher proliferative activity in the samples obtained from the sternum could contribute to the circadian rhythm observed, and further to see if the rhythm at the two localizations was the same, we pooled all the samples taken from the sternum

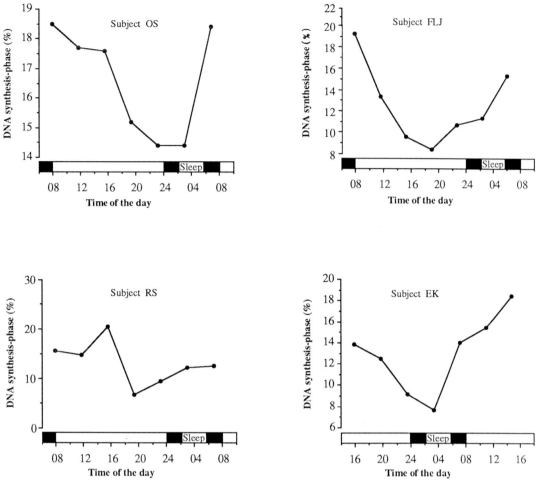

**Fig. 2.** DNA synthesis variation along the 24-h time span in four different subjects, sampling of bone marrow being done every 4 h, seven times in total for each individual

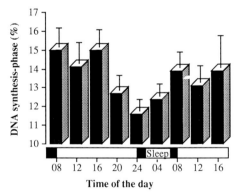

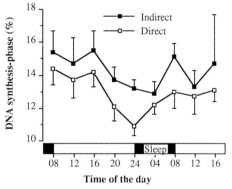

**Fig. 3.** The mean S-phase values at each time point for twenty-one 24-h periods. The time scale extends for over 32 h because of the different time of starting each individual study (see text). The lowest proliferative activity (DNA synthesis) is observed during late evening and night, while the highest DNA synthesis is during daytime

**Fig. 4.** A comparison of the S-phase of the bone marrow samples prepared using the direct and indirect staining methods (see text). Nearly the same circadian pattern is seen

and iliac crests at each time point and analyzed them separately. We found a statistically significant circadian rhythm for samples harvested from the sternum both for the absolute values and for the normalized data ($P < 0.05$), with acrophase at the same time as all the S-phase data. There was a similar circadian rhythm in the S-phase of the samples obtained from the iliac crests with peak values at the same time of day, but the rhythm was not statistically significant ($P = 0.20$). For this localization the range of change in S-phase was not as large as for the sternum, and the scatter of data about each time point therefore prevented rhythm detection below the 5% level.

Theoretically, a possible stress-related influence on DNA synthesis could not be ruled out. We therefore compared the cortisol levels 24 h apart; no statistically significant difference was observed ($P = 0.93$). There was also no statistically significant difference in percentage of cells in DNA synthesis sampled 24 h apart ($P = 0.17$).

The lowest value of the $G_2$/M phase was found at about 0400 hours, i.e., about 4 h after the trough value of the S-phase. The curve of the $G_2$/M phase showed a sinusoidal shape with highest value at 1200 hours. However, preliminary analyses of the $G_2$/M phase at 0400 hours were not statistically lower than the day values.

## Discussion

Although there were interindividual differences in circadian phasing, a well known phenomenon for most physiological parameters, a consistent and statistically significant pattern was manifested when pooling the S-phase for all subjects. The period of lowest DNA synthesis was found to be around midnight, and the highest DNA synthesis was found to be during the day.

A potential problem is that DNA synthesis may vary as a function of the location in the bone marrow from which the sample is taken. Conceptually one would regard the total red bone marrow as one organ, being affected by the same endogenous physiological and hormonal factors, making site-dependent variations less important. This is supported by earlier-reported data from Dosik et al. [44], who demonstrated a very close correlation between DNA synthesis in bone marrow samples obtained simultaneously by biopsies from right and left iliac crests. Good reproducibility, although with larger individual variations, was also demonstrated by simultaneous bilateral aspirations. In the present study, sampling was also done from the sternum, in addition to the left and right iliac crests. In the protocol for the reported study we tried to minimize this potential site-related problem by harvesting bone marrow from different sites (i.e., sternum and iliac crests) at the same time points for different individuals according to a randomized schedule. In agreement with the results of Dosik et al. [44], we found no statistical difference of the S-phase between the left and right iliac crests, being $12.1 \pm 0.5$ and $11.6 \pm 0.5$, respectively ($P = 0.47$).

DNA analyses from samples from both the sternum and iliac crests demonstrated the same circadian pattern of the S-phase. This finding rules out the possibility of different sampling sites being the reason for the observed circadian stage dependence of DNA synthesis. It further contradicts the possibility that the overall circadian rhythm detected could be attributed to a difference in level of S-phase dependent on the sampling site.

The finding of nearly the same mean value of DNA synthesis from samples of the left and right iliac crests strongly indicates that the total red bone marrow must be looked upon as a functional entity. The demonstration of the same circadian variation in the bone marrow of the sternum as in the iliac crests further corroborates this functional homogeneity. The slightly higher DNA synthesis in the sternum is most likely due to less blood contamination of these samples.

The possibility that multiple punctures into the bone could have an impact on DNA synthesis by virtue of stress has to be considered. We therefore compared the level of the stress-related hormone cortisol at the start and at the end of the sampling procedure, i.e., 24 h apart. No statistically significant difference was observed ($P = 0.93$). In addition, the possible stress-related effect on the result of the pooled data was already in the study protocol, aimed at being minimized by starting sampling at three different times of day. Neither did we find any statistically different level of DNA synthesis 24 h apart ($P = 0.17$). These three factors should contradict the possibility of a stress-induced circadian rhythm of the DNA synthesis.

Our data for the DNA synthesis are in agreement with those of Mauer, who found that the [$^3$H]TdR-labeled cells of the myeloid lineage in three of four individuals was clearly higher during the day than at midnight. In addition, the observed trough value of the $G_2$/M phase at 0400 hours in our study corresponds very well with the observed trough of mitosis in the study of Mauer. The nonsignificant difference between the 0400-hour value when compared to the day values for the pooled data indicates that there are greater circadian variations of the S-phase than of the

$G_2/M$ phase, or that there are greater interindividual differences in the $G_2/M$ phase [45].

The potential importance of the data presented here is underlined by the results recently published by Lévi for murine bone marrow [46], demonstrating a corresponding circadian variation in colony-forming unit-granulocyte macrophage (CFU-GM) and DNA synthesis, with highest and lowest values of these parameters in the activity and rest spans, respectively. The potential use of taking such rhythms into consideration was demonstrated in the same study with a circadian toxicity rhythm of the anticancer agent $4'$-$O$- tetrahydropyranyl adriamycin showing the lowest toxicity when the CFU-GM and DNA synthesis were at their lowest.

By taking circadian stage-dependent variations in DNA synthesis into account it may therefore be possible to reduce bone marrow toxicity of cytotoxic drugs by administering the drugs or the major dose of a continuous drug infusion during the time of lowest proliferative activity, i.e., in the late evening or at night. Cells in the S-phase will then be less susceptible, and cells in the $G_0/G_1$ phase will have more time for repairing damage before entering into the S-phase.

A potentially important consequence of these findings is that it may be possible to increase the effect of biological response modifiers, such as granulocyte-macrophage-colony-stimulating factor (GM-CSF) and granulocyte-colony-stimulating factor (G-CSF), and regulatory peptides [47], by administering the optimal dose at the time of greatest responsiveness of the bone marrow. This may increase the usefulness and effect of these biological substances, and thereby also possibly reduce potential side effects. In addition, the data suggest that the rate of success of bone marrow auto- or allografting may improve with careful selection of the sampling time of the bone marrow from the donor.

# Myeloid Progenitor Cells in Human Bone Marrow

## Introduction

With the introduction of assay methods for the various classes of stem and progenitor cells, data have become available not only from the stem cell proliferation of animals, but also from humans. Several authors have demonstrated circadian and seasonal variations of multipotent and committed stem cells in mice [48–52]. In addition, the circadian rhythms of multipotent stem cells, myelopoietic progenitor cells (CFU-GM), and recognizable myelopoietic cells have been measured in parallel in female C3H mice. An interesting finding was that the rhythms were partly synchronized [53].

However, data on biological rhythms in stem and progenitor cell activity of human bone marrow have so far been sparse, mainly due to the practical limitations of obtaining multiple samples. Since these cells are also in the circulation in small numbers, they have been subject to such investigations by sampling of peripheral blood. Early erythroid progenitor cells (burst-forming unit erythroid, BFU-E) were shown not to exhibit circadian variations in peripheral blood [54], and in one study the same applied to CFU-GM [55], which showed only some variations and differences due to age and sex. On the other hand, Ross et al. [56] found significant variations in the numbers of CFU-GM in blood, with a maximum at 0900 hours. Verma et al. also demonstrated a circadian-stage-dependent variation in circulating myeloid progenitor cells, with a significant elevation in the afternoon as compared with the morning levels in normal human volunteers [57].

It should be kept in mind that age-related changes in physiological variations, including hematopoiesis, may be another modifying factor as to circadian-stage-dependent effects of cytotoxic therapy. Reports on hematopoiesis in aging mice are inconsistent, however, both with regard to multipotent stem cells (colony-forming unit-spleen, CFU-S) [58–60] and myelopoietic progenitor cells (CFU-GM) [61, 62]. In our laboratory we have demonstrated that both the 24-h mean values and the amplitudes of the circadian variation were declining in aging mice. In addition, a phase shift of the CFU-GM peak was observed, indicating that equal timing of chemotherapy in young and old mice might have different effects [63].

It is generally accepted that the volume of active hematopoietic tissue in the human bone marrow decreases in old age [64–67]. In addition, reduced numbers of both myelopoietic and erythropoietic progenitor cells have been reported [68]. However, Corberand et al. have reported that blood cell parameters do not change during physiological human aging [69], and Resnitzky et al. demonstrated an inverse correlation between marrow cellularity and the myeloid progenitor cell numbers [70]. Thus, previous studies are inconclusive with regard to age-related proliferative activity.

No extensive studies of CFU-GM in humans along the 24-h time span by sampling of bone marrow have previously been done. When sampling bone marrow for study of the cell cycle distribution, we also harvested bone marrow for soft agar culture to look for a possible circadian and seasonal dependent variation in myeloid progenitors (CFU-GM). In addition, it was of great interest to study how this most important cell lineage relative to cancer chemotherapy correlated to the total proliferative activity of the bone marrow.

## Materials and Methods

After local anesthesia of the periost, as described earlier, and after the sampling of bone marrow for flow cytometry analyses, about 2 ml of marrow was drawn through an 18-gauge needle into a syringe filled with 5 ml 0.9% sodium chloride and 10% ethylene-diamine-tetraacetate (EDTA). Bone marrow smears were also prepared.

*CFU-GM assay.* Myelopoietic colonies were cultured using the single layer method described by Burgess et al. [71]. Human placenta conditioned medium (HPCM) was used as a CSF, according to the method of Schlunk and Schleyer [72]. Bone marrow cells at a concentration of $1 \times 10^6$/ml (0.5 ml) were mixed with 1.0 ml fetal calf serum (Flow, Scotland), 0.5 ml HPCM, 0.5 ml of cell suspension, and 0.5 ml of 3% agar. With eight dishes per time point, 1 ml of this mixture, containing $1 \times 10^5$ cells, was then plated per dish. The dishes were incubated for 14 days at 37°C in 5.0% $CO_2$. A colony was defined as more than 50 cells; all colonies were counted on coded specimens.

## Results

A large circadian variation of CFU-GM was seen in all nineteen 24-h periods (two 24-h periods lost), with a difference between the lowest and highest colony number ranging from 47.1% to 233.0% (mean difference = 136.8%) as compared with the mean colony number of each volunteer. Three examples of individual variations along the circadian scale are shown in Fig. 5.

When evaluating six time points along the circadian scale, comparing 0800, 1200, and 1600 hours (day) with 2000, 2400, and 0400 hours (evening/ night), 13 volunteers were found to have their highest colony number in daytime, while 6 had their highest colony number in the evening or at night. In addition, 5 volunteers had their lowest colony number in daytime,

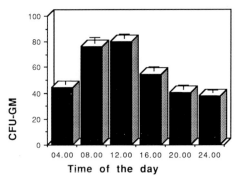

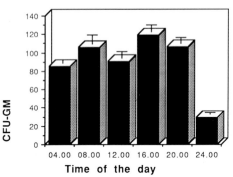

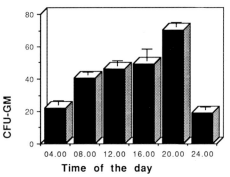

**Fig. 5.** Three examples of individual variations of myeloid progenitors (CFU-GM; mean ± SE) according to the time of day. Eight parallel samples were counted for each time point

while 14 had their lowest colony number at night ($x^2 = 5.3$, $P = 0.02$; $P = 0.05$ with continuity correction). We then normalized the CFU-GM values due to large interindividual mean values and compared the values during the day with the values in the evening or at night. In this way, a significant difference was also demonstrated between day and evening or night values: mean, 88.6% ± 6.8% (SE) and 110.8% ± 7.2%, respectively ($P = 0.01$, Mann-Whitney U test).

For the pooled data there was a significantly lower colony number at 2400 hours compared with daytime, i.e., at 0800 hours, 1200 hours, and 1600 hours (ANOVA, $P < 0.05$), both for original values and for normalized data.

By comparing the CFU-GM values obtained during the light season (April–September) with those of winter (October–March), a significantly higher colony number was found during summer, $68.2 \pm 7.9$ (SE) and $45.3 \pm 6.8$, respectively ($P = 0.02$, Mann-Whitney U test).

## Discussion

In our study a large circadian variation of CFU-GM was seen in all nineteen 24-h periods. As for the S-phase when considering the pooled data, the differences became smaller due to interindividual differences in phasing. However, the variations were to a large extent predictable, with the lowest number of CFUs during nighttime and the highest during daytime. We also found a seasonal difference in the level of CFU-GM, with a higher colony number during the light season than in winter.

These data are in good agreement with those recently published by Lévi for murine bone marrow, where the highest and lowest numbers of CFU-GM were found during the activity and rest span, respectively [46].

By taking these circadian and circannual variations of hematopoietic cell proliferation into consideration, it may be possible to optimize the therapeutic index of cytotoxic drugs.

As discussed previously in relation to the cell cycle distribution, our data may also indicate an optimal time for administering biological response modifiers like GM-CSF and G-CSF by administering the highest dose at the time of greatest responsiveness of the bone marrow. This may increase the usefulness and effect of these biological substances, and thereby possibly reduce potential side effects. The temporal variation of CFU-GM also implies a possibility of optimizing the time of harvesting bone marrow cells for bone marrow transplantation by taking biological rhythms into consideration.

# Glutathione Content in Human Bone Marrow

## Introduction

Glutathione (GSH) is a cysteine-containing tripeptide which has been assigned an important role in the cellular defence against free radicals and oxidative in-jury, detoxification processes, and in the protection of the cell against radiation damage [23–26]. It is the most abundant intracellular nonprotein thiol, and the average cellular content amounts to 0.5–10 mmol/liter [73, 74]. It has been demonstrated that depletion of intracellular GSH in vitro enhances the cytotoxicity of several chemotherapeutic agents such as melphalan (L-PAM) [26, 75–77], Adriamycin [24, 78], cis-platinum [79, 80], mitomycin C [81], and cyclophosphamide [81, 82, 83]. In addition, GSH has a radioprotective effect [23, 84, 85].

It has also been shown that tumor cells obtained from a patient with ovarian adenocarcinoma after the onset of resistance had a significantly higher level of GSH as compared with the level before onset of resistance [30]. In vitro studies have further demonstrated that resistant cells can be sensitized by depleting GSH using the metabolic inhibitor buthionine sulfoximine (BSO) [74, 76, 79, 86]. Thus, pharmacological modulation of cellular GSH content may alter the sensitivity towards several cytotoxic drugs, pointing to the possibility that time-dependent physiological changes in cellular GSH content could play a role determining the susceptibility of both normal and malignant cells to such agents. It was, therefore, of interest to measure the level of GSH in human bone marrow and its possible circadian variation in order to evaluate its detoxifying capacity and relate it to the proliferative activity, i. e., the S-phase.

## Materials and methods

The GSH content and its circadian stage dependent variation in human bone marrow samples from ten healthy males were investigated. From each individual, bone marrow samples were collected every 4 h during a 24-h period, seven samples altogether. The bone marrow sample was immediately put into liquid nitrogen for the determination of reduced (GSH) and oxidized (GSSG) glutathione.

Two droplets of bone marrow (in liquid nitrogen) were extracted within 3 days of sampling with 1 ml of ice-cold 5% sulfosalicylic acid containing $50 \, \mu M$ DTE, and the precipitated protein removed by centrifugation. This protein precipitate was retained for subsequent protein determination. GSH was determined in the acid extract by a published method [87], slightly modified by us [88]. Briefly, free sulfhydryl groups were derivatized with monobromobimane, and the GSH-bimane derivative was quantitated by chromatography on a 3-µm ODS Hypersil column, which was equilibrated and eluted with 14.2% meth-

**Table 1.** Circadian variation in glutathione content in human bone marrow and results of cosinor analysis

| Subject | Age, y | N* | Range† | | ROC. ≠ | Results of fit of 24-h cosine (single cosinor) | | | | |
| --- | --- | --- | --- | --- | --- | --- | --- | --- | --- | --- |
| | | | Lowest | Highest | % | p | Mesor§ ± SE | Amplitude† | Acrophase ‖ | Trough ‖ |
| AA | 19 | 7 | 1.94 | 2.80 | 44.3 | .51 | 2.63 ± 0.12 | | | |
| BS | 33 | 7 | 1.55 | 2.40 | 54.8 | .03 | 1.94 ± 0.04 | 0.27 | 0738 | 1938 |
| GW | 39 | 7 | 2.75 | 3.75 | 36.4 | .80 | 3.27 ± 0.13 | | | |
| IK | 31 | 7 | 2.13 | 3.28 | 54.0 | .44 | 2.75 ± 0.16 | | | |
| KL | 43 | 7 | 1.80 | 2.38 | 32.2 | .71 | 2.06 ± 0.09 | | | |
| MJ | 24 | 7 | 1.53 | 2.80 | 83.0 | .09 | 2.26 ± 0.10 | 0.46 | 1238 | 0038 |
| OH | 35 | 7 | 2.03 | 2.66 | 31.0 | .25 | 2.23 ± 0.07 | | | |
| RK | 25 | 7 | 2.58 | 3.63 | 40.7 | .53 | 2.93 ± 0.17 | | | |
| EK | 28 | 7 | 1.74 | 3.31 | 90.2 | .28 | 2.78 ± 0.19 | | | |
| SF | 30 | 7 | 2.23 | 3.28 | 47.1 | .59 | 2.68 ± 0.13 | | | |

| Group rhythm summary by single cosinor analysis of the population mean according to circadian stage | | | | | |
| --- | --- | --- | --- | --- | --- |
| No. of subjects | p | Mesor§ ± SE | Amplitude† | Acrophase ‖ | Trough ‖ |
| 10 | .055 | 2.54 ± 0.01 | 0.07 | 0835 | 035 |

*N = No. of sampling procedures during the 24-h time span
† nmol/mg protein
≠ ROC = range of change from lowest to highest value
§Mesor = rhythm-adjusted mean (average)
‖ An hour and minute specification

anol in 43.5 m$M$ sodium acetate, pH 3.9. The column was washed by increasing the methanol concentration to 90%. We measured both GSH and GSSG in the bone marrow samples. The oxidized form accounted for less than 10% of total glutathione, i.e., reduced plus oxidized form. GSSG plus GSSR was determined by a procedure which has recently been developed in our laboratory [88].

## Results

The mean GSH content during the sample periods was low in accordance with previous reports of murine bone marrow and lower compared with other murine normal and tumor tissues, varying from 1.94 to 3.27 nmol/mg protein between the subjects, i.e., a difference of 68.6% between the lowest and highest individual average GSH content. The mean values for all samples ($n = 70$) was 2.54 ± 0.06 nmol/mg protein. The single lowest and highest GSH content measured were 1.53 nmol/mg protein and 3.75 nmol/mg protein, respectively, i.e., a difference of 145.1%.

There was a marked circadian stage dependent variation (31.0%–90.2%) of GSH content in all individuals, but with considerable interindividual variation with regard to time of acrophase or trough. For the individual subjects we observed a circadian stage dependent variation in reduced GSH from 31.0% to 90.2%, as compared with the lowest value (Table 1).

Only one subject had a significant circadian rhythm by single cosinor analysis ($P = 0.03$), while another individual had a marginally significant rhythm ($P = 0.09$). When the data from all subjects were pooled there was a trend toward increasing GSH content from midnight till 1200 hours. By single cosinor analysis of the population mean this circadian variation was marginally statistically significant ($P = 0.06$) with the time of highest concentration (acrophase) found to be at 0835 hours, and the time of lowest concentration (trough) at 2035 hours.

For all ten subjects there seemed to be a covariation between GSH concentration and S-phase according to circadian stage, either demonstrating a near-identical covariation or a slightly phase-shifted near-identical covariation. This is shown in Fig. 6 for four individuals.

When adjusting for the phase difference in five of the ten subjects, we found a close correlation between each subject's GSH content and S-phase, with a statistically significant correlation for the pooled data ($r = 0.40$, $P = 0.002$, $n = 60$).

The mean and median $r$-values for the individual covariation between GSH and DNA synthesis were 0.77 and 0.82, respectively. The pooled GSH values showed a highly significant correlation with the S-phase when discarding one highly deviating time point in six of ten individuals ($P < 0.0001$, $n = 54$).

Single cosinor analysis for DNA synthesis resulted in a circadian rhythm-detection for the DNA syn-

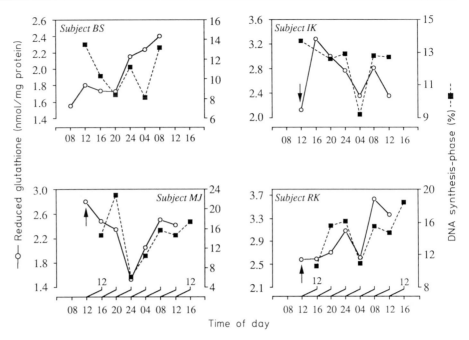

**Fig. 6.** Circadian stage dependent covariation between DNA synthesis and reduced GSH content in four individuals. The S-phase values of subjects MJ *(bottom left)* and RK *(bottom right)* are phase-shifted 4 h forward in time. An *arrow* indicates the start time for the experiment. The *upper x-axis* denotes time when DNA synthesis is phase-shifted

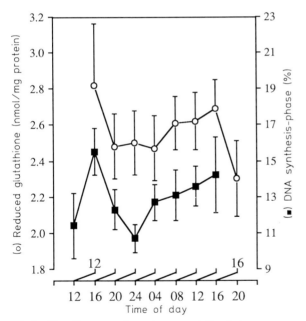

**Fig. 7.** Circadian stage dependent covariation between mean values of DNA synthesis and reduced GSH for pooled data of ten individuals when GSH is phase-shifted 4 h forward in time

thesis, with acrophase at 1317 hours and trough at 0117 hours ($P = 0.055$), i.e., the highest GSH content preceding that of DNA synthesis by about 4.5 h.

In accordance with this, when directly correlating each time point for both parameters according to circadian stage in a chronogram, a close circadian stage

dependent covariation between mean GSH content and S-phase was demonstrated when GSH content was phase-shifted 4 h forward in time (Fig. 7).

## Discussion

The low mean GSH content of 2.54 nmol/mg protein in human bone marrow found from repeated measurements over a 24-h period is in accordance with murine data reported by others [24, 83, 86].

The highest values of GSH content in our study were still lower than the average GSH content of other normal and tumor tissues. In rodents, five to more than ten times higher levels of GSH have been measured in many normal organs than in bone marrow [24, 83, 86, 89], the difference being even greater when compared with malignant tumors [90, 91]. In addition, some human tumor cells, and especially cells with acquired or de novo multidrug resistance, contain very high concentrations of GSH [25, 74, 85, 92].

The low level of GSH content in the bone marrow may suggest a limited capacity of the GSH-dependent detoxification mechanisms in this tissue, which should be related to the high sensitivity of human bone marrow to many cytotoxic drugs.

A possible relation between GSH content and DNA synthesis should therefore be considered in bone marrow cells, especially since an elevation of intracellular GSH has been suggested to be an event associated with proliferation in a wide variety of cell types [91, 93]. Our data suggest a close circadian stage dependent covariation of GSH content and fraction of cells in S-phase in the human bone marrow within the same subjects, either with the same phasing or being slightly phase-shifted. This is reflected in a statistically significant correlation between GSH content and fraction of cells in DNA synthesis when half of the cases are adjusted for the phase difference observed. These findings are thus in agreement with earlier findings demonstrating a relation between GSH and DNA synthesis when cells are activated into cell cycle progression or are in exponential growth [91, 93]. Since the timing was not always identical for the two parameters and the maximal GSH content for the pooled data preceded the DNA synthesis by about 4.5 h, the relation may be indirect and not necessarily causal.

## Examples of Peripheral Blood Parameters that Can Be Related to Bone Marrow Proliferation

### Introduction

Due to the observed interindividual differences in the DNA synthesis (and in CFU-GM), which also means differences in phasing of the different phases of the cell cycle, a more exact determination of the individual cell cycle distribution according to circadian stage would be desirable. This might be achieved by individual monitoring of the bone marrow proliferative parameters, or by monitoring of a parameter in peripheral blood that has a certain phase relation to the bone marrow cell proliferation. This phase relationship does not need to be causal, but the actual peripheral parameter could serve as a marker rhythm. We have so far looked at cortisol and peripheral granulocytes and their relation to DNA synthesis of total bone marrow cells and myeloid progenitor cells (CFU-GM), respectively.

### Cortisol and DNA Synthesis

Cortisol has a strong endogenous rhythm with large-amplitude variations during the 24-h cycle, performing an important function in the homeostasis of many organ functions. Therefore, we related this rhythmic endocrine parameter to the fraction of bone marrow cells in the S-phase.

The bone marrow sampling was performed as outlined earlier. At each sampling time serum was obtained for cortisol determination, and the usual circadian pattern for all individuals, i.e., high morning levels and low evening levels, was demonstrated.

We found that in most of the subjects there seemed to be a circadian stage dependent covariation between cortisol and DNA synthesis, but with a phase difference in several of the individuals. To quantitate the relationship between the two parameters, the measurements of each parameter were normalized relative to the mean value, and the difference between them squared and summed, either for corresponding time points or with cortisol phase-shifted 4 and 8 h forward. The lowest sum would then indicate the extent of phase difference between cortisol and DNA synthesis. Five subjects had the lowest sum when the circadian stage dependent variation of cortisol was not phase-shifted, while two sets of eight subjects had the lowest sum when phase-shifted 4 or 8 h, respectively, i. e., a phase delay in DNA synthesis as compared to cortisol of 4 or 8 h in 16 of 21 subjects. This corresponded very well to the best correlation between the two parameters when these were phase-shifted either 0, 4, or 8 h. Thirteen of 21 individuals had a correlation coefficient greater than 0.50 for the five to seven time points measured (range 0.08–0.96; mean $r = 0.55 \pm 0.23$ SD, median $r = 0.60$). A statistically significant correlation for the pooled data was demonstrated when cortisol (phase-shifted forwards 0, 4, or 8 h) was related to DNA synthesis (Spearman correlation test; $P < 0.0001$). Individual examples of covariation are shown in Fig. 8 (chronogram) and Fig. 9 (correlation).

By cosinor analysis of pooled data from all individuals for each parameter, a statistically significant circadian rhythm was found for cortisol ($P = 0.009$) and also a statistically significant rhythm for DNA synthesis ($P = 0.018$), with acrophase at 0850 hours and 1316 hours, respectively; while the trough was at 2050 hours and at 0116 hours, respectively. Thus, the cosinor analysis was in accordance with the sums of squares analysis of the two parameters, showing that the majority fo the individuals had the closest covariation when cortisol was phase-shifted forward 4–8 h.

These findings may suggest that DNA synthesis of bone marrow cells is directly or indirectly related to the cortisol level, with a phase delay of 4–5 h. That there might be a causal relationship is supported by findings of glucocorticoids having a stimulatory effect on the proliferation of normal human granulocyte-

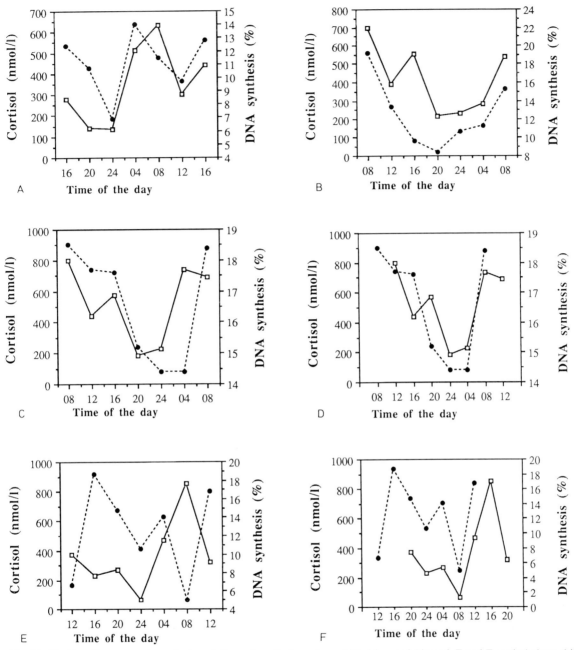

**Fig. 8 A–F.** Circadian stage dependent variation of cortisol (nmol/liter) and DNA synthesis (%) in human bone marrow in four individuals: **A** and **B** not phase-shifted; **C** and **D** cortisol phase-shifted forward 4 h; and **E** and **F** cortisol phase-shifted forward 8 h. — □ — cortisol; --- ● --- DNA synthesis

macrophage progenitors in the long-term liquid system of bone marrow cultures [94, 95]. An alternative hypothesis may be that the two parameters are not causally interrelated, but that the cortisol level of the individual person may be a marker rhythm to be taken advantage of concerning determination of the individual bone marrow proliferation state.

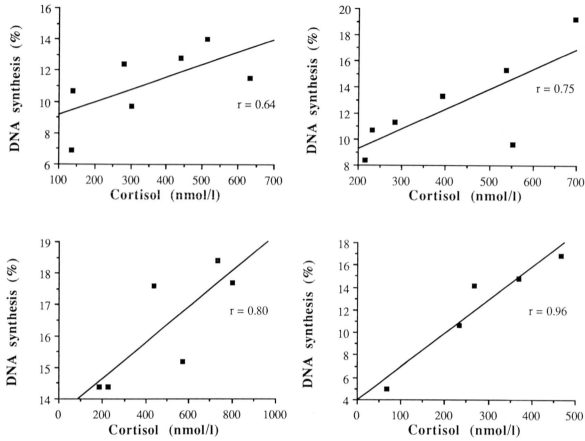

**Fig. 9.** Linear correlation between cortisol (nmol/liter) and DNA synthesis (%) in human bone marrow for the same subjects as in Fig. 8 when cortisol is phase-shifted forward 0, 4 or 8 h

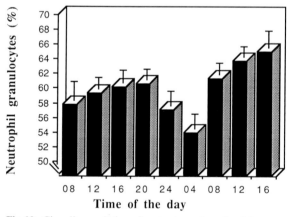

**Fig. 10.** Circadian variation of percentage of neutrophil granulocytes in peripheral blood during twenty-one 24-h periods. (See text)

## Granulocytes and Myeloid Progenitors (CFU-GM)

Because granulocytes in the blood represent the mature form of one of the two cell populations cultured in the CFU-GM assay, we also correlated CFU-GM and granulocytes to each other, granulocytes measured both as absolute and relative numbers.

Figure 10 shows the circadian variation of percentage of neutrophil granulocytes in peripheral blood during twenty-one 24-h periods, demonstrating a significant circadian variation. There was a tendency toward increasing values in the last part of the 24-h period. Analysis of the cortisol levels does not so far explain this increase.

Three examples of correlation between the percentage of neutrophil granulocytes in peripheral blood and CFU-GM are shown in Fig. 11. For the first individual there is a close covariation. For the second and third individuals a close covariation is also seen, but slightly phase-shifted, with the acrophase for the

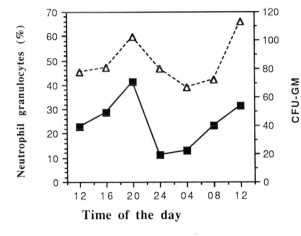

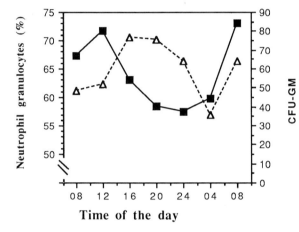

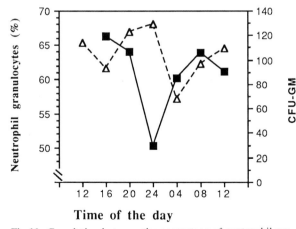

**Fig. 11.** Correlation between the percentage of neutrophil granulocytes in peripheral blood and CFU-GM in three individuals (see text). —■— CFU-GM; ---△--- Neutrophil granulocytes

CFU-GM occurring about 4 h before the acrophase of the neutrophil count, and with the trough of the CFU-GM occurring about 4 h before the trough of the neutrophils.

When pooling the data for CFU-GM and percentage of peripheral granulocytes, a very close and slightly phase-shifted covariation was demonstrated between these two parameters when measured consecutively.

The highest colony numbers were seen during early day, while the percentage of neutrophil granulocytes was also high and increased during the day, with the acrophase occurring about 4 h later than the CFU-GM. In addition, the trough value of neutrophils occurred about 4 h later (0400 hours) than the trough of the CFU-GM (2400 hours). This slightly phase-shifted covariation for the pooled data, as well as for most of the individuals, could imply that the percentage of peripheral granuloytes could serve as a marker rhythm for the production of myeloid progenitors in the bone marrow.

## Regulatory Aspects

During the last few years there has been great progress in the understanding of regulatory mechanisms of the bone marrow. Several stimulators of proliferation and maturation of bone marrow cells have been identified, such as the colony-stimulating factors (CSFs). This is a class of glycoproteins that mainly act on myelomonopoiesis and have pronounced effects at the stem and progenitor cell levels. Multi-CSF or interleukin-3 (IL-3) stimulates formation of colonies of all types of hematopoietic cell from the stem cell level. G-CSF stimulates formation of granulocytes, macrophage-CSF (M-CSF) of macrophages, and GM-CSF of both. It has been found that the receptors of these growth factors interact with each other, and that the simultaneous presence of several of the factors greatly enhances the effect. The factors have now been cloned and are available by use of recombinant DNA techniques [96–98]. In addition, erythropoietin, necessary for normal erythropoiesis, has been characterized [99]. The proteins of the CSF class seem to be necessary for normal growth and maturation of the different hematopoietic cell lineages. These factors are now currently being evaluated in phase I–II trials. Although the use of G-CSF and GM-CSF has not eliminated chemotherapy-associated neutropenia, they have reduced the period of dangerous neutropenia by accelerating hemato-poietic recovery. However, G-CSF and GM-CSF are much less effective in patients with reduced bone marrow reserve [100]. By optimizing the administration schedule of cytotoxic drugs and the hemato-poietic growth factors according to circadian stage dependent variation in bone marrow cell proliferative activity, one could possibly, to a larger extent, protect the pluripotent and committed stem cells.

In addition, data indicate that stem cell proliferation is controlled by one or more negative proliferative inhibitors, and several negative regulators of peptide nature have been identified and chemically characterized [47, 101–105]. In fact, pluripotent and committed stem cells seem to be under the control of competing inhibitory and stimulatory factors. Therefore, in some situations the inhibitors block entry from $G_0/G_1$ phase to DNA synthesis and accumulating cells in the stationary $G_0$ phase. On the other hand, stimulating factors will trigger cells from $G_0$ rapidly into DNA synthesis. By alternating the application of stimulator and inhibitor, it is possible to switch CFU-S (pluripotent stem cells) into and out of the cell cycle via a $G_0$ S switch [106]. This labile and sensitive system may very likely also react to time-dependent variations of different endogenous physiological/biochemical regulators, which may have their origin outside the bone marrow as well as in the bone marrow itself, including stromal cells [106]. However, at present it is not known the extent to which these factors are responsible for the feedback system that maintains a steady state.

## General Discussion

The sensitivity of both normal and malignant cells to cytotoxic agents is partly related to a high proliferation rate [3, 4]. On the other hand, GSH and GSH-dependent enzymes are now established as important factors in the defense of several normal and malignant cell types against cytotoxic agents [24, 30]. The myelosuppressive effect of many chemotherapeutic agents assumed to be detoxified by GSH-dependent mechanism(s) should therefore be considered in the light of the low GSH content in the human bone marrow as demonstrated here, and also its circadian stage relation to DNA synthesis. Thus, both proliferation rate and GSH content may play a role in the response of the human bone marrow to cytotoxic drugs.

The larger circadian stage dependent variation seen in the DNA synthesis and CFU-GM, indicates that the proliferative status of the bone marrow may be the most important single factor of these parameters to be taken into account for drugs mostly affecting the DNA synthesis when administering according to a circadian schedule. Flow cytometry methods are now available that allow rapid multiparameter analysis of single cells. Hematopoietic cells of different lineages and maturity can then be recognized, both in the bone marrow and peripheral blood, and their phenotype characterized by the combined use of flow cytometry and monoclonal antibodies to different subpopulation antigens [107]. This opens the possibility of studying specific subpopulations' rhythmicity according to proliferative activity and function.

Two main approaches can be followed when exploring effects of the underlying rhythmicity in bone marrow cell proliferation as to the response to cytotoxic chemotherapy or radiotherapy. First, one may minimize damage to the cells by giving bolus infusions or constant infusion with varying infusion rate according to the circadian-stage-dependent proliferation pattern of the cells. Another approach would be to focus on the rhythms or lack of rhythms in the tumors themselves. By using the first alternative, one may achieve a "shielding in time" of the normal tissue, while still attacking the tumor.

The possibility of protecting the bone marrow against cytotoxic damage by taking circadian variations into account is well documented in animal studies. Clinical studies have indicated that this is also possible in the human [18, 19]. These human studies are mainly based on empirical data from animal studies, finding the least toxic schedule along the 24-h time scale with extrapolation to the human activity-rest schedule. In the studies reported here we have measured proliferation parameters of the human bone marrow, as well as GSH, in order to devise a more direct and rational way to schedule cytotoxic treatment. This can thus be done by taking into consideration the bone marrow's circadian stage dependent proliferation pattern and GSH content. It is important to underline that the mechanism(s) of effect of the actual drugs must be considered relative to these parameters.

It has also been the purpose of these studies to relate parameters in the bone marrow to easily accessible and measurable parameters in the peripheral blood. In this way it may be possible to optimize chronotherapy further by taking into account the individual phasing of the relevant parameter in the bone marrow of the patient to be treated.

*Acknowledgments.* We are indebted to the volunteer subjects in the study. The authors gratefully acknowledge analyses of GSH by Asbjørn Svardal and the skillful technical assistance of Gro Olderøy, Dagny Ann Sandnes, and Jan Solsvik. The work was supported by the Norwegian Cancer Society.

# References

1. Laerum OD, Smaaland R, Sletvold O (1989) Rhythms in blood and bone marrow: Potential therapeutic implications. In: Lemmer B (ed) Chronopharmacology. Cellular and biochemical interactions. Dekker, New York
2. Gordon MY, Barrett AJ, Gordon-Smith EC (1985) Bone marrow disorders. The biological basis of clinical problems. Blackwell Scientific, Oxford
3. Lohrman H-P, Schreml W (1982) Cytotoxic drugs and the granulopoietic system. Springer, Berlin Heidelberg New York (Recent results in cancer research, vol 81)
4. Pollak MN, Brennan LV, Antman K, Elias A, Cannistra SA, Socinsky MA, Schnipper LE, Frei E, Griffin JD (1989) Recombinant GM-CSF in myelosuppression of chemotherapy. N Engl J Med 320: 253–254
5. Frei E et al. (1980) Dose: a critical factor in cancer chemotherapy. Am J Med 69: 585–593
6. Hryniuk W, Bush H (1984) The importance of dose intensity in chemotherapy of metastatic breast cancer. J Clin Oncol 2: 1281–1288
7. de Vita VTJ (1986) Dose-response is alive and well (editorial). J Clin Oncol 4: 1157–1159
8. Hryniuk W, Figueredo A, Goodyear M (1987) Applications of dose intensity to problems in chemotherapy of breast and colorectal cancer. Semin Oncol 14: (Suppl 4) 3–11
9. Cardoso SS, Scheving LE, Halberg F (1970) 1. Mortality of mice as influenced by the hour of day of drug (ara-C) administration. Pharmacologist 12: 302
10. Haus E, Halberg F, Scheving LE, Pauly JE, Cardoso S, Kuhl JFW, Sothern RB, Shiotsuka RN, Hwang DS (1972) Increased tolerance of leukemic mice to arabinosyl cytosine with schedule adjusted to circadian system. Science 177: 80–82
11. Scheving LE, Haus E, Kühl JFW, Pauly JE, Halberg F, Cardoso SS (1976) Close reproduction by different laboratories of characteristics of circadian rhythms in 1-$\beta$-D-arabinofuranosylcytosine tolerance by mice. Cancer Res 36: 1133–1137
12. Scheving LE, Pauly JE, Tsai TH, Scheving LA (1983) Chronobiology of cell proliferation. Implication for cancer chemotherapy. In: Reinberg A, Smolensky MH (eds) Biological rhythms and medicine. Cellular, metabolic, physiopathologic, and pharmacologic aspects. Springer, Berlin Heidelberg New York
13. Kühl JFK, Haus E, Halberg F, Scheving LE, Pauly JE, Cardoso SS, Rosene G (1974) Experimental chronotherapy with ara-C; comparison of murine ara-C tolerance on differently timed treatment schedules. Chronobiologia 1: 316–317
14. Scheving LE, Burns ER, Halberg F, Pauly JE (1980) Combined chronochemotherapy of L1210 leukemic mice using 1-$\beta$-D-arabino-furanosylcytosine, cyclophosphamide, vincristine, methylprednisolone, and cis-platinum. Chronobiologia 17: 33–40
15. Scheving LE, Burns ER, Pauly JE, Halberg F (1980) Circadian bioperiodic response of mice bearing advanced L1210 leukemia to combination therapy with adriamycin and cyclophosphamide. Cancer Res 40: 1511–1515
16. Roemeling R, Hrushesky WJM (1990) Determination of the therapeutic index of floxuridine by its circadian infusion pattern. JNCI 82: 386–393
17. Scheving LE, Burns ER, Pauly JE, Halberg F, Haus E (1977) Survival and cure of leukemic mice after optimization of cancer treatment with cyclophosphamide and ara-C. Cancer Res 37: 3648–3655
18. Hrushesky WJM (1985) Circadian timing of cancer chemotherapy. Science 228: 73–75
19. Kerr DJ, Lewis C, O'Neill B, Lawson N, Blackie RG, Newell DR, Boxall F, Cox J, Rankin EM, Kaye SB (1990) The myelotoxicity of carboplatin is influenced by the time of its administration. Hematol Oncol 8: 59–63
20. Hrushesky WJM, von Roemeling R, Sothern RB (1990) Preclinical and clinical cancer chronotherapy. In: Arendt J, Minors DS, Waterhouse JM (eds) Biological rhythms in clinical practice. Wright, London
21. Rivard GE, Infante-Rivard C, Hoyoux C, Champagne J (1985) Maintenance chemotherapy for childhood acute lymphoblastic leukemia: better in the evening. Lancet 2: 1264–1266
22. Nienhuis AW (1988) Hematopoietic growth factors. Biologic complexity and clinical promise. N Engl J Med 318: 916–918
23. Dethmers JK, Meister A (1981) Glutathione export by human lymphoid cells: depletion of glutathione by inhibition of its synthesis decreases export and increases sensitivity to irradiation. Proc Natl Acad Sci USA 78: 7492–7496
24. Lee FYF, Allalunis-Turner MJ, Siemann DW (1987) Depletion of tumor versus normal tissue glutathione by buthionine sulfoximine. Br J Cancer 56: 33–38
25. Lee FYF, Siemann DW, Allalunis-Turner MJ, Keng PC (1988) Glutathione contents in human and rodent tumor cells in various phases of the cell cycle. Cancer Res 48: 3661–3665
26. Friedman HS, Colvin OM, Griffith OW, Lippitz B, Elion GB, Schold JSC, Hilton J, Bigner DD (1989) Increased melphalan activity in intracranial human medulloblastoma and glioma xenografts following buthionine sulfoximine-mediated glutathione depletion. JNCI 81: 524–527
27. Hrushesky WJM (1987) The rationale for non-zero order drug delivery using automatic, computer based drug delivery systems (chronotherapy). J Biol Response Mod 6: 587–598
28. Arendt J, Minors DS, Waterhouse JM (1989) Basic concepts and implications. In: Arendt J, Minors DS, Waterhouse JM (eds) Biological rhythms in clinical practice. Wright, London
29. Butcher EC (1990) Cellular and molecular mechanisms that direct leukocyte traffic. Am J Pathol 136: 3–11
30. Lewis AD, Hayes JD, Wolf CR (1988) Glutathione and glutathione-dependent enzymes in ovarian adenocarcinoma cell lines derived from a patient before and after the onset of drug resistance: intrinsic differences and cell cycle effects. Carcinogenesis 9: 1283–1287

31. Pizzarello DJ, Witcofski RL (1970) A possible link between diurnal variations in radiation sensitivity and cell division in bone marrow of male mice. Radiology 97: 165–167

32. Burns ER (1981) Circadian rhythmicity in DNA synthesis in untreated and saline-treated mice as a basis for improved chronochemotherapy. Cancer Res 41: 2795–2802

33. Scheving LE, Pauly JE (1973) Cellular mechanisms involving biorhythms with emphasis on those rhythms associated with the S and M stages of the cell cycle. Int J Chronobiol 1: 269–286

34. Scheving LE, Burns ER, Pauly JE, Tsai TH (1978) Circadian variation in cell division of the mouse alimentary tract, bone marrow, and corneal epithelium, and its possible implication in cell kinetics and cancer chemotherapy. Anat Res 191: 479–486

35. Sharkis SJ, LoBue J, Alexander PJ, Rakowitz F, Weitz-Hamburger A, Gordon AS (1971) Circadian variations in mouse hematopoiesis. II. Sex differences in mitotic indices of femoral diaphyseal marrow cells. Proc Soc Exp Biol Med 138: 494–496

36. Sharkis SJ, Palmer JD, Goodenough J, LoBue J, Gordon AS (1974) Daily variations of marrow and splenic erythropoiesis, pinna epidermal cell mitosis and physical activity in C57B 1 + 6J mice. Cell Tissue Kinet 7: 381–387

37. Moskalik KG (1976) Diurnal rhythm of mitotic activity, DNA synthesis, and duration of mitoses in mouse bone marrow cells. Biull Eksp Biol Med 81: 594

38. Dörmer P, Schmolke W, Muschalik P, Brinkman W (1970) Die DNS-Synthesegeschwindigkeit im Verlaufe der DNS-Synthesephase von Erythroblasten der Maus in vivo. Beitr Pathol 141: 174–186

39. Mauer AM (1965) Diurnal variation of proliferative activity in the human bone marrow. Blood 26: 1–7

40. Killmann S-Å, Cronkite EP, Fliedner TM, Bond VP (1962) Mitotic indices of human bone marrow cells. I. Number and cytologic distribution of mitosis. Blood 19: 743–750

41. Vindeløv LL (1977) Flow microfluorometric analysis of nuclear DNA in cells from solid tumors and cell suspensions. Virch Arch [B] 24: 227–242

42. Gray JW, Dean PN (1980) Display and analysis of flow cytometric data. Ann Rev Biophys Bioeng 9: 509–539

43. Ginsberg L, Ludman PF, Anderson JV, Burrin JM, Joplin GF (1988) Does stressful venepuncture explain increased midnight serum cortisol concentration? Lancet II 8622: 1257

44. Dosik GM, Barlogie B, Göhde W, Johnston D, Tekell JL, Drewinko B (1980) Flow cytometry of DNA content in human bone marrow: a critical reappraisal. Blood 55: 734–740

45. Smaaland R, Lote K, Sletvold O, Bjerknes R, Laerum OD (1989) Circadian stage dependent variations in the DNA synthesis-phase and G2/M-phase of human bone marrow. Proc Am Assoc Cancer Res 30: 35

46. Lévi F, Blazcek I, Ferlé-Vidovic A (1988) Circadian and seasonal changes in murine bone marrow colony forming cells affect tolerance for 4′tetrahydropyranyladriamycin. Exp Hematol 16: 696–701

47. Paukovits WR, Guigon M, Binder KA, Hergl A, Laerum OD, Schulte-Hermann R (1990) Prevention of hematotoxic side effects of cytostatic drugs in mice by a synthetic hemoregulatory peptide. Cancer Res 50: 328–332

48. Stoney PJ, Halberg F, Simpson HW (1975) Circadian variation in colony-forming ability of presumable intact murine bone marrow cells. Chronobiologia 2: 319

49. Laerum OD, Aardal NP (1981) Chronobiological aspects of bone marrow and blood cells. In: Mayersbach, Scheving, Pauly (eds) 11th International congress of anatomy, part C, Biological rhythms in structure and function, pp 87–97. Liss, New York

50. Bartlett P, Haus E, Tuason T, Sackett-Lundeen L, Lakatua D (1982) Circadian rhythm in number of erythroid and granulocytic colony forming units in culture (ECFU-C and GSFU-C) in bone marrow of BDF1 male mice. In: Haus E, Kabat HF (eds) Proc. 15th international conference on chronobiology. Karger, Basel

51. Aardal NP, Laerum OD, Paukovits WR (1982) Biological properties of partially purified granulocyte extract (chalone) assayed in soft agar culture. Virch Arch [B] 38: 253–261

52. Aardal NP (1984) Circannual variations of circadian periodicity in murine colony-forming cells (CFU-C). Exp Hematol 12: 61–67

53. Aardal NP, Laerum OD (1983) Circadian variations in mouse bone marrow. Exp Hematol 11: 792–801

54. Meytes DMA, Powell WB, Ortega JA, Shore NA, Dukes PP (1980) Constancy of erythroid burst forming unit (BFU-E) levels in the blood of hematologically normal individuals. Exp Hematol 8: 641–644

55. Ponassi A, Morra L, Bonanni F, Molinari A, Gigli G, Mercelli M, Sachetti C (1979) Normal range of blood colony-forming cells (CFU-C) in humans: influence of experimental conditions, age, sex and diurnal variations. Blut 39: 257–263

56. Ross DD, Pollack A, Akman SA, Bachur NR (1980) Diurnal viaration of circulating human myeloid progenitor cells. Exp Hematol 8: 954–960

57. Verma DS, Fisher R, Spitzer G, Zander AR, McCredie KB, Dicke KA (1980) Diurnal changes in circulating myeloid progenitor cells in man. Am J Hematol 9: 185–192

58. Harrison DE (1979) Proliferative capacity of erythropoietic stem cell lines and aging: an overview. Mech Ageing Dev 9: 409–426

59. Williams LH, Udupa KB, Lipschitz DA (1986) Evaluation of the effect of age on hemopoiesis in young and old mice. Exp Hematol 14: 827–832

60. Sletvold O, Laerum OD (1988) Multipotent stem cell (CFU-S) numbers and circadian variations in aging mice. Eur J Haematol 41: 230–236

61. Metcalf D, Stevens S (1972) Influence of age and antigenic stimulation on granulocyte and macrophage progenitor cells in the mouse spleen. Cell Tissue Kinet 5: 433–446

62. Akagawa T, Onari KJPW, Makinodan T (1984) Differential effect on mitotically active and inactive bone marrow stem cells and splenic stem T cells in mice. Cell Immunol 86: 53–63

63. Sletvold O, Laerum OD, Riise T (1988) Age-related differences and circadian and seasonal variations in myelopoietic progenitor cell (CFU-GM) numbers in mice. Eur J Haematol 40: 42–49

64. Custer RP, Ahlfeldt FE (1932) Studies on the structure and function of bone marrow. II. Variations in cellularity in various bones with advancing years of life and their relative response to stimuli. J Lab Clin Med 17: 960–962

65. Hartsock RJ, Smith EB, Petty CS (1965) Normal variations with aging of the amount of hematopoietic tissue in the

bone marrow of the anterior iliac crest. Am J Clin Pathol 43: 326–331

66. Dunhill MS, Anderson JA, Whitehead R (1967) Quantitative histologic studies on the age changes in bone. J Pathol Bacteriol 94: 275–291

67. Schroder U, Tougaard L (1977) Age changes in the quality of hematopoietic tissue. Acta Pathol Microbiol Scand Sect A 85: 559–560

68. Lipschitz DA, Udupa KB, Milton KY, Thompson CO (1984) Effect of age on hemopoiesis in man. Blood 63: 502–509

69. Corberand JP, Laharrague P, Fillola G (1987) Blood cell parameters do not change during physiological aging. Gerontology 33: 72–76

70. Resnitzky P, Segal M, Barak Y, Dassa C (1987) Granulopoiesis in aged people: inverse correlation between bone marrow cellularity and myeloid progenitor cell number. Gerontology 33: 109–114

71. Burgess AW, Wilson EMA, Metcalf D (1977) Stimulation by human placental conditioned medium of hemopoietic colony formation by human marrow cells. Blood 49: 573–583

72. Schlunk T, Schlunk M (1980) The influence of culture conditions on the production of colony-stimulating activity by human placenta. Exp Hematol 8: 179–184

73. Meister A, Anderson ME (1983) Glutathione. Annu Rev Biochem 52: 711–760

74. Dusre L, Mimnaugh EG, Myers CE, Sinha BK (1989) Potentiation of doxorubicin cytotoxicity by buthionine sulfoximine in multidrug-resistant human breast tumor cells. Cancer Res 49: 511–515

75. Suzukake K, Petro BJ, Vistica DT (1982) Reduction in glutathione content of L-PAM-resistant L1210 cells confers drug sensitivity. Biochem Pharmacol 31: 121–124

76. Green JA, Vistica DT, Young RC, Hamilton TC, Rogan AM, Ozols RF (1984) Potentiation of melphalan cytotoxicity in human ovarian cancer cell lines by glutathione depletion. Cancer Res 44: 5427–5431

77. Kramer RA, Schuller HM, Smith AC, Boyd MR (1985) Effects of buthionine sulfoximine on the nephrotoxicity of 1-(2-chloroethyl)-3-(trans-4- methylcyclohexyl)-1-nitrosurea (MeCCNU). J Pharmacol Exp Ther 234: 498–506

78. Russo A, Mitchell JB (1985) Potentiation and protection of doxorubicin cytotoxicity by cellular glutathione modulation. Cancer Treat Rep 69: 1293–1296

79. Hamilton TC, Winker MA, Louie KG, Batist G, Behrens BC, Tsuruo T, Grotzinger KR, McKoy WM, Young RC, Ozols RF (1985) Augmentation of adriamycin, melphalan and cisplatin cytotoxicity in drug-resistant and -sensitive human ovarian carcinoma cell lines by buthionine sulfoximine mediated glutathione depletion. Biochem Pharmacol 34: 2583–2586

80. Andrews PA, Schiefer MA, Murphy MP, Howell SB (1988) Enhanced potentiation of cisplatin cytotoxicity in human ovarian carcinoma cells by prolonged gutathione depletion. Chem Biol Interact 65: 51–58

81. Ono K, Shrieve DC (1986) Enhancement of EMT6/SF tumor cell killing by mitomycin C and cyclophosphamide following in vivo administration of buthionine sulfoximine. Int J Radiat Oncol Biol Phys 12: 1175–1178

82. Crook TR, Souhami RI, Whyman GD, McLean AEM (1986) Glutathione depletion as a determinant of sensitivity of human leukemia cells to cyclophosphamide. Cancer Res 46: 5035–5038

83. Tsutsui K, Komuro C, Ono K, Nishidai T, Shibamato Y, Takahashi M, Abe M (1986) Chemosenzitation by buthionine sulfoximine in vivo. Int J Radiat Oncol Biol Phys 12: 1183–1186

84. Bump EA, Yu NY, Brown JM (1982) The use of drugs which deplete intracellular glutathione in hypoxic cell radiosensitization. Int J Radiat Oncol Biol Phys 8: 439–442

85. Biaglow JE, Varnes ME, Clark EP, Epp EP (1983) The role of thiols in cellular response to radiation and drugs. Radiat Res 94: 437–455

86. Somfai-Relle S, Suzukake K, Vistica BP, Vistica DT (1984) Reduction in cellular glutathione by buthioninesulfoximine and sensitization of murine tumor cells resistant to L-phenylalanine mustard. Biochem Pharmacol 33: 485–490

87. Anderson M (1985) Determination of glutathione and glutathione disulfide in biological samples. Methods Enzymol 113: 548–555

88. Svardal AM, Mansoor MA, Ueland PM (1990) Determination of reduced, oxidized, and protein-bound glutathione in human plasma with precolumn derivatization with monobromobimane and liquid chromatography. Anal Biochem 184: 338–346

89. Jaeschke H, Wendel A (1985) Diurnal fluctuation and pharmacological alteration of mouse organ glutathione content. Biochem Pharmacol 34: 1029–1033

90. Somfai-Relle S, Suzukake K, Vistica BP, Vistica DTG (1984) Glutathione conferred resistance to antineoplastics: approaches toward its reduction. Cancer Treat Rev 11: (Suppl A) 43–54

91. Lee FYF, Vessey A, Rofstad E, Siemann DW, Sutherland RM (1989) Heterogeneity of glutathione content in human ovarian cancer. Cancer Res 49: 5244–5248

92. Kramer RA, Zakher J, Kim G (1988) Role of glutathione redox cycle in acquired and de novo multidrug resistance. Science 241: 694–697

93. Shaw JP, Chou IN (1986) Elevation of intracellular glutathione content associated with mitogenic stimulation of quiescent fibroblasts. J Cell Physiol 129: 193–198

94. Suda T, Dexter T (1981) Effect of hydrocortisone on long-term human bone marrow cultures. Br J Haematol 48: 661–664

95. Pasquale D, Chikkappa G, Wang G, Santella D (1989) Hydrocortisone promotes survival and proliferation of granulocyte macrophage progenitors via monocytes/macrophages. Exp Hematol 17: 1110–1115

96. Metcalf D (1986) Annotation. Haematopietic growth factors now cloned. Br J Haematol 62: 409–412

97. Bronchud MH, Scarfe JH, Thatcher N et al. (1987) Phase I/II sudy of recombinant human granulocyte colony-stimulating factor in patients receiving intensive chemotherapy for small cell lung cancer. Br J Cancer 56: 809–813

98. Antman KS, Griffin JD, Elians A et al. (1988) Effect of recombinant human granulocyte-macrophage colony stimulating factor on chemotherapy-induced myelosuppression. N Engl J Med 319: 593–598

99. Graber SE, Krantz SB (1989) Erythropoietin: biology and clinical use. Hematol Oncol Clin North Am 3: 369–400

100. Demetri GD, Griffin JD (1990) Hematopoietic growth factors and high-dose chemotherapy: will grams succeed where milligrams fail? J Clin Oncol 8: 761–764

101. Lord BI, Mori KJ, Wright EG, Lajtha LG (1976) An inhibitor of stem cell proliferation in normal bone marrow. Br J Haematol 34: 441–445

102. Frindel E, Guigon M (1977) Inhibition of CFU-entry into cycle by a bone marrow extract. Exp Hematol 5: 74–76

103. Laerum OD, Paukovits WR (1984) Inhibitory effects of a synthetic pentapeptide on hemopoietic stem cells in vitro and in vivo. Exp Hematol 12: 7–17

104. Laerum OD, Paukovits WR (1989) Biological and chemical properties of the hemoregulatory peptide and possibilities for clinical applications. Pharmacol Ther 44: 335–349

105. Paukovits WR, Elgjo K, Laerum OD (1990) Pentapeptide growth inhibitors. In: Sporn MB, Roberts AB (eds) Peptide growth factors and their receptors II. Springer, Berlin Heidelberg New York (Handbook of experimental pharmacology, vol 95)

106. Lord BI, Testa NG (1988) The hemopoietic system. Structure and regulation. In: Testa NG, Gale RP (eds) Hematopoiesis. Long-term effects of chemotherapy and radiation. Dekker, New York

107. Lund-Johansen F, Bjerknes R, Laerum OD (1990) Flow cytometric assay for the measurement of human bone marrow phenotype, function and cell cycle. Cytometry 11: 610–616

# Chronobiology of Polymorphonuclear Leukocyte Migration

J. P. Bureau and L. Garrelly

## Introduction

The inflammatory reaction is characterized by a multifrequency time structure with prominent circadian, ultradian and circannual rhythms in cell proliferation and cell function. The chemotaxis of polymorphonuclear neutrophil leukocytes (PMN) appears to be an important effector mechanism in the defenses against external antigenic agents. Granulocytes are present in large numbers in inflammatory reactions and their numbers are also increased in chronic inflammatory diseases. Numerous models have been used for analyzing the mechanisms of inflammation because any single one cannot express the whole inflammatory process of disease. Many pharmacology studies indicate circadian and circannual changes in drug effectiveness and toxicity. The mechanisms underlying these changes are not fully understood yet, but there is evidence indicating that time-dependent variations in the metabolism of drugs and in drug pharmacokinetics may partly explain them.

This review will describe the acute phase of inflammation based on the PMN accumulation at an inflammatory site, and the chronomodulation of the inflammatory responses.

## Kinetics of the Inflammatory Response

### Inflammatory Processes

The inflammatory reaction is characterized first by an increase of vascular permeability with diffusion of fluid rich in serum proteins, which forms the exudate, and second by a cell migration into the injured site. Labrecque et al. (1984) found a circadian variation in the amount of plasma exudation and in the formation of edema in carrageenan paw edema in the rat. The highest paw volumes were found in animals on LD 12:12 (light on: 0700 to 1900 hours), between 1900 and 0100 hours, and the minimal values were obtained at 1600 hours (during the rest span).

In human subjects and animals circadian and circannual rhythms have been documented for numerous blood proteins involved in inflammatory processes (Reinberg et al. 1977; Haus et al. 1983; Bruguerolle et al. 1986). In healthy human subjects maximal values of blood proteins were found around 1200 hours. Comparable variations were observed in the concentration of serum complement components (C3 and C4) (Haus et al. 1983).

Circadian variations have been described in the concentration of hematological, chemical, [e.g., histamine, serotonine (Wisser et al. 1981), prostaglandin E (PGE) (Abe et al. 1981; Bowden et al. 1977)] and endocrinological parameters, [e.g., adrenocorticotropic hormone (ACTH) (Desir et al. 1980; Reinberg et al. 1983), corticosterone (Szafarczyk et al. 1979), catecholamines (Wisser et al. 1981), etc.]. The number of circulating blood cells shows also highly reproducible circadian rhythms in human subjects and in animals (Reinberg et al. 1977; Haus et al. 1983). The human circadian rhythm in the number of circulating PMN shows its acrophase during the late afternoon and early evening hours. In mice the acrophase occurs during the late night and early morning.

## Time Course of Cell Migration in Mice

*Cell Migration Induced by Antigens.* The migration of PMN was studied using an in vivo migration test (Senelar and Bureau 1979a; Bureau et al. 1980, 1981). A disk of sterilized mesh material made of rayon was implanted subcutaneously in Swiss male mice, 8–12 weeks of age, housed in an air-conditioned room on an L:D of 12:12 (light on: 0700–1900 hours). Just before grafting, 10µl of either saline, or $4–6 \times 10^6$ viable BCG/ml suspension, or 2% of carrageenan

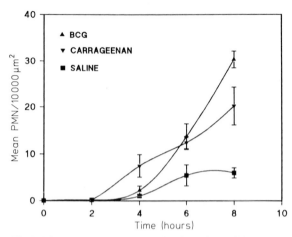

**Fig.1.** Time course of cell migration in saline, BCG, and carrageenan cell traps implanted at 1300 hours in Swiss male mice, 8–12 weeks of age (L:D 12:12 with L from 0700 to 1900 hours). Results expressed in mean number of cells/$10000\ \mu m^2 \pm SE$

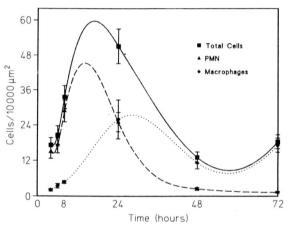

**Fig.2.** Time course of PMN and macrophage migration in BCG-induced granulomas in mice. Experimental design as in Fig.1. Results expressed in number of cells/$10000\ \mu m^2 \pm SE$

solution were added onto the disk. The implant was inserted beneath the skin at 1300 hours and removed 2, 4, 6, or 8 h later, fixed in Bouin-Holland's solution, processed for 5 µm paraffin sections and studied with Groat's hematoxylin. Five sections of each disk were counted with an Artek cell counter. The results were expressed as the mean number of PMN ± SE/10000 µm² of the original section.

A significant increase in cell numbers was found with carrageenan and BCG 480 min after implantation, but not when saline was added to the cell trap (Fig.1).

*Study of PMN and Macrophage Migration Induced by BCG.* The time course of migration of PMN and

macrophages produced by the implantation of the cell trap during 72 h is presented in Fig.2, which shows an increase in the number of PMN during the first 12 h. The increase of macrophage migration begins only 8 h after cell trap implantation. The time course of the cell infiltration of the trap explains the choice of the sampling time of 480 min after each of the time-specific implantations.

## Circadian Rhythm

### Time Course of PMN Migration Induced by Saline and BCG

The time course of the migration of PMN produced by the implantation of the disk at six different times of the day is presented in Fig.3 (Bureau et al. 1986b). The experiments were carried out in autumn, under the same conditions as described above. When saline is added to the cell trap the number of cells found in the trap at 0500 hours out of the six times of experimentation is high after 480 min implantation. In the BCG-treated mice maximal PMN migration is obtained at 0500 and 1700 hours. These data suggest that the 1700-hours peak is due to the effect of BCG itself, whereas the nonspecific stimulating effect of the cell trap explains the 0500-hours peak.

### Circadian Rhythm of Cell Migration Induced by Different Antigens

*BCG-Induced Cell Migration.* The experiments shown in Figs.4 and 5 demonstrate that the migration of PMN induced by BCG is not constant throughout the year. In Fig.4, the experiments were carried out in November showing two peaks in the number of PMN, at 0500 and 1700 hours. When experiments were carried out in March (Fig.5) the leukocyte peak was found at 0500 hours and the trough occurred at 17 hours (Bureau et al. 1984a, 1986a). These findings could be related to circadian and circannual variations of plasma corticosterone and the well-known inverse relationship between plasma corticosterone levels and the mechanism underlaying the circadian variation in the effect of BCG on PMN migration as will be discussed elsewhere.

*Carrageenan-Induced Cell Migration.* The circadian rhythm observed with carrageenan when experi-

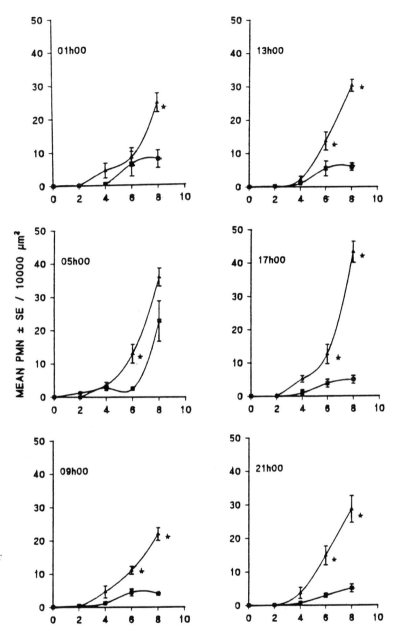

**Fig.3.** Time course of the migration in vivo of PMN in response to saline and BCG after implantation six times of the day in mice on L:D 12:12 with L from 0700 to 1900 hours. Results expressed as mean PMN/10000 $\mu m^2$ ± SE. The *asterisk* denotes $P < 0.05$ (Student's *t* test)

ments were carried out in November (Fig.4) is similar to the one induced by BCG.

*Lipopolysaccharide-Induced Cell Migration.* The experiment with lipopolysaccharide (LPS) was performed in March and the results compared with PMN migration induced by BCG (Fig.5). Maximal LPS-induced migration occurred at 9000 hours and the trough was observed at 1700 hours. Although the immunological responses are different when induced by LPS as compared with those obtained by BCG, there is no difference between the circadian vari-

ations of PMN migration induced by the two antigens.

## Circannual Rhythm

### BCG-Induced PMN Migration

The PMN migration induced by live BCG in mice is not constant throughout the year (Bureau et al. 1985) (Fig.6). The mean PMN values are higher when cell

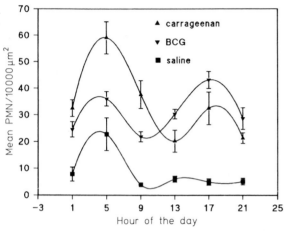

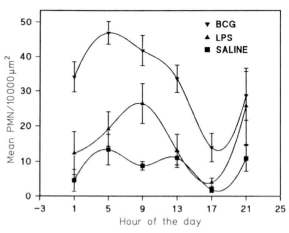

**Fig. 4.** Circadian rhythms of PMN migration induced by saline, carrageenan, and BCG in mice. The experimental design is summarized in the legend of Fig. 3. Experiment carried out in November

**Fig. 5.** Circadian rhythms of PMN migration induced by saline, LPS, and BCG in mice. The experimental design is summarized in the legend of Fig. 3. Experiment carried out in March

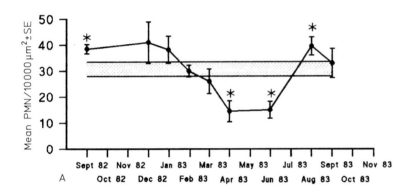

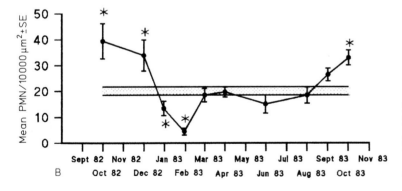

**Fig. 6 A, B.** Circannual variations in BCG-induced migration of PMN in mice **A** at 0900 hours and **B** at 1700 hours. Results expressed as mean PMN/10000 μm² ± SE. The asterisk denotes $P < 0.05$ (Student's $t$ test)

traps are implanted at 0900 hours as compared with the number of PMN obtained at 1700 hours. The PMN migration is significantly higher during autumn than at any other time of the year. Minimal PMN migration is found in April when the serum corticosterone levels were reported to be low, whereas the migration is very high in November (Haus et al. 1970; Reinberg et al. 1983). This result shows that other factors or hormones are involved in PMN migration (Senelar et al. 1979a).

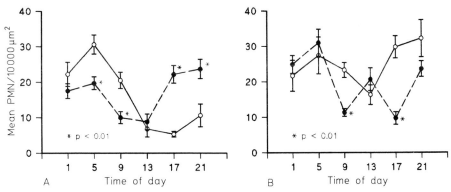

**Fig. 7 A, B.** Chronopharmacology of corticosterone on BCG-induced granulocyte migration in normal and adrenalectomized mice. **A** Presents the number of PMN in intact mice injected with either 0.2 ml saline (○—○) or corticosterone, 2 mg/kg (●—●). **B** Illustrates the granulocyte migration observed in adrenalectomized mice injected with saline (○—○) or corticosterone, 2 mg/kg (●—●)

## Other Circannual Variations

Circannual variations in carrageenan-induced paw edema in the rat has been described by Labrecque et al. (1982). These investigators found that the paw edema volume was minimal when carrageenan was injected during the winter and maximal when carrageenan was injected in spring.

Haus et al. (1983) pointed out the circannual changes can be observed in the immune system. It was shown that the total number of circulating lymphocytes and of individual cell types was much higher in autumn and winter as compared to spring and summer.

## Chronomodulation of PMN Migration

### Steroid Hormones

#### Adrenal Gland

*Effect of Adrenalectomy.* When adrenalectomy is performed in mice during winter, the mean values of PMN migration induced by BCG are significantly higher in the adrenalectomized than in normal mice (Figs. 7 and 8). Compared with normal mice, the peak-PMN number is found at 0500 and 1700 hours; the lowest values are obtained at 0100 and 0900 hours. The circadian rhythm detected in normal mice becomes an ultradian rhythm in adrenalectomized animals (Bureau et al. 1984b).

*Corticosterone effect.* When 2 mg/kg corticosterone are injected IP in mice immediately before implantation of BCG-impregnated disks at six different times of the day, the data show that corticosterone can either inhibit or stimulate PMN migration. When the endogenous plasma corticosterone level is known to be low the cell migration is high (Fig. 7 A). The injected corticosterone stimulates or inhibits the migration of PMN induced by BCG depending on the time of its administration. It is interesting to note that Walker et al. (1985) found a similar biphasic effect for ACTH 1-17 which stimulated or inhibited DNA synthesis in murine metaphyseal bone as a function of

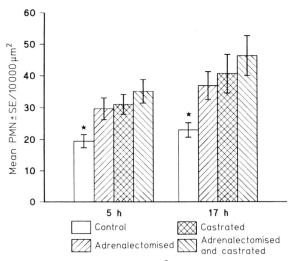

**Fig. 8.** Number of cells/10000 μm² in cell traps implanted at 0500 and 1700 hours in mice after castration, adrenalectomy, and castration with adrenalectomy. The asterisk denotes $P < 0.05$ (Student's *t* test)

the time of drug administration. In adrenalectomized mice the biphasic effect of corticosterone is virtually absent (Fig. 7 B). The mechanism of the biphasic effect of corticosterone could be related to receptors mediating the effects of glucocorticoids on inflammation (Laue et al. 1988) or to different endogenous substances such as prostaglandins or leukotrienes (Rigas and Levine 1983; Dore et al. 1985).

### Gonads

*Gonadectomy.* It is known that there are interactions between gonadal steroids and the immune system (Castro 1976; Grosmann 1985; Lebreton et al. 1988). Castration increases the PMN migration induced by BCG in mice (Fig. 8). The PMN level is higher when the cell trap is implanted at 1700 rather than at 0500 hours. When gonadectomy is added to adrenalectomy the increase of the PMN number is more prominent at 1700 than at 0500 hours.

### Other Inflammatory Modulators

*Peptides.* The hemoregulatory peptide extracted from mature granulocytes inhibits PMN migration induced by BCG (Bureau et al. 1990). This inhibition effect is dose- and time-dependent. Also other peptides are involved in inflammatory response such as substance P (Gibson et al. 1989), tumor necrosis factor (TNF) (Harmsen et al. 1990; Cybulsky et al. 1988; Visner et al. 1990), and interleukin-1 (IL-1) (Vargaftig 1977; Vincent et al. 1978; Rubin et al. 1988; Visner et al. 1990). Circadian variations have been described for some of them.

*Lipid Mediators.* Chronoperiodicity is described in the hypotensive effect of $PGE_2$ in rats (Dore et al. 1985). In human subjects there is a circadian variation in the salivary concentration of $PGE_2$ and prostaglandin F ($PGF_2$) (Rigas and Levine 1983). The inhibition of the PMN movement is related to the elevation of the cyclic AMP levels by prostaglandins (Rivkin et al. 1975; Rosbash and Hall 1989). The circadian rhythmicity in neutrophil chemotaxis (Muniain et al. 1988) is related to the circadian variations in the production of superoxide (Muniain et al. 1989; Symons et al. 1988).

*Anti-Inflammatory Drugs.* There is a circannual variation of phenylbutazone effects on carrageenan-induced paw edema in the rat (Labrecque et al. 1982).

Inhibition of leukocyte migration by salicylates is more effective in winter than in summer (Warne and West 1978).

## Conclusion

The experimental data and the different investigations reported in this chapter show ultradian, circadian, and/or circannual rhythms in the inflammatory response. These rhythms are related to many hematological, chemical, hormonal, and cellular mediators, most of which show periodic variations. The chronobiological approach to the study of the biological events taking place in the course of the process of inflammation will lead to a better understanding of related physiological processes. Moreover, its application to the treatment of inflammatory diseases will improve the efficacy of medications.

## References

Abe K, Sato M Kasai Y, Haruyana T, Sato K, Miyasaki S, Imai Y, Hiwatari M, Itoh T, Sakurai Y, Goto T, Tajima J, Seino M, Yoshinaga K (1981) A circadian variation in the excretion of urinary kinin, kallikrein and prostaglandin E in normal volunteers. Jpn Circ J 45: 1098–1103

Bongrand P, Bouvenot G, Bartolin R, Tatossian J, Bruguerolle B (1988) Are there ciradian variations of polymorphonuclear phagocytosis in man? Chronobiol Int 5: 81–83

Bowden RE, Ware JH, Demets DL, Keiser HR (1977) Urinary excretion of immunoreactive E: a circadian rhythm and the effect of posture. Prostaglandins 14: 151–161

Bruguerolle B, Levi F, Arnaud C, Bouvenot G, Mechkouri N, Vannetzel J, Touitou Y (1986) Alteration of physiologic circadian time structure of six plasma proteins in patients with advanced cancer. Annu Rev Chronopharmacol 32: 207–210

Bureau JP, Senelar R, Cupissol D (1980) Restoration by Ketoprofen of defective neutrophil granulocyte migration induced in guinea pigs by plasma from cancer patients. Br J Exp Pathol 61: 479–485

Bureau JP, Senelar R, Cupissol D (1981) Plasma transferable inhibition of BCG induced subcutaneous inflammation in human cancer. J Pathol 133: 215–227

Bureau JP, Coupe M, Garrelly L, Labrecque G (1984a) Circannual rhythm of PMN migration induced by BCG in intact mice. Annu Rev Chronopharmacol 1: 337–340

Bureau JP, Garrelly L, Coupe M, Labrecque G (1984b) Circadian rhythm studies on BCG-induced migration of PMN in normal and in adrenalectomized rats. Annu Rev Chronopharmacol 1: 331–336

Bureau JP, Labrecque G, Garrelly L, Coupe M (1985) Monthly changes in the effect of BCG on the migration of polymorphonuclear leukocytes. Res Commun Chem Pathol Pharmacol 49: 317–320

Bureau JP, Coupe M, Labrecque G (1986a) Chronopharmacological study of the effect of corticosterone on BCG induced granulocytose migration in normal and in adrenalectomized mice. Annu Rev Chronopharmacol 3: 309–312

Bureau JP, Labrecque G, Coupe M, Garrelly L (1986b) Influence of BCG administration time on the in vivo migration of leukocytes. Chronobiol Int 3: 23–28

Bureau JP, Garrelly L, Vago P, Bayle S, Labrecque G, Laerum OD (1990) Circadian variations of hemoregulatory peptide effect on PMN migration in mice. Annu Rev Chronopharmacol 7: 205–208

Castro JE (1976) Orchidectomy and immune response. Ann R Coll Surg Engl 58: 359–367

Cybulsky MI, McComb DJ, Rovat HZ (1988) Neutrophil leukocyte emigration induced by endotoxin. Mediator roles of interleukin 1 and tumor necrosis factor alpha 1. J Immunol 140: 3144–3149

Desir D, Caliter EV, Golstein J, Fang S, Leclerq R, Refetoff S, Copinschi G (1980) Circadian and ultradian variations of ACTH and cortisol secretion. Horm Res 13: 302–316

Dore F, Labrecque G, Belanger PM, D'Auteuil C (1985) Chronobiological studies on the hypotensive effect of prostaglandin E2 and arachidonic acid in the rats. Chronobiol Int 1: 273–278

Gibson SJ, Andrews PV, Helme RD (1989) Photoperiodic variation in substance P induced plasma extravasation in the rat. Neuropharmacology 28: 889–892

Grosmann CJ (1985) Interactions between the gonadal steroïds and the immune system. Science 227: 257–261

Harmsen AG, Havell EA (1990) Roles of tumor necrosis factor and macrophages in lipolysaccharide-induced accumulation of neutrophils in cutaneous air pouches. Infect Immun 58: 297–302

Haus E, Halberg F (1970) Circannual rhythm in level and timing of serum corticosteroid in standardized inbred C-mice. Environ Res 3: 81–106

Haus E, Lakatua DJ, Swoyer J, Sackett-Lundeen L (1983) Chronobiology in hematology and immunology. Am J Anat 168: 467–517

Hollwich F, Dieckhues B (1989) Effect of light on the eye on metabolism and hormones. Klin Monatsbl Augenheilkd 195: 284–290

Ishibashi Y, Nagaoka I, Yamashita T (1988) Studies on inflammation in mice: dynamics of inflammatory response induced by extract of Escherichia Coli and influence of inflammatory drugs. Int J Tissue React 10: 7–16

Labrecque G, Dore F, Belanger PM (1981) Circadian variation of carrageenan-paw oedema in the rat. Life Sci 28: 1337–1343

Labrecque G, Belanger PM, Dore F (1982) Circannual variations in carrageenan-induced paw oedema and in the antiinflammatory effect of phenylbutazone in the rat. Pharmacology 24: 169–174

Labrecque G, Dore FM, Belanger PM, Carter V (1984) Chronobiological study of plasma exudation in carrageenan-paw oedema in the rat. Agents Actions 14: 719–722

Laue L, Kawai S, Brandon DD, Brightwell D, Barnes K, Knazek RA, Loriaux DL, Chrousos GP (1988) Receptor mediated effects of glucocorticoids on inflammation: enhancement of the inflammatory response with a glucocorticoid antagonist. J Steroid Biochem 29: 591–598

Lebreton JP, Hiron M, Biou D, Daveau M (1988) Regulation of 1-acid glycoprotein plasma concentration by sex steroids and

adrenal-cortical hormones during experimental inflammation in the rat. Inflammation 12: 413–424

Muniain MA, Mata R, Pozuelo F, Rodriguez C, Rodriguez D, Romero A, Trueba A, Spilberg I (1988) Circadian rhythmicity in neutrophil chemotaxis. J Rheumatol 15: 1044–1045

Muniain MA, Rodriguez MD, Mata R, Pozuelo F, Gimenez MJ, Romero A (1989) Circadian variations on the production of superoxide and liberation of lysosomal enzymes of neutrophils in peripheral blood. Rev Clin Esp 184: 20–23

Perper RJ, Sandar M, Stretcher VJ, Oronsky AL (1975) Physiological and pharmacological alterations of rat leukocyte chemotaxis in vivo. Ann NY Acad Sci 256: 190–195

Pownall R, Kabler PA, Knapp MS (1979) The time of the day of antigen encounter influences the magnitude of the immune response. Clin Exp Immunol 36: 347–354

Reinberg A, Sidi E, Ghata J (1965) Circadian reactivity rhythms of human skin to histamine or allergen and the adrenal cycle. J Allergy 36: 273–283

Reinberg A, Schuller E, Delasnerie N, Clench J, Helary M (1977) Rythmes circadiens et circannuels de leucocytes, protéines totales, immunoglobulines A, G, M. Etude chez 9 adultes jeunes et sains. Nouv Presse Med 6: 819–823

Reinberg A, Touitou Y, Levi F, Nicolai A (1983) Circadian and seasonal changes in ACTH-induced effects in healthy young men. Eur J Clin Pharmacol 25: 657–665

Rigas B, Levine L (1983) Human salivary eicosanoids: circadian variation. Biochem Biophys Res Commun 115: 201–215

Rivkin I, Rozenblatt J, Becker EL (1975) The role of cyclic AMP in the chemotactic responsiveness and spontaneous mobility of rabbit peritoneal exudate neutrophils. The inhibition of neutrophil movement and the elevation of cyclic AMP levels by catecholamins, prostaglandins, theophylline and choleratoxin. J Immunol 115: 1126–1132

Rosbash M, Hall JC (1989) The molecular biology of circadian rhythms. Neuron 3: 387–398

Rubin RM, Rosenbaum MD, Rosenbaum JT (1988) A platelet-activating factor antagonist inhibits interleukin I-induced inflammation. Biochem Biophys Res Commun 154 (1): 429–436

Senelar R, Bureau JP (1979a) In vivo effects of human chorionic gonadotropin on the migration of inflammatory cells in intact or castrated male and female guinea-pigs. A quantitative histological study. II. Study of castrated males and females. Br J Exp Pathol 60: 489–492

Senelar R, Bureau JP (1979b) Inhibitory effects of pregnancy on the migration of the inflammatory cells: a quantitative histological study. Br J Exp Pathol 60: 286–293

Symons AM, Dowling EJ, Parke DV (1988) Lipid peroxydation, free radicals and experimental inflammation. Basic Life Sci 49: 987–990

Szafarczyk A, Ixart G, Malaval F, Nouguier-Soule J, Assenmacher I (1979) Effects of lesions of the suprachiasthmatic nuclei and of p-chlorophenylalanine on the circadian rhythms of adrenocorticotrophic hormone and corticosterone in the plasma and on locomotor activity of rats. J Endocrinol 83: 1–16

Vargaftig BB (1977) Involvement of mediators in the interaction of platelets and carrageenan. Agents Actions 2 (suppl): 9–39

Vincent JE, Bonta IL, Zijlstra FJ (1978) Accumulation of blood platelets in carrageenan paw oedema: possible role in the inflammatory response. Agents Actions 8: 291–293

Visner GA, Dougall WL, Wilson JM, Burr IA, Nick HS (1990) Regulation of manganese superoxide dismutase by lipopolysaccharide, interleukine 1 and tumor necrosis factor. Role in the acute inflammatory response. J Biol Chem 265: 2856–2864

Walker VW, Russell JE, Simmons DJ, Scheving LE, Cornelissen G, Halberg F (1985) Effects of an adrenocorticotrophin analogue, ACTH 1-17, on DNA synthesis in murine metaphyseal bone. Biochem Pharmacol 34: 1191–1196

Ward PA, Remold HG, David JR (1969) Leukotactic factor produced by sensitized lymphocytes. Science 163: 1079–1080

Warne PJ, West GB (1978) Inhibition of leucocyte migration by salicylates and indomethacin. J Pharm Pharmacol 30: 783–785

Warne PJ, West GB (1980) Seasonal variations in drug action and animal responses in models of inflammation. Int Arch Allergy Appl Immunol 61: 11–13

Wisser H, Breuer H (1981) Circadian changes of clinical chemical and endocrinological parameters. J Clin Chem Clin Biochem 19: 323–338

# Chronobiology in Hemostastis

H. Decousus

## Introduction

Arterial and venous thromboembolic disorders remain the leading cause of death in most of the develop countries. A thrombotic occlusion of a coronary or cerebral artery, already damaged by atherosclerosis, appears to be a common and essential link in the onset of most of the cases of such vascular events and induces myocardial or cerebral infarctions [1, 2]. It is not surprising, therefore, that these vascular disorders and the hemostatic system are correlated. For example, high levels of plasma fibrinogen and factor VII coagulant activity were associated with an increased risk of death from cardiovascular disease [3]. High plasma levels of the fast-acting plasminogen activator inhibitor (PAI-1) seem also to be correlated with acute coronary thrombosis [4]. In venous thrombosis the relationship between the disease and the hemostatic system is even more patent. Thus, biological rhythms which modulate the hemostatic system are expected to relate to temporal changes in the incidence of thromboembolic disorders. Such rhythms may also induce temporal variations in the biological effect of antithrombotic agents. Possible therapeutic implications of these observations for the treatment of thromboembolic disorders deserve to be considered and evaluated.

## Chronophysiology of Hemostastis

### Animal Studies

Cohen et al. [5] showed that, in rats, bleeding time had a peak at night during their activity span and a trough at noon during their rest span. There was an opposite variation for platelet concentrations. Scheving and Pauly [6] showed, also in rats, a nocturnal

peak and a morning trough for blood clotting time. The day-night difference reached 50%. This circadian rhythm was not related in its timing to the circadian rhythm in motor activity. Total adrenalectomy and adrenal medullectomy did not abolish this rhythm but minimized its amplitude. Recently Soulban and Labrecque [7] confirmed these results, in rats again, and found similar variations for blood clotting time. Coagulation factors II, VII, and X had an opposite time course with a morning peak and a nocturnal trough. To our knowledge no study has been performed on the circadian variation of fibrinolysis in animals. To summarize, a morning "hypercoagulability" and a nocturnal "hypocoagulability" was observed in rats.

## Human Studies

Circadian Rhythm of Platelet Functions

*Platelet Aggregability.* Most of the studies performed in health [8–12] and disease, e.g., studies in diabetic patients [12] and in patients with stable coronary disease [13], showed a morning peak and a nocturnal trough for platelet aggregability in response to epinephrine and adenosine diphosphate (ADP) (Fig. 1). The discovery of the same pattern in platelet binding affinity to $\alpha_2$-adrenoreceptor might be one of the explanations of this circadian variation [10]. Brezinski et al. [9] found that the morning increase in platelet aggregability was correlated with increased levels of endogenous catecholamines released during the assumption of the upright posture. In this experiment, the circadian variation in platelet aggregability was abolished when the subjects remained recumbent. Another study [12] showed that this variation was also abolished in fasting subjects. Thus, exogenous factors might be a major determinant for circadian variations in platelet aggregability. In a recent study on healthy volunteers, Haus et al. [14] showed an op-

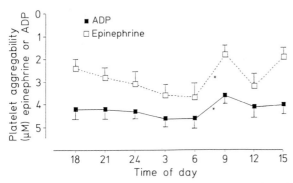

**Fig. 1.** Circadian changes in the minimum concentration of ADP and epinephrine (mean ± SEM) required to produce biphasic aggregation in ten healthy volunteers. The asterisks denote $p < 0.01$ for mean values at 0600 versus 0900 hours. The scale of the ordinate is inverted so that increasing aggregability is represented by an upward-sloping line. (From [8])

posite time course for platelet aggregability with a nocturnal peak and a trough in the afternoon. This discrepancy might be explained by the differences in the experimental design since, in this experiment, the subjects remained recumbent for at least 30 min before each sampling.

*Platelet Adhesiveness.* To our knowledge, only one study has been performed [14], in which platelet adhesiveness, as measured by the retention of platelets in a glass bead column, showed a morning peak and a nocturnal trough.

## Circadian Rhythm of Blood Coagulation

Seven studies in healthy volunteers [11, 12, 14–18] and two studies in patients with disease, one in diabetic patients [19] and one in arterial thrombosis [20], have been carried out to evaluate the circadian variation of blood coagulation processess.

*Plasma Fibrinogen Level.* Four studies [11, 14, 15, 18] showed a peak in plasma fibrinogen level in the morning and a trough at night. The peak-trough difference was large, reaching 20 % (Fig. 2). However, a lack of circadian variation was described in another study [16]. This difference may be in relation to a different technique of measurement.

*Factor VIII Activity.* A peak in the morning and troughs in the late evening [14] or at night [15, 19] have been described for factor VIII activity, with a peak-trough difference of about 30 %. A circadian

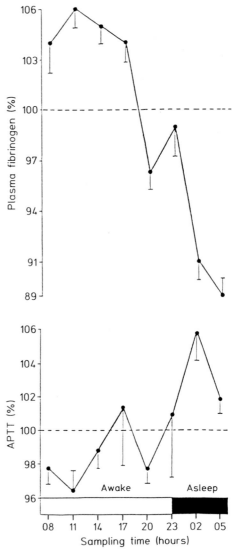

**Fig. 2.** Circadian variation of plasma fibrinogen levels and activated partial thromboplastin time *(APTT)* expressed in percentage of the individual 24-h means in eight healthy subjects. *Vertical bars* represent mean (SEM). Significant circadian (24 h) and ultradian (8 and 12 h) rhythms were validated by cosinor analysis for these two parameters ($p < 0.05$). (From [15])

variation was, however, not validated in the study of Fornasari et al. [16].

*Von Willebrand Factor Activity.* In diabetic patients [19] the peak in von Willebrand factor activity was found in the morning and low values at night, a finding which may be correlated with the nocturnal peak of growth hormone [21].

*Antithrombin III Activity.* Four studies [11, 14, 16, 20] using a functional method with a chromogenic substrate described a lack of circadian variation in anti-thrombin III activity whereas one study [17] showed a circadian variation with a small amplitude (less than 10%) and an acrophase in the afternoon.

*Activated Partial Thromboplastin Time.* Three experiments [14–16] reported a morning trough for APTT and a peak either in the late evening [14] or at night [15, 16]. In all cases the amplitude of this rhythm (Fig. 2) was small (about 10%). In contrast Petralito et al. [11] reported a morning peak and a nocturnal trough for APTT. The reason for this discrepancy is not clear.

*Thrombin Time.* Two studies [11, 14] reported a small-amplitude circadian rhythm for thrombin time (TT) with a peak at night and a trough in the morning [11] or in the afternoon [14].

*Prothrombin time.* Contradictory results were reported in two studies on prothrombin time (PT). Haus et al. [14] found nocturnal and morning troughs and a peak in the afternoon, whereas Weniger and Panzram [12] reported lower PT values at night. Fornasari et al. [16] was not able to demonstrate a circadian rhythm.

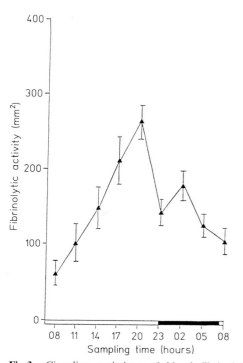

**Fig. 3.** Circadian variations of blood fibrinolytic activity measured by fibrin plate assay (mean ± SEM) in nine healthy nonfasting subjects. The peak-trough difference was highly significance (*p* < 0.001). (From [26])

*Other Parameters.* Haus et al. [14] did not find a circadian rhythm for factor V, plasminogen, $\alpha_2$-macroglobulin, and fibrinogen degradation products. To our knowledge no study is available on the possible circadian rhythm of other parameters such as factor VII coagulant, protein C or protein S.

The time course of several but not all coagulation parameters most often suggests a transient state of hypercoagulability during the morning (high values in plasma figrinogen levels and factor VIII and von Willebrand's factor activities, and low values in APTT and TT). However, the amplitude of these circadian rhythms is small, except for plasma fibrinogen levels (a factor particularly important in the thrombosis processes) and for factor VIII activity. To our knowledge no study has been carried out to evaluate the influence of plasma catecholamines and exogenous factors (e.g., fasting-nonfasting, rest-activity cycles) on these variations. Further investigations are needed to evaluate these influences and to document the potential circadian variations of some coagulation parameters particularly important in the thrombosis processes (factor VII coagulant, protein C and S in particular). More studies in patients with thrombosis are also needed.

## Circadian Rhythm of Blood Fibrinolysis

Circadian variations in blood fibrinolytic activity were described in healthy volunteers as far back as 1957 [22]. Many other experiments in healthy volunteers [23–28] as well as in patients with deep vein thrombosis [29] or patients with arterial disease [30] confirmed these preliminary results. In all these studies fibrinolytic activity, measured by global methods (fibrin plate assays and/or euglobulin clot lysis time), was lower in the early morning and higher in the late afternoon, with a peak-trough difference of about 300% (Fig. 3). The fibrinolytic response to different standardized exercises followed the same circadian pattern with a lower response in the morning than in the evening [26].

Recently the mechanism of the circadian variation in this activity has been analyzed by some of its components. The total fibrinolytic activity is the resultant of several factors and chain reactions. Fibrinolysis in blood is mediated by plasmin which is formed from plasminogen through the action of plasminogen activators. The two major plasminogen activators in plasma are tissue-type plasminogen activator (t-PA) and urokinase (UK) regulated by PAI-1 [31]. No circadian changes were found for plasminogen in two studies performed in healthy volunteers [14, 26]. Only minor circadian variations were observed in healthy volunteers for UK components [28, 32]. On the contrary, the circadian variations of t-PA activity were similar to the circadian rhythm of fibrinolysis assessed by global methods: Four studies carried out in healthy volunteers [28, 32–34] showed an early morning trough and a peak in the late afternoon with a peak-trough difference around 150 %. This circadian rhythm cannot be explained by the circadian variation of t-PA antigen which was found to decrease from the morning to the evening [28, 32–34]. The circadian variation of overall PAI-1 activity showed an opposite time course to that of t-PA activity with an early morning peak and a trough in the later afternoon (peak-trough difference around 200 %), whether in healthy volunteers [28, 32–35] or in patients with deep vein thrombosis [29] or acute coronary thrombosis [36, 37]. Finally the circadian variation of PAI-1 antigen concentration was found to be similar to the circadian variation of PAI-1 activity, a threefold increase in the early morning when compared with the evening value [28], and is thought to be the main explanation of the circadian variation in fibrinolysis.

Circadian variations in total fibrinolytic activity were found in fasting and nonfasting subjects whether ambulatory or at bedrest [22, 26]. In the same way, the circadian variation of PAI-1 activity was not influenced by the assumption of the upright posture [35]. Thus, endogenous rhythms could be a determinant factor for the circadian variations in fibrinolysis. Among these rhythms the circadian rhythm of catecholamines [8] does not seem to be important since the circadian variation in fibrinolytic activity is apparently not influenced by either $\alpha$- or $\beta$-adrenergic-receptor blockade [30].

Circadian Rhythm of Blood Viscosity

Two studies performed in healthy volunteers [38, 39] have demonstrated a circadian rhythm in hematocrit, protein concentration, blood and plasma viscosity with a morning peak and a nocturnal trough. This rhythm is consistent with the circadian rhythm of plasma fibrinogen levels. The influence of exogenous factors on this rhythm has not been evaluated.

To summarize, an increase in platelet aggregability, in platelet adhesiveness in the plasma levels of several coagulant factors and of blood viscosity, and a decrease in fibrinolytic activity, which could produce a transient state of hypercoagulability, are observed in the morning hours. This seems to be the resultant of both endogenous and exogenous factors. The magnitude of the circadian rhythms of two parameters (plasma fibrinogen and PAI-1) directly involved in the occurrence of coronary thrombosis must be emphasized [3, 4]. Such rhythms might result in circadian changes in the incidence of thromboembolic disorders. One study [36] likely confirms this hypothesis: in 63 patients with unstable angina pectoris, the circadian variation in PAI-1 activity was correlated directly with the onset of ischemic symptoms. In each patient in whom acute myocardial infarction developed, PAI-1 activity was signficantly increased immediately prior to the acute event. Other studies are needed to confirm these preliminary results.

Circannual Rhythm

Two studies have been performed in healthy volunteers [40, 41]. A positive correlation was found between increased antithrombin III concentrations and high temperatures [40]. This is countered, however, by an inverse relationship between fibrinolytic activity and temperature, i.e., increased fibrinolysis at low temperatures [40]. In the second study increases in platelet and red cell counts and in whole blood viscosity were found at low temperatures [41]. These data suggest an increased tendency to thrombosis in winter when temperatures are low.

# Chronopathology of Thromboembolic Disorders

## Circadian Rhythm of Myocardial Infarction

Fifteen epidemiologic studies have demonstrated a circadian rhythm in the onset of myocardial infarction [42–56]. They all showed a morning peak between 0600 and 1200 hours and most often a nocturnal trough. In two studies, circadian data were obtained from clinical trials: the Multicenter Investigation of Limitation of Infarct Size (MILIS) [49] and the Intravenous Streptokinase in Acute Myocardial Infarction (ISAM) study [56], Fig. 4. Both investigations estimated the time of the beginning of myocardial infarction by the onset of clinical symptoms and objectively by the evaluation of plasma creatine kinase curves. Myocardial infarction occurred three times more often in the morning than in the late eve-

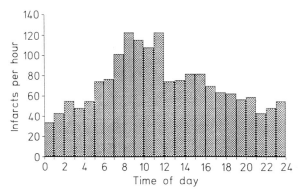

**Fig. 4.** Incidence of myocardial infarction in 1741 patients of the ISAM study. A marked circadian variation ($p < 0.001$) occurs with a peak during the morning hours and a trough during the night. (From [56])

ning. In these two studies, the group of patients receiving $\beta$-adrenergic blocking therapy before the event did not show an increased morning incidence of myocardial infarction.

### Circadian Rhythm of Sudden Cardiac Death

Evidence of a circadian variation in the incidence of myocardial infarction led to looking for a similar pattern in the incidence of sudden cardiac death. Such an objective raised several methodological problems, reported by Muller et al. [57]. These problems were solved in studies of three large data bases, mortality records for Massachusetts [57], the Framinghan Heart Studies [58] and the Beta-Blocker Heart Attack trial [59]. All showed a prominent circadian pattern similar to that observed in myocardial infarction, with a morning peak and a nocturnal trough. Again

no increased morning incidence of sudden cardiac death could be found in patients receiving beta-adrenergic-blocking therapy [59]. In addition, a recent epidemiologic study [60] reported a circadian rhythm in fatal pulmonary embolism with a morning peak which certainly may contribute to the circadian rhythm of sudden cardiac death.

### Circadian Rhythm of Thrombotic Stroke

The circadian variation in the onset of thrombotic strokes has been evaluated in seven studies [61–67]. Five of them [61–65] showed a morning peak of incidence which was similar to that observed in myocardial infarction and sudden cardiac death. In all of these studies, stroke was less likely to occur in the late evening (Fig. 5), and one of them [62] showed a dramatic morning peak in the onset of progression of stroke for 283 patients whose strokes worsened while in the hospital, suggesting a circadian variation in recurrence of thrombosis.

### Circadian Rhythm of Arterial Embolism

A study performed by Dewar and Weightman [68] showed a morning peak in the onset of arterial embolism in patients with mitral valve disease and atrial fibrillation.

These circadian variations in the hemostatic system and in cardiovascular and/or cerebrovascular events seem to correlate directly suggesting a cause-effect relationship. However, as pointed out by Muller et al. [69], many other factors, such as arterial blood pressure, coronary blood flow, plasma catecholamines,

**Fig. 5.** Frequency, in 2-h intervals, of onset of ischemic stroke for 1167 patients. Hypothesis for uniform distribution of time of onset was rejected ($p < 0.005$). (From [62])

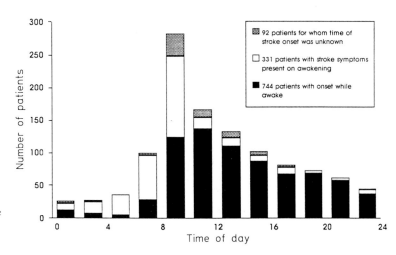

and plasma cortisol, may participate in the morning increase of the onset of these vascular events. It is difficult to determine the respective roles of endogenous rhythms and of the rest-activity cycle in the genesis of this chronopathology. However, the study of Goldberg et al. [70] suggests that the rest-activity cycle plays a preponderant role in myocardial infarction. Further epidemiologic studies are needed to evaluate the real contribution of awakening and rising time as triggering processes for acute vascular events.

## Circannual Rhythm

A peak in winter was found for the incidence of arterial embolism [71] and deep vein thrombosis [72]. It was also observed by Reinberg et al. [73] for the mortality from myocardial infarction and thrombotic stroke. A comparable seasonal variation has been found for the mortality from pulmonary embolism [74–76]. These seasonal variations may be partly induced by those factors which affect the hemostatic

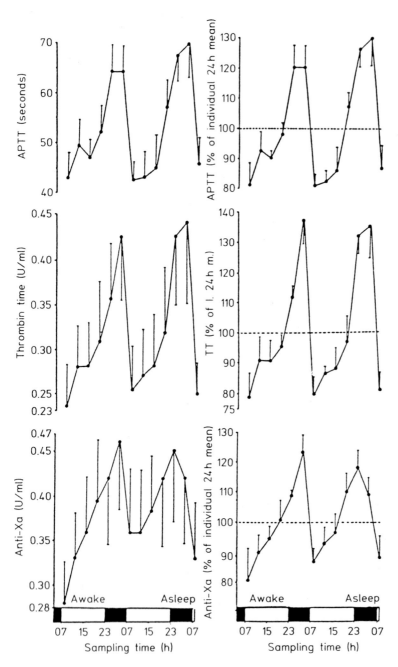

**Fig. 6.** Time course of activated partial thromboplastin time *(APTT)*, thrombin time, and factor Xa inhibition assay in six patients with deep venous thrombosis receiving continuous intravenous heparin treatment. *Vertical bars* represent mean (SEM). Significant circadian and ultradian rhythms were validated by cosinor analysis $(p < 0.05)$. (From [77])

system [40, 41], and environmental conditions, especially ambient temperature [40, 41, 72].

# Chronopharmacology
# of Antithrombotic Agents

## Heparin

In patients with deep vein thrombosis the anticoagulant effect of standard heparin was found less pronounced in the morning than during the night (difference of about 50%). This circadian variation, which occurred in spite of a continuous intravenous infusion ([77], Fig. 6), was manifest in all three coagulation tests used (APTT, TT and factor Xa inhibition assay) and reproducible. The same variation was observed when heparin was administered as a bolus subcutaneously in high doses to patients with a deep vein thrombosis or in arteritis of the lower extremities [78]. Schved et al. [79] confirmed these findings with standard heparin given to postoperative patients as subcutaneous bolus in small doses. In these three studies considerable interindividual differences in the amplitude of the circadian variation were observed. Toulon et al. [20] could not find a circadian variation in heparin effect in hospitalized patients (intensive care units) who were treated by standard heparin administered in continuous venous infusion. In this study the test used to evaluate the heparin effect was different from those used in the three previous studies (factor Xa inhibition assay measured by a chromogenic method). Fagrell et al. [80] did not find a circadian variation after a subcutaneous bolus injection or a continuous venous infusion of standard heparin to patients with deep vein thrombosis. The reasons for these divergent findings are probably multiple and presumably may be found both in the difference of the coagulation tests used and of the population studied. Also a seasonal variation and important intraindividual differences may play a role. There may be an influence of external factors such as associated medications, diet, rest-activity and sleep-wakefulness cycles, etc. For example, the rising of the patient in the morning seems to be essential for the circadian rhythm of platelet aggregation [9] and possibly also for that of myocardial infarction [70]. It is possible that this activity pattern represents a determinant element for the circadian rhythm of heparin. On the other hand, Schved et al. [79] showed that the peak of the nocturnal increase in heparin activity was

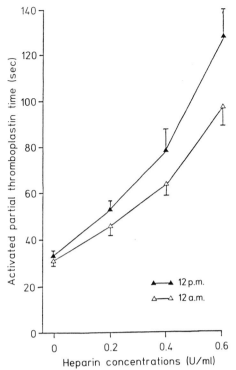

**Fig. 7.** Circadian changes in heparin sensitivity (tested in vitro) of human plasma sampled at 1200 and 2400 hours in ten subjects. The heparin sensitivity was lower in the diurnal samples than in the nocturnal ones ($p < 0.01$). (From [83])

correlated with the time of beginning of sleep. Studies are presently being carried on to elucidate these different aspects. Furthermore, several chronopharmacologic studies using the new low molecular weight heparins (LMWH) which will very probably replace the regular heparin in the future are in progress [81]. To document the mechanisms of the circadian variation of the anticoagulant effect of heparin, several investigations were carried out. A circadian variation of the pharmacokinetic of labeled heparin was found in healthy volunteers [82]. The in vitro sensitivity to standard heparin was examined in human plasma obtained at 1200 and at 2400 hours in subjects free of any antithrombotic treatment ([83], Fig. 7). No matter what dose of heparin was used, the anticoagulant effects obtained in vitro were less pronounced for plasma obtained during the day than during the night. These results suggest that the spontaneous circadian rhythm of blood coagulation determines the circadian rhythm in heparin effect which, under the circumstances, only amplifies a physiologic circadian variation.

## Thrombolytic Therapy

A study has been carried out to investigate the existence of a circadian variation in thrombolytic response to either recombinant t-PA (rtPA) or streptokinase in acute myocardial infarction. No circadian variation was found for streptokinase. But patients treated with rtPA [84] between 2400 and 1200 hours were four times less likely to achieve coronary reperfusion than patients treated during other time intervals. However this difference seems above all to be accounted for by those patients who received rtPA between 0600 and 1200 hours.

## Possible Therapeutic Implications

### Treatment of Arterial Thrombosis

Prophylactic

For the prevention of arterial thrombosis (myocardial infarction and thrombotic stroke) it appears logical to reinforce the activity of the antithrombotic treatment in the morning at the time of the maximal risk. The drugs of value for this indication are the platelet antiaggregation agents, mainly aspirin and ticlopidine. These drugs have a prolonged antiaggregation effect which in a single daily dose covers essentially the entire 24-h span. It therefore appears likely (although no study has been conducted on the subject) that the time of administration of these agents will not substantially influence their antithrombotic effect. However, the gastric tolerance of the nonsteroidal anti-inflammatory agents in general, and aspirin in particular, as shown by endoscopic studies [85] and by two controlled clinical studies [86, 87] is better in the evening. This better tolerance of the gastric mucosa to aspirin may be explained by a nocturnal peak in prostaglandins which exert a protecting effect against the physiologic nocturnal hyperacidity. The improved evening tolerance can also be explained on the basis of the circadian variations in the pharmacokinetics of the nonsteroidal anti-inflammatory agents. The peak is less pronounced in the serum after the evening administration than after the morning administration [88, 89]. Administration of aspirin in the evening therefore appears preferable in order to improve its long-term tolerance.

No data have been published in the literature which would allow a recommendation of a precise time of administration for ticlopidine.

Curative

The fibrinolytic agents have now become a proven treatment of myocardial infarctions during their acute stage. The time of injection of the fibrinolytic agents may influence their effect on the mortality (lesser effect during the morning?) and/or the risk of hemorrhages (more pronounced during the night?). The first answer to this question will be available when the European Myocardial Infarction Project (EMIP) is completed. This study will register in 10000 patients the exact hour of the injection of the fibrinolytic agent tested in order to determine whether the time of administration may modulate the incidence of mortality and hemorrhagic accidents.

### Treatment of Venous Thrombosis

*Prophylactic Treatment.* For the prevention of deep vein thrombosis the question arises of the time of injection of the LMWH which are of value when given once during 24 h [81]. Again the epidemiologic data appear to be controversial. The objective would be to obtain the peak in plasma heparin concentration in the morning when the highest risk of fatal embolism occurs. The peak of plasma heparin concentratin occurs on average 3 h after injection which implies an injection at 0400 hours. This awaits further confirmation.

*Curative Treatments.* It does not appear, at this time, reasonable to recommend a modification of the classic scheme of the diurnal dose of heparin as a function of APTT. In contrast, the subcutaneous administration (two s.c. injections/24 h) with injections at 0400 hours (to obtain a morning peak of blood heparin levels and to limit the risk of an underdosage and rethrombosis) and at 1600 hours (to avoid a nocturnal peak of the blood heparin levels and limit the risk of a possible nocturnal overdose) seems a better schedule. However, further studies are needed to confirm the clinical pertinence of such a schedule.

# Conclusion

The circadian variations of blood coagulation and of the fibrinolytic system suggest a transient state of hypercoagulability in the morning hours (high concentrations and/or activity of plasma fibrinogen, factor VIII and von Willebrand factor activities and low values in APTT and TT, high values in platelet aggregation and retention). A circadian variation in fibrinolytic activity (lower activity in the early morning and higher activity in the late afternoon) added to the early morning risk state concur for the occurrence of thromboembolic phenomena. This transient risk state involving both the blood coagulation and the fibrinolytic system may explain the increased incidence in the early morning hours of pulmonary embolism, myocardial infarction, sudden cardiac death, thrombotic stroke and some forms of arterial embolism.

Seasonal variations in the incidence of arterial embolism, deep vein thrombosis, myocardial infarction and thrombotic stroke, with a peak in winter, may also be related in part to variations in the hemostatic/fibrinolytic system.

The chronopharmacologic studies of antithrombotic agents most often show a circadian variation of heparin effect which appears to be less pronounced in the morning than during the night. An adaptation of the timing of heparin administration according to the expected peak in hypercoagulability in the morning appears legitimate, with subcutaneous injections given in hospitalized patients approximately at 0400 hours and again at 1600 hours in order to avoid a nocturnal elevation of the blood heparin levels with increased bleeding tendency. Other thrombolytic therapy may in the long term benefit from timed treatment. However, other studies are needed to assess this presumption.

# References

1. Davies MJ, Thomas AC (1984) Thrombosis and acute coronary-artery lesions in sudden cardiac ischemic death. N Engl J Med 310: 1137–1140
2. De Wood MA, Spores J, Notske R, Mouser LT, Burroughs R, Golden MS, Lang HT (1980) Prevalence of total coronary occlusion during the early hours of transmural myocardial infarction. N Engl J Med 303: 897–902
3. Meade TW, Brozovic M Chakrabarti RR, Haines AP, Imeson JD, Mellows S, Miller GJ, North WRS (1986) Haemostatic function and ischaemic heart disease: principal results of the Northwick Park Study. Lancet 2: 533–537
4. Hamsten A, Waldius G, Szamosi A (1987) Plasminogen activator inhibitor in plasma: risk factor for recurrent myocardial infarction. Lancet 2: 3–9
5. Cohen M, Simmons DJ, Joist JH (1978) Diurnal hemostatic changes in the rat. Thromb Res 12: 965–971
6. Scheving LE, Pauly JE (1967) Daily rhythmic variations in blood coagulation times in rats. Anat Rec 157: 657–666
7. Soulban G, Labrecque G (1989) Circadian rhythms of blood clotting time and coagulation factors II, VII, IX and X in rats. Life Sci 45: 2485–2489
8. Tofler GH, Brezinski D, Schafer AI, Czeiser CA, Rutherford JD, Willich SN, Glaeson RE, Williams GH, Muller JE (1987) Concurrent morning increase in platelet aggregability and the risk of myocardial infarction and sudden cardiac death. N Engl J Med 316: 1514–1518
9. Brezinski D, Tofler GH, Muller JE (1988) Morning increase in platelet aggregability association with assumption of upright posture. Circulation 78: 35–40
10. Mehta JL, Lawson D, Mehta P (1987) Circadian variation in platelet aggregation and alpha 2 adrenoceptor binding affinity. Circulation 76 (suppl IV): 364
11. Petralito A, Gibbino S, Maino MF (1982) Daily modification of plasma fibrinogen, platelet aggregation, Howell's time, PTT, TT and antithrombin III in normal subjects and in patients with Valsalvar disease. Chronobiologia 9: 195–201
12. Weniger J, Panzram G (1985) Untersuchungen über das zirkadiane Verhalten hämostaseologischer Parameter bei Diabetikern und Stoffwechselgesunden. Z Gesamte Inn Med 40: 489–492
13. Willich SN, Pohjolva-Sintonen S, Bhatia SJS, Shook TL, Tofler GH, Muller JE, Curtis DG, Williams GH, Stone PH (1989) Suppression of silent ischemia by metoprolol without alteration of the morning increase of platelet aggregability in patients with stable coronary artery disease. Circulation 79: 557–565
14. Haus E, Cusulos M, Sackett-Lundeen L, Swoyer J (1990) Circadian variations in platelet functions and coagulation parameters. Annu Rev Chronopharmacol 7: 153–156
15. Conchonnet P, Decousus H, Boissier C, Perpoint B, Reynaud J, Mismetti P, Tardy B, Queneau P (1990) Morning hypercoagulability in man. Annu Rev Chronopharmacol 7: 165–168
16. Fornasari P, Gratton L, Dolci D, Gamba G, Ascari E, Montalbetti N, Halberg F (1977) Circadian rhythms of clotting, fibrinolytic activators and inhibitors. In: Halberg F (ed) Proceedings of the XIII international conference of the International Society for Chronobiology. Il Ponte, Milan, pp 155–158
17. Casale G, Butte M, Pasotti C, Ravecca D, de Nicola P (1983) Antithrombin III and circadian rhythms in the aged and in myocardial infarction. Haematologica 68: 615–619
18. Hajjar GC, Whissen NC, Moser K (1961) Diurnal variations in plasma euglobulin activity and fibrinogen levels. Angiology 12: 160
19. Porta M, Maneschi F, White MC, Kohner E (1981) Twenty-four hour variations of von Willebrand factor and factor VIII. Releated antigen in diabetic retinopathy. Metabolism 30: 695–699
20. Toulon P, Vitoux JF, Leroy C, Lecomte TH, Roncato M, Motobashi Y, Aiach M, Fiessinger JN (1987) Circulating activities during constant infusion of heparin or a low molecu-

lar weight derivative (Enoxaparine): failure to demonstrate any circadian variations. Thromb Haemost 58: 1068–1072

21. Sarji KE, Levine JH, Nair RMG, Sagel J, Colwell JA (1977) Relation between growth hormone levels and von Willebrand factor activity. J Clin Endocrinol Metab 45: 853

22. Fearnley GR, Balmforth G, Fearnley E (1957) Evidence of a diurnal fibrinolytic rhythm with a simple method of measuring natural fibrinolysis. Clin Sci 16: 645

23. Buckell M, Elliott FA (1959) Diurnal fluctuation of plasma fibrinolytic activity in normal males. Lancet 1: 660

24. Kowarzyk H, Kaniak J, Kotschym M (1960) Diurnal fluctuations of plasma fibrinolytic activity. Lancet 1: 176

25. Cepelak V, Barcal R, Celepakova H, Mayer O (1978) Circadian rhythm of fibrinolysis. In: Davidson JF et al. (eds) Progress in chemical fibrinolysis and thrombolysis, vol 3. Raven, New York, pp 571–578

26. Rosing DR, Brakman P, Redwood DR (1970) Blood fibrinolytic activity in man: diurnal variation and the response to varying intensities of exercise. Circ Res 27: 171–184

27. Grimaudo V, Omri A, Kruithof EKO, Hauert J, Bachman F (1988) Fibrinolytic and anticoagulant activity after a single subcutaneous administration of a low dose of heparin or a low molecular weight heparin-dihydroergotamine combination. Thromb Haemost 59: 388–391

28. Grimaudo V, Hauert J, Bachmann F, Kruithof EKO (1988) Diurnal variation of the fibrinolytic system. Thromb Haemost 59: 495–499

29. Haglund O, Wibell L, Saldeen T (1985) Plasminogen activators and inhibitors in patients with deep venous thrombosis (DVT). Thromb Haemost 54: 271

30. Hafenberg J, Weber E, Spohr V, Morl H (1980) Is the diurnal increase in fibrinolytic activity influenced by alpha-adrenergic or beta-adrenergic blockade? Blut 41: 455–458

31. Sprengers ED, Kluft C (1987) Plasminogen activator inhibitors. Blood 69: 381–387

32. Andreotti F, Davies GJ, Hackett D, Khan MI, de Bart A, Dooijewaard G, Maseri A, Kluft C (1988) Circadian variation of fibrinolytic factors in normal human plasma. Fibrinolysis 2 (suppl): 90–92

33. Kluft C, Jie AFH, Rijken DC, Verheijen JH (1988) Daytime fluctuations in blood of tissue-type plasminogen activator (t-PA) and its fast-acting inhibitor (PAI-1). Thromb Haemost 59: 329–332

34. Köhler M, Miyashita C (1988) Probleme bei der Messung von Parametern des fibrinolytischen Systems: circadiane Rhythmik von Gewebe-Plasminogen-Aktivator und Plasminogen-Activator-Inhibitor. Klin Wochensch 66 (suppl XII): 62–67

35. Huber K, Beckmann R, Lang I, Schuster E, Binder BR (1989) Circadian fluctuations of plasma levels of tissue plasminogen activator antigen and plasminogen activator inhibitor activity. Fibrinolysis 3: 41–43

36. Huber K, Resch I, Rosc D (1987) Thrombotic complications in acute CAD can be correlated with elevated plasminogen activator inhibitor levels in plasma. Circulation 76 (suppl IV): IO/O

37. Angleton P, Chandler WL, Schmer G (1987) Diurnal variation in tissue plasminogen activator and its rapid inhibitor. Circulation 76 (suppl IV): 339

38. Seaman GVF, Engel R, Swank RL, Hissen W (1965) Circadian periodicity in some physicochemical parameters of circulating blood. Nature 4999: 833–835

39. Ehrly AM, Jung G (1973) Circadian rhythm of human blood viscosity. Biorheology 10: 557–583

40. Bull GM, Brozonic M, Chakrabarti R, Meade TW, Horton J, North WRS, Stirling Y (1979) Relationship of air temperature to various chemical haematological and haemostatic variables. J Clin Pathol 32: 16–20

41. Keatinge WR, Coleshaw SRK, Cotter F, Mottoie M, Murphy M, Chelliah R (1984) Increases in platelet and red cell counts, blood viscosity and arterial pressure during mild surface cooling: factors in mortality from coronary and cerebral thrombosis in winter. Br Med J 289: 1405–1408

42. Churina SK, Ganelina JE, Volpert EL (1975) On the distribution of the incidence of acute myocardial infarction within a 24-hour period. Kardiologiia 15: 115–119

43. Dimitrov L, Khadzhikhristev A (1983) Dynamics of the incidence of myocardial infarction in Smoljan District for the period 1965–1979. Vutr Boles 22: 40–46

44. Ganelina IE, Burisova IY (1983) Circadian rhythm of working capacity sympathicoadrenal activity, and myocardial infarction. Hum Physiol 9: 113–120

45. Gyarfas I, Csukas A, Horath U, Gaudi I (1976) Analysis of the diurnal periodicity of acute myocardial infarction attacks. Sante Publique (Bucur) 19: 77–84

46. Johansson BW (1972) Myocardial infarction in Malmo (1960–1968). Acta Med Scand 191: 505–515

47. Kaufmann MW, Gottlieb G, Kahaner K (1981) Circadian rhythm and myocardial infarct: a preliminary study. IRCS J Med Sci 9: 557

48. Master AM (1960) The role of effort and occupation (including physicians) in coronary occlusion. JAMA 174: 942–948

49. Muller JE, Stone PH, Turi SG, Czeisler C (1985) Circadian variation in the frequency of onset of acute myocardial infarction. N Engl J Med 313: 1315–1322

50. Myers A, Dewar HA (1975) Circumstances attending 100 sudden deaths from coronary artery disease with coroner's necropsies. Br Heart J 37: 1133–1143

51. WHO (1976) Myocardial infarction community registers. Results of a WHO international collaborative study coordinated by the regional office for Europe. In: Public Health in Europe, no 5, Copenhagen: Regional office for Europe (World Health Organization) 1976: 1–232

52. Pedoe HT, Clayton D, Morris JN, Bridgen W, Mac Donald L (1975) Coronary heart attacks in East London. Lancet 2: 833–838

53. Pell S, D'Alonzo CA (1963) Acute myocardial infarction in a large industrial population: report of a 6 years study of 1356 cases. JAMA 185: 831–838

54. Smolensky M (1983) Human chronopathology. In: Reinberg A, Smolensky M (eds) Biological rhythms and medicine: cellular, metabolic, physiopathologic and pharmacologic aspects. Springer, Berlin Heidelberg New York, p 209 (Topics in environmental physiology and medicine)

55. Thompson DR, Blandford RL, Sutton TW, Marchant PF (1985) Time of onset of chest pain in acute myocardial infarction. Int J Cardiol 7: 1439–1467

56. Willich SN, Linderer T, Wegscheider K, Leizorovicz A, Alamercery Y, Schröder R (1989) Increased morning incidence of myocardial infarction in the ISAM study: absence with prior beta-adrenergic blockade. Circulation 80: 853–858

57. Muller JE, Ludmer PL, Willich SN, Tofler GH, Aylmer G, Klangos I, Stone PH (1987) Circadian variation in the frequency of sudden cardiac death. Circulation 75: 131–138

58. Willich SN, Levy D, Rocco MB, Tofler GH, Stone PH, Muller JE (1987) Circadian variation in the incidence of sudden

cardiac death in the framingham heart study population. Am J Cardiol 60: 801–806

59. Peters RW, Muller JE, Goldstein S, Byington R, Friedman LM (1989) Propanolol and the morning increase in the frequency of sudden cardiac death (BHAT study). Am J Cardiol 63: 1518–1520

60. Colantinio D, Casale R, Abruzzo B, Lorenzetti G, Pasqualetti P (1989) Circadian distribution in fatal pulmonary thromboembolism. Am J Cardiol 64: 403–404

61. Agnoli A, Manfredi M, Mossuto L, Piccinelli A (1975) Rapport entre les rythmes héméronyctaux de la tension artérielle et sa pathogénie de l'insuffisance vasculaire cérébrale. Rev Neurol (Paris) 131: 597–606

62. Marler JR, Price TR, Clark GL, Muller JE, Robertson T, Mohr JP, Hier DB, Wolf PA, Caplan LR, Foulkes MA (1989) Morning increase in onset of ischemic stroke. Stroke 20: 473–476

63. Jovicic A (1983) Bioritam i shemieni cerebrovaskularni poremecaji, Vojnosanit Pregl 40: 347–351

64. Tsementzis SA, Gill JS, Hitchcock ER (1985) Diurnal variation of and activity during the onset of stroke. Neurosurgery 17: 901–904

65. Kaps M, Busse O, Hofmann O (1983) Zur circadianen Häufigkeitsverteilung ischämischer Insulte. Nervenarzt 54: 655–657

66. Marshall J (1977) Diurnal variation of occurrence of strokes. Stroke 8: 230–231

67. Hossmann V (1971) Circadian changes of blood pressure and stroke. In: Zulch KJ (ed) Cerebral circulation and stroke. Springer, Berlin Heidelberg New York, pp 203–208

68. Dewar HA, Weightman D (1983) A study of embolism in mitral valve disease and atrial fibrillation. Br Heart J 49: 133–140

69. Muller JE, Tofler GH, Stone PH (1989) Circadian variation and triggers of onset of acute cardiovascular disease. Circulation 79: 733–743

70. Goldberg R, Brady P, Chen Z, Gore J, Flessas AK, Greenberg J, Thedosiou G, Dalen J, Muller JE (1989) Time of onset of acute myocardial infarction after awakening. J Am Coll Cardiol 13: 133 A

71. Clark CV (1978) Seasonal variation in incidence of brachial and femoral emboli. Br Med J 287: 1109

72. Lawrence JC, Xabregas A, Gray L, Jam JL (1977) Seasonal variation in the incidence of deep vein thrombosis. Br J Surg 64: 777–780

73. Reinberg A, Gervais P, Halberg F, Gauthier M (1973) Mortalité des adultes: rythmes circadiens et circannuels dans un hôpital parisien et en France. Presse Méd 2: 289–294

74. Feinleib M (1972) Venous thrombosis in relation to cigarette smoking, physical activity and seasonal factors. Milbank Mem Fund Q 50 (suppl 2): 123–141

75. Colantonio D, Casale R, Natali G, Pasqualetti P (1990) Seasonal periodicity in fatal pulmonary thromboembolism. Lancet 1: 56

76. Wroblewski BM, Siney P, White R (1990) Seasonal variation in fatal pulmonary embolism after hip arthroplasty. Lancet 1: 56–57

77. Decousus H, Croze M, Levi F, Perpoint B, Jaubert J, Bonadona JF, Reinberg A, Queneau P (1985) Circadian changes in anticoagulant effect of heparin infused at a constant rate. Br Med J 290: 341–344

78. Decousus H, Scully MF, Reynaud J (1987) Circadian changes in anticoagulant effect of heparin given by subcutaneous bolus. Thromb Haemost 58: 1376

79. Schved JF, Gris JC, Eledjam JJ (1985) Circadian changes in anticoagulant effect of heparin infused at a constant rate. Br Med J 290: 1286

80. Fagrell B, Arver S, Intaglietta M, Tsai AG (1989) Changes of activated partial thromboplastin time during constant intravenous and fixed intermittent subcutaneous administration of heparin. J Int Med 225: 257–260

81. Kher A, Bara L, Samama M (1986) Les héparines de bas poids moléculaire. Pathol Biol 34: 61–69

82. Docousus M, Gremillet E, Decousus H, Champailler A, Housard D, Perpoint B, Jaubert J (1985) Nycthemeral variations of 99 Tc-labelled heparin-pharmacokinetic parameters. Nucl Med Commun 6: 633–640

83. Scully MF, Decousus H, Ellis C, Girard P, Parker C, Kakkar VV (1987) Measurement of heparin in plasma: influence of inter-subject and circadian variability in heparin sensitivity according to method. Thromb Res 46: 447–455

84. Becker RC, Carrao JM, Baker SP, Gore JM, Alpert JS (1988) Circadian variation in thrombolytic response to recombinant tissue-type plasminogen activator in acute myocardial infarction. J Appl Cardiol 3: 213–221

85. Moore JG, Goo H (1987) Day and night aspirin induced gastric mucosal damage and protection by ranitidine in man. Chronobiol Int 4: 111–116

86. Levi F, Le Louarn G, Reinberg A (1985) Timing optimizes sustained-release indomethacin treatment of osteoarthritis. Clin Pharmacol Ther 37: 77–84

87. Boissier C, Decousus H, Perpoint B, Laporte S, Mismetti P, Hocquart J, Gayet JL, Queneau P (1990) Timing optimizes sustained-release ketoprofen treatment of osteoarthritis. Annu Rev Chronopharmacol 7: 289–292

88. Clench J, Reinberg A, Dziewanowska Z, Ghata J, Smolensky MH (1981) Circadian changes in the bioavailability and effects of indomethacin in healthy subjects. Eur J Clin Pharmacol 20: 359–369

89. Ollagnier M, Decousus H, Cherrah Y, Levi F, Mechkouri M, Queneau P, Reinberg A (1987) Circadian changes in the pharmacokinetics of oral ketoprofen. Clin Pharmacokinet 12: 367–378

# Normal and Abnormal Cell Proliferation in Mice Especially as It Relates to Cancer

L. E. Scheving, T.-H. Tsai, L. A. Scheving, R. J. Feuers, and E. L. Kanabrocki

## History of Research on Cell Division Rhythms

A great deal of confusion has prevailed as to whether cell proliferation undergoes circadian change. Because of this confusion, a brief history at the beginning of this Chapter seems relevant.

### Plants

One of the first reports suggesting a circadian rhythm in cell division in plants was by Killicott (1904). He found that the maximum mitotic index in the *Allium* root tip occurred at 2300 hours and the minimum at 0700 hours. Karsten (1918) reported a similar phenomenon for the mitotic index of *Spirogyra*, with highest activity during midday, and for *Zea mays* where the highest mitotic index occurred during the night. Stalfelt (1921) reported that the highest mitotic index in *Pisum sativum* occurred at midday and the minimum during the morning. With these and other early studies, botanists pioneered in exploring rhythms in cell division as well as in other diverse rhythmic variables (Bünning 1958).

### Animals

Fortuyn-van-Leyden (1917, 1926) was the first to report on mitotic index rhythms in several tissues of the cat. Most of these studies involved once-a-day and once-a-night sampling, and small numbers of animals were used; consequently, her findings in some cases were contradictory from one study to another. However, in spite of confusing data, she proved to be correct in suggesting that cell division in the cat was periodic and in this respect it was not unlike cell division in plants.

In the 1930s and 1940s there was a renewed interest in this phenomenon, and several papers appeared. Carleton (1934), Ortiz-Picón (1934), and Cooper and Franklin (1940) all reported a mitotic rhythm in the epidermis of the mouse. These studies were very limited as to sampling intervals and number of animals used; generally mitotic activity was found to be highest about midday and lowest at night. The work of Ortiz-Picón (1934) was, up to that time, the most extensive as he sampled at 1200, 1900, 2000, and 2400 hours and used six mice per time point. In reality, the above represented only three circadian stages since the 1900- and 2000-hour time points were so close together.

More extensive, and thus more reliable, investigations were made on rhythms of cell division in the rat by Blumenfeld (1939, 1943). He found that the maximum mitotic index in the abdominal epidermis occurred between 0800 and 1000 hours and the minimum occurred between 2000 and 2400 hours. In these same rats, he determined the mitotic index in the submandibular gland and kidney. Although the times of peak mitotic index were different for these three diverse tissues, it was of interest to find that the time of the lowest mitotic index was the same for all. Blumenfeld discussed at length the differences in the peak timing of the rhythms, but did not discuss similarity of phasing of the lowest values. We frequently have found that the timing of the lowest data points from one study to another of many rhythmic variables show far less variation than that of the peak of the rhythm. We believe that this is an important observation, but one that rarely has been commented on in the literature over the years. The emphasis is usually on the peak or "acrophase" of the rhythm.

With sampling at 1-h intervals, it was found that the peak mitotic index in the epidermis of the pinna of the rat occurred somewhat earlier than that reported by Blumenfeld (1939) (Scheving and Pauly 1960). We later confirmed and extended our own findings (Scheving and Pauly 1967) (Fig. 1). It should be noted that nothing was reported about the light-

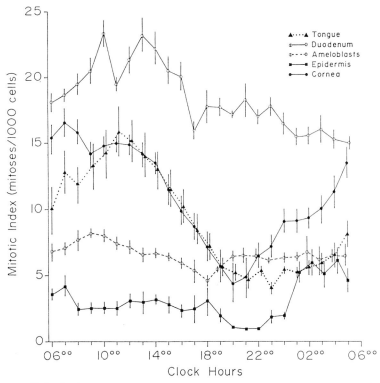

**Fig. 1.** The mitotic indices of five different tissues, all obtained from the same animals (rats; $n = 5$) on the same date (7 August 1965). The trough levels in all tissues occur between 1800 and 2200 hours. Lights were on from 0600 to 1800 hours and off from 1800 to 0600 hours; the animals were fed ad libitum. It should be noted that the evaluation of the mitotic indices was made by five different technicians. The ameloblasts had the lowest amplitude rhythm. Rhythmicity was statistically documented for all tissues by cosinor and variance analysis. (Scheving et al. 1972).

| p | Mesor ± SE | Amplitude ± SE | Acrophase ∅ ± CI |
|---|---|---|---|
| < 0.001 | 9.1 ± 0.2 | 5.2 ± 0.2 | − 156° ( − 161° to − 171°) |
| < 0.001 | 18.5 ± 0.2 | 2.8 ± 0.2 | − 196° ( − 187° to − 206°) |
| < 0.01 | 6.5 ± 0.1 | 0.8 ± 0.2 | − 125° ( − 103° to − 147°) |
| < 0.01 | 3.1 ± 0.2 | 1.2 ± 0.3 | − 82° ( − 50° to − 115°) |
| < 0.01 | 11.0 ± 0.4 | 5.2 ± 0.6 | − 134° ( − 128° to − 140°) |

dark cycle to which Blumenfeld's animals were subjected; we can only assume that it was the natural light-dark cycle. Tvermyr (1969), in an extensive study on hairless mice, also reported a rhythm in the mitotic index of the epidermis with a phasing similar to what we have reported. In addition to the epidermis, we carried out extensive studies on the mitotic indices of the cornea, tongue, ameloblasts, and duodenum in rats (Fig. 1), all of which demonstrated overt circadian rhythms (Scheving and Pauly 1967; Gasser et al. 1972a, b; Scheving and Pauly 1977). A review article summarized time series studies that had been published up to that time (1973) on rhythmic variation in the mitotic indices of 15 different tissues of the rodent; the interested reader is referred to this article (Scheving and Pauly 1973). As documented, many of these earlier studies were based on data obtained from two or only a few sampling times along a 24-h time scale.

## Man

Cooper and Schiff (1938) analyzed mitoses in the prepuce of 8-day-old infants and concluded that the highest mitotic index occurred at 0900 hours and the lowest at 2200 hours. Only 13 specimens were used to evaluate the entire 24-h span, and no samplings were performed between 1245 and 0730 hours. A similar study followed in 1939 when Broders and Dublin examined 14 infant foreskins and concluded that the highest mitotic activity occurred at night. No samples were obtained between 1200 and 0730 hours. These two studies are still cited as evidence for rhythmicity in the human epidermis with no comment on the scanty sampling (Broders and Dublin 1939).

It was against the above background that our interest in research on rhythmicity of cell division in humans arose. Anatomists and pathologists had for a long time been bewildered by the sparse number of mitotic figures that frequently were found in histological preparations. This was especially puzzling in the case of the epidermis, which was known to rapidly shed and renew. The question often asked was whether the few mitotic figures frequently encountered in biopsy specimens (which were, more often that not, obtained in the morning), could account for the renewal of all the skin that was known to be lost during the course of a single day. Some said "no", and several investigators, including Frieboes (1920), Bostroem (1928), Cameron (1936), Levander (1950), and others, even postulated that epidermal cells might have a mesodermal origin. It was in 1949 that a report by Andrew and Andrew appeared introducing the hypothesis that in the human epidermis migrating lymphocytes gradually became transformed into "clear cells" which later differentiated

into typical epidermal cells. Andrew [a histologist] also believed that this was the reason why normally one did not find mitotic figures in skin specimens found in histology or pathology slide sets used for teaching students (Andrew and Andrew 1949).

This explanation was difficult to accept and it seemed more logical in light of the reports on rhythmicity in plants and animals that the paucity of mitotic figures might well be explained by the fact that mitosis occurs in daily cycles. Therefore, a study was designed to further test this hypothesis and evaluate for evidence of lymphocyte transformation into epithelial cells.

In retrospect, it was unfortunate that initially Scheving selected the newborn infant prepuce to investigate for circadian rhythmicity. This selection was prompted principally by the availability of this tissue

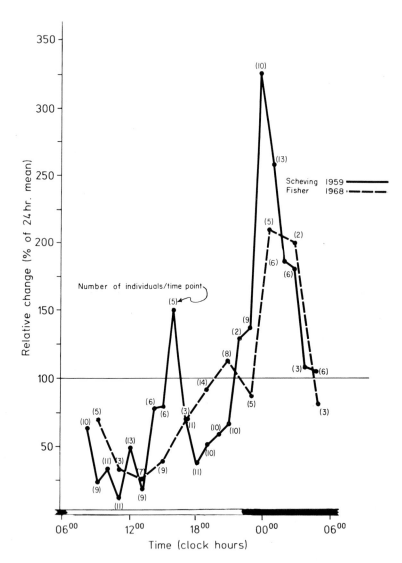

**Fig. 2.** The rhythm of mitotic index in the adult human epidermis. The data show that a majority of the cells divide during a predictable stage of the circadian system, between midnight and 0400 hours. Remarkable reproducibility has been demonstrated in studies done miles (London-Chicago) and many years apart. Data from *Fisher* is given as the mean number of mitoses per 1000 cells, data from *Scheving*, as mean mitoses per 1000 cells (2500–3000 cells per specimen). (Scheving 1959; Fisher 1968)

and also because the authors of the two earlier studies on infant prepuce had based their conclusion on far too few specimens. We began collecting foreskins from newborn infants at different hours of the day and night from local Chicago area hospitals. Thus, with the help of a colleague, over 150 foreskins were ultimately collected, which represented circumcisions performed within the course of about 1 and ½ years, with specimens spanning the entire 24-h time scale.

The end result of this Herculean task was that no evidence of a circadian rhythm of the mitotic index was found; at least, it was not evident from the plotting of the data as a chronogram or after analyzing using conventional statistics. Moreover, no compelling evidence of lymphocyte transformation into epidermal cells was found. What appeared in the data was the suggestion of an ultradian rhythm as there were multiple peaks and troughs in the mitotic index along the 24-h time span. In short, the claim of a circadian rhythm in the infant prepuce by Schiff and Cooper (1938) and Broders and Dublin (1939) was not confirmed by us. This was rather discouraging; what would seem to be "negative" data could not be easily published since they tended to refute the two earlier published works, albeit that in both of the earlier studies the data were very scanty. Moreover, in those days the concept of rhythmicity was not readily accepted by many editors. As a consequence, our data were put in a desk drawer where they would remain for a couple of years. Later, Scheving moved to another city and the foreskin slides and data sheets were stored in a box in the crawl space beneath the newly built house. Shortly thereafter, this space was inundated with several feet of flood water, and the slides, labels, and data sheets became wet and impossible to read and everything had to be thrown out. It is hoped that in the future someone will repeat this study on newborn infants because it is now known that the circadian rhythm of the mitotic index in the epithelium of the rodent does not manifest itself until about the 15th day of postnatal life (Goodrun et al. 1974).

During the early 1950s the lymphocyte hypothesis of replacement of skin by Andrew became more and more discussed and popular; because of this, it was decided to do a preliminary study on Scheving's skin. Limited skin samples were removed by punch biopsy from his shoulders. Samples of skin were obtained during the morning and after midnight on several occasions. Clearly, there was a remarkable difference in the number of mitotic figures encountered between the different time points, with the highest mitotic activity occurring shortly after midnight.

A more extensive investigation prompted by the above preliminary studies found that the mitotic index rhythm in the epidermis of the adult male clearly was circadian-stage-dependent with a great bulk of the cell division taking place after midnight (Scheving and Gatz 1955; Scheving 1959). These data are illustrated in Fig. 2, which also illustrates and compares data obtained almost 10 years later by Fisher (1968). It is of interest that this finding was again confirmed 20 years later by Zugula-Mally et al. (1979). As of 1991 the controversy that generated this study seems to have been forgotten by most.

In addition to the skin studies and the extensive documentation of the remarkable circadian variation found in human circulating blood cells (covered in the chapter by Haus) there are, to the best of our knowledge, two other reports in the literature show-

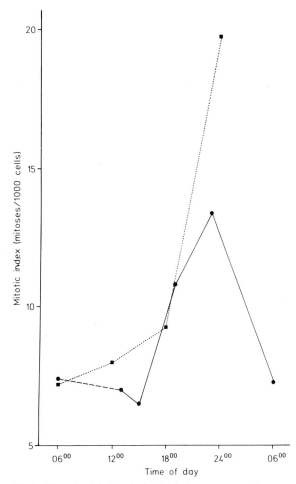

**Fig. 3.** The mitotic index rhythm in the bone marrow of a group of young men *(solid line)* and the mitotic rhythm in bone marrow obtained from a single individual *(dotted line)* along the 24-h domain. (For details see References)

ing a mitotic index rhythm in the bone marrow of man (Killman et al. 1962; Mauer 1965). The data from these two studies are illustrated in Fig. 3. Although data of this kind are scanty, it must be kept in mind that the taking of serial bone marrow biopsies were difficult to ethically justify or obtain from healthy subjects. It should be mentioned that very recent studies in humans by another group has shown that cell proliferation in the rectum of man also is highly circadian-stage-dependent, this data only recently has been submitted for publication. We have seen the data and are convinced of its validity, but do not feel free to further comment on it until we know that it is accepted for publication.

It is emphasized that man is a very rhythmic creature. Not only are there rhythms in cell division of the above tissues and circulating levels of blood cells. Man also shows a remarkably rhythmic time structure for vital signs and performance in a host of metabolic variables found in blood, urine, and saliva; evidence of this is amply documented throughout this book.

## More Recent Studies on Cell Proliferation in Animals

Up until the 1960s most studies on the rhythmic nature of cell proliferation were based on mitotic counts; these were very time-consuming and laborious. For example, to prepare one tissue properly and to evaluate the mitotic activity with a adequate number of specimens at frequent enough intervals of sampling along a single 24-h time scale could require the labor of one technician for the better part of a year. Naturally, there were attempts to make this task easier; one involved arresting mitotic figures with colchicine over a fixed span of time. However, such techniques were sometimes fraught with pitfalls, which have been discussed earlier in detail (Scheving 1980; Scheving and Pauly 1973, 1977; Scheving et al. 1974b, c, 1983a, b). In our opinion, the practice of sampling at too-infrequent intervals, to reduce the volume of work involved, also was a major source of the ambiguity and confusion that arose during that time as to the nature of the rhythmicity in many tissues (Scheving 1980a).

With the advent of radioisotopes, such as tritiated thymidine ([$^3$H]TdR), it became possible to study cell proliferation by measuring the uptake of [$^3$H]TdR into DNA, which supposedly reflected the synthesis (S)-phase of the cell cycle. This could be done either by preparing the tissue for autoradiography and then counting the labeled cells, number of grains per nucleus or by scintillation counting, which was less time-consuming (Fig. 4) (Scheving and Pauly 1967). Only more recently has flow cytometry (FCM) made the task of evaluating cell proliferation easier. However, at present this technique works only for determining rhythmicity accurately for a limited number of tissues because it requires a homogenous single-cell suspension; this frequently is difficult or even impossible to obtain for many heterogenous tissues and is particularly difficult when solid tumors are involved. In this section we shall discuss briefly studies on cell proliferation using all the different techniques.

The mitotic activity in the rodent corneal epithelium has been the most extensively investigated tissue over the past 25 years simply because it can easily be prepared as a whole mount and thus reliable counts of mitotic figures are made more easily than from tissues that first have to be sectioned and stained. The rhythm of the mitotic index in the cornea exhibits a large amplitude change over the 24-h span, as is illustrated in Fig. 5. Often the peak in the mitotic index is more than 20 times the trough value. The trough value frequently is zero or almost zero; thus we found what seems to be a "shutting off" of cell proliferation at one circadian stage. From the literature relative to rhythmic cell proliferation in unicellular organisms and insects, such a phenomenon is common and has been called "gating". Edmunds (1971) presents much evidence for such a gate in the mitosis of *Euglena*. Are we observing a similar phenomenon for cell proliferation in the cornea of rodents to that in *Euglena* or insects? Rubin (1981) suggested, from a series of radiation studies using the corneal epithelium as a target organ, that gating could help explain her interesting results pertaining to recovery of the mitotic index in the cornea subsequent to radiation; the interested reader is referred to her papers. The wide fluctuation around the 24-h mean illustrates the hazard of sampling this tissue at only one or two random times during the 24-h cycle. The maximum and minimum values clearly do not occur at random, but are very predictable from one experiment to another as long as the animals used have been kept under the same standardized conditions. It should be emphasized that the once very popular colchicine technique for evaluating mitotic figures did not always work properly as no rhythm was found in this tissue of the rat in some earlier studies by others. Also, to the best of our knowledge no one has been able to show this same rhythm in the cornea of rodents using flow cytometry, yet we know that a number of investigators

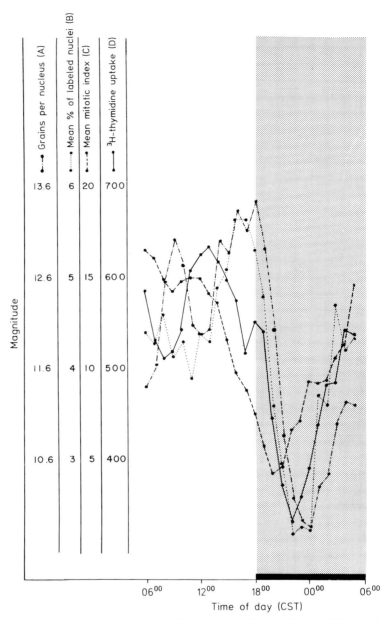

**Fig. 4.** Significant differences in the ³H-thymidine uptake *(D)*, mean percentage of labeled cells *(B)*, numbers of grains per labeled nucleus *(A)*, and mitotic rate in rat corneal epithelium (mean mitotic index, *C*). These were demonstrated by injecting subgroups of five animals every hour during a 24-h period with ³H-thymidine, killing them 2 h later, and analyzing the corneal epithelium by scintillation counting and radioautographic techniques. The individual plotted values, calculated as 3-h moving averages, are the means. The data are plotted as kill-time. In all cases, *n* = 5. The rhythmometric summary of the data is shown. The standard errors were left off to avoid a too-cluttered figure.

For a more detailed account of the conditions of this experiment, see Scheving and Pauly (1967)

|  | P | VR | Mesor ± SE | Amplitude ± SE | Acrophase ± SE |
|---|---|---|---|---|---|
| A = | 0.01 | 0.49 | 11.8 ± 0.2 | 1.3 ± 0.3 | − 201 ± 12 |
| B = | 0.11 | 0.19 | 4.2 ± 0.2 | 0.7 ± 0.3 | − 194 ± 25 |
| C = | 0.01 | 0.80 | 11.0 ± 0.4 | 5.2 ± 0.6 | − 134 ± 6 |
| D = | 0.01 | 0.49 | 518.0 ± 16 | 102.0 ± 22.7 | − 176 ± 12 |

Phase reference = local midnight

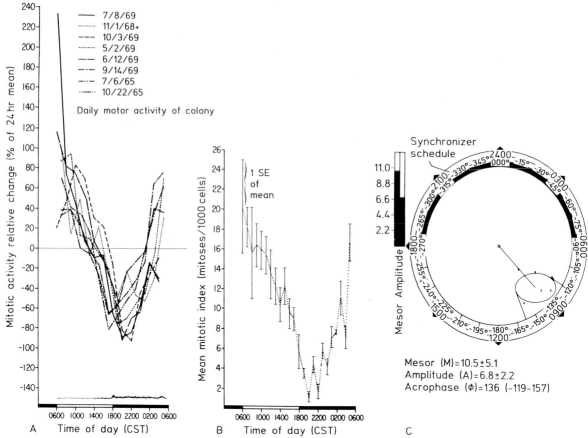

**Fig. 5. A** Conventional chronograms showing the circadian variation in the mitotic index of the epithelium of the rat cornea on eight different dates. Also shown is the daily motor activity of the colony based on noises emanating from it.

**B** Chronogram representing the summary of all eight studies expressed as absolute values ± SE. The dates of study and the number of animals used are given in the first two columns of the rhythmometric summary displayed *above* the chronograms.

The data are expressed as percent-change from the 24-h mean, to facilitate comparison.

**C** The cosinor plot is a display of the data obtained by a fit of the eight sets of time series data (shown in the tabular rhythmometric summary) to a 24-h cosine curve (courtesy of Chronobiology Laboratories, University of Minnesota). (Scheving et al. 1983)

| Condition of experiment | N of rats | P | SE/C | Mesor ± SE | Amplitude ± SE | Acrophase ∅ ± SE |
|---|---|---|---|---|---|---|
| $^L06\text{-}18^D18\text{-}06$ | | | | | | |
| 07-06-65 | 10 | 0.001 | 0.09 | 14.9 ± 0.6 | 9.81 ± 0.09 | −138 (−128 to −149) |
| 09-14-65 | 14 | 0.001 | 0.10 | 12.4 ± 0.6 | 7.72 ± 0.10 | −119 (−108 to −131) |
| 10-22-65 | 8 | 0.001 | 0.15 | 11.6 ± 0.6 | 5.64 ± 0.15 | −179 (−161 to −197) |
| 11-01-68 | 7 | 0.010 | 0.25 | 7.5 ± 0.7 | 3.79 ± 0.25 | −140 (−111 to −168) |
| 05-02-69 | 8 | 0.001 | 0.12 | 7.4 ± 0.5 | 5.58 ± 0.12 | −125 (−110 to −140) |
| 06-12-69 | 8 | 0.001 | 0.13 | 9.1 ± 0.8 | 7.61 ± 0.13 | −135 (−119 to −151) |
| 07-08-69 | 8 | 0.020 | 0.27 | 8.5 ± 1.6 | 7.96 ± 0.27 | −120 (−89 to −151) |
| 10-03-69 | 8 | 0.001 | 0.09 | 10.4 ± 0.5 | 7.57 ± 0.09 | −155 (−146 to −166) |

have tried. It is interesting to note that in mice bearing L1210 leukemia we found that the rhythm in the corneal epithelium persisted until the day of death from the disease (Scheving et al. 1983a). Moreover (Fig. 6), it persists in rats who had been hypophysectomized and adrenalectomized (Scheving and Pauly 1973).

We conclude that two types of cell proliferation are encountered among different tissues of rodents. One is a high-amplitude rhythm where very low levels of activity, or even none, are encountered at certain circadian stages. The other type of rhythm, as exemplified by the uptake of [³H]TdR into DNA in the duodenum,

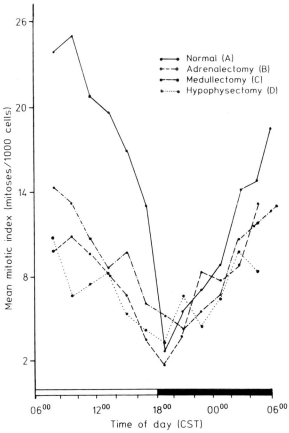

**Fig. 6.** Comparison of mitotic index rhythm in corneal epithelium of normal rats *(A)* with that of rats sampled 3 weeks after adrenalectomy *(B)*, adrenomedulectomy *(C)*, or hypophysectomy *(D)*. Adrenalectomized rats died when deprived of physiological saline, and this was taken as evidence of complete adrenalectomy. For a more detailed description, see Scheving and Pauly (1967)

|   | P | SE/C | Mesor ± SE | Amplitude ± SE | Acrophase ⊘ ± SE |
|---|---|------|------------|----------------|-------------------|
| A = | 0.001 | 0.12 | 7.4 ± 0.51 | 5.5 ± 0.15 | − 125 ( − 110  − 140) |
| B = | 0.001 | 0.15 | 7.7 ± 0.46 | 4.1 ± 0.65 | − 108 ( −  90 − 125) |
| C = | 0.001 | 0.11 | 9.0 ± 0.36 | 4.4 ± 0.50 | − 128 ( − 115 − 141) |
| D = | 0.010 | 0.25 | 6.8 ± 0.44 | 2.4 ± 0.62 | − 108 ( −  79 − 136) |

Phase reference = local midnight

is characterized by relatively high levels of activity over the entire 24-h time scale, but a low-amplitude rhythm is still present. Using data from the rat duodenum and mouse ovary as examples of low-amplitude rhythms, the two types of rhythms are illustrated in Fig. 7. The tongue and esophagus represent examples of high-amplitude rhythms. In our opinion the majority of the tissues in the body exhibit a high-amplitude rhythm, but most do not show the almost complete absence of cell proliferation at certain circadian stages as seen in the corneal epithelium (Scheving 1981).

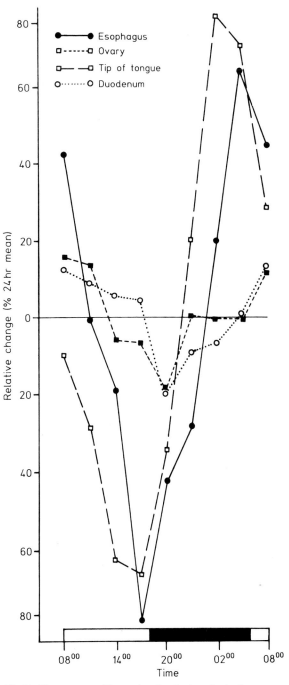

**Fig. 7.** Chronograms illustrating the rhythms in the incorporation of [³H] TdR into DNA of the esophagus, ovary, tip of the tongue, and duodenum of 7 CDF, female mice. To facilitate comparison, all data have been converted into "percentage of 24-h mean." (Scheving 1981)

## Intestinal Tract, Bone Marrow, Spleen, and Thymus

Since rhythmic behavior in cell proliferation of the intestinal tract, bone marrow, spleen, and thymus are more relevant to this section of the volume than rhythms in some other tissues, we present some examples of each. The greatest emphasis in this section will be on the intestinal tract, since damage to this organ system is among the hardest to manage and control in cancer chemotherapy or radiotherapy. Moreover, the studies on cell proliferation in the in-

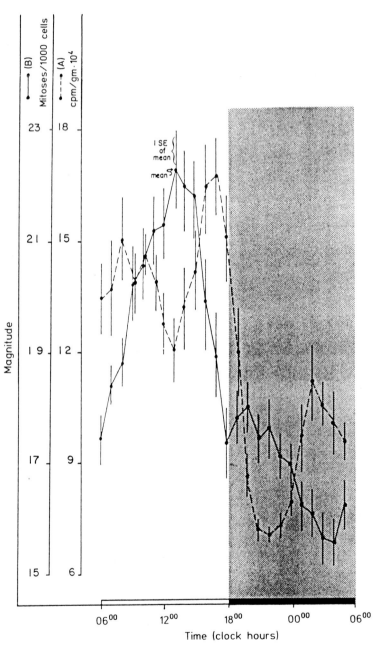

**Fig. 8.** Chronograms illustrating the mitotic index and uptake of [$^3$H]TdR into DNA of the duodenal epithelium of the rat. Each time point is the mean and standard error of seven animals $n = 7$ per time point. The animals had been standardized for at least 2 weeks prior to the experiment. (Scheving et al. 1972)

|     | P     | SE/C | Mesor ± SE   | Amplitude ± SE | Acrophase ∅ ± SE     |
|-----|-------|------|--------------|----------------|----------------------|
| A = | 0.001 | 0.18 | 11.9 ± 0.4   | 3.2 ± 0.5      | −181 (−161, −202)    |
| B = | 0.001 | 0.08 | 18.5 ± 0.17  | 2.8 ± 0.2      | −196 (−187, −206)    |

testinal tract have had a long history and have generated more confusion in the past than any other tissue. It is important to understand why this came about since many researchers still (1991) ignore its rhythmicity. As early as 1947, Klein and Geisel reported a rhythmicity for the mitotic rate in the duodenal epithelium of rodents; Bullough (1947) and Raitsina (1961) reported the same phenomenon, and Gololobova (1958) described the same thing in an undefined region of the gut. Bullough (1947) and Liosner et al. (1962) claimed periodicity in cell proliferation in the esophagus of the rodent and it was also documented for the gastric epithelium of the rat by Sinha (1960); Clark and Baker (1963) demonstrated that this rhythm persisted in hypophysectomized rats. Sigdestad et al. (1969) and Sigdestad and Lesher (1970) reported a circadian rhythm in mitosis and DNA synthesis in the jejunum of the mouse. In spite of such evidence, reports persisted claiming that there was no rhythm in cell proliferation in the gut epithelium. Leblond and Stevens (1948), Stevens and Leblond (1953), Bertalanfly (1960), Hunt (1952), Muhlemann et al. (1956), and Pilgrim et al. (1963) all reported that they found no evidence of rhythmicity for the specific region of the gut each group investigated. Unfortunately, for a long time the latter negative reports had a great deal of influence on many scientists, because they were the most frequently cited, often without reference to any of the positive findings mentioned above. This controversy has been discussed in detail previously and will not be further discussed here because of a lack of space (Scheving et al. 1972, 1978).

Our interest in rhythms in the digestive tract dates back to 1961 (Chiakulas and Scheving 1966). By sampling at 2-h intervals along a 24-h span we reported a high-amplitude rhythm in the mitotic index of the gastric mucosa epithelium of the salamander *Amblystoma punctuatum*. Later, the same phenomenon was more intensely studied by us in the digestive tract of the rodent. Our interest then was prompted by the erroneous widespread assumption, despite considerable evidence to the contrary, that cell proliferation in the gut was random; or if there was any circadian rhythmicity, it merely represented a minor fluctuation around a 24-h mean and was of little consequence. It was really investigations into cell proliferation in the small intestine that generated much of the confusion as to whether cell proliferation was rhythmic or random. As briefly mentioned above, in 1967 Scheving and Pauly reported that there was a circadian rhythm in both the mitotic index and synthesis phase of the cell cycle in the rat duodenum (Fig. 8). This investigation involved sampling at 2-h intervals, and therefore

it represented a far more extensive study than any reported previously. Thus it extended and confirmed work of others who noted at least some evidence of circadian variation, especially that involving the jejunum (Sigdestad et al. 1969; Sigdestad and Lesher 1970), where the animals used were standardized in a manner similar to our own. The duodenum, which up to that time was the most extensively studied region of the gut from the viewpoint of cell kinetics and regeneration times, showed a rhythm in cell proliferation. In several subsequent investigations, we confirmed these findings and Fig. 9 represents such data (Scheving et al. 1974a, 1978). Although such investigations were not carried out simply to test for reproducibility, they do point out retrospectively that, despite its comparatively low amplitude, a rhythm in cell proliferation exists in the duodenum. The percentage change form the lowest to highest group mean value in the duodenum was 30%–60%, depending on the particular study. The data are best illustrated as a percentage change of the 24-h mean, because the overall mean levels are so different. These differences in the overall mean levels may result from many things, including different specific activities of the radioisotopes injected and different ages of the animals. What is relevant to this discussion and what logically can be compared is the phasing of the rhythms.

Earlier it was demonstrated that the rhythm was important, because a reported mitotic stimulant, isoproterenol (IPR), when given as a single intraperitoneal injection, would bring about an increase in DNA synthesis in the duodenum at one circadian stage, a decrease at another stage, while at other stages there was no difference between the responses to IPR and to saline (Scheving et al. 1972). Moreover, IPR caused a phase shift in the duodenal rhythm (Burns et al. 1972). About this same time, there was great interest in IPR because of its reported stimulatory effect on cell proliferation in the parotid gland (Serota and Baserga 1970). Obviously, if such diverse effects could be produced subsequent to perturbation of a rhythm by a single dose of IPR, a naturally occurring oscillation should be an important variable to consider, irrespective of whether its amplitude was small or large. We emphasize this point since some investigators still select the small intestine for study because it has a small-amplitude rhythm which they believe is of little consequence!

The practice heretofore usually had been to study only one region of the gut because of the tremendous amount of work and time involved in evaluating cell proliferation; frequently, it was the duodenum that was studied, primarily because of its comparatively

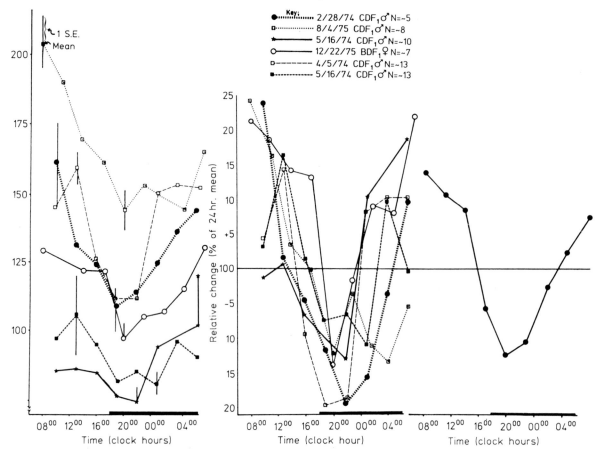

**Fig. 9.** A composite of chronograms for DNA synthesis in the duodenum of mice based on [³H]TdR incorporation into DNA. Tissues were obtained from control animals (*n* = 248) used in a variety of experiments in our laboratory on different dates. Although the study was not designed to test for this, there clearly was a high degree of reproducibility in the phasing of the different rhythms. The mesor varied greatly as did the amplitude (A), and these can be explained. The point to emphasize is that al-though the A may be small, there is a clear-cut rhythm in DNA synthesis in the duodenum. For the sake of clarity, standard errors are shown only for the high and low points for each chronogram. In general, from one study to another, the trough is the most fixed phase of a rhythm; this is clearly evident in the series of data and the summary of the six studies *(right).* (Scheving et al. 1978)

high rate of cell proliferation as well as the reason cited above. Later, cell proliferation was examined in several regions in the digestive tract (Scheving et al. 1978). Five regions were evaluated (esophagus, stomach, duodenum, jejunum, and rectum) along with the tongue; moreover, they were monitored along a 48-h span. These data were significant since they systematically documented for the first time, in the same animals, the dramatic variation encountered in cell proliferation from one region to another over a 2-day span (Fig. 10). The major variations seen were in the amplitude of the rhythms and in the 24-h overall means. The phasing of the rhythms in the different regions of the gut were remarkable similarly. In these animals, which were standardized to 12 h of light (0600 to 1800 hours) alternating with 12 h of darkness, it ap-

peared that the peak in DNA synthesis occurred around the time of transition from dark to light or even just before this; the trough occurred about the time of transition from light to dark. The rhythm in the duodenum, as illustrated by the chronogram in Fig. 10, was not very impressive, especially when compared to the rhythms in the other regions; but this was due, in part, to the scale on which the data were plotted.

## Reproducibility from One Study to Another in Intestinal Rhythm Using the Same and Different Techniques

Further evidence shows that such rhythms are reproducible from one study to another. This is shown in the data plotted in Fig. 11 for the epithelia of both tongue

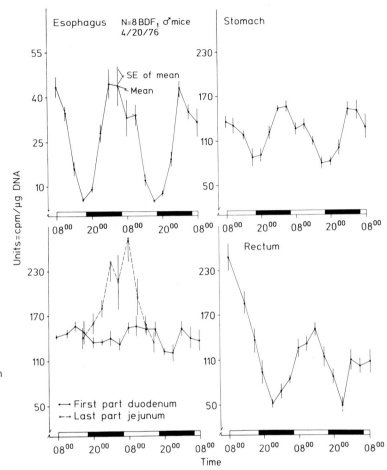

**Fig. 10.** Chronograms illustrating the rhythmic patterns of [³H]TdR-incorporation into DNA over a 48-h span in five different regions of the alimentary canal of male CD2F₁ mice. The *abscissa* represents the light-dark cycle to which the mice were standardized: L 0600–1800 hours. The lowest-amplitude rhythm was found in the duodenum. (Scheving et al. 1978)

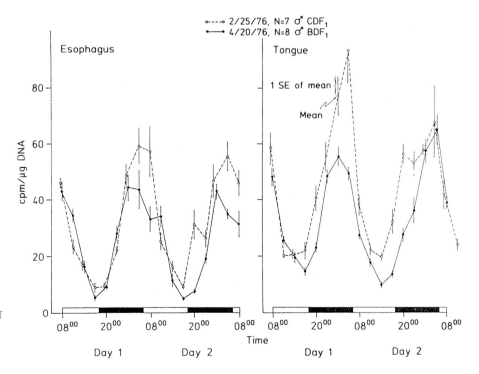

**Fig. 11.** High-amplitude rhythms in the incorporation of [³H]TdR into DNA in the tongue and esophagus of two species of mice monitored over a 48-h span. These rhythms are remarkably similar in the two strains of mice, especially regarding their phasing. (From Scheving et al. 1983a)

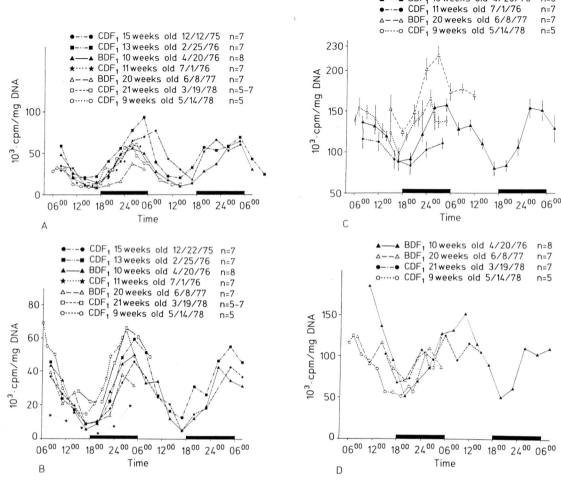

**Fig. 12 A–D.** Remarkable reproducibility of the phasing of the rhythm in DNA synthesis in the mouse tongue (**A**) esophagus (**B**) stomach (**C**) and rectum (**D**) in relation to the light-dark cycle from one study to another. It is important to bear in mind that data in this figure represent a retrospective approach in that data were obtained from animals that had served as controls for various experiments over a 5-year span. The minor differences seen could be due to many factors, including the specific activity of the isotope, and the season. Standard errors were purposely omitted to avoid an overly cluttered graph; however, analysis of variance and cosinor analysis for each series of data showed high statistically significant changes ($P < 0.05$ in all cases). (Scheving et al. 1983)

and esophagus (Scheving 1980b, c); the phasing was remarkably similar in two strains of mice. The data in Fig. 12 shows similar reproducibility, specifically for the rhythms of the tongue, esophagus, glandular stomach, and rectum. These data were obtained from control animals used in several different studies carried out for diverse reasons on different dates in our laboratory; they did not result from studies done simply to test for reproducibility; the data in Fig. 11 were, however, carried out to specifically test for reproducibility. The differences in the mean and amplitude of the chronograms in Fig. 12, as well as the smaller differences in the phasing, can be attributed to many things, including the specific activity of the radioiso-

topes and age. The data unequivocally demonstrates that cell proliferation is rhythmic throughout the gut and that the once-prevailing view of randomness must be abandoned. Of course, the same conclusion applies to all tissues that undergo cell proliferation. Such fluctuations are evident whether mitotic figures are counted, the incorporation of [3H]TdR into DNA is measured, or the cytofluorometric technique (FCM) of analyzing cell proliferation is utilized (Thorud et al. 1978; Møller et al. 1974; Clausen et al. 1979; Laerum 1981; Rubin 1981; Rubin et al. 1983a, b).

It should also be pointed out that the biochemical technique used to measure [3H]TdR uptake into DNA has been confirmed by the FCM technique to

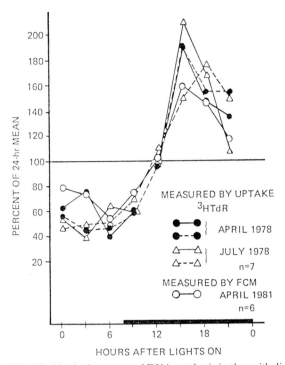

**Fig. 13.** Rhythmic pattern of DNA synthesis in the epithelium of the tongue as determined by two different methods: incorporation of [³H]TdR into DNA, and FCM. The "counts per minute per microgram" and "percentage of cells" have been converted into "percentage of 24-h mean" to facilitate comparison of data. Note the remarkable reproducibility between the two methods as far as the phasing of the rhythms is concerned. (Rubin et al. 1983b)

be a highly reliable method for analyzing circadian variation in vivo, but only in those tissues that reliably can be analyzed by the FCM technique (Fig. 13). Such a demonstration was important, because in the past the reliability of biochemical techniques, especially in vitro, had been questioned (Maurer 1981). Rubin et al. (1983b) addressed this problem by comparing the two methods in vivo and concluded that the FCM, although a rapid and elegant technique, is suitable only for a small part of the gut. Essentially they are those regions, such as the esophagus and tongue, where the epithelial cells can be easily separated into a homogeneous population. The stomach is one region where the characteristic rhythm, often documented by mitotic counts and [³H]TdR uptake, cannot be demonstrated by FCM (at least to date). Rubin et al. (1981, 1983a, b) discussed the problem in detail and explained why the FCM technique, at least at present, cannot always replace the more tedious mitotic counts or biochemical analysis. This limitation of the FCM does not seem to be appreciated by all those involved in cell-kinetic re-

search. We mention this because our own work was recently criticized by a referee because we did not use the FCM technique. Obviously, he/she, as well as the editor, was not aware of its limitations.

## Differences in Rhythm Characteristics Between Various Regions of the Gut

### Amplitude and Mesor

As mentioned above, available data indicate that all tissues that undergo proliferation in the intestinal tract do so in a circadian manner, but the amplitude of the rhythms vary in different regions. It also is apparent that large variations exist in the mesors between tissues of different organs and/or sites within the same organ. Fig. 14 shows the overall 24-h levels in different regions of the gut in graphic form (Scheving et al. 1979, 1980c, 1982). Although some variation may be found from one study to another, we believe,

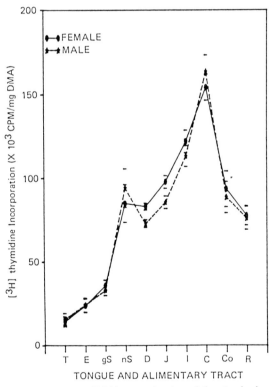

**Fig. 14.** The overall 24-h mean levels and the standard errors (mesor) in the incorporation of [³H]TdR into DNA for each region of the intestinal tract of CD2F₁ mice. *T*, tongue; *E*, esophagus; *gS*, glandular stomach; *nS*, nonglandular stomach; *D*, duodenum; *J*, jejunum; *I*, ileum; *C*, cecum; *Co*, colon; *R*, rectum. The highest levels are found in the cecum, the lowest in the tongue and esophagus. (Scheving et al. 1984)

based on data from a number of studies, that the data presented in Fig. 14 are typical for mice that have been standardized to 12 h of light alternating with 12 h of darkness.

Variations in amplitude of rhythms could reflect either regional differences in the extent and cellular composition of the mucosa or circadian variations in the distribution of positive or negative growth-controlling signals. The tissues with the larger amplitudes in DNA synthesis in the digestive tract such as the tongue, esophagus, stomach, colon, and rectum are not necessarily those with the overall-highest levels of DNA synthesis, but they are the ones most prone to cancer; the significance, if any, of the latter observation has not yet been determined. Interestingly, these tissues also exhibit the greatest and most consistent growth response to epidermal growth factor (EGF), a polypeptide isolated from the submandibular salivary glands; this molecule plays an important role in the control of growth in the digestive tract. The role that EGF plays in cell proliferation has been discussed elsewhere (Scheving LA et al. 1979, 1980) and will only be briefly commented on below.

## Phase

Figure 15 is a map that compares the acrophase of the circadian rhythm in the incorporation of [³H]TdR into DNA in many tissues of the mouse as well as the acrophase of the mitotic index of the corneal epithelium for ten different tissues studied in one investigation with 13 tissues in another (Scheving 1981). Note the reproducibility in these two different studies; however, it should be pointed out that neither study was specifically designed to test for reproducibility. It is of interest that there was approximately a 1- to 2-h delay in the acrophase of the rectum compared to that of the esophagus (Scheving et al. 1981 a).

## Waveform

The data from studies of cell proliferation also reveal variation in the waveform of the rhythm. In the digestive tract, the rhythm in DNA synthesis tends, from one study to another, to be strongly monophasic and sinusoidal in the oral, esophageal, stom-

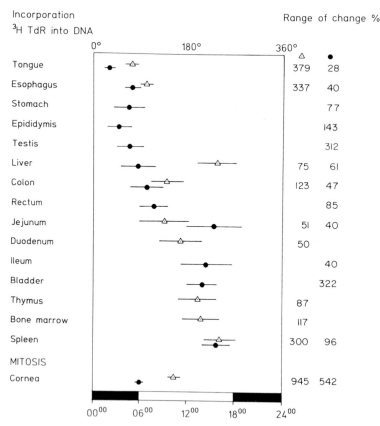

**Fig. 15.** Acrophase map showing the approximate peak of the circadian cycle in either the incorporation of [³H]TdR into DNA of 15 different tissues or in the mitotic index of the corneal epithelium of mice. The acrophase is represented by a *dot* or *triangle,* and the *bars* extending from the *dots* or *triangles* represent the confidence limits. The *abscissa* shows the environmental light-dark cycle. The percentage rate reflects the change from the lowest to the highest time-point means, with the lowest value equal to 100%; values are rounded to nearest integer. (Scheving 1981)

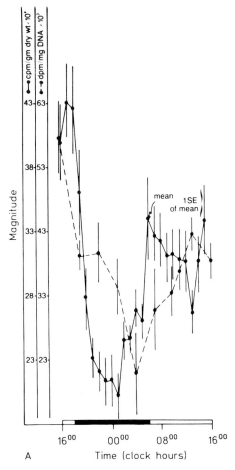

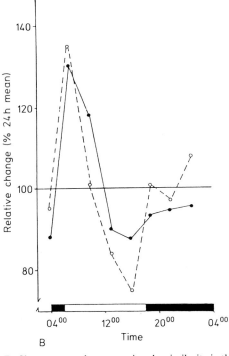

**Fig.16. A** Chronograms showing the rhythmic pattern in DNA synthesis in the bone marrow of rats. The bone marrow generally peaks around the transition from light to dark. It has a tendency to begin its daily climb later in the dark span than that which is seen in the intestinal tract. The standard errors were intentionally omitted to avoid an overly cluttered figure. The changes are highly statistically significant. (Scheving and Pauly 1973)

**B** Chronograms demonstrating the similarity in the phasing of the rhythm in DNA synthesis in the thymus and spleen of CD2F₁ mice. They do have a tendency to peak later in the dark span or even at the transition from dark to light compared to the intestinal tract, and this can be better appreciated by comparing with the acrophase map in Fig.15. The standard errors were intentionally omitted to avoid an overly cluttered figure. The data were highly statistically significant

ach, and anal regions. Occasionally it is irregularly monophasic or even multiphasic in the regions between the stomach and the rectum; it should be emphasized, however, that more frequently it is the monophasic rhythm that is encountered throughout the hindgut. Such occasional variation is mentioned because it suggests that the cells dividing within the tongue, esophagus, stomach, and rectum progress through the cell cycle with a greater degree of synchrony that do those in the small intestine, cecum, and colon. This technique of whole organ analysis of DNA synthesis does not give the best resolution of the inherent rhythmicity in cell proliferation within an organ, because it fails to discriminate between the

rhythmic tendencies of regionally or functionally different subpopulations of cells. For example, the technique fails to discriminate between the lymphocytic and epithelial components of the ileum. In other words, a multiphasic rhythm may represent a combination of the monophasic rhythms of several different populations of cells.

Figure 16a illustrates the rhythmic pattern in DNA synthesis in the bone marrow of the rat and Fig.16b illustrates the same for the spleen and thymus of CD2F₁ mice. We shall not discuss these in the same detail as we have in the rhythm in the cornea and intestinal tract, but they both undergo robust cyclic variation (Scheving and Pauly 1973).

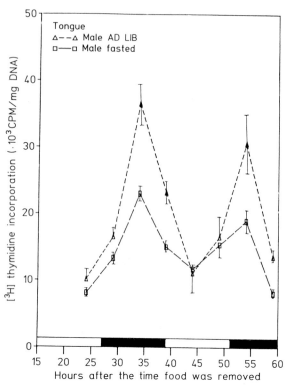

**Fig. 17.** Chronograms of data over a 59-h span of fasting are compared with the rhythm in ad libitum-fed mice. Each time point is the mean of five mice plus the standard error. The mice were 8-week-old CD2F₁ male mice standardized to 12 h of light alternating with 12 h of darkness. Clearly, as reported earlier, the rhythm persisted in the fasting state for 59 h. (Scheving 1987)

## Do Such Rhythms Persist in Fasted Animals?

This is an important question and the answer generally is "yes" for the intestinal tract and cornea which have been the tissues most extensively studied in the fasting state. However, the response is somewhat different in different regions, but space does not permit adequate discussion. Earlier we dealt with this question using male mice who had been fasted from 16 to 24 h, which is the traditional time for researchers in gastroenterology to fast their animals (Scheving et al. 1984a). Ostensibly one reason (but not the only one) for doing this has been to avoid rhythmic fluctuation. In the foregut, stomach, and hindgut there was a generalized decrease in the overall level and amplitude of the rhythms, but the phasing persisted in the above-mentioned report for 34 h, which was the duration of the study. More recently it was found that the rhythm would persist over a much longer span of fasting and this is illustrated using the tongue epithelium as an example (Fig. 17). In the latter study the mice were fasted for 59 h, at which time the monitoring was terminated because some of the mice were beginning to die from starvation. Burholt et al. (1985) reported similar findings, but they fasted their mice for 72 h. Such data challenge the strongly held dogma (not based on adequate data) that "daily rhythms" are no more than responses to food intake. Certainly, food intake may affect the mesor and amplitude, but the timing mechanism must remain under endogenous control.

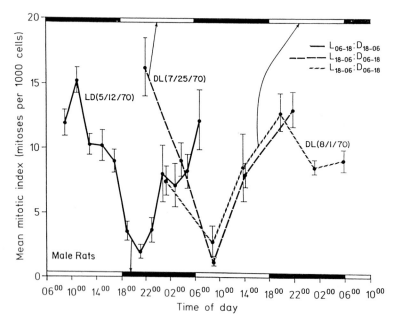

**Fig. 18.** The *solid curve* illustrates a typical mitotic index rhythm in rats standardized at 12 h of light alternating with 12 h of darkness, shown on the *bottom abscissa*. The *broken line (right)* represents the inverted pattern 7 days subsequent to inverting the light-dark cycle (see *horizontal axis* at the *top*). The *slanted broken line* accompanying represents another study conducted a week later showing that the rhythm was still "locked on" to the light-dark cycle. In all cases *n* = 5. (From Scheving and Pauly 1973)

## Summary of Some Properties of Rhythms

### Synchronization of Rhythms

To a Light-Dark Cycle

All rhythms in cell division are synchronized to the light-dark cycle if the animals are fed ad libitum. Figure 18 illustrates that the mitotic index rhythm in corneal epithelium can be inverted within seven days or less simply by inverting the ambient light-dark cycle 180°. The rhythms in cell division in other tissues may take longer to invert; for example, the rhythm in incorporation of [3H]TdR into DNA in the spleen took 3 weeks to invert (unpublished results by us). For more extensive treatment of synchronization of rhythms to the light-dark cycle see Scheving and Pauly (1973) and Scheving et al. (1981 a). Figure 19 illustrates how variation in the duration of the light or dark spans may alter the waveform of the rhythm.

To Meal Timing

Some rhythms in cell proliferation, such as those in the bone marrow, spleen, and intestinal tract can be manipulated by restricting feeding to a particular time of the day. For such tissues, restricted meal timing is capable, with time, of partially or even completely overriding the light-dark cycle (Scheving et al. 1976). The same has been reported for the circulating eosinophils in these same mice (Pauly et al. 1975). Other tissues such as the corneal epithelium (Scheving et al. 1974a) cannot easily be manipulated by restricted feeding indicating that the light-dark cycle is far more dominant in the synchronizing of cell division rhythms in such tissues than it is for the gut, bone marrow, spleen, or eosinophils. Phillipens et al. 1977, confirmed our finding for the cornea of the rat and also carried out additional studies on the effect of meal timing on a host of biochemical variables. Data along this line have been extensively reviewed by Scheving (1981). The data obtained from the many studies performed by ourselves and many others supports the following generalizations, not only for rhythms in cell proliferation, but for rhythmic variables in general:

1. The effect of synchronization by the light-dark cycle or restriction of food to specific circadian stages cannot be generalized with respect to all variables within an organism's circadian system.
2. Food is managed differently when given at different circadian stages; it simply is not enough to give

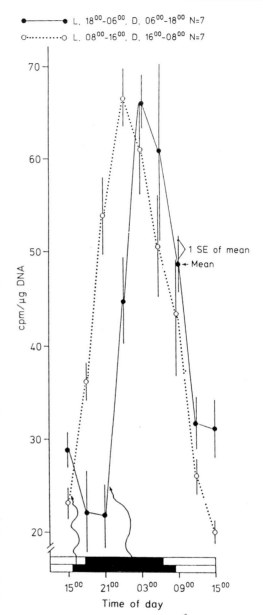

**Fig. 19.** Rhythm in incorporation of [3H]TdR into DNA in the esophagus of CD2F₁ male mice synchronized to two different light-dark cycles. Note that the waveform of the rhythm in the animals subjected to the longer dark span (16 h) is more sinusoidal than that of the rhythm in the animals subjected to the shorter (12 h) dark span; this is not surprising because the ascending limb and peak of the rhythm in this variable occur within the dark span, and there is simply more time in the 16-h dark span. The amplitudes and overall 24-h means are remarkably similar. When comparing data drom different studies, it is important to take into consideration the ratio of light to dark. (From Scheving et al. 1983)

the meal at a single stage of the organism's circadian system, as frequently has been done, and assume that a particular response would be the same if the same meal was presented at a different stage.

3. Restricted feeding in spite of the diverse responses seen in different regions of the digestive tract might prove to be a very useful tool to facilitate investigation of the control mechanisms into cell division or to improve cancer chemotherapy. This hypothesis has been supported by one study on experimental chemotherapy in the rodent (Nelson et al. 1974), but it has not been tested in the clinics.

## Phase Shifting of Rhythms

It has been shown repeatedly that stimuli such as drugs and physical agents can alter the rhythm of the mitotic index (Scheving and Pauly 1973); Burns and Scheving 1973). Figure 20 illustrates the results of an early investigation performed on adult mice standardized for 7 days prior to the beginning of the study to artificial light from 0600 to 1800 hours, alternating with darkness from 1800 to 0600 hours; food and water were available ad libitum. Forty BD2F$_1$ mice were injected intraperitoneally with 2.6/mg of cytosine arabinoside (ara-C) and 40 with 0.2 ml saline. Two h after the injection, sub-

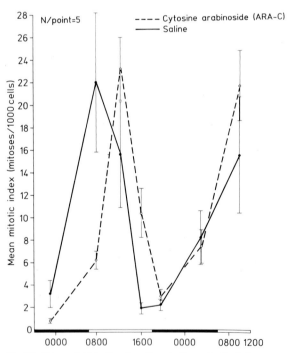

**Fig. 20.** Phase shift in the circadian rhythm of mitosis in mouse corneal epithelium induced by ara-C. For further explanation see text, as well as Scheving and Pauly (1973)

groups of five mice from both the control and experimental groups were killed at intervals indicated in Fig. 20 for 36 h. The mitotic rate of the corneal epithelium and the saline-injected animals showed the expected circadian rhythm, with phasing corresponding to that previously found in similarly standardized rats. In the ara-C-injected animals a phase delay of about 4 h was apparent on the first day after administration of the drug. On the second day after treatment the usual phase relationship was reestablished between the mitotic rhythm and the synchronizing cycle of the alternating light-dark cycle (Scheving and Pauly 1973).

The rationale at the time of performing the experiment dealing with the effect of ara-C on the mitotic rate in the corneal epithelium stemmed from a series of earlier unpublished experiments done in 1971. It had been found that when ara-C was injected during the circadian stage at which the rate of corneal epidermal mitosis was usually highest, no statistically significant decreases in the mitotic count could be detected 12 or 24 h later. The mitotic peak occurred 24 h later at the expected time. When the same dose of ara-C was injected at the circadian stage when the corneal mitotic rate was lowest there usually was a significant decrease in mitotic rate the next day or 12 h later, at the time of expected circadian "high". The question asked was whether this apparent decrease was the result of a shift in the phasing of the rhythm, or whether it actually represented a real depression of the mitotic rate? From examination of Fig. 20 it appears that the results obtained were due to phase delay of about 4 h. Single time point sampling could be quite misleading because in this situation the data obtained might suggest that a depression of mitosis had taken place when actually there was a phase delay in the rhythm. Of course, it is recognized that ara-C can depress mitosis in corneal epithelium if given sufficiently high doses, for a longer duration of time, or perhaps even at a different stage in the circadian system. The data in Fig. 20 clearly points out the complexity of evaluating the effect of a drug on mitotic activity, moreover, they demonstrate the pitfalls awaiting those who would ignore the organism's time structure in circadian and other frequency ranges.

## Cell Proliferation Rhythms in Tumors

A number of workers have reported evidence for and against circadian variation in the mitotic index of several different tumor types when studied at different times along the 24-h domain in both animals and human. We have previously reviewed this subject matter and will only summarize herein (Scheving et al. 1983 a).

## Animal Models

Among the first adequately designed studies for rhythm exploration in animals were those of Blumenfeld (1943) who demonstrated the occurrence of mitotic rhythms in various healthy tissues of rabbits and mice. He then compared the mitotic activity of the presumably normal epidermis with that of a carcinoma induced in the same animals by the topical application of 3-methylcholanthrene. Unlike the normal epidermis, the induced epidermal carcinoma did not show a 24-h synchronized circadian rhythm (Fig. 21 f). The interpretation of this finding was that the malignancy represented an "escape" of the tumor from physiologic control of cell proliferation. A series of studies by Bertalanffy and Lau (1962 b), Bertalanffy (1963 a, b), and Bertalanffy and McAskill (1964) failed to detect any circadian variation in several tumor models, among them being spontaneous mammary gland adenocarcinoma in C3H/HeJ mice. As mentioned earlier, these workers have often failed to detect any circadian variation even in normal tissues. Nash and Echave Llanos (1971) reported a circadian rhythm in DNA synthesis of slow-growing hepatoma (SS1H), but the variation was not as prominent as in a fast-growing hepatoma (SS1K) (Fig. 21 d). Moreover, Badran and Echave Llanos (1965) reported the persistence of circadian rhythmicity in the mitotic activity of mammary carcinoma of C3H/Mza female mice and found that this tumor persisted over 35 transplant generations carried out over a 3-year span (Fig. 21 b). Rosene and Halberg (1970) demonstrated the presence of a high-amplitude change along the 24-h scale in mammary carcinoma in albino mice (Fig. 21 a). We believe that this well-designed study was especially relevant because the tumors were spontaneous in origin. Burns et al. (1979) reported statistically significant decreases in the uptake of [³H]TdR into DNA in 10-day and 14-day transplanted Lewis lung carcinoma. The differences observed along the 24-h scale were as great as 320%. There was, however, no evidence from their limited data of any circadian pattern.

We also have carried out repeated studies on three other experimental mouse tumor models; they include: (1) the transplanted L1210 leukemia, (2) the Ehrlich ascites tumor, and (3) the mammary adenocarcinoma. An impression from studying the data of our collected studies is that when the tumor cells of different models are transplanted while still in the experimental growth phase, they are very likely to exhibit more of an ultradian (high-frequency) than a circadian variation. We have found that occasionally, when the data of particular study are plotted over a single 24-h span, a typical circadian pattern may emerge. But our impression is that this is not likely to be predictable from one study to another, especially in the transplanted tumor model. Evidence to support such a view comes from Rosene and Halberg (1970) who, as indicated above, reported the absence of a circadian rhythm in transplanted Ehrlich ascites carcinoma in adult albino mice, while Rubin et al. (1983 a) found rhythms in the $G_1$, S, and $G_2$ phases of the cell cycle in male Swiss mice using the FCM technique. Figure 21 illustrates some animal tumor models (Scheving et al. 1983 a). In our opinion the question of circadian variation in tumors is questionable in transplanted tumors.

## Human Models

Among the first such studies in humans was that of Dublin et al. (1940), who studied the mitotic activity of a human carcinoma in the large intestine. They simply removed specimens once during the day (1000–1200 hours) and once during the night (2000–2400 hours) and, finding little difference, concluded that no circadian variation was present.

Voutilainen (1953) carried out a series of studies on 21 patients with malignant tumors. Repeated biopsies were taken from the tumors of these patients along the 24-h time scale (longitudinal sampling). The patients were then irradiated (each one at a different time after the first sampling), and another series of biopsies were carried out on all the patients at 2-h intervals over a single 24-h span, just subsequent to radiation. Thus, Voutilainen was able to compare the pattern of mitotic activity in subjects who had been sampled sufficiently far in advance of the radiation with the pattern in the same subjects immediately after radiation.

Following this, another Herculean study was carried out by Tähti (1956) on 20 patients with exophytic ulcerated malignant tumors. A biopsy was taken from each tumor at 2-h intervals during a 24-h span immediately preceding exposure of the patients to X-rays and again during another 24-h span immediately after radiotherapy. In all, 24 samples were obtained from the tumor of each patient during a single 48-h span. The major difference between the two studies was that Voutilainen did not sample the tumors immedi-

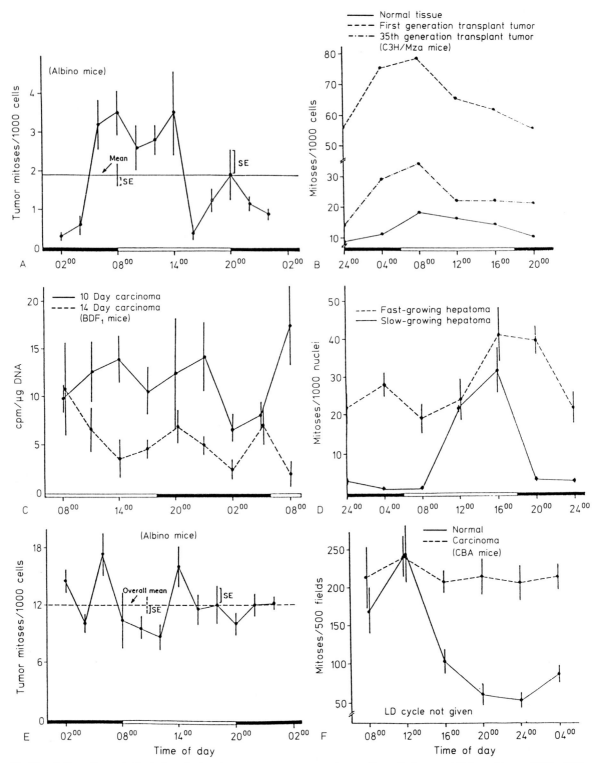

**Fig. 21. A** Circadian mitotic index rhythm in spontaneous mammary tumors in mice; 96 "A"-strain mice were used (from Rosene and Halberg 1970). **B** Mitotic rhythm in normal mammary epithelium in mice in the first generation of a transplanted mammary carcinoma and the 35th transplant generation of the same tumor (from Badran and Echave Llanos 1965). **C** The pattern of variation along the 24-h scale in 10- and 14-day-old Lewis lung carcinoma in BD2F₁ mice. Although there is little evidence of a circadian rhythm, there is as much as 320% variation (from Burns et al. 1979). **D** Daily fluctuation in the mit-

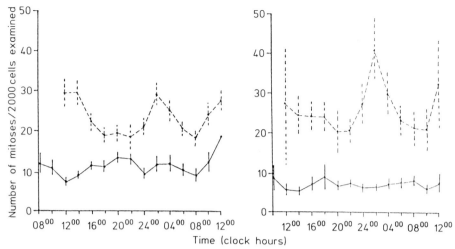

**Fig. 22.** Mean mitotic counts on serial biopsies from human malignancies, before *(left)* and after *(right)* radiotherapy. In the Tähti (1956) series of data (10 patients, *solid lines*) a biopsy was taken from each tumor at 2-h intervals during the 24-h span preceding exposure to X-rays and for 24-h immediately after radiotherapy. Thus, 24 samples were obtained from each patient during a 48-h span. Voutilainen's (1953) data *(dashed line)* were obtained from 21 patients with malignant tumors. The studies were similar, except that Voutilainen did not sample from the same tumor immediately before irradiation as did Tähti. These were tumors of several types (squamous cell, basal cell, and mammary carcinomas). When Garcia-Sainz and Halberg (1966) further analyzed these same data using the cosinor, a clear-cut rhythm could be detected for the data on the mammary carcinoma but not for the squamous or basal cell carcinomas. (Tähti 1958; Voutilainen 1953)

ately before the irradiation as did Tähti. The conclusion of both authors, based upon examination of the data which are displayed in Fig. 22, was that prior to irradiation they were dealing with a bimodal rhythm.

Garcia-Sainz and Halberg (1966), using the cosinor program for analyzing time series data, reevaluated the data from both the studies. Although originally the three kinds of tumors were considered all together, in the reanalyses, the different tumors were separated into (1) squamous, (2) basal cell, and (3) mammary carcinoma types and tested separately for the presence of circadian variation. The conclusion from the statistical evaluation used by Garcia-Sainz and Halberg was that a statistically significant circadian rhythm could be demonstrated only for the mammary carcinoma data.

It is of interest to us that in both of the above studies, but especially in that of Voutilainen, maximum mitotic activity took place around midnight, which is the time when we and others had reported the occurrence of the highest mitotic activity in the normal human epidermis, as shown in Fig. 2 (Scheving and Gatz 1955; Scheving 1959; Fisher 1968; Zagula-Mally et al. 1979). This also was the time when Killman et al. (1962) and Mauer (1965), while evaluating human bone marrow for circadian variation, found the highest mitotic index. It is our opinion that this similarity in phasing is not likely to be a coincidence, but of course this has not been proven.

## Influence of Some Hormones and Growth Factors on Normal Rhythms

### Epidermal Growth Factor

EGF was originally isolated from the submandibular gland of the adult mouse and shown to stimulate cell proliferation in the epidermis. Since its original dis-

otic activity of an SS1K and an SS1H. Significantly higher mean values of mitotic activity are observed during light (L) than during darkness. The SS1K presents higher values than the SS1H during darkness, but the values for the two hepatomas are not different from one another during part of the L span. The circadian amplitude is higher in the SS1H and approached that found in the normal immature liver (from Echave Llanos and Nash 1970). **E** Variation along the 24-h scale in an Ehrlich as-

cites tumor transplanted into "A" mice; no circadian variation is evident (Rosene and Halberg 1970). **F** Mitotic activity in an epidermal carcinoma compared with that in the normal epidermis of the rat (sampling at 4-h intervals for a period of 24 h. Note that there is no circadian variation in the tumor, whereas there is a characteristic circadian variation in the normal epidermis. The author failed to give the LD cycle by which the animals were standardized

covery in 1962 EGF has been shown to play a role in the growth control of many different cell types in vitro and in vivo.

Since 1978, we have carried out a comprehensive investigation of the effects brought about by EGF on cell proliferation in a number of different organ systems in intact adult animals. We also documented a more rapid action than previously based on in vitro studies. Our results demonstrated that EGF had a pronounced stimulatory effect on DNA synthesis in a number of tissues. The greatest effect was observed in the fore- and hindgut of the digestive tract. Fig-

ure 23 illustrates the response of the esophagus and rectum to EGF at 4, 8, and 12 h after intraperitoneal injection. Indeed EGF by itself was found to be capable of phase shifting the prominent circadian rhythm of DNA synthesis in the esophagus of the adult mouse (Scheving et al. 1983b).

Interestingly, in two of the tissues which are very responsive to EGF, the cornea and the esophagus, the phasing of the DNA-synthesis rhythms can be dissociated in phase from each other by changing the lighting schedules and feeding times, suggesting that a general systemic release of EGF was insufficient by

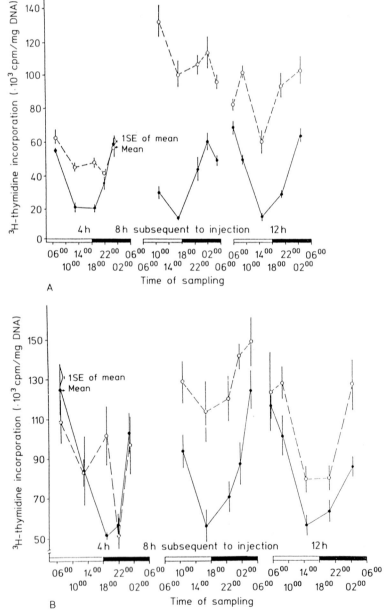

**Fig. 23 A, B.** Effect of EGF on [³H]TdR incorporation into the **A** esophagus and **B** rectum is plotted versus the time of death for the different injection times. Mice were killed at either 4, 8, or 12 h after each time of injection. (Scheving LA et al. 1979)

itself to account for rhythmic variation in these tissues (LA Scheving et al. 1980). The data indicate that EGF also inhibited DNA synthesis at certain circadian stages in several tissues including the thymus, bone marrow, and spleen.

We often have observed that food accumulates in the stomachs of EGF-treated mice resulting in a marked enlargement in stomach size. EGF is secreted into the saliva of mice and into both saliva and duodenal fluid of man. In mice, based on the circadian rhythm of EGF concentration in the submandibular gland, the secretion of EGF appears to coincide with the span of time when there is increased feeding. These results suggest that oral and intestinal mechanisms involving EGF may exist which regulate stomach emptying to allow for maximal duodenal absorption of food. Clinical support for the role of saliva in gastric emptying was provided by Malhotra (1967), who showed that human gastric emptying was delayed when food was premixed with saliva. Others have recently reported that sheep continually infused with EGF exhibited decreased feeding behavior. Our results suggest that the effects of EGF on stomach emptying, perhaps as a part of a safety mechanism, may be involved in food rejection. Finally, the ability of transformed cells to produce EGF-like molecules that can interact with this hormone's cell-surface receptor in vitro raises the possibility that ectopic EGF-like molecules may play a role in the anorexia observed in some cancers. Such factors have recently been reported to be elevated in the urine of a number of cancer patients.

Our laboratory has found very little evidence of a stimulatory effect of EGF on DNA synthesis in the duodenal mucosa of mice. Dembinski et al. (1982) found that EGF stimulated DNA synthesis in the duodenum, and Al-Nafussi and Wright (1982) reported that EGF stimulated cell proliferation in duodenal and ileal crypts. Goodlad et al. (1988) claimed stimulatory effects of urogastrone-EGF on stomach and colon DNA synthesis and to a lesser degree in the small intestine. Moreover, in a recent study, we found that exogenous EGF did not stimulate DNA synthesis in the duodenum or proximal part of the jejunum, but did so in the distal jejunum and in the ileum, although still to a much lesser degree than that of other parts of the digestive tract (unpublished results). Clearly, at this time, a controversy does exist as to what effect exogenous EGF has on cell proliferation in the small intestine. This could be due to the presence of an endogenous EGF in the small intestine in at least some species which may mask any effect that exogenous EGF might be capable of bringing about.

In recent studies we and others have carried out many investigations related to EGF and its role in the small intestine. This was promoted in part by its reported cytoprotective role in gastric acid secretion (Konturek et al. 1981). EGF has been localized by immunocytochemical techniques to sites within the small intestine including Paneth's cells at the base of the intestinal crypts, villous goblet cells, and the duodenal glands of Brunner (Poulsen et al. 1986; LA Scheving et al. 1989a). Moreover, the affinity and capacity of EGF receptors in the liver of the mouse recently have been reported to be very circadian stage dependent in both ad libitum fed and fasted $CD_2F_1$ mice (LA Scheving et al. 1989b).

The administration of EGF reportedly reverses the atrophy observed in the alimentary canal during total parental nutrition (Goodlad et al. 1987, 1988) and increases the rate of closure of intestinal wounds. Indeed, mucosal glands capable of producing EGF have been shown to arise at sites close to cellular injury. Together these findings point to EGF as being a major hormone in the control of growth and differentiation in the entire gut.

EGF inhibits gastric acid secretion. Konturek et al. (1981) reported that EGF when given in non-antisecretory doses prevented dose-dependent aspirin-induced gastric ulcers. The formation of aspirin-induced ulcers resulted in a significant reduction of DNA synthesis, and this was restored to normal levels by the administration of EGF. Kirkegaard et al. (1983) reported that the exocrine secretion of EGF from the submandibular gland in rats has a cytoprotective function in that it inhibited the formation of cysteamine-induced duodenal ulcers in rats. They do not attribute this action to a decrease in gastric acid secretion and they question the conclusion that cell proliferation accounts for this protection.

## Insulin and Glucagon

Our interest in insulin and glucagon arose from the demonstration by Bucher et al. (1978) of the importance of EGF, insulin, and glucagon in supporting cell proliferation in the regeneration of liver in rodents.

*Insulin.* In the second study, female Balb/Cann mice, which had been standardized to 12 h of light and 12 h of dark, were divided into four groups; and each group was injected with insulin at one of four different circadian stages. These stages represented 1 and 9 h after the lights went on, and 1 and 9 h after they went off. From each of these four groups, subgroups of mice were killed at 4, 8, 12 and 18 h after injection. A conclusion from data of this study was that insulin did affect, in many parts of the alimentary tract, the incorporation of [³H]TdR into DNA, but its effect was very circadian-stage-dependent. Further comments and specific examples will be given below; see text and reference for more details.

*Glucagon.* While doing the insulin study, a similar investigation was carried out, but in this case glucagon was administered instead of insulin. It was found that glucagon also had an effect on the incorporation of [³H]TdR into DNA. Further comment and specific examples will be cited below.

In summary, the data from these studies demonstrated for the first time that both glucagon and insulin affected the incorporation of [³H]TdR into DNA in all the organs examined, but in different ways at different circadian stages. The effects of these hormones were complex, but several generalizations may be made:

1. Insulin tended to increase the rate of incorporation of [³H]TdR into DNA in the organs examined, whereas glucagon decreased it, but only at certain circadian stages.
2. Insulin was more effective in stimulating incorporation of [³H]TdR into DNA when injected at either the end of the dark span or the beginning of the light span, as opposed to the end of the light span or beginning of the dark span.
3. Insulin had its greatest effect on [³H]TdR incorporation into DNA in the glandular stomach and rectum, whereas glucagon had its greatest effect on the colon and spleen.
4. The effects of both insulin and glucagon were different from those from EGF, as revealed in a similar study done by us.

In general, the results suggest that insulin, glucagon, and EGF all play important roles in the control of cell proliferation of various endodermally derived organs.

## Why is Circadian Variation Important in Experimental Design?

It should be evident from the plethora of data presented above and elsewhere that cell proliferation oscillates. A consideration of this oscillation will be essential if one ultimately expects to fully understand the mechanism of normal or abnormal cell division.

We simply cannot assume that such in vivo oscillation is of little consequence and that it simply can be avoided by sampling at the same time of day or night or that it realistically can be avoided simply by doing in vitro studies. It must be realized that cells grown in culture are freed from many selective pressures that affect them in the whole animal. As mentioned above it was found that IPR did nothing to cell proliferation in the duodenum if it was administered at one circadian stage, but had inhibitory or stimulatory effects if administered at other stages, and that it could even cause a phase shift in the rhythm. Since then, numerous other examples of such circadian-dependent responses in vivo to different hormonal, pharmacological, biochemical, and physical stimuli have been demonstrated for many metabolic and physiological variables by us and others.

## Adrenocorticotropin 1-17

From among the many examples for which data now exist, we shall cite, in addition to IPR, three examples picked more or less at random from data generated in our own laboratory using adrenocorticotropin (ACTH) 1-17, insulin, and glucagon. This is done to further emphasize that temporal organization is a most important factor to consider in experimental design.

Figure 24 represents data from a study that was designed to test the effect of a synthetic adrenocorticotropin analog, ACTH-17, on DNA synthesis in murine bone. The detailed design and results of the study have been published (Walker et al. 1985). Essentially, we wanted to know if this molecule stimulated or inhibited DNA synthesis, how long it took to act, the duration of its effectiveness and whether its effect was circadian-stage-dependent. Two doses were used and the two sets of data were compared with the data obtained from control mice who were injected only with the vehicle in which the ACTH-17 was suspended (we refer to this as a placebo group). Fig. 24 illustrates the data from the larger dose only (simply to avoid an overly cluttered Fig.). The following is evident: (1) When ACTH-17 was injected 2 h after "lights-on" (2 HALO), there was stimulation of DNA synthesis for 24 h; if administered at 14 HALO (or 2 h into the dark) there was inhibition for 24 h. If administered at 10 HALO, there followed inhibition for about 12 h, only to be followed by stimulation for about 12 h. When administered at 18 HALO there was stimulation for about 12 h, followed by inhibition for about the next 12 h. Clearly, the diverse responses seen were dramatically circadian stage dependent.

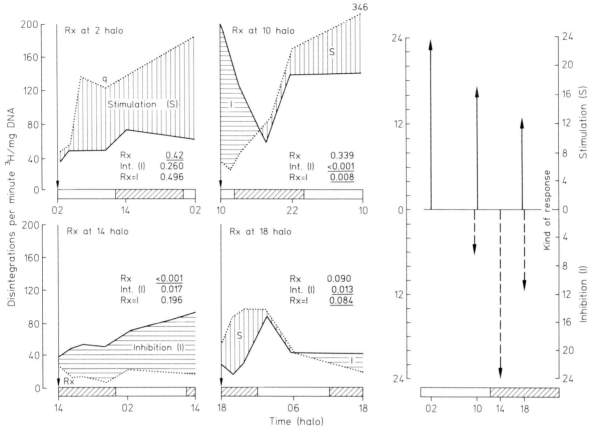

**Fig. 24.** DNA labeling in metaphyseal bone of CD2F$_1$ mice during a 24-h span after treatment with either a placebo (saline) or ACTH 1–17 (on the *left*) and duration of response (h) based on the same data (right). The statistical significance of the findings was established by three-way ANOVA of all data and by two-way ANOVA for data at each treatment time. The latter yielded the *P* values for the effect of treatment (R$_x$), those of the time interval elapsed since treatment (Int), and for interaction. See text for further explanation. Timing does make a difference! (From Walker et al. 1985)
___ Placebo     ..... 20 IU/kg ACTH 1–17

## Insulin

Two additional examples are mentioned using insulin and glucagon just in case one might have the idea that somehow ACTH-17 is a unique molecule when it comes to bringing about such diverse responses as a function of time. When 1.6 mg/g of pork insulin was administered intraperitoneally to a group of mice at 1 HALO there were statistically significant increases in DNA synthesis in the mouse cecum at both 8 and 12 h postinjection. However, if the identical dose was administered at 13 HALO (1 h into the dark span), there were no statistically significant increases at time points sampled after treatment; thus we found no effect when insulin was administered at one circadian stage, whereas if it were given at another stage there was a rather dramatic effect (Fig. 25 a, b). Such findings have been reported for many regions of the digestive tract. Thus we found, as cited above, that exogenous insulin can bring about an effect on DNA synthesis in the intestine, but it is likely to be stimulatory and circadian stage dependent (Scheving et al. 1982).

## Glucagon

Another example involving glucagon, also illustrated in Fig. 25 c, d, shows that when 1.0 mg/g of glucagon was administered intraperitoneally to mice at 1 HALO there was a dramatic decrease in DNA synthesis in the incorporation of [$^3$H]TdR into DNA in the colon of mice at 4 h after treatment which did not persist to the 8th h. However, if the same dose was administered at 9 HALO (still in the light span), statistically significant decreases persisted for at least 8 h. The latter example is given simply to point out that

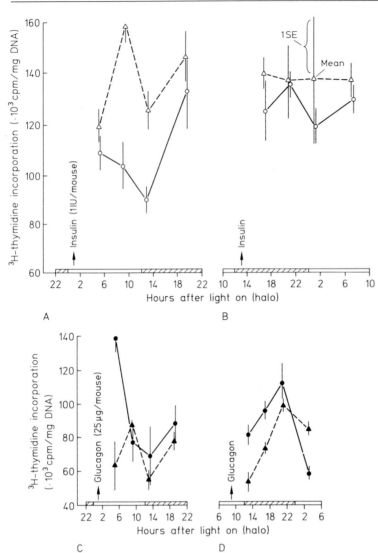

**Fig. 25 A–D.** When insulin was administered to different subgroups of mice at 1 h into the light (**A**) or 1 h into the dark (**B**), the pattern of response in DNA synthesis of the cecum was different. This shows that difference in response is not simply a light-dark response since both injections were administered in the light span. (Scheving LA et al. 1982) When glucagon was administered 1 h into the light (**C**), statistically significant decreases in DNA synthesis in the colon followed within 4 h. However, when similarly administered at 9 h into the light (**D**), statistically significant decreases were recorded for both time points. (Scheving LA et al. 1982)

variation in the response seen is not simply a day-night (light-dark) difference since both treatments were administered during the light span.

We could continue to cite numerous similar findings from our own data showing that other molecules such as somatostatin (Scheving 1984), pentagastrin, and more recently, interleukin-2 do influence, in diverse ways, cell proliferation in the intestinal tract (Scheving et al. 1980c). Undoubtedly, there are many more, including corticosterone and others still yet to be discovered.

The data clearly demonstrate that when experimental design is being considered, the investigator must be sensitive to temporal organization of the experimental animal. In the case of cell proliferation the different molecules are all playing a role in its

regulation, and under normal conditions they are doing so in a finely orchestrated time-dependent manner. Recognizing this could aid in unraveling the yet unexplained (as of 1991) mechanism of normal and abnormal cell division!

Another example of where failure to recognize rhythms created much confusion in the 1960s and 1970s involved the determination of cell cycle time. Determination of cell cycle time and duration of various phases of the cell cycle in many tissues were made using the once-popular frequency of labeled mitosis (FLM) method which originally was designed by Quastler and Sherman (1959). In fact, the intestinal epithelium was selected by Quastler and Sherman to design this technique because they believed that intestinal cells divided randomly. Reasons for this erro-

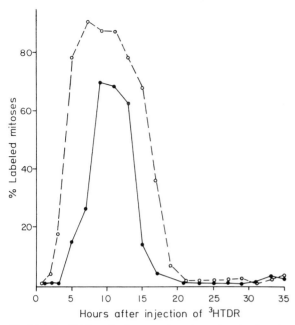

**Fig. 26.** The FLM method was used to determine the duration of the S phase of the cell cycle in the epithelium of the mouse cornea. The *dashed line* represents the data obtained from mice that were injected with [³H]TdR at 0900 hours and killed at frequent intervals thereafter. The *solid line* represents data obtained from mice that were injected with [³H]TdR at 2100 hours and killed at frequent intervals thereafter. (For complete details see Burns and Scheving 1975 and the text)

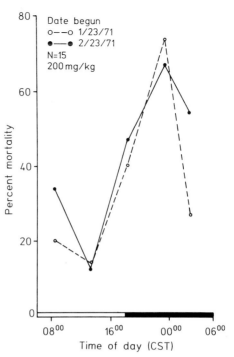

**Fig. 27.** Circadian susceptibility rhythm of BD2F$_1$ mice to ara-C given on five consecutive days at single, defined, circadian stages. See text for further information. (Scheving et al. 1974a)

neous view have been discussed (Burns and Scheving 1975; Scheving and Burns 1977). Obviously, this method no longer can be considered reliable for investigating cell kinetics in the in vivo system. The data in Fig. 26 illustrate why this statement can be made. We have shown that if one injects [³H]TdR when the mitotic index is lowest in the corneal epithelium, about 2100, and analyzes the cell division in this tissue in a conventional manner, then $G_2 + 1/2 M = 8$ h and $T_s = 12.2$ h (Burns and Scheving 1975; Scheving and Burns 1977). Tvermyr (1972) and Møller et al. (1974) have reported similar findings for the epidermis of the hairless mouse and the hamster cheek pouch, respectively. Unfortunately the FLM method generated unreliable data, and much research time and money was wasted in the 1960s and 1970s.

## Host Susceptibility Rhythms

Because there is a rhythm in cell proliferation and other metabolic events, it should not be surprising

that there also is a rhythm in host toxicity to any drug which has as its target any specific phase of the cell cycle. Such a rhythm is illustrated in Fig. 27; 15% of the animals treated succumbed to the toxic effects of ara-C administered at one circadian stage, whereas approximately 74% of the animals died after being given the same injection of the drug at another circadian stage (Cardoso et al. 1970; Scheving et al. 1974b). When one compares this variation in drug toxicity to the rhythm in DNA synthesis in the duodenum, it is seen that the highest mortality occurs at the time when DNA synthesis was just beginning its daily upswing (Fig. 8) (Scheving et al. 1977b). Of course, we cannot be sure that this is the cause of the variation in toxicity, but it is recognized that the bone marrow and gut are two of the primary target organs of this agent.

Because the biological system is rhythmically changing, it follows that the organism is biochemically a different entity at different circadian stages; consequently, as illustrated for ara-C, it reacts differently to the same stimulus when applied at different times. This differential response to an identical stimulus repeatedly has been confirmed for a variety of stimuli; these include: drugs, poisons, chemical substances, physical agents such as noise and X-ir-

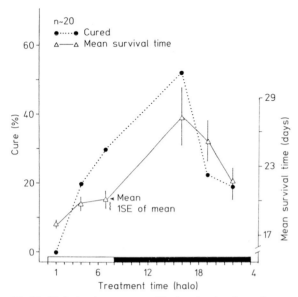

**Fig. 28.** Variation in cure rate of leukemic mice depending on the timing of combined treatment with CTX and ADR. Study 1 *(solid line)* on male mice in LD 12:12; study 2 *(dotted line)* on female mice in LD 8:16. See text and reference cited for further explanation. (Scheving et al. 1980)

radiation, and biological agents such as endotoxins. (For reviews see Reinberg and Halberg 1971; Scheving et al. 1974c; Moore 1973)

We and others have carried out a series of studies which suggest that such variation in susceptibility to several anticancer agents which are cell cycle-specific can be optimized in experimental cancer chemotherapy (Haus et al. 1972; Kühl et al. 1974; Scheving et al. 1977). For a review, see Scheving (1980a) and Levi et al. (1980). Figure 28 presents one such example. In this particular case, cyclophosphamide (CTX; 100 mg/kg) and adriamycin (ADR; 5 mg/kg) were found to be synergistic in treating mice that had been inoculated with $1 \times 10^5$ L1210 leukemia cells 4 days prior to treatment. With only one course of treatment, there was a dramatic circadian variation in response, as monitored by mean survival time and cure rate. The variation in cure rate (mice alive and apparently free of disease 75 days post-tumor inoculation) as a function of treatment timing ranged from 8% to 68% in male animals standardized to 12 h of light alternating with 12 h of darkness. Similarly, in female mice standardized to 8 h of light alternating with 16 h of darkness, the cure rate ranged from 0% to 56% depending upon when the drugs were injected during the 24-h span. No cures were obtained with either drug used alone (Scheving et al. 1980a). The maximum cure rate was recorded when the two drugs

were administered in the early part of the dark span of the light-dark cycle (whether 12 h of light and 12 h of darkness, or 8 h of light and 16 h of darkness); maximum mortality occurred following treatment in the L span. The data also document that maximal therapeutic advantage was obtained when the two drugs were separated by 1- or 3-h intervals and that this effect of drug sequencing was strongly circadian stage dependent (Scheving et al. 1980). We believe that the variation in toxicity to cell cycle-specific agents is due to a "shielding in time" of the normal tissues such as the intestinal tract, while still killing cancer cells which in our experience generally are found to divide in a noncircadian manner. (For further discussion see Scheving 1981.)

The above data from the experimental models demonstrate clearly why one should not ignore circadian variation in tolerance when dealing with chemotherapy or sequencing studies. The evidence is compelling that the temporal organization in general carries with it significant implications, not only for cancer chemotherapy and radiotherapy, but equally important for basic research on normal as well as abnormal growth (Scheving 1980a; Haus et al. 1974).

## Conclusions

The time has now arrived when most scientists accept without question the existence of and the potential importance of such rhythms, as considered above, even though it still remains a mystery why some still ignore them in their experimental design. Those responsible for regulatory control of toxicity such as the Food and Drug Administration must become more sensitive to such data than they have been! Based on a critical mass of data investigators must completely abandon the erroneous concept that somehow sampling at the same time of day takes care of the rhythm problem (Scheving 1980a). Moreover, it cannot be assumed that rhythmic variation involves only minor change around a 24-h mean and consequently is of no consequence.

Circadian cytokinetics, pharmacokinetics, endocrinology, immunology, and toxicology no longer can be ignored, but must be used to treat human cancer more effectively and eventually to prevent some varieties of malignancies. A major short-coming of classical noncircadian cytokinetics has been the inability to directly translate in vitro advances from the petri dish to the cancer patient. It has been thought that this has been because of the recognition that the in vitro synchronized cell population behaves differently than the in vivo populations which classically have been assumed to be nonrhythmic with regard to cytokinetic parameters. The realization that there is a relative degree of synchrony along the circadian time scale within all normal in vivo cell populations is the only solution for cytokinetics as a practical, clinically relevant tool for the treatment of cancer patients. If the basic rhythmicity inherent in the organism is properly considered, we believe this will greatly benefit cancer patients.

Perhaps we can come to better grips with the mechanism of such diseases as cancer if we pay more attention to its rhythmic nature. As of 1992 the mechanisms of normal or abnormal cell division are not known!

# References

Al-Nafussi AL, Wright NA (1982) The effect of epidermal growth factor (EGF) on cell proliferation of the gastrointestinal mucosa in rodents. Virchows Arch [B] 40: 71–79

Andrew W, Andrew NB (1949) Lymphocytes in the normal epidermis of the rat and man. Anat Rec 104: 217–241

Badran AF, Echave Llanos JM (1965) Persistence of mitotic circadian rhythm of a transplantable mammary carcinoma after 35 generations: its bearing on the success of treatment with endoxan. J Natl Cancer Inst 35: 285–290

Barnum CP, Jardetzky CD, Halberg F (1957) Nucleic acid synthesis in regenerating liver. Tex Rep Biol Med 15: 134–147

Bertalanffy FD (1960) Mitotic rates and renewal times of the digestive tract epithelia in the rat. Acta Anat 40: 130–148

Bertalanffy FD (1962) Cell renewal in the gastrointestinal tract of man. Gastroenterology 43: 472–473

Bertalanffy FD (1963a) Mitotic rate of methylcholanthrene induced subcutaneous fibrosarcoma. Naturwissenschaften 50: 648–659

Bertalanffy FD (1963b) Mitotic rate of spontaneous mammary gland adenocarcinoma in C3H/HeJ mice. Nature 198: 496–497

Bertalanffy FD, McAskill C (1964) Rate of cell division of malignant mouse melanoma. J Natl Cancer Inst 32: 535–545

Blumenfeld CM (1939) Periodic mitotic activity in the epidermis of the albino rat. Science 90: 446–447

Blumenfeld CM (1943) Studies of normal and abnormal mitotic activity. II. The rate and periodicity of the mitotic activity of experimental epidermoid carcinoma in mice. Arch Pathol 35: 667–673

Bostroem E (1928) "Der Krebs des Menschen". Thieme, Leipzig

Broders AD, Dublin WB (1939) Rhythmicity of mitosis in the epidermis of human beings. Proc Staff Meet Mayo Clin 14: 423–425

Bucher NLR, Patel U, Cohen S (1978) Hormonal factors concerned with liver regeneration. In: Hepatotrophic factors, a Ciba symposium, p 95

Bullough WS (1947) Mitotic activity in the adult male mouse, Mus musculus L. The diurnal cycles and their relation to waking and sleeping. Proc R Soc Lond [Biol] 135: 212–242

Bünning E (1958) Das Weiterlaufen der "Physiologischen Uhr" im Zuckerdarm ohne zentrale Steuerung. Naturwissenschaften 45: 68

Burns ER, Scheving LE (1973) Isoproterenol-induced phase shifts in circadian rhythm mitosis in murine corneal epithelium. J Cell Biol 56: 605–608

Burns ER, Scheving LE (1975) Circadian influence of the waveform of the frequency of labeled mitosis and mouse corneal epithelium. Cell Tissue Kinet 8: 61

Burns ER, Scheving LE, Tsai TH (1972) Circadian rhythm in uptake of triatiated thymidine by kidney, parotid and duodenum of isoproterenol-treated mice. Science 175: 71–73

Burns ER, Scheving LE, Tsai TH (1979) Circadian rhythms in DNA synthesis and mitosis in normal mice and in mice bearing a Lewis Lung Carcinoma. Eur J Cancer 15: 233–242

Cardoso SS, Scheving LE, Halberg F (1970) Mortality of mice are influenced by the hour of the day of drug (ARAC) administration. Pharmacology 12: 302

Carleton A (1934) A rhythmic periodicity in the mitotic division of animal cells. J Anat 68: 251–263

Cameron JH (1936) The origin of new epidermal cells in the skin of normal and x-rayed frogs. J Morphol 59: 327–350

Chabot JG, Payet N, Hugon JS (1983) Effects of epidermal growth factor (EGF) on adult mouse small intestine in vivo and in organ culture. Comp Biochem Physiol 74 (2): 247–252

Cherington PV, Smith BL, Pardee AB (1979) Loss of epidermal growth factor requirement and malignant transformation. Proc Natl Acad Sci USA 70: 3937

Chiakulas JJ, Scheving LE (1966) Circadian phase relationships of $^3$H-thymidine uptake, numbers of labeled nuclei, grain counts and cell division in the larval urodele epidermis. Exp Cell Res 44: 256–262

Clark RH, Baker BL (1963) Effect of hypophysectomy on mitotic proliferation in gastric epithelium. Am J Physiol 204: 1018–1022

Clausen OBF, Thorud E, Bjerknes R, Elgjo K (1979) Circadian rhythms in mouse epidermal basal cell proliferation. Variations in compartment size, flux and phase duration. Cell Tissue Kinet 12: 319–337

Cooper ZK, Franklin HC (1940) Mitotic rhythms in the epidermis of the mouse. Anat Rec 78: 1–8

Cooper ZK, Schiff A (1938) Mitotic rhythms in the human epidermis. Proc Soc Exp Biol Med 39: 323–324

Dembinski A, Gregory H, Konturek SJ, Polanski M (1982) Trophic action of epidermal growth factor on the pancreas and gastrointestinal mucosa in rats. J Physiol (Lond) 325: 35–42

Dublin WB, Gregg RO, Broders AC (1940) Mitosis in specimens removed during the day and night from carcinoma of large intestine. Arch Pathol 30: 893–895

Echave Llanos JM, Nash RE (1970) Mitotic circadian rhythm in a fast growing and slow growing hepatoma. Mitotic rhythm in hepatomas. JNCI 44: 581–586

Edmunds LN Jr (1971) Persistent circadian rhythm of cell division in Euglena: some theoretical considerations and the problem of intracellular communication. In: Menaker NM (ed) Biochemistry. National Academy of Sciences, Washington, DC

Fisher LB (1968) The diurnal mitotic rhythm in the human epidermis. Br J Dermatol 180: 75–80

Forgensen PE, Poulsen SS, Nexo E (1988) Distribution of i.v.

administered epidermal growth factor in the rat. Regul Peptides 23: 161–169

Fortuyn-van-Leyden CED (1917) Some observations of periodic nuclear division in the cat. Proc K Ned Akad Wet 38: 44

Fortuyn-van-Leyden CED (1926) Day and night period in nuclear division. Proc K Ned Akad Wet 29: 979–988

Frieboes W (1920) Beiträge zur Anatomie und Biologie der Haut. Dermatol Z 31: 57–83

Gallo-Paget N, Pothier P, Hugon JS (1987) Ontogeny of EGF receptors during postnatal development of mouse small intestine. J Pediatr Gastroenterol Nutr 6: 114–20

Garcia-Sainz M, Halberg F (1966) Mitotic rhythms in human cancer, reevaluated by electronic computer programs – evidence for chronopathology. J Natl Cancer Inst 37: 279–292

Gasser RF, Scheving LE, Pauly JE (1972a) Circadian rhythm in the mitotic index and in the uptake of $^3$H-thymidine by the tongue of the rat. J Cell Physiol 80: 437–442

Gasser RF, Scheving LE, Pauly JE (1972b) Circadian rhythm in the cell division rate of the inner enamel epithelium and in the uptake of $^3$H-thymidine by the root tip of rat incisors. J Dent Res 51: 740–746

Gololobova MT (1958) Changes in mitotic activity in rats in relation to the time of day or night. Bull Exp Biol Med 40: 1143–1146

Goodlad RA, Wilson TJ, Canton W, Gregory H, McCullagh KG, Wright NA (1987) Proliferative effects of urogasterone – EGF on the intestinal epithelium. Gut 28 [Suppl]: 37–43

Goodlad RA, Savage AR, Lenton W, Ghatei MA, Gregory H, Bloom SR, Wright NA (1988) Does resection enhance the response of the intestine to urogastrone – epidermal growth factor in the rat? Clin Sci 75: 121–126

Goodrum PJ, Sowall JG, Cardoso SS (1974) Characterization of the circadian rhythm of mitosis in the corneal epithelium of the immature rat. In: Scheving LE, Halberg F, Pauly JE (eds) Chronobiology. Igaku Shoin, Tokyo, p 29–32

Gupta BD, Deka AC, Halberg E (1975) Application of chronobiology to radiotherapy of tumor of the oral cavity. Chronobiologia 2, Suppl 1: 125

Halberg F (1964) Grundlagenforschung zur Aetiologie des Karzinoms. Monatskurse Aerztl Fortbild 14: 67–77

Halberg F, Halberg E, Barnum CP, Bittner JJ (1959) Physiologic 24-hour periodicity in human beings and mice, the lighting regimen and daily routine. In: Withrow RB (ed) The cellular aspects of biorhythms. Springer, Berlin Heidelberg New York, p 20

Halberg F, Gupta BD, Haus E, Halberg E, Deka AC, Nelson W, Sothern RB, Cornelissen G, Lee JK, Lakatua DJ, Scheving LE, Burns ER (1977) Steps toward cancer chronopolytherapy. In: Proceedings of XIVth International Congress of Therapeutics. L'Expansion Scientifique Francaise, Montpellier, p 151–196

Haus E, Halberg F, Scheving LE, Cardoso S, Kühl A, Sothern R, Shiotsuka R, Hwang DS, Pauly JE (1972) Increased tolerance of leukemic mice to arabinosyl cytosine with schedule adjusted to circadian system. Science 77: 80–82

Haus E, Halberg F, Loken MK, Kim YS (1974) Circadian rhythmometry of mammalian radiosensitivity. In: Tobias CA, Todd P (eds) Space radiation biology and related topics. Academic, New York, p 435

Hrushesky W, Levi F, Halberg F, Haus E, Scheving LE, Sanchez S, Medini E, Brown H, Kennedy BJ (1980) Clinical chronooncology. In: Scheving LE, Halberg F (eds) Chronobiology: principles and applications to shifts in schedules. Sijthoff and Noordhoff, Alphen aan den Rijn, p 513

Hrushesky W, Halberg F, Heinlen T, Murray C, Kennedy BJ (1981) Human bone marrow toxicity of adriamycin (AD) reduced by optimal circadian timing. Cancer Res to be published

Hunt TE (1952) Mitotic activity in the gastric glands of the rat. Anat Rec 112: 346

Kanabrocki EL, Scheving LE, Halberg F, Brewer RL, Bird TJ (1974) Circadian variation in presumable healthy young soldiers. US Department of Commerce, Doc No PB228427, PO Box 51553, Springfield, Virginia 22151

Karsten G (1918) Über Tagesperiode der Kern- und Zellteilung. Z Bot 10: 1–20a

Killicott WE (1904) The daily periodicity of cell division and the elongation in the root of *Allium*. Bull Torrey Bot Club 31: 529

Killman SA, Cronkite EP, Fliedner TM, Bond VT (1962) Mitotic indices of human bone marrow cells. I. Number and cytologic distribution of mitoses. Blood 19: 743–750

Kirkegaard P, Skov-Olsen P, Poulsen SS, Ebba Nexø (1983) Epidermal growth factor inhibits cysteamine-induced duodenal ulcers. Gastroenterology 85: 1277

Klein H, Geisel H (1947) Zum Nachweis eines 24-Stundenrhythmus der Mitosen bei Ratte und Maus. Klin Wochenschr 25: 662–663

Konturek SJ, Radecki T, Brozozowski T, Piastuchi I, Dembinski A, Dembinsha-Kiec A, Zumuda A, Gryglewski R, Gregory H (1981) Gastric cytoprotection by epidermal growth factor. Gastroenterology 81: 438

Krieger DT, Hauser H, Liotta A, Zelenetz A (1976) Circadian periodicity of epidermal growth factor ad its abolition by superior cervical ganglionectomy. Endocrinology 99: 1589–1596

Kühl JFK, Haus E, Halberg F, Scheving LE, Pauly JE, Cardoso SS, Rosene G (1974) Experimental chronotherapy with ara-C; comparison of murine ara-C tolerance on differently timed treatment schedules. Chronobiologia 1: 316–317

Laerum OD, Aardal NP (1981) Chronobiological aspects of bone marrow and blood cells. In: Scheving LE, Pauly JE (eds) Biological rhythms in structure and function. Liss, New York, p 87–97

Leblond CP, Stevens CE (1948) The constant renewal of the intestinal epithelium in the albino rat. Anat Rec 100: 357–377

Levander J (1950) On the epithelium-regeneration in the healing wounds. Acta Chir Scand 100: 637–649

Levi F, Hrushesky W, Haus E, Halberg F, Scheving LE, Kennedy BJ (1980) Experimental chono-oncology. Principles and applications to shifts and schedules. Sijthoff and Noordhoff, Alphen aan den Rijn, p 481

Li AK, Schatternkerk ME, de Vries JE, Ford WDA, Malt RA (1980) Submandibular sialoadenectomy retards dimethylhydrazine induced colonic carcinogenesis. Gastroenterology 78: 1207

Liosner LD, Artemieva NS, Babaeva EG, Romanova LK, Ryabinina ZA, Sidorva VF, Kharlova GV (1962) On the level and 24-hour rhythm of mitotic activity in hypophysectomized rats. Eksp Biol Med 54: 77–81

Malhotra SL (1967) Effect of saliva on gastric emptying. Am J Physiol 213: 169

Mauer AM (1965) Diurnal variation of proliferative activity in the human bone marrow. Blood 26: 1–6

Maurer HR (1981) Potential pitfalls of [3H]-thymidine techniques to measure cell proliferation. Cell Tissue Kinet 14: 111–120

Miettinen PJ, Perheentupa J, Otonkoski T, Lahteenmaki A, Panula P (1989) EGF- and TGF-alpha-like peptides in human fetal gut. Pediatr Res 26: 25–30

Møller U, Larsen JK, Faber M (1974) The influence of injected triatiated thymidine on the mitotic circadian rhythm in the epithelium of the hamster cheek pouch. Cell Tissue Kinet 7: 1007

Moore EM (1973) Circadian rhythms of drug effectiveness and toxicity. Clin Pharmacol Ther 14: 925

Muhlemann HR, Marthaler TM, Rateitschak KH (1956) Mitosenperiodik in der Nebennierenrinde, Schilddrüse, im Duodenal- und Mundhohlenepithel der Ratte. Acta Anat 28: 331–341

Müller O (1974) Circadian rhythmicity and response to barbiturates. In: Scheving LE, Halberg F, Pauly JE (eds) Chronobiology. Igaku Shoin, Tokyo, p187

Nash RE, Echave Llanos JM (1971) Twenty-four-hour variations in DNA synthesis of a fast-growing and slow growing hepatoma: DNA synthesis rhythm in hepatoma. J Natl Cancer Inst 47: 1007–1012

Nelson W, Zinneman H, Selden JA, Schaber K, Halberg F, Bazin H (1974) Circadian rhythm in Bence-Jones protein excretion by LOU rats bearing a transplantable immunocytoma, responsive to adriamycin treatment. Int J Chronobiol 2: 359–366

Ortiz-Picón JM (1934) Über Zellteilungsfrequenz und Zellteilungsrhythmus in der Epidermis der Maus. Z Zellforsch Mikrosk Anat 19: 488–509

Pauly JE (1981) An introduction to chronobiology. In: Scheving LE, Pauly JE (eds) Biological rhythms in structure and function. Liss, New York, p1

Pauly JE, Scheving LE (1964) Temporal variations in the susceptibility of white rats to pentobarbital sodium and tremorine. Int J Neuropharmacol 3: 651

Pauly JE, Burns ER, Halberg F, Tsai S, Betterton HO, Scheving LE (1975) Meal-timing dominates lighting regimen as a synchronizer of the eosinophil rhythm in mice. Acta Anat 93: 60

Phillipens KMH, v Mayersbach H, Scheving LE (1977) Effects of the scheduling of meal-feeding at different phases of the circadian system in rats. J Nutr 107: 176–193

Pilgrim C, Erb W, Mauer W (1963) Diurnal fluctuations in the numbers of DNA synthesizing nuclei in various mouse tissues. Nature 199: 863

Poulsen SS, Nexo E, Olsen TS, Hess J, Kirkegaard P (1986) Immunochemical localization and epidermal growth factor in rat and man. Histochemistry 85: 389–394

Powell EW, Pasley JN, Scheving LE, Halberg F (1980) Amplitude-reduction and acrophase advance of circadian mitotic rhythm in corneal epithelium of mice with bilaterally lesioned suprachiasmatic nuclei. Anat Rec 197: 227–281

Quastler H, Sherman FG (1959) Cell population kinetics in the intestinal epithelium of the mouse. Exp Cell Res 17: 420–438

Raitsina SS (1961) Physiological regeneration in the mucosa of the small intestine of hypophysectomized rats. Bull Exp Biol Med 51: 494–497

Reinberg A, Halberg F (1971) Circadian chronopharmacology. Annu Rev Pharmacol 2: 455

Reinberg A, Reinberg MA (1977) Circadian changes of duration of action of local anaesthetic agents. Naunyn Schmiedeberg's Arch Pharmacol 297: 149

Roberts ML, Friston JA, Reade PC (1976) Suppression of immune responsiveness by a submandibular salivary gland factor. Immunology 30: 811

Rosene GL, Halberg F (1970) Circadian and ultradian rhythms in liver, breast cancer and Ehrlich ascites tumor of mice. Bull All India Inst Med Sci 4: 77–94

Rubin NH (1981) Circadian stage dependence in radiation: response of dividing cells in vivo. In: Scheving LE, Pauly JE (eds) Biological rhythms in structure and function. Liss, New York, pp151–196

Rubin NH, Hokanson JA, Bodgon G (1983a) Circadian rhythms in phases of the cell cycle as demonstrated by flow cytometry. Cell Tissue Kinet 16: 115–123

Rubin NH, Hokanson JA, Mayshack JW, Tsai TH, Barranco SC, Scheving LE (1983b) Several cytokinetic methods for showing circadian variation in normal murine tissue and in a tumor. Am J Anat 168: 15–26

Scheving LA, Yeh YC, Tsai TH, Scheving LE (1979) Circadian phase dependent stimulatory effects of epidermal growth factor on DNA synthesis in the tongue, esophagus and stomach of the adult male mouse. Endocrinology 105: 1475

Scheving LA, Yeh YC, Tsai TH, Scheving LE (1980) Circadian-phase dependent stimulatory effects of epidermal growth factor on DNA synthesis in the duodenum, jejunum, ileum, caecum, colon and rectum of the adult male mouse. Endocrinology 106: 1498–1503

Scheving LA, Scheving LE, Tsai TH, Pauly JE (1982) Circadian stage-dependent effects of insulin and glucagon on deoxyribonucleic acid synthesis in the esophagus, stomach, duodenum, jejunum, ileum, caecum, rectum and spleen of the adult female mouse. Endocrinology 111: 308–315

Scheving LA, Shiurba RA, Nguyen TD, Gray GM (1989a) Epidermal growth factor receptor of intestinal enterocyte localization to laterobasal but not brush border membrane. J Biol Chem 264 (3): 1735–1741

Scheving LA, Tsai TH, Cornett LE, Feuer RJ, Scheving LE (1989b) Circadian variation in binding and affinity of epidermal growth factor to its receptors mouse liver. Anat Res 224: 559–565

Scheving LE (1959) Mitotic activity in the human epidermis. Anat Rec 135: 7–15

Scheving LE (1976) The dimension of time in biology and medicine-chronobiology. Endeavour 35: 66–76

Scheving LE (1980a) Chronotoxicology in general and experimental chronotherapeutics of cancer. In: Scheving LE, Halberg F (eds) Chronobiology: principles and applications of shifts and schedules. Sijthoff and Noordhoff, Alphen aan den Rijn, p455–479 (NATO Advanced Study Inst Ser)

Scheving LE (1980b) Chronobiology: a new perspective for biology and medicine. In: E.Saletu (ed) Proceedings of the eleventh Collegium International Neuro-Psychopharmacologicum (CINP) congress Vienna, Austria, 9–14 July, 1978. Pergamon, Oxford, pp629–646

Scheving LE (1981) Circadian rhythms of cell proliferation: their importance when investigating the basic mechanism of normal vs. abnormal growth. In: v.Mayersbach H, Scheving LE, Pauly JE (eds) Biological rhythms in structure and function. Liss, New York, pp39–79

Scheving LE (1984) Chronobiology of cell proliferation in mammals – implications for basic research and cancer chemotherapy. In: Edwards LN (ed) Cell cycle clocks. Dekker, New York, pp455–500

Scheving LE, Burns ER (1977) Some evidence that a consideration of the chronobiology of a cell cycle may improve chemotherapy. In: Scharf JH, v. Mayersbach H (eds) Die Zeit und das Leben. Nova Acta Leopold 46: 277

Scheving LE, Gatz AJ (1955) Mitotic activity in the human epidermis. Anat Rec 121: 363

Scheving LE, Pauly JE (1960) Daily mitotic fluctuation in the epidermis of the rat and their relation to variations in spontaneous activity and rectal temperature. Acta Anat 43: 337

Scheving LE, Pauly JE (1967) Circadian phase relationship of thymidine-H[3] uptake, labeled nuclei, grain counts and cell division rate in rat and corneal epithelium. J Cell Biol 32: 677–683

Scheving LE, Pauly JE (1971) Pitfalls awaiting those who ignore circadian time structure in experimental design: a conclusion based on experiments to test the effect of cytosine arabinoside on mitosis. Anat Rec 169: 419

Scheving LE, Pauly JE (1973) Cellular mechanisms involving biorhythms with emphasis on those rhythms associated with the S and M stages of the cell cycle. Int J Chronobiol 1: 269–283

Scheving LE, Pauly JE (1977) Several problems associated with the conduct of chronobiological research. In: Scharf KH, v. Mayersbach H (eds) Die Zeit und das Leben. Nova Acta Leopold 46: 237–258

Scheving LE, Vedral E (1966) Circadian variation of susceptibility of rat to several different pharmacological agents. Anat Rec 154: 417

Scheving LE, Vedral DF, Pauly JE (1968) Daily circadian rhythm in rats to D-amphetamine sulphate: effect of blinding and continuous illumination on the rhythm. Nature 219: 621

Scheving LE, Burns ER, Pauly JE (1972) Circadian rhythms in the mitotic activity and [3]H-thymidine uptake in the duodenum; effect of isoproterenol on the mitotic rhythm. Am J Anat 135: 311–317

Scheving LE, Pauly JE, Burns ER, Halberg F, Tsai TH, Betterton HO (1974a) Lighting regimen dominates interacting meal schedules and synchronizes mitotic rhythms in the mouse corneal epithelium. Anat Rec 180: 447

Scheving LE, Cardoso SS, Pauly JE, Halberg F, Haus E (1974b) Variation in susceptibility of mice to the carcinostatic agent arabinosyl cytosine. In: Scheving LE, Halberg F, Pauly JE (eds) Chronobiology. Igaku Shoin, Tokyo, pp 213–217

Scheving LE, v. Mayersbach H, Pauly JE (1974c) An overview of chronopharmacology (a general review). J Eur Toxicol 7: 203–227

Scheving LE, Enna CC, Halberg F, Jacobsen RR, Mather A, Pauly JE (1975) Mean circadian cosinors of vital signs, performance and of blood and unnary constituents in patients with leprosy. Int J Lepr 43: 364

Scheving LE, Burns ER, Pauly JE, Tsai TH, Betterton HO, Halberg F (1976) Meal scheduling, cellular rhythms and the chronotherapy of cancer. In: Kioshi H (ed) Nutrition. Victory-sha Kyoto, p 141

Scheving LE, Burns ER, Pauly JE (1977b) Can chronobiology be ignored when considering the cancer problem? In: Nieburgs HE (ed) Prevention and Detection of Cancer. Pt 1, prevention, vol 1, etiology. Dekker M, New York, p 1063

Scheving LE, Burns ER, Pauly JE, Halberg F, Haus E (1977c) Survival and cure of leukemic mice after optimization of cancer treatment with cycleophosphamide and arabinosyl cytosine. Cancer Res 37: 3648

Scheving LE, Burns ER, Pauly JE, Tsai TH (1978) Circadian variation and cell division of the mouse alimentary tract, bone marrow and corneal epithelium. Anat Rec 191: 479–485

Scheving LE, Burns ER, Pauly JE, Halberg H (1980b) Circadian bioperiodic response of mice bearing a L1210 leukemia to combination therapy with adriamycin and cyclophosphamide. Cancer Res 40: 1511

Scheving LE, Pauly JE, Scheving LA (1981a) Circadian rhythms at the cellular level. In: v. Mayersbach H, Reinberg A, Smolensky M (eds) Chronobiology in pharmacology and nutrition. Springer, Berlin Heidelberg New York

Scheving LE, Pauly JE, Tsai TH, Scheving LA (1981b) Effect of insulin on DNA labelling in gastric mucosa depends on circadian stage of mouse when drug is administered. Anat Rec 199: 226 A

Scheving LE, Pauly JE, Tsai TH, Scheving LA (1983a) Chronobiology of cellular proliferation: implication for cancer chemotherapy. In: Reinberg A, Smolensky M (eds) Topics in environmental physiology and medicine. Springer, Berlin Heidelberg New York

Scheving LE, Tsai TH, Powell EW, Pasley JN, Halberg F (1983b) Bilateral lesions of suiprachiasmatic nuclei affect circadian rhythms in [3H]-thymidine incorporation into deoxyribonucleic acid in mouse intestinal tract, in mitotic index of corneal epithelium, and in serum corticosterone. Anat Rec 205: 239–249

Scheving LE, Tsai TH, Pauly JE, Halberg F (1983c) Circadian effect of ACTH 1-17 on mitotic Index of the corneal epithelium of Balb/c mice. Peptides 4: 183–190

Scheving LE, Tsai TH, Pauly JE, Scheving LA, Feuers RJ, Kanabrocki EL, Lucas EA (1989b) Temporal organization some reasons why it should not be ignored by biological researchers as well as practicing oncologists. Chronobiology 16: 307–329

Scheving LE, Tsai TH, Scheving LA (1989a) Rhythmic behavior in the gastrointestinal tract. In: Szabo S, Pfeiffer CJ (eds) Ulcer disease: new aspects of pathogenesis and pharmacology. Chap 21. CRC, Boca Raton, pp 239–270

Serota FT, Baserga R (1970) Polyinosinic acid-polycytidylic acid: inhibition of DNA synthesis by isoperoternol. Science 167: 1379–1380

Sigdesdad CP, Lesher S (1970) Further studies on the circadian rhythm in the proliferative activity of the mouse intestinal epithelium. Experientia 26: 1321–1322

Sigdesdad CP, Bauman J, Lesher S (1969) Diurnal fluctuation in the number of cells in mitosis and DNA synthesis in the jejunum of the mouse. Exp Cell Res 58: 159–162

Sinna HP (1960) Observations on the mitotic activity in the esophageal epithelium of the rat and its relationship the blood-sugar level. Patna J Med 34: 301–304

Stalfelt MG (1921) Studie über die Periodizität der Zellteilung und sich daran anschließende Erscheinungen. Kunst Svenska Vetensh Hand 62: 1–14

Stevens CE, Leblond CP (1953) Renewal of gastric mucosa cells. Anat Rec 115: 231–243

Sturtevant RP, Sturtevant FM, Pauly JE, Scheving LE (1978) Chronopharmacokinetics of ethanol. III. Circadian variations in rate of ethanolemia decay in human subjects. Int J Clin Pharmacol Biopharm 16: 594

Tähti E (1956) Studies of the effect of x-radiation on 24-hour variations in the mitotic activity of human malignant tumors. Acta Pathol Microbiol Scand Suppl 117: 166

Thorud E, Clausen OPF, Bjerknes R, Elgjo K, Lindmo T (1978) Circadian rhythms in the proliferative compartment of epidermis in the hairless mouse demonstrated by pulse cytophotometry, autoradiography and a statmokinetic (colcemid). In: Lutz D (ed) Pulse, cytophotometry, vol 3. European, Ghent, pp 359–364

Tsai TH, Scheving LE, Pauly JE (1970) Circadian rhythms in plasma inorganic phosphorus and sulphur of the rat, also in susceptibility to strychnine. Jap J Physiol 20: 12

Tvermyr EM (1969) Circadian rhythms in epidermal mitotic activity. Diurnal variations of the mitotic index, the mitotic rate and the mitotic duration. Virchows Arch [B] 2: 318–325

Tvermyr EMF (1972) Circadian rhythms in hairless mouse epidermal DNA-synthesis as measured by double labelling with $^3$H-thymidine [($^3$H)TdR]. Virchows Arch [B] 11: 43

Voutilainen A (1953) Über die 24-Stundenrhythmik der Mitosenfrequenz in malignen Tumoren. Acta Pathol Microbiol Scand Suppl 99: 1–104

Walker WV, Russell JE, Simmons DJ, Scheving LE, Cornelissen G, Halberg F (1985) Effect of a synthetic adrenocorticotropin analogue ACTH on DNA synthesis in metaphyseal bone. Biochem Pharmacol 34: 1191–1196

Wright NA, Pike C, Elia G (1990) Induction of a novel epidermal growth factor secreting cell lineage by mucosed ulceration in human gastrointestinal stem cells. Nature 343: 82–85

Yeh YC, Scheving LA, Tsai TH, Scheving LE (1981) Circadian-phase dependent effects of epidermal growth factor on deoxyribonucleic acid synthesis in ten different organs of the adult male mouse. Endocrinology 109: 644–645

Zugula-Mally ZW, Cardoso SS, Williams D, Simpson H, Reinberg A (1979) Time point differences in skin mitotic activity of actinic keratoses and skin cancers. Circadian reference: plasma cortisol. In: Reinberg A, Halberg F (eds) Chronopharmacology. Pergamon, New York, pp 399–402

# Chronobiology of Endocrine and Endocrine-Responsive Tumors

R. von Roemeling

## Introduction

### Temporal Organization of the Endocrine System

A prominent feature of the endocrine system is its high degree of temporal organization, which was recently reviewed by van Cauter [1]. Far from obeying the concept of constancy of the internal environment, circulating hormone levels spontaneously undergo pronounced oscillations. Plasma levels of adrenocorticotropin (ACTH), growth hormone (GH) and prolactin (PRL) follow a circadian pattern which repeats itself day after day. PRL levels decrease rapidly after morning awakening, a time when ACTH release is close to its maximum and GH secretion is generally quiescent. Both PRL and GH increase rapidly after sleep onset, a time when ACTH levels are essentially suppressed. Under normal conditions, the 24-h profile of plasma PRL levels follows a bimodal pattern, with minimal concentrations around noon, and afternoon phase of augmented secretion and a major nocturnal elevation starting shortly after sleep onset and culminating around mid-sleep. Thus, the release of these three hormones by the pituitary follows a highly coordinated temporal program. Pulsatile secretion is evident throughout the 24-h cycle for ACTH and PRL (Fig. 1). In contrast, pulses of GH secretion occur less frequently and are often confined to the early part of sleep. In addition to the circadian and pulsatile variations, other ultradian and infradian rhythms also occur. The ultradian range includes a pulsatile release of pituitary and pituitary-dependent hormones and ultrafast fluctuations with periods of reoccurrence in the range of minutes. In man, hormonal pulses generally recur at intervals ranging between 1 and 2 h (e.g., "hourly"), but this may vary greatly because the pulse frequence of a given hormone may be modulated by numerous factors, including levels of other hormones, age, circadian rhythmicity, and sleep. When intensified rates of blood sampling are used and hormonal levels are measured at 1–4 min intervals, ultrafast fluctuations of low amplitude may appear superimposed on the larger hourly pulses. Rapid oscillations also appear to characterize beta cell function of the islets of Langerhans as oscillations with periods in the 10–14 min range have been observed for both insulin and glucagon. The infradian range of human en-

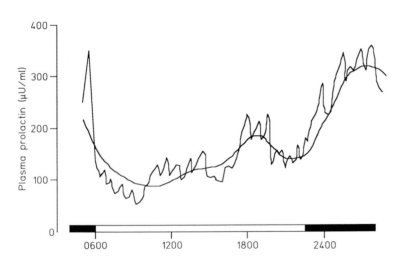

**Fig. 1.** Typical profile (24 h) of plasma prolactin. Pulsatile secretion plus circadian rhythm with bimodal pattern; augmented secretion in the afternoon with major nocturnal elevation after sleep onset

docrine rhythms include the menstrual cycle and seasonal variations. The latter are still poorly defined and appear to affect primarily the reproductive axis.

The interest in endocrine rhythms has greatly increased since the recognition of their essential role in maintaining normal endocrine function. Evidence showing that pathological states are associated with abnormal rhythms has accumulated, and therapies taking into account the temporal organization of endocrine release are being developed.

## Zeitgeber for Circadian and Seasonal Rhythms

The light-dark cycle is the major zeitgeber for both circadian and seasonal rhythms in most species [2]. A functional pineal gland is essential for the perception of changing day length in a number of photoperiodic species. The pineal hormone, melatonin, is the photoperiodic messenger molecule. Removal of the daily melatonin rhythm by pinealectomy or by sympathetic denervation of the gland abolishes the appropriate seasonal response to applied artificial photoperiod. Suitable administration of melatonin in physiological quantities to pinealectomized animals replicates the effects of varying photoperiod. Essentially, the pineal gland, through melatonin, appears to act as a synchronizer of circadian and seasonal cycles. How it achieves this is at present under intense investigation [3].

Melatonin is normally only secreted in high levels at night, and light serves both to suppress and entrain the endogenous rhythm of secretion. It thus serves as a "marker" molecule for the effects of light on circadian rhythms and provides an index of light sensitivity via light suppression of its nighttime secretion. The duration of melatonin secretion, which reflects the length of the dark phase, is the critical parameter in the induction of photoperiodic responses. The effects of melatonin may also depend on an underlying daily rhythm of sensitivity to the hormone [3, 4]. Melatonin is a methoxyindole secreted by the pineal gland; a secondary source is the retina, although whether its contribution is other than local is unclear. The final pathway controlling pineal function, insofar as indoleamines are concerned, lies in its sympathetic innervation from the superior cervical ganglion.

## Genetic Control of Circadian Rhythmicity

It is clear from a cursory examination of the biology of circadian rhythms that a wide variety of processes are being controlled, directly or indirectly, by genetic clocks. One important aspect of this temporal control within a rhythmic cell's metabolism network is the daily control of gene expression. Experiments by Loros and coworkers [5] have identified three genes in Neurospora which are regulated by light changes. Clock control of the genes identified could be exerted either by changes in initiation of gene expression or by changes in the stability of existing messages that are always transcribed at constant rates. In either case, the present data suggest that direct or indirect clock control of messenger ribonucleic acid (mRNA) levels may be an important aspect of developmental and environmental control of gene expression.

## Rhythm Disturbances May Promote Carcinogenesis

Tissue growth and function are, to some extent, controlled by local hormones (chalones and growth factors) and hormones that are elaborated some distance from that tissue (growth hormone, insulin, thyroid hormones, glucocorticoids, sex hormones, and many others). The endocrine tissues responsible for the release of these hormones are under similar local and distant controls, but are also subject to tight control by more or less specific trophic hormones usually elaborated from within the central nervous system. These pineal-hypothalamic-pituitary-endocrine-target tissue networks interact with one another and the local growth controls to coordinate cell turnover, as well as hormone production, and metabolic response hormones. A complex, rhythmic pattern characterizes healthy endocrinologic function and healthy target tissue response.

Hormonal influences upon cancer development have been extensively investigated. Endogenous hormones have been implicated as possible contributing causes of a number of tumors [6, 7]. These include, but are not limited to, cancers of the thyroid [8], endometrium [9], breast [10], prostate [11, 12], ovaries [13], and testis [14].

Barbason and coworkers [15, 16] have studied the correlation of liver growth and function during liver regeneration and hepatocarcinogenesis. It is generally observed, in different biological systems, that cell division and specific differentiated function present a mutually exclusive relationship. A hepatocyte, for instance, cannot be both dividing and synthesizing a specific enzyme. Under normal condition, most of the hepatocytes are in $G_0$ phase of the cell cycle. The authors studied the normal liver function of young

Wistar rats (weighing 150–180 g) by measuring cholesterol-7-alpha-hydroxylase activity. This enzyme is synthesized during the $G_0$ phase of the cell cycle, and its activity has a very short half-life. In addition, the mitotic index and labeling index were determined. Both enzyme activity and mitotic index showed strong circadian rhythmicity with opposite peaks and troughs. Several regulating factors could be involved in triggering this rhythmic activity. It is well known that the adrenal hormones act upon the two functions studied. Corticosterone inhibits cell division and stimulates the synthesis of a number of specific hepatic enzymes. However, glucocorticosteroids may require the intervention of another regulating factor. Subsequent disturbance of the regulatory mechanism was achieved by long-term administration of the hepatotoxic carcinogen diethylnitrosamine (DENA); initially, the drug induced necroses, and subsequently, increased proliferation (phase I; regulatory mechanism still intact). This was followed by a progressive loss of the temporal pattern of cell division and tissue function, during which preneoplastic areas emerged (phase II). Finally, frank neoplastic foci developed after 60 days of drug exposure, resulting in a 100% incidence of liver cancers. DENA exposure over at least 4 weeks was necessary to induce liver cancers. At this time a "point of no return" was passed, which coincided with the regulatory breakdown.

Liver generation after two-thirds hepatectomy was also studied. Partial hepatectomy triggered circadian mitotic waves presenting peaks in the morning. It did not promote neoplastic transformation, but the addition of phenobarbital after DENA accelerated the pathological evolution and increased the tumor incidence. The promoting effect was associated with the induction of chronic cell proliferation, the inhibition of the rapid response to the two-thirds partial hepatectomy, and the suppression of the mitotic circadian rhythm normally present during liver regeneration.

In a subsequent study, Mormont et al. [17] described that phenobarbital altered the circadian corticosterone rhythm, but pentobarbital did not. It is known that pentobarbital has no tumor promotion effects in contrast to phenobarbital.

Hrushesky et al. [18] used a different model to test the relationship between circadian stage and malignant growth. They investigated the effects of circadian timing, inoculum size, and location of tumor cell implantation upon the frequency of tumor take and subsequent tumor growth rate in a total of 96 young female $C_3$H/HeJ mice. These mice were adapted to a synchronizing schedule of 12 h light alternating with 12 h dark. Fibrosarcoma cells were injected subcutaneously in one of eight different doses, at one of eight permutated anatomic sites, and at one of six equispaced circadian times. Regardless of tumor cell dose and inoculum location, tumor take strongly depended on the circadian stage of inoculation, with injections of live tumor cells early in the animal's activity span resulting in the lowest incidence of tumor take. Injection site did not affect take frequency, but affected subsequent tumor growth rate. The lag time of tumor nodules of equal size also varied directly with the initial number of cells inoculated. Mice inoculated during mid-light (resting span) lived significantly longer than those inoculated during mid-dark (activity span). These data suggested that the factors involved in first-line tumor defense exhibit marked circadian rhythmicity. The tumor growth rate was dependent on the anatomic region of inoculation and the initial tumor cell dosage.

Natural killer (NK) cells are generally considered to be one of the effector systems of the immunosurveillance network against malignant tumors. Gatti et al. [19] had previously documented circadian changes in the number and activity of circulating NK cells in humans. They also determined that NK cell responsiveness to positive and negative modulation by immune interferon (IFN-gamma) and cortisol is also circadian-stage-dependent [20]. Spontaneous NK cytotoxicity was maximal in the morning at 8 h and then declined in the nocturnal hours. Maximum enhancement by IFN-gamma was attained between midnight and early morning at 4 h. Susceptibility to cortisol inhibition was maximal from evening to midnight. These results suggested that chronobiology of the immune system should be considered in immunotherapy schedules. They were confirmed by independent studies reported by Levi et al. [21] and Hrushesky et al. [22].

## Seasonality of Tumor Symptomatology

Annual cycles in many functions have been described in practically all vertebrates. In humans, the incidence of general mortality, suicide, and birth rates exhibit seasonal fluctuations. Significant seasonal variations in testicular size, testosterone levels, male fertility, and progesterone levels in women have been found in vertebrates, which may be the cause of a seasonal increase of cellular proliferation, preceding the development of neoplasia. Hrushesky et al. [23] hypothesized that this may also be reflected by a seasonality in the occurrence of testicular seminoma. Initially, they demonstrated in male $BDF_1$ mice that testicular cell proliferation gauged by tritiated thymidine uptake varied markedly as a function of season, with highest activity during the winter. Secondly, they retrospectively analyzed the onset of symptoms and the date of orchiectomy for seminoma in 185 patients and found that more than twice as many cases were diagnosed during the winter months as opposed to the summer. This coincidence of winter peaks of testicular proliferation and the circannual variation in the seminoma incidence suggested a periodic control mechanism for cellular growth in normal and malignant testicular tissues.

Seasonal variation in malignant growth has also been shown for other neoplasias. In a study reported by Boon et al. [24], a retrospective analysis of the incidence of squamous and cyclindrical epithelial neoplasia during the various months of the year was performed. In a series of 144,018 cervical smears obtained from a routine cytology laboratory in the Netherlands, during the period October 1982 to May 1985, infections and epithelial neoplasia were scored as incidence per thousand smears per month. Different significant circannual rhythms were found. Condylomata acuminata had a peak incidence in summer, whereas the chlamydia and trichomonas infections mostly occurred around autumn and winter. Neoplasia scores of squamous, ectocervical cells peaked around August. Cyclindrical, endocervical malignancies, however, occurred most frequently in autumn (October to December). This rhythmicity of the described abnormalities detection scores should be considered in the evaluation and planning of screening programs.

## Circadian Cytokinetics in Ovarian Cancer

Because of the difficulty in obtaining repeated biopsies of in vivo human cancers, most studies of tumor kinetic perturbations caused by surgery have been performed either with transplantable animal tumors or in in vitro tissue culture systems; these results have limited applicability for improving the treatment of human cancers. However, Klevecz et al. [25–27] were able to repeatedly collect human malignant cells during an intraperitoneal chemotherapy program, which required the surgical implantation of peritoneal dialysis catheters, followed by frequent irrigation of the peritoneal cavity in an attempt to decrease adhesion formation and maintain catheter patency. Tumor and normal host cells from these washings were analyzed by flow cytometry for DNA

content and cytokinetic parameters. When analyzed over time, proliferation dynamics including the period and amplitude of S-phase waves and the time of day when peak proliferation occurs could be determined. Thirty-one patients with advanced gynecologic malignancies were studied with serial, around-the-clock measurements, and of these 28 showed rhythmic changes in the fraction of cells in S phase. Eight of these patients were studied with serial analyses for 3 or more days postoperatively and seven showed a superimposed marked increase (two- to five fold) in the fraction of cells in S phase, with the maximum peak of proliferation occurring between days 4 and 7 postoperatively. These findings could be correlated to clinically relevant factors, including histologic type and grade of tumor, previous therapy, amount of tumor debulked, and residual tumor. Such information may also be useful in modifying the timing of postoperative chemotherapy in patients who have undergone surgical debulking, because cell cycle-specific drugs are more effective at times of maximal mitotic activity.

## Circadian Rhythms of Tumor Markers

Little has been published in the literature on the circadian rhythm of tumor marker antigens, although rhythmic sources of marker variability are of interest to interpret adequately laboratory values. Cancer patients with normal levels of biological markers in the morning at the usual time of sampling, may actually have circadian desynchronization or alteration of the pattern of the marker rhythm. Touitou et al. [28, 29] recently documented the circadian variations of plasma cortisol, total proteins, carcinoembryonic antigen (CEA) and breast cancer associated antigen, CA 15-3, two markers widely used in the follow up of breast cancer patients. All patients in this study had elevated levels of CA 15-3 and CEA. A small-amplitude circadian rhythm was validated in the patients before chemotherapy with doxorubicin by infusion. Treatment resulted in an impairment of circadian rhythmicity of all documented variables except cortisol. Whether this discrepancy is in relation to the small number of patients ($n = 6$) or with tumor burden remains to be shown.

Diurnal changes have also been reported by Hallek and Emmerich [30] for the serum concentrations of such markers as alpha-fetoprotein (AFP), prostatic acid phosphatase (PAP), lactate dehydrogenase (LDH) and thymidine kinase (TK). The clinical value of these observations is still hypothetical: Peaks of the serum levels of these markers might indicate situations of an increased tumor proliferation in a given patient and thus allow a temporally optimized application of anticancer treatment at that time. Moreover, repeated determinations may help to avoid misinterpretations of high tumor marker concentrations which might simply be due to rapid time-related changes.

## Chronotoxicity of Chemotherapy and Its Manipulation by Hormones

English et al. [4] have demonstrated a circadian variation in the toxicity of methotrexate in the rat. This variation can be modulated by abolition of the circadian rhythm in corticosterone production, by continuous light, by timed melatonin administration, or by melatonin implantation for 3 weeks prior to methotrexate injection.

Glucocorticoids and other hormones may interact with the toxicity and the efficacy of antineoplastic agents in either a favorable or an unfavorable way. This aspect is particularly relevant since hormonal therapy is often administered concurrently with cancer chemotherapy. Moreover, high doses of glucocorticosteroids are often used as antiemetics, mostly before and after cisplatin administration. Preclinical studies suggest that concurrent high-dose methylprednisolone and cisplatin administration may reduce the therapeutic index of the latter [31–33]. Clearly, further work is needed to define the temporal aspects of drug interactions and the effects on the time structures of the treatment recipient.

Both Hrushesky [34, 35] and Lévi et al. [36] have independently demonstrated in randomized clinical studies that the therapeutic index of cytotoxic chemotherapy in patients with advanced ovarian cancer is circadian-stage-dependent. The administration of anthracyclines during the early morning hours and cisplatin in the afternoon resulted in significantly reduced toxicity and improved therapeutic outcome. Underlying causes for these observations include circadian rhythmic changes in drug pharmacokinetics, tumor cell susceptibility including repair mechanisms, and susceptibility of normal host tissues which experience dose-limiting toxicity from the treatment.

## Breast Cancer: A Case in Point

### Murine Studies

Hormone rhythm alterations have had profound effects upon the incidence of murine breast cancer. Estrogen and prolactin are the hormones which have been most frequently implicated as contributing to the development of experimental murine and perhaps human breast cancer. Estrogens, if given by implantation, are more often carcinogenic to murine breast tissue than if similar total doses are given intermittently by injection [37]. Subcutaneous transplantation of syngeneic pituitary gland results in a disruption of the circadian rhythms in hormone release. When hypothalamus and pituitary are transplanted subcutaneously, hormonal rhythm characteristics are less radically altered.

Tumor Growth Control by Melatonin

In a study by Haus et al. [38] the incidence of breast cancer in Balb/C female mice which received subcutaneous muscle transplants was less than 1% during 800 days of subsequent observation. When pituitary gland was transplanted subcutaneously with muscle, 50% of these animals developed breast cancer during the more than 2-year follow-up period. When hypothalamus was transplanted with pituitary to another group, fewer breast cancers were found than in the group receiving pituitary only. These data indicated that breast cancer incidence in this strain of mouse is dependent upon a factor elaborated from the pituitary, which is modulated by the concurrent presence of hypothalamic control. Unfortunately, in these animals the 24-h average levels of serum corticosterone did not differ in both groups, but the circadian profiles were not given, and estrogen and prolactin rhythms were not measured.

A possible role for the pineal and melatonin in malignant disease was also suggested by studies of hormone-dependent tumors reviewed by Tamarkin et al. [39]. Melatonin protected against, while pinealectomy enhanced 7,12-dimethylbenz(alpha)-anthracene-induced mammary tumors in rats. Melatonin may protect against tumors through a suppressive effect on prolactin secretion or an action on estrogen receptors. It can change the concentration of estrogen receptors in hamsters and in a human breast cancer cell line, hereby inhibiting estrogen-stimulated growth.

The spontaneous breast cancer incidence was compared by Wrba et al. [40, 41] in more than 500 $C_3C$ $F_1$ mice, receiving daily injections of melatonin or of a vehicle or no treatment at one of 6 different circadian stages. Results were analyzed when 33% of all animals had developed breast cancer. Vehicle treatment increased tumor incidence, but melatonin reduced it circadian-stage-dependently.

Hormone Responsivity

Receptors are mediators of target tissue hormone responsivity. The significance of hormone receptors for breast cancer prognosis has been reviewed by Manni [42]. Estrogen and progesterone receptor status are important prognostic factors predicting increased disease-free survival and overall survival. Mammary tumors with positive estrogen and progesterone receptors, have a better prognosis than their hormone-independent counterparts. Autocrine/paracrine mechanisms may play an important role in breast cancer growth. A significant fraction of human breast tumors contain receptors for epidermal growth factor (EGF) as well as insulin-like growth factor (IGF-1). EGF receptors but not IGF-1 receptors have been consistently found to be negatively correlated with the estrogen and progesterone receptor status of the tumor. Furthermore, presence of EGF receptor has been found to be associated with a shorter relapse-free and overall survival. Recent preliminary data indicate that EGF receptor status may divide the estrogen receptor-negative population into good and poor prognosis subgroups. Preliminary data suggest that the relapse-free survival and overall survival of "double negative" patients may be as good as those observed in estrogen receptor-positive patients. These findings are especially interesting, because EGF may have a profound, circadian-stage-dependent effect on cell proliferation, including carcinogenesis [43].

Receptor circadian periodicity in noncancerous murine breast tissue was first reported by Labrosse et al. [44]. Pituitary isografting increased receptor circadian mean levels and decreased the daily amplitude of these levels. These data describe a coordination of hormone-hormone response, which is rhythmic in many time scales in health and if disturbed results in the development of breast cancer. Others have extensively documented monthly and circannual receptor periodicity in normal breast, oviduct, and endometrial tissue.

Data reported by Kiang et al. [45] addressed the cyclic biologic properties of the breast cancer of Balb/C GR-$F_1$ mice. These investigators studied tumor growth rate, tumor cell progesterone receptor content, the frequency of polyploidy in tumor cell karyotypes, and tumor cell thymidine kinase activity over 22 approximately monthly tumor cell transplantation generations. Visual inspection of these data revealed a high degree of variability of each of the studied endpoints. Rhythmometric analysis of these data over a 1-year period showed a 6-month periodicity for tumor polyploidy and tumor thymidine kinase levels. Tumor cell progesterone receptor concentrations were inversely related to tumor cell ploidy and thymidine kinase activity.

Menstrual Stage and Metastatic Potential

Ratajczak and coworkers [46] studied the effect of estrous stage, as reflected by vaginal cellularity, at the time of surgical resection of an estrogen receptor-bearing mammary adenocarcinoma upon the metastatic potential of that tumor in the $C_3$HeB/FeJ mouse. Presence of the tumor prolonged the length of the estrous cycle by about 25% and removal of the tumor returned the cycle to its usual duration. Neither estrous stage at tumor implant nor size of tumor at resection (within a small range) had significant independent effects upon differences observed in the incidence of subsequent pulmonary metastases. However, estrous stage at time of surgical removal of the tumor, as reflected by cell types in vaginal smear, markedly affected whether or not metastases ultimately appeared.

Hrushesky et al. [22] investigated the potential role of the immune response for the metastatic potential of malignancy by measuring splenic NK cell activity and interleukin-2 (IL-2) production in syngeneic tumor-free mice of two age groups at each of two circadian times and in each of four estrous stages. Mice from the younger age group were found to have eightfold higher NK activity and 35% greater IL-2 production (as gauged by CTLL-2 cell line bioassay). After normalization of NK and IL-2 values for age, a highly significant difference in NK activity was found among the four estrous and between the two circadian stages. NK activity was greater during the daily resting span across every estrous stage. IL-2 values were highest in diestrus and proestrus when sampled in the light span and in estrus-metestrus when sampled in the dark. The stages within the fertility cycle associated with lowest metastatic potential (proestrus/estrus) corresponded precisely with those of highest splenocyte NK activity. These results indicated that an important component of the cellular immune response varies rhythmically both during the fertility and circadian cycles of the host and many in fact control metastatic potential of malignancy.

## Human Studies

In a clinical study testing the same hypothesis, Hrushesky et al. [47] retrospectively examined the incidence of breast cancer recurrence, disease-free interval, and survival duration of 44 premenopausal women with 5- to 12-year follow-up (7-year median) after primary breast resection. As extrapolated from the above murine studies, each patient was assigned to two menstrual cycle categories based upon the number of days between the first day of the last menstrual period and the date of operation: 7–20 days, midcycle, and 0–6 or 21–36 days, perimenstrual. Patients resected near to menses had a more than fourfold higher risk of recurrence and death than did those resected near midcycle. These provocative results, which were predicted by the murine experimental model, suggest that the endocrine milieu at the time of resection impacts upon the eventual outcome of primary breast cancer. They await confimation from data gathered by major cooperative study groups.

### Breast Cancer Incidence and Mortality

Using data from the New York State Cancer Control Bureau, Jacobson and Janerich [48] described a seasonal variation in the diagnosis of breast cancer during a 3-year span between 1970 and 1973. Breast cancer was found to be most common in spring and autumn with the highest monthly incidence in May.

A nationwide study of all 3183 female patients with breast cancer in Israel, diagnosed over a 7-year period (1960–1966), was conducted by Cohen et al. [49]. Monthly series analysis showed a seasonal pattern in the symptomatology of the disease, which was most pronounced in patients younger than 55 years. It was seen in all ethnic groups and was mainly confined to cases with a nonlocalized tumor at diagnosis. Peaks occurred during spring, and troughs appeared during autumn. This pattern was believed to be of endogenous, probably hormonal nature, although environmental factors could not be excluded.

Langlands and associates [50] noted that the mortality of breast cancer in several series of patients treated for breast cancer by operation and/or local irradiation in the United Kingdom before the availability of chemotherapy was highly seasonal. Timing of peak breast cancer mortality for pre- and postmenopausal patients was opposite. Premenopausal patients more often succumbed to their disease in autumn while this was the season of lowest breast cancer mortality in the postmenopausal patients suffering from this disease. These seasonal differences in mortality suggest a circannual bioperiodicity of endocrine malignancy.

### Human Breast Cancer Risk: Prolactin and Melatonin Rhythms

Circadian and circannual patterns of putatively important hormones were studied in Japanese and American individuals at risk for the development of breast cancer by Halberg et al. [51]. It was hypothesized that intact circadian and circannual rhythms are important in maintaining health, and that statistically significant risk-dependent differences in the circadian and circannual patterns of these hormone levels would be discovered. The circadian characteristics of prolactin in plasma differed in certain seasons. The circadian rhythm-qualified plasma prolactin mean of the low-risk population was higher than that of the high-risk population. An even greater difference in circadian amplitude characterized the risk groups with a high circadian amplitude of plasma prolactin concentration seen in the low-risk group. Year-long sampling allowed evaluation of circannual plasma prolactin patterns. The seasonal means of plasma prolactin from women at high and low risk for the development of breast cancer differed little in summer and autumn but were statistically different in winter, when low-risk women had higher 24-h mean levels and a circannual rhythm with a higher amplitude. These results suggest that discrimination of populations at varying risk for the development of breast cancer is possible by prolactin rhythm analysis. The high-risk profile shows flat rhythm curves on circadian and circannual time scales. The low-risk profile is characterized by higher amplitude rhythms.

Tarquini et al. [52] compared the circadian and circannual rhythm characteristics of plasma prolactin concentrations of healthy Italian women with those of women with fibrocystic breast disease, who have a higher risk of developing breast cancer. Stratified by menstrual stage, both groups had similar circadian rhythms, but the mean level was higher in patients with fibrocystic mastopathy. The circannual rhythm amplitude was higher in the healthy population with a yearly variation of 112% of the mean value. The circannual rhythm seasonal amplitude was reduced in the population with fibrocystic breast disease. The observation of flattening or disappearance of the circannual rhythm curve in high-risk patients is consistent with Halberg's data [51].

A possible relationship between melatonin and estrogen receptors was examined in women with breast cancer [39]. An inverse correlation was observed between the estrogen receptor concentration in each patient's tumor and her plasma melatonin level. However, the hypothesis that a more robust daily melatonin rhythm offers greater protection against breast cancer is not easily tested, as shown in a study of women at high risk for developing breast cancer. These women had daily melatonin profiles that did not differ from those of a normal population. Thus, the use of plasma melatonin as a screening tool for breast cancer risk is not reasonable; however the basic and clinical data argue that the role of melatonin in endocrine-related cancers requires more extensive clinical investigation.

## Altered 17-Hydroxycorticosteroid Rhythm in Breast Cancer Patients

Alterations in corticosteroid metabolism have been reported in patients with breast cancer by Singh et al. [53]. A marked circadian rhythm in plasma 17-hydroxycorticosteroid (17-OHCS) was seen in normal controls with a maximum level at 0800 hours. While all breast cancer patients had increased levels, this circadian rhythm was found to be deranged in six of ten patients. It was postulated that altered rhythms may be associated with an inability to cope with stress, such as surgical procedures.

## Evidence for Residual Hormonal Control of Malignant Tissue of Endocrine Origin

Hughes and coworkers [54, 55] analyzed primary breast cancer samples obtained by biopsy or mastectomy from 98 consecutive postmenopausal patients between 0700 and 0900 hours over the span of 5 years for annual rhythmicity of estrogen receptor activity. Fourteen samples in which no estrogen receptor activity could be detected were removed from their analysis. The monthly means of the remaining 84 samples were not randomly distributed. The highest values were seen in autumn and spring and the lowest in summer and winter.

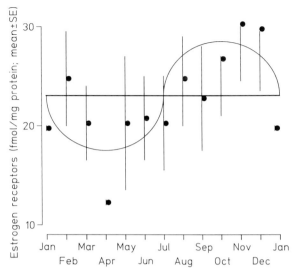

**Fig. 2.** Seasonality of estrogen receptor content in 797 breast cancer biopsies. Cosinor $p < 0.05$. (Hrushesky et al., 1979)

Hrushesky et al. [56] reported a larger series, in which estrogen receptor concentrations were determined by competitive binding (sucrose gradient) technique with radioactive estrogen in 797 consecutive specimens of primary breast cancer. Surgery was performed during the morning hours between 0800 hours and noon over a 4-year span from November 1974 to November 1978. A statistically significant circannual rhythm in estrogen receptor activity was found with a predicted difference between the lowest value in spring and highest value in autumn of 10.4 fmol; the mean value was 22.8 fmol (Fig. 2). This low estrogen receptor value in spring may determine an escape of breast cancer from hormonal growth control mechanism, explaining the higher circannual incidence during this season. However, this conclusion is inconsistent with the data of Hughes et al. [54], even considering that the data from the studies cited above were gathered from patients of different geographical hemispheres, in which seasonal environmental influences may be different.

## Circadian Temperature Rhythms and Breast Cancer: Early Detection?

As reviewed by Smolensky [57], the circadian temperature rhythm of a cancerous breast in comparison with that of the contralateral healthy breast becomes altered. Spectral analysis of continuous recordings of either surface or deep breast temperature has shown that noncancerous breasts exhibit rhythms with predominant periods of 24 h and 7 days. A tumorous breast, on the other hand, often exhibits non-24-h rhythms, i.e., predominant periodicities of approximately 20, 40, and 80 h. In other words, the temperature pattern of the cancerous breast undergoes a transformation from a circadian to an (about 20-h) ultradian organization. A corresponding swith from circadian to ultradian rhythmicity occurs in the mitotic rhythm of cells of cancerous breasts. Changes in the surface temperature rhythm of cancerous breasts appear to have application in clinical oncology as a diagnostic tool. Continuous monitoring of the breast surface temperature is currently being evaluated as a means of detecting breast cancer and of rapidly evaluating the efficacy of its treatment [58]. For these purposes, Simpson and Griffith [59] have developed a "thermobra" which allows the automatic and continuous recording of breast surface temperature.

## Outlook

Our knowledge of multiple high-amplitude interacting rhythms in hormone concentrations and rhythm abnormalities in patients with endocrine responsive tumors may help in understanding the etiology of cancer of the endocrine organs. Malignant transformation of tissues derived from endocrine target organs may result from aberrations in the temporal patterns of hormone concentrations. This information could be used for tumor prevention, early detection, treatment, and surveillance. High-risk patients for the development of malignancies may be identified by studying abnormalities of rhythmic patterns. Those individuals would be candidates for more intensive cancer screening efforts. Prognostic factors may be found which help to determine the need for adjuvant treatment after curative intervention. Restoration of abnormal rhythm patterns may be effective in adjuvant cancer therapy.

However, although these concepts appear intriguing, there has been mixed enthusiasm, at least by medical oncologists, to incorporate chronobiological principles into daily practice. Underlying reasons include reluctant acceptance of dynamic model systems rather than homeostasis, the complexity of chronobiological experiments, and the difficulty in conducting well-controlled prospective clinical studies in order to reproduce and expand upon the above observations. This difficulty relates in part to the necessity of frequent sampling and the expenses for serial determinations of biochemical and other endpoints. Many researchers are also unfamiliar with the basic terminology and analytical techniques used for the detection of rhythmic phenomena. Finally, automatic monitoring devices and drug administration systems have only recently become available. This will hopefully facilitate accelerate the pace of chronobiology research.

## Conclusion

The endocrine system has a high degree of temporal organization. Disturbance of this complex regulatory network causes disease, including an increased risk of malignancy. Endocrine tumors may still respond to the complex rhythmic control mechanisms, and altering the hormonal milieu may cause disease arrest or progression. Our knowledge of the chronobiology of host and tumor and their interaction is fragmentary. However, the available data suggest that the exploration of temporal regulatory patterns may be helpful in risk factor assessment for malignancy, the development of more effective tumor prevention measures, early detection, treatment with improved therapeutic index, and follow-up observation with tumor markers. Most of the chronobiologic work on endocrine and endocrine-reponsive tumors has been performed in breast cancer. This review has decribed some of the emerging puzzle pieces from preclinical and clinical research and suggests possible relationships.

## References

1. van Cauter E (1989) Endocrine rhythms. In: Arendt J, Minors DS, Waterhouse JM (eds) Biological rhythms in clinical practice. Wright, London, pp 23–50
2. Arendt J, Broadway J (1987) Light and melatonin as zeitgebers in man. Chronobiol Int 4: 273–282
3. English J, Arendt J, Symons AM, Poulton AL, Tobler I (1988) Pineal and ovarian response to 22- and 14-h days in the ewe. Biol Reprod 39: 9–18
4. English J, Aherne GW, Arendt J (1988) Modulation of the circadian rhythm in methotrexate toxicity in the rat by melatonin and photoperiod. In: Reinberg A, Smolensky M, Labrecque G (eds) Annual review of chronopharmacology, vol 5. Pergamon, Oxford, p 359
5. Loros JJ, Denome SA, Dunlap JC (1989) Molecular cloning of genes under control of the circadian clock in Neurospora. Science 243: 385–388
6. Henderson BE, Ross RK, Pike MC, Casagrande JT (1982) Endogenous hormones as a major factor in human cancer. Cancer Res 42: 3232–3239

7. Huggins C (1967) Endocrine-induced regression of cancers. Science 156: 1050–1054

8. Crile G (1966) Endocrine dependency of papillary carcinomas of the thyroid. J Am Med Assoc 195: 721–724

9. Sommers SC, Meissner WA (1957) Endocrine abnormalities accompanying human endometrial cancer. Cancer 10: 516–521

10. McMahon B, Cole P, Brown J (1973) Etiology of human breast cancer: a review. J Natl Cancer Inst 50: 21–42

11. Tarquini B, Halberg F, Seal US, Benvenuti M, Cagnoni M (1981) Circadian aspects of serum prolactin and TSH lowering by bromocriptine in patients with prostatic hypertrophy. Prostate 2: 269–279

12. Crawford ED, Blumenstein BA, Goodman PJ, Davis MA, Eisenberger MA, McLeod DG, Spaulding JT, Benson R, Dorr FA (1990) Leuprolide with and without flutamide in advanced prostate cancer. Cancer 66 (Suppl 5): 1039–1044

13. Joly DJ, Lilienfeld AM, Diamond EL, Bross IDJ (1974) An epidemiologic study of the relationship of reproductive experience to cancer of the ovary. Am J Epidemiol 99: 190–209

14. van Vliet G, Canfriez A, Robyn C, Wolter R (1980) Plasma gonadotropin values in prepubertal cryptorchid boys: similar increase of FSH secretion in uni- and bilateral cases. J Pediatr 39: 253–255

15. Barbason H, Smoliar V, van Cantfort J (1979) Correlation of liver growth and function during liver regeneration and hepatocarcinogenesis. Arch Toxicol Suppl 2: 157–169

16. Barbason H, Rassenfosse C, Betz EH (1983) Promotion mechanism of phenobarbital and partial hepatectomy in DENA hepatocarcinogeneses cell kinetcs effect. Br J Cancer 47: 517–525

17. Mormont MC, Rabatin J, Lakatua D, Sothern R, Roemeling R, Hrushesky WJM (1987) Phenobarbital tumor promotion is related to an alteration of the circadian corticosterone rhythm. Proc Am Assoc Cancer Res 28: 170

18. Hrushesky WJM, Sothern RB, Levi FA, Olshefsky R, Lannin D, Berestka JS, Gruber SA (1989) Circadian timing, anatomic location and dose of tumor cell inoculum each affect tumor biology. Proc Am Assoc Cancer Res 30: 82

19. Gatti G, Cavallo R, Del Ponte D, Sartori M, Masera R, Carignola R, Carandente F, Angeli A (1986) Circadian changes of human natural killer (NK) cells and their in vitro susceptibility to cortisol inhibition. In: Reinberg A, Smolensky M, Labrecque G (eds) Annual review of chronopharmacology, vol 3. Pergamon, Oxford, pp 75–78

20. Gatti G, Sartori ML, Cavallo R, Del Ponte D, Carignola R, Salvadori A, Masera R, Angeli A (1987) Circadian variation of human natural killer (NK) cell response to positive modulation by immune interferon and negative modulation by cortisol. Chronobiologia 14: 178

21. Lévi FA, Canon C, Touitou Y, Reinberg A, Mathé G (1988) Seasonal modulation of the circadian time structure of circulating T and natural killer lymphocyte subsets from healthy subjects. J Clin Invest 81: 407–413

22. Hrushesky WJM, Gruber SA, Sothern RB, Hoffman RA, Lakatua D, Carlson A, Cerra F, Simmons RL (1988) Natural killer cell activity: age, estrous- and circadian-stage dependence and inverse correlation with metastatic potential. J Natl Cancer Inst 80: 1232–1237

23. Hrushesky WJM, Haus E, Lakatua D, Vogelzang N, Kennedy BJ (1983) Seasonality in testicular cell proliferation and seminoma incidence. Proc Am Soc Clin Oncol 24: 18

24. Boon ME, Rietveld PEM, Rietveld WJ, Sothern RB, Hrushesky WJM (1987) Seasonal rhythmicity in incidence of cervical neoplasia and infections in Dutch women. Chronobiologia 14: 154

25. Klevecz RR, Braly PS (1986) Synchronous waves of proliferation in human ovarian cancers. In: Reinberg A, Smolensky M, Labrecque G (eds) Annual review of chronopharmacology, vol 5. Pergamon, Oxford, pp 175–178

26. Klevecz RR, Shymko RM, Blumenfeld D, Braly PS (1987) Circadian gating of S phase in human ovarian cancer. Cancer Res 47: 6267–6271

27. Braly PS, Klevecz RR (1987) Cell kinetic measurements in "in vivo" human cancers – the effect of surgical debulking. Chronobiologia 14: 154

28. Touitou Y, Lévi F, Ferment O, Bailleul F, Bogdan A, Auzeby A, Chevelle C, Lesaunier F (1987) CA 125 rhythmicity in ovarian cancer patients. The effect of chemotherapy. Chronobiologia 14: 250

29. Touitou Y, Bailleul F, Lévi F, Bogdan A, Touitou C, Metzger G, Mechkouri M (1990) Circadian rhythms of tumor markers in breast cancer patients. In: Hayes DK, Pauly JE, Reiter RJ (eds) Chronobiology: its role in clinical medicine, general biology, and agriculture, part A. Wiley-Liss, New York, pp 59–66

30. Hallek M, Emmerich B (1990) Biological rhythms of tumor markers in cancer patients. Cancer Res Clin Oncol 116 (Suppl 1): 1073

31. Lévi F, Halberg F, Haus E, Sanchez de la Peña S, Sothern RB, Halberg E, Hrushesky W, Brown H, Scheving LE, Kennedy BJ (1980) Synthetic adrenocorticotropin for optimizing murine circadian chronotolerance for Adriamycin. Chronobiologia 7: 227–244

32. Kodama M, Kodama T (1982) Influence of corticosteroid hormones on the therapeutic efficacy of cyclophosphamide. Gann 73: 661–666

33. Halberg F, Sanchez S, Brown H, Haus E, Melby J, Wilson T, Sothern R, Berg H, Scheving LE (1980) Pretreatment with time-dependent active short-chain adrenocorticotropin (ACTH-17); HOE 433) for convenient optimization of murine Adiamycin chronotolerance. Am Assoc Cancer Res 21: 307

34. Hrushesky WJM (1985) Circadian timing of cancer chemotherapy. Science 228: 73–75

35. Hrushesky WJM (1987) Circadian scheduling of chemotherapy increases ovarian patient survival and cancer responses significantly. Proc Am Soc Clin Oncol 6: 120

36. Lévi F, Benavides M, Chevelle C, Le Saunier F, Bailleul F, Misset JL, Regensberg C, Vannetzel JM, Reinberg A, Mathé G (1990) Chemotherapy of advanced ovarian cancer with 4'-0-tetrahydropyranyl doxorubicin and cisplatin: a randomized phase II trial with an evaluation of circadian timing and dose-intensity. J Clin Oncol 8: 705–714

37. Mühlbock O (1958) The hormonal genesis of mammary cancer. J Endocrinol 17: vii–xv

38. Haus E, Halberg F (1962) Adrenocortical changes after supplementary heterotopic pituitary isografting in ovariectomized C-mice. Endocrinology 70: 837–841

39. Tamarkin L, Almeida OFX, Danforth DN Jr (1985) Melatonin and malignant disease. In: Ciba Foundation Symposium 117: Photoperiodism, melatonin, and the pineal. Pitman, London, pp 284–299

40. Wrba H, Halberg F, Dutter A (1986) Melatonin circadian stage-dependently delays breast tumor development in mice injected daily for several months. Chronobiologia 13: 123–128

41. Wrba H, Dutter A, Sanchez de la Peña S, Wu J, Carandente

F, Cornélissen G, Halberg F (1989) Secular or circannual effects of placebo and melatonin on murine breast cancer. In: Hayes DK, Pauly JE, Reiter RJ (eds) Chronobiology: its role in clinical medicine, general biology, and agriculture, part A. Wiley-Liss, New York, pp 31–40

42. Manni A (1989) Endocrine therapy of breast and prostate cancer. Endocrinology and metabolism. Clin N Am 18: 569–592

43. Scheving LE, Pauly JE, Tsai TH, Scheving LA (1986) Rhythms in enzymes of brain and liver; cell proliferation in various tissues and the effect of epidermal growth factor (EGF), insulin and glucagon on these rhythms. In: Rietveld WJ (ed) Clinical aspects of chronobiology. Cip-Gegevens, Den Haag, pp 105–126

44. Labrosse K, Haus E, Lakatua DJ, Halberg F (1978) Circadian rhythm in mammary cytoplasmic estrogen receptor content in mice with and without pituitary isografts and demonstrated differences in breast cancer risk. Proc Endocrine Soc, 60th annu meeting, p 407

45. Kiang DT, King M, Zhang HJ, Kennedy BJ (1982) Cyclic biological expression in mouse mammary tumors. Science 216: 68–70

46. Ratajczak HV, Sothern RB, Hrushesky WJM (1988) Estrous influence on surgical cure of a mouse breast cancer. J Exp Med 168: 73–83

47. Hrushesky WJM, Bluming AZ, Gruber SA, Sothern RB (1989) Menstrual influence on surgical cure for breast cancer. The Lancet II: 949–952

48. Jacobson HI, Janerich DT (1977) Seasonal variation in the diagnosis of breast cancer. Proc Am Assoc Cancer Res 18: 93

49. Cohen P, Wax Y, Modan B (1983) Seasonality in the occurrence of breast cancer. Cancer Res 43: 892–896

50. Langlands AO, Simpson H, Sothern RB, Halberg F (1977) Different timing of circannual rhythm in mortality of women with breast cancer diagnosed before and after menopause. Proceedings of the 8th international scientific meeting of the International Epidemiological Association. San Juan, Puerto Rico, 9/17–23, 1977

51. Halberg F, Cornélissen G, Sothern RB, Wallach LA, Halberg E, Ahlgren A, Kuzel M, Radke A, Barbosa J, Goetz F, Buckley J, Mandel J, Schuman L, Haus E, Lakatua D, Sackett L, Berg H, Wendt HW, Kawasaki T, Ueno M, Uezono K, Matsuoka M, Omae T, Tarquini B, Cagnoni M, Sainz MG, Vega EP, Wilson D, Griffith K, Donati L, Tatti P, Vasta M, Locatelli J, Camagna A, Lauro R, Tritsch G, Wetterberg L (1981) International geographic studies of oncological interest on chronobiological variables chap 37. In: Kieser H (ed) Neoplasms – comparative pathology of growth in animals, plants, and man. Williams and Wilkins, Baltimore, pp 553–596

52. Tarquini B, Gheri R, Romano S, Costa A, Cagnoni M, Lee JK, Halberg F (1979) Circadian mesor-hyperprolactinemia in fibrocystic mastopathy. Am J Med 66: 229–237

53. Singh RK, Singh S, Razdan JL (1987) Circadian periodicity of plasma 17-hydroxycorticosteroids in advanced breast cancer. In: Pauly JE, Scheving LE (eds) Advances in chronobiology, part B. Progress in clinical and biological research, vol 227 B. Liss, New York pp 335–342

54. Hughes A, Jacobson HI, Wagner RK, Jungblut PW (1976) Ovarian independent fluctuations of estradiol receptor levels in mammalian tissues. Mol Cell Endocrinol 5: 379–398

55. Jacobson HI, Janerich DT (1980) Is seasonality in human reproduction related to seasonality in tissue levels of estrogen receptor? In: Mahesh VB, Muldoon TG, Saxena BB, Sadler WA (eds) Functional correlates of hormone receptors in reproduction. Elsevier North Holland, New York, pp 573–578

56. Hrushesky WJM, Teslow I, Halberg F, Kiang D, Kennedy BJ (1979) Temporal components of predictable variability along the 1-year scale in estrogen receptor concentration of primary human breast cancer. Proc, Am Soc Clin Oncol 20: 331

57. Smolensky MH (1983) Aspects of human chronopathology. In: Reinberg A, Smolensky MH (eds) Biological rhythms and medicine. Cellular, metabolic, physiopathologic, and pharmacologic aspects. Springer, Berlin Heidelberg New York, pp 131–209

58. Simpson HW, Pauson A, Cornélissen G (1989) The chronopathology of breast cancer. Chronobiologia 16: 365–372

59. Simpson HW, Griffith K (1989) The diagnosis of breast precancer by the chronobra. J. Background review. II. The breast pre-cancer test. Chronobiol Int 6: 355–393

# Chronochemotherapy of Malignant Tumors: Temporal Aspects of Antineoplastic Drug Toxicity

W. J. M. Hrushesky and W. J. März

## Introduction: Chrono-oncology

Chronobiology defines and quantifies the temporal organization of life on every level of structural and functional complexity with a focus on exogenous and endogenous periodic biophysical phenomena called biological rhythms. The effect of time can be considered as a systematic and, to a certain degree, predictive factor rather than a chaotic error term in the quantitative assessment of biological data [1–3]. Thus, chronobiologists are utilizing the understanding of such phenomena in the temporal design of laboratory experiments and clinical observations to improve on the precision of the answers of scientific experiments and epidemiological and clinical studies.

The chrono-oncologist is primarily concerned with temporal aspects of cancer chemotherapy to improve on the effectiveness of such therapy in favor of the patients general condition by minimizing systemic toxicity and to extend the survival time of cancer patients.

In general, drugs are given at a time considered most suitable for the ward routine and the doctor and nursing staff administering them. Clinicians rarely consider the administration time of a drug as an important factor of pharmacotherapy effectiveness. A growing body of chronopharmacological data on animals and humans, however, supports the hypothesis that the therapeutic effect can be maximized, the drug toxicity can be minimized, and thus the therapeutic index can be optimized if drugs are given at carefully selected times of the day. Optimization of the therapeutic ratio of anticancer drugs bears potential for improved cure rates in some, if not all, cancers in the future.

## Temporal Structure of Host-Tumor Systems

Both host and tumor show an intricate multifrequency time structure of parameters of importance for tumor growth host defense which manifests itself in marked temporal changes in the effects of anticancer treatments and in the resistance of the tumor and of host tissues to such treatment. Some of the evidence pertaining to the role of biological rhythms in oncology are summarized below.

## Circadian Cytokinetics of Host and Tumor Cells

*There is ample evidence of circadian modulation in cell growth and cell division in normal tissues of animals and humans [2–8] as well as in tumor tissue.* Bone marrow in normal human beings exhibits circadian rhythms in DNA synthesis and the mitotic index [9, 10]. Polyamines, organic anions involved in the regulation of nucleic acid biosynthesis [11–13], are a marker of cell growth and exhibit a circadian rhythm as determined by the N1/N8-acetylspermidine urinary ratio (Fig. 1). This provides indirect evidence for circadian synchrony in the cytokinetic activity of normal human tissues. The circadian rhythm of polyamine excretion may be disturbed in patients with cancer.

*There is also evidence that the temporal organization of tumor tissue proliferation kinetics is a marker of degree of the malignancy.* Since cell proliferation in host tissues and in the majority of tumors occurs in a circadian periodic pattern, the growth fraction of a host tissue or a tumor will vary predictably as a function of time. Mitotic index and/or DNA synthesis have been used for evaluating the proliferative activity of laboratory tumor models. Comparative study of several experimental tumor models supports the hypothesis that well-differentiated slow growing tumors may retain a circadian time structure, whereas poorly differentiated fast growing tumors may tend to lose it

[14]. Loss of circadian organization of cell division may also be acquired along the course of tumor growth. Data on fast or slowly growing hepatomas illustrate the fact that circadian organization of tumor cell divisions is present, though dependent upon the stage of tumor growth. That is, the early stage of tumor tissue formation (prior to the development of necrosis) is more tightly tied to host circadian time structure, than later stages when partial necrosis of tumor tissue occurs [15].

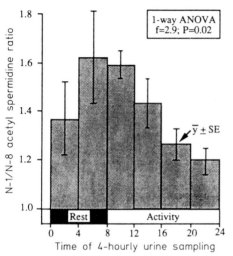

**Fig. 1.** Circadian rhythm in $N_1N_8$ acetyl spermidine ratio in the urine of apparently healthy volunteers sampled approximately 4-hourly

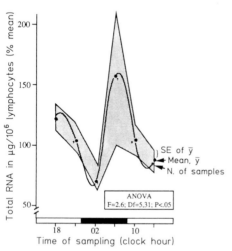

**Fig. 2.** Circadian pattern of lymphocyte total RNA concentration in five apparently healthy volunteers sampled in March–April 1980. Two peaks are noted allowing for the possibility of a single lymphocyte population with 12 h rhythmicity or two populations each with circadian rhythms in metabolic activity with opposite timings

*There is evidence of a periodic temporal variation of the mammalian immune system with a likely effect on host-tumor balance.* Circadian rhythms in all blood cell types are well-documented in both experimental animals and human beings [16, 17]. Numbers of total lymphocytes, B and T lymphocytes and natural killer cells, demonstrate circadian periodicity [18]. Circadian stage dependency of certain functional tests of the immune system was documented in vivo and in vitro. Tuberculin skin test reacitivity and incidence of human kidney rejections are circadian stage dependent [19] Tuberculin-, pokeweed-, and PHA-induced human lymphocyte transformation are circadian stage-dependent in antiphase to the "immunosuppressive" serum cortisol rhythm [20]. The plaque-forming cell response of spleens from mice immunized with sheep red blood cells has a marked circadian stage dependency [21, 22]. Total RNA content of circulating lymphocytes exhibits a rhythm with two peaks during a circadian cycle at 9 h (circadian peak of circulating steroid level) and 18 h after sleep onset reflecting either two subpopulations of lymphocytes with their RNA synthesis out of phase or a bimodal RNA distribution within lymphocytes of healthy indi-

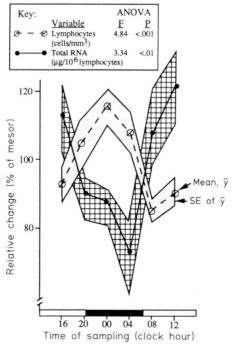

**Fig. 3.** Circadian rhythms in lymphocyte number and total lymphocyte RNA are precisely out of phase in our patients with metastatic malignancy receiving intermittent chemotherapy. Data obtained from 19 circadian profiles in 10 women with advanced cancer sampled 1 month after chemotherapy with adriamycin and cisplatin. This population has a single peak perhaps indicating a deficienay of one lymphocyte population.

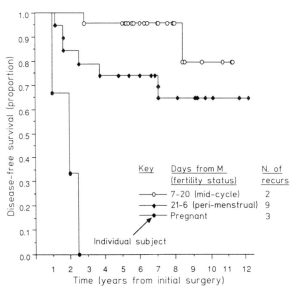

**Fig. 4.** The pattern of distant tumor recurrence-free survival for 44 women following breast cancer surgery performed within 7–20 days (*n* = 22) or within 21–26 days (*n* = 19) of the first day of the last menstrual period (LMP) or during pregnancy (*n* = 3). Each *open* or *closed symbol* represents an individual patient. Outcome was different depending upon the timing of surgery, from life table comparison of perimenstrual vs midcycle patients (*w* = 1.98, *p* < 0.05). Outcome of both of the groups also differed from that of pregnant women (*w* = 4.24 and 2.42, *p* = 0.002 and 0.016 of midcycle and perimenstrual groups, respectively)

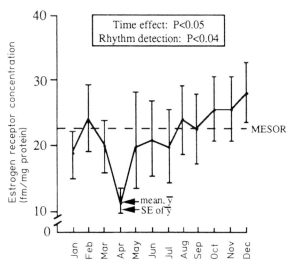

**Fig. 5.** Seasonal pattern of estrogen receptor concentration in 797 serially independent histologically validated primary breast cancer specimens obtained over a 3.5 year span. An early spring trough and late autumn peak characterize these data

viduals (Fig. 2). A circadian rhythm in RNA content of leukocytes was present in five healthy volunteers [23]. With a peak at 11:15, a circadian rhythm in total RNA content of lymphocytes was also present in cancer patient (*n* = 10, metastatic ovarian cancer) with a peak 11 h after sleep onset (Fig. 3). These data establish a molecular basis for predictable differences in lymphocyte receptivity to therapeutic manipulation by drugs affecting RNA synthesis.

## Infradian Structure of Host-Tumor Balance

*There is evidence of menstrual cycle changes in the balance between host and cancer.* Animal experiments and observations in human subjects suggest that trauma, e.g., surgical procedures, may adversely affect the balance between cancer and host and, for example, favor the metastatic spread of breast cancer. The observation that subgroups of younger women with early stage breast cancer are either cured by local resection or develop widespread metastatic disease has been ascribed to the "biological heterogeneiety" of breast cancer. It appears that the endocrine

milieu may affect the balance between a woman host and her breast cancer at the time of surgery. High amplitude rhythms of follicle stimulating hormone (FSH), luteinizing hormone (LH), prolactin, estrogens, and progesterone associated with the menstrual cycle might have some influence on the metastatic potential of tumor cells spread during tissue trauma associated with surgical resection. The natural killer cell activity is estrus stage-dependent [24], suppressed by surgical stress [25], and coincidentally inversely correlated with metastatic potential.

Laboratory experiments with a C3HeB/FeJ female murine transplanted mammary cancer model support this hypothesis, since surgical cure was 2.5-fold more frequent, if the tumor was resected near mid-cycle (proestrus/estrus) than in the other half of the cycle (metestrous/diestrous) [26]. Concurrently, it was noted that splenocyte natural killer cell activity was greatest at the time of highest surgical curability.

Epidemiological retrospective analysis (Cox proportional hazard model) of frequency and rapidity of cancer recurrence and death from breast cancer in 41 nonpregnant premenopausal women followed for 5–12 years after surgery revealed that surgical timing and lymph node status were the only two prognostic variables which independently predicted recurrence and death. Women having operations near the menstrual period had a quadrupled risk of relapse (Fig. 4) and death from breast cancer, indicating that timing of surgery may be more influential on outcome than adjuvant chemotherapy [27].

The mechanisms of this effect are not precisely known but both hormone-dependent temporal variations of anticancer efficiency of immune defense networks and autocrine growth factors within tumor cells or paracrine functions (e.g., angiogenetic growth factors) of tissue, harboring nascent metastatic deposits of the primary tumor, may be involved.

*There is evidence of circannual (seasonal) changes in the balance between host and cancer.* The marked circannual rhythm in breast cancer incidence attest to the role of known circannual rhythms of the endocrine system in human cancer development [28, 29]. This is supported by the seasonal fluctuation of estrogen receptor concentration in human breast cancer tissue [30, 31] (Fig. 5) and seasonal dependency of breast cancer mortality [32]. The seasonality of deaths from breast cancer is opposite for pre- and postmenopausal women and should not be disregarded in the design of hormonal therapy protocols.

**Table 1.** Chronotoxicity of anticancer agents in murine models (From Levi et al. 1987 [132])

| Agent | Optimal circadian time (HALO) | Target |
|---|---|---|
| Ara-C, Daunorubicin, Doxorubicin | 8–10 | Bone marrow |
| Etoposide, Fluorouracil, L-PAM, THP | | Bone marrow |
| Cisplatin, MTX, Vinblastine, MMC | 17–19 | Bone marrow |
| Cisplatin, MTX | 18–19 | Kidney |
| Doxorubicin | 14–16 | Heart |
| FUDR, MTX | 18–19 | Liver |
| Fluorouracil, Cisplatin | 18–19 | GI tract |
| Cyclophosphamide | 11 and 23 | Bladder |

HALO, hours after light onset; Ara-C, cytosine arabinoside; L-PAM, melphalan; THP, 4'tetrahydropyranyl adriamycin, MTX, methotrexate; FUDR, 5-fluoro-2'-deoxyuridine

## Temporal Factors in Cancer Etiology and Natural History

Differences in incidence of cancer in laboratory animals exposed to carcinogens at different specified times is evidence that carcinogenesis is dependent on the timing of carcinogen exposure.

The observation that individuals at high risk for the development of breast cancer exhibit a lower circadian and circannual amplitude of the prolactin rhythm suggests that a disruption of the pineal-hypothalamic-pituitary temporal balance may affect the incidence of breast cancer.

The circannual rhythm of mortality from breast cancer and estrogen receptor content of breast cancer tissue is evidence of a low frequency rhythmic balance between host and cancer. Also, circannual rhythm in the incidence of endocrine malignancy provides evidence of neurohumoral control of cancer development and growth [29].

mal experiments have indicated that the lethal doses of a cytotoxic drug is highly dependent on the time of administration. This phenomenon of chronotolerance was first described in 1972, when the survival rate of animals exposed to cytosine arabinoside (ara-C) was found to be predictably time-dependent [33, 34].

Evidence is accumulating that proliferating tissue "in vivo" is more sensitive to toxic drug effects at one time of the day than at another time of the same day. Table 1 lists a variety of cytotoxic agents, the target tissue of the drug's toxicity, and the times within the sleep/wake cycle in hours after light onset (HALO) when the drug has the least toxicity. Also, the antineoplastic effects of some of these drugs are time-dependent in rats and mice, in agreement with the finding, that some though not all tumor tissues exhibit patterns of time-dependent susceptibility to cytotoxic drugs, similar to the tissues of the host. These susceptibility rhythms may be exploited to optimize tumor cell damage.

## Time Dependency of Host Tolerance of Cytotoxic Drug Effects

### Chronotolerance

Not only the balance between host and tumor is predictable in time, but also the reaction of the host-tumor system to the exposure to cytotoxic drugs. Ani-

## Host Tissue Properties of Drug Detoxification

Reduced glutathione (GSH) content of heart muscle cells, a determinant of redox potential and thus salvage from free oxygen radicals, maintains a circadian rhythm [35]. Kidney function – in itself a determinant of renal toxicity – exhibits a circadian rhythm [36, 37].

The mitotic index and DNA synthesis of rapidly regenerating tissue, e.g., rat and mouse GI tract mucosa [38, 39], bone marrow, and the regeneration of the liver parenchyma after experimental partial hepatectomy in rats [40], show a circadian rhythmicity.

In the mouse and rat, liver DNA synthesis, RNA synthesis, RNA translational activity, mitotic index, weight, glucagon, content, and activity of numerous enzyme systems are circadian periodic. The liver detoxification potential of several drugs, e.g., paraoxone, nicotine, antimycin-A, phenobarbital, hexobarbital, and ara-C, is circadian stage-dependent, thus affecting drug pharmacokinetics [41–43]. Circadian rhythms of pharmacokinetics were described for 5-fluorouracil (5-FU), cis-diaminedichloroplatinum-II (cisplatin), oxalipatin, methotrexate, 6-mercapto-purine, and doxorubicin ([44–48]; R. von Roemeling, unpublished results).

A precondition of the improvement of the therapeutic index by optimal circadian drug timing is the detection and quantification of biological rhythms acting upon chemotherapy pharmacodynamics in regard to unwanted toxicity and wanted antitumor activity of the drugs [49].

**Table 2.** Circadian rhythmic changes in normal human tissues

| Tissue | Parameter | Time of peak | Change found (%)[a] | References |
|--------|-----------|--------------|---------------------|------------|
| PB | Granulocyte count | 1900 | 20 | [55] |
| PB | Lymphocyte count | 0030 | 40 | [56] |
| PB | Lymphocyte count | 0100 | 60 | [54] |
| PB | NK cell activity | 0800 | 76 | [57] |
| PB | T4/T8 lymphocyte ratio | 0630 | >200 | [58] |
| PB | CFU-C | 0900 | 100 | [59] |
| PB | CFU-C | 1500 | >100 | [60] |
| BM | Mitotic index | 2300 | 100 | [61] |
| BM | Mitotic index | 2400 | >100 | [62] |
| BM | Nonprotein sulfhydryl level | 0800 | 140 | [63] |
| BM | S phase DNA | 2000 | ? | [9] |
| Colon rectum) | S phase DNA | 0700 | 34 | [64, 65] |
| Jejunum | Mitotic index | 0600 | ? | [66] |

PB, peripheral blood; BM, bone marrow; CFU-C, colony forming units-culture
[a] Peak – trough differences as % of the mean

## Rhythms of Cell Proliferation as Marker Rhythms of Chronotolerance

The effect of antineoplastic drugs differentially active during certain cell cycle stages can be predicted if the timing of a significant rhythm in cell proliferation is known. Optimization of the time of drug exposure within a given host-tumor-drug system may then serve to improve the therapeutic ratio of the drug of choice.

Several investigators have demonstrated cytokinetic circadian periodicity in non-human systems, probably pertinent to the treatment of human beings with anticancer drugs in regard to efficacy and toxicity [2–8, 10, 50–53]. Data on rhythms of cell proliferation in normal (Table 2) and neoplastic human tissues are available (Table 3), but are rather scarce due to the ethical problems involved in frequent tissue sampling. However, the rhythmic fashion of physiological proliferative tissue activity can be directly or indirectly measured by using less or non-invasive laboratory test parameters which express circadian rhythmicity [9, 54–66] with a strong correlation to cell proliferation (Table 2).

As in experimental oncology, the assessment of mitotic index, DNA synthesis, and local tissue temperature in clinical oncology may serve as a marker of predictable periodic change of proliferation indices in normal tissue and in spontaneous human malignancies. Pronounced population rhythms of cell kinetics can be found in human tissues. However, related to tumor stage, grade, and the condition of the host, large interindividual and intertumor differences in circadian organization and chronopharmacological effect can be present [67]. Thus, circadian monitoring of selected marker variables may be advised to further define an individual's circadian stage and constitutes a tool for individualization of timed cancer chronotherapy [68–70].

Apart from cytokinetic rhythms of the tumor tissue and pharmacokinetic rhythms of the drug of choice, other body functions that vary predictably with time, e.g., endocrinologic and immunological

**Table 3.** Rhythmic changes in human tumors

| Tumors | Rhythm parameters | Time(s) of maxima | Change (%) | Frequency (h) | References |
|--------|-------------------|-------------------|------------|---------------|------------|
| Ovarian carcinoma | S phase DNA | 0800 and 2000 | 200 | 12 | [68] |
| Malignant cells | | 0800 | | | |
| Nonmalignant cells | | 2000 | | | |
| Mammary carcinoma | Mitotic index | 1330 | ? | 24 | [69] |
| Squamous cell carinoma | Mitotic index | – | – | None | [69] |

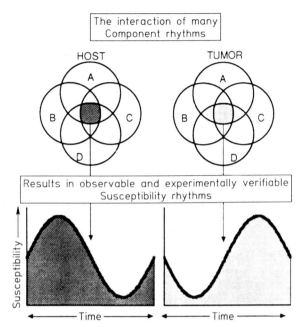

Fig.6. Host and tumor susceptibility rhythms. The figure shows the interaction of many component rhythms. *A*, cell cycle rhythm; *B*, drug absorption, activation, cellular uptake, metabolism, and exeretion; *C*, overlying hormonal rhythm affecting drug toxicity; *D*, immunological rhythm

rhythms, may influence the pharmacodynamic rhythm of the drug with regards both to host and tumor tissues [71–74] (Fig.6).

All the above mentioned "marker rhythms" may provide substantial information and help predict both desired and undesired pharmacodynamic effects and the amount of drug that may be given safely at a given circadian stage. On the other hand, time of day plus or minus 2–4 hours has already demonstrated significant and easily obtainable advantage.

## Experimental Design
## of Chronotoxicological Studies

Chronotoxicological experiments investigate whether mice or rats tolerate the same dose of the same anticancer drug differently as a function of the time of administration throughout the 24 h span. Animals of the same strain, sex, and age are housed in an enclosed environment with defined lighting regimen, e.g., consisting of an alternation of 12 h of light alternating with darkness (LD 12:12) and thus modeling a 24 h solar day. These animals are exposed to presumably toxic doses of a drug at one of several well-

defined time points, i.e., circadian stages throughout the 24 h light-dark cycle.

The effect of timing of drug exposure upon drug action and toxicity is examined with survival rate and mean survival time or organ-specific measures of lethal and sublethal toxicity as end points. The behavior of the rhythmic variations encountered after temporal variation of the lighting regimen and food exposure serves conventionally as evidence of their endogenous nature. Changing the lighting regimen and/or the feeding pattern of the animals can predictively shift, but never eliminate, cell proliferation rhythms and thus the susceptibility rhythms to cancer chemotherapeutic agents.

## Development of Chronotherapeutic
## Treatment Schedules in Clinical Oncology.
## Drug Pharmacokinetics and Pharmacodynamics

Prediction of the time structure of cell cytokinetics and drug pharmacokinetics is a prerequisite for the prediction of the chronopharmacodynamics of chemotherapeutic drugs chosen for the development of optimized chronotherapeutic treatment schedules. The circadian stage dependency of the antitumor activity of commonly used chemotherapeutic agents and combination chemotherapy has been documented for several experimental tumor types [67, 75–77]. Table 4 lists the optimal circadian timing for single or combination chemotherapy with respect to long-term cure or improved survival rate.

Initial clinical chronotherapy trials have focused upon dosing time-related differences of drug toxicity in cancer patients. Clinical protocols were guided by results from animal chronotolerance studies and produced preliminary evidence of time effects on pharmacodynamics: chronotherapy of FIGO III (Federation Internationale de Gynecologie et Obstet) and IV ovarian adenocarcinoma was tested with a combination of 4' tetrahydropyranyl adriamycin (THP) (50 mg/m$^2$ i.v. bolus) and cisplatin (100 mg/m$^2$ i.v. over 4 h) [78].

Comparison of two schedules (THP at 0600 followed by cisplatin at 1600–2000 vs THP at 1800 hours followed by cisplatin at 0400–800) revealed a marked difference in complete remission rate (73% vs 38%) and mean survival time (20 months vs 13 months) in favor of giving THP in the morning followed by cisplatin in the evening.

Complex chemotherapy with metothrexate, 5-FU, vinblastine, and cyclophosphamide was demonstrated to be time dependent ($n = 68,58\%$ vs 23%

**Table 4.** Chronoeffectiveness of anticancer agents in murine tumor systems (From Levi et al. 1987 [68])

| Agent | Tumor | Optimal Circadian Time (HALO) | Index of Effectiveness | Differences due to dosing Time(s) |
|---|---|---|---|---|
| CY | Mammary adenocarcinoma | 08 | PR | ? |
|  | T9 + T10 sarcoma | 02 | CR | 14% |
|  | L1210 leukemia | 12 | CR | 27% |
|  | Ehrlich's ascites | 04 | CR | 13% |
| CY + Ara-C | L1210 leukemia | 10 | CR | 20–50% |
| CY + D | L1210 leukemia | 13 | CR | 60% |
| Ara-C | L1210 leukemia | 08 | CR | 18% |
| D | Immunocytoma | 10 | PR | ? |
| L-PAM + D | Mammary Adenocarcinoma | 10 | CR | 38% |
| Cisplatin + D | Immunocytoma | 18 | CR | 40% |
| FUDR infusion | Mammary Adendocarcinoma | 22–04 | PR | 35% |

HALO, hours after light onset; CY, cyclophosphamide; Ara-C, cytosine arabinoside; D, doxorubicin; L-PAM, melphalan; FUDR, 5-fluoro-2'-deoxyuridine; PR, pertial response; CR, complete response

partial response rate) [79]. The 5 year disease-free survival of children with acute lymphoblastic leukemia was found to be different depending upon the timing of their maintenance chemotherapy with 6-mercaptopurine and methotrexate, with 80% after evening therapy vs 40% after morning therapy [78].

## Factors Potentially Modulating Chronotolerance and Therapeutic Index in Time

*Seasonal Effects (Circannual Rhythms).* Mortality from doxorubicin, daunomycin, and cisplatin is influenced by a nontrivial seasonality, even if the animals are kept in a laboratory under constant conditions. For cisplatin, the extent of predictable seasonal variation in the mean survival after drug injection was about 40% for rats kept under constant laboratory conditions, indicating that the "in vivo" effect of circannual rhythms may be even greater [80].

*Administration Pattern.* The therapeutic index may be improved by a sinusoidal pattern of drug administration. A circadian sinusoidal modulation (every 3 h) of ara-C administration, instead of a constant dose, could double the cure rate after inoculation of L1220 leukemia cells into mice and increased the survival time by 60% when the highest doses were given near the time of best drug tolerance [33, 81].

*Therapeutic Synergism of Combination Chemotherapy.* Drug resistance may be circumvented by combination chemotherapy. Circadian stage dependency was demonstrated for several combination chemotherapy regimens in a murine L1220 leukemia model, e.g., ara-C + cyclophosphamide [82], doxorubicin + cyclophosphamide, and ara-C + 1-[2 chlor-

ethyl] 3-cyclohexyl-1-nitroso-urea (CCNU) [83]. A rat immunocytoma model received combination chronotherapy with doxorubicin + cisplatin [84] and a rat mammary adenocarcinoma model was treated with doxorubicin + *L*-phenylanaline mustard.

## Development of Chronotherapeutic Treatment Schedules in Oncology

The attempt of a systematic design of clinical chronochemotherapy guided by experimental animal data can be illustrated for the therapy with doxorubicin + cisplatin and floxuridine.

### Chronochemotherapy with Doxorubicin and Cisplatin

Conventional Therapy of Ovarian and Transitional Cell Bladder Cancer with Doxorubicin + Cisplatin

Doxorubicin and cisplatin are the most active drugs in treatment of several cancer types, including transitional cell cancer and ovarian cancer. In ovarian cancer, the combination of these drugs is superior to single agent chemotherapy in respect to response rate and survival. The achieved dose intensity determines tumor control [85]. However, only a third of patients with advanced disease (FIGO stage III, IV) will have a complete clinical response and an additional third a partial response. Only 20% of patients with complete clinical response do not have any microscopic residual disease at the occasion of a second-look staging procedure. Consequently, most patients relapse and only have median survival times between 10 and 36 months [86]. Obviously, optimization of the therapeutic index is very desirable for this disease.

Transitional cell carcinoma presently has an even lesser response rate to chemotherapy than ovarian carcinoma. Newer protocols, including the combination of doxorubicin and cisplatin, have resulted in long lasting complete responses in some patients; however, the prognosis of the entire patient population remained poor. Higher doses in conventional therapeutic trials were associated with better response rates but also with substantial toxicity. Adjuvant chemotherapy after surgical resection improved the length of disease-free survival [87].

## Experimental Animal Studies on Doxorubicin and Cisplatin

To improve this therapy by a chronotherapeutic approach, the first steps consisted of a series of animal experiments which were systematically designed to lead to a chemotherapeutic treatment schedule applicable to patients. The issues to be considered in the development of such a schedule were as follows:

1. Exploration of mechanisms of circadian differences in cisplatin toxicity. In a series of 11 experiments, rats were exposed to toxic doses of cisplatin (11 mg/kg) at six different circadian stages. A pronounced circadian rhythm of cisplatin lethal toxicity was observed. Cisplatin was tolerated better when given late in the animal's active phase (Fig. 7) [88]. Lethal toxicity of cisplatin resulted from its nephrotoxic effects most extensively affecting the proximal convoluted tubules. A marker of cisplatin nephrotoxicity was the urinary release of $\beta$-$N$-acetylglucosaminidase (NAG), a brush border lysosomal enzyme.

Both the normal kidney function [37] and cisplatin pharmacokinetics, e.g. cisplatin plasma protein binding [89], are circadian stage dependent. The NAG urinary excretion in normal animals was found to show a high amplitude circadian rhythm. When cisplatin was given at its most unfavorable time of day, the circadian rhythm of NAG excretion was maintained with a five-fold increase of mean level and circadian amplitude in direct proportion to a subsequent rise of BUN. When cisplatin was given at a favorable circadian stage, a smaller rise in mean NAG, little histologic renal damage, and only a small rise in BUN were observed (Fig. 8).

A marked circadian rhythm in the kidney protective effect of fluid loads was noted when an intraperi-

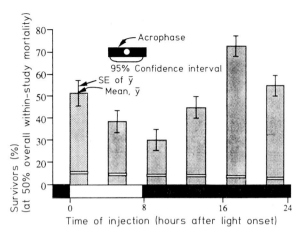

**Fig. 7.** The circadian stage-dependent pattern of cisplatin lethal toxicity demonstrates least toxicity when drug is given late in the activity span. Data tased on single i. p. injection of 11 mg/kg cisplatin to 1265 female Fischer rats; 6 injection times in each of 9 studies; $p$ (from $\chi^2$) < .05 in each study; zero amplitude test, $p < .001$

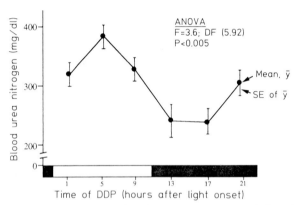

**Fig. 8.** Organ-specific toxicity is dependent upon the timing of cisplatin administration. The pattern for blood urea nitrogen (BUN) deviation is dependent upon when the cisplatin was given. If it was given late in the daily activity span the risk in BUN was much less dramatic

toneal fluid load of 3 % body weight was given or withheld from animals concurrently treated with cisplatin at six different times of day. A highly protective effect of hydration was found late in the activity span as compared to the early stage of the activity span, indicating that not only lethal nephrotoxicity of cisplatin is circadian stage-dependent, but also the ability of hydration to reduce cisplatin nephrotoxicity (Fig. 9) [90].

2. Mechanisms of circadian differences in doxorubicin toxicity. In a series of six experiments, the lethal toxicity of doxorubicin was also found to be circadian

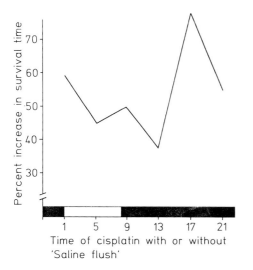

**Fig. 9.** Serially independent blood samples were obtained on day 4.5 after a i. p. saline hydration to 98 female F344 Fischer rats at one of six different circadian stages. Mean ± SE of 10 control rats: 19 ± 1 mg/dl. The degree of protection conferred by an intraperitoneal saline load upon animals given cisplatin alone or cisplatin plus concurrent 3 % body weight saline is markedly time dependent. The most protection occurs at the best time for cisplatin

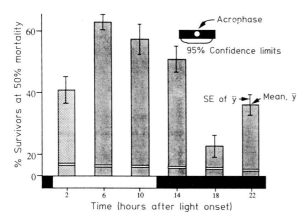

**Fig. 10.** Doxorubicin lethality is dependent to a large extent upon when in the day the drug is given. The safest time for intraperitoneal doxourbicin is just before the animals usually awaken, late in the resting phase. Data based on single i. p. injection of 18 mg/kg adriamycin to a total of 690 CDF and BDF male mice; 6 injection times in each of 6 studies

stage-dependent. The circadian stage of maximum doxorubicin tolerance occurred shortly before awakening, thus 12 h apart from the optimal time for cisplatin administration (Fig. 10). Doxorubicin-related heart muscle toxicity is believed to be secondary to its activation into a free radical intermediate. The availability of the main free radical scavenger, GSH is circadian stage-dependent, with a peak at the time of

lowest drug toxicity [35, 67]. Doxorubicin antineoplastic activity is exerted by drug binding to DNA (intercalation) and single strand break. Circadian changes in cell cytokinetics, (e. g., the rate of cells in S phase), tissue-specific differences between tumor and normal organs, and circadian changes in drug pharmacokinetics, may have determined the observed circadian differences in doxorubicin therapeutic index.

3. Pharmacodynamic effects of doxorubicin and cisplatin. The first chronotherapy studies using doxorubicin, performed in 1977, revealed that the rate of tumor shrinkage following doxorubicin treatment of a transplanted plasmocytoma of rats is dependent upon the time of drug administration, i. e., fastest shrinkage of the tumor occurred when animals were exposed to doxorubicin at the end of their daily resting span, just prior to awakening [91–94]. Later, time-dependent synergistic effects of doxorubicin and cisplatin were demonstrated in tumor bearing rats, since reduction in tumor size and the quantity of renal excretion of the paraneoplastic Bence-Jones protein varied predictably, depending on one of two selected administration times (late light or late dark).

4. Therapeutic trials with intention to cure. Therapeutic trials were repeated with less toxic doses of doxorubicin alone at six different times of day or doxorubicin given at its optimal circadian time combined with cisplatin given at six different times of day. LOU rats bearing a plasmocytoma showed the most favorable response to treatment, i. e., maximal tumor reduction and minimal weight loss at 10 HALO on a LD12:12 schedule [94, 95]. If doxorubicin was given at its optimal time at 10 HALO, additional treatment with cisplatin resulted in a statistical prolongation of life only if given at a certain time of day, i. e., at 18 HALO (mid-dark). These studies were repeated with a lower total dose to avoid obscuring of the circadian stage-dependent survival advantage by lethal treatment toxicity. The best therapeutic index was confirmed with a regimen of doxorubicin at 10 HALO and cisplatin at 18 HALO, with confirmation of a statistically significant rhythm in rate of complete remission as a function of cisplatin treatment time.

Chronotherapy of Ovarian and Transitional Cell Bladder Cancer with Doxorubicin + Cisplatin

Faced with the poor prognosis of ovarian and transitional cell carcinoma, our goal was to reduce treatment-related toxicity and complication rates by opti-

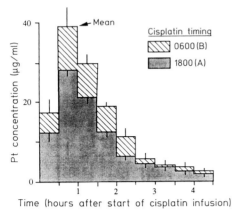

**Fig. 11.** Each *bar* represents the concentration of cisplatin + SE in urine excreted in 30 min samples after evening *(shaded)* and morning *(hatched)* cisplatin ($f = 7.4$, $p < 0.01$)

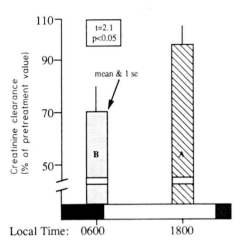

**Fig. 12.** Patients treated with cisplatin in the morning had a statistically significant greater fall in creatinine clearance than did those who received cisplatin in the evening

mal circadian drug timing, allowing high dose intensity treatment to be administered safely and most effectively. We expected to achieve improved tumor control and patient survival in both diseases with optimal drug timing.

According to laboratory animal experimental treatment trials, the best therapeutic index of doxorubicin-cisplatin combination chemotherapy is achieved by administration of doxorubicin in the early morning (e. g., 0600) and cisplatin in the evening (e. g., 1800). The therapeutic effect is related to the circadian stage and needs to be adjusted if a patient is on a different sleeping routine. The monitoring of marker rhythms, e.g., temperature and urinary potassium, may help to define the circadian phases and the degree of environmental synchronization for an individual patient.

Based upon the results of the studies in animals the following protocols were designed for a toxicity study with a crossover design to be used in patients: doxorubicin 60 mg/m$^2$ followed by cisplatin 60 mg/m$^2$ 12 h later. The timing of doxorubicin was alternated between 0600 (referred to as schedule A) and 1800 hours (referred to as schedule B) for each subsequent cycle.

In 23 patients treated with these chronotherapeutic schedules, the following results were obtained: urinary cisplatin kinetics were predictably different depending upon when the drug was infused and reached significantly higher concentrations in the urinary tract following morning administration (Fig. 11) [45, 46]. Cisplatin-induced nephrotoxicity, measured as a drop in creatinine clearance, was significantly greater in the morning than in the evening (Fig. 12). Subsequent to the first course of cisplatin, there was either a permanent 30% decline or no change in creatinine clearance depending on treatment time.

Bone marrow toxicity, measured as the degree of neutropenia and thrombocytopenia, was less if doxorubicin was given at 0600 (schedule A) than when given at 1800 hours (schedule B). The morning schedule resulted in a higher nadir count and full recovery of all counts to pretreatment levels within 21 days, whereas the evening schedule led to less than full recovery even after 28 days following therapy. This phenomenon was reproducible for individuals on alternating timing of doxorubicin (Fig. 13).

Cisplatin-induced nausea and vomiting, the most common reason for patients to refuse further therapy, cannot be completely controlled by antiemetic therapy and seems to be related to the administration time, since patients who received cisplatin at 0600 (schedule B) reported significantly more vomiting episodes which tended to begin sooner and last longer [96]. Neurotoxicity and chronic anemia were also statistically significantly different in favor of morning doxorubicin and evening cisplatin [75].

In a randomized noncrossover study of cumulative drug toxicity and efficacy, the protocol used consisted of nine courses of doxorubicin 60 mg/m$^2$ followed by cisplatin 60 mg/m$^2$ 12 h later, starting always at the same circadian stage with doxorubicin either at 0600 or at 1800.

## Improvement of Drug Toxicity by Chronotherapeutic Treatment Schedules

*Bone Marrow Toxicity.* The circadian stage of drug administration determined whether this combination chemotherapy induced bone marrow suppression or

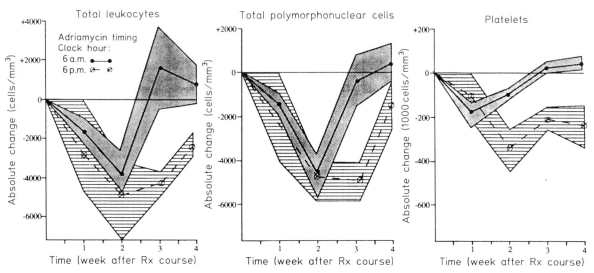

**Fig. 13.** When treated with doxorubicin in the evening, this patient had a greater falloff and a slower recovery for total leuko- cytes, PMNs, and platelets than when treated with doxorubicin in the morning

not in 37 patients that completed nine courses of chemotherapy. The majority of patients receiving doxorubicin in the evening followed by cisplatin in the morning (schedule B) had to have a greater than 33% doxorubicin dose reduction and many of them had to have treatment delays of greater than 2 weeks as opposed to patients receiving doxorubicin in the morning and cisplatin in the evening (schedule A). Assessment of individual white blood cell decrease and recovery after 7, 14, and 28 days after treatment revealed more cumulative bone marrow toxicity for patients treated with doxorubicin in the evening despite substanial dose reductions.

*Nephrotoxicity.* Renal function of 43 patients was studied prior to 295 separate treatment courses. The creatinine clearance decline 4 weeks after treatment was then assessed as a function of treatment timing relative to the patient's peak of circadian rhythm of potassium excretion, which was used as a marker rhythm of the circadian change of renal function. Patients who were treated between 3 hours before and after their peak potassium excretion suffered no subsequent loss of renal function, while those patients receiving cisplatin farthest away from the time of highest potassium excretion had an average loss of 8 ml/min in creatinine clearance per treatment course. Since the protocol consisted of nine courses of therapy, inopportune timing of repeated cisplatin administration resulted in a substantial and preventable loss of kidney function in more than 50%.

Dose and schedule modification forced by drug toxicity represent an index for the frequency of side effects of the drugs. Doxorubicin dose modifications (25% reduction of dose) or schedule delays (1 week, then re-evaluation) were forced by a recovery absolute granulocyte count below 1500/μl 28 days after treatment, a recovery platelet count under 100000/vl, or interim infection or bleeding. Cisplatin was dis-

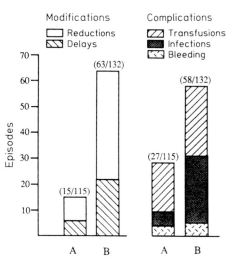

**Fig. 14.** A total of 247 courses of doxorubicin-cisplatin therapy were administered; 115 on schedule A and 132 on schedule B (see text). Dose reductions were three times more frequent when the drugs were given on schedule B (31% vs 9%, $X_2 = 17.4$, $p < 0.01$). Despite schedule B dose reductions and treatment delays, 44% of schedule B treatments were associated with bleeding, infection, or transfusion requirement, while only 23% of treatments given on schedule A were associated with these complications ($X_2 = 10.5$, $p < 0.01$)

continued if creatinine clearance fell below 30 ml/min.

The incidence of treatment complications represents another index in the evaluation of drug toxicity. Complications were defined as interim clinical infections that required oral or parenteral antibiotics; interim bleeding episodes of any kind whether or not platelet transfusions were administered; and anemia requiring a transfusion, usually two or three units of packed red cells, Figure 14 demonstrates the marked difference of required modifications and complications in favor of the morning doxorubicin followed by evening cisplatin schedule.

Tumor Response and Patient Survival
in Ovarian Cancer

Four comparable subgroups of patients with FIGO III and FIGO IV ovarian cancer were treated with a combination of cisplatin and doxorubicin on four different circadian schedules: (1) unspecified time of treatment with no consistent sequence or interval between drugs, (2) doxorubicin at 0600 followed by cisplatin at 1800, (3) doxorubicin at 1800 followed by cisplatin at 0600, or (4) alteration between morning and evening administration of doxorubicin followed by cisplatin 12 h later. Circadian scheduling significantly increased clinical complete response rates, pathological complete response rate, and survival after a median follow-up of 67 (16–105) months. All patients treated without regard to drug timing died within 3 years (Fig. 15).

Patients receiving doxorubicin in the morning and cisplatin in the evening had a 44% 5 year survival, exceeded even by the survival rate of patients on alternating schedule, while patients receiving doxorubicin in the evening and cisplatin in the morning had a 5 year survival rate of 11%. Patients receiving the alternative circadian timed regimen also had excellent survivel. Part of this schedule-dependent difference may be attributed to differences in average dose intensity (% of planned dose intensity) since patients on morning doxorubicin followed by evening cisplatin received 94% of the planned doxorubicin dose and 103% of the planned cisplatin dose, whereas patients on evening doxorubicin followed by morning cisplatin received only 80% of the planned dose of doxorubicin and 82% of the planned dose of cisplatin.

Patients receiving unsystematic therapy, with the drugs given without regard to timing, received 94% of the planned doxorubicin dose and 103% of the planned cisplatin dose however still failed fairly rapidly, i.e., had received less than four of the nine planned courses at the time of failure; patients on timed treatment were able to complete nine courses as a rule. A difference of 10% in dose intensity may not appear important, but, due to the nonlinear relationship between dose intensity and disease control, a 10% dose intensity reduction may cause a 50% drop in disease control [86, 87] due to failure of drug administration at the optimal circadian time. This imperfect association between dose intensity and outcome indicates that dose intensity is only part of the story.

Tumor Response and Patient Survival
for Transitional Cell Bladder cancer

Two comparable subgroups of patients with widely metastatic transitional cell carcinoma of the bladder received up to nine courses of a timed combination of doxorubicin 60 mg/m$^2$ followed by cisplatin 60 mg/m$^2$ 12 h later at one of two circadian stages (0600 vs

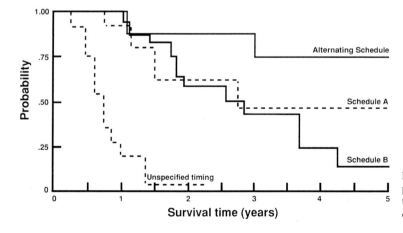

**Fig. 15.** Survival time of ovarian cancer patients varied depending on the circadian timing of treatment. This was only in part due to variations in dose intensity

1800). This was followed by a maintenance therapy for up to 2 years with a combination of cyclophosphamide, 5-FU, and cisplatin [97]. The response rate for the 35 evaluable patients was 57% with a complete response seen in 23%. Median survival after the first course of therapy was more than 1 year for partial responders and more than 2 years for complete responders, with three of the complete resonders alive without evidence of cancer more than 2 years after their last course of therapy. In addition, circadian optimized therapy has also been used for the adjuvant treatment of patients with advanced bladder cancer following resection. There were 5 patients with stage C and 11 patients with stage D1 transitional carcinoma of the bladder who received timed adjuvant combination therapy with doxorubicin followed by cisplatin 12 h later. Of these patients 11 showed no recurrence of the disease after a median follow-up time of 3.5 years (1–5.5 years). Two of the nonresponders had unusually late local recurrence after 37 and 42 months posttreatment respectively [85]. Similar differences in schedule-dependent toxicity were observed as in the ovarian cancer trials; however, the small size of the treatment groups did not allow interpretation of schedule-dependent differences in drug efficacy. The potential for chemotherapeutic curability of metastatic transitional cell cancer and the delay of local recurrence was demonstrated for the first time in this series and is possibly attibutable to sequence and timing of the drugs given.

## Chronotherapy with Fluoropyrimidines

Chemotherapy with fluoropyrimidines (5-FU; 5-fluoro-2'-deoxyuridine, FUDR) is important as a monotherapy and in combination with other antineoplastic agents for many malignancies. The efficacy of the drug may be further increased by prolonged tumor cell exposure to fluoropyrimidines resulting from continuous infusion schedules and by the possibility of synergistic effects from the addition of drugs modulating FU/FUDR interaction with biochemical targets (including tetrahydrofolic cid, methotrexate, cisplatin, and dipyridamol) [98]. A steep dose intensity-response relationship for fluoropyrimidines is documented, and this indicates the potential importance of any method that can safely increase fluoropyrimidine dose intensity [99].

### Experimental Animal Studies on Fluoropyrimidine Toxicity

*Experimental Animal Studies with 5-FU Bolus Injection.* According to clinical observation, toxicity patterns of bolus injection or long-term infusion of fluoropyrimidine are different, but most of the toxicity studies to date have been done with bolus administration. Several murine studies have demonstrated a circadian stage dependency of a 5-FU bolus injection, with the best drug tolerance at the mid-rest phase of the animals [100, 102] and several hours before their usual awakening [102] (Fig. 16).

The time dependency of the hematotoxic effect of fluoropyrimidine bolus therapy has also been documented, since leukopenia was observed at 12 HALO but not at 2 HALO in a colon-38 murine tumor model. In that experiment increased tumor activity was also observed at 2 HALO as compared to 12 HALO [103]. The time dependency of intestinal toxicity of fluoropyrimidine bolus therapy was assessed in vivo as a function of incidence of diarrhea and weight loss in Wistar rats and in vitro as a function of water absorption of isolated small intestine [104]. Impairment of water absorption, incidence of diarrhea, and weight loss were minimal after treatment in the center of the activity span coincidental with the circadian peak of mucosa cell DNA content of control animals. At that time the maximum number of cells are in the postmitotic resting phase (G1) and accordingly less susceptible to antimetabolic ef-

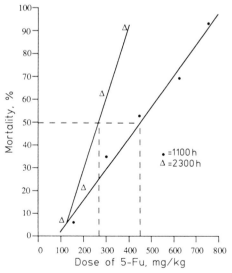

**Fig. 16.** Toxicity of FU. This figure demonstrates that the LD50 for 5FU is very different depending upon when in "the day" it is given. (From Burns and Beland 1984 [100])

fects of FU. It is noteworthy that diurnal variations in various absorptive, enzymatic, and cytokinetic activities of the intestine are related to feeding time, with the lowest DNA synthesis rate apparent at or after feeding. This mid-activity optimum for the intestine to tolerate bolus FU is in ad libitum fed animals several hours later in the day than the time optimum for whole animal toxicity and lethality. This finding raises the possibility that optimal times for toxic drug exposure may be different for different target tissues.

*Experimental Animal Studies woth FUDR Bolus Injection.* Circadian stage dependency of FUDR bolus injection toxicity (1000–2000 mg/kg) was tested by our group in a series of murine experiments. A time-dependent variation of survival was found to be 50% in favor of drug exposure during the mid-to-late activity span (18–20 HALO). On autopsy of the rats exposed to lethal doses of FUDR, pathological changes were found in both small and large bowel, as an explanation for diarrhea and dehydration prior to death, and in liver (diffuse liver necrosis) and bone marrow (aplasia), thus in organs identified as anatomical targets of lethal FUDR toxicity. A time dependency of FUDR bolus toxicity was confirmed [102] by similar experiments with a low mortality (20%) at 18 HALO, but in one strain of mice [Balb C] a minimal mortality [13%] was seen at 2 HALO, 6 h off the optimal treatment time in the experiments described above.

*Circadian "Shaping" of Continuous FUDR Infusion and Related Toxicity* Effect of circadian shaping of continuous FUDR infusion toxicity was tested in murine experiments using a quasi-sinusoidal infusion pattern with a peak at one of six different circadian time points [105]. At a dose level resulting in 50% overall mortality, lethal toxicity differed substantially depending upon the circadian stage of maximum drug delivery. Variable rate infusions showed increased, decreased, or equal toxicity compared to constant rate infusions depending on the circadian stage of maximum drug flow, with the least mortality during late activity/early rest span of the recipients (Fig.17). The best time of drug tolerance was slightly later than the results of bolus studies.

*Fluoropyrimidine Effectiveness in Experimental Cancer Chronochemotherapy: Therapeutic Index.* Effect of circadian shaping of continuous FUDR infusion on anticancer efficacy was tested on a rat adenocarcinoma model. At a therapeutic dose level and with identical dose intensity, the sinusoidal infusion pattern with a peak during late activity/early rest

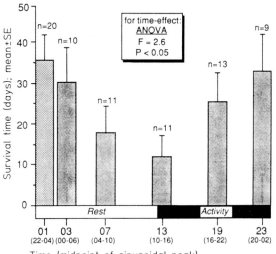

**Fig.17.** Survival time of female Fischer rats following 48 h of six equal doses (100 mg/kg/48 h) of sinusoidal FUDR infusions

phase resulted in significantly greater tumor growth delay than that observed with flat infusion or other peak times of the sinusoidal infusion and actually was the only infusion pattern associated with significant objective tumor shrinkage.

In summary, as the preliminary animal experiments indicated, FUDR pharmacodynamics, determined as toxicity and antitumor activity, is dependent on the circadian timing of drug exposure rather than just on quasi-intermittency of drug exposure, since, given with the "wrong" peak, the sinusoidal infusion was more toxic than the constant rate infusion. The optimal pharmacodynamic exposure time to FUDR (late acitivity/early sleep) in this rat model is different from that of 5-FU as determined by murine experiments [103].

## Clinical Results with 5-FU Chronotherapy

Limited clinical experience is available to date with circadian stage specified 5-FU administration. In a dose escalation study 34 patients with colon cancer were treated with external ambulatory infusion pumps set to deliver a sinusoidal infusion of 5-FU with a peak at 0400 [106] every 3 weeks over a period of 5 days. Intrapatient dose escalation by 1 g/m²/course was planned from the starting dose of 4 g/m²/course. Dose limiting toxicity (WHO grade II or higher) included stomatitis (9%), diarrhea (5%), neutropenia (4%), hand-foot syndrome (3%), anemia (2%), and cardiac toxicity (0%). Median maximal tolerated dose per course was 8 g/m$^2$ in patients with good performance status and 6.5 g/m$^2$ in patients with impaired performance status (bed-ridden for 50% or more of the day). The partial response rate was 30% and the stabilization rate was 44%. In previously untreated patients ($n = 22$) the partial response rate was 36% and the stabilization rate was 41%. Median survival was more than 20 months. Ambulatory circadian chronotherapy permitted an increase of dose intensity by approximately 75% with fewer side effects retrospectively than in a control group of patients on ambulatory chemotherapy with a constant infusion pattern over 5 days. Systemic FUDR Chronotherapy. Continuous long-term FUDR infusion frequently causes severe and dose limiting gastrointestinal toxicity when given by constant rate at commonly prescribed dose levels. Animal data allow the assumption that FUDR would be better tolerated when most of the daily infusion is given during the evening. A phase I randomized trial on 54 patients with metastatic cancer [107] compared a circadian shaped FUDR infusion with a peak flow rate during late afternoon/early evening to a constant rate infusion administered by a programmable implanted infusion pump (Synchromed, Medtronic Inc., Minneapolis, MN USA). Patients with metastatic cancer treated with equal dose intensities experienced less frequent and less severe diarrhea, nausea, and vomiting on circadian timed treatment. Dose intensity escalation studies with a circadian shaped infusion pattern revealed on the average a 1.45-fold increased drug tolerance with minimal toxicity encountered. The 29% objective response rate observed in the trial suggest that increased dose intensity achieved by optimal circadian shaping may improve the therapeutic index of infusional FUDR and may help control malignancies to date refractory to conventional chemotherapy, e.g., renal cell carcinoma [108].

Recently, [109] a phase II trial reported the highest objective response rate achieved to date (66%) in patients with metastatic colorectal cancer treated with i.v. time specified delivery of 5-FU, tetrahydrofolate (Leucovorin, LV) and oxaliplatin (I-OHP). In a 5 day chronotherapy protocol, the infusion rates of 5-FU, LV, and I-OHP were automatically increased and decreased along the 24 h scale by a multichannel pump (Intelliject-Aguettat, France). Daily doses were: FU, 500 mg/m$^2$; LV, 300 mg/m$^2$; and I-OHP, 25 mg/$^2$. A mixture of FU and LV was infused in a bell shaped pattern between 2200 and 1000. I-OHP was infused in a bell shaped pattern from 1000 to 2200. Results from 32 surviving patients receiving a total of 120 cycles of therapy are available.

Dose limiting toxicity included vomiting ( > 2 episodes/day despite 10 mg/day alizapride) and diarrhea ( ≥ 5 stools/day for > 7 days despite 10 mg/day loperamide) and reversible neuropathy (observed in 3 patients after ≥ 8 courses). The disease progressed in 6 patients (19%), was stabilized in 5 patients (16%), and regressed objectively by more than 50% as determined by computer assisted tomography in 21 patients (66%, 57% of pretreated patients, 72% of untreated patients).

*Regional FUDR Chronotherapy.* Liver metastases from colorectal cancer can be treated effectively with prolonged intra-arterial infusion of FUDR, although frequently severe and sometimes fatal hepatocellular necrosis is encountered. In a therapeutic trial, 50 patients with liver metastases received ambulatory intra-arterial chemotherapy of FUDR (0.1–0.3 mg/kg per day for 14 days every 4 weeks) via implanted infusion pump devices [110]. Patients recieved either a constant rate infusion ($n = 24$) or a circadian shaped infusion ($n = 26$). FUDR attributable cholestasis-related jaundice was less frequent, less severe, and delayed when a circadian shaped infusion was employed.

Alkaline phosphatase elevation occurred in 65% vs 44% and clinically apparent jaundice in 26% vs 4% of patients when flat vs circadian shaped infusion was compared. Gastrointestinal symptoms (nausea, vomiting, diarrhea) (20%) and gastric ulcers (12%) were equally encountered in both arms of the study. Patients receiving circadian shaped infusion tolerated a 70% higher average dose intensity of FUDR. Still, in spite of this difference in dose intensity, 52% of patients on a circadian shaped infusion regimen experienced no obvious toxicity, whereas only 22% of patients on a constant infusion regimen were free of toxicity. The objective response rates were comparable, at 35% and 32% for flat and circadian infusion, respectively. The trial demonstrated that the maximum tolerated dose intensity of intrahepatic arterial FUDR may be increased substantially by circadian modification of the continuous infusion rate and still lead to a significant reduction in hepatotoxic side effects.

Possible Mechanisms of FUDR
Chronopharmacodynamic Properties

The main organ for fluoropyrimidine metabolism is the liver. There is mounting empirical information that FU catabolism is highly coordinated in circadian time, although the circadian stage-dependent steps of the catabolic pathways have not been identified. Among the metabolic and catabolic events which may play a role in the time dependence of fluoropyrimidine toxicity are, for FUDR: (a) entry into cells using the pyrimidine nucleotide transport system [111], (b) activation by enzymatic conversion [112], (c) retention in the cells with a pharmacokinetic profile dependent on individual tissue characteristics [113], (d) competitive enzyme blockage with interference with the (circadian periodic) DNA synthesis and interaction in this respect with a number of enzymatic and nonenzymatic factors [114], and (e) its enzymatic breakdown [115, 116]. Similarly, 5-FU is activated [115, 117] and catabolized [118] by a multistep process with involvement of numerous enzyme systems. The elimination of 5-FU and FUDR is a rapid process with conversion largely into carbon dioxide, urinary excretion of less than 20%, and conjugation with bile acids with a probable impact on subsequent cholestasis and cholangitis after hepatic arterial infusion of fluoropyrimidines [116]. Since bile acid production has a marked non-meal-dependent circadian stage dependency, the conjugation-dependent drug inactivation can be expected to be also circadian periodic.

Dynamics of drug administration pattern determine FUDR metabolic pathways. The clearance of an equivalent of FUDR is much slower if given as i.v. bolus than as a constant 24 h infusion, the determinant being saturation of the degrading enzyme DPD [118]. FUDR given as bolus injection is primarily split into FU and deoxyribose-1-phosphate. The dose limiting toxic effect of a bolus injection is bone marrow depression, partly correlated with direct FU incorporation into bone marrow cell RNA via FUTP, blockage of DNA synthesis via FdUMP, and impaired RNA synthesis via FUTP [119].

FUDR given at low dose continuously over a longer period of time undergoes primarily "intracellular salvage" and is converted into FdUMP. As a result the toxicity pattern is profoundly changing with an increased infusion duration of an equivalent of FUDR [120]. Now the mucosa cell of the gastrointestinal tract is the dose limiting target, since ulceration and profuse fluid loss occur in the absence of clinically significant bone marrow depression. This may be explained by the relative accumulation of FdUMP in the gastrointestinal mucosa cell compared to the bone marrow cell for a given plasma level of FUDR [121].

Circadian stage dependency of FUDR/FU interconversion was studied by drug monitoring during a constant infusion of FUDR (350 mg/kg over 6 h i.v.) in a rat model (female F344 Fischer rats) exposed to

**Table 5.** FUDR and FU plasma levels following continuous FUDR infusion

| Drug | Treatment time | $C_{max}$ (mg/ml) | $AUC_{0-6.5}$ $\times 1000$ | $t_{1/2}$ (h) |
|------|------|------|------|------|
| FUDR | 22-04 | $17.1 \pm 6.0$ | $64.1 \pm 8.6$ | $0.10 \pm 0.03$ |
| FUDR | 10-16 | $14.4 \pm 1.0$ | $51.8 \pm 3.7$ | $0.25 \pm 0.14$ |
| FU | 22-04 | $5.1 \pm 0.3$ | $16.3 \pm 2.6$ | $0.11 \pm 0.06$ |
| FU | 10-16 | $10.6 \pm 4.1$ | $25.0 \pm 5.7$ | $0.13 \pm 0.02$ |

Mean $\pm$ SE; $AUC_{0-6.5}$, area under the plasma concentration curve from 0 to 6.5 h of treatment start

the infusion at two different circadian stages (Table 5). The conversion of FUDR into FU was greater after FUDR infusion between 10 HALO and 16 HALO suggesting that the activity of the converting enzyme PDP is circadian stage dependent. The half life ($t_{1/2}$) of FUDR appears prolonged at 16 HALO though not statistically significant. Small sample size ($n = 60$) and limitation to two circadian stages may have limited the results of this pilot study.

A significant circadian variation of elimination of FU and FU catabolites with a 40%–60% variation around the mean was shown in a model of isolated perfused rat livers, with a peak elimination rate of FU at 19 HALO and a reciprocal peak for FU catabolites at 7 HALO [122].

The circadian stage dependency of dihydropyrimidine dehydrogenase (DPD), an enzyme participating in fluoropyrimidine catabolism, was documented in a rat model [123] with a peak activity at 10 HALO (2.96 nmol/min/mg) and a trough activity at 22 HALO (0.40 nmol/min/mg) revealing a sevenfold difference in activity as a function of time.

The specific activity of DPD in cytosolic preparations of human mononuclear cells assayed by high-performance liquid chromatography (HPLC) and radioactivity flow monitoring was circadian stage dependent in seven healthy volunteers on a usual daily sleep-wakefulness pattern and a regular diet [124]. The peak enzyme activity was consistently located around 0000 (2200–0200) and the variation around the individual mean DPD activity was between 20% and 60.8%. This finding was confirmed in cancer patients ($n = 7$) on continuous FU infusion (300 mg/ kg/day) chemotherapy for several weeks [125] with the peak of the circadian DPD rhythm at 0100. A circadian rhythm of FU plasma level was found in the same study with a peak value ($27.4 \pm 1.3$ ng/ml) at 1100 and a trough value ($5.6 \pm 1.3$ ng/ml) at 2300 hour, this ratio being almost fivefold (Table 6). An inverse relationship between DPD activity in peripheral blood mononuclear cells and plasma FU concentration was found by linear regression between the two parameters in all

**Table 6.** Circadian changes of DPD activity and FU plasma levels in humans

|  | Dose (mg/m²/day) | Peak (HALO) | Trough (HALO) | Max/min | Reference |
|---|---|---|---|---|---|
| DPD activity[a] | - | 0100 | 1300 | 1.5 | [124] |
| DPD activity | - | 0100 | 1300 | 1.7 | [123] |
| FU level | 300 | 1100 | 2300 | 4.9 | [123] |
| FU level after pretreatment with cisplatin | 450–966 | 0100 | 1300 | 2.3 | [126] |

[a] In mononuclear blood cells

patients ($r = 0.63$) and within patients ($-0.98 < r < -0.74$). Finally, a circadian variation of FU plasma level was also found in patients receiving combination chemotherapy [126].

The significant differences in circadian stage of optimal drug exposure between 5-FU and FUDR are of interest. Such differences in the optimal timing of two closely related drugs are not uncommon. Despite the chemical similarity between FUDR and FU and the fact that both drugs are interconvertible, the circadian time of best drug tolerance may be different (late activity/early rest phase for FUDR and mid-to-late rest phase for 5-FU). We have learned that some apparently very closely related drugs (e.g., anthracyclines including doxorubicin, daunomycin, and epirubicin) can have individually optimal circadian times [94], while other classes of related drugs (cisplatin, carboplatin, I-OHP) have very similar or identical best and worst circadian times [127, 128].

The differential, intraindividual, time-variant tissue susceptibility to FUDR may be explained by differential tissue specific accumulation of the active metabolite FdUMP, which is found at relatively higher levels in epithelial cells of the gastrointestinal tract than in bone marrow cells for the same plasma levels of FUDR [120]. It may further be explained by the circadian dependence of FUDR and FU interconversion catabolism in rodents and humans as described above. However, we did not find a correlation between a spot determination of DPD activity and FUDR toxicity in patients receiving the drug by low dose long-term infusion [129]. We may have sampled DPD activity at a suboptimal diagnostic circadian stage or the catabolic rate may be high enough to prevent drug accumulation at any circadian stage and then other factors may be determinants of the circadian stage-dependent toxicity pattern.

In the complicated metabolic pathways of activation and catabolism, only a small fraction of potentially relevant end points have been studied. Currently one can only speculate on the determinant mechanisms of circadian FUDR host and tumor susceptibility rhythms. These mechanisms are probably drug-specific including multifactoral temporal changes in the biochemical drug pathways and time-dependent patterns in host and tumor cell cytokinetics which determine the susceptibility to cytotoxic agents and deserve urgent attention. 5-FU/FUDR toxicity is clearly circadian stage dependent in animals. Divergent findings about the optimal circadian exposure to 5-FU/FUDR in animal experiments may be due to a lack of standardization of animal model characteristics and treatment plan. Still for the clinical utilization of the chronopharmacological observations one has to rely on animal data in the design of human therapeutic trials, since clinical trials testing six different circadian stages of drug exposure in a randomized fashion are not feasible.

In vitro studies are not useful for testing chronopharmacodynamics because the underlying biological rhythms may be altered or destroyed outside the intact organism. Still human trials reproduced the observation of chronotoxicity and produced information that allowed a markedly higher dose intensity to be given safely at well-defined times of the day. Evidence of improved antitumor activity is still pending. The ability to predict an individual's drug susceptibility pattern can be improved as many treatment-related variables, recipient-related factors, and environmental and seasonal factors that have a potential for modulation of the pharmacodynamic rhythm are quantitatively defined in time and taken into account when formulating a prediction model. Periodic changes in hormonal and immunological functions may also have modulating effects upon the ultimate treatment outcome [130, 131].

## Chronotherapy of Biological Response Modifiers

It has been adequately demonstrated that the therapeutic ratio of anticancer drugs is potentially dependent upon the timing of drug delivery – relative

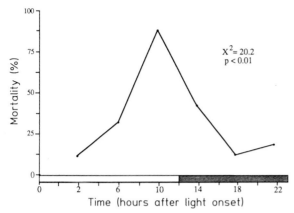

**Fig. 18.** The circadian stage-dependent pattern of human recombinant tumor necrosis factor (rHuTNF) demonstrates greatest toxicity when given in the late sleeping span. Single intravenous injection of 750 and 1000 mg/kg to a total of 60 Balb C mice; five mice/time point/dose

to circadian stage, time between doses, and sequencing of drugs in combination chemotherapy. However the time factor is even much more relevant to the effective use of biological response modifiers. The time of day when interferons, leukotrienes, tumor necrosis factors, interleukin-2, LAK cells, growth factors, e.g., epidermal growth factor [50, 51], or monoclonal antibodies are given, with respect to the patient's circadian cycle, as well as repeat dosages, their sequences vis a vis one another, and standard cytotoxic treatment will ultimately determine how effective these factors are.

The relationship between host and tumor is extremely complex. The standard phase I and phase II approaches to the study of biological agents will therefore not adequately define their activity or even their relevant toxicities. Biological agents have profound effects in picogram quantities at the level of the cell. Giving milligram quantities without regard to this complexity will only result in great expense, great toxicity, and finally great frustration.

The lethal toxicity of human recombinant tumor necrosis factor has now been demonstrated to be circadian stage dependent with a tenfold time-dependent difference in lethal toxicity (Fig. 18) [132]. Once the time structure of the pharmacodynamic effects of biological agents is well-defined, the use of infusion devices able to stipulate sequence, interval, circadian stage and infradian pattern of immune modulation will be a sine qua non to optimal biotherapy [133].

## State of the Art Drug Delivery Systems for Use in Chronotherapy

The wider application of optimally timed therapy requires technological means for automated drug delivery. The technology necessary to accomplish these goals is widely available for cost-effective and easy use [134]. Complex drug timing schedules (timed bolus, constant infusion, or circadian shaped infusion pattern) are only feasible in a clinical or ambulatory setting if automatic programmable delivery systems are used. Such devices with one or multiple reservoirs are now available [135, 136]. There is a choice between external wearable devices hooked up to a port system and implantable devices, the latter only available as single channel device. The following instruments are currently available:

- Synchromed pump (Medtronic Inc. Minneapolis, MN USA): an implantable device, filled and programmed transcutaneously by telemetry after surgical implantation. The system can deliver variable rate infusion cycles or timed bolus injections intravenously, intra-arterially, or intrathecally. The pump reservoir has a maximum usable capacity of 18 ml. This system can administer either a single drug or a mixture of drugs which are compatible with one another while stored in the reservoir at body core temperature and with the pump components and catheters.
- Parker-Hannifin Biomedical Products Division (Irvine, CA USA) produces the only single channel device with a clock and calendar. The device is capable of increasing the drug flow over time in a limited circadian cycle.
- Intelligent Medicine Inc. (Englewood, CO USA) has produced a four channel flexibly programmable device which can deliver optimal time specified treatments with up to four drugs with a maximum infusate of 30 ml per channel. The device is fairly bulky and programmed by an add-on microchip that needs to be replaced if program modifications are necessary.

Time-based infusion delivery systems currently in development include a single channel device with up to 24 steps per program cycle (Parker micropump) and a four channel wearable device for polychemotherapy (I-Flow Co., Irvine, CA USA).

Future generations of automated drug delivery devices will all be multichannel devices with each channel independently programmable. Eventually, closed loop systems now under development will become available in which the delivery of drugs and hormones is determined by an array of physiological signals sensed by the device and fed through algorithms that determine the appropriate dose, dose timing, sequence of agents, and time span between pulses or patterns of these agents. Implantable systems will be used for locoregional therapy, for quasi-continuous systemic cell cycle synchronization, and for long-term immunomodulation. Extracorporeal devices will complement these instruments for short-term therapy in the same patients.

The outlook for future instrumentation applications involves numerous areas in clinical medicine, many of which will require timed treatment schedules. *Modern oncology* will demand circadian and infradian timing-stipulated and complex sequential multidrug regimens, biological therapies, and hybrid chemobiotherapies. Moreover, with the advent of multichannel delivery devices complex protocol design will no longer be constrained by questions of protocol compliance. In *endocrinology* today, ultradian pulsatile and circadian patterned hormone therapy is essential in treating infertility and growth hormone disorders, and this therapy will soon be applied to alleviate sleep and emotional disorders. Tomorrow, endocrinology will rely upon implanted, closed loop, artificial hypothalamus, pituitary, adrenal, and pancreas technologies. In *gastroenterology,* delivery of $H_2$-blocking agents at the time of day associated with highest basal and stimulated acid secretion. In *hematology,* treatment of iron overload by desferrioxamine chelation may be improved by automated systems. In *cardiology,* chronotherapy of cardiac arrhythmias and hypertension by circadian-modified transcutaneous drug delivery is possible today, and will be further improved if closed loop automated systems become available. Finally, *intensive care units* will probably be the first site to use polypharmacy in a closed loop manner for treatment of patients with multiple organ failure.

The adoption of "intelligent", automatic, programmable, drug delivery devices will make medicine both intrinsically more complex and extrinsically more simple. It will make delivery of complex drug and biological regimens safer, less error prone, and less expensive. The therapy of serious diseases will be more standardized, with protocols developed in clinical research centers widely available to all medical practitioners.

The adoption of programmed automatic drug delivery will bring attention to temporal chronobiological questions of drug treatment which have not yet been solved. This attention will turn chronobiology into what it truly is – a multidimensional and dynamic perspective on life science. This medical movement toward temporal considerations will abolish the separate science of chronobiology and ultimately convert all biologists and physicians to chronobiologists.

# Conclusions

To date, only very few solid tumors can be cured by chemotherapy alone. Most malignancies, especially the more frequent cancer types, are primarily chemotherapy resistant or develop resistance. The chronobiological approach in experimental and clinical oncology contributed to the definition of a complex and predictable time structure of host-tumor interaction with periodic changes in several frequency domains.

Timing of drug exposure to cytotoxic drugs with known time dependency of pharmacokinetic and pharmacodynamic effects allows chronotherapeutic optimization of dose intensity tolerated by the host and also improvement of the therapeutic index according to empirical findings in several host-tumor systems. Reproducible circadian differences of drug pharmacokinetics and pharmacodynamics in human beings have been found for several anticancer drugs, including fluoropyrimidines, cisplatin, I-OHP carboplatin, doxorubicin, 6-mercapto-purine, and methotrexate [46, 48, 129, 137–139].

The chronotherapeutic models tested to date are simple. However, even the most crude specification of a chronotherapy schedule, namely, treating each individual at a time of day determined on the basis of animal experiments and/or the known periodicity of vital tissues of the host involved in the tolerance or in the toxic side effects of the drug(s), clearly results in a measurable overall advantage of tumor control.

Against this background there is a strong argument for mandatory preclinical chronotoxicology studies of all new chemotherapeutic agents and for rigorous, circadian, time-qualified, phase I studies performed on subsets of patients at one of several well-defined time points of the day (and possibly of other frequency ranges, i.e., the week and/or the month of the year). Monitoring of physiological marker variables is further advised for individualization of chronotherapy, since significant interindividual differences in circadian time structure can exist.

The development of biological monitoring capability to assign a treatment time relevant to an individual's internal circadian time and the availability of automatic programmable drug delivery devices to execute a complex time-specified polychemotherapy protocol may be expected to add further to the improvement of therapeutic ratio of conventional chemotherapy.

Chronotherapy must also include the timing of surgical intervention, the timing of radiotherapy, and the timing of manipulation of the host immunological and endocrinological functions, all known to be significantly and predictably time dependent, in order to improve the clinical effectiveness of systemic antineoplastic chronotherapy.

# References

1. Halberg F, Nelson W, Cornélissen G, Haus E, Scheving LE, Good RA (1979) On methods for testing and achieving cancer chronotherapy. Cancer Treat Rep 63: 1428–1430
2. Burns ER (1981) A critique of the practice of comparing control data obtained at a single time point to experimental data obtained at multiple time points. Cell Tissue Kinet 14: 219–224
3. Burns ER (1982) A critique of the practice of plotting data obtained in two in vivo on an 'hours after treatment' format. Oncology 39: 250–254
4. Scheving LE (1981) 11th International Congress of Anatomy: biological rhythms in structure and function. Liss, New York, pp 39–79
5. Thorud E, Clausen OPF, Bjerknes R, Aarnaes E (1980) The stathmokinetic method in vivo time-response with special reference to circadian variationsin epidermal cell proliferation in the hairless mouse. Cell Tissue Kinet 13: 625–634
6. Rubin NH (1982) Influence of the circadian rhythm in cell division on radiation-induced mitotic delay in vivo. Radio Res 89: 65–76
7. Mamontov SG (1968) Diurnal rhythm of mitoses in the epithelium of the mouse tongue. Bull Exp Biol Med 66: 1277–1278
8. Izquierdo JN, Gibbs SJ (1974) Turnover of cell-renewing populations undergoing circadian rhythms in cell proliferation. Cell Tissue Kinet 7: 99–111
9. Smaaland R, Sletvold O, Bjerknes R, Lote K, Laerum OD (1987) Circadian variations of cell cycle distribution in human bone marrow. Chronobiologia 14: 239
10. Laerum OD, Aardal NP (1981) Chronobiological aspects of bone marrow and blood cells. In: Acosta EV, Mayersbach H, Scheving LE et al. (eds) Eleventh International Congress of Anatomy: Biological Rhythms in Structure and Function. Liss, New York, pp 87–97
11. Durie BGM, Salmon SE, Russell DH (1977) Polyamines as markers of response and disease activity in cancer chemotherapy. Cancer Res 36: 214–221
12. Janne J, Poso H, Raina A (1978) Polyamines in rapid growth and cancer. Biochem Biophys Acta 473: 241–243
13. Hrushesky W, Merdink J, Abdel-Monem M (1983) Circadian rhythmicity characterizes monacetyl polyamine urinary excretion. Cancer Res 43: 3944–3947
14. Burns, ER, Scheving LE, Tsai TH (1979) Circadian rhythms in DNA synthesis and mitosis in normal mice and in mice bearing the Lewis lung carcinoma. Eur J Cancer 15: 233–242
15. Nash RE, Echave Llanos JM (1971) Circadian variations in DNA synthesis of a fast-growing and a slow-growing hepatoma: DNA synthesis rhythm in hepatoma. J Natl Cancer Inst 47: 1007–1012
16. Brown HE, Dougherty TF (1956) The diurnal variations of blood leucocytes in normal and adrenalectomized mice. Endocrinology 58: 365–375
17. Kanabrocki EL, Scheving LE, Halberg F et al. (1975) Circadian variation in presumably healthy young soldiers. Department of the Army, Document PF 228437, National Technical Information Service, US Department of Commerce, PO Box 1553, Springfield, Virginia
18. Abo T, Kumagai K (1978) Studies of surface immunoglobulins on human b lymphocytes. III: Physiologic variations of sig + cells in peripheral blood. Clin Exp Immunol 33: 441–452
19. Cove-Smith JR, Pownall R, Kabler TA et al. (1979) Chronopharmacology. In: Reinberg A, Halbert G (eds) International Congress of Pharmacology. Pergamon, Oxford, pp 369–373
20. Tavadia H, Fleming K, Hume P, Simpson HW (1975) Circadian rhythmicity of human plasma cortisol and PHA-induced lymphocyte transformation. Clin Exp Immunol 22: 190–193
21. Fernandes G, Halberg F, Yunis E, Good RA (1976) Circadian rhythmic plaque-forming cell response of spleens from mice immunized with SRBC. J Immunol 117: 962–966
22. Fernandes G, Carandente F, Halberg E, Halberg F, Good RA (1979) Circadian rhythm in activity of lymphocytic natural killer cells from spleens of Fischer rats. J Immunol 123: 622–625
23. Kohler WC, Karacan I, Rennert OM (1972) Circadian variation of RNA in human leucocytes. Nature 238: 94–96
24. Hrushesky WJM, Gruber SA, Sothern RB, Hoffman RA, Lakatua D, Carlson A, Cessa F, Simmons RL (1988) Natural killer cell activity is age, estrous and circadian-stage dependent and correlates inversely with metastatic potential. J Natl Cancer Inst 80: 1232–1237
25. Pollock RE, Babcock GF, Romsdahl M, Nishioka K (1984) Surgical stress-mediated suppression of murine natural killer cell cytotoxicity. Cancer Res 44: 3888–3891
26. Ratajczak HV, Sothern RB, Hrushesky WJM (1988) Estrous influence on surgical cure of a mouse breast cancer. J Exp Med 168: 88–96
27. Gruber SA, Nichol KL, Sothern RB, Malone ME, Potter JD, Lakatua D, Hrushesky WS (1989) Menstrual history and breast cancer. Breast Cancer Res Treat 13: 278
28. Cohen P, Wax Y, Modan B (1983) Seasonality in the occurrence of breast cancer. Cancer Res 43: 892–896
29. Jacobson H, Janerich DT, Nasca P, Langevin T, Steiner B, Hrushesky WSM (1983) Circannual rhythmicity in the incidence of endocrine malignancy: evidence for neurohumoral control of cancer (ca) development and growth, abstracted. Chronobiol. 10: 135
30. Hrushesky W, Teslow T, Halberg F et al. (1979) Temporal components of predictable variability along the 1-year scale in estrogen receptor concentration of primary human breast cancer, abstracted. Proc Am Soc Clin Oncol 20: 331

31. Hughes A, Jacobson HI, Wagner RK et al. (1976) Ovarian independent fluctuations of estradiol receptor levels in mammalian tissues. Mol Cell Endocrinol 5: 379–388

32. Langlands AO, Simpson H, Sothern R et al. (1977) Different timing of circannual rhythms in mortality of women with breast cancer diagnosed before and after menopause. In: Proceedings of the 8th International Scientific Meeting of the International Epidemiological Association, San Juan, Puerto Rico, 17–23 Sept 1977

33. Haus E, Halberg F, Scheving LE, Cordoso S, Kühl A, Sothern R, Shiotsuka R, Hwang DS, Pauly JE (1972) Increased tolerance of leukemic mice to arabinosyl cytosine with schedule adjusted to circadian system. Science 77: 80–82

34. Scheving LE, Haus E, Kuhl JFW, Pauly JE, Halberg F, Cardoso S (1976) Close reproduction by different laboratories of characteristics of circadian rhythm in 1-$\beta$-D-arabinofuranosylcystosine tolerance by mice. Cancer Res 36: 1133–1137

35. Hrushesky WJM, Dell I, Eaton J, Halberg F (1982) Circadian-stage-dependent effect of doxorubicin upon reduced glutathione in the murine heart. Proc Am Assoc Cancer Res 23:12

36. Hrushesky WJM (1983) The clinical application of chronobiology to oncology. Am J Anat 168: 519–542

37. Wesson LG (1964) Electrolyte excretion in relation to diurnal cycles of renal function. Medicine (Baltimore) 43: 547–592

38. Scheving LE (1981) Circadian rhythms in cell proliferation: their importance when investigating the basic mechanism of normal versus abnormal growth. In: Acosta EV, Mayersbach H, Scheving LE et al. (eds) 11th International Congress of Anatomy, Biological Rhythms in Structure and Function. Liss, New York, pp 39–79

39. Burns RE (1981) Circadian rhythmicity in DNA synthesis in untreated mice as a basis for improved chemotherapy. Cancer Res 41: 2795–2802

40. LaBrecque DR, Feigenbaum A, Bachur NR (1978) Diurnal rhythm: effects on hepatic regeneration and hepatic regenerative stimulator substance. Science 199: 1082–1084

41. Scheving LE (1984) Chronobiology of cell proliferation in mammals: implications for basic research and cancer chemotherapy. In: Edmunds LN (ed) Cell cycle clocks. Dekker, New York, pp 455–499

42. Mayersbach HV (1978) Die Zeitstruktur des Organismus. Auswirkungen auf zellulaere Leistungsfähigkeit und Medikamentenempfindlichkeit. Arzneimittelforsch/Drug Res 28: 1824–1836

43. Kinlaw WB, Fish LH, Schwartz HL, Oppenheimer JH (1987) Diurnal variation in hepatic expression of the rat S14 gene is synchronized by the photoperiod. Endocrinology 120: 1563–1567

44. Petit E, Milano G, Levi F, Thyss A, Bailleul F, Schneider M (1987) Circadian rhythm in 5-FU pharmacokinetics during 5-day continuous infusion. Satellite Symposia on the Proceedings of the European Conference on Clinical Oncology, vol 4, p 293

45. Levi F, Hrushesky WJM, Borch RF, Pleasants ME, Kennedy BJ, Halberg F (1982) Cisplatin urinary pharmacokinetics and nephrotoxicity: a common circadian mechanism. Cancer Treat Rep 66: 1933–1938

46. Hrushesky WJM, Borsch R, Levi F (1982) Circadian time dependence of cisplatin urinary kinetics. Clin Pharmacol Ther 32: 330–339

47. Boughattas AN, Levi F, Roulon A, Mechkouri M, Lernaigre G, Cal JC, Camber J, Reinberg A, Mathe G (1987) Similar circadian rhythm in murine host tolerance for two platinum analogs: carboplatin (CBDCA) and oxaliplatin (I-OHP). Proc Am Assoc Cancer Res 28: 1788

48. Aherne GW, English J, Burton N, Arendt J, Marks V (1987) Chronopharmacokinetics and their relationship to toxicity and effect with reference to methotrexate, 6-mercaptopurine and morphine. Satellite Symposia on the Proceedings of the European Conference on Clinical Oncology, vol 4, p 40

49. Halberg F, Haus E, Cardoso SS, Scheving LE, Kuhl JFW, Shiotsuka R, Rosene G, Pauly JE, Runge W, Spalding JF, Lee JK, Good RA (1972) Toward a chronotherapy of neoplasia: tolerance of treatment depends upon host rhythms. Experientia 29: 909–934

50. Scheving LA, Yeh YC, Tsai T, Scheving LE (1980) Circadian phase-dependent stimulatory effects of epidermal growth factor on deoxyribonucleic acid synthesis in the duodenum, jejunum, ileun, caecum, colon, and rectum of the adult male mouse. Endocrinology 106: 1498–1503

51. Scheving LA, Yeh YC, Tsai TH, Scheving LE (1979) Circadian phase-dependent stimulatory effects of epidermal growth factor on deoxyribonucleic acid synthesis in the tongue, esophagus, and stomach of the adult male mouse. Endocrinology 105: 1475–1480

52. Halberg F, Haus E, Scheving LE (1978) Sampling of biologic rhythms, chronocytokinetics and experimental oncology. In: Valleron AJ, Macdonald PDM (eds) Biomathematics and cell kinetics. Elsevier North-Holland, New York, pp 175–190

53. Clausen OPF, Thorud E, Bjerknes R, Elojor K (1979) Circadian rhythms in mouse epidermal basal cell proliferation. Variations in compartment size, flux, and phase duration. Cell Tissue Kinet 12: 319–337

54. Haus E, Lakatua DJ, Swoyer J, Sackett-Lundeen L (1983) Chronobiology in hematology and immunology. Am J Anat 168: 467

55. Swoyer J, Haus E, Lakatua D, Sackett-Lundeen L, Thompson M (1984) Chronobiology in the clinical laboratory. In: Haus E, Kabat HF (eds) Chronobiology 1982–1983. Karger, New York, pp 533–543

56. Levi FA, Canon C, Blum JP, Mechkouri M, Reinberg A, Mathe G (1985) Circadian and/or circahemidian rhythms in nine lymphocyte-related variables from peripheral blood of health subjects. J Immunol 134: 217–220

57. Gatti G, Cavallo R, Del Ponte D, Sartori M, Masera R, Carignola R, Carandente F, Angeli A (1986) Circadian changes of human natural killer (NK) cells and their in vitro susceptibility to cortisol inhibition. Annu Rev Chronopharmacol 3: 75–78

58. Levi, F, Canon C, Blum JP, Reinberg A, Mathes G (1983) Large-amplitude circadian rhythm in help: suppressor ratio of peripheral blood lymphocytes. Lancet ii: 462–463

59. Ross DD, Pollak A, Akman SA, Bachur NR (1980) Diurnal variation of circulating human myeloid progenitor cells. Exp Hematol 8: 954–960

60. Verma DS, Fisher R, Spitzer G, Zander AR, McCredie KB, Dicke KA (1980) Diurnal changes in circulating myeloid progenitor cells in man. Am J Hematol 9: 185–192

61. Killmann SA, Cronkite EP, Fliedner TM, Bond VP (1962) Mitotic indices of human bone marrow cells. I. Number and cytologic distribution of mitoses. Blood 19: 743–750

62. Mauer AM (1965) Diurnal variation of proliferative activity in the human bone marrow. Blood 26: 1–7

63. Bellamy WT, Alberts DS, Dorr RT (1988) Circadian variation in non-protein sulfhydryl levels of human bone marrow. Cancer Res Eur J Cancer Clin Oncol 24: 1759–1762

64. Buchi KN, Moore JG, Rubin NH (1987) Circadian cellular proliferation measurements in human rectal mucosa. Chronobiologia 14: 155–156

65. Buchi KN, Rubin N, Moore JG (1988) Circadian rhythm of cellular proliferation in the human rectal mucosa. Annu Rev Chronopharmacol 5: 355

66. Markiewicz A, Lelek A, Panz B, Wagiel J, Boldys H, Hartleb M, Kaminski M (1987) Chronomorphology of jejunum in man. Chronobiologia 14: 202

67. Levi F (1991) Chronopharmacology of anticancer agents and cancer chronotherapy. In: Kummerle H (ed) International handbook of clinical pharmacology. Ecomed, Landsberg (in press)

68. Klevecz RR, Shymko RM, Blumenfeld D, Braly PS (1987) Circadian gating of S phase in human ovarian cancer. Cancer Res 47: 6267–6271

69. Garcia Sainz M, Halberg F (1966) Mitotic rhythm in human cancer, reevaluated by electronic computer programs- evidence for chronopathology. J Natl Cancer Inst 37: 279–292

70. Hrushesky WJM, Haus E, Lakatua DJ, Halberg F, Langevin T, Kennedy BJ (1985) Marker rhythms for cancer chrono-chemotherapy. In: Haus E, Kabat HF (eds) Chronobiology 1981–1983. Karger, New York, pp 493–499

71. Halberg F (1974) From aniatoxicosis and aniatrosepses toward chronotherapy. Introductory remarks to the 1974 Capri Symposium on timing and toxicity: the necessity for relating treatment to bodily rhythms. Tempus non solum dosis venenum facit. In: Aschoff J, Ceresa F, Halberg F (eds) Chronobiological Aspects if Endocrinology. Stuttgart, F.K.Schattauer Verlag, pp 1–34

72. Derer I (1960) Rhythm and proliferation with special reference to the six day rhythm of blood leukocyte count. Neoplasma 7: 117–134

73. Ashkenazi IE, Hartman H, Strulovitz B et al. (1975) Activity rhythms of enzymes in human red blood cell suspensions. J Interdisc Cycle Res 6: 291–301

74. Haus E, Halberg F, Loken MD, Kim YS (1973) Circadian rhythmometry of mammalian radiosensitivity. In: Tobias A, Todd P (eds) Space radiation biology and related topics. Academic, London, pp 435–474

75. Hrushesky WJM (1985) Circadian timing of cancer chemotherapy. Science 228: 73–75

76. Roemeling R, Mormont M-C, Walker K, Olshefski R, Langevin T, Rabatin J, Wick M, Hrushesky W (1987) Cancer control depends upon the circadian shape of continuous FUDR infusion. Proc Am Assoc Cancer Res 28: 1293

77. Peters GJ, Van Dijk J, Nadal JC, Van Groeningen CJ, Lankelma J, Pinedo HM (1987) Diurnal variation in the therapeutic efficacy of 5-fluorouracil against murine colon cancer. In Vivo 1: 113–118

78. Rivard G, Infante-Rivard C, Hoyeux C, Champagne J (1985) Maintenance chemotherapy for childhood acute lymphoblastic leukemia: better in the evening. Lancet ii: 1264–1266

79. Focan C (1979) Sequential chemotherapy and circadian rhythm in human solid tumors. Cancer Chemother Pharmacol 3: 197–202

80. Halberg F (1964) Medical aspects of stress in the miliary climate. In: Walter Reed Army Institute of Research Symposium. US Government Printing Office, Washington DC, pp 1–36

81. Halberg F, Haus E, Cardoso SS, Scheving LE, Kühl JF, Shiotsuka R, Rosene G, Pauly JE, Runge W, Spelding JF, Lee JK, Good RA (1973) Toward a chronotherapy of neoplasia: tolerance of treatment depends upon host rhythms. Experientia 29: 909–1044

82. Scheving LE, Burns ER, Pauly JE, Halberg F, Haus E (1977) Survival and care of leukemic mice after circadian optimization of treatment with cyclophosphamide and 1-$\beta$-D-arabinofuranosylcytosine. Cancer Res 37: 3648–3655

83. Rugh R, Castro V, Balter S et al. (1963) X-rays: are there cyclic variations in radiosensitivity? Science 142: 53–56

84. Sothern RB, Halberg F, Halberg E, Zinneman HH, Kennedy BJ (1981) Circadian and methodologic aspects of toxicity from cis-diamminedichloroplatinum, adriamycin and methylprednisolone interaction in rats with immunocytoma. In: Walker CA, Winget CM, Soliman KFA (eds) Chronopharmacology and Chronotherapeutics, Florida State University Institute, Tallahassee, pp 247–256

85. Roemeling R von, Hrushesky WJM (1986) Advanced transitional cell bladder cancer: a treatable disease. Semin Surg Oncol 2: 76–89

86. Hryniuk WM, Levine MN, Levin L (1986) Analysis of dose intensity for chemotherapy in early (stage II) and advanced breast cancer. NCI Monogr 1: 87–94

87. Levin L, Hryniuk W (1987) The use of dose intensity (DI) to solve problems in gynecologic oncology. Proc Am Soc Clin Oncol 6: 119

88. Hrushesky WJM, Levi F, Halberg F, Kennedy BJ (1982) Circadian stage dependence of cisdiamminedichloroplatinum lethal toxicity in rats. Cancer Res 42: 945–949

89. Hecquet B, Meynadier J, Bonneterre J, Adenis L, Demaille A (1985) Time dependency in plasmatic protein binding of cisplatin. Cancer Treat Rep 69: 79–82

90. Levi FA, Hrushesky WJM, Halberg F, Langevin TR, Haus E, Kennedy BJ (1982) Lethal nephrotoxicity and hematologic toxicity of cisdiamminedichloroplatinum ameliorated by optimal circadian timing and hydration. Eur J Cancer Clin Oncol 18: 471–477

91. Good RA, Sothern RB, Stoney PJ, Simpson HW, Halberg E, Halberg F (1977) Circadian state dependence of adriamycin-induced tumor regression and recurrence rates in immunocytomabearing LOU rats. Chronobiologia 4: 174

92. Halberg F, Gupta BD, Haus E, Halberg E, Deka AC, Nelson W, Sothern RB, Cornelissen G, Lee JK, Lakatua DJ, Scheving LE, Burns ER (1977) Steps toward a cancer chronopolytherapy. In: Proceedings of the XIVth International Congress of Therapeutics, Montpellier. L'Expansion Scientifique Française, Paris, pp 151–196

93. Sothern RB, Nelson WL, Halberg F (1977) A circadian rhythm in susceptibility of mice to the anticancer drug, adriamycin. In: Proceedings of the XII International Conference of the International Society for Chronobiology, Washington DC II Ponte, Milano, pp 433–438

94. Sothern RB, Halberg F, Good RA, Simpson HW, Grage TB (1981) Difference in timing of circadian susceptibility rhythm in murine tolerance of chemically-related antimalignant antibiotics: adriamycin and daumomycin. In: Walker CA, Winget CM, Soliman KFA (eds) Chronopharmacology and Chronotherapeutics. Florida A&M University Foundation Tallahassee, pp 257–268

95. Mormont MC, Roemeling R von, Sothern RB, Berestka JS, Langevin TR, Olshefski R, Wick M, Hrushesky WJM (1988) Circadian rhythm and seasonal dependence in tolerance of mice to 4'-epidoxorubicin. Invest New Drugs 6: 273–283

96. Roemeling R von, Christiansen NP, Hrushesky WJM (1985) Lack of antiemetic effect of high dose metoclopramide. J Clin Oncol 3: 1273–1276

97. Roemeling R von, Hrushesky WJM, Fraley E (1987) Long-term control of locally advanced transitional cell bladder cancer (TCCB) by high-dose intensity, circadian-based adjuvant chemotherapy. In: Salmon SE (ed) Adjuvant treatment of cancer, vol 5. Grune and Stratton Orlando, pp 571–580

98. Pinedo HM, Peters GJ (1988) Fluorouracil: biochemistry and pharmacology. J Clin Oncol 6: 1653–1664

99. Hryniuk WM, Figueredo A, Goodyear M (1987) Applications of dose intensity to problems in chemotherapy of breast and colorectal cancer. Semin Oncol 14 [Suppl 4]: 3–11

100. Burns ER, Beland SS (1984) Effect of biological time on the determination of the $LD_{50}$ of 5-fluorouracil in mice. Pharmacology 28: 296–300

101. Popovic P, Popovic V, Baughman J (1982) Circadian rhythm and 5-fluorouracil toxicity in $C_3H$ mice. Biomed Therm 25: 185–187

102. Gonzales JL, Sothern RB, Thatcher G, Nguyen N, Hrushesky WJM (1989) Substantial difference in timing of murine circadian susceptibility to 5-fluorouracil and FUDR. Proc Am Assoc Cancer Res 30: 616

103. Peters GJ, Van Dijk J, Nadal JC, Van Groeningen CJ, Lankelma J, Pinedo HM (1987) Diurnal variation in the therapeutic efficacy of 5-fluorouracil against murine colon cancer. In Vivo 1: 113–118

104. Gardner MLG, Plumb JA (1981) Diurnal variation in the intestinal toxicity of 5-fluorouracil in the rat. Clin Sci 61: 717–722

105. Roemeling R von, Hrushesky WJM (1990) Circadian FUDR infusion pattern determines its therapeutic index. J Natl Cancer Inst 82: 386–393

106. Levi F, Soussan A, Adam R, Caussanel JP, Metzger G, Misset JL, Descorps-Decleres, Kustlinger F, Lorphelin D, Jasmin C, Bismuth H, Reinberg A, Mathe G (1989) Programmable-in-time pumps for chronotherapy of patients (pts) with colorectal cancer with 5-day circadian modulated venous infusion of 5-fluorouracil (CVI-5FUra). Proc Am Soc Clin Oncol 8: 111 1989 (abstr)

107. Roemeling R von, Hrushesky WJM (1989) Circadian patterning of continuous FUDR infusion reduces toxicity and allows higher dose intensity. J Clin Oncol 7: 1710–1719

108. Vugrin D (1988) Systemic therapy of metastatic renal cell carcinoma. Semin Nephrol 7: 155–162.

109. Levi F (1989) Ambulatory chronotherapy of colorectal cancer with 5-fluorouracil, folinic acid and oxaliplatinum via a multichannel programmable pump. Results of a phase II trial. Proceedings of the 5th European Conference on Clinical Oncology (ECCO), London, 9 March 1989. (abst)

110. Wesen C, Roemeling R von, Lanning R, Grage T, Olson G, Hrushesky WJM (1992) The effect of circadian modification of intra-arterial FUDR infusion rate upon hepatic toxicity. J Clin Oncol (in press)

111. Chabner BA (1982) Pyrimidine antagonists. In: Chabner BA (ed) Pharmacologic principles of cancer treatment. Saunders, Philadelphia, pp 183–212

112. Reichard P, Skold O, Klein G (1959) Possible enzymatic mechanisms for development of resistance against fluorouracil in ascites tumors. Nature 183: 939–941

113. Klubes P, Connelly K, Cerna I (1978) Effects of 5-fluorouracil on 5-fluorodeoxyuridine 5'-monophosphate and 2-deoxyuridine 5'-monophosphate pools and DNA synthesis in solid mouse L1210 and rat Walker 256 tumors. Cancer Res 38: 2325–2331

114. Evans RM, Laskin JD, Hakala MT (1981) Effect of excess folates and deoxyinosine on the activity and site of action of 5-fluorouracil. Cancer Res 41: 3288

115. Peters GJ, Laurensse E, Leyva A, Lankelma J, Pinedo HM (1986) Sensitivity to human, murine and rat cells to 5-fluorouracil and 5'-deoxy-5-fluorouridine in relation to drug-metabolizing enzymes. Cancer Res 46: 20–28

116. Sweeny DJ, Barnes S, Heggie GD, Diasio RB (1987) Metabolism of 5-fluorouracil to an N-cholyl-2-fluoro-$\beta$-alanine conjugate: previously unrecognized role for bile acids in drug conjugation. Proc Natl Acad Sci USA 84: 5439–5443

117. Caradonna SJ, Cheng Y-C (1980) The role of deoxyuridine triphosphate nucleotidohydrolase, uracil-DNA glycosylase and DNA polymerase alpha in the metabolism of FUDR in human tumor cells. Mol Pharmacol 18: 513–520

118. Tuchman M, Stoeckeler JS, Kiang DT, O'Dea RF, Ramnaraine ML, Mirkin BL (1985) Familial pyrimidinemia and pyrimidinuria associated with severe fluorouracil toxicity. N Engl J Med 313: 245–249

119. Schuetz JD, Wallace HJ, Diasio RB (1984) 5-Fluorouracil incorporation into DNA of CF-1 mouse bone marrow cells as a possible mechanism of toxicity. Cancer Res 44: 1358–1363

120. Fraile RJ, Baker LH, Buroker TR (1980) Pharmacokinetics of 5-fluorouracil administered orally, by rapid intravenous and by slow infusion. Cancer Res 40: 2223–2228

121. Danhauser LL, Rustum YM (1979) A method for continuous drug infusion in unrestrained rats: its application in evaluating the toxicity of 5-fluorouracil/thymidine combinations. J Lab Clin Med 93: 1047–1053

122. Harris BE, Song R, Soong SJ, Diasio RB (1989) Circadian variation of 5-fluorouracil catabolism in isolated perfused rat liver. Cancer Res 19: 6610–6614

123. Harris BE, Song R, He Y-J, Soong S-J, Diasio RB (1988) Circadian rhythm of rat liver dihydropyrimidine dehydrogenase. Possible relevance to fluoropyrimidine chemotherapy. Biochem Pharmacol 37: 4759–4762

124. Tuchman M, Roemeling R von, Lanning R, Sothern RB, Hrushesky WJM (1988) Source of variability of dihydropyrimidine dehydrogenase (DPD) activity in human blood mononuclear cells. Annu Rev Chronopharmacol 5: 399–402

125. Harris BE, Song R, Diasio RB (1989) Circadian variation (CV) of dihydropyrimidine dehydrogenase (DPD) activity and plasma 5-fluorouracil (FUra) levels in cancer patients receiving FUra by protracted continuous infusion. Proc Am Assoc Cancer Res 30: 247

126. Petit E, Milano G, Levi F, Thyss A, Bailleul F, Schneider M (1988) Circadian rhythm-varying plasma concentration of 5-fluorouracil during a five-day continuous venous infusion at a constant rate in cancer patients. Cancer Res 48: 1676–1679

127. Boughattas NA, Levi F, Hecquet B, Lemaigre G, Roulon A, Fournier C, Reinberg A (1988) Circadian time depen-

634    W.J.M.Hrushesky and W.J.März

dence of murine tolerance for carboplatin. Toxicol Appl Pharmacol 96: 233–247

128. Boughattas NA, Levi F, Fournier C, Lemaigre G, Roulon A, Hecquet B, Mathe G, Reinberg A (1989) Circadian rhythm in toxicities and tissue uptake of 1,2-diammino- cyclohexane (trans-1) oxalatoplatinum (II) in mice. Cancer Res 49: 3362–3368

129. Tuchman M, Roemeling RV, Hrushesky WJM, O'Dea RF (1989) Dihydropyrimidine dehydrogenase activity in human blood mononuclear cells. Enzyme 42: 15–24

130. Lévi F, Halberg F, Haus E, Sanchez de la Pena S, Sothern R, Halberg E, Hrushesky W, Brown H, Scheving L, Kennedy BJ (1980) Synthetic adrenocorticotropin for optimizing murine circadian chronotolerance for Adriamycin. Chronobiologia 7: 227–244

131. Levi FA, Canon C, Touitou Y, Reinberg A, Mathé G (1988) Seasonal modulation of the circadian time structure of circulating T and natural killer lymphocyte subsets from health subjects. J Clin Invest 81: 407–413

132. Levi F, Bailleul F, Chevelle C, Benavides M, Missett JL, Le Saunier F, Despax R, Ribaud P, Machover D, Jasmin C, Regensberg C, Reinberg A, Mathe G (1987) Chronotherapy of ovarian cancer with 4'tetrahydropyranyl adriamycin (THP) and cisdichlorodiammine platinum (CDDP). Proc Am Soc Clin Oncol 6: 119

133. Burns RE (1981) Circadian rhythmicity in DNA synthesis in untreated mice as a basis for improved chemotherapy. Cancer Res 41: 2795–2802

134. Hrushesky WJM (1987) The rationale for non-zero order drug delivery using automatic, computer based drug delivery systems (chronotherapy). J Biol Res Modif 6: 587–598

135. Roemeling R, Hrushesky WJM (1987) Circadian shaping of FUDR infusion reduces toxicity even at high-dose intensity. Proc Am Soc Clin Oncol 6: 293

136. Roemeling R, Hrushesky WJM, Kennedy BJ, Buchwald H (1987) Programmed automatic FUDR chronotherapy improves therapeutic index. Surg Forum 27: 401–402

137. Edmunds LN (1978) clocked cell cycle clocks: implications toward chronopharmacology and aging. In: Samis HV Jr, Capoblanco S (eds) Aging and biological rhythms. Plenum, New York, pp 125–184

138. Haen E, Golly I (1986) Circadian variation in the cytochrome P-450 system of rat liver. Annu Rev Chronopharmacol 3: 357–360

139. Langevin T, Young J, Waler K, Roemeling R, Nygaard S, Hrushesky WJM (1987) The toxicity of tumor necrosis factor (TNF) is reproducibly different at specific times of the day. Proc Am Assoc Cancer Res 28: 281

# Immune System in Relation to Cancer

C. Canon and F. Lévi

## Introduction

Through its continuous interactions with the environment, the immune system constitutes one of the major sources of external information of the organism.

*Accessory cells* are those immune cells which capture antigens. They mostly consist of monocytes and macrophages. These cells are present in all tissues, although in a greater concentration in lymphoid organs such as lymph nodes and spleen, where encounters between antigens and reactive lymphocytes take place. Furthermore, skin, gastrointestinal and respiratory tracts have their own specialized accessory cells and lymphocytes to capture antigens.

Molecular mechanisms of antigen recognition involve specific receptors on the surface of lymphocytes and antibody production. The immune system reacts with chemical agents and assesses whether these molecules belong to "self" or to "non self" (immune surveillance) [1].

Lymphocytes are formed in the primary lymphoid organs: B lymphocytes in the bone marrow and T lymphocytes in the thymus. During the ontogeny of T and B cell lines the expression of receptors at the surface of these cells are antigen-independent. Once lymphocytes have been activated with an antigen in the secondary lymphoid organs, they proliferate then differentiate into effector cells with specific receptors.

An antigen can be recognized directly by B cells through their immunoglobulin receptors. However, antigen recognition by T cells through their antigen receptor requires antigen association with self molecules. These are glycoproteins coded by the major histocompatibility complex (MHC). There are two different classes: class I molecules are restricted to accessory cells and to CD8 suppressor/cytotoxic T lymphocytes. Class II molecules (Ia-like) are restricted to CD4 helper/inducer T lymphocytes. After antigen encounter and recognition, lymphocytes become activated. As a consequence of this activation they synthesize a variety of molecules: cytokines are released, whereas other molecules become components of the cell membrane; activated lymphocytes proliferate then differentiate into cytotoxic effector T cells. NK cells constitute another class of cytotoxic effectors. They are MHC-independent and lyse certain tumor or virally-infected cells without prior sensitization.

In several instances, quantitative differences in the amount of antigens present at the cell surface were found between the cancerous and the homologous healthy tissues, but the concept of specific antigens on human tumor has failed. Nonetheless, immunological approaches to cancer diagnosis and therapy have proved useful. The immune system is thus indispensable for the individual's life and its appropriate activation may be required for achieving cancer cures. The present chapter indicates that the temporal organization of immunologic functions along several time scales constitutes an important feature of the immune system. Endogenous and genetically based rhythms are thought to result from an adaptive process of living species to environmental cycles [2].

The understanding of the physiology of the immune system needs to take into account the coordination of the several immune functions along the 24-h scale. Immune functions of living beings are highly and predictably organized in time. As a result: (1) the immune response to antigen presentation differs both quantitatively and qualitatively according to time of exposure; (2) the pharmacodynamic effects of substances which affect immunity, "biological response modifiers" (BRM), strongly and predictably depend upon dosing time (clock hour, stage of the menstrual cycle, season); (3) chronotherapy (a strategy aimed at delivering the amount of drug needed at appropriate times) should become a necessary step for fully exploiting the pharmacologic properties of BRM. In the present chapter, clinical data will be complemented by results from animal experiments whenever needed.

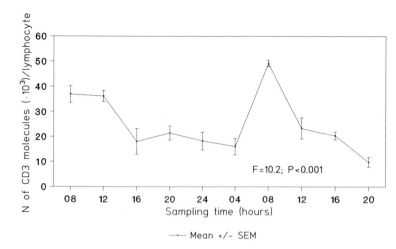

**Fig. 1.** Circadian rhythm in the density of CD3 molecules on human lymphocyte surface. Data obtained from five healthy subjects, in March 1987, active from 0700 hours to 2300 hours and recumbent at night. Blood samples were obtained every 4 h. The density of these molecules was assessed by flow cytometry. (After Canon et al. [6])

## The Immune Clock

### Circadian Rhythms

Knowledge of the temporal circadian organization of the immune system seems to be necessary to the understanding of its physiology and its involvment in the etiology and control of several diseases, including cancer.

Several steps are necessary to make the immune system able to eliminate a foreign antigen: encounter, recognition, activation, proliferation, differentiation and destruction of the antigen, as well as regulation of the immune system [3].

#### Encounter with Antigen

In living beings, encounter with environmental foreign antigens takes place mostly during the activity span rather than during rest.

#### Antigen Recognition

From an adaptative viewpoint, the identification of a foreign substance and its subsequent destruction and recognition constitutes the task of the immune system. Human data on antigen recognition are scarce. Antigen-specific receptors on lymphocytes and antibody molecules formed recognize and combine with particular antigens.

### T Lymphocytes

The antigen is usually presented by macrophages to T lymphocytes. T lymphocytes recognize antigens only after these are expressed on the surface of an antigen-presenting cell and in physical association with an MHC molecule. The T cell receptor consists of a clonotypic Ti $\alpha\beta$ heterodimer and four monomorphic T3 molecules ($\gamma$, $\delta$, $\varepsilon$ and $\xi$). The $\alpha$ and $\beta$ proteins contain variable (V) and constant (C) regions similar to immunoglobulins on B cells. The T $\alpha\beta$ T3 (T3Ti) molecules constitute the binding site for antigen and MHC complex. The T3 units are used as a signal transduction function [4]. T lymphocytes (CD3$^+$) are recognized with anti-CD3 monoclonal antibodies.

The presence of CD4 molecules on the surface of T (CD3$^+$) lymphocytes makes these cells able to further amplify the immune response – helper T lymphocytes, T (CD4$^+$) lymphocytes [5].

The density of CD3 molecules at the surface of T-lymphocytes in peripheral blood of five healthy male subjects was two- to threefold higher in the morning than in the evening or early night (Fig. 1). Similar results were obtained for the density of CD4 molecules on the surface of T (CD4$^+$) lymphocytes [6]. This indeed suggests that the availability of T cell receptors for antigens is increased at that time, so that the efficiency of T cell response may be enhanced then. Such response may subsequently be primarily B- or T-mediated.

### Antibody Molecules

The main function of B cells is to produce antibodies against microorganism antigens or harmful toxins.

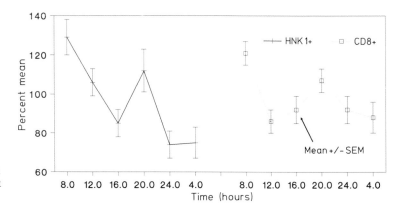

**Fig. 2.** Circadian rhythms in circulating natural killer (HNK1[+] on *left*) and suppressor-cytotoxic (CD8[+], on *right*) mononuclear cells of healthy young adults. Peak occurred at 0800 hours with a marked depression in the early night. (After Lévi et al. [15, 48])

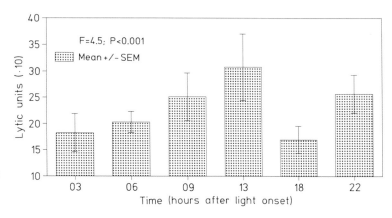

**Fig. 3.** Circadian rhythm in NK cell activity of murine splenocytes (300 B6D2F$_1$ male mice aged 9–10 weeks). NK activity, as gauged against YAC-1 tumor cells, was twice as high at the beginning of the active span of mice (13 h after light onset, HALO) as compared to mid-activity (18 HALO) or early rest (3 HALO)

IgM antibodies are more effective to stimulate the complement cascade and, thus, promote phagocytosis and killing of foreign microorganisms. IgG antibodies facilitate phagocytic destruction of microorganisms. IgA antibodies are transported across mucous membranes along the intestinal and respiratory tracts to protect them against invasion of the body. A circadian rhythm was documented for circulating immunoglobulins IgM, IgG, IgA and fragment C3 of complement in man [7–10]. Maximal values were located in the early afternoon. This time span may correspond to a more efficient antigen recognition.

Natural Killer Cells

Natural killer (NK) cells are not immunologically specific, are non-MHC-dependent, and do not usually express CD3 antigen. Interleukin-2 (IL-2) stimulates growth and differentiation of NK cells and generates lymphokine-activated killer cells (LAK). In mice, interleukin-2 and gamma interferon (IFN-γ) are synergistic with regard to peritoneal NK cell ac-

tivity [11]. These activated killer cells are thought to be a potential cancer therapy.

Figure 2 displays the circadian rhythms which characterize these cells [12–16]. Murine data indeed indicate that the organism is more resistant to tumor cells in the middle to late rest span [17, 18] when an increased number of NK lymphocytes are circulating ([19] and own data – Fig. 3). Results shown in Figs. 2 and 3 emphasize that immune data obtained in nocturnally active laboratory rodents such as mice or rats usually correspond well to human results if referred to the species-specific rest-activity cycle.

In response to foreign antigens, the reactivity of the immune system was enhanced. Thus, in several studies, the reactivity of this system depended on the circadian time of antigen exposure.

The skin response of sensitized subjects to tuberculin challenge was 2.5-fold larger following antigen exposure at 0700 hours as compared to 2200 hours [20] and episodes of rejection of kidney allograft were estimated to start at 0600 hours in patients bearing a renal transplant [21].

In nocturnally active mice, macrophage reactivity to a phagocytic stimulus as gauged by zymosan-induced chemiluminescence, was greatest in the middle to late rest span [22–24] or early activity span [24]. Other phagocytic cells, such as polymorphonuclears, had increased migratory activity at these times [25].

## Lymphocyte Activation

As a consequence of activation, T lymphocytes synthesize a variety of proteins such as IL-1, -2, -3, -4, . . . and IFN-$\gamma$. . .

Resting T cells do not express receptors for IL-2. Following activation, the number of surface CD3 and CD4 molecules decrease and that of IL-2 receptors increases a few hours later. Subsequently, IL-2 is secreted, then binds to IL-2 receptors, and DNA synthesis occurs. Stimulation of T cells trough the T3-Ti complex leads to a large increase in cytoplasmic free calcium and potassium and in intracellular pH, an activation of membrane methyltransferases, adenylate and guanylate cyclases, an enhanced membrane fluidity and a triggering of RNA synthesis. This latter signal permits cell division. Macrophage-derived products, including IL-1, must be present. In the absence of antigenic stimulation, there is a reexpression of T3-T1 receptor complex and decrease in the number of surface IL-2 receptors.

Small virgin or memory resting B cells are converted to large proliferating B cells following binding of antigen to surface immunoglobulins. These activated B cells require lymphokines proliferating (clonal expansion) and differentiating into antibody-secreting plasma cells.

A physiologic state of latent activation seems to characterize T lymphocytes between 1200 and 1800 hours. Thus energy metabolism, as gauged by several enzyme activities, was highest at 1600 hours in circulating lymphocytes [26]. A peak in RNA content of total lymphocytes was observed at 1800 hours, another one occuring at 0600 hours [Sanchez de la Pena S, Hrushesky W, Levi F, unpublished results]. High affinity IL-2 receptors are constituted with both p55 and p70 proteins. p55 Protein is recognized with monoclonal antibodies to the Tac antigens [27]. p55 Proteins tended to be expressed in a significantly larger number of circulating lymphocytes (CD25$^+$) in the early afternoon, in healthy subjects. However, spontaneous expression of these receptors was very low, as expected [6].

In human erythrocytes, the fluidity of cell membrane was increased at 1800 hours, the activity of methyltransferase I was highest at 1800–2400 hours, and intracellular potassium concentration was great-

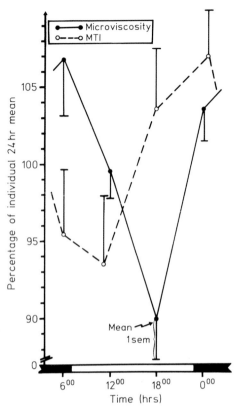

**Fig. 4.** Circadian rhythms in the erythrocyte membrane: Blood was obtained every 6 h in seven diurnally active healthy subjects. Membrane fluidity was highest at 1800 hours ($p < 0.001$) and negatively correlated with the activity of membrane-bound methyltransferase I (MTI) ($p < 0.001$). (After Lévi et al. [28])

est at 1800 hours (Fig. 4) [28]. The same circadian mechanisms might govern these rhythms in different blood cells.

## Lymphocyte Proliferation

After recognition of an antigen and activation, lymphocytes amplify and distribute defense functions through proliferation and differentiation into cytotoxic T cells and immunoglobulin production.

Lectins are cell recognition molecules [29], which are nonspecific inducers of lymphoblastic transformation and proliferation. The response of lymphocytes to phytohemagglutinin (PHA), closely resembles that elicited by antigen presentation, and only involves T cells. Proliferation of T cells was twice as large following PHA exposure of separated lymphocytes in the early morning (0600–1000 hours) as compared to late evening (1800–2200 hours) [30, 31]. When whole blood, rather than mononuclear cells, was exposed to PHA, greatest lymphoblastic transformation occured at 2000 hours [32, 33].

An explanation for such discrepancy has been provided [33]. T lymphocytes were more prone to proliferate following contact with autologous PHA-stimulated T cells at 0800 hours than at 1200 or at 2000 hours. The reverse was true if T lymphocytes were exposed to non-T cells. This indeed documents that a rather precise coordination in time characterizes these different stages of immune defenses.

Proliferation of immune cells primarily occurs in lymphoid organs. B cells are located mostly in the bone marrow but also in the germinal centers of lymph nodes and in the spleen; T cells proliferate and differentiate mostly in the thymus during the early stages of life, then do so in the deep cortex of lymph nodes and in the spleen.

Cell proliferation, as gauged by DNA synthesis, was highest near 1600–2000 hours in human bone marrow [34]. The proliferative ability of bone marrow granulo- and monocytic precursors was also highest in the second half of the activity span both in man [34] and in mice [35]. These cells give rise to phagocytic cells, namely polymorphonuclear cells, mono-cytes, and macrophages. Maximal thymic DNA synthesis also occurred in the late acticity span of mice [36, 37] or guinea pigs [38]. Nonetheless, results obtained in mice may differ from one strain to another since these circadian rhythms have a genetic basis [39, 40]. Human peripheral blood lymphocytes also exhibited a spontaneous circadian variation in DNA synthetic activity, with two peaks, one in the morning (0800–1000 hours), the other one in the evening (2000–2400 hours) with a marked depression at 0400 hours [32–42].

### Lymphocyte Blood Release

Lymphoid organs release lymphocytes into the peripheral blood; the counts of circulating total lymphocytes are 50%–100% higher during the early night (midnight – 0400 hours) than in the morning (0800–1200 hours) [2, 12, 14–16, 30–33, 41–44]. B lymphocytes are released between 2000 and 2400 hours, a few hours before T cells (between 2400 and 0400 hours).

Mature B lymphocytes can be characterized by surface Ig and identified by a monoclonal antibody

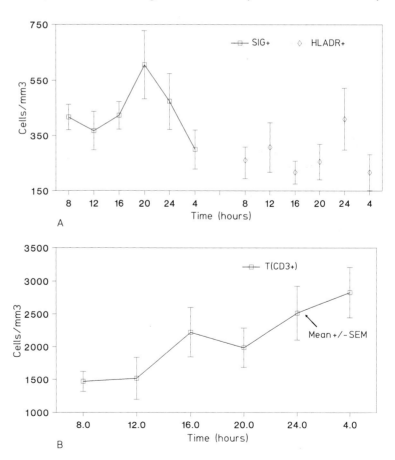

**Fig. 5.** Circadian changes in circulating B lymphocytes (SIg$^+$, *left* and HLA-DR$^+$, right in **A**) T lymphocytes CD3$^+$, in **B**) in seven healthy young men. (After Levi et al. [42])

directed against Ig (SIg$^+$). The circulating count of B cells doubled between 0800 hours and 2000 or 2200 hours [12, 42] (Fig. 5). Different laboratory techniques and study designs and/or infradian periodicity may account for discrepant results in one study [31], as discussed elsewhere [43]. The late evening peak in B cell count is close to the time of maximal skin or bronchial reactivity to antigens such as house dust and/or penicillin in allergic patients [2, 45–47]. B cells take part in this reaction.

The circulating count of T lymphocytes, as identified with a variety of techniques, doubled between morning (0800–1200 hours) and early night (midnight – 0400 hours) (Fig. 5) [12, 14, 31, 42, 43, 48]. A similar circadian pattern characterized the T-helper subset (T (CD4$^+$)) [2, 43, 48].

In mice, the count of circulating T lymphocytes also peaked in the early rest span [49]. Such increases result from a release of T lymphocytes at this circadian stage, as supported by:
– A circadian rhythm in circulating thymosin-$\alpha_1$, a thymic hormone involved in T cell maturation (and release?) which peaked in the early rest span in mice [50]
– The appearance of presumably immature T cells in peripheral blood in the early night span in healthy subjects [51]

Potential pitfalls of such models of immune surveillance result from the scarcity of *human* data on circadian rhythms in T and B intraorgan proliferation, and on phagocyte activity, among other functions. The relationship of such circadian immune clocks with other oscillators will deserve further studies. For instance, the normal circadian rhythms in cortisol or testosterone secretion play a minor role, if any, on lymphocyte circulation [48], but NK cells may exhibit a circadian-dependent susceptibility to cortisol [52].

## Circannual Time Structure of Immunity

Immune defenses are also organized along other time scales. These are reviewed elsewhere in detail [43].

Table 1 indeed documents that winter depression of T cell immunity may characterize most living beings including man. In these studies, such circannual rhythms remained similar despite ambiant temperature, and photoperiod duration was kept constant throughout the year [53–63]. The circannual time structure of immunity is thus endogenous.

Furthermore, circannual rhythms may modulate the circadian time structure of some immune func-

**Table 1.** Winter Depression of T cell Immunity

| Species | Environment | | Author | Year |
|---|---|---|---|---|
| | Light-Dark | Temperature | | |
| Lizard | Natural | Natural | Hussein | 1979 |
| Rainbow trout | Natural | Constant | Yamaguchi | 1981 |
| Guinea pig | LD12:12 | 22 ± 1°C | Godfrey | 1975 |
| | | | Hildeman | 1980 |
| Ground squirrel | Natural | 23 ± 1°C | Sidky | 1972 |
| Mice | LD12:12 | 23 ± 1°C | Brock | 1983 |
| | | | Pati | 1988 |
| Beagle dog | Natural | Natural | Shifrine | 1980 |
| Man | Natural | Natural | | |

tions. Thus, the circadian rhythms in peripheral counts of T(CD3$^+$) and T(CD4$^+$) lymphocytes were not detected at a group level in March or in June, but had a large amplitude in August, November, and January [15]. The lack of a circadian rhythm at a group level may result from interindividual desynchronization, for example, individual circadian rhythms may exhibit period lengths differing from precisely 24 h. This phenomenon might be more likely to occur at specific times of the year, as shown for testosterone and lutropin secretions as discussed elsewhere [15, 43].

The prominent group-synchronized circadian rhythm in the density of CD3 molecules on the surface of T lymphocytes (Fig. 1) was observed in March, when the peripheral count of T(CD3$^+$) lymphocytes lacked any circadian rhythm at a group level. This indeed indicates that both systems may be "physiologically" desynchronized.

These results may account for the increased spring incidence of some infectious diseases and certain cancers (Hodgkin's lymphomas and breast cancer).

## Other Infradian Immune Rhythms

Other periodicities – circamenstrual (about 30 days) and circaseptan (about 7 days) – also modulate immune defenses. An extensive literature review led us to hypothesize that once an immune response had been triggered, its various cellular and humoral components became organized along multiples of 7 days as shown for the incidence of allograft rejection, both in laboratory rodents and in transplant patients [64].

NK cell activity was shown to predictably double along the estrous cycle in female mice. NK cell activity was highest near the middle of the estrous cycle (proestrus) in female mice. This rhythm appeared to influence the metastatic potential of a transplanted mammary tumor [65]. Similar results were reported in premenopausal women with breast cancer: the risk of recurrence was four times larger in women who had tumor removed around menstruation as compared to ovulation [66].

The knowledge of the time structure of immune defenses may be essential for understanding physiopathological processes. For instance, experiments in mice have indicated the role of time of antigen exposure upon the generation of an allergic process.

## Hormones and Immunity

Exogenous glucocorticoids and adrenalectomy, respectively, decrease or enhance immune responses, both in laboratory animals and in man. Migration of T immunocompetent lymphocytes into the bone marrow is triggered by glucocorticoids. Sex hormones may also be involved since a sexual dimorphism characterizes immune responses [67]. More recently ACTH secretion was found to be regulated by thymic peptides, such as thymosin fraction 5, thymopoietin, IL-1. The interaction between neuroendocrine factors and their receptors result in the activation of second messengers such as cAMP in immunocompetent cells [68].

Neuroendocrine factors may also modulate the immune response through the modulation of lymphokine secretion. Glucocorticoids and ACTH depress antibody production, NK cell activity, and cytokine production. Sex hormones suppress or enhance lymphocyte transformation and mixed lymphocyte cultures depending upon concentration. Beta-endorphin enhances antibody synthesis and macrophage and T cell activation.

Circadian rhythms are exhibited by the free total and free plasma cortisol, with an acrophase at the beginning of the activity phase and, for testosterone, at the end of the morning, but without correlation with lymphocyte subpopulation rhythms, in peripheral blood [43] of healthy male subjects. Circadian time structure of T and B lymphocyte subsets was altered in male patients infected with human immunodeficiency virus (HIV) [69]. Conversely, cortisol, ACTH, beta-endorphin, testosterone, dehydroepiandrosterone (DHEA) and its sulfate (DHEA-S) exhibited significant circadian rhyhtms in patients suffering from either asymptomatic disease or acquired immunodeficiency syndrome (AIDS). Acrophases were similar in patients and in controls. Mesors were

higher for cortisol and lower for DHEA, DHEA-S and ACTH in all HIV-infected patients as compared to healthy controls. The plasma testosterone mesor was similar in controls and in asymptomatic subjects but decreased in AIDS patients.

The pineal gland is an important transducer of light-dark information to circadian rhythms [70] and plays a role in immunity, too. Primary antibody synthesis and mixed lymphocyte reaction are altered by pinealectomy or in animals exposed to continuous light. In mice, administration of melatonin in the evening reverses these activities, enhances antibody formation and antagonizes immunosuppressive effects of corticosterone.

NK cells are involved in immunosurveillance of tumor cells. Their activity is inhibited by glucocorticoids. Their lytic activity is enhanced at the end of the night, after the peak of melatonin. In vivo, melatonin administration in the afternoon induces a slight increase of NK activity and enhances the responsiveness of NK cell to IFN-$\gamma$. Chronic administration of melatonin versus placebo enhanced significantly spontaneous NK cytotoxicity. However melatonin exerted no effect upon NK cell activity in vitro [52].

## Cancer-Associated Alterations of the Immunologic Time Structure

Alterations of circadian rhythms in lymphocyte subpopulations of patients with hematologic malignancies were investigated in 12 patients at an early stage (five with chronic lymphocytic leukemia-CLL) or in clinical complete remission (five after Hodgkin's lymphoma and two after non-Hodgkin's lymphoma). Such patients are prone to develop infectious diseases. No subject had received any treatment for the 6 months preceding the study. Venous blood was obtained every 3–4 h for 12–48 h from all subjects for determination of circulating counts in total lymphocytes and in T (CD3, CD4, CD8) lymphocytes. No circadian rhythm was found in the group of patients with CLL (Fig.6) or with non-Hodgkin's lymphoma in clinical remission, whereas patients with Hodgkin's disease in complete remission exhibited a 24-h rhythm in total and T lymphocytes similar to that of controls. A profound alteration of the circadian rhythm in circulating T helper lymphocytes was documented in all three groups of patients [71].

Likewise, the circadian rhythmicity in the activity of several enzymes from circulating lymphocytes was

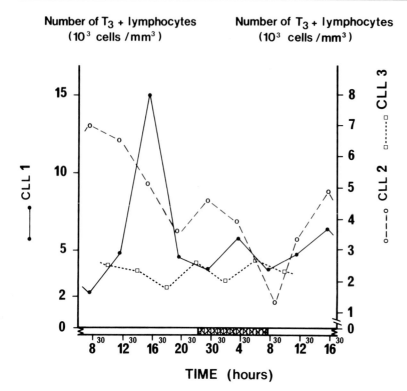

**Fig. 6.** Chronogram of total T (T3, CD3) circulating lymphocytes in three patients with CLL (CLL1, CLL2 and CLL3). Blood samples were obtained from each patient every 4 h for 36 h. (After Canon et al. [71])

found to differ between healthy subjects and patients with CLL [26].

The circadian rhythm of the metabolic activity of lymphocytes differed between healthy subjects and women with advanced ovarian tumors [72].

Furthermore, circadian rhythms in circulating B and T (CD3, CD4, CD8) lymphocytes were studied in 13 HIV-infected men [73]. Circadian rhythms in B and T lymphocyte subsets (CD3, CD4, CD8) were largely altered in 13 patients with asymptomatic or acquired immuno deficiency syndrome (AIDS). In HIV-infected subjects, both mesors and amplitudes of B and T (CD4) lymphocytes were decreased. The normal circadian rhythm in circulating CD8 lymphocytes disappeared in early-stage disease (asymptomatic infection). It was reexpressed at a later stage (AIDS), but with a timing different from the normal one.

## Chronotherapy with BRM

Antigen presentation to T cells, T cell recognition, activation and interactions with endothelial cells are involved in the identification of tumor antigens by immune cells. These different steps are organized

along the circadian time scale. When there is alteration of immunological rhythms, restoration or such rhythmicity may constitute a goal for therapy with BRM. Present cancer immunotherapy has mostly involved the administration of interferons and/or lymphokines such as IL-2 alone or associated with LAK cells [11, 74, 75]. Conflicting results have often been reported, yet little attention has been paid to differences in time effects in understanding these discrepancies.

Lentinan, an immunostimulant, was administered to rats, prior to tumor inoculation. The tumor grew faster or more slowly according to both time of administration and season of treatment [76]. Major differences in the toxicity of tumor necrosis factor [77], a cytokine used for treating cancer, were found according to dosing time in mice.

## Cyclosporin A

Cyclosporin A (CYA) is an immunosuppressive drug used in kidney, liver, heart transplantation and bone marrow allografts. Furthermore, it reverses resistance to anticancerous drugs.

Several studies indicate that CYA interferes with $T\alpha\beta$ receptors on T cells. Mice treated with CYA exhibit abnormal thymic ontogenesis, with absence of mature T cells in thymus and bone marrow. T (CD4$^+$, CD8$^+$) corticothymocytes were blocked in their maturation, a step at which T$\alpha\beta$ receptor selection usually occurs. In chickens, CYA blocked the activation of resting T cells but not that of proliferative T cells expressing IL-2 receptors. In some cases, CYA induced T cell autoimmunity. CYA binds to cytosolic proteins [78]. Other molecular mechanisms of CYA effects have remained poorly understood.

The dosing time of cyclosporin A indeed influenced the extent of its immunosuppressive effect in mice [79]. Major differences in the toxicity of CYA were found according to dosing time in mice [79, 80].

## Interferon

Preclinical and early clinical chronopharmacologic studies have been performed with (IFN-$\alpha$) and will be reported.

There are two types of interferon with different biochemical characteristics: type I, IFN-$\alpha$ and -$\beta$ and type II, IFN-$\gamma$. The latter is produced only by T lymphocytes and large granular lymphocytes (LGL). On the other hand, IFN-$\alpha$ and -$\beta$ are potentially produced by all cells and are in competition for the same receptors. These differ from those of IFN-$\gamma$. IFN has antiviral, antiproliferative, cytotoxic, immunodulating, gene activating and differentiating effects on cells involved in the immune response.

IFN-$\alpha$ has proved effective against metastatic renal cell carcinoma, among other malignancies. Its antitumor effectiveness was clearly dose-related [81, 82]. Toxicities of IFN-$\alpha$ may be severe and are dose-limiting. They consist of fever, chills, malaise, somnolence and asthenia, as well as neutropenia, thrombocytopenia and liver or renal damage. Various schedules and routes of administration have been used. This drug is usually injected intramuscularly three times per week at doses ranging from 3 to 10 MU/m$^2$ per day. Doses of 3–5 MU/m$^2$ per day were recommended for continuous venous infusion of this drug [83]. Evening intramuscular injections were found less toxic than morning ones in patients [84]. In a crossover study, IFN-$\alpha$ was administered either at 0800 h or at 2000 h to healthy subjects; plasma cortisol and circulating T and NK lymphocyte subsets were measured every 3–6 h for 24 h. Circadian rhythms in these variables were similar both in the absence of IFN injection and following drug administration at 2000 h. IFN-$\alpha$ dosing at 0800 h resulted in a profound alteration of these physiologic rhythms, which likely relates to the increased toxicity of morning IFN-$\alpha$ [85].

## Preclinical Findings

The ability of IFN-$\alpha$ to stimulate NK cell activity was examined in mice at different circadian stages. Male B6D2F$_1$ mice were randomly allocated to one of six groups of ten animals. Each group was housed on a different shelf of an autonomous chronobiologic facility (ESI-Flufrance, Arcueil, France). Such a facility allows one to synchronize mice in six different lighting schedules consisting of an alternation of 12 h light (L) and 12 h dark (D). The phasing of circadian rhythms can be shifted by altering the LD cycle. Food and water were provided ad libitum. Light onset was staggered by 4 h 15 min in each shelf, with temperature being automatically controlled (23°–25°C). Light intensity ranged from 600 to 800 lux at 30 cm from the source. After 3 weeks of synchronization, the mice had adjusted to the altered lighting schedule and six circadian stages could be explored at the same clock hour. Mice from each circadian group were killed at 3, 6, 10, 13, 18, or 22 h after light onset (HALO). Spleens were removed. Three suspensions per timepoint (each corresponding to three or four spleens) were prepared in RPMI 1640 culture medium by gently teasing apart the organs between the frosted ends of two microscope slides. After filtration through a nylon gauze, the suspensions were centrifuged at 300 g for 7 min at 4°C. Cells were resuspended into medium and their viability was assessed by the trypan blue dye exclusion test. NK cell activity of splenocytes was assessed from $^{51}$Cr-release by YAC-1 lymphoma cells, following 4-h exposure, as previously described [49]. A circadian rhythm characterized splenic NK cell activity with maximal value in the early activity span of mice, similar to that shown in Fig. 4. In vitro exposure of splenocytes to mouse IFN-$\alpha\beta$ (1000 U/ml for 1 h) resulted in an increase of NK cell activity which was fourfold greater in the second half of the active span of mice as compared to exposure in the late rest or early activity spans (Fig. 7).

## Clinical Study

We hypothesized that IFN-$\alpha$ would be more effective and less toxic in the late evening in man. A phase I clinical trial was undertaken in patients with metastatic carcinoma. IFN-$\alpha$ was continuously infused for

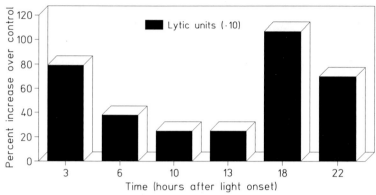

**Fig. 7.** In vitro stimulation of murine splenocyte NK activity with alpha-beta interferon (IFN-$\alpha\beta$) as a function of circadian time of exposure. Splenocytes sampled in the mid to late activity or early rest span of 60 male B6D2F$_1$ mice (18,22 or 3 HALO) and subsequently exposed to IFN-$\alpha\beta$ exhibited greatest increase in NK cell activity. Multivariant analyses were performed on NK activity according to different effector to target ratio (e:t ratio) in the presence or absence of IFN. Results indicated statistically significant differences according to circadian time ($F = 116$, $p < 0.001$), e:t ratio ($F = 927$, $p < 0.001$), IFN-$\alpha\beta$ exposure ($F = 512$, $p < 0.001$), and a significant interaction between circadian time and IFN-$\alpha\beta$ exposure ($F = 6,4$; $p < 0.03$). These results indicate that (1) a circadian rhythm characterized NK cell activity; (2) IFN-$\alpha\beta$ exposure affected the 24-h-mean level of this variable; (3) the extent of the pharmacodynamic effect of IFN-$\alpha\beta$ upon NK cell activity varied from 20% to 100% according to the circadian time of exposure as depicted in this figure

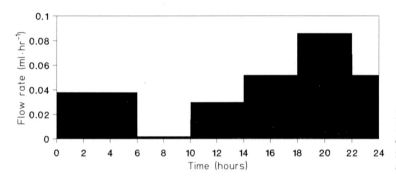

**Fig. 8.** Schedule of intravenous delivery of IFN-$\alpha$ over 24 h in patients with renal cell carcinoma. Automatized circadian changes in flow rate were performed with a small ambulatory pump (Minimed, SP 404)

21-day courses every month (10 days' rest), at doses ranging from 15 to 20 MU/m$^2$/day according to an intrapatient dose-escalation protocol. Drug delivery was automatically modulated along the 24 h scale from 0.002 ml/h between 0600 hours and 1000 hours up to 0.09 ml/h between 1800 hours and 2200 hours, using a small ambulatory pump (SP 404 pump, Minimed, France) (Fig. 8). This device has a 3-ml reservoir which was connected to the central venous system via an implanted access port. Syringe changes were required every 3 days. IFN-$\alpha$ was concentrated up to 30 MU/ml.

All patients tolerated the first dose level for the whole 21-day course. Five patients have received at least three courses at present. These very high dosages were safely infused to four of them, for 3–6 months. All remained ambulatory and two of them continued their professional activities during infusional chronotherapy. These preliminary results suggest that chronotherapy with BRM may profoundly increase dose intensity and desired immunopharmacodynamic effect. Programmable-in-time pumps are now available for such chronotherapy with high drug doses and minimal side effects to be performed in ambulatory patients.

## Conclusions

The activities of the several immune cells are normally programmed along both 24-h and yearly time scales in living beings. Such physiologic temporal organization, although spread all over the body, may be hypothesized as a coordinated clock, periodically reset by cyclic antigen exposure. Circadian immune rhythms may be altered in cancer patients and the restoration of normal periodicity may become an endpoint for chronoimmunotherapy. Chronoimmunopharmacologic investigations have successfully led to optimizing the therapeutic index of cyclosporin A and IFN-$\alpha$ among other BRM in mice.

A clinical phase I trial has been conducted with IFN-$\alpha$ in ten patients with metastatic renal cell carcinoma. Continuous intravenous infusion of IFN-$\alpha$ was performed at circadian-modulated rate, using a small ambulatory programmable-in-time pump. Chronotherapy allowed a two-to threefold increase in tolerable doses of IFN-$\alpha$, as compared to the standard administration schedule.

Experimental chronopharmacology of BRM should guide the development of original timed schedules of drug delivery, with enhanced desired effects on immune functions. Such chronotherapy can now be implemented on a large scale thanks to ambulatory programmable-in-time pumps. Benefits from such strategy will be assessed in ongoing phase II clinical trials.

Acknowledgements. The autors are indebted to A. Reinberg for critical discussions and permanent encouragements which helped design the present model of circadian-adapted immune surveillance.
We are indebted to G. Debotte, M. Mechkouri and A. Nicolai for technical assistance.

# References

1. Bach JF, Lesavre P (1989) Immunologie. Flammarion, Paris
2. Reinberg A, Smolensky M (1983) Biological rhythms and medicine. Cellular, metabolic physiopathologic and pharmacologic aspects. Springer, Berlin Heidelberg New York
3. Nossal GJV (1987) Current concepts: immunology. The basic components of the immune system. N Engl J Med 21: 1320–1325
4. Alcover A, Ramarli D, Richardson NE, Chang HC, Reinherz EL (1987) Functional and molecular aspects of human T lymphocytes activation via T3–T1 and T11 pathways. Immunol Rev 95: 5–36
5. Takada S, Engleman EG (1987) Evidence for an association between CD8 molecules and the T cell receptor complex on cytotoxic T cells. J Immunol 139: 3231–3235
6. Canon C, Lévi F, Reinberg A (1990) Rythme circadien des lymphocytes circulants de la densité des épitopes de surface (CD3 et CD4) chez l'homme adulte sain (to be published)
7. Reinberg A, Schuller E, Delasnerie N, Clench J, Helary M (1977) Rythmes circadiens et circannuels des leucocytes, protéines totales, immunoglobulines A, G et M: etude chez 9 adultes jeunes et sains. Nouv Presse Med 6: 3819–3823
8. Halberg F, Duffert D, von Mayersbach H (1977) Circadian

rhythm in serum immunoglobulins of clinically healthy young men. Chronobiologia 4: 114
9. Pallansch M, Kim Y, Halberg E et al. (1979) Circadian rhythm in several components of the complement cascade in healthy women. Chronobiologia 6: 139–140
10. Kim Y, Pallansch M, Carandente F, Reissman G, Halberg E, Halberg F (1980) Circadian and circannual aspects of the complement cascade. New and old differing in specificity. Chronobiologia 7: 189–204
11. Hercend T, Schmidt RE (1988) Characteristics and uses of natural killer cells. Immunol Today 10: 291–293
12. Abo T, Kawate K, Itoh K, Kumagai K (1981) Studies on the bioperiodicity of the immune response. I. Circadian rhythms of human T, B and K cell traffic in the peripheral blood. J Immunol 126: 1360–1363
13. Williams R, Kraus L, Inbar M, Dubey D, Yunis E, Halberg F (1980) Circadian bioperiodicity of natural killer cell activity in human blood (individually-assessed). Chronobiologia 6: 172
14. Ritchie WS, Oswald I, Micklem HS, Boyd JE, Elton RA, Jazwinska E, James K (1983) Circadian variation of lymphocyte subpopulations: a study with monoclonal antibodies. Br Med J 286: 1773–1775
15. Lévi F, Canon C, Touitou Y, Reinberg A, Mathé G (1988) Seasonal modulation of the circadian time structure of circulating T and NK lymphocyte subsets from healthy subjects. J Clin Invest 81: 407–413
16. Gatti G, Masera R, Cavallo R, Delponte D, Sartori ML, Salvadori A, Carignola R, Angeli A (1988) Circadian variations of interferon-induced enhancement of human natural killer (NK) cell activity. Cancer Detect Prev 12: 431–438
17. Fernandes G, Carandente F, Halberg F, Halberg E (1979) Circadian rhythm in activity of lympholytic natural killer cells from spleens of Fisher rats. J Immunol 123: 622–625
18. Hrushesky W, Gruber S, Sothern R, Olshefski R, Beretska J, Lévi F, Lannin D (1990) Circadian timing, anatomic location and dose of tumor cell inoculum predictably affect murine tumor biology. (to be published)
19. Tsai TH, Burns ER, Scheving E (1979) Circadian influence on the immunization of mice with live bacillus Calmette-Cuérin (BCG) and subsequent challenge with Ehrlich ascites carcinoma. Chronobiologia 6: 187–201
20. Cove-Smith MS, Kabler PA, Pownall R, Knapp MS (1978) Circadian variation in an immune response in man. Br Med J 11: 253
21. Knapp MS, Cove-Smith MS, Dugdale R, MacKenzie N, Pownall R (1979) Possible effect of time on renal allograft rejection. Br Med J 75: 75–77
22. Knyszynski A, Fischer H (1981) Circadian fluctuations in the activity of phagocytic cells in blood, spleen and peritoneal cavity of mice as measured by zymosan-induced chemiluminescence. J Immunol 127: 2508–2511
23. Szabo I, Kovats T, Halberg F (1978) Circadian rhythms in murine reticuloendothelial function. Chronobiologia 5: 137–143
24. Carrere V, Dorfman P, Bastide M (1988) Evaluation of various factors influencing the action of mouse-interferon on the chemiluminescence of mouse peritoneal macrophages. Annu Rev Chronopharmacol 5: 9–12
25. Bureau JP, Garelly L, Coupé M, Labrecque G (1985) Circadian rhythm studies on BCG-induced migration of PMN in normal and adrenalectomized mice. Annu Rev Chronopharmacol 1: 333–336
26. Ramot B, Brok-Simoni F, Chiveidman E, Ashkenazi YE

(1976) Blood Leucocyte enzymes. III. Diurnal rhythm of activity in isolated lymphocytes of normal subjects and chronic lymphatic leukemia patients. Br J Haematol 34: 79–85

27. Siegel JP, Sharon M, Smith PL, Leonard WJ (1987) The IL2 receptor $\beta$ chain (p70): role in mediating signals for LAK, NK and proliferative activities. Science 238: 75–78

28. Lévi F, Benavidès J, Touitou Y, Quarteronnet D, Canton T, Uzan A, Auzeby A, Gueremy C, Sulon J, Le Fur G, Reinberg A (1987) Circadian rhythm in the membrane of circulating human blood cells: microviscosity and number of benzodiazepine binding sites. A search for regulation by plasma ions, nucleosides, proteins or hormones. Chronobiol Int 4: 235–243

29. Nathan S, Halina L (1989) Lectins as recognition molecules. Science 246: 227–234

30. Tavadia HB, Fleming KA, Hume PD, Simpson HW (1972) Circadian rhythmicity of plasma cortisol and PHA-induced lymphocyte transformation. Clin Exp Immunol 22: 190–193

31. Haus E, Lakatua DJ, Swoyer J, Sackett-Lundeen L (1983) Chronobiology in hematology and immunology. Am J Anat 168: 467–517

32. Eskola J, Frey H, Molnar G, Soppi E (1976) Biological rhythm of cell-mediated immunity in man. Clin Exp Immunol 26: 253–257

33. Indiveri F, Pierri I, Rogna S, Poggi P, Montaldo P, Romano R, Pende A, Morgano A, Barabino A, Ferrone S (1985) Circadian variations of autologous mixed lymphocyte reactions and endogenous cortisol. J Immunol Methods 82: 17–24

34. Smaaland R, Lote K, Sletwold O, Kamp D, Wiedemann G, Laerum O (1989) Rhythmen in Knochenmark und Blut: Unterschiede wie Tag und Nacht. Dtsch Med Wochenschr 114: 845–849

35. Lévi F, Blazscek I, Ferlé-Vidovic A (1988) Circadian and seasonal rhythms in murine bone marrow colony forming cells affect tolerance for the anticancer agent 4'-O-tetrahydropyranyladriamycin (THP). Exp Hematol 16: 696–701

36. Haus E, Taddeini L, Larson K, Bartlett P, Sackett-Lundeen L (1984) Circadian rhythm in spontaneous 3H-thymidine uptake and in PHA-response of splenic cells of BDF1 male mice in vitro. Phase relations to hematologic rhythms in vivo. In: Haus E, Kabat H (eds) Chronobiology 1982–1983. Karger, Basel, pp 178–182

37. Kirk H (1972) Mitotic activity and cell degeneration in the mouse thymus over a period of 24 hours. Z Zellforsch 129: 188–195

38. Dlouhy W, Sawicki W (1969) Diurnal fluctuations in the numbers of mitotic and labelled with tritiated thymidine cells of colonic lymphatic noduli of guinea pig. Bull Acad Pol Sci CI, VI, 17: 517–521

39. Scheving LE, Pauly J, Tsai T, Scheving LA (1983) Chronobiology of cell proliferation. Implications for cancer chemotherapy. In: Reinberg A, Smolensky M (eds) Biological rhythms and medicine. Springer, Berlin Heidelberg New York, pp 79–130

40. Peleg L, Nesbitt M, Ashkenazi IE (1984) Genetic variation in a circadian rhythm in mouse thymus. In: Haus E, Kabat H (eds) Chronobiology. Karger, Basel, pp 155–159

41. Carter JB, Barr GD, Levin AS, Byers VS, Ponce B, Fudenberg H (1975) Standardization of tissue culture conditions for spontaneous thymidine 14C incorporation by unstimulated normal human peripheral lymphocytes: circadian rhythm of DNA synthesis. J Allergy Clin Immunol 56: 191–205

42. Lévi F, Canon C, Blum JP, Mechkouri M, Reinberg A, Mathé G (1985) Circadian and/or circahemidian rhythms in nine lymphocyte-related variables from peripheral blood of healthy subjects. J Immunol 134: 217–225

43. Lévi F, Reinberg A, Canon C (1989) Clinical immunology and allergy. In: Arendt J (ed) Biological rhythms in clinical practice. Butterworth, Guildford, pp 99–135

44. Sabin FR, Cunningham RS, Doan CA, Kindwall JA (1925) The normal rhythm of the blood cells. Bull John Hopkins Hosp 37: 14–67

45. Gervais P, Reinberg A, Gervais C, Smolensky MH, Defrance O (1977) Twenty-four hour rhythm in the bronchial hyperreactivity to house dust in asthmatics. J Allergy Clin Immunol 59: 207–213

46. Reinberg A, Sidi E, Ghata J (1965) Circadian rhythms of human skin to histamine or allergen and the adrenal cycle. J Allergy Clin Immunol 36: 279–283

47. Reinberg A, Zagula-Mally Z, Ghata J, Halberg F (1969) Circadian reactivity rhythm of human skin to house dust, penicillin and histamine. J Allergy Clin Immunol 44: 292–306

48. Lévi F, Canon C, Touitou Y, Sulon J, Demey-Ponsard R, Mechkouri M, Mowrowicz I, Touboul JP, Reinberg A, Mathé G (1988) Circadian rhythms in circulating T lymphocyte subsets, plasma total and free cortisol and testosterone in healthy men. Clin Exp Immunol 71: 329–335

49. Kawate T, Abo T, Hinuma S, Kumagai K (1981) Studies on the bioperiodicity of the immune response. II. Covariations of murine T and B cells and a role of corticosteroid. J Immunol 126: 1364–1367

50. McGillis J, Hall N, Goldstein A (1983) Circadian rhythms of thymosin-$\alpha$ 1 in normal and thymectomized mice. J Immunol 131: 148–151

51. Canon C, Lévi F, Reinberg A, Mathé G (1985) Circulating CALLA positive lymphocytes exhibit circadian rhythms in man. Leukemia Res 9: 1539–1546

52. Gatti G, Carignola R, Masera R, Sartori ML, Salvadori A, Magro E, Angeli A (1989) Circadian-stage-specified effects of melatonin on human natural killer (NK) cell activity: in vitro and in vivo studies. Annu Rev Chronopharmacol 5: 25–29

53. Sidky Y, Hayward J, Ruth R (1972) Seasonal variations of the immune response of ground squirrels kept at 22–24°C. Can J Physiol Pharmacol 50: 203–206

54. Shifrine M, Taylor N, Rosenblatt L, Wilson F (1980) Seasonal variation in cell mediated immunity of clinically normal dogs. Exp Hematol 8: 318–326

55. Brock M (1983) Seasonal rhytmicity in lymphocyte blastogenic responses of mice persist in a constant environment. J Immunol 130: 2586–2588

56. Godfrey H (1975) Seasonal variation of induction of contact sensitivity and of lymph node T lymphocytes in guinea pigs. Int Arch Allergy Appl Immunol 49: 411–414

57. Patti A, Florentin I, Chung V, De Sousa M, Lévi F, Mathé G (1987) Circannual rhythm in natural killer cell activity and mitogen responsiveness of murine splenocytes. Cell Immunol 108: 227–234

58. Ratjczak H, Thomas P, Vollmuth T, Heck D, Sothern R, Hrushesky W (1988) Seasonal variation of host resistance and in vitro antibody formation of spleen cells from the B6C3F1 mouse. Annu Rev Chronopharmacol 5: 49–52

59. Sinha A, Linscombe A, Gollapudi B, Jersey G, Flake R (1986) Cytogenic variability of lymphocytes from phenotypically normal men. Influence of smoking, age, season and sample storage. J Toxicol Environ Health 17: 325–345

60. Shifrine M, Garsd A, Rosenblatt LS (1982) Seasonal variation in immunity of humans. J Interdiscip Cycle Res 13: 157–165

61. Bratescu A, Teodorescu M (1981) Circannual variation in the B -T cell ratio in normal human peripheral blood. J Allergy Clin Immunol 68: 273–280

62. Paigen BE, Ward E, Reilly A, Houten L, Gurtoo H, Minowada J, Steenland K, Harens MB, Sartori P (1981) Seasonal variation of arylhydrocarbon hydroxylase activity in human lymphocytes. Cancer Res 41: 2757–2761

63. Sofronov B, Nazarov P, Purin V (1976) Some characteristics of human lymphocyte responsiveness to stimulation in vitro. Allerg Immunol 22: 383–386

64. Lévi F, Halberg F (1981) Circaseptan (about 7-days) bioperiodicity – spontaneous and reactive – and the search for pacemakers. La Ricerca Clin Lab 12: 323–370

65. Hrushesky W, Gruber S, Sothern R, Hoffman R, Lakatua D, Carlson A, Cerra F, Simmons R (1988) Natural killer cell activity: age, estrous- and circadian stage dependence and inverse correlation with metastatic potential. J Natl Cancer Inst 80: 1232–1237

66. Hrushesky W, Bluming A, Gruber S, Sothern R (1989) Menstrual influence on surgical cure of breast cancer. Lancet ii: 949–952

67. Grossman CJ (1985) Interactions between the gonadal steroids and the immune system. Science 227: 257–260

68. Khansari DN, Margo AJ, Faith RE (1990) Effects of stress on the immune system. Immunol Today 11: 170–175

69. Villette JM, Bourin P, Doinel C, Mansour I, Fiet J, Boudou P, Dreux C, Roue E, Debord M, Lévi F (1990) Circadian variations in plasma levels of hypophyseal adrenocortical and testicular hormones in men infected with human immunodeficiency virus. J Clin Endocrinol Metab 70: 572–577

70. Arendt J (1989) Melatonin and the pineal gland. In: Arendt J, Minors DS, Waterhouse JM (eds) Biological rhythms in clinical practice. Butterworth, Guildford, pp 184–206

71. Canon C, Lévi F, Bennaceur M, Touboul JP, Patti A, Reinberg A, Mathé G (1986) Alterations of circadian rhythms in lymphocyte subpopulations of patients with hematologic malignancies. Proceedings of the 12th Annual Meeting of the European Society of Medical Oncology, Nice, 28–30 Nov 1986. Cancer Chemother Pharmacol 18 (suppl 1)

72. Hrushesky W (1983) The clinical application of chronobiology to oncology. Am J Anat 168: 519–542

73. Bourin P, Mansour I, Lévi F, Villette JM, Roué R, Fiet J, Rouger P, Doisnel C (1989) Perturbations précoces des rythmes circadiens des lymphocytes T et B au cours de l'infection par le virus de l'immunodéficience humaine (VIH). C R Acad Sci 308: 431–436

74. Lotze MT, Finn OJ (1990) Recent advances in cellular immunology: implications for immunity to cancer. Immunol Today 6: 190–193

75. Orband ME, Ross S (1990) Problems in the investigational study and clinical use of cancer immunotherapy. Immunol Today 6: 193–195

76. Lévi F, Halberg F, Chihara G, Byram J (1982) Chronoimmunomodulation circadian circaseptan and circannual aspects of immuno-potentiation or suppression with lentinan. In: Takahashi R, Halberg F, Walker CA (eds) Toward chronopharmacology. Pergamon, Oxford, pp 289–311

77. Hrushesky W, Langevin T, Nygaard S, Young J, Roemeling R (1987) Circadian stipulation required for reduction of variability in TNF toxicity/efficacity. Proceedings of the International conference TNF and related cytotoxins, 14–18 Sept 1987, Heidelberg

78. Bucy RP, Li J, Xu XY, Char D, Chen CLH (1990) Effect of cyclosporin A on the ontogeny of different T cell sublineages in chickens. J Immunol 144: 3257–3265

79. Pati A, Florentin I, Lemaigre G, Mechkouri M, Lévi F (1988) Chronopharmacologic optimization of oral ciclosporin A in mice: a search for a compromise between least renal toxicity and highest immunosuppressive effects. Annu Rev Chronopharmacol 5: 43–44

80. Magnus G, Cavallini M, Halberg F, Cornelissen G, Sutherland D, Najarian J, Hrushesky W (1985) Circadian toxicology of cyclosporin. Toxicol Appl Pharmacol 77: 181–185

81. Rosenberg SA, Longo DL, Lotze MT (1989) Principles and applications of biologic therapy. In: De Vita VT, Hellman S, Rosenberg SA (eds) Principles and practice of oncology. Lippincott, Philadelphia, pp 301–347

82. Quesada J, Rios A, Swanson D, Trown P, Gutterman J (1985) Antitumor activity of recombinant-derived interferon alpha in metastatic renal cell carcinoma. J Clin Oncol 3: 1522–1528

83. Smith D, Wagstaff J, Thatcher N, Scarffa H (1987) A phase I study of rDNA alpha 2b interferon as a 6-week continuous intravenous infusion. Cancer Chemother Pharmacol 20: 327–331

84. Abrams P, McClamrock E, Foon K (1985) Evening administration of alpha-interferon. N Engl J Med 312: 443–444

85. Indiveri F, Puppo F (1989) Neuroendocrine effects of biologic response modifiers. Proceedings of the 16th International Congress of Chemotherapy, 11–16 June 1989. Jérusalem, p 324

# Rhythms in Tumor Markers

Y. Touitou and C. Focan

## Introduction

Tumor markers, especially oncofetal or tumor-associated antigens as well as hormones or enzymes, have been documented in a large number of human malignancies. They have been used clinically to aid in the differential diagnosis, to provide information on the stage of the disease, on response of the tumor to therapy and on recurrence. The specificity of tumor markers in cancer is poor since a large number of pathophysiological events, e.g., biological factors or non cancer related diseases can induce significant variations in the plasma concentration of tumor markers (Touitou and Bogdan 1988a, Touitou et al. 1989a). Besides these classical sources of variations, rhythmicity of biological markers of cancer, if found, may be of great interest, although data in the literature are scarce. The incidence of several of the tumor markers is very different depending upon the origin of the cancer. For example, 70%–80% of patients with colorectal and 10%–30% of women with ovarian and gynecologic malignancies have elevated levels of carcinoembryonic antigen (CEA). Similarly, 60%–80% of patients with metastatic breast cancer have elevated levels of the tumor marker antigen CA 15-3. Thus, a large number of cancer patients present normal levels (so called "false negative" results) in their plasma concentrations of tumor markers (Touitou and Bogdan 1988b). These cancer patients with presumably normal blood levels of biological markers in the morning, at the usual time of sampling, may actually present a marker rhythm showing higher levels at other times along the 24-h scale. Circadian variations of tumor markers are therefore worth ascertaining to improve both treatment and follow-up of cancer patients. Indeed, in many instances serial blood samples or some time-specified sampling in false negative patients would improve the clinical usefulness of these tests.

## Cortisol as a Rhythm Marker in Cancer Patients

Cortisol shows a high-amplitude circadian rhythm and is therefore suitable as marker of the circadian rhythmicity in man (Touitou et al. 1982). Changes in the secretory pattern of plasma cortisol have been found by Touitou et al. (1990b) in 7 out of 13 patients with metastatic breast cancer and in 15 out of 20 patients with ovarian cancer (stage IC-IV). Abnormalities in cortisol secretory pattern consisted of flattened profiles, decrease in amplitude, and/or shift in the peak or trough time, and/or plateau with high values in the morning (Fig. 1). These abnormalities were mostly observed in patients with a poor performance status ($\geq 2$) which is a clinical estimate of the patient's level of physical activity. All these changes are compatible with the hypothesis of a rhythm desynchronization in these patients and fits well with data reported by Bailleul et al. (1986) on the hematologic time structure in cancer patients.

## Biochemical Markers for Hematologic Proliferation

In acute lymphoblastic leukemia and in lymphosarcoma, Makhonova et al. (1978) described a significant shift in the acrophase of circulating granulo- and lymphocytes peaking at 12 h intervals from circulating blastic cells. In both types of cells, Kachergene et al. (1971) described also evening elevated levels of lactic dehydrogenase enzymatic activity. These authors also showed that normal rhythmicity could be rapidly restored after effective chemotherapy.

In multiple myeloma, (MM), circadian rhythm varying according to the individuals have been scarcely studied both for host rhythms and urinary

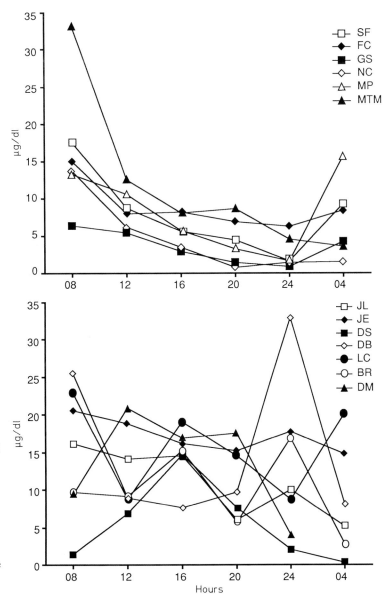

**Fig. 1.** Individual pattern of plasma cortisol in 13 patients with advanced breast cancer. *Top:* six patients *(SF, FC, GS, NC, MP,* and *MTM)* had a normal general secretory pattern, though one patient *(GS)* had abnormally low levels around the day and one other *(MTM)* had abnormally high values at the time of peak. *Bottom:* seven patients *(JL, JE, DS, DB, LC, BR,* and *DM)* had abnormal profiles of the hormone and most often bad performance status ( ≥ 2). (From Touitou et al. 1990a)

excretion of Bence Jones protein (Zinneman et al. 1974).

Ferritin can occasionnally by synthesized in high amounts by tumor cells and it has been proposed as a nonspecific tumor marker for both hematologic and solid proliferations (Touitou et al. 1985). Of interest, at low plasma levels in normal subjects or in benign monoclonal gammopathies, no circadian rhythm could be detected, whereas, in MM, levels of ferritin are usually high and a nice circadian rhythm with a peak during the afternoon appeared. Therefore the circadian rhythm in serum ferritin levels would represent a simple discriminant test between benign and tumoral gammopathies (Neri et al. 1983a).

Phosphohexoisomerase (PGI) glycolytic enzymatic activity is present both in normal and tumoral tissues. Neri et al. (1981, 1983b) have reported circadian rhythm for PGI serum activity with peaks around 1600 hours. However, the rhythm characteristics of PGI were different in MM (lower amplitude, higher levels); furthermore, after effective chemotherapy, serum levels of PGI rapidly dropped while the circadian rhythm became indetectable.

# Biochemical Markers
# for Human Solid Tumors

## Carcinoembryonic antigen

About 75% of patients with ovarian cancer (Malkin et al. 1978; van Nagell et al. 1978) and 50% of patients with bladder cancer (Guinan et al. 1978) have plasma CEA levels within the usual range when sam-

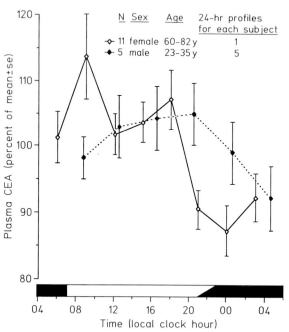

**Fig. 2.** Circadian variation in plasma CEA for groups of healthy males and females. The individual series were expressed as percentage of mean before group summary. (From Touitou et al. 1988)

pled during the usual morning sampling span. Whether serial blood samples or some time-specified sampling in these false negative patients would improve the clinical usefulness of this test was an hypothesis worth examining. A circadian rhythm of plasma CEA was found in separate groups of healthy male and female subjects by Focan et al. (1986 a, b, c) and Touitou et al. (1986) with higher levels during the afternoon and lower levels during sleep (Fig. 2). The total circadian variability ranged between 5% and 38% according to the subjects, which represents a much greater variability than can be ascribed to the daily change in plasma volume as reflected by hemoglobin, hematocrit, and erythrocyte counts (Touitou 1986). A circannual rhythmicity was also shown by Touitou et al. (1988) in four out of five healthy young men with a large total variability (from 19% to 60% according to the subjects) but with very disparate individual circannual peak times (Fig. 3). In contrast to the clinically healthy subjects, a circadian rhythm of plasma CEA could not be described in patients with ovarian ($n = 15$) or bladder ($n = 10$) cancer prior to the onset of monthly courses of chemotherapeutic treatment with doxorubicin and cisplatin. However the circadian rhythm was restored by chemotherapy in general and restored even sooner by four courses of effective chronotherapy with doxorubicin and cisplatin (Fig. 4) (Touitou et al. 1988). This may reflect a synchronizing effect of the timed chemotherapy and/or the hospital regime and may imply that the rhythm in plasma CEA in some cancer patients may be desynchronized from the usual day-night synchronization of the host. The reestablishment of a circadian rhythm in plasma CEA may be associated with a significant decrease in tumor burden. The same disappearance of the circadian rhythmicity of

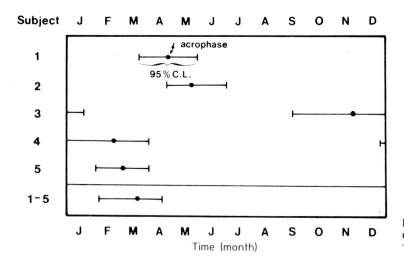

**Fig. 3.** Circannual rhythmicity of plasma CEA concentration in healthy men. (From Touitou et al. 1988)

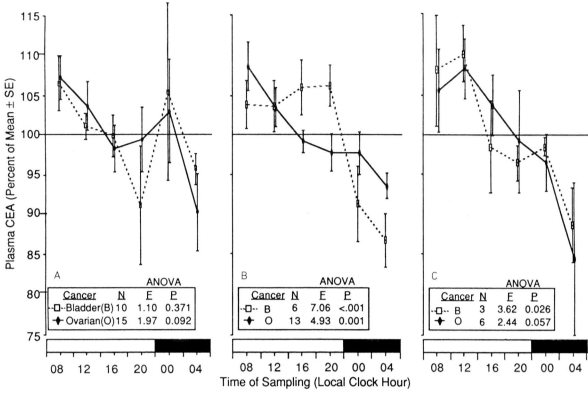

**Fig. 4A–C.** Circadian waveform of plasma CEA in patients with advanced ovarian and bladder cancer before (**A**) and after four (**B**) or eight (**C**) monthly courses of chemotherapy with adriamycin and cisplatin. Individual series were expressed as percentage of mean before group summary. (From Touitou et al. 1988)

plasma CEA was also found by Focan et al. (1986 a, b, c) in patients suffering from lung and gastrointestinal malignancies.

In a recent study on 13 patients with advanced breast cancer circadian variations of CEA, total proteins, $\gamma$-glutamyl transferase, and alkaline phosphatase were validated as a group phenomenon. However, a large variability in the individual patterns of the tumor marker was found. The amplitude of the variations during the 48 h of sampling averaged 10%–15% but reached 30%–50% in some patients (Touitou et al., to be published). Jäger et al. (1987) also found in metastatic breast cancer patients rhythmic fluctuations of CEA serum levels, which seemed to be independent of the localization of metastases and kind of treatment.

## CA 15-3 Antigen

Ca 15-3 is a carbohydrate antigen mainly used as a marker in the follow-up of breast cancer patients. Abnormally high plasma levels of this marker are found in 25%–50% of these patients and up to 60%–80% in the case of metatatic carcinomas (Hayes et al. 1986). In a study of patients with advanced breast cancer, CA 15-3 fluctuations were determined over a 48-h span before and after treatment with doxorubicin. A circadian rhythm of small amplitude (15% of total variability) was found only before treatment in the group of patients with high levels of CA 15-3 (Touitou et al. 1990a). However, this amplitude could reach 50% in individual patients (Fig. 5), which is important to consider in regard to the follow-up to these patients since these circadian variations were not found to be associated with changes in the clinical status (Touitou et al., to be published).

## CA 125 Antigen

CA 125 is a carbohydrate antigen mainly used in the investigation of patients with non-mucinous ovarian cancer (Kawabat et al. 1983). About 80% of patients with stage III and IV have elevated plasma levels of this marker. In a study of 20 patients with ovarian

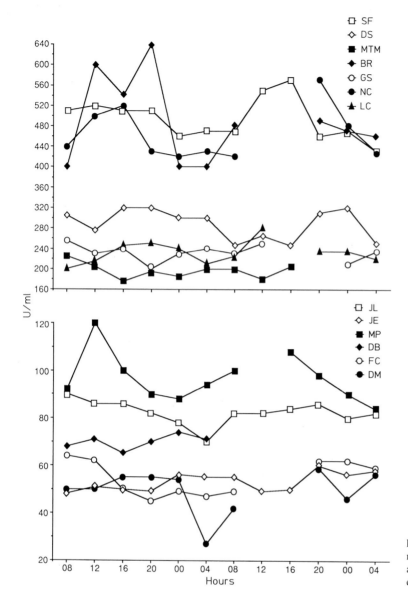

**Fig. 5.** Individual patterns of the tumor marker antigen CA 15-3 in 13 patients with advanced breast cancer. See Fig. 1. (Touitou et al., 1992, submitted)

cancer (stage IC-IV) both ANOVA and cosinor analysis validated the effect of time upon plasma concentrations of CA 125 in data expressed as percentages of the mean, but could not detect it on the raw data due to large interindividual differences. The total variability of the group was around 15%, but could reach 200% in some patients. The plasma CA 125 patterns was inconstant from one patient to another, e.g., in five patients the peak concentration was observed at 0800 hours, whereas in three other patients the 0800-hour concentration was their lowest one. The large variations within the 24 h were not related to apparent modifications of the patient's clinical state (Touitou et al., to be published).

## Alpha-Fetoprotein

The oncofetal marker, AFP, has been shown to exhibit a circadian rhythm in experimental hepatomas (Romanov et al. 1982). A circadian rhythm of low amplitude (8%) has been documented in control groups with a peak around 1100 hours as well as in gastrointestinal tract cancer patients with a somewhat different peak time (1530 hours) as shown in Fig. 6 (Focan et al. 1986a, b, 1987).

However, despite the fact that individual circadian rhythm could be recorded in cancer patients, no population rhythmicity could be ascertained due to the variability of individual rhythm characteristics,

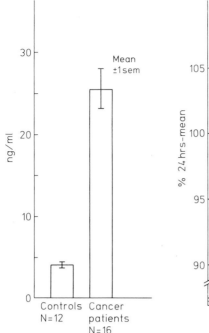

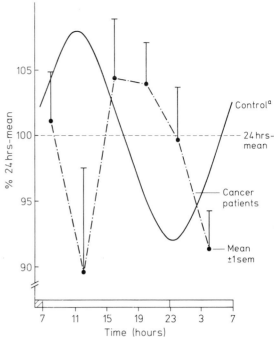

**Fig. 6.** Cancer-associated alteration of circadian rhythm in serum concentration of AFP. Cosine curve fitted to the data with $p < 0.01$. (From Focan et al. 1986b)

e. g., various peaks, amplitude and acrophases (Focan et al. 1986a, b, Focan 1987, Focan et al. 1987).

## Amines

Polyamines (putrescine, spermidine) are natural components playing an essential role in the regulation of RNA dependent protein synthesis and in cell proliferation (Dreyfuss et al. 1975; Durie et al. 1977). In control subjects, and in a case of retinoblastoma, circadian rhythm of urinary excretion of monoacetyl spermidine (peak at 1100 hours) has been reported (Hrushesky et al. 1983).

## Hormones

Some hormone-secreting experimental tumors have been shown to retain circadian rhythms of cell proliferation (Focan et al. 1982). In man, despite the fact that, classically, the disappearance of circadian hormonal secretions could be the rule when malignant degeneration appears, some exceptions have been reported in the literature. In some Cushing syndromes related to adrenal adenomas, as well as in juxtaglomerular tumors, hormonal release could present its usual circadian peak at 0800 hours. However, characteristics of these rhythms are markedly different

from those of controls while hormonal reactivity to physiological stimuli is markedly impaired (Ambrosi et al. 1985; Conn et al. 1972). In Cushing syndrome, circadian rhythm of serum cortisol peaks can also be shifted (Olsen et al. 1978).

Circadian rhythm in urinary excretion of norepinephrine has been described in a case of neuroblastoma (Focan 1987a). Diurnal variability has also been described for calcitonin serum levels in medullary carcinoma with a peak around midday (Wahl et al. 1980). More recently, we have shown diurnal rhythmicity in serum levels of parathyroid hormone in two cases of primitive hyperparathyroidy and in one case of small-cell lung carcinoma, of calcitonin in four cases of tumor ectopic secretion, and of gastrine in two cases of polyendocrinopathy (Focan, unpublished).

## Prostatic Acid Phosphatase

The specificity and the relevance of prostatic acid phosphatase (PAP) for the clinical workup of prostatic cancer has been well established for more than 4 decades. A circadian rhythm has been looked for using colorimetric methods (Doe and Mellinger 1964), but such analyses must be scrutinized with scepticism due to the numerous possible biases of the methodology. More recently, relevant enzyme-immuno and radio-immunoassays have allowed relevant titrations, but most of the analyses were performed either in a limited number of subjects or without any statistical rhythmometric tools (Mannini et al. 1988).

In three groups of old subjects (controls, patients with prostate adenoma, and patients with advanced prostate cancer), Focan et al. (1986b) were unable to describe any group rhythmicity. However, untreated cancer patients with high levels of circulating PAP had a circadian rhythmicity with a 16.4% amplitude and a peak around 1600 hours. These results are to be compared with those of Benvenuti et al. (1985), who reported a circadian rhythm in nine old control subjects (peak at 1500 hours) as well as in six patients with nondisseminated prostate cancer (acrophase disrupted to 0200 hours). These authors suggested also the disappearance of PAP circadian rhythm after effective treatment. Finally, in young adult males, a circadian rhythm of PAP was also present (Benvenuti et al. 1985). Other authors, however, more recently concluded that there is spontaneous diurnal variability of PAP (Dejter et al. 1988; Mannini et al. 1988).

### Prostate-Specific Antigen

Prostate-specific antigen (PSA), as a specific tumor marker, has been proposed to improve the clinical workup of prostatic cancer. Touitou et al. (unpublished data) have documented circadian rhythmicity in serum plasma levels only in prostate cancer patients after effective treatment (peak 1346 hours) amplitude 13.3%); significant circadian rhythm could be found in old male controls for patients suffering from a prostate adenoma or for cancer patients before treatment. Others reported a spontaneous, unpredictable variability from −72% to +190% of mean PSA values, but failed to sustain their assertions by exploration of at least one complete 24-h period (Dejter et al. 1988; Mannini et al. 1988).

### Thymidine Kinase

Elevated levels of thymidine kinase have been reported in viral disease, vitamin $B_{12}$ deficiency, and malignant diseases including hematologic neoplasms, lung and brain tumors. Large interindividual fluctuations were described in cancer patients sometimes retaining a circadian pattern, whereas no circadian rhythm could be shown at a group level (Hallek et al. 1988).

## Ultradian and Infradian Rhythms

Besides the circadian rhythms of tumor markers, some authors have described other rhythm periods. Ultradian rhythms are present in cell division and circulating levels of nucleated blood cells in chronic myelogenous leukemia (CML) (Lévi et al., unpublished). Focan et al. (1986a) reported 8-h period rhythmicity in 6 of 49 cancer patients among whom 5 of 6 were suffering from squamous cell carcinoma. Twelve-hour periods were also noted in two cases of hepatoma.

Infradian rhythms of circulating tumor cells were described in chronic lymphocytic leukemia (Lévi et al., unpublished), as well as in hormone release (i.e., ACTH) from pituitary adenomas or lung adenocarcinomas with periods ranging from 11 to 86 days (Bailey et al. 1971; Liberman et al. 1976).

Finally, as mentioned above, a circannual rhythmicity of CEA, peaking during springtime, has been described in adult healthy males (Touitou et al. 1988).

## Evolution of Tumor Markers' Rhythms Under Treatment

Only few reports dealt with the evolution of tumor marker rhythms after anticancer treatment. According to some authors (Kachergene et al. 1971; Makhonova et al. 1978), the sequential analysis of rhythmicity of serum levels of dehydrogenase lactic enzymes could allow to predict the outcome of acute leukemia: a rapid restoration of physiological rhythmicity under effective chemotherapy was noted in future responders to treatment.

The prevention of infradian rhythmicity of circulating nucleated blood cells in CML has been described after sequential time-scheduled chemotherapy.

In juxtaglomerular renin-secreting tumors (Focan et al. 1987), normal renin circadian rhythms quickly reappeared after a transient trough after surgical removal of the primary tumor in some cases, i.e., in a norepinephrine-secreting neuroblastoma, the surgical treatment abolished the circadian release of the marker.

In multiple myeloma, after active chemotherapy, the circadian rhythms of PGI disappeared. The evolution of the circadian rhythmicity of other markers remained either unreported (i.e., ferritin), or unexploitable due to large individual fluctuations (Neri et al. 1981, 1983 a, b).

In solid tumors, the reappearance (or the restoration) after chemotherapy of a circadian rhythm comparable to that of the control group, has been reported for CEA in homogenous groups of bladder or ovarian cancers (Touitou et al. 1988).

Similarly, a circadian rhythm reappeared for CA 125 after two courses of chemotherapy for advanced ovarian carcinoma. In breast cancer, circadian fluctuations with limited amplitudes remained unchanged under chemotherapy for both CEA and CA 15-3 tumor markers (Touitou et al. 1989 b, 1990 a).

## Conclusions

Different studies have shown that tumor markers (CEA, CA 15-3, CA 125) exhibit oscillations within the 24-h scale. In some patients the amplitude (peak-trough difference) of these variations is very large, up to 200%. Therefore a single determination may not be adequate to arrive at a clinical judgement. The variations within the 24-h scale, which are not always rhythmic and which can be different from one patient to another, e.g., in terms of plasma levels, amplitude, peak and trough location, etc., do not seem to be dependent upon changes in the clinical status of the patient. They could be related, e.g., to changes in the secretory activity of cells producing the marker and (or) to changes in its metabolism and elimination (among other hypotheses). These changes, together with those of plasma cortisol as found in breast and ovarian cancer, strongly suggest that some cancer patients present an external and/or internal desynchronization. These data are important to take into account to avoid possible errors in the therapeutic or surgical follow-up of cancer.

From the side of the host, desynchronization of circadian rhythmicity may lead to changes in the time of maximal and/or minimal sensitivity of critical organ systems (e.g., the bone marrow) to chemotherapeutic agents or radiation. Furthermore, if some of the tumor markers represent metabolic or secretory products of the tumor cells they may indicate time-dependent differences in tumor metabolism and possibly a circadian rhythmicity in the tumor. If this should be the case, new perspectives could be opened for the timed treatment (chronotherapy) of malignant growth not only in relation to host-rhythm but to rhythmic and thus in their timing predictable metabolic differences in the tumor.

At the individual level, it has become evident that the characteristics of the tumor-related rhythms can vary significantly from one subject to another; this has been clearly documented for light-chain urinary excretion in multiple myeloma and for CEA, AFP, PAP, PSA, CA 125, and CA 15-3 in different cancers.

At a group level circadian rhythmicity of some markers did appear to be a criterion of malignancy. As an example, the circadian rhythms of ferritin and PGI allowed distinction between benign and malignant gammopathies (Neri et al. 1981, 1983 a, b). It is worth noting that one marker exhibits circadian rhythm in normal subjects (ferritin) while the other did not (PGI). Similarly, disappearance of "normal" circadian variations in serum CEA titration appears to represent a discriminant criterion of malignancy (Focan et al. 1986 a; Focan 1987); indeed, patients with cancer have lost their circadian rhythm, which remained unaltered in noncancer patients with elevated non-timed qualified CEA levels.

It is striking to state that in the presence of cancer, some biochemical markers loose their circadian rhythm or change the pattern of their circadian rhythm (i.e., leukemia, PGI, CEA) while other markers develop de novo a circadian rhythm (i.e., ferritin). Similarly, the disappearance of a "normal" rhythm could be seen in some malignant tumors, but not in others (i.e., AFP, CEA, CA 15-3).

Most often however, the loss of the characteristics of circadian synchronization was observed for the majority of serum tumor markers. These markers are polypeptides or small serum proteins. In young or old normal subjects, circadian rhythms characterized by a nocturnal dip have been described for circulating plasma protein levels (Touitou et al. 1986). The peaks regularly reported for tumor markers in serum either in control groups or in some cancer subpopulations coincide precisely with day time, i.e., 11–16 h. Profound alteration of the physiological time structure has been described for plasmatic proteins in advanced cancer (Bruguerolle et al. 1986). Similarly, modifications of the rhythmicity of physiological cortisol or hematologic parameters have been reported (Bailleul et al. 1986, Touitou et al. 1989 b, 1990 b). These modifications, not clearly related to the stage of cancer, but also not influenced by renal function alteration or hepatic metastases, could indicate modifications of general metabolic activities when cancer develops. These phenomena could also reflect an internal desynchronization as suggested from individual shifts in acrophases of tumor markers or alterations of the time structure of cortisol as well as of most tumor marker serum levels. The question of how to relate such desynchronization to circadian cell division activity remains putative. Nevertheless, a progressive disruption of circadian waves of metabolic and cell division activities has been described in rats when nitrosamine-induced hepatomas developed (Barbason 1985; Focan 1985, 1987).

Some information available regarding the evolution of tumor marker rhythms under and/or after treatment also merits discussion. In some cases, this evolution could be related to the disease outcome. No general rule, however, could be obtained as sometimes circadian rhythms either disappear, such as ferritin in MM (Neri et al. 1983a) and norepinephrine in neuroblastoma (Focan 1987a); become "normal", such as PGI in MM (Neri et al. 1981, 1983b), lacticodehydrogenase in blastic and normal cells in leukemia (Kachergene et al. 1971; Makhonova et al. 1978), and renin in juxtaglomerular tumors (Conn et al. 1972); reappear, such as CEA or CA 125 in ovarian cancer and CEA in bladder cancer (Touitou et al. 1988); remain unchanged, such as CEA and CA 15-3 in breast cancer (Touitou et al. 1989b, 1990a); or present unpredictable fluctuations impossible to integrate into the clinical situation, e. g., the light-chain excretion in MM (Zinneman et al. 1974). The question of whether reappearance of a normal circadian rhythm in some tumor marker serum titrations (e. g., CEA, CA 125) represents an independent, good prognostic variable remains, at present, unanswered.

In conclusion, biochemical tumor markers reflect the rhythmic activity of the general metabolism of the host and perhaps of cell division of tumors. Their practical use for the follow-up and management of cancer patients or even for the choice of individualized time-scheduled chemotherapies remain, at present, purely speculative.

# References

Ambrosi B, Riva E, Faglia G (1985) Persistence of a circadian rhythmicity of glucocorticoid secretion in a patient with Cushing's syndrome: study before and after unilateral adrenalectomy. J Endocrinol Invest 8: 363–367

Bailey RE (1971) Periodic hormonogenesis. A new phenomenon. Periodicity in function of a hormone producing tumor in man. J Clin Endocrinol 32: 317–327

Bailleul F, Levi F, Reinberg A, Mathé G (1986) Interindividual differences in the circadian hematologic time structure of cancer patients. Chronobiol Int 3: 47–54

Barbason H (1985) Contribution à l'étude du contrôle de la prolifération cellulaire dans le foie de rat normal et précancéreux. Analyse de la cinétique cellulaire. Thèse d'Agrégation Enseignement Supérieur, Université de Liège

Benvenuti M, Legnaioli M, Melone F, Taddei I, Tarquini B (1985) Circadian rhythm in prostatic acid phosphatase: a potential tumor marker in prostatic cancer. Chronobiologia 10: 383–386

Bruguerolle B, Levi F, Arnaud C, Bouvenot G, Mechkouri M, Vannetzel JM, Touitou Y (1986) Alterations of physiologic circadian time structure of six plasma proteins in patients with advanced cancer. Annu Rev Chronopharmacol 3: 207–210

Conn JW, Cohen EL, Lucas CP, McDonald WJ, Mayor GH, Blough WM, Eveland WC, Bookstein JJ, Lapides J, Arbor A (1972) Primary reninism. Hypertension, hyperreninemia, and secondary aldosteronism due to renin-producing juxtaglomerular cell tumors. Arch Intern Med 130: 682–696

Dejter SW Jr, Martin JS, McPherson RA, Lynch JH (1988) Daily variability in human serum prostate-specific antigen and prostatic acid phosphatase: a comparative evaluation. Urology 32: 288–292

Doe RP, Mellinger GT (1964) Circadian variation of serum acid phosphatase in prostatic cancer. Metabolism 13: 445–452

Dreyfuss F, Chayen R, Dreyfuss G, Doir R, Ratan JP (1975) Polyamine excretion in the urine of cancer patients. Isr J Med Sci 11: 785–795

Durie BGM, Salmon SB, Russel DH (1977) Polyamines as markers of response and disease activity in cancer chemotherapy. Cancer Res 37: 217–221

Focan C (1985) Le rhythme nycthemeral de la prolifération tumorale. Aspects expérimentaux et cliniques. Implications pour la chimiothérapie oncolytique. Ph D Thesis, University Liege

Focan C (1987) Chronobiologie et marqueurs biochimiques du cancer humain. Pathol Biol 35: 951–959

Focan C, Schyns-Mosen, Luyckx A, Lefebvre P (1982) Circadian variability of kinetics in a transplantable islet-cell tumor of the golden hamster. J Cancer Res Clin Oncol 104: 197–206

Focan C, Collette J, Levi F, Hrushesky W, Touitou Y, Franchimont P (1986a) Circadian rhythms in human carcinoembryonic antigen (CEA), alphafoeto protein (AFP) and prostatic acid phosphatase (PAP). Evidence of chronopathology in cancer. Annu Rev Chronopharmacol 3: 417–420

Focan C, Focan-Henrard D, Collette J, Mechkouri M, Levi F, Hrushesky W, Touitou Y, Franchimont P (1986b) Cancer-associated alteration of circadian rhythms in carcinoembryonic antigen (CEA) and alpha foetoprotein (AFP) in humans. Anticancer Res 6: 1137–1144

Focan C, Focan-Henrard D, Frere MH, Hung SL, Castronovo V, Collette J, Franchimont P, Touitou Y, Lévi F, Roemeling R, Hrushesky WJM (1986c) Circadian CEA variability: when to sample. J Clin Oncol 4: 607–608

Focan C, Collette J, Levi F, Touitou Y, Hrushesky W, Franchimont P (1987) Rythmes biologiques et marqueurs tumoraux chez l'homme. Référence particulière à l'antigène carcinoembryonnaire (ACE) et à l'alphafoetoprotéine (AFP) dans les tumeurs malignes. Acta Gastroenterol Belg 50: 12–21

Guinan PD, McKiel C, Sundar B, Veith R, Dubin A, Ablin RJ (1978) The carcinoembryonic antigen test in urologic cancer. Natl Cancer Inst Monogr 49: 225–229

Hallek M, Emmerich B, Reichle A, Schick HD, Senekowitsch R, Reinberg A (1988) Observation of circadian variations of serum thymidine kinase levels in control and tumor patients. Annu Rev Chronopharmacol 5: 361–364

Hayes DF, Zurawski VR, Kufe W (1986) Comparison of circulating CA 15-3 and carcinoembryonic antigen levels in patients with breast cancer. J Clin Oncol 4: 1542–1550

Hrushesky WJM, Merkink J, Abdel-Monen MM (1983) Circadian rhythmicity of polyamine urinary excretion. Cancer Res 43: 3944–3947

Jäger W, Diedrich M, Sauerbrei W, Wildt L (1987) Rhythmic fluctuations of CEA levels in breast cancer patients. Anticancer Res 7: 711–716

Kachergene NB, Voshol IV, Narisyssov RP (1971) Circadian rhythm of dehydrogenase activity in the blood cells in acute leukemia in children. Bull Acad Sci USSR 9: 81:85 (in Russian)

Kawabat SE, Bast RC, Welch WR, Knapp RC, Colvin RB (1983) Immunopathologic characterization of a monoclonal antibody that recognizes common surface antigens of human ovarian tumor of serous, endometroid, and clear cell types. Am J Clin Pathol 79: 98–104

Le Thi Huong Du, Mohattane H, Piette JC, Bogdan A, Auzeby A, Touitou Y, Godeau P (1988) Specificité du marqueur tumoral CA 125. Etude de 328 observations de médecine interne. Presse Med 17: 2287–91

Liberman B, Wajchenberg BL, Tambascia MA, Mesquita CH (1976) Periodic remission in Cushing's disease with paradoxal dexamethasone response: an expression of periodic hormonogenesis. J Clin Endocrinol Metab 43: 913–918

Makhonova LA, Peterson IS, Mayakova SA, Berezin AA, Gubarev KM (1978) Investigation of circadian biorhythms of peripheral circulation in children with acute leukemia and lymphosarcoma. Bull Acad Sci USSR 23: 32–36

Malkin A, Kellen JA, Lickrish GM, Bush RS (1978) Carcinoembryonic antigen (CEA) and other tumor markers in ovarian and cervical cancer. Cancer 42: 1452–1456

Mannini D, Maver P, Aiello E, Corrado G, Vecchi F, Bellanova B, Marengo M (1988) Spontaneous circadian fluctuations of prostate specific antigen and prostatic acid phosphatase serum activities in patients with prostatic cancer. Urol Res 16: 9–12

Neri B, Ciapini A, Comparini T, Guidi G, Guisi S, Lupi R, Romano S (1981) Circadian variations of phosphohexoseisomerase (PGI) in multiple myeloma. Eur J Cancer Clin Oncol 17: 1177–1181

Neri B, Comparini Y, Guidi S, Pieri A, Tommasi M (1983a) Ferritin as a chronobiological marker in immunoproliferative diseases. Oncology 40: 111–114

Neri B, Necuccia A, Ciapini A, Comparini T, Guidi G, Guidi S (1983b) Chronobiological aspects of phosphohexoseisomerase in monitoring multiple myeloma. Oncology 40: 332–335

Olsen NJ, Fang VS, Degroot LJ (1978) Cushing's syndrome due to adrenal adenoma with persistent diurnal cortisol secretory rhythm. Metabolism 27: 695–700

Romanov YA, Plaksina TI, Filippovich SS (1982) Effect of cyclophosphamide injection time on changes in rhythms of serum alpha-fetoprotein levels in mice with hepatoma 22a and of tumor cell proliferation. Bull Exp Biol Med 93: 475–478

Touitou Y, Bogdan A (1988a) Tumor markers in non-malignant diseases. Eur J Cancer Clin Oncol 24: 1083–1091

Touitou Y, Bogdan A (1988b) Etude critique des marqueurs tumoraux récents. Bull Cancer 75: 247–262

Touitou Y, Sulon J, Bogdan A, Touitou C, Reinberg A, Beck H, Sodoyez JC, Van Cauwenberg H (1982) Adrenal circadian system in young and elderly human subjects: a comparative study. J Endocrinol 93: 201–210

Touitou Y, Proust J, Carayon A, Klinger E, Nakache JP, Huard D, Sachet A (1985) Plasma ferritin in old age. Influence of biological and pathological factors in a large elderly population. Clin Chim Acta 149: 37–45

Touitou Y, Touitou C, Bogdan A, Reinberg A, Auzeby A, Beck H, Guillet P (1986) Differences between young and elderly subjects in seasonal and circadian variations of total plasma proteins and blood volume as reflected by hemoglobin, hematocrit and erythrocyte counts. Clin Chem 32: 801–804

Touitou Y, Darbois Y, Bogdan A, Auzeby A, Keusseoglou S (1989a) Tumour markers antigens during menses and pregnancy. Br J Cancer 60: 419–420

Touitou Y, Lévi F, Bailleul F, Bogdan A, Touitou C (1989b) Circadian rhythmicity of CA 15.3, CEA and cortisol in breast cancer patients. Chronobiologia 16: 191

Touitou Y, Sothern RB, Lévi F, Focan C, Bogdan A, Auzéby A, Franchimont P, Roemeling RW, Hrushesky WJM (1988) Sources of predictable tumor variation within the so-called normal range: circadian and circannual aspects of plasma carcinoembryonic antigen (CEA) in health and cancer. J Tumor Marker Oncol 4: 351–359

Touitou Y, Bailleul F, Lévi F, Bogdan A, Touitou C, Metzger G Mechkouri M (1990a) Circadian rhythms of tumor markers in breast cancer patients. In: Hayes D, Pauly J, Reiter R (eds) The clinical and agricultural applications of chronobiology. Liss, New York

Touitou Y, Lévi F, Bogdan A, Bruguerolle B (1990b) Abnormal patterns of plasma cortisol in breast cancer patients. Annu Rev Chronopharmacol 7: 245–248

Van Nagel JR Jr, Donaldson ES, Wood EG, Goldenberg DM (1978) The clinical significance of carcinoembryonic antigen in the plasma and tumors of patients with gynecologic malignancies. Cancer 42: 1527–1532

Wahl HM, Schmidt-Gayk H, Wahl R, Cordes D, Limba-Cach HJ (1980) Serum calcitonin in medullary thyroid carcinoma: circadian variation and response to pentagastrin. Tumor Diagnostik 2: 85–88

Zinneman HH, Halberg F, Haus E, Kaplan M (1974) Circadian rhythms in urinary light chains, serum iron and other variables of multiple myeloma patients. Int J Chronobiol 2: 3–16

# Chronoepidemiology: Chronobiology and Epidemiology

M. H. Smolensky

## Introduction

Epidemiology as a science is concerned with the occurrence and determinants of diseases in human populations (Lilienfeld and Lilienfeld 1980; Barker 1982). Although epidemiologists are interested in temporal factors relating to the occurrence and clustering of disease, the interest is primarily with the *prevalence* of disease, that is the number of individuals exhibiting a given disease at a specific time relative to the number of individuals making up the population at that point in time. Too, they are concerned with the *incidence rate* of a disease, that is the number of cases of a given disease occurring in a population during a span of time in relation to the length of time at which the population is at risk of experiencing the disease. Essentially, prevalence is the proportion of the population that exhibits a particular disease at a certain point in time. Thus, the major concern of epidemiologists regarding the variable of time, per se, is in terms of the time interval used in the denominator, often expressed as "person-years," "person-time" or "risk-time." For each individual in the study population the time at risk is that span during which the individual remains free from disease and therefore is at risk to it. Collectively, for a given study group or cohort, the time period of risk represents the summation of disease-free durations (Barker 1982; Ahlbom and Norell 1984).

Epidemiologists also are interested in the clustering of particular diseases in time. For example, greater than expected morbidity and/or mortality of a group with respect to a given interval of time usually implies to classically trained epidemiologists the likelihood of an environmental etiology or cause and effect relationship. Even if the clustering of events tends to be confined to a particular span of day or season of year, the focus of epidemiologic investigations is to determine causality by examining environmental factors which are external to the body. The assumption of constancy, or homeostasis, of bodily functions by epidemiologists and other environmental scientists precludes the consideration of possible chronobiologic (biological rhythm-dependent) determinants of the observed temporal patterns in disease. Even when epidemiologists note clear temporal features in morbidity and mortality, such as in the case of the seasonality of infectious diseases, their perspective is mainly in terms of the incubation time of the illnesses and/or those temporal factors in specific vectors giving rise to seasonal differences in such diseases (Lilienfeld and Lilienfeld 1980).

Chronobiologists, as compared to those trained traditionally in biology with the point of view of constancy (homeostasis) in biological functions, have a vastly different perspective about the significance of time with regard to the temporal patterning in the occurrence of human diseases. In particular, chronobiologists are concerned with *predictable* variability in biological functions and processes over time (bioperiodicities), including biological rhythm-related patterns in the susceptibility/resistance to disease as well as in the manifestation of illnesses and their symptoms. Obviously, if the symptoms and markers of illness vary in a predictable manner over 24 h, the menstrual cycle, year, etc., due to rhythmic processes, the time when certain types of epidemiologic assessments or surveys are conducted, such as during the morning versus the evening or summer versus the winter, may result in very different findings. An epidemiologic survey of hypertension by means of single blood pressure measurements made late in the day is likely to result in a higher estimate of the prevalence or incidence of disease than if the measurements are done earlier in the day (Smolensky et al. 1976). Similarly, *when* pulmonary assessments are performed, morning vs evening, may strongly influence the findings of industrial or community surveys designed to evaluate the effect of work place or environmental contaminants on airways function and status (Smolensky and Halberg 1977; Gervais et al. 1979;

Smolensky et al. 1980). Indeed, reliance upon single once-a-day determinations and measurements to document the presence or absence of disease is likely to be inappropriate especially in the case when the amplitude (peak-trough variation) of the selected study variables may be large. Although epidemiologists have neglected possible contributions of chronobiology to their activities, chronobiologists as well have been negligent in not directing their research efforts toward solving medical and public health matters. Many chronobiologists have been overly consumed only with the detection and description of biological rhythms, rather.than investigating their pertinence or application to public health problems. The purpose of this chapter is to illustrate how the findings and methods of chronobiology are of critical importance to epidemiologic considerations and to the field of public health in general.

## Philosophy and Methods of Chronobiology

Research in the medical sciences, including epidemiology, is strongly dominated by the prevailing theory of homeostasis. According to this hypothesis, a set of specific mechanisms maintains the *milieu interieur* relatively constant. The working model (Fig.1) of epidemiologists, whether overtly stated or not, assumes the capability of human beings to respond to environmental challenges, medications, toxic agents, etc., is comparable no matter the hour of the day, day

Homeostatic model

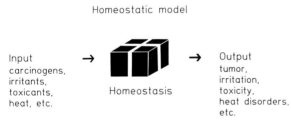

**Fig. 1.** The central importance ascribed to the hypothesis of homeostasis in conceptualizing and investigating the causality of disease. Biological functions of the body (denoted by the *black box*) are presumed to be relatively constant. Accordingly, nonvarying or cyclic inputs from the ambient or occupational environments are attended to by homeostatic processes. Explanation of temporal (e.g., 24-h and seasonal) patterns in the symptoms, signs, and frank manifestations of disease (in the figure termed *output*) in populations or in single individuals is sought by the search for cyclic alterations in the external environment exclusively. The contribution of biological rhythms to the observed temporal patterns is not considered. (From Smolensky 1983)

of the week, or month of the year. Thus, observed cyclic day-night and seasonal patterns in events or measurements are interpreted as being the direct result of temporally varying environmental phenomena. This perspective is prevalent even though numerous published reports (for overviews see Arendt et al. 1989; Halberg 1969; Moore-Ede et al. 1982; Reinberg and Smolensky 1983a) document the existence of large-amplitude biological rhythms with periods of 24 h (circadian), 7 days (circaseptan), 28–30 days (circamensual), and 1 year (circannual), among others, which are endogenous in origin (Barcal et al. 1968; Halberg 1969; Konopka and Beuzer 1971; Reinberg et al. 1985). Although the expression of biological rhythms may be influenced by the environment, they are not caused by environmental periodicities.

The detection and quantification of rhythms are accomplished through the application of special time series analyses (De Prins et al. 1986; Halberg et al. 1972; Reinberg and Smolensky 1983b). One method, the Cosinor (Halberg et al. 1972), enables the objective detection of statistically significant bioperiodicities and in addition their description by specific characteristics, i.e., the period length, time series mean (also referred to as *mesor* – a middle value around which predictable variation is exhibited), *amplitude* (a measure of the variability due to rhythmicity), and *acrophase* (crest time expressed relative to a designated reference marker) (De Prins et al. 1984; Halberg et al. 1972; Reinberg and Smolensky 1983b). The staging (occurrence of peak and trough values) of human circadian rhythms with respect to clock time under usual conditions of life is affected primarily by the temporal pattern of rest and activity (Halberg and Simpson 1967). For example, the peak and trough times of circadian rhythms in persons chronobiologically adjusted to night work differ with respect to clock hour from those of persons adhering to a schedule of day work alternating with night sleep (Reinberg 1979; Johnson et al. 1981).

Nearly all biological processes exhibit rhythms. Bioperiodicity has been shown to be so important that the ability to withstand and survive an environmental challenge can depend literally on the time with regard to biological rhythms when it is encountered. This important fact was initially demonstrated by Halberg (1960) in laboratory studies on rodents. Halberg termed these predictable cyclic differences in the ability of organisms to tolerate physical, chemical, and bacterial challenges susceptibility/resistance rhythms. Herein, this concept is extended to epidemiologic concerns. Different examples have been selected to illustrate the scope of chronobiologic findings to the field of epidemiology and to public health issues, in general.

## The Chronoepidemiology of Human Disease

A majority of epidemiologists are engaged in the study of patterns in human morbidity and mortality as well as their etiology. Ordinarily, this necessitates the identification of those individuals who exhibit the signs and symptoms of disease or those who have died due to specific diseases or causes with respect to selected time scales. The latter is investigated by the review of death certificates for cause, time, and place of death. The former is accomplished either by surveys which examine persons for the symptoms and signs of illness and/or the use of clinical data arising from specific measurements and tests. From an epidemiological point of view, *symptoms* refer to manifestations of disease that only the examined person (or patient) may know; this is the case of dyspnea, pain, nausea, etc. In contrast, *signs* refer to manifestations of illness which may be observed by an independent examiner, typically a professional health worker; these include the interpretation of lung or heart sounds or radiographic films (X-rays), among others. *Tests* refer to special techniques or instruments which document the presence of abnormal function and/or disease; these include clinical laboratory procedures, electrocardiograms, etc. Epidemiologic concerns about the specificity and reliability of techniques to accurately assess disease through the evaluation of symptoms, signs, and tests have seldom included a chronobiologic perspective, i.e., that the biological time of such events could well be a very significant variable. Indeed, *when* during the 24 h, week, menstrual cycle, or year persons are evaluated for the symptoms and sings of disease or *when* blood samples or other instrument-dependent assessments are conducted can make a profound difference (Bartter et al. 1976; Smolensky and Halberg 1977; Reinberg and Smolensky 1983a; Haus et al. 1984).

When diseases cluster in time and/or in a given region, epidemiologists generally turn their attention to aspects of the surrounding environment for clues to their etiology. Both epidemiologists and chronobiologists consider the occurrence of human diseases and their symptoms in time not to be random. However, to epidemiologists the clustering of illnesses in time suggests causality due to specific, perhaps cyclic, en-

viromental factors. In contrast, to chronobiologists the clustering of illnesses in time infers the possible contribution of biological rhythmic influences in the occurrence or exacerbation of human disease or the contribution of both exogenous and endogenous cyclic phenomena. Actually, the occurrence or exacerbation of many human diseases can be shown to vary in a predictably cyclic manner over 24 h, a week, month, and year. This is exemplified by the morbidity and mortality of various cardiovascular diseases (Lemmer 1989b; Marshall 1977; Kuroiwa 1978; Muller et al. 1985), asthma (Smolensky et al. 1981a; Dethlefsen and Repges 1985; Turner-Warwick 1988; Smolensky et al. 1986a; Smolensky et al. 1986b) and arthritis (Kowanko et al. 1982; Labrecque and Reinberg 1989), among many others (Smolensky 1983).

The investigation and documentation of temporal patterns in the *signs* of certain diseases in individuals and in groups of persons at risk can be done by the continuous monitoring of selected variables over consecutive 24-h periods using suitable ambulatory instrumentation or self-assessment techniques. Thus, for chronic disorders, such as asthma, angina, arthritis, and epilepsy, which are typically non-life threatening, their signs and/or symptoms tend to be recurrent in time describing a definite pattern over 24 h, not only in individuals but in groups of individuals suffering the same disease. In contrast, the occurrence of serious, life-threatening morbid events in individuals is rather infrequent during one's lifetime. This is the situation for stroke and heart attack, for example. For these, one most rely on data obtained from studies of large groups of persons or data available from special patient registries.

Some of the points discussed above are illustrated in Fig. 2, which presents the 24-h pattern in the signs and events of several diseases of the cardiovascular system. Myocardial infarction (MI), cerebral infarction (CI), and cerebral hemorrhage (CH) all constitute the type of diseases for which the occurrence is rather rare during one's lifetime. Thus, the only way to assess circadian patterns is by the pooling of data for the clock time of the event from a relatively large number of persons so affected. The temporal pattern in ST-segment anomaly and pain of angina, which is recurrent from one day to the next in the same person, can be determined by the investigation of either separate individuals or the pooling of data from groups of persons who suffer from the same disorder.

As a group phenomenon in diurnally active persons at risk, MI exhibits a major peak in cases around 0900–1100 hours. The prevailing theory held by the majority of physicians and epidemiologists is that MI

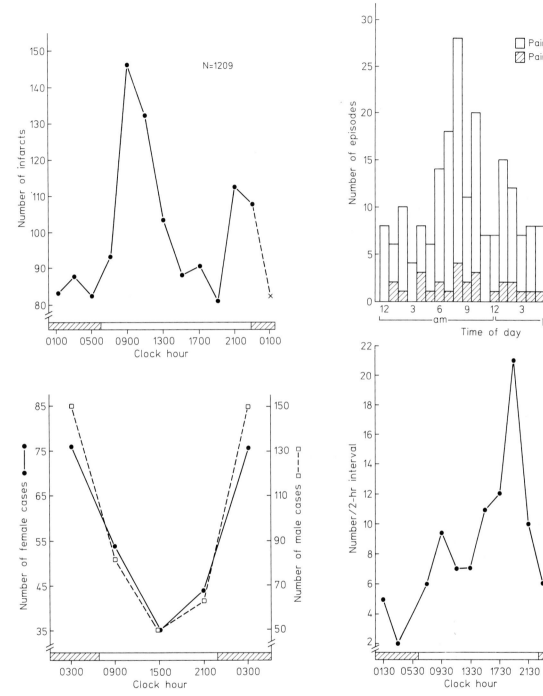

**Fig. 2.** Chronoepidemiology of selected diseases of the cardiovascular system. The occurrence in groups of persons, presumably adhering to a life routine of diurnal activity alternating with nighttime sleep, of the morbid events of myocardial infarction *(top left)*, cerebral infarction *(bottom left)*, ST-segment anomaly and pain of regular angina *(top right)*, and spontaneous intracerbral hemorrhage *(bottom right)* is strongly variable during 24 h. Individual graphs or data obtained from Smolensky (1983), Marshall (1977), and Lemmer (1989 b)

is more common during and following strenuous activity and work or in the evening after a heavy meal (Master 1960); however, the distribution of attacks over the 24 h does not support this hypothesis. CI, in comparison, appears to be most frequent during the night, especially around 0300. On the other hand, CH is most frequent early in the evening, around 1930. Finally, the pooled data regarding ST-segment anomaly of regular angina indicates this is most frequent during the first few hours after the commencement of the daily activity span, while for Prinzmetal patients it is most frequent during nocturnal sleep (see Smolensky 1983; Lemmer 1989 a, b).

From a methodological point of view the use of population-derived data poses some special problems for chronoepidemiologists. First, in most cases the sleep-wake synchronizer schedule of the persons from which the data are obtained is typically unknown. Often, one must *assume* that all the data are representative of diurnally active persons. Second, in the case of elderly persons, in particular those who are likely to be under treatment with medications for a variety of chronic ailments, the interpretation of any detected temporal pattern in the occurrence of disease is confounded. One does not know if a day-night pattern in the disease arises from the type and dosing schedule of medications, the circadian time structure, itself, and/or environmental influences. In most cases, relevant information addressing these matters ordinarily is not incorporated into patient registries from which epidemiologists obtain data, thus confounding the investigation and/or interpretation of temporal patterns.

Being so strongly influenced by schooling in the prevailing theory of homeostasis, epidemiologists (as do most environmental scientists) preclude the possible contribution of endogenous biological rhythms to the etiology of human diseases. This in part explains why the focus of epidemiologic studies is exclusively upon the external environment for explanation of temporal patterns in symptoms or manifestations of illness. Although the underlying mechanisms responsible for the temporal patterns in the occurrence and signs of diseases presented in Fig. 2 have yet to be identified, the timing of the peak (between late afternoon and early evening) and trough (around mid-sleep) of the circadian rhythm of blood pressure with regard to the occurrence of peak numbers of CH and CI attacks seems more than a coincidence. Too, the circadian rhythm in ST-segment anomaly as well as MI, CI, and CH may result, in part, from the staging of various 24-h rhythms, such as those in catecholamines, cardiac physiology,

coronary artery patency, and blood coagulation (Smolensky et al. 1976; Smolensky 1983; Decousus et al. 1985; Muller et al. 1985; Haus et al. 1990; Tofler et al. 1987). These, as well as other relevant biological rhythms as discussed elsewhere in this volume, conceivably contribute to the circadian variation in the susceptibility/resistance of human beings to the manifestation and signs of human disease.

Asthma is one disease for which chronobiologic factors have been rather well researched (Smolensky et al. 1981 a; Barnes et al. 1980; Smolensky et al. 1986 a, 1986 b). Asthmatic patients exhibit marked seasonal as well as day-night differences in the occurrence of their symptoms. As detailed in another contribution to this volume, the manifestation and/or the severity of asthma in day-active patients is primarily a nocturnal event due to at least in part a host of circadian processes which affect airways patency (Reinberg et al. 1963; Prevost et al. 1980; Smolensky 1983; Turner-Warwick 1988; Dethlefsen and Repges 1985).

It is pertinent to discuss here the chronoepidemiology of so-called recurrent nocturnal asthma, which arises from the hypersensitization of the airways to antigens oftentimes experienced in the work place (Siracusa et al. 1978; Cockcroft et al. 1984; Mapp et al. 1986). Figure 3 provides an example of the phenomenon. The figure illustrates the effect of a single daytime (around 1100) exposure to a triggering agent of asthma on airways status in a former pigeon breeder who had became sensitized to an antigen excreted by the birds. Airways patency ($FEV_{1.0}$) was monitored frequently throughout a span of several consecutive days before and after the exposure study. The baseline level of airways patency is within the normal range prior to the 10-min daytime handling of the pigeons in the special study unit of the hospital. During the initial few hours after the daytime exposure, airways patency declines, indicative of a mild, immediate asthmatic reaction. However, it is not until midnight of the first night that asthma is fully manifested in terms of a precipitous drop in airways patency requiring emergency treatment with a potent beta-agonist bronchodilator medication. It is noteworthy that airways status during the daytime on the subsequent 3 days, although somewhat compromised with reference to the level observed during the baseline condition of day 1, is considerably better than it is at 0000 for each of the postexposure nights.

This example of recurrent nocturnal asthma demonstrates two issues of importance in regard to chronoepidemiologic considerations. First, a daytime exposure of some asthmatics to triggering agents may not necessarily provoke asthma immediately, but may

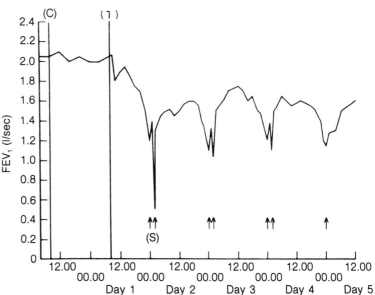

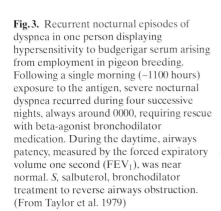

**Fig. 3.** Recurrent nocturnal episodes of dyspnea in one person displaying hypersensitivity to budgerigar serum arising from employment in pigeon breeding. Following a single morning (~1100 hours) exposure to the antigen, severe nocturnal dyspnea recurred during four successive nights, always around 0000, requiring rescue with beta-agonist bronchodilator medication. During the daytime, airways patency, measured by the forced expiratory volume one second (FEV$_1$), was near normal. *S*, salbuterol, bronchodilator treatment to reverse airways obstruction. (From Taylor et al. 1979)

do so in a very severe manner late at night, well after the exposure time. Second, this example shows that a single daytime exposure to a specific triggering agent is capable of provoking asthma, not only during the first night subsequent to the exposure, but for several consecutive nights even though there is no additional contact with the offending agent. This example illustrates the fact that the etiology of nocturnal asthma can be particularly difficult to determine by epidemiologists when it is suspected that the disease is exacerbated or provoked by antigens encountered in the ambient or work place environment during the daytime. The nighttime manifestation of asthma and perhaps other disease represents the time required for the development of the pathophysiologic processes underlying the symptoms observed as well as the programmed-in-time, 24-h alterations in the one's biology, which in turn gives rise to circadian rhythmicity in the resistance/susceptibility in the threshold for the breakthrough of symptoms or frank manifestation of disease itself.

It is important that epidemiologists and clinicians recognize that the effect of a deleterious exposure during daytime activity need not result in the symptoms of disease immediately. Moreover, it is important to note that symptoms which develop at home many hours after the termination of a work shift may be occupationally induced. Recently, chronobiologic methods have proved useful in defining the etiology of certain job-related illnesses. For example, Gervais et al. (1979), by means of employee self-measurement of bronchial patency (peak expiratory flow,

PEF) several times daily during the workweek, and vacations, demonstrated the etiology of one worker's asthma to be of occupational origin. Too, Randem et al. (1987) showed the persistence of compromised airways status due to worker sensitization to colophony in employees of an electronics industry in Great Britain. The same PEF self-assessment method also has been proven sufficiently sensitive to quantify the deleterious effect of environmental air pollutants on the airways status of asthmatic patients (Gervais et al. 1979; Smolensky et al. 1980). Oftentimes, epidemiologic surveys involving the spirometric assessment of airways function are conducted without regard to time of day. In persons without lung disease, the circadian change in airways patency is nil, amounting to about 5% of the 24-h mean level. In healthy persons the time when airways assessments are conducted by means of field surveys is unlikely to have a significant effect on the findings. In contrast, in asthmatic persons a 25%–50% circadian variation is not atypical. The airways status of diurnally active asthmatics is best around midday and afternoon and worse overnight and especially around the time of awaking (Hetzel and Clark 1980; Smolensky and Halberg 1977; Smolensky et al. 1986b). Thus, in air pollution surveys which focus on asthmatic participants, the time when assessments of airways status are conducted with reference to the circadian staging will greatly influence the findings. For this reason, many chronobiologists (see Gervais et al. 1977) recommend the use of ambulatory peak flow meters several times daily to quantify the effect of ambient air quality by the rhythm endpoints of mesor (24-h mean), amplitude, and acrophase using cosinor or other time series analyses.

## Chronoepidemiology, Sexually Deviant Behavior, Sexually Transmitted Disease and Congenital Malformations

Seasonal variation in human reproduction and reproductive physiology has long been of interest to scientists of many disciplines (see Smolensky and Sargent 1972). As shown in Fig. 4 (Smolensky et al. 1981b), the circannual rhythm in human birth peaks during the later part of summer. The staging of the circannual rhythm in human sexual activity (using data from couples or unmarried males or by examining data on rape) is greatest between late summer and early autumn and appears to be closely associated in time with the circannual peak of serum testosterone (Reinberg and Lagoguey 1978). Although the occurrence of menarche in young girls (Engle and Shelesnyak 1934; Grimm 1952; Bojlen and Bentzon 1974; Smolensky 1983; Albright et al. 1990) is more common in the autumn than in other seasons, little as yet is known about seasonal variation in the reproductive physiology and endocrinology of adult women.

Even though a satisfactory explanation for the seasonal variation in human birth continues to be explored (Roenneberg and Aschoff 1990a, 1990b), the information summarized in Fig. 4 appears to be enlightening for more pertinent reasons. From a chronobiological perspective the seasonal pattern of rape, an unacceptable behavior toward women by men, being greatest during the summer and autumn and coinciding closely with the circannual peak of serum testosterone and sexual activity of males and couples, suggests a hypothesis which could include an endogenous biological component as discussed in greater detail elsewhere (Smolensky et al. 1981b; Bicakova-Rocher et al. 1985). Admittedly, this perspective is at odds with the prevailing behavioral hypothesis that views rape to be a crime of pure violence. Although speculation at this time, one wonders if the epidemiology of this unacceptable behavior toward women by men when timed during the summer and autumn months includes the chronobiologically modulated factors of heightened sexual and/or aggressive drive related to elevated testosterone levels at this time of the year. Too, one wonders if rapes which are committed during the spring have a different etiology, mainly one related to criminal violence per se. In this regard, it seems pertinent that a circannual rhythm in violent crime has been documented with a springtime peak (Smolensky et al. 1981b). Although these circumstantial facts and speculations are of curious interest, they do not unquestioningly define the epidemiology and etiology of rape; however, the circannual time structure of males does suggest that the seasonality of rape may not be simply a response to heat and the wearing of revealing apparel per se during the summer (Michael and Zumpe 1983).

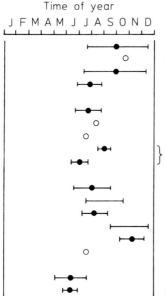

**Fig. 4.** Circannual acrophase (peak time, ∅, with respect to month of the year) along with 95% confidence intervals of variables thought to be directly or indirectly associated with or related to the circannual pattern of human sexual activity. *T*, trichomonas; *G*, gonorrhea; *PS*, primary syphilis; ⊢●⊣, annual acrophase and 95% confidence limits, cosinor method; ○ and ⊢●⊣, annual peak time or peak time zone in reported raw data. (From Smolensky et al. 1981b)

| Variables | Time of year<br>J F M A M J J A S O N D | Authors |
|---|---|---|
| Pl. testosterone | | Reinberg & Lagoguey (1975) |
| Pl. testosterone | | Smals (1976) |
| Sexual ⎰ young males | | Reinberg & Lagoguey (1978) |
| outlet ⎱ couples | | Udry, Morris (1967) |
| Reported rapes | | |
| Great Britain 1880-4 | | Leffingwell (1892) |
| USA Oakland 1970 | | Amir (1971) |
| USA Memphis 1973 | | Brown (1974) |
| USA Houston 1974-5 | | ⎫ |
| France, Paris 1973-8 | | ⎬ This study |
| Sex. trans. diseases | | |
| W. Germany (T) | | Hildebrandt (1962) |
| USA Denver (G) | | Wright & Judson (1978) |
| USA Houston (G) | | This study |
| USA Denver (P.S.) | | Wright & Judson (1978) |
| USA Houston (P.S.) | | This study |
| Sales of contraceptives | | Parkes (1968) |
| Human birth | | |
| Belgium | | van Cauter (1973) |
| USA | | Batschelet (1973) |

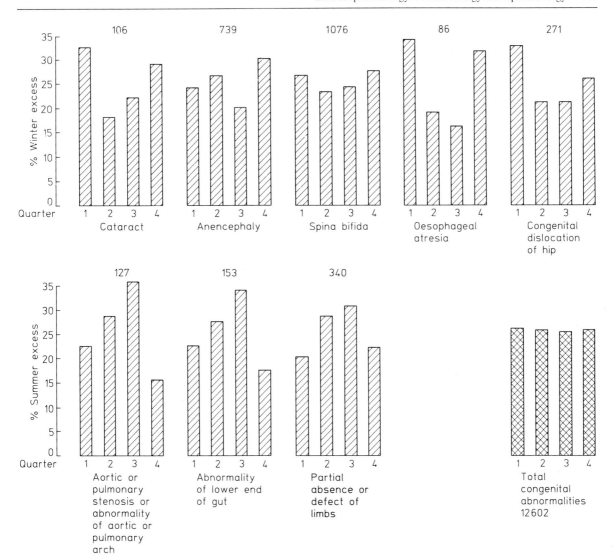

**Fig. 5.** Quarterly distribution of births associated with congenital malformations. The *upper* portion of the figure presents those congenital malformations which are more common in winter births than in summer births. The *lower* portion of the figure depicts those which are more common in births during the summer. (From Slater et al. 1964)

The circannual rhythm in human sexual activity may have relevance from another perspective – the epidemiology of the seasonality of sexually transmitted diseases (STD). Figure 6 (Smolensky 1983) shows that the occurrence of reported cases of gonorrhea is circannually variable. Regarding gonorrhea, for which the symptoms are rather easily noted and immediate in development, the peak occurs late in summer or early in autumn, not too long delayed from the circannual peak in sexual activity as reported by several investigators (Fig. 4). Although the seasonality of this STD could represent seasonal differences in the reporting of cases, this hypothesis seems untenable. The circannual staging of this primarily adult infectious disease differs from that of several mainly childhood infectious diseases – mumps, chicken pox, rubella, and rubeola – which are more frequent during the spring (Smolensky 1983). The circannual peak of these morbidity rhythms differs somewhat from that of upper respiratory infections, cold, and influenza, which occurs during late winter, both in children and adults (Smolensky and Sargent 1972; Smolensky 1983). While many epidemiologists cite the importance of the crowding of children in school as a major factor in the timing of the peak numbers of infectious diseases in this age group (Lilienfeld and Lilienfeld 1980), this explanation may be incomplete. Since in the United

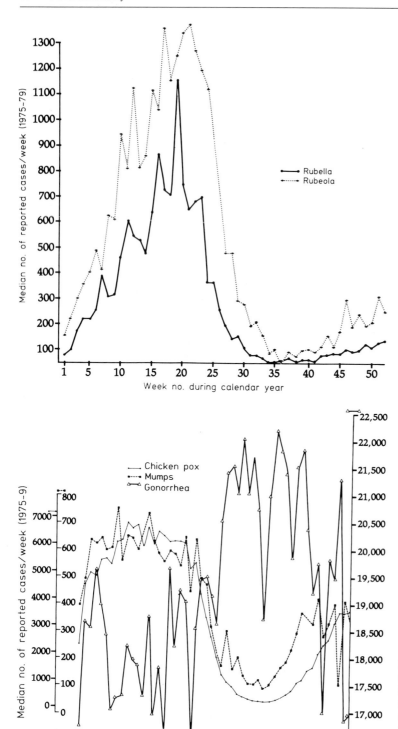

**Fig.6.** Circannual rhythms in adult and childhood infectious diseases. Using data from the United States, chicken pox and mumps in populations of youngsters are most frequent during the spring or early summer. In contrast, the occurrence of the primarily adult infectious disease, gonorrhea, is most frequent during the autumn. (From Smolensky 1983)

States school attendance resumes in the autumn of each year after summertime vacation, it is difficult to understand why the epidemic of these infectious diseases does not occur earlier during the fall. Undoubtedly, seasonal differences in environmental conditions conducive to the virulence and spread of infectious agents are critical; however, since biological rhythms in human immune function and in immunosurveillance have been identified (Lu et al. 1980; Pownall et al. 1979; Bratescu and Teodorescu 1981; Shifrine et al. 1982; MacMurry et al. 1983; Levi et al. 1984, 1985, 1989), it appears that the etiology of these seasonal patterns in infections is multifactorial, involving both exogenous and endogenous cyclic phenomena.

A number of epidemiologists have been concerned with the etiology of congenital malformations. In this regard, seasonal patterns have been identified in cataract (Slater et al. 1964), anencephaly (McKeown and Record 1951; Edwards 1961; Slater et al. 1964), spina bifida (Slater et al. 1964), coarctation of the aorta (Miettiner et al. 1970), ventricular septal defect (Rothman and Flyer 1974; Rosenberg and Heinonen 1974; Rothman and Flyer 1976), and congenital dislocation of the hip (Cohen 1970; Cohen 1971; Slater et al. 1964), to mention just a few. The data obtained by the epidemiologic study of Slater and colleagues (1964) reveal that while the total number of congenital malformations is rather constant over the months of the year, the distribution of the specific malformations does vary according to the season of birth. As shown in Fig. 5, the congenital malformations of cataract, anencephaly, spina bifida, esophageal atresia, and dislocation of the hip are more common in births occurring during the winter than the summer (Slater et al. 1964). In contrast, aortic and pulmonary stenosis, abnormalities of the aorta, pulmonary arch, and gastrointestinal tract, and limb defects are more common in births occurring during the summer than the winter (Slater et al. 1964).

The majority of the epidemiologists have hypothesized that the seasonality of congenital malformations represents the seasonality of certain common viral infections as contracted during pregnancy (Edwards 1961; Slater et al. 1964). However, some (Stoller and Collmann 1965; Cohen 1971; Smolensky 1983) suggest that circannual differences in gestational physiology may well be contributory. While it is well known that biological and endocrine rhythms of high amplitude characterize the physicochemical processes of the body, there is no evidence as yet that these are directly related to the etiology of the described seasonal differences in specific congenital malformations. Thus, in spite of the fact that circannual variation in teratospermia (percent abnormal sperm/total sperm count) and necrospermia (percent nonliving sperm/total sperm count) (Pinatel et al. 1981, 1982) is known, the nature (endogenous circannual rhythmicities and/or exogenous cycles of infections) of the causal factors involved in the etiology of seasonal patterns in congenital abnormalities remains to be thoroughly researched.

## Chronoepidemiology and Community-Based Medication Trials

Chronopharmacology is the study of biological rhythm influences on medications (Reinberg et al. 1986). The pharmacokinetics, metabolism, and desired and undesired effects of medications can vary markedly according to their timing in relation to rhythms during 24 h and perhaps other time scales (Reinberg and Smolenksy 1982; Reinberg 1983; Reinberg et al. 1986). In certain cases the importance of rhythms can be so great that *when* a given medication is timed determines whether it is effective or not and/or whether it is dangerous or safe! This is true for many classes and types of medications and whether delivered by means of oral tablets, capsules, or elixirs or by other routes including a parenteral one. Trials involving the epidemiology of drug safety and efficacy are based on homeostatic tenets. In other words, the timing of medications is defined in terms of the physician's insert regarding their recommended scheduling to maintain constant drug levels in the blood. Typically, the epidemiologists involved in community drug trials have no knowledge or recognition of information pertaining to the chronopharmacology of the medications under study. Moreover, the biological endpoints and measurements selected for evaluating the effects of the medications are believed to be nonvarying over 24 h such that the time of conducting tests or measurements to assess efficacy is not considered of importance in the epidemiologic trials.

Such assumptions are unfounded based on the tremendous amount of reference information regarding the chronopharmacology of medications and also the circadian features of human biological function. For example, rather large scale studies have been conducted on the administration-time dependency of the safety and efficacy of nonsteroidal anti-inflammatory medications for the management of arthritic disease. The timing of such medications as a once-daily administration has been shown to differ greatly according to when it is taken, in the morning upon arising versus before bedtime (Labrecque and Reinberg 1989). Information of this nature is known from trials involving antihistamines for allergic rhinitis, bronchodilators for asthma, synthetic corticosteroids for inflammatory diseases, $H_2$-receptor antagonist drugs for the treatment of ulcer disease, and antitumor agents used to treat human cancers to mention a few (Reinberg 1989; Reinberg et al. 1988; Smolensky et al. 1986a, 1987; Levi 1987; Hrushesky 1985, 1991; Merki et al. 1987).

Today, in the United States and other nations very large epidemiologic drug trials involving antihypertensive medications are in progress. As of this writing, none have incorporated a chronopharmacologic perspective into the study methods. Failure to evaluate the role of drug administration schedule in relation to circadian time and the role of the latter on assessment of drug effect in these epidemiologic investigations is likely to result in inaccurate or even misleading findings. One can only hope that future epidemiologic investigations involving medical interventions will entail chronobiologic principles and

concepts to achieve a much more thorough and accurate assessment of the medication under evaluation.

## Discussion

Herein, the topic of chronobiology has been discussed in relation to common epidemiologic concerns. Chronoepidemiology has been discussed with regard to several human illnesses, reproductive behavior, and rape as well as the conduct of community drug trials. The design of epidemiologic studies must take into consideration chronobiologic concepts and methods. The latter entails as a minimum: (1) the knowledge of the sleep-activity schedule of the persons under evaluation; usually most people are active between 0600 to 2300 hours and at rest from 2300 to 0600, (2) the use of instrumentation and/or visual analogue scales to obtain data at frequent intervals at least during the waking span to supplement, for example, the clinical evaluation of symptoms and

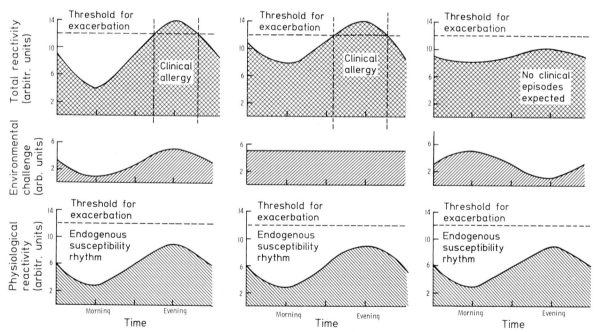

**Fig. 7.** Interaction of the circadian susceptibility/resistance rhythm of the airways of asthmatic patients to illustrated cyclic or noncyclic changes in air quality over 24 h constituting a challenge to the well-being (asthma) of human populations. The susceptibility/resistance rhythm of asthmatic persons to asthma (dyspnea) is based on the documented circadian vulnerability of the airways both to chemical agents and antigenic substances.

The considered environmental patterns in the concentration of air pollutants is based on documented findings of air quality investigations. The likelihood of dyspnea varies according to the level of air pollution to which one is exposed and *when* the exposure occurs relative to the endogenously changing circadian difference in airways vulnerability (see text)

signs of a given disease or drug effects, and (3) the analysis of the resulting data not only by statistical techniques common to epidemiology but when appropriate by quantitative time series analyses.

In this paper, the discussion of chronobiology and epidemiology has centered around the concept of susceptibility/resistance rhythms. Most epidemiologists presume the homeostatic model in devising epidemiologic investigations (Fig. 1); however, a chronobiologic model seems more appropriate in most instances. To illustrate this concept, Fig. 7 presents as an example the circadian rhythm in the physiologic circadian vulnerability of the airways of asthmatic patients to the environmental challenge of air pollutants (Halberg et al. 1977; Smolensky 1983; Gervais et al. 1979). The figure defines three possibilities. The one shown to the left exemplifies the situation that an environmental challenge exhibits cyclic variation of 24 h with its staging being identical to that of the endogenous circadian rhythm in airways susceptibility as verified by several patient studies entailing selected agents (Gervais et al. 1977, Reinberg et al. 1971; DeVries et al. 1962; Sly and Landau 1986). Under this condition, when the challenge is moderate the threshold for dyspnea (asthma) is exceeded only if and when exposure occurs during the evening. During the daytime the biological hyperreactivity of the airways is reduced as is the magnitude of environmental challenge resulting in a very low probability of dyspnea. The example illustrated in the middle depicts the situation in which the environmental challenge is nonvarying over 24 h. Again, only during a relatively short interval of time, during the evening, does the exposure to the environmental agent result in dyspnea. Finally, the example depicted to the right presents the situation in which the environmental challenge is cyclic but differing by 12 h in its staging from the circadian rhythm in airways susceptibility. When the challenge is moderate, dyspnea is unlikely at anytime during the 24 h.

The concept of susceptibility/resistance rhythms can be extended to other scales of time, such as the week, menstrual cycle, or year, for particular biological indicators of disease, effects of medications (positive or negative), industrial accidents (Novak et al. 1990) and air pollutants, among others. With regard to human morbidity and mortality, the extent to which an environmental challenge is harmful to human health and well-being depends on many variables, one of which appears to be *when* it occurs with regard to the staging of certain critical rhythms.

The concept of susceptibility/resistance rhythmicity seems to add a new dimension to the epidemiologic concept of relative risk (Ahlbom and Norell 1984; Lilienfeld and Lilienfeld 1980). For example, in epidemiology, relative risk refers to the occurrence of disease in one population relative to another *at a given point in time*. Data on relative risk motivates investigative efforts to elucidate the difference between populations and/or to initiate meaningful interventions. Since many diseases in susceptible or exposed populations exhibit high-amplitude temporal patterns *over specific time domains*, e.g., 24 h and year, the meaning of relative risk to most chronobiologists is different than it is to epidemiologists. Chronobiologists commonly compare the occurrence of an illness at different times, such as during 24 h, a year, or stages of the menstrual cycle. This has been done, for example, for the occurrence of asthma and cardiovascular diseases during 24 h and a year, to mention but a few examples (Smolensky and Sargent 1972; Smolensky 1983). The evaluation of the relative risk of populations both in terms of epidemiologic criteria at a specific point in time and in terms of chronobiologic criteria will provide a more complete perspective about associated or causal factors underlying the disease itself and its periodicity.

## Conclusion

Epidemiology and chronobiology are two emerging scientific disciplines of the twentieth century. In the United States and certain European and Asian countries, epidemiology is a well-developed science; however, in many countries this field is not yet well practiced. In several European countries and certain centers in the United States, chronobiology is quite well known and respected as a science. Yet, in many academic centers this field is just now being recognized. Epidemiologists are concerned with temporal patterns in illnesses, primarily in terms of the prevalence and incidence of human disease. For the most part, epidemiologists do not consider biological rhythmic phenomena relevant to their pursuits. The symptoms, signs, and tests which document illness and determine prevalence and incidence all may be influenced by *when* evaluations are conducted with respect to the staging of circadian and other period rhythms. Chronobiologists, on the other hand, too often have been overly concerned with the elucidation and quantification of human bioperiodicities without regard to their relevance to the etiology of temporal patterns in the occurrence or exacerbation of illness, matters of particular importance to epidemiologists. Herein, the topic of epidemiology and chronobiology has been discussed in terms of the relevance of each field to the other. The concept of susceptibility/resistance rhythmicities, with respect to patterns in illness, medication effects, and behavior, is central to elucidating the etiology of observed phenomena in terms of cyclic exogenous and endogenous factors.

# References

Ahlbom A, Norell S (1984) Introduction to modern epidemiology. Epidemiology Resources, Inc, Chestnut Hill

Albright D, Voda A, Smolensky MH, Hsi B, Decker M (1990) Seasonal characteristics of and age at menarche. Chronobiol Int 7: 251–258

Arendt J, Minors DS, Waterhouse JM (1989) Biological rhythms in clinical practice. Wright, London

Barcal R, Sova J, Krizanovska M, Levy J, Matousek J (1968) Genetic background of circadian rhythms. Nature 220: 1128–1131

Barnes PJ, FitzGerald G, Brown M, Dollery C (1980) Nocturnal asthma and changes in circulating epinephrine, histamine and cortisol. N Engl J Med 303: 263–267

Barker DJP (1982) Practical epidemiology. Churchill Livingstone, Edinburgh

Bartter FC, Delea CS, Baker W, Halberg F, Lee Jung-Juen (1976) Chronobiology in the diagnosis and treatment of mesor-hypertension. Chronobiologia 3: 199–213

Bicakova-Rocher A, Smolensky MH, Reinberg A, De Prins J (1985) Seasonal variations in socially and legally unacceptable sexual behavior. Chronobiol Int 2: 203–208

Bojlen K, Bentzon MW (1974) Seasonal variation in the occurrence of menarche. Dan Med Bull 21: 116–168

Bratescu A, Teodorescu M (1981) Circannual variation in the B cell/T cell ratio in normal human peripheral blood. J Allergy Clin Immunol 68: 273–280

Cockcroft DW, Hoeppner VH, Werner GD (1984) Recurrent nocturnal asthma after bronchoprovocation with Western Red Cedar sawdust: association with acute increase in nonallergic bronchial responsiveness. Clin Allergy 14: 61–68

Cohen P (1970) Seasonal variations of congenital dislocation of the hip. J Interdisp Cycle Res 2: 417–425

Cohen P (1971) Seasonal variation of congenital malformation. J Interdiscipl Cycle Res 3: 271-274

Decousus HA, Croze M, Levi FA, Perpoint B, Jaubert J, Bormadonna JF, Reinberg A, Queneau P (1985) Circadian changes in anti-coagulant effect of heparin infused at a constant rate. Br Med J 290: 341–344

De Prins J, Cornelissen G, Malbecq W (1986) Statistical procedures in chronobiology and chronopharmacology. Annu Rev Chronopharmacol 2: 27–141

Dethlefsen U, Repges R (1985) Ein neues Therapieprinzip bei nächtlichem Asthma. Klin Med: 44–47

Devries K, Goei JT, Booy-Noord H, Orie NG (1962) Changes during 24 hours in the lung function and histamine hyperreactivity of the bronchial tree in asthmatic and bronchitic patients. Int Arch Allergy 20: 93–101

Edwards JH (1961) Seasonal incidence of congenital disease in Birmingham. Ann Hum Genet 25: 89–93

Engle ET, Shelesnyak MG (1934) First menstruation and subsequent menstrual cycles of pubertal girls. Human Biol 6: 431–453

Gervais P, Reinberg A, Gervais C, De France O, Smolensky MH (1977) Twenty-four-hour rhythm in bronchial hyperreactivity to house dust in asthmatics. J Clin Allergy Immunol 59: 207–214

Gervais P, Reinberg A, Fraboulet G, Abulker C, Vignaud D, Delcourt MER (1979) Circadian changes in peak expiratory flow of subjects suffering from allergic asthma documented in areas of high and low air pollution. In: Reinberg A, Halberg F (eds) Chronopharmacology. Pergamon, Oxford, pp 203–212

Grimm H (1952) Über jahreszeitliche Schwankungen in Eintritt der Menarche. Zentralbl Gynäkol 74: 1577–1581

Halberg F (1960) Temporal coordination of physiologic function. Cold Spring Harb Symp Quant Biol 25: 289–310

Halberg F (1969) Chronobiology. Annu Rev Physiol 31: 675–725

Halberg F, Simpson H (1967) Circadian acrophases of human 17-hydroxycorticosteroid excretion referred to midsleep rather than midnight. Hum Biol 39: 405–413

Halberg G, Johnson EA, Nelson W, Runge W, Sothern R (1972) Autorhythmometry. Procedures for physiologic self-measurements and their analysis. Physiol Teacher 1: 1–11

Halberg F, Reinberg A, Reinberg Agnes (1977) Chronobiologic serial sections gauge circadian rhythm adjustments following transmeridian flight and life in a novel environment. Waking and Sleeping 1: 259–279

Haus E, Lakatua DJ, Sackett-Lundeen LL, Swoyer J (1984) Chronobiology in laboratory medicine. In: Reitveld WJ (ed) Clinical aspects of chronobiology. Cip-Gegevens Koninklijke Bibliotheek, The Haag, pp 13–84

Haus E, Cusulos M, Sackett-Lundeen L, Swoyer J (1990) Circadian variations in blood coagulation parameters, alpha-antitrypsin antigen and platelet aggregation and retention in clinically healthy subjects. Chronobiol Int 7: 203–216

Hetzel MR, Clark TJH (1980) Comparison of normal and asthmatic circadian rhythms in peak expiratory flow rate. Thorax 35: 732–738

Hrushesky WJM (1985) Circadian timing of cancer chemotherapy. Science 228: 73–75

Hrushesky WJM (1991) The multifrequency (circadian, fertility cycle and season) balance between host and cancer. In: Hrushesky WJM, Langer R, Theeuwes F (eds) Temporal control of drug delivery. Ann NY Acad Sci 618: 228–256

Johnson LC, Tepas DI, Colquhoun WP, Colligan MJ (eds) (1981) Biological rhythms, sleep and shift work. SP Medical and Scientific Books, New York (Advances in sleep research, vol 7)

Konopka RJ, Beuzer S (1971) Clock mutants of drosophila melanogaster. Proc Natl Acad Sci USA 58: 2112–2116

Kowanko IC, Knapp MS, Pownall R, Swannell AJ (1982) Domiciliary self-measurement in rheumatoid arthritis and the demonstration of circadian rhythmicity. Ann Rheum Dis 41: 453–455

Kuroiwa A (1978) Symptomatology of variant angina. Jpn Circ J 42: 459–476

Labrecque G, Reinberg AE (1989) Chronopharmacology of non-steroid anti-inflammatory durgs. In: Lemmer B (ed) Chronopharmacology. Dekker, New York, pp 545–579

Lemmer B (1989 a) Circadian rhythms in the cardiovascular

system. In: Arendt J, Minors D, Waterhouse J (eds) Biological rhythms in clinical practice. Wright, London, pp 51–70

Lemmer B (1989 b) Temporal aspects of the cardiovascular system in humans. In: Lemmer B (ed) Chronopharmacology. Dekker, New York, pp 525–541

Levi F (1987) Chronobiology in oncology. Pathol Biol (Paris) 35: 960–968

Levi F, Cannon C, Blum JP, Misset JL, Mechkouri M, Bennaceur M, Mathe G (1984) Circadian rhythms in 6 circulating lymphocyte subtypes in healthy men: 100-fold variation may be physiologc. Ann Rev Chronopharmacol 1: 131

Levi F, Cannon C, Blum J-P, Mechkouri M, Reinberg A, Mathe G (1985) Circadian and/or circahemidian rhythms in nine lymphocyte-related variables from peripheral blood of healthy subjecs. J Immunol 134: 217–222

Levi F, Reinberg A, Cannon C (1989) Clinical immunology and allergy. In: Arendt J, Minors D, Waterhouse JM (eds) Biological rhythms in clinical practice. Wrigth, London, pp 99–135

Lilienfeld AM, Lilienfeld DE (1980) Foundations of epidemiology. Oxford University Press, New York

Lu M, Smolensky MH, Hsi B, McGovern JP (1980) Seasonal changes in immunoglobulin and complement levels in atopic and non-atopic persons. In: Reinberg A, Smolensky MH, McGovern JP (eds) Recent advances in the chronobiology of allergy and immunology. Pergamon, New York, pp 261–273

MacMurry JP, Barker JP, Armstrong JD, Bozzetti LP, Kuhn IN (1983) Circannual changes in immune function. Life Sci 32: 2363–2370

Mapp CE, Digiacomo GR, Ominic et al. (1986) Late, but not early, asthmatic reactions induced by toluene-diisocyanate are associated with increased airways responsiveness to methacholine. Eur J Respir Dis 69: 276–284

Marshall J (1977) Diurnal variation in occurrence of strokes. Stroke 8: 230–231

Master AM (1960) The role of effort and occupation (including physicians) in coronary occlusion. JAMA 174: 942–948

McKeown T, Record RG (1951) Seasonal incidence of congenital malformations of the central nervous system. Lancet I: 192–196

Merki H, Witzel L, Hane K, Scheule E, Neumann J, Rohmel J (1987) Single dose treatment with $H_2$-receptor antagonists: Is bedtime administration too late? Gut 28: 451–457

Michael RP, Zumpe D (1983) Sexual violence in the United States and the role of season. Am J Psychiatry 140: 883–886

Miettiner OS, Reiner ML, Nadas AS (1970) Seasonal incidence of coarctation of the aorta. Br Heart J 32: 103–107

Moore-Ede MC, Sulzman FM, Fuller MH (1982) The clocks that time us. Harvard University Press, Cambridge

Muller JE, Stone PH, Turi ZG, Rutherford JD, Czeisler CA, Partker C, Poole K, Passamani E, Roberts R, Robertson T, Sobel BE, Willerson JT, Braunwald E, Miles Study Group (1985) Circadian variation in the frequency of onset of acute myocardial infarction. N Engl J Med 313: 1315–1322

Novak RD, Smolensky MH, Fairchild EJ, Reves RR (1990) Shiftwork and industrial injuries at a chemical plant in southeast Texas. Chronobiol Int 7: 155–164

Pinatel MC, Souchier C, Croze JP, Czyba JC (1981) Seasonal variation of necrospermia in man. J Interdiscipl Cycle Res 12: 225–235

Pinatel MC, Czyba JC, Souchier C (1982) Seasonal changes in sexual hormones secretion, sexual behavior and sperm production in man. Int J Androl Suppl 5: 183–190

Pownall R, Kabler PA, Knapp MS (1979) The time of day of

antigen encounter influences the magnitude of the immune response. Clin Exp Immunol 36: 347–354

Prevost RJ, Smolensky MH, Reinberg A, Raymer WJ, McGovern JP (1980) Circadian rhythm of respiratory distress in asthmatic, bronchitic and emphysemic patients. In: Reinberg A, Smolensky MH, McGovern JP (eds) Recent advances in the chronobiology of allergy and immunology. Pergamon, New York, pp 237–250

Randem B, Smolensky MH, Hsi B, Albright D, Burge S (1987) Field survey of circadian rhythm in PEF of electronics workers suffering from colophony-induced asthma. Chronobiol Int 4: 263–272

Reinberg A (1979) Chronobiological field studies of oil refinery shift-workers. Chronobiologia 6 (Suppl 1)

Reinberg A (1983) Clinical chronopharmacology: an experimental basis for chronotherapy. In: Reinberg A, Smolensky MH (eds) Biological rhythms and medicine. Springer, Berlin Heidelberg New York, pp 211–263

Reinberg A (1989) Chronopharmacology of corticosteroids and ACTH. In: Lemmer B (ed) Chronopharmacology. Dekker, New York, pp 137–167

Reinberg A, Lagoguey M (1978) Circadian and circannual rhythms in sexual activity and plasma hormones (FSH, LH, testosterone) of five human males. Arch Sex Behav 7: 13–30

Reinberg A, Smolensky MH (1982) Circadian changes in drug disposition in man. J Clin Pharmacokin 7: 401–420

Reinberg A, Smolensky MH (1983 a) Biological rhythms and medicine. Springer, Berlin Heidelberg New York

Reinberg A, Smolensky MH (1983 b) Investigative methodology for chronobiology. In: Reinberg A, Smolensky MH (eds) Biological rhythms and medicine. Springer, Berlin Heidelberg New York, pp 23–46

Reinberg A, Ghata J, Sidi E (1963) Nocturnal asthma attacks; their relationship to the circadian adrenal cycle. J Allergy 34: 323–330

Reinberg A, Gervais P, Morin M, Abulker C (1971) Rythme circadien humain du seuil de la response bronchique a l'acetylcholine. C R Acad Sci 272: 1879–1881

Reinberg A, Touitou Y, Restoin A, Migraine C, Levi F, Montagner H (1985) The genetic background of circadian and ultradian rhythm patterns of 17-hydroxycorticosteroids: a cross-twin study. J Endocrinol 105: 247–253

Reinberg A, Smolensky MH, Labrecque G (1986) New aspects in chronopharmacology. Ann Rev Chronopharmacol 2: 3–26

Reinberg A, Gervais P, Levi F, Smolensky MH, Del Cerro L, Ugolini C (1988) Circadian and circannual rhythms of allergic rhinits. A chronoepidemiology study. J Allergy Clin Immunol 81: 51–62

Roenneberg T, Aschoff J (1990 a) Annual rhythm of human reproduction. I. Biology, sociology, or both? J Biol Rhythms 5: 195–216

Roenneberg T, Aschoff J (1990 b) Annual rhythm of human reproduction. II. Environmental correlations. J Biol Rhythms 5: 217–240

Rosenberg LA, Heinonen OP (1974) Seasonal occurrence of ventricular septal defect. Lancet II: 903–904

Rothman KJ, Fyler DC (1974) Seasonal occurrence of complex ventricular septal defect. Lancet II: 193–197

Rothman KJ, Flyer DC (1976) Association of congenital heart defects with season and population density. Teratology 13: 29–34

Shifrine M, Garsd A, Rosenblatt LS (1982) Seasonal variation in immunuty of humans. J Interdiscipl Cycle Res 13: 157–165

Siracusa A, Curradi F, Abbritti G (1978) Recurrent nocturnal asthma due to toluene di-isocyanate: a case report. Clin Allergy 8: 195–201

Slater BCS, Watson GI, McDonald JC (1964) Seasonal variation in congenital abnormalities. Preliminary report of a survey conducted by the research committee of the council of the college of general practitioners. Br J Prev Soc Med 18: 1–7

Slatis HM, Decloux RJ (1967) Seasonal variation in stillbirth frequencies. Hum Biol 39: 284–294

Sly PD, Landau LI (1986) Diurnal variation in bronchial responsiveness in asthmatic children. Pediatr Pulmonol 2: 344–352

Smolensky MH (1983) Aspects of human chronopathology. In: Reinberg A, Smolensky MH (eds) Biological rhythms and medicine. Springer, Berlin Heidelberg New York, pp 131–209

Smolensky MH (1984) Establishing time-qualified reference values for Spacelab 1, hematology experiment INS 103. Report for NASA Contract NAS 9-16911, Houston, Texas

Smolensky MH, Halberg F (1977) Circadian rhythm in airway patency and lung volumes. In: Reinberg A, Smolensky MH, McGovern JP (eds) Chronobiology in allergy and immunology. Thomas, Springfield, pp 117–138

Smolensky MH, Sargent FS II (1972) Chronobiology of the life sequence. In: Itoh S, Ogata K, Yoshimura H (eds) Advances in climatic physiology. Igaku Shoin, Tokyo, pp 281–318

Smolensky MH, Tatar SE, Bergman SA, Losman JG, Barnard CN, Cacso CC, Kraft IA (1976) Circadian rhythmic aspects of human cardiovascular function: a review by chronobiologic statistical methods. Chronobiologia 3: 337–371

Smolensky MH, Reinberg A, Prevost RJ, McGovern JP, Gervais P (1980) The application of chronobiological findings and methods to the epidemiological investigations of the health effects of air pollutants on sentinel patients. In: Reinberg A, Smolensky MH, McGovern JP (eds) Recent advances in the chronobiology of allergy and immunology. Pergamon, New York, pp 211–236

Smolensky MH, Reinberg A, Queng J (1981 a) Chronobiology and chronopharmacology of allergy. Ann Allergy 47: 234–252

Smolensky MH, Bicakova-Rocher A, Reinberg A, Sanford J (1981 b) Chronoepidemiological search for circannual changes in sexual activity of human males. Chronobiologia 8: 217–230

Smolensky MH, Scott PH, Barnes PJ, Jonkman JHG (1986 a) The chronopharmacology and chronotherapy of asthma. Annu Rev Chronopharmacol 2: 229–273

Smolensky MH, Barnes PJ, Reinberg A, McGovern JP (1986 b) Chronobiology and asthma. I. Day-night differences in bronchial patency and dyspnea and circadian rhythm dependencies. J Asthma 23: 321–343

Smolensky MH, McGovern JP, Scott PH, Reinberg A (1987) Chronobiology and asthma: II. Body-time-dependent differences in the kinetics and effects of bronchodilator medications. J Asthma 24: 91–123

Stoller A, Collmann RD (1965) Incidence of infective hepatitis followed by Down's syndrome nine months later. Lancet I: 1221–1223

Taylor AM, Davies RJ, Hendrick DJ, Pepys J (1979) Recurrent nocturnal asthmatic reactions to bronchial provocation tests. Clin Allergy 9: 213–219

Tofler GH, Brezinski D, Schafter AI, Czeisler CA, Rutherford JD (1987) Concurrent morning increase in platelet aggregability and the risk of myocardial infarction and sudden death. N Engl J Med 316: 1514–1516

Turner-Warwick M (1988) Epidemiology of nocturnal asthma. Am J Med 85 (Suppl 1 B): 6–8

# Chronobiology in Laboratory Medicine

E. Haus and Y. Touitou

## Introduction

The multifrequency human time structure represents a challenge for sampling and interpreting laboratory measurements. However, it also represents an opportunity for refining the diagnostic value of the measurement of many variables showing high amplitude rhythms and opens a new field in laboratory diagnosis. Statistically quantified rhythm parameters, and their relation to astronomic time and to each other, can serve as new end points in defining normality and in recognizing deviations from time-qualified reference values, including reference values to be established for the rhythm parameters themselves.

A thorough knowledge of the principles of chronobiology has become essential for data collection and interpretation of laboratory values in order to provide meaningful information for the clinical questions asked. The temporal behavior of the human organism has to be kept in mind in preparing the patient for sample collection, in the collection process itself, in the sampling procedure, and in the timing, the number, and the intervals of samples to be collected. Most laboratory parameters are subject to rhythmic variations in not one but several frequency ranges and the time(s) of sampling and the interpretation of the results have to be adjusted accordingly. A physiologic measurement (or laboratory value) which represents the result of a spotcheck, e.g., a blood drawing, may in part be determined by the interaction of several biologic rhythms in different frequencies, by trends occurring during a lifetime, and by the effects of random and nonrandom environmental stimuli acting all upon the same parameter. The result of a measurement obtained at one astronomic time may, therefore, represent an entirely different functional state of the organism than an identical result of the same parameter obtained at another time, depending on the stage of one or more rhythmic functions at the moment of sampling.

## Chronobiologic Reference Values

In the establishment of chronobiologic reference values, numerous factors of biologic and environmental variation have to be considered which are similar to those for reference values in laboratory medicine in general (Solberg 1987; Petit Clerc and Solberg 1987). However, some of these factors are especially important in regard to chronobiologic investigations since they may alter the rhythm under study. Also, many of the environmental factors act differently depending on the stage of the rhythm in which they are applied (Haus 1964; Haus et al. 1974a,b; Moore-Ede et al. 1983; Wever 1983).

Laboratory measurements for chronobiologic observations have to be obtained either at a certain defined biologic time of the individual or, preferably, have to be adequate in sample density and length of the sampling span to provide statistically meaningful rhythm characteristics and their variance estimate (Haus 1987). After these laboratory measurements have been obtained, they have to be compared with pertinent chronobiologic reference values derived from clinically healthy subjects comparable in their population characteristics with the subjects or patients to be evaluated and obtained under comparable conditions. Time-qualified reference ranges, so-called chronodesms (Halberg et al. 1978), can be developed for single individuals by repeated measurement of the same subject over numerous periods or they can be determined for groups of subjects by measurements over a single or a limited number of periods. While longitudinal observations and individual chronodesms are preferable for studying a single individual, especially trends seen during aging, such studies require a long-term follow-up of an individual which is seldom feasible. In using peer groups for the establishment of a group chronodesm, the choice of the peer population and the conditions of the study will determine the validity of the chronodesm for a given individual or a group of subjects.

## Collection of Data
## for Time-Qualified Reference Values

Factors to be taken into account and specified when establishing reference values in variables showing rhythmic variations in one or several frequency ranges are described below.

### Characteristics of the Reference Population

The characteristics of the reference population have to be defined in respect to age, sex, weight, ethnic-geographic background, climatic conditions, eating habits, time of rising and retiring, and, if feasible, as to social background, lifestyle, and economic factors.

The definition of health in a reference population entails several problems. No definition of health appears to be completely satisfactory, including the definition presented in the constitution of the World Health Organization describing health as "a state of complete physical, mental, and social well-being and not merely the absence of disease or infirmity." Health is conceptually different in different countries and in the same country at different times and in the same individual at different ages. Thus it is a relative and not an absolute state. Ideally individuals should be required to satisfy rigid criteria including a thorough assessment of their health by the use of questionnaires and by clinical and laboratory investigations (Grassbeck and Alstrom 1981). This is seldom feasible and for practical purposes an approximation to this ideal state has to be accepted.

### Size of Reference Population

The reference sample has to consist of an adequate number of individuals selected to represent the reference population. This number will vary from variable to variable with the regularity of the rhythm encountered, the more or less close comparability of the reference group to the subject(s) to be studied, and with the degree of statistical power of decision making desired (Haus 1987; De Prins and Hecquet, this volume).

### Conditions of Sampling

The physiologic and environmental conditions have to be defined under which the reference group is studied as has the way the measurements are obtained. The description of the sampling conditions has to include the astronomic and the biologic time at sample collection. In chronobiologic measurements, the astronomic time, i.e., the clock hour of the measurements, the day of the week, and the season alone are not adequate to define the biologic time of an individual. The rhythmic variations encountered under the sampling conditions, usual in clinical medicine, frequently include the manifestation of an endogenous and apparently genetically fixed rhythm, which may be altered in a free-living population exposed to everyday environmental stimuli, including behavioral differences. Under certain circumstances, the endogenous rhythms may free run from the astronomic and/or social environmental counterpart. Thus, in chronobiologic studies, the determination of the time of measurement has to take into account to whatever extent possible, not only the astronomic time, but also the environmental synchronizers known to act upon a rhythm under study. Depending upon the frequency studied this may include but not be limited to the day-night cycle, the activity-rest pattern of the individuals and changes in activity along both the weekly and seasonal scales (e.g., observation of weekends, vacations, etc.) (Halberg and Reinberg 1967; Haus et al. 1980a, 1984, 1988a; Touitou et al. 1986c, 1989). Also the times of food uptake are of importance for certain rhythms but not for others (Haus et al. 1984, 1988a). In the collection of reference values (and in the sampling of patient material and its comparison with those reference values) fasting and nonfasting conditions should be indicated. In women, the stage of the menstrual cycle or premenarchal, peri-, and post-menstrual and menopausal conditions are important. For some variables, the observation of posture (including time spent in this position), smoking, degree of obesity, etc., may be of interest. Physical exercise, psychologic stress, alcohol, and smoking have to be kept in mind (for a review of sources of variation in laboratory values, apart from chronobiologic investigations, see Statland and Winkel 1977; Statland 1979).

Information on the conditions of sampling is important not only in the study of circadian and higher frequency rhythms, but also in the study of lower frequencies such as circaseptan, menstrual, and seasonal variations. Differences in the timing of, e. g., circadian rhythm, may lead to apparent differences in lower frequency rhythms, i.e., when only limited sampling (e.g., once a day) can be applied. In some parameters the lower frequency rhythms are also known to alter the parameters of superimposed higher frequency rhythms (Haus et al. 1988a; Nicolau et al. 1984; Touitou et al. 1983, 1984, 1986c, 1989).

The criteria used for including or excluding individuals from the reference sample group have to be clearly defined. In chronobiologic studies, these include the avoidance for at least 2–3 weeks prior to study, of changes in schedule like night or shift work, transmeridian flights, heavy environmental loads, certain types of medication (e. g., corticosteroids), or other treatment potentially changing circadian or other synchronization. Little is known about synchronization in frequency ranges other than the circadian one and about the time required for adaptation and stabilization of infradian rhythms, e. g., after the change of a weekend-weekday working schedule. To be meaningful, the description of a reference group should indicate the month and the year of the study, the location where it was performed, and any events which may have influenced the results (including environmental temperature, relative humidity, schedule of meals, etc.).

## Time Qualified Reference Ranges

Time-qualified reference ranges (chronodesms) can be established by different methods, the choice of which will depend upon the parameters studied, the amplitude of the rhythm, and upon the shape of the rhythmic variation encountered, i.e., the Gaussian or non-Gaussian distribution of the measurements at different time points. If the function under study is Gaussian and its distribution can be approximated by a mathematical model, for example, by a cosine curve, a reference range can be calculated based upon this model (Halberg et al. 1978). In this case, the chronobiologic reference values should also include reference values for the rhythm parameters, rhythm adjusted mean (or MESOR), amplitude, and acrophase. Alternatively, when a non-Gaussian distribution is found, nonparametric statistics, such as percentiles, can be used (Fig. 1) (Haus et al. 1983; Swoyer et al. 1984). The use of percentiles has the advantage of greatest simplicity, but in order to be meaningful requires a sizable number of subjects. This, however, is the case for any reference range and statistical complexity and sophistication is no substitute for adequate data.

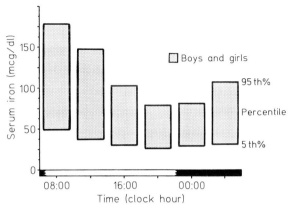

**Fig. 1.** Circadian time-qualified reference range (circadian chronodesm) of serum iron in prepubertal children (86 boys and 108 girls, 11 ± 1.5 years of age) delineated by 5th and 95th percentile of the data. (After Nicolau et al. 1987b)

Indication of peak and trough time of the measured values and the peak-trough difference (expressed in the units of measurement, as percent of the mean value of all measurements, or as percent of the lowest or highest value) may serve as end points of clinical interest, but these end points have to be based upon large enough groups of subjects to provide meaningful variance estimates.

Although a certain amount of information on chronobiologic reference values in different frequency ranges has been published (Reinberg et al. 1977, 1978; Touitou 1982; Touitou et al. 1981, 1982, 1986b,c, 1989; Nicolau et al. 1982, 1983b, 1984, 1985a; Haus et al. 1983, 1984, 1988a; Montalbetti and Halberg 1983; Halberg et al. 1966, 1969, 1977, 1981a, 1983), these data are still far too limited to provide a valid and meaningful chronobiologic reference for each patient to be examined. The availability of those widely applicable reference values will depend on worldwide cooperative studies with collection of the material in a computerized data bank. At the present time, a physician investigating a possible temporal abnormality in a patient will have to choose from the presently available reference material those groups of subjects which are most closely comparable to the specific patient and are most pertinent for the patient's problems.

## Factors To Be Considered in the Study of Rhythmic Variables

### Ethnic-Geographic Differences

Ethnic-geographic differences in biologic rhythms of human subjects have been reported (Haus et al. 1980b; Wetterberg et al. 1986; Kawasaki et al. 1983; Reinberg et al. 1985; Halberg et al. 1981a, b; Lakatua et al. 1987; Sensi et al. 1984; Sasaki 1977). These differences may involve the rhythm parameters of several frequencies and may be detectable at some time points but not at others (Fig. 2). The distinction between ethnic and geographic factors is often difficult. The latter include differences in climate, diet, social customs, etc. Among dietary factors, prolonged deficiencies of certain vital components, like iron, have been reported to lead to alterations of circadian rhythms in human subjects (Stramba-Badiale et al. 1981).

### Interactions of Rhythms

Among the factors which may lead to apparent differences in the mammalian time structure are the superimposed higher and lower frequency rhythms. While higher frequency rhythms can lead to spuriously elevated or lowered values if relatively infrequent measurements are used, lower frequency rhythms may be mistaken for trends (depending upon their stage sometimes in one or sometimes in the other direction). Some lower frequency rhythms may be accompanied by a shift in acrophase of the rhythm under study, e.g., circadian acrophase shifts have been observed in the course of the menstrual cycle (Reinberg and Smolensky 1974; Simpson and Halberg 1974) and in the course of circannual rhythms (Haus and Halberg 1970; Lagoguey and Reinberg 1981; Touitou et al. 1986c; Nicolau et al. 1984, 1987a; Reinberg et al. 1977, 1978; Haus et al. 1988a). In the measurement of circadian rhythms we have to take into account, in many variables, superimposed ultradian rhythms or episodic variations. If circadian rhythms are studied by sampling at 3 or 4 h

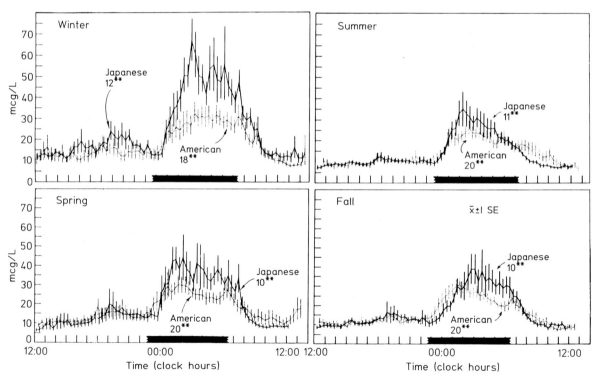

**Fig. 2.** Plasma prolactin in clinically healthy American and Japanese women sampled at 20 min intervals over a 24 h span during each season. The data show a circadian and a circannual rhythm, both with higher amplitude in the Japanese. A group difference is detectable only at certain clock hours (i.e., the time of circadian peak) during certain seasons (i.e., at the time of the circannual peak during winter and spring). Depending upon the time of sampling, the plasma prolactin concentrations in the Japanese can be found to be higher, the same, or lower than in the Americans. **, The number of subjects in each season. (After Haus et al. 1980b)

intervals, as is frequently the case, superimposed high amplitude ultradian rhythms or pulsatile secretions can lead to aliasing. To obtain meaningful data with such sampling schedules in parameters showing high amplitude ultradian variations, either a large number of cycles has to be studied in an individual or a large enough group of subjects has to be followed over one or over a few cycles to obtain statistically meaningful results.

## Circadian Rhythm Alterations by Drugs and Other Stimuli

Certain pharmacologic agents, including some drugs used widely in clinical medicine, may present problems for the use of time-qualified reference values, since they may alter the phase and/or the amplitude of circadian rhythms (Reinberg 1983). Agents in question are, among others, some neurotransmitters and/or their agonists or antagonists (Eskin et al. 1982; Eskin and Takahashi 1983), many psychotropic drugs, i.e., benzodiazepines (Childs and Redfern 1981; Ralph and Menaker 1986; Turek and Losee-Olson 1986), gamma-aminobutyric acid (GABA) antagonists (Ralph and Menaker 1983, 1985), reserpine (Halberg 1963), certain beta-blockers (Abetel et al. 1986), and others. Also, the administration of hormones which usually are secreted in a rhythmic fashion, e.g., ACTH and corticosteroids, or substances which interfere with their secretion, e.g., metyrapone, can alter the rhythms of their target organs and/or may change rhythms of other variables through feedback effects. In the case of metyrapone, the drug-induced interference with the prominent circadian rhythms of plasma and urinary cortisol is accompanied in human subjects during prolonged administration of the drug by the appearance of a high amplitude circadian rhythm of compound S and of tetrahydrocompound S, which without this treatment is minimal in MESOR and amplitude (Touitou et al. 1976, 1977).

Another example of an originally unexpected drug-induced rhythm alteration is the effect of an ACTH analogue (ACTH 1-17) upon human growth hormone (GH) secretion. Without involving GH releasing hormone, ACTH 1-17 leads to a circadian phase-dependent increase in plasma GH concentration with a higher response at 1400 or 2100 than at 0700 (Touitou et al. 1990). Beta-blockers, even in relatively low doses and given to clinically healthy subjects in the morning, lead to a marked decrease in the nighttime rise in melatonin (Wetterberg 1978;

Cowen et al. 1983; Arendt et al. 1985), presumably indirectly by suppressing catecholamine secretion. The effect of these agents on circadian rhythms may vary with the stage of the rhythm and the dose in which they are given. Although many observations on the action of drugs upon biologic rhythms are based upon animal experiments and the observations in human subjects are thus far limited, it has to be expected that, if given at a critical time and in an effective dose, many pharmacologic agents may affect the patient's time structure in the circadian or in other frequency ranges. This has to be taken into account in the selection of individuals to establish reference values and also in the comparison of an individual's data against a set of established reference values.

Elderly patients who received different medications, including barbiturates, phenothiazines, reserpine, and tranquilizers, in doses as they are used by physicians in a gerontologic practice, did show certain rhythm alterations (Haus et al. 1988b; Lakatua et al. 1990). The design of the study, however, did not allow one to distinguish between rhythm alterations due to the drugs and/or due to the underlying condition for which the drugs were given. In this study, the most frequent and pronounced changes involved the MESOR and in some instances the amplitude but only in very few of a total of 17 endocrine functions studied was the acrophase involved. Circadian rhythm alterations were found most frequently in plasma prolactin and urinary epinephrine. There was a change in acrophase in ACTH in subjects treated with reserpine and in prolactin in patients treated with phenothiazines. These data indicate that relatively few of the drugs which show circadian rhythm alterations in the laboratory did so when encountered under field conditions and given in a dosage customary in geriatric practice.

Exposure to environmental contaminants and pollutants such as herbicides, insecticides, and fungicides was shown to lead to rhythm alterations in mammalian nontarget organisms in doses which did not cause recognizable clinical disease during the time span of the studies (Nicolau 1984). No human data on rhythm alterations due to environmental pollutants have thus far been reported, but in view of the wide exposure of humans and animals, further investigations of environmentally induced dysrhythmias and possibly chronopathology seem to be of considerable interest.

Rhythm alterations and internal and/or external desynchronization have been reported to occur, e.g., during treatment of human malignancies with ionizing radiation (Charyulu et al. 1974).

## Rhythm Alteration by Alcohol, Drug Abuse, and Smoking Habits

Cigarette smoking was shown to alter certain hormonal rhythms (Haus et al. 1986). In both men and in women smokers, the circadian MESOR of norepinephrine was elevated. An elevation of MESOR and amplitude of the circadian rhythm in epinephrine excretion was found in women only. In men, the circadian rhythms of ACTH and of the related adrenal steroids were unchanged. In women, there were changes in the amplitude and in the acrophase of cortisol and a change in the amplitude of aldosterone and 17-hydroxyprogesterone. Dehydroepiandrosterone-sulfate showed in women smokers an increase in MESOR and amplitude and a change in acrophase. In men, the circadian mean in FSH and LH was lower in the smokers. In contrast, the testosterone MESOR was significantly higher in the male smokers than in the nonsmokers. Total $T_3$ and $T_4$ were statistically significantly higher in women smokers than in the nonsmokers. No such differences were noted in the men. It is difficult to ascribe all changes observed to the specific effects of nicotine, since the possibility of a "smoker personality" cannot be excluded. Of interest and importance, however, is the observation of marked sex differences in response to an environmental stimulus, which may have to be kept in mind in the evaluation of such studies.

Short- and long-term ingestion of ethanol has been reported to lead to rhythm disturbances in, among others, the circadian rhythms of cortisol and testosterone, which in turn may alter the circadian rhythmicity of related biochemical functions (Angeli et al. 1981; Bertello et al. 1982; Rosman et al. 1982; Minors and Waterhouse 1980; Reinberg et al. 1975; for a review on alcohol effects on biologic rhythms see Paille et al. 1988). Patients taking different and multiple drugs of abuse were reported to show frequently abnormal circadian variation in melatonin, cortisol, growth hormone, and thyroid stimulating hormone although the few data available do not allow a reliable statistical characterization of these changes (Veit et al. 1986).

## Induction of Circaseptan Rhythms by Environmental Stimuli

Suggestive evidence for a drug-induced rhythm alteration in the circaseptan frequency range was presented for the rhythm of 17-ketosteroid excretion in a subject treated with testosterone (Halberg et al. 1965). In the study of circaseptan rhythms, their induction by environmental, including immunologic stimuli, has to be considered. In these instances, the organism responds in an infradian rhythmic fashion to a single (e.g., introduction of an antigen), repeated, or persistent (e.g., high salt diet) stimulation. The exact length of the period, the amplitude, and the acrophase may vary with the time of exposure and with the nature and severity of the stimulus (Lévi and Halberg 1982; Uezono et al. 1984). Circaseptan rhythms and their synchronization are thus, in many instances, not related to the calendar week and their phase can not necessarily be equated with a certain day of the week.

## Sampling for Chronobiologic Studies

Sampling schedules for chronobiologic investigations in clinical medicine have to be based on some knowledge of the mammalian time structure in general and on the rhythm to be studied in particular. Statistical rhythmometry can evaluate and quantify the physiologic information obtained and, in the noisy time series obtained in clinical medicine, can help to separate information on rhythmic variables from the noise of a biologic time series (De Prins and Hecquet, this volume). However, to be valid rhythmometry requires an adequate amount of properly collected data and statistical rhythmometry is no substitute for poor data. If rhythmic functions are to be explored, the experimental design should be based on as much chronobiologic information as can be obtained which is pertinent to a given problem (e.g., on the frequencies and stages anticipated).

The "right" time (e.g., clock hour, day of the week, or month of the year) for sample collection may vary both with the biologic timing of the individual(s) and with the specific problem to be investigated. A group difference or an abnormality in an individual may be detectable at a certain circadian, circaseptan, or circannual time but not at others. This is now generally recognized for patients with Cushing's syndrome, whose plasma corticosteroid concentrations may be well within the usual range of a peer group in the early morning but will most likely be above that range in the afternoon and evening. In contrast, in patients with Addison's disease, the plasma corticosteroid concentration will often be within the usual range during the evening, which at this time in healthy subjects falls to very low values (Krieger 1979). The selection of the "right time" for sampling may involve several rhythms of different frequencies as shown for the detection of differences in plasma prolactin concentration between a population of women with high and a population of women with low risk to develop breast cancer (Fig. 2) (Haus et al. 1980 a, b; Halberg et al. 1981 a). Sampling at different circadian and also circannual times can, in the comparison of these two groups of women, lead to different and even opposite conclusions. The choice of a given convenient fixed time of day, and season and even of a similar stage of the menstrual cycle may not be adequate to "control" a rhythmic variation, since superimposed lower frequencies may or may not induce phase-shifts and free-running of cycles may lead to erroneous results.

While relatively small sample sizes may easily detect large differences, small differences will usually require a large amount of data. The length of the observation span required depends on the spectral region and on the characteristics of the rhythm to be studied.

## Longitudinal vs Transverse Study Designs

There are basically three different ways of collecting data as a function of time:
1. A single subject is followed over a long time span, i.e., over many periods of the rhythm in a so-called longitudinal study design.
2. Alternatively, numerous individuals, in whom on the basis of additional information, one can expect a comparable rhythm of the same variable, may be studied over a single period or over a few periods in a so-called transverse study design.
3. Limited numbers of subjects may be followed over several periods in a so-called hybrid (between the longitudinal and the transverse) design. The latter represents a compromise solution which is often practiced in human chronobiology.

Longitudinal sampling is the prefered approach when information regarding a given individual has to be obtained. In order to obtain representative and statistically meaningful rhythm parameters with their variance estimate as end points, a large enough number of cycles has to be studied. An estimate of the number of periods needed or the number of subjects to be investigated is possible only after the characteristics of a rhythm have once been rigorously established (as for prominence, amplitude, and amount of noise in a given function).

## Sampling Interval – Sampling Density

Sampling should follow a schedule which takes into account the regularity of each rhythm to be studied and the relative importance of the rhythm in relation to superimposed higher frequencies and to lower frequencies upon which the rhythm may be superimposed. The number of samples required to allow characterization of the parameters of a rhythm by rhythmometric statistical procedures will vary from one variable to the other. In many chronobiologic designs, at least six measurements per cycle spread over the length of the expected period are advocated when no prior information is available concerning the timing of the rhythm under study. If a rhythm is to be described on individual subjects, sampling often will have to occur more frequently during one cycle and/or will have to be extended over several cycles in order to obtain statistically meaningful results. Achievement of a reproducible and statistically meaningful description of a rhythm, with a clinically feasible sampling density, often requires groups of subjects, the size of which will vary with the variable studied.

The interval between consecutive observations must be short and frequent enough to allow for a reliable resolution of the rhythm to be studied. In variables with marked and rapid response to a variety of environmental stimuli or showing superimposed higher frequencies, considerably more measurements may be necessary. In choosing the sampling interval for ultradian or pulsatile variations, the half-life time of a hormone or other solute should be considered; however, this also varies as a function of time (chronopharmacokinetics). In many instances, the assessment of certain frequencies may have to be omitted from evaluation if the sampling requirements cannot be met. In such a case one may have to choose the period of main interest for the problem to be studied and sample either at intervals sufficiently short to resolve the rhythms with high frequencies or sample for a long enough time span to yield valid information on the components with lower frequencies.

## Sampling Requirements

Only very little generally applicable information is available on the sampling requirements for obtaining statistically meaningful rhythmometric results for common laboratory functions in individual subjects.

In ultradian variations, with periods occurring at more than 1 h, as seen in most pituitary and steroid

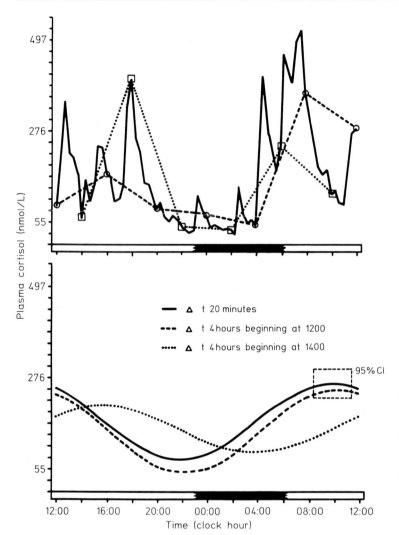

Fig. 3. Circadian variations in plasma cortisol; pitfalls of infrequent sampling. *Top*, chronogram of plasma cortisol concentrations with sampling at 4 min intervals. The two chronograms were obtained by using only plasma concentrations found every 4th hour in the same set of data. Aliasing leads to apparent differences in the rhythm's timing if the onset of the 4 h sampling is chosen to begin at 1200 or at 1400 hours. *Bottom*, Cosine curves best fitting to the selected 4-hourly measurements suggest substantially different acrophases in values obtained from same set of data. (After Haus 1987)

hormones, sampling intervals of 10–15 min will be adequate (Veldhuis and Johnson 1990; Veldhuis et al. 1987; see also Veldhuis et al., this volume). Higher sampling frequencies may yield additional information in some relatively high frequency variables, e.g., LH. Methods for the identification of ultradian peaks have been described (see below and Veldhuis, this volume). The sensitivity of "rhythm detection" and the requirements for the determination of peak and/or trough by these methods can be modified by the investigator.

In the evaluation of circadian rhythms, the superimposed ultradians can lead to a considerable degree of aliasing if longer sampling intervals are used (Fig. 3). The ultradian rhythms or (if a nonrandom nature of these variations cannot be demonstrated) the episodic secretions superimposed upon a circadian rhythm require for many parameters caution

in the interpretation of "single point" measurements, even at defined circadian stages. In such variables, results obtained from a single blood sample may be rather difficult to interpret and occasionally may be misleading, since samples taken only a few minutes earlier or later may show substantially different values. A more meaningful "integrated single sample" can be derived from prolonged continuous blood aspiration (e.g., with a withdrawal pump) or from several blood collections obtained through an intravenous line and spread over a certain time span (e.g., three or four samples obtained at 15 min intervals). These samples can then be pooled resulting in a single (more cost-effective) chemical determination. The time span for withdrawal or the sampling interval for such an integrated sample depends upon the parameter to be studied, i.e., the information available on its "usual" periodic behavior and upon its half-life time in the circulation. Most suitable for the measurement of high frequency variations would be either continuous or frequent short interval monitoring over a time span of several periods of the rhythm under study, preferably with a noninvasive or only minimally invasive sensor with computer compatible output. Such automated instrumentation also would minimize the interference of the investigator with a subject.

Rhythmometric procedures like the cosinor (Nelson et al. 1979) are often used to "detect" rhythms in noisy time series of laboratory values containing superimposed rhythms of several frequencies masking effects and noise. The incidence of "rhythm detection" by cosinor in a reference population can provide an estimate about the sampling frequency or, alternatively, about the size of the group of subjects needed to obtain results which allow rhythm description by this procedure. If, quite empirically and with all qualifications for the use of this procedure in mind, "rhythm detection" by cosinor at the $p \leq 0.05$ value is taken as end point, the percentage of subjects showing a detectable rhythm in individual profiles of 13 hormones measured in serum or plasma is shown in Fig. 4. The differences in the percentage of time series allowing "rhythm detection" by cosinor varies as a function of the hormone studied and for prolactin and cortisol, in which 20 min and 100 min sampling is available, as a function of the sampling interval. At the far left of Fig. 4, prolactin and cortisol are shown with sampling at 20 min intervals. In almost all instances, "rhythm detection" can be achieved with this sampling schedule in these two variables in a single subject studied over a single 24 h span. The same parameters are shown again farther to the right of the figure together with the other variables which were studied at 100 min intervals only. If the sampling frequency for cortisol and prolactin is reduced to 100 min intervals, still a large percentage but not all subjects show "rhythm detection" for cortisol and even fewer for prolactin. The decrease in sampling density from 20 to 100 min leads to a certain but not dramatic loss of information. Comparison of the 13 hormonal variables sampled at 100 min intervals over one 24 h span shows that there are considerable differences from one variable to the other in the number of individuals in whom "rhythm detection" was significant by the cosinor analysis of each subject's data. Apart from cortisol and prolactin, a large percentage of subjects showed a circadian rhythm characterized by cosinor analysis in DHEA-S, while LH, a predominantly menstrually and ultradian cycling variable, was found to be at the very low end of the circadian spectrum.

If different sampling intervals in cortisol were studied in the same subjects (every second, third, fourth,

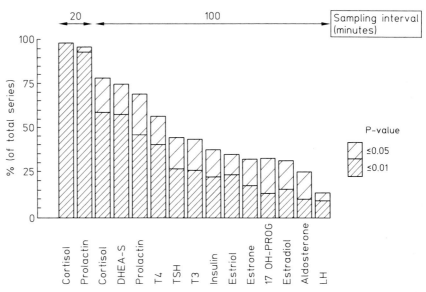

**Fig. 4.** Rhythm detection by single cosinor analysis in per cent of 24 h profiles of 13 hormonal variables studied at 100 min intervals over one 24 h span (cortisol and prolactin also at 20 min intervals), 141 profiles in 41 subjects (except in estriol 84 profiles in 35 subjects). In cortisol and prolactin 20 min sampling *(left)* allows "rhythm detection" in 95%–100% of individual profiles. Decrease in sampling density to 100 min lowers the percentage of subjects in whom a rhythm can be described by single cosinor to 75%. Lesser rhythm detection rate with same sampling density in most other hormonal variables. (After Halberg et al. 1981a, Haus et al. 1988a)

or fifth 20 min measurement used as input for a cosinor analysis), it was shown that, with a sampling interval of 120 min, still approximately 77% of subjects showed "rhythm detection by cosinor," but with a drop in sampling density to 180 min, only about 35%. With a decrease in sampling density to 240 min (an interval which is used often in clinical studies), only about 25% of the subjects showed rhythm detection in a single subject sampled over a single 24 h span (Haus 1987). In prolactin, after 120 min, approximately 68% of profiles allowed rhythm detection, but with a decrease in sampling density to 180 min, this number dropped to 22% and at 240 min was approximately 17% (Haus 1987).

The sampling requirements will be considerably greater for demonstrating a rhythm of unknown period, especially if it is unexpected. In such a case, the research for unknown bioperiodicities may be aided by the knowledge of an environmental rhythmic input (such as the day/night cycle) likely to synchronize the biologic rhythm. The appearance of an unexpected "candidate component" in a rhythmometric analysis (based, e.g., upon the fit of a series of cosine curves as periods covering a certain frequency range) requires confirmation by different techniques and/or separate information on potential synchronizers or environmental factors which might induce masking of the rhythm under study, exposure to drugs, etc. Statistical procedures for determining the minimal sample size and experimental designs based on a single cosinor were presented by Bingham et al. (1982).

## Data Acquisition for Clinical Studies

For the clinical application of chronobiology, it is necessary to obtain meaningful chronobiologic measurements in individual subjects. For some purposes this can be accomplished with single samples if reference values from the same subject (individual chronodesm) or from a representative and comparable peer population (group chronodesm) are available. In this case, a single measurement can be extrapolated against a time-qualified usual range (chronodesm) sampled at the same biologic time (Figs. 1, 5). For this approach it is essential that the biologic time of the individual is known, which can be achieved to a certain degree by observing simple reference functions like the habitual time of rising or time of retiring, major meal times, etc. Such reference

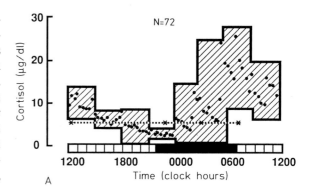

A

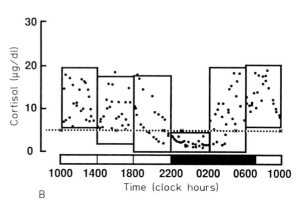

B

**Fig. 5. A** Individual circadian chronodesm of plasma cortisol in clinically healthy young adult women sampled at 20 min intervals over a 24 h span (72 samples); calculated as tolerance interval (tolerance intervals determined separately for 3 h spans and indicate limits within which 90% of measurements would be expected to fall with 90% confidence). **B** Group circadian chronodesm of plasma cortisol (obtained from a group of 15–21 year old young adult women sampled at 20 min intervals over a 24 h span) shows a wider range than the individual chronodesm. In the individual and in the group chronodesm, both the same value of plasma cortisol obtained at different times and evaluated without regard to the time of sampling could be below, within, or above the "usual range." (After Halberg et al. 1981 a)

functions establish the probability of a certain timing of the subject's circadian system, although certain phase differences and free-running of a rhythm from its usual synchronizers may remain undetected.

If rhythm parameters are to be used as end points, the measurements have to be adequate in sampling density and the length of the sampling span has to be adequate to allow a quantitative, inferential, statistical description of the rhythms and of their parameters with variance estimates for each. However, repeated measurement with invasive techniques such as blood drawing can lead to responses by the subject to the sampling procedure which may alter ("mask") the rhythm under study. It is essential, therefore, to minimize interference with the subject during chronobiologic studies and create conditions under which the interference by the investigator will not bias the results of the examination.

## Blood, Plasma, or Serum Collection

The collection of blood is an invasive procedure, the psychologic effect of which can lead to masking of a rhythm. The ease of venipuncture and the skill of the venipuncturist are essential factors in the study of many blood parameters, i.e., in rapidly stress responsive variables such as plasma catecholamines, the number of circulating granulocytes, etc. For example, the measurement of plasma catecholamines is substantially altered by the procedure of a venipuncture and reliable measurements reflecting the circadian rhythm can be obtained only through an indwelling catheter with sampling at least 30 min after the venipuncture. The personality and bedside manner of the venipuncturist can lead to differences in the results. Also prolonged tourniquet action can lead to changes in some blood parameters and has to be avoided (see review by Statland and Winkel 1977). Establishment of an intravenous line prior to sampling can avoid some masking effects. For prolonged sampling at short intervals (e.g., 15 min intervals over 24 h) an intravenous line is essential. With less frequent sampling, the benefit of the line has to be weighed against the liabilities due to relative immobilization of the patient, the need for heparin flush and/or drip, the need to obtain blood free from infusion fluid (with some associated blood loss), etc.

Sampling for parameters showing high frequency variations which are unpredictable in their timing has to occur at short enough intervals and over a long enough time span to allow to measure these variations. If sampling at frequent intervals has to be extended over a prolonged time span (e.g., 24–48 h) the amount of blood to be removed has to be considered. In children and in elderly subjects, the amount of blood required may limit chronobiologic studies. A hemoglobin or hematocrit determination should precede any study involving repeated and/or prolonged blood sampling. Pre-existing anemia will be a cause for exclusion for some of these studies.

Shielding of the investigator from the subject is desirable if the studies are extended, e.g., over 24 h including a dark span during which the subject is expected to continue a quasi-usual sleep-wakefulness pattern. Connection of an IV catheter to a blood drawing line extending into a neighboring chamber may be recommendable to avoid, as far as possible, interference with the subject under study. In order to obtain meaningful and statistically valid results in single subjects available only over a limited time span, frequent sampling will be necessary in the study of most variables. In using invasive methods, a sampling frequency or length of sampling span may have to be followed which may often be difficult to accomplish or not be feasible at all. For this purpose, techniques and/or materials for study will have to be sought which provide for reliable sampling by noninvasive or minimally invasive methods. Among the body fluids which can be obtained by noninvasive methods are urine, saliva and sweat.

## Urine Sampling

Urine sampling can be obtained in babies by collection vessels fixed to the skin, and in children and in young and middle aged adults by spontaneous voiding. Urine collection by spontaneous voiding has to be regarded with caution in elderly subjects who frequently have residual urine. An advantage of urine collection in the study of, e.g., circadian periodicity, is the integration of the high frequency variations of serum components over a certain length of time, thus avoiding some spuriously high or low results. For the same reason, however, urine sampling by spontaneous voiding is not suitable for the study of higher frequency rhythms. Also, on the negative side is the not always linear relationship between serum concentrations and urinary excretion, with the urinary rhythm not necessarily representing the rhythm in plasma concentration. Rhythms in renal excretion of some solutes often show a phase difference in comparison to the plasma concentrations. In some instances, the renal production and a nephrogenic rhythm in the excretion of solutes may add a renal component in comparison to the rhythm observed in plasma. This has been shown, e.g., for glucosaminidase (Lakatua et al. 1982) and other enzymes (Jung et al. 1986) and certain hormones (Gowenlock 1988).

## Collection of Saliva

Saliva is easily obtainable by noninvasive techniques even in babies and during sleep. Numerous solutes in saliva mirror the concentration of these solutes in the plasma (see below) while others are secretory products of the salivary glands.

There are, however, certain peculiarities in the collection of saliva for chronobiologic studies which have to be understood to obtain meaningful results and avoid pitfalls. These include, among others, the type of saliva, the flow rate, duration of stimulation, nature of the stimulus, frequency of collection, relation of the collection period to meals, and the storage of the saliva prior to analysis. With knowledge of its physiologic behavior and of the sampling procedures to be employed, saliva can serve as a useful medium to study marker rhythms for the monitoring of circadian and lower frequency rhythms. For solutes in saliva which show a direct correlation with the concentrations found in plasma, salivary measurements may provide information on the rhythmic (and other) variations of these compounds. Due to the ease and noninvasive nature of sampling, we expect that salivary rhythms will be used more widely in chronobiologic studies in the future.

Saliva can be sampled easily as mixed saliva including the secretory material of the parotid, the submandibular, sublingual, and other salivary glands. Saliva can be collected with an attempt at minimal stimulation and will then be designated as "unstimulated" or "resting saliva". However, there is some question to which degree uniform collection procedures can be established in chronobiologic studies requiring frequent sampling in order to obtain truly "unstimulated saliva" comparable from one subject to the other and from one time point to the other. The psychologic interference with the process of sample collection may not allow a uniform degree of seclusion of the subject from environmental stimuli leading to salivation.

Whole ("mixed") saliva is a complex mixture from four different types of salivary glands, the relative contribution of which varies with the total flow rate. For example, the contribution of the parotid varies from 23% in unstimulated saliva to 50% after maximal gustatory stimulation (for review see Dawes 1974). Whole saliva also contains a variety of nonsalivary components such as gingival crevicular fluid, leukocytes, epithelial cells, bacteria, and occasionally plaque and food debris. The insoluble components of whole saliva may be centrifuged down and the supernatant removed. To avoid bacterial or enzymatic degradation of organic components of whole saliva, saliva should either be stored frozen or analyzed immediately after collection.

For sampling of unstimulated whole saliva, the subject should rinse his mouth with lukewarm water, approximately 20–30 min prior to collection. Saliva is then allowed to drip from the lower lip into a funnel and a graduated tube in cracked ice. For many applications direct salivation into a wide-mouthed collection tube may be satisfactory. It is important to avoid movement of the oral musculature which has been shown to cause an increase in salivary flow rate. The minimum time for collection of a sufficient volume for an accurate estimation of the flow rate is probably about 5 min. About 3 ml of "unstimulated" mixed saliva can usually be obtained within 10 min.

A collection of stimulated whole saliva uses chewing of inert materials such as paraffin or rubber bands. Relatively large volumes of stimulated saliva can usually be collected over a time span of a few minutes. Gustatory stimuli cannot be used since material eliciting such a stimulus would dissolve in the saliva.

Saliva can be collected from individual glands by use of a cannula which fits over the duct orifices. A device for the collection of submandibular and sublingual saliva has been described by Truelove et al. (1967). Shannon and Chauncey (1967) devised a cannula for collection of parotid saliva. When the individual gland is cannulated, gustatory stimuli can be used for sample collection (e.g., sour lemon drops). The gustatory stimulus can be kept relatively constant throughout the collection period and at different collection times, although constancy of the stimulus does not imply constancy of the response.

Factors Influencing Salivary Composition

The different salivary glands show quantitative differences in the relative proportion of proteins and electrolytes excreted. Thus, any factor (e.g., a change in total flow rate) which causes a change in the relative proportions of the individual secretions will influence the composition of whole saliva.

The flow rate has a very marked influence on some salivary solutes. In general, if the flow rate is increased only very slightly above the unstimulated rate, sodium and bicarbonate concentration and pH increase, whereas potassium, calcium, phosphate, chloride, and protein concentrations decrease. At higher flow rates, sodium, calcium, chloride, bicarbonate, and protein concentrations and pH increase, while the phosphate concentration decreases and the potassium concentration shows little further change (Shannon and Prigmore 1960; Dawes 1969, 1974).

The salivary flow rate shows a circadian rhythm which influences salivary composition (Ferguson and Botchway 1979, 1980). In order to measure this rhythm, it is important to standardize as many other variables as possible. Apart from the intensity of gustatory or pharmacologic stimulation, factors that can stimulate flow rate are olfactory stimuli and smoking, which should be avoided prior to collection times.

The duration of stimulaton influences the composition of saliva, even if the flow rate of stimulated parotid or submandibular saliva remains constant. Although almost all components of saliva show a change in concentration during the first 1 or 2 min of stimulation, the concentrations of total protein, calcium, bicarbonate, and chloride and the pH are particularly affected and may not achieve steady state levels even after 15 min of stimulation (Dawes 1974). Thus, it is important to control in the collection of stimulated saliva the duration of stimulation. A compromise procedure suitable for most applications would be to discard the saliva collected during the first 5 min of stimulation, since during this time span most constituents of saliva show marked changes in composition due to transition from the unstimulated to the stimulated condition. The saliva to be analyzed should then be collected for a further fixed time span (e.g., 5 min).

The study of salivary rhythms requires multiple collections. If these are made too frequently, there is a possibility that a collection procedure at one sampling time may influence the flow rate or composition of saliva collected at a later time. In the study of parotid saliva, Dawes and Chebib (1972) found that this effect could be avoided if the collections were separated from other collections or meals by 1 or 2 h, depending upon the component of interest.

Present Use of Saliva in Chronobiology

For many components of saliva collected under standardized conditions, a positive correlation was found between the concentration in plasma and in saliva (e.g., for potassium, calcium, urea, and uric acid). In contrast, the phosphate concentration in human saliva is relatively independent of that in plasma.

The saliva/plasma ratios for numerous drugs, including anticonvulsants, have been established and saliva is used widely for therapeutic drug monitoring (Horning et al. 1977).

Measurement of steroid hormones has been practiced extensively in saliva, and reports summarizing the experience with numerous hormones have been published (Read et al. 1984; Touitou et al. 1986a; Walker et al. 1990; Sufi et al. 1985). Steroid concentrations in saliva do not seem to be flow-dependent and the neutral steroids in stimulated and unstimulated mixed saliva were found to be almost identical (Read et al. 1983). The concentration of steroids in saliva has been thought to approximate the free, nonprotein bound, presumably biologically active fraction in plasma (Read et al. 1983). A certain problem in the use of saliva for steroid determination is the low concentration of steroids found in saliva (about 5 % of that in plasma), which requires highly sensitive techniques to obtain reliable results.

Numerous matched plasma and saliva samples of steroid hormones, including cortisol, aldosterone, progesterone, 17-hydroxyprogesterone and others, collected at different circadian stages, under conditions of stimulation and suppression, show a remarkably good correlation over a wide range of concentrations (Read et al. 1983, 1984; Touitou et al. 1986a). Also, other (nonprotein) hormonal variables which may be of interest for chronobiology have been measured in saliva, e.g., melatonin (Miles et al. 1987) or $T_4$ (Vining et al. 1983).

## Statistical Evaluation of Rhythms in Laboratory Data

The application of chronobiology to clinical medicine depends upon the objective and quantitative evaluation of data collected as a function of time. To be useful, the time-dependent, i.e., rhythmic changes, observed have to be quantitated and, whenever feasible measurable numerical end points should be established for rhythm parameters such as phase and amplitude. Since biologic rhythms are not as precise as their counterparts in physics, their rhythm parameters cannot be expressed by a single number but have to be seen always in the context of their variance estimate established by an appropriate inferential statistical procedure. The problems encountered in clinical chronobiology are varied and complex and require a variety of methods for data collection and analysis. The physician treating patients is often limited in the amount of data which can be obtained, i.e., in regard to the length of the observation span, the number of measurements, the regularity or irregularity of sampling, etc. In view of the different problems encountered and the different experimental or clinical conditions, an array of methods each with its own domain of applicability has to be considered. The choice of the analytical procedure will depend upon the questions to be examined and the kind and amount of data available. No single procedure of analysis is applicable to all situations (Halberg 1965; Halberg et al. 1967; Halberg and Panofsky 1961; Haus and Halberg 1980; Nelson et al. 1979; Panofsky and Halberg 1961; De Prins et al. 1986; Van Cauter 1979; Winfree 1980; Minors and Waterhouse 1989). The larger the amplitude of a rhythm and the lesser the interference by superimposed higher frequencies and random or nonrandom noise, the lesser will be the sampling requirements and the simpler will be the techniques which can be used for the evaluation of the rhythm (De Prins et al. 1986; Haus and Halberg 1980). Some of the aspects of rhythmometric analysis are presented in more detail in the chapter by De Prins and Hecquet in this volume.

## Observation of the Raw Data (Chronogram and Plexogram)

The first step in the evaluation of biologic rhythms is the inspection of the raw data plotted as a function of time (the "chronogram"). Obvious rhythmicity may easily be recognized and an impression concerning the shape of the waveform can be gained which in part may determine the statistical method to be used for its quantitation. The existence of outliers and the need for data transformation prior to further analysis can also be determined.

If a rhythm is relatively regular and prominent and enough data are available, even the computation of means and standard errors of data obtained at different sampling times can give reproducible curves on a chronogram and allow tentative rhythm detection.

If data extending over several cycles are available these data can be assembled in the time frame of a single period, irrespective of the sequence in which the measurements were collected, and presented in this form as a "plexogram".

Assembled as chronogram or plexogram the data can be evaluated by conventional statistical procedures like analysis of variance (ANOVA) or $t$-test. These methods will provide information as to whether time can be considered a statistically significant source of variation or will determine the statistical significance of peak-trough differences.

In the analysis of the data by one factor (time) analysis of vaiance (ANOVA), the procedure indicates whether the variance between time points is significantly greater than the variation within them. The times of measurements do not have to be equally spaced and the shape of the rhythm does not affect the statistical outcome. Although the ANOVA may

establish that a time-dependent variation exists, it gives no information as to its period, amplitude, or phase.

## Periodogram and Variance Spectra

Biologic data obtained in healthy subjects on usual living routines and in patients are usually rather noisy, and the question has to be examined whether the variations found may be random, caused by exogenous stimuli, or actually represent biologic rhythmicity. For a physiologic function to be considered rhythmic, the event, the intervals, and the sequences need not to be precisely repetitive, but they must show some regularity of repetition. To qualify as a rhythm, the regularity must have an identifiable pattern, and it must be shown that its occurrence as a matter of chance is unlikely. Especially the assessment of rhythmicity when the period is not known or cannot be assumed requires a considerable amount of data and more complex methods of time series analysis (Enright 1965; Rummel et al. 1974; De Prins et al. 1986; Fookson et al. 1984).

Fourier or harmonic analysis was used originally to find periodic components of known periods (for a review see Bloomfield 1976). However, when Fourier analysis was used to search for periodic components of unknown period in empirical data, the results were misleading. Schuster (1898) then introduced the periodogram and applied the theory of probability to the solution of the question whether the value of any particular Fourier coefficient indicates a true periodicity or may be accounted for by purely accidental causes. The periodogram is an appropriate means for the detection of periodicity in sufficiently long time series sampled at equidistant intervals with a consistent rhythm overlayered by noise. Spurious peaks may, however, be found because of the large statistical variability of the estimates. Fisher (1929) provided a significance test appropriate for independent errors. This test has been modified to allow for correlated series and has been extended to multiple times series (Hanson 1970; MacNeill 1974). A method for characterization of the circadian variations of blood components, in relatively short time series from data collected in equal intervals and based upon periodogram analysis, was presented by Van Cauter (1979).

The problem of estimating the spectrum of a time series of measurements led to the smoothing of the periodogram and the concept of the power (variance) spectrum, which is more reliable for testing unknown periodicities if one is concerned about falsely alleging rhythms when none occur (Blackman and Tukey 1954; Bendat and Piersol 1971). Halberg and Panofsky (1961) and Panofsky and Halberg (1961) adapted variance spectra for use in biology and medicine, notably under conditions when the occurrence of a rhythm cannot be postulated *a priori,* on the basis of concomitantly obtained marker rhythms, or on other prior information. When no prior information on a given bioperiodicity is available, variance spectra can be used to detect rhythms in noisy time series and to evaluate their statistical significance. However, the method requires a considerable amount of data and remains limited to equidistant data while the phase information is lost. Moreover, the amplitude information is not directly attainable since the procedure usually refers to the variance per unit frequency vs frequency.

## Curve Fitting by Least Squares Techniques – The Cosinor Procedure

In order to apply the study of biologic rhythms of laboratory parameters to medical practice, statistically meaningful end points have to be reached in individual subjects and the statistical methods have to be applicable to relatively short time series obtained by measurements in unequal intervals. The sampling requirement for the statistical description and quantification of rhythm parameters in single subjects may be substantial, i.e., if rhythm parameters such as phase and amplitude and their alterations are to be used as end points (Haus 1987). Least squares techniques are useful for curve fitting when it is desirable to obtain a functional form that best fits a given set of measurements. The goodness of fit is indicated by minimizing the sum of squares of differences between the actual measurement and the "estimated" functional form or "best fitting" curve. The least squares methods of fitting relatively simple models is especially attractive since it allows the analysis of nonequidistant observations. However, the least squares methods are sensitive to outliers and one or a few "bad data" can markedly influence the estimation of the rhythm parameters and the confidence intervals (De Prins et al. 1986).

The Single Cosinor Procedure

The methods presently in use include the different variants of the cosinor procedure introduced and widely used by Halberg (Halberg et al. 1965, 1967;

Nelson et al. 1979). The single cosinor procedure (Halberg et al. 1972; Nelson et al. 1979) was designed for the detection of periodic components in short and noisy time series as they are usually presented in clinical situations involving patients. For the cosinor analysis the period to be examined is chosen by the investigator and has to be known to avoid misleading results. The cosine curve best fitting to the data is determined by the method of least squares. In order to test the presence of a rhythm, the total sum of squares is partitioned into the sum of squares due to the regression model and the residual sum of squares. The sum of squares due to the regression is the amount of variability accounted for by the fit of the model to the data. The residual sum of squares indicates the difference between the data and the fitted model. "Rhythm detection" is sought by testing the zero amplitude hypothesis with an $F$ test.

For rhythms "detected," the cosinor procedure yields the rhythm adjusted mean (the so-called MESOR), the amplitude (the distance from the MESOR to the peak or trough of the cosine curve best fitting to the data), and, as indicator of the timing of the rhythm, the acrophase (the peak time of the cosine curve best fitting to the data), each with the their variance estimate. The method also provides a confidence region for phase and amplitude, in which, in the graphic representation of this procedure, the "polar cosinor plot" usually is presented in the form of an ellipse (for more information see Haus and Touitou, this volume).

Cosinor analysis requires as a major assumption that the data obtained can be reasonably well represented by a cosine curve and nonsinusoidality limits the applicability of the method. For a meaningful rhythmometric analysis by cosinor and a correct interpretation of the results, it is important to determine the approximate sinusoidality of the data. This requires at least the inspection of the chronogram before application of the rhythmometric procedure and/or a mathematical test for sinusoidality which may be incorporated into the computer program containing the cosinor. Unfortunately, few biologic rhythms of laboratory function are ideally sinusoidal. Rhythms vary greatly in shape and for a particular variable that shape may be quite characteristic. Depending on the shape of the curve of the actual data observed, the maxima and the minima of the fitted function may be quite different from those of the periodic variable. A small irregularity of the waveform of the order of a few percent may result in an apparent shift of the maximum of the fitted curve which may be misleading. Thus the acrophase and its confidence interval provided by the cosinor method have to be interpreted with some caution (De Prins et al. 1986). Similarly, one has to be aware that the amplitude provides information on the extent of the variation of the best fitting cosine curve and does not necessarily represent the range between the highest and lowest values actually measured, especially if the distribution of the data is less than ideally sinusoidal. If the form of the curve is other than sinusoidal, more complex curve fitting procedures may have to be used (Halberg et al. 1977).

The rejection of the zero amplitude assumption ("rhythm detection") by single cosinor refers to the given data set and does not allow extrapolation to a population as a whole. However, the rhythm parameters obtained by the single cosinor procedure in different subjects may serve as input for a population mean cosinor for further quantitation.

### The Population Mean Cosinor Procedure

In order to summarize results obtained for different individuals belonging to the same population, the rhythm characteristics obtained by the single cosinor can be further analyzed by the population mean cosinor (Halberg et al. 1967; Nelson et al. 1979). The rhythm characteristics obtained by the single cosinor are then considered as imputations or as a first order statistic. The population mean cosinor, in turn, contributes a second order statistic applied to derive confidence regions relating to the whole population. The parameter estimates are based on the means of estimates obtained from the individuals in the sample. The confidence regions for the true parameters depend on the variability among individual parameter estimates.

If the sample is to be characterized without further inference to others in the population, a single cosinor is much more efficient. However, when inference is to be drawn on the basis of the sample for the population as a whole, the population mean cosinor is indicated (Haus 1987; Haus and Halberg 1980; Nelson et al. 1979). Use of one or of the other of these procedures depends on the purpose and on the design of a given study. In the use and interpretation of the cosinor, one has to be aware of its limitations and avoid the associated pitfalls. Unsatisfactory results of rhythmometric procedures frequently are due to the inadequacy of the data available for analysis (which of course is not the fault of the procedure). If all necessary qualifications are kept in mind, the cosinor procedure with its variants can provide meaningful stat-

istical end points for the detection and quantification of rhythms and can represent a useful base for the comparison of rhythms and their parameters between different subjects and different populations.

### Comparison of Rhythm Parameters Obtained by Cosinor Models

Methods for the comparison of the parameters of curves fitted to different data sets were developed by Bingham et al. (1982) and are useful in the determination of the statistical significance of differences between MESOR, amplitude, and acrophase of two or more time series.

### The Single Cosinor Window ("Chronobiologic Window")

When the period length of a biologic rhythm is unknown but is assumed to lie in a certain range, a "single cosinor window" can be applied. The method consists of applying the single cosinor procedure not only at a single fixed period, but to a set of trial periods in a given range. In doing so, one is usually interested in determining the period length within that range for which the residual sum of squares is minimal. Rhythm detection, especially of rhythms of unknown or unexpected frequencies, requires a considerable amount of data collected over numerous periods of the rhythm(s) in question. Since 95% probability is usually regarded as the limit of significance for rhythm detection, 5 out of 100 cosine curves fitted to a set of noisy data will falsely indicate a rhythm. In the case of an unknown or unexpected frequency, therefore, it is necessary to confirm the candidate rhythm by different techniques (e. g., an analysis of variance applied to the data assembled as plexogram in the candidate period) and/or in different time series of data of the same or of different subjects.

### Nonlinear Least Squares Rhythmometry and Deviation Spectrum

Another possible approach to this problem is to use nonlinear least squares methods to estimate both the period and the amplitude and acrophase. This method allows one to fit concomitantly several components. A combination of computer programs has been used for this purpose (Marquardt 1963; Rummel et al. 1974). Linear least squares rhythmometry is first done, separate components being successively fitted at each single frequency in the spectral region of biologic interest. Nonlinear least squares rhythmometry follows, with the concomitant fit of candidate periods and their characteristics (obtained by linear least squares) used as initial estimates. After an appropriate number of iterations, final parameter estimates are obtained. Further resolution may be achieved with a deviation spectrum based on these two procedures (Halberg et al. 1977).

## Multicomponent Models

The most frequent problem in the use (and abuse) of cosinor techniques is the nonsinusoidality of biologic rhythms. This problem can partially be overcome by the fitting of curves of several periods to the data. With multiple component models, information concerning the waveform of the rhythm is provided. The amplitudes and acrophases of each component may, however, no longer be easily interpreted, and new parameters needed to be introduced. Of interest is the so called orthophase representing the timing of the maximum of this more complex model, given a proper reference time and time scale (Halberg et al. 1977).

Similarly, for short and sparse equally spaced series, a reconstruction of the continuous function from the samples can be computed by harmonic interpolation (De Prins et al. 1981). The paraphase or lag from zero time to the crest time in the reconstructed function may then be defined and used in the same way as an orthophase or an acrophase.

## The Chronobiologic "Serial Section"

Rhythm characteristics undergo changes spontaneously (e.g., as a function of a rhythm interaction such as circadian vs menstrual or circannual) or following a change in routine (e.g., after a transmeridian flight or in isolation without environmental time cues); however, changes in rhythm parameters may also be found in some clinical conditions. It is important to detect and determine these changes. The chronobiologic serial section was developed for that purpose (Arbogast et al. 1983). A data section of fixed length (interval) is first defined which may consist of one or, in order to achieve statistical significance, more often of several periods to be examined and is then displaced in increments throughout the time

series. A single cosine with a constant period is fitted to the data in each interval, providing rhythm parameter estimates which may be graphically displayed in parallel with the original data. This procedure is particularly useful in showing the occurrence of desynchronization. It has the effect of smoothing any day to day changes and shows the general trend of a phase-shift or free-running rhythm. However, the procedure misses changes in the rhythm's shape which tend to occur during phase shift. Also, if the interval extends over several (and in some instances numerous) periods of the rhythms examined, the procedure will not allow one to pinpoint when a change has taken place nor its extent at a given time.

Many of these procedures are now available for use with the small computers found in most clinical laboratories. Depending upon the nature and the extent of the measurement available, the clinical evaluation of data obtained at different times or deliberately collected as a function of time can be very simple, e.g., by observation of the chronogram or by comparison with time-qualified reference ranges (chronodesms). If the data are adequate, statistical methods are now available for rhythmometric analyses which allow one to take advantage of quantitatively assessed rhythm parameters as new end points in physiology and medicine. With the wider availability of monitoring devices, including those in use today for the electrocardiogram and the assessment of blood pressure, activity, temperature, and other physiologic variables and those in development in the fields of clinical chemistry and endocrinology (for review see Haus 1987), more data suitable for rhythmometric analysis will be obtained. As with any new methodology, however, the clinical investigator has to be aware not only of its possibilities, but also of its limitations. If used critically, chronobiologic information derived from appropriately designed laboratory testing and analysis can open a new dimension for the interpretation and use of laboratory data.

# Biologic Rhythms in Hematology and Immunology

The number of circulating formed elements in the peripheral blood shows circadian rhythms in all cell lines. While those in the red cells and their related parameters are of low amplitude, the circadian rhythms of circulating lymphocytes and granulocytes are high amplitude rhythms which may in certain instances be of diagnostic importance. The circadian rhythms of the formed elements in the circulating blood are of complex nature. Although distribution between compartments (e.g., circulating vs marginated) seems to play a major role, other factors such as marrow release and cell removal may be involved. Therefore, extrapolation from the circadian rhythm in the peripheral blood to that in the bone marrow has to be approached with much caution.

Circadian rhythms and seasonal variations (or circannual rhythms if endogenous in nature) have been documented for several end points either of or related to cell proliferation in the human and animal bone marrow (see chapter by Smaaland and Laerum) and may be at least partly responsible for the circadian (and possibly other) changes in sensitivity and/or resistance of this organ to many of the agents used in cancer chemotherapy.

Higher and lower frequency rhythms other than circadian have been described in the number of circulating human blood cells, but have not been uniformly reproducible in clinically healthy human subjects.

Chronopathology is found, e.g., in the number of circulating lymphocytes and granulocytes in HIV infected patients with and without AIDS (Swoyer et al. 1990), in the number of circulating lymphocytes in some patients with solid lymphomas, in cyclic neutropenia and thrombocytopenia, and in a proportion of patients with chronic myelocytic leukemia. Details on the importance of chronobiology for laboratory hematology and hematopathology are presented in the chapters by Haus, by Smaaland and Laerum and by Canon and Lévi in this volume.

In immunology, circadian rhythms have been shown for circulating lymphocytes and their subtypes and for their responsiveness to pokeweed mitogen (PWM), phytohemagglutin (PHA), the mixed lymphocyte reaction (MLR), and for the development of cellular and humoral immunity. Of interest are the circaseptan rhythms in antibody formation in numerous animal models. The circaseptan rhythms in immunity are usually independent from the calendar week and seem to be triggered by the time of the introduction of the antigen. Applied to transplantation biology, the occurrence of rejection reaction of kidney, pancreas, and heart transplants shows predictable circaseptan periods suggesting transient states of elevated risk to develop rejection (Lévi and Halberg 1982). Critical circadian times for the occurrence of rejection reaction have also been described (Knapp et al. 1967, 1979). The multifrequency rhythmic structure of the immune system is of special interest for immune stimulation, immune response, and immune suppression. The chronobiology of the immune system is discussed in detail in the chapter by Fernandes and by Canon and Lévi in this volume.

## Biologic Rhythms in Clinical Chemistry

In clinical chemistry, circadian, circaseptan, circavigintan, circatrigintan, and circannual rhythms of plasma and/or urinary solutes have been described. In most functions the circadian and/or the circannual rhythms show the highest amplitudes. Only in a few parameters, however, is the amplitude of the rhythms of plasma solutes large enough to pose diagnostic problems (Nicolau et al. 1987b; Haus et al. 1988a; Touitou et al. 1979, 1986c, 1989). Age, sex, ethnic-geographic factors, etc., may modulate the rhythms observed.

The establishment of reference values for plasma urinary and salivary solutes in clinical chemistry has only begun and will require cooperation and use of similar techniques of subject selection and standardization, sample collection, and chemical and statistical analyses in many laboratories in many parts of the world. Few such comparable series are available. In the following figures and tables, we present some reference values of subjects sampled at different geographic locations, with cooperation by the investigators, using comparable sampling techniques. All chemical and statistical analyses were done in the same laboratory (St. Paul-Ramsey Medical Center, St. Paul, MN USA). All subjects from which these data were obtained were clinically healthy, diurnally active, ambulatory, and on a regular three meal schedule of their local cooking. The average time of rising was 0630–0700 and average time of retiring 2100–2300 local time. The daily routine of the subjects was maintained during the studies as far as feasible. Evaluation of the data was done by the population mean cosinor procedure.

The timing of the circadian rhythms observed in the different groups of subjects is presented in Fig. 6–10 as acrophase charts indicating the acrophase provided by the cosine curve best fitting to the data with its 95% confidence interval. The number of subjects in each group and their age, sex, and geographic location are indicated. The acrophase of the parameters in which a circadian rhythm was verified and characterized by cosinor analysis ($p < 0.05$) is shown with its 95% confidence interval as a horizontal bar. If the rhythm detection by cosinor was between $p = 0.05$

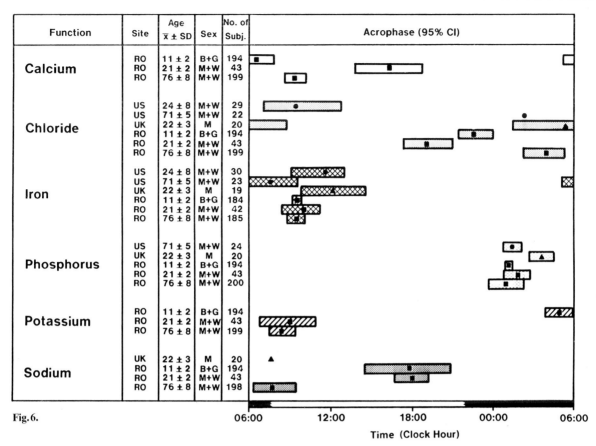

Fig. 6.

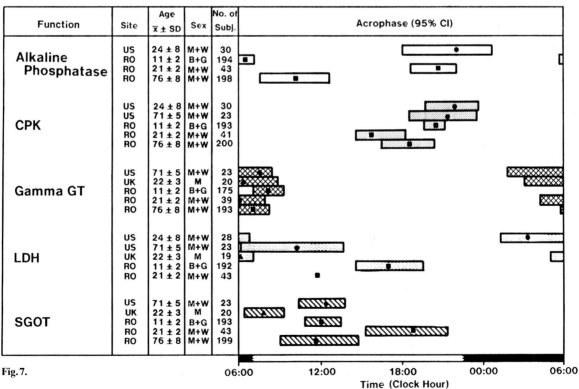

Fig. 7.

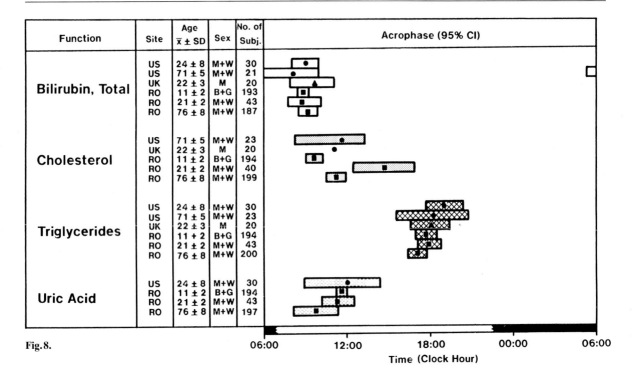

Fig. 8.

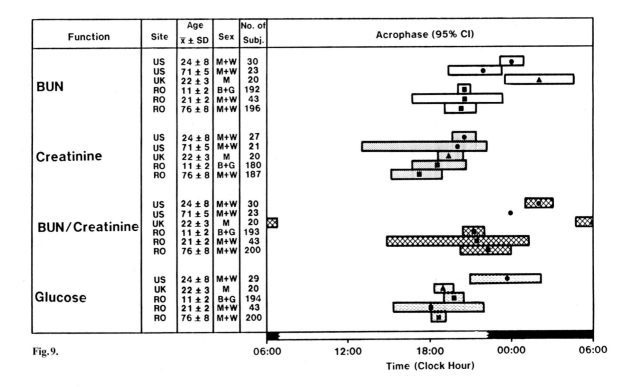

Fig. 9.

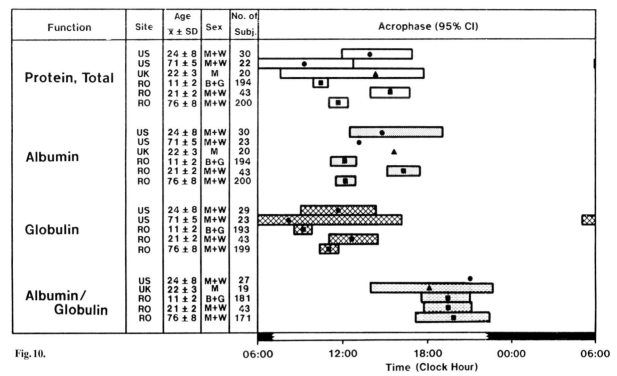

**Fig. 10.**

| Function | Site | Age $\bar{x} \pm SD$ | Sex | No. of Subj. |
|---|---|---|---|---|
| Protein, Total | US | 24 ± 8 | M+W | 30 |
| | US | 71 ± 5 | M+W | 22 |
| | UK | 22 ± 3 | M | 20 |
| | RO | 11 ± 2 | B+G | 194 |
| | RO | 21 ± 2 | M+W | 43 |
| | RO | 76 ± 8 | M+W | 200 |
| Albumin | US | 24 ± 8 | M+W | 30 |
| | US | 71 ± 5 | M+W | 23 |
| | UK | 22 ± 3 | M | 20 |
| | RO | 11 ± 2 | B+G | 194 |
| | RO | 21 ± 2 | M+W | 43 |
| | RO | 76 ± 8 | M+W | 200 |
| Globulin | US | 24 ± 8 | M+W | 29 |
| | US | 71 ± 5 | M+W | 23 |
| | RO | 11 ± 2 | B+G | 193 |
| | RO | 21 ± 2 | M+W | 43 |
| | RO | 76 ± 8 | M+W | 199 |
| Albumin/ Globulin | US | 24 ± 8 | M+W | 27 |
| | UK | 22 ± 3 | M | 19 |
| | RO | 11 ± 2 | B+G | 181 |
| | RO | 21 ± 2 | M+W | 43 |
| | RO | 76 ± 8 | M+W | 171 |

**Figs. 6–10.** Timing of circadian rhythms of biochemical serum parameters in subjects of different ages studied at different geographic locations: USA *(US)*; United Kingdom *(UK)*; Romania *(RO)*. Acrophase as determined by population mean cosinor analyses (*p* < 0.05) shown with its 95% confidence interval *(horizontal bar)*. If "rhythm detection" by cosinor *p* > 0.05 but < 0.15, the tentative acrophase is shown without confidence interval. The average activity-rest schedule is indicated as *white* and *black bar* at the bottom above clock hour (in local time). (After Haus et al. 1988a, 1990)

and *p* < 0.15, the tentative acrophase is indicated without a confidence interval. The average activity-rest schedule of the subjects is shown as a white and black bar at the bottom of the figures above the clock hours in local time.

Some biochemical serum parameters show a very similar timing irrespective of geographic location and age differences, e. g., gamma glutamyl transpeptidase (GGT), total bilirubin, and triglycerides (Figs. 7–9). Other parameters show a much greater variability, e. g., alkaline phosphatase and LDH (Fig. 7).

The extent of the circadian rhythms is shown in Table 1 as "range of change" expressed by the difference between the highest value and the lowest value observed in each time series in percent of the lowest value:

$$(\% \text{ROC} = \frac{\text{highest value} - \text{lowest value}}{\text{lowest value}} \times 100)$$

In Table 1 the "range of change" is shown for functions exhibiting rhythmic variations verified by cosinor analysis, although the range of change expressed in this manner is not dependent upon the presence of

a detectable rhythm. The range of change during the 24 h span is very similar and reproducible in different age groups studied at different geographic locations, e. g., in the low amplitudes of serum sodium, chloride and serum proteins, but is considerably more variable in others indicating the need for the wider and more refined study of reference populations.

Circannual rhythms were also found in some of the groups which were examined during all seasons of the year (Nicolau et al. 1986; Reilly et al. 1987; Haus et al. 1988a).

Circaseptan rhythms have been observed for some biochemical variables, the timing of some of which is shown in Fig. 14.

## Biologic Rhythms in Endocrinology

The endocrine system shows a complex time structure with a wide spectrum of rhythms ranging in period from a few minutes to a year. Ultradian

**Table 1.** Extent of circadian variation of biochemical serum parameters. (After Haus et al. 1990)

| Variables | RO 11 ± 2 M + F | RO 21 ± 2 M + F | UK 22 ± 3 M | US 24 ± 8 M + F | US 71 ± 5 M + F | RO 76 ± 8 M + F | Site Age Sex |
|---|---|---|---|---|---|---|---|
| Alkaline P'tase | 8 | 10 | | 12 | | 7 | |
| CPK | 28 | 30 | | 40 | 18 | 18 | |
| Gamma GT | 50 | 53 | 37 | | 34 | 15 | |
| LDH | 5 | 17 | 118 | 23 | 22 | | |
| SGOT | 11 | 15 | 28 | | 15 | 9 | |
| Bilirubin, Total | 81 | 65 | 67 | 75 | 52 | 42 | |
| Cholesterol | 7 | 6 | 4 | | 4 | 9 | |
| Triglyceride | 82 | 129 | 78 | 73 | 57 | 38 | |
| Uric Acid | 14 | 10 | | 6 | | 4 | |
| BUN | 30 | 10 | 11 | 34 | 13 | 9 | |
| Creatinine | 6 | | 29 | 22 | 14 | 7 | |
| BUN/Creatinine | 22 | 11 | 32 | 25 | 9 | 9 | |
| Glucose | 21 | 23 | 56 | 14 | | 48 | |
| Calcium | 3 | 4 | | | | 4 | |
| Chloride | 1 | 3 | 2 | 1 | 1 | 2 | |
| Iron | 106 | 54 | 66 | 49 | 60 | 48 | |
| Phosphorus | 24 | 18 | 28 | 33 | | 7 | |
| Potassium | 8 | 8 | | | | 8 | |
| Sodium | 1 | 3 | 1 | | | 2 | |
| Protein, Total | 5 | 7 | 5 | 6 | 5 | 8 | |
| Albumin | 5 | 7 | 6 | 6 | 3 | 7 | |
| Globulin | 8 | 6 | | 6 | 9 | 9 | |
| Albumin/Globulin | 5 | 6 | 6 | 4 | | 3 | |

Difference between highest and lowest value encountered in groups of subjects expressed as percent of the lowest value. Same subjects as shown in Fig. 6–10. RO = Romania; UK = United Kingdom; US = USA

rhyhtms are prominent in most hormonal variables measured in serum or plasma. Secretory episodes occurring in about 1–4 h intervals are superimposed upon circadian rhythms. In some variables, lower frequency rhythms are found with periods of about a week (circaseptan), about 20 days (circavigintan), or about 30 days (circatrigintan), the latter especially but not exclusively in women during the reproductive years. In many variables there are also seasonal variations or circannual rhythms.

The circadian rhythms are environmentally synchronized by our social routine and the light-dark cycle and some of them may be influenced to some extent by the timing of the main meal. The environmental time information through light stimuli is relayed from the retina to the suprachiasmatic nuclei in the hypothalamus, which act as pacemaker for a number of endocrine and metabolic rhythms. Light stimuli are relayed also to the pineal and determine the timing of the circadian rhythms of this organ which in turn modulates other circadian rhythms, i.e., of the gonads and adrenals.

Also, some circaseptan rhythms appear to be environmentally synchronized presumably by our social weekly routine. Circaseptan rhythmic response patterns may also be environmentally triggered, e.g., by a metabolic or antigenic stimulus, and then may proceed in a rhythmic fashion without regard to the calendar week. Some seasonal variations seem to be induced and maintained by environmental stimuli, including temperature or the length and/or timing of the daily photoperiod. Other rhythmic changes with a period of about 1 year are endogenous in nature and continue under constant laboratory conditions and/or have been found free-running from the calendar year. An anatomic structure which may serve as pacemaker for these rhythms is not known at this time.

The ultradian rhythms (or episodic or pulsatile secretions) and the circatrigintan and circavigintan rhythms are also expressions of internal timekeeping systems which have no known synchronizers. Although these rhyhtms may be modulated and in turn may modulate some of the circadian rhythms, these frequencies show no fixed time relation to the environmentally synchronized rhythms in the circadian, circaseptan, and circannual frequency ranges.

The rhythmicity of the endocrine system is found at all levels of organization from the superimposed central nervous system controls to the receptor and the metabolic response in the individual target cell. Thus the responsiveness of the neuroendocrine and endocrine structures and of the peripheral target cells will vary rhythmically and to some extent predictably as a function of time. In consequence, a certain hormone concentration measured in the clinical laboratory will not always have the same biologic significance and a certain hormonal or pharmacologic agent used for either stimulation or suppression testing may elicit substantially different responses if given at different biologic times (see chapter by Touitou in this volume). The chronophysiology and related pathobiology of the individual components of the endocrine system are discussed in detail in several chapters in this volume.

## Ultradian Rhythms – Pulsatile (Episodic) Variations

Most hormones are secreted in pulses rather than continuously and their physiologic effects seem to depend upon the alteration between peaks and troughs of hormone concentration. The rhythmic (ultradian) or nonrhythmic timing of these pulsatile episodes still have to be explored with respect to mechanism and interactions with endocrine rhythms of other frequencies. Alterations in the rhythmic patterns of the secretion of some hormones are associated with certain endocrine disorders and may actually be involved in their etiology.

The extent and frequency of the hormonal secretory pulses are modulated by the circadian system (Van Cauter and Honinckx 1985). Therefore, the study of pulsatile hormone secretion may have to cover an entire 24 h span. If only a limited time span of sampling is feasible, the selection of this time span has to take into consideration the specific pulsatile pattern of the variable to be studied and the circadian synchronization of the subject. While episodic pulses occur throughout the 24 h span for cortisol and prolactin, the profile of plasma GH consists of only a few pulses timed after onset of the daily sleep span.

The selection of an optimal sampling protocol depends upon the phenomenon under study, the secretion pattern of the hormone, and its kinetics in the circulation (i.e., its half-life). Taking both the expected frequency and the kinetics of the secreted hormone into account, a sampling interval has been proposed of two to three samples per half-life of the hormone under study or five to six times the expected interval between peaks, whichever is more frequent

(Merriam and Wachter 1984). For most hormonal variables a protocol with sampling every 15 min for 24 h may be adequate for the study of pulsatile hormone secretions and may allow one to obtain statistically meaningful circadian rhythm description by rhythmometric techniques (Van Cauter and Honinckx 1985). For LH, which shows the highest frequency of major secretory pulses, sampling intervals of 5–10 min were recommended (Filicori et al. 1987; Reame et al. 1984; Veldhuis et al. 1987). More frequent sampling (e.g., 1–4 min) will allow recognition of more rapid episodic variations in hormones such as ACTH, LH, TSH, and GH. These high frequency variations, however, seem to be different from the larger amplitude ultradian (pulsatile) secretions occurring at intervals of 1 h or more, and their physiologic significance is at this time unclear. However, longer sampling intervals of 20–30 min will only detect major pulses lasting more than 1 h. In establishing a sampling protocol, the amount of plasma needed for each determination and the total amount of blood to be withdrawn have to be considered.

In the study of pulsatile variations, i.e., if they are nonperiodic in nature and recur at irregular (unpredictable) intervals, the recognition of a secretory pulse and its distinction from variations due to the imprecision of the chemical method used are essential. The first methods of pulse recognition proposed were based upon a simple percentage increase of a peak over a preceding trough value (Santen and Bardin 1973) and, in later versions, on the comparison of a peak with multiples of the intra-assay coefficient of variation of the method used (Ross et al. 1984; Merriam and Wachter 1984; Veldhuis et al. 1985, 1986). Computer programs for pulse detection based on these general principles were developed, e.g., the Ultra program presented by Van Cauter (1981) and the Pulsar program by Merriam and Wachter (1982). In the Ultra algorithm both the increment and decrement of each peak are taken into account. In the Pulsar algorithm both the height and the width of the peaks are included in the evaluation. Later developments of pulse detection and evaluation procedures include the Cluster program by Veldhuis and Johnson (1986), the Detect program by Oester et al. (1986), and, most recently, methods of deconvolution analysis (Veldhuis and Johnson 1990; Veldhuis et al. 1990).

The Cluster algorithm (Veldhuis and Johnson 1986) defines peaks and trough of plasma concentration as "clusters" of two, three, or more data points rather than as single time points with the number of data points forming a "cluster" selected.

The Detect algorithm (Oester et al. 1986) was developed from programs originally designed for the analysis of more or less Gaussian peaks, as found in electrophoresis and chromatography, and provides peak detection based on analysis of first derivatives with logic which has been optimized for asymmetrical peaks with exponential decays. In spite of the differences between the more recent programs of pulse detection and quantitative evaluation, the results obtained in cross-validation studies between the Ultra, Cluster, and Detect algorithms on time series of hormonal profiles were found to be similar (Urban et al. 1988).

The methods of deconvolution analysis which have been recently applied to biologic time series attempt to estimate hormone secretion and/or clearance based on the serial circulating hormone concentration measurements and include both waveform-defined procedures and waveform-independent algorithms (Veldhuis and Johnson 1990). The application of some of these procedures to pituitary hormones is discussed in the chapter by Veldhuis et al. in this volume.

## Circadian Endocrine Rhythms

Circadian rhythms are an ubiquitous finding in metabolizing structures and thus occur at all levels of organization of the endocrine system. In the clinical laboratory the circadian variations in hormone concentrations in plasma, saliva, and urine have to be considered in sampling for diagnostic purposes and for the delineation of usual values in different populations. In addition, circadian rhythm alterations are thought to be characteristic for some disorders and may serve as new end points for study. Finally, the circadian changes in responsiveness at all levels of the endocrine system, from the central nervous system structures to the peripheral target cells, lead to circadian changes in the results of endocrine testing by stimulation or inhibition. The circadian rhythms in the response of endocrine structures and target tissues to stimulation and suppression are of diagnostic significance and are discussed in the chapter by Touitou and Bogdan in this volume.

Circadian endocrine rhythms usually maintain certain characteristic phase relations to their environmental synchronizers and among each other as part of the complex endocrine time structure of the organism. Timing and amplitude are characteristic for each circadian endocrine rhythm, but may be different in populations of different ethnic-geographic back-ground (Halberg et al. 1981a; Ghata et al. 1977; Haus et al. 1980b, 1988a) and may be modified by the stage of circaseptan, circavigintan, circatrigintan (including menstrual), and seasonal or circannual rhythms at the time of sampling (Touitou et al. 1986b, 1989; Ghata et al. 1977; Reinberg et al. 1978; Halberg et al. 1983; Haus et al. 1980b, 1988a; Nicolau et al. 1984).

The circadian rhythms in hormone concentration are often characterized by the timing, frequency, and/or amplitude of the pulsatile ultradian peaks. However, these often rapidly alternating peaks and troughs do represent a substantial problem in the measurement of circadian rhytms by a limited number of (strategically placed) samples. In spot-checks of hormone concentrations "extended sampling," e.g., with a continuous withdrawal pump over a 30–45 min span or three samples in 15 min intervals may be helpful for those variables in which rapid pulsatile secretory episodes are expected.

The choice of the (biologic) time of sampling is essential for the diagnostic value of a spot-check of hormones showing high amplitude circadian rhythms. In Addisonian patients, for example, only the early morning value of plasma cortisol may be abnormally low while an equally low evening value may well fall within the low usual range found at that time. In contrast the plasma cortisol concentration in a group of patients with Cushing's syndrome showed a wide overlap with those of normal subjects except during a time span of about 4 h centered around midnight when plasma cortisol offers an excellent index of discrimination between normals and patients with hypercortisolism (Van Cauter and Aschoff 1989; Zadik et al. 1982).

Similar time-dependent group differences have been observed among many others, e. g., for aging effects upon serum testosterone, with lower values in the aged found only during the early morning hours (Bremner et al. 1983); the aging effects upon plasma TSH (see chapter by Nicolau and Haus in this volume); for ethnic-geographic differences in catecholamine excretion (Lakatua et al. 1987); urinary melatonin excretion (Wetterberg et al. 1986); and plasma prolactin concentrations (Haus et al. 1980b). In plasma prolactin concentrations the ethnic-geographic differences between American and Japanese women (Fig. 2) are obvious during winter if sampling is done during the night hours but are absent during summer and during most of the daytime during all seasons. At some time points the plasma prolactin concentrations in the Japanese were statistically significantly lower than those in the Americans. These differences are due to higher circadian and circannual amplitudes of plasma prolactin in the Japanese and the interaction between the circadian and circannual rhythms of this hormone. Sampling at single time points chosen without knowledge of the multifrequency time structure of this variable may thus be grossly misleading and even lead to opposite conclusions. Seasonal differences in prolactin concentrations have been reported in relation to patients with breast cancer (Hermida et al. 1982, 1984) and fibrocystic mastopathy (Tarquini et al. 1979) and for TSH in relation to prostate cancer (Tarquini et al. 1981). Circadian and circannual stage-dependent differences have been reported for aldosterone and TSH in relation to hypertension (Hermida and Halberg 1986; Hermida et al. 1987) and for dehydroepiandrosterone-sulfate (DHEA-S) in relation to personality trends favoring alcoholism and drug addiction (Hermida et al. 1982). The latter findings are of considerable pathobiologic interest as group phenomena but will need confirmation in larger groups of patients. Their applicability to individual subjects may be limited due to the wide range of individual variability and the numerous other factors which act upon hormone concentrations. Also the sampling requirements to establish a circannual amplitude with any degree of statistical probability are prohibitive for application of some of these findings to single individuals.

Endocrine circadian rhythm alterations have been described in some but not in all patients with Cushing's syndrome (Krieger 1973; Kreiger and Allen 1975; Glass et al. 1984; Refetoff et al. 1985), in patients with chronic adrenal insufficiency (Halberg et al. 1951), hypothalamic diseases (Krieger 1973, 1979), and sleep disturbances (Weitzman et al. 1981). Free-running circadian rhythms have been found in some blind patients but occasionally also in subjects without clinically recognizable disorders or other alterations in their relationship to their circadian periodic environment (Moore-Ede et al. 1983).

Of considerable interest are apparent rhythm alterations of the hypothalamic-pituitary-adrenal and thyroid axes in affective disorders. The findings reported, however, are quite complex and the role of the endocrine rhythm abnormalities found (i.e., the abnormal dexamethasone suppression test and TRH/TSH abnormalities) in the etiology and pathogenicity of the disorder is regarded by many as unresolved (for a detailed review see the chapter by Kripke in this volume).

The large amplitude circadian rhythms in numerous endocrine functions require the establishment of time-qualified reference ranges. In following single subjects longitudinally, reference ranges established for individual subjects are relatively narrow and would be most meaningful in discovering already minor changes early in their occurrence (Fig. 5). Unfortunately, individual reference ranges are seldom available and the clinician has to rely on reference ranges derived from comparable peer populations which are usually much wider.

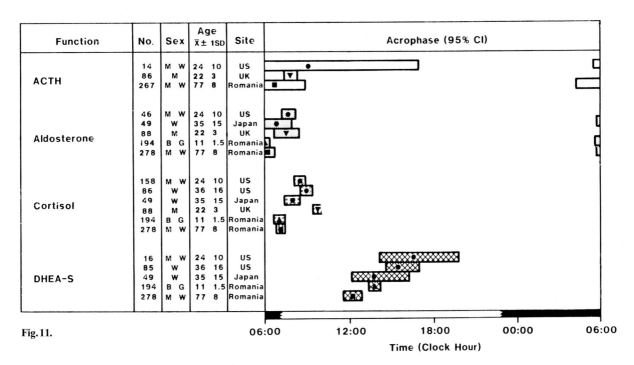

**Fig. 11.**

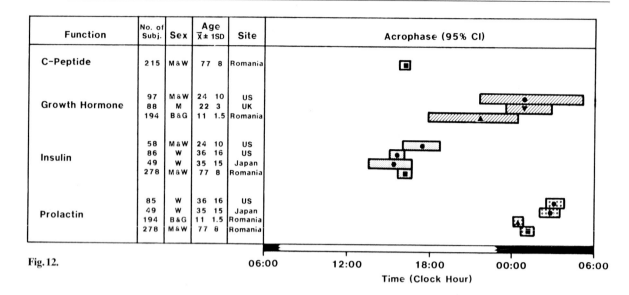

Fig. 12.

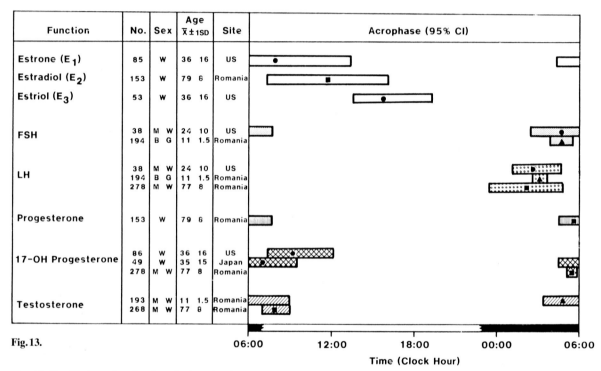

Fig. 13.

**Figs. 11–13.** Timing of endocrine rhythms in the plasma or serum of subjects of different ages studied at different geographic locations. Acrophase as determined by cosinor analysis ($p < 0.05$) shown with its 95% confidence interval. Average activity-rest schedule indicated as *white* and *black bar* at the bottom above clock hour (in local time). For timing of thyroid and related parameters see chapter by Nicolau and Haus. (After Haus et al. 1988a, 1990)

The timing of the circadian rhythms of a number of endocrine variables, as studied by comparable collection techniques in diurnally active populations of different ages living at different geographic locations, is shown in Figs. 11–13 (the thyroid and thyroid related variables are shown in the chapter by Nicolau and Haus). All chemical analyses of the data shown were done in the same laboratory (Haus et al. 1988a, 1990; Nicolau et al. 1984). The extent of the circadian variation is shown as "range of change" (difference between highest and lowest value measured expressed in percent of the lowest value) in Table 2.

**Table 2.** Extent of circadian variation of endocrine parameters. (After Haus et al. 1990)

| Variables | RO 11 ± 2 M + F | UK 22 ± 3 M | US 24 ± 10 M + F | JAPAN 35 ± 15 F | US 36 ± 16 F | RO 77 ± 8 M + F | Site Age Sex |
|---|---|---|---|---|---|---|---|
| ACTH | | 356 | 63 | | | 83 | |
| Aldosterone | 383 | 241 | 195 | 127 | | 109 | |
| Cortisol | 165 | 634 | 182 | 327 | 261 | 140 | |
| DHEA-S | 26 | | 21 | | | | |
| DHEA-S – M | | | | | | 20 | |
| DHEA-S – F | | | | 38 | 39 | 24 | |
| C-Peptide | | | | | | 144 | |
| Growth Hormone | 107 | 152 | 74 | | | | |
| Insulin | | | 119 | 234 | 366 | 158 | |
| Prolactin | 174 | | | 350 | 169 | 69 | |
| Estrone | | | | | 12 | | |
| Estradiol – F | | | | | | 23 | |
| Estriol | | | | | 59 | | |
| FSH | 28 | | 7 | | | | |
| LH | 51 | | 38 | | | | |
| LH – M | | | | | | 7 | |
| LH – F | | | | | | 11 | |
| Progesterone – F | | | | | | 76 | |
| 17-OH Prog – M | | | | | | 77 | |
| 17-OH Prog – F | | | | 66 | 34 | 271 | |
| Testosterone – M | 128 | | | | | 30 | |
| Testosterone – F | 105 | | | | | 32 | |

Difference between highest and lowest value encountered in groups of subjects expressed as percent of the lowest value. Same subjects as shown in Figs. 11–13. RO = Romania; UK = United Kingdom; US = USA

The timing of circaseptan rhythms of hematologic, chemical, and endocrine variables studied in one of our laboratories (St. Paul-Ramsey Medical Center) is shown in Fig. 14. The amplitudes of the circaseptan rhythms are invariably less than those of the circadian rhythms. The number of subjects studied, however, is still too small to provide reference values.

Circannual rhythms of endocrine parameters have been widely reported by numerous investigators (Halberg et al. 1965, 1981 a, b, 1983; Nicolau et al. 1983 a, 1984, 1985 b, 1987 a; Haus et al. 1987, 1980 b, 1988 a; Reinberg et al. 1978; Touitou et al. 1983, 1984) and those of the thyroid hormones and related variables are presented in the chapter by Nicolau and Haus. The timing of many of the circannual rhythms, however, is less consistent than that of the circadian rhythms. Some reviews on circannual rhythms of endocrine functions in human subjects have been presented (Halberg et al. 1983; Nicolau et al. 1984; Haus et al. 1988 a; Reinberg et al. 1978, etc.). Due to the multitude of factors, rhythmic and non-rhythmic, determining plasma hormone concentrations and their rhythm parameters, truly pertinent peer populations for a specific problem are difficult to find and will depend ultimately upon the cooperative efforts of many laboratories around the world and upon the availability of a computerized data bank.

## Current Perspectives of Chronobiology in Laboratory Medicine

The data accumulated over the last two decades allow definition of areas in which chronobiological methods are essential in the study of laboratory parameters to obtain meaningful results. In addition, there are numerous areas in which chronobiologic study design and analysis may open new avenues in the exploration of disease states and their diagnoses. A review of the present role and potential application of chronobiology in laboratory medicine leads to a number of observations and conclusions.

Almost all laboratory parameters studied are periodic in one or several frequency ranges. Rhythmic variations with large amplitudes may be detectable in single individuals with relatively few appropriately timed measurements. Rhythms with small amplitude and/or high variability may require extensive sampling in order to reach statistical significance in a single

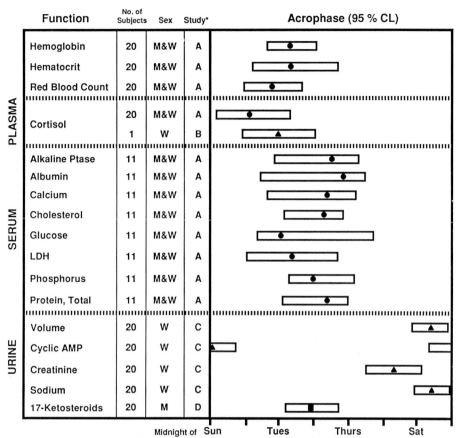

**Fig. 14.** Circaseptan acrophase chart of blood and urinary functions in clinically healthy diurnally active subjects. Study *A*, Samples 3 times/week (usually Mon., Wed., Fri.) between 07:30 and 08:00 from Nov.–Feb. (approximately 90 days); pooled data of two groups of subjects 21–46 years of age. *B*, Sampled 3 times/week (Mon., Wed., Fri.) between 08:50 and 09:00 from Sept. 3–Oct. 22 in a 31-year-old woman. *C*, Collection span: Sept. 8 to Oct. 25 during the waking span in a 31-year-old woman. *D*, Collection span: Aug. 3, 1963 to April 18, 1964 at approximately 4 h intervals in a 37-year-old man. (After Haus et al. 1988 a)

subject or may become manifest only as a group phenomenon.

The length of the observation span and the sampling intervals required for statistical rhythm detection depend on the spectral region and on the characteristics of the rhythm to be studied. An estimate of the number of periods needed for detection of a rhythm of known frequency in an individual or of the number of subjects needed for the detection of such a rhythm as a group phenomenon is possible only after the characteristics of the rhythm have been established and the amount of noise to be expected is approximately known. The sampling requirements will be considerably greater for demonstrating a rhythm of unknown period, especially if it is unexpected.

In many but by far not in all functions of clinical interest, the circadian frequency component seems to have the highest amplitude. This is not the case, however, for, e.g., hormones related to reproduction, which in women during the reproductive years have a higher circatrigintan (menstrual) rhythm amplitude and which over the 24 h span are secreted predominantly in a pulsatile fashion. In certain immunologic functions (e.g., in transplantation biology) a prominent circaseptan periodicity seems to be operational in determining in part the response of the host's immune system to the introduced antigen (e.g., allograft). In some variables, e.g. DHEA-S, the circannual variation may be more important than the circadian amplitude (Nicolau 1984).

Circadian rhythms are modulated by ultradian rhythms with a higher frequency and by rhythms with lower frequencies, i.e., in the circaseptan, circatrigintan, and circannual frequency range. The higher frequencies may lead to spurious results and to aliasing in (infrequent) spot-checks in the study of rhythms with longer periods. The lower frequencies found in many parameters may alter MESOR, amplitude, and acrophase of a higher frequency rhythm and have to be kept in mind in experimental designs and in clinical studies.

There are a limited number of laboratory functions showing high amplitude circadian rhythms. These are predominantly found in endocrinology and in hematology with a few also in clinical chemistry. Rhythms of relatively large amplitude require time-qualified reference ranges for the interpretation of individual measurements. The determination of reference limits, irrespective of time, may be too broad and hence not discriminating. High amplitude rhythms of different frequencies characterize variables of vital interest in biology and medicine and may contribute a large fraction of the variability seen in clinical data. Sampling at time points chosen (on the basis of previous chronobiologic information) by pertinence rather than only by convenience is essential to obtain meaningful and comparable results in parameters showing high amplitude rhythms. In some of these, the same value, when checked against an appropriate chronodesm, may in fact be too low at one time, quite normal at another time, and even too high at yet another time (Haus et al. 1980a; Haus and Halberg 1980). If measurements are performed near the acrophase of the rhythms and compared to a non-time-qualified usual range, a normal value may well appear to be elevated (a false positive) or an abnormally low value may be considered normal (false negative). Similarly, if a measurement is made near the bathyphase (trough) of the rhythm, a normal value may be misinterpreted as abnormally low (a false positive) or an abnormally high value may be considered normal (false negative).

For some parameters which show marked interindividual variations, even the time-qualified reference ranges obtained from well-selected and apparently appropriate peer groups may be quite large. In such cases it is essential in the evaluation of a laboratory test in a given clinical situation to keep in mind the rhythm parameters of rhythm adjusted level (MESOR), the timing (acrophase), and the extent of the rhythm (as characterized either by its amplitude and/or by the range of the variation) and to evaluate consecutive measurements in a given individual against this background.

Time-qualified reference ranges (chronodesms) are desirable for high amplitude rhythms in all frequency ranges. The choice of the method chosen for establishing these reference ranges will depend on the shape of the rhythmic variation, the distribution of the measurements, and the number of measurements available. In the circadian range, the nonsinusoidal shape of many of the high amplitude rhythms of laboratory functions and the marked differences in variance occurring characteristically at different times along the 24 h scale have led us to prefer a time-qualified usual range calculated for each time point of sampling over the fitting of reference ranges according to sinusoidal models. The non-Gaussian distribution of the values found in many laboratory functions often requires the use of nonparametric statistics (i.e., percentiles) to delineate the usual range in a chronodesm.

The circadian rhythms in most laboratory functions can be detected in diurnally active, clinically healthy, ambulatory subjects following their habitual daily routine. For most functions standardization beyond that used for blood drawing in the laboratory did not appear essential for rhythm detection (some exceptions, e.g., in plasma catecholamines, plasma renin, etc., are obvious). It has to be realized that under the conditions encountered in a free-living population and in most clinical situations, the circadian variation measured is likely to consist of an endogenous rhythm component which may be modified in both timing and amplitude by environmental factors which may act as synchronizers and/or as masking agents.

It has to be emphasized that a given clock hour, day of the week, or month of the year is not necessarily representative of the biologic time of an organism. For example, circadian periodic changes must not be misinterpreted as time of day effects. The same time of day may have an entirely different meaning for two individuals on two different working schedules. Phase-shifts, phase-drift, and within group desynchronization have to be kept in mind in the interpretation of experimental or laboratory results.

In the study of human subjects or populations, chronobiologic sampling should be accomplished against the background of the subject's rest-activity and sleep-wakefulness patterns, working schedules, and dietary habits, including the times of the main meals. Inquiry into such simple biologic reference functions provides some information on the subject's synchronization and may be critical if sampling has to be limited to a single or to a few time points. Other potential reference functions or marker rhythms in the circadian domain are body temperature, urinary volume, urinary, salivary and sweat electrolytes, notably sodium and potassium, and corticosteroids. Many variables can serve as potential reference functions. The pertinence of a reference function for the rhythm under study will have to be explored in each instance. Preferably, a reference function should be longitudinally measurable at frequent intervals by noninvasive methods. Variables measurable by physical or physicochemical end points that do not involve expensive and/or lengthy chemical analyses and which can be recorded automatically in a computer compatible form are particularly attractive.

Telemetering of some biochemical or physical reference functions will be of crucial importance and will – in the long run – greatly reduce the cost of chronobiologic studies. Automated monitoring will allow one to limit invasive sample collection for more costly procedures, e. g., for chemical analysis, to a few well-defined time points chosen for their pertinence rather than convenience. The practical application of human chronobiology to environmental and work physiology and to clinical medicine will depend critically on the availability of instrumentation for the noninvasive or minimally invasive study of marker rhythms.

Many circadian (and other) rhythms encountered are of low amplitude and do not pose a diagnostic problem in today's clinical practice. They may, however, gain in importance if the sampling requirements for statistically valid evaluation of low amplitude rhythms can be met (e. g., frequent noninvasive sampling over a prolonged time span by automatic miniaturized instrumentation preferably interfaced with a data processor or computer). The low amplitude rhythms indicate the usual sequence of many metabolic events and by phase and/or amplitude alterations may indicate early pathology. Temporal abnormalities in low amplitude circadian rhythm parameters have, in some instances, been found to precede the relatively coarse abnormalities regarded as clinically important at this time. The circadian rhythm in serum bilirubin, for example, has too small an amplitude to present a problem for the diagnosis of jaundice. However, it may be of considerable biologic interest since changes in the circadian rhythm of serum bilirubin have been reported as one of the earliest signs of liver damage (Ferrari et al. 1981). Rhythm alterations may indicate early pathology by revealing temporal changes of laboratory values which, by a conventional approach, would be regarded as within the "usual range."

Time-dependent changes in drug kinetics are reviewed in the chapter in this volume by Bruguerolle and will have to be applied to the evaluation of clinical drug level studies.

## Conclusions

Chronobiology plays an important role in laboratory medicine which will undoubtedly expand in the future. A critical amount of chronobiologic information in parameters measured in the clinical laboratory is now available which indicate that:

1. The laboratory values regarded as usual under conditions of health of a biologic system cannot be expressed in terms of time-unqualified measurements of clinical laboratory functions. We have to take into account the time structure of the mammalian organism, consisting of a wide spectrum of rhythms superimposed upon each other, modulating each other, and often showing a certain relation in phase, frequency, and/or amplitude which may be critical for normal function.

2. Time-qualified reference values (chronodesms) are essential for a definition of "usual values" of certain large amplitude laboratory parameters.

3. Statistically quantified rhythm parameters have become new end points for laboratory measurements. These include a rhythm's frequency or period, a rhythm adjusted mean (the MESOR), the extent of a rhythm (its amplitude or its range of change), and its timing (e. g., the acrophase). In the ultradian range, the description of the rhythmic or pulsatile variations also includes number, height, width, and distribution of the episodic peaks and troughs (e. g., of polypeptide and steroid hormone concentrations). These new end points may not only serve to characterize normalcy, but in their alterations may indicate risk states and disease. Some of the rhythmic and/or episodic variations may be essential for certain body functions and if found to be abnormal require correction and/or substitution (e. g., in the field of fertility and reproduction).

4. The task to obtain an adequate data base for the chronobiologic evaluation of laboratory results in the several pertinent frequencies encountered in most parameters is still incomplete. However, after the background information on a biologic rhythm has been obtained by more extended studies of representative populations, a much more limited approach will yield meaningful and cost-effective results in samples obtained at a time pertinent for the problem to be investigated. The introduction of the time domain into the practice of laboratory medicine will greatly enlarge its scope and is expected to lead in the future to a large number of useful applications.

5. The future application of chronobiology in laboratory medicine will depend critically on advances in the field of data collection and data analysis. The technology for automated sample collection including *in vivo* measurements of biochemical end points as such is available but has not been refined and miniaturized for routine chronobiologic sampling in biology and medicine. Also, the technology for direct data transfer to the computer for rhythmometric analysis is available in specialized laboratories and will have to be adapted for wider cost-effective use in clinical routines.

# References

Abetel G, Karly M, Genoud G (1986) Evaluation of the anti-hypertensive effect of a beta-blocker with the aid of daily blood pressure profiles. J Cardiovasc Pharmacol 8 [Suppl 6]: 77–79

Angeli A, Agrimonti F, Frairia R, Vilino PL, Barbadoro E, Ceresa F (1981) Circadian patterns of plasma cortisol and testosterone in chronic male alcoholics. Int J Chronobiol 7: 199 (abstr)

Arbogast B, Lubanovic W, Halberg F, Cornelissen G, Bingham C (1983) Chronobiologic serial sections of several orders. Chronobiologia 10: 59–68

Arendt J, Bojkowski C, Franey C, Wright J, Marks V (1985) Immunoassay of 6-hydroxymelatonin sulfate in human plasma and urine: abolition of the urinary 24-hour rhythm with atenolol. J Clin Endocrinol Metab 60: 1166–1173

Bendat JS, Piersol AG (1971) Random data-analysis and measurement procedures. Wiley Interscience, New York

Bertello P, Agrimonti F, Gurioli L, Frairia R, Fornaro D, Angeli A (1982) Circadian patterns of plasma cortisol and testosterone in chronic male alcoholics. Alcohol Clin Exp Res 6: 475–481

Bingham C, Arbogast B, Cornelissen-Guillaume G, Lee JK, Halberg F (1982) Inferential statistical methods for estimating and comparing cosinor parameters. Chronobiologia 9: 397–439

Blackman RB, Tukey JW (1954) The measurement of power spectra, from the point of view of communications engineering. Dover, New York

Bloomfield P (1976) Fourier analysis of time series: an introduction. Wiley, New York

Bremner WJ, Vitiello MV, Prinz PN (1983) Loss of circadian rhythmicity in blood testosterone levels with aging in normal men. J Clin Endocrinol Metab 56: 1278–1281

Charyulu K, Halberg F, Reeker E, Haus E, Buchwald H (1974) Autorhythmometry in relation to radiotherapy: case report as tentative feasibility check. In: Scheving LE, Halberg F, Pauly JE (eds) Chronobiology. Igaku Shoin, Tokyo, pp 265–272

Childs G, Redfern PH (1981) A circadian rhythm in passive avoidance behaviour: the effect of phase shift and the benzodiazepines. Neuropharmacology 20: 1365–1366

Cowen PJ, Fraser S, Sammons R, Green AR (1983) Atenolol reduces plasma melatonin concentrations in man. Br J Clin Pharmacol 15: 579–581

Dawes C (1969) The effects of flow rate and duration of stimulation on the concentrations of protein and the main electrolytes in human parotid saliva. Arch Oral Biol 14: 277–294

Dawes C (1974) Rhythms in salivary flow rate and composition. Int J Chronobiol 2: 253–279

Dawes C, Chebib FS (1972) The influence of previous stimulation and the day of the week on the concentrations of protein and the main electrolytes in human parotid saliva. Arch Oral Biol 17: 1289–1301

De Prins J, Cornelissen G, Hillman D, Halberg F, Van Dijck C (1981) Harmonic interpolation yields paraphases and orthophases for biologic rhythms. In: Hayes DK, Halberg F, Scheving LE (eds) Proceedings of the XIII International Conference of the Society of Chronobiology, pp 333–334

De Prins J, Cornelissen G, Malberg W (1986) Statistical procedures in Chronobiology and Chronopharmacology. In: Reinberg A, Smolensky M, Labreque G (eds) Annual Review of Chronopharmacology, vol 2. Pergamon, New York, pp 27–141

Enright JT (1965) The search for rhythmicity in biological time series. J Theor Biol 3: 426–468

Eskin A, Takahashi JS (1983) Adenylate cyclase activation shifts the phase of a circadian pacemaker. Science 220: 82–84

Eskin A, Corrent G, Lin CY, McAdoo DJ (1982) Mechanism for shifting the phase of a circadian rhythm by serotonin: involvement of CAMP. Proc Natl Acad Sci USA 79: 660–664

Ferguson DB, Botchway CA (1979) Circadian variations in flow rate and composition of human stimulated submandibular saliva. Arch Oral Biol 24: 433–437

Ferguson DB, Botchway CA (1980) A comparison of circadian variation in the flow rate and composition of stimulated human parotid, submandibular and whole salivas from the same individuals. Arch Oral Biol 25: 559–568

Ferrari E, Bossolo PA, Daguati M, Canepari C, Ficara S (1981) Lack of adrenocortical rhythmicity in liver cirrhosis. Int J Chronobiol 7: 239 (abstr)

Filicori M, Flamigni C, Crowley WF Jr (1987) The critical role of blood sampling frequency in the estimation of episodic luteinizing hormone secretion in normal woman. In: Crowley WF Jr, Hofler JG (eds) The episodic secretion of hormones. Wiley, New York, pp 5–13

Fisher RA (1929) Tests of significance in harmonic analysis. R Soc Lond Proc Ser A 125: 54–59

Fookson JE, Weitzman ED, Cseisler CA, Zimmerman JC, Ronda J (1984) Development of mathematical techniques to describe chronophysiological rhythms in man during temporal isolation. In: Haus E, Kabat H (eds) Chronobiology 1981–1983. Karger, New York, pp 73–82

Ghata J, Reinberg A, Lagoguey M, Touitou Y (1977) Human circadian rhythms documented in May–June from three groups of young, healthy males living respectively in Paris, Colombo and Sydney. Chronobiologia 4: 181–190

Glass AR, Zavadie AP III, Halberg F, Cornelissen G, Schaaf M (1984) Circadian rhythm of serum cortisol in Cushing's disease. J Clin Endocrinol Metab 59: 161–165

Gowenlock AE (ed) (1988) Practical clinical biochemistry. CRC, Boca Raton, pp 808–856

Grassbeck R, Alstrom T (eds) (1981) Reference values in laboratory medicine. Wiley, New York

Halberg F (1963) Circadian (about 24-hour) rhythms in experimental medicine. Proc R Soc Med 56: 253–257

Halberg F (1965) Some aspects of biologic data analysis; longitudinal and transverse profiles of rhythms. In: Aschoff J (ed) Circadian clocks. Proceedings of the Feldafing Summer School. North-Holland, Amsterdam, pp 13–22

Halberg F, Panofsky H (1961) I. Thermo-variance spectra; method and clinical illustrations. Exp Med Surg 19: 284–309

Halberg F, Reinberg A (1967) Rythmes circadiens et rythmes de basses frequences en physiologie humaine. J Physiol (Paris) 59: 117–200

Halberg F, Cohen SL, Flink EB (1951) Two new tools for the diagnosis of adrenal dysfunction. J Lab Clin Med 38: 817–825

Halberg F, Engeli M, Hamburger C, Hillman D (1965) Spectral resolution of low-frequency, small amplitude rhythms in excreted ketosteroids; probable androgen-induced circaseptan desynchronization. Acta Endocrinol [Suppl] (Copenh) 103: 5–54

Halberg F, Engel R, Swank R, Seaman G, Hissen W (1966) Cosinor Auswertung circadianer Rhythmen mit niedriger Amplitude im menschlichen Blut. Phys Med Rehabil 7: 1–7

Halberg F, Tong YL, Johnson EA (1967) Circadian system phase – an aspect of temporal morphology; procedures and illustrative examples. In: von Mayersbach H (ed) The cellular aspects of biorhythms. Springer, Berlin, Heidelberg New York, pp 20–48

Halberg F, Reinhardt J, Bartter FC, Delea C, Gordon R, Reinberg A, Ghata J, Hofmann H, Halhuber M, Gunther R, Knapp E, Pena JC, Garcia Sainz M (1969) Agreement in endpoints from circadian rhythmometry on healthy human beings living on different continents. Experientia 25: 107–112

Halberg F, Johnson EA, Nelson W, Runge W, Sothern R (1972) Autorhythmometry procedures for physiologic self-measurements and their analysis. Physiol Teach 1: 1–11

Halberg F, Carandente F, Cornelissen G, Katinas GS (1977) Glossary of chronobiology. Chronobiologia 4 [Suppl 1]: 1–189

Halberg F, Lee JK, Nelson WL (1978) Time-qualified reference intervals chronodesms. Experientia 34: 713–716

Halberg F, Cornelissen G, Sothern RB, Wallach LA, Halberg E, Ahlgren A, Kuzel M, Radke A, Barbosa J, Goetz F, Buckley J, Mandel J, Schuman L, Haus E, Lakatua D, Sackett L, Berg H, Kawasaki T, Ueno M, Uezono K, Matsuoka M, Omae T, Tarquini B, Cagnoni M, Garcia Sainz M, Vega EP, Griffiths K, Wilson D, Wetterberg L, Donati L, Tatti P, Vasta M, Locatelli I, Camagna A, Lauro R, Tritsch G, Wendt HW (1981a) International geographic studies of oncological interest on chronobiologic variables. In: Kaiser HE (ed) Neoplasms – comparative pathology of growth in animals, plants, and man. Williams and Wilkins, Baltimore, pp 553–596

Halberg F, Tarquini B, Lakatua D, Halberg E, Seal U, Haus E, Cagnoni M (1981b) Circadian and circannual plasma TSH rhythms, human mammary and prostatic cancer and steps toward chrono-oncoprevention. Lab J Res Lab Med 8: 251–257

Halberg F, Lagoguey M, Reinberg A (1983) Human circannual rhythms over a broad spectrum of physiological processes. Int J Chronobiol 8: 225–268

Hanson EJ (1970) Multiple time series. Wiley, New York

Haus E (1964) Periodicity in response and susceptibility to environmental agents. Ann NY Acad Sci 117: 281–291

Haus E (1987) Requirements for chronobiotechnology and chronobiologic engineering in laboratory medicine. In: Scheving LE, Halberg F, Ehret CF (eds) Chronobiotechnology and chronobiological engineering. Kluwer Academic, Netherlands, pp 331–372 (NATO ASI series, series E: applied sciences, no 120)

Haus E, Halberg F (1970) Circannual rhythm in level and timing of serum corticosterone in standardized inbred mature C-mice. Environ Res 3: 81–106

Haus E, Halberg F (1980) The circadian time structure. In: Scheving LE, Halberg F (eds) Chronobiology-principles and applications to shifts and schedules. Sijthoff, Leiden, pp 47–94 (NATO advanced study institute series D)

Haus E, Halberg F, Kuhl JWF, Lakatua DJ (1974a) Chronopharmacology in animals. Chronobiologia [Suppl 1]: 122–156

Haus E, Halberg F, Loken MK, Kim YS (1974b) Circadian rhythmometry of mammalian radiosensitivity. In: Tobias CA, Todd P (eds) Space radiation biology. Academic, New York, pp 435–474

Haus E, Cornelissen G, Halberg F (1980a) Introduction to chronobiology. In: Scheving LE, Halberg F (eds) Chronobiology-principles and applications to shifts and schedules. Sijthoff, Leiden, pp 1–32 (NATO advanced study institute series D)

Haus E, Lakatua DJ, Halberg F, Halberg E, Cornelissen G, Sackett L, Berg H, Kawasaki T, Ueno M, Uezono K, Matsouka M, Omae T (1980b) Chronobiological studies of plasma prolactin in woman in Kyushu, Japan and Minnesota, USA. J Clin Endocrinol Metab 51: 632–640

Haus E, Lakatua D, Swoyer J, Sackett-Lundeen L (1983) Chronobiology in hematology and immunology. Am J Anat 168: 467–517

Haus E, Lakatua DJ, Sackett-Lundeen L (1984) Chronobiology in laboratory medicine. In: Reitveld WT (ed) Clinical aspects of chronobiology. Baarn-Bakker, The Netherlands, pp 13–83

Haus E, Nicolau GY, Lakatua DJ, Sackett-Lundeen L, Bogdan C, Petrescu E (1986) Circadian endocrine rhythm alterations in elderly cigarette smokers. Annu Rev Chronopharmacol 3: 115–118

Haus E, Nicolau GY, Lakatua DJ, Jachimowicz E, Plinga E, Sackett-Lundeen L, Petrescu E, Ungureanu E (1987) Circannual variations in blood pressure, urinary catecholamine excretion, plasma aldosterone and serum sodium, potassium, calcium and magnesium in children 11 ± 1.5 years of age. In: Advances in chronobiology, part B. Liss, New York, pp 3–19

Haus E, Nicolau GY, Lakatua D, Sackett-Lundeen L (1988a) Reference values for chronopharmacology. Annu Rev Chronopharmacol 4: 333–424

Haus E, Nicolau GY, Lakatua DJ, Bogdan C, Popescu M, Sackett-Lundeen L, Fraboni A, Petrescu E (1988b) Circadian rhythm parameters of clinical and endocrine functions in elderly subjects under treatment with various commonly used drugs. Annu Rev Chronopharmacol 5: 77–80

Haus E, Nicolau GY, Lakatua DJ, Sackett-Lundeen L, Swoyer J (1990) Circadian rhythms in laboratory medicine. In: Fanfani M, Tarquini B (eds) Reference values and chronobiology. Arand and Brent, Florence, pp 21-32

Hermida RC, Halberg F (1986) Bootstrapping and added data discriminate, at low blood pressure, neuroendocrine risk of developing mesor hypertension. Chronobiologia 13: 29–36

Hermida RC, Halberg F, Del Pozo F, Haus E (1982) Toward a chronobiologic pattern of the risk of breast cancer and other diseases. Rev Esp Oncol 29: 199–207

Hermida RC, Halberg F, Del Pozo F, Chavarria F (1984) Pattern discrimination and the risk to develop breast cancer. In: Haus E, Kabat H (eds) Chronobiology 1982–1983. Karger, New York, pp 399–412

Hermida RC, Bingham C, Halberg F, Del Pozo F (1987) Bootstrapped potential circadian harbingers if not determinants of cardiovascular risk. In: Pauly JE, Scheving LE (eds) Advances in chronobiology, part B. Liss, New York, pp 571–583 (Progress in clinical and biological research, vol 227B)

Horning MG, Brown L, Nowlin J, Lertratanangkoon K, Kellaway P, Zion TE (1977) Use of saliva in therapeutic drug monitoring. Clin Chem 23: 157–164

Jung K, Schulze G, Reinholdt C (1986) Different diuresis-dependent excretions of urinary enzymes: $N$-acetyl-$\beta$-D-glutaminidase, alanine aminopeptidase, alkaline phosphatase, and glutamyl transferase. Clin Chem 32: 529–532

Kawasaki T, Uezono K, Ueno M, Omae T, Matsuoka M, Haus E, Halberg F (1983) Comparison of circadian rhythms of the renin-angiotensin-aldosterone system and electrolytes in clinically healthy young women in Fukuoka (Japan) and Minnesota (USA). Acta Endocrinol (Copenh) 102: 246–251

Knapp MS, Keane PM, Wright JG (1967) Circadian rhythm of plasma 11-hydroxycorticosteroids in depressive illness, congestive heart failure, and Cushing's syndrome. Br Med J 2: 27–30

Knapp MS, Cove-Smith JR, Dugdale R, MacKenzie N, Pownall R (1979) Possible effect of time on renal allograft rejection. Br Med J 1: 75–77

Krieger DT (1973) Pathophysiology of central nervous system regulation of anterior pituitary function. In: Pathophysiology of central nervous system regulation of anterior pituitary function. In: Biology of brain dysfunction, vol 2. Plenum, New York, pp 351

Krieger DT (1979) Rhythms in CRF, ACTH, and corticosteroids. In: Krieger DT (ed) Endocrine rhythms. Raven, New York, pp 123–142

Krieger DT, Allen W (1975) Relationship of bioassayable and immunoassayable ACTH and cortisol concentrations in normal subjects and in patients with Cushing's disease. J Clin Endocrinol Metab 40: 675–687

Lagoguey M, Reinberg A (1981) Circadian and circannual changes of pituitary and other hormones in healthy human males: their relationship with gonadal activity. In: Van Cauter E, Copinschi C (eds) Human pituitary hormones. Nijhoff, The Hague, pp 261–278

Lakatua DJ, Blomquist CH, Haus E, Sackett-Lundeen L, Berg H, Swoyer J (1982) Circadian rhythm in urinary $N$-acetylglucosaminidase (NAG) of clinically healthy subjects: timing and phase relation to other urinary circadian rhythms. Am J Clin Pathol 78: 69–77

Lakatua DJ, Nicolau GY, Bogdan C, Plinga L, Jachimowicz A, Sackett-Lundeen L, Petrescu E, Ungureanu E, Haus E (1987) Chronobiology of catecholamine excretion in different age groups. In: Pauly JE, Scheving LE (eds) Advances in chronobiology, part B. Liss, New York, pp 31–50 (Progress in clinical and biological research, vol 227B)

Lakatua D, Nicolau GY, Haus E, Sackett-Lundeen L, Petrescu E (1990) Circadian rhythm parameters of seventeen hormonal variables in plasma and of urinary catecholamines in elderly subjects treated with cyclobarbital. Annu Rev Chronopharmacol 7: 269–272

Levi F, Halberg F (1982) Circaseptan (about 7-day) bioperiodicity – spontaneous and reactive – and the search for pacemakers. La Ricerca Clin Lab 12: 323–370

MacNeill IB (1974) Tests for periodic components in multiple time series. Biometrika 61: 57–70

Marquardt DW (1963) An algorithm for least squares estimation of nonlinear parameters. J Soc Ind Appl Math 11: 431–441

Merriam GH, Wachter DW (1982) Algorithms for the study of episodic hormone secretion. Am J Physiol 243: E 310–318

Merriam GR, Wachter KW (1984) Measurement and analysis of episodic hormone secretion. In: Rodbard D, Forti G (eds) Computers in endocrinology. Raven, New York, pp 325–346

Miles A, Philbrick DRS, Thomas DR, Grey J (1987) Diagnostic and clinical implications of plasma and salivary melatonin assay. Clin Chem 33: 1295–1297 (lett)

Minors DS, Waterhouse JM (1980) Aspects of chronopharmacokinetics and chronergy of ethanol in healthy man. Chronobiologia 7: 465–480

Minors DS, Waterhouse JM (1989) Mathematical and statistical analysis of circadian rhythms. Psychoneuroendocrinology 13: 443–464

Montalbetti N, Halberg F (1983) Cronopatologia clinica. In: Baserga A (ed) Patologia clinica: a cura di F. Corso. Masson, Milan, pp 73–85

Moore-Ede MC, Czeisler CA, Richardson GS (1983) Circadian time keeping in health and disease. N Engl J Med 309: 469–476

Nelson WL, Tong YL, Lee JK, Halberg F (1979) Methods for cosinor rhythmometry. Chronobiologia 6: 305–323

Nicolau GY (1984) Environmental chronopathology: circadian dyschronisms as early indicator of toxicity in chronic pesticide exposure. In: Haus E, Kabat H (eds) Chronobiology 1982–1983. Karger, New York, pp 379–389

Nicolau GY, Haus E, Lakatua D, Bogdan C, Petrescu E, Sackett-Lundeen L, Berg H, Ioanitiu D, Popescu M, Chiopan C, Milcu SM (1982) Endocrine circadian time structure in the aged. Rev Roum Med Endocrinol 20: 165–176

Nicolau GY, Haus E, Lakatua DJ, Bogdan C, Popescu M, Petrescu E, Sackett-Lundeen L, Ioanitiu D (1983a) Circadian and circannual variations in plasma immunoreactive insulin (IRI) and C-peptide concentrations in elderly subjects. Rev Roum Med Endocrinol 21: 243–255

Nicolau GY, Haus E, Lakatua DJ, Bogdan C, Popescu M, Petrescu E, Sackett-Lundeen L, Swoyer J, Adderley J (1983b) Circadian periodicity of the results of frequently used laboratory tests in elderly subjects. Rev Roum Med Endocrinol 21: 3–21

Nicolau GY, Lakatua D, Sackett-Lundeen L, Haus E (1984) Circadian and circannual rhythms of hormonal variables in elderly men and women. Chronobiol Int 1: 301–319

Nicolau GY, Haus E, Lakatua D, Bogdan C, Petrescu E, Robu E, Sackett-Lundeen L, Swoyer J (1985a) Chronobiologic observations of calcium and magnesium in the elderly. Rev Roum Med Endocrinol 23: 39–53

Nicolau GY, Haus E, Lakatua DJ, Bogdan C, Sackett-Lundeen L, Popescu M, Berg H, Petrescu E, Robu E (1985b) Circadian and circannual variations of FSH, LH, testosterone, dehydroepiandrosterone-sulfate (DHEA-S) and 17-hydroxy progesterone (17 OH-PROG) in elderly men and women. Rev Roum Med Endocrinol 23: 223–246

Nicolau GY, Haus E, Lakatua DJ, Bogdan C, Sackett-Lundeen L, Petrescu E, Reilly C (1986) Circannual rhythms of laboratory parameters in serum of elderly subjects – evaluation by cosinor analysis. Rev Roum Med Endocrinol 24: 281–292

Nicolau GY, Dumitriu L, Plinga L, Petrescu E, Sackett-Lundeen L, Lakatua DJ, Haus E (1987a) Circadian and circannual variations of thyroid function in children 11 ±1.5 years of age with and without endemic goiter. In: Pauly JE, Scheving LE (eds) Advances in chronobiology, part B. Liss, New York, pp 229–247 (Progress in clinical and biological research, vol 227B)

Nicolau GY, Haus E, Lakatua DJ, Bogdan C, Plinga L, Irvine P, Petrescu E, Sackett-Lundeen L, Swoyer J (1987b) Chronobiology of serum iron concentrations in subjects of different ages at different geographic locations. Rev Roum Med Endocrinol 25: 63–82

Oester KE, Guardabasso V, Rodbard D (1986) Detection and characterization of peaks and estimation of instantaneous secretory rate for episodic pulsatile hormone secretion. Comput Biomed Res 19: 170–191

Paille F, Royer-Morrot MJ, Trechot P, Royer RJ (1988) Alcohol and biological rhythms. Ann Med Int (Paris) 139: 213–218

Panofsky H, Halberg F (1961) II. Thermo-variance spectra: simplified computation computational example and other methodology. Exp Med Surg 19: 323–338

Petit Clerc C, Solberg H (1987) Approved recommendation (1987) on the theory of reference values, part 2. Selection of individuals for the production of reference values. Report of expert panel on theory of reference values (EPTRV) of the international federation of clinical chemistry (IFCC). J Clin Chem Clin Biochem 25: 639–644

Ralph MR, Menaker M (1983) Bicuculline selectively blocks light induced phase delays in the circadian rhythms of hamsters. Soc Neurosci Abstr 9: 1073

Ralph MR, Menaker M (1985) Bicuculline blocks circadian phase delays but not advances. Brain Res 325: 362–365

Ralph MR, Menaker M (1986) Effects of diazepam on circadian phase advances and delays. Brain Res 372: 405–408

Read CF, Riad-Fahmy D, Wilson DW, Griffiths K (1983) A new approach to breast cancer research: assays for steroids in saliva. In: Bulbrook RD, Taylor JD (eds) Commentaries on research in breast disease, vol 3. Liss, New York, pp 61–92

Read CF, Riad-Fahmy D, Walker RF, Griffiths K (eds) (1984) Proceedings of the 9th Tenovus workshop: immunoassays of steroids in saliva. Alpha Omega, Cardiff, pp 1–346

Reame N, Sauder SE, Kelch RP, Marshall JC (1984) Pulsatile gonadotropin secretion during the human menstrual cycle: evidence for altered frequency of gonadotropin releasing hormone secretion. J Clin Endocrinol Metab 59: 328–337

Refetoff S, Van Cauter E, Fang VS, Laderman C, Graybeal ML, Landau RL (1985) The effect of dexamethasone on the 24 hour profiles of adrenocorticotropin and cortisol in Cushing's syndrome. J Clin Endocrinol Metab 60: 527–535

Reilly C, Nicolau GY, Lakatua DJ, Bogdan C, Sackett-Lundeen L, Petrescu E, Haus E (1987) Circannual rhythms of laboratory parameters in serum of elderly subjects. In: Pauly JE, Scheving LE (eds) Advances in chronobiology, part B. Liss, New York, pp 51–72 (Progress in clinical and biological research, vol 227B)

Reinberg A (1983) Clinical chronopharmacology: an experimental basis for chemotherapy. In: Reinberg A, Smolensky MH (eds) Biological rhythms and medicine: cellular, metabolic, physiopathologic and pharmacologic aspects. Springer, Berlin Heidelberg New York, pp 211–263

Reinberg A, Smolensky M (1974) Circatrigintan secondary rhythms related to hormonal changes in the menstrual cycle: general considerations. In: Anderson JA (ed) Biorhythms and human reproduction – seminars in human reproduction. Wiley, New York, p 241–258

Reinberg A, Clench J, Aymard N (1975) Variations circadiennes des effets de l'éthanol et de l'éthanolémie chez l'homme adulte sain. Etude chronopharmacologique. J Physiol (Paris) 70: 435–456

Reinberg A, Schuller E, Delasuerie N, Clench J, Helary M (1977) Rythmes circadiens et circannuels des leucocytes, proteines totales, immunoglobulines A, G et M; Etude chez 9 adultes jeunes et sains. Nouv Presse Med 6: 3819–3823

Reinberg A, Lagoguey M, Cesselin F, Touitou Y, Legrand JC, De La Salle A, Antreassian J, Lagoguey A (1978) Circadian and circannual rhythms in plasma hormone and other variables of five healthy young human males. Acta Endocrinol (Copenh) 88: 417–427

Reinberg A, Touitou Y, Restoin A, Migraine C, Levi F, Montagner H (1985) The genetic background of circadian and ultradian rhythm patterns of 17 hydroxycorticosteroids: a cross-twin study. J Endocrinol 105: 247–253

Rosman PM, Farag A, Benn R, Tito J, Mishik A, Wallace EZ (1982) Modulation of pituitary-adrenocortical function: decreased secretory episodes and blunted circadian rhythmicity patients with alcoholic liver disease. J Clin Endocrinol Metab 55: 709–717

Ross JL, Barnes KM, Brody S, Merriam GR, Loriaux DL, Cutler GB Jr (1984) A comparison of two methods for detecting hormone peaks: the effect of sampling interval on gonadotropin peak frequency. J Clin Endocrinol Metab 59: 1159–1163

Rummel J, Lee JK, Halberg F (1974) Combined linear-nonlinear chronobiologic windows by least squares resolve neighboring components in a physiologic rhythm spectrum. In: Ferrin M, Halberg F, Richart RM (eds) Biorhythms and human reproduction. Wiley, New York, pp 53–82

Santen RJ, Bardin CW (1973) Episodic luteinizing hormone secretion in man: pulse analysis, clinical interpretation, physiologic mechanisms. J Clin Invest 52: 2617–2628

Sasaki T (1977) Basal metabolism. A recent trend of seasonal variations in basal metabolism. Bull Inst Const Med Kumamoto Univ 27: 2–7

Schuster A (1898) On the investigation of hidden periodicities with application to a supposed 26 day period of meteorological phenomena. Terr Magn 3: 13–41

Sensi S, Haus E, Nicolau GY, Halberg F, Lakatua DJ, Del Ponte A, Guagnano MT (1984) Circannual variation of insulin secretion in clinically healthy subjects in Italy, Romania and the USA. Riv Ital Biol Med 4: 1–8

Shannon IL, Chauncey HH (1967) A parotid fluid collection device with improved stability characteristics. J Oral Ther Pharmacol 4: 93–97

Shannon IL, Prigmore JR (1960) Parotid fluid flow rate. Its relationship to pH and chemical composition. Oral Surg, Oral Med, Oral Pathol 13: 1488–1500

Simpson HW, Halberg EA (1974) Menstrual changes of the circadian temperature rhythm in women. In: Anderson JA (ed) Biorhythms and human reproduction – seminars in human reproduction. Wiley, New York, pp 549–556

Solberg HE (1987) Approved recommendation (1986) on the theory of reference values, part 1. The concept of reference values. Report of expert panel on theory of reference values (EPTRV) of the international federation of clinical chemistry (IFCC). Clin Chim Acta 165: 111–118

Statland BE (1979) Fundamental issues in clinical chemistry. Am J Pathol 95: 243–272

Statland BE, Winkel P (1977) Effects of preanalytic factors on the intraindividual variation of analytes in the blood of healthy subjects: consideration of preparation of the subject and time of venipuncture. CRC Crit Rev Lab Sci 8: 105–144

Stramba-Badiale M, Castoldi C, Ceretti A (1981) Mantenimento del bioritmo del ferro nel longero e sua alterazione in presenza di anemia sideropenica. Minerva Med 72: 2173–2174

Sufi SB, Donaldson A, Gandy SG, Jeffcoate SL, Chearskul S, Goh H, Hazra D, Romero C, Wang HZ (1985) Multicenter evaluation of assays for estradiol and progesterone in saliva. Clin Chem 31: 101–103

Swoyer J, Haus E, Lakatua D, Sackett-Lundeen L, Thompson M (1984) Chronobiology in the clinical laboratory. In: Haus E, Kabat H (eds) Chronobiology 1981–1983. Karger, New York, pp 533–543

Swoyer J, Rhame F, Hrushesky W, Sackett-Lundeen L, Sothern R, Gale H, Haus E (1990) Circadian rhythm alterations in HIV infected subjects. In: Hayes DK, Pauly JE, Reiter RJ (eds) Chronobiology: its role in clinical medicine, general biology, and agriculture, part A. Wiley/Liss, New York, pp 437–449

Tarquini B, Gheri R, Romano SR, Costa A, Cagnoni M, Lee JK, Halberg F (1979) Circadian mesor-hyperprolactinemia in fibrocystic mastopathy. Am J Med 66: 229–237

Tarquini B, Halberg F, Seal US, Benvenuti M, Cagnoni M (1981) Circadian aspects of serum prolactin and TSH lowering by bromocriptine in patients with prostatic hypertrophy. Prostate 2: 269–279

Touitou Y (1982) Some aspects of the circadian time structure in the elderly. Gerontology 28: 53–67

Touitou Y, Limal JM, Bogdan A, Reinberg A (1976) Circadian rhythms in adrenocortical activity during and after a 36 hour 4-hourly-sustained administration of metyrapone in humans. J Steroid Biochem 7: 517–520

Touitou Y, Bogdan A, Limal JM, Touitou C, Reinberg A (1977) Circadian rhythm in urinary steroids in response to a 36-hour sustained metyrapone administration in 8 young men. Horm Metab Res 9: 314–321

Touitou Y, Touitou C, Bogdan A, Chasselut J, Beck H, Reinberg A (1979) Circadian rhythm in blood variables of elderly subjects. In: Reinberg A, Halberg F (eds) Chronopharmacology. Pergamon, New York, pp 283–290

Touitou Y, Fevre M, Lagoguey M, Carayon A, Bogdan A, Reinberg A, Beck H, Cesselin F, Touitou C (1981) Age and mental health-related circadian rhythm of plasma levels of melatonin, prolactin, luteinizing hormone and follicle-stimulating hormone in man. J Endocrinol 91: 467–475

Touitou Y, Sulon J, Bogdan A, Touitou C, Reinberg A, Beck H, Sodoyez JC, Van Cauwenberge H (1982) Adrenal circadian system in young and elderly human subjects: a comparative study. J Endocrinol 93: 201–210

Touitou Y, Carayon A, Reinberg A, Bogdan A, Beck H (1983) Differences in the seasonal rhythmicity of plasma prolactin in elderly human subjects. Detection in women but not in men. J Endocrinol 96: 65–71

Touitou Y, Fevre M, Bogdan A, Reinberg A, De Prins J, Beck H, Touitou C (1984) Patterns of plasma melatonin with aging and mental condition: stability of nyctohemeral rhythms and differences in seasonal variation. Acta Endocrinol (Copenh) 106: 145–151

Touitou Y, Motohashi Y, Pati A, Lévi F, Reinberg A, Ferment O (1986a) Comparison of cortical circadian rhythms documented in samples of saliva, capillary (fingertips) and venous blood from healthy subjects. Annu Rev Chronopharmacol 3: 297–299

Touitou Y, Reinberg A, Bogdan A, Auzeby A, Beck H, Touitou C (1986b) Age-related changes in both circadian and seasonal rhythms of rectal temperature with special reference to senile dementia of Alzheimer type. Gerontology 32: 110–118

Touitou Y, Touitou C, Bogdan A, Reinberg A, Auzeby A, Beck H, Guillet PH (1986c) Differences between young and elderly subjects in seasonal and circadian variations of total

plasma proteins and blood volume as reflected by hemoglobin, hematocrit and erythrocyte counts. Clin Chem 32: 801–804

Touitou Y, Touitou C, Bogdan A, Reinberg A, Motohashi Y, Auzeby A, Beck H (1989) Circadian and seasonal variations of electrolytes in aging humans. Clin Chim Acta 180: 245–254

Touitou Y, Garnier P, Reinberg A, Castagno L, Donnadieu M, Bogdan A, Motohashi Y (1990) Growth hormone (GH) and GH-releasing hormone response to ACTH 1-17 at different times of day. Evidence of a circadian stage dependence. Eur J Clin Pharmacol 38: 149–152

Truelove EL, Bixler D, Merritt AD (1967) Simplified method for collection of pure submandibular saliva in large volumes. J Dent Res 46: 1400–1403

Turek FW, Losee-Olson S (1986) A benzodiazepine used in the treatment of insomnia phase-shifts in the mammalian circadian clock. Nature 321: 167–168

Uezono K, Haus E, Swoyer J, Kawasaki T (1984) Circaseptan rhythms in clinically healthy subjects. In: Haus E, Kabat H (eds) Chronobiology 1982–1983. Karger, New York, pp 257–262

Urban RJ, Kaiser DL, Van Cauter E, Johnson ML, Veldhuis JD (1988) Comparative assessments of objective peak-detection algorithms. II. Studies in men. Am J Physiol 1 (1): E113–E119

Van Cauter E (1979) Method for characterization of 24-h temporal variations of blood components. Am J Physiol 237: E255–E264

Van Cauter E (1981) Quantitative methods for the analysis of circadian and episodic hormone fluctuations. In: Van Cauter E, Copinschi G (eds) Human pituitary hormones: circadian and episodic variations. Nijhoff, The Hague, p 1

Van Cauter E, Aschoff J (1989) Endocrine and other biologic rhythms. In: De Groot LJ (ed) Endocrinology. Saunders, Pennsylvania, pp 2658–2705

Van Cauter E, Honinckx E (1985) The pulsatility of pituitary hormones. In: Schulz H, Lavie P (eds) Ultradian rhythms in physiology and behavior. Springer, Berlin, Heidelberg, New York, pp 41–60

Veit I, Dietzel M, Lesch DM, Hermann P, Bipsak L, Reschenhofer E (1986) Polytoxicomane Patienten im neuroendokrinologischen Tagesprofil. Wien Med Wochenschr 19/20, pp 500–504

Veldhuis JD, Johnson ML (1986) Cluster analysis: a simple, versatile, and robust algorithm for endocrine pulse detection. Am J Physiol 250: E486–E493

Veldhuis JD, Johnson ML (1990) New methodological aspects of evaluating episodic neuroendocrine signals. In: Yen SSC, Vale W (eds) Advances in neuroendocrine regulation of reproduction. Plenum, New York, pp 123–139

Veldhuis JD, Rogol AD, Johnson ML (1985) Minimizing false-positive errors in hormonal pulse detection. Am J Physiol 248: E475–481

Veldhuis JD, Weiss J, Mauras N, Rogol AD (1986) Appraising endocrine pulse signals at low circulating hormone concentrations: use of regional coefficients of variation in the experimental series to analyze pulsatile luteinizing hormone release. Pediatr Res 20: 632–637

Veldhuis JD, Evans WS, Rogol AD, Thorner MO, Johnson ML (1987) Influence of rapid and extended venous sampling paradigms on the detection of gonadotropin pulses in men. In: Crowley WF Jr, Hofler JG (eds) The episodic secretion of hormones. Wiley, New York, pp 15–31

Veldhuis JD, Johnson ML, Ivanmanesh A, Lizarralde G (1990)

Temporal structure of *in vivo* secretory activity estimated by deconvolution analysis. J Biol Rhythms 5: 247–255

Vining RF, McGinley RA, Symons RG (1983) Hormones in saliva: mode of entry and consequent implications for clinical interpretation. Clin Chem 29: 1752–1756

Walker RF, Wilson DW, Riad-Fahmy D, Griffith K (1990) Chronobiology in laboratory medicine: Principles and clinical applications illustrated from measurements of neutral steroids in saliva. In: Hayes DK, Pauly JE, Reiter RE (eds) Chronobiology: its role in clinical medicine, general biology and agriculture, part A. Wiley/Liss, New York, pp 105–117

Weitzman ED, Czeisler CA, Zimmerman JC, Moore-Ede MC (1981) Biological rhythms in man: relationship of sleep-wake, cortisol, growth-hormone, and temperature during temporal isolation. Adv Biochem Psychopharmacol 28: 475–499

Wetterberg L (1978) Melatonin in humans. Physiological and clinical studies. J Neural Transm Suppl 13: 289–310

Wetterberg L, Halberg F, Halberg E, Haus E, Kawasaki T, Ueno, M Uezono K, Cornelissen G, Matsuoka M; Omae T (1986) Circadian characteristics of urinary melatonin from clinically healthy young women at different civilization disease risks. Acta Med Scand 220: 71–81

Wever R (1983) Bright light affects human circadian rhythm. Pflügers Arch 396: 85–87

Winfree AT (1980) The geometry of biological time. Springer, Berlin, Heidelberg, New York (Biomathematics, vol 8)

Zadik Z, de Lacerda L, Kowarski AA (1982) Evaluation of the 6-hour integrated concentration of cortisol as a diagnostic procedure for Cushing Syndrome. J Clin Endocrinol Metab 54: 1072–1074

# Glossary

Like any other discipline in biology and medicine, chronobiology has developed its own nomenclature. In relation to rhythms numerous terms describing their properties and parameters were borrowed from physics. Biologic rhythms, however, do not show the same precision as their counterparts in physics and thus cannot be characterized by point estimates. Every parameter of a biologic rhythm is a statistical entity which always has to be viewed with its variance estimate. This qualification has to be kept in mind if terms used in physics are applied to biologic rhythms and is expressed in the term "circa" which is used to describe frequencies which are known to change their cycle length under certain conditions. Although the terms used in physics are mostly well-defined and described in mathematical terms, their adaptation to chronobiology has led to some differences in their use by different investigators.

Other terms were adopted or coined (often derived from Latin or Greek) to describe aspects of biologic rhythms for which no suitable term was available and/or in order to avoid a lengthy descriptive phrase, the frequent use of which in a chronobiologic text could be quite cumbersome. Many of these terms were necessary and have been introduced similar to every other subspecialty of medicine and have been widely accepted by the specialists in the field. However, since chronobiology is a new and rapidly developing specialty, many of the terms introduced recently are still unknown to many investigators and physicians who might benefit from the application of chronobiologic principles and findings to their work. We have tried to help the reader of this book by presenting many of the more widely used terms with a definition which is kept as simple and generally understandable as possible. We are aware that some chronobiologists may be using more complicated definitions which may have some merit as such but often make the understanding of chronobiologic texts rather difficult for the nonspecialized reader. We have not included terms which have not yet been widely accepted and which are not generally used in the field. It is certain that, like in any living branch of science, some of these terms will be accepted while others will disappear as unnecessary or cumbersome. The following glossary is thus a snapshot of terms widely used today in chronobiology in a presentation aimed at being short and understandable (although sometimes perhaps simplified) and does not make any claim to completeness.

*Acrophase (Φ).* Measure of timing of a rhythm in relation to a defined reference timepoint selected by the investigator (e.g., local midnight for circadian rhythms); used for data which can be described by the fitting of a mathematical model, e.g., a cosine curve, and represents the crest time of the cosine curve best fitting to the data; may be expressed in (negative) degrees as the lag from the acrophase reference (360°C = 1 period) or in calendar time units (e.g., hours and minutes for circadian rhythms, days or months for infradian rhythms).

*Amplitude (A).* The measure of one half of the extent of the rhythmic change estimated by the mathematical model (e.g., cosine curve) best fitting to the data (e.g., the difference between the maximum and the rhythm-adjusted mean (MESOR) of the best-fitting curve).

*Autorhythmometry.* Self-measurement of biologic rhythms by the subject examined.

*Bathyphase.* The time of the lowest point of a mathematical model (e.g., cosine curve) fitted to a time series and describing a rhythm. If a sine or a cosine curve is fitted, the bathyphase will differ 180° from the acrophase, measured in relation to a defined reference time point selected by the investigator (e.g., local midnight for circadian rhythms); may be expressed in degrees as the lag from the phase reference (360° = 1 period) or in calendar time units (e.g., hours and minutes for circadian rhythms, days or months for infradian rhythm).

*Biologic time structure.* The sum of nonrandom time-dependent biologic changes, including growth, development, and aging, and a spectrum of rhythms with different frequencies.

**Biological clocks.** Self-sustained oscillators which generate biologic rhythms in absence of external periodic input (e. g., at the gene level in individual cells)

**Biologic rhythm.** A regularly recurring (periodic) component in a series of measurements of a biologic variable obtained as a function of time.

**Chronergy.** Represents the rhythmic change of the response of the organism to a drug (its total effect) according to its chronokinetics and its chronesthesy (see below).

**Chronesthesy.** Rhythmic (thus predictable-in-time) changes in the susceptibility or sensitivity of a target biosystem (cell or organism) to an agent. May be caused by temporal changes in receptors of target cells or organs, membrane permeability, etc.

**Chronobiology.** The science of investigating and objectively quantifying phenomena and mechanisms of the biologic time structure, including the rhythmic manifestations of life. Term derived from: Chronos (time), bios (life), and logos (science).

**Chronobiotic.** An agent capable of influencing biologic rhythm parameters (e.g., the phase setting).

**Chronodesm.** Time-qualified reference intervals. Reference intervals constructed along the time scale by Gaussian or non-Gaussian methods. Include time-qualified prediction and tolerance intervals.

**Chronogram.** Display of data as a function of time.

**Chronopathology.** Changes in an individual's biologic time structure preceding, coincident or following functional disorders or organic disease and/or time-dependent manifestation of disease.

**Chronopharmacodynamics.** Temporal variations in the mode of action of a drug.

**Chronopharmacokinetics.** The study of the temporal changes in absorption, distribution, metabolism, and elimination of a drug. Describes the influence of the time of administration of a drug on the mathematical parameters which describe these processes in terms of absorption rate, peak drug concentration ($C_{max}$), time-to-peak drug concentration ($E_{max}$), area under the concentration time curve (AUC), half-life ($t_{1/2}$), etc.

**Chronotherapy.** Use of treatment timed according to the stages in the sensitivity-resistance cycles of target (or nontarget) tissues and organs (or of the organism as a whole) to enhance the desired pharmacologic effect and/or reduce undesirable side effects of drugs or other therapeutic agents.

**Chronotolerance.** Time-dependent tolerance of an organism to environmental stimuli and xenobiotics.

**Chronotoxicology.** Time-dependent variation in toxicity.

**Circadian.** About 24 h. The term describes rhythms with an about 24-h ( >20 to <28 h) cycle length whether they are synchronized with a 24-h periodic surrounding or not.

**Circadiseptan.** A rhythm with a period of about 14 ( ±3) days.

**Circannual.** A rhythm with a period of about 1 year ( ±2 months), synchronized with or desynchronized from the calendar year.

**Circaseptan.** A rhythm with a period of about 7 ( ±3) days, which may or may not be synchronized with the calendar week.

**Circatrigintan.** A rhythm with a period of about 30 ( ±5) days. Includes, in mature women during the time of ovarian activity, the menstrual cycle. The term is preferred to the term "menstrual" because rhythms of this frequency are found in premenarchal girls, postmenopausal women and in men.

**Circavigintan.** A rhythm with a period of about 20 ( ±3 days).

**Clinospectrometry.** Resolving of a spectrum of rhythms and trends (cline) by (computer-implemented) time series collection and analysis. With rhythms quantified as algorithmically formulated phenomena validated in inferential statistical terms.

**Cosinor procedure.** A mathematical-statistical method of describing a rhythm by determining by least squares technique the cosine curve best fitting to the data and exploring the presence of a rhythm by examining the null hypothesis for amplitude in an F-test. If a rhythm can be described by this procedure the cosinor yields a rhythm-adjusted mean (MESOR), an amplitude as measure of the extent of the rhythm, and an acrophase as indication of its timing with variance estimates for each.

**Cosinor.** *Single cosinor* – a cosinor procedure applicable to single biologic time series.

*Population mean cosinor* – the cosinor procedure applicable to parameter estimates from three or more biologic time series for assessing the rhythm characteristics of a population. The parameter estimates are based on the means of estimates obtained from individuals in the samples.

**Daily.** Occurring every day.

**Dampened oscillation.** Oscillation decreasing (dampened) in amplitude due to inevitable loss of energy.

**Desynchronization.** State of two or more previously synchronized rhythmic variables that have ceased to exhibit the same frequency and/or the same acrophase relationships and show different than usual and/or changing time relations.

**Diurnal.** Day related (in contrast to nocturnal), e.g., diurnal (vs nocturnal) activity pattern.

**Endogenous rhythm.** Presumably genetically fixed biologic rhythm, persisting in an environment without outside time cues.

**Entrainment.** Coupling of two rhythms of the same frequency to one of them (the entraining agent or synchronizer) determining the phase of the other. e.g., Coupling of endogenous rhythms to environmental oscillator of the same frequency and/or determination of the phase of biologic rhythms by an internal pacemaker.

**Episodic variation.** Apparently irregular (nonrhythmic) variation of a biologic variable, e.g., episodic secretion of certain hormones (used by some as synonymous with "pulsatile").

**External desynchronization.** Desynchronization of a biologic rhythm from an environmental cycle.

**Feedsideward coordination.** Interaction of several rhythms (multifrequency coordination). Involves rhythmic and to that extent predictable sequences of effects depending upon the phase of each of the rhythms involved. "Feedsideward" may manifest itself as rhythmic alteration of stimulation, no effect or inhibition by an action of a rhythmic entity upon two other interacting entities.

**Free running.** Continuance of an endogenous bioperiodicity at least slightly but consistently different from any known environmental schedule, i.e., from its usual synchronizer or usual pacemaker rhythm.

**Frequency (f).** The number of cycles occurring per time unit; f is the reciprocal of the period (t).

**Frequency ranges.** Groups of frequencies (or periods) frequently encountered in biologic rhythms. (Circadian frequency range: rhythm with periods of about one day, i.e., by definition $> 20$ to $< 28$ h).

**Infradian rhythm.** Rhythm with a period longer (by definition $> 28$ h) than the circadian range; the term includes circaseptan, circatrigintan, circannual, and other rhythms of lower frequency.

**Internal desynchronization.** State in which two or more previously synchronized variables within the same organism have ceased to exhibit the same frequency and/or the same acrophase relationships and show different than usual and/or changing time relations.

**Jet lag.** Desynchronization and its clinical effect after rapid movement over several time zones (after transmeridian flights).

**Lighting regimen.** The light-dark cycle (LD), or constant light (LL), or constant dark (DD) conditions used for chronobiologic studies.

**Longitudinal sampling.** Study of the same subject or of a group of subjects over numerous cycles.

**Longitudinal study.** Study of the same individual over a prolonged time span (e.g., aging).

**Marker rhythm.** Rhythm of use in monitoring an organism's biologic timing and/or the timing of a related rhythm showing a fixed time relation to the rhythm used as "marker". Can be used where appropriate for decision-making in applied or basic physiologic or pharmacologic work, e.g., for time of sampling, timing of therapy, or for assessing therapeutic response (without any implication of causal relations between the rhythmic process and its marker). See also "reference rhythm".

**Masking of a rhythm.** Alteration of the usual shape and/or parameters of a rhythm due to random or nonrandom environmental stimuli, persisting for the duration of the stimulus only (without persistent alteration of endogenous rhythm components). e.g., change in body temperature after a hot bath.

**MESOR.** Midline Estimating Statistic of Rhythm. The value midway between the highest and the lowest values of the (cosine) function best fitting to the data. The "M" is equal to the arithmetic mean only for equidistant data covering an integral number of cycles.

**Pacemaker.** A functional entity capable of self-sustaining oscillations which synchronize other rhythms (e.g., the suprachiasmatic nucleus in man).

**Peak.** The highest point in a series of measurements obtained as a function of time.

**Period (τ).** Duration of one complete cycle in a rhythmic variation.

**Phase.** The value of a rhythmic biological variable at a certain time. Each instantaneous state of an oscillation represents a phase.

**Phase advance.** Involves the earlier occurrence of a rhythm's phase, usually the acrophase (denoted by a plus sign).

**Phase delay.** Involves the later occurrence of a rhythm's (acro)phase (denoted by a minus sign).

**Phase drift.** During free running of an endogenous rhythm with a period slightly but consistently different from its usual environmental synchronizer, the rhythm's acrophase will occur during every synchronizer cycle at a different time (e.g., clock hour in the case of circadian rhythms) in relation to the phase reference.

**Phase reference.** Time point chosen by the investigator as reference for the estimation of the timing of a rhythm (e.g., local midnight for circadian rhythms).

**Phase response curve.** Graphical plot indicating how the amount and the duration of a phase shift, induced by a single stimulus, depend upon the rhythm's stage at which the stimulus is applied.

**Phase shift.** Single relatively abrupt or gradual change in the timing of a rhythm (completed within a finite time span) and described by the difference between the initial and final (acro)phase.

**Photoperiod.** In a light-dark regimen the duration of the light span (e.g., in light-dark = LD 12:12 h, the photoperiod L = 12 h).

**Plexogram.** Display of original data covering spans longer than the period of a rhythm investigated along an abscissa of a single period (irrespective of time order of data collection).

**Pulsatile variation.** Variation of a biologic function with an irregular period higher than circadian of which a regular recurrence (rhythm) cannot be documented. May be the result of circadian-ultradian interactions or of other rhythmic and/or nonrhythmic mechanisms (used by some as synonymous with "episodic").

**Reference rhythm.** A rhythm in one variable used as a time reference for other rhythms, events, or actions. (See also "marker rhythms").

**Rhythm.** A regularly recurring and thus, to a certain degree, predictable (periodic) component of a (biologic) time series, demonstrated by inferential statistical means.

**Scotoperiod.** In a light-dark regimen the duration of the dark span (e. g., light-dark = LD 12:12 h the scotoperiod D = 12 h).

**Seasonal variation.** Change in a biologic system brought about by seasonal changes of temperature, light-span, etc, and not observed in the absence of such changes.

**Self-sustained oscillation.** System that can make use of a constant source of energy (to counteract energy losses) and is able to continue to oscillate without outside energy input.

**Shift work.** Transient or permanent change in work schedule in relation to the social surroundings (e. g., $3 \times 8$-h work shifts).

**Suprachiasmatic nucleus.** Group of hypothalamic neurons situated above the optic chiasm exhibiting an endogenous circadian oscillation acting as circadian pacemaker, receiving external phase information via the retina.

**Synchronization.** State of a system when two or more variables exhibit periodicity with the same frequency and specifiable acrophase and phase relation.

**Synchronizer.** Environmental periodicity determining the temporal placement of a biologic rhythm along an appropriate time scale. Synonyms: entraining agent, time giver, Zeitgeber.

**Synchronizing agent.** See Synchronizer.

**Time giver.** See Synchronizer.

**Time series.** A series of measurements obtained as a function of time.

**Transmeridian flight.** Movement over time zones (see Jet lag).

**Transverse sampling.** Sampling of a group of subjects over one cycle of a rhythm.

**Transverse study.** Comparison of two groups differing by a parameter (e. g., age, sex, etc.) studied at one time (e. g., over one cycle).

**Trough.** The lowest point in a series of measurements obtained as a function of time.

**Ultradian rhythm.** Biologic rhythm with a period shorter than circadian (less than 20 h).

**Zeitgeber.** See Synchronizer. It has to be understood that the "Zeitgeber" does not "give time" (does not induce a rhythm) but determines its arrangement in time (synchronizes).

# Subject Index